T0252189

THE THEORY OF ATOMIC STRUCTURE AND SPECTRA

LOS ALAMOS SERIES IN BASIC AND APPLIED SCIENCES
David H. Sharp and L. M. Simmons, Jr., editors

DETONATION
Wildon Fickett and William C. Davis

NUMERICAL MODELING OF DETONATIONS
Charles L. Mader

THE THEORY OF ATOMIC STRUCTURE AND SPECTRA
Robert D. Cowan

THE THEORY OF ATOMIC STRUCTURE AND SPECTRA

Robert D. Cowan

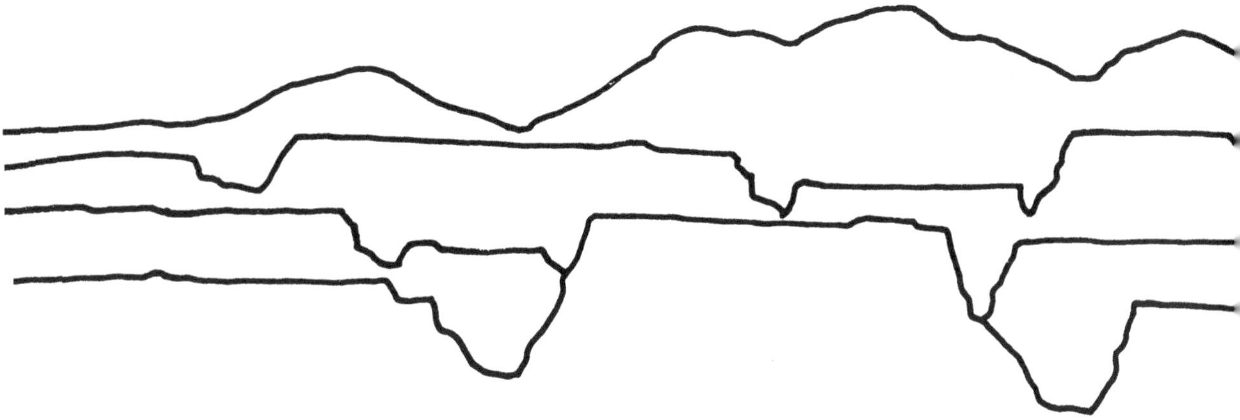

UNIVERSITY OF CALIFORNIA PRESS
Berkeley · Los Angeles · London

During its cataclysmic eruptions over a million years ago, the great Jemez volcano deposited enormous quantities of ash, pumice, and other debris. Compacted to form a soft rock, and extending to depths of more than a thousand feet, this is the material of the Pajarito Plateau in north-central New Mexico. The Los Alamos Scientific Laboratory is situated here, at an elevation of 7300 feet, on the eastern edge of the Jemez range. The plateau is deeply cut by numerous canyons which create sheer walls and mesas of striking beauty. The motifs on the jacket and title page are sketches of this terrain, as viewed from the valley of the Rio Grande.

University of California Press
Berkeley and Los Angeles, California

University of California Press, Ltd.
London, England

© 1981 by
The Regents of the University of California

Library of Congress Cataloging in Publication Data

Cowan, Robert Duane, 1919–
The theory of atomic structure and spectra.

(Los Alamos series in basic and applied sciences)
Bibliography: p.
Includes indexes.
1. Atomic structure. 2. Atomic spectra.
I. Title. II. Series.
QC173.C647 539′.14 81-4578
ISBN 978-0-520-03821-9 AACR2

12 11 10 09
12 11 10 9 8

To the memory of G. H. Dieke

Contents

PREFACE xv

CHAPTER 1. EXPERIMENTAL BACKGROUND AND BASIC CONCEPTS 1

1-1. Line Spectra 2
1-2. Characteristic Spectra of Atoms and Ions 4
1-3. Isoelectronic Sequences 6
1-4. Wavelength and Wavenumber Units 6
1-5. Energy Levels 8
1-6. Units for Energy Levels 9
1-7. Energy Level Diagrams 10
1-8. The Ritz Combination Principle; Grotrian Diagrams 16
*1-9. The Profiles of Spectrum Lines 17
*1-10. Wavelength Range and Accuracy 24
1-11. Empirical Spectrum Analysis 25
1-12. The Role of Theory 27

CHAPTER 2. ANGULAR-MOMENTUM PROPERTIES OF WAVEFUNCTIONS 33

2-1. Introduction 33
2-2. Angular-Momentum Operators 34
2-3. Eigenvalues of Angular-Momentum Operators 35
2-4. Step-up and Step-down Operators 38
2-5. Orbital Angular Momentum 39
2-6. The Addition Theorem of Spherical Harmonics 44
2-7. Electron Spin 49
2-8. Addition of Two Angular Momenta 50
2-9. The Vector Model 53
2-10. Coupled Wavefunctions 53
2-11. Coupling Schemes; LS Coupling 56

2-12. Notation for Angular-Momentum States 58
2-13. Parity 61
2-14. Parity and Angular-Momentum Selection Rules 62
*2-15. Appendix 64

CHAPTER 3. ONE-ELECTRON ATOMS 67
3-1. The Schrödinger Equation 67
3-2. Central-field Problems 68
3-3. Analytical Solution of the Radial Equation 70
3-4. Numerical Solution of the Radial Equation 72
3-5. Electron Probability Density 78
3-6. Energy Levels and Wavelengths 79
3-7. Relativistic Corrections 81
*3-8. Appendix: The Virial Theorem 88
3-9. Appendix: The Spin-Orbit Interaction 91

CHAPTER 4. COMPLEX ATOMS—THE VECTOR MODEL 93
4-1. The Schrödinger Equation 93
4-2. The Matrix Method 94
4-3. The Central-Field Model 97
4-4. Product Wavefunctions 98
4-5. Antisymmetrization; Determinantal Functions 99
4-6. Coupling of Antisymmetrized Wavefunctions 102
4-7. Electron Configurations 103
4-8. Equivalent Electrons; Closed Subshells 106
4-9. Permitted LS Terms for Equivalent Electrons 108
4-10. Configurations with Several Open Subshells 109
4-11. Configuration-Average Energies 112
4-12. Relative Energies of Configurations 113
4-13. The Periodic System 115
4-14. Variation of Ionization Energy with Z 120
4-15. Level Structure under LS-Coupling Conditions 122
4-16. Hund's Rule 124
4-17. jj Coupling 125
4-18. Pair Coupling 128
4-19. Example: Si I 3pns 131
4-20. Other Coupling Schemes 134
4-21. Statistical Weights (Hydrogenic Atoms) 134
4-22. Statistical Weights (Complex Atoms) 136
4-23. Quantitative Calculation of Level Structures 138

CHAPTER 5. THE 3n-j SYMBOLS 142
5-1. The 3-j Symbol 142
5-2. The 6-j Symbol 147

5-3. The 9-j Symbol 150
*5-4. Graphical Methods 152

CHAPTER 6. CONFIGURATION-AVERAGE ENERGIES 156
6-1. Diagonal Matrix Elements of Symmetric Operators 156
6-2. One-Electron and Total-Atom Binding Energies 160
6-3. Ionization Energy and One-Electron Binding Energies 168
6-4. Numerical Example 171

CHAPTER 7. RADIAL WAVE EQUATIONS 176
7-1. The Variational Principle 176
7-2. The Hartree-Fock Equations 178
7-3. The Classical Potential Energy 181
7-4. The Exchange Potential Energy 183
7-5. Solution of the Hartree-Fock Equations 184
*7-6. Complications and Instabilities 185
7-7. Homogeneous-Equation (Local-Potential) Methods 190
7-8. The Thomas-Fermi (TF) and Thomas-Fermi-Dirac (TFD) Methods 191
*7-9. The Parametric Potential Method 193
7-10. The Hartree Method (H) 194
7-11. The Hartree-Fock-Slater Method (HFS) 194
7-12. The Hartree-plus-Statistical-Exchange Method (HX) 197
*7-13. The Hartree-Slater Method (HS) 199
7-14. Relativistic Corrections 200
7-15. Correlation Corrections 202
*7-16. Appendix: The Thomas-Fermi and Thomas-Fermi-Dirac Atoms 206
*7-17. Appendix: Small-r Solution for the HXR Method 212

CHAPTER 8. RADIAL WAVEFUNCTIONS AND RADIAL INTEGRALS 214
8-1. Computer Calculation of Radial Functions 214
8-2. Comparison of Methods 218
8-3. Accuracy of Computed Configuration-Average Energies 219
8-4. Relativistic Effects 223
8-5. Variation with the Principal Quantum Number 224
8-6. Variation with Z of One-Electron Binding Energies 229
8-7. Variation with Z of Coulomb and Spin-Orbit Integrals 236

CHAPTER 9. COUPLED ANTISYMMETRIC BASIS FUNCTIONS 240
9-1. Coupling of Two Angular Momenta 240
9-2. Recoupling of Three Angular Momenta 245
9-3. Transformations Between Coupling Schemes 248
9-4. Antisymmetrization Difficulties for Equivalent Electrons 250
9-5. Coefficients of Fractional Parentage 255
9-6. Coefficients of Fractional Grandparentage 259

9-7. The Seniority Number 261
9-8. Antisymmetrized Functions for an Arbitrary Configuration 265
9-9. Single-Configuration Matrix Elements of Symmetric Operators 267
9-10. Uncoupling of Spectator Subshells 270
*9-11. Appendix: Vector-Coupling Coefficients 271

CHAPTER 10. ENERGY LEVEL STRUCTURE (SIMPLE CONFIGURATIONS) 276
10-1. Kinetic and Electron-Nuclear Energies 276
10-2. Effects of Closed Subshells 278
10-3. One-Electron Configurations 279
10-4. Two-Electron Configurations; Spin-Orbit Matrix Elements 280
10-5. Two-Electron Configurations; Coulomb Matrix Elements 282
10-6. Intermediate Coupling 288
10-7. Eigenvector Purities 291
10-8. Examples: ps and p^5s 295
10-9. Pair Coupling 298

CHAPTER 11. THE ALGEBRA OF IRREDUCIBLE TENSOR OPERATORS 303
11-1. Calculation of Complex Matrix Elements 303
11-2. Racah Algebra 304
11-3. Irreducible Tensor Operators 305
11-4. The Wigner-Eckart Theorem 307
11-5. Matrix Elements of the Product of Two Operators 310
11-6. Tensor Product of Two Tensor Operators 311
11-7. Uncoupling Formulae for Reduced Matrix Elements 313
11-8. Scalar Product of Two Tensor Operators 314
11-9. Unit Tensor Operators 316
11-10. Double Tensor Operators 319
11-11. Spin-Orbit Matrix Elements (One-Electron Configurations) 321
11-12. Direct Coulomb Matrix Elements (Two-Electron Configurations) 322
11-13. Exchange Coulomb Matrix Elements (Two-Electron Configurations) 323
11-14. The Effective Operator for Exchange Matrix Elements 323
11-15. Summary 326

CHAPTER 12. ENERGY LEVEL STRUCTURE (COMPLEX CONFIGURATIONS) 327
12-1. Introduction 327
12-2. Coulomb Matrix Elements: Equivalent Electrons 328
12-3. Non-Equivalent Electrons: Direct Integral 331
12-4. Non-Equivalent Electrons: Exchange Integral 333
12-5. Spin-Orbit Matrix Elements 335
12-6. Example: f^3sd^2 337

12-7. Level Structure: Spin-Orbit Effects under Near-𝔏𝔖 Coupling Conditions 339
12-8. Level Structure: Configurations l^w 341
12-9. Level Structure: Configurations $l_1^{w1}l_2$ 346
*12-10. Level Structure: More-Complex Configurations 356
*12-11. Appendix: Center-of-Gravity Relations 357

CHAPTER 13. CONFIGURATION INTERACTION 358
13-1. General Remarks 358
*13-2. Wavefunction Orthogonality Problems 361
13-3. Two-Electron Configurations 366
13-4. Single-Configuration-Like Interactions 368
13-5. Rydberg-Series Interactions 369
13-6. One-Electron Configurations 371
13-7. Brillouin's Theorem 372
*13-8. Arbitrary Configuration Interactions 373
*13-9. Matrix Elements of Symmetric Operators 375
*13-10. Coulomb Matrix Elements; General Case 378
*13-11. One-Electron Matrix Elements; General Case 390

CHAPTER 14. RADIATIVE TRANSITIONS (E1) 395
14-1. The Einstein Transition Probabilities 395
14-2. Electric Dipole Radiation (Classical) 398
14-3. Electric Dipole Radiation (Quantum Mechanical) 400
14-4. Selection Rules (Electric Dipole Radiation) 402
14-5. The Dipole Line Strength 402
14-6. Oscillator Strengths 404
14-7. Theoretical Calculation of Line Strengths 405
14-8. Selection Rules and Relative Intensities for LS Coupling 406
14-9. One-Electron Configurations; The Radial Dipole Integral 410
14-10. Transitions Involving an Electron in Singly Occupied Subshells 412
*14-11. Line Strengths for General Transition Arrays 417
14-12. Selection-Rule Violations 421
*14-13. Line Strength Sum Rules 422
*14-14. Oscillator Strength Sum Rules 424
14-15. Cancellation Effects 432
14-16. Experimental Measurement of Oscillator Strengths 434
*14-17. Systematics of Oscillator Strengths 436
*14-18. Radiative Decay in Low-Density Plasmas 440

CHAPTER 15. RADIATIVE TRANSITIONS (M1 AND E2) 442
15-1. Magnetic Dipole Radiation (M1) 442
15-2. Electric Quadrupole Radiation (E2) 445
*15-3. Interference Between M1 and E2 Radiation 449
15-4. Examples of Forbidden Transitions 450

CHAPTER 16. NUMERICAL CALCULATION OF ENERGY LEVELS AND SPECTRA 456
16-1. Computer Programs 456
16-2. *Ab Initio* Calculations 461
16-3. Least-Squares Calculations 465
16-4. Basic Least-Squares Method 468
*16-5. Modifications of the Basic Method 473
16-6. Phase of the Configuration-Interaction Parameters 477
*16-7. Effective Operators 477
*16-8. LS-Dependent Hartree-Fock Calculations 481
*16-9. Highly Excited Configurations 483

CHAPTER 17. EXTERNAL FIELDS AND NUCLEAR EFFECTS 485
17-1. The Zeeman Effect 485
17-2. Matrix Elements of the Magnetic-Energy Operator 486
17-3. The Weak-Field Limit 488
17-4. Weak-Field Zeeman Patterns 492
*17-5. Strong Magnetic Fields; The Paschen-Back Effect 494
*17-6. Intermediate Magnetic Fields 497
*17-7. The Stark Effect 498
*17-8. Isotope Shifts 505
*17-9. Hyperfine Structure 506

CHAPTER 18. CONTINUUM STATES; IONIZATION AND RECOMBINATION 512
18-1. Introduction 512
18-2. One-Electron Continuum Basis Functions 515
18-3. Normalization of the Radial Function 516
18-4. The Energy Dimensions of $P_{\varepsilon l}$ 521
18-5. Relation Between Radial Integrals for Discrete and Continuum States ... 522
18-6. Photoionization 523
18-7. Interaction of a Discrete State with a Single Continuum; Autoionization 526
*18-8. Pseudo-Discrete Treatment of Continuum Problems 535
*18-9. Ionization Equilibrium 544
*18-10. Radiative Recombination 547
*18-11. Dielectronic Recombination 549
*18-12. Generalized Oscillator Strengths 563
*18-13. Plane-Wave-Born Collision Strengths 567

CHAPTER 19. HIGHLY IONIZED ATOMS 570
19-1. Binding Energies in Isoelectronic Sequences 570
19-2. Energy-Level Structures 573
19-3. Configuration Interactions 578

19-4. Radial Multipole Integrals 580
19-5. Spectra 581
19-6. X-Ray Spectra 583
19-7. Comparison of Highly Ionized and X-Ray Spectra 584
*19-8. Autoionization and Dielectronic Recombination 589
*19-9. Inner-Subshell Excitations 590
*19-10. Temperature and Density Diagnostics 590
*19-11. Iron K_α Diagnostics 593

CHAPTER 20. RARE-EARTH AND TRANSITION ELEMENTS 598
20-1. Lanthanide Configurations 598
20-2. Level Structure; General Remarks 600
20-3. Low-Level Structures and Coupling Conditions 603
20-4. Spectra 604
20-5. Aids to Empirical Spectrum Analysis 606
20-6. Ions 610
*20-7. Actinides 612
20-8. Transition Elements 613

CHAPTER 21. STATISTICAL DISTRIBUTIONS 616
*21-1. Thermodynamic Functions of Atomic Gases 616
*21-2. Opacities of Thick Plasmas 617
*21-3. Energy Distribution of Levels of a Configuration 618
*21-4. Wavelength and Oscillator-Strength Distributions Within a Transition Array 625

Appendix A. Physical Constants, Units, and Conversion Factors 632
Appendix B. Conversion Between Vacuum and Air Wavelengths 634
Appendix C. 3-j Symbols 635
Appendix D. 6-j Symbols 646
Appendix E. One-Electron and Total-Atom Energies 665
Appendix F. Basis Functions, Matrix Elements, and Racah Algebra 668
Appendix G. Matrix Elements of Spherical Harmonics 677
Appendix H. Coefficients of Fractional Parentage, $U^{(k)}$, $V^{(k1)}$ 679
Appendix I. Relative Line Strengths Within an LS Multiplet 694

Bibliography 702

Name Index 709
Subject Index 717

*The starred sections may be skipped with little loss of continuity.

Preface

During the past half century an enormous amount of effort has been expended on the important task of deriving energy levels of atoms and ions from their observed optical spectra. An equally enormous amount of work remains to be done, particularly on the very complex rare-earth (lanthanide and actinide) spectra, and on highly ionized atoms, where serious experimental difficulties exist. In addition, much of the older work on relatively simple spectra needs to be redone because of errors arising from the inadequate experimental apparatus and theoretical knowledge available at that time.

Essential both to persons doing basic research of this type and to those interested in astrophysical, plasma-physics, and other applications of spectroscopy is an adequate knowledge of the theory of atomic structure and spectra, and the ability to use modern electronic computers for the calculation of energy levels and wavefunctions. The basic theoretical principles were established within a decade after the invention of quantum mechanics, and were well summarized in the classic book by Condon and Shortley in 1935; mathematical techniques adequate to apply the theory to the complex atomic systems of current interest became available within another decade, as a result of the work of Racah and others. In spite of the thirty-odd years that have passed since, and the appearance during that time of several books dealing with the theory of atomic spectra, many aspects of the theory and its application remain widely scattered in the literature.

The present book is an outgrowth of theoretical work in which I have been engaged at the Los Alamos Scientific Laboratory over the past two decades; the topics emphasized most strongly are those that I have had occasion to investigate myself, and many of the illustrative examples are drawn from this work. However, I have tried to present a coherent and reasonably complete treatment of the Slater-Condon theory, in a form adapted particularly to the use of digital computers as an aid to the interpretation of observed complex spectra. For this purpose, calculations—though generally very involved and lengthy—usually need give results of only moderate accuracy. Consequently, there is no mention in this book of the highly accurate calculational methods that are available for helium and other systems containing only a very few electrons. The hydrogen-atom problem is discussed only in the detail required as a basis for the treatment of many-electron atoms; there is no discussion of the Lamb shift and similar fine points. There is also no consideration of accurate methods of calculating correlation energies in many-electron systems.

Early drafts of the book were prepared in connection with courses in atomic structure that I taught at Purdue University in the spring of 1971 and at the Los Alamos Graduate Center of the University of New Mexico in the spring of 1972. These drafts have been considerably expanded, and the present version contains far more material than can comfortably be included in a typical one-semester, three-hour course, even though several tedious mathematical derivations have been omitted, and other detailed and expendable sections have been indicated by asterisks. The text can be further shortened by omission of various obvious topics in Chapters 17-21, depending on the particular interests of the user. For those who wish to truncate the book even further, or who may wish to skim through portions of the book in a preliminary fashion in order to obtain a relatively brief overview of the subject, two outlines are suggested—one for those whose interest lies in the mathematical theory and numerical calculations, and one for those interested primarily in the qualitative features of atomic structure:

Theory	Common	Qualitative
	Secs. 1-1 to 1-8	
	1-11 to 1-12	
	2-1 to 2-3	
	2-5	
	2-7 to 2-14	
	3-1 to 3-7	
	4-1 to 4-12	
Secs. 4-21 to 4-23		Secs. 4-13 to 4-19
5-1 to 5-3		
	Appendix E	
	Secs. 7-1 to 7-2	
	7-7	
	7-14 to 7-15	
	8-1	
Appendix F		
		Secs. 8-6
Secs. 12-1 to 12-6		10-6 to 10-9
13-1		12-7 to 12-9
	14-1 to 14-10	
16-1 to 16-4		14-12

Finally, for anyone wishing a super-short outline of the Slater-Condon method of calculating energy levels and spectra, with only references to the pertinent equations, I suggest Secs. 4-1 to 4-6, 4-23, 7-1, 7-2, 7-7, 7-14, 7-15, 8-1, 16-1, and 16-2. Even those working their way through the book in detail may find it helpful, after finishing Chapter 4, to skim through Appendix F in order to obtain an outline of the general mathematical method, and to refer to this appendix occasionally by way of review while wading through Chapters 9-11. This appendix also contains a summary of the basic equations of Racah algebra for convenient reference in connection with Chapters 12-15, or independent evaluations of other matrix elements.

Particularly for those interested in the theoretical details, it is assumed that readers have a good working knowledge of elementary quantum mechanics, through the hydrogen atom and first-order perturbation theory, and some familiarity with matrix theory (eigenvalues and eigenvectors, and orthogonal transformations). Some previous acquaintance with the qualitative elements of atomic spectra and structure (the vector model of the atom and the Zeeman effect) would be helpful, but should not be essential; persons having a strong background in this subject can omit or skim through much of the first four chapters. Although an extensive knowledge of group theory can be used to greatly shorten some otherwise tedious derivations, and can lead to a deeper understanding of many aspects of the theory, I have preferred to follow the approach of Condon and Shortley, and base the discussion of wavefunction symmetries on more physical angular-momentum arguments. An effort has been made to present all material in the most elementary and unsophisticated manner possible consistent with the aim of being able to treat atoms of arbitrary complexity. The use of coupled wavefunctions, 3n-j symbols, and Racah algebra is unavoidable; however, these topics are discussed thoroughly from an elementary point of view, and no previous acquaintance with them should be necessary. Wherever feasible, I have included sample numerical calculations to illustrate how the theory is applied in practice, and have included comparisons of theoretical results with experimental data to show the degree of accuracy that can be expected of the theory. The limited number of homework problems included are mainly intended for this same purpose.

Phase and normalization conventions in atomic structure theory are far from standardized, forming repeated pitfalls for the unwary. I have chosen to employ the conventions of Condon and Shortley, except to use LS coupling ($\mathbf{L} + \mathbf{S} = \mathbf{J}$) instead of SL coupling ($\mathbf{S} + \mathbf{L} = \mathbf{J}$). The latter is more commonly used, and is more in line with the standard Russell-Saunders LS-coupling [sic!] level notation $^{2S+1}L_J$, but the former fits in better with the progression of couplings LS $\rightarrow$ LK $\rightarrow$ jK $\rightarrow$ jj. In any event, the difference between LS and SL is only a phase factor $(-1)^{L+S-J}$ in each wavefunction. I have tried at all appropriate points in the text to call attention to the most common phase and normalization differences in the literature.

It is no exaggeration to say that this book would never have been written without the inspiration and encouragement of K. L. Andrew. Many of the topics included are ones that I have investigated jointly with him and his students—especially D. C. Griffin—and his laboratory has provided much of the experimental data against which theoretical methods have been tested. N. J. Peacock was responsible for my initial interest in the calculation of atomic spectra by *ab initio* methods (as opposed to least-squares fitting of experimental energy levels). For suggestions, data, and general encouragement, I am greatly indebted also to J. Blaise, P. G. Burkhalter, C. Corliss, H. M. Crosswhite, G. A. Doschek, B. Edlén, J. O. Ekberg, B. C. Fawcett, U. Feldman, M. Fred, J. E. Hansen, W. F. Huebner, R. C. Isler, V. Kaufman, N. H. Magee, Jr., J. B. Mann, W. C. Martin, A. L. Merts, L.

Minnhagen, L. J. Radziemski, Jr., J. Reader, W. D. Robb, D. W. Steinhaus, J. Sugar, L. Å. Svensson, S. von Goeler, K. G. Widing, W. L. Wiese, and many others—especially M. Wilson, who read preliminary drafts of the entire manuscript, catching innumerable errors, and offering many suggestions for improvement.

My deep thanks are due to Martha Wooten and Christella Secker, who typed a long and difficult manuscript, and to members of the LASL photocomposition group who converted it beautifully to type—particularly to Katherine Valdez, head compositor and layout artist, and to Mary Louise Garcia and Samia L. Davis, compositors. The University of California Press editorial and production staffs headed by Grant Barnes and Chet Grycz were always friendly and helpful. And finally, I am much indebted to my wife, Dot, who sympathetically endured countless evenings and weekends alone, and also helped with proofreading and matters of grammar and style.

Robert D. Cowan
Los Alamos, New Mexico, March 1981

1

EXPERIMENTAL BACKGROUND AND BASIC CONCEPTS

Spectroscopic studies of the light emitted or absorbed by atoms and ions date from the early nineteenth century. From these studies, it gradually became clear that the particular wavelengths of light associated with atoms of a given element are characteristic of that element, and that spectral information must therefore provide clues to the internal structure of the atom. During the last quarter of the century, important regularities were discovered among the wavelengths of hydrogen and other comparatively simple spectra. However, efforts to interpret these regularities by means of classical models of the atom all ended—despite some apparent initial successes—in dismal failure. The first real success appeared only in 1913 with Bohr's theory of the hydrogen atom. Progress in the understanding of multi-electron atoms had to await the invention of quantum mechanics a decade later, but thereafter was extremely rapid.

It is not necessary for our purposes to describe the slow and painful pre-quantum-mechanical steps that led to our present understanding of the relationship between atomic structure and atomic spectra. The interested reader may find brief outlines in the introductory chapters of the books by Sawyer[1] and by White;[2] a more detailed account of the earlier phases has been given by McGucken.[3] English translations of some important early research papers are included in Hindmarsh's source book.[4]

There is likewise no need here to include an extensive description of spectroscopic apparatus and techniques; these are discussed in detail by Sawyer[1] and by Harrison, et al.,[5] though their books are somewhat obsolescent.

[1]Ralph A. Sawyer, *Experimental Spectroscopy* (Dover Publications, New York, 1963), 3rd ed. (For convenience of later reference, Sawyer's text and other major items are listed in a bibliography at the end of this book.)

[2]Harvey E. White, *Introduction to Atomic Spectra* (McGraw-Hill, New York, 1934).

[3]William McGucken, *Nineteenth-Century Spectroscopy; Development of the Understanding of Spectra, 1802-1897* (The Johns Hopkins Press, Baltimore, 1969).

[4]W. R. Hindmarsh, *Atomic Spectra* (Pergamon, Oxford, 1967).

[5]George R. Harrison, Richard C. Lord, and John R. Loofbourow, *Practical Spectroscopy* (Prentice-Hall, Englewood Cliffs, N. J., 1948). See also additional references listed in Sec. I of the bibliography.

It is anticipated that most readers of the present book will have some familiarity with the qualitative aspects of spectra and atomic structure, as presented in the recent books by Woodgate[6] and Kuhn,[7] or in the older but still useful texts of White[2] and Herzberg.[8] Nonetheless, it is deemed appropriate in this chapter to review some of the basic facts and terminology of spectroscopy and atomic structure, as a background for the more highly theoretical material to follow. Numerous references to recent spectroscopic techniques and observations are included to supplement the older works cited above.

1-1. LINE SPECTRA

Information concerning the electronic structure of an isolated atom (or ion[9]) can be inferred from a variety of different types of experimental information—for example, data on elastic and inelastic scattering of electrons, ions, or x-rays by the atom, and data on the energies of photoelectrons ejected from the atom. By far the most important source of high-accuracy information, however, comes from the spectroscopic study of light radiated by, or absorbed by, the atom.

The spectrum of the light radiated by atoms in an appropriate light source may be examined with the aid of a simple prism spectrograph as illustrated in Fig. 1-1. Light from the source is focused on a narrow entrance slit that is oriented parallel to the dispersing faces of the prism. The light passing through the slit is dispersed according to wavelength in passing through the prism. For each wavelength of light present, a camera lens forms an image of the slit on a photographic plate; the various images for the different wavelengths are displaced from one another in a direction perpendicular to the length of each slit image. If the source is such (for example, an incandescent solid) as to radiate light of all wavelengths, the spectrographic plate records a continuous succession of overlapping slit images, and the spectrum is correspondingly known as a continuous spectrum. However, if the light from the source is radiated by a collection of isolated atoms, it is found to consist of a greater or smaller number of isolated wavelengths: if a very narrow spectrograph slit is used, the spectrum appears on the photographic plate as a set of isolated parallel lines. Such spectra are accordingly referred to as line spectra, to distinguish them from the continuous spectra of solids and the "band" spectra of molecules; examples of all three types of spectra are shown in Fig. 1-2.

Evidently, the line-image form of the observed spectra of atoms is a characteristic of the shape of the entrance slit of the spectrograph. For modern research work, the prism spectrograph has been almost entirely replaced by spectrographs using a diffraction grating[10]

[6]G. K. Woodgate, *Elementary Atomic Structure* (McGraw-Hill, London, 1970).

[7]H. G. Kuhn, *Atomic Spectra* (Longmans, Green, London, 1969), 2nd ed.

[8]Gerhard Herzberg, *Atomic Spectra and Atomic Structure* (Dover Publications, New York, 1944), 2nd ed.

[9]For brevity, the unqualified term "atom" will generally be used to mean either a neutral atom (a nucleus of charge $+Ze$ surrounded by a number of electrons N equal to Z), or a positively or negatively charged ion ($N < Z$ or $N > Z$, respectively).

[10]The term "diffraction grating" is a universally used misnomer for what should more appropriately be called an interference grating.

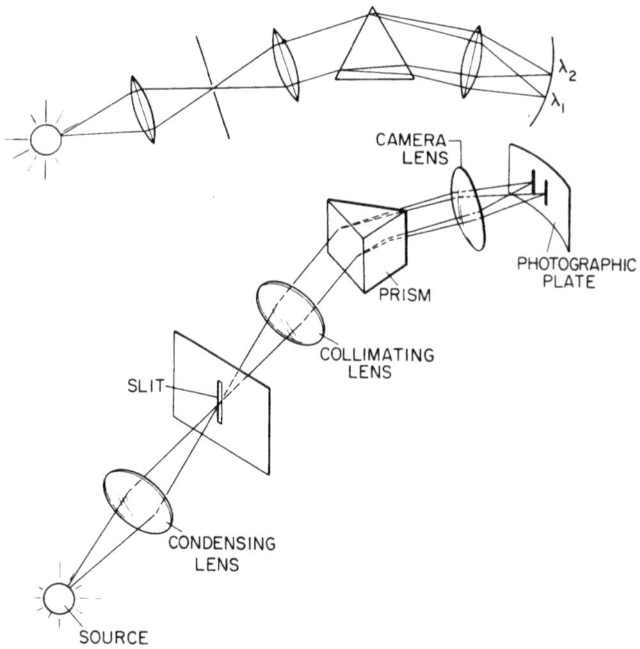

Fig. 1-1. Simple prism spectrograph, showing the separation of two spectrum lines. Top, plan; bottom, perspective view.

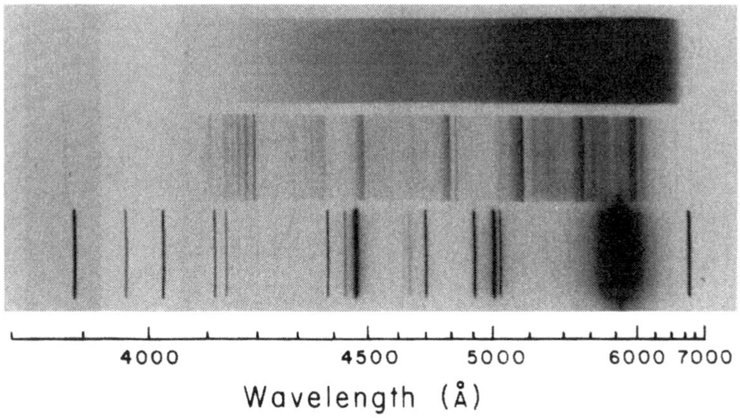

Fig. 1-2. Typical emission spectra, taken with a low-dispersion prism spectrograph. Top: Continuous spectrum of an incandescent filament. Middle: Band spectrum of the CO molecule. Bottom: The line spectrum of the neutral helium atom in a Geissler discharge. (Photographic negatives, courtesy of K. L. Andrew.)

for the dispersing element, but a narrow entrance slit is normally still used, and a line spectrum still results. For ultra-high resolution work, Fabry-Perot and Michelson interferometers may be used; these produce interference fringes in the form of concentric circles rather than straight lines (Fig. 1-3). Even with a grating spectrograph, the instrument is sometimes used without a slit (or, rather, with an extremely wide slit), and the spectrum then appears as a set of (usually overlapping) images of the source, one image for each wavelength present (Fig. 1-4). Nonetheless, because of the appearance of the spectrum produced by the classic narrow-slit instrument, each essentially monochromatic wavelength present in the light emitted by an atom is called a *spectrum line.*

1-2. CHARACTERISTIC SPECTRA OF ATOMS AND IONS

Each atom or ion emits a line spectrum which is characteristic of that atom or ion. This provides the basis for spectrographic methods of the analysis of the composition of (vaporized) alloys or other mixtures of atoms; indeed, several elements (among them helium, rubidium, cesium, gallium, indium, and thallium) were discovered spectroscopically.[5]

The spectrum emitted by neutral atoms of a given element is called the first spectrum of that element, and is denoted by the Roman numeral I; the spectrum emitted by singly ionized atoms is called the second spectrum and is denoted by Roman numeral II; etc. The first spectrum is sometimes also called the *arc* spectrum, because it is the one that is most prominent in the spectrum of a low-current d-c arc. The second spectrum is called the (first) *spark* spectrum because it becomes strong in a high-voltage (low-current) spark

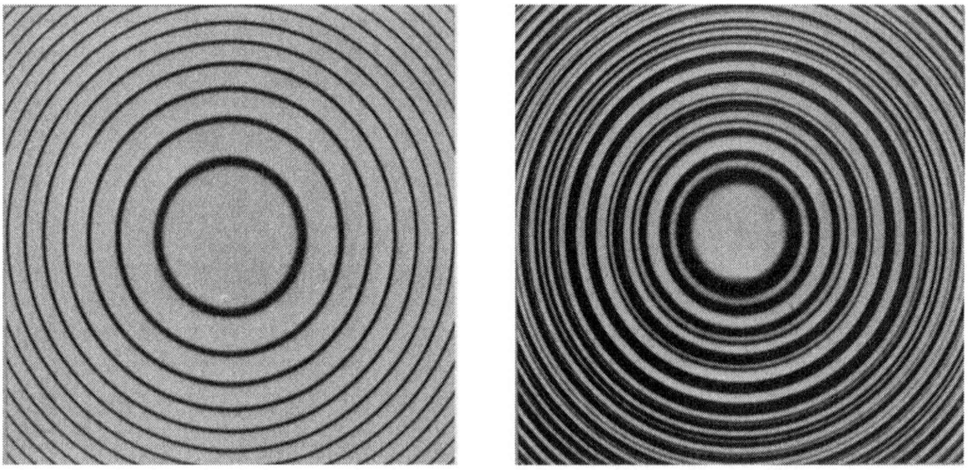

Fig. 1-3. Circular fringes produced by a Fabry-Perot interferometer for the red cadmium line at 6438.47 Å (left) and for a superposition of all strong visible lines of neutral cadmium (right). (Photographic negatives, courtesy of K. L. Andrew.)

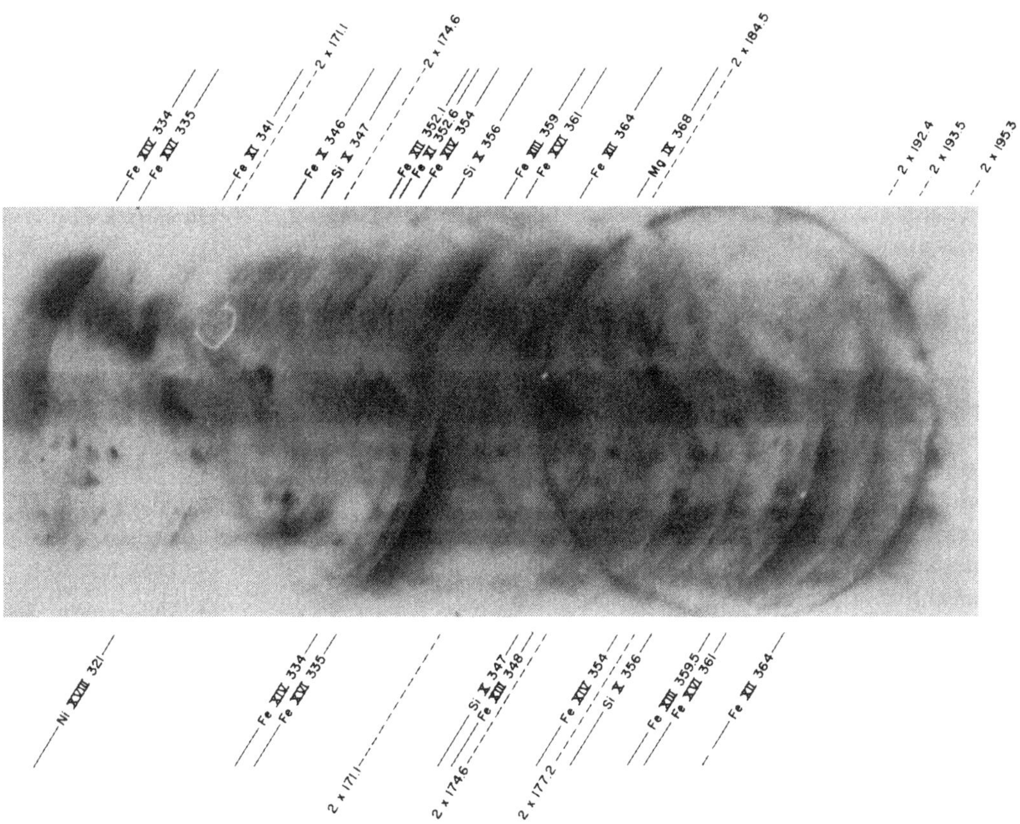

Fig. 1-4. Vacuum ultraviolet slitless spectrogram of the sun, showing images of the solar limb in light of wavelengths corresponding to spectrum lines of highly ionized Mg, Si, Fe, and Ni. [Photographic negative, from K. G. Widing, G. D. Sandlin, and R. D. Cowan, Astrophys. J. **169**, 405 (1971).]

discharge.[11] Spectra of still higher ionization stages can readily be produced by very high-voltage and/or high-current sparks,[12] in theta-pinch and tokamak discharges,[13,14] or by vaporizing materials with the beam from a very high energy laser pulse.[15]

[11]For research purposes, arc and spark sources have been largely replaced by hollow-cathode sources and microwave-excited electrode-less discharges. See, for example, H. M. Crosswhite, G. H. Dieke, and C. S. Legagneur, J. Opt. Soc. Am. **45**, 270 (1955); C. H. Corliss, W. R. Bozman, and F. O. Westfall, J. Opt. Soc. Am. **43**, 398 (1953).

[12]B. Edlén, Rep. Prog. Phys. **26**, 181 (1963); L. Minnhagen, J. Res. Natl. Bur. Stand. **68C**, 237 (1964); U. Feldman, M. Swartz, and L. Cohen, Rev. Sci. Instr. **38**, 1372 (1967).

[13]B. C. Fawcett, B. B. Jones, and R. Wilson, Proc. Phys. Soc. (London) **78**, 1223 (1961); D. D. Burgess, Space Sci. Rev. **13**, 493 (1972); J. P. Connerade, N. J. Peacock, and R. J. Speer, Solar Phys. **14**, 159 (1970).

[14]E. Hinnov, Phys. Rev. A **14**, 1533 (1976); Equipe TFR, Nucl. Fusion **15**, 1053 (1975).

[15]B. C. Fawcett, A. H. Gabriel, F. E. Irons, N. J. Peacock, and P. A. H. Saunders, Proc. Phys. Soc. (London) **88**, 1051 (1966); B. C. Fawcett, J. Phys. B **3**, 1152, 1732 (1970).

Obviously the number of different possible line spectra of an element is equal to its atomic number Z, so that there exist the spectra H I, He I, He II, Li I, Li II, Li III, etc.[16] Such high-ionization-stage spectra as Fe XVIII-XXVI, Mo XXXI-XXXIII, and Au LII have been produced in the laboratory,[15,17] and lines of Fe XXVI (hydrogen-like iron) may have been seen in spectra of solar flares.[18]

By extension, the Roman numeral notation is used loosely to refer also to the ion that produced the spectrum; thus C I refers to the neutral carbon atom, C II refers to the singly ionized atom C^+, C IV refers to the triply ionized atom C^{3+}, etc.[19]

1-3. ISOELECTRONIC SEQUENCES

Spectra of ions of different elements having the same number of electrons N tend to be very similar in general structure, especially for highly ionized atoms. A sequence of ions having fixed N, or the corresponding sequence of spectra, is called an isoelectronic sequence. A sequence is usually denoted by its first (neutral-atom) member; for example, the Ar I sequence consists of Ar I, K II, Ca III, Sc IV, $\cdots$. However, if emphasis is on a particular highly ionized member of a sequence, the sequence might be denoted by this member; for example, the Ar I sequence might also be called the Fe IX sequence, and thought of as the sequence $\cdots$ Cr VII, Mn VIII, Fe IX, Co X, Ni XI, $\cdots$.

1-4. WAVELENGTH AND WAVENUMBER UNITS

The wavelengths of spectrum lines are most commonly given in Ångströms[20] ($1\ \text{Å} = 10^{-8}$ cm $= 10^{-10}$ m), but sometimes in microns or micrometers ($1\mu = 10^{-6}$ m $= 10000$ Å), in millimicrons or nanometers ($1\text{m}\mu = 1$ nm $= 10$ Å), or in X units (1 X-unit $\cong$ 0.00100208 Å).

From a theoretical point of view, it is much more convenient to describe a spectrum line not in terms of its wavelength but rather in terms of its frequency or some quantity proportional thereto. For example, the line can be described by the energy of one photon of wavelength λ:

$$E = h\nu = hc/\lambda_{vac} \; ; \tag{1.1}$$

[16]Astronomers speak of H II, meaning the continuous spectrum produced by the scattering of free electrons from hydrogen nuclei. However, we shall not treat free-free transitions, and shall therefore consider H II, He III, etc., to be nonexistent.

[17]U. Feldman and L. Cohen, Astrophys. J. **151**, L55 (1968); L. Cohen, U. Feldman, M. Swartz, and J. H. Underwood, J. Opt. Soc. Am. **58**, 843 (1968); E. Hinnov, ref. 14; P. G. Burkhalter, C. M. Dozier, and D. J. Nagel, Phys. Rev. A **15**, 700 (1977).

[18]W. M. Neupert, W. Gates, M. Swartz, and R. Young, Astrophys. J. **149**, L79 (1967).

[19]This spectroscopic usage must not be confused with a notation sometimes used for chemical oxidation states, where for example Fe (III) refers to Fe^{3+}.

[20]After the Swedish physicist A. J. Ångström (1814-74), who made extensive studies of the spectra of gases, metallic vapors, the sun, and the aurora, and pioneered in the accurate measurement of spectrum lines to establish wavelength standards.

for wavelengths (in vacuo) measured in Å,

$$E = \frac{12398.5}{\lambda_{vac}} \quad eV \tag{1.2}$$

$$= \frac{911.27}{\lambda_{vac}} \quad Ry , \tag{1.3}$$

where the rydberg [≅ 13.6058 eV, see (1.10) or Appendix A] is the unit of energy that we shall use throughout most of this book.

Most commonly, however, spectrum lines are described simply in terms of the wavenumber σ, or the number of wavelengths (in vacuo) per unit of length (usually per centimeter):

$$\sigma \equiv \frac{1}{\lambda_{vac}} = \frac{E}{hc} . \tag{1.4}$$

With λ in Å,

$$\sigma = \frac{10^8}{\lambda_{vac}} \quad cm^{-1} . \tag{1.5}$$

The unit cm^{-1} is frequently called a kayser (K).[21] The kilokayser (kK) is convenient for lines in the extreme vacuum ultraviolet (XVUV), a wavelength of 100 Å corresponding to a wavenumber of 10^6 K or 1000 kK. The millikayser (mK) is convenient for the discussion of isotope shift, hyperfine structure, and other small effects.

Wavelengths of observed spectrum lines of most elements are available in numerous tabulations, particularly the MIT Tables and the *Handbook of Chemistry and Physics* for low ionization stages,[22] Outred's tables for the infrared region,[23] and the tables of Kelly and Palumbo for the vacuum ultraviolet.[24] For additional references, see Sec. VI of the bibliography at the end of this book.

Below 2000 Å, wavelengths in vacuum are always tabulated because air is opaque shortly below that point, and so vacuum spectrographs must be used. Above that point,

[21]The name "kayser" (after the German physicist and spectroscopist H. G. J. Kayser, 1853-1940) has not been officially approved by the International Union of Pure and Applied Physics nor other members of the International Council of Scientific Unions [cf. J. Opt. Soc. Am. **43**, 410 (1953), **47**, 1035 (1957)], but is much shorter to read than "centimeters to the minus one" or "inverse centimeters." (These last long phrases are sometimes avoided by reading "cm^{-1}" as "wavenumbers," but this is as incorrect as is reading "100 Å" as "one hundred wavelengths.")

[22]G. R. Harrison, *Massachusetts Institute of Technology Wavelength Tables* (The M.I.T. Press, Cambridge, Mass., 1969), revision of 1939 edition; I-II spectra, 2000-10000 Å. R. C. Weast, ed., *Handbook of Chemistry and Physics* (CRC Press, West Palm Beach, Fla.), 59th (1978-79) and later editions; I-V spectra.

[23]M. Outred, J. Phys. Chem. Ref. Data **7**, 1-262 (1978); 10000-40000 Å.

[24]R. L. Kelly and L. J. Palumbo, *Atomic and Ionic Emission Lines below 2000 Å—Hydrogen through Krypton*, Naval Research Laboratory Report NRL-7599 (U. S. Govt. Printing Off., Washington, D. C., 1973).

however, tabulated wavelengths are almost always values in standard air. The relation between the two is

$$\lambda_{vac} = n\lambda_{air} , \tag{1.6}$$

where the index of refraction of standard air (dry air containing 0.03% CO_2 by volume at normal pressure and T = 15°C) is given by[25]

$$n = 1 + 8342.13 \cdot 10^{-8} + \frac{2406030}{130 \cdot 10^8 - \sigma^2} + \frac{15997}{38.9 \cdot 10^8 - \sigma^2} . \tag{1.7}$$

A brief conversion table between λ_{air} and λ_{vac} is given in Appendix B. (Inasmuch as spectrum lines are now always measured by comparison with known standard lines, whose wavelengths in turn are usually measured with an evacuated Fabry-Perot interferometer, there would be no point in working with λ_{air} at all except for the fact that most spectrographs working above 2000 Å are not evacuated, and it is usually necessary to intercompare wavelengths in different spectral orders.)

1-5. ENERGY LEVELS

The fact that light is radiated by an atom at only certain discrete wavelengths is associated with the fact that the atom can exist only in stationary states having certain discrete values of internal energy E. The existence of such discrete energy states is confirmed by the observation that bombardment of an atom with mono-energetic x-rays produces photoelectrons of only certain discrete kinetic energies, and by the observation that low-energy electrons scattered inelastically from an atom at a given angle can lose only certain discrete energies.

The various possible energies of the atom are called *energy levels*. The lowest possible energy is called the *ground level*, and each quantum state of the atom having this energy (there may be more than one such state) is called a ground state. All other levels are called *excited levels*, and the corresponding quantum states are called *excited states*.

We shall be interested primarily in energy levels that correspond to excitation of one of the more loosely bound electrons of the atom. These levels are inferred from observed line spectra in the manner to be described in Sec. 1-11. Known energy levels of most atoms and many ions are tabulated in AEL[26] and its revision;[27] bibliographies of more recent results

[25] B. Edlén, Metrologia **2**, 71 (1966). See also J. C. Owens, Appl. Opt. **6**, 51 (1967), and E. R. Peck and K. Reeder, J. Opt. Soc. Am. **62**, 958 (1972). Tabular values of n and σ vs. λ_{air} can be found in C. D. Coleman, W. R. Bozman, and W. F. Meggers, *Table of Wavenumbers*, U. S. NBS Monograph 3 (U. S. Govt. Printing Off., Washington, D. C., 1960), 2 vols.

[26] C. E. Moore, *Atomic Energy Levels, as Derived from the Analyses of Optical Spectra*, U. S. Natl. Bur. Stand. Circ. 467 (U. S. Govt. Printing Off., Washington, D. C.), Vols. I-III, 1949-58 [reissued 1971 as NSRDS-NBS 35, Vols. I-III]; W. C. Martin, R. Zalubas, and L. Hagan, *Atomic Energy Levels—The Rare Earth Elements*, NSRDS-NBS 60 (U. S. Govt. Printing Off., Washington, D. C., 1978). These volumes are commonly referred to by spectroscopists as "AEL," and by astrophysicists as "Moore's tables."

[27] C. E. Moore, *Selected Tables of Atomic Spectra*, NSRDS-NBS 3 (U. S. Govt. Printing Off., Washington, D. C.), Secs. 1-8, 1965-79.

are available.[28] For additional references, see Secs. VII and VIII of the bibliography at the end of this book.

Levels corresponding to tightly bound (inner-shell) electrons can be determined through a study of photoelectrons produced by absorption of x-rays,[29] directly through a study of emission and absorption x-ray spectra,[30] through line absorption spectra using synchrotron radiation as a continuous background source,[31] etc. Tables of such levels are available.[29,30,32]

1-6. UNITS FOR ENERGY LEVELS

The energies of the various levels of an atom are most commonly specified in terms of the excitation energy above the ground state. A frequently used unit, especially for levels corresponding to excitation of inner-shell electrons, is the electron-volt:

$$1 \text{ ev} = 10^8 e/c = 1.60219 \cdot 10^{-12} \text{ ergs} \ . \tag{1.8}$$

For levels corresponding to excitation of the loosely bound "valence" electrons, the pseudo-energy unit

$$E/hc \tag{1.9}$$

is used almost exclusively, and is measured in cm^{-1} (kaysers) or (for highly ionized atoms) sometimes in kilokaysers; cf. Eqs. (1.4) and (1.5).

For theoretical purposes, the most convenient unit of energy is the rydberg:[33]

$$1 \text{ Ry} = \frac{me^4}{2h^2} = \frac{e^2}{2a_0} = 2.17991 \cdot 10^{-11} \text{ ergs} = 13.6058 \text{ eV} \ , \tag{1.10}$$

[28]C. E. Moore, *Bibliography on the Analyses of Optical Atomic Spectra*, U. S. Natl. Bur. Stand. Special Publ. 306 (U. S. Govt. Printing Off., Washington D. C.), Secs. 1-4, 1968-69; L. Hagan and W. C. Martin, *Bibliography on Atomic Energy Levels and Spectra*, July 1968 through June 1971, U. S. Natl. Bur. Stand. Special Publ. 363 (U. S. Govt. Printing Off., Washington, D. C., 1972); L. Hagan, Supplement 1 to Special Publ. 363, July 1971 through June 1975 (1977).

[29]K. Siegbahn, et al., *ESCA—Atomic, Molecular, and Solid State Structure Studied by Means of Electron Spectroscopy* (Almqvist and Wiksells, Uppsala, 1967).

[30]J. A. Bearden and A. F. Burr, "X-ray Wavelengths and X-Ray Atomic Energy Levels," Rev. Mod. Phys. **39**, 78, 125 (1967); also, reprinted as NSRDS-NBS 14 by U. S. Govt. Printing Off., Washington, D. C., 1967. More detailed tables appear in *Atomic Energy Levels*, Report NYO-2543-1 (1965).

[31]R. P. Madden, D. L. Ederer, and K. Codling, Appl. Opt. **6**, 31 (1967); K. Codling, Rep. Prog. Phys. **36**, 541 (1973).

[32]J. C. Slater, "One-Electron Energies of Atoms, Molecules, and Solids," Phys. Rev. **98**, 1039 (1955); W. Lotz, J. Opt. Soc. Am. **58**, 915 (1968), **60**, 206 (1970); D. A. Shirley, R. L. Martin, S. P. Kowalczyk, F. R. McFeely, and L. Ley, Phys. Rev. B **15**, 544 (1977).

[33]After the Swedish physicist J. R. Rydberg (1854-1919), who found important mathematical relations connecting the wavenumbers of series of related spectrum lines; see, for example, W. McGucken, ref. 3, and N. Bohr, "Rydberg's Discovery of the Spectral Laws," Lunds University Årsskr., Avd. 2, **50**, 15 (1955).

where $a_0 = \hbar^2/me^2 = 0.529177 \cdot 10^{-8}$ cm is the radius of the first Bohr orbit of hydrogen. The hartree $(= e^2/a_0 = 2$ Ry) is also frequently employed, but we shall use only the rydberg in this book.

The equivalences between the energy and pseudo-energy units are:

$$1 \text{ Ry} = 109737.3 \text{ cm}^{-1} \ , \tag{1.11}$$

$$1 \text{ eV} = 8065.479 \text{ cm}^{-1} \ . \tag{1.12}$$

The first of these two numbers is known as the Rydberg constant, R_∞, for infinitely large nuclear mass.

1-7. ENERGY LEVEL DIAGRAMS

The possible energy levels of an atom are conveniently depicted by means of a diagram in which each level is represented by a short horizontal line placed at the appropriate point along a vertical energy scale. Each level is usually placed in one of several different columns, according to certain properties of the quantum state(s) that correspond to that level.

As a simple example, some of the known levels[26,27] of C II are shown (slightly simplified) in Fig. 1-5. Each level has been labeled with appropriate values of quantum numbers n*l*, which will be defined in detail in Chapter 3. For the present, suffice it to say that in each column (corresponding to a specific value of *l*), the various levels (differing only in the value of the quantum number n) form a regular sequence of energies known as a *Rydberg series*;[33] as n tends to ∞, each series tends to a limiting energy known as the series limit. The limit is also called an ionization limit, as it represents the physical state in which an electron has been removed to an infinite distance from the ion (with zero kinetic energy).

In general, there are many possible ionization limits (only one of which is shown in the figure), corresponding to different possible energy states of the residual ion. The lowest possible limit corresponds to the case in which the residual ion lies in its ground state, and is known as the first ionization limit, or simply as *the* ionization limit.

The energy difference between the ground state of an atom (or ion) and the first ionization limit (i.e., between the ground state of an ion and the ground state of the next-higher-stage ion) is called the ***ionization energy*** of the atom (or ion).[34] The ***ionization potential*** is a potential (in volts) numerically equal to the ionization energy in electron volts. Experimental (and some theoretical) ionization energies are given in Table 1-1.[35]

If I_m is the ionization energy of the mth ionization stage, $1 \leq m \leq Z$, then the total energy required to remove all $Z - m + 1$ electrons is

[34]An infinity of other ionization energies can be defined involving the energy differences between an excited level of the atom and/or an excited level of the next higher ion, but in such cases the excited levels must be stated explicitly.

[35]C. E. Moore, *Ionization Potentials and Ionization Limits Derived from the Analyses of Optical Spectra*, NSRDS-NBS 34 (U. S. Govt. Printing Off., Washington, D. C., 1970), and other references given in Sec. IX of the bibliography.

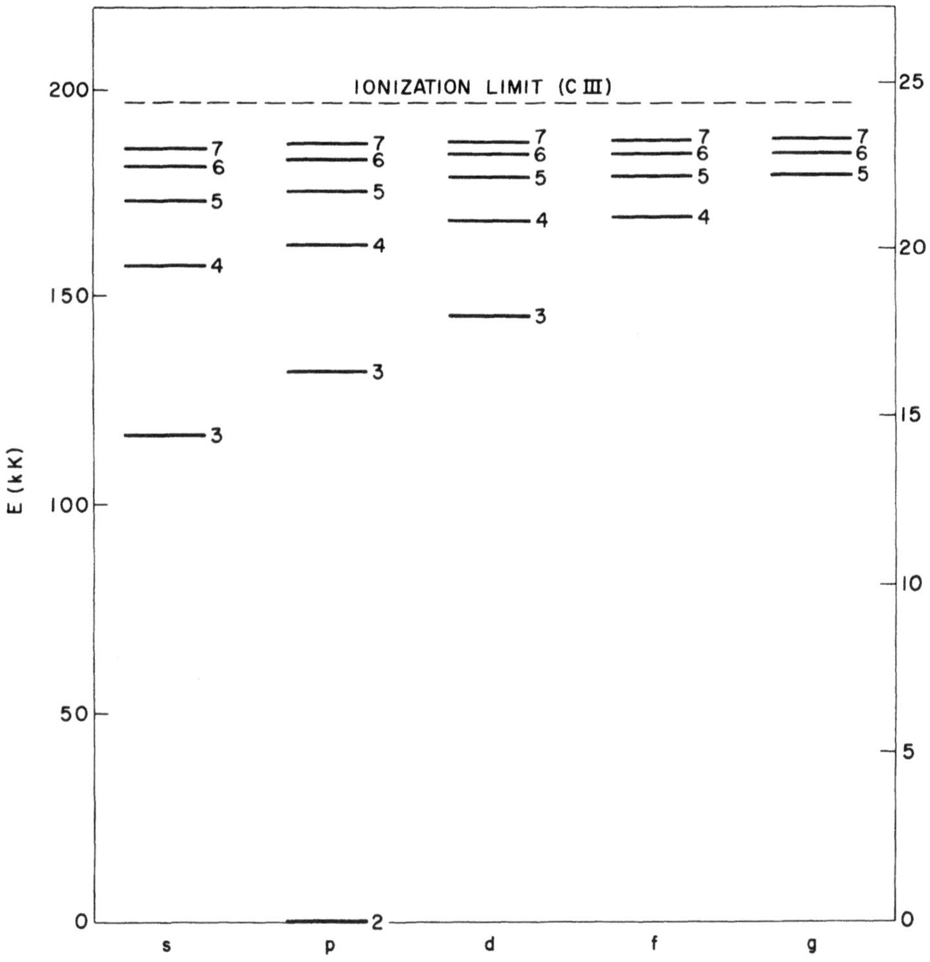

Fig. 1-5. Energy level diagram of some of the known levels of C II (slightly simplified). The labels s, p, d, f, g represent values of the orbital-angular-momentum quantum number $l = 0, 1, 2, 3, 4$, respectively, and the number beside each level is the principal quantum number n.

$$\sum_{j=m}^{Z} I_j \; .$$

For theoretical work, the zero of energy is usually taken to be the energy of the state in which the nucleus and all electrons are at rest infinitely far apart from each other; with respect to this zero, the total electronic energy of an ion in the ground state of the mth ionization stage is

$$E = -\sum_{j=m}^{Z} I_j \; . \tag{1.13}$$

TABLE 1-1. IONIZATION ENERGIES (eV)[a]

		I	II	III	IV	V	VI	VII	VIII	IX	X	XI	XII	XIII	XIV	XV	XVI	XVII	XVIII
1	H	13.599																	
2	He	24.588	54.418																
3	Li	5.392	75.641	122.455															
4	Be	9.323	18.211	153.896	217.720														
5	B	8.298	25.155	37.931	259.374	340.228													
6	C	11.260	24.384	47.888	64.494	392.091	490.00												
7	N	14.534	29.602	47.450	77.474	97.891	552.12	667.05											
8	O	13.618	35.118	54.936	77.414	113.900	138.12	739.30	871.42										
9	F	17.423	34.971	62.709	87.141	114.244	157.17	185.19	953.94	1103.13									
10	Ne	21.565	40.963	63.46	97.12	126.22	157.93	207.3	239.09	1195.9	1362.21								
11	Na	5.139	47.287	71.621	98.92	138.39	172.15	208.48	264.19	299.88	1465.14	1648.71							
12	Mg	7.646	15.035	80.144	109.266	141.27	186.51	224.95	265.96	328.24	367.54	1761.86	1962.68						
13	Al	5.986	18.829	28.448	119.994	153.72	190.48	241.44	284.60	330.11	399.37	442.08	2086.05	2304.16					
14	Si	8.152	16.346	33.493	45.142	166.769	205.06	246.53	303.18	351.11	401.38	476.08	523.52	2437.76	2673.20				
15	P	10.487	19.726	30.203	51.444	65.026	220.43	263.23	309.42	371.74	424.51	479.59	560.43	611.87	2817.04	3069.87			
16	S	10.360	23.33	34.83	47.31	72.68	88.054	280.94	328.24	379.11	447.10	504.79	564.67	651.65	707.16	3223.95	3494.22		
17	Cl	12.968	23.814	39.61	53.47	67.8	97.03	114.20	348.29	400.06	455.63	529.28	591.99	656.71	749.76	809.41	3658.55	3946.33	
18	Ar	15.760	27.630	40.74	59.81	75.02	91.01	124.32	143.46	422.45	478.69	538.96	618.26	686.11	755.75	854.78	918	4120.92	4426.26
19	K	4.341	31.626	45.73	60.91	82.66	100.0	117.56	154.71	175.82	503.45	564.15	629.11	714.04	787.16	861.80	968	1034.87	4611.11
20	Ca	6.113	11.872	50.914	67.10	84.41	108.78	127.7	147.24	188.35	211.28	591.27	656.41	726.06	816.64	895.15	974	1087	1158.1
21	Sc	6.54	12.80	24.757	73.669	91.66	111.1	138.0	158.7	180.03	225.11	249.84	685.91	755.49	829.82	926.03	1009	1094	1213
22	Ti	6.82	13.58	27.492	43.267	99.30	119.5	140.8	170.4	192.1	215.92	265.0	291.50	787.84	863.1	941.9	1044	1131	1221
23	V	6.74	14.66	29.311	46.709	65.282	128.13	150.17	173.7	205.8	230.5	255.05	308.26	336.28	896.0	974	1060	1168	1260
24	Cr	6.766	16.50	30.96	49.1	69.46	90.64	160.18	184.7	209.3	244.4	270.8	298.0	354.8	384.17	1010.6	1097	1185	1299
25	Mn	7.437	15.640	33.668	51.2	72.4	95	119.27	194.5	221.8	248.3	286.0	314.4	343.6	403.0	435.6	1133.1	1244	1317
26	Fe	7.871	16.183	30.652	54.8	75.0	99.1	125	151.06	233.6	262.1	290.3	330.8	361.0	392.2	457.0	489.26	1262	1358
27	Co	7.86	17.083	33.50	51.3	79.5	102	129	157	186.14	276	305	336	379	411	444	512	546.8	1397
28	Ni	7.635	18.169	35.17	54.9	75.5	108	133	162	193	224.5	321.2	352	384	430	464	499	571	607.2
29	Cu	7.726	20.293	36.84	55.2	79.9	103	139	166	199	232	266	368.9	401	435	484	520	557	633
30	Zn	9.394	17.965	39.724	59.4	82.6	108	134	174	203	238	274	310.8	419.7	454	490	542	579	619
31	Ga	5.999	20.51	30.71	64														
32	Ge	7.900	15.935	34.22	45.71	93.5													
33	As	9.81	18.589	28.352	50.14	62.63	127.6												
34	Se	9.752	21.19	30.821	42.945	68.3	81.81	155.4											
35	Br	11.814	21.8	36	47.3	59.7	88.6	103.0	192.8										
36	Kr	14.000	24.360	36.95	52.5	64.7	78.5	111.0	125.94	230.9	[268]	[318]	[367]	[416]	[465]	[515]	[565]	(591)	(641)

TABLE 1-1 (cont)

		I	II	III	IV	V	VI	VII	VIII	IX	X	XI	XII	XIII	XIV	XV	XVI	XVII	XVIII
37	Rb	4.177	26.050	39.02	52.6	71.0	84.4	99.2	136	150	277.1								
38	Sr	5.695	11.030	42.884	56.28	71.6	90.8	106	122.3	162	177	324.1							
39	Y	6.38	12.24	20.525	60.60	75.0	89.26	116	129	146.2	191	205.82	374.0						
40	Zr	6.84	13.13	22.99	34.412	80.35							236.25						
41	Nb	6.88	14.32	25.05	38.3	50.55	102.06	118.7						268.7					
42	Mo	7.099	16.16	27.17	46.4	61.2	68	125.67	144.0	[167]	[189]	[217]	[239]	(278)	302.6	(543)	[571]	[636]	[702]
43	Tc	7.28	15.26	29.55															
44	Ru	7.37	16.76	28.47						178.41									
45	Rh	7.46	18.08	31.06							207.51								
46	Pd	8.34	19.43	32.93								238.57							
47	Ag	7.576	21.484	34.83									271.47						
48	Cd	8.994	16.908	37.48															
49	In	5.786	18.870	28.044	54														
50	Sn	7.344	14.632	30.503	40.735	72.28													
51	Sb	8.642	16.53	25.3	44.2	56	99												
52	Te	9.010	18.6	27.96	37.42	58.76	70.7	124											
53	I	10.451	19.131	33.			75.76	87.53											
54	Xe	12.130	21.21	32.1	[46.7]	[59.7]	[71.8]	92.1	105.9	[171]	[202]	[233]	[263]	[294]	[325]	[358]	[390]	[421]	[452]
55	Cs	3.894	23.14	33.38															
56	Ba	5.212	10.004	35.844	47.1														
57	La	5.577	11.06	19.177	49.95	61.6													
58	Ce	5.539	10.85	20.198	36.76	65.55	77.6												
59	Pr	5.473	10.55	21.624	38.98	57.53													
60	Nd	5.525	10.73	22.1	40.41	60.00													
61	Pm	5.582	10.90	22.3	41.1	61.69													
62	Sm	5.644	11.07	23.4	41.4	62.66													
63	Eu	5.670	11.24	24.92	42.6	63.23													
64	Gd	6.150	12.09	20.63	44.0	64.76													
65	Tb	5.864	11.52	21.91	39.79	66.46													
66	Dy	5.939	11.67	22.8	41.47	62.08													
67	Ho	6.022	11.80	22.84	42.5	63.93													
68	Er	6.108	11.93	22.74	42.65	65.10													
69	Tm	6.184	12.05	23.68	42.69	65.42													
70	Yb	6.254	12.176	25.05	43.74	65.58													
71	Lu	5.426	13.9	20.955	45.250	66.47													
72	Hf	6.65	14.9	23.3	33.37	68.38													

TABLE 1-1 (cont)

		I	II	III	IV	V	VI	VII	VIII	IX	X	XI	XII	XIII	XIV	XV	XVI	XVII	XVIII
73	Ta	7.89				48.27													
74	W	7.98					64.77	122.01	(140)	(158)	(177)	(207)	(231)	(256)	(289)	(324)	(360)	(387)	(419)
75	Re	7.88						82.71	144.4										
76	Os	8.7							102.0	168.8									
77	Ir	9.1								122.8	194.7								
78	Pt	9.0	18.563								145.0	220.4							
79	Au	9.226	20.5									168.2	248.0						
80	Hg	10.438	18.756	34.2									192.7	276.8					
81	Tl	6.108	20.428	29.83										218.4	306.9				
82	Pb	7.417	15.032	31.938	42.32	68.8									245.2	338.1			
83	Bi	7.289	16.69	25.56	45.3	56.0	88.3									272.7	370.2		
84	Po	8.42																	
85	At																		
86	Rn	10.749																	
87	Fr																		
88	Ra	5.279	10.147																
89	Ac	5.17	12.1																
90	Th	6.08	11.5	20.0	28.8														
91	Pa	5.89																	
92	U	6.194																	
93	Np	6.266																	
94	Pu	6.06																	
95	Am	5.99																	
96	Cm	6.02																	
97	Bk	6.23																	
98	Cf	6.30																	
99	Es	6.42																	
100	Fm	6.50																	
101	Md	6.58																	
102	No	6.65																	

TABLE 1-1 (cont)

		XIX	XX	XXI	XXII	XXIII	XXIV	XXV	XXVI	XXVII	XXVIII	XXIX	XXX	XXXI	XXXII	XXXIII	XXXIV	XXXV	XXXVI
19	K	4934.10																	
20	Ca	5129.22	5469.92																
21	Sc	1288.4	5675	6034															
22	Ti	1346	1425.9	6249.4	6626														
23	V	1355	1486	1570.3	6851	7246													
24	Cr	1396	1496	1634	1721.4	7482	7894.8												
25	Mn	1437	1539	1644	1788	1880	8141.4	8671.9											
26	Fe	1456	1582	1689	1799	1950	2045	8828	9277.7										
27	Co	1500	1603	1735	1846	1962	2119	2218	9544										
28	Ni	1540	1648	1756	1894	2011	2131	2295	2399										
29	Cu	671	1690	1804	1916	2060	2182	2308	2478										
30	Zn	698	737	1846	1966	2084	2234	2361	2493										
36	Kr	(786)	(833)	(884)	(936)	(997)	(1050)	(1152)	(1205)	(2928)	(3070)	(3227)	(3381)	(3594)	(3760)	(3966)	(4112)	(17310)	(17940)
42	Mo	|767|	|833|	|902|	|968|	(1020)	(1083)	(1264)	(1323)	(1387)	(1449)	(1535)	(1601)	(1726)	1789	(4257)	(4430)	(4617)	(4798)
54	Xe	(549)	(583)	(618)	(651)	(701)	(737)	(819)	(857)	(1495)	|1491|	|1587|	|1684|	|1781|	|1877|	|1987|	|2085|	(2211)	(2302)
74	W	(463)	(506)	(549)	(584)	(635)	(685)	(733)	(783)	(832)	(881)	(1129)	(1177)	(1228)	(1280)	(1335)	(1388)	(1450)	(1510)

		XXXVII	XXXVIII	XXXIX	XL	XLI	XLII	XLIII	XLIV	XLV	XLVI	XLVII	XLVIII	XLIX	L	LI	LII	LIII	LIV
42	Mo	(5099)	(5296)	(5546)	(5716)	(23840)	(24570)												
54	Xe	(2554)	(2639)	(2728)	(2812)	(2979)	(3071)	(3245)	(3334)	(7663)	(7893)	(8143)	(8381)	(8987)	(9257)	(9582)	(9813)	(40245)	(41211)
74	W	(1566)	(1620)	(1827)	(1881)	(1938)	(1993)	(2148)	(2209)	(2354)	(2415)	(4055)	(4184)	(4314)	(4447)	(4584)	(4721)	(4896)	(5060)

[a]Values in \|\| or () are *ab initio* theoretical values, the former from T. A. Carlson, et al, Atomic Data **2**, 63 (1970) accurate to about 5 percent, and the latter from R. D. Cowan, Los Alamos Scientific Laboratory Report LA-6679-MS (January 1977) accurate to about one percent. Other values are based on experimental spectrum analyses, though in many cases (especially those values given only to the nearest eV) interpolation or extrapolation along an isoelectronic sequence is involved.

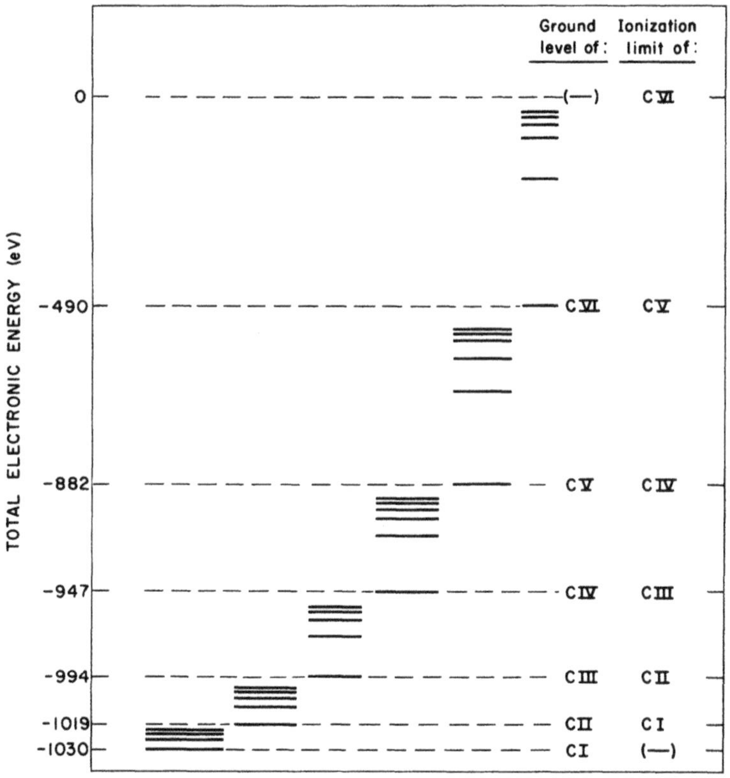

Fig. 1-6. Energy level diagram for all ionization stages of carbon, with all energies based on an energy zero for the completely stripped carbon atom. (Highly schematic, and not to scale.)

An illustrative example is given in Fig. 1-6. The energy in any excited state of the mth ionization stage is of course found by adding the excitation energy of that state to the ground-state energy (1.13).

1-8. THE RITZ COMBINATION PRINCIPLE; GROTRIAN DIAGRAMS

If an atom exists in an excited state of energy E_2 it may spontaneously decay to some other state of lower energy E_1, the energy difference appearing as a photon[36] of energy $E_2 - E_1$. This photon corresponds to emitted radiation of frequency ν, vacuum wavelength λ, and wavenumber σ given by

$$E_2 - E_1 = h\nu = \frac{hc}{\lambda} = hc\sigma \;, \tag{1.14}$$

[36]With very low probability, the atom may decay with emission of two or more photons of total energy $E_2 - E_1$; see M. Goeppert Mayer, Ann. Physik **9**, 273 (1931); H. R. Reiss, Phys. Rev. Lett. **25**, 1149 (1970). Such multi-photon decay modes will not be discussed in this book.

or if the energy levels are given in the pseudo-energy units E/hc

$$\frac{E_2}{hc} - \frac{E_1}{hc} = \frac{1}{\lambda} = \sigma \ . \tag{1.15}$$

Conversely, if the atom exists in the state 1 and lies in a radiation field that includes radiation of wavenumber σ given by (1.15), it may be excited to the state 2 by absorption of one photon of energy $E_2 - E_1$.

The Ritz combination principle[37] states that any two energy levels of the atom can combine in the above manner to give rise to a spectrum line. In principle this is true, but in practice most pairs of levels give rise to negligibly weak spectrum lines, as the result of various *selection rules* that will be discussed in Sec. 2-14 and in Chapters 14-15 on multipole spectra.

In general, the different observed lines of a spectrum cover a very wide range of intensities. It has been estimated that in favorable cases (low-background sources and detectors) an intensity range as great as 10^8:1 can be observed.[38] This intensity range may be due either to inherent differences in line strength, or to excitation conditions (greatly different numbers of atoms in the initial states of different lines)—usually to both.

Frequently, the strongest and most important spectrum lines are shown in an energy-level diagram in the form of vertical or diagonal lines connecting the appropriate pairs of energy levels, with the corresponding wavelength (in Å) written beside each line. The diagram is then known as a *Grotrian diagram*;[39] an example, from Moore and Merrill,[40] is shown in Fig. 1-7.

Lines that involve the ground level are called *resonance lines*. That resonance line (of appreciable strength) which involves the lowest-lying upper level is called the principal resonance line, or simply *the* resonance line; in Fig. 1-7, this is the line with wavelength 1816 Å.

*1-9. THE PROFILES OF SPECTRUM LINES

For most purposes we shall consider the energy of each level of an atom to be perfectly definite in value; i.e., to have zero energy width. The wavenumber or wavelength given by (1.15) is then also definite, and each spectrum line is monochromatic, with zero wavenumber width.

[37]W. Ritz, Astrophys. J. **28**, 237 (1908).

[38]D. R. Wood, K. L. Andrew, A. Giacchetti, and R. D. Cowan, J. Opt. Soc. Am. **58**, 830 (1968).

[39]W. Grotrian, *Graphische Darstellung der Spektren von Atomen und Ionen mit Ein, Zwei, und Drei Valenzelektronen* (Julius Springer, Berlin, 1928; reprinted by J. W. Edwards, Ann Arbor, Mich., 1946).

[40]C. E. Moore and P. W. Merrill, *Partial Grotrian Diagrams of Astrophysical Interest*, NSRDS-NBS 23 (U. S. Govt. Printing Off., Washington, D. C., 1968). See also S. Bashkin and J. O. Stoner, Jr., *Atomic Energy Level and Grotrian Diagrams* (North-Holland Publ. Co., Amsterdam), Vol. I, H I - P XV (1975), Vol. I addenda (1978). Vol. II, S I - Ti XXII (1978).

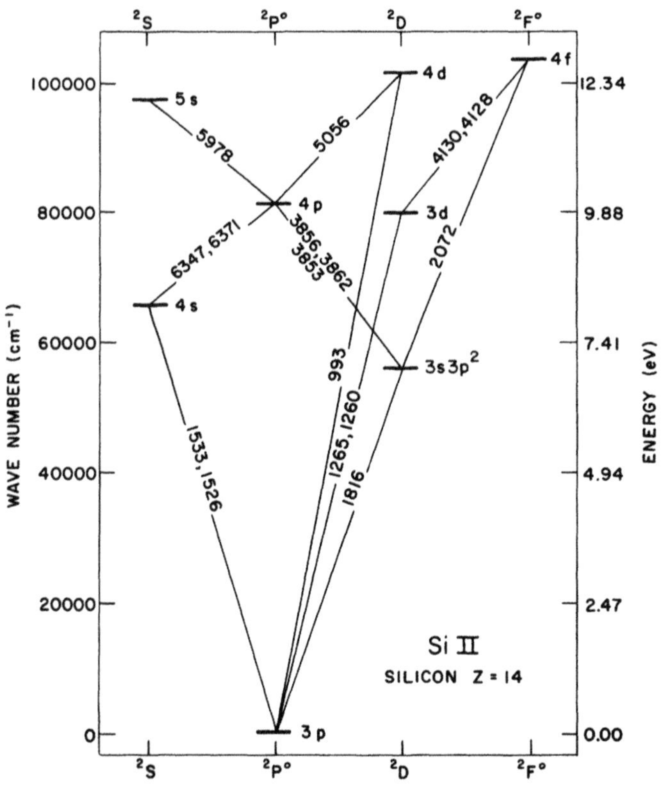

Fig. 1-7. Grotrian diagram showing transitions among the low-lying levels of Si II; wavelengths are in Å. (From Moore and Merrill, ref. 40.)

In reality, spectrum lines are never monochromatic, but show intensities distributed over a finite wavenumber range; this may be described by an intensity distribution function $I(\sigma)$, which we assume to be normalized:

$$\int I(\sigma)d\sigma = 1 \ . \tag{1.16}$$

This intensity distribution may arise from a number of different causes,[41] which we shall describe only briefly.

Natural width. The fact that an atom in an excited state may spontaneously decay to a lower state implies that the former has a finite natural lifetime. The Heisenberg uncertainty principle then implies an uncertainty $\Delta E \cong \hbar/\Delta t$ in the energy of the upper state. In neutral

[41]See, for example, H. R. Griem, *Plasma Spectroscopy* (McGraw-Hill, New York, 1964) or C. R. Cowley, *The Theory of Stellar Spectra* (Gordon and Breach, New York, 1970).

atoms, Δt is no smaller than about 10^{-8} sec, so that in wavenumber units the energy uncertainty is

$$\frac{\Delta E}{hc} \cong \frac{1}{2\pi c\ \Delta t} \leqq \frac{1}{6\cdot 3\cdot 10^{10}\cdot 10^{-8}} = 5\cdot 10^{-4}\ \mathrm{cm}^{-1}\ . \tag{1.17}$$

Corresponding to this uncertainty of the energy of any excited level, there follows from (1.15) an uncertainty in the wavenumber of each spectrum line. It can be shown quantum mechanically[42] that the resulting intensity distribution has the resonance form

$$I(\sigma) = \frac{\Gamma/\pi}{(\sigma - \sigma_0)^2 + \Gamma^2}\ , \tag{1.18}$$

where σ_0 is the wavenumber at the center of the spectrum line and Γ is the half-width of the line at half the maximum amplitude. This natural line width of $\Delta\sigma = 2\Gamma \leqq 5\cdot 10^{-4}\ \mathrm{cm}^{-1}$ corresponds at $\lambda_0 = 5000$ Å ($\sigma_0 = 20000\ \mathrm{cm}^{-1}$) to a wavelength width of

$$\Delta\lambda = \lambda_0 \frac{\Delta\sigma}{\sigma_0} \leqq \sim 10^{-4}\text{Å}\ , \tag{1.19}$$

which is usually completely negligible by comparison with other line-broadening effects discussed below.

A semiclassical derivation of (1.18) can be made by considering the atom to radiate at the fixed wavenumber σ_0 with an amplitude that decreases exponentially with time as the atom loses energy:

$$e^{-t/(\Delta t)} \cos(2\pi c\sigma_0 t) \tag{1.20}$$

[so that the energy decreases as $e^{-2t/(\Delta t)}$]. A Fourier analysis of (1.20) leads to exactly the result (1.18) with $\Gamma = (2\pi c\ \Delta t)^{-1}$.

Resonance and collision broadening. In any actual light source, atoms are not isolated, but are coupled together by the radiation field: a photon emitted by one atom may be absorbed by some other atom lying in the appropriate lower level, or it may induce the emission of a like photon by another atom in the appropriate upper level. This induced absorption and emission decreases the lifetimes of the atoms below the natural lifetimes, resulting in a corresponding resonance broadening of the spectrum lines.

Collisional excitation and de-excitation by electrons or by other atoms in the light source may also greatly reduce the lifetime of each state, leading to still further broadening of the energy levels and spectrum lines. From a classical point of view, the collisions may be considered as interrupting the radiation wave train (1.20), thereby resulting in a broadening of the Fourier spectrum.

Both resonance and collision broadening leave the resonance line profile (1.18) essentially unchanged in form, merely increasing the half-width Γ by one or two (or even more) orders of magnitude over the natural half-width.

[42]V. Weisskopf and E. Wigner, Z. Physik **63**, 54 (1930); F. Hoyt, Phys. Rev. **36**, 860 (1930).

Doppler broadening. If a radiating atom has a translational motion with a component parallel to the observer's line of sight, the apparent wavenumber of the observed radiation is Doppler-shifted from the actual wavenumber σ_o in the atom's frame of reference by an amount

$$\sigma - \sigma_o = \sigma_o \frac{v_\parallel}{c} \qquad (v_\parallel \ll c) \ . \tag{1.21}$$

If there exist atoms having a continuous distribution of translational velocities, the result is a continuous distribution of observed wavenumbers σ and a corresponding apparent broadening of the spectrum line. In the absence of other line-width effects, a Maxwellian velocity distribution at absolute temperature T would produce the Doppler line shape

$$I(\sigma) = \left(\frac{\ln 2}{\pi\Gamma^2}\right)^{1/2} e^{-(\ln 2)(\sigma - \sigma_o)^2/\Gamma^2} \ ; \tag{1.22}$$

the half-width at half maximum is given by

$$\Gamma = [2kT(\ln 2)/Mc^2]^{1/2}\sigma_o = 3.581 \cdot 10^{-7}(T/A)^{1/2}\sigma_o \ , \tag{1.23}$$

where M is the mass of the atom and A is its atomic weight. For the international-length-standard Kr line at 6057.8 Å in a source with T = 4000°K, this gives a full line width of

$$\Delta\sigma = 2\Gamma = 7.162 \cdot 10^{-7}(4000/83.8)^{1/2}\sigma_o = 4.948 \cdot 10^{-6}\sigma_o = 0.0817 \text{ cm}^{-1} \tag{1.24}$$

or

$$\Delta\lambda = 4.948 \cdot 10^{-6}\lambda_o = 0.0300 \text{ Å} \ , \tag{1.25}$$

which is much greater than the natural width (1.19). Doppler widths are even greater for lighter elements, but can be reduced by using atomic-beam sources.[43]

Voigt profiles. The relative shapes of the resonance and Doppler distributions are compared in Fig. 1-8. In general, both forms of broadening are present, and the line shape is then a convolution of the two distribution functions; this leads to a shape known as a Voigt profile.[44] The detailed form of this profile depends on the relative values of the resonance and Doppler half-widths; however, it is clear from the figure that the Doppler form tends to be dominant near the center of the line, whereas the resonance form is controlling in the "wings" of the line.

External fields. The presence of an external electric or magnetic field may result in either a slight decrease or a slight increase in the energy of an atom. In an energy-level diagram, each possible level is in general split into several possible *sublevels*, the magnitude of the

[43]R. W. Stanley, J. Opt. Soc. Am. **56**, 350 (1966).

[44]W. Voigt, Sitzber. Akad. Wiss. München (1912), p. 603; B. H. Armstrong, J. Quant. Spectrosc. Radiat. Transfer **7**, 61 (1967); J. F. Kielkopf, J. Opt. Soc. Am. **63**, 987 (1973).

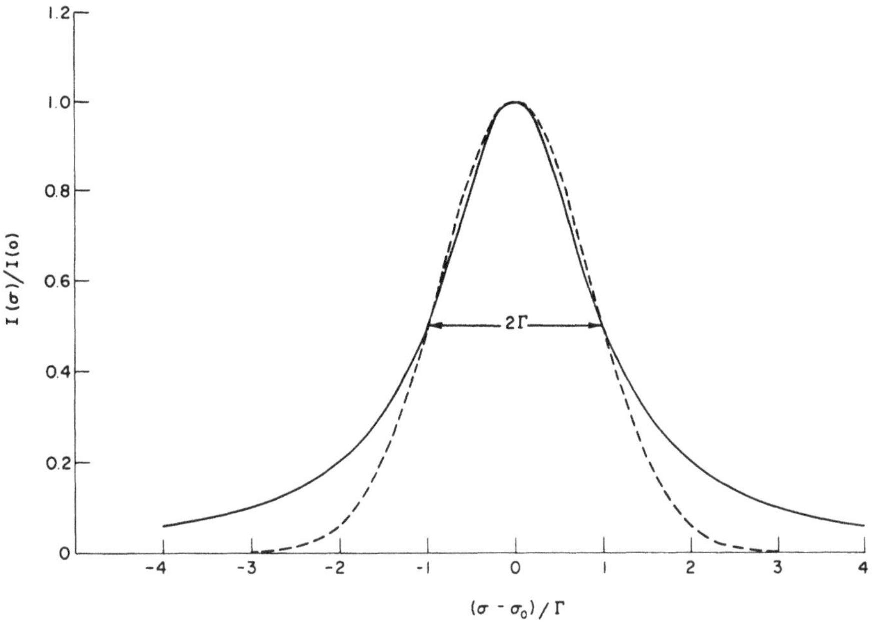

Fig. 1-8. Comparison of resonance (solid) and Doppler (dashed) line profiles.

splitting depending on the strength of the field. (If no energy degeneracy remains, a sublevel is also called a *state*). Correspondingly, each spectrum line is in general split into several ***components***; this is known as the Stark effect[45] in the case of an electric field, and as the Zeeman effect[46] for a magnetic field.

Stray electric and magnetic fields in a light source result in small and variable line splittings, which on a time average result in a broadening of the spectrum lines. Stark broadening is usually asymmetrical, and may thus involve a wavenumber shift of the line center. Stark-effect broadening is particularly noticeable in the high-current sparks required to produce high ionization stages. Zeeman-effect broadening is usually not important in laboratory sources, but is significant in spectra of certain stars; in some stars, the magnetic field strengths are sufficiently uniform to produce resolved Zeeman patterns rather than Zeeman broadening.[47]

Hyperfine structure and isotope shift. In addition to the splittings produced by external fields, each level may be split into a number of sublevels as a result of hyperfine-structure splitting (caused primarily by magnetic interactions of the electrons with the nuclear magnetic moment). In heavy elements, these splittings are commonly large enough to produce distinctly separate (non-overlapping) energy levels; correspondingly, each spectrum line is split into a number of separate hyperfine-structure (hfs) components. In light

[45]J. Stark, Sitzber. preuss. Akad. Wiss. Berlin (1913), p. 932.

[46]P. Zeeman, Phil. Mag. **43**, 226 (1897); **44**, 55, 255 (1897).

[47]G. W. Preston, Astrophys. J. **160**, 1059 (1970); Publ. Astron. Soc. Pacific **83**, 571 (1971).

elements, however, the level splitting is very small and the line components overlap to a large extent, so that the effect may be only a (usually asymmetrical) broadening of the spectrum line.

The energy levels of an atom depend on the nucleus not only through its spin, but also through its mass (reduced-mass effects) and its shape (nuclear quadrupole-moment effects). These effects result in slightly different wavenumbers for a given spectrum line in different isotopes of the same element. In light sources containing more than one isotope, this isotope shift may result in distinctly different lines or only in the apparent broadening of a single line. Nuclear and external-field effects are discussed in greater detail in Chapter 17.

Self absorption. A photon emitted by an atom at one point in a light source may be absorbed by a different atom before it has a chance to escape from the source; this photon may be lost as a contribution to the original spectrum line either as a result of radiative decay to a different lower level or through collisional de-excitation of the absorbing atom. In the same way that the probability of emission was greatest at the center of the line, the probability of absorption is also greatest at the center. Thus the result of this re-absorption within the light source (self absorption) is to reduce the intensity of the spectrum line proportionately more in the center of the line than elsewhere, altering the line shape and making the line appear broader.

If the excitation conditions within the source are uniform throughout, then under extreme conditions (an optically thick source provided by large physical dimensions, high atom density, and/or strong spectrum lines involving highly populated levels), the spectrum line acquires a flat top at an intensity level equal to that of a black body radiating thermally at the temperature of the source; indeed the atomic source *is* a black body over the limited wavelength range of the flat top.[48]

If the self absorption is strong and the outer portion of the source is more weakly excited than the inner portion, then an intensity minimum develops in the center of the spectrum line, and the line is said to be self reversed.[48,49] It is sometimes difficult to distinguish a self-reversed line from two closely spaced lines of (more or less) equal intensity.

When conditions are not so extreme as to produce self-reversed or obviously flat-topped lines, it is not easy to ascertain the degree of self absorption existing in the light source. The absorption may nevertheless be sufficiently large to greatly alter the observed relative intensities of spectrum lines with respect to what would be seen with an optically thin source.

The effects of (symmetrical) self absorption on observed line shapes are illustrated in Fig. 1-9. Observed self reversals are frequently somewhat asymmetrical because of different values of σ_0 in the emitting and absorbing portions of the light source.

Instrumental effects. Finite slit widths, diffraction effects, optical aberrations, and finite resolving power of the detector (photographic grain, finite exit-slit widths for electronic detectors, etc.) may all serve to alter the apparent shapes and widths of spectrum lines. A wide slit may even cause a self-reversed line to appear as *three* overlapping lines.[48]

[48]R. D. Cowan and G. H. Dieke, Rev. Mod. Phys. **20**, 418 (1948).

[49]In wavelength tables, lines that tend to show strong self-reversal are usually labeled R, and those tending to show less extreme reversal are labeled r.

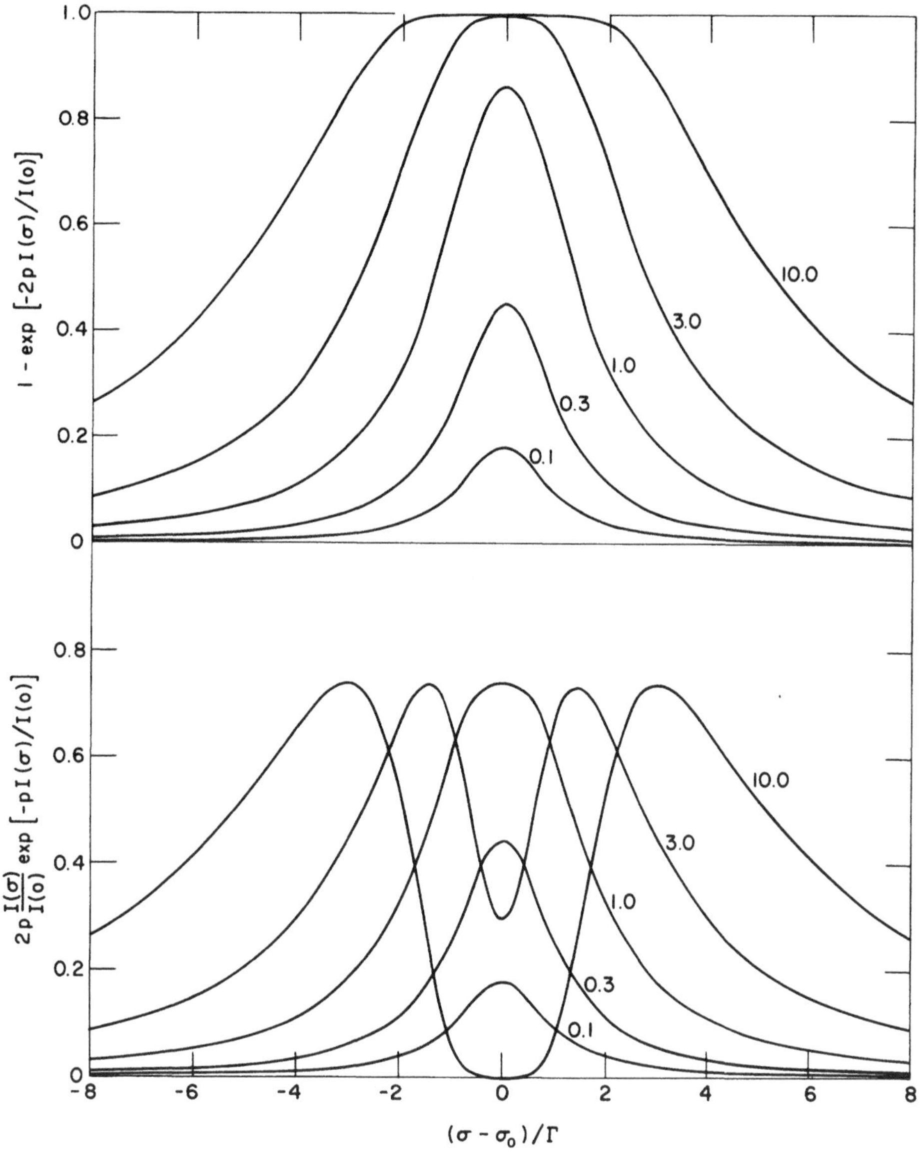

Fig. 1-9. Theoretical shapes of self-absorbed spectrum lines for a resonance profile I(σ). Top: uniformly excited source; bottom: source with all absorption between the observer and the emitting atoms, showing self reversal for $p > 1$. The value of p indicated on each curve is proportional to the intrinsic strength of the unabsorbed line: $p = (\pi e^2 f/mc^2)\, I(\sigma_0) N_a$, where N_a is the number of atoms in the lower level of the line, per unit area integrated along the line of sight from the center of the source, and f is the oscillator strength of the line (Sec. 14-6).

*1-10. WAVELENGTH RANGE AND ACCURACY

From Fig. 1-5 it may be inferred that energy-level differences in any atom range from zero up to the ionization energy (for excitation of the outermost electron). From the combination principle (1.14) and the Table 1-1 of ionization energies, this implies photon energies between 0 and 4 to 25 eV for neutral atoms, or—from (1.2)—wavelengths anywhere from infinity down to about 500 Å. For highly ionized atoms, the ionization energies are much greater, and wavelengths range down to 1 Å and less.

In practice, experimental work in atomic emission spectroscopy is limited at the long-wavelength end by the limits of sensitivity of the photographic plate (13000 Å = 1.3 μ) and of cooled PbS detectors (4 μ)[50] and InSb detectors (5.5 μ); beyond 2.5 μ there may also be complications due to absorption by atmospheric water vapor and carbon dioxide unless work is done in a dry nitrogen atmosphere or a vacuum.

Below about 1850 Å all work must be done in a vacuum because of very strong atmospheric absorption, principally by molecular oxygen. For spectroscopic observation of astronomical sources, the cutoff occurs at 2900 Å because of absorption by ozone in the upper atmosphere; all work below this wavelength is done with rocket- or satellite-borne instruments.

Diffraction-grating instruments are usable at all wavelengths down to about 1 Å, though below about 300 Å the gratings must be used at grazing incidence because of the low reflectivity of all materials at soft-x-ray wavelengths.[51] Below 20 Å, crystal spectrometers with photoelectric detection are also used.[52]

Above about 2000 Å, higher resolution than that provided by grating spectrographs can be obtained by using an evacuated Fabry-Perot interferometer in combination with a spectrograph,[53] or by doing Fourier-transform spectroscopy with a Michelson interferometer.[54]

The wavelengths of spectrum lines can in general be measured with very high accuracy. Errors are nevertheless nonnegligible, partly because of the finite line widths and wavelength shifts discussed in the previous section. Typical accuracies obtainable in the study of atomic emission spectra are indicated in Table 1-2.

It is evident that the percent accuracy deteriorates greatly below 2000 Å. For this reason, higher accuracy wavelengths of some lines (for use as wavelength standards for

[50]C. J. Humphreys, J. Opt. Soc. Am. **50**, 1171 (1960); C. J. Humphreys and E. Paul, Jr., Appl. Opt. **2**, 691 (1963).

[51]R. J. Speer, N. J. Peacock, W. A. Waller, and P. J. H. Osborne, J. Phys. E. **3**, 143 (1970).

[52]J. F. Meekins, R. W. Kreplin, T. A. Chubb, and H. Friedman, Science **162**, 891 (1968).

[53]S. Tolansky, *An Introduction to Interferometry* (John Wiley, New York, 1973), 2nd ed.; R. W. Stanley and W. F. Meggers, J. Res. Natl. Bur. Stand. **58**, 41 (1957).

[54]P. Connes, in J. Home Dickson, ed., *Optical Instruments and Techniques 1969* (Oriel Press, Newcastle upon Tyne, 1970); R. J. Bell, *Introductory Fourier Transform Spectroscopy* (Academic Press, New York 1972).

TABLE 1-2. APPROXIMATE WAVELENGTH AND WAVENUMBER ACCURACIES IN VARIOUS SPECTRAL RANGES

		Grating spectrograph		Fourier spectrometer or Fabry-Perot etalon	
		Δλ (Å)	Δσ (cm^{-1})	Δλ (Å)	Δσ (cm^{-1})
PbS region	25000-40000 Å	1.0	0.1[a]	0.02	0.002[h,c]
	13000-25000	0.2	0.05[b]	0.01	0.002[h,c]
Photographic IR	7800-13000	0.003	0.003[c]	0.002	0.002[i]
Visible	3800-7800	0.001	0.003[c]	0.0005	0.002[i]
Ultraviolet	2000-3800	0.0002	0.003[c]	0.0001	0.002[i]
Near vacuum UV	1000-2000	0.002[d]	0.1	---	---
Far vacuum UV	100-1000	0.004[e]	4	---	---
Extreme VUV	10-100	0.003[f]	300	---	---
X-ray	1-10	0.001[g]	10000	---	---

[a]C. J. Humphreys et al., J. Opt. Soc. Am. **57**, 855 (1967), **61**, 110 (1971).
[b]E. B. M. Steers, Spectrochim. Acta **23B**, 135 (1967).
[c]J. Blaise and L. J. Radziemski, Jr., J. Opt. Soc. Am. **66**, 644 (1976).
[d]V. Kaufman, L. J. Radziemski, Jr., and K. L. Andrew, J. Opt. Soc. Am. **56**, 911 (1966).
[e]L. Å. Svensson and J. O. Ekberg, Ark. Fys. **37**, 65 (1968).
[f]S. Hoory et al., J. Opt. Soc. Am. **60**, 1449 (1970).
[g]R. J. Speer et al., J. Phys. E **3**, 143 (1970).
[h]J. Connes, P. Connes et al., Nouv. Rev. d'Opt. Appl. **1**, 3 (1970).
[i]A. Giacchetti, R. W. Stanley, and R. Zalubas, J. Opt. Soc. Am. **60**, 474 (1970).

other measurments) are derived through use of the Ritz combination principle: Once an accurate energy-level scheme has been derived from longer-wavelength lines (by methods discussed in the next section), accurate wavenumbers of short-wavelength lines follow directly from (1.15). Wavelengths obtained in this way are known as Ritz standards.[55]

1-11. EMPIRICAL SPECTRUM ANALYSIS

The classical problem of experimental spectroscopy is to infer the possible energy levels of an atom or ion from the observed wavenumbers of its spectrum lines, thereby obtaining a Grotrian diagram similar to (but more extensive and detailed than) Fig. 1-7. Basically, the procedure is to calculate the wavenumber difference

$$\Delta_{ij} = |\sigma_i - \sigma_j|$$

for each pair (i,j) of spectrum lines, and to look for values of Δ that (within some small margin of error) appear several times; each such value of Δ is assumed to represent the

[55]F. Paschen, Ann. Physik **35**, 860 (1911), **42**, 840 (1913); B. Edlén, Rep. Prog. Phys. **26**, 181 (1963).

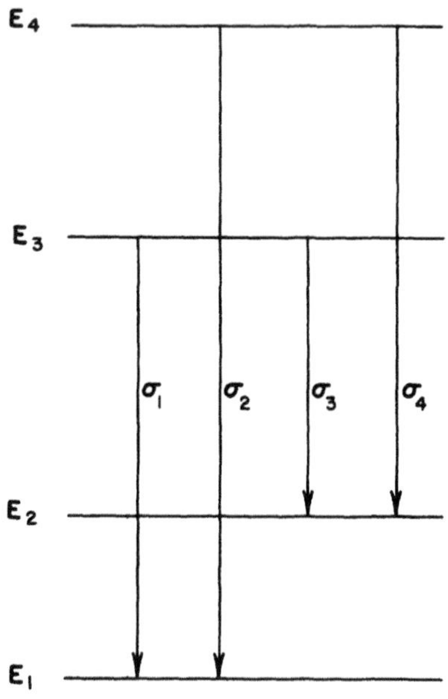

Fig. 1-10. Grotrian diagram showing four spectrum lines produced by transitions among four energy levels of a hypothetical atom.

energy difference between a pair of energy levels. In Fig. 1-10, for example, the wavenumber differences

$$\sigma_2 - \sigma_1 = \sigma_4 - \sigma_3$$

correspond to the energy difference between the levels E_4 and E_3, and the differences

$$\sigma_1 - \sigma_3 = \sigma_2 - \sigma_4$$

correspond to the energy difference of the levels E_2 and E_1.

Only the magnitudes of energy differences are obtained in this way—there is no indication as to which level is the higher and which is the lower of a pair. However, once a fairly extensive array of levels has been derived, clues as to which extreme of the array is the bottom are provided by the fact that the strongest lines and those tending to show self reversal in dense sources usually involve the ground level or other low-lying levels (because these are the most highly populated levels),[56] by the fact that the closely spaced levels tend to be the high-energy levels (cf. Fig. 1-5), and by various theoretical generalizations regarding energy-level structures, which we shall discuss at various points in this book.

In comparatively simple atoms, there is in principle no real problem in making the above type of purely empirical spectrum analysis. In practice, there can be serious difficulties in carrying the procedure beyond the determination of some of the lower-lying levels, especially in the more complex atoms: (1) In spite of the very low percentage error in measured wavenumbers, the finite errors indicated in Table 1-2 are frequently great enough

[56]For neutral atoms, at least, similar information can be obtained by observing a continuous background spectrum through a cool vapor of these atoms, to see which spectrum lines appear in absorption.

to fortuitously produce spuriously equal values of Δ_{ij} (within the experimental error); (2) Critically important spectrum lines may not have been observed because they are weaker than the ever-present continuous background, because they are masked by stronger lines, or because they lie outside the accessible wavelength range; (3) The task may be confused by the presence of spurious lines arising from impurity elements or (particularly when investigating an ion) other ionization stages.

The problem is rather like that of trying to put together the pieces of a complicated jigsaw puzzle when the pieces never fit exactly, some pieces fit spuriously, some critical pieces are missing, and there are pieces present that belong to one or more entirely different puzzles. The task may be aided somewhat by the knowledge of not only the *shape* of each piece (the wavenumber of the line), but also of an obscure section of picture on some or all of the pieces in the form of self-reversal, Zeeman-effect, or isotope-shift information; however, these small portions of the overall picture are helpful in putting the puzzle together only if the key to the total picture is available in the form of a knowledge of the theory of atomic structure.

1-12. THE ROLE OF THEORY

The Grotrian-diagram puzzle, put together either entirely empirically on the basis of the combination principle or with the addition of some theoretical help, is of direct interest for a considerable variety of purposes. For example, low-lying levels and associated transitions are of interest for laser materials; a knowledge of the energy levels associated with each spectrum line can help in the choice of lines that are relatively free from self-absorption, for use in quantitative spectrochemical analysis; if the energy-level diagram is sufficiently complete, ionization potentials and thermodynamic functions can be computed; if intrinsic line strengths are determined either experimentally or theoretically, these can be combined with the level energies to estimate temperatures and compositions of stellar atmospheres.

Although energy levels and spectra are thus of direct practical utility, knowledge of the level structure of an atom is not a final goal, by any means: energy levels and line strengths provide the main clues to a fundamental understanding of the electronic structure of the atom. These clues, however, can be interpreted meaningfully only in terms of a theoretical model. Classical attempts to find a suitable model were of course failures, and an adequate theory of multi-electron atoms had to await the development of quantum mechanics in the middle and late nineteen-twenties. Indeed, development of the understanding of atomic (and molecular) structure and development of the principles of quantum theory went hand in hand—each rapidly providing greater and greater insight into the other.

By the early thirties, the principles of atomic structure were well understood, and a quantum mechanical interpretation could be given to each observed energy level; that is, the qualitative nature of the corresponding wavefunction could be specified, and somewhat rough quantitative calculations of level energies and line strengths could be made for comparatively simple cases. The state of the art by 1935 was well summarized in Condon and Shortley's classic work.[57] This book still provides a basically correct and rather complete account of the principles of atomic structure, but the mathematical apparatus available at

[57]E. U. Condon and G. H. Shortley, *The Theory of Atomic Spectra* (University Press, Cambridge, 1935).

that time was not adequate to handle very complex atoms, such as the transition and rare-earth elements.

A very powerful new mathematical technique—the algebra of irreducible tensor operators—was developed by Racah[58] in the early forties. However, by this time the forefront of physics had passed from the electronic structure of atoms to the structure of nuclei; the exigencies of wartime had also siphoned much effort away from basic physical research. As a consequence, Racah's work lay almost unnoticed for more than a decade.

By about 1960, several developments had combined to produce a renaissance of activity in the fields of atomic spectra and structure. The use of spectroscopy as a diagnostic tool in plasma-physics research, and the observation of solar spectra from rockets fired above the earth's atmosphere produced renewed interest in the vacuum ultraviolet spectra of highly ionized atoms, a field that had been largely untouched since the pioneering work of Edlén and others in the mid-thirties.[59] There were increasing demands for energy-level and line-strength data for applications to both astrophysical and laser problems; the development of nanosecond and picosecond electronic-detection circuits[60] and of beam-foil spectroscopy[61] made possible the determination of line strengths through the direct measurement of the lifetimes of excited states. Efforts were being renewed to unravel the extremely complex spectra of the lanthanide and actinide elements, a field in which only meager progress had been made before the war; higher quality diffraction gratings and improved spectrograph designs made possible the higher accuracy measurements required for this work. This improved equipment, and the availability of PbS detectors for infrared work, provided the incentive to revise and extend earlier work on simpler atoms. And last but not least, the development of large digital computers, and the techniques of Racah algebra, made feasible theoretical calculations that were of great practical utility to the empirical spectroscopist.

The above brief historical outline is illustrated by Table 1-3, which summarizes the progress made in the investigation and analysis of atomic spectra. Table 1-4 gives a detailed breakdown of the state of analysis of the spectra of all atoms and ions, as of 1969 and 1979, graded from A (practically complete) down to F (fragmentary).[62] It is evident that a great deal of work remains to be done before the analyses of all spectra of importance will have been carried to a satisfactory stage. The theory of atomic structure can be used to aid this work in three different ways:

(1) Without making any detailed calculations, the theory can be used in a largely qualitative fashion to predict the numbers and types of energy levels to be expected for any given atom, and also (to a limited extent) to predict the approximate relative positions of

[58] G. Racah, "Theory of Atomic Spectra," II. Phys. Rev. **62**, 438 (1942); III. Phys. Rev. **63**, 367 (1943); IV. Phys. Rev. **76**, 1352 (1949).

[59] B. Edlén, Z. Physik **100**, 621, 726 (1936); **101**, 206 (1936); **103**, 536 (1936).

[60] D. R. Shoffstall and D. G. Ellis, J. Opt. Soc. Am. **60**, 894 (1970).

[61] L. Kay, Phys. Lett. **5**, 36 (1963); Stanley Bashkin ed., *Beam-Foil Spectroscopy* (Gordon and Breach, New York, 1968); S. Bashkin, Nucl. Instrum. Methods **90**, 3 (1970); I. Martinson, Nucl. Instrum. Methods **110**, 1 (1973); I. A. Sellin and D. J. Pegg, eds., *Beam-Foil Spectroscopy* (Plenum, New York, 1976), 2 vols.

[62] In a few cases, the 1979 grade has been lowered from the 1969 value because of errors discovered in the analysis or a change in grading criteria.

TABLE 1-3. PROGRESS IN SPECTRUM ANALYSIS[a]

Year	Partial Analyses Available for:[b]
1922	38 spectra of 30 elements
1932	231 spectra of 69 elements
1939	400 spectra of 77 elements
1946	445 spectra of 82 elements
1951	504 spectra of 84 elements
1959	511 spectra of 87 elements
1969	648 spectra of 95 elements
1979	1002 spectra of 99 elements

[a]For 104 elements, there are a total of 5460 spectra.

[b]A. Fowler, *Report on Series in Line Spectra* (Fleetway Press, London, 1922); F. Paschen and R. Götze, *Seriengesetze der Linienspektren* (Julius Springer, Berlin, 1922). R. F. Bacher and S. Goudsmit, *Atomic Energy States* (McGraw-Hill, New York, 1932). A. G. Shenstone, Rep. Prog. Phys. V, 210 (1939). W. F. Meggers, J. Opt. Soc. Am. **36**, 431 (1946) and **41**, 143 (1951); Rev. Univers. Mines **XV**, 230 (1959). C. E. Moore Sitterly, Opt. Pura y Apl. **2**, 103 (1969). J. Blaise, private communication, J.-F. Wyart, J. Opt. Soc. Am. **68**, 197 (1978), and literature search.

the levels. These aspects of the theory are largely represented by the vector model of the atom, which will be discussed in Chapter 4.

(2) The theory can be used in a semi-empirical parameterized form, in which the values of certain theoretical parameters are determined by least-squares fitting of empirically determined energy levels; this is the time-honored method in which the quantitative theory has been mainly employed since the nineteen-thirties.[57,63] This procedure can of course be utilized only after a number of levels have been found by empirical spectrum analysis, but it is then very useful for identifying the nature of the corresponding quantum states and for predicting approximate energies of unknown levels. These predictions may then make it possible to find the missing levels by means of more detailed wavenumber searches and/or to substantiate levels that have been tentatively identified through a single observed spectrum line.[64]

(3) The theory can also be used in a completely *ab initio* fashion by computing theoretical values for the parameters mentioned in (2). The accuracy of computed results is limited, especially for very complex neutral atoms, but the predicted energy level structures can still provide a very helpful guide for getting a start to an empirical spectrum analysis. Particularly for not-too-complex spectra of multiply ionized atoms, *ab initio* calculations

[63]A thorough discussion of energy-level structures from the point of view of the semi-empirical theory has been given by B. Edlén, "Atomic Spectra," in S. Flügge, ed., *Handbuch der Physik* (Springer-Verlag, Berlin, 1964), Vol. XXVII.

[64]See, for example, L. J. Radziemski, Jr. and K. L. Andrew, J. Opt. Soc. Am. **55**, 474 (1965).

TABLE 1-4. GRADES OF ANALYSIS OF ATOMIC SPECTRA (CAPITAL LETTERS, NOVEMBER 1979; LOWER CASE, APRIL 1969)

	1	2	3	4	5	6	7	8	9	10	11	12	13	14	15	16	17	18	19	20	21	22	23	24	25	26	27	28	29	30	31	32	33	34	35	36	37	38	39	40	41	42
1 H	aA																																									
2 He	aA	aA																																								
3 Li	aA	aA	aA																																							
4 Be	aA	aA	bA	aA																																						
5 B	bA	bA	bA	bA	aA																																					
6 C	aA	aA	aA	aA	bB	aA																																				
7 N	aA	aA	aA	aA	aA	bB	aA																																			
8 O	aA	aA	aA	aA	aA	bB	bB	aA																																		
9 F	aA	aA	aA	bB	bB	bB	bB	bB	bB																																	
10 Ne	aA	aA	cC	bB	bB	bB	bB	cB	cC	bB																																
11 Na	aA	bB	bB	cB	cC	bB	bB	bB	bB	dD	bB																															
12 Mg	aA	aA	bB	bB	bB	bB	bB	bB	bB	cB	cC	bB																														
13 Al	aA	aA	aA	bA	bB	bB	bB	bB	bB	cB	dB	cC	bB																													
14 Si	aA	aA	aA	aA	cA	bB	bB	bB	bB	cB	cB	dB	dD	bB																												
15 P	aA	aA	bA	cA	cA	cB	cC	cC	cC	cC	cC	cC	dC	-D	bB																											
16 S	aA	bB	bB	cC	dC	cA	cC	cC	dC	cC	cC	eC	eC	eC	-D	bB																										
17 Cl	aA	aA	bB	cC	cC	cC	cC	cC	cC	cC	cC	cC	eC	eD	eD	-E	bB																									
18 Ar	aA	aA	cC	cC	cC	cB	dB	cA	cA	cB	cC	cC	dD	eD	eE	-E	dD	bB																								
19 K	aA	cB	cC	cC	dC	dC	dB	dB	cA	dD	dC	dC	dC	eD	eD	eE	-C	eE	bB																							
20 Ca	aA	aA	aA	cC	cC	dC	cC	cB	cB	cA	cC	dC	dC	dC	dD	-D	eD	-C	eD	bB																						
21 Sc	aA	bA	cA	dA	dC	cC	dC	dC	dB	dB	cB	dC	cC	eC	eC	-C	-D	-E	-C	eE	---																					
22 Ti	aA	aA	aA	aA	aA	aB	aB	bB	bB	bB	bB	bA	dC	cC	eC	-C	-C	-C	-C	-B	dD	-E																				
23 V	aA	aA	aA	aA	bA	dB	cC	cC	dC	dC	dC	cC	bA	cC	cC	-C	-C	-C	-D	-C	-B	dC	---																			
24 Cr	aA	aA	bB	cB	cB	cB	dB	dC	dC	eC	eC	dC	cC	cA	dC	cC	-C	-C	-D	-E	-D	-C	dD	---																		
25 Mn	aA	aA	bA	bB	cB	eC	cC	dD	cC	dC	eC	eC	eC	dB	cA	dC	cC	-C	-D	-D	-E	-C	-C	dD	---																	
26 Fe	aA	aA	aA	bB	cB	cB	cB	cC	dC	cC	cC	dC	dC	dC	cB	cB	dC	cB	dB	dC	dC	dC	eC	dC	dD	dD																
27 Co	aA	bA	aA	eD	eC	cC	cC	cC	cC	dD	dD	eD	-D	eD	eC	dC	cB	dC	cC	-E	-E	-E	-E	-C		dD	---															
28 Ni	aA	aA	aA	cB	eB	eE	dD	cC	cC	cC	dD	dD	eD	eD	eD	eC	dC	cB	dC	cC	-C	-E	-E	-E	-D	-D	dD	---														
29 Cu	aA	aA	aA	cB	-B	-D	eE			cC	dC	dD	-E		-E	-D	-E	-C	cB	dC	cC	-D	-E	-E		-E		eD	---													
30 Zn	aA	bB	bB	cB	-E	-E		-E			cC	dD	eE	-E	-E	-E	-D	-E	-C	cB	dC	-C	-E	-E			-C		eE	---												
31 Ga	aA	bB	dC	dC	-C	-E	-E					-E	-E	-E	-E	-E	-E	-E	-E	-C	-B	-D	-D	-E							---											
32 Ge	bA	aA	cC	cC	dC	-D							-E	-E	-E	-E	-E	-E	-E	-E	-C	-B	-C	-C	-E				-E			---										
33 As	aA	bA	cC	cC	dD	dD	-C							-E	-E	-E	-E	-E			-E	-C	-B	-D	-D								---									
34 Se	bA	bB	cC	cC	cC	dB	cC	-C							-E	-E		-E	-E			-E	-D	-C	-C	-C	-E				-D			---								
35 Br	aA	bB	cC	dD	dD	dD	eE	cC	dD															-E	-C	-D	-D								---							
36 Kr	aA	aA	bB	dD	-E	dD	eD	eB	cC	dD															-E	-E	-D	-D					-E	-E		---						
37 Rb	aA	cA	cA	-D				-E	eE	cC	dD	eE	eE	eE														-D	-E								---					
38 Sr	aA	aA	-A	dA	-D				-E	-E	cC	dD	eE	eE													-E	-E	-D	-E								---				
39 Y	aA	aA	cA	-B	eC	-C	-B	-C	-E	-E	-A	cD	dD	eE	eE													-E	-E	-D	-E								---			
40 Zr	aA	aA	aA	aA	bA	bB	bB	-C	-C	-E	-D	-A	-E	dD	-D														-E	-E	-D									---		
41 Nb	aA	aA	aA	bB	dD	bB	bB	bB	-C	-C	-E	-E	-E	-E	-E															-E	-E	-D									---	
42 Mo	aA	aA	dD	dD	cC	cC	bB	bB	bB	-C	-C	-E	-D	-A	-E	eC	-D	-F	-F	-F	-F	-F	-F	-E	-E		-F	-F	-F	-D	-C	-B	-D	E						-E	-D	-D
43 Tc	bB	cC																																								
44 Ru	aA	aA	cC					-C	-D	-D					-E	-E																	-E	-E								
45 Rh	bB	bB	aA						-C	-D	-D																							-E	-E							
46 Pd	aA	aA	aA							-C	-D	-D					-E	eE		eE																-E						
47 Ag	aA	aA	bB								-E	-E					-E	-E	eE		eE																	-E				
48 Cd	aA	aA	aA	bB															-E	eE		eE																				
49 In	aA	aA	cB	dB	dC	-D														-E	eE		eE																			
50 Sn	aA	aA	cC	cC	dC	dC															-E	eD	-E	eD	-F	-F	-F	-F														

TABLE 1-4 (cont)

Z	Ion	1	2	3	4	5	6	7	8	9	10	11	12	13	14	15	16	17	18	19	20	21
51	Sb	aA	cC	dD	cC	dD	dD	·C														
52	Te	cA	bB	cC	cC	cC	dD	dD	·C													
53	I	aA	aA	eE	eE	·E	eD	eD	eD	·C												
54	Xe	aA	aA	bB	cC			dC	dB	eE												
55	Cs	aA	bB	dC						·E	·E											
56	Ba	aA	aA	·A	dD						·E	·E										
57	La	aA	aA	aA	·B	·C						·E	·E									
58	Ce	bB	bA	aA	cB	·E	·E						·E	·E								
59	Pr	dC	bB	bB	bB	bB								·D	·E							
60	Nd	bB	bB	·D	·E										·E	·E						
61	Pm	cC	cC																			
62	Sm	bB	bB	cC	eE												·E	·E				
63	Eu	aB	bB	dB	eE													·E	·E			
64	Gd	bB	bB	cC	dD														·E	·E		
65	Tb	cB	dC	·B	·C															·E	·E	
66	Dy	eB	cB		·E																·E	·E
67	Ho	eB	eC	cB	·E																	·E
68	Er	bB	bB	·C	cE																	
69	Tm	aB	aB	dB																		
70	Yb	aA	aA	bB	dD																	
71	Lu	bA	aA	dA	·B	·C																
72	Hf	bA	aA	cC	dB	·B																
73	Ta	aA	bB		·B	·B	·B															
74	W	bB	bB				·B	·B														
75	Re	bB	bB					·C	·C													
76	Os	bB	cC						·C	·C												
77	Ir	aA	cC							·C	·D											
78	Pt	bB	aA								·E	·E										
79	Au	aA	bB	aB								·E	·E									
80	Hg	aA	aA	cC	dD								·E	·E								
81	Tl	aA	aA	cA	dD									·E	·E							
82	Pb	aA	aA	bB	bA	bB									·E	·E						
83	Bi	bB	cC	cC	cC	dD	bB									·E	·E					
84	Po	cC																				
85	At	eE																				
86	Rn	aA	eE																			
87	Fr	·E																				
88	Ra	aA	aA																			
89	Ac	bB	aA	cC																		
90	Th	bA	bA	bB	cC																	
91	Pa	cB	cC																			
92	U	bA	bB				·C															
93	Np	dB	·D																			
94	Pu	cA	cB																			
95	Am	cC	cC																			
96	Cm	cB	cB																			
97	Bk	·C	·D																			
98	Cf	·C	·D																			
99	Es	·D	·D																			
100	Fm																					

Z	Ion	22	23	24	25	26	27	28	29	30	31	32	33	34	35	36	37	38	39	40	41	42	43	44	45	46	47	51	52
51	Sb	·E	eE		eE																								
54	Xe				·E	·E																							
56	Ba						·E	·E																					
57	La							·E	·E																				
60	Nd										·E	·E																	
62	Sm												·E	·E	·E														
64	Gd														·E	·E	·E	·F	·F										
66	Dy																·E	·E	·E										
68	Er																		·E	·E									
70	Yb																				·E	·E							
72	Hf																								·D				
73	Ta																							·E	·E	·D	·E		
74	W								·F	·F	·F	·F	·F												·E	·E	·D		
78	Pt																											·E	
79	Au													·F	·F	·F	·F	·F											·E

Key

F: No individual lines identified—only unresolved transition arrays
E: Fragmentary analysis—2 to 5 energy levels
D: For moderately complex spectra, about a dozen levels known
C: Two dozen or so levels known
B: All of lowest 3 or 4 configurations and portions of some higher ones known
A: Analysis essentially complete—further progress would be extremely difficult

Note

The grades for hydrogenic ions beyond $Z = 8$ reflect mainly theoretical calculations, which are now available for all Z (as well as to $Z = 42$ for helium-like ions).

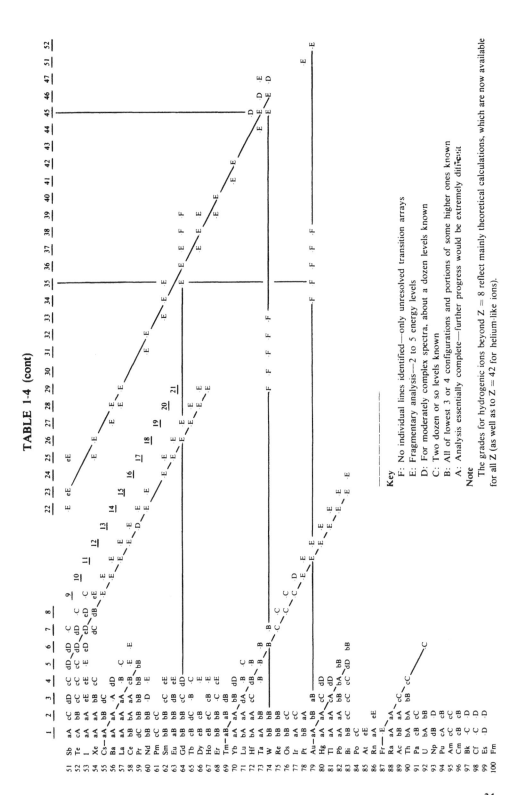

may be adequate in accuracy to clarify dubious points of an analysis[65] or even to identify the transitions producing spectrum lines when the empirical information alone is quite inadequate.[66]

In the *ab initio* form of the theory (and to a certain extent in the semi-empirical form also), one obtains not only level energies but also the wavefunctions of the corresponding quantum states. From these wavefunctions one can (at least in principle) compute any desired properties of the atom—of which the most important for our purposes are the strengths of spectrum lines.[67]

Our principal aim in the remainder of this book is to develop the theory of atomic structure, using Racah's methods, in as elementary a fashion as possible, but in a form suitable for the calculation of energy levels and line strengths for all atoms from the simplest to the most complex. The emphasis is on the use of computers in making both *ab initio* and least-squares calculations, but for illustrative purposes simple hand calculations and numerical comparisons with experiment are given wherever feasible.

[65]L. Å. Svensson and J. O. Ekberg, Ark. Fys. **37**, 65 (1968).

[66]R. D. Cowan and N. J. Peacock, Astrophys. J. **142**, 390 (1965), **143**, 283 (1966); R. D. Cowan, Astrophys. J. **147**, 377 (1967); B. C. Fawcett, R. D. Cowan, E. Y. Kononov, and R. W. Hayes, J. Phys. B **5**, 1255 (1972).

[67]In the semi-empirical method, information is obtained on only the angular (not radial) portion of the wavefunctions, and on only relative (not absolute) line strengths.

2

ANGULAR-MOMENTUM PROPERTIES OF WAVEFUNCTIONS

2-1. INTRODUCTION

The fundamental problem in the theory of atomic structure is the calculation of the wavefunction for each quantum state of interest. For practical purposes, it is convenient to be able to describe each wavefunction briefly in terms of characteristic properties of the function; these properties then serve also to provide an informative and useful designation for the corresponding energy level. The properties in question consist primarily of certain symmetry properties of the wavefunction. Symmetries are appropriately discussed in group theoretical terminology, and a number of books[1-5] are available which treat atomic structure from that point of view.

For an adequate treatment of atoms in molecules or crystals, formal use of group theory is indispensable. For free atoms, however, which show basically a spherical symmetry, the pertinent full rotation group is closely related to the angular momentum properties of the atom. Since angular momentum terminology is always used anyway in describing atomic wavefunctions, we prefer to avoid any discussion of the rotation group and to work directly in terms of the quantum mechanical theory of angular momentum. (In addition to the continuous rotation group, several finite groups are useful in discussing certain theoretical aspects of wavefunctions;[4,5] however, they are not essential in practical applications, and we shall by-pass group theoretical methods almost entirely.)

[1]E. P. Wigner, *Group Theory and Its Application to the Quantum Mechanics of Atomic Spectra*, transl. by J. J. Griffin (Academic Press, New York, 1959).

[2]M. Hamermesh, *Group Theory and Its Application to Physical Problems* (Addison-Wesley, Reading, Mass., 1962).

[3]M. Tinkham, *Group Theory and Quantum Mechanics* (McGraw-Hill, New York, 1964).

[4]B. R. Judd, *Operator Techniques in Atomic Spectroscopy* (McGraw-Hill, New York, 1963).

[5]B. G. Wybourne, *Symmetry Principles and Atomic Spectroscopy* (Wiley-Interscience, New York, 1970).

2-2. ANGULAR-MOMENTUM OPERATORS

In classical mechanics, the orbital angular momentum (about some given reference point) of a particle moving with linear momentum **p** at a position **r** with respect to that point is defined as

$$\mathbf{L} = \mathbf{r} \times \mathbf{p} , \tag{2.1}$$

or

$$\begin{aligned} L_x &= yp_z - zp_y \\ L_y &= zp_x - xp_z \\ L_z &= xp_y - yp_x . \end{aligned} \tag{2.2}$$

The corresponding quantum-mechanical operators, obtained as usual by replacing p_x with $-i\hbar\partial/\partial x$, etc., are

$$\begin{aligned} L_x &= - i\hbar\left(y \frac{\partial}{\partial z} - z \frac{\partial}{\partial y}\right) \\ L_y &= - i\hbar \left(z \frac{\partial}{\partial x} - x \frac{\partial}{\partial z}\right) \\ L_z &= - i\hbar \left(x \frac{\partial}{\partial y} - y \frac{\partial}{\partial x}\right) . \end{aligned} \tag{2.3}$$

From these expressions we find by straightforward differentiation that

$$\begin{aligned} L_xL_y - L_yL_x &= - \hbar^2 \left(y \frac{\partial}{\partial z} - z \frac{\partial}{\partial y}\right)\left(z \frac{\partial}{\partial x} - x \frac{\partial}{\partial z}\right) \\ &\quad + \hbar^2 \left(z \frac{\partial}{\partial x} - x \frac{\partial}{\partial z}\right)\left(y \frac{\partial}{\partial z} - z \frac{\partial}{\partial y}\right) \\ &= - \hbar^2 \left(y \frac{\partial}{\partial x} + yz \frac{\partial^2}{\partial z \partial x} - yx \frac{\partial^2}{\partial z^2} - z^2 \frac{\partial^2}{\partial y \partial x} + zx \frac{\partial^2}{\partial y \partial z}\right) \\ &\quad + \hbar^2 \left(zy \frac{\partial^2}{\partial x \partial z} - z^2 \frac{\partial^2}{\partial x \partial y} - xy \frac{\partial^2}{\partial z^2} + x \frac{\partial}{\partial y} + xz \frac{\partial^2}{\partial z \partial y}\right) \\ &= \hbar^2 \left(x \frac{\partial}{\partial y} - y \frac{\partial}{\partial x}\right) = i\hbar L_z . \end{aligned}$$

The commutation relation

$$L_xL_y - L_yL_x = i\hbar L_z$$

and the two analogous expressions obtained by cyclic permutation of the subscripts are more general than the partial-derivative expressions (2.3), and are applicable to electron spin **S** as well as to orbital angular momentum **L**. We accordingly write the commutation relations for a general angular-momentum operator **J**:

$$J_x J_y - J_y J_x = ihJ_z$$
$$J_y J_z - J_z J_y = ihJ_x \qquad (2.4)$$
$$J_z J_x - J_x J_z = ihJ_y \ .$$

2-3. EIGENVALUES OF ANGULAR-MOMENTUM OPERATORS

The commutation relations show that it is impossible to find an eigenfunction of the angular-momentum operator $\mathbf{J}$ with non-zero eigenvalue—i.e., to find an eigenfunction of all three operators J_x, J_y, J_z, with at least one of the three eigenvalues m_x, m_y, m_z different from zero. For, if we assume such an eigenfunction f to exist with m_z (say) not equal to zero, then the first of the relations (2.4) leads to

$$\begin{aligned}
ihJ_z f = ihm_z f &= (J_x J_y - J_y J_x)f \\
&= (J_x m_y - J_y m_x)f \\
&= (m_y J_x - m_x J_y)f \\
&= (m_y m_x - m_x m_y)f = 0 \ ,
\end{aligned}$$

and this is impossible unless $f \equiv 0$.

Although it is not possible to find eigenfunctions of $\mathbf{J}$, it *is* possible to find functions that are not only eigenfunctions of *one* component of $\mathbf{J}$ (usually taken to be J_z) but are also eigenfunctions of

$$\mathbf{J}^2 \equiv \mathbf{J} \cdot \mathbf{J} = J_x^2 + J_y^2 + J_z^2 \ . \qquad (2.5)$$

This is possible because $\mathbf{J}^2$ and J_z commute, as can be easily seen from (2.4):

$$\begin{aligned}
\mathbf{J}^2 J_z - J_z \mathbf{J}^2 &= J_x^2 J_z - J_z J_x^2 + J_y^2 J_z - J_z J_y^2 + J_z^3 - J_z^3 \\
&= J_x(J_x J_z) - (J_z J_x)J_x + J_y(J_y J_z) - (J_z J_y)J_y \\
&= J_x(J_z J_x - ihJ_y) - (J_x J_z + ihJ_y)J_x \\
&\quad + J_y(J_z J_y + ihJ_x) - (J_y J_z - ihJ_x)J_y \\
&= 0 \ . \qquad (2.6)
\end{aligned}$$

Let ψ_{bc} be such an eigenfunction of $\mathbf{J}^2$ and J_z with eigenvalues b and c, respectively. Then by definition

$$(J_x^2 + J_y^2 + J_z^2)\psi_{bc} = b\psi_{bc}$$

and

$$J_z \psi_{bc} = c\psi_{bc} \ ,$$

from which it follows that

$$J_z^2\psi_{bc} = J_z c\psi_{bc} = cJ_z\psi_{bc} = c^2\psi_{bc}$$

and

$$(J_x^2 + J_y^2)\psi_{bc} = (b - c^2)\psi_{bc} .$$

Since J_x and J_y are real observables, $J_x^2 + J_y^2$ is an essentially positive operator and the eigenvalue $b - c^2$ must be non-negative, so that

$$|c| \leqq b^{1/2} . \tag{2.7}$$

From the commutation relations (2.4) it follows that

$$J_z(J_x \pm iJ_y) = J_xJ_z + i\hbar J_y \pm i(J_yJ_z - i\hbar J_x) = (J_x \pm iJ_y)(J_z \pm \hbar) .$$

Operating with this equality on ψ_{bc} we obtain

$$J_z(J_x \pm iJ_y)\psi_{bc} = (J_x \pm iJ_y)(J_z \pm \hbar)\psi_{bc} = (c \pm \hbar)(J_x \pm iJ_y)\psi_{bc} ;$$

this shows that if the function

$$(J_x \pm iJ_y)\psi_{bc} \tag{2.8}$$

is not identically zero, then it is an eigenfunction of J_z with eigenvalue

$$c \pm \hbar .$$

Moreover, J_x and J_y commute with $\mathbf{J}^2$ [cf. (2.6)] so that

$$\mathbf{J}^2(J_x \pm iJ_y)\psi_{bc} = (J_x \pm iJ_y)\mathbf{J}^2\psi_{bc} = b(J_x \pm iJ_y)\psi_{bc} ,$$

and (2.8) is (like ψ_{bc} itself) an eigenfunction of $\mathbf{J}^2$ with eigenvalue b. Repeating this argument for the function (2.8) in place of ψ_{bc}, we find that unless

$$(J_x \pm iJ_y)(J_x \pm iJ_y)\psi_{bc}$$

vanishes, it is an eigenfunction of $\mathbf{J}^2$ with eigenvalue b and an eigenfunction of J_z with eigenvalue

$$c \pm 2\hbar .$$

Thus for a given eigenvalue b of $\mathbf{J}^2$, there in general exists a whole sequence of possible eigenvalues of J_z:

$$c = \cdots , c' - 2\hbar, c' - \hbar, c', c' + \hbar, c' + 2\hbar, \cdots , \tag{2.9}$$

where c' is any specific eigenvalue. But from (2.7) this sequence is bounded, so that there must exist a smallest member c_n and a largest member c_x. The corresponding eigenfunctions must satisfy the relations

$$(J_x - iJ_y)\psi_{bc_n} = 0 \tag{2.10}$$

$$(J_x + iJ_y)\psi_{bc_x} = 0 , \tag{2.11}$$

for otherwise the left-hand sides of these equations would be eigenfunctions of $c_n - \hbar$ and $c_x + \hbar$, respectively. But from the commutation relations (2.4)

$$(J_x + iJ_y)(J_x - iJ_y) = J_x^2 + J_y^2 + \hbar J_z \tag{2.12}$$

$$(J_x - iJ_y)(J_x + iJ_y) = J_x^2 + J_y^2 - \hbar J_z , \tag{2.13}$$

so that from (2.10) and (2.11) we find

$$(J_x^2 + J_y^2 + \hbar J_z)\psi_{bc_n} = (b - c_n^2 + \hbar c_n)\psi_{bc_n} = 0$$

$$(J_x^2 + J_y^2 - \hbar J_z)\psi_{bc_x} = (b - c_x^2 - \hbar c_x)\psi_{bc_x} = 0 .$$

Since ψ_{bc_n} and ψ_{bc_x} are non-zero by hypothesis,

$$b - c_n^2 + \hbar c_n = b - c_x^2 - \hbar c_x = 0 , \tag{2.14}$$

or

$$(c_x + c_n)(c_x - c_n + \hbar) = 0 .$$

The second factor on the left is greater than zero and hence

$$c_n = - c_x ,$$

so that the sequence of values (2.9) is symmetrical about zero.

The quantity $(c_x - c_n)/\hbar$ is obviously a non-negative integer, which we denote by 2j. Thus $c_n = -j\hbar$ and $c_x = +j\hbar$, and from (2.14) we find the possible eigenvalues b of $\mathbf{J}^2$ to be

$$j(j + 1)\hbar^2 \qquad (j = 0, 1/2, 1, 3/2, 2, \cdots) ; \tag{2.15}$$

for any given value of j, the possible eigenvalues c of J_z are[6]

$$m\hbar \qquad (m = -j, -j + 1, -j + 2, \cdots j - 1, j) . \tag{2.16}$$

[6]The notation m_j is sometimes used in place of m to distinguish this quantum number from similar quantum numbers m_l and m_s used in the cases of orbital and spin momenta, respectively.

These results correspond to a semiclassical picture of angular momentum as a vector **J** with magnitude

$$[j(j + 1)]^{1/2}\hbar , \tag{2.17}$$

quantized in such a way that the component of **J** along the z-axis can have only the values (2.16), as sketched in Fig. 2-1: note that the maximum component along the z axis can never be as large as the magnitude (2.17). In the limit of very large j, this picture tends to the classical result in which the z-component of **J** can be as great as $|\mathbf{J}| = j\hbar$, and can vary in infinitesimally small fractions of $|\mathbf{J}|$.

2-4. STEP-UP AND STEP-DOWN OPERATORS

The operators $J_x + iJ_y$ and $J_x - iJ_y$ that we used in the preceding section are called step-up and step-down operators, respectively, because when they operate on an eigenfunction of J_z they produce new eigenfunctions with m increased and decreased by unity, respectively:

$$(J_x \pm iJ_y)\psi_{jm} = N_\pm \psi_{j,m\pm1} ,$$

where the notation has been altered by replacing the eigenvalue subscripts b and c by the corresponding quantum numbers j and m.

If we assume both ψ_{jm} and $\psi_{j,m\pm1}$ to be normalized in the usual manner,

$$\langle\psi_{jm}|\psi_{jm}\rangle = 1 , \tag{2.18}$$

then the multiplicative constant $N_\pm$ can be easily found with the aid of (2.12)-(2.16):

$$\langle N_\pm\psi_{j,m\pm1}|N_\pm\psi_{j,m\pm1}\rangle = \langle(J_x \pm iJ_y)\psi_{jm}|(J_x \pm iJ_y)\psi_{jm}\rangle ,$$

or

$$\begin{aligned} |N_\pm|^2 &= \langle\psi_{jm}|(J_x \mp iJ_y)(J_x \pm iJ_y)|\psi_{jm}\rangle \\ &= \langle\psi_{jm}|(J_x^2 + J_y^2 \mp \hbar J_z)|\psi_{jm}\rangle \\ &= \langle\psi_{jm}|j(j + 1)\hbar^2 - m^2\hbar^2 \mp m\hbar^2|\psi_{jm}\rangle \\ &= \hbar^2[j(j + 1) - m^2 \mp m]\langle\psi_{jm}|\psi_{jm}\rangle \\ &= \hbar^2[(j \mp m)(j \pm m + 1)] . \end{aligned}$$

Thus

$$N_\pm = e^{i\delta}\,\hbar[(j \mp m)(j \pm m + 1)]^{1/2} ,$$

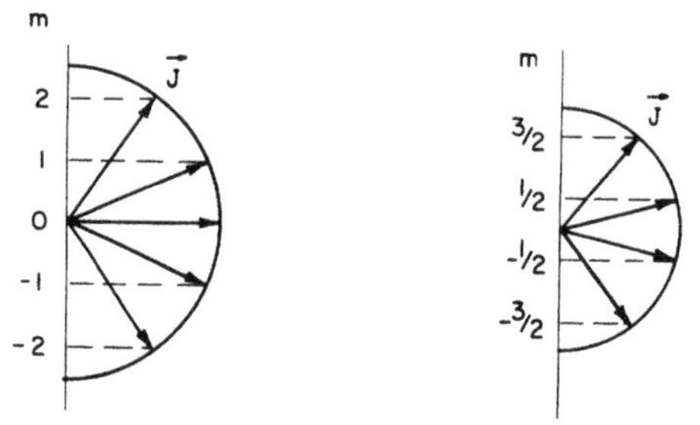

Fig. 2-1. Semiclassical picture of the possible orientations of the angular-momentum vector **J** for integral (j=2, left) and half-integral (j=3/2, right) values of the quantum numbers; the radius of the semicircle (in units of $\hbar$) is $[j(j+1)]^{1/2}$ in each case. Note: the plane of the figure may be any plane containing the z-axis, as only the z-component of **J** is known.

where δ is a real number. The phase factor $e^{i\delta}$ is completely arbitrary, because ψ_{jm} and $\psi_{j,m\pm1}$ are completely independent wavefunctions. We shall always choose $\delta = 0$, so that

$$J_\pm\psi_{jm} \equiv (J_x \pm iJ_y)\psi_{jm} = \hbar[(j \mp m)(j \pm m + 1)]^{1/2}\psi_{j,m\pm1}$$

$$= \hbar[j(j + 1) - m(m \pm 1)]^{1/2}\psi_{j,m\pm1} \ . \qquad (2.19)$$

2-5. ORBITAL ANGULAR MOMENTUM

In working with the orbital angular momentum of an electron about the nucleus, it is convenient, because of the spherical symmetry of the problem, to transform to the spherical coordinates illustrated in Fig. 2-2 and defined by:

$$\begin{aligned} x &= r \sin\theta \cos\phi \\ y &= r \sin\theta \sin\phi \\ z &= r \cos\theta \ , \end{aligned} \qquad (2.20)$$

or

$$\begin{aligned} r &= (x^2 + y^2 + z^2)^{1/2} \\ \cos\theta &= \frac{z}{(x^2 + y^2 + z^2)^{1/2}} \\ \tan\phi &= \frac{y}{x} \ . \end{aligned}$$

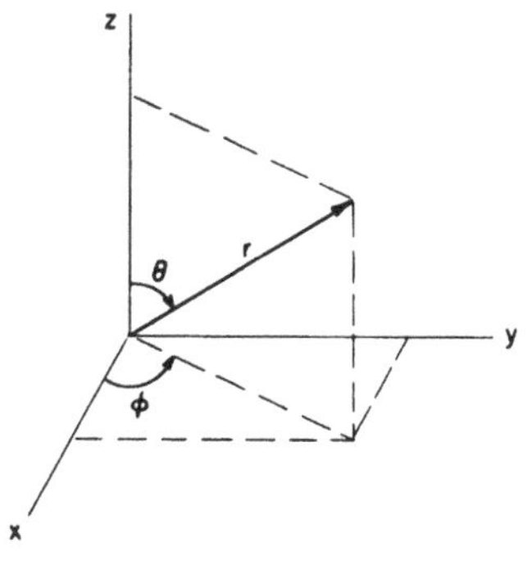

Fig. 2-2. Spherical coordinates (r,θ,ϕ).

Differentiation of these latter expressions gives

$$\frac{\partial r}{\partial x} = \frac{x}{(x^2 + y^2 + z^2)^{1/2}} = \sin\theta\cos\phi$$

$$-\sin\theta\,\frac{\partial\theta}{\partial x} = -\frac{xz}{(x^2 + y^2 + z^2)^{3/2}} = -\frac{1}{r}\sin\theta\cos\theta\cos\phi$$

$$\frac{1}{\cos^2\phi}\,\frac{\partial\phi}{\partial x} = -\frac{y}{x^2} = \frac{-\sin\phi}{r\sin\theta\cos^2\phi},$$

etc., from which

$$\begin{aligned}
\frac{\partial}{\partial x} &= \sin\theta\cos\phi\,\frac{\partial}{\partial r} + \frac{1}{r}\cos\theta\cos\phi\,\frac{\partial}{\partial\theta} - \frac{\sin\phi}{r\sin\theta}\,\frac{\partial}{\partial\phi} \\
\frac{\partial}{\partial y} &= \sin\theta\sin\phi\,\frac{\partial}{\partial r} + \frac{1}{r}\cos\theta\sin\phi\,\frac{\partial}{\partial\theta} + \frac{\cos\phi}{r\sin\theta}\,\frac{\partial}{\partial\phi} \qquad (2.21)\\
\frac{\partial}{\partial z} &= \cos\theta\,\frac{\partial}{\partial r} - \frac{\sin\theta}{r}\,\frac{\partial}{\partial\theta}\,.
\end{aligned}$$

By substituting (2.20) and (2.21) into (2.3), we easily find the following expressions for the Cartesian components of the orbital angular momentum in terms of (r,θ,ϕ):

$$\begin{aligned}
L_x &= i\hbar\left(\sin\phi\frac{\partial}{\partial\theta} + \cot\theta\cos\phi\,\frac{\partial}{\partial\phi}\right) \\
L_y &= i\hbar\left(-\cos\phi\,\frac{\partial}{\partial\theta} + \cot\theta\sin\phi\,\frac{\partial}{\partial\phi}\right)
\end{aligned} \qquad (2.22)$$

$$L_z = -i\hbar\,\frac{\partial}{\partial\phi}\,. \qquad (2.23)$$

By combining the two expressions (2.22), the step-up and step-down operators are found to be

$$L_x \pm iL_y = \hbar e^{\pm i\phi}\left[\pm\frac{\partial}{\partial\theta} + i\cot\theta\,\frac{\partial}{\partial\phi}\right]. \qquad (2.24)$$

Straightforward partial differentiation gives the result

$$\begin{aligned}
(L_x - iL_y)(L_x + iL_y) &= \hbar^2 e^{-i\phi}\left[-\frac{\partial}{\partial\theta} + i\cot\theta\frac{\partial}{\partial\phi}\right]e^{i\phi}\left[\frac{\partial}{\partial\theta} + i\cot\theta\frac{\partial}{\partial\phi}\right] \\
&= -\hbar^2\left[\frac{\partial^2}{\partial\theta^2} + \cot\theta\,\frac{\partial}{\partial\theta} - i\,\frac{\partial}{\partial\phi} + \cot^2\theta\,\frac{\partial^2}{\partial\phi^2}\right],
\end{aligned}$$

so that from (2.13) and (2.23) we obtain for the operator $\mathbf{L}^2$:

$$\mathbf{L}^2 \equiv L_x^2 + L_y^2 + L_z^2$$

$$= -\hbar^2\left[\frac{\partial^2}{\partial\theta^2} + \cot\theta\,\frac{\partial}{\partial\theta} + \frac{1}{\sin^2\theta}\,\frac{\partial^2}{\partial\phi^2}\right]$$

$$= -\hbar^2\left[\frac{1}{\sin\theta}\,\frac{\partial}{\partial\theta}\left(\sin\theta\,\frac{\partial}{\partial\theta}\right) + \frac{1}{\sin^2\theta}\,\frac{\partial^2}{\partial\phi^2}\right] . \tag{2.25}$$

We now wish to determine the form of functions of (θ,ϕ) that are simultaneous eigenfunctions of L_z and $\mathbf{L}^2$, together with the corresponding eigenvalues.

From (2.16) we know that any eigenvalue of L_z is of the form $m_l\hbar$, and from (2.23) we see that the corresponding eigenfunction must satisfy

$$L_z\psi = -i\hbar\,\frac{\partial\psi}{\partial\phi} = m_l\hbar\psi .$$

The only solutions of this differential equation are of the form

$$\psi(\theta,\phi) = \Theta(\theta)\cdot\Phi_{m_l}(\phi) = \Theta(\theta)(2\pi)^{-1/2}e^{im_l\phi} . \tag{2.26}$$

In order that this be a single-valued function of ϕ, the quantum number m_l must be integral. The factor $(2\pi)^{-1/2}$ has been included in order to make Φ orthonormal in the sense

$$\langle\Phi_{m_l}|\Phi_{m_l'}\rangle \equiv \int_0^{2\pi} \Phi^*_{m_l}(\phi)\cdot\Phi_{m_l'}(\phi)\,d\phi$$

$$= \frac{1}{2\pi}\int_0^{2\pi} e^{i(m_l'-m_l)\phi}\,d\phi = \delta_{m_l m_l'} . \tag{2.27}$$

Since m_l is integral, it follows from (2.15) and (2.16) that all eigenvalues of $\mathbf{L}^2$ must be of the form

$$l(l+1)\hbar^2 \qquad (l \geqq |m_l|) , \tag{2.28}$$

with the quantum number l also integral. The form of the function Θ in (2.26) can be found directly from the eigenvalue equation

$$\mathbf{L}^2\psi = l(l+1)\hbar^2\psi , \tag{2.29}$$

by use of the differential form (2.25) for the operator $\mathbf{L}^2$. However, it is more convenient to find Θ through the medium of the step-up and step-down operators.

Starting with the eigenfunction ψ_{lm_l} [Eq. (2.26)] for minimum m_l (= $-l$), we find from (2.10) that

$$(L_x - iL_y)\psi_{l,-l}(\theta,\phi) = (L_x - iL_y)\Theta_{l,-l}(\theta)\cdot\phi_{-l}(\phi) = 0 ,$$

which from (2.24) gives

$$-\frac{d}{d\theta}\,\Theta_{l,-l}(\theta) + l(\cot\theta)\Theta_{l,-l}(\theta) = 0 .$$

The integral of this differential equation is

$$\Theta_{l,-l} = \left[\frac{(2l+1)!}{2}\right]^{1/2}(2^l l!)^{-1}\sin^l\theta ; \tag{2.30}$$

the coefficient of $\sin^l\theta$ has been chosen such that

$$\langle\Theta_{l,-l}|\Theta_{l,-l}\rangle \equiv \int_0^\pi \Theta^*_{l,-l}\Theta_{l,-l}\sin\theta\, d\theta = 1 , \tag{2.31}$$

as proved in the appendix to this chapter, with the arbitrary phase factor[7] chosen to be +1.

Normalized functions[8] Θ_{lm} for all $m > -l$ can be found by using (2.19):

$$\Theta_{l,m+1}\Phi_{m+1} = \hbar^{-1}[(l-m)(l+m+1)]^{-1/2}(L_x + iL_y)\Theta_{lm}\Phi_m ,$$

which from (2.24) and (2.26) may be written

$$\Theta_{l,m+1} = [(l-m)(l+m+1)]^{-1/2}\left[\frac{\partial}{\partial\theta} - m\cot\theta\right]\Theta_{lm} . \tag{2.32}$$

The solution to this recursion relation is

$$\Theta_{lm} = \frac{(-1)^{l+m}}{2^l l!}\left[\frac{(2l+1)(l-m)!}{2(l+m)!}\right]^{1/2}\sin^m\theta\,\frac{d^{l+m}}{d(\cos\theta)^{l+m}}\sin^{2l}\theta , \tag{2.33}$$

as it is seen to satisfy (2.32) and to reduce to (2.30) when $m = -l$. For $m = 0$, (2.33) may be written

$$\Theta_{l0} = \frac{1}{2^l l!}\left[\frac{1}{2}(2l+1)\right]^{1/2}\frac{d^l}{d(\cos\theta)^l}(\cos^2\theta - 1)^l$$
$$= \left[\frac{1}{2}(2l+1)\right]^{1/2} P_l(\cos\theta) , \tag{2.34}$$

where $P_l(\cos\theta)$ is the lth Legendre polynomial; for $m = +l$, (2.33) becomes

[7]Our choice of phase agrees with that used by Condon and Shortley, by Racah in his original papers, and by A. R. Edmonds, *Angular Momentum in Quantum Mechanics* (Princeton University Press, 1960). The only other common phase convention (used especially by the Israeli group) is that of U. Fano and G. Racah, *Irreducible Tensorial Sets* (Academic Press, New York, 1959), who use a phase factor $i^l = (-1)^{l/2}$.

[8]We temporarily drop the subscript from m_l for brevity.

$$\Theta_{ll} = \frac{(-1)^l}{2^l l!}\left[\frac{(2l+1)}{2(2l)!}\right]^{1/2} \sin^l\theta \frac{d^{2l}}{d(\cos\theta)^{2l}}(\cos^2\theta - 1)^l$$

$$= \frac{(-1)^l}{2^l l!}\left[\frac{1}{2}(2l+1)!\right]^{1/2} \sin^l\theta \ . \tag{2.35}$$

[Note that $\Theta_{l,l+1} = 0$, in agreement with (2.11) and (2.16).]

An alternative expression for Θ_{lm} can be obtained by starting with (2.35) as a solution to (2.11), and then applying the step-down relation [(2.19) and (2.24), with lower signs] $l - m$ times. The result is

$$\Theta_{lm} = \frac{(-1)^l}{2^l l!}\left[\frac{(2l+1)}{2}\frac{(l+m)!}{(l-m)!}\right]^{1/2} \sin^{-m}\theta \frac{d^{l-m}}{d(\cos\theta)^{l-m}} \sin^{2l}\theta \tag{2.36}$$

$$= \frac{1}{2^l l!}\left[\frac{(2l+1)}{2}\frac{(l+m)!}{(l-m)!}\right]^{1/2} \sin^{-m}\theta \frac{d^{l-m}}{d(\cos\theta)^{l-m}}(\cos^2\theta - 1)^l \ . \tag{2.37}$$

For $m = 0$ and $m = -l$, this reduces to (2.34) and (2.30), respectively, and for $m = -l - 1$ it gives zero, in agreement with (2.10) and (2.16).

By replacing m with $-m$ in (2.36) and comparing with (2.33), we see that

$$\Theta_{l,-m} = (-1)^m \Theta_{lm} \ , \tag{2.38}$$

and also that we may write still a third expression for Θ_{lm}:

$$\Theta_{lm} = (-1)^{(m+|m|)/2}\left[\frac{(2l+1)(l-|m|)!}{2(l+|m|)!}\right]^{1/2} \frac{\sin^{|m|}\theta}{2^l l!} \frac{d^{l+|m|}}{d(\cos\theta)^{l+|m|}}(\cos^2\theta - 1)^l$$

$$\equiv (-1)^{(m+|m|)/2}\left[\frac{(2l+1)(l-|m|)!}{2(l+|m|)!}\right]^{1/2} P_l^{|m|}(\cos\theta) \ , \tag{2.39}$$

where $P_l^{|m|}(\cos\theta)$ is the associated Legendre function. Although (2.39) is the most commonly quoted expression, actual computation of Θ_{lm} in terms of $\sin\theta$ and $\cos\theta$ is simpler using (2.37) and (2.38) with $m > 0$ because fewer differentiations are required.

The final expression for eigenfunctions of $\mathbf{L}^2$ and L_z is, from (2.26) and (2.39),

$$Y_{lm}(\theta,\phi) = \Theta_{lm}(\theta)\cdot\Phi_m(\phi)$$

$$= (-1)^{(m+|m|)/2}\left[\frac{(2l+1)(l-|m|)!}{4\pi(l+|m|)!}\right]^{1/2} P_l^{|m|}(\cos\theta)e^{im\phi} \ . \tag{2.40}$$

It may be readily verified from (2.34) and from (2.39)-(2.40) that

$$P_l(1) \equiv P_l^0(1) = 1 \ , \qquad Y_{lm}(0,\phi) = \left(\frac{2l+1}{4\pi}\right)^{1/2}\delta_{m0} \ . \tag{2.41}$$

The functions Y_{lm} are called spherical harmonics; that they indeed satisfy the eigenvalue equation (2.29) with $\mathbf{L}^2$ given by (2.25) is shown directly in the appendix to this chapter. They are orthonormal in the sense

$$\langle Y_{lm}|Y_{l'm'}\rangle \equiv \int_0^{2\pi}\int_0^{\pi} Y_{lm}^{*}Y_{l'm'} \sin\theta \, d\theta \, d\phi = \delta_{ll'}\delta_{mm'} \,. \qquad (2.42)$$

Orthogonality for $m \neq m'$ follows from (2.27); if $m = m'$ but $l \neq l'$ then orthogonality follows from general theoretical principles,[9] and also is demonstrated directly in the appendix. Normalization for $m = m'$ and $l = l'$ is assured because of (2.27), (2.31), and the use of the normalized relation (2.19) in deriving the expressions (2.33)-(2.39) for Θ.

Explicit expressions for the first few spherical harmonics are given in Table 2-1. Angular distributions of the quantity

$$|Y_{lm}(\theta,\phi)|^2 = [\Theta_{lm}(\theta)]^2 \,, \qquad (2.43)$$

which is proportional to electron density, are shown in Fig. 2-3 in the form of polar diagrams in the y-z plane; three-dimensional angular distributions are obtained by rotating these figures about the z axis. Note that for $m = l$ and large l, the electron density tends to be concentrated near the x-y plane; this is consistent with the classical picture of an electron with $L_z = \pm |\mathbf{L}|$, which must be moving about the nucleus in an orbit in the x-y plane. For $m = 0$, the electron density is concentrated along the z axis, which is of course consistent with $L_z = 0$. For intermediate values of m ($0 < m < l$), the density is concentrated not near a plane making some angle θ with the z axis, as in the classical case, but near the surface of a double-ended cone (of half-angle θ); this cone represents the locus of that portion of the classical orbit-plane which lies in the **L**-z plane, as **L** precesses about the z axis. (It is to be recalled from Sec. 2-3 that only $|\mathbf{L}|$ and L_z can be known quantum mechanically, and not the vector **L**.)

2-6. THE ADDITION THEOREM OF SPHERICAL HARMONICS

In later chapters, we shall wish to evaluate the electrostatic interaction energy between two electrons at positions defined by radius vectors $\mathbf{r}_1$ and $\mathbf{r}_2$, separated by an angle ω (Fig. 2-4). For this purpose, it proves to be necessary to express the Legendre polynomial P_l ($\cos\omega$) in terms of the angles θ_1, ϕ_1, θ_2, ϕ_2. This can be done as follows.[10]

Setting $m = 0$ in (2.40) we obtain

$$P_l(\cos\omega) = \left(\frac{4\pi}{2l+1}\right)^{1/2} Y_{l0}(\omega,\alpha) \,, \qquad (2.44)$$

[9] Namely, Sturm-Liouville theory: see, for example, H. Margenau and G. M. Murphy, *The Mathematics of Physics and Chemistry* (D. Van Nostrand, New York, 1956), 2nd ed., p. 267.

[10] E. U. Condon and G. H. Shortley, *The Theory of Atomic Spectra* (University Press, Cambridge, 1935), p. 53.

TABLE 2.1 SPHERICAL HARMONICS

$l = 0$

$$Y_{00} = \frac{1}{\sqrt{4\pi}}$$

$l = 1$

$$Y_{1,\pm 1} = \mp \sqrt{\frac{3}{8\pi}} \sin\theta\, e^{\pm i\phi}$$

$$Y_{10} = \sqrt{\frac{3}{4\pi}} \cos\theta$$

$l = 2$

$$Y_{2,\pm 2} = \sqrt{\frac{15}{32\pi}} \sin^2\theta\, e^{\pm 2i\phi}$$

$$Y_{2,\pm 1} = \mp \sqrt{\frac{15}{8\pi}} \sin\theta \cos\theta\, e^{\pm i\phi}$$

$$Y_{20} = \sqrt{\frac{5}{16\pi}} (2\cos^2\theta - \sin^2\theta)$$

$l = 3$

$$Y_{3,\pm 3} = \mp \sqrt{\frac{35}{64\pi}} \sin^3\theta\, e^{\pm 3i\phi}$$

$$Y_{3,\pm 2} = \sqrt{\frac{105}{32\pi}} \sin^2\theta \cos\theta\, e^{\pm 2i\phi}$$

$$Y_{3,\pm 1} = \mp \sqrt{\frac{21}{64\pi}} (4\cos^2\theta \sin\theta - \sin^3\theta)\, e^{\pm i\phi}$$

$$Y_{30} = \sqrt{\frac{7}{16\pi}} (2\cos^3\theta - 3\cos\theta \sin^2\theta)$$

where ω and α are spherical-coordinate angles expressing the direction of $\mathbf{r}_2$ in a coordinate system (x′,y′,z′) having the z′ axis along the direction of $\mathbf{r}_1$. Since (2.44) is an eigenfunction of $\mathbf{L}^2$ belonging to the quantum number l, it may be expanded in terms of the set of eigenfunctions $Y_{lm}(\theta_2,\phi_2)$ referred to the coordinate system (x,y,z):

$$Y_{l0}(\omega,\alpha) = \sum_{m=-l}^{l} A_{lm} Y_{lm}(\theta_2,\phi_2) , \tag{2.45}$$

where the expansion coefficients A_{lm} will of course depend parametrically on θ_1 and ϕ_1.

Given an angular momentum vector $\mathbf{L}$ with components (L_x,L_y,L_z) in the (x,y,z) coordinate system, the component of $\mathbf{L}$ along the z′ axis is

$$\begin{aligned} L_{z'} &= \cos\theta_1\, L_z + \sin\theta_1 \cos\phi_1\, L_x + \sin\theta_1 \sin\phi_1\, L_y \\ &= \cos\theta_1\, L_z + \frac{1}{2}\sin\theta_1\, e^{-i\phi_1}(L_x + iL_y) + \frac{1}{2}\sin\theta_1\, e^{i\phi_1}(L_x - iL_y) . \end{aligned}$$

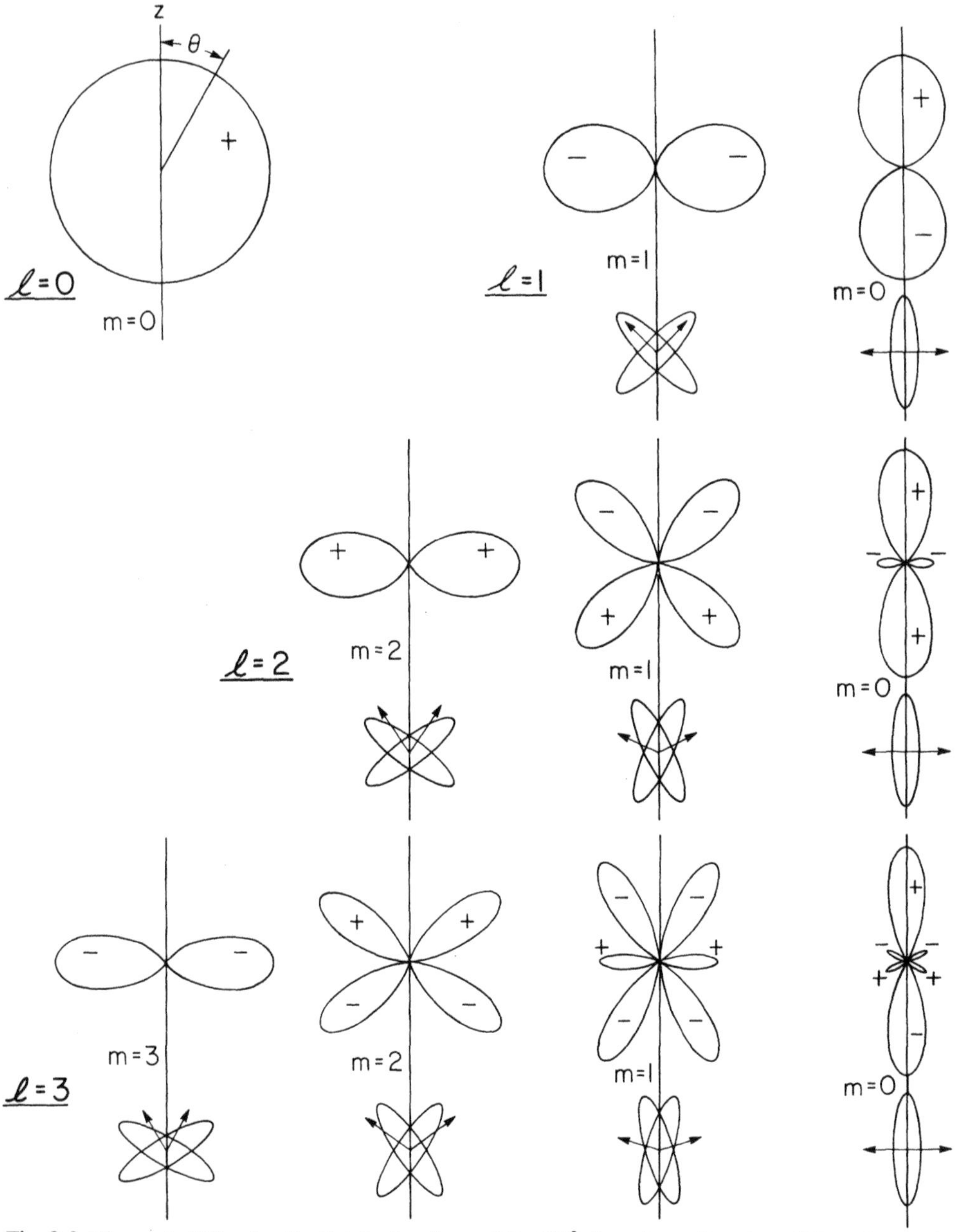

Fig. 2-3. The probability-density distribution factor $[\Theta_{lm}(\theta)]^2$ plotted as a function of the polar angle θ (as defined in Fig. 2-2) for s, p, d, and f electrons. Each figure is rotationally symmetric about the vertical z-axis, and the various lobes are marked + or − according as the sign of $\Theta_{l|m|}$ is positive or negative, respectively. [From (2.38), all signs are changed for m negative and odd.] The classical oriented orbit for each state is given below each figure, tilted slightly out of the normal plane to show an orbit rather than a straight line. [After H. E. White, Introduction to Atomic Spectra (McGraw-Hill, New York, 1934).]

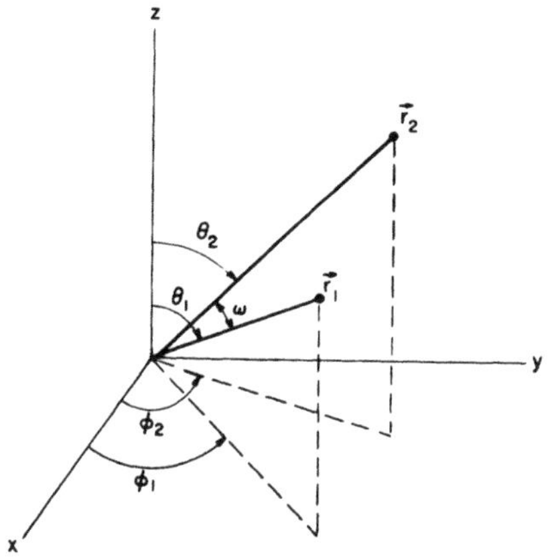

Fig. 2-4. Spherical coordinates of two points in space, and the angel ω between the two radius vectors $\mathbf{r}_1$ and $\mathbf{r}_2$.

Operating with either side of this expression on the corresponding side of (2.45), we obtain with the aid of (2.16) and (2.19)

$$0 = \sum_{m=-l}^{l} A_{lm} \Big\{ m\hbar \cos\theta_1 \, Y_{lm}(\theta_2,\phi_2) + \frac{1}{2}\sin\theta_1 \, e^{-i\phi_1}\hbar[(l-m)(l+m+1)]^{1/2} Y_{l,m+1}(\theta_2,\phi_2) + \frac{1}{2}\sin\theta_1 \, e^{i\phi_1}\hbar[(l+m)(l-m+1)]^{1/2} Y_{l,m-1}(\theta_2,\phi_2) \Big\} .$$

Since the Y_{lm} are linearly independent functions, this expression can be zero for all (θ_2,ϕ_2) only if the coefficient of each Y_{lm} is zero:

$$m\hbar \cos\theta_1 \, A_{lm} + \frac{1}{2}\sin\theta_1 \, e^{-i\phi_1}\hbar[(l-m+1)(l+m)]^{1/2} A_{l,m-1} + \frac{1}{2}\sin\theta_1 \, e^{i\phi_1}\hbar[(l+m+1)(l-m)]^{1/2} A_{l,m+1} = 0 . \tag{2.46}$$

This requirement is satisfied, for any value of (θ_1,ϕ_1), by letting

$$A_{lm} = f(\theta_1,\phi_1)\, Y^{*}_{lm}(\theta_1,\phi_1) = (-1)^m f(\theta_1,\phi_1)\, Y_{l,-m}(\theta_1,\phi_1) , \tag{2.47}$$

where $f(\theta_1,\phi_1)$ is an arbitrary function; for then the left-hand side of (2.46) becomes, upon use of (2.19) and (2.24),

$$(-1)^m f(\theta_1,\phi_1) \Big\{ m\hbar \cos\theta_1 \, Y_{l,-m}(\theta_1,\phi_1) - \frac{1}{2}\sin\theta_1 \, e^{-i\phi_1}\hbar[(l-m+1)(l+m)]^{1/2} Y_{l,-m+1}(\theta_1,\phi_1) - \frac{1}{2}\sin\theta_1 \, e^{i\phi_1}\hbar[(l+m+1)(l-m)]^{1/2} Y_{l,-m-1}(\theta_1,\phi_1) \Big\}$$

$$= (-1)^m f(\theta_1,\phi_1)\Big\{ m\hbar \cos\theta_1 - \frac{1}{2}\sin\theta_1\, e^{-i\phi_1}(L_x + iL_y) - \frac{1}{2}\sin\theta_1\, e^{i\phi_1}(L_x - iL_y)\Big\} Y_{l,-m}(\theta_1,\phi_1)$$

$$= (-1)^m f(\theta_1,\phi_1)\Big\{ m\hbar \cos\theta_1 - i\hbar\cos\theta_1 \frac{\partial}{\partial\phi_1}\Big\} Y_{l,-m}(\theta_1,\phi_1) \,,$$

and this is zero because

$$\frac{\partial}{\partial\phi_1} Y_{l,-m} = -\, im\, Y_{l,-m} \,.$$

Thus, substituting the expression (2.47) into (2.45) we obtain

$$Y_{l0}(\omega,\alpha) = f(\theta_1,\phi_1) \sum_{m=-l}^{l} (-1)^m Y_{l,-m}(\theta_1,\phi_1)\, Y_{lm}(\theta_2,\phi_2) \,. \tag{2.48}$$

But it is obvious that we could have interchanged the roles of $\mathbf{r}_1$ and $\mathbf{r}_2$ throughout the above derivation, and we would then have obtained

$$Y_{l0}(\omega,\alpha) = f(\theta_2,\phi_2) \sum_{m=-l}^{l} (-1)^m Y_{l,-m}(\theta_2,\phi_2)\, Y_{lm}(\theta_1,\phi_1)$$

$$= f(\theta_2,\phi_2) \sum_{m=l}^{-l} (-1)^{-m} Y_{lm}(\theta_2,\phi_2)\, Y_{l,-m}(\theta_1,\phi_1) \,.$$

This can be equal to (2.48), for all $(\theta_1,\phi_1,\theta_2,\phi_2)$, only if f is a constant. Taking $\theta_1 = \theta_2 = \omega = 0$, we find from (2.41) and (2.48) that the value of this constant is $f = [4\pi/(2l+1)]^{1/2}$. Substituting (2.48) into (2.44) we thus finally obtain the spherical-harmonic addition theorem

$$P_l(\cos\omega) = \frac{4\pi}{2l+1} \sum_{m=-l}^{l} (-1)^m Y_{l,-m}(\theta_1,\phi_1)\, Y_{lm}(\theta_2,\phi_2)$$

$$= \frac{4\pi}{2l+1} \sum_{m=-l}^{l} Y^*_{lm}(\theta_1,\phi_1)\, Y_{lm}(\theta_2,\phi_2) \,. \tag{2.49}$$

As an immediate corollary of this result we find, by setting $\theta_1 = \theta_2$ and $\phi_1 = \phi_2$ (whence $\omega = 0$) and using (2.41), that

$$\sum_{m=-l}^{l} |Y_{lm}(\theta,\phi)|^2 = \frac{2l+1}{4\pi} \,. \tag{2.50}$$

Thus the electron density distribution (2.43), summed over all possible values of m, is spherically symmetrical.

2-7. ELECTRON SPIN

An electron is known from both empirical evidence and the relativistic Dirac theory to possess an intrinsic angular momentum corresponding to the value $j = 1/2$. For electron spin, the general symbols **J**, j, and m are customarily replaced by **S**, s ($\equiv 1/2$), and m_s. For a single electron, the only possible eigenvalue of $\mathbf{S}^2$ is

$$s(s + 1)\,\hbar^2 = \frac{3}{4}\,\hbar^2 , \qquad (2.51)$$

and the possible eigenvalues of S_z are

$$m_s\hbar \qquad \left(m_s = -\frac{1}{2} \text{ or } +\frac{1}{2}\right) . \qquad (2.52)$$

For our purposes, an adequate one-electron spin eigenfunction is[11]

$$\sigma_{m_s}(s_z) \equiv \delta_{m_s s_z} , \qquad (2.53)$$

where s_z is the z-component of the spin vector **S** (in units of $\hbar$), and $\delta_{m_s s_z}$ is a Kronecker delta. Thus $\sigma_{1/2}$ is zero unless the spin s is oriented "parallel" to the z axis ($s_z = +1/2$), and $\sigma_{-1/2}$ is zero unless the spin is "antiparallel" to the z axis ($s_z = -1/2$). Orthonormalization of σ is to be defined in terms of a summation over the two possible values of s_z:[12]

$$\langle \sigma_{m_s}(s_z) | \sigma_{m_s'}(s_z) \rangle \equiv \sum_{s_z=-1/2}^{+1/2} \sigma_{m_s}(s_z)\sigma_{m_s'}(s_z)$$

$$= \sum_{s_z} \delta_{m_s s_z}\delta_{m_s' s_z} = \delta_{m_s m_s'} . \qquad (2.54)$$

It is to be understood implicitly that

$$(S_x \pm iS_y)\sigma_{m_s} = \hbar\left[\left(\frac{1}{2} \mp m_s\right)\left(\frac{1}{2} \pm m_s + 1\right)\right]^{1/2} \sigma_{m_s \pm 1}$$

$$= \hbar\left(\frac{1}{2} \mp m_s\right)\sigma_{m_s \pm 1} , \qquad (2.55)$$

in line with the general relationships (2.10), (2.11), and (2.19).

The complete eigenfunction of all four operators $\mathbf{L}^2$, L_z, $\mathbf{S}^2$, and S_z is just the product

$$Y_{l m_l}(\theta,\phi)\cdot\sigma_{m_s}(s_z) . \qquad (2.56)$$

[11]The symbol σ used here for a wavefunction is of course not to be confused with that used in (1.4) for the wavenumber of a spectrum line.

[12]Alternatively, σ could have been defined in terms of the square root of a Dirac delta function:

$$\sigma_{m_s}(s) = \delta^{1/2}(m_s - s_z) ,$$

and orthonormalization then defined in terms of integration over the orientation angles of **S**.

The orthonormalization "integral"

$$\langle Y_{lm_l}\sigma_{m_s}|Y_{l'm'_l}\sigma_{m'_s}\rangle = \delta_{ll'}\delta_{m_lm'_l}\delta_{m_sm'_s} \tag{2.57}$$

of course involves summation over s_z as well as integration over θ and ϕ.

2-8. ADDITION OF TWO ANGULAR MOMENTA

So far we have discussed separately the orbital angular momentum and the spin of an electron. The question now is, how do we describe the total angular momentum $\mathbf{L} + \mathbf{S}$ of the electron? Once we know how to add two angular momenta, then by extension we will be able to add any number of angular momenta together to obtain, for example, the total angular momentum of a multi-electron atom.

To state the problem more precisely, consider a system involving two angular-momentum operators $\mathbf{J}_1$ and $\mathbf{J}_2$ that commute with each other. (That is, each component of $\mathbf{J}_1$ must commute with each component of $\mathbf{J}_2$. As an example, $\mathbf{L}$ and $\mathbf{S}$ commute with each other because they operate on two entirely different sets of coordinates; the same is true of the angular momenta of two different electrons.) If $\psi_{j_1m_1}(q_1)$ is an eigenfunction of $\mathbf{J}_1^2$ and J_{1z} with eigenvalues $j_1(j_1 + 1)\hbar^2$ and $m_1\hbar$, respectively, and if $\psi_{j_2m_2}(q_2)$ is an eigenfunction of $\mathbf{J}_2^2$ and J_{2z} with eigenvalues $j_2(j_2 + 1)\hbar^2$ and $m_2\hbar$, then

$$\psi_{j_1j_2m_1m_2}(q_1,q_2) \equiv \psi_{j_1m_1}(q_1)\cdot\psi_{j_2m_2}(q_2) \tag{2.58}$$

is clearly an eigenfunction of all four operators $\mathbf{J}_1^2$, J_{1z}, $\mathbf{J}_2^2$, and J_{2z}; an example that we have already encountered is the wavefunction (2.56). But

$$\mathbf{J} = \mathbf{J}_1 + \mathbf{J}_2 \tag{2.59}$$

is also an angular-momentum operator, and so it must be possible to find functions

$$\psi_{jm}(q_1,q_2) \tag{2.60}$$

that are eigenfunctions of $\mathbf{J}^2$ and J_z with eigenvalues $j(j + 1)\hbar^2$ and $m\hbar$. The questions are: How are the quantum numbers j and m related to j_1, j_2, m_1, and m_2, and how are the eigenfunctions (2.60) related to the functions (2.58)?

We show first that it is possible to find functions (2.60) that are eigenfunctions of $\mathbf{J}_1^2$ and $\mathbf{J}_2^2$ as well as of $\mathbf{J}^2$. To do this, we note from (2.6) that J_{1z} commutes with $\mathbf{J}_1^2$, and that $J_z \equiv J_{1z} + J_{2z}$ therefore commutes with $\mathbf{J}_1^2$. Similarly, J_x and J_y commute with $\mathbf{J}_1^2$, and likewise all components of $\mathbf{J}$ commute with $\mathbf{J}_2^2$. Therefore

$$\mathbf{J}^2 \equiv \mathbf{J}\cdot\mathbf{J} = J_x^2 + J_y^2 + J_z^2 \tag{2.61}$$

(as well as J_z) commutes with both $\mathbf{J}_1^2$ and $\mathbf{J}_2^2$, which is sufficient to prove the possibility of simultaneous eigenfunctions of the four operators $\mathbf{J}_1^2$, $\mathbf{J}_2^2$, $\mathbf{J}^2$, J_z. [However, these functions cannot also be eigenfunctions of J_{1z} and J_{2z} because, for example, J_{1z} does not commute with

$$J_x^2 = (J_{1x} + J_{2x})^2 = J_{1x}^2 + 2J_{1x}J_{2x} + J_{2x}^2$$

because it does not commute with J_{1x}.]

Eigenfunctions ψ_{jm} that are also eigenfunctions of $\mathbf{J}_1^2$ and $\mathbf{J}_2^2$ we shall denote by

$$\psi_{j_1j_2jm} \equiv \psi_{j_1j_2jm}(q_1,q_2) \ . \tag{2.62}$$

For given values of j_1 and j_2, the set of functions (2.62) for all possible values of j and m must be a complete set spanning the same subspace as does the complete set of functions (2.58) for all possible values of m_1 and m_2. Thus the one set of functions must be expressible as an orthogonal set of linear combinations of the other set of functions. From (2.16) there are $k = (2j_1 + 1)(2j_2 + 1)$ different functions in the set (2.58), and there must be the same number k of different functions in the set (2.62). We shall leave till Chapter 9 the problem of actually calculating the orthogonal transformation connecting the two sets of functions, and here determine only the nature of the k sets of values (jm).

Though the simple product functions (2.58) are in general not eigenfunctions of $\mathbf{J}^2$, they *are* already eigenfunctions of J_z because

$$\begin{aligned}(J_{1z} + J_{2z})\psi_{j_1m_1}\psi_{j_2m_2} &= \psi_{j_2m_2}J_{1z}\psi_{j_1m_1} + \psi_{j_1m_1}J_{2z}\psi_{j_2m_2} \\ &= (m_1 + m_2)\hbar\psi_{j_1m_1}\psi_{j_2m_2} \ .\end{aligned} \tag{2.63}$$

This means that the set of possible values of m must be identical with the set of possible values of $m_1 + m_2$; in particular, for some given value of m, the number of different values of j ($\geqq$ m) must be the same as the number of ways of choosing m_1 and m_2 such that $m_1 + m_2 = m$.

It follows from (2.16) that the maximum value of m is $j_1 + j_2$, and that therefore the maximum value of j is also equal to $j_1 + j_2$. There is only one pair of values of (m_1,m_2) for which $m_1 + m_2$ is equal to $j_1 + j_2$, and consequently only one eigenfunction having j (and m) equal to $j_1 + j_2$. However, there are two pairs of values of (m_1, m_2)—namely, (j_1,j_2-1) and (j_1-1,j_2)—for which $m_1 + m_2 = j_1 + j_2 - 1$. Consequently, there must be two eigenfunctions $\psi_{j_1j_2jm}$ with $m = j_1 + j_2 - 1$; one of these is a function with the value $j = j_1 + j_2$ that we have already found, and the other must be a function with a new value $j = j_1 + j_2 - 1$. Similarly, there are (if j_1 and j_2 are sufficiently large) three pairs of values of (m_1,m_2) for which $m_1 + m_2 = j_1 + j_2 - 2$—

$$(j_1,\ j_2-2), \qquad (j_1-1,\ j_2-1), \qquad (j_1-2,\ j_2)$$

—and consequently $j_1 + j_2 - 2$ must also be a possible value of j. This argument can be continued until we have encountered the value $m_1 = -j_1$ or $m_2 = -j_2$, whichever occurs first; beyond this point the number of pairs of values of (m_1,m_2) for fixed m no longer increases, and so we obtain no new values of j.

This completes the task of finding the possible sets of quantum numbers for the eigenfunctions $\psi_{j_1j_2jm}$: For given j_1 and j_2, the possible values of j are

$$j = j_1 + j_2,\ \ j_1 + j_2 - 1,\ \ j_1 + j_2 - 2,\ \cdots,\ |j_1 - j_2| \ ; \tag{2.64}$$

and for each j the values of m run as usual [see (2.16)] from $-j$ to $+j$ in integral steps; a numerical example is given in Table 2-2. The number of different possible values of j is $2j_1 + 1$ or $2j_2 + 1$, whichever is smaller. The total number of states (different states being specified by different pairs of values jm) is

$$\sum_{j=|j_1-j_2|}^{j_1+j_2} (2j + 1) = \frac{1}{2}\left[2(j_1 + j_2) + 2|j_1 - j_2| + 2\right]\left[(j_1 + j_2) - |j_1 - j_2| + 1\right]$$

$$= (j_1 + j_2 + 1)^2 - (j_1 - j_2)^2$$

$$= 4j_1j_2 + 2j_1 + 2j_2 + 1$$

$$= (2j_1 + 1)(2j_2 + 1) \ , \qquad (2.65)$$

TABLE 2-2. MAGNETIC QUANTUM NUMBERS FOR THE UNCOUPLED AND COUPLED ORBITAL-ANGULAR-MOMENTUM STATES OF AN ELECTRON WITH $l = 2$ PLUS AN ELECTRON WITH $l = 1$

l_1	m_{l_1}	l_2	m_{l_2}	m_l	No. of states for given m_l
2	2	1	1	3	x
	2		0	2	x x
	1		1	2	
	2		−1	1	x x x
	1		0	1	
	0		1	1	
	1		−1	0	x x x
	0		0	0	
	−1		1	0	
	0		−1	−1	x x x
	−1		0	−1	
	−2		1	−1	
	−1		−1	−2	x x
	−2		0	−2	
	−2		−1	−3	x
					↑ ↑ ↑
				L =	3 2 1

The numbers of states for the various values of m_l are sufficient to corresppond to values of L (=quantum number for L^2, where $L = l_1 + l_2$) of 3, 2, and 1.

which is indeed the same as the number k of different states $\psi_{j_1j_2m_1m_2}$.

A specific example of great importance is that of a single electron in a state with orbital angular momentum quantum number l. The states in which $\boldsymbol{l}$ and $\boldsymbol{s}$ have been added together to give states of definite total angular momentum **J** have quantum number $j = l + 1/2$ or $l - 1/2$ (except that only $j = l + 1/2$ is possible if $l = 0$). The total number of possible states computed in the jm scheme,

$$\left[2\left(l + \frac{1}{2}\right) + 1\right] + \left[2\left(l - \frac{1}{2}\right) + 1\right] = 4l + 2 ,$$

is the same as that found in the m_1m_2 scheme,

$$(2l + 1)(2s + 1) = (2l + 1)\cdot 2 = 4l + 2 . \tag{2.66}$$

2-9. THE VECTOR MODEL

The result (2.64) can be visualized graphically in the manner illustrated in Fig. 2-5(a) by adding two vectors of lengths j_1 and j_2 at different relative angles, starting with the parallel case which gives a resultant vector of length $j_{max} = j_1 + j_2$ and taking all possible resultant lengths that differ from this by integers, down to $j_{min} = |j_1 - j_2|$ for the antiparallel case.

However, this semiclassical vector model of the addition of two angular-momentum vectors must not be used for any purpose other than visualizing the range of possible values of j. It must be remembered from (2.17) that the magnitudes of the angular-momentum vectors (in units of $\hbar$) are greater than the corresponding quantum numbers. Thus if, in the case $j = j_{max} = j_1 + j_2$, the two vectors $\mathbf{J}_1$ and $\mathbf{J}_2$ were actually parallel the square of the magnitude of the vector sum would be

$$\left\{[j_1(j_1 + 1)]^{1/2} + [j_2(j_2 + 1)]^{1/2}\right\}^2 = j_1^2 + j_2^2 + j_1 + j_2 + 2[(j_1^2 + j_1)(j_2^2 + j_2)]^{1/2} ;$$

this is greater than the squared magnitude of the vector **J** with quantum number j_{max},

$$(j_1 + j_2)(j_1 + j_2 + 1) = j_1^2 + j_2^2 + j_1 + j_2 + 2j_1j_2 .$$

Thus in the more accurate quantum-mechanical picture, the vectors $\mathbf{J}_1$, $\mathbf{J}_2$, $\mathbf{J}_{max}$ cannot form a degenerate triangle; this triangle—and, similarly, that formed by $\mathbf{J}_1$,$\mathbf{J}_2$,$\mathbf{J}_{min}$—must be non-degenerate, as shown in Fig. 2-5(b).

2-10. COUPLED WAVEFUNCTIONS

The wavefunctions

$$\psi_{j_1j_2m_1m_2} \equiv \psi_{j_1m_1}\cdot\psi_{j_2m_2} \tag{2.67}$$

are called uncoupled functions. The functions

$$\psi_{j_1j_2jm} \tag{2.68}$$

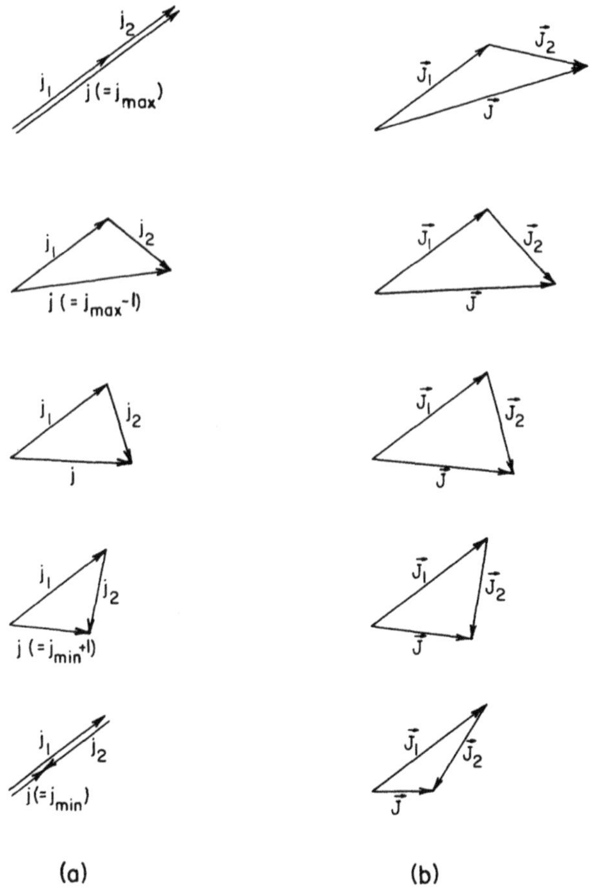

Fig. 2-5. (a) Vector-model picture of the possible values of the quantum number j corresponding to the resultant **J** of two angular-momentum vectors $\mathbf{J}_1$ and $\mathbf{J}_2$ with quantum numbers j_1 and j_2. (b) Quantum-mechanical relationship of the three vectors with magnitudes $|j_1(j_1+1)|^{1/2}$, $|j_2(j_2+1)|^{1/2}$, and $|j(j+1)|^{1/2}$.

are called coupled functions, because they represent states in which the angular-momentum vectors $\mathbf{J}_1$ and $\mathbf{J}_2$ have been coupled together in such a way as to give a definite resultant vector **J**. Either set of functions forms a complete set of orthonormal functions spanning the same quantum space, and consequently each of the functions (2.68) can be written as a linear combination of the functions (2.67):

$$\psi_{j_1j_2jm} = \sum_{m_1=-j_1}^{j_1} \sum_{m_2=-j_2}^{j_2} C(j_1j_2m_1m_2;jm)\psi_{j_1m_1}\psi_{j_2m_2} \ . \tag{2.69}$$

From the discussion of Sec. 2-8 [especially Eq. (2.63)], it is evident that the summation can actually include only those terms for which $m_1 + m_2 = m$; thus each coefficient C must contain a factor $\delta_{m_2,m-m_1}$, and (2.69) simplifies to

$$\psi_{j_1j_2jm} = \sum_{m_1} C(j_1j_2m_1,m-m_1;jm)\psi_{j_1m_1}\psi_{j_2,m-m_1} \; . \tag{2.70}$$

The coefficients are called vector-addition or Clebsch-Gordon coefficients; we shall postpone till Chapter 9 the problem of evaluating them in the general case, and here merely consider a specific example, corresponding to the second and third lines of Table 2-2: For $j_1 = 2$ and $j_2 = 1$, the two coupled functions for $m = 2$ are

$$j = 3: \qquad \psi_{2132} = (1/3)^{1/2}\psi_{22}(1)\psi_{10}(2) + (2/3)^{1/2}\psi_{21}(1)\psi_{11}(2) \; , \tag{2.71}$$

$$j = 2: \qquad \psi_{2122} = (2/3)^{1/2}\psi_{22}(1)\psi_{10}(2) - (1/3)^{1/2}\psi_{21}(1)\psi_{11}(2) \; , \tag{2.72}$$

where the numbers in parentheses indicate the coordinates of which each ψ is a function. It is easily verified from the assumed orthonormality of the functions $\psi_{j_im_i}$ that (2.71) and (2.72) constitute a pair of orthogonal normalized functions. That they are eigenfunctions of J_z with an eigenvalue corresponding to $m = 2$ is easily seen from (2.63), and that they are eigenfunctions of $\mathbf{J}^2$ corresponding to $j = 3$ and $j = 2$, respectively, can be verified in the manner outlined in the problem at the end of this section.

The physical reason for our interest in coupled wavefunctions is that there usually exists some sort of interaction between the two subsystems to which the angular momenta $\mathbf{J}_1$ and $\mathbf{J}_2$ pertain. For example, in the case of two electrons in an atom moving with orbital angular momenta $\boldsymbol{l}_1$ and $\boldsymbol{l}_2$, there exists the Coulomb repulsion of either electron by the other; in the case of a single electron, there exists a magnetic interaction between the magnetic dipole associated with the spin angular momentum $\mathbf{s}$ and that associated with the orbital angular momentum $\boldsymbol{l}$. As a consequence of such interactions, each subsystem exerts a torque on the other so that neither $\mathbf{J}_1$ nor $\mathbf{J}_2$ is a constant of the motion; however, if no *external* torque acts on the combined system $(\mathbf{J}_1,\mathbf{J}_2)$, then the total angular momentum $\mathbf{J}$ does remain constant. The internal torques cause $\mathbf{J}_1$ and $\mathbf{J}_2$ to precess about the resultant $\mathbf{J}$; since $\mathbf{J}$ cannot lie exactly along the z axis (Fig. 2-1), the z-components of $\mathbf{J}_1$ and $\mathbf{J}_2$ are not constant, and m_1 and m_2 are therefore not "good" (i.e., not physically significant) quantum numbers—only the sum $J_{1z} + J_{2z}$ remains constant during the precession, and so only $m_1 + m_2$ is a good quantum number. The coupled wavefunctions (2.70)-(2.72) reflect this physical situation exactly: each corresponds to definite values of the quantum numbers j and m, in line with the fact that the magnitude and orientation of $\mathbf{J}$ are constant in time, but consists of a mixture of uncoupled functions for all possible pairs of values (m_1,m_2) such that $m_1 + m_2$ is equal to m.

If the interaction between $\mathbf{J}_1$ and $\mathbf{J}_2$ is small in magnitude, the rate of precession of $\mathbf{J}_1$ and $\mathbf{J}_2$ is small, and these angular momenta remain well-defined in magnitude even though not in direction. This is reflected in that the coupled wavefunctions, as well as the uncoupled ones, correspond to definite values of the quantum numbers j_1 and j_2. If the interaction is very strong, the precession becomes so fast that the individual angular momenta cannot really be defined even as to magnitude; correspondingly, the quantum-mechanical state of the system must be described by a mixture of the coupled wavefunctions (2.70) for all possible pairs of values of $(j_1 j_2)$ that can combine according to the vector model to give the appropriate value of the quantum number j; thus only j and m remain as good quantum numbers.

If an external torque acts on the total angular momentum **J**, owing for example to an external magnetic field oriented in the direction of the z-axis, then **J** is no longer fixed in direction, but precesses about the z-axis. If the precession is sufficiently fast (because of the external field being very strong) then the magnitude of **J** (but not the z-component) loses its significance, and the wavefunction must be written as a superposition of functions of fixed m but of all possible values of j ($\geqq$m), so that only m remains as a good quantum number.

Wavefunction mixings of the above types occur almost universally in atomic structure calculations, and many examples will be encountered in the following chapters; their importance can hardly be overemphasized.

Problem

2-10(1). (a) Show that

$$\mathbf{J}_1^2 + \mathbf{J}_2^2 + 2J_{1z}J_{2z} + (J_{1x} + iJ_{1y})(J_{2x} - iJ_{2y}) + (J_{1x} - iJ_{1y})(J_{2x} + iJ_{2y})$$
$$= (J_{1x} + J_{2x})^2 + (J_{1y} + J_{2y})^2 + (J_{1z} + J_{2z})^2 \equiv \mathbf{J}^2 .$$

(b) Using this result and (2.19) show that

$$\mathbf{J}^2\psi_{j_1m_1}\psi_{j_2m_2} = [j_1(j_1 + 1) + j_2(j_2 + 1) + 2m_1m_2]\hbar^2\psi_{j_1m_1}\psi_{j_2m_2}$$
$$+ [(j_1 - m_1)(j_1 + m_1 + 1)(j_2 + m_2)(j_2 - m_2 + 1)]^{1/2}\hbar^2\psi_{j_1,m_1+1}\psi_{j_2,m_2-1}$$
$$+ [(j_1 + m_1)(j_1 - m_1 + 1)(j_2 - m_2)(j_2 + m_2 + 1)]^{1/2}\hbar^2\psi_{j_1,m_1-1}\psi_{j_2,m_2+1} \cdot$$

(c) Verify with the aid of this expression that (2.71) and (2.72) are eigenfunctions of $\mathbf{J}^2$ corresponding to j = 3 and 2, respectively.

2-11. COUPLING SCHEMES; LS COUPLING

In an atom containing N electrons with N > 1, there are more than 2 elementary angular momenta—N orbital angular momenta $\boldsymbol{l}_i$, and N spin momenta $\mathbf{s}_i$—and the total angular momentum of the atom is

$$\mathbf{J} = \sum_{i=1}^{N} (\boldsymbol{l}_i + \mathbf{s}_i) . \qquad (2.73)$$

Starting with product wavefunctions of the form

$$\prod_{i=1}^{N} Y_{l_i m_{l_i}}(\theta_i,\phi_i)\cdot\sigma_{m_{s_i}}(s_{iz}) , \qquad (2.74)$$

completely coupled wavefunctions that are eigenfunctions of $\mathbf{J}^2$ and J_z can be obtained by $2N - 1$ successive applications of the procedure described in the preceding section; for example, by coupling a third angular momentum onto the resultant of the first two, a fourth momentum onto the resultant of these three, etc.

Evidently the order in which the momenta are coupled together can be chosen in many different ways; any specific choice is known as a *coupling scheme.* The individual wavefunctions will correspond most closely to the various physical states of the atom if the coupling scheme used corresponds to the coupling of successive momenta in the order of decreasing strength of the various interactions.

Usually, the strongest interactions among the electrons of an atom are their mutual Coulomb repulsions. These repulsions affect only the orbital angular momenta, and not the spins. It is thus most appropriate to first couple together all the orbital angular momenta (in some, for the present, unspecified order) to give eigenfunctions of $\mathbf{L}^2$ and L_z, where

$$\mathbf{L} = \sum_i \boldsymbol{l}_i \tag{2.75}$$

is the total orbital angular momentum of the atom, and similarly to couple together all spins[13] to give eigenfunctions also of $\mathbf{S}^2$ and S_z, where

$$\mathbf{S} = \sum_i \mathbf{s}_i \tag{2.76}$$

is the total spin angular momentum; **L** and **S** are then coupled together to give eigenfunctions of $\mathbf{J}^2$ and J_z, where

$$\mathbf{J} = \mathbf{L} + \mathbf{S} \;. \tag{2.77}$$

This scheme is known as the[14] LS or Russell-Saunders[15] coupling scheme.

If the physical coupling conditions existing in the atom closely approximate LS-coupling conditions, then the individual $\boldsymbol{l}_i$ are to be thought of as precessing rapidly in some complicated (but unspecified) manner about the resultant **L**, and the individual $\mathbf{s}_i$ similarly as precessing about their resultant **S**. If all interactions other than the Coulomb interactions are neglected, then **L** and **S** are constants of the motion, the angular-momentum coupling procedure can be ended with (2.75) and (2.76), and the wavefunctions and quantum states can be described in terms of the four good quantum numbers L, S, M_L, M_S. Physically, however, the vectors **L** and **S** are always coupled together by means of non-zero (even if weak) spin-orbit interactions and the vectors **L** and **S** precess about **J**—sufficiently slowly

[13]The (principal) interactions that couple the spins together are also electrostatic in origin, but are quantum mechanical in nature and have no classical analog.

[14]Or, rather, as *an* LS-coupling scheme, since there is more than one possible order of coupling the various $\boldsymbol{l}_i$ if $N > 2$. Condon and Shortley and most other spectroscopy texts use a coupling scheme that corresponds to $\mathbf{J} = \mathbf{S} + \mathbf{L}$; this is usually also referred to as LS coupling, though we shall call it SL coupling to distinguish it from the coupling scheme (2.77). The reason for our choice of the LS scheme will become clear when we discuss LK coupling in Secs. 4-18 and 9-3.

[15]H. N. Russell and F. A. Saunders, Astrophys. J. **61**, 38 (1925).

that the magnitudes of **L** and **S** (described by the quantum numbers L and S) are fairly well-defined, but such that only the sum of the z components ($M = M_L + M_S$) is defined; the quantum states are then described in terms of the four quantum numbers L, S, J, M.

The LS scheme is physically the most important single coupling scheme, and is also usually the most convenient one to use for practical calculations. There are, however, many instances in which the coupling conditions in an atom correspond more or less closely to various other coupling schemes, which will be described in Chapters 4 and 9. There also exist innumerable cases in which the coupling conditions do not closely approximate any conceivable coupling scheme. These cases are referred to generally as intermediate-coupling cases. The wavefunctions of the various quantum states can then be expressed only as mixtures of pure-coupling functions of the sort described in the preceding section; the only really "good" quantum numbers may be the quantum numbers J and M that correspond to the total angular momentum **J** of the atom.

2-12. NOTATION FOR ANGULAR-MOMENTUM STATES

The customary notation that we have employed in this chapter for the discussion of orbital and spin angular momenta of electrons and atoms, together with the possible values of the various quantum numbers, is summarized in Table 2-3 for convenient reference.

It is confusing to have to designate a specific angular-momentum state in terms of a string of numerical values of quantum numbers [note, for example, Eqs. (2.71) and (2.72)], and so an alphabetic notation has been developed for the designation of orbital angular-momentum states. For historical reasons having to do with properties of certain series of observed spectral lines of the alkali and akaline-earth elements (namely, sharp, principal, diffuse, and fundamental series),[16] orbital angular-momentum states corresponding to $l =$ 0, 1, 2, and 3 are called s, p, d, and f states, respectively. For orbital-angular momentum states of the atom as a whole (or for states of certain significant subsets of atomic electrons), analogous capital-letter notation is used. For values of l (or L) greater than 3, successive letters of the alphabet following f are used (with omission of j, which is reserved for other purposes, and omission of s, p, and d which have already been used):[17]

l or L:	0	1	2	3	4	5	6	7	8	9
Symbols:	s	p	d	f	g	h	i	k	l	m
	S	P	D	F	G	H	I	K	L	M
l or L:	10	11	12	13	14	15	16	17	18	19
Symbols:	n	o	q	r	t	u	v	w	x	y
	N	O	Q	R	T	U	V	W	X	Y

[16]See, for example, H. G. Kuhn, *Atomic Spectra* (Longmans, Green, London, 1969), 2nd ed., Chap I.

[17]The symbols s, S and l, L used here for states with l or L = 0 and l or L = 8 can be distinguished from the letters used to designate the general spin and orbital angular-momentum quantum numbers (Table 2-3) only through the context in which they are used.

TABLE 2-3. ANGULAR MOMENTUM NOTATION

	General	Angular Momentum of One Electron			Angular Momentum of N-Electron Atom		
		Orbital	Spin	Total	Orbital	Spin	Total
Vector	$\mathbf{J}$	$\boldsymbol{l}$	$\mathbf{s}$	$\mathbf{j} \equiv \boldsymbol{l} + \mathbf{s}$	$\mathbf{L} \equiv \sum_{i=1}^{N} \boldsymbol{l}_i$	$\mathbf{S} \equiv \sum_{i=1}^{N} \mathbf{s}_i$	$\mathbf{J} \equiv \sum_{i=1}^{N} (\boldsymbol{l}_i + \mathbf{s}_i)$
Eigenfunction[a]	ψ_{jm} (or ψ_{jm_j})	$\psi_{lm_l}(\theta,\phi)$	$\sigma_{m_s}(s_z)$	$\psi_{ljm}(\theta,\phi,s_z)$	ψ_{LM_L}	ψ_{SM_S}	ψ_{JM} (for LS coupling, ψ_{LSJM})
Operator	$\mathbf{J}^2$	$\boldsymbol{l}^2$	$\mathbf{s}^2$	$\mathbf{j}^2$	$\mathbf{L}^2$	$\mathbf{S}^2$	$\mathbf{J}^2$
Eigenvalue	$j(j+1)\hbar^2$	$l(l+1)\hbar^2$	$s(s+1)\hbar^2$	$j(j+1)\hbar^2$	$L(L+1)\hbar^2$	$S(S+1)\hbar^2$	$J(J+1)\hbar^2$
Possible values of quantum numbers	$j = 0,1,2,3,\cdots$ or $j = \frac{1}{2},\frac{3}{2},\frac{5}{2},\frac{7}{2},\cdots$	$l = 0,1,2,3,\cdots$	$s = \frac{1}{2}$ only	$j = l - \frac{1}{2}$ $(l > 0)$ and $j = l + \frac{1}{2}$	L integral ($L_{max} = \Sigma l_i$)[b]	S integral (N even), or S half-integral (N odd) ($S_{max} = \Sigma s_i = N/2$)[b]	J integral (N even), or J half-integral (N odd) ($J_{max} = \Sigma l_i + \Sigma s_i$)[b] (for LS, $J = \|L-S\|,\cdots,L+S$)
Operator	J_z	l_z	s_z	j_z	L_z	S_z	J_z
Eigenvalue	$m\hbar$ (or $m_j\hbar$)	$m_l\hbar$	$m_s\hbar$	$m\hbar$	$M_L\hbar$	$M_S\hbar$	$M\hbar$
Possible values of quantum numbers	$m = -j,-j+1,\cdots,j-1,j$	$m_l = -l,-l+1,\cdots,l-1,l$	$m_s = \pm\frac{1}{2}$	$m = -j,-j+1,\cdots,j-1,j$	$M_L = -L,-L+1,\cdots,L-1,L$	$M_S = -S,-S+1,\cdots,S-1,S$	$M = -J,-J+1,\cdots,J-1,J$

[a] For a multi-electron atom, the eigenfunction is labeled not only by the quantum numbers shown (LM_L,SM_S,JM), but also by all l_i and (if $N > 2$) by other quantum numbers as well.

[b] Subject to limitations resulting from the Pauli exclusion principle, as discussed in Chapter 4.

Single-electron states with values of l greater than 6 (i states) have not yet been identified experimentally with certainty,[18] but in complex rare-earth atoms K, L, M, and N states are encountered frequently.

For LS-coupled functions, the shorthand notation introduced by Russell and Saunders[15,19] is universally used—specifically,

$$^{2S+1}L_j \; , \tag{2.78}$$

where numerical values are to be substituted for $2S + 1$ (S = total spin quantum number) and J, and the appropriate letter symbol is to be used for the total orbital quantum number L; except when discussing the Zeeman or Stark effect, there is usually no need to specify the value of M. In the special case of a one-electron atom, S is necessarily equal to s ($=1/2$) and L is necessarily equal to l; the possible values of (2.78) are then readily seen from (2.77) and the vector model (2.64) to be[20]

$$^2S_{1/2}, \, ^2P_{1/2}, \, ^2P_{3/2}, \, ^2D_{3/2}, \, ^2D_{5/2}, \; \cdots \; . \tag{2.79}$$

For the case of a d plus a p electron used in Table 2-2, the possibilities are seen to be

$$^1P_1, \, ^3P_{0,1,2}, \, ^1D_2, \, ^3D_{1,2,3}, \, ^1F_3, \, ^3F_{2,3,4} \; .$$

When the coupling conditions within an atom correspond closely to pure LS-coupling conditions, then the quantum states of the atom can be accurately described in terms of LS-coupling quantum numbers. Giving values of L and S specifies a *term*;[21] giving values of L, S, and J specifies a *level*; giving values of L, S, J, and M specifies a *state*. The value of $2S + 1$ is called the *multiplicity* of the term, even though the number of levels in the term is equal to $2L + 1$ rather than to $2S + 1$ when $L < S$.

Problem

2-12(1). Use the vector model to show that for a three-electron system consisting of an f, a d, and an s electron, the possible terms are

$$^{2,2,4}PDFGH \; ;$$

i.e., two doublets and a quartet for each value $L = 1, 2, 3, 4$, and 5.

[18]k states have been produced by laser excitation: W. E. Cooke, T. F. Gallagher, S. A. Edelstein, and R. M. Hill, Phys. Rev. Lett. **40**, 178 (1978).

[19]H. N. Russell, A. G. Shenstone, and L. A. Turner, Phys. Rev. **33**, 900 (1929).

[20]Read "doublet S one-half," "doublet D five-halves," etc. The multiplicities 1, 2, 3, 4, 5, 6, 7, 8, 9, $\cdots$ are read singlet, doublet, triplet, quartet, quintet, sextet, septet, octet, nonet, $\cdots$.

[21]Or, more precisely, an "LS term," because one may also refer to "terms" of a different sort when discussing other coupling schemes. (In order to completely specify a term it is necessary to give not only values of L and S, but also values of all lower-order quantum numbers, such as the $n_i l_i$.)

2-13. PARITY

In addition to the classification of wavefunctions by means of their angular-momentum properties, they can be further described in terms of the important concept of parity, which is defined in the following manner.

The total wavefunction of an atom is a function of the N positional-coordinate vectors and N spin angular momenta of the electrons, which we shall indicate collectively by $\mathbf{r}_i,\mathbf{s}_i$. The Hamiltonian of the system is a function of the $\mathbf{r}_i$, the corresponding linear momenta $\mathbf{p}_i$, and (if magnetic interactions are included) the various angular momenta $\boldsymbol{l}_i$ and $\mathbf{s}_i$. The wavefunction satisfies the Schrödinger equation

$$H(\mathbf{r}_i,\mathbf{p}_i,\boldsymbol{l}_i,\mathbf{s}_i)\psi(\mathbf{r}_i,\mathbf{s}_i) = E\psi(\mathbf{r}_i,\mathbf{s}_i) \tag{2.80}$$

for all values of the coordinates $\mathbf{r}_i$; in particular, it is satisfied if all $\mathbf{r}_i$ are replaced by $-\mathbf{r}_i$ (and thereby all $\mathbf{p}_i$ are replaced by $-\mathbf{p}_i$, since $p_x = -i\hbar\partial/\partial x$):

$$H(-\mathbf{r}_i,-\mathbf{p}_i,\boldsymbol{l}_i,\mathbf{s}_i)\psi(-\mathbf{r}_i,\mathbf{s}_i) = E\psi(-\mathbf{r}_i,\mathbf{s}_i) \ . \tag{2.81}$$

(The spin vectors are of course unaffected by this substitution, as are the orbital angular-momentum vectors $\boldsymbol{l}_i = \mathbf{r}_i \times \mathbf{p}_i$.)

For an isolated atom in field-free space

$$H(-\mathbf{r}_i,-\mathbf{p}_i,\boldsymbol{l}_i,\mathbf{s}_i) = H(\mathbf{r}_i,\mathbf{p}_i,\boldsymbol{l}_i,\mathbf{s}_i) \ , \tag{2.82}$$

because neither kinetic energies nor the interparticle distances involved in Coulomb interactions are affected by changing the signs of all the $\mathbf{r}_i$ and $\mathbf{p}_i$. Thus we see that $\psi(\mathbf{r}_i,\mathbf{s}_i)$ and $\psi(-\mathbf{r}_i,\mathbf{s}_i)$ are solutions of the same Schrödinger equation for the same eigenvalue E. If the energy state E is non-degenerate,[22] this can be true only if the one function is a constant multiple of the other:

$$\psi(-\mathbf{r}_i,\mathbf{s}_i) = c\psi(\mathbf{r}_i,\mathbf{s}_i) \ . \tag{2.83}$$

Since this relation holds for all $\mathbf{r}_i$, it must hold if the signs of all the $\mathbf{r}_i$ are changed:

$$\begin{aligned}\psi(\mathbf{r}_i,\mathbf{s}_i) &= c\psi(-\mathbf{r}_i,\mathbf{s}_i)\\ &= c^2\psi(\mathbf{r}_i,\mathbf{s}_i) \ .\end{aligned}$$

Thus the only possible values of c are ± 1; ψ is said to have even or odd parity according as the value of c is $+1$ or -1, respectively.

From Fig. 2-2 it is easily seen that in spherical coordinates, inversion of a coordinate vector $\mathbf{r}$ means

[22]Degeneracy with respect to the quantum number M, which specifies the orientation of the total angular momentum $\mathbf{J}$, can be removed by placing the atom in an external magnetic field. This does not invalidate (2.82), because the magnetic energy of interaction with the field depends only on the $\boldsymbol{l}_i$ and $\mathbf{s}_i$.

$$\begin{aligned} r &\rightarrow r, \\ \theta &\rightarrow \pi - \theta , \\ \phi &\rightarrow \pi + \phi . \end{aligned} \tag{2.84}$$

Under such an inversion, $\cos\theta \rightarrow (-\cos\theta)$, $\sin\theta \rightarrow \sin\theta$, and $e^{im\phi} \rightarrow (-1)^m e^{im\phi}$, and it is easily seen from (2.39) and (2.40) that

$$R(r)\cdot Y_{lm}(\theta,\phi)\cdot\sigma_{m_s}(s_z) \rightarrow (-1)^l R(r)\cdot Y_{lm}(\theta,\phi)\cdot\sigma_{m_s}(s_z) , \tag{2.85}$$

where $R(r)$ is any function of $|\mathbf{r}|$. Thus product functions of the form (2.67), except containing N factors instead of only two, have a parity given by

$$p = (-1)^{\Sigma l_i} , \tag{2.86}$$

where the sum is over all N values of i. The parity of coupled wavefunctions is given by this same expression, because the coupling of angular momenta (2.69) involves summations over magnetic quantum numbers for fixed values of l_i (and s_i).

In the Russell-Saunders notation (2.78) for LS-coupled functions, odd terms are indicated[19] by a superscript $^\circ$ appended to the letter symbol for L. Including this parity symbol, the possible levels (2.79) of a one-electron atom are:[23]

$$^2S_{1/2},\ ^2P^\circ_{1/2},\ ^2P^\circ_{3/2},\ ^2D_{3/2},\ ^2D_{5/2},\ ^2F^\circ_{5/2},\ \cdots .$$

All the other examples given in the preceding section have odd parity:

$$\begin{aligned} &dp\ \ ^{1,3}P^\circ,\ ^{1,3}D^\circ,\ ^{1,3}F^\circ\ ; \\ &fds\ \ ^{2,2,4}P^\circ D^\circ F^\circ G^\circ H^\circ\ . \end{aligned}$$

2-14. PARITY AND ANGULAR-MOMENTUM SELECTION RULES

The theory of atomic structure and spectra that will be developed in the following chapters is based almost entirely on matrix methods. Consequently, we shall repeatedly be involved in evaluating matrix elements of the form

$$\langle\psi|O|\psi'\rangle \equiv \int\!\!\int\!\!\int_{-\infty}^{\infty} \psi^*(\mathbf{r}_i)O(\mathbf{r}_i)\psi'(\mathbf{r}_i)\,dx_i\,dy_i\,dz_i , \tag{2.87}$$

where ψ and ψ' are wavefunctions for the atomic system (or a portion thereof) and O is some operator. (Possible functional dependence of ψ and O on angular momenta has been omitted for brevity, as have summations over the spin directions.) In many cases, it is possible to say that the values of certain matrix elements are zero simply on the basis of the symmetry properties of the wavefunctions and the operator. The restrictions thereby placed on the values of the quantum numbers of ψ and ψ' in order that (2.87) be non-zero are called *selection rules*.

[23]Read "doublet S one-half," "doublet P odd one-half," etc.

All operators in which we shall be interested have a definite parity p_o. If the parities of the wavefunctions are p and p', then changing the signs of the integration variables ($\mathbf{r} \rightarrow -\mathbf{r}$) in (2.87) we find (since renaming of integration variables cannot change the value of an integral) that

$$\langle\psi|O|\psi'\rangle = \int\int\int_{\infty}^{-\infty} \psi^*(-\mathbf{r}_i)O(-\mathbf{r}_i)\psi'(-\mathbf{r}_i)(-dx_i)(-dy_i)(-dz_i)$$

$$= (-1)^{p+p_o+p'} \int\int\int_{-\infty}^{\infty} \psi^*(\mathbf{r}_i)O(\mathbf{r}_i)\psi'(\mathbf{r}_i)\ dx_i\ dy_i\ dz_i$$

$$= (-1)^{p+p_o+p'}\langle\psi|O|\psi'\rangle \ . \tag{2.88}$$

We thus obtain the selection rule that ψ must have the same or the opposite parity as ψ' according as O is of even or odd parity, respectively.

For example, the Hamiltonian operator for an isolated field-free atom has even parity [Eq. (2.82)], and so connects two functions ψ and ψ' only if these functions have a common parity. Likewise, the energy of the atom in an external magnetic field, $-\mathbf{B}\cdot\mu$ where μ is the magnetic moment of the atom, is of even parity[22] and so matrix elements of this operator are non-zero only between wavefunctions of the same parity. On the other hand, the energy of interaction $\mathfrak{E}\cdot(\Sigma_i -e\mathbf{r}_i)$ between an external electric field and the electric dipole moment $\Sigma(-e)\mathbf{r}_i$ of the atom is obviously of odd parity, so that this operator connects only wavefunctions of opposite parity. This is true whether the electric field is static (Stark effect), quasi-static (Stark-effect collisional line broadening), or associated with an electromagnetic field (radiative electric-dipole transitions).

Some important selection rules involving angular-momentum quantum numbers are also easily discussed using arguments similar to those employed in Secs. 2-10 and 2-11:

Any operator O related to forces internal to the atom cannot alter the total angular momentum **J** of the atom. Consequently, for fully coupled functions $\psi_{JM} \equiv |JM\rangle$ the matrix elements

$$\langle JM|O|J'M'\rangle \tag{2.89}$$

are zero unless $J = J'$ and $M = M'$; moreover, the values of the matrix elements must be independent of M, because they cannot depend on the arbitrarily chosen orientation of the coordinate system. In the case of the Coulomb interaction between two electrons i and j, $O = e^2/r_{ij} \equiv e^2/|\mathbf{r}_i - \mathbf{r}_j|$, which acts only on the orbital angular momenta and not on the spins, **L** and **S** as well as **J** are constants of the motion; therefore, using LS-coupled functions, the matrix elements

$$\langle LSJM|e^2/r_{ij}|L'S'J'M'\rangle$$

must be zero unless $L = L'$, $S = S'$, $J = J'$, and $M = M'$, and must be independent of J as well as of M.

If incompletely coupled LS functions are used, then the matrix with elements

$$\langle LSM_LM_S|e^2/r_{ij}|L'S'M'_LM'_S\rangle$$

must be diagonal in all four quantum numbers, and independent of M_L and M_S.

The above selection rules ($\Delta J = 0$, etc.) are mathematically a result of the fact that the operators involved are scalars (independent of the orientation of the coordinate system) and are therefore proportional to the spherical harmonic Y_{00} of degree zero. In contrast, the electric dipole operator $\mathbf{D}_e \equiv \Sigma(-e)\mathbf{r}_i$ involved in the calculation of radiative transitions is a vector quantity, related to the spherical harmonics Y_{lm} with $l = 1$ [cf. (2.20) with Table 2-1]; this corresponds physically to the radiated photon carrying off one unit of angular momentum. The law of conservation of angular momentum applied to the matrix element (2.89) then requires that **J**, **J**′, and the photon angular momentum form a vector triangle similar to those in Fig. 2-5; the vector model thus indicates the selection rule

$$J' = J + 1,\ J,\ \text{or}\ J - 1, \qquad (J' = J = 0\ \text{not allowed})\ . \tag{2.90}$$

In addition, the electric dipole operator involves only the orbital and not the spin coordinates, so that S is not involved in the radiation process; thus for matrix elements between LS-coupled functions, we have the additional selection rules

$$\begin{aligned} S' &= S, \\ L' &= L + 1,\ L,\ L - 1, \qquad (L' = L = 0\ \text{not allowed})\ . \end{aligned} \tag{2.91}$$

The simplifications in the evaluation of Coulomb and electric-dipole matrix elements associated with the above selection rules make LS-coupled functions the most convenient for use in numerical calculations, even when the coupling conditions in the atom are not close to pure LS coupling.

*2-15. APPENDIX

We now give proofs of certain relationships omitted in Sec. 2-5.

First we wish to verify the normalization of the function $\Theta_{l,-l}$ given in (2.30). From the differential formula

$$\frac{d}{d\theta}(\sin^{n-1}\theta \cos\theta) = (n-1)\sin^{n-2}\theta\cos^2\theta - \sin^n\theta = (n-1)\sin^{n-2}\theta - n\sin^n\theta\ ,$$

it follows that

$$\int_0^\pi \sin^n\theta\, d\theta = -\frac{1}{n}\Big[\sin^{n-1}\theta\cos\theta\Big]_0^\pi + \frac{(n-1)}{n}\int_0^\pi \sin^{n-2}\theta\, d\theta\ ,$$

where the first term on the right is zero for $n > 1$. Using this result repeatedly we obtain

$$\int_0^\pi \sin^{2l+1}\theta \, d\theta = \frac{(2l)(2l-2)(2l-4)\cdots 2}{(2l+1)(2l-1)(2l-3)\cdots 3}\int_0^\pi \sin\theta \, d\theta$$
$$= \frac{[2^l \, l!]^2}{(2l+1)!} \cdot 2 \, , \tag{2.92}$$

whence (2.31) follows immediately.

Secondly, we wish to verify that the spherical harmonics (2.40) satisfy (2.29) with $\mathbf{L}^2$ given by (2.25). Carrying out the ϕ differentiation shows that we need only verify that Θ_{lm} satisfy

$$\left[-\frac{1}{\sin\theta}\frac{d}{d\theta}\left(\sin\theta\frac{d}{d\theta}\right) + \frac{m^2}{\sin^2\theta}\right]\Theta_{lm} = l(l+1)\Theta_{lm} \, , \tag{2.93}$$

or with the substitution $x = \cos\theta$

$$\left[-\frac{d}{dx}(1-x^2)\frac{d}{dx} + \frac{m^2}{1-x^2}\right]\Theta_{lm} = l(l+1)\Theta_{lm} \, . \tag{2.94}$$

To do this we first define

$$v^n = \frac{d^n}{dx^n}(x^2-1)^l \tag{2.95}$$

and prove that

$$(x^2-1)v^{n+2} + 2(n-l+1)xv^{n+1} + (n-2l)(n+1)v^n = 0 \, . \tag{2.96}$$

The correctness of (2.96) can be established by induction: direct differentiation of (2.95) shows (2.96) to be true for $n = 0$; a single differentiation of (2.96) gives the identical expression over again except with n replaced by $n + 1$, thus showing (2.96) to be true for all $n \geqq 0$. From (2.33), we may write

$$\Theta_{lm} = (1-x^2)^{m/2}v^{l+m}$$

except for constant factors that are of no importance in establishing (2.94). Straightforward differentiation of this expression gives

$$-\frac{\partial}{\partial x}(1-x^2)\frac{\partial\Theta_{lm}}{\partial x} = (1-x^2)^{m/2}\Big[(x^2-1)v^{l+m+2}$$
$$+ 2(m+1)xv^{l+m+1} + \left(m + m^2 - \frac{m^2}{1-x^2}\right)v^{l+m}\Big] \, .$$

Use of (2.96) for $n = l + m$ then gives

$$-\frac{\partial}{\partial x}(1-x^2)\frac{\partial\Theta_{lm}}{\partial x} = \left[(l-m)(l+m+1) + m + m^2 - \frac{m^2}{1-x^2}\right]\Theta_{lm} \, ,$$

which reduces directly to (2.94).

Finally, we wish to prove directly the orthogonality relation

$$\int_0^\pi \Theta_{lm}\Theta_{l'm} \sin\theta \, d\theta = 0 , \qquad l \neq l' . \tag{2.97}$$

Without loss of generality we may assume $l > l'$; using the form (2.36) for Θ_{lm} and the expression (2.33) for $\Theta_{l'm}$, we have only to prove that

$$\int_{-1}^{1} \left[\frac{d^{l-m}}{dx^{l-m}}(x^2 - 1)^l\right]\left[\frac{d^{l'+m}}{dx^{l'+m}}(x^2 - 1)^{l'}\right] dx = 0 . \tag{2.98}$$

Assuming that $m < l$, integration by parts gives for this integral

$$\left[\left\{\frac{d^{l-m-1}}{dx^{l-m-1}}(x^2 - 1)^l\right\}\left\{\frac{d^{l'+m}}{dx^{l'+m}}(x^2 - 1)^{l'}\right\}\right]_{-1}^{1}$$
$$- \int_{-1}^{1} \left[\frac{d^{l-m-1}}{dx^{l-m-1}}(x^2 - 1)^l\right]\left[\frac{d^{l'+m+1}}{dx^{l'+m+1}}(x^2 - 1)^{l'}\right] dx .$$

The first term in this result is zero because the lowest powers of $(x^2 - 1)$ in the two factors in braces are

$\max(0, m + 1)$

and

$\max(0, -m)$,

one of which is certainly greater than zero so that the corresponding factor is zero at both $x = -1$ and $x = +1$. Successive additional integrations by parts for a total of $l - m$ times give for the integral in (2.98) the result

$$(-1)^{l-m}\int_{-1}^{1} (x^2 - 1)^l\left[\frac{d^{l'+l}}{dx^{l'+l}}(x^2 - 1)^{l'}\right] dx .$$

But the remaining derivative is certainly identically zero because of the assumption $l > l'$, and this then proves (2.98).

3

ONE-ELECTRON ATOMS

3-1. THE SCHRÖDINGER EQUATION

The preceding chapter has discussed in a general way the nature of the angular portions of wavefunctions of atoms containing one or more electrons. We have yet to discuss the radial dependence of wavefunctions; i.e., the dependence on the distances $r_i \equiv |\mathbf{r}_i|$ of the electrons from the nucleus. It is convenient to introduce this subject with a review of the one-electron-atom problem—either neutral hydrogen, or He^+, Li^{++}, Be^{3+}, etc.

The Schrödinger equation for the single electron of mass m moving in the electrostatic field of the (stationary) nucleus with a potential energy $V(\mathbf{r})$ is

$$H\psi = E\psi \, , \tag{3.1}$$

where

$$H = -\frac{\hbar^2}{2m}\nabla^2 + V(\mathbf{r}) \, . \tag{3.2}$$

The Laplacian operator

$$\nabla^2 = \frac{\partial^2}{\partial x^2} + \frac{\partial^2}{\partial y^2} + \frac{\partial^2}{\partial z^2}$$

can be converted to spherical coordinates by means of the relations (2.21) to give

$$H = -\frac{\hbar^2}{2m}\left[\frac{1}{r}\frac{\partial^2}{\partial r^2}r + \frac{1}{r^2 \sin\theta}\left(\frac{\partial}{\partial\theta}\sin\theta\frac{\partial}{\partial\theta}\right) + \frac{1}{r^2\sin^2\theta}\frac{\partial^2}{\partial\phi^2}\right] + V(\mathbf{r}) \, . \tag{3.3}$$

It is worth noting that this result follows from the classical kinetic energy in spherical coordinates,

$$\frac{p^2}{2m} = \frac{p_r^2}{2m} + \frac{L^2}{2mr^2} \, ,$$

if the operator (2.25) is used for $\mathbf{L}^2$, and if p_r^2 is replaced not by the expression

$$\left(-\,i\hbar\frac{\partial}{\partial r}\right)^2 = -\,\hbar^2\frac{\partial^2}{\partial r^2}$$

appropriate for Cartesian coordinates but rather by[1]

$$\left(-\,i\hbar\frac{1}{r}\frac{\partial}{\partial r}r\right)^2 = -\,\hbar^2\frac{1}{r}\frac{\partial^2}{\partial r^2}r\ .$$

In all theoretical work in this book we shall measure distances in units of the Bohr radius[2]

$$a_0 = \frac{\hbar^2}{me^2} = 0.529177\,\text{Å}\ , \tag{3.4}$$

energies in units of the rydberg[2]

$$\frac{\hbar^2}{2ma_0^2} = \frac{e^2}{2a_0} = 13.6058\ \text{eV}\ , \tag{3.5}$$

and angular momenta in units of $\hbar$. With it understood that r is measured in Bohr units and $\mathbf{L}^2$ in units of $\hbar^2$, the Schrödinger equation becomes

$$\frac{\hbar^2}{2ma_0^2}\left[-\,\frac{1}{r}\frac{\partial^2}{\partial r^2}r\psi + \frac{\mathbf{L}^2}{r^2}\psi\right] + V(\mathbf{r})\psi = E\psi\ ,$$

or, with $V(\mathbf{r})$ and E measured in rydbergs,

$$\left[-\,\frac{1}{r}\frac{\partial^2}{\partial r^2}r + \frac{1}{r^2}\mathbf{L}^2 + V(\mathbf{r})\right]\psi = E\psi\ . \tag{3.6}$$

3-2. CENTRAL-FIELD PROBLEMS

In the case of one-electron atoms, the potential energy $V(\mathbf{r})$ is a function only of the magnitude of $\mathbf{r}$, and not of the angles θ and ϕ. In any such central-field problem, the angular momentum $\mathbf{L}$ is classically a constant of the motion; quantum mechanically, we

[1]F. S. Crawford, Jr., Am. J. Phys. **32**, 611 (1964).

[2]In the one-electron problem, a correction for the motion of the nucleus of finite mass M can be made by replacing m in (3.3) by the reduced mass $\mu \equiv mM/(m + M)$ of the system; i.e., by measuring distances in units of $a_0 m/\mu$ and energies in units of μ/m rydbergs. In multi-electron atoms we need not be concerned with such a correction because the magnitude of the effect is much smaller than in hydrogen, and because the inherent accuracy of numerical calculations is comparatively low.

may expect (from Sec. 2-3) that ψ should be an eigenfunction of $\mathbf{L}^2$ and L_z. Indeed, (3.6) may be written in the form

$$\mathbf{L}^2\psi = \left[r\frac{\partial^2}{\partial r^2}r - r^2V(r) + r^2E \right] \psi \; ; \tag{3.7}$$

then since the coefficient of ψ on the right side of this equation is constant so far as θ and ϕ are concerned, (3.7) is in the form of the eigenvalue equation (2.29). Thus ψ can only be some function of r times a spherical harmonic Y_{lm} (times a spin function, if desired), and the quantity in brackets in (3.7) must be equal to $l(l+1)$. [The factor $\hbar^2$ in (2.29) is no longer present because we are now measuring $\mathbf{L}^2$ in units of $\hbar^2$.] It is convenient to write the radial function with an explicit factor r^{-1}, so that from (2.56) the total wavefunction is of the form

$$\psi_{nlm_lm_s}(r,\theta,\phi,s_z) = \frac{1}{r} P_{nl}(r)\cdot Y_{lm_l}(\theta,\phi)\cdot\sigma_{m_s}(s_z) \; , \tag{3.8}$$

where n is a new quantum number whose significance will become clear later on.

Substituting this into

$$\left[r\frac{\partial^2}{\partial r^2}r - r^2V(r) + r^2E \right] \psi = l(l+1)\psi$$

and dividing out the factor $Y\sigma$, we obtain

$$\left[r\frac{\partial^2}{\partial r^2} - rV(r) + rE \right] P_{nl}(r) = \frac{l(l+1)}{r} P_{nl}(r) \; .$$

Rearranging terms back into the form (3.6) we obtain the differential equation

$$\left[-\frac{d^2}{dr^2} + \frac{l(l+1)}{r^2} + V(r) \right] P_{nl}(r) = EP_{nl}(r) \; , \tag{3.9}$$

which must be satisfied by the radial function P_{nl}.

Inasmuch as the wavefunction (3.8) must be everywhere finite (and in fact must tend to zero as r tends to infinity if ψ represents a bound electronic state), P_{nl} must satisfy the boundary conditions

$$P_{nl}(0) = 0 \; , \tag{3.10}$$

$$P_{nl}(\infty) = 0 \; . \tag{3.11}$$

Moreover, if ψ is to be normalized,

$$\langle\psi|\psi\rangle \equiv \sum_{s_z} \int \psi^*\psi \, r^2 \sin\theta \, dr \, d\theta \, d\phi = 1 \; , \tag{3.12}$$

then it follows from (2.57) that P_{nl} must be normalized in the sense

$$\int_0^\infty P_{nl}^*(r)P_{nl}(r)\,dr = 1 , \tag{3.13}$$

where dr is (as usual) measured in Bohr units. [Of the volume-element weighting factor r^2 sin θ, the factor sin θ goes into the θ integral (2.31), and the factor r^2 cancels out the explicit r^{-1} in (3.8).]

For a system consisting of a single electron of charge −e lying at a distance r from a nucleus with charge +Ze, the potential energy V(r) (in rydbergs) is equal from (3.5) to

$$V(r) = -\frac{Ze^2}{r} \div \frac{e^2}{2a_o} = -\frac{2Z}{(r/a_o)} , \tag{3.14}$$

so that with r measured in Bohr units, as required in (3.9), the differential equation to be solved is

$$\left[-\frac{d^2}{dr^2} + \frac{l(l+1)}{r^2} - \frac{2Z}{r}\right]P_{nl}(r) = EP_{nl}(r) . \tag{3.15}$$

3-3. ANALYTICAL SOLUTION OF THE RADIAL EQUATION

In order to find solutions of (3.15), it is convenient to make the substitutions

$$\rho = \frac{2Zr}{n} , \qquad E = -\frac{Z^2}{n^2} , \tag{3.16}$$

where n is a constant whose value remains to be determined. Then (3.15) reduces to

$$\left[\frac{d^2}{d\rho^2} - \frac{1}{4} + \frac{n}{\rho} - \frac{l(l+1)}{\rho^2}\right]P_{nl} = 0 . \tag{3.17}$$

For very large ρ, the solution is clearly of the form $e^{\pm\rho/2}$; the negative sign must be chosen if the boundary condition (3.11) is to be satisfied. For very small ρ, (3.17) reduces to a form having solutions ρ^{l+1} and ρ^{-l}; only the former satisfies the boundary condition (3.10), so that we must have

$$P_{nl}(\rho) = b_o\rho^{l+1} , \qquad (r \to 0) , \tag{3.18}$$

where b_o is some constant. This suggests the substitution

$$P_{nl}(\rho) = \rho^{l+1}e^{-\rho/2}f(\rho) , \tag{3.19}$$

from which we find that f(ρ) must satisfy the equation

$$f'' + \left[\frac{2l + 2}{\rho} - 1\right]f' + \frac{n - l - 1}{\rho}f = 0 .$$

If a solution f(ρ) in the form of a power series in ρ is sought, it is found that for non-integral n the series becomes infinite like $e^{+\rho}$ as ρ tends to infinity, which together with (3.19) violates the boundary condition (3.11). On the other hand, if n is an integer greater than l, the power series breaks off as a polynomial of degree $n - l - 1$, thereby giving a solution (3.19) that satisfies the boundary condition at ∞, as well as that at $\rho = 0$. This polynomial proves to be the associated Laguerre polynomial

$$L_{n+l}^{2l+1}(\rho) = - [(n + l)!]^2 \sum_{k=0}^{n-l-1} \frac{(-\rho)^k}{k!(n - l - 1 - k)!(2l + 1 + k)!} , \tag{3.20}$$

and the final form of the radial function P_{nl}, properly normalized according to (3.13), is

$$P_{nl}(r) = - \left[\frac{Z(n - l - 1)!}{n^2[(n + l)!]^3}\right]^{1/2} \rho^{l+1}e^{-\rho/2}L_{n+l}^{2l+1}(\rho) , \qquad \rho = \frac{2Zr}{n} . \tag{3.21}$$

We have chosen to include a negative sign in (3.21) to cancel the one in (3.20), and to thus provide the phase convention

$$P_{nl}(r) > 0 , \qquad r \to 0 . \tag{3.22}$$

The integer n which has appeared in the course of finding solutions of (3.15) that satisfied the boundary conditions (3.10)-(3.11) is called the principal quantum number, because the energy E of the atom is seen from (3.16) to depend only on n, and not (in the present non-relativistic approximation) on any of the other three quantum numbers l, m_l, m_s. The radial function P_{nl} depends on the angular-momentum quantum number l as well as on n, but not on the magnetic quantum numbers m_l and m_s. As can be seen from (3.20), the possible l values are restricted to the range

$$0 \leqq l < n ; \tag{3.23}$$

they are usually denoted by means of the letter code s, p, d, f ··· given in Sec. 2-12, so that we have the possible radial functions

$$P_{1s}, P_{2s}, P_{2p}, P_{3s}, P_{3p}, P_{3d}, P_{4s}, \cdots .$$

The detailed algebraic expressions for these functions (for $n \leqq 4$) are given in Table 3-1. Graphs of some of the functions will be described shortly. Analytical expressions for various integrals of the radial functions can be derived using (3.21); Table 3-2 gives some results for expectation values of r^k,

$$\langle r^k \rangle \equiv \int_0^\infty P_{nl}^*(r)\, r^k\, P_{nl}(r)\, dr \ , \qquad (3.24)$$

in units of a_o^k.

3-4. NUMERICAL SOLUTION OF THE RADIAL EQUATION

The analytical expression (3.21) for P_{nl} provides a complete solution to the non-relativistic hydrogenic problem, and is usually the most convenient form to use for practical purposes. However, in dealing with atoms containing more than one electron, differential equations similar in form to (3.9) will be encountered; the potential energy function V(r) in these cases has no simply expressible form, so that analytical solution is not possible. It is then necessary to obtain some sort of numerical solution—for example, by

TABLE 3-1. NORMALIZED RADIAL FUNCTIONS FOR Z = 1[a]

$$P_{1s} = 2re^{-r}$$

$$P_{2s} = \frac{1}{\sqrt{2}}\, re^{-r/2}\left(1 - \frac{1}{2}r\right)$$

$$P_{2p} = \frac{1}{2\sqrt{6}}\, r^2e^{-r/2}$$

$$P_{3s} = \frac{2}{3\sqrt{3}}\, re^{-r/3}\left(1 - \frac{2}{3}r + \frac{2}{27}r^2\right)$$

$$P_{3p} = \frac{8}{27\sqrt{6}}\, r^2e^{-r/3}\left(1 - \frac{1}{6}r\right)$$

$$P_{3d} = \frac{4}{81\sqrt{30}}\, r^3e^{-r/3}$$

$$P_{4s} = \frac{1}{4}\, re^{-r/4}\left(1 - \frac{3}{4}r + \frac{1}{8}r^2 - \frac{1}{192}r^3\right)$$

$$P_{4p} = \frac{\sqrt{5}}{16\sqrt{3}}\, r^2e^{-r/4}\left(1 - \frac{1}{4}r + \frac{1}{80}r^2\right)$$

$$P_{4d} = \frac{1}{64\sqrt{5}}\, r^3e^{-r/4}\left(1 - \frac{1}{12}r\right)$$

$$P_{4f} = \frac{1}{768\sqrt{35}}\, r^4e^{-r/4}$$

[a]Radial functions for $Z \neq 1$ may be obtained by multiplying the above functions by $\sqrt{Z}$ and replacing r by Zr; in any case, r is measured in Bohr units $a_o = \hbar^2/me^2$.

TABLE 3-2.[a] EXPECTATION VALUES of r^k,

$$\langle r^k \rangle = \int_0^\infty r^k P_{nl}^2(r)\, dr$$

1	$\frac{1}{2Z}[3n^2 - l(l+1)]$
2	$\frac{n^2}{2Z^2}[5n^2 + 1 - 3l(l+1)]$
3	$\frac{n^2}{8Z^3}[35n^2(n^2-1) - 30n^2(l+2)(l-1) + 3(l+2)(l+1)l(l-1)]$
4	$\frac{n^4}{8Z^4}[63n^4 - 35n^2(2l^2+2l-3) + 5l(l+1)(3l^2+3l-10) + 12]$
−1	$\frac{Z}{n^2}$
−2	$\frac{Z^2}{n^3(l+1/2)}$
−3	$\frac{Z^3}{n^3(l+1)(l+1/2)l}$
−4	$\frac{Z^4[3n^2 - l(l+1)]}{2n^5(l+3/2)(l+1)(l+1/2)l(l-1/2)}$

[a]K. Bockasten, Phys. Rev. A 9, 1087 (1974).

straightforward numerical integration. We shall discuss this numerical procedure here for the simple Coulomb-potential case because it helps to illuminate the role of the boundary conditions (3.10)-(3.11) in leading to the quantized energy levels of the atom,

$$E = -\frac{Z^2}{n^2}$$

with n an integer, and it also helps to illuminate the nature of the principal quantum number n.

It is convenient to write (3.15)—or, more generally, (3.9)—in the form

$$\frac{d^2P_{nl}(r)}{dr^2} = \left[V_{eff}(r) - E\right]P_{nl}(r)\ . \tag{3.25}$$

The function

$$V_{eff}(r) = V(r) + \frac{l(l+1)}{r^2} \tag{3.26}$$

is an effective potential-energy function for purely radial motion, consisting of the true three-dimensional potential energy V(r), plus a term that actually comes from the angular-momentum portion of the kinetic energy, but that may be thought of as a radial potential energy resulting from work done against centrifugal forces. The form of $V_{eff}(r)$ is shown in Fig. 3-1 for a Coulomb potential $-2Z/r$ with $Z = 1$, and for $l = 0, 1, 2, 3$. For s states ($l = 0$), V_{eff} is equal to V and increases monotonically from $-\infty$ at $r = 0$ to zero at $r = \infty$. For all other states ($l > 0$), the centrifugal term in (3.26) goes to $+\infty$ (as r tends to zero) faster than V goes to $-\infty$; the result is that V_{eff} is positive at small r, and shows a potential-well minimum at some finite (non-zero) radius.

Classically, an electron can exist only at radii where the total energy E is greater than V_{eff} (i.e., where the "radial portion" of the kinetic energy is non-negative). Thus, for example, a p electron ($l = 1$) with a total energy of -0.4 Ry may move only within the range $1.4a_0 \leqq r \leqq 3.6a_0$; the two limits represent the perihelion and aphelion distances, respectively, for the elliptical motion of the electron about the nucleus. If the p electron has an energy of -0.5 Ry, it can have only the radius $r = 2a_0$; i.e., it moves in a circular orbit. For $E < -0.5$, p electrons cannot exist at all. Only s electrons can reach the position of the nucleus ($r = 0$), but they may have any energy whatever.

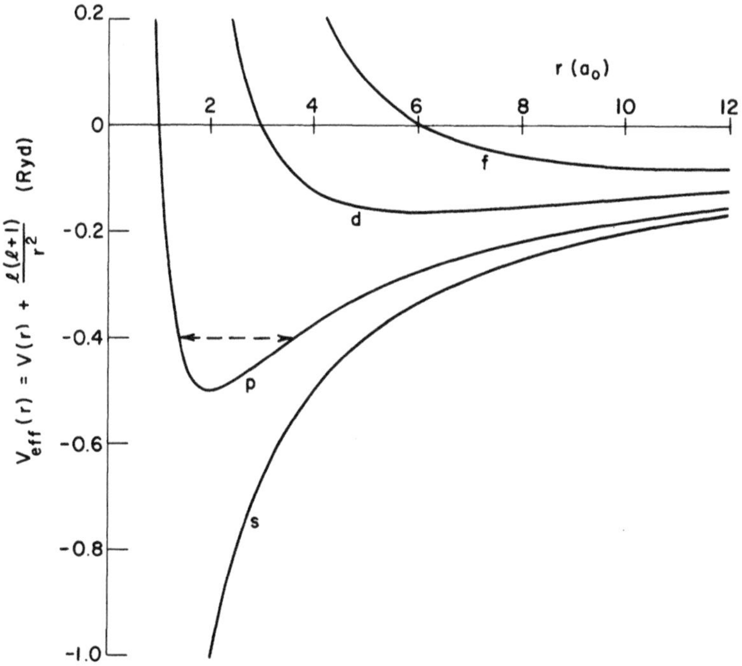

Fig. 3-1. Effective potential energy curves for s, p, d, and f electrons in the hydrogen atom, $V(r) = -2/r$. The classical motion of a p electron with total energy -0.4 Ry would be restricted to the range indicated by the dashed line.

Quantum mechanically, an electron can penetrate outside the classically allowed region. In particular, an electron of any l can penetrate to arbitrarily small radii. In this region, a series expansion beginning with the term (3.18)—which satisfies the boundary condition (3.10) at r = 0—can be used to calculate values of P_{nl} at several values of r in order to provide numbers with which to start the numerical integration of the differential equation (3.25); the constant b_0 should be positive so as to satisfy the convention (3.22), but may have an arbitrary magnitude. Integration of (3.25) is then carried forward to larger r by standard numerical techniques,[3] for some fixed value of the energy E, to give a tentative solution P(r).

The general shape of this tentative solution can be easily described. In the classically forbidden region ($V_{eff} > E$), it follows from (3.25) that P″ has the same sign as P, so that a graph of P(r) must continuously increase in magnitude, and so can never satisfy the boundary condition (3.11) at infinity; such a value of E is thus forbidden quantum mechanically as well as classically. It is evident that E must be greater than the minimum value of V_{eff}, so that there exists a region in which $V_{eff} < E$ and in which P is concave *toward* the horizontal axis—i.e., in which P tends to be oscillatory in form. Moreover, E must be such that the degree of oscillatory character is exactly right; that is, such that in the large-radius region P_{nl} decays asymptotically toward the horizontal axis, thereby satisfying (3.11).

All of the above is illustrated in Fig. 3-2, which shows six selected frames from a computer-produced movie in which the form of the integral P(r) of the differential equation (for $l = 0$) is shown as a function of E. Within the energy range shown, only the value E = − 1/4 results in an integral that satisfies the boundary condition at infinity. The numerical procedure for solving the radial differential equation thus consists in adjusting the value of E until the corresponding integral P(r) tends to zero at infinity.[4] Since the differential equation (3.25) is linear, any multiple of P(r) is a solution of (3.25) that satisfies the boundary conditions (3.10)-(3.11). We can thus evaluate numerically the integral

$$C^2 \equiv \int_0^\infty [P(r)]^2\, dr \ ,$$

[3]See, for example, D. R. Hartree, *The Calculation of Atomic Structures* (John Wiley, New York, 1957). A commonly used method is that due to B. Numerov, Publ. Observ. Central Astrophys. Russ. 2, 188 (1933): If the differential equation is written $y'' = f(r)y + g(r)$ (where $g \equiv 0$ in the present case of interest) and if y is already known at the points r_j and r_{j-1} on a mesh in which $h = r_i - r_{i-1}$ (all i), then the value of y at r_{j+1} is given by

$$Y_{j+1} = 2Y_j - Y_{j-1} + h^2\left[f_j y_j + \frac{1}{12}(g_{j+1} + 10g_j + g_{j-1})\right] ,$$

where

$$Y = y\left(1 - \frac{1}{12}h^2 f\right) .$$

[4]In practice, numerical instabilities dictate that the outward integration be carried only to some matching radius a short distance beyond the classically-allowed region, that an inward integration then be performed starting with $P(r) \propto \exp\left[-r(-E)^{1/2}\right]$ at some appropriately large radius, and that E be adjusted until, at the matching radius, the outward and inward integrations give functions with both equal values and equal slopes; see, for example, ref. 3.

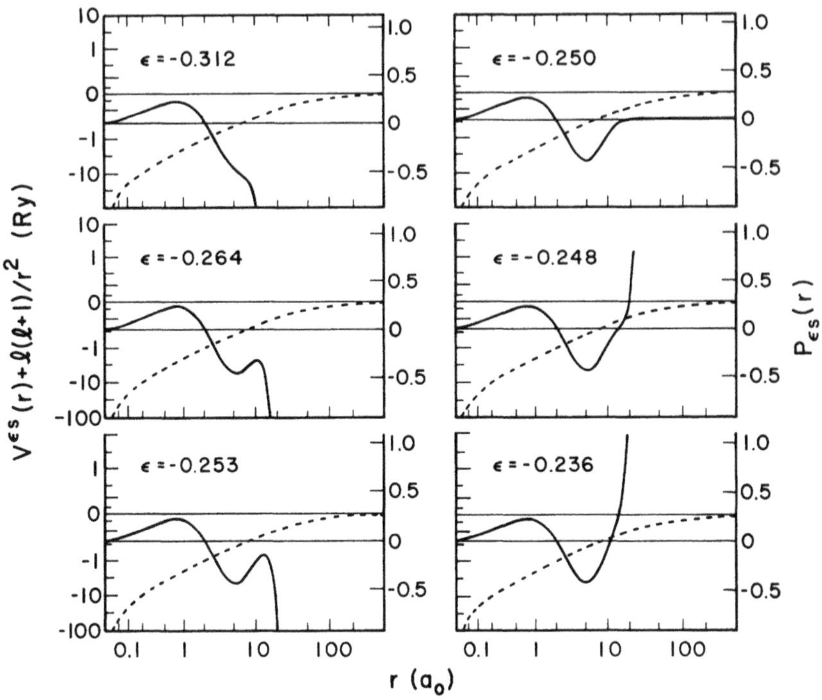

Fig. 3-2. The form of the integral P(r) of the differential equation (3.25), as a function of the assumed value of the energy parameter E, for an s electron in the field of a hydrogen nucleus. The dotted curve is the effective potential energy (3.26)(in this case, $-2/r$), plotted on the nonlinear energy scale shown (in rydberg units). The solid curve is P(r), plotted on a linear vertical scale with zero at the position $\varepsilon \equiv E$ on the energy scale. (The apparent concave-upward form of P at small r would actually be concave-downward if a linear r scale had been used.)

and P(r)/C is then a solution of (3.25) that satisfies also the normalization condition (3.13).

It is evident that (for any given value of l, including $l = 0$), there exists a lowest value of E giving an acceptable (normalized) radial function P(r). There are, however, many higher eigenvalues—and, in fact, a denumerably infinite number, which can therefore be denoted by integral serial numbers n. For an s electron, the lowest-energy solution ($E = -1$ Ry for hydrogen) is given the serial number $n = 1$, the next lowest ($E = -1/4$) is denoted by $n = 2$, etc., in line with the analytical result for hydrogenic atoms,

$$E_n = -\frac{Z^2}{n^2} ; \tag{3.27}$$

the function $P_{ns}(r)$ is plotted for several values of n in Fig. 3-3. For a p electron, the lowest-energy solution in hydrogen occurs for $E = -1/4$; this solution is accordingly given the serial number 2 rather than 1, in line with (3.27) (cf. Fig. 3-4). In general, the minimum value of n is chosen to satisfy

$$n \geqq l + 1 \,, \tag{3.28}$$

which is of course identical with the result (3.23) obtained from analytical solution of the radial equation.

It may be seen from the above discussion and the examples in Figs. 3-3 and 3-4 that for each l, the radial function $P_{nl}(r)$ of lowest energy has no nodes (other than those at $r = 0, \infty$) and one antinode (one point of maximum magnitude), and that each successively higher-energy eigenfunction has one additional node and one additional antinode. In general,

$$\begin{aligned} &\text{number of nodes} = n - l - 1 \,, \\ &\text{number of antinodes} = n - l \,. \end{aligned} \tag{3.29}$$

When numerical integration of the differential equation (3.25) is carried out to find a radial function for a non-Coulomb potential V(r) [where the relation (3.27) does not hold], node counting provides the means of assigning a value of n to the function P(r) and to the corresponding eigenvalue E.

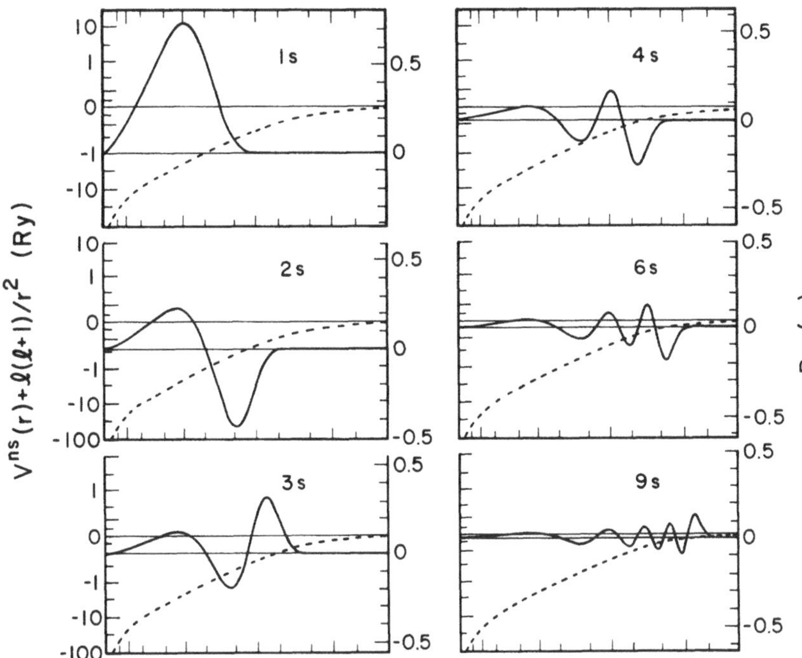

Fig. 3-3. The form of some of the integrals P(r) (for s electrons in hydrogen) that satisfy the boundary condition $P(\infty) = 0$.

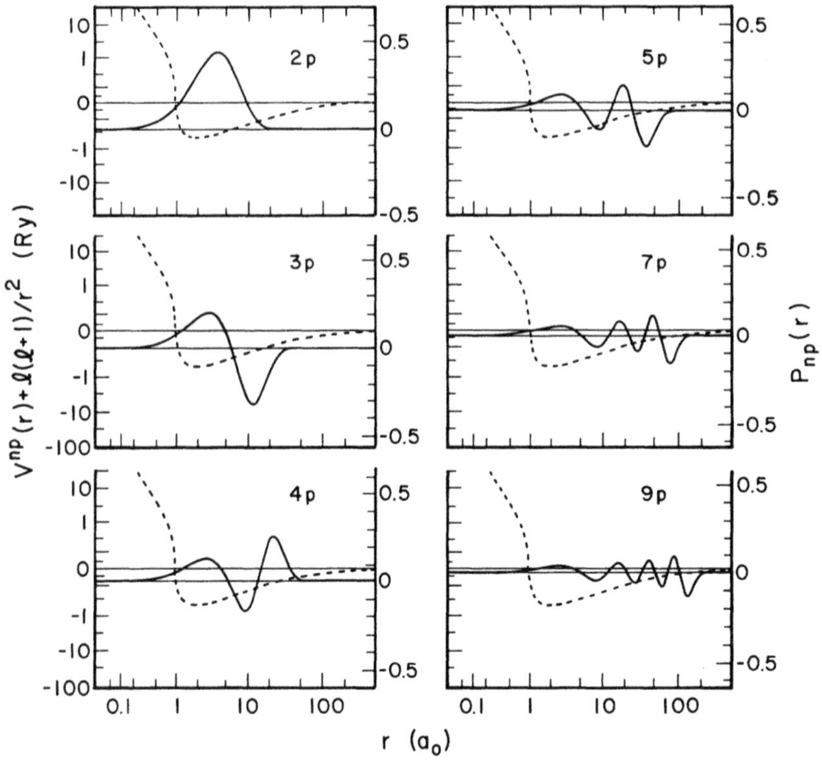

Fig. 3-4. Same as Fig. 3-3, except for p electrons. Note the different form of V_{eff} at small r, and the resulting zero slope of P(r) at the origin.

3-5. ELECTRON PROBABILITY DENSITY

With P_{nl} known (either analytically or numerically), the probability-density distribution of the electron position (assuming $s_z = m_s$) is given by (3.8) and (2.40) as:

$$|\psi_{nlm_lm_s}|^2 = \frac{1}{2\pi r^2}|P_{nl}(r)|^2 \cdot |\Theta_{lm_l}(\theta)|^2 \ . \tag{3.30}$$

"Photographs" of this electron-cloud distribution are given by White[5] and by Herzberg.[6]

The fraction of the total electronic charge lying in a spherical shell of radius r and thickness dr is from (2.42)

$$\int_0^{2\pi}\int_0^{\pi} |\psi|^2 \, r^2 \sin\theta \, dr \, d\theta \, d\phi = |P_{nl}(r)|^2 \, dr \ . \tag{3.31}$$

[5]H. E. White, *Introduction to Atomic Spectra* (McGraw-Hill, New York, 1934).

[6]G. Herzberg, *Atomic Spectra and Atomic Structure* (Dover Publications, New York, 1944), 2nd ed.

The radial probability distribution function[7] P_{nl}^2 for each state of hydrogen with $n \leqq 4$ is shown in Fig. 3-5. It is evident that, in each case, most of the charge is concentrated in the vicinity of the outermost antinode, and thus within a spherical shell of moderate thickness. With increasing n, the radius of this shell increases, and the amplitudes of the inner antinodes decrease; in the limit $n = \infty$, all electronic charge lies at infinity, corresponding to ionization of the atom.

For given n, Fig. 3-5 and the expressions for $\langle r \rangle$ and $\langle r^2 \rangle$ in Table 3-2 show that most of the charge lies at greater radii the smaller the value of l. This is in line with the semiclassical Bohr-orbit picture (Fig. 3-6), in which smaller angular momentum means an elliptical orbit of greater eccentricity; this implies (since the length of the major axis of the ellipse depends only on energy) a larger aphelion distance, and (since the orbital speed is smallest at aphelion) a greater mean distance from the nucleus.

At the same time, Fig. 3-5 and the expressions for $\langle r^{-2} \rangle$ and $\langle r^{-3} \rangle$ show that the probability of the electron lying close to the nucleus is also greatest for small l. Again, this is consistent with the Bohr-orbit picture, as the highly eccentric small-angular-momentum orbit is the one that has the small perihelion distance. Indeed, only for zero angular momentum can the point $r = 0$ be reached classically; correspondingly, the quantum-mechanical density for small r (averaged over all angles θ, ϕ) is given from (3.31) and (3.18) by

$$\frac{1}{4\pi r^2}|P_{nl}|^2 \propto r^{2l} , \qquad (3.32)$$

and this is non-zero at the origin only for $l = 0$. [As a consequence, only s electrons show sizable interactions with the nucleus and therefore appreciable isotope shifts (Sec. 17-8)].

3-6. ENERGY LEVELS AND WAVELENGTHS

The possible energies of a hydrogenic atom, relative to the ionization limit as zero, are given by (3.16) as

$$E_n = -\frac{Z^2}{n^2}\ \text{Ry} \qquad (n = 1, 2, 3, \cdots) . \qquad (3.33)$$

It is worth noting that this expression is consistent with the virial theorem which, for Coulomb forces, states that (see the appendix to this chapter, Sec. 3-8):

$$\langle E_{kin} \rangle = -\frac{1}{2}\langle E_{pot} \rangle ,$$

[7]As defined in Secs. 3-3 and 3-4, the function P_{nl} is always real; however, in what follows the unessential asterisk in P_{nl}^* will sometimes be retained to make it clear which radial factors arose from bra functions and which from ket functions.

or

$$\langle E\rangle = \langle E_{kin}\rangle + \langle E_{pot}\rangle = \frac{1}{2}\langle E_{pot}\rangle \ . \tag{3.34}$$

From (3.14), this becomes for hydrogenic atoms

$$\langle E\rangle = \frac{1}{2}\langle -\frac{2Z}{r}\rangle = - Z\langle r^{-1}\rangle \quad Ry \ , \tag{3.35}$$

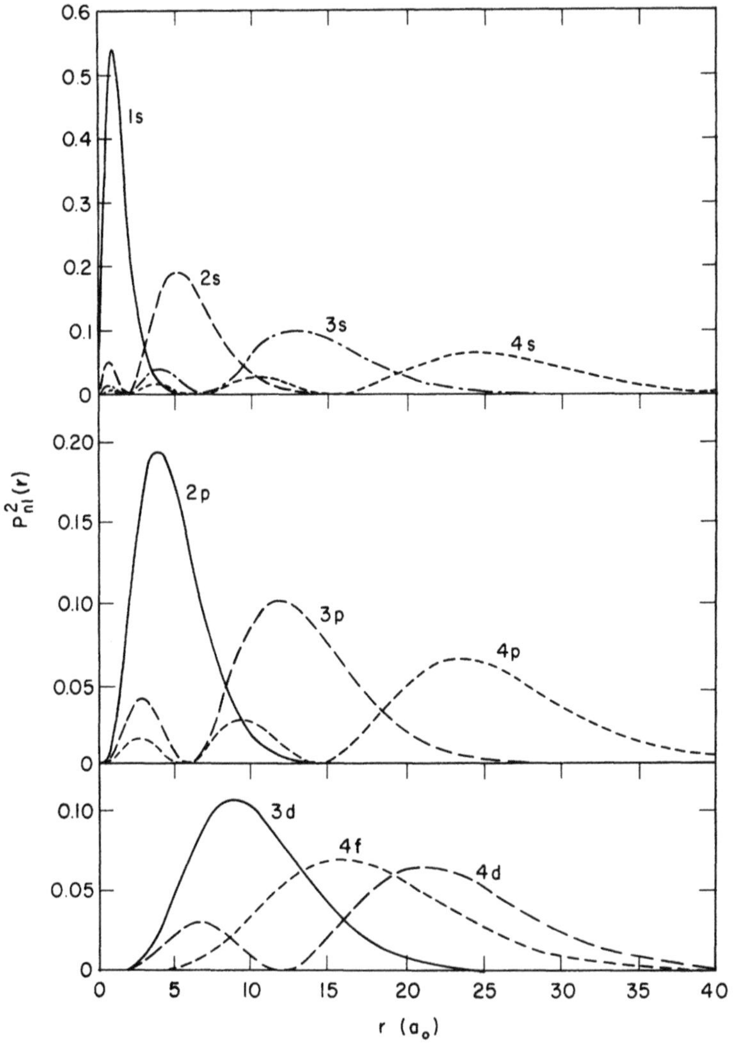

Fig. 3-5. Radial probability distribution function $[P_{nl}(r)]^2$ for the lowest-energy states of hydrogen.

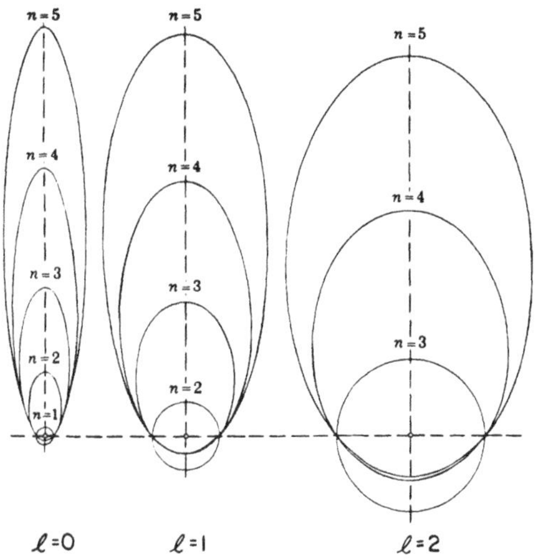

Fig. 3-6. Elliptical Bohr-Sommerfeld orbits for the hydrogen atom. (After G. Herzberg, *Atomic Spectra and Atom Structure*, Dover Publications, New York, 1944.)

and since the expectation value of r^{-1} is Z/n^2 (see Table 3-2), this result is identical with (3.33).

The energy levels E_n depend only on n and not on *l*, and are shown for $Z = 1$ in the form of a Grotrian diagram in Fig. 3-7, which depicts also some of the possible radiative transitions. The wavenumbers of the possible transitions are given, from the Ritz combination principle (1.15) and the value (1.11) for the Rydberg constant, as[8]

$$\sigma_{nk} = \frac{E_k}{hc} - \frac{E_n}{hc} = 109667.6\left(\frac{1}{n^2} - \frac{1}{k^2}\right) \quad cm^{-1}\ , \qquad k > n\ . \tag{3.36}$$

For a fixed value of n and successively larger values of k, there corresponds a series of regularly spaced spectrum lines whose wavenumbers or wavelengths converge to a series limit as k tends to ∞. The series with $n = 2$ (Fig. 3-8) lies in the easily accessible visible and near-ultraviolet region, and is named after the Swiss mathematician and physicist J. J. Balmer, who in 1885 found empirically the equivalent of the formula (3.36), thereby providing the first important step toward a theoretical understanding of atomic structure. Other observed series of lines in the spectrum of hydrogen (named after their discoverers) are listed in Table 3-3. Lines of the Balmer and Lyman series are commonly denoted H_α, H_β, H_γ, $\cdots$ and Ly_α, Ly_β, Ly_γ, $\cdots$, respectively.

3-7. RELATIVISTIC CORRECTIONS

In order to take relativistic effects into account properly, we should solve the Dirac equation rather than the Schrödinger equation (3.1)-(3.2). The former is, however, much

[8]The value of the Rydberg constant given in (1.11) is that for an infinitely heavy nucleus. In (3.36) we have made the correction for reduced-mass effects, $R_M = R_\infty M/(m + M)$, which was discussed in fn. 2. Experimental determination of the Rydberg constant is discussed by G. W. Series, Contemp. Phys. 15, 49 (1974).

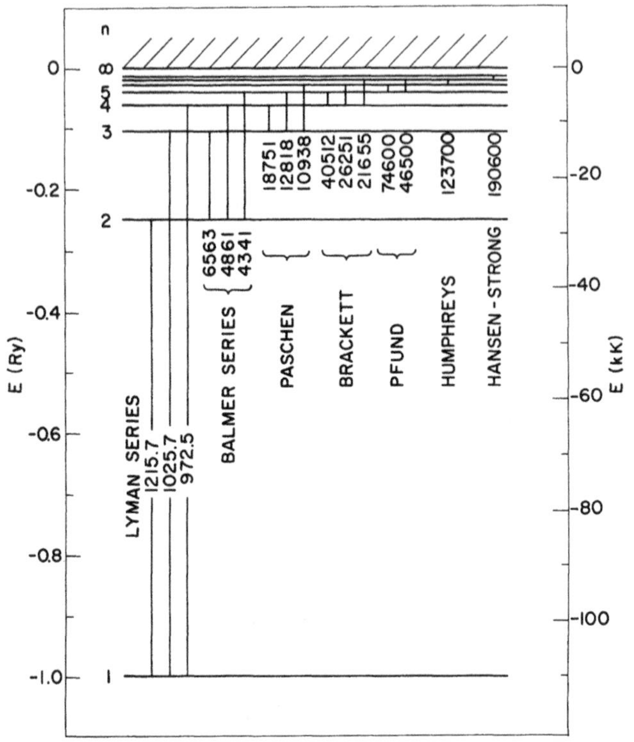

Fig. 3-7. Diagram of the non-relativistic energy level structure of hydrogenic atoms. (The energy scales must be multiplied by Z^2 if $Z > 1$.) Also shown are transitions giving rise to the first few members of the seven observed series of lines in the spectrum of hydrogen, with wavelengths in Ångströms.

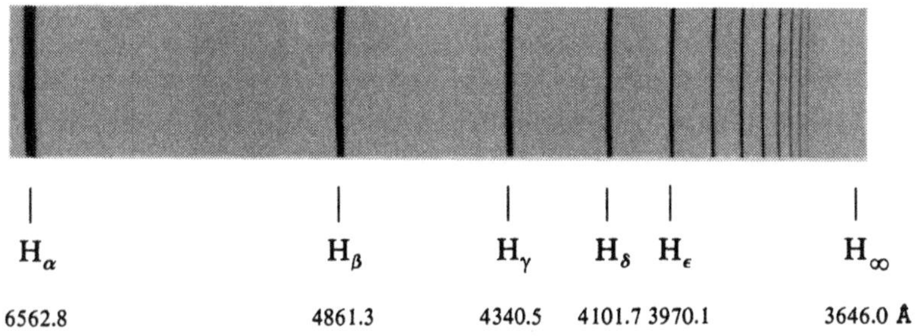

Fig. 3-8. Balmer-series lines in the visible and near ultraviolet portion of the hydrogen spectrum. [Photographic negative, from G. Herzberg, *Atomic Spectra and Atomic Structure* (Dover Publications, New York, 1944), courtesy of the author and publisher.]

TABLE 3-3. SERIES IN THE SPECTRUM OF HYDROGEN (WAVENUMBER AND WAVELENGTH OF THE FIRST MEMBER AND OF THE SERIES LIMIT)

n	Name[a]	$\sigma_{n,n+1}$	$\sigma_{n\infty}$	$\lambda_{n,n+1}$	$\lambda_{n\infty}$
1	Lyman	82259 K	109679 K	1215.67Å	911.75Å
2	Balmer	15233	27420	6562.8	3646
3	Paschen	5331.6	12186	18751	8204
4	Brackett	2467.8	6854.9	40512	14584
5	Pfund	1340.5	4387.1	74578	22788
6	Humphreys	808.3	3046.6	123690	32814
7	Hansen-Strong	524.6	2238.3	190570	44664

[a]T. Lyman, Astrophys. J. **23**, 181 (1906), Nature **93**, 241 (1914).
J. J. Balmer, Ann. Physik **25**, 80 (1885).
F. Paschen, Ann. Physik **27**, 537 (1908).
F. S. Brackett, Astrophys. J. **56**, 154 (1922).
A. H. Pfund, J. Opt. Soc. Am. **9**, 193 (1924).
C. J. Humphreys, J. Opt. Soc. Am. **42**, 432 (1952).
P. Hansen and J. Strong, Appl. Opt. **12**, 429 (1973).

more complex than the latter—involving, for example, the calculation of four different radial functions (for given nl) instead of only one. Fortunately, for most problems of interest to us (involving primarily the outer valence electrons of complex atoms), it is sufficient to consider only first-order relativistic corrections to the Schrödinger equation. These corrections are derived from the Dirac equation by assuming that the velocity of the electron v is always small with respect to the velocity of light, expanding in powers of $(v/c)^2$, and neglecting all terms that involve powers of v/c higher than the second.

The result, for a central field potential V(r), is the same as the Schrödinger equation (3.1) except for the addition of three new terms to the Hamiltonian (3.2). In cgs units these terms are[9]

$$-\frac{1}{2mc^2}(E - V)^2 - \frac{\hbar^2}{4m^2c^2}\left(\frac{dV}{dr}\right)\left(\frac{\partial}{\partial r} - \frac{2}{r}\boldsymbol{l}\cdot\mathbf{s}\right) , \tag{3.37}$$

where E is the eigenvalue of the Schrödinger equation (now also appearing in the Hamiltonian itself), and $\boldsymbol{l}$ and s are orbital and spin angular-momentum operators, in units of ħ. The first of these new terms is called the mass-velocity term because it arises from the relativistic variation of mass with velocity; note that E − V is just the kinetic energy of the electron, so that this term represents a correction to the kinetic-energy term in (3.2) consisting essentially of $(-E_{kin})\cdot(E_{kin}/2mc^2)$. The third term is called the spin-orbit term, as it represents the magnetic interaction energy between the electron's spin magnetic moment

[9]H. A. Bethe and E. E. Salpeter, *Quantum Mechanics of One- and Two-Electron Atoms* (Springer-Verlag, Berlin, 1957), or the corresponding article in S. Flügge, ed., *Handbuch der Physik* (Springer-Verlag, Berlin, 1957), Vol. XXXV, especially Sec. 13.

and the magnetic field that the electron sees as a result of its orbital motion through the electric field of the nucleus; this interaction is extremely important in the calculation of energy-level structures of complex atoms, and a classical derivation of this term is given in an appendix (Sec. 3-9). The middle term is called the Darwin term; it may be thought of as arising from a relativistically-induced electric moment of the electron, or (better) from the relativistic non-localizability of the electron.[10]

If distances are measured in Bohr units a_o and all energies (including the Hamiltonian itself) are measured in Rydberg units $= \hbar^2/2ma_o^2$, then both coefficients in (3.37) become equal to

$$\frac{\hbar^2}{4m^2c^2a_o^2} = \frac{\alpha^2}{4} ,$$

where

$$\alpha = \frac{e^2}{\hbar c} = \frac{\hbar}{mca_o} = \frac{1}{137.036} \tag{3.38}$$

is the fine-structure constant. The complete Hamiltonian thus becomes

$$H = -\nabla^2 + V - \frac{\alpha^2}{4}(E - V)^2 - \frac{\alpha^2}{4}\left(\frac{dV}{dr}\right)\frac{\partial}{\partial r} + \frac{\alpha^2}{2}\frac{1}{r}\left(\frac{dV}{dr}\right)(\boldsymbol{l}\cdot\mathbf{s}) . \tag{3.39}$$

To solve the Schrödinger equation with this Hamiltonian, we first define a new angular-momentum operator

$$\mathbf{j} = \boldsymbol{l} + \mathbf{s} . \tag{3.40}$$

The operators $\boldsymbol{l}$ and s commute with each other because they act on entirely different sets of coordinates; it follows from Sec. 2-8 that we can find a function F_{lsjm} that is an eigenfunction of the four operators $\boldsymbol{l}^2$, s^2, j^2, and j_z with eigenvalues (in our present units) of $l(l + 1)$, $s(s + 1) = 3/4$, $j(j + 1)$, and m, respectively. But

$$\mathbf{j}^2 \equiv \mathbf{j}\cdot\mathbf{j} = (\boldsymbol{l} + \mathbf{s})\cdot(\boldsymbol{l} + \mathbf{s}) = \boldsymbol{l}^2 + \boldsymbol{l}\cdot\mathbf{s} + \mathbf{s}\cdot\boldsymbol{l} + \mathbf{s}^2 ,$$

so that F_{lsjm} is also an eigenfunction of

$$\boldsymbol{l}\cdot\mathbf{s} = \frac{1}{2}(\mathbf{j}^2 - \boldsymbol{l}^2 - \mathbf{s}^2) \tag{3.41}$$

with eigenvalue

$$X \equiv \frac{1}{2}[j(j + 1) - l(l + 1) - s(s + 1)] . \tag{3.42}$$

[10]C. G. Darwin, Proc. Roy. Soc. (London) **120**, 621 (1928); V. B. Berestetskii, E. M. Lifshitz, and L. P. Pitaevskii, *Relativistic Quantum Theory*, transl. J. B. Sykes and J. S. Bell (Pergamon, Oxford, 1971), p. 2; J. D. Bjorken and S. D. Drell, *Relativistic Quantum Mechanics* (McGraw-Hill, New York, 1964), p. 52.

Thus if, analogously to (3.8), we seek a solution of the Schrödinger equation (3.1) with Hamiltonian (3.39) having the form[11]

$$\psi_{nljm} = \frac{1}{r} P_{nlj}(r) \cdot F_{lsjm} , \tag{3.43}$$

then

$$(l \cdot s)\psi_{nljm} = X\psi_{nljm} ,$$

and the Schrödinger equation can be reduced to a radial equation for P_{nlj} analogous to (3.9):

$$\left[-\frac{d^2}{dr^2} + \frac{l(l+1)}{r^2} + V - \frac{\alpha^2}{4}(E - V)^2 - \frac{\alpha^2}{4}\left(\frac{dV}{dr}\right) r \frac{d}{dr} r^{-1} + \frac{\alpha^2}{2}\frac{X}{r}\left(\frac{dV}{dr}\right)\right] P_{nlj}(r) = EP_{nlj}(r) , \tag{3.44}$$

where $V = V(r)$ is as before equal to $-2Z/r$ for hydrogenic atoms.

Rather than attempt to solve this equation for modified radial functions $P_{nlj}(r)$, we content ourselves with the solutions $P_{nl}(r)$ already found for the non-relativistic equation (3.15), and merely calculate first-order perturbation corrections to the non-relativistic eigenvalues (3.33). The result is just

$$E_{nlj} = -\frac{Z^2}{n^2} + E_m + E_D + E_{so} , \tag{3.45}$$

where

$$E_m = -\frac{\alpha^2}{4}\int_0^\infty P_{nl}^*(E_n - V)^2 P_{nl}\, dr , \tag{3.46}$$

$$E_D = -\frac{\alpha^2}{4}\int_0^\infty P_{nl}^* \left(\frac{dV}{dr}\right) r\left(\frac{dr^{-1}P_{nl}}{dr}\right) dr , \tag{3.47}$$

$$E_{so} = \frac{\alpha^2 X}{2}\int_0^\infty P_{nl}^* \left(\frac{1}{r}\frac{dV}{dr}\right) P_{nl}\, dr . \tag{3.48}$$

For a Coulomb potential $V = -2Z/r$, the correction terms are easily evaluated with the aid of the expressions for $\langle r^k \rangle$ given in Table 3-2. For the mass-velocity correction we find

[11]For brevity, the quantum number s has been omitted from the subscripts on ψ and P because its value is always unambiguously 1/2.

$$E_m = -\frac{\alpha^2}{4}\left[E_n^2 + \left\langle\frac{4Z}{r}\right\rangle E_n + \left\langle\frac{4Z^2}{r^2}\right\rangle\right]$$

$$= -\frac{\alpha^2}{4}\left[\frac{Z^4}{n^4} - \frac{4Z^4}{n^4} + \frac{4Z^4}{n^3(l + 1/2)}\right]$$

$$= -\frac{\alpha^2 Z^4}{4n^4}\left[\frac{4n}{l + 1/2} - 3\right] \quad \text{Ry} . \tag{3.49}$$

For the Darwin correction, we obtain with the aid of (3.11), (3.20), and (3.21)

$$E_D = -\frac{\alpha^2 Z}{2}\int_0^\infty (r^{-1} P_{nl})\, d(r^{-1} P_{nl})$$

$$= \frac{\alpha^2 Z}{4}[(r^{-1} P_{nl})^2_{r=0} - (r^{-1} P_{nl})^2_{r=\infty}]$$

$$= \delta_{l0}\frac{\alpha^2 Z}{4}\left[\frac{Z(n + l)!}{n^2(n - l - 1)![(2l + 1)!]^2}\right]\left(\frac{\rho}{r}\right)^2$$

$$= \delta_{l0}\frac{\alpha^2 Z^4}{n^3} \quad \text{Ry} . \tag{3.50}$$

Finally, for the spin-orbit correction we obtain from (3.42) and Table 3-2

$$E_{so} = \alpha^2 Z X\langle r^{-3}\rangle \tag{3.51}$$

$$= (1 - \delta_{l0})\frac{\alpha^2 Z^4}{n^3 l(l + 1)(2l + 1)}[j(j + 1) - l(l + 1) - s(s + 1)] \quad \text{Ry} . \tag{3.52}$$

The derivation of (3.52) is straightforward except in the case $l = 0$, where $X = 0$ and $\langle r^{-3}\rangle = \infty$, so that (3.51) is indeterminate; as pointed out by Bethe and Salpeter[9] the proper result in this case[12] is $E_{so} = 0$, and this we have indicated by the explicit factor $(1 - \delta_{l0})$ in (3.52).

From the vector model (Secs. 2-8, 9), the possible values of j are $l \pm 1/2$ (except $l + 1/2$ only, if $l = 0$). It is easily seen that the factor in brackets in (3.52) is equal to

$$\begin{aligned} &l, && j = l + \frac{1}{2} \\ &-(l + 1), && j = l - \frac{1}{2} \end{aligned} \tag{3.53}$$

[12]The difficulty arises from the fact that $2mc^2 + E - V(r)$ has been approximated by $2mc^2$ in the denominator of the last two terms of (3.37) and by the fact that for $l = 0$ and small r, $P_{nl} \propto r$ from (3.18), so that the integral of $r^{-3}P^2 \propto r^{-1}$ diverges at $r = 0$. If we use the more accurate expression $2mc^2 + E - V = 2mc^2(1 + \alpha^2 E/4 + Z\alpha^2/2r)$, then r^{-1} is replaced by the quantity $(r + \alpha^2 rE/4 + Z\alpha^2/2)^{-1}$ which is finite at $r = 0$, so that in place of $\langle r^{-3}\rangle$ we obtain a finite result.

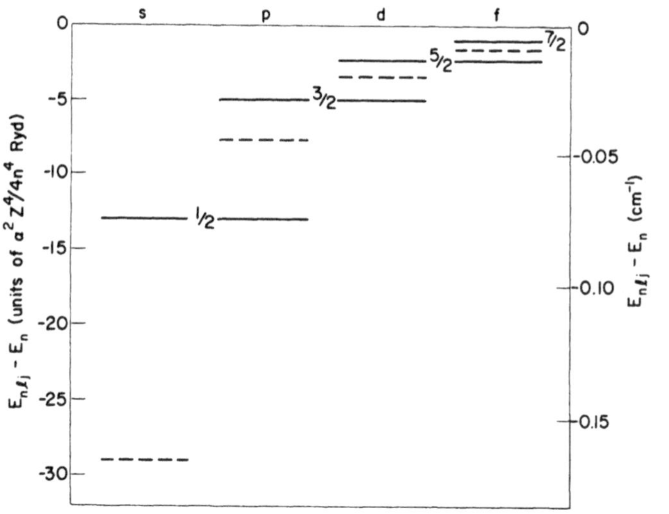

Fig. 3-9. Relativistic corrections to the n = 4 energy levels of hydrogenic atoms. (The cm^{-1} scale must be multiplied by Z^4 if $Z > 1$.) The dashed lines show the corrections arising from the mass-velocity term (E_m) only.

and that (3.45) and (3.49)-(3.52) may be represented in all cases by the single expression

$$E_{nlj} = -\frac{Z^2}{n^2} - \frac{\alpha^2 Z^4}{4n^4}\left[\frac{4n}{j + 1/2} - 3\right] \text{ Ry} . \tag{3.54}$$

From (3.28), the maximum value of $j + 1/2 = l + 1$ is n, so that the total relativistic correction is always negative. In contrast to the non-relativistic result $-Z^2/n^2$ which depends only on n, the mass-velocity correction (3.49) has removed the degeneracy with respect to l; the spin-orbit term has then removed the degeneracy with respect to j [cf. (3.53)] but (together with the Darwin term) has left the two states with equal j (and $l = j \pm 1/2$) exactly degenerate; this is illustrated in Fig. 3-9. Additional correction terms arising from the finite size of the nucleus and from quantum-electrodynamical effects remove the remaining degeneracy, leaving (for given j) the level of lower l slightly higher in energy;[9] the effect is greatest for $j = 1/2$, where the energy difference $^2S_{1/2} - {}^2P^{\circ}_{1/2}$ is known as the Lamb shift.[13] These correction terms have been included in theoretically calculated energy levels for hydrogenic atoms;[14] we shall not discuss the details here, as these effects do not contribute importantly to the energy-level structure of multi-electron atoms.

[13]W. E. Lamb, Jr. and R. C. Retherford, Phys. Rev. **72**, 241 (1947); N. M. Kroll and W. E. Lamb, Jr., Phys. Rev. **75**, 388 (1949).

[14]J. D. Garcia and J. E. Mack, J. Opt. Soc. Am. **55**, 654 (1965); G. W. Erickson, J. Phys. Chem. Ref. Data **6**, 831 (1977).

As a consequence of the relativistic splitting of energy levels, it is obvious that each of the spectrum lines indicated in Fig. 3-7 is split into a number of closely-spaced line-structure components. Because of the parity and angular-momentum selection rules discussed in Sec. 2-14, however, not all pairs of energy levels give rise to observable (electric-dipole) lines. For example, the possible transitions for each of the Paschen-series lines [remembering the restriction $l < n$, Eq. (3.28)] are those listed in the first column of Table 3-4.[15] For low-Z elements, the magnitude of the fine-structure splitting is very small, as can be seen from the example of the He II line 4686 Å included in the table. The effects increase very rapidly with Z, however, and in multi-electron atoms the spin-orbit interactions are indispensable in the discussion of energy-level structures. Indeed, there are many cases with Z as small as ten in which the spin-orbit terms are responsible for the gross, rather than the fine, features of the energy-level structure; some examples will be discussed in the following chapter.

Problem

3-7(1). For the hydrogenic atom $_{18}$Ar XVIII (= Ar^{17+}):
(a) Calculate the energies of the levels

$$1s\ ^2S_{1/2},\ 2p\ ^2P^{\circ}_{1/2},\ \text{and}\ 2p\ ^2P^{\circ}_{3/2}$$

in the non-relativistic approximation, Eq. (3.33).
(b) Calculate the above energies including relativistic and spin-orbit corrections, Eq. (3.54).
(c) Using the reduced-mass-corrected value 109735.8 cm^{-1} for the Rydberg constant, calculate the wavenumbers (and the wavelengths, in Å) of the transitions

$$1s\ ^2S_{1/2} - 2p\ ^2P^{\circ}_{1/2,3/2}$$

both without and with the corrections in (b). Compare these with the more accurate theoretical wavenumbers computed by Garcia and Mack;[14] to what degree do our corrections approximate the more inclusive corrections? Compare also with the wavelength of a feature observed in a solar flare by G. A. Doschek, et al., Astrophys. J. **170,** 573 (1971).

*3-8. APPENDIX: THE VIRIAL THEOREM

We shall here prove the virial theorem in classical mechanics, but it is valid in quantum mechanics as well.[16]

Let us consider a closed system made up of a collection of N particles contained in a rectangular box (with sides of length a, b, c in the x, y, z directions, respectively) that is

[15] By convention, atomic transitions are always indicated by listing the lower energy level first and the upper level second, no matter whether one is referring to absorption or to emission of a photon. In hydrogenic atoms, l is necessarily equal to L and the former is commonly deleted from the notation. Similarly, helium-like cases $1snl\ ^{1,3}L$ are frequently abbreviated to $n\ ^{1,3}L$.

[16] J. O. Hirschfelder, C. F. Curtiss, and R. B. Bird, *Molecular Theory of Gases and Liquids* (John Wiley, New York, 1954), pp. 41 and 68; A. G. McLellan, Am. J. Phys. **42,** 239 (1974).

TABLE 3-4. WAVENUMBERS OF THE FINE-STRUCTURE COMPONENTS: first member of the Paschen series (n = 3 to n = 4) for He II (λ = 4686 Å), relative to the 3 $^2D_{5/2}$ − 4 $^2F^o_{7/2}$ component at σ_1 = 21335.08205 K

Transition	σ (theory[a])	σ − σ_1 (theory[a])	Exp.[b] − theory
3 $^2S_{1/2}$ − 4 $^2P^o_{1/2}$	21336.15701 K	1074.96 mK	1.1 mK
− 4 $^2P^o_{3/2}$	6.88912	1807.07	0.9
3 $^2P^o_{1/2}$ − 4 $^2S_{1/2}$	6.35552	1273.47	0.3
− 4 $^2D_{3/2}$	7.02736	1945.31	−0.1
3 $^2P^o_{3/2}$ − 4 $^2S_{1/2}$	4.62013	−461.92	0.2
− 4 $^2D_{3/2}$	5.29197	209.92*	1 *
− 4 $^2D_{5/2}$	5.53599	453.94**	
3 $^2D_{3/2}$ − 4 $^2P^o_{1/2}$	4.56400	−518.05	−18
− 4 $^2P^o_{3/2}$	5.29611	214.06*	
− 4 $^2F^o_{5/2}$	5.53845	456.40**	0.6**
3 $^2D_{5/2}$ − 4 $^2P^o_{3/2}$	4.71770	−364.35	14
− 4 $^2F^o_{5/2}$	4.96004	−122.01	14
− 4 $^2F^o_{7/2}$	5.08205	0	0

[a]Ref. 14.
[b]H. P. Larson and R. W. Stanley, J. Opt. Soc. Am. **57**, 1439 (1967).
*The asterisks indicate experimentally unresolved pairs.

centered at the origin. If the ith particle has mass m_i, and at time t is located at the point $\mathbf{r}_i = (x_i, y_i, z_i)$ with velocity $\dot{\mathbf{r}}_i$ and is acted on by a force $\mathbf{F}^i$, then

$$\frac{d}{dt}(m_i x_i \dot{x}_i) = m_i \dot{x}_i^2 + m_i x_i \ddot{x}_i = m_i \dot{x}_i^2 + x_i F_x^i . \qquad (3.55)$$

The left-hand side vanishes if averaged over a sufficiently long time interval τ:

$$\left\langle \frac{d}{dt}(m_i x_i \dot{x}_i) \right\rangle = \frac{1}{\tau}\int_0^\tau \frac{d}{dt}(m_i x_i \dot{x}_i)\, dt = \frac{1}{\tau}\left[(m_i x_i \dot{x}_i)_{t=\tau} - (m_i x_i \dot{x}_i)_{t=0} \right] ,$$

and this tends to zero as $\tau \to \infty$ because $x_i \dot{x}_i$ remains bounded. Thus the time average of (3.55) leads to

$$-\frac{1}{2}\left\langle x_i F_x^i \right\rangle = \frac{1}{2}\left\langle m_i \dot{x}_i^2 \right\rangle .$$

Adding similar expressions for the y and z directions and summing over all particles, we may write

$$\mathcal{V} \equiv -\frac{1}{2}\sum_i \langle \mathbf{r}_i \cdot \mathbf{F}^i \rangle = \langle K \rangle , \qquad (3.56)$$

where $\langle K \rangle$ is the time average of the total kinetic energy of the system. $\mathcal{V}$ is called the virial of the system.

The forces $\mathbf{F}^i$ may be of two types. There are first the forces exerted on the particles by the walls of the container. For the walls that are perpendicular to the x axis, each of area A = bc, we have[17]

$$\sum_i \langle \mathbf{r}_i \cdot \mathbf{F}^i \rangle = \left(-\frac{a}{2}\right)(pA) + \left(\frac{a}{2}\right)(-pA) = -pV ,$$

where p is the pressure and V the volume of the container. There are identical contributions from the other two pairs of walls, giving a total contribution to $\mathcal{V}$ of

$$-\frac{1}{2}\sum_i \langle \mathbf{r}_i \cdot \mathbf{F}^i \rangle = \frac{3}{2} pV . \tag{3.57}$$

Secondly, there may be forces between pairs of particles. We assume these forces to be conservative, and therefore expressible in terms of a potential-energy function $U = U(\mathbf{r}_1, \mathbf{r}_2, \cdots \mathbf{r}_N)$, such that

$$\mathbf{F}^i = -\frac{\partial U}{\partial \mathbf{r}_i} .$$

If U is a homogeneous function of the coordinates of degree n, then by Euler's theorem[18]

$$\sum_i \mathbf{r}_i \cdot \mathbf{F}^i = -\sum_i \mathbf{r}_i \cdot \frac{\partial U}{\partial \mathbf{r}_i} = -nU . \tag{3.58}$$

Combining (3.57) and (3.58) to obtain the total virial, we find from (3.56) that the virial theorem becomes

$$\frac{3}{2} pV = \langle K \rangle - \frac{n}{2}\langle U \rangle . \tag{3.59}$$

As a simple example, the potential energy for an ideal gas of point particles is zero and the kinetic energy under conditions of thermal equilibrium is (3/2)NkT, so that the virial theorem reduces to the ideal gas law

$$pV = NkT . \tag{3.60}$$

An isolated atom at rest constitutes a system of charged particles acted on by zero external forces ($p = 0$), and interacting with each other via Coulomb forces for which $n = -1$. The virial theorem then reduces to

[17]Components of $\mathbf{F}^i$ parallel to the walls average to zero.

[18]See, for example, J. S. R. Chisholm and R. M. Morris, *Mathematical Methods in Physics* (North-Holland Publ. Co., Amsterdam, 1966), 2nd ed. Crudely, if the potential energy of a particle at r due to a particle at the origin is $U = kr^n$, then $rF = -rnkr^{n-1} = -nU$.

$$\langle K \rangle = -\frac{1}{2}\langle U \rangle , \tag{3.61}$$

as used in (3.34).

3-9. APPENDIX: THE SPIN-ORBIT INTERACTION

The spin-orbit interaction term in (3.39) was obtained by a consideration of relativistic corrections to the non-relativistic Hamiltonian. However, the concept of electron spin is predicted not only relativistically, but also by non-relativistic theories (Biedenharn and Louck, *Angular Momentum in Quantum Physics,* pp. 5, 331. In any case, if the electron spin (or, more accurately, the intrinsic magnetic moment of the electron) is accepted as an experimental fact,[19] then the spin-orbit *interaction energy* can be viewed as being essentially non-relativistic in origin. This energy cannot be correctly calculated as the energy of interaction of the electron's magnetic moment with the magnetic field that is produced by the apparent motion of the nucleus about the electron; such a picture involves a noninertial reference frame in which the electron is considered to be at rest, and results in a computed energy that is too large by a factor of two.[20] This same error is present even if the interaction energy is pictured as that of the electron magnetic moment in the magnetic field which the electron sees as the result of its motion through the electrostatic field of the nucleus.

The correct result can be obtained non-relativistically only by computing the energy of the (essentially) stationary nucleus; i.e., the *electrostatic* potential energy of the nuclear charge q_N in the electric field produced by the moving electron magnetic moment μ. This can be computed as follows:[21] If **r** is the position of the electron with respect to the nucleus, then $\mathbf{A} = -\mu \times \mathbf{r}/r^3$ is the vector potential associated with the magnetic field of the dipole μ, and the electric field at the position of the nucleus is, in cgs units,

$$-(1/c)\partial \mathbf{A}/\partial t = (1/c)[(\dot{\mu} \times \mathbf{r}/r^3) + (\mu \times \mathbf{v}/r^3) - (3v\,\mu \times \mathbf{r}/r^4)] . \tag{3.62}$$

The potential energy of the nuclear charge in this changing electric field will be the same as in a static field given by the above expression at any given instant of time; this energy may readily be evaluated by imagining the nuclear charge to be moved from infinity to the position of the nucleus along a straight-line path directed toward the electron. The vector directions of the first and last terms of (3.62) contribute components to the electrostatic force **F** on the nucleus that are perpendicular to the path of integration, and so

$$E_{so} = -\int_{\infty}^{r} \mathbf{F}\cdot d\mathbf{r} = -(q_N/c)\int_{\infty}^{r} (\mu \times \mathbf{v})\cdot d\mathbf{r}/r^3$$

$$= (q_N/c)(\mu \times \mathbf{v})\cdot \mathbf{r}/2r^3 = -(q_N/2r^3c)(\mathbf{r} \times \mathbf{v})\cdot \mu .$$

[19]G. E. Uhlenbeck and S. Goudsmit, Naturwissenschaften **13**, 953 (1925), Nature **117**, 264 (1926).

[20]L. H. Thomas, Nature **117**, 514 (1926).

[21]W. T. Dixon, Am. J. Phys. **38**, 1162 (1970).

But $\mathbf{r} \times \mathbf{v} = l\hbar/m$, where l is the angular momentum of the electron about the nucleus, in units of $\hbar$; also, for an electron spin $\mathbf{s}$ of magnitude 1/2 (in units of $\hbar$) we may write[22] $\mu = -e\hbar\mathbf{s}/mc$ because μ has a magnitude of one Bohr magneton ($= e\hbar/2mc$) and a direction opposite to that of $\mathbf{s}$. Finally, the potential energy of the electron in the electrostatic field of the nucleus is $V(r) = -q_N e/r$, and so we may write

$$E_{so} = \frac{\hbar^2}{2m^2c^2}\left(\frac{1}{r}\frac{dV}{dr}\right)(\boldsymbol{l}\cdot\mathbf{s})$$

in cgs units, which is identical with the spin-orbit portion of (3.37).

[22]Quantum-electrodynamical corrections result in the gyromagnetic ratio for the electron (the ratio of $|\mu|$ in Bohr magnetons to $|s|$ in units of $\hbar$) being slightly greater than two:

$$g_e = 2\left(1 + \frac{\alpha}{2\pi} - \frac{0.328\alpha^2}{\pi^2} + \cdots\right) \cong 2.002319 \ .$$

This small correction need not concern us here because of the relatively low accuracy with which spin-orbit effects can be computed in complex atoms.

4

COMPLEX ATOMS—THE VECTOR MODEL

4-1. THE SCHRÖDINGER EQUATION

Theoretical treatment of an atom containing N electrons (N greater than one, but not necessarily equal to the atomic number Z of the nucleus) requires first of all knowledge of a suitable Hamiltonian operator. An appropriate operator may be obtained by summing the one-electron operator (3.39) over all N electrons, and adding a term for the electrostatic Coulomb interactions among the electrons:

$$H = H_{kin} + H_{elec\text{-}nucl} + H_{elec\text{-}elec} + H_{s-o}$$

$$= -\sum_i \nabla_i^2 - \sum_i \frac{2Z}{r_i} + \sum_{i>j}\sum \frac{2}{r_{ij}} + \sum_i \xi_i(r_i)(\boldsymbol{l}_i \cdot \mathbf{s}_i) \; . \tag{4.1}$$

Here $r_i = |\mathbf{r}_i|$ is the distance of the ith electron from the nucleus, $r_{ij} = |\mathbf{r}_i - \mathbf{r}_j|$ is the distance between the ith and j^{th} electrons, and the summation over $i > j$ is over all pairs of electrons. As before, we are measuring distances in Bohr units (a_o) and energies in rydbergs. For brevity, we have omitted the mass-velocity and Darwin terms. These terms depend only on $|\mathbf{r}_i|$ and have the effect only of shifting the absolute energies of a group of related levels, without affecting the energy differences among these levels; we shall discuss their effects in greater detail in Chapters 7 and 8.

The final term of the Hamiltonian represents the sum over all electrons of the magnetic interaction energy between the spin of an electron and its own orbital motion, represented by a quantity proportional to the scalar product of the orbital- and spin-angular-momentum operators. Unlike the mass-velocity and Darwin terms, the spin-orbit interaction involves the angular portion of the wavefunction through the operators $\boldsymbol{l}$ and $\mathbf{s}$, and has a pronounced effect on energy-level structures (i.e., on energy differences within a group of related levels); it is therefore necessary to retain it explicitly in the Hamiltonian. The proportionality factor ξ_i is measured in rydbergs (with $\boldsymbol{l}$ and $\mathbf{s}$ in units of $\hbar$) and has here been left unspecified in form, though we shall mostly use the expression

$$\xi(r) = \frac{\alpha^2}{2}\frac{1}{r}\left(\frac{dV}{dr}\right) \tag{4.2}$$

that is applicable for hydrogenic atoms when we have defined an appropriate potential-energy function V(r) for an electron in a multi-electron atom.[1] Other magnetic interactions may be considered—orbit-orbit ($\boldsymbol{l}_i \cdot \boldsymbol{l}_j$), spin-spin ($\mathbf{s}_i \cdot \mathbf{s}_j$), and spin-other-orbit ($\boldsymbol{l}_i \cdot \mathbf{s}_j$)—but are usually much less important than the spin-orbit term, and will be neglected throughout this book.

Our task is to solve the Schrödinger equation

$$H\Psi^k = E^k\Psi^k \tag{4.3}$$

to obtain the wavefunction Ψ^k and the energy E^k of the atom for every stationary quantum state k of interest. However, the wavefunction is a function of 4N variables (three space and one spin coordinate for each electron), and the quantum mechanical problem is extremely complex; for $N > 1$, exact solutions cannot be found at all and approximations of one sort or another are required. The usual approximation is to assume some form of wavefunction that contains several adjustable parameters, and to vary the values of these parameters so as to obtain the best possible function, as judged by some appropriate criterion. If the interelectronic distances r_{ij} are included explicitly in the wavefunction so as to properly take account of correlations among positions of the various electrons owing to their mutual Coulomb repulsions, the problem becomes prohibitively complex even for $N = 3$ or 4. Even accurate perturbation calculations of correlation energies (using simpler zero-order wavefunctions) are extremely lengthy for N greater than about 6.

For spectroscopic purposes, we are interested in tens, hundreds, or even thousands of different levels of each atom and ion, containing usually 10 to 100 (or even 150) electrons. It is obvious that we can attack the problem only by making very drastic approximations.

4-2. THE MATRIX METHOD

The approximation method that we shall use was first developed by Slater,[2] and later extended by Condon and Shortley[3] and many others; it is commonly known as the Slater-Condon theory. The basic procedure consists of expanding the unknown wavefunction Ψ^k in terms of a set of known basis functions Ψ_b:

$$\Psi^k = \sum_b y_b^k \Psi_b \, . \tag{4.4}$$

The basis functions are assumed to be members of a complete set of orthonormal functions,

$$\langle \Psi_b | \Psi_{b'} \rangle = \delta_{bb'} \, . \tag{4.5}$$

[1]An alternative formulation of the operator ξ has been given by M. Blume and R. E. Watson, Proc. Roy. Soc. (London) **A270**, 127 (1962), **A271**, 565 (1963), which is much more accurate than (4.2) for Z less than 15 or 20. However, the operator is rather complex, and gives results in no better agreement with experiment than does (4.2) when Z is greater than about 30—which is where the spin-orbit effects are most important—unless relativistic corrections to the radial wavefunctions are taken into account (see Table 19-2).

[2]J. C. Slater, Phys. Rev. **34**, 1293 (1929).

[3]E. U. Condon and G. H. Shortley, *The Theory of Atomic Spectra* (University Press, Cambridge, 1935).

In general this set has an infinite number of members, and so in principle (4.4) represents an infinite series. In practice, it is necessary to truncate the series to a finite number of terms; it is then essential that basis functions be of an appropriate type, and that the particular functions to be included in (4.4) in any given case be chosen judiciously.

For the present, let us assume that a set of M suitable basis functions has been chosen, and consider the problem of determining the values of the M expansion coefficients y_b $(1 \leqq b \leqq M)$. Substitution of (4.4) into the Schrödinger equation (4.3) gives

$$\sum_{b'=1}^{M} H\, y^k_{b'} \Psi_{b'} = E^k \sum_{b'=1}^{M} y^k{}_{b'} \Psi_{b'} \; .$$

Multiplying this from the left by any one of the basis functions Ψ_b, and integrating over all 3N space coordinates, and summing over both possible directions of each of the N spins [cf. (2.54)], we obtain from (4.5)

$$\sum_{b'=1}^{M} H_{bb'} y^k_{b'} = E^k \sum_{b'=1}^{M} y^k_{b'} \langle \Psi_b | \Psi_{b'} \rangle$$

$$= E^k y^k_b \; , \qquad 1 \leqq b \leqq M \; , \tag{4.6}$$

where

$$H_{bb'} \equiv \langle \Psi_b | H | \Psi_{b'} \rangle \tag{4.7}$$

is the matrix element of the Hamiltonian operator (4.1) between the basis functions b and b′. As in most quantum mechanical problems, the Hamiltonian matrix $H \equiv (H_{bb'})$ is Hermitian ($H_{b'b} = H^*_{bb'}$); in our case, it will always prove to be a real symmetric matrix ($H_{b'b} = H_{bb'}$).

The relations (4.6) comprise a set of M simultaneous linear homogeneous equations in the M unknowns $y^k_{b'}$. This set of equations has a non-trivial solution (at least one $y^k_{b'} \neq 0$) only if the determinant of the matrix $(H_{bb'} - E^k \delta_{bb'})$ is zero:

$$|H - E^k I| = 0 \; , \tag{4.8}$$

where I is the unit matrix. One possible procedure is to expand this determinant into a polynomial of degree M in E^k, the zeroes of which represent M different possible energy levels of the atom. Each of these values of E^k, substituted back into (4.6), gives M − 1 independent equations for the M − 1 ratios

$$y^k_b / y^k_i \; , \qquad b \neq i \; ;$$

the value of y^k_i is then chosen such that[4]

[4]Since H is real symmetric, the coefficients y^k_b can be chosen to be all real and the absolute-value signs are unnecessary.

$$\sum_{b=1}^{M} |y_b^k|^2 = 1 \ , \tag{4.9}$$

so that Ψ^k is normalized:

$$\begin{aligned}\langle\Psi^k|\Psi^k\rangle &= \langle\sum_b y_b^k\Psi_b|\sum_{b'} y_{b'}^k\Psi_{b'}\rangle \\ &= \sum_b\sum_{b'}(y_b^k)^* y_{b'}^k\langle\Psi_b|\Psi_{b'}\rangle \\ &= \sum_b |y_b^k|^2\end{aligned}$$

from (4.5).

For M no greater than 2 or 3, the above procedure is feasible, either numerically or analytically. For appreciably larger M, the only practical procedure is to numerically diagonalize the Hamiltonian matrix $H \equiv (H_{bb'})$. That is, if we write the set of expansion coefficients in the form of a column vector

$$\mathbf{Y}^k = \begin{pmatrix} y_1^k \\ y_2^k \\ y_3^k \\ \cdot \\ \cdot \\ \cdot \end{pmatrix} , \tag{4.10}$$

then equations (4.6) may be written as the single matrix equation

$$H\mathbf{Y}^k = E^k\mathbf{Y}^k \ , \tag{4.11}$$

and the problem is to find the M eigenvalues E^k of the matrix H, together with the corresponding eigenvectors $\mathbf{Y}^k$. This is accomplished by supplying numerical values of the matrix elements $H_{bb'}$ to a digital computer, and finding by standard techniques[5] the matrix T that diagonalizes H. The kth diagonal element of the diagonalized Hamiltonian matrix is the eigenvalue E^k,

$$T^{-1}HT = (E^k\delta_{kb}) \ , \tag{4.12}$$

[5]For example, by first tri-diagonalizing H via a sequence of Householder transformations, and then using the QR algorithm with origin shifts: J. H. Wilkinson, *The Algebraic Eigenvalue Problem* (Clarendon Press, Oxford, 1965), p. 290; Linear Algebra and Its Appls. 1, 409 (1968). G. W. Stewart, *Introduction to Matrix Computations* (Academic Press, New York, 1973). The order of the columns of T is not unique; for the sake of definiteness we shall usually assume that the columns are arranged in the order of increasing value of the corresponding eigenvalues, so that $E^1 \leqq E^2 \leqq E^3 \cdots$.

and the kth column of T represents the corresponding eigenvector $\mathbf{Y}^k$; this is true because if we multiply both sides of (4.12) from the left by T we obtain

$$HT = T(E^k\delta_{kb}) , \tag{4.13}$$

and it is easily seen that the kth column of the matrix $T(E^k\delta_{kb})$ is equal to E^k times the kth column of T, which is just (4.11).

As is well known, the eigenvectors $\mathbf{Y}^k$ are automatically mutually orthogonal if they belong to non-degenerate eigenvalues, and they may be chosen orthogonal even if the eigenvalues are degenerate; we assume that this orthogonalization has been caried out if necessary, and that each vector has been normalized according to (4.9). The eigenvectors then form an orthonormal set:

$$(\mathbf{Y}^k)^\dagger \mathbf{Y}^{k'} = \delta_{kk'} , \tag{4.14}$$

where $\mathbf{Y}^\dagger$ is the complex conjugate of the transpose of $\mathbf{Y}$. This result implies in turn that

$$T^\dagger T = I , \tag{4.15}$$

where I is the M $\times$ M identity matrix, and that therefore the eigenvector matrix T is unitary ($T^{-1} = T^\dagger$); actually, all matrix elements are real, so that T is an orthogonal matrix

$$T^{-1} = \tilde{T} , \tag{4.16}$$

where $\tilde{T}$ is the transpose of T. The matrix is actually the transformation matrix from the basis-function representation b to the representation k formed by the eigenfunctions Ψ^k.

In summary, once we have computed the Hamiltonian matrix elements $H_{bb'}$, the calculation of energy levels and eigenfunctions of the atom is essentially trivial—all we have to do is find the eigenvalues of the matrix H. The difficulty lies in computing H; i.e., in setting up appropriate basis functions and in using them to evaluate the matrix elements. This is a long and involved task that will occupy us for some chapters to come. In the remainder of this chapter we shall discuss semi-quantitative aspects of basis functions, and some qualitative aspects of energy-level structures that can be inferred from the so-called vector model of the atom.

4-3. THE CENTRAL-FIELD MODEL

The N electrons in the atom interact with each other through their mutual Coulomb repulsions, and the resulting correlation between the positions of the various electrons must be reflected in our basis functions if we are to obtain reasonably accurate results. Nevertheless, we begin with the central-field model of the atom: we make the approximation that any given electron i moves independently of the others in the electrostatic field of the nucleus (assumed stationary) and the other N − 1 electrons; this field is assumed to be time-averaged over the motion of the N − 1 electrons, and therefore (neglecting correlation with the position of the ith electron) to be spherically symmetric. In this central field, the probability distribution of electron i will be described by a one-electron wavefunction (or "spin-orbital") identical in form with (3.8):

$$\varphi_i(\mathbf{r}_i) = \frac{1}{r}P_{n_il_i}(r_i)\cdot Y_{l_im_{l_i}}(\theta_i,\phi_i)\cdot\sigma_{m_{s_i}}(s_{i_z}) , \tag{4.17}$$

where $\mathbf{r}_i$ denotes position (r,θ,ϕ) with respect to the nucleus and also the spin orientation $\mathbf{s}$.

In line with the fact that the angular momentum of electron i (moving in our postulated central field) is a constant of the motion, the function (4.17) is an eigenfunction of the one-electron angular-momentum operators l_i^2, l_{z_i}, s_i^2, and s_{z_i}, with eigenvalues $l_i(l_i + 1)$, m_{l_i}, $s_i(s_i + 1) \equiv 3/4$, and m_{s_i}, respectively. The one difference from (3.8) lies in the quantitative form of the radial factor $P_{nl}(r)$. We still expect this to be a solution of a differential equation similar to (3.9), except that the potential energy V(r) is no longer a simple Coulomb function, $-2Z/r$; exact analytical solution of the differential equation is therefore not possible, and we have to calculate P_{nl} by a numerical procedure such as that described in Sec. 3-4. Methods of determining V(r)—and thereby P_{nl}—will be discussed in Chapter 7. For the present we can leave the detailed form of P_{nl} unspecified, except to note that: (a) it will depend on the quantum number l (which will appear just as before in the radial differential equation); (b) for bound electrons, only certain discrete solutions of the differential equation will satisfy the boundary and normalization conditions (3.10)-(3.13)—which fact we indicate by means of the serial quantum number n, assigned according to the number of nodes of P_{nl} as specified in (3.29); and (c) we make the phase of φ_i definite by choosing $P_{nl}(r)$ to be positive at small r, in line with the convention (3.22).

We shall assume that two radial functions of the same l but different n are orthogonal:

$$\int_0^\infty P_{nl}(r)P_{n'l}(r)\, dr = \delta_{nn'} . \tag{4.18}$$

Such will always be the case if the function V(r) is chosen to be independent of n,[6] but we shall see in Chapter 7 that (4.18) can be satisfied under considerably less restrictive conditions. It then follows from (2.57) that

$$\langle\varphi_{nlm_lm_s}|\varphi_{n'l'm_l'm_s'}\rangle = \delta_{nn'}\delta_{ll'}\delta_{m_lm'_l}\delta_{m_sm'_s} , \tag{4.19}$$

or in abbreviated notation

$$\langle\varphi_i|\varphi_j\rangle = \delta_{ij} . \tag{4.20}$$

4-4. PRODUCT WAVEFUNCTIONS

We now need to construct a basis function for the entire atom from the one-electron spin-orbitals φ_i. Since the probability that electron i lies at position $\mathbf{r}_i$ is given by $|\varphi_i(\mathbf{r}_i)|^2$, then the probability that electron i lies at $\mathbf{r}_i$ and at the same time electron j lies at $\mathbf{r}_j$ is (neglecting any correlation between the positions of electrons i and j)

[6]See ref. 9 of Chap. 2.

$$|\varphi_i(\mathbf{r}_i)|^2 \cdot |\varphi_j(\mathbf{r}_j)|^2 \ .$$

The probability distribution function for all N electrons is thus

$$\prod_{i=1}^{N} |\varphi_i(\mathbf{r}_i)|^2 \ ,$$

and we are led to take as a basis function the simple product of spin-orbitals

$$\psi = \varphi_1(\mathbf{r}_1)\varphi_2(\mathbf{r}_2)\varphi_3(\mathbf{r}_3) \cdots \varphi_N(\mathbf{r}_N) \ . \tag{4.21}$$

In this expression, each subscript i is an abbreviation for the four one-electron quantum numbers $n_i l_i m_{l_i} m_{s_i}$, which for brevity may be written in the more compact form (omitting the i's)

$$n l^{m_s}_{m_l} \ . \tag{4.22}$$

Keeping in mind the limitations (2.28) and (2.52) on the possible values of the magnetic quantum numbers, the first few possible sets of quantum numbers (4.22), arranged in a standard speedometer order, are

$$1s^-_0,\ 1s^+_0,\ 2s^-_0,\ 2s^+_0,\ 2p^-_{-1},\ 2p^+_{-1},\ 2p^-_0,\ 2p^+_0,\ 2p^-_1,\ 2p^+_1,\ 3s^-_0,\ 3s^+_0,\ 3p^-_{-1},\ 3p^+_{-1},\ \cdots \ . \tag{4.23}$$

For simplicity, we shall (in this section only) assume that a product function is always written as in (4.21), with the factors arranged in the order of increasing coordinate subscript. Then for a specified form of the radial functions $P_{nl}(r)$, a product function is completely specified by an ordered list of the N sets of quantum numbers (4.22). For example, in a three-electron atom we could form the product functions

$$\begin{array}{ll} (1s^-_0)(1s^+_0)(2s^-_0), & (1s^+_0)(1s^-_0)(2s^-_0) \ , \\ (1s^-_0)(2s^+_0)(2p^+_1), & (2s^+_0)(2p^+_1)(1s^-_0) \ , \\ (2p^+_{-1})(2p^-_0)(3d^+_2), & (3d^+_2)(2p^-_0)(2p^+_{-1}) \ , \end{array}$$

together with a multifold infinity of other possibilities.

The product functions are clearly orthonormal because of (4.20):

$$\langle\psi|\psi'\rangle = \langle\varphi_1|\varphi_{1'}\rangle\langle\varphi_2|\varphi_{2'}\rangle \cdots \langle\varphi_N|\varphi_{N'}\rangle = \delta_{11'}\delta_{22'} \cdots \delta_{NN'} = \delta_{\psi\psi'} \ . \tag{4.24}$$

4-5. ANTISYMMETRIZATION; DETERMINANTAL FUNCTIONS

The trouble with the product function (4.21) is that it does not reflect the physical indistinguishability of electrons, which requires that when two electrons are interchanged—for example, when we go from

$$\psi_c = \varphi_1(\mathbf{r}_1)\varphi_2(\mathbf{r}_2)\varphi_3(\mathbf{r}_3)\varphi_4(\mathbf{r}_4) \cdots$$

to

$$\psi_d = \varphi_1(\mathbf{r}_2)\varphi_2(\mathbf{r}_1)\varphi_3(\mathbf{r}_3)\varphi_4(\mathbf{r}_4) \cdots$$

—the probability density should be unchanged. That is, ψ_c and ψ_d should be such that $|\psi_d|^2 = |\psi_c|^2$ and hence $\psi_d = k\psi_c$, where $|k|^2 = 1$; in fact, interchanging the two coordinates a second time brings one back to the original function, so that $\psi_c = k\psi_d = k^2\psi_c$, and $k = \pm 1$. It is a fundamental postulate of quantum mechanics that ψ be antisymmetric upon interchange of two electrons, so that

$$\psi_d = -\psi_c \ . \tag{4.25}$$

A wavefunction that is antisymmetric upon interchange of any two electron coordinates can be formed by taking the following linear combination of product functions.[7]

$$\Psi = (N!)^{-1/2}\sum_P (-1)^p \varphi_1(\mathbf{r}_{j_1})\varphi_2(\mathbf{r}_{j_2})\varphi_3(\mathbf{r}_{j_3})\cdots\varphi_N(\mathbf{r}_{j_N}) \ . \tag{4.26}$$

In each product function, the same set of one-electron quantum numbers is arranged in the same order [usually in the standard order (4.23)], but the electron coordinates $\mathbf{r}_1, \mathbf{r}_2, \mathbf{r}_3, \cdots$ have been rearranged into some new order $\mathbf{r}_{j_1}, \mathbf{r}_{j_2}, \mathbf{r}_{j_3}, \cdots$. The summation is over all N! possible permutations $P = j_1 j_2 j_3 \cdots j_N$ of the normal coordinate ordering $1\ 2\ 3 \cdots N$, and p is the parity of the permutation P ($p = 0$ if P is obtained from the normal ordering by an even number of interchanges, and $p = 1$ if an odd number of interchanges is involved). The antisymmetry of (4.26) can be established as follows: For any specific ordering of the coordinates $\mathbf{r}_3, \mathbf{r}_4, \cdots \mathbf{r}_N$, there are in Σ_P two terms

$$\pm \cdots \varphi_i(\mathbf{r}_1) \cdots \varphi_k(\mathbf{r}_2) \cdots \mp \cdots \varphi_i(\mathbf{r}_2) \cdots \varphi_k(\mathbf{r}_1) \cdots , \tag{4.27}$$

which carry opposite signs because the two permutations are of opposite parity. When the coordinates $\mathbf{r}_1$ and $\mathbf{r}_2$ are interchanged, the sum of these two terms changes sign, as does the sum of all other similar pairs of terms. The same argument is valid for the interchange of any other two electron coordinates. For example, in the case $N = 3$ the function (4.26) would be

$$\Psi = \frac{1}{\sqrt{6}}\Big[\varphi_1(\mathbf{r}_1)\varphi_2(\mathbf{r}_2)\varphi_3(\mathbf{r}_3) - \varphi_1(\mathbf{r}_2)\varphi_2(\mathbf{r}_1)\varphi_3(\mathbf{r}_3) + \varphi_1(\mathbf{r}_3)\varphi_2(\mathbf{r}_1)\varphi_3(\mathbf{r}_2) - \varphi_1(\mathbf{r}_1)\varphi_2(\mathbf{r}_3)\varphi_3(\mathbf{r}_2) + \varphi_1(\mathbf{r}_2)\varphi_2(\mathbf{r}_3)\varphi_3(\mathbf{r}_1) - \varphi_1(\mathbf{r}_3)\varphi_2(\mathbf{r}_2)\varphi_3(\mathbf{r}_1)\Big] ,$$

and it is easily verified that this example is antisymmetric upon interchange of $\mathbf{r}_1$ and $\mathbf{r}_2$, of $\mathbf{r}_1$ and $\mathbf{r}_3$, or of $\mathbf{r}_2$ and $\mathbf{r}_3$.

[7]In all that follows, we shall use the symbol Ψ for a completely antisymmetric function, and ψ for a function that is not antisymmetric with respect to interchange of at least one pair of coordinates. Antisymmetrization could be effected by summing over permutations of the quantum numbers instead of the electron coordinates, but the former procedure leads to complications when forming coupled wavefunctions, as will be seen from Secs. 4-6 and 9-1.

The antisymmetrized function (4.26) has the property that if any two orbitals are identical, then Ψ is identically zero. For example, for any specified coordinate ordering $\mathbf{r}_3$, $\mathbf{r}_4$, $\cdots$ $\mathbf{r}_N$, the two terms (4.27) in Σ_P add to zero if the two sets of quantum numbers $i \equiv n_i l_i m_{l_i} m_{s_i}$ and $k \equiv n_k l_k m_{l_k} m_{s_k}$ are identical; similarly, *all* terms of (4.26) cancel in pairs. The antisymmetrized function Ψ therefore satisfies the Pauli exclusion principle, which states that[8]

No two electrons can occupy the same spin-orbital. (4.28)

The antisymmetrized functions have a second very important property: if any two electrons lie at the same position, for example $\mathbf{r}_1 = \mathbf{r}_2$ (remember that this includes the statement that the electrons have parallel spin, $\mathbf{s}_1 = \mathbf{s}_2$), then $\Psi \equiv 0$ because (4.27) is then zero even if the orbitals are all different. Because the functions φ are continuous in the spatial variables (r,θ,ϕ), it follows that $|\Psi|$ must be unusually small whenever two electrons having parallel spin are close together. Thus unlike the product function (4.21), the antisymmetrized function (4.26) shows a certain degree of electron correlation. This correlation is incomplete—it arises by virtue of the Pauli exclusion principle rather than as a result of electrostatic repulsion, and there is no correlation at all between two electrons having antiparallel spins. Nevertheless, energies computed using antisymmetrized basis functions show decidedly better agreement with experiment than do those computed from product functions.

The factor $(N!)^{-1/2}$ is required in (4.26) in order to give a properly normalized function:

$$\langle\Psi|\Psi'\rangle = (N!)^{-1}\sum_{P}\sum_{P'}(-1)^{p+p'}\langle\varphi_1(\mathbf{r}_{j_1})\varphi_2(\mathbf{r}_{j_2})\cdots|\varphi_{1'}(\mathbf{r}_{j'_1})\varphi_{2'}(\mathbf{r}_{j'_2})\cdots\rangle \ .$$

For any given permutation P, all of the terms in the sum over P′ will be zero unless Ψ and Ψ' have identical sets of one-electron quantum numbers [cf. (4.24)]; even then the only non-zero term will be that for which $P' = P$ because, by (4.28), no two of the spin-orbitals are identical. Thus

$$\langle\Psi|\Psi'\rangle = (N!)^{-1}\sum_{P}(-1)^{2p}\delta_{\Psi\Psi'} = \delta_{\Psi\Psi'} \ , \tag{4.29}$$

since there are N! terms in the summation over P.

The antisymmetrized function (4.26) may be written in the form of a determinant

[8] W. Pauli, Z. Physik **31**, 765 (1925). Actually, unlike the situation for the product function (4.21), with an antisymmetrized function we can no longer say that any given electron occupies a specific orbital, because the electron with coordinates $\mathbf{r}_i$ is distributed uniformly among all N different orbitals; the statement (4.28) of the Pauli principle is just a convenient semiclassical way of saying that no antisymmetric wavefunction describing an N-electron atom can be made up of one-electron functions, two of which have the same set of one-electron quantum numbers.

$$\Psi = \frac{1}{N!^{1/2}} \begin{vmatrix} \varphi_1(\mathbf{r}_1) & \varphi_1(\mathbf{r}_2) & \varphi_1(\mathbf{r}_3) & \cdots \\ \varphi_2(\mathbf{r}_1) & \varphi_2(\mathbf{r}_2) & \varphi_2(\mathbf{r}_3) & \cdots \\ \varphi_3(\mathbf{r}_1) & \varphi_3(\mathbf{r}_2) & \varphi_3(\mathbf{r}_3) & \cdots \\ \cdot & \cdot & \cdot & \\ \cdot & \cdot & \cdot & \\ \cdot & \cdot & \cdot & \end{vmatrix} , \qquad (4.30)$$

and is therefore referred to as a determinantal function or a Slater determinant.[2] The antisymmetrization, Pauli principle, and correlation effects then follow immediately from well-known properties of determinants: (1) interchanging the coordinates of two electrons is equivalent to interchanging two columns of the determinant, which changes its sign; (2) if two orbitals have the same quantum numbers, then two rows of the determinant are identical and the determinant is therefore zero; (3) if two electrons have the same coordinates, then two columns are identical and the determinant is zero. Although the determinantal form thus exhibits the properties of Ψ very neatly, it is simpler to use the form (4.26) for actual calculations.

4-6. COUPLING OF ANTISYMMETRIZED WAVEFUNCTIONS

As already mentioned in Sec. 4-2, it is desirable that the number of terms which must be retained in the basis-function expansion (4.4) be as small as possible consistent with the desired accuracy. Toward this end, it is important that each basis function Ψ_b be as close as practicable to one of the actual eigenfunctions Ψ^k of the atom. The one property of an isolated atom (in field-free space) of which we can be certain, other than the total energy, is its total angular momentum; thus it is desirable that Ψ_b be an eigenfunction of $\mathbf{J}^2 \equiv \mathbf{J}\cdot\mathbf{J}$ and of J_z, where $\mathbf{J}$ is the total angular-momentum operator for the system of N electrons:

$$\mathbf{J} = \sum_{i=1}^{N} (\boldsymbol{l}_i + \mathbf{s}_i) . \qquad (4.31)$$

Since each term of the antisymmetrized function (4.26) involves the same N sets of one-electron quantum numbers, it is evident that such a function is already an eigenfunction of J_z with eigenvalue

$$M = \sum_{i=1}^{N} (m_{l_i} + m_{s_i}) , \qquad (4.32)$$

just as was the case with the simple product functions considered in Secs. 2-8 to 2-11. Similarly, it may be anticipated that an eigenfunction of $\mathbf{J}^2$ can be formed in much the same way as for product functions, by taking an appropriate linear combination of determinantal functions—each determinant involving the same set of quantum numbers $n_i l_i$ ($1 \leqq i \leqq N$), but different sets of values of m_{l_i} and m_{s_i} subject to the restriction that the sum (4.32) of all magnetic quantum numbers have the same value M for each determinant.

It may be noted that antisymmetrization involves summations over permutations of electron coordinates, whereas coupling involves summations over functions with different values of the magnetic quantum numbers. Since these are independent finite summations, the order in which they are performed is interchangeable—that is, we can in principle either antisymmetrize first and couple second, or vice versa. In practice, the Pauli principle (4.28) restricts the possible values of the m_{l_i} and m_{s_i} whenever two or more spin-orbitals have the same values of $n_i l_i$, and as a result of these complications it proves to be more convenient to couple first and antisymmetrize second (with the antisymmetrization not always being done by means of coordinate permutations). The qualitative vector-model aspects of these Pauli-principle complications will be discussed in Secs. 9 and 17 of this chapter and the quantitative mathematical aspects will be dealt with in Chapter 9.

4-7. ELECTRON CONFIGURATIONS

As indicated in the preceding section, a coupled basis function is formed by taking an appropriate linear combination of uncoupled (product or determinantal) functions, all of which are formed from spin-orbitals having the same set of N values of $n_i l_i$. For any given fixed set of $n_i l_i$ values, the number of different possible coupled functions that can be formed (described by different sets of the coupled quantum numbers $\cdots$JM) is equal to the number of different possible uncoupled functions (described by different sets of the one-electron magnetic quantum numbers $m_{l_i} m_{s_i}$); this follows directly from the considerations of Sec. 2-8, especially Eq. (2.65), and an example is given in Table 4-1. The set of coupled

TABLE 4-1. UNCOUPLED AND LS-COUPLED BASIS FUNCTIONS OF THE CONFIGURATION ps

Uncoupled	M	LS-Coupled
$p_{-1}^{-}\ s_0^{-}$	−2	ps $^3P_2^o$
$p_{-1}^{-}\ s_0^{+}$ $p_{-1}^{+}\ s_0^{-}$ $p_0^{-}\ s_0^{-}$	−1	ps $^3P_2^o$ ps $^3P_1^o$ ps $^1P_1^o$
$p_{-1}^{+}\ s_0^{+}$ $p_0^{-}\ s_0^{+}$ $p_0^{+}\ s_0^{-}$ $p_1^{-}\ s_0^{-}$	0	ps $^3P_2^o$ ps $^3P_1^o$ ps $^3P_0^o$ ps $^1P_1^o$
$p_0^{+}\ s_0^{+}$ $p_1^{-}\ s_0^{+}$ $p_1^{+}\ s_0^{-}$	1	ps $^3P_2^o$ ps $^3P_1^o$ ps $^1P_1^o$
$p_1^{+}\ s_0^{+}$	2	ps $^3P_2^o$

functions is related to the set of uncoupled functions by an orthogonal linear transformation,[9] and each set spans the same angular-momentum subspace as does the other.

The list of N pairs of quantum numbers $n_i l_i$ that defines such a set of functions (either coupled or uncoupled) is called an *electron configuration*, or simply a *configuration*. A configuration may also be thought of as the set of functions itself. In general there may be more than one spin-orbital with a given value of $n_i l_i$, and so the list is written by means of the condensed notation

$$(n_1 l_1)^{w_1}(n_2 l_2)^{w_2} \cdots (n_q l_q)^{w_q} , \qquad \text{where } \sum_{j=1}^{q} w_q = N . \tag{4.33}$$

An *occupation number* w_j that is equal to unity is usually omitted from the notation; an example is the configuration

$$1s^2\ 2s\ 2p^3 ,$$

which may be described semiclassically as the collection of all basis functions that can be formed from two 1s electrons, one 2s electron, and three 2p electrons.[10] The angular momentum properties of the basis functions of a configuration do not depend explicitly on the values of the principal quantum numbers n_j. Consequently, these values are sometimes left unspecified; examples are the configuration ps in Table 4-1 and the configurations dp and fds discussed in Sec. 2-12.

The concept of a configuration is of physical as well as mathematical significance. It is quite commonly found that in an energy-level diagram, such as that for Ne I in Fig. 4-1, the lower energy levels tend to appear in closely spaced groups. In any one group the number of levels, and the values of J observed to be associated with these levels,[11] are found to correlate exactly with the corresponding values for the coupled basis functions of an electron configuration appropriate to the number of electrons in the atom in question. Conversely, if an energy-level calculation is made by the procedure outlined in Sec. 4-2, using as a basis in (4.4) the set of functions that constitute an appropriate configuration (the *single-configuration approximation*), then the computed energy levels (and their J values) correlate straightforwardly with the observed levels of one of these groups. In the sense of this correlation, each group of levels is also referred to as an electron configuration, and is labeled by means of the notation (4.33). The configuration that includes the ground (lowest-energy) level of an atom is called the *ground configuration* of that atom; all other configurations are called *excited configurations*.

[9]Actually, by a *set* of orthogonal linear transformations, one transformation for each value of M.

[10]Quantum-mechanically we cannot really talk about a "1s electron," etc. The above terminology is a loose way of referring to basis functions formed from two 1s orbitals, one 2s orbital, and three 2p orbitals (cf. fn. 8). It is such a convenient manner of speaking that we shall use it often, but the reader should always remember the quantum-mechanical reservations.

[11]The J values may most easily be determined experimentally via the Zeeman effect: when the atom is placed in an external magnetic field, each level is split into 2J + 1 *sublevels* or *states* according to the 2J + 1 possible values of M; cf. Fig. 4-5.

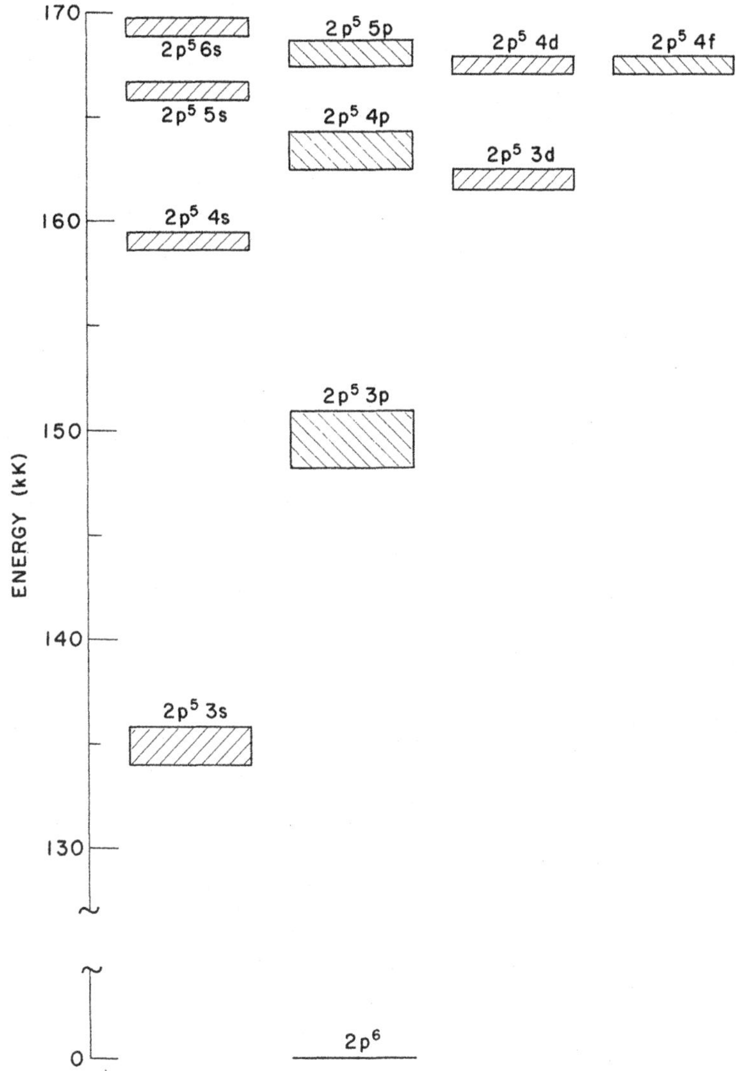

Fig. 4-1. Block diagram of the lowest configurations of Ne I. The levels of each configuration lie within the limited energy range shown by the corresponding shaded block. (There is one level in $2p^6$, and there are four levels in each p^5s configuration, ten levels in each p^5p, and twelve levels in each p^5d or p^5f configuration.)

It may be true, particularly in the case of highly excited configurations which overlap each other in energy, that accurate calculations cannot be made within the limitations of a single-configuration basis set. It is then necessary to use a *multi-configuration* approximation, in which the basis set includes functions from two or more configurations. Each computed eigenfunction Ψ will in general then be a mixture of basis functions from all configurations included in the calculation—a result referred to as *configuration mixing.*

Correspondingly, the computed energies will be different from the values that would have been given by a set of single-configuration calculations—a result referred to as *configuration-interaction perturbations.* The wavefunction-mixing and energy-level-perturbation aspects jointly are referred to simply as *configuration interaction.*

Whenever configuration mixing of basis functions is pronounced, it is clear that each eigenfunction Ψ (and its corresponding energy level) cannot unambiguously be said to belong to a specified configuration. Such cases are certainly not uncommon, but even so they do not occur with nearly the frequency that one might anticipate. In part, this is due to the fact that eigenfunctions of a field-free atom must correspond to a definite parity and a definite value of J (cf. Secs. 2-13 and 2-10). Thus in Fig. 4-1, for example, mixing can occur only among states belonging to configurations in the first and third columns (odd parity) *or* among states of the configurations in the second and fourth columns (even parity), and then only among states that have the same J value. It is therefore usually possible to assign most observed levels, even of rather high energy, to definite configurations.

Configurations that differ only in the principal quantum number of an excited electron—for example, the configurations $2p^5 3p$, $2p^5 4p$, $2p^5 5p$, $\cdots$ in the second column of Fig. 4-1—are called a *Rydberg series of configurations.* Corresponding levels (having the same J, etc.), one from each configuration of a Rydberg series, are called a *Rydberg series of levels.*

4-8. EQUIVALENT ELECTRONS; CLOSED SUBSHELLS

One-electron spin-orbitals having the same value of *nl* are called *equivalent orbitals.* Correspondingly, we may speak loosely of electrons having the same values of *nl*, and refer to them as *equivalent electrons.* A set of equivalent electrons

$$(nl)^w$$

is called a *subshell*, or sometimes simply a *shell*—though the latter term is also used to refer to the set of all possible subshells with given n:

K shell:	n = 1 (1s)
L shell:	n = 2 (2s + 2p)
M shell:	n = 3 (3s + 3p + 3d)
N shell:	n = 4 (4s + 4p + 4d + 4f)
O shell:	n = 5 (5s + 5p + 5d + 5f + 5g)
P shell:	n = 6 (6s + 6p + 6d + 6f + 6g + 6h)
etc.	

For given l, a spin-orbital may from (2.16) have one of $(2l + 1)$ different values of m_l; in each case the value of m_s may be either $+1/2$ or $-1/2$. As a consequence of the Pauli exclusion principle (4.28), the maximum number of electrons that can exist in a given subshell is $4l + 2$. A subshell occupied by the maximum number of electrons is called a *filled* or *closed subshell*; half-filled subshells $(w = 2l + 1)$ are also of special interest:

l	Half-filled subshell	Closed subshell
0	s^1	s^2
1	p^3	p^6
2	d^5	d^{10}
3	f^7	f^{14}
4	g^9	g^{18}

For every spin-orbital with quantum numbers $m_l m_s$ in a filled subshell there exists another orbital $-m_l\ -m_s$; the only possible value of $M_L = \Sigma m_l$ and of $M_S = \Sigma m_s$ is therefore zero. The only possible LS-coupled function (identical, therefore, with the determinantal function) is consequently one with $L = S = J = 0$; i.e., a $^1S_0$ function. For a single s electron ($l = 0$) the orbital is independent of the spatial angles θ,ϕ (Table 2-1); correspondingly, we may anticipate that a filled-shell ^1S function is also spherically symmetric. Indeed, in the approximation in which we consider the electrons as independent particles, uncorrelated in position, the total probability density of electrons of both spins at a point (r,θ,ϕ) is given from (3.30) and (4.17) as

$$\frac{1}{2\pi r^2} \sum_i \sum_{s_{i_z}} |P_{n_i l_i}(r)|^2 \cdot |Y_{l_i m_{l_i}}(\theta,\phi)|^2 \cdot |\sigma_{m_{s_i}}(s_{i_z})|^2$$

$$= \frac{1}{2\pi r^2} \sum_i |P_{n_i l_i}(r)|^2 \cdot |Y_{l_i m_{l_i}}(\theta,\phi)|^2 \ . \tag{4.34}$$

For the $4l + 2$ electrons in a filled subshell this becomes from (2.50)

$$\frac{1}{\pi r^2} |P_{n_i l_i}(r)|^2 \sum_{m_{l_i}=-l_i}^{l_i} |Y_{l_i m_{l_i}}(\theta,\phi)|^2 = \frac{2l_i + 1}{4\pi^2 r^2} |P_{n_i l_i}(r)|^2 \ , \tag{4.35}$$

which is independent of the angular variables—a result known as Unsöld's theorem.[12]

It is evident that any filled subshell k, having $L_k = S_k = J_k = 0$, contributes nothing to the possible values of the total L, S, and J for states of a configuration (4.33), and likewise contributes nothing to the corresponding energy-level structure of the atom. Consequently, filled subshells are commonly omitted from the configuration notation. For example, the configuration

Ne I $1s^2 2s^2 2p^5 3s$

is usually abbreviated to simply

Ne I $2p^5 3s$,

as indicated in Fig. 4-1.

[12]A. Unsöld, Ann. Physik **82**, 355 (1927).

4-9. PERMITTED LS TERMS FOR EQUIVALENT ELECTRONS

For a set of equivalent electrons $(nl)^w$, the LS terms predicted by the vector model are not all permitted, because of limitations on the permissible values of the $m_{l_i} m_{s_i}$ imposed by the Pauli principle; an example has already been seen, in that the only allowed term for a filled subshell is ^1S. The terms allowed in any given case may be determined by a procedure[13] similar to that used in Table 2-2, if we take care to include only those combinations of values m_{l_i} and m_{s_i} allowed by the exclusion principle. As an example we consider the case p^2. We first find the possible (M_L, M_S) states:

$m_{l_1}^{m_{s_1}}$	$m_{l_2}^{m_{s_2}}$	M_L	M_S	$M = M_L + M_S$
-1^-	-1^+	-2	0	-2
	0^-	-1	-1	-2
	0^+	-1	0	-1
	1^-	0	-1	-1
	1^+	0	0	0
-1^+	0^-	-1	0	-1
	0^+	-1	1	0
	1^-	0	0	0
	1^+	0	1	1
0^-	0^+	0	0	0
	1^-	1	-1	0
	1^+	1	0	1
0^+	1^-	1	0	1
	1^+	1	1	2
1^-	1^+	2	0	2

We next write the *number* of (M_L, M_S) states in an array in which M_S is tabulated against M_L:

	M_S		
M_L	-1	0	$+1$
2		1	
1	1	2	1
0	1	3	1
-1	1	2	1
-2		1	

It is evident that this array can be decomposed into an array for which $|M_L| \leqq 2$, $M_S = 0$, an array for which $|M_L| \leqq 1$, $|M_S| \leqq 1$, and an array $M_L = M_S = 0$; these correspond

[13]Ref. 2, or J. C. Slater, *Quantum Theory of Atomic Structure* (McGraw-Hill, New York, 1960), Chap. 13.

respectively to a ^1D, a ^3P, and a ^1S term. (As a check, the values of M in the first table correspond to two J = 2 levels, one J = 1 level, and two J = 0 levels, which is what we should have for $^1D_2$, $^3P_{012}$, and $^1S_0$.)

If in setting up the first table we had not excluded those combinations $(m_{l_1}^{m_{s1}}, m_{l_2}^{m_{s2}})$ forbidden by the Pauli principle, we would have found the six terms ^1S, ^3S, ^1P, ^3P, ^1D, ^3D, which are just the ones to be expected from the vector model for a pair of non-equivalent p electrons (i.e., a pp′ configuration). Thus the effect of the Pauli principle in the LS-coupled scheme for p^2 has been to delete the terms ^3S, ^1P, ^3D. In general, the only allowed terms of l^2 are those for which L + S is even, as we shall show in Sec. 9-4.

In the above manner it is possible to determine the permitted LS terms for any set of equivalent electrons l^w. Such a calculation has to be performed only for $w \leqq 2l + 1$, because the allowed terms for l^w are the same as those for the conjugate configuration l^{4l+2-w}; this is true because the only possible term of l^{4l+2} is ^1S, and the only way we can obtain a ^1S term by the vector-model combination

$$l^w\, L_1S_1 + l^{4l+2-w}\, L_2S_2 \rightarrow l^{4l+2}\ {}^1S \tag{4.36}$$

is by having $L_1S_1 = L_2S_2$. Even so, this method is very laborious in complex cases, and group theoretical methods have been developed.[14]

The permitted LS terms are given in Table 4-2 for p, d, and f electrons; results for g^w, h^w, and i^w are tabulated by Shudeman.[15] It will be noted that in many cases with $l > 1$ and $2 < w < 4l$ there exist 2 or more terms with given values of LS. In such cases, additional quantum numbers are required to distinguish one term from another. These quantum numbers we denote by α; their nature will be discussed in Sec. 9-7.

Problem

4-9(1). By the method used above for p^2, verify that the allowed terms of p^3, d^2, and d^3 are as listed in Table 4-2.

4-10. CONFIGURATIONS WITH SEVERAL OPEN SUBSHELLS

Many configurations of practical interest (especially ground configurations) include only one open (i.e., non-filled) subshell; in such cases, the possible LS terms of the configuration are the same as those of the open subshell l^w. More often, however, a configuration will contain two or more open subshells;[16] for example,

[14]R. F. Curl, Jr. and J. E. Kilpatrick, Am. J. Phys. **28**, 357 (1960); N. Karayianis, J. Math. Phys. **6**, 1204 (1965); B. G. Wybourne, J. Chem. Phys. **45**, 1100 (1966); P. E. S. Wormer, Chem. Phys. Lett. **5**, 355 (1970).

[15]C. L. B. Shudeman, J. Franklin Inst. **224**, 501 (1937).

[16]Some authors refer to a configuration containing at most one open subshell as a pure configuration, and to a configuration containing more than one open subshell as a mixed configuration. We shall not use these terms, as they are too easily confused with the absence or presence, respectively, of configuration-interaction mixing, which is an entirely different concept.

TABLE 4-2. PERMITTED LS TERMS FOR s^w, p^w, d^w, and f^w SUBSHELLS[a]. The subscripts indicate the number of different terms having the same values of LS. (Odd-parity superscripts omitted for brevity.)

					Total Number
s	^2S				1
s^2	^1S				1
p, p^5	^2P				1
p^2, p^4	$^1(SD)$	^3P			3
p^3	$^2(PD)$	^4S			3
d, d^9	^2D				1
d^2, d^8	$^1(SDG)$	$^3(PF)$			5
d^3, d^7	$^2(PD_2FGH)$	$^4(PF)$			8
d^4, d^6	$^1(S_2D_2FG_2I)$	$^3(P_2DF_2GH)$	^5D		16
d^5	$^2(SPD_3F_2G_2HI)$	$^4(PDFG)$	^6S		16
f, f^{13}	^2F				1
f^2, f^{12}	$^1(SDGI)$	$^3(PFH)$			7
f^3, f^{11}	$^2(PD_2F_2G_2H_2IKL)$	$^4(SDFGI)$			17
f^4, f^{10}	$^1(S_2D_4FG_4H_2I_3KL_2N)$	$^3(P_3D_2F_4G_3H_4I_2K_2LM)$	$^5(SDFGI)$		47
f^5, f^9	$^2(P_4D_5F_7G_6H_7I_5K_5L_3M_2NO)$	$^4(SP_2D_3F_4G_4H_3I_3K_2LM)$	$^6(PFH)$		73
f^6, f^8	$^1(S_4PD_6F_4G_8H_4I_7K_3L_4M_2N_2Q)$	$^3(P_6D_5F_9G_7H_9I_6K_6L_3M_3NO)$	$^5(SPD_3F_2G_3H_2I_2KL)$	^7F	119
f^7	$^2(S_2P_5D_7F_{10}G_{10}H_9I_9K_7L_5M_4N_2OQ)$	$^4(S_2P_2D_6F_5G_7H_5I_5K_3L_3MN)$	$^6(PDFGHI)$	^8S	119

[a]H. N. Russell, Phys. Rev. **29**, 782 (1927); R. C. Gibbs, D. T. Wilber, and H. E. White, Phys. Rev. **29**, 790 (1927).

C I	$1s^2 2s 2p^3$,
Al I	$1s^2 2s^2 2p^6 3s 3p 3d$,
Ti I	$\cdots 3p^6 3d^2 4p^2$.

In this case, the overall quantum numbers for LS-coupled functions may be found by applying the vector model to the vector sums

$$\mathbf{L} = [(\mathbf{L}_1 + \mathbf{L}_2) + \mathbf{L}_3] + \cdots ,$$
$$\mathbf{S} = [(\mathbf{S}_1 + \mathbf{S}_2) + \mathbf{S}_3] + \cdots ,$$
$$\mathbf{J} = \mathbf{L} + \mathbf{S} ,$$

where L_iS_i represents any one of the terms listed in Table 4-2 for the ith open subshell. (Closed subshells may be omitted since the corresponding values of L_i and S_i are zero.) If the various intermediate (and final) quantum numbers are denoted by script letters, the quantum numbers and coupling scheme can both be indicated by the convenient notation

$$\{[((L_1,L_2)\mathfrak{L}_2,L_3)\mathfrak{L}_3, \cdots]\mathfrak{L}_q, [((S_1,S_2)\mathfrak{S}_2,S_3)\mathfrak{S}_3, \cdots]\mathfrak{S}_q\}\mathfrak{J}_q\mathfrak{M}_q .$$

A more compact notation, which does not separate the related quantum numbers L_i and S_i, and which also includes specification of each subshell, is

$$\{[(l_1^{w_1}\alpha_1L_1S_1,l_2^{w_2}\alpha_2L_2S_2)\mathfrak{L}_2\mathfrak{S}_2, l_3^{w_3}\alpha_3L_3S_3]\mathfrak{L}_3\mathfrak{S}_3, \cdots\}\mathfrak{L}_q\mathfrak{S}_q\mathfrak{J}_q\mathfrak{M}_q . \tag{4.37}$$

All terms predicted by the vector model are allowed because we are coupling vectors associated with electrons having different values of n_il_i, so that the Pauli principle imposes no limitations on the values of the magnetic quantum numbers. As an example, in the configuration d^2p^2 we obtain the following 49 different terms:

	p^2		
d^2	^1S	^3P	^1D
^1S	^1S	^3P	^1D
^3P	^3P	$^{1,3,5}SPD$	^3PDF
^1D	^1D	^3PDF	^1SPDFG
^3F	^3F	$^{1,3,5}DFG$	^3PDFGH
^1G	^1G	^3FGH	^1DFGHI

It is clear that the number of LS terms, and therefore the number of energy levels, can be very large, especially in configurations of the lanthanide and actinide elements (Z = 57-70 and 89-102), such as

Gd I $4f^7 5d^2 6s$;

correspondingly, the spectra of these elements are exceedingly complex, and comparatively little progress has yet been made in their analysis (cf. Table 1-4).

For any given configuration, the values obtained for the quantum numbers $\mathfrak{L}_q\mathfrak{S}_q$ are independent of the order of coupling together the various subshells. However, if the genealogical coupling order (4.37) is used, the term $\mathfrak{L}_{q-1}\mathfrak{S}_{q-1}$ is called the *parent* of the final term $\mathfrak{L}_q\mathfrak{S}_q$, $\mathfrak{L}_{q-2}\mathfrak{S}_{q-2}$ is called its *grandparent*, etc. Conversely, $\mathfrak{L}_q\mathfrak{S}_q$ is a *daughter* of $\mathfrak{L}_{q-1}\mathfrak{S}_{q-1}$, a *granddaughter* of $\mathfrak{L}_{q-2}\mathfrak{S}_{q-2}$, etc. For example, in the configuration d^2p, the parent term $d^2(^1G)$ leads to three daughter terms

$$d^2(^1G)p\ \ ^2F^\circ,^2G^\circ,^2H^\circ\ ;$$

such a set of all possible terms based on a single parent is sometimes called a *subconfiguration.*

Problem

4-10(1). Using Table 4-2 and the vector model, derive the 38 LS terms of the configuration d^2sp. How many energy levels are there in this configuration for each possible value of J?

4-11. CONFIGURATION-AVERAGE ENERGIES

It is convenient to be able to describe the energy of a configuration in terms of an appropriate mean value

$$E_{av} = \langle b|H|b\rangle_{av}\ ,$$

where the average is to be carried out over all basis functions b belonging to the configuration in question. The appropriate form of this average is the simple one in which all basis functions are given equal weight:

$$E_{av} = \frac{\sum_b \langle b|H|b\rangle}{\text{number of basis functions}}\ ; \tag{4.38}$$

the appropriateness of this choice lies in the fact that the numerator of (4.38) is the trace of the Hamiltonian matrix, and the trace is invariant under the orthogonal transformation that connects one set of basis functions with any other set. If uncoupled basis functions are used, then the evaluation of E_{av} amounts to averaging over all possible sets of values of the one-electron magnetic quantum numbers $m_{l_i}m_{s_i}$. From Unsöld's theorem (4.35), such an average is equivalent to performing a spherically symmetrized average over the angular distribution of the electrons in the atom; this is consistent with the central-field model of the atom which we assumed in Sec. 4-3.

Since the trace of a matrix is invariant under all orthogonal transformations, and in particular under that transformation (4.12) which diagonalizes H, (4.38) may also be written in the form

$$E_{av} = \frac{\sum\limits_{\text{states}} E^k}{\text{number of states}}\ , \tag{4.39}$$

where the summation is over all eigenstates k of the configuration, having corresponding eigenvalues E^k. If the atom lies in field-free space, each state is an eigenstate of the total-angular-momentum operators $\mathbf{J}^2$ and J_z described by quantum numbers JM, and the eigenvalues are independent of M. Since each energy level thus has a (2J + 1)-fold degeneracy, we may also write

$$E_{av} = \frac{\sum_{\text{levels}} (2J + 1)E^J}{\sum_{\text{levels}} (2J + 1)} . \tag{4.40}$$

This expression may be used to calculate an experimental value of E_{av}, using the observed energies of the levels belonging to the configuration in question; for example, the configuration p^2 has LS terms ^3P, ^1D, and ^1S, and so

$$E_{av} = \frac{1}{15} [E(^3P_0) + 3E(^3P_1) + 5E(^3P_2) + 5E(^1D_2) + E(^1S_0)] .$$

This quantity is called the center-of-gravity or baricenter energy of the levels of the configuration; its value is of course significant only to the extent that configuration-interaction perturbations are small.

4-12. RELATIVE ENERGIES OF CONFIGURATIONS

It is readily possible to make some general qualitative statements about the relative values of E_{av} for different configurations, based on the concept of the (spherically averaged) binding energies of electrons in the various subshells $n_i l_i$.

The magnitude of the binding energy of an electron in a complex atom will be smaller the greater the value of n, just as in a hydrogenic atom, only more so: An electron in the K shell (n = 1) will see almost the full nuclear charge Z, and so will be very tightly bound. But an electron in the L shell (n = 2) is partially shielded from the nucleus by the electrons in the K shell, and is thus bound by an effective nuclear charge less than that for the K electrons. Similarly, electrons with n = 3, 4, $\cdots$ see successively smaller effective nuclear charges, and thus the dependence of binding energy on n is considerably stronger (for given l) than the hydrogenic $1/n^2$ dependence of (3.27).

For a one-electron atom, we have seen in Sec. 3-3 that the (non-relativistic) binding energy $|E| = Z^2/n^2$ depends only on n and not on l; however, this is a peculiarity of a Coulomb field only. In a multi-electron atom the energy depends strongly on l, for reasons very similar to those just discussed: just as in hydrogen (Sec. 3-5), the probability that an electron in a multi-electron atom lies close to the nucleus is greater the smaller the value of l; but the closer it comes to the nucleus, the greater the effective nuclear charge seen by the electron, and hence the more tightly it is bound. Associated with this greater nuclear attraction for small-angular-momentum electrons is the result that for given n the mean value of r tends to be smallest for small l, rather than for large l as in hydrogenic atoms.

This dependence of energy on n and l may be seen quite clearly in Fig. 1-5, which depicts the energies of the C II configurations[17]

$1s^2 2s^2 2p$,
$1s^2 2s^2 3s$,
$1s^2 2s^2 3p$,
$1s^2 2s^2 3d$,
$1s^2 2s^2 4s$, etc.

The greater the values of n and l for the excited outer electron, the farther this electron is from the nucleus, and the less it penetrates inside the $1s^2 2s^2$ electron "core"—consequently, the more nearly independent of l is the effective nuclear charge, and therefore also the binding energy. Because the Pauli principle limits occupation of an s subshell to at most two electrons (Sec. 4-8), $1s^2 2s^2 2p$ is the ground configuration.

A second example may be seen in Fig. 1-7, which again shows that the energies of the configurations

Si II $1s^2 2s^2 2p^6 3s^2 3p$,
$1s^2 2s^2 2p^6 3s^2 3d$,
$1s^2 2s^2 2p^6 3s^2 4s$, etc.

depend strongly on both n and l of the outer electron. In addition, it may be seen that the configuration[18]

$1s^2 2s^2 2p^6 3s 3p^2$

lies higher than does

$1s^2 2s^2 2p^6 3s^2 3p$,

in a manner completely consistent with the above binding-energy arguments.

There is, however, in Si II a complication that did not appear in C II: the dependence of energy on l has become so strong (because of the greater number of core electrons) that the configuration $3s^2 4s$ lies lower than does $3s^2 3d$—the smaller angular momentum of 4s (by two units) more than compensating for the fact that n is greater by one. Similarly, in Fig. 4-1 the configuration $2p^5 4s$ of Ne I lies lower than $2p^5 3d$, and $2p^5 5s$ lies lower than $2p^5 4d$ and $2p^5 4f$. Such crossovers become more and more common the greater the value of Z,[19]

[17]In each configuration $1s^2 2s^2 nl$, there is only a single term ^2L, with $L = l$. For each term except ns ^2S, there are two levels $^2L_{l-1/2}$ and $^2L_{l+1/2}$, but the spin-orbit splitting is too small for these levels to be plotted separately; the figure therefore shows only the value of E_{av} for each configuration.

[18]Actually, only the term $3s3p^2\ ^2D$ is plotted in the figure, but the other terms (^2S, ^2P, ^4P) of $3s3p^2$ also lie much higher than the lone $^2P^o$ term of the ground configuration $3s^2 3p$.

[19]This remark applies only to the outer two or three shells of neutral or a few-fold-ionized atoms. For highly ionized atoms or for shells in the inner part of the core, the nuclear attraction overshadows the electron-electron repulsions sufficiently that all nl electrons are more tightly bound than any (n + 1)l' electrons.

but the extent of such non-hydrogenic relationships in any particular case cannot be predicted on the basis of qualitative arguments alone.

4-13. THE PERIODIC SYSTEM

The arguments of the preceding section go a long way toward predicting the ground configurations of neutral atoms, and thereby the nature of the periodic system: For the most part, we may expect the ground configuration (the configuration to which the ground level belongs) to be the configuration with the lowest average total energy. We would expect that this lowest-energy configuration would involve no more than one open subshell, and that the magnitude of the one-electron binding energy for this subshell would be smaller than that for each filled subshell, but greater than that for each unoccupied subshell.

Thus in the ground configuration of neutral carbon, for example, we would expect two of the six electrons to fill the very tightly bound 1s subshell and two more electrons to fill the 2s subshell, with two electrons left to occupy the third-most-tightly-bound 2p subshell. This $1s^2 2s^2 2p^2$ configuration has the same LS terms as a p^2 subshell alone, which from Table 4-2 should be a ^3P, a ^1D, and a ^1S term. These expectations are confirmed by analysis of the observed C I spectrum: the lowest energy levels (derived empirically as described in Sec. 1-11) consist of three closely spaced levels with $J = 0, 1, 2$ (indicating a ^3P term),[11] a $J = 2$ level ($^1D_2$) somewhat higher, and a $J = 0$ level ($^1S_0$) still higher—all other levels being much higher yet.[20]

With each unit increase in the nuclear charge, we would expect the added electron to enter the partly filled subshell if one exists or, otherwise, to enter that unoccupied subshell for which its binding energy is the greatest. This is known as the building-up principle, or aufbauprinzip.[21] Taking into account non-hydrogenic irregularities such as the 4s-3d ordering discussed in the preceding section, the basic structure of the periodic system[22] is as depicted schematically in Fig. 4-2. Filling of the 3d and 4d subshells takes place after completion of the 4s and 5s subshells, respectively, forming the first two d transition series of elements from $Z = 21$ to 30 and from $Z = 39$ to 48. Upon completion of the 6s subshell, filling of the 4f subshell takes place before the 5d subshell (with two or three exceptions) producing the 4f lanthanide series and then the third d transition series, which together extend from $Z = 57$ to 80; similarly, filling of the 5f and 6d subshells produces the actinide and the fourth d transition series, extending from $Z = 89$ to (presumably) $Z = 112$. Hypothetical super-heavy elements of still higher Z are predicted theoretically to include,

[20] C. E. Moore, *Atomic Energy Levels*, Vol. I (1949) or NSRDS-NBS 3, Sec. 3 (1970).

[21] N. Bohr, Z. Physik **9**, 1 (1922); E. C. Stoner, Phil. Mag. **48**, 719 (1924).

[22] Numerous schemes for displaying this structure have been suggested; see, for example, C. Janet, *La structure du noyau de l'atome, considérée dans la classification périodique, des éléments chimiques* (Imprimerie départmentale de l'Oise, Beauvais, 1927); S. A. Goudsmit and P. I. Richards, Proc. Natl. Acad. Sci. **51**, 664 (1964); D. Neubert, Z. Naturforsch. **25a**, 210 (1970).

n(s + p)			
1	1 (1s) 2		
2	3 (2s) 4		5 (2p) 10
3	11 (3s) 12		13 (3p) 18
4	19 (4s) 20	21 (3d) 30	31 (4p) 36
5	37 (5s) 38	39 (4d) 48	49 (5p) 54
6	55 (6s) 56	57 (4f) – (5d) 80	81 (6p) 86
7	87 (7s) 88	89 (5f) – (6d) 112	113 (7p) 118
8	119 (8s) 120	121 (5g) – (6f) – (7d) 162	163 (8p) 168

n + ℓ				
1				1 (1s) 2
2				3 (2s) 4
3			5 (2p) 10	11 (3s) 12
4			13 (3p) 18	19 (4s) 20
5		21 (3d) 30	31 (4p) 36	37 (5s) 38
6		39 (4d) 48	49 (5p) 54	55 (6s) 56
7	57 (4f) – (5d) 80		81 (6p) 86	87 (7s) 88
8	89 (5f) – (6d) 112		113 (7p) 118	119 (8s) 120
9	121 (5g) – (6f) – (7d) 162		163 (8p) 168	169 (9s) 170

Fig. 4-2. Schematic form of the periodic system, indicating the valence-electron subshells for the ground configurations of the neutral free atoms: (upper) a common tabular arrangement; and (lower) Janet's form, with rows arranged according to Madelung's rule of constant n + *l*. Extrapolation beyond the list of presently known elements is not straightforward: because of strong relativistic effects, $8p_{1/2}$ electrons probably are present as early as Z = 121, and the 9s and $9p_{1/2}$ subshells are probably filled before the $8p_{3/2}$ subshell.

after filling of the 7p and 8s subshells, a "super-actinide" series having complicated ground configurations involving 5g, 6f, 7d, and 8p electrons.[23]

In addition to the gross non-hydrogenic irregularities associated with the positions of the transition and rare-earth series of elements, there are also minor irregularities within each series, as shown in Table 4-3. For example, at $_{24}$Cr and $_{29}$Cu, half-filled and filled 3d subshells, respectively, form at the expense of a 4s electron, and the ground configurations of $_{57}$La, $_{58}$Ce, and $_{64}$Gd each contain one 5d electron at the expense of a 4f electron. It is important to keep in mind that we are working with a one-electron model, but *not* with an *independent*-electron model. The electrons interact strongly with each other via Coulomb

[23] G. T. Seaborg, Ann. Rev. Nucl. Sci. **18**, 53 (1968); J. B. Mann and J. T. Waber, J. Chem. Phys. **53**, 2397 (1970); R. D. Cowan and J. B. Mann, in G. K. Woodgate and P. G. H. Sandars, eds., *Atomic Physics 2* (Plenum, London, 1971), p. 215.

TABLE 4-3. EXPERIMENTALLY OBSERVED GROUND CONFIGURATIONS AND GROUND LS TERMS OF FREE ATOMS[a]

Element	Neutral		Singly-Ionized		Doubly-Ionized	
$_{1}$H	1s	^2S				
$_{2}$He	1s^2	^1S	1s	^2S		
$_{3}$Li	2s	^2S	1s^2	^1S	1s	^2S
$_{4}$Be	2s^2	^1S	2s	^2S	1s^2	^1S
$_{5}$B	2p	$^2P^\circ$	2s^2	^1S	2s	^2S
$_{6}$C	2p^2	^3P	2p	$^2P^\circ$	2s^2	^1S
$_{7}$N	2p^3	$^4S^\circ$	2p^2	^3P	2p	$^2P^\circ$
$_{8}$O	2p^4	^3P	2p^3	$^4S^\circ$	2p^2	^3P
$_{9}$F	2p^5	$^2P^\circ$	2p^4	^3P	2p^3	$^4S^\circ$
$_{10}$Ne	2p^6	^1S	2p^5	$^2P^\circ$	2p^4	^3P
$_{11}$Na	3s	^2S	2p^6	^1S	2p^5	$^2P^\circ$
$_{12}$Mg	3s^2	^1S	3s	^2S	2p^6	^1S
$_{13}$Al	3p	$^2P^\circ$	3s^2	^1S	3s	^2S
$_{14}$Si	3p^2	^3P	3p	$^2P^\circ$	3s^2	^1S
$_{15}$P	3p^3	$^4S^\circ$	3p^2	^3P	3p	$^2P^\circ$
$_{16}$S	3p^4	^3P	3p^3	$^4S^\circ$	3p^2	^3P
$_{17}$Cl	3p^5	$^2P^\circ$	3p^4	^3P	3p^3	$^4S^\circ$
$_{18}$Ar	3p^6	^1S	3p^5	$^2P^\circ$	3p^4	^3P
$_{19}$K	4s	^2S	3p^6	^1S	3p^5	$^2P^\circ$
$_{20}$Ca	4s^2	^1S	4s	^2S	3p^6	^1S
$_{21}$Sc	3d 4s^2	^2D	3d 4s	^3D	3d	^2D
$_{22}$Ti	3d^{2}4s^2	^3F	3d^{2}4s	^4F	3d^2	^3F
$_{23}$V	3d^{3}4s^2	^4F	3d^4	^5D	3d^3	^4F
$_{24}$Cr	3d^{5}4s	^7S	3d^5	^6S	3d^4	^5D
$_{25}$Mn	3d^{5}4s^2	^6S	3d^{5}4s	^7S	3d^5	^6S
$_{26}$Fe	3d^{6}4s^2	^5D	3d^{6}4s	^6D	3d^6	^5D
$_{27}$Co	3d^{7}4s^2	^4F	3d^8	^3F	3d^7	^4F
$_{28}$Ni	3d^{8}4s^2	^3F	3d^9	^2D	3d^8	^3F
$_{29}$Cu	3d^{10}4s	^2S	3d^{10}	^1S	3d^9	^2D
$_{30}$Zn	3d^{10}4s^2	^1S	3d^{10}4s	^2S	3d^{10}	^1S
$_{31}$Ga	4p	$^2P^\circ$	4s^2	^1S	4s	^2S
$_{32}$Ge	4p^2	^3P	4p	$^2P^\circ$	4s^2	^1S
$_{33}$As	4p^3	$^4S^\circ$	4p^2	^3P	4p	$^2P^\circ$
$_{34}$Se	4p^4	^3P	4p^3	$^4S^\circ$	4p^2	^3P
$_{35}$Br	4p^5	$^2P^\circ$	4p^4	^3P	4p^3	$^4S^\circ$
$_{36}$Kr	4p^6	^1S	4p^5	$^2P^\circ$	4p^4	^3P

TABLE 4-3 (cont)

Element	Neutral		Singly-Ionized		Doubly-Ionized	
${}_{37}$Rb	$5s$	^2S	$4p^6$	^1S	$4p^5$	$^2P^\circ$
${}_{38}$Sr	$5s^2$	^1S	$5s$	^2S	$4p^6$	^1S
${}_{39}$Y	$4d\,5s^2$	^2D	$5s^2$	^1S	$5s$	^2S
${}_{40}$Zr	$4d^25s^2$	^3F	$4d^25s$	^4F	$4d^2$	^3F
${}_{41}$Nb	$4d^45s$	^6D	$4d^4$	^5D	$4d^3$	^4F
${}_{42}$Mo	$4d^55s$	^7S	$4d^5$	^6S	$4d^4$	^5D
${}_{43}$Tc	$4d^55s^2$	^6S	$4d^55s$	^7S	$4d^5$	^6S
${}_{44}$Ru	$4d^75s$	^5F	$4d^7$	^4F	$4d^6$	^5D
${}_{45}$Rh	$4d^85s$	^4F	$4d^8$	^3F	$4d^7$	^4F
${}_{46}$Pd	$4d^{10}$	^1S	$4d^9$	^2D	$4d^8$	^3F
${}_{47}$Ag	$4d^{10}5s$	^2S	$4d^{10}$	^1S	$4d^9$	^2D
${}_{48}$Cd	$4d^{10}5s^2$	^1S	$4d^{10}5s$	^2S	$4d^{10}$	^1S
${}_{49}$In	$5p$	$^2P^\circ$	$5s^2$	^1S	$5s$	^2S
${}_{50}$Sn	$5p^2$	^3P	$5p$	$^2P^\circ$	$5s^2$	^1S
${}_{51}$Sb	$5p^3$	$^4S^\circ$	$5p^2$	^3P	$5p$	$^2P^\circ$
${}_{52}$Te	$5p^4$	^3P	$5p^3$	$^4S^\circ$	$5p^2$	^3P
${}_{53}$I	$5p^5$	$^2P^\circ$	$5p^4$	^3P	$5p^3$	$^4S^\circ$
${}_{54}$Xe	$5p^6$	^1S	$5p^5$	$^2P^\circ$	$5p^4$	^3P
${}_{55}$Cs	$6s$	^2S	$5p^6$	^1S	$5p^5$	$^2P^\circ$
${}_{56}$Ba	$6s^2$	^1S	$6s$	^2S	$5p^6$	^1S
${}_{57}$La	$5d6s^2$	^2D	$5d^2$	^3F	$5d$	^2D
${}_{58}$Ce	$4f5d6s^2$	^1G	$4f5d^2$	^4H	$4f^2$	^3H
${}_{59}$Pr	$4f^3\,6s^2$	$^4I^\circ$	$4f^3\,6s$	$^5I^\circ$	$4f^3$	$^4I^\circ$
${}_{60}$Nd	$4f^4\,6s^2$	^5I	$4f^4\,6s$	^6I	$4f^4$	^5I
${}_{61}$Pm	$4f^5\,6s^2$	$^6H^\circ$	$4f^5\,6s$	$^7H^\circ$	($4f^5$	$^6H^\circ$)
${}_{62}$Sm	$4f^6\,6s^2$	^7F	$4f^6\,6s$	^8F	$4f^6$	^7F
${}_{63}$Eu	$4f^7\,6s^2$	$^8S^\circ$	$4f^7\,6s$	$^9S^\circ$	$4f^7$	$^8S^\circ$
${}_{64}$Gd	$4f^75d6s^2$	$^9D^\circ$	$4f^75d6s$	$^{10}D^\circ$	$4f^75d$	$^9D^\circ$
${}_{65}$Tb	$4f^9\,6s^2$	$^6H^\circ$	($4f^9\,6s$	$^7H^\circ$)	$4f^9$	$^6H^\circ$
${}_{66}$Dy	$4f^{10}6s^2$	^5I	$4f^{10}6s$	^6I	($4f^{10}$	^5I)
${}_{67}$Ho	$4f^{11}6s^2$	$^4I^\circ$	$4f^{11}6s$	$^5I^\circ$	$4f^{11}$	$^4I^\circ$
${}_{68}$Er	$4f^{12}6s^2$	^3H	$4f^{12}6s$	^4H	$4f^{12}$	^3H
${}_{69}$Tm	$4f^{13}6s^2$	$^2F^\circ$	$4f^{13}6s$	$^3F^\circ$	$4f^{13}$	$^2F^\circ$
${}_{70}$Yb	$4f^{14}6s^2$	^1S	$4f^{14}6s$	^2S	$4f^{14}$	^1S

TABLE 4-3 (cont)

Element	Neutral		Singly-Ionized		Doubly-Ionized	
$_{71}$Lu	5d $6s^2$	^2D	$6s^2$	^1S	6s	^2S
$_{72}$Hf	$5d^26s^2$	^3F	5d $6s^2$	^2D	$5d^2$	^3F
$_{73}$Ta	$5d^36s^2$	^4F	$5d^36s$	^5F		
$_{74}$W	$5d^46s^2$	^5D	$5d^46s$	^6D		
$_{75}$Re	$5d^56s^2$	^6S	$5d^56s$	^7S		
$_{76}$Os	$5d^66s^2$	^5D	$5d^66s$	^6D		
$_{77}$Ir	$5d^76s^2$	^4F				
$_{78}$Pt	$5d^96s$	^3D	$5d^9$	^2D		
$_{79}$Au	$5d^{10}6s$	^2S	$5d^{10}$	^1S		
$_{80}$Hg	$5d^{10}6s^2$	^1S	$5d^{10}6s$	^2S	$5d^{10}$	^1S
$_{81}$Tl	6p	$^2P^\circ$	$6s^2$	^1S	6s	^2S
$_{82}$Pb	$6p^2$	^3P	6p	$^2P^\circ$	$6s^2$	^1S
$_{83}$Bi	$6p^3$	$^4S^\circ$	$6p^2$	^3P	6p	$^2P^\circ$
$_{84}$Po	$6p^4$	^3P				
$_{85}$At						
$_{86}$Rn	$6p^6$	^1S				
$_{87}$Fr						
$_{88}$Ra	$7s^2$	^1S	7s	^2S		
$_{89}$Ac	6d $7s^2$	^2D	$7s^2$	^1S	7s	^2S
$_{90}$Th	$6d^27s^2$	^3F	6d $7s^2$	$^2D^{(b)}$	5f6d	$^3H^\circ$
$_{91}$Pa	$5f^26d7s^2$	^4K	$5f^2$ $7s^2$	^3H		
$_{92}$U	$5f^36d7s^2$	$^5L^\circ$	$5f^3$ $7s^2$	$^4I^\circ$		
$_{93}$Np	$5f^46d7s^2$	^6L				
$_{94}$Pu	$5f^6$ $7s^2$	^7F	$5f^6$ 7s	^8F		
$_{95}$Am	$5f^7$ $7s^2$	$^8S^\circ$	$5f^7$ 7s	$^9S^\circ$	$5f^7$	$^8S^\circ$
$_{96}$Cm	$5f^76d7s^2$	$^9D^\circ$	$5f^7$ $7s^2$	$^8S^\circ$		
$_{97}$Bk	$5f^9$ $7s^2$	$^6H^\circ$	$5f^9$ 7s	$^7H^\circ$		
$_{98}$Cf	$5f^{10}7s^2$	^5I	$5f^{10}7s$	^6I		
$_{99}$Es	$5f^{11}7s^2$	$^4I^\circ$	$5f^{11}$ 7s	$^5I^\circ$	$5f^{11}$	$^4I^\circ$
$_{100}$Fm	$5f^{12}7s^2$	^3H	($5f^{12}$ 7s	^4H)		
$_{101}$Md	($5f^{13}7s^2$	$^2F^\circ$)	($5f^{13}$ 7s	$^3F^\circ$)		
$_{102}$No	($5f^{14}7s^2$	^1S)	($5f^{14}$ 7s	^2S)	($5f^{14}$	^1S)

[a]C. E. Moore, *Ionization Potentials and Ionization Limits Derived from the Analyses of Optical Spectra*, NSRDS-NBS 34 (U.S. Govt. Printing Off., Washington, D.C., 1970), and references in Sec. IX of the bibliography. (Entries in parentheses have not been experimentally determined, but are reasonably certain.)
[b]Strongly mixed with $6d^27s$ ^4F.

forces, and consequently the binding energy of a given electron depends not only on Z but also on the number of other electrons present and on the subshells in which they lie—i.e., on the electron configuration. In the first transition series, for example, the binding energy of a 3d electron depends on how many 4s (and other 3d) electrons are present, and vice versa. It is because of this fact that—contrary to our original expectations—we find ground configurations that involve more than one open subshell; to some extent, it is also this fact that is responsible for the variable number of s electrons in the neutral and singly ionized transition elements, and the variable number of d and s electrons in the rare-earth elements.

The minor irregularities in the periodic table can thus be understood qualitatively, but they cannot be predicted in detail except by quantitative numerical calculations. In fact, even this cannot always be done reliably; in the case of $_{41}$Nb I, the lowest level of $4d^3 5s^2$ lies only 1143 cm^{-1} above the ground level (of $4d^4 4s$), and this is a factor two or more smaller than the accuracy of theoretical calculations. (Actually, the center of gravity of $4d^3 5s^2$ lies well below the center of gravity of $4d^4 5s$, so that Nb I is not really out of line with expectations based on spherically averaged one-electron binding energies; however, many of the minor irregularities remain even when one looks at configuration-average energies rather than lowest-level energies.)

It may be seen from Table 4-3 that most irregularities have disappeared in twofold ionized ions. This is the result of the effective nuclear charge for the outer-shell electrons being large enough to make the n-dependence of binding energy tend to again overshadow the l-dependence (cf. footnote 19), and also to make the binding energy of an nl electron less dependent on the configuration of the other electrons.

4-14. VARIATION OF IONIZATION ENERGY WITH Z

The qualitative variation with Z of the ionization energy of neutral atoms (Table 1-1 and Fig. 4-3) follows directly from the shell structure of atoms as reflected in the periodic system. Consider, for example, the filling of the 2p subshell which takes place in the interval $5 \leq Z \leq 10$. In $_5$B, there is only a single 2p electron, which for the most part lies outside the $1s^2 2s^2$ core; it is therefore fairly well shielded from the nucleus by the other four electrons and sees an effective nuclear charge not much greater than unity, and so is not too tightly bound. In $_6$C, on the other hand, the two 2p electrons only partially shield each other from the added unit of nuclear charge, so that each is more tightly bound than the single 2p electron of B; the ionization energy of C is therefore greater than that of B. Similarly, for N, O, F, and Ne, the 2p electrons become successively more tightly bound because there is an increasingly greater number of equivalent 2p electrons, each of which provides incomplete shielding from the corresponding unit of nuclear charge; consequently, the ionization potential of neutral atoms increases from B to Ne.[24]

[24]A monotonic increase with Z exists only for the spherically averaged ionization energy. When detailed level structures are taken into account, the ionization energy shows a local maximum at nitrogen (Table 1-1), where the half-filled $2p^3$ shell results in the maximum possible number of parallel spins and therefore the maximum Pauli-principle correlation effects.

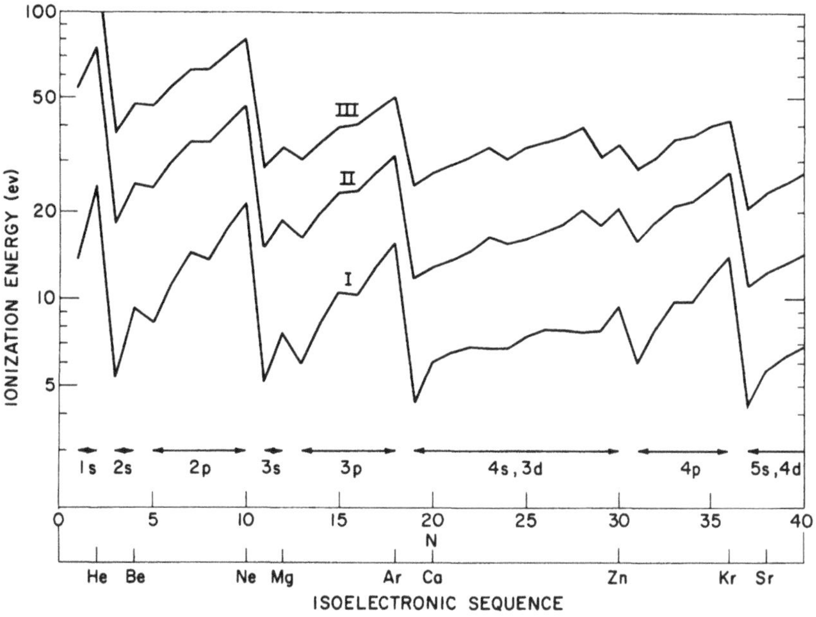

Fig. 4-3. Ionization energies of neutral (I), singly ionized (II), and doubly ionized (III) atoms as a function of the number of electrons N. The range of values of N for which ionization involves removal of an electron from a given subshell (1s, 2s, 2p, etc.) is indicated by the arrows. (The electron comes from the 4s subshell for N = 19—30 in the case of neutral atoms; for N = 19—21, 24—25, 29, 30 in the case of singly-ionized atoms; and for N = 29, 30 in doubly-ionized atoms.)

At Ne the 2p shell is filled, and so in Na the added electron has to go into the 3s shell, where it not only lies farther from the nucleus but also (in spite of its appreciable core penetration) is much more thoroughly shielded from the nucleus than was each 2p electron in Ne; consequently, the ionization energy of Na is much less than that of Ne.

The gradual increase in ionization energy while a given subshell is being filled, and the sudden drop upon starting a new subshell, may be readily observed as far as neutral calcium. At scandium, the added electron starts a new subshell (3d), but this electron is more tightly bound than the 4s electrons, so that ionization consists of removal of a 4s electron, just as in K and Ca. The ionization energy therefore does not show a drop at Sc, but continues to increase throughout most of the transition series. Actually, the binding energies of 3d and 4s electrons are not greatly different, and in V I, Co I, and Ni I ionization involves not only removal of a 4s electron but also rearrangement of the ion core, with the second 4s electron dropping down to the 3d subshell. Similar irregularities can be seen in Table 4-3 in the other transition-element and the rare-earth series.

Problems

4-14(1). Using ground-configuration energy levels from AEL,[20] verify that neutral-atom center-of-gravity ionization energies increase monotonically and smoothly from carbon through neon.

4-14(2). Using data from AEL, show that the spherical-average binding energy of a 4s electron in the ground configuration of neutral scandium, E_{av}(Sc II 3d4s) – E_{av}(Sc I 3d4s^2), is less than the binding energy of a 3d electron, E_{av}(Sc II 4s^2) – E_{av}(Sc I 3d4s^2). Why then is the ground configuration not 3d^{2}4s (or even 3d^3)? Hint: Compare the average-energy differences Sc II 3d^2 ^{3}F – Sc I 3d^2(^{3}F)4s $\cong$ E_{av}(3d^2) – E_{av}(3d^{2}4s) and Sc II 3d4s – Sc I 3d^2(^{3}F)4s > E_{av}(3d4s) – E_{av}(3d^{2}4s).

4-15. LEVEL STRUCTURE UNDER LS-COUPLING CONDITIONS

Having discussed qualitatively the relative energies of different configurations, we turn now to a qualitative discussion of the relative order and spacing of energy levels within a configuration. More specifically, we wish to discuss how the level structure of a configuration arises from the various interactions within the atom, and thereby the manner in which the level structure depends on the relative magnitudes of the different interactions—i.e., on the coupling conditions within the atom. For the most part we shall consider only the simple case of a configuration containing just two non-equivalent electrons outside of closed shells.

In light atoms especially, spin-orbit interactions tend to be small compared with the electrostatic interactions between electrons. To the extent that spin-orbit effects can be neglected, the remaining terms of the Hamiltonian (4.1) are independent of the total spin **S**, which must therefore be a constant of the motion. Since the total angular momentum **J** is constant, it follows that $\mathbf{L} = \mathbf{J} - \mathbf{S}$ must also be constant. This implies that the eigenfunction Ψ^k must be an eigenfunction of $\mathbf{L}^2$ and $\mathbf{S}^2$ as well as of $\mathbf{J}^2$ (i.e., L and S are "good" quantum numbers because there is no interaction that couples **L** and **S** together), and we therefore have a close approximation to pure LS coupling.

The relative energies of the various LS terms can be qualitatively predicted in simple cases by the following semiclassical argument based on the vector model.[25] For each electron with orbital angular momentum $\boldsymbol{l}_i$, the electronic charge cloud is concentrated near an orbital plane perpendicular to $\boldsymbol{l}_i$ (Fig. 2-3). For either maximum L ($\boldsymbol{l}_1$ parallel to $\boldsymbol{l}_2$) or minimum L ($\boldsymbol{l}_1$ antiparallel to $\boldsymbol{l}_2$), the two orbital planes more or less coincide; the two charge clouds thus overlap strongly and there is a high repulsive energy. For intermediate L, on the other hand, the orbital planes lie at an angle to each other, the charge clouds are more separated, and the energy is lower. Thus in a pd configuration, for example, we would expect the D levels to lie lower than the P and F levels; this is seen to be true for the experimentally observed levels of 4p3d in Ti III, shown in Fig. 4-4.[20]

It will be observed from the figure that for given L the energies of the ^{1}L and ^{3}L terms differ. This arises through the partial correlation of electron position (and therefore lower repulsion energy) that exists for parallel spins (see Sec. 4-5): if $\mathbf{s}_1$ is parallel to $\mathbf{s}_2$, then we must necessarily have $S = 1$ (rather than zero) and therefore a triplet state. The figure shows that the triplet levels indeed lie lower in energy than the singlets, but not for every L—only on an average. Figure 4-4 also shows the effect of the small spin-orbit interactions in producing a small energy dependence on the total angular-momentum quantum number J, thus splitting each triplet term into three levels.

[25]H. G. Kuhn, *Atomic Spectra* (Longmans, Green, London, 1969), 2nd ed., pp. 254, 267-69.

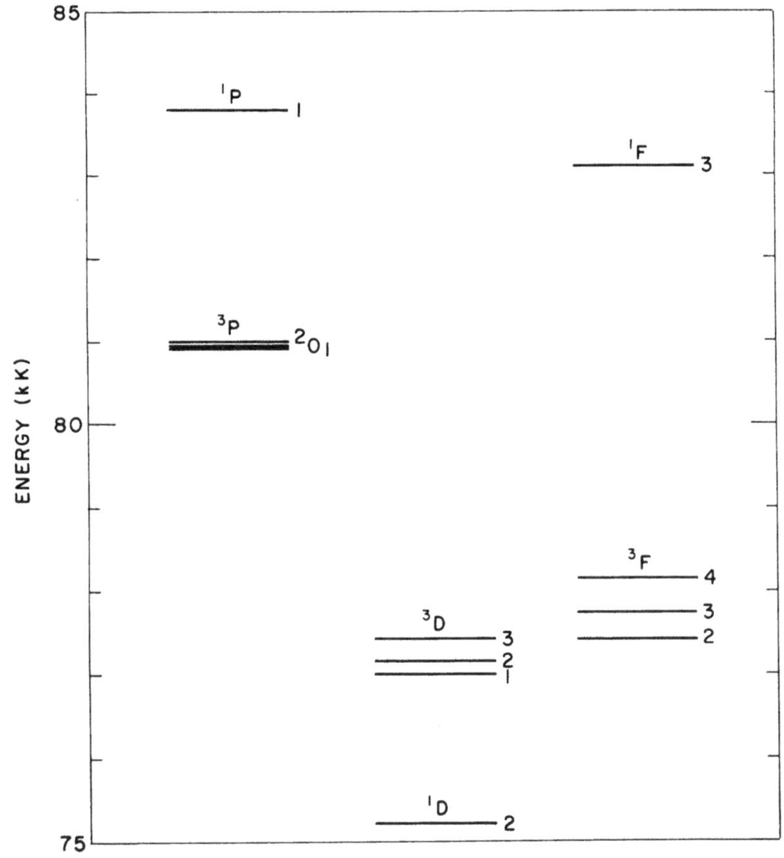

Fig. 4-4. The observed energy levels of the configuration Ti III 3d4p. Note that with increasing L, the ^{1}L term lies alternately above and below the ^{3}L term. The number at the right of each level is the value of the total angular-momentum quantum number J.

The various interactions contributing to the overall energy-level structure are depicted schematically in Fig. 4-5: We see (a) the average energy for the spherically symmetrized atom; (b) splitting according to L value resulting from the electron-electron Coulomb interactions, neglecting the Pauli-principle correlations (the so-called "direct" interaction); (c) splitting according to S value resulting from the Pauli-principle ("exchange") effects; (d) splitting by J value caused by spin-orbit interactions. Also shown is (e) splitting according to M value, caused by an external magnetic field (Zeeman effect); this magnetic field splitting of each energy level into $2J + 1$ sublevels provides an experimental method of determining the value of J for each level. It is worth noting that if the level splittings produced by a given interaction are small compared with the level separations that have resulted from the stronger interactions, then the center of gravity of each group of split levels is identical with the position of the unsplit parent level; this will be proved mathematically in Sec. 12-11.

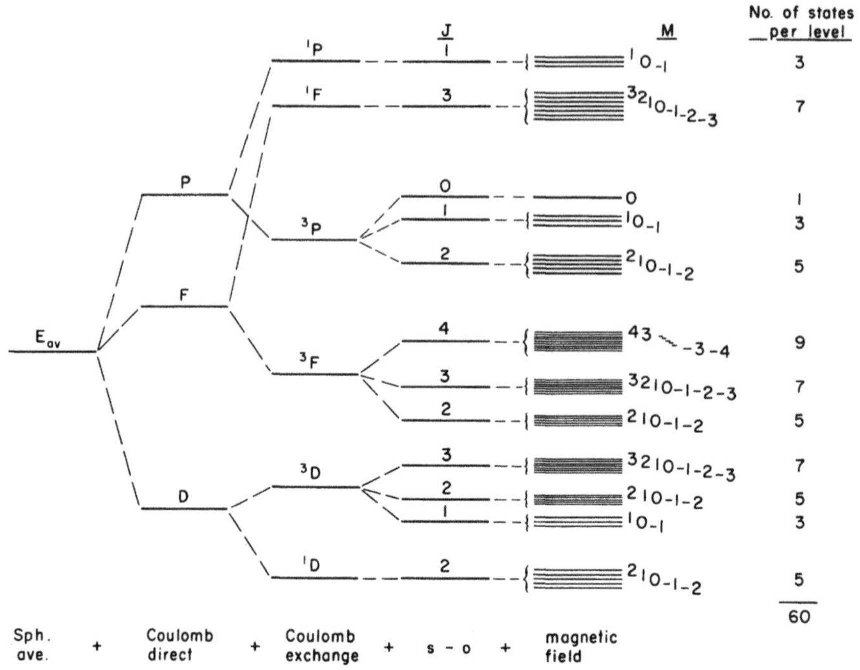

Fig. 4-5. Schematic drawing of the development of the energy-level structure of a pd configuration under LS-coupling conditions—starting with the spherically averaged central-field energy, and then taking into account successively the large Coulomb interaction energies (first for product, and then for determinantal functions), the small spin-orbit effects, and finally a splitting of levels into sublevels by an external magnetic field. (The s-o and external-field splittings have been greatly exaggerated for clarity.)

4-16. HUND'S RULE

Not a great deal can be said in general, on the basis of qualitative arguments, about the energy-level structure of configurations containing more than two electrons. Strictly on the basis of empirical evidence, Hund[26] proposed the following relations applicable for LS-coupling conditions:

(1) Within any given configuration, of all those terms LS having a given value of L, the lowest in energy will be that which has the greatest S;

(2) Among all of the above lowest terms (one for each value of L), the lowest of all will be that with greatest L.

With the limited amount of experimental evidence available at that time, Hund thought the above relations to be quite general. Although they are now known to be misleading more often than not (note the two points of disagreement in Fig. 4-4), they are still sometimes invoked in an attempt to predict the lowest term of a complex configuration.[27]

[26]F. Hund, *Linienspektren und periodisches System der Elemente* (Julius Springer, Berlin, 1927), p. 124.

[27]See, for example, L. Brewer, J. Opt. Soc. Am. **61**, 1101, 1666 (1971).

Hund's rule can safely be applied only to configurations with a single open subshell or with one subshell plus an s electron, and then only in the restricted form:

The lowest-energy term of a configuration l^w or of $l^w s$ *is that term of maximum S which has the largest value of L.*

In Table 4-2, the Hund's-rule term is the rightmost term in each line.

The restricted form of Hund's rule has been well verified experimentally (compare Table 4-3 with Table 4-2); that it should be true can be made plausible for l^w by the following argument:

(1) With determinantal wave functions, we have seen that partial correlations in position exist between electrons that have parallel spins. The resulting decrease in electron-electron repulsive energy should give the lowest total energy for the atom when all electrons have parallel spin, to the extent permitted by the Pauli principle; this implies maximum S.

(2) So far as L is concerned, we saw in the preceding section that for non-equivalent electrons, the highest-L terms were *not* those of lowest energy. However, with equivalent electrons, the largest L value predicted by the vector model is (for maximum S) forbidden by the Pauli principle; thus it is not unreasonable that the highest *allowed* value of L (for maximum S) should indeed have the lowest energy.

The addition of an s electron to l^w in forming an $l^w s$ configuration does not change the possible values of L; of the two resultant spin values $S - 1/2$ and $S + 1/2$, the latter always has the lower energy by the argument used above.

4-17. jj COUPLING

So far, we have considered only LS-coupling conditions, in which the electrostatic interactions between electrons are much stronger than the interaction between the spin of an electron and its own orbital motion. With increasing Z, the spin-orbit interactions become increasingly more important; in the limit in which these interactions become much stronger than the Coulomb terms, the coupling conditions approach pure jj coupling.

In the jj-coupling scheme, basis functions are formed by first coupling the spin of each electron to its own orbital angular momentum, and then coupling together the various resultants $\mathbf{j}_i$ in some arbitrary order to obtain the total angular momentum $\mathbf{J}$:

$$\mathbf{l}_i + \mathbf{s}_i = \mathbf{j}_i \,, \qquad \text{(each i);}$$

$$\mathbf{j}_1 + \mathbf{j}_2 = \mathbf{J}_2 \,,$$
$$\mathbf{J}_2 + \mathbf{j}_3 = \mathbf{J}_3 \,,$$
$$\vdots$$
$$\mathbf{J}_{N-1} + \mathbf{j}_N = \mathbf{J}_N \equiv \mathbf{J} \,.$$

For two-electron configurations, this coupling scheme may be described by the condensed notation

$$[(l_1,s_1)j_1, (l_2,s_2)j_2]JM \tag{4.41}$$

and the usual jj-coupling notation for energy levels [analogous to the Russell-Saunders notation ${}^{2S+1}L_J$] is

$$(j_1, j_2)_J \ . \tag{4.42}$$

In jj coupling, the Pauli principle takes the form that no two electrons may have the same set of one-electron quantum numbers

$$n_i l_i j_i m_{j_i} \ .$$

For equivalent electrons with equal values of j_i, the Pauli-principle limitations on the possible values of m_{j_i} prohibit some of the values of J that would be predicted by the vector model; the allowed values may be found by examining the permitted values of M. For example, in the two-electron case p^2, we have the possibilities listed in Table 4-4. The allowed states $(j_1 j_2)_J$ are evidently $(1/2,1/2)_0$, $(1/2,3/2)_{1,2}$, $(3/2,3/2)_{0,2}$; the states forbidden by the Pauli principle are $(1/2,1/2)_1$, $(3/2,1/2)_{1,2}$, and $(3/2,3/2)_{1,3}$. [More correctly, the antisymmetric linear combination of the states $(1/2,3/2)_J$ and $(3/2,1/2)_J$ is allowed, and the symmetric combination is forbidden.] The general result is that for any two-equivalent-electron case $(nl)^2$, only the even-J states are allowed when $j_1 = j_2$, and only half of the $j_1 \neq j_2$ states are allowed.

For cases $(nl)^w$ with $w > 2$, the allowed states can be found in the manner used above. However, it is simpler to consider the subcases $(nlj)^w$, with w restricted by the Pauli principle to be no greater than $2j + 1$. It is necessary to make calculations only for $w \leq j + 1/2$,

TABLE 4-4. PERMITTED VALUES OF jj-COUPLED QUANTUM NUMBERS FOR A p^2 CONFIGURATION

$j_1 m_1$		$j_2 m_2$		M	J
$\frac{1}{2}$	$-\frac{1}{2}$	$\frac{1}{2}$	$\frac{1}{2}$	0	0
$\frac{1}{2}$	$-\frac{1}{2}$	$\frac{3}{2}$	$-\frac{3}{2}$	−2	
			$-\frac{1}{2}$	−1	
			$\frac{1}{2}$	0	
			$\frac{3}{2}$	1	1,2
$\frac{1}{2}$	$\frac{1}{2}$	$\frac{3}{2}$	$-\frac{3}{2}$	−1	
			$-\frac{1}{2}$	0	
			$\frac{1}{2}$	1	
			$\frac{3}{2}$	2	
$\frac{3}{2}$	$-\frac{3}{2}$	$\frac{3}{2}$	$-\frac{1}{2}$	−2	
			$\frac{1}{2}$	−1	
			$\frac{3}{2}$	0	0,2
$\frac{3}{2}$	$-\frac{1}{2}$	$\frac{3}{2}$	$\frac{1}{2}$	0	
			$\frac{3}{2}$	1	
$\frac{3}{2}$	$\frac{1}{2}$	$\frac{3}{2}$	$\frac{3}{2}$	2	

TABLE 4-5. ALLOWED STATES of $(lj)^w$

l	j	w	J
s,p	$\frac{1}{2}$	0, 2	0
		1	$\frac{1}{2}$
p,d	$\frac{3}{2}$	0, 4	0
		1, 3	$\frac{3}{2}$
		2	0, 2
d,f	$\frac{5}{2}$	0, 6	0
		1, 5	$\frac{5}{2}$
		2, 4	0, 2, 4
		3	$\frac{3}{2}, \frac{5}{2}, \frac{9}{2}$
f,g	$\frac{7}{2}$	0, 8	0
		1, 7	$\frac{7}{2}$
		2, 6	0, 2, 4, 6
		3, 5	$\frac{3}{2}, \frac{5}{2}, \frac{7}{2}, \frac{9}{2}, \frac{11}{2}, \frac{15}{2}$
		4	0, 2, 2, 4, 4, 5, 6, 8
g,h	$\frac{9}{2}$	0, 10	0
		1, 9	$\frac{9}{2}$
		2, 8	0, 2, 4, 6, 8
		3, 7	$\frac{3}{2}, \frac{5}{2}, \frac{7}{2}, \frac{9}{2}, \frac{9}{2}, \frac{11}{2}, \frac{13}{2}, \frac{15}{2}, \frac{17}{2}, \frac{21}{2}$
		4, 6	0, 0, 2, 2, 3, 4, 4, 4, 5, 6, 6, 6, 7, 8, 8, 9, 10, 12
		5	$\frac{1}{2}, \frac{3}{2}, \frac{5}{2}, \frac{5}{2}, \frac{7}{2}, \frac{7}{2}, \frac{9}{2}, \frac{9}{2}, \frac{9}{2}, \frac{11}{2}, \frac{11}{2}, \frac{13}{2}, \frac{13}{2}, \frac{15}{2}, \frac{15}{2}, \frac{17}{2}, \frac{17}{2}, \frac{19}{2}, \frac{21}{2}, \frac{25}{2}$

because it is easily seen that the only allowed value is $J = 0$ when $w = 2j + 1$, and that the allowed values of J for $(nlj)^{2j+1-w}$ are the same as for $(nlj)^w$ [compare the corresponding relation (4.36) for the LS-coupling case]. The results for $j \leqq 9/2$ are given in Table 4-5.[28] The allowed states of

$$(nlj_-)^{w-}(nlj_+)^{w+} , \tag{4.43}$$

[28]See, for example, A. de-Shalit and I. Talmi, *Nuclear Shell Theory* (Academic Press, New York, 1963).

where $j_{\pm} = l \pm 1/2$, may be found from Table 4-5 and the vector model, with no further restrictions arising from the Pauli principle.

Equivalent-particle jj-coupled basis states are used extensively in nuclear physics.[28] However, for atomic-electron subshells l^w, the coupling conditions are usually closer to the LS than to the jj scheme, and the latter is seldom even considered except in a very few high-Z cases such as $_{82}$Pb $6p^2$, $_{83}$Bi $6p^3$, and $_{84}$Po $6p^4$. At the same time, the coupling between *different* subshells may lie close to pure jj conditions, and in these cases the most appropriate basis states will be

$$\{[(l_1^{w_1}\alpha_1L_1S_1)J_1,\ (l_2^{w_2}\alpha_2L_2S_2)J_2]\mathfrak{J}_2,\ \cdots\ (l_q^{w_q}\alpha_qL_qS_q)J_q\}\mathfrak{J}_q\mathfrak{M}_q\ . \tag{4.44}$$

In excited configurations with $w_q = 1$, the coupling of the excited electron with the core is often close to jj because the excited electron is automatically removed from the vicinity of the other electrons, thereby reducing the strength of the Coulomb interaction between the excited electron and the core electrons.

Figure 4-6 shows schematically the development of the energy level structure of the two-electron pd configuration under jj-coupling conditions. The characteristic feature is the presence of four groups of levels, corresponding to the four possible pairs of values $(j_1 j_2)$. A secondary characteristic, not predictable on the basis of qualitative considerations, is illustrated by the two upper groups of levels. If the energies of $l^w l'$ are plotted (on a vertical scale) against J on a horizontal scale, the result is usually either a concave-upward (bowl-shaped) pattern or a concave-downward (umbrella-shaped) pattern; for $w = 1$, only bowl patterns occur, whereas for $w = 4l + 1$ one sees only umbrella patterns. (See also Sec. 20-3.)

Problem

4-17(1). Show, both without and with the use of Table 4-5, that the allowed jj-coupled states of p^3 are

$$\left(\frac{1}{2}\frac{1}{2}\frac{3}{2}\right)^{\circ}_{3/2}, \qquad \left(\frac{1}{2}\frac{3}{2}\frac{3}{2}\right)^{\circ}_{1/2,3/2,5/2}, \qquad \text{and} \qquad \left(\frac{3}{2}\frac{3}{2}\frac{3}{2}\right)^{\circ}_{3/2}.$$

4-18. PAIR COUPLING

More common than jj coupling, though less well known, are coupling conditions under which the energy levels tend to appear in pairs.[29] These conditions occur for excited configurations in which the energy depends only slightly on the spin **s** of the excited electron; the level pairs correspond to the two possible values of J that are obtained when **s** is added to the resultant, **K**, of all other angular momenta. Pair-coupling conditions occur mainly when the excited electron has large angular momentum (an f or g electron) because such an electron tends not to penetrate the core (Figs. 3-5 and 3-6) and thus experiences only a

[29]Such a paired structure was first noted by A. G. Shenstone, Phil. Trans. Roy. Soc. London **A235**, 195 (1936) in the configurations $3d^9nf$ and $3d^9ng$ of Cu II, but has since been observed in many other cases; see footnote 30 for further references.

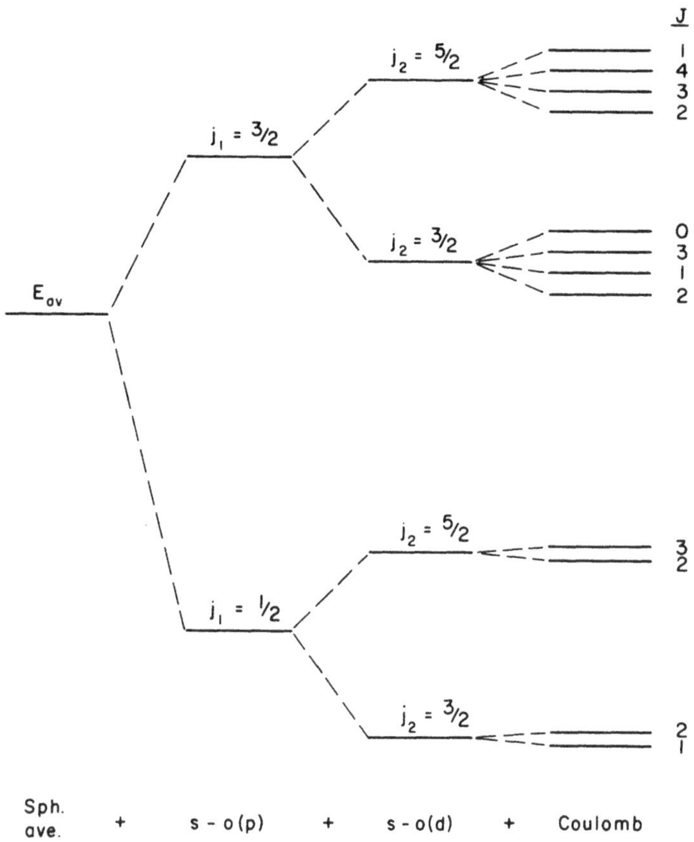

Fig. 4-6. Schematic illustration of the development of the energy-level structure of a pd configuration under jj-coupling conditions. The two strong spin-orbit interactions result in four different energies, corresponding to the four possible pairs of values of $j_1 j_2$; the weak Coulomb interactions then produce small splittings according to the possible values of J.

small spin-dependent (exchange) Coulomb interaction, and its spin-orbit interaction is likewise small [being roughly proportional to l^{-3}, Eq. (3.52)]. For simplicity, we discuss special cases of pair coupling for configurations containing only two electrons outside of closed subshells.[30]

The more common limiting type of pair coupling, jK coupling,[31] occurs when the strongest interaction is the spin-orbit interaction of the more tightly bound electron, and the next strongest interaction is the spin-independent (direct) portion of the Coulomb interaction between the two electrons. The corresponding angular-momentum coupling scheme is

[30]R. D. Cowan and K. L. Andrew, J. Opt. Soc. Am. **55**, 502 (1965).

[31]Also called *jl* coupling; G. Racah, Phys. Rev. **61**, 537 (1942).

$$\mathbf{l}_1 + \mathbf{s}_1 = \mathbf{j}_1 \,,$$
$$\mathbf{j}_1 + \mathbf{l}_2 = \mathbf{K} \,,$$
$$\mathbf{K} + \mathbf{s}_2 = \mathbf{J},$$

or

$$\{[(l_1,s_1)j_1,l_2]K,s_2\}\ JM \,, \tag{4.45}$$

and the standard energy-level notation is

$$j_1[K]_J \,. \tag{4.46}$$

This type coupling occurs particularly in excited configurations of the noble gases (Ne, Ar, Kr, Xe, Rn) and of the carbon-group elements (C, Si, Ge, Sn, Pb), but also in many other cases.[30]

The other limiting form of pair coupling is called LK (or Ls) coupling. In two-electron configurations, it corresponds to the case in which the direct Coulomb interaction is greater than the spin-orbit interaction of either electron, and the spin-orbit interaction of the inner electron is next most important. The coupling scheme is[32]

$$\mathbf{l}_1 + \mathbf{l}_2 = \mathbf{L} \,,$$
$$\mathbf{L} + \mathbf{s}_1 = \mathbf{K} \,,$$
$$\mathbf{K} + \mathbf{s}_2 = \mathbf{J} \,,$$

or

$$\{[(l_1,l_2)L,s_1]K,s_2\}\ JM \,, \tag{4.47}$$

and the standard level notation is

$$L[K]_J \,, \tag{4.48}$$

where L is the usual letter symbol (S, P, D, F, ⋯). LK-coupling conditions occur rather infrequently, notably in the configurations N II 2p4f and P II 3p4f.

The developments of the energy-level structures of a pd configuration under jK- and LK-coupling conditions are shown in Figs. 4-7 and 4-8, respectively. In the jK case, the various pairs of levels are grouped according to the two possible values of j_1; in the LK case, they are grouped according to the three possible values of L. The general case of intermediate pair coupling corresponds to coupling conditions

$$\{(l_1,l_2,s_1)K,s_2\}\ JM \tag{4.49}$$

in which the spin $\mathbf{s}_2$ is only weakly coupled to **K**, but in which the other three vectors are all mutually coupled together via strong interactions. The various pairs of levels are the same as under jK- and LK-coupling conditions, but are grouped neither according to j_1 nor according to L (which are of course not good quantum numbers).

[32] K. B. S. Eriksson, Phys. Rev. **102**, 102 (1956). Note that the relationship between SL and LK coupling is much less straightforward than that between LS and LK coupling (cf. fn. 14 of Chap. 2).

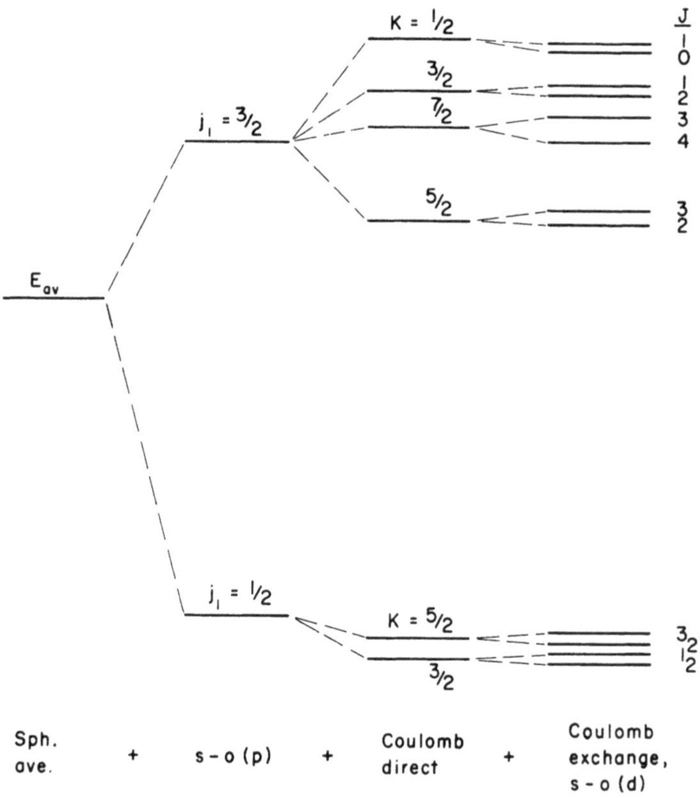

Fig. 4-7. Schematic illustration of the development of the energy-level structure of a pd configuration under jK-coupling conditions. The large spin-orbit interaction of the p electron produces the major energy separation according to the two possible values of j for this electron. Further separation according to the possible values of K is produced by the moderate spin-independent part of the Coulomb interaction between electrons. Finally, the small Coulomb exchange interaction and the small spin-orbit interaction of the d electron produce a small splitting into level pairs according to the two possible values $J = K \pm 1/2$.

4-19. EXAMPLE: Si I 3pns

Within a given atom or ion, the coupling conditions usually vary greatly from one configuration to another, for reasons alluded to in the preceding section. For example, the electrons in the configuration $n_1l_1^2$ may be very close to purely LS coupled, but in a Rydberg series of excited configurations $n_1l_1\ n_2l_2$ ($n_2 \rightarrow \infty$) the Coulomb interaction between the two electrons decreases toward zero with increasing n_2; at some point this interaction must become weaker than the spin-orbit interaction of the electron n_1l_1 (which is approximately independent of n_2), with the result that the coupling conditions must tend toward j_1j_2 or j_1K

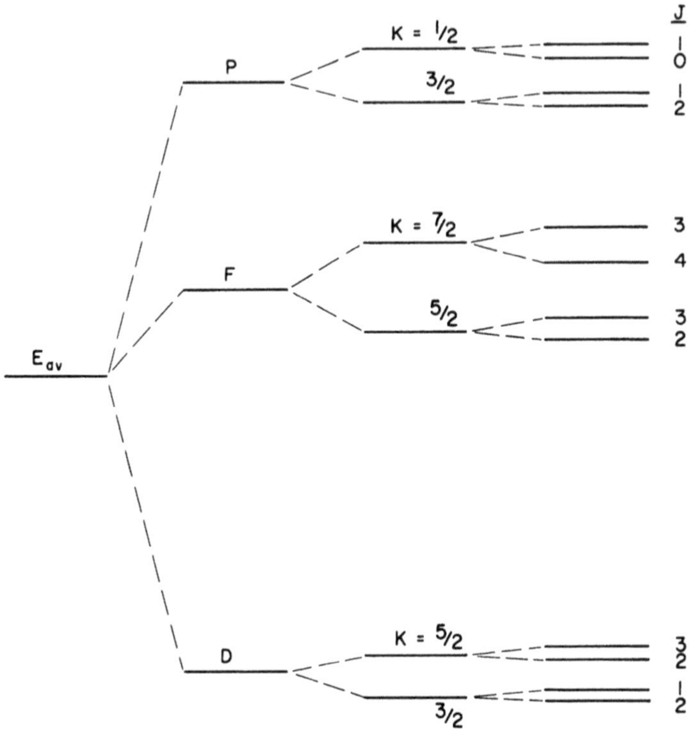

Fig. 4-8. Schematic illustration of the development of the energy-level structure of a pd configuration under LK-coupling conditions. The largest interaction is the spin-independent portion of the Coulomb interaction, and produces three different energies according to the possible values of L, just as in LS coupling (Fig. 4-5). The spin-orbit interaction of the p electron is next most important, and produces a separation according to the two possible values $K = L \pm 1/2$. The remaining small interactions produce splittings according to the two possible values $J = K \pm 1/2$. In contrast to the LS-coupling case, which for each L gives a singlet and a triplet, LK coupling results in two pairs of levels.

or similar coupling. This is well illustrated by the 3pns configurations of neutral silicon (Fig. 4-9),[33] in which the level structure clearly changes from an LS (singlet-triplet) structure in 3p4s to a paired structure in 3p7s; the energy difference between the two pairs obviously corresponds to the spin-orbit interaction of the 3p electron, as it is equal to the separation of the $^2P^\circ_{1/2}$ and $^2P^\circ_{3/2}$ levels of Si II 3p.

[33]L. J. Radziemski, Jr. and K. L. Andrew, J. Opt. Soc. Am. **55**, 474 (1965).

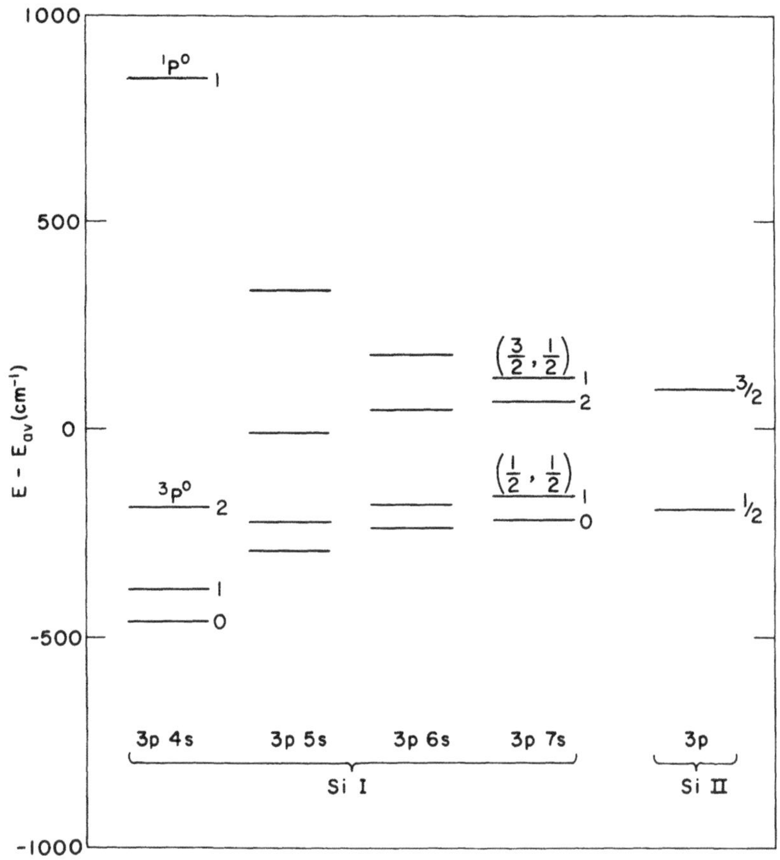

Fig. 4-9. The observed energy-level structures of the four lowest 3pns configurations of Si I and of the Si II ground configuration, plotted relative to the respective centers of gravity and showing a rapid change from LS- to pair-coupling conditions.

Problems

4-19(1). Show from (4.41), (4.45), and (4.47) that the configuration ps is a degenerate case in which jj, jK, and LK couplings are mutually equivalent.

4-19(2). Compare the coupling conditions shown in Fig. 4-9 for Si I 3pns with those for C I 2pns, Ge I 4pns, Sn I 5pns, and Pb I 6pns. [AEL, ref. 20; also L. Johansson, Ark. Fys. **31**, 201 (1966); K. L. Andrew and K. W. Meissner, J. Opt. Soc. Am. **49**, 146 (1959); D. R. Wood and K. L. Andrew, J. Opt. Soc. Am. **58**, 818 (1968).]

4-20. OTHER COUPLING SCHEMES

For a two-electron configuration $l_1 l_2$, various pure-coupling schemes other than those already discussed can obviously be mathematically defined—for example,

$$\{[(l_1,s_1),s_2],l_2\} \text{ or } \{(l_1,s_2),(l_2,s_1)\} \ .$$

These are of no physical interest because of the small magnitude of the spin-spin and spin-other-orbit interactions, and level structures have never been experimentally observed that could be attributed to pure-coupling schemes other than the four already discussed (LS, LK, jK, jj).[34] Frequently, however, the coupling conditions do not lie particularly close even to one of these four cases (as, for example, in Si I 3p5s and 3p6s); such a situation is referred to as *intermediate coupling*. The energy-level structure then does not lie close to any of the limiting forms illustrated by the examples in Figs. 4-5 to 4-8, and the energy levels can only be labeled in terms of the least objectionable of the four pure-coupling schemes—with the understanding that these labels may give a very poor description of the true angular-momentum properties of the corresponding quantum states.

In more complicated configurations involving more than two electrons in unfilled shells, the coupling conditions may lie close to one of a number of different hybrids of the four simple schemes;[35,36] for example, in $l_1^{w_1} l_2$ the most appropriate coupling scheme may be

$$\{[(l_1^{w_1}\alpha_1 L_1 S_1)J_1,l_2]K,s_2\} \ JM \ , \tag{4.50}$$

and in $l_1^{w_1} l_2 l_3$ the coupling conditions may be closest to

$$\{(l_1^{w_1}\alpha_1 L_1 S_1)J_1,[(l_2 l_3)L_3,(s_2 s_3)S_3]J_3\} \ JM \ . \tag{4.51}$$

In many cases, however, the coupling conditions are so hopelessly far from any pure-coupling scheme that it is meaningless to do anything more than label the energy levels and quantum states by means of serial numbers or some similar arbitrary device,[35,37] or to list the values of the largest few eigenvector components (or the squares thereof) in the expansion (4.4).

4-21. STATISTICAL WEIGHTS (HYDROGENIC ATOMS)

In a one-electron atom the possible wavefunctions (ignoring the spin-orbit interaction) were labeled in Secs. 3-2 to 3-4 by the four quantum numbers

$$n l m_l m_s \ , \tag{4.52}$$

[34]For two equivalent electrons (l^2) only the LS and jj coupling schemes are physically significant, because the other two schemes do not treat the two electrons equivalently.

[35]G. Racah, J. Opt. Soc. Am. **50**, 408 (1960).

[36]A. M. Gutman and I. B. Levinson, Astron. Zh. **37**, 86 (1960) [English transl: Soviet Astron.-AJ **4**, 83 (1960)].

[37]O. Laporte and J. E. Mack, Phys. Rev. **63**, 246 (1943).

corresponding to the four variables r,θ,ϕ,s_z required to specify the classical position and spin orientation of the electron. For given values of n and l, the number of different possible quantum states of the atom was the number of different pairs of values that $m_l m_s$ could assume—namely, two different values of m_s for each of the $2l + 1$ possible values of m_l, or a total of $4l + 2$ different quantum states.

When we took the spin-orbit interaction into account in Sec. 3-7, we switched to coupled wavefunctions that were eigenfunctions of the operators $\mathbf{J}^2$ and J_z corresponding to the total angular momentum **J**. The wavefunctions were still labeled by four quantum numbers,

$$nljm \ . \tag{4.53}$$

The number of quantum states (for given nl) was the number $(2j + 1)$ of possible values of m, summed over the two possible values $j = l + 1/2$ and $j = l - 1/2$, or $2(l + 1/2) + 1 + 2(l - 1/2) + 1 = 4l + 2$. This is of course the same number as for the uncoupled functions, the coupled functions being nothing more than $4l + 2$ independent linear combinations of the $4l + 2$ linearly-independent uncoupled functions.

It is a well substantiated postulate of atomic theory that, for statistical purposes, the atom must be considered to exist in any given stationary quantum state with the same *a priori* probability as for any other quantum state. Thus for an ensemble of atoms of given Z and N under conditions of complete thermal equilibrium at temperature T, the (time-averaged) number of atoms that exist in a quantum state of energy E depends only on that energy, through the Boltzmann factor

$$e^{-E/kT} \ , \tag{4.54}$$

and not on any other property of the state. This postulate we express by assigning to each possible quantum state a "statistical weight" g of unity.

An energy level of a hydrogenic atom that corresponds to a total-angular-momentum quantum number j then possesses a statistical weight of

$$g = 2j + 1 \ , \tag{4.55}$$

corresponding to the $2j + 1$ different quantum states (4.53) that differ only in their values of the magnetic quantum number m $(= -j, -j + 1, \cdots j - 1, j)$.

For statistical calculations involving the Boltzmann factor (4.54), it is frequently convenient to lump together all levels of nearly equal energy, and to treat each group as a single level with an appropriate total statistical weight. We saw in Sec. 3-7 that the energy levels of a hydrogenic atom are nearly independent of everything except n (provided $\alpha^2 Z^2 \ll 1$). It is therefore of interest to consider the total statistical weight of all quantum states having given values of n and l, which is

$$g = \sum_{j=l-1/2}^{l+1/2} (2j + 1) = 4l + 2 \ , \tag{4.56}$$

and the total statistical weight of all states of given n, which is

$$g = \sum_{l=0}^{n-1} (4l + 2) = \frac{1}{2} [(2) + (4n - 2)] \cdot n = 2n^2 . \qquad (4.57)$$

4-22. STATISTICAL WEIGHTS (COMPLEX ATOMS)

For an N-electron atom, the problem of evaluating statistical weights is complicated by the fact that the basis functions which we have considered do not necessarily correspond even approximately to actual quantum states of the atom.

We began with uncoupled (product or determinantal) functions defined by the 4N quantum numbers

$$n_i l_i m_{l_i} m_{s_i} \qquad (1 \leqq i \leqq N) ,$$

corresponding to the 4N variables $r_i, \theta_i, \phi_i, s_{z_i}$. In order to obtain eigenfunctions of the total-angular-momentum operators $\mathbf{J}^2$ and J_z, we considered sets of coupled functions constructed according to various possible coupling schemes—LS, LK, jK, jj, etc. In all cases, each coupled function was still defined by a set of 4N quantum numbers, because each coupling resulted in the replacement of two magnetic quantum numbers by a new intermediate quantum number (L_i,j_i,K,etc.) and its corresponding magnetic quantum number.

In any given scheme (coupled or uncoupled), the number of different basis functions corresponding to a given configuration is equal to the number of different sets of values that can be assumed by the 2N quantum numbers other than $n_i l_i$ (making due allowance for the restrictions imposed by the Pauli exclusion principle in the case of equivalent electrons). In all schemes the number of different functions is the same, each coupling scheme simply producing a different set of k linear combinations of the k uncoupled functions.

Not only is the total number of basis functions fixed, but also the number of functions having a given value of the quantum number J is found to be the same, no matter what coupling scheme is employed. It has been well verified empirically that the number of observed energy levels of each J is never in conflict with the number of basis functions predicted by the vector model. If this were not the case, the theory would of course be in deep trouble; the fact that it *is* true implies that though the individual pure-coupling basis functions for any given coupling scheme may not even approximately reflect the ***nature*** of the various quantum states of the atom (other than the properties specified by J and M), the number of basis functions predicted by the theory does give precisely the correct ***number*** of possible quantum states of the atom, and therefore also correct values of statistical weights.

For any given energy level (defined by a fixed set of values of all quantum numbers except M), the number of quantum states is equal to the number of possible values of M (= $-J, -J + 1, \cdots J - 1, J$). Thus the statistical weight of a level is

$$g = 2J + 1 ; \qquad (4.58)$$

this result has already been encountered in the expression (4.40) for the center-of-gravity energy of a configuration.

The total statistical weight of any group of closely-spaced levels is found by summing 2J + 1 over all those levels. When the coupling conditions in an atom closely approximate some pure-coupling scheme, each group of related levels consists of those levels J that can be formed through the vector addition of two component quantum numbers, j_1 and j_2. Equation (2.65) shows the total statistical weight of the group to be just

$$g = \sum_{J=|j_1-j_2|}^{j_1+j_2} (2J + 1) = (2j_1 + 1)(2j_2 + 1) . \tag{4.59}$$

For example, under LS-coupling conditions, the total statistical weight of all levels (L,S)J belonging to a given term LS is

$$g = (2L + 1)(2S + 1) ; \tag{4.60}$$

examples may be seen in Fig. 4-5. Under pair-coupling conditions, the statistical weight of a pair (K,s)J is

$$g = (2K + 1)(2s + 1) = 4K + 2 ; \tag{4.61}$$

and under jj-coupling conditions, the statistical weight of all levels $(j_1 j_2)J$ for fixed $(j_1 j_2)$ is given by (4.59), provided j_1 and j_2 refer to non-equivalent electrons or $j_1 \neq j_2$.

In some cases, the complete set of levels belonging to a configuration extend over a rather limited energy range; this is true, for example, of the configurations of Ne I shown in Fig. 4-1. It may then be of interest to consider the total statistical weight of all levels of the configuration. This total may be obtained by considering any specific coupling scheme, and successively "uncoupling" the various quantum numbers through repeated application of (2.65). For example, using the jK scheme we find the total statistical weight of a two-electron configuration $l_1 l_2$ to be

$$\begin{aligned} g &= \sum_{j_1} \sum_{K} \sum_{J} (2J + 1) \\ &= \sum_{j_1} \sum_{K} (2K + 1)(2s_2 + 1) \\ &= \sum_{j_1} (2j_1 + 1)(2l_2 + 1)(2s_2 + 1) \\ &= (2l_1 + 1)(2s_1 + 1)(2l_2 + 1)(2s_2 + 1) \\ &= (4l_1 + 2)(4l_2 + 2) . \end{aligned} \tag{4.62}$$

This same result is obtained using the LS scheme and summing (4.60) over L and S, or using the LK or jj schemes and summing over the pertinent intermediate quantum numbers; it is equal to the product of the statistical weight (4.56) for the two singly occupied subshells l_1 and l_2.

For a general configuration of the form (4.33), the total statistical weight is clearly

$$g = \prod_{i=1}^{q} g_i , \tag{4.63}$$

where

$$g_i = \sum_{L_jS_j} (2L_j + 1)(2S_j + 1) \tag{4.64}$$

is the total statistical weight for all allowed terms L_jS_j of the subshell of equivalent electrons $l_i^{w_i}$. The summation in (4.64) cannot be evaluated through use of (2.65) because (for $2 \leqq w_i \leqq 4l_i + 2$) some LS terms indicated by the vector model are prohibited by the Pauli exclusion principle. However, g_i can be easily evaluated in the uncoupled scheme as the number of ways in which we can choose w_i different pairs of one-electron quantum numbers $m_l m_s$ from the total of $4l_i+2$ possible pairs:

$$\begin{aligned} g_i &= \frac{(4l_i + 2)(4l_i + 1) \cdots (4l_i + 3 - w_i)}{w_i!} \\ &= \frac{(4l_i + 2)!}{w_i!(4l_i + 2 - w_i)!} = \binom{4l_i+2}{w_i} . \end{aligned} \tag{4.65}$$

For either an empty ($w_i = 0$) or a filled ($w_i = 4l_i + 2$) subshell this expression reduces to the expected value unity, and for a subshell containing a single electron or a single hole ($w_i = 1$ or $4l_i + 1$) it reduces to $4l_i + 2$.

For a subshell of equivalent jj-coupled electrons $(nlj)^w$, the statistical weight is similarly found to be

$$\begin{aligned} g &= \frac{(2j + 1)(2j) \cdots (2j + 2 - w)}{w!} \\ &= \frac{(2j + 1)!}{w!(2j + 1 - w)!} = \binom{2j+1}{w} . \end{aligned} \tag{4.66}$$

The total statistical weight of a subshell $(nlj_-)^{w_-}(nlj_+)^{w_+}$ containing electrons with both possible values $j_\pm = l \pm 1/2$, summed over all possible values $w_\pm$ satisfying $w_- + w_+ = w$ is[38]

$$\sum_{w_-=0}^{w} \binom{2l}{w_-}\binom{2l+2}{w-w_-} = \binom{4l+2}{w} , \tag{4.67}$$

which is of course identical with (4.65).

4-23. QUANTITATIVE CALCULATION OF LEVEL STRUCTURES

In the preceding sections we have discussed a number of qualitative aspects of the energy-level structure of complex atoms. Arguments of the type that we used there can be

[38]See Eq. (9.92).

applied to help classify and correlate much empirical data on atomic energy levels,[39] but little more can be said from a theoretical point of view without going into the quantitative details of the theory.

The quantitative calculation of wavefunctions and energy levels involves two distinct stages:

(1) Determination of the detailed shape of the functions $P_{nl}(r)$ that form the radial factors of the one-electron spin-orbitals (4.17).

(2) Calculation of the energy matrix elements $H_{bb'} \equiv \langle b|H|b'\rangle$ [using as basis functions Ψ_b either the uncoupled determinantal functions (4.26) or functions constructed according to any desired coupling scheme] and diagonalization of the energy matrix, as discussed in Sec. 4-2.

These two stages will be discussed in turn in the following chapters; as both problems are rather lengthy, we here outline each briefly.

Calculation of Radial Functions. In any determinantal basis function (4.26) corresponding to a configuration (4.33) there are q different radial functions $P_{n_j l_j}(r)$, one for each subshell of equivalent electrons $n_j l_j^{w_j}$ $(1 \leqq j \leqq q)$. The matrix elements $H_{bb'}$, and therefore the eigenvalues E^k, will depend on the detailed shapes of these functions. We determine the $P_{n_j l_j}(r)$ by the criterion that they should be such as to minimize the calculated energy of the atom, within the limitations set by the orthonormalization conditions (4.18). In principle, one might think that a different set of functions should be found for each different energy level of the atom (i.e., for each different eigenvalue E^k), or at least for each set of related closely-spaced levels (e.g., under LS-coupling conditions, for each different LS term; cf. Sec. 16-8). In practice, this would involve a great deal of effort, and we shall usually content ourselves with a single set of functions to be used for all energy levels of the configuration. This set of functions we choose to be the set that minimizes the configuration-average energy (4.38) of all quantum states of the configuration,

$$E_{av} = \frac{\sum_b H_{bb}}{\text{number of states}}, \tag{4.68}$$

where H_{bb} is a diagonal element of the energy matrix in any desired representation, and where the summation is over all basis functions b of the configuration. As discussed in Sec. 4-11, this average is equivalent to the energy of the spherically symmetrized atom. Thus determination of the $P_{n_j l_j}$ by minimizing E_{av} is consistent with the fact that the form of the spin-orbitals (4.17) was determined in the first place on the basis of the central field model of the atom.

Minimization of E_{av} with respect to variations in the form of the $P_{n_j l_j}$ leads to a set of coupled differential equations (one for each j), similar in form to the hydrogenic radial equation (3.9), known as the spherically averaged Hartree-Fock (HF) equations. Numerical solution of these equations in a manner similar to that described in Sec. 3-4

[39]See, for example, G. K. Woodgate, *Elementary Atomic Structure* (McGraw-Hill, London, 1970); H. G. Kuhn, *Atomic Spectra* (Longmans, Green, London, 1969), 2nd ed.; C. Candler, *Atomic Spectra and the Vector Model* (D. Van Nostrand, Princeton, N. J., 1964), 2nd ed.

provides us with the desired functions $P_{n_j l_j}(r)$ for the configuration in question. Separate calculations of this type have to be made for each configuration of interest.

Calculation of Energy Level Structures. With the radial functions $P_{nl}(r)$ determined by solution of the HF equations (or simpler approximations thereto), the determinantal functions (4.26) are completely known. These functions may be used as basis functions for evaluation of the matrix elements $H_{bb'}$, and thereby calculation of the energy levels of the atom; this is the basic procedure followed in Slater's original treatment[2] and also in other more recent works.[3,13,40] There are, however, important calculational advantages in the use of coupled basis functions, having to do with their higher symmetry properties. We shall illustrate these advantages by means of the pd example of Fig. 4-5.

There are a total of 60 different possible sets of quantum numbers in this example, and consequently a total of 60 basis functions, either in the coupled scheme (Fig. 4-5), or in the uncoupled scheme (each of the 15 cases of Table 2-2, with any of the four spin cases $\uparrow\uparrow$, $\uparrow\downarrow$, $\downarrow\uparrow$, $\downarrow\downarrow$). However, this does not mean that we have to compute matrix elements for, and diagonalize, a 60×60 matrix. Even if the atom lies in a magnetic field (oriented parallel to the z-axis), the component of **J** along the magnetic field is a constant of the motion. This means that M is a good quantum number and that the wavefunction expansion (4.4) for any given state k can involve basis functions for only a single value of M. The energy matrix [for a Hamiltonian involving a magnetic-field term, in addition to the terms shown in (4.1)] can therefore have no non-zero off-diagonal elements connecting basis functions of different M, and the 60×60 matrix breaks up into smaller blocks—one for each value of M. The size of each matrix is shown in Table 4-6.

For an atom in field-free space, the matrix elements (and the eigenvalues) are independent of M, and it is necessary to diagonalize only the $M = 0$ matrix to obtain all 12 energy levels of the configuration. (Diagonalization of the $M = \pm 1$ matrices just gives over again the energies of all levels except the $J = 0$ level, the $M = \pm 2$ matrices give again the energies of all levels except those with $J = 0$ and 1, etc.)

If coupled functions are used, additional simplification is possible. The angular momentum **J** is a constant of the motion, and therefore in the wavefunction expansion (4.4) the Hamiltonian operator cannot mix basis functions that are eigenfunctions corresponding to different values of the quantum number J. The 12×12 matrix thus breaks down into smaller matrices, one for each value of J, as shown in the table, and it is necessary to diagonalize only one 4×4 matrix and two 3×3 matrices (plus two trivial 1×1 matrices) instead of a 12×12 matrix.

In addition to the fact that smaller matrices are obtained, there are other advantages to the use of coupled functions. For example, it is immediately apparent which eigenvalues (and eigenvectors) correspond to states of any given value of J. Moreover, if the coupled functions have been constructed according to the coupling scheme that corresponds most closely to the coupling conditions of the atomic configuration in question, then each eigenvector will be fairly close to a pure basis vector and it is immediately possible to label the eigenvalues with the proper values of other quantum numbers (for example L and S, or j_1

[40]H. H. Theissing and P. J. Caplan, *Spectroscopic Calculations for a Multielectron Ion* (Interscience, New York, 1966).

TABLE 4-6. MATRIX SIZES FOR THE CONFIGURATION pd

Interaction	No. of Basis Functions	Matrix:	Size
External Magnetic Field	60	M= −4:	1 × 1
		M= −3:	4 × 4
		M= −2:	8 × 8
		M= −1:	11 × 11
		M=0:	12 × 12
		M=1:	11 × 11
		M=2:	8 × 8
		M=3:	4 × 4
		M=4:	1 × 1
Coulomb + Spin-Orbit			
Uncoupled functions	12	M=0:	12 × 12
Coupled functions	12	J=0:	1 × 1
		J=1:	3 × 3
		J=2:	4 × 4
		J=3:	3 × 3
		J=4:	1 × 1

and K) in addition to J. (Alternatively, the diagonalization of H can be done in any convenient basis representation, and the resulting eigenvectors then transformed into the desired representation with the aid of the appropriate transformation matrix.)

If in using coupled functions it were necessary to explicitly write these functions down as specific linear combinations of determinantal functions, and if it were necessary to evaluate matrix elements between coupled functions as linear combinations of matrix elements between determinantal functions, then much of the above advantage would be counterbalanced by the additional complexity of evaluating these linear combinations. However, the mathematical techniques developed by Racah[41] make it possible to completely bypass explicit use of determinantal functions, and to write down formulae for the direct evaluation of matrix elements for coupled basis functions. The mathematical background required to use Racah's techniques is rather extensive and will require several chapters for its development;[42] however, once this facility has been acquired, it becomes possible to calculate matrix elements with a small fraction of the effort otherwise required, and to deal with complex configurations that it would be completely impractical to handle by means of determinantal functions. Essential to these mathematical methods are the Wigner 3n-j symbols, which we discuss in the next chapter.

[41]G. Racah, (II) Phys. Rev. **62**, 438 (1942); (III) Phys. Rev. **63**, 367 (1943); (IV) Phys. Rev. **76**, 1352 (1949).

[42]A summary of the Racah method is given in Appendix F. The reader may find it helpful to refer to this appendix occasionally by way of review.

5

THE 3n-j SYMBOLS

5-1. THE 3-j SYMBOL

The Wigner 3n-j symbols,[1,2] or their close relatives the Clebsch-Gordon and Racah coefficients (for n = 1 and 2, respectively), are practically indispensable for quantitative calculations of atomic structure and spectra. We here indicate a few of their properties that are important for our purposes; the discussion of most applications will be deferred till the need arises.

The 3-j symbol is an algebraic function of six arguments that may be defined by the expression

$$\begin{pmatrix} j_1 & j_2 & j_3 \\ m_1 & m_2 & m_3 \end{pmatrix} \equiv \delta_{m_1+m_2+m_3,0}(-1)^{j_1-j_2-m_3}$$

$$\times \left[\frac{(j_1+j_2-j_3)!(j_1-j_2+j_3)!(-j_1+j_2+j_3)!(j_1-m_1)!(j_1+m_1)!(j_2-m_2)!(j_2+m_2)!(j_3-m_3)!(j_3+m_3)!}{(j_1+j_2+j_3+1)!}\right]^{1/2}$$

$$\times \sum_k \frac{(-1)^k}{k!(j_1+j_2-j_3-k)!(j_1-m_1-k)!(j_2+m_2-k)!(j_3-j_2+m_1+k)!(j_3-j_1-m_2+k)!} \;. \qquad (5.1)$$

This function is defined (i.e., is non-zero) only for values of j_i and m_i such that the arguments of all factorials are non-negative integers. It follows that j_i and m_i must both be either integral or half-integral, with $j_i \geqq |m_i| \geqq 0$ (each i), that $j_1 + j_2 + j_3$ and $m_1 + m_2 + m_3$ must be integral, and that $j_1 - j_2 - m_3$ is integral so that the 3-j symbol is real. Also, the three j_i must satisfy the three inequalities

$$\begin{aligned} j_1 + j_2 &\geqq j_3 \,, \\ j_2 + j_3 &\geqq j_1 \,, \\ j_3 + j_1 &\geqq j_2 \,; \end{aligned} \qquad (5.2)$$

[1]E. P. Wigner, in L. C. Biedenharn and H. van Dam, eds., *Quantum Theory of Angular Momentum* (Academic Press, New York, 1965).

[2]A. R. Edmonds, *Angular Momentum in Quantum Mechanics* (Princeton University Press, Princeton, N.J., 1960), 2nd ed.

these inequalities together with the integral-sum restriction are referred to as the *triangle relations*. We shall use the symbol $\delta(j_1j_2j_3)$ to represent +1 if the triangle relations are satisfied, and to represent zero otherwise. In practice, each j_i is an angular-momentum quantum number, m_i is the corresponding magnetic quantum number, and the j_i correspond to three angular-momentum operators of which one is the vector sum of the other two;[3] all of the above restrictions are thus familiar from the vector model of the addition of angular momenta. The summation in (5.1) is finite, being over those integral values of k that satisfy

$$\max(0,\ j_2-j_3-m_1,\ j_1-j_3+m_2) \leqq k \leqq \min(j_1+j_2-j_3,\ j_1-m_1,\ j_2+m_2) \ . \tag{5.3}$$

Symmetry properties. It may readily be seen from the definition (5.1) that the 3-j symbols satisfy the symmetry properties

$$\begin{pmatrix} j_2 & j_1 & j_3 \\ m_2 & m_1 & m_3 \end{pmatrix} = \begin{pmatrix} j_1 & j_3 & j_2 \\ m_1 & m_3 & m_2 \end{pmatrix} = (-1)^{j_1+j_2+j_3}\begin{pmatrix} j_1 & j_2 & j_3 \\ m_1 & m_2 & m_3 \end{pmatrix} \ . \tag{5.4}$$

From these two relations it follows that

$$\begin{pmatrix} j_i & j_k & j_n \\ m_i & m_k & m_n \end{pmatrix} = \varepsilon\begin{pmatrix} j_1 & j_2 & j_3 \\ m_1 & m_2 & m_3 \end{pmatrix} , \tag{5.5}$$

where ε is +1 or $(-1)^{j_1+j_2+j_3}$ according as (ikn) is an even or an odd permutation of (123). It can also be easily seen with the aid of (5.4) that

$$\begin{pmatrix} j_1 & j_2 & j_3 \\ -m_1 & -m_2 & -m_3 \end{pmatrix} = (-1)^{j_1+j_2+j_3}\begin{pmatrix} j_1 & j_2 & j_3 \\ m_1 & m_2 & m_3 \end{pmatrix} \ . \tag{5.6}$$

These symmetry properties make it possible to greatly shorten tables of numerical values of the 3-j symbols.[4] An abbreviated table, for argument values up to 4, is given in Appendix C.

Problems

5-1(1). Verify the symmetry relation (5.4) for interchange of the first two columns of the 3-j symbol by making the substitution $k = j_1 + j_2 - j_3 - l$ in (5.1); similarly, verify the effect of interchange of the second and third columns by means of the substitution $k = j_1 - m_1 - l$. Verify (5.6) by changing the signs of all m_i in (5.1) and then using (5.4). [Hint:

[3]Actually, the sign of one of the m_i has been reversed, so that the sum of all three m_i is zero rather than one of them being equal to the sum of the other two.

[4]M. Rotenberg, R. Bivins, N. Metropolis, and J. K. Wooten, Jr., *The 3-j and 6-j Symbols* (The Technology Press, Massachusetts Institute of Technology, Cambridge, 1959).

Remember that $(-1)^{4j} = +1$ for any j (integral or half-integral), and that $(-1)^{2j} = +1$ if j is integral.]

5-1(2). Use the symmetry properties (5.5)-(5.6) and the tabulated values in Appendix C to obtain the results

$$\begin{pmatrix} \frac{1}{2} & 1 & \frac{3}{2} \\ -\frac{1}{2} & 1 & -\frac{1}{2} \end{pmatrix} = -\begin{pmatrix} \frac{3}{2} & 1 & \frac{1}{2} \\ -\frac{1}{2} & 1 & -\frac{1}{2} \end{pmatrix} = \begin{pmatrix} \frac{3}{2} & 1 & \frac{1}{2} \\ \frac{1}{2} & -1 & \frac{1}{2} \end{pmatrix} = 3^{1/2}/6 ,$$

$$\begin{pmatrix} 1 & 2 & 2 \\ 1 & 0 & -1 \end{pmatrix} = -\begin{pmatrix} 2 & 2 & 1 \\ -1 & 0 & 1 \end{pmatrix} = \begin{pmatrix} 2 & 2 & 1 \\ 1 & 0 & -1 \end{pmatrix} = 10^{-1/2} ,$$

$$\begin{pmatrix} 1 & 1 & 2 \\ 0 & 1 & -1 \end{pmatrix} = \begin{pmatrix} 2 & 1 & 1 \\ -1 & 0 & 1 \end{pmatrix} = \begin{pmatrix} 2 & 1 & 1 \\ 1 & 0 & -1 \end{pmatrix} = -10^{-1/2} .$$

Notation. The following abbreviated notation will be used extensively throughout the remainder of this book:

$$[j] \equiv 2j + 1 ,$$
$$[j_1, j_2, \cdots] \equiv (2j_1 + 1)(2j_2 + 1)\cdots . \tag{5.7}$$

(This notation is to be inferred only when the quantities within the brackets are simple angular momentum quantum numbers, not when they are complex expressions.)

A variety of notations have been introduced for various quantities related to the 3-j symbol; for convenience of cross reference, we mention those used by Racah[5]

$$V(j_1 j_2 j_3; m_1 m_2 m_3) = (-1)^{j_1 - j_2 - j_3} \begin{pmatrix} j_1 & j_2 & j_3 \\ m_1 & m_2 & m_3 \end{pmatrix} , \tag{5.8}$$

by Fano and Racah[6]

$$\bar{V}\begin{pmatrix} j_1 & j_2 & j_3 \\ m_1 & m_2 & m_3 \end{pmatrix} = (-1)^{2(j_2 + j_3)} V(j_1 j_2 j_3; m_1 m_2 m_3) = (-1)^{j_1 + j_2 + j_3} \begin{pmatrix} j_1 & j_2 & j_3 \\ m_1 & m_2 & m_3 \end{pmatrix} , \tag{5.9}$$

and by Condon and Shortley[7]

$$(j_1 j_2 m_1 m_2 | j_1 j_2 j_3 m_3) = (-1)^{j_1 - j_2 + m_3} [j_3]^{1/2} \begin{pmatrix} j_1 & j_2 & j_3 \\ m_1 & m_2 & -m_3 \end{pmatrix} ; \tag{5.10}$$

[5]G. Racah, II, Phys. Rev. **62**, 438 (1942).

[6]U. Fano and G. Racah, *Irreducible Tensorial Sets* (Academic Press, New York, 1959).

[7]E. U. Condon and G. H. Shortley, *The Theory of Atomic Spectra* (University Press, Cambridge, 1935), p. 75.

this last form is also known as a vector-coupling or Clebsch-Gordon coefficient.

Special values. Some special 3-j symbols that are frequently encountered are

$$\begin{pmatrix} j & j & 0 \\ m & -m & 0 \end{pmatrix} = (-1)^{j-m}[j]^{-1/2} , \tag{5.11}$$

$$\begin{pmatrix} j & j-\frac{1}{2} & \frac{1}{2} \\ m & -m-\frac{1}{2} & \frac{1}{2} \end{pmatrix} = (-1)^{j-m-1}\left[\frac{j-m}{2j(2j+1)}\right]^{1/2} , \tag{5.12}$$

and

$$\begin{pmatrix} j_1 & j_2 & j_3 \\ 0 & 0 & 0 \end{pmatrix} = (-1)^J\left[\frac{(2J-2j_1)!(2J-2j_2)!(2J-2j_3)!}{(2J+1)!}\right]^{1/2}\frac{J!}{(J-j_1)!(J-j_2)!(J-j_3)!} , \tag{5.13}$$

where $2J = j_1 + j_2 + j_3$ must be even; if 2J is odd, then (5.13) is zero from (5.6). The first two of the above expressions can be readily derived directly from the definition (5.1); the third requires a somewhat lengthy calculation.[5]

Orthogonality relations and sum rules. The following additional properties of 3-j symbols we state without proof:

$$\sum_{j_3}\sum_{m_3}[j_3]\begin{pmatrix} j_1 & j_2 & j_3 \\ m_1 & m_2 & m_3 \end{pmatrix}\begin{pmatrix} j_1 & j_2 & j_3 \\ m_1' & m_2' & m_3 \end{pmatrix} = \delta_{m_1m_1'}\delta_{m_2m_2'} , \tag{5.14}$$

$$\sum_{m_1}\sum_{m_2}[j_3]\begin{pmatrix} j_1 & j_2 & j_3 \\ m_1 & m_2 & m_3 \end{pmatrix}\begin{pmatrix} j_1 & j_2 & j_3' \\ m_1 & m_2 & m_3' \end{pmatrix} = \delta_{j_3j_3'}\delta_{m_3m_3'}\delta(j_1j_2j_3) , \tag{5.15}$$

$$\sum_{m}(-1)^{j-m}\begin{pmatrix} j & j & j_3 \\ m & -m & 0 \end{pmatrix} = \delta_{j_30}[j]^{1/2} . \tag{5.16}$$

In the first two relations, the usual restrictions $j_i \geqq |m_i|$ apply. In the special case $j_3 = 0$, the result (5.16) follows directly from (5.11).

From (5.14)-(5.15) it follows that the quantities

$$[j_3]^{1/2}\begin{pmatrix} j_1 & j_2 & j_3 \\ m_1 & m_2 & m_3 \end{pmatrix}$$

form the elements of an orthogonal matrix in which the rows are labeled by (m_1m_2) and the columns are labeled by (j_3m_3).

The product of two spherical harmonics. The 3-j symbols provide a convenient means of expressing the coefficients in the spherical-harmonic expansion of a product of two spherical harmonics:

$$\left(\frac{4\pi}{[k]}\right)^{1/2} Y_{kq}(\theta,\phi) Y_{l'm'}(\theta,\phi)$$

$$= [l']^{1/2} \sum_{j} \sum_{m_j} (-1)^{-m_j} [j]^{1/2} \begin{pmatrix} j & k & l' \\ 0 & 0 & 0 \end{pmatrix} \begin{pmatrix} j & k & l' \\ -m_j & q & m' \end{pmatrix} Y_{jm_j}(\theta,\phi) \ . \qquad (5.17)$$

This result is most easily proved by group theoretical methods.[8] It may be noted that the only non-zero terms in the double sum are those with $m_j = q + m'$, with $j \geq |m_j|$, and—because of (5.13)—with $j + k + l'$ even.

Matrix elements of spherical harmonics. The renormalized spherical harmonic $(4\pi/[k])^{1/2} Y_{kq}$ will be encountered sufficiently frequently that it is convenient to abbreviate it by a special symbol,

$$C_q^{(k)} \equiv \left(\frac{4\pi}{[k]}\right)^{1/2} Y_{kq} \ . \qquad (5.18)$$

The angular matrix element of this quantity over two spherical-harmonic wavefunctions is easily evaluated with the aid of (5.17) and the orthonormality property (2.42):

$$\langle lm | C_q^{(k)} | l'm' \rangle \equiv \left(\frac{4\pi}{[k]}\right)^{1/2} \int_0^{2\pi} \int_0^{\pi} Y^*_{lm} Y_{kq} Y_{l'm'} \sin\theta \, d\theta \, d\phi$$

$$= (-1)^{-m} [l,l']^{1/2} \begin{pmatrix} l & k & l' \\ 0 & 0 & 0 \end{pmatrix} \begin{pmatrix} l & k & l' \\ -m & q & m' \end{pmatrix}$$

$$\equiv \delta_{q,m-m'} c^k(lm,l'm') \ . \qquad (5.19)$$

This result may also be obtained by a direct algebraic evaluation of the integral, without recourse to (5.17); the integration over ϕ is trivial and leads to the δ-factor, but evaluation of the θ integral is a lengthy process[5,9] and will not be given here. In order that the θ integral $c^k(lm,l'm')$ be non-zero, it follows from (5.13) that k must be even or odd according as $l + l'$ is even or odd; from the triangle relation (5.2) we may then have only

$$k = |l-l'|, \ |l-l'|+2, \ |l-l'|+4, \ \cdots, \ l+l'-2, \ l+l' \ . \qquad (5.20)$$

It follows from (5.5) that

$$c^k(l'm',lm) = (-1)^{m-m'} c^k(lm,l'm') \ , \qquad (5.21)$$

and from (5.11) that

[8]See, for example, ref. 2, Eq. (4.6.5).

[9]J. A. Gaunt, Phil. Trans. Roy. Soc. London **A228**, 192 (1929).

$$c^0(lm,lm) = (-1)^m[l]\begin{pmatrix} l & 0 & l \\ 0 & 0 & 0 \end{pmatrix}\begin{pmatrix} l & 0 & l \\ -m & 0 & m \end{pmatrix} = 1 \; . \tag{5.22}$$

[Because $C_0^{(0)} = (4\pi)^{1/2}Y_{00}$ is unity (Table 2-1), the result (5.22) represents simply the normalization property of the spherical harmonics.]

Tables of values of the c^k are given by Condon and Shortley[10] and by Slater[11] for s, p, d, and f electrons, and by Shortley and Fried[12] for g electrons; however, we shall have no need for these quantities in this book.

5-2. THE 6-j SYMBOL

The 6-j symbol is a function of six arguments that may be defined by the expression[2,4]

$$\begin{Bmatrix} j_1 & j_2 & j_3 \\ l_1 & l_2 & l_3 \end{Bmatrix} \equiv \Delta(j_1j_2j_3)\,\Delta(j_1l_2l_3)\,\Delta(l_1j_2l_3)\,\Delta(l_1l_2j_3)$$

$$\times \sum_k \left[\frac{(-1)^k(k+1)!}{(k-j_1-j_2-j_3)!(k-j_1-l_2-l_3)!(k-l_1-j_2-l_3)!(k-l_1-l_2-j_3)!} \right.$$

$$\left. \times \frac{1}{(j_1+j_2+l_1+l_2-k)!(j_2+j_3+l_2+l_3-k)!(j_3+j_1+l_3+l_1-k)!} \right] \; , \tag{5.23}$$

where

$$\Delta(abc) \equiv \left[\frac{(a+b-c)!(a-b+c)!(-a+b+c)!}{(a+b+c+1)!} \right]^{1/2} . \tag{5.24}$$

In atomic-structure theory, the 6-j symbol usually arises as a five-fold summation over a product of four 3-j symbols:[5]

$$\begin{Bmatrix} j_1 & j_2 & j_3 \\ l_1 & l_2 & l_3 \end{Bmatrix} = [j_3] \sum_{\substack{m_1m_2 \\ n_1n_2n_3}} (-1)^S \begin{pmatrix} j_1 & j_2 & j_3 \\ m_1 & m_2 & m_3 \end{pmatrix}\begin{pmatrix} j_1 & l_2 & l_3 \\ m_1 & n_2 & -n_3 \end{pmatrix}\begin{pmatrix} l_1 & j_2 & l_3 \\ -n_1 & m_2 & n_3 \end{pmatrix}\begin{pmatrix} l_1 & l_2 & j_3 \\ n_1 & -n_2 & m_3 \end{pmatrix} , \tag{5.25}$$

where

$$S = l_1 + l_2 + l_3 + n_1 + n_2 + n_3 \; .$$

[10]E. U. Condon and G. H. Shortley, *The Theory of Atomic Spectra* (University Press, Cambridge, 1935), pp. 178-80. (Beware the decimal points, which are printed English style as center-dots!)

[11]J. C. Slater, *Quantum Theory of Atomic Structure* (McGraw-Hill, New York, 1960), Vol. II, App. 20.

[12]G. H. Shortley and B. Fried, Phys. Rev. 54, 739 (1938).

It should be noted that the result of this multiple summation is independent of m_3; the physical reasons for this will be indicated in Sec. 10-5. Clearly, each argument of the 6-j symbol has the nature of an angular-momentum quantum number (*not* a magnetic quantum number).

The 6-j symbol is considered to be zero unless the argument of each factorial in (5.23), (5.24) is a non-negative integer. This means that each of the arguments of the 6-j symbol must be a non-negative integer or half-integer, and that all four of the triangle relations

$$\delta(j_1j_2j_3), \quad \delta(j_1l_2l_3), \quad \delta(l_1j_2l_3), \quad \delta(l_1l_2j_3) \tag{5.26}$$

must be satisfied; all of these restrictions follow also from the restrictions on the upper-row arguments of the 3-j symbols in (5.25). The summation in (5.23) is finite, and covers integral values of k in the range

$$\max(j_1+j_2+j_3,\ j_1+l_2+l_3,\ l_1+j_2+l_3,\ l_1+l_2+j_3)$$
$$\leqq k \leqq \min(j_1+j_2+l_1+l_2,\ j_2+j_3+l_2+l_3,\ j_3+j_1+l_3+l_1)\ . \tag{5.27}$$

Symmetry properties. It may be seen directly from the definition (5.23) that the value of the 6-j symbol is unchanged if any two columns are interchanged, or if any *two* numbers in the bottom row are interchanged with the *corresponding* two numbers in the top row. These symmetry properties make possible a great reduction in the size of tables of values of the 6-j symbols;[4] an abbreviated table, for argument values up to 4, is given in Appendix D.

Problem

5-2(1). Use the symmetry properties of the 6-j symbol and the tables of Appendix D to obtain the results

$$\begin{Bmatrix} \frac{5}{2} & 1 & \frac{3}{2} \\ 2 & \frac{1}{2} & 3 \end{Bmatrix} = \begin{Bmatrix} \frac{5}{2} & \frac{1}{2} & 3 \\ 2 & 1 & \frac{3}{2} \end{Bmatrix} = \begin{Bmatrix} 3 & \frac{5}{2} & \frac{1}{2} \\ \frac{3}{2} & 2 & 1 \end{Bmatrix} = 30^{-1/2}\ ,$$

$$\begin{Bmatrix} 1 & 2 & 2 \\ 2 & 3 & 2 \end{Bmatrix} = \begin{Bmatrix} 2 & 3 & 2 \\ 1 & 2 & 2 \end{Bmatrix} = \begin{Bmatrix} 3 & 2 & 2 \\ 2 & 2 & 1 \end{Bmatrix} = -\frac{14^{1/2}}{35}\ .$$

Notation. A variety of notations are used for various quantities related to the 6-j symbol, but the only one much used in the literature of atomic structure is the Racah coefficient[5]

$$W(j_1j_2l_2l_1;j_3l_3) = (-1)^{j_1+j_2+l_1+l_2}\begin{Bmatrix} j_1 & j_2 & j_3 \\ l_1 & l_2 & l_3 \end{Bmatrix}\ . \tag{5.28}$$

Racah's notation has typographical advantages but the much higher symmetry of the 6-j symbol makes it vastly simpler for use with numerical tables, and we shall use only the 6-j form. Fano and Racah[6] use the notation

$$\overline{W}\begin{pmatrix} j_1 & j_2 & j_3 \\ l_1 & l_2 & l_3 \end{pmatrix}$$

to represent the 6-j symbol.

Special value. When one argument (e.g., l_3) of the 6-j symbol is zero, the definition (5.23) is easily seen to reduce to the simple expression

$$\begin{Bmatrix} j_1 & j_2 & j_3 \\ l_1 & l_2 & 0 \end{Bmatrix} = \delta(j_1 j_2 j_3)\delta_{j_1 l_2}\delta_{j_2 l_1}\frac{(-1)^{j_1+j_2+j_3}}{[j_1,j_2]^{1/2}} . \tag{5.29}$$

Orthogonality relations and sum rules. It can be shown[5] that the quantities

$$[j_3,l_3]^{1/2}\begin{Bmatrix} j_1 & j_2 & j_3 \\ l_1 & l_2 & l_3 \end{Bmatrix} \tag{5.30}$$

form the elements of a real orthogonal matrix, where j_3 and l_3 label the matrix rows and columns, respectively; thus the orthonormality relation

$$\sum_{l_3}[j_3,l_3]\begin{Bmatrix} j_1 & j_2 & j_3 \\ l_1 & l_2 & l_3 \end{Bmatrix}\begin{Bmatrix} j_1 & j_2 & j_3' \\ l_1 & l_2 & l_3 \end{Bmatrix} = \delta_{j_3 j_3'} \tag{5.31}$$

holds (provided that the four triangle relations are satisfied). This result will be proved in Sec. 9-2.

Two additional sum rules of considerable utility have been given by Racah[5] and by Biedenharn and Elliott,[13] respectively:

$$\sum_{l_3}(-1)^{j_3+j+l_3}[l_3]\begin{Bmatrix} j_1 & j_2 & j_3 \\ l_1 & l_2 & l_3 \end{Bmatrix}\begin{Bmatrix} j_1 & l_1 & j \\ j_2 & l_2 & l_3 \end{Bmatrix} = \begin{Bmatrix} j_1 & j_2 & j_3 \\ l_2 & l_1 & j \end{Bmatrix} , \tag{5.32}$$

$$\sum_{k}(-1)^{S+k}[k]\begin{Bmatrix} l_1 & j_2 & l_3 \\ l_3' & l_2' & k \end{Bmatrix}\begin{Bmatrix} j_2 & j_3 & j_1 \\ l_1' & l_3' & k \end{Bmatrix}\begin{Bmatrix} l_1 & j_3 & l_2 \\ l_1' & l_2' & k \end{Bmatrix} = \begin{Bmatrix} j_1 & j_2 & j_3 \\ l_1 & l_2 & l_3 \end{Bmatrix}\begin{Bmatrix} l_3 & j_1 & l_2 \\ l_1' & l_2' & l_3' \end{Bmatrix} , \tag{5.33}$$

where

$$S = j_1 + j_2 + j_3 + l_1 + l_2 + l_3 + l_1' + l_2' + l_3' .$$

[13]L. C. Biedenharn, J. Math. and Phys. 31, 287 (1952); J. P. Elliott, Proc. Roy. Soc. (London) A218, 345 (1953).

Substituting $j_3' = 0$ into (5.31) and using (5.29), we obtain

$$\sum_k (-1)^{j+l+k}[k]\begin{Bmatrix} j & j & j' \\ l & l & k \end{Bmatrix} = [j,l]^{1/2}\delta_{j'0} \ . \tag{5.34}$$

Similarly, substitution of $j = 0$ into (5.32) leads to

$$\sum_k (-1)^{2k}[k]\begin{Bmatrix} j & l & j' \\ j & l & k \end{Bmatrix} = \delta(jlj') \ , \tag{5.35}$$

which may also be written

$$\sum_k [k]\begin{Bmatrix} j & l & j' \\ j & l & k \end{Bmatrix} = (-1)^{2j+2l}\delta(jlj') \ . \tag{5.36}$$

5-3. THE 9-j SYMBOL

The 9-j symbol is defined as[2,4,14]

$$\begin{Bmatrix} j_{11} & j_{12} & j_{13} \\ j_{21} & j_{22} & j_{23} \\ j_{31} & j_{32} & j_{33} \end{Bmatrix} = \sum_j (-1)^{2j}[j]\begin{Bmatrix} j_{11} & j_{21} & j_{31} \\ j_{32} & j_{33} & j \end{Bmatrix}\begin{Bmatrix} j_{12} & j_{22} & j_{32} \\ j_{21} & j & j_{23} \end{Bmatrix}\begin{Bmatrix} j_{13} & j_{23} & j_{33} \\ j & j_{11} & j_{12} \end{Bmatrix}$$

$$= \sum_{\text{all } m} \begin{pmatrix} j_{11} & j_{12} & j_{13} \\ m_{11} & m_{12} & m_{13} \end{pmatrix}\begin{pmatrix} j_{21} & j_{22} & j_{23} \\ m_{21} & m_{22} & m_{23} \end{pmatrix}\begin{pmatrix} j_{31} & j_{32} & j_{33} \\ m_{31} & m_{32} & m_{33} \end{pmatrix}$$

$$\times \begin{pmatrix} j_{11} & j_{21} & j_{31} \\ m_{11} & m_{21} & m_{31} \end{pmatrix}\begin{pmatrix} j_{12} & j_{22} & j_{32} \\ m_{12} & m_{22} & m_{32} \end{pmatrix}\begin{pmatrix} j_{13} & j_{23} & j_{33} \\ m_{13} & m_{23} & m_{33} \end{pmatrix} , \tag{5.37}$$

where all nine arguments are obviously angular-momentum quantum numbers, and hence non-negative integers or half-integers. From the properties of the 6-j symbols it follows that the 9-j symbol is zero unless the arguments in each row and in each column satisfy the triangle relations.

Symmetry properties. From the 3-j-symbol form of (5.37), it may be easily seen that the value of the 9-j symbol is unchanged upon reflection about either diagonal; upon permutation of the rows or of the columns, the value of the 9-j symbol is unchanged or is multiplied by $(-1)^S$, where S is the sum of all nine arguments, according as the permutation is even or odd respectively.

[14]Tables of numerical values are given by K. Smith and J. W. Stephenson, Argonne National Laboratory Reports ANL-5776 (1957) and ANL-5860 (1958), and by S. H. Koozekanani, *9j Symbols* (Engineering Publications, Ohio State University, Columbus, 1972).

Special values. Using (5.29) one finds from (5.37) that

$$\begin{Bmatrix} a & b & e \\ c & d & e \\ f & f & 0 \end{Bmatrix} = (-1)^{b+c+e+f}[e,f]^{-1/2}\begin{Bmatrix} a & b & e \\ d & c & f \end{Bmatrix} \tag{5.38}$$

and that

$$\begin{Bmatrix} a & a & 0 \\ c & c & 0 \\ f & f & 0 \end{Bmatrix} = [a,c,f]^{-1/2}\delta(acf) \ . \tag{5.39}$$

Orthogonality. It is easily proved directly from the definition (5.37) and the orthonormality property (5.31) of the 6-j symbol that

$$\sum_i \sum_k [i,k,s,t] \begin{Bmatrix} A & B & s \\ C & D & t \\ i & k & E \end{Bmatrix}\begin{Bmatrix} A & B & s' \\ C & D & t' \\ i & k & E \end{Bmatrix} = \delta_{ss'}\delta_{tt'}\delta(ABs)\delta(CDt)\delta(stE) \ . \tag{5.40}$$

From the symmetry properties of the 9-j symbol, it then follows that

$$\sum_s \sum_t [i,k,s,t] \begin{Bmatrix} A & B & s \\ C & D & t \\ i & k & E \end{Bmatrix}\begin{Bmatrix} A & B & s \\ C & D & t \\ i' & k' & E \end{Bmatrix} = \delta_{ii'}\delta_{kk'}\delta(ACi)\delta(BDk)\delta(ikE) \ . \tag{5.41}$$

Thus the quantities

$$[i,k,s,t]^{1/2}\begin{Bmatrix} A & B & s \\ C & D & t \\ i & k & E \end{Bmatrix} \tag{5.42}$$

form the elements of an orthogonal matrix in which the rows are labeled by values of (ik) and the columns are labeled by values of (st).

Higher-order 3n-j symbols can be defined, though not uniquely—there being two essentially different 12-j symbols, five different 15-j symbols, and eighteen different 18-j symbols, for example.[15,16] We shall have no need for any of these. Indeed, even the 9-j symbol is mainly of notational utility; for calculational purposes, it can be evaluated only in terms of 6-j symbols via the expansion (5.37).

[15]A. P. Yutsis, I. B. Levinson, and V. V. Vanagas (A. Jucys, J. Levinsonas, and V. Vanagas), *Mathematical Apparatus of the Theory of Angular Momentum* (English translation of 1960 Russian edition, available from National Technical Information Service, Springfield, Va., 1962).

[16]E. El-Baz and B. Castel, *Graphical Methods of Spin Algebras in Atomic, Nuclear, and Particle Physics* (Marcel Dekker, New York, 1972).

*5-4. GRAPHICAL METHODS

Graphical methods have been developed for representing the 3n-j symbols and relationships among them. Contrary to what one might anticipate, the 3-j symbol is represented not by a closed triangle, but by three lines mutually joined at a single node (Fig. 5-1a); the lengths of the lines and the angles between pairs of lines have no quantitative significance.

The LS coupling of two electrons, $[(l_1l_2)L, (s_1s_2)S]J$, may thus be represented by a diagram such as Fig. 5-1b, and the LK coupling of the same angular momenta may be represented by Fig. 5-1c. If we delete the node (l_1l_2), which is common to the two diagrams, the remaining portions of the figures can be combined into the single figure 5-1d. This combined diagram represents the 6-j symbol

$$\begin{Bmatrix} L & S & J \\ s_2 & K & s_1 \end{Bmatrix}, \tag{5.43}$$

as can be inferred from the fact that the four nodes of the diagram correspond to the four triangle conditions (5.26); also, the symmetry properties of the 6-j symbol may be seen in the topological equivalence of the diagram under various distortions (Fig. 5-2).

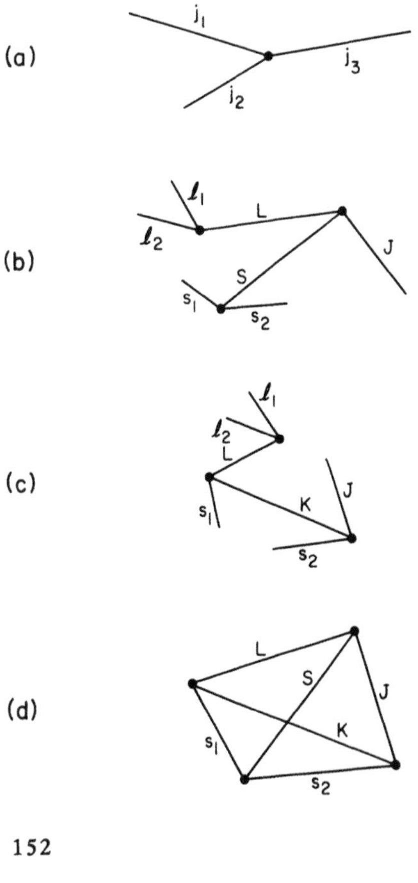

Fig. 5-1. Graphical representations of (a) a 3-j symbol representing the Clebsch-Gordon coupling of two angular-momentum vectors to give a resultant third vector, (b) LS coupling of two electrons, (c) LK coupling, and (d) a 6-j symbol associated with the relation between LS and LK coupling.

$$\begin{Bmatrix} L & S & J \\ s_2 & K & s_1 \end{Bmatrix}$$

$$\begin{Bmatrix} s_2 & K & J \\ L & S & s_1 \end{Bmatrix}$$

$$\begin{Bmatrix} J & K & s_2 \\ s_1 & S & L \end{Bmatrix}$$

$$\begin{Bmatrix} K & J & s_2 \\ S & s_1 & L \end{Bmatrix}$$

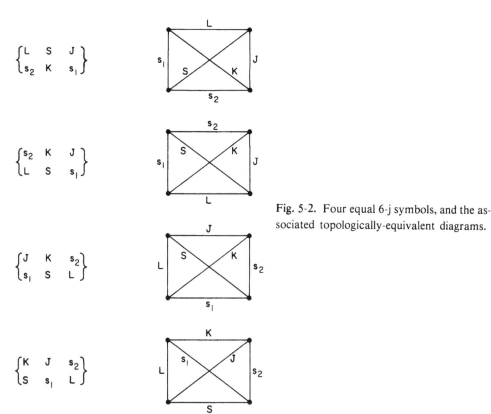

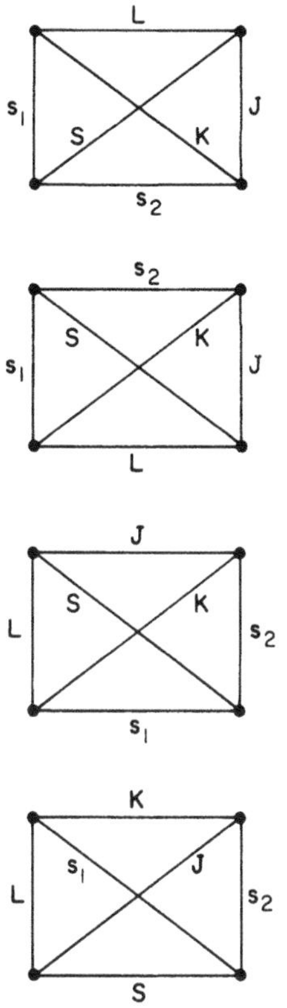

Fig. 5-2. Four equal 6-j symbols, and the associated topologically-equivalent diagrams.

Inasmuch as Fig. 5-1d represents some sort of relationship between the LS- and LK-coupling diagrams, we may suspect that the 6-j symbol (5.43) is closely related to the transformation between LS- and LK-coupled basis functions. Since this transformation must be an orthogonal one (if each basis set consists of orthonormal functions), then from the orthonormality relation (5.31) for the 6-j symbol we may infer that the transformation-matrix elements are given by

$$T_{LS,LK} = (-1)^m [S,K]^{1/2} \begin{Bmatrix} L & J & S \\ s_2 & s_1 & K \end{Bmatrix}, \tag{5.44}$$

to within an unknown phase factor $(-1)^m$.

Similarly, we show in Fig. 5-3 diagrams for the LS and the jj coupling of two electrons, both separately and combined into a single diagram; the latter represents (to within a phase factor) the 9-j symbol

$$\begin{Bmatrix} l_1 & l_2 & L \\ s_1 & s_2 & S \\ j_1 & j_2 & J \end{Bmatrix},$$

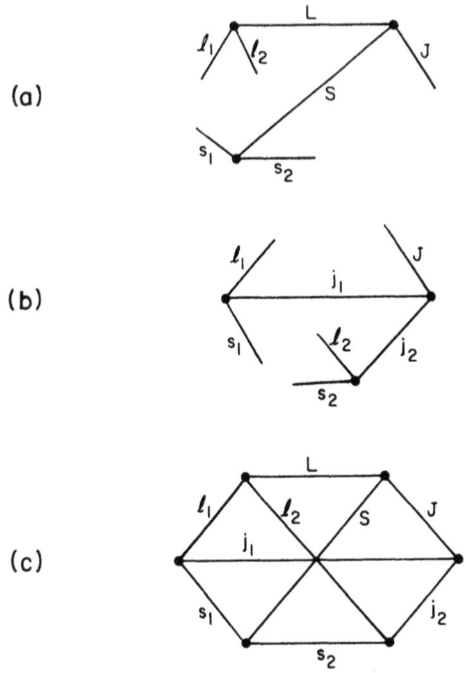

Fig. 5-3. Graphical representations of (a) the LS and (b) the jj coupling of two electrons, and (c) the 9-j symbol associated with the relation between these two couplings.

because the diagram's six nodes correspond to the six triangle conditions $\delta(l_1 l_2 L)$, $\delta(l_1 s_1 j_1)$, etc., and its topological properties correspond to the symmetry properties of the 9-j symbol. From the orthonormality properties (5.40)-(5.41) of the 9-j symbol, we thus infer that the matrix elements of the LS-jj transformation are equal to

$$T_{LS,jj} = [L,S,j_1,j_2]^{1/2} \begin{Bmatrix} l_1 & l_2 & L \\ s_1 & s_2 & S \\ j_1 & j_2 & J \end{Bmatrix}, \tag{5.45}$$

at least to within a phase factor. The conclusions (5.44) and (5.45) will be verified algebraically in Sec. 9-3.

The diagrammatic methods that we have outlined very briefly can be extended to include indications of phase factors and multiplicative constants, and can be very useful in the simplification of complicated angular-momentum and angular-matrix-element relationships. We shall not go into the details here, as we shall be able to algebraically derive all necessary results directly in the simplest possible form; those interested in further discussion of graphical methods are referred to several available papers and monographs.[15-21] Bordarier[19] has described not only the basic principles, but also computer methods of representing and simplifying the graphs. Angular-momentum graphs are closely related to Feynman diagrams.[22]

Problems

5-4(1). By the graphical method, obtain the magnitudes of the transformation matrix elements from LK to jK coupling and from jK to jj coupling.

5-4(2). Check the above results algebraically by means of the expansion (5.37) for the 9-j symbol involved in the LS-jj transformation matrix element (5.45). [Hint: Note that the LS-jj transformation matrix is the matrix product of the LS-LK, LK-jK, and jK-jj transformation matrices. The resulting double summation of a product of the three matrix elements reduces to a single sum over the intermediate quantum number K.]

[17]B. R. Judd, *Operator Techniques in Atomic Spectroscopy* (McGraw-Hill, New York, 1963).

[18]J.-N. Massot, E. El-Baz, and J. Lafoucrière, Rev. Mod. Phys. **39**, 288 (1967).

[19]Y. Bordarier, "Contribution à l'emploi de méthodes graphiques en spectroscopie atomique" (Thesis, Univ. of Paris, Orsay, 1970).

[20]E. El-Baz and B. Castel, Am. J. Phys. **39**, 868 (1971).

[21]J. S. Briggs, Rev. Mod. Phys. **43**, 189 (1971).

[22]See, for example, P. G. H. Sandars, Adv. Chem. Phys. **XIV**, 365 (1969); B. R. Judd, *Second Quantization and Atomic Spectroscopy* (The Johns Hopkins Press, Baltimore, 1967).

6

CONFIGURATION-AVERAGE ENERGIES

6-1. DIAGONAL MATRIX ELEMENTS OF SYMMETRIC OPERATORS

As discussed in Sec. 4-11, the energy of the spherically averaged atom is given by

$$E_{av} = \frac{\sum_b \langle b|H|b\rangle}{\text{number of basis functions}} , \tag{6.1}$$

where the sum is over all basis functions belonging to the electron configuration of interest. The Hamiltonian operator in the approximation that we shall use for the present is given by (4.1):

$$H = -\sum_i \nabla_i^2 - \sum_i \frac{2Z}{r_i} + \sum_i \xi_i(r_i)(\boldsymbol{l}_i \cdot \mathbf{s}_i) + \sum_{i>j}\sum \frac{2}{r_{ij}} , \tag{6.2}$$

where distances are measured in Bohr units and energies in rydbergs. The first three terms of H are one-electron operators of the form

$$\sum_{i=1}^{N} f_i \equiv \sum_i f(\mathbf{r}_i) \tag{6.3}$$

symmetric in the spatial-plus-spin coordinates $\mathbf{r}_i$ of all N electrons, and the last term is a two-electron operator of the form

$$\sum_{i=2}^{N}\sum_{j=1}^{i-1} g_{ij} \equiv \sum_{i>j}\sum g(\mathbf{r}_i,\mathbf{r}_j) , \tag{6.4}$$

symmetric in all N(N − 1)/2 pairs of coordinates. Because of the indistinguishability of electrons, all physically significant operators must be symmetric in the above fashion,[1] and

[1]The spin-other-orbit operator $\boldsymbol{l}_i \cdot \mathbf{s}_j$ is not inherently symmetric, but can be written in the symmetrized form $g_{ij} = (\boldsymbol{l}_i \cdot \mathbf{s}_j + \boldsymbol{l}_j \cdot \mathbf{s}_i)/2$.

we first consider the calculation of diagonal matrix elements of the general operators (6.3) and (6.4), for uncoupled basis functions (which are the simplest type to use for the evaluation of E_{av}).

Product functions. For a product function (4.21), the diagonal matrix element of the one-electron symmetric operator is easily simplified with the aid of the orthonormality property (4.20) of the one-electron spin-orbitals φ_i:

$$\langle\psi|\sum_i f_i|\psi\rangle = \sum_i \langle\varphi_1(\mathbf{r}_1)\varphi_2(\mathbf{r}_2)\cdots|f_i|\varphi_1(\mathbf{r}_1)\varphi_2(\mathbf{r}_2)\cdots\rangle$$

$$= \sum_i \langle\varphi_i(\mathbf{r}_i)|f_i|\varphi_i(\mathbf{r}_i)\rangle \equiv \sum_i \langle\varphi_i|f|\varphi_i\rangle \equiv \sum_i \langle i|f|i\rangle . \tag{6.5}$$

In the last two steps the notation has been condensed by first deleting the integration variable as unnecessary, and then by using simply $|i\rangle$ to represent a spin-orbital φ_i with quantum numbers $n_i l_i m_{l_i} m_{s_i}$. Diagonal matrix elements of the two-electron operator simplify similarly:

$$\langle\psi|\sum_{i>j}\sum g_{ij}|\psi\rangle = \sum_{i>j}\sum \langle\varphi_i(\mathbf{r}_i)\varphi_j(\mathbf{r}_j)|g_{ij}|\varphi_i(\mathbf{r}_i)\varphi_j(\mathbf{r}_j)\rangle$$

$$= \sum_{i>j}\sum \langle ij|g|ij\rangle . \tag{6.6}$$

It should be noted that in the operators (6.3) and (6.4) the summations are over electron coordinates; however, in the matrix elements (6.5) and (6.6) the coordinates have disappeared upon integration, and the summations are over the various sets of one-electron quantum numbers involved in the product function.

Determinantal functions. Simplification of matrix elements for determinantal functions is somewhat more complicated than for product functions. We first prove that because the equivalence of electron coordinates is now reflected in the wavefunction as well as in the operator, the value of the matrix element of f_i is the same as that of f_j: In the matrix element

$$\langle\Psi|f_i|\Psi\rangle ,$$

the coordinates appear only as integration variables, and changing the names of these variables does not affect the value of the integral. Thus if we change the name of the integration variable $\mathbf{r}_i$ to $\mathbf{r}_j$ and simultaneously change the name of the variable $\mathbf{r}_j$ to $\mathbf{r}_i$, we obtain

$$\langle\Psi|f_i|\Psi\rangle = \langle\Psi'|f_j|\Psi'\rangle ,$$

where Ψ' is the same as Ψ except that the names of the variables $\mathbf{r}_i$ and $\mathbf{r}_j$ have been interchanged. But because of the antisymmetry of Ψ, Ψ' is just the negative of Ψ. It follows that

$$\langle\Psi|f_i|\Psi\rangle = \langle\Psi|f_j|\Psi\rangle , \tag{6.7}$$

and therefore that

$$\langle\Psi|\sum_i f_i|\Psi\rangle = N\langle\Psi|f_1|\Psi\rangle . \tag{6.8}$$

In order to simplify this result further, we rewrite the determinantal function (4.26) by arranging the spin-orbitals in each term in the order of the electron-coordinate subscripts instead of the quantum-number subscripts:

$$\Psi = (N!)^{-1/2}\sum_P(-1)^p\varphi_{k_1}(\mathbf{r}_1)\varphi_{k_2}(\mathbf{r}_2)\cdots\varphi_{k_N}(\mathbf{r}_N) ; \tag{6.9}$$

the summation is then over all N! permutations of the quantum-number subscripts. Substituting this into (6.8) and factoring out the summations, we obtain

$$\langle\Psi|\sum_i f_i|\Psi\rangle = \frac{1}{(N-1)!}\sum_P\sum_{P'}(-1)^{p+p'}\langle\varphi_{k_1}(1)\varphi_{k_2}(2)\cdots|f_1|\varphi_{k'_1}(1)\varphi_{k'_2}(2)\cdots\rangle$$

Let us consider a specific permutation P; then because of the orthogonality of the φ_i, the matrix element on the right is zero except for permutations P′ for which

$$k'_n = k_n , \qquad n \geqq 2 .$$

But because of the Pauli principle (4.28), each set of one-electron quantum numbers k_n is different from every other set, and therefore the only such permutation P′ is that which is identical with P—for which we additionally have $k'_1 = k_1$, so that

$$\langle\Psi|\sum_i f_i|\Psi\rangle = \frac{1}{(N-1)!}\sum_P(-1)^{2p}\langle\varphi_{k_1}(1)|f_1|\varphi_{k_1}(1)\rangle = \frac{1}{(N-1)!}\sum_P\langle k_1|f|k_1\rangle .$$

Now for any given value of k_1, there are in the P summation (N − 1)! different terms, each of which has of course the same value of $\langle k_1|f|k_1\rangle$. Thus we obtain finally

$$\langle\Psi|\sum_i f_i|\Psi\rangle = \sum_{k_1}\langle k_1|f|k_1\rangle = \sum_i\langle i|f|i\rangle ; \tag{6.10}$$

this is exactly the same as the result that we obtained with product functions.

For the two-electron operator (6.4), each term g_{ij} gives the same contribution to a matrix element as does any other term g_{kl}; this can be proved in the same manner as used for (6.7) except that two coordinate interchanges $i \leftrightarrow k$ and $j \leftrightarrow l$ are now involved. The double summation over $i > j$ contains N(N − 1)/2 terms, and therefore

$$\langle\Psi|\sum_{i>j}\sum g_{ij}|\Psi\rangle = \frac{1}{2(N-2)!}\sum_P\sum_{P'}(-1)^{p+p'}\langle\varphi_{k_1}(1)\varphi_{k_2}(2)\cdots|g_{12}|\varphi_{k'_1}(1)\varphi_{k'_2}(2)\cdots\rangle .$$

For a given permutation P, the only permutations P′ for which the matrix element on the right is non-zero are those for which

$$k_n' = k_n \,, \qquad n \geqq 3 \,.$$

There are two such permutations: (1) that with $k_1' = k_1$ and $k_2' = k_2$ (for which $p' = p$), and (2) that with $k_1' = k_2$ and $k_2' = k_1$ (for which $p' = p \pm 1$). Thus

$$\langle\Psi|\sum_{i>j}\sum g_{ij}|\Psi\rangle = \frac{1}{2(N-2)!}\sum_{P}[\langle k_1k_2|g|k_1k_2\rangle - \langle k_1k_2|g|k_2k_1\rangle] \,,$$

where this abbreviated notation is understood to imply that in both the bra and the ket function, the spin-orbital written first is a function of $\mathbf{r}_1$ and that written second is a function of $\mathbf{r}_2$. Now for any given values of k_1 and k_2, there are $(N-2)!$ different permutations of the other $N-2$ quantum-number sets k_n, $n > 2$. Also, there are two permutations of k_1 and k_2; because of the symmetry $g_{ij} = g_{ji}$ of the two-electron operator,[1] it follows that

$$\langle j_1(1)j_2(2)|g_{12}|j_1(1)j_2(2)\rangle = \langle j_1(1)j_2(2)|g_{21}|j_1(1)j_2(2)\rangle = \langle j_2(1)j_1(2)|g_{12}|j_2(1)j_1(2)\rangle$$

(where the last step follows from simply renaming the two integration variables and then interchanging the order of the two factors in each product function), and similarly that

$$\langle j_1j_2|g|j_2j_1\rangle = \langle j_2j_1|g|j_1j_2\rangle \,.$$

Thus we obtain finally

$$\langle\Psi|\sum_{i>j}\sum g_{ij}|\Psi\rangle = \sum_{i>j}\sum[\langle ij|g|ij\rangle - \langle ij|g|ji\rangle] \,. \tag{6.11}$$

Comparing this result with (6.6), we see that for every "direct" term which we obtained using product functions, we have here obtained an additional "exchange" term in which the spin-orbitals of the two-electron ket function have been interchanged. These exchange terms are thus a consequence of having used an antisymmetrized wavefunction to reflect the quantum-mechanical indistinguishability of electrons. For the case of the electron-electron Coulomb interaction, $g_{ij} = 2/r_{ij}$, we shall see that both the direct and exchange matrix elements are positive (when averaged over all basis functions of a configuration); the physical interpretation is therefore as follows: The direct terms represent the positive energy of mutual electrostatic repulsion for an uncorrelated spatial distribution of the electrons; the exchange terms represent the decrease in this energy that results when one includes the positional correlation of parallel-spin electrons which we discussed earlier for determinantal functions (Sec. 4-5).

6-2. ONE-ELECTRON AND TOTAL-ATOM BINDING ENERGIES

With the aid of the above results for determinantal basis functions, we now derive expressions for the center-of-gravity energy (6.1) of atomic states corresponding to an arbitrary electron configuration.

For the spin-orbit term of the Hamiltonian (6.2), the diagonal matrix element is

$$\langle\Psi|\sum_i \xi_i(l_i\cdot s_i)|\Psi\rangle = \sum_i \langle n_i l_i m_{l_i} m_{s_i}|\xi(l\cdot s)|n_i l_i m_{l_i} m_{s_i}\rangle ,$$

where it is to be remembered that calculation of the matrix element involves summation over both possible values of s_z, as well as integration over the spatial coordinates. When this expression is averaged over all basis functions, there will—for any specific value of $n_i l_i m_{l_i}$—be one-electron matrix elements with m_{s_i} equal to both $+1/2$ and $-1/2$. Because the spin appears in the operator in the form of a scalar product with l and because the value of $l_i m_{l_i}$ is the same in both matrix elements, it is clear that these two matrix elements will be equal in magnitude but opposite in sign; thus the spin-orbit contribution to E_{av} is zero.

There remains, then,

$$E_{av} = \sum_i \langle i|-\nabla^2|i\rangle_{av} + \sum_i \langle i|-2Z/r_1|i\rangle_{av} + \sum\sum_{i>j} [\langle ij|2/r_{12}|ij\rangle_{av} - \langle ij|2/r_{12}|ji\rangle_{av}] . \tag{6.12}$$

This expression is of such a form that we can think of it in essentially classical terms: The configuration-average binding energy of an electron in an orbital $n_i l_i$ is[2]

$$E^i = E_k^{\ i} + E_n^{\ i} + \sum_{j\neq i} E^{ij} , \tag{6.13}$$

and the configuration-average total binding energy[2] of all N electrons may be written in any of the equivalent forms

$$E_{av} = \sum_i E_k^{\ i} + \sum_i E_n^{\ i} + \sum\sum_{i>j} E^{ij} \tag{6.14}$$

$$= \sum_i \left(E_k^{\ i} + E_n^{\ i} + \frac{1}{2}\sum_{j\neq i} E^{ij}\right) \tag{6.15}$$

$$= \sum_i \left(E^i - \frac{1}{2}\sum_{j\neq i} E^{ij}\right) \tag{6.16}$$

$$= \frac{1}{2}\sum_i (E_k^{\ i} + E_n^{\ i} + E^i) . \tag{6.17}$$

[2]The quantities E^i and E_{av} are negative for bound states, and the term "binding energy" should more properly be applied to $-E^i$ and $-E_{av}$.

That is, the average one-electron binding energy of an electron in orbital i is equal to its kinetic energy and potential energy of interaction with the nucleus, plus its (spherically averaged) energy of interaction with the $N - 1$ other electrons in the atom; the average total binding energy of the atom is the sum of all kinetic and electron-nuclear energies, plus the (averaged) electron-electron Coulomb interactions summed over all electron pairs. Note that E_{av} is *not* equal to $\sum E^i$, because the latter quantity counts all electron pairs twice.

We now evaluate each of the quantities $E_k{}^i$, $E_n{}^i$, and E^{ij} in turn.

Kinetic energy. Just as in the one-electron problem (Secs. 3-1 and 3-2), we may write

$$-\nabla^2\varphi = -\frac{1}{r}\frac{\partial^2 r\varphi}{\partial r^2} + \frac{1}{r^2}\mathbf{L}^2\varphi = -\frac{1}{r}\frac{\partial^2 r\varphi}{\partial r^2} + \frac{l(l+1)}{r^2}\varphi ,$$

because the spin-orbital φ defined by (4.17) is an eigenfunction of $\mathbf{L}^2$ with eigenvalue $l(l + 1)$. Thus

$$\langle i|-\nabla^2|i\rangle$$

$$= \sum_{s_{iz}} \int_0^\infty \int_0^{2\pi} \int_0^\pi \frac{1}{r} P_i{}^* Y_i{}^* \sigma_i \left\{ \frac{1}{r}\left[-\frac{\partial^2}{\partial r^2} + \frac{l_i(l_i+1)}{r^2} \right] P_i Y_i \sigma_i \right\} \sin\theta \, d\theta \, d\phi \, r^2 dr .$$

The spin summation and angular integrations can be carried out immediately, being simply the normalization integral (2.57). The result is a radial integral that involves only the quantum numbers $n_i l_i$; there is thus no necessity for averaging over $m_{l_i} m_{s_i}$, and we obtain

$$E_k{}^i \equiv \langle i|-\nabla^2|i\rangle_{av} = \int_0^\infty P^*_{n_i l_i}(r)\left[-\frac{d^2}{dr^2} + \frac{l_i(l_i+1)}{r^2} \right] P_{n_i l_i}(r)\, dr . \qquad (6.18)$$

If the radial function $P_i(r)$ is known, this integral can readily be evaluated by numerical quadrature. In practice, however, the necessity for this can be avoided by obtaining the value of $E_k{}^i$ from the eigenvalue of the differential equation for P_i, in a manner to be described in Sec. 8-1.

Electron-nuclear energy. Since the pertinent Hamiltonian operator is a function of r only, the angular integration and summation over spins can be carried out immediately and there is no necessity for averaging over $m_{l_i} m_{s_i}$; we obtain

$$E_n{}^i \equiv \langle i|-2Z/r|i\rangle = \int_0^\infty (-2Z/r)|P_i(r)|^2 dr . \qquad (6.19)$$

Electron-electron Coulomb energy. Evaluation of the matrix elements of the electron-electron electrostatic interaction is a somewhat lengthy process. The first step is to make a multipole expansion of $2/r_{12}$ by applying the cosine law to the triangle formed by the three vectors $\mathbf{r}_1$, $\mathbf{r}_2$, $\mathbf{r}_{12}$ (Fig. 2-4):

$$2/r_{12} = 2[r_1^2 + r_2^2 - 2r_1r_2\cos\omega]^{-1/2}$$
$$= (2/r_>)[1 + (r_</r_>)^2 - 2(r_</r_>)\cos\omega]^{-1/2}$$
$$= \sum_{k=0}^{\infty} \frac{2r_<^k}{r_>^{k+1}} P_k(\cos\omega) \ . \tag{6.20}$$

Here $r_<$ and $r_>$ are respectively the lesser and greater of the distances r_1 and r_2 of the electrons from the nucleus, and ω is the angle between the two radius vectors. The final step in (6.20) is accomplished by expanding the quantity in brackets via the binomial theorem, collecting terms in like powers of $(r_</r_>)$, and showing that the coefficient of the kth power is just the kth Legendre polynomial.[3] Using the spherical-harmonic addition theorem (2.49) and the abbreviation (5.18), we may expand further to

$$\frac{2}{r_{12}} = \sum_{k=0}^{\infty} \frac{2r_<^k}{r_>^{k+1}} \sum_{q=-k}^{k} (-1)^q C_{-q}^{(k)}(\theta_1,\phi_1) C_q^{(k)}(\theta_2,\phi_2) \ , \tag{6.21}$$

thus obtaining an expression that not only effects a separation of radial from angular factors but also a separation of the angular variables of the individual electrons.

With the aid of (5.19), we can readily calculate the matrix element of (6.21) between two two-electron product functions, the result being

$$\langle ij|2/r_{12}|tu\rangle = \delta_{m_{s_i},m_{s_t}} \delta_{m_{s_j},m_{s_u}} \sum_{k=0}^{\infty} R^k(ij,tu)$$
$$\times \sum_{q=-k}^{k} \delta_{q,m_{l_t}-m_{l_i}} \delta_{q,m_{l_j}-m_{l_u}} (-1)^q c^k(l_i m_{l_i}, l_t m_{l_t}) c^k(l_j m_{l_j}, l_u m_{l_u}) \ , \tag{6.22}$$

where the first two δ-factors arise from summations over the spin variables [cf. (2.54)], and where

$$R^k(ij,tu) \equiv \int_0^\infty \int_0^\infty \frac{2r_<^k}{r_>^{k+1}} P_i^*(r_1) P_j^*(r_2) P_t(r_1) P_u(r_2)\, dr_1\, dr_2$$
$$= \int_0^\infty \left\{ \frac{2}{r_2^{k+1}} \int_0^{r_2} r_1^k P_i^* P_t\, dr_1 + r_2^k \int_{r_2}^\infty \frac{2}{r_1^{k+1}} P_i^* P_t\, dr_1 \right\} P_j^* P_u\, dr_2 \ . \tag{6.23}$$

In (6.23), the abbreviations ijtu refer only to the corresponding quantum numbers nl (not $m_l m_s$) pertinent to the radial factors P_{nl} of the spin-orbitals.

Note that the matrix elements (6.22) are zero unless

$$q = m_{l_t} - m_{l_i} = m_{l_j} - m_{l_u} \ . \tag{6.24}$$

[3]See, for example, H. Margenau and G. M. Murphy, *The Mathematics of Physics and Chemistry* (D. Van Nostrand, Princeton, N.J., 1956), 2nd ed., Sec. 3.3.

Mathematically, this means that there is at most one non-zero term in the summation over q. Rewriting (6.24) in the form

$$m_{l_i} + m_{l_j} = m_{l_t} + m_{l_u} , \tag{6.25}$$

we see that this is just a reflection of the law of conservation of angular momentum: the electrostatic interaction between two electrons cannot change the total orbital angular momentum of the two electrons, nor the z-component thereof. The situation for the spins is even more restricted, the δ-factors giving

$$m_{s_i} = m_{s_t} , \qquad m_{s_j} = m_{s_u} ; \tag{6.26}$$

since the electrostatic interaction does not operate on the electron spins, not only is the *total* spin angular momentum conserved, but so also is the spin of each electron separately.

The electron-electron matrix elements involved in the expression (6.12) for E_{av} follow immediately as special cases of (6.22). For the direct contribution we obtain

$$\langle ij|2/r_{12}|ij\rangle = \sum_{k=0}^{\infty} F^k(ij)\, c^k(l_i m_{l_i}, l_i m_{l_i})\, c^k(l_j m_{l_j}, l_j m_{l_j}) , \tag{6.27}$$

where

$$F^k(ij) \equiv R^k(ij,ij) = \int_0^\infty \int_0^\infty \frac{2r_<^k}{r_>^{k+1}} |P_i(r_1)|^2 |P_j(r_2)|^2 \, dr_1\, dr_2 ; \tag{6.28}$$

the exchange contribution simplifies with the aid of (5.21) to

$$-\langle ij|2/r_{12}|ji\rangle = -\delta_{m_{s_i} m_{s_j}} \sum_{k=0}^{\infty} G^k(ij)\, [c^k(l_i m_{l_i}, l_j m_{l_j})]^2 , \tag{6.29}$$

where

$$G^k(ij) \equiv R^k(ij,ji) = \int_0^\infty \int_0^\infty \frac{2r_<^k}{r_>^{k+1}} P_i^*(r_1) P_j^*(r_2) P_j(r_1) P_i(r_2) \, dr_1\, dr_2 . \tag{6.30}$$

The radial integrals F^k and G^k (or more generally, R^k) are frequently referred to as Slater integrals.[4] Since the integrand of (6.28) is everywhere positive and for every (r_1,r_2) is smaller the larger the value of k, it follows that

$$F^0 > F^1 > F^2 > \cdots > 0 . \tag{6.31}$$

[4]J. C. Slater, Phys. Rev. 34, 1293 (1929).

Although it is not immediately obvious from the definition (6.30), it can be proved[5] that G^k is also positive, and that

$$\frac{G^k}{[k]} > \frac{G^{k+1}}{[k+1]} > 0 \; ; \qquad (6.32)$$

for realistic forms of the radial functions P(r), the exchange integrals usually (though not always) satisfy the more restricted inequalities

$$G^0 > G^1 > G^2 > \cdots > 0 \, . \qquad (6.33)$$

From (5.20), $c^k(lm,l'm')$ is zero unless the triangle relation $\delta(lkl')$ is satisfied and $l + k + l'$ is even; thus the summations in (6.27) and (6.29) involve only a small number of non-zero terms, and the only radial integrals required are

$$F^k(ij), \qquad k = 0, 2, 4, \cdots, \min(2l_i, 2l_j) \qquad (6.34)$$

and

$$G^k(ij), \qquad k = |l_i - l_j|, \; |l_i - l_j|+2, \; |l_i - l_j|+4, \cdots, l_i + l_j \, . \qquad (6.35)$$

There still remains the problem of averaging over all permissible values of the four magnetic quantum numbers; to accomplish this we first sum over all permitted pairs of values of the two quantum numbers $m_{l_j}m_{s_j}$, and divide by the number of such pairs. If i and j are non-equivalent electrons ($n_il_i \neq n_jl_j$), then there are $2[l_j]$ permitted pairs—$[l_j]$ different values of m_{l_j} for each of the two values of m_{s_j}. For equivalent electrons ($n_il_i = n_jl_j$), the case $m_{l_j}m_{s_j} = m_{l_i}m_{s_i}$ is not allowed by the Pauli exclusion principle. We can nevertheless still formally sum over all $2[l_j]$ pairs, because when i = j the direct and exchange contributions cancel $[\langle ij|2/r_{12}|ij\rangle - \langle ij|2/r_{12}|ji\rangle = 0]$; however, we must divide by the number $4l_j + 1$ of allowed pairs rather than by $4l_j + 2$.

Using (5.19), (5.16), (5.11), and (5.22), we easily find for the sum of the angular integrals in (6.27)

$$c^k(i,i) \sum_{m_{l_j}m_{s_j}} c^k(j,j) = c^k(i,i)2[l_j]\begin{pmatrix} l_j & k & l_j \\ 0 & 0 & 0 \end{pmatrix} \sum_{m_{l_j}} (-1)^{-m_{l_j}} \begin{pmatrix} l_j & l_j & k \\ m_{l_j} & -m_{l_j} & 0 \end{pmatrix}$$
$$= c^k(i,i)2[l_j]^{3/2}\begin{pmatrix} l_j & k & l_j \\ 0 & 0 & 0 \end{pmatrix} \delta_{k0}(-1)^{-l_j} = 2[l_j]\,\delta_{k0} \, . \qquad (6.36)$$

In summing the exchange term (6.29) over m_{s_j} we will not get a factor 2 as we did in the direct term, because of the δ-factor; using (5.19) and (5.15), we obtain

[5]G. Racah II, Phys. Rev. 62, 460 (1942).

$$-\sum_{m_{l_j}m_{s_j}} \delta_{m_{s_i}m_{s_j}}[c^k(i,j)]^2 = -[l_j]\begin{pmatrix} l_i & k & l_j \\ 0 & 0 & 0 \end{pmatrix}^2 \sum_{m_{l_j}}[l_i]\begin{pmatrix} k & l_j & l_i \\ m_{l_i}-m_{l_j} & m_{l_j} & -m_{l_i} \end{pmatrix}^2$$
$$= -[l_j]\begin{pmatrix} l_i & k & l_j \\ 0 & 0 & 0 \end{pmatrix}^2 . \qquad (6.37)$$

Both (6.36) and (6.37) are independent of $m_{l_i}m_{s_i}$ and so there is no need to average over these quantum numbers. Dividing by $2[l_j]$ we find from (6.27) and (6.29) that for non-equivalent electrons

$$E^{ij} \equiv \langle ij|2/r_{12}|ij\rangle_{av} - \langle ij|2/r_{12}|ji\rangle_{av}$$
$$= F^0(ij) - \frac{1}{2}\sum_k \begin{pmatrix} l_i & k & l_j \\ 0 & 0 & 0 \end{pmatrix}^2 G^k(ij) . \qquad (6.38)$$

For equivalent electrons, we divide by $4l_j + 1$ $(= 4l_i + 1)$, note from (6.28) and (6.30) that $G^k(ii) \equiv F^k(ii)$, and obtain with the aid of (5.11)

$$E^{ii} = \frac{1}{4l_i+1}\left[2[l_i]F^0(ii) - F^0(ii) - [l_i]\sum_{k>0}\begin{pmatrix} l_i & k & l_i \\ 0 & 0 & 0 \end{pmatrix}^2 F^k(ii)\right]$$
$$= F^0(ii) - \frac{2l_i+1}{4l_i+1}\sum_{k>0}\begin{pmatrix} l_i & k & l_i \\ 0 & 0 & 0 \end{pmatrix}^2 F^k(ii) . \qquad (6.39)$$

Use of values of the 3-j symbols from Appendix C leads to the specific expressions for E^{ij} given in Table 6-1 and the more extensive table in Appendix E.

TABLE 6-1. AVERAGE COULOMB ENERGY OF ELECTRON PAIRS

Equivalent Electrons	*Non-equivalent Electrons*
$E^{ss} = F^0$	$E^{ss'} = F^0 - \frac{1}{2}G^0$
$E^{pp} = F^0 - \frac{2}{25}F^2$	$E^{sp} = F^0 - \frac{1}{6}G^1$
$E^{dd} = F^0 - \frac{2}{63}F^2 - \frac{2}{63}F^4$	$E^{sd} = F^0 - \frac{1}{10}G^2$
	$E^{pp'} = F^0 - \frac{1}{6}G^0 - \frac{1}{15}G^2$
	$E^{pd} = F^0 - \frac{1}{15}G^1 - \frac{3}{70}G^3$
	$E^{dd'} = F^0 - \frac{1}{10}G^0 - \frac{1}{35}G^2 - \frac{1}{35}G^4$

If we had used product functions rather than antisymmetrized (determinantal) functions, we would from (6.6) have obtained only the direct matrix elements and would have found for either equivalent or non-equivalent electrons that

$$E^{ij} = \langle ij | 2/r_{12} | ij \rangle_{av} = F^0(ij) \; ; \tag{6.40}$$

this is positive, as is to be expected classically for the Coulomb interaction energy of two negative charges. Since the F^k and G^k are all positive, the exchange contribution $-\langle ij|2/r_{12}|ji\rangle_{av}$ to E_{ij} is always negative, and represents the energy decrease arising from the partial positional correlation of electrons that we anticipated in Sec. 6-1.

Problems

6-2(1). Using the semiclassical expression $dq = |P(r)|^2 dr$, Eq. (3.31), for the fractional electronic charge lying in a spherical shell of radius r and thickness dr, show that the direct part $F^0(ij)$ of E^{ij} is just the classical electrostatic potential energy of two concentric spherically symmetric charge distributions q_i and q_j.

6-2(2). The term F^0 arose from the monopole term of (6.20). Why do the higher-order multipoles all give zero contribution to the direct part of E^{ij}? Why do some of them give non-zero contributions to the exchange part of E^{ij}?

Notation. The coefficients of the Slater integrals F^k and G^k in (6.27), (6.29) are rational fractions. In order to be able to write these expressions in a form involving integral coefficients, one sometimes uses subscripted quantities defined by

$$F_k(ij) = F^k(ij)/D_k \; , \qquad G_k(ij) = G^k(ij)/D_k \; , \tag{6.41}$$

where the constants D_k are the least common denominators (for different $m_{l_i} m_{l_j}$) of the coefficients of the corresponding F^k, G^k. However, use of these subscripted quantities does not in general lead to integral coefficients in expressions for E_{av} nor for off-diagonal matrix elements; moreover, integral values are of no particular advantage when using digital computers to handle complex cases. In this book we shall therefore use only the theoretically more fundamental superscripted quantities. Values[6] of the D_k are given in Tables 6-2 and 6-3 for convenience of intercomparison of the two types of quantities, both of which are used in the literature.

For the interaction of two equivalent electrons within a subshell l^w, various linear combinations of the F^k (or F_k) are sometimes used in order to simplify the appearance of energy-matrix-element expressions. For d^w, Racah[7] defined the quantities

$$\begin{aligned} A &= F_0 - 49F_4 = F^0 - F^4/9 \; , \\ B &= F_2 - 5F_4 = (9F^2 - 5F^4)/441 \; , \\ C &= 35F_4 = 5F^4/63 \; ; \end{aligned} \tag{6.42}$$

[6]See refs. 10-12 of Chap. 5.

[7]G. Racah II, Phys. Rev. 62, 438 (1942).

TABLE 6-2. VALUES OF D_k FOR $F_k = F^k/D_k$
($D_0 = 1$ in all cases)

$l_i l_j$	k = 2	k = 4	k = 6
pp	25		
pd	35		
pf	75		
dd	49	441	
df	105	693	
ff	225	1089	7361.64

TABLE 6-3. VALUES OF D_k FOR $G_k = G^k/D_k$
($D_0 = 1$ in all cases)

$l_i l_j$	k = 1	k = 3	k = 5	$l_i l_j$	k = 2	k = 4	k = 6
sp	3			sd	5		
sf		7		pp	25		
pd	15	245[a]		pf	175[b]	189	
df	35	315	1524.6	dd	49	441	
				ff	225	1089	7361.64

[a]The value 81 2/3 is also used.
[b]The value 58 1/3 is also used.

and for f^w he defined

$$A = F_0 - 21F_4 - 468F_6 , \qquad B = (5F_2 + 6F_4 - 91F_6)/5 ,$$
$$C = 7(F_4 - 6F_6)/5 , \qquad D = 462F_6 . \tag{6.43}$$

More commonly used for f^w are the linear combinations[8]

$$\begin{aligned} E^0 &= F_0 - 10F_2 - 33F_4 - 286F_6 , \\ E^1 &= (70F_2 + 231F_4 + 2002F_6)/9 , \\ E^2 &= (F_2 - 3F_4 + 7F_6)/9 , \\ E^3 &= (5F_2 + 6F_4 - 91F_6)/3 ; \end{aligned} \tag{6.44}$$

these expressions can be inverted to give equations for calculating F^k from values of E^k tabulated in the literature:

[8]G. Racah IV, Phys. Rev. 76, 1352 (1949); arrived at by consideration of the group-theoretical properties of LS-coupled basis functions.

$$
\begin{aligned}
F^0 &= (7E^0 + 9E^1)/7 \ , \\
F^2 &= 225(E^1 + 143E^2 + 11E^3)/42 \ , \\
F^4 &= 1089(E^1 - 130E^2 + 4E^3)/77 \ , \\
F^6 &= 7361.64(E^1 + 35E^2 - 7E^3)/462 \ .
\end{aligned}
\tag{6.45}
$$

For calculations using digital computers, there is little advantage in using any of these linear combinations, and in this book we shall use only the F^k.

6-3. IONIZATION ENERGY AND ONE-ELECTRON BINDING ENERGIES

If the radial functions $P_i(r) \equiv P_{n_il_i}(r)$ are known for each subshell of an electron configuration

$$
(n_1l_1)^{w_1}(n_2l_2)^{w_2}\cdots(n_ql_q)^{w_q} \ , \qquad \sum_{i=1}^{q} w_i = N \ , \tag{6.46}
$$

then Eqs. (6.13), (6.17)-(6.19), (6.28), (6.30), and (6.38)-(6.39) provide everything needed to calculate numerically the one-electron (spherically averaged) binding energy E^i of an electron in the subshell $(n_il_i)^{w_i}$, and the total (averaged) energy E_{av} for the atom.

We wish now to consider the subject of ionization energies. As normally defined, the ionization energy of an atom is the energy required to remove an electron from the ground (lowest-energy) state of an atom, leaving a system that consists of an ion in its ground state plus a free electron at infinity with zero kinetic and potential energy. We cannot yet compute such a quantity because we have not yet derived expressions for the energies of individual states—only energies representing configuration centers-of-gravity. However, we can compute (and compare with experiment) what may be called a *configuration-average ionization energy*—the energy required to go from the center of gravity of some configuration of the atom to the center of gravity of some configuration of the ion. The closest analog to the usual definition of ionization energy arises when the configurations involved are the ground configurations of the atom and ion; in this case, the distinction between the standard ionization energy and the configuration-average ionization energy ΔE_{av} is illustrated in Fig. 6-1.

In some cases, ionization (as usually defined) does not correspond simply to the removal of an electron from some orbital n_il_i. For example, the ground configurations of Ni I and Ni II are respectively $3d^84s^2$ and $3d^9$, so that the ionization process consists of removal of a 4s electron[9] (leaving Ni II in the excited configuration $3d^84s$), followed by (or accompanied by) de-excitation of the remaining 4s electron to a 3d orbital. Usually, however, ionization consists of simple removal of an electron from the outermost (least tightly bound) orbital; this is true, for example, in the ionization of C I $1s^22s^22p^2$ to C II $1s^22s^22p$ by removal of a 2p electron, as illustrated in Fig. 6-1.

[9]Because of the indistinguishability of electrons, we should not really speak of "removal of a 4s electron." This phrase is shorthand for the more nearly correct expression "removal of an electron, leaving the ion in a configuration involving one fewer 4s orbital than the configuration of the atom." (Cf. fns. 8 and 10 of Chap. 4).

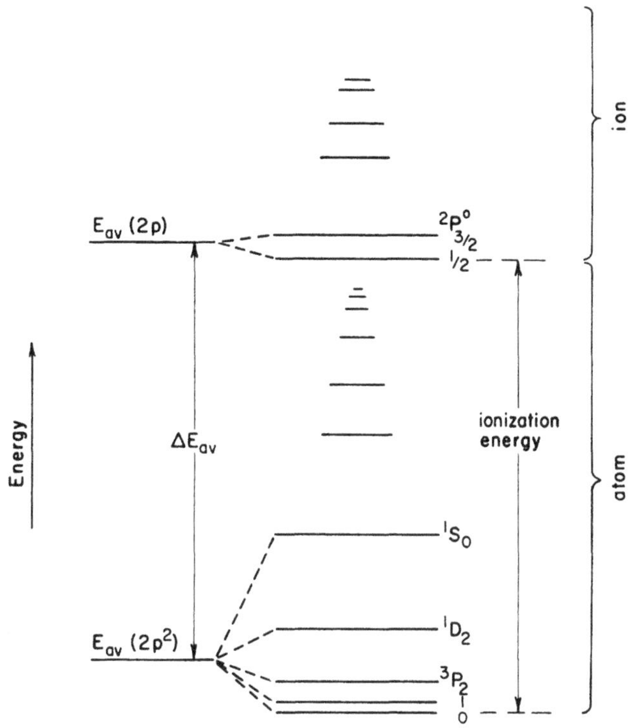

Fig. 6-1. Schematic energy-level diagrams of the neutral and singly ionized carbon atoms, illustrating the difference between the true ionization energy and the configuration-average ionization energy.

Let us consider ionization of this simple type, involving just the removal of an electron from the Nth orbital of the atom, with no quantum number changes of the other $N - 1$ orbitals. Then from (6.14)

$$\begin{aligned} \Delta E_{av} &= (E_{av})_{ion} - (E_{av})_{atom} \\ &= \left(\sum_{i=1}^{N-1} E_k{}^i + \sum_{i=1}^{N-1} E_n{}^i + \sum_{i=2}^{N-1} \sum_{j=1}^{i-1} E^{ij} \right)_{ion} \\ &\quad - \left(\sum_{i=1}^{N} E_k{}^i + \sum_{i=1}^{N} E_n{}^i + \sum_{i=2}^{N} \sum_{j=1}^{i-1} E^{ij} \right)_{atom} . \end{aligned} \tag{6.47}$$

Now, removal of the Nth electron will result in some minor redistribution of the remaining electronic charge—not quantum-number changes, but just readjustments in the form of the radial functions $P_{n_i l_i}$. Because of these changes, the values of $E_k{}^i$, $E_n{}^i$, and E^{ij} for the ion will differ somewhat from the corresponding values for the atom. If we neglect these differences and assume $(E_k{}^i)_{ion} = (E_k{}^i)_{atom}$, etc., then (6.47) simplifies to

$$\Delta E_{av} \cong -\left(E_k^N + E_n^N + \sum_{j=1}^{N-1} E^{Nj}\right)_{atom} = -(E^N)_{atom} . \qquad (6.48)$$

Thus the one-electron (spherically averaged) binding energy E^N has the physical significance (as its name implies) of being just the negative of the configuration-average ionization energy, in the case that removal of the electron N produces no change in the quantum numbers nl of the other N − 1 orbitals, and in the approximation that it produces no quantitative change in the form of these orbitals (the frozen-core approximation).

In the above discussion, we have for convenience spoken of "atom" and "ion," but similar results are of course applicable to the removal of an electron from an ion to produce a new ion which is one-fold more highly ionized. Likewise, the expression (6.48) is not restricted to the removal of an electron N from the most loosely bound subshell. This result is applicable to ionization from an inner subshell as well, though the approximation will probably be somewhat poorer because removal of an electron from an inner shell usually results in larger changes in the forms of the radial functions $P_i(r)$ for the remaining electrons than does removal of an electron from an outer subshell. Even so, the approximation (6.48) is normally rather good, the disagreement of $|E^N|$ with experimental configuration-average ionization energies being usually only 1 to 5 percent, and only infrequently rising to 10 percent.

The reason why this approximation is so good can be understood qualitatively from Fig. 6-2.[10] The horizontal axis of the figure represents in a highly schematic one-dimensional fashion the infinite-dimensional variation in the forms of the radial functions $P_i(r)$ of the atom or ion [subject of course to the normalization and node-count conditions (3.13), (3.29)]. The two solid curves represent the centers of gravity of the ground configurations of the atom and ion, computed from (6.17) as functions of the variations in the P_i. Physically, we expect the electrons in the atom or ion to distribute themselves in such a manner as to have the lowest possible total energy. Mathematically we thus expect the best computed value of E_{av} to be the lowest value that can be obtained upon variation of the P_i—that is, the energies A and C for atom and ion, respectively. The configuration-average ionization energy ΔE_{av} is then the vertical distance AC. On the other hand, the one-electron binding energy $|E^N|$, computed using the P_i for the atom, is the vertical distance AB, and this is obviously greater than ΔE_{av}. Nonetheless, $|E^N|$ is frequently closer than ΔE_{av} to the experimental ionization energy for the following reason:

The expression that we have derived for E_{av} includes only a partial allowance for electron correlation—the exchange terms that arose from the use of antisymmetrized wavefunctions. Suppose that we were to include further energy corrections for the remaining correlation effects, thereby obtaining lower curves (E'_{av}) than before, with minima at some points D and F. The energy decrease DA for the atom will be greater in magnitude than the decrease FC for the ion because there is one more electron to experience correlation effects. Thus the corrected ionization energy $\Delta E'_{av}$ is, like $|E^N|$, greater than ΔE_{av}.

[10] A. Rosén and I. Lindgren, Phys. Rev. 176, 114 (1968).

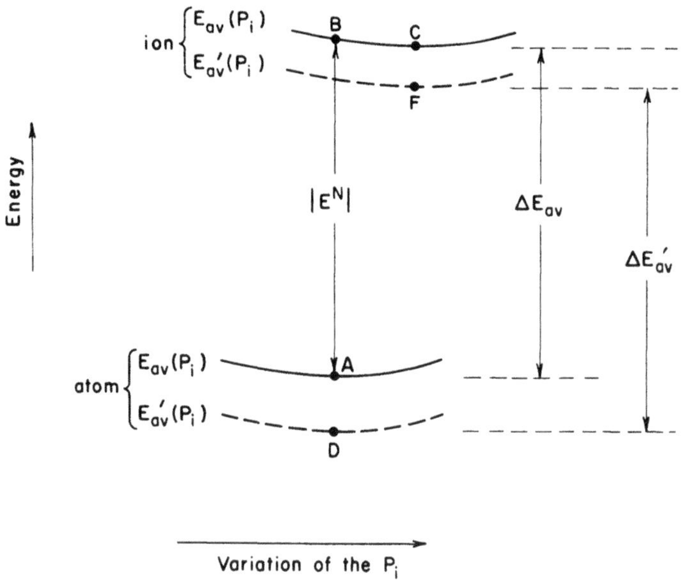

Fig. 6-2. Neglecting correlation effects, the configuration-average ionization energy ΔE_{av} is equal to the one-electron binding energy $|E^N|$ minus the relaxation energy upon ionization (vertical distance BC). The effect of electron correlation (dashed curves) is to increase the ionization energy.

Moreover, the energy difference $|E^N| - \Delta E_{av} = BC$ actually has something of the nature of a correlation energy: BC is the energy change that results when we allow the electrons of the ion to redistribute themselves so as to take account of the absence of electron N.

Thus it is not surprising that $|E^N|_{atom}$ may provide a better approximation to the ionization energy than does the uncorrected value ΔE_{av}. In Sec. 7-15 we shall discuss a way of approximating the correlation energies DA and FC so as to obtain $\Delta E'_{av}$; this value usually is more accurate than $|E^N|$. (Note also that it may be a mistake to try to add a correlation correction to E^N itself).

6-4. NUMERICAL EXAMPLE

As an example illustrating the above equations and discussion, let us consider the ground configurations N I $2p^3$ and N II $2p^2$, for which Table 6-4 gives theoretical values of the various radial integrals (computed from radial functions P_i obtained by a method to be described in Sec. 7-12). From Table 6-1, we find the various electron-electron potential energies to be (values for N II in parentheses):

TABLE 6-4. THEORETICAL (HX) VALUES OF RADIAL INTEGRALS (in rydbergs) FOR (a) N I $2p^3$, (b) N I $2p^2 3d$, AND (c) N II $2p^2$

i	Config.	E_k^i	E_n^i	$E_k^i + E_n^i$
1s	a	44.007	−92.848	−48.841
	b,c	44.027	−92.870	−48.843
2s	a	4.791	−15.442	−10.651
	b	5.373	−16.330	−10.957
	c	5.374	−16.331	−10.957
2p	a	3.763	−13.377	−9.614
	b	4.553	−14.830	−10.277
	c	4.554	−14.831	−10.278
3d	b	0.116	−1.593	−1.477

i	j	Config.	$F^0(ij)$	$F^2(ij)$	$G^0(ij)$	$G^1(ij)$	$G^2(ij)$	$G^3(ij)$
1s	1s	a	8.219					
		b,c	8.222					
1s	2s	a	1.963		0.149			
		b,c	2.062		0.166			
1s	2p	a	1.889			0.180		
		b,c	2.091			0.225		
1s	3d	b	0.228				0.000	
2s	2s	a	1.373					
		b,c	1.451					
2s	2p	a	1.330			0.807		
		b,c	1.456			0.887		
2s	3d	b	0.228				0.001	
2p	2p	a	1.296	0.576				
		b,c	1.472	0.673				
2p	3d	b	0.228	0.011		0.002		0.001

		(n,n') :	(1,1)	(1,2)	(2,2)
$E^{ns,ns}$	$= F^0$	=	8.219 (8.222)		1.373 (1.451)
$E^{ns,n's}$	$= F^0 - \frac{1}{2} G^0$	=		1.889 (1.979)	
$E^{ns,n'p}$	$= F^0 - \frac{1}{6} G^1$	=		1.859 (2.054)	1.195 (1.308)
$E^{np,np}$	$= F^0 - \frac{2}{25} F^2$	=			1.250 (1.418)

From (6.13) the spherically averaged one-electron binding energies are

$$E^{1s} = E_k^{1s} + E_n^{1s} + E^{1s,1s} + 2E^{1s,2s} + \left\{ \begin{matrix} 3 \\ (2) \end{matrix} \right\} E^{1s,2p} = \left\{ \begin{matrix} -31.266 \\ (-32.556) \end{matrix} \right. ,$$

$$E^{2s} = E_k^{2s} + E_n^{2s} + 2E^{1s,2s} + E^{2s,2s} + \left\{ \begin{matrix} 3 \\ (2) \end{matrix} \right\} E^{2s,2p} = \left\{ \begin{matrix} -1.914 \\ (-2.933) \end{matrix} \right. ,$$

$$E^{2p} = E_k^{2p} + E_n^{2p} + 2E^{1s,2p} + 2E^{2s,2p} + \left\{ \begin{matrix} 2 \\ (1) \end{matrix} \right\} E^{2p,2p} = \left\{ \begin{matrix} -1.004 \\ (-2.137) \end{matrix} \right. .$$

Finally, the configuration-average total binding energies are found from (6.17) to be

$$\text{NI } 2p^3\text{:} \quad \frac{1}{2}[2\cdot(-80.107) + 2\cdot(-12.565) + 3\cdot(-10.618)] = -108.599 \text{ Ry} ,$$

$$\text{NII } 2p^2\text{:} \quad \frac{1}{2}[2\cdot(-81.399) + 2\cdot(-13.890) + 2\cdot(-12.415)] = -107.705 \text{ Ry} .$$

These results, together with values of E_{av} calculated similarly for all higher ionization stages, are compared with experimental values in Table 6-5. We see that in all cases the computed energies are less negative than the observed values, owing to the neglect of correlation effects (and, to a lesser extent, of relativistic effects); the error becomes increasingly greater with decreasing ionization stage because of the greater number of electrons that contribute to the correlation effects. In the final column of the table, the theoretical values ΔE_{av}^t include relativistic and correlation corrections computed by methods to be described in Chapter 7; the agreement with experiment is thereby greatly improved.

In Table 6-6, computed ionization energies are compared with experiment. It may be seen that values of $|E^i|$ are greater than ΔE_{av} and tend to agree better with experiment (for the reasons discussed in connection with Fig. 6-2), and that they agree rather well with experiment even for the inner-shell electrons. However, the best results are obtained by differencing values of E_{av}^t, which include correlation and relativistic corrections.

TABLE 6-5. COMPARISON OF THEORETICAL AND EXPERIMENTAL VALUES OF E_{av} (rydbergs)

Config.		*Experiment*			*Theory*	
		Ioniz. Energy[a]	E (ground level)	E_{av}[a,b]	E_{av}	E_{av}^{t}
N VIII	---	0	0	0	0	0
N VII	1s	49.027	−49.027	−49.027	−49.000	−49.032
N VI	$1s^2$	40.580	−89.604	−89.604	−89.472	−89.607
N V	2s	7.195	−96.799	−96.799	−96.662	−96.842
N IV	$2s^2$	5.694	−102.493	−102.493	−102.180	−102.429
N III	2p	3.487	−105.980	−105.979	−105.643	−105.959
N II	$2p^2$	2.176	−108.156	−108.089	−107.705	−108.086
N I	$2p^3$	1.068	−109.224	−109.058	−108.599	−109.041

[a]C. E. Moore, *Ionization Potentials and Ionization Limits Derived from the Analyses of Optical Spectra*, NSRDS-NBS 34 (1970).
[b]See Eq. (4.40).

TABLE 6-6. COMPARISON OF THEORETICAL AND EXPERIMENTAL IONIZATION ENERGIES (rydbergs)

Config.	i	*Theory*			*Experiment*[a]			
		$-E^i$	ΔE_{av}	ΔE_{av}^{t}	Moore	Slater	Bearden	Siegbahn
N II $2p^2$	2p	2.137	2.062	2.127	2.110			
N I $2p^3$	2p	1.004	0.894	0.956	0.969	0.95	0.68	
	2s	1.914	1.797	1.862	1.879	1.88		
	1s	31.27	30.26	30.34		30.0	29.52	29.6

[a]C. E. Moore, *Ionization Potentials and Ionization Limits, and Atomic Energy Levels*; J. C. Slater, Phys. Rev. **98**, 1039 (1955); J. A. Bearden and A. F. Burr, Rev. Mod. Phys. **39**, 78, 125 (1967); K. Siegbahn et al., *ESCA—Atomic, Molecular, and Solid State Structure Studied by Means of Electron Spectroscopy* (Almqvist and Wiksells, Uppsala, 1967), App. 2.

Problems

6-4(1). For N I $2p^3$, the value of $\sum E^i = -69.372$ Ry differs greatly from the value $E_{av} = -108.599$. Discuss the reasons why this is true even though the value of $|E^i|$ agrees with the experimental binding energy to within 5 percent for each of the three orbitals 1s, 2s, and 2p.

6-4(2). Using values of the radial integrals from Table 6-4, calculate E^i and E_{av} for N I $2p^2 3d$. How does the theoretical value of E^{3d} compare with the energy of H I 3d, and why? Calculate (from E_{av}) the excitation energy of N I $2p^2 3d$ relative to N I $2p^3$, and the configuration-average ionization energy from N I $2p^2 3d$ to N II $2p^2$; compare each with experimental values (from Moore, *Atomic Energy Levels*). Which value agrees better with experiment, and why?

6-4(3). From the entries in Table 6-4, calculate the total kinetic energy of N I $2p^3$ and check the accuracy with which the virial theorem (3.34) is satisfied.

7

RADIAL WAVE EQUATIONS

7-1. THE VARIATIONAL PRINCIPLE

We now take up the problem of determining the quantitative form of the radial factors P(r) appearing in the spin-orbitals

$$\varphi_i(r) = \frac{1}{r} P_{n_i l_i}(r) \cdot Y_{l_i m_{l_i}}(\theta,\phi) \cdot \sigma_{m_{s_i}}(s_z) , \qquad 1 \leqq i \leqq N . \tag{7.1}$$

As indicated by the notation, the radial factors $P_i \equiv P_{n_i l_i}$ involved in the basis functions of a configuration

$$(n_1 l_1)^{w_1} (n_2 l_2)^{w_2} \cdots (n_q l_q)^{w_q} , \qquad \sum_{i=1}^{q} w_i = N , \tag{7.2}$$

depend only on n_i and l_i, and so there are q different such factors,[1] one for each subshell of equivalent electrons $(n_i l_i)^{w_i}$. When dealing with the P_i, it will therefore be convenient to replace summations over the N spin-orbitals by summations over the q subshells, with a weighting factor w_i for the ith subshell.

As discussed in Sec. 4-23, the P_i will be chosen so as to minimize the center-of-gravity energy for the configuration (7.2). That is, the $P_i(r)$ are assumed to be such that for any set of small variations $\delta P_i(r)$ which do not violate the orthonormalization conditions (4.18),

$$\int_0^\infty P^*_{n_i l_i}(r_1) P_{n_j l_i}(r_1)\, dr_1 = \delta_{n_i n_j} , \tag{7.3}$$

[1]The theory can be extended so as to employ more than q different radial factors. See, for example, D. R. Hartree and W. Hartree, Proc. Roy. Soc. (London) **A154**, 595 (1936); G. W. Pratt, Jr., Phys. Rev. **102**, 1303 (1956); A. P. Jucys and V. A. Kaminskas, Adv. Chem. Phys. **XIV**, 207 (1969). The increase in accuracy is, however, small compared with the resulting mathematical complexities.

the resulting variation δE_{av} will be zero. The restrictions on the permissible forms of the δP_i posed by (7.3) are mathematically very awkward, but can be removed by the usual method of Lagrangian multipliers.[2] These multipliers we shall write in the form $-\varepsilon_{ij}w_iw_j$ for the orthogonality restriction between P_i and P_j ($l_i = l_j$), and $-\varepsilon_{jj}w_j$ (or simply $-\varepsilon_j w_j$) for the normalization restriction on P_j.[3] The condition for minimum E_{av} may then be written

$$\delta\left\{E_{av} - \sum_{j=1}^{q} \varepsilon_j w_j \int_0^\infty P_j^* P_j \, dr_1 - \sum_{j=1}^{q}\sum_{t\neq j}^{q} \delta_{l_j l_t} \varepsilon_{jt} w_j w_t \int_0^\infty P_j^* P_t \, dr_1\right\} = 0 . \qquad (7.4)$$

The variations δP_i employed in this expression may now be completely arbitrary (except that boundary conditions on P_i require $\delta P_i = 0$ at $r = 0$ and ∞); it is necessary only that the values of ε_j and ε_{jt} be so chosen (by procedures outlined in Sec. 7-5) that the functions P_j deduced from (7.4) indeed satisfy the corresponding normalization and orthogonality conditions.

The expression (6.15) for E_{av} may be written in the form

$$E_{av} = \sum_{j=1}^{N} \left\{E_k^j + E_n^j + \frac{1}{2}\sum_{t\neq j}^{N} E^{jt}\right\}$$

$$= \sum_{j=1}^{q} w_j\left\{E_k^j + E_n^j + \frac{1}{2}(w_j - 1)E^{jj} + \frac{1}{2}\sum_{t\neq j}^{q} w_t E^{jt}\right\} ,$$

from which the variation of E_{av} due to a variation of P_i *only* is

$$\delta_i E_{av} = w_i\left\{\delta_i E_k^i + \delta_i E_n^i + \frac{1}{2}(w_i - 1)\delta_i E^{ii} + \frac{1}{2}\sum_{t\neq i} w_t \delta_i E^{it} + \frac{1}{2}\sum_{j\neq i} w_j \delta_i E^{ji}\right\} ,$$

where the two summations are identical since $E^{ij} = E^{ji}$. Substituting this into (7.4) and dividing out w_i we obtain

$$\delta_i E_k^i + \delta_i E_n^i + \frac{1}{2}(w_i - 1)\delta_i E^{ii} + \sum_{j\neq i}^{q} w_j \delta_i E^{ij}$$

$$= \varepsilon_i \delta_i \int_0^\infty P_i^* P_i \, dr_1 + \sum_{j\neq i}^{q} \delta_{l_j l_i} w_j \left\{\varepsilon_{ij}\delta_i \int_0^\infty P_i^* P_j \, dr_1 + \varepsilon_{ji}\delta_i \int_0^\infty P_j^* P_i \, dr_1\right\} . \qquad (7.5)$$

[2]See, for example, H. Margenau and G. M. Murphy, *The Mathematics of Physics and Chemistry (D. Van Nostrand, Princeton, N.J., 1956)*, 2nd ed., Sec. 6.5.

[3]With this definition, the matrix of "energy parameters" is symmetric; $\varepsilon_{ij} = \varepsilon_{ji}$. Alternatively, the Lagrangian multipliers are frequently taken to be $+\varepsilon_{ij}w_i = +\varepsilon_{ji}w_j$ (and $+\varepsilon_{jj}w_j$): Douglas R. Hartree, *The Calculation of Atomic Structures* (John Wiley, New York, 1957), pp. 54, 58; C. Froese [Fischer], Can. J. Phys. **45**, 7 (1967).

7-2. THE HARTREE-FOCK EQUATIONS

If we substitute into (7.5) the expressions (6.18), (6.19), (6.38), and (6.39) that we derived for E_k^i, E_n^i, and E^{ij} using determinantal wavefunctions, we arrive ultimately at a set of q equations (one for each value of i) known as the Hartree-Fock (HF) equations—or more specifically as the (restricted[4]) HF equations for the configuration-average energy. Before going into the details, we first note that up to this point we have written all expressions as though the $P_i(r)$ were complex functions. There was some advantage in this, because in expressions such as (6.30) it helps one keep track of which radial function came from the bra wavefunction and which from the ket. Actually, however, there is no need for P_i to be complex, and to continue this fiction in the present derivation complicates the mathematics appreciably. From now on, therefore, we shall assume all P_i to be real. The two orthogonality integrals on the right side of (7.5) are then identical, and it follows that we should take $\varepsilon_{ij} = \varepsilon_{ji}$.

Let us consider first the variations of the integrals on the right-hand side of (7.5), which can be simplified to

$$\delta_i \int_0^\infty P_i^2 \, dr_1 = 2\int_0^\infty (\delta P_i)P_i \, dr_1 \tag{7.6}$$

and

$$2\varepsilon_{ij}\delta_i \int_0^\infty P_i P_j \, dr_1 = 2\varepsilon_{ij}\int_0^\infty (\delta P_i)P_j \, dr_1 \ . \tag{7.7}$$

Next, in dealing with the integral (6.18) for E_k^i we can integrate by parts twice to give

$$\begin{aligned}\int_0^\infty P_i\left(-\frac{d^2}{dr_1^2}\right)(\delta P_i)\,dr_1 &= \left[P_i\left(-\frac{d}{dr_1}\right)(\delta P_i)\right]_0^\infty - \int_0^\infty \left(\frac{dP_i}{dr_1}\right)\left(-\frac{d}{dr_1}\right)(\delta P_i)\,dr_1 \\ &= -\left[\left(\frac{dP_i}{dr_1}\right)(-\delta P_i)\right]_0^\infty + \int_0^\infty \left(\frac{d^2P_i}{dr_1^2}\right)(-\delta P_i)\,dr_1 \\ &= \int_0^\infty (\delta P_i)\left(-\frac{d^2P_i}{dr_1^2}\right)dr_1 \ ;\end{aligned}$$

in this derivation the quantities in brackets vanish at both limits because P_i (and therefore also δP_i) must be zero at $r_1 = 0$ and $r_1 = \infty$ from the boundary conditions (3.10)-(3.11). With this result we readily obtain

[4]That is, all spin-orbitals for the ith subshell are restricted to having the same radial function $P_i(r)$, as opposed to the spin-polarized HF method in which different $P_i(r)$ are assumed for different spin orientations, and to the unrestricted HF method in which different $P_i(r)$ are assumed for different m_l as well as different m_s; cf. fn. 1.

$$\delta_i E_k^{\,i} = 2\int_0^\infty (\delta P_i)\left[-\frac{d^2}{dr_1^{\,2}} + \frac{l_i(l_i+1)}{r_1^{\,2}}\right] P_i\, dr_1 \ . \tag{7.8}$$

Finally, the variations of E_n^i and of $F^k(ij)$, $i \neq j$, are trivial, and the variations of $F^k(ii)$ and $G^k(ij)$ involve only minor complications: From (6.28) and (6.30) we find, since the names used for integration variables are immaterial, that

$$\begin{aligned}
\delta F^k(ii) &= 2\int_0^\infty\!\!\int_0^\infty \frac{2r_<^k}{r_>^{k+1}} \{\delta P_i(r_1)\}\, P_i(r_1) P_i^2(r_2)\, dr_1\, dr_2 \\
&\quad + 2\int_0^\infty\!\!\int_0^\infty \frac{2r_<^k}{r_>^{k+1}} P_i^2(r_1)\{\delta P_i(r_2)\}\, P_i(r_2)\, dr_1\, dr_2 \\
&= 4\int_0^\infty\!\!\int_0^\infty \frac{2r_<^k}{r_>^{k+1}} \{\delta P_i(r_1)\}\, P_i(r_1) P_i^2(r_2)\, dr_1\, dr_2 \ ,
\end{aligned} \tag{7.9}$$

and

$$\begin{aligned}
\delta G^k(ij) &= \int_0^\infty\!\!\int_0^\infty \frac{2r_<^k}{r_>^{k+1}} \{\delta P_i(r_1)\}\, P_j(r_2) P_j(r_1) P_i(r_2)\, dr_1\, dr_2 \\
&\quad + \int_0^\infty\!\!\int_0^\infty \frac{2r_<^k}{r_>^{k+1}} P_i(r_1) P_j(r_2) P_j(r_1)\{\delta P_i(r_2)\}\, dr_1\, dr_2 \\
&= 2\int_0^\infty\!\!\int_0^\infty \frac{2r_<^k}{r_>^{k+1}} \{\delta P_i(r_1)\}\, P_j(r_2) P_j(r_1) P_i(r_2)\, dr_1\, dr_2 \ .
\end{aligned} \tag{7.10}$$

Since the form of the variation $\delta P_i(r_1)$ is completely arbitrary (except for the requirement that it be zero at $r_1 = 0$ and $r_1 = \infty$), we may take it to be zero everywhere except in the immediate vicinity of some point $r_1 = r$. Then (7.6) can be written

$$2\int_0^\infty \{\delta P_i(r_1)\}\, P_i(r_1)\, dr_1 = 2P_i(r)\int_0^\infty \{\delta P_i(r_1)\}\, dr_1 \ ,$$

and all other integrals (7.7)-(7.10) simplify similarly. Substituting everything into (7.5) and dividing out a factor

$$2\int_0^\infty \{\delta P_i(r_1)\}\, dr_1 \ ,$$

we obtain finally

$$\left[-\frac{d^2}{dr^2} + \frac{l_i(l_i+1)}{r^2} - \frac{2Z}{r} + \sum_{j=1}^{q}(w_j - \delta_{ij})\int_0^\infty \frac{2}{r_>}P_j^2(r_2)\,dr_2 - (w_i - 1)A_i(r)\right]P_i(r)$$

$$= \varepsilon_i P_i(r) + \sum_{j(\neq i)=1}^{q} w_j[\delta_{l_i l_j}\varepsilon_{ij} + B_{ij}(r)]P_j(r)\ , \qquad (7.11)$$

where

$$A_i(r) = \frac{2l_i+1}{4l_i+1}\sum_{k>0}\begin{pmatrix} l_i & k & l_i \\ 0 & 0 & 0 \end{pmatrix}^2 \int_0^\infty \frac{2r_<^k}{r_>^{k+1}}P_i^2(r_2)\,dr_2 \qquad (7.12)$$

and

$$B_{ij}(r) = \frac{1}{2}\sum_{k}\begin{pmatrix} l_i & k & l_j \\ 0 & 0 & 0 \end{pmatrix}^2 \int_0^\infty \frac{2r_<^k}{r_>^{k+1}}P_j(r_2)P_i(r_2)\,dr_2\ . \qquad (7.13)$$

In all integrals, $r_<$ and $r_>$ represent respectively the lesser and the greater of r and r_2. The set of q equations (7.11)—one for each subshell n_il_i—are the Hartree-Fock equations[5] for the spherically averaged atom. [If the variational principle is applied to the energy of a specific LS state rather than to E_{av}, then HF equations are obtained that (in simple cases) are of the same form as (7.11), but with expressions for A_i and B_{ij} that differ from (7.12) and (7.13). We will discuss these modifications in Sec. 16-8, after we have derived expressions for the energies of specific LS terms.]

Let us review the origin of the various terms in (7.11). The first two terms, $-d^2/dr^2 + l_i(l_i+1)/r^2$, arose from variation of the kinetic energy E_k^i, and the third term came from the nuclear potential energy E_n^i. The next term, which could be written as a summation over all electrons other than i, came from the direct portion F^0 of the electron-electron interactions E^{ij} for electrons both equivalent and non-equivalent to i; the physical significance of this term will be discussed in the next section. The terms involving A_i and B_{ij} came from the exchange portions of E^{ij}, for electrons equivalent and non-equivalent to i, respectively.

The terms involving the Lagrangian multipliers ε_{ij} of course arose from the orthogonality requirement. If they were omitted from the HF equations the solutions P_i and P_j for $l_i = l_j$ but $n_i \neq n_j$ (for example, P_{2p} and P_{3p}) would not in general be exactly orthogonal; inclusion of the ε_{ij} (and suitable adjustment of their numerical values by more or less trial-and-error methods) in effect produces solutions that are orthogonal linear combinations of the non-orthogonal solutions. It may be shown[6] that in the special case in

[5]V. Fock, Z. Physik **61**, 126 (1930), **62**, 795 (1930); V. Fock and M. J. Petrashen, Physik. Z. Sowjetunion **6**, 368 (1934).

[6]D. R. Hartree, *The Calculation of Atomic Structures* (John Wiley, New York, 1957). C. Froese Fischer, *The Hartree-Fock Method for Atoms* (John Wiley, New York, 1977).

which the subshells i and j are both closed, P_i and P_j are automatically orthogonal and so we may take $\varepsilon_{ij} = \varepsilon_{ji} = 0$.

The Lagrangian multipliers ε_i were introduced in order to take account of the normalization condition on the P_i; correspondingly, solution of the ith HF equation proceeds by trial-and-error adjustment of ε_i to find that value which results in a normalized solution P_i, as we shall discuss in detail in Secs. 7-5 and 7-6. Once this value of ε_i has been found, it has a direct physical significance. To see this, we multiply all terms of (7.11) from the left by $P_i^*(r)$ and integrate with respect to r. The first three terms give just $E_k^i + E_n^i$ from (6.18) and (6.19). The remaining terms on the left-hand side of (7.11), together with the terms in B_{ij} (transferred to the left-hand side) are easily seen to give just

$$\sum_{j(\neq i)=1}^{N} E^{ij}$$

from (6.38) and (6.39). Because of the orthonormality of the P_j, the remaining terms give simply ε_i. Thus we see from (6.13) that

$$\varepsilon_i = E_k^i + E_n^i + \sum_{j(\neq i)=1}^{N} E^{ij} = E^i , \tag{7.14}$$

or that ε_i is equal to the binding energy[7] of an electron in the subshell $n_i l_i$. This, together with the result (6.48) leads to *Koopmans' theorem*:[8] The negative of the eigenvalue ε_i of the HF equation (for the spherically averaged atom) is equal to the configuration-average ionization energy for an electron in subshell $n_i l_i$, in the approximation in which the orbitals for the ion are taken to be the same as those for the atom.

7-3. THE CLASSICAL POTENTIAL ENERGY

In looking for the physical significance of the terms in the HF equations that came from E^{ij}, let us recall that we started out with the central-field model, in which we considered the motion of one electron in the field of the nucleus and the spherically averaged field of the other N−1 electrons. Let us calculate the potential energy of this electron semiclassically.

We may write the normalization condition (7.3) for a radial wavefunction P_j in the form

$$\int_0^\infty P_j^2(r)\,dr = \int_0^\infty \frac{P_j^2(r)}{4\pi r^2}\cdot 4\pi r^2\,dr = 1 ,$$

from which it follows that

[7]With our sign convention, ε_i is negative for bound electrons and positive for free electrons. The opposite sign convention is frequently used (see fn. 3).

[8]T. Koopmans, Physica 1, 104 (1934).

$$\rho_j(r) = \frac{P_j^2(r)}{4\pi r^2} \tag{7.15}$$

is the spherically averaged electron probability density distribution of one electron in subshell j [in units of electrons/(Bohr unit)^3].

Consider first a nucleus of charge Z units, surrounded by one electron j with a smeared-out spherically symmetrical charge distribution given by (7.15). Then consider a test-charge electron localized at a distance r_1 from the nucleus. That portion of the charge distribution ρ_j lying outside a sphere of radius r_1 exerts no force on the test electron, and the portion lying inside the sphere exerts the same force as though it lay at the center of the sphere. Therefore, the force on the test electron is

$$F = r_1^{-2}\left[-2Z + 2\int_0^{r_1} \rho_j(r_2) 4\pi r_2^2 \, dr_2\right] ;$$

the factor 2 in front of the integral arises by virtue of the units that we are using—it is analogous to the factor 2 in front of the Z. Using (7.15) and integrating by parts, we find the potential energy of the test electron at a distance r from the nucleus to be

$$\begin{aligned}
V(r) &= \int_r^\infty F \, dr_1 = -\int_r^\infty \frac{2Z}{r_1^2} dr_1 + \int_r^\infty \frac{2}{r_1^2}\int_0^{r_1} P_j^2(r_2)\, dr_2 \, dr_1 \\
&= \left[\frac{2Z}{r_1}\right]_r^\infty + \left[-\frac{2}{r_1}\int_0^{r_1} P_j^2(r_2)\, dr_2\right]_r^\infty + \int_r^\infty \frac{2}{r_1} P_j^2(r_1)\, dr_1 \\
&= -\frac{2Z}{r} + \frac{2}{r}\int_0^r P_j^2(r_2)\, dr_2 + \int_r^\infty \frac{2}{r_2} P_j^2(r_2)\, dr_2 \\
&= -\frac{2Z}{r} + \int_0^\infty \frac{2}{r_>} P_j^2(r_2)\, dr_2 ,
\end{aligned} \tag{7.16}$$

where $r_>$ is the larger of r and r_2.

The first term in this expression is the potential energy of the test electron due to the nucleus, and the integral is the potential energy due to the density distribution of electron j. Thus

$$V_c(r) \equiv \sum_{j=1}^{q} w_j \int_0^\infty \frac{2}{r_>} P_j^2(r_2)\, dr_2 \tag{7.17}$$

is the classical potential energy of a test electron in the spherically averaged field of all N electrons of the atom, and the term

$$V_H(r) \equiv \sum_{j=1}^{q} (w_j - \delta_{ij}) \int_0^\infty \frac{2}{r_>} P_j^2(r_2)\, dr_2 \qquad (7.18)$$

that appears in the HF equation (7.11) is the potential energy of the ith electron of the atom in the (averaged) field of the other N−1 electrons—the energy that Hartree[9] used in his early calculations, and that he arrived at by exactly the above semiclassical argument.

7-4. THE EXCHANGE POTENTIAL ENERGY

The quantity

$$-(w_i - 1)A_i(r) \qquad (7.19)$$

in (7.11) arose from the exchange portion of the interaction energy E^{ii} between the electron i in question and each of the other $(w_i - 1)$ electrons equivalent to it. This quantity can evidently be thought of as an exchange correction to the Hartree potential-energy operator (7.18); this correction corresponds to the partial positional correlation among electrons of parallel spin.

The terms that arose from the exchange portion of E^{ij} $(j \neq i)$ can be considered to contribute an additional potential-energy-operator correction to V_H of the form

$$\frac{-\sum_{j \neq i} w_j B_{ij}(r) P_j(r)}{P_i(r)} \,. \qquad (7.20)$$

For radial wavefunctions P_i with $n_i = l_i + 1$ (e.g., 1s, 2p, 3d, 4f), which have no nodes other than at $r = 0$ and ∞, this function is well-behaved. However, for cases in which $n_i > l_i + 1$, the nodes in $P_i(r)$ produce singularities in (7.20), this function going to $+\infty$ on one side of each node, and to $-\infty$ on the other side of the node. It is then not very useful to picture (7.20) as a contribution to a radial potential-energy function.

Problems

7-4(1). Simplify the HF equation for any configuration of a one-electron atom, and show that it reduces to the hydrogenic Schrödinger equation (3.15), with the Lagrange multiplier ε taking the place of the total binding energy E of the hydrogenic atom. Is this consistent with the physical significance of ε, as given by Koopmans' theorem?

7-4(2). Simplify the HF equations for the ground configuration of Ne IX. What are the fundamental reasons that are responsible for these particular simplifications?

7-4(3). Show that at large r the HF equation (7.11) reduces to the form

$$\left[-\frac{d^2}{dr^2} - \frac{2(Z-N+1)}{r}\right] P_i(r) = \varepsilon_i P_i(r) + \sum_{j \neq i} w_j \delta_{l_i l_j} \varepsilon_{ij} P_j(r) \,.$$

[9]D. R. Hartree, Proc. Cambridge Phil. Soc. **24**, 111 (1928).

[Hint: $B_{ij}(r)$ contains a k = 0 term only if $l_i = l_j$, in which case P_i and P_j are orthogonal.] Show that if the terms in $\varepsilon_{ij}P_j$ are negligible compared with $\varepsilon_i P_i$, then at large r, $P_i(r) \propto \exp[-(-\varepsilon_i)^{1/2}r]$ for a bound wavefunction ($\varepsilon_i < 0$).

7-5. SOLUTION OF THE HARTREE-FOCK EQUATIONS

Since the ith HF equation involves the P_j for all j, we have a set of q coupled integro-differential equations that (if q is greater than 1) can be solved only by an iterative procedure, which we now describe briefly. Each iterative cycle consists of:

(1) assuming a set of trial functions $P_j(r)$, $1 \leqq j \leqq q$;

(2) for each i, computing V_H, A_i, and B_{ij}, and estimating the ε_{ij} ;

(3) solving the ith HF equation for a new $P_i(r)$, each i.

These three steps are repeated, using a new set of trial functions each time, until the output functions obtained in step (3) are identical with the functions assumed in step (1), and all functions with given l are mutually orthogonal, within the desired tolerances. The output functions are then self-consistent with the trial input functions used in computing the central field in step (2), and so this procedure is called a self-consistent-field (SCF) method.

In the above procedure, the values of ε_{ij} required for orthogonality can be estimated each cycle from equations obtained by multiplying (7.11) from the left by P_j ($l_j = l_i$) and integrating, the various integrals being evaluated numerically by using the trial functions P_j. Appropriate trial input functions for the $(m + 1)^{st}$ iteration cycle can usually be obtained by taking (normalized) linear combinations of the input and output functions from the preceding cycle:

$$P_i^{(m+1)}(\text{input}) = c\ P_i^{(m)}(\text{output}) + (1 - c)\ P_i^{(m)}(\text{input}) \ . \tag{7.21}$$

The value of c is chosen by trial and error so as to give the maximum speed of convergence; usually this value is about 0.5, but it may be anywhere from 0.05 to 1.1. In some cases of excited d and f orbitals, the above procedure does not lead to stable convergence of the SCF interation, as will be discussed in the next section.

There are several different procedures[6] that can be used in step (3) in order to find a new output function $P_i(r)$ from the differential equation (7.11), but the most straightforward one is the following. As a result of steps (1) and (2) we consider everything in the HF equation (7.11) to be known except for the desired function $P_i(r)$ and the unknown parameter ε_i. We start with the boundary condition

$$P_i(0) = 0 \ , \tag{7.22}$$

which is required to keep the electron density (7.15) finite at r = 0. For small r, all electron-electron terms in (7.11) are negligible compared with the kinetic and nuclear terms, and the differential equation reduces to that for a hydrogenic atom, for which we know that $P_i \propto r^{l_i+1}$ at small r [Eq. (3.18)]. We thus assume some arbitrary value $a_o^{(o)}$ of the "initial slope"

$$a_o \equiv \left[\frac{P_i(r)}{r^{l_i+1}}\right]_{r\to 0} \ . \tag{7.23}$$

(By convention, the trial functions P_j are always assumed to have positive a_0, and hence the self-consistent functions must also have $a_0 > 0$.)

Beginning with (7.22) and (7.23), and assuming some value of ε_i, we then numerically integrate[6,10] the differential equation (7.11) out to some suitably large value of r to give a particular integral $P^I(r)$ of the inhomogeneous differential equation. With the same starting conditions, we also numerically integrate the homogeneous equation obtained from (7.11) by setting all ε_{ij} and $B_{ij}(r)$ to zero, thereby obtaining a function $P^H(r)$. It is well-known that any integral of (7.11) satisfying (7.22) can be written as

$$P_i(r) = P^I(r) + \alpha P^H(r) \ , \tag{7.24}$$

where α is an appropriate constant. We then choose α such that P_i satisfies the boundary condition

$$\lim_{r \to \infty} P_i(r) = 0 \tag{7.25}$$

required for the wavefunction of any bound electron; the value of the "initial slope" of P_i is obviously

$$a_0 = (1 + \alpha) a_0^{(o)} \ , \tag{7.26}$$

and we compute the norm of P_i from the definition

$$\|P_i\| = \left[\int_0^\infty P_i^2(r)\, dr \right]^{1/2} .$$

Both a_0 and $\|P_i\|$ depend on the assumed value of ε_i. The parameter ε_i was originally introduced as the Lagrangian multiplier associated with the normalization condition on $P_i(r)$. It must therefore be chosen (by means of a secondary iterative procedure) such that $\|P_i\| = 1$. The additional requirements that a_0 must be positive and that $P_i(r)$ have $n - l - 1$ nodes[11] [by analogy with the hydrogenic result (3.29)] lead to a unique integral $P_i(r)$ of the HF equation (7.11).

Extensive numerical results for (primarily) ground configurations of neutral and few-fold ionized atoms are tabulated in the references listed in Sec. XIV of the bibliography.

*7-6. COMPLICATIONS AND INSTABILITIES

The iteration on ε_i required to arrive at the desired integral of the inhomogeneous HF equation is considerably more complicated than the procedure involved in the solution of the homogeneous differential equation (3.9) for a hydrogenic atom.

[10]Using, for example, the Numerov method described in fn. 3 of Chap. 3.

[11]This excludes "spurious" nodes that may be introduced into the tail of $P_i(r)$ by the inhomogeneous terms of (7.11) involving $P_j(r)$, $j \neq i$.

In the homogeneous case, we saw in Sec. 3-4 that there existed only a discrete set of values of ε_i at which the boundary condition (7.25) was satisfied; it was necessary only to find that element of the set for which $P_i(r)$ had the desired number of nodes, and then to scale $P_i(r)$ by a constant such that $\|P_i\| = 1$ and $a_o > 0$.

In the HF case, integrals $P_i(r)$ of the inhomogeneous differential equation can be found satisfying (7.25) for all ε_i *except* at a discrete set of values. If $P_i(r)$ is such an integral (for given ε_i), then $cP_i(r)$ is an integral only for $c = +1$; i.e., $P_i(r)$ is unique and so is its norm. The variation of $\|P_i\|$ with ε_i is qualitatively of the form shown in the upper portion of Fig. 7-1, as may be seen from the following argument: The values $\varepsilon_i = \varepsilon_1{}^H, \varepsilon_2{}^H, \varepsilon_3{}^H, \cdots$ are values for which the homogeneous integral $P^H(r)$ tends to zero as $r \to \infty$; at each of these points there exists no finite value of α that can force (7.25) to be true [unless $P^I(\infty) = 0$

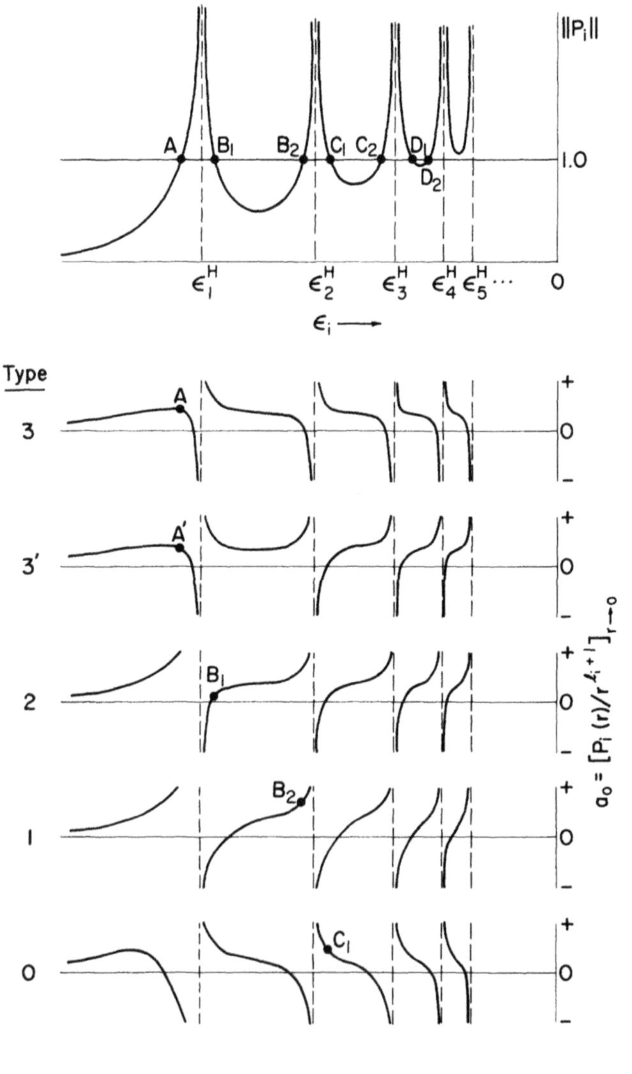

Fig. 7-1. Top: The variation with ε_i of the norm of the integral $P_i(r)$ of the HF equation that satisfies the boundary conditions $P_i(0) = P_i(\infty) = 0$. The singularities occur at values of ε_i equal to the eigenvalues of the associated homogeneous equation; $\varepsilon_k{}^H$ corresponds to the eigenfunction having $k - 1$ nodes. Bottom: The variation with ε_i of the "initial slope" a_o of $P_i(r)$, for different cases of type 0, 1, 2, and 3, in which the positive-a_o, one-node, unit-norm solution $P_i(r)$ occurs at C_1, B_2, B_1, and A, respectively. Type 3′ is a hybrid case in which a_o varies as in type 3 for large $|\varepsilon_i|$, and as in type 2 for small $|\varepsilon_i|$.

for all values of a_0, which is impossible]. When ε_i is nearly equal to one of the $\varepsilon_m{}^H$, P^H almost satisfies $P^H(\infty) = 0$ and therefore $|\alpha|$ has to be very large to force the condition (7.25); this means that both a_0 and $\|P_i\|$ show singularities at $\varepsilon_k{}^H$, as shown in the figure.

In the illustrative example of Fig. 7-1 (top) there are seven values of ε that give integrals P(r) having unit norm, corresponding to solutions at A, B_1, B_2, C_1, C_2, D_1, and D_2. Most commonly, the variation of a_0 with ε_i is of the type 1 shown in the lower portion of the figure; the integrals $P_i(r)$ at B_1, C_1, and D_1 then have negative values of a_0, and are therefore not close to the self-consistent function for which we are looking. The integrals at A, B_2, C_2, and D_2 have positive a_0, and are functions having respectively 0, 1, 2, and 3 nodes. By analogy with the radial wavefunctions for hydrogen, it is the integral A that we want if we are seeking a 1s, 2p, 3d, 4f, $\cdots$ solution [according as $l_i = 0, 1, 2, 3, \cdots$ in (7.11)]; we wish integral B_2 for the cases 2s, 3p, 4d, 5f $\cdots$; we need C_2 for 3s, 4p, 5d, 6f, $\cdots$; etc. In the section of Fig. 7-1 that is labeled "type 1," it has been assumed that we are seeking a 4d or 5f radial function, so that the integral B_2 is the desired solution.

There exist in certain cases (especially, excited d and f orbitals in configurations of transition and rare-earth elements, respectively) a number of variations from the above situation, which greatly complicate the task of arriving at the proper self-consistent solution.[12] Suppose, for example, that we are looking for a 4d or a 5f function, having one node, which in the common case (type 1) is the solution B_2. In one variation (type 2), the integrals at both B_1 and B_2 have positive a_0, but it is only choice of the integral B_1 that will ultimately lead to a self-consistent solution; the only certain *a priori* way to detect this situation is to observe that continued choice of the integral B_2 leads inexorably to divergence of the SCF iterative process. In other cases, there is only one normalized integral having positive a_0 and one node, but this is the integral A (type 3) or C_1 (type 0) rather than B_2.[13] In these cases, the number of nodes in $P_i(r)$ — and sometimes also the value of a_0—varies with ε_i in a somewhat complicated manner, as illustrated in Fig. 7-2 (and in Fig. 7-1, type 3′); it takes careful computer programming to recognize these situations so that the program will change the trial values of ε_i in the proper direction to arrive at that integral P_i which has positive a_0 and the desired number of nodes. Moreover, in the cases where the final self-consistent integral lies at A or B_1, the method (7.21) for obtaining a new trial function will result in unconditional divergence of the SCF iteration if the normalized integral A or B_1 is chosen each cycle and a positive value of c is used; two procedures for obtaining convergence are to choose an appropriate *unnormalized* integral near A or B_1 (on the basis of some complex criterion), or to repeatedly use a small negative value of c for several cycles and then c = 1 for one cycle.[12] Finally, if the trial functions are too far from the final self-consistent solutions, the minimum in the appropriate segment of the norm curve may lie above unity, so that there exist no normalized integrals at all having the

[12] D. C. Griffin, Thesis, Purdue Univ., Lafayette, Ind., 1970; D. C. Griffin, R. D. Cowan, and K. L. Andrew, Phys. Rev. A 3, 1233 (1971).

[13] When seeking a function having two nodes (e.g., P_{5d} or P_{6f}), the desired integrals in cases of type 0, 1, 2, 3 occur at D_1, C_2, C_1, and B_2, respectively; etc. Cases of type 0 are believed to occur only when seeking solutions of the HF equations for a specified LS term (Sec. 16-8), and not with the spherically averaged HF equations.

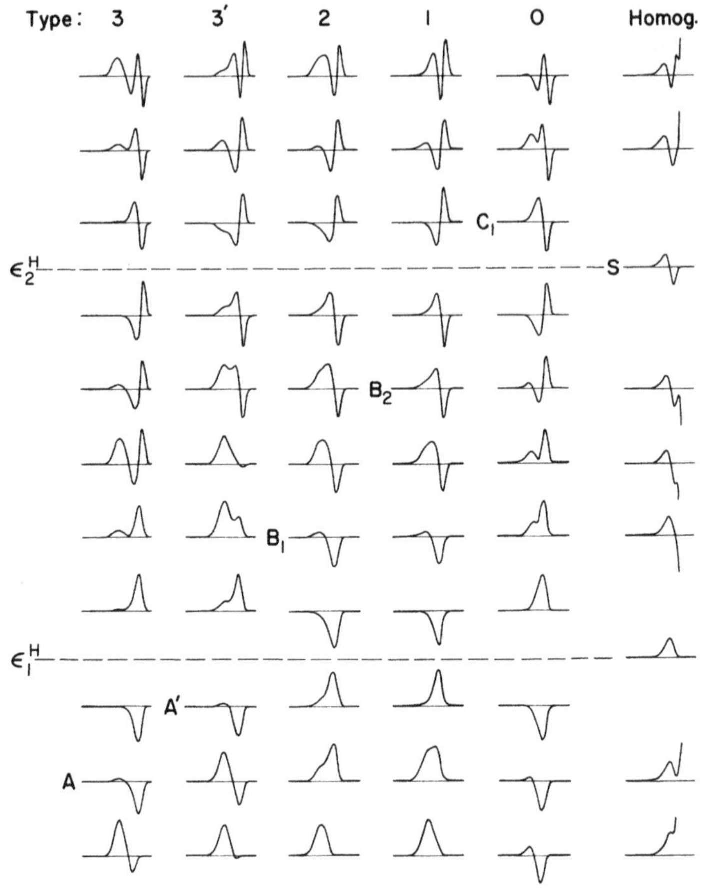

Fig. 7-2. The variation with ε_l of the form of the function $P_l(r)$ for the SCF HF equation corresponding to the 5f or 4d orbital of: type 3, Pr I $4f^2 6s^2 5f$; type 3′, Pr II $4f^2 6s5f$; type 2, Ti I $3d^2 4s4d$; type 1, K I 4d; type 0, P II 3p4d $^1P^\circ$. The self-consistent, positive-a_0, one-node, unit-norm function is that at A, A′, B_1, B_2, and C_1, respectively. (For the 5f orbital of Pr II $4f^3 5f$, the HF integrals are of type 2, and the self-consistent function corresponds to B_1.) For comparison, the right-hand column shows the variation with ε_l of the integral of a homogeneous differential equation; the one-node function *always* appears at S (i.e., at $\varepsilon_l = \varepsilon_2^H$). (Note: All functions are plotted on a logarithmic horizontal r scale, which is the same for all segments of a given column. The vertical P scales are linear but differ from one segment to another, so that the variation of $\|P_l\|$ with ε_l is not reflected in this figure.)

desired number of nodes—as is the case in Fig. 7-1, for example, if one is looking for an integral with more than three nodes.[14]

[14]Even if we use the final SCF functions, the form of Fig. 7-1 is such as to frustrate all attempts to find normalized radial functions for highly excited "unoccupied" orbitals.

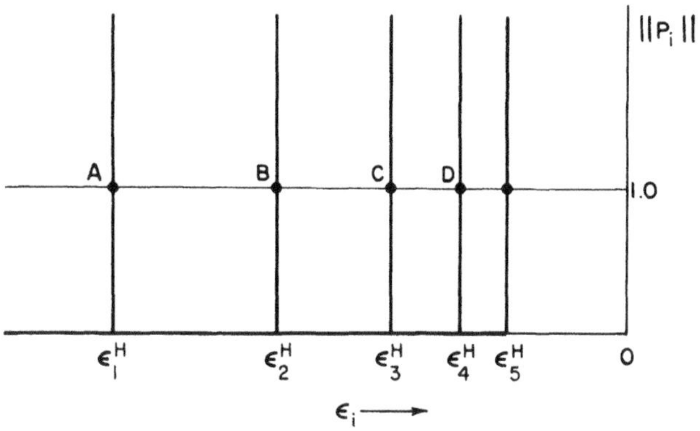

Fig. 7-3. Variation with ε_i of the norm of the integral of a homogeneous equation, under the restrictions $P_i(0) = P_i(\infty) = 0$. The norm is zero everywhere except at the infinite set of discrete points $\varepsilon_k{}^H$, where it can have any non-negative value.

All of the above complications result from the large effect of the inhomogeneous terms in the differential equation. When these terms are small, the parabola-like sections of the norm curve in Fig. 7-1 lie much lower, and approach as a limit the rectangular wells formed by the ε axis and the vertical dashed lines at the points $\varepsilon = \varepsilon_k{}^H$ (Fig. 7-3). The solutions are always of type 1, and the value of ε_i for the required integral $P_i(r)$ lies only very slightly to the left of the corresponding $\varepsilon_i{}^H$ (e.g., for a one-node integral, B_2 is almost coincident with $\varepsilon_2{}^H$).

This suggests the following very simple method of eliminating all difficulties.[15] We abbreviate the HF equation (7.11) to

$$P'' + fP = g \ ,$$

where $g = g(r)$ represents the sum of all the inhomogeneous terms. We then rewrite this equation in the form

$$P'' + \left(f - \frac{bg}{P}\right)P = (1-b)g \ ,$$

or approximately

$$P'' + \left(f - \frac{bg}{P(\text{input})}\right)P \cong (1-b)g \ , \tag{7.27}$$

where P(input) is the trial input function for the current SCF iteration cycle. When self-consistency has been reached, P will be identical with P(input), and the approximation becomes exact. By choosing the value of b to be nearly unity (e.g., $b = 0.95$), we produce a differential equation that in effect has only a very small inhomogeneous term.

It should be emphasized that the above modified methods are called for only in certain cases involving singly occupied outer d and f (and sometimes p) orbitals of neutral and singly ionized atoms—all other cases converge straightforwardly using the simple method of (7.21).

[15]R. D. Cowan and J. B. Mann, J. Comput. Phys. **16**, 160 (1974). Minor modifications of (7.27) are required to avoid the introduction of singularities at the positions of the nodes of P(input).

7-7. HOMOGENEOUS-EQUATION (LOCAL-POTENTIAL) METHODS

Even without the complications discussed in the preceding section, the Hartree-Fock equations are numerically complex and their solution is time consuming. In addition, we saw from (7.20) that the HF method does not in general lead to the definition of a smooth central potential-energy function V(r). This complicates the calculation of relativistic corrections and spin-orbit parameters from the simple formulae (3.46), (3.47), and (4.2), and prevents the easy discussion of certain properties of the radial functions $P_i(r)$, in the manner to be used in Chapter 8. For these reasons, a number of simpler methods have been developed for the calculation of approximate radial functions.

In all common approximations to the HF equations, each differential equation to be solved is of the form

$$\left[-\frac{d^2}{dr^2} + \frac{l_i(l_i+1)}{r^2} + V^i(r)\right]P_i(r) = \varepsilon_i P_i(r) , \tag{7.28}$$

where $V^i(r)$ is some assumed potential-energy function for the field in which the electron i moves. This is a homogeneous equation and is free of all the complexities present with inhomogeneous equations: (1) It is not necessary to integrate both an inhomogeneous and a homogeneous equation, and take the linear combination (7.24), and this alone cuts the integration time in half. (2) The sometimes erratic effects of the off-diagonal Lagrangian multipliers ε_{ij} are not present. (3) The parabolic-like segments of the norm curve, $\|P_i\|$ vs. ε_i, degenerate into rectangular wells, with $\|P_i\| = 0$ everywhere except at the set of discrete points $\varepsilon_i = \varepsilon_1^H, \varepsilon_2^H, \varepsilon_3^H \cdots$, as illustrated in Fig. 7-3. There is no possibility of existence of either two or zero positive-slope solutions with the desired number of nodes: there is always exactly one such solution—which is found not by adjusting ε_i to obtain a normalized solution, but by adjusting ε_i to give a solution satisfying the boundary condition at ∞, (7.25), and then multiplying this solution with whatever constant is required to give a normalized function with positive initial slope. The contrast between the situations for homogeneous and inhomogeneous equations may be seen by comparing the right-hand column of Fig. 7-2 with the other five columns.

In short, the differential equations (7.28) each have exactly the same form as the hydrogenic Schrödinger equation (3.9), and the method of solution is entirely analogous to that discussed in Sec. 3-4. The only complication is that if $V^i(r)$ depends functionally on the $P_j(r)$, then a SCF iteration is required; however, this iteration always converges straightforwardly by use of (7.21), with positive c.

In the remainder of this chapter, we describe several possible forms of the potential energy function V^i. In the following chapter, we discuss further details of the SCF solution of homogeneous equations, and compare results obtained by the various methods.

7-8. THE THOMAS-FERMI (TF) AND THOMAS-FERMI-DIRAC (TFD) METHODS

The simplest approximation assumes for V^i the Thomas-Fermi potential,[16] which is derived from a statistical (or semi-free-electron) theory of the atom.[17] A detailed derivation is given in an appendix to this chapter; here we give only a brief summary of the method and the results.

The N electrons in the atom are assumed to be described by a spherically symmetrical number-density distribution function $\rho(r)$, within a sphere of radius r_0 (which may be infinite). The kinetic-energy density at a distance r from the nucleus is assumed to be that appropriate to a zero-temperature free-electron gas of density $\rho(r)$,

$$dE_k/d\Omega = (3/5)(3\pi^2)^{2/3}\,\rho(r)^{5/3} \;, \tag{7.29}$$

where as usual E is measured in rydbergs and distances in Bohr units, so that the volume element $d\Omega$ is in units of a_0^3 and ρ is in electrons/a_0^3. The potential energy is computed classically from the continuous charge distribution $-e\rho(r)$, and the form of the function $\rho(r)$ is adjusted so as to minimize the total kinetic plus potential energy of the atom. The potential-energy function is then found by integrating Laplace's equation, and the result is found to be

$$V_{TF}(r) = -\frac{2Z}{r}\varphi(x) - \frac{2}{r_0}(Z - N) \;, \tag{7.30}$$

where

$$x = r/\mu, \qquad \mu = (1/4)(9\pi^2/2Z)^{1/3} \;, \tag{7.31}$$

and $\varphi(x)$ is that solution of

$$\varphi''(x) = \varphi^{3/2}/x^{1/2} \tag{7.32}$$

which satisfies the boundary conditions

$$\varphi(0) = 1 \tag{7.33}$$

and

$$\varphi(x_0) = 0 \;, \qquad x_0\varphi'(x_0) = -(Z - N)/Z \;. \tag{7.34}$$

Solution of these equations is particularly simple for neutral atoms. The boundary condition (7.34) is then independent of Z, and one universal solution $\varphi(x)$ (for which $r_0 = \mu x_0$ proves to be infinite) can be scaled to give $V_{TF}(r)$ for any desired element. For (positive)

[16] L. H. Thomas, Proc. Cambridge Phil. Soc. **23**, 542 (1927); E. Fermi, Atti accad. Lincei **6**, 602 (1927), Z. Physik **48**, 73 (1928).

[17] P. Gombás, *Die Statistische Theorie des Atoms und ihre Anwendungen* (Springer-Verlag, Wien, 1949), *Pseudopotentiale* (Springer-Verlag, Wien, 1967).

ions, a separate function $\varphi(x)$ is required for each value of $(Z - N)/Z$, but these functions may still be found by a comparatively simple calculation.

Use of the TF potential-energy function for $V^i(r)$ in the differential equation (7.28) makes determination of the radial functions $P_i(r)$ comparatively easy: no SCF iteration is involved, and the only iteration is that required to find the eigenvalue ε_i for each subshell i. Since V^i is the same for all i, the various P_i for a given value of l are automatically mutually orthogonal, even though no Lagrange multipliers ε_{ij} are involved. The drawback is that the form of $V_{TF}(r)$ is rather poor. This is due in part to the fact[17] that at small r, $\rho(r)$ becomes infinite as $r^{-3/2}$ rather than remaining finite, and at large r it tends to zero as r^{-6} for neutral atoms (or becomes zero at a finite radius for positive ions) rather than going exponentially to zero. However, the primary difficulty is that V_{TF} has been calculated classically for a completely smeared charge distribution, and it therefore contains the self-interaction energy of the electron i. Thus at large r, rV_{TF} tends to the limiting value $-2(Z - N)$ [as does $-2Z + rV_c$, Eq. (7.17)], rather than to the correct value $-2(Z - N + 1)$ [as does the Hartree potential, $-2Z + rV_H$, Eq. (7.18)]; thus an excited electron in a neutral atom sees almost zero electric field, and is hardly bound at all. Various modifications of the TF potential have been proposed in order to remedy this defect,[17-20] of which the simplest is that used by Latter:

$$V^i(r) = \min\{V_{TF}(r), -2(Z - N + 1)/r\} \ . \tag{7.35}$$

Although this modification greatly improves computed binding energies of excited electrons, overall results are still comparatively poor because of the basically unsatisfactory form of the TF potential.

The Thomas-Fermi-Dirac model of the atom[21] provides some improvement over the TF model by including an exchange contribution to the electron potential energy. In the free-electron-gas approximation, the exchange-energy density is shown in an appendix to this chapter to be

$$\frac{dE_{ex}}{d\Omega} = -\frac{3}{2}\left(\frac{3}{\pi}\right)^{1/3} \rho(r)^{4/3} \ . \tag{7.36}$$

If the volume integral of this quantity is included in the total energy of the atom, application of the variational principle leads to the TFD potential-energy function

$$V_{TFD}(r) = -\frac{2Z}{r}\psi(x) + \frac{1}{16\pi^2} - \frac{2}{r_0}(Z - N) \ , \tag{7.37}$$

where x is defined by (7.31) and $\psi(x)$ is that solution of

[18]E. Fermi and E. Amaldi, Mem. Acc. Italia **6**, 119 (1934).

[19]C. A. Coulson and C. S. Sharma, Proc. Phys. Soc. (London) **79**, 920 (1962).

[20]R. Latter, Phys. Rev. **99**, 510 (1955).

[21]P. A. M. Dirac, Proc. Cambridge Phil. Soc. **26**, 376 (1930).

$$\psi''(x) = x\left\{\left(\frac{\psi}{x}\right)^{1/2} + \beta_o\right\}^3 \tag{7.38}$$

which satisfies the boundary conditions

$$\psi(0) = 1$$

and

$$\psi(x_o) = \beta_o^2 x_o/16, \qquad x_o\psi'(x_o) - \psi(x_o) = -(Z - N)/Z\ , \tag{7.39}$$

where

$$\beta_o = (\mu/2\pi^2 Z)^{1/2} = (3/2^5\pi^2 Z^2)^{1/3}\ . \tag{7.40}$$

The quantity β_0 and the term $1/16\pi^2$ in (7.37) arose from the exchange energy; if these are deleted, everything of course reduces to the TF results. Since β_0 depends on Z, the function ψ is different for each element (and r_0 is finite) even for neutral atoms. As we shall see in Table 8-8, the TFD potential is appreciably better than V_{TF} provided a tail cutoff analogous to (7.35) is used, but it is still greatly inferior to SCF methods.

The TF (or the TFD) method can be significantly improved by altering the radial scale until computed eigenvalues agree with observed energy levels.[22] However, the method then becomes a semi-empirical rather than *ab initio* one, and is of no use for problems in which experimental data are unavailable.

*7-9. THE PARAMETRIC POTENTIAL METHOD

Intermediate in complexity between the P_j-independent TF and TFD potentials and the SCF potentials to be discussed in the following sections is an assumed analytical potential of some convenient form such as[23]

$$V(r) = -\frac{2}{r}\{(N - 1)e^{-\alpha_1 r} + \alpha_2 r e^{-\alpha_3 r} + \cdots + \alpha_{n-1} r^k e^{-\alpha_n r} + Z - N + 1\}$$

which depends on several parameters α_i. Most commonly the values of these parameters are determined so as to give the best possible agreement (least-squares-wise) between computed and experimental energies; the method is then a semi-empirical one, and is not appropriate for our purposes.

Alternatively, the parameters can be determined by minimization of the computed energies of the ground and perhaps a few excited configurations, and the method then becomes a purely *ab initio* one. Applications of the parametric potential method have been quite successful, especially for highly ionized atoms.

[22] J. C. Stewart and M. Rotenberg, Phys. Rev. **140**, A1508 (1965).

[23] M. Klapisch, C. R. Acad. Sci. B **265**, 914 (1967), Comput. Phys. Commun. **2**, 239 (1971); M. Klapisch, et al., Phys. Lett. **62A**, 85 (1977); E. Luc-Koenig, Physica **62**, 393 (1972). See also P. P. Szydlik and A. E. S. Green, Phys. Rev. A **9**, 1885 (1974); J. D. Talman and W. F. Shadwick, Phys. Rev. A **14**, 36 (1976).

7-10. THE HARTREE METHOD (H)

Hartree[9] originally set up his differential equation of the form (7.28) by intuitive arguments, using a potential-energy function

$$V^i(r) = -\frac{2Z}{r} + V_H(r) \tag{7.41}$$

in which the electron-electron portion V_H [Eq. (7.18)] was derived by the semiclassical argument used in Sec. 7-3. However, it is clear by comparison of expressions (6.6) and (6.11) that we would have obtained the Hartree result (7.28), (7.41) in place of the HF result (7.11) if we had applied the variational principle to the energy expression for product wavefunctions instead of determinantal wavefunctions, and had omitted the orthogonality restrictions that brought in the ε_{ij}.

The Hartree method is the simplest of all SCF methods. However, since the exchange terms in the potential have been omitted, the potential function (7.41) is not sufficiently strongly attractive, the radial wave functions P_i extend too far outward from the nucleus, and the computed one-electron and total binding energies are too small in magnitude and do not agree well with experiment. Moreover, since the ε_{ij} have been omitted and the V^i are different for different subshells, the P_i for given l are not necessarily orthogonal—and in fact the overlap integrals

$$\int_0^\infty P_{n_il_i}P_{n_jl_i}\,dr\ , \qquad n_i \neq n_j\ , \tag{7.42}$$

are sometimes as large as 0.2 or 0.25.

7-11. THE HARTREE-FOCK-SLATER METHOD (HFS)

The Hartree-Fock-Slater method (sometimes called simply Hartree-Slater) is an approximation to HF consisting of a sort of hybrid of the Hartree and TFD schemes. It is a SCF method in which the direct terms of the electron-electron potential energy are computed from the $P_j^2(r)$ as in the HF method, but in which the exchange terms are approximated by a statistical free-electron expression similar to that used in the TFD theory of the atom. The exchange approximation is obtained by dividing (7.36) by the number-density ρ to obtain for the average exchange energy per electron of a free-electron gas

$$\frac{dE_{ex}}{dN} \equiv \frac{1}{2N}\sum_i\sum_j E_{ex}^{ij} = -\frac{3}{2}\left(\frac{3\rho}{\pi}\right)^{1/3} . \tag{7.43}$$

The exchange energy of electron i in a free-electron gas of density ρ (averaged over all possible momenta of this electron) is just

$$\langle E_{ex}^i\rangle_{av} = \langle\sum_j E_{ex}^{ij}\rangle_{av} = \frac{1}{N}\sum_i\sum_j E_{ex}^{ij} = -\frac{3}{2}\left(\frac{24\rho}{\pi}\right)^{1/3}\ \text{Ry}\ . \tag{7.44}$$

Using this expression, Slater[24] introduced the HFS potential-energy function

$$V^i(r) = -\frac{2Z}{r} + V_c(r) - \frac{3}{2}\left(\frac{24\rho}{\pi}\right)^{1/3} \quad \text{Ry} , \tag{7.45}$$

where ρ is the total spherically averaged number density of electrons at a distance r from the nucleus, given from (7.15) by

$$\rho(r) = \sum_{j=1}^{q} w_j \rho_j(r) = \frac{1}{4\pi r^2} \sum_{j=1}^{q} w_j P_j^2(r) . \tag{7.46}$$

It is important to notice that Slater has used the total classical potential energy V_c given by (7.17), which includes the self-interaction of the electron i, rather than using the Hartree energy V_H, (7.18). The reason for this is partly that ρ is the density of *all* N electrons, and therefore includes the self-exchange energy of electron i which (approximately) cancels out the direct self-energy term. [This is analogous to our including the self-energy term $\langle ii|2/r_{12}|ii\rangle$ in both the direct and exchange sum when we were averaging the expressions (6.27) and (6.29) to obtain the result (6.39) for equivalent electrons.] The real advantage is that both V_c and ρ include *all* electrons symmetrically; therefore V^i is the same for all i, which saves some computational effort and (as in the TF methods) automatically produces orthogonal functions $P_i(r)$.

It may be noted that the exchange contribution to the HFS eigenvalue ε_i [evaluated from (7.45) similarly to the derivation of (7.14)] is

$$\int_0^\infty \left[-\frac{3}{2}\left(\frac{24\rho}{\pi}\right)^{1/3}\right] P_i^2(r)\, dr . \tag{7.47}$$

From (7.44), this is just the expectation value of the exchange energy of an electron in orbital i; that is, Koopmans' theorem holds, and the eigenvalue still has the physical significance of a binding energy, in the approximation that the exchange contribution is calculated for a free-electron gas instead of from the correct quantum-mechanical expression.

There is, however, an alternative method of deriving an expression for the exchange contribution to $V_i(r)$.[25,26] If we use the volume integral of (7.36) for the total exchange energy of the atom and apply the variational principle, we obtain from (7.46) a term

$$\begin{aligned}\delta_i \int_0^\infty \left(-\frac{3}{2}\right)\left(\frac{3}{\pi}\right)^{1/3} \rho^{4/3} 4\pi r_1^2\, dr_1 &= -\int_0^\infty \left(\frac{24\rho}{\pi}\right)^{1/3} (\delta_i\rho) 4\pi r_1^2\, dr_1 \\ &= -2w_i \int_0^\infty \left(\frac{24\rho}{\pi}\right)^{1/3} P_i\,(\delta P_i)\, dr_1 .\end{aligned}$$

[24] J. C. Slater, Phys. Rev. **81**, 385 (1951).

[25] R. Gáspár, Acta Phys. Hung. **3**, 263 (1954).

[26] W. Kohn and L. J. Sham, Phys. Rev. **140**, A1133 (1965).

If, as in the derivation of Eqs. (7.5) and (7.11), we take $\delta P_i(r_1)$ to be zero everywhere except near the point $r_1 = r$ and divide out a factor

$$2w_i \int_0^\infty \{\delta P_i(r_1)\}\, dr_1 \ ,$$

we are left with

$$-\left(\frac{24\rho}{\pi}\right)^{1/3} P_i(r)$$

and thus are led to a potential-energy function

$$V^i(r) = -\frac{2Z}{r} + V_c(r) - \alpha\cdot\frac{3}{2}\left[\frac{24}{\pi}\rho(r)\right]^{1/3} \tag{7.48}$$

with $\alpha = 2/3$, instead of $\alpha = 1$ as we had in (7.45). There is some indication that this smaller coefficient of the exchange term gives radial functions $P_i(r)$ that are in better agreement with HF,[27-29] and that still better agreement results from using some value of α between 2/3 and 1 [determined such as to give a minimum value for the total energy E calculated from the correct quantum-mechanical expression (6.14)-(6.39)—the so-called Xα method].[30] It is to be noted, however, that with any value of α other than unity, the eigenvalue ε_i includes α times the expectation value (7.47) of the exchange energy, and so no longer has the physical significance of a binding energy; it is therefore not surprising that the magnitude of ε_i is found to differ considerably from the corresponding HF value.[29-31]

Regardless of the value used for α, however, the principal difficulty with the HFS method is that the $\rho^{1/3}$ exchange approximation only *very* roughly cancels out the direct self-interaction which is included in $V_c(r)$: at large r, ρ tends exponentially to zero, whereas the correct self-interaction term is of the form

$$\int_0^\infty \frac{2}{r_>} P_i^2(r_2)\, dr_2 \rightarrow \frac{2}{r}\int_0^\infty P_i^2(r_2)\, dr_2 = \frac{2}{r} \ .$$

Thus, at large r, the HFS function (7.48) reduces from (7.17) to

$$-\frac{2}{r}(Z - N) \ , \tag{7.49}$$

[27]D. R. Hartree, ref. 6, p. 61.

[28]B. Y. Tong and L. J. Sham, Phys. Rev. **144**, 1 (1966).

[29]R. D. Cowan, A. C. Larson, D. Liberman, J. B. Mann, and J. Waber, Phys. Rev. **144**, 5 (1966).

[30]J. C. Slater, T. M. Wilson, and J. H. Wood, Phys. Rev. **179**, 28 (1969). See also R. W. Harrison and L. C. Cusachs, Int. J. Quantum Chem., Symp. **6**, 217 (1972).

[31]R. D. Cowan, Phys. Rev. **163**, 54 (1967).

rather than to the value

$$-\frac{2}{r}(Z - N + 1) \tag{7.50}$$

shown by either the Hartree function (7.41), (7.18) or the HF potential [Prob. 7-4(3)]; the value (7.50) is what we would expect physically for a highly excited electron far away from the nucleus and the other N−1 electrons. This defect can be remedied as in (7.35) by taking a potential equal to either (7.48) or (7.50), whichever is the more negative,[32] but the results are still in poor agreement with both HF and experiment. In particular, the radius at which the change-over from (7.48) to (7.50) occurs is different in an ion from what it is in an atom, because of the different value of N in (7.50); this results in very poor ionization energies when computed by differencing values of E_{av} as in (6.47).[31]

7-12. THE HARTREE-PLUS-STATISTICAL-EXCHANGE METHOD (HX)

The major difficulties with the HFS method arise from the fact that the exchange approximation used in (7.48) does not accurately account for the most important quantity, the self-interaction energy of the electron i. The obvious solution is to leave this self-energy out in the first place by using V_H instead of V_c, and then to use some statistically based approximation for only the non-self-exchange terms:

$$V^i(r) = -\frac{2Z}{r} + V_H(r) + V_x(r) \; . \tag{7.51}$$

The approximate function $V_x(r)$ should be chosen such that as nearly as possible it should have the following properties of the exact HF terms (involving A_i, B_{ij}, and ε_{ij}) that it replaces:

(1) V_x should vanish for any single-electron configuration, and for any two-electron configuration ns^2 [since exchange terms arise only between electrons of parallel spin—Eq. (6.29)];

(2) for each subshell, the eigenvalue ε_i of the homogeneous differential equation (7.28) should be equal [see (7.14)] to the one-electron binding energy E^i computed quantum mechanically from the solutions P_j of the equations (7.28);

(3) radial functions for equal l but different n should be orthogonal.

Hopefully, these properties would suffice to give eigenvalues and radial wavefunctions (and quantities evaluated from the latter, such as F^k, G^k, E^i, E_{av}, $\langle r^n \rangle_i$, etc.) in fairly good agreement with the corresponding HF quantities.

In attempting to find a suitable function $V_x(r)$ we start from the free-electron expression used by Slater in (7.45), using however a modified number density $\rho'(r)$ that excludes the self-exchange effects of electron i so as to satisfy property (1) above. Since the exchange energy really depends not on ρ but on the density $\rho/2$ of electrons having spins parallel to that of i, we assume tentatively

[32]F. Herman and S. Skillman, *Atomic Structure Calculations* (Prentice-Hall, Englewood Cliffs, N.J., 1963).

$$\frac{1}{2}\rho' = \frac{1}{2}\rho - \rho_i$$

where ρ_i is the probability density of electron i [cf. (7.46)]. However, if $w_i = 1$ then $\rho/2$ cannot include more than $\rho_i/2$ (statistics are not very good for only one electron!), and so we have to modify the above expression to

$$\rho'(r) = \rho(r) - [\min(2,w_i)]\rho_i(r) \ . \tag{7.52}$$

We might then expect the desired function to be

$$V_x(r) = -\frac{3}{2}\left(\frac{24\rho'}{\pi}\right)^{1/3} = -\frac{3}{2}\left(\frac{\rho'}{\rho}\right)^{1/3}\left(\frac{24\rho}{\pi}\right)^{1/3} \quad \text{Ry} \ .$$

However, trial calculations show that the coefficient 3/2 is too large to give the desired property (2) above: the eigenvalue ε_i is too negative, the wavefunctions therefore too tightly bound, the electron-electron repulsions are too large, and E^i therefore not negative enough. Reducing the value of the coefficient makes ε_i less negative and E^i more negative, bringing the two more nearly into equality and (very satisfyingly) bringing both into better agreement with the corresponding HF value.

Calculations for a variety of cases show that there is nothing magic about using the cube root of ρ'/ρ—the first power is simpler, and leads to somewhat better agreement with HF. Further calculations suggest two additional correction factors, giving finally for the HX method[31]

$$\left[-\frac{d^2}{dr^2} + \frac{l_i(l_i+1)}{r^2} + V^i(r)\right]P_i(r) = \varepsilon_i P_i(r) \tag{7.53}$$

with

$$V^i(r) = -\frac{2Z}{r} + \sum_{j=1}^{q}(w_j - \delta_{ij})\int_0^{\infty}\frac{2}{r_>}P_j^2(r_2)\,dr_2$$
$$-k_x f(r)\left[\frac{\rho'}{\rho' + 0.5/(n_i - l_i)}\right]\left(\frac{\rho'}{\rho}\right)\left(\frac{24\rho}{\pi}\right)^{1/3} , \tag{7.54}$$

where sample calculations indicate that $k_x = 0.65$ gives the best agreement with HF. (It may be noted that for $k_x = 0$, HX reduces to the Hartree method.)

The factor in brackets in (7.54) is a minor correction factor to decrease the magnitude of the exchange term at large r (or, really, at small ρ'), introduced to improve agreement with HF values for excited configurations.

The remaining modification factor f(r) is unity, except in certain cases where it is necessary to slightly increase the magnitude of the exchange term at small r in order to obtain functions that are sufficiently close to being orthogonal [property (3) above]; this factor thus takes the place of the omitted Lagrangian multipliers ε_{ij}. It is defined in the following manner: Arrange the subshells of the configuration in the order of decreasing magnitude of eigenvalue ε_i. Then $f(r) \equiv 1$ for the ith subshell except when (a) $l_{i-1} = l_i > 1$, or (b) $l_{i-2} = l_i > 1$ and $w_{i-1} = 1$. In either of these two cases,

$$f(r) = \begin{cases} 1 , & r \geqq r_o , \\ 1 + 0.7(1 - r/r_o) , & r < r_o . \end{cases} \tag{7.55}$$

Here r_o is the location of the kth node of the radial function $P_{n_i l_i}(r)$, where k is the number of subshells having $l = l_i$ and $n < n_i$. [As we shall see in Sec. 8-6, this k^{th} node is the one that lies in the region of the effective-potential barrier between the inner and outer potential wells; the effect of f(r) is thus to make the inner well deeper.]

The HX method is reasonably simple to use, and gives results in rather good agreement[31] with HF except for the inner portions of the $P_i(r)$, which are relatively unimportant for all purposes except the calculation of interactions with the nucleus. In particular, values of E_{av} are quite good because the HF value of E_{av} represents a variational minimum. The overlap integrals (7.42) between HX radial functions of the same l but different n are not exactly zero, but are usually no larger than 0.02 to 0.05. This is sufficiently small that we can simply consider the functions to be exactly orthogonal without making any significant error (see Sec. 13-2).

Except as specifically indicated to the contrary, all theoretical results given in this book have been computed using radial funtions $P_i(r)$ obtained by the HX method.

*7-13. THE HARTREE-SLATER METHOD (HS)

An approach similar to the HX method has been suggested by Lindgren and Rosén,[33] which starts from the Hartree potential but uses the expression

$$V_x(r) = -(24/\pi)^{1/3}[\rho_s^{1/3} - (2\rho_i)^{1/3}] \text{ Ry} \tag{7.56}$$

for the exchange term in (7.51). Here ρ_i is again the probability density of one electron in orbital i, and $\rho_s = \rho$ unless $w_i = 1$, in which case $\rho_s = \rho + \rho_i$; note that the constant coefficient corresponds to the Kohn-Sham value $\alpha = 2/3$ in (7.48).

The Lindgren-Rosén expression for V_x is both much simpler and more firmly founded theoretically than its HX analog [the final term of (7.54)], and contains no empirically adjusted parameters. It satisfies the three requirements listed after (7.51) just about as well as does the HX expression, and for the most part gives results in about equally good agreement with HF and with experiment. However there exist a number of cases (near the beginnings of the transition and rare-earth series) in which the HS method gives poor results; in Ar I $3p^5 3d$ and Ba I 6s4f, for example, HS gives values of F^k and G^k for the open shells that are two to nine times as large as HF values, whereas HX is fairly close to both HF and experiment. [Indeed, it was just such difficulties that dictated the inclusion of the two correction factors in the HX expressions (7.54).]

[33] I. Lindgren, Int. J. Quantum Chem., Symp. **5**, 411 (1971); A. Rosén and I. Lindgren, Phys. Scr. **6**, 109 (1972). See also M. S. Gopinathan, Phys. Rev. A **15**, 2135 (1977).

7-14. RELATIVISTIC CORRECTIONS

Relativistic effects on total binding energies and radial wavefunctions become appreciable for Z as small as 10 and 30, respectively. Proper treatment of these effects leads to what may be called the Dirac-Fock (DF) or Dirac-Hartree-Fock (DHF) equations.[34] However, these are considerably more complex than the HF equations, there being nearly four times as many radial functions to compute—one "large" and one "small" component for each set of quantum numbers nlj, with $j = l + 1/2$ and (except for s orbitals) $j = l - 1/2$. Moreover, a relativistic configuration is defined by a set of subshells

$$\left\{(n_i l_i j_i)^{w_i}\right\} ,$$

and so in general one non-relativistic configuration is made up of several relativistic ones. Thus the problem of calculating energy-level structures becomes much more complicated: where there previously was one value of E_{av} to be found, there now are several—and except in the comparatively infrequent case of nearly pure jj-coupling conditions, these values of E_{av} have little physical significance.

Thus for our purposes it is desirable to incorporate the major relativistic effects within the format of the non-relativistic approach. In many cases, particularly when only the outermost subshells of the atom are concerned, it is sufficient to calculate relativistic energy corrections by the perturbation methods used for one-electron atoms in Sec. 3-7. We then take as relativistic corrections E_r^i and E_r to the one-electron and total binding energies,

$$E_r = \sum_{i=1}^{N} E_r^{\,i} = \sum_{i=1}^{N} (E_m^{\,i} + E_D^{\,i}) \; . \tag{7.57}$$

The mass-velocity and Darwin contributions are given by

$$E_m^{\,i} = -\frac{\alpha^2}{4}\int_0^\infty P_i(r)[\varepsilon_i - V^i(r)]^2 P_i(r)\, dr \tag{7.58}$$

and

$$E_D^{\,i} = -\delta_{l_i 0}\frac{\alpha^2}{4}\int_0^\infty P_i(r)\left[\frac{dV^i(r)}{dr}\right]\left[r\frac{dr^{-1}P_i(r)}{dr}\right] dr \; , \tag{7.59}$$

where $\alpha = 1/137.036$ is the fine-structure constant and all energies are in rydbergs.[35] These expressions involve the central-field potential-energy function $V^i(r)$. Since the HF

[34] I. P. Grant, Adv. Phys. 19, 747 (1970); J. B. Mann and J. T. Waber, J. Chem. Phys. 53, 2397 (1970), At. Data 5, 201 (1973); J. P. Desclaux, Comput. Phys. Commun. 9, 31 (1975), At. Data Nucl. Data Tables 12, 311 (1973).

[35] For a Coulomb potential, $V^i \propto r^{-1}$, we saw in Eq. (3.50) that the Darwin term is zero for $l_i \neq 0$. In multi-electron atoms, calculations show the integral in (7.59) to be very small for $l_i > 0$ even with non-Coulomb potentials.

function (7.20) usually contains one or more singularities, it is difficult to calculate relativistic corrections from (7.58)-(7.59) if the HF method is used, as we have already indicated. There is, however, no problem when using the HX potential-energy function (7.54) or any similar local potential.

For high Z, or when inner subshells of moderate-Z atoms are involved (as in x-ray transitions), energy perturbation corrections may be inadequate and it becomes desirable to include relativistic corrections in the radial functions $P_{nl}(r)$. This can be done rather satisfactorily by using a Pauli-type approximation to the DHF equations:

Local-potential approximations to the DHF equations are of the form (Desclaux[34])

$$P_\kappa' = -\frac{\kappa}{r}P_\kappa + \frac{\alpha}{2}\left(\varepsilon_i - V^i + \frac{4}{\alpha^2}\right)Q_\kappa ,$$

$$Q_\kappa' = \frac{\alpha}{2}(V^i - \varepsilon_i)P_\kappa + \frac{\kappa}{r}Q_\kappa ,$$

where P_κ and Q_κ are respectively the large- and small-component radial functions, ε_i and V^i are measured in rydbergs (not hartrees), and

$$\kappa = \begin{cases} l_i , & j = l_i - 1/2 , \\ -l_i - 1 , & j = l_i + 1/2 . \end{cases}$$

By solving the first equation for Q_κ and substituting into the second, and noting that $\kappa(\kappa + 1) = l_i(l_i + 1)$, we obtain a differential equation for P_κ that involves κ in only one term, which is proportional to κ/r. We replace this coefficient by its $(2j + 1)$-weighted average

$$\frac{2l_i \cdot l_i/r - (2l_i + 2)\cdot(l_i + 1)/r}{(4l_i + 2)} = -\frac{1}{r} ,$$

and thereby obtain the following differential equation for a j-independent radial wavefunction:[36]

$$\left\{-\frac{d^2}{dr^2} + \frac{l_i(l_i + 1)}{r^2} + V^i(r) - \frac{\alpha^2}{4}[\varepsilon_i - V^i(r)]^2 \right.$$
$$\left. -\delta_{l_i 0}\frac{\alpha^2}{4}\left[1 + \frac{\alpha^2}{4}\left(\varepsilon_i - V^i(r)\right)\right]^{-1}\left(\frac{dV^i}{dr}\right)\left(\frac{dP_i/dr}{P_i} - \frac{1}{r}\right)\right\} P_i(r) = \varepsilon_i P_i(r) . \qquad (7.60)$$

It may be seen that this result could have been obtained simply by adding the mass-velocity and Darwin operators of (7.58)-(7.59) into the non-relativistic differential equation (7.53), much as in the one-electron case (3.44). The only difference is inclusion in (7.60) of the factor $[\ \ ]^{-1}$ that was discussed earlier for the spin-orbit term (see fn. 12 of Chapter 3);

[36]R. D. Cowan and D. C. Griffin, J. Opt. Soc. Am. **66**, 1010 (1976); D. D. Koelling and B. N. Harmon, J. Phys. C **10**, 3107 (1977). A delta-factor has been added to the Darwin term similarly to (7.59); it would probably have been better omitted, but the effect is small.

inclusion here is important because it keeps the singularity in the Darwin term at $r = 0$ of order r^{-2} instead of r^{-3}. (Even so, there are some complications in deriving a satisfactory series expansion for P_i at small r to provide starting values for the numerical integration; these are discussed in an appendix to this chapter.)

It is important to remember that we have purposely omitted the spin-orbit term from (7.60) in order to obtain radial functions that are independent of κ (or j), and thereby make possible retention of the vastly simpler non-relativistic formalism of atomic structure. What we are in effect obtaining from (7.60) is a $(2j + 1)$-weighted average of the two relativistic functions P_κ; this is the proper average to use for a closed subshell, or for a spherically averaged open subshell. (We ignore the small components Q_κ completely, and normalize P_i in the usual manner.)

Compared with the non-relativistic differential operator in (7.53), the factor in braces on the left hand side of (7.60) now not only includes P_i implicitly within V^i, but also explicitly involves both P_i and ε_i. However, the effects of the relativistic terms are small enough (even for $Z = 125$) that values of these quantities may be taken from the preceding cycle of the SCF iteration, without complication and without unduly slowing the convergence. Thus the procedure for solution is identical with that used in the non-relativistic case.

Use of the differential equation (7.60) with V^i given by the HX potential-energy function (7.54) will be referred to as the HXR method. Similarly, addition of the two relativistic terms to the HF equations (7.11) — using for V^i in these terms the HX function (7.54) evaluated with the Hartree-Fock radial functions—gives what we shall call the HFR method. Numerical results will be compared with DHF values in Chapter 8.

7-15. CORRELATION CORRECTIONS

In the HF method (or any of the approximations thereto) correlations among the positions of the various electrons are only partially taken into account—somewhat accidentally through the action of the Pauli exclusion principle, as discussed in Sec. 4-5. As a consequence, HF values of the total binding energy E_{av} are not as negative as those observed experimentally. The additional binding energy (after allowing for relativistic effects) is defined as the so-called "correlation energy":

$$E_c \equiv E_{av}^{exp} - (E_{av}^{HF} + E_r) \ . \tag{7.61}$$

Several methods have been developed for the theoretical calculation of correlation energies;[37] however, each of these involves a major computational effort for each configuration of interest. We shall here restrict ourselves to a rather rough, but comparatively simple, calculational method which (like the HFS, HX, and HS exchange approximations) is based on the free-electron-gas model.

It is customary to discuss this subject not in terms of the electron density of the gas, but rather in terms of the radius r_s (in Bohr units) of a sphere whose volume is the average volume per electron:

[37]R. Lefebvre and C. Moser, eds., *Correlation Effects in Atoms and Molecules*, Adv. Chem. Phys. **XIV** (1969); O. Sinanoğlu and K. A. Brueckner, *Three Approaches to Electron Correlation in Atoms* (Yale University Press, New Haven, Conn., 1970); A. Hibbert, Rep. Prog. Phys. **38**, 1217 (1975); A. W. Weiss, Adv. Atomic Mol. Phys. **9**, 1 (1973).

$$\rho = \frac{3}{4\pi r_s^3} \; . \tag{7.62}$$

The electron gas is assumed to be neutralized by a uniform positive background charge. The classical potential energy is then zero, and in the HF approximation the only energies are the kinetic and exchange energies. From (7.29) and (7.36) the average HF energy per electron is

$$\begin{aligned} \bar{E}_{av}^{HF} = \bar{E}_k^i + \bar{E}_{ex}^i &= \frac{3}{5}(3\pi^2\rho)^{2/3} - \frac{3}{2\pi}(3\pi^2\rho)^{1/3} \\ &= \frac{3}{5}\left(\frac{9\pi}{4r_s^3}\right)^{2/3} - \frac{3}{2\pi}\left(\frac{9\pi}{4r_s^3}\right)^{1/3} \\ &= \frac{2.210}{r_s^2} - \frac{0.916}{r_s} \quad \text{Ry} \; . \end{aligned} \tag{7.63}$$

A number of calculations of the non-relativistic energy of a uniform free-electron gas have been made by more accurate methods, using approximations valid in either the high-density[38] or the low-density[39] limit. The difference between any one of these results and the HF expression (7.63) then gives a theoretical value for the average correlation energy per electron, $\bar{e}_c(r_s)$, in a uniform free-electron gas. These estimates are shown by the plotted points in Fig. 7-4. A weighted average of all points is represented by the dashed line; it has the form

$$\bar{e}_c = -\,(1.142 r_s)^{-1} \tag{7.64}$$

for large r_s, and shows a comparatively weak logarithmic dependence for small r_s.

Electrons in atoms cannot move around as readily as free electrons because they are localized by their attraction to the nucleus; we would therefore expect correlation energies in atoms to be smaller in magnitude than those in a free-electron gas, especially at high densities. This is born out by the empirical observation[40] that in atoms

$$\bar{e}_c \equiv \frac{E_{av}^{\,exp} - (E_{av}^{\,HF} + E_r)}{N} \cong -\,0.08 \text{ Ry/electron} \; , \tag{7.65}$$

even though the values of r_s involved are mostly less than 0.5. Accordingly, we estimate the function $\bar{e}_c(r_s)$ for electrons in atoms to be something like the solid curve in the figure.

In calculating the total correlation energy of an atom, we must be careful not to include any self-correlation energy. (The self-correlation energy of one electron out of 10^{23} in a free-electron gas is negligible, but this is not true of a one-electron atom!) One method of accomplishing this[31] is to calculate the expectation value of $\bar{e}_c$ for each electron,

[38]M. Gell-Mann and K. A. Brueckner, Phys. Rev. **106**, 364 (1957); W. J. Carr, Jr. and A. A. Maradudin, Phys. Rev. **133**, A371 (1964); L. Hedin, Phys. Rev. **139**, A796 (1965).

[39]D. Pines, Solid State Physics **I**, 375 (1955); R. D. Cowan and J. G. Kirkwood, Phys. Rev. **111**, 1460 (1958); W. J. Carr, Jr., R. A. Coldwell-Horsfall and A. E. Fein, Phys. Rev. **124**, 747 (1961).

[40]E. Clementi, J. Chem. Phys. **38**, 2248 (1963), **39**, 175 (1963), **42**, 2783 (1965).

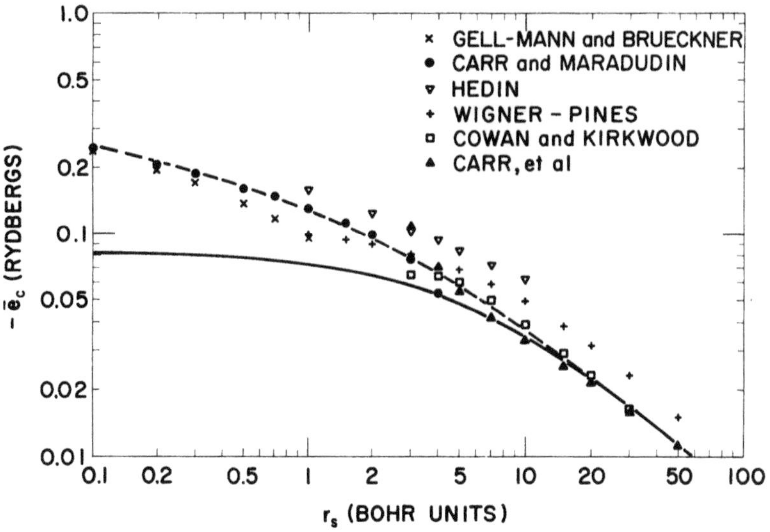

Fig. 7-4. Dotted curve: The average correlation energy per electron in a uniform-density, zero-temperature free electron gas, computed in various approximations. Solid curve: The correlation energy, modified at high density to fit the empirical value $\bar{e}_c \cong -0.08$ Ry/electron for bound electrons in an atom. The independent variable r_s is the radius of a sphere whose volume is the (local) average volume per electron.

$$E_c^{\,i} = 4\pi \int_0^\infty \rho_i(r)\, \bar{e}_c(r_s)\, r^2\, dr = \int_0^\infty P_i^{\,2}(r)\, \bar{e}_c(r_s)\, dr \ , \tag{7.66}$$

with r_s obtained from (7.62) using for $\rho(r)$ the density of all $N-1$ electrons other than i, and then summing (7.66) over all electrons:

$$E_c = \sum_{i=1}^{N} E_c^{\,i} = \sum_{j=1}^{q} w_j E_c^{\,j} \ . \tag{7.67}$$

This certainly gives $E_c = 0$ for a one-electron atom, but it is very poor for a two-electron atom (or indeed for the contribution from any doubly occupied subshell far removed from other electrons in the atom—e.g., both $1s^2$ and $2s^2$ in Be I $1s^22s^2$, or $1s^2$ and $2p^2$ in the excited configuration Be I $1s^22p^2$); this is because correlation energies are a two-or-more electron effect, and because (7.67) tends to count pairs twice. More precisely, the number of pair-correlation terms that contribute to $\bar{e}_c$ in a free-electron gas is roughly proportional to the number of close-neighbor electrons (which is perhaps a dozen); when we have only one close neighbor in an atom, $\bar{e}_c$ greatly over-estimates the correlation energy.

Instead of the above procedure, we here compute E_c as though we were adding electrons to the atom one at a time (in the order from most strongly bound to least strongly bound—i.e., 1s, 2s, 2p, $\cdots$) and compute the correlation energy of the ith electron not from the average free-electron value $\bar{e}_c$, but from the incremental value e_c upon adding one more electron.

The incremental correlation energy per electron is

$$e_c = \left[\frac{d(N\bar{e}_c)}{dN}\right]_{vol.} = \left[\frac{d(\rho\bar{e}_c)}{d\rho}\right]_{vol.} = \frac{d(r_s^{-3}\bar{e}_c)}{d(r_s^{-3})} = -\frac{r_s^4}{3}\frac{d(r_s^{-3}\bar{e}_c)}{dr_s} . \tag{7.68}$$

In the low-density limit where $\bar{e}_c \propto r_s^{-1}$ this gives

$$e_c = \frac{4}{3}\bar{e}_c , \tag{7.69}$$

and in the high-density limit $\bar{e}_c \cong$ const we have

$$e_c = \bar{e}_c . \tag{7.70}$$

As a semi-empirical interpolation formula between (7.69), (7.64) and the high-density limit (7.70), (7.65), we assume

$$e_c(r_s) = -\left[4(r_s + 9)^{1/2} + \frac{3}{4}1.142r_s\right]^{-1} . \tag{7.71}$$

We then take the correlation correction to the total binding energy E_{av} to be

$$E_c = \sum_{i=2}^{N}\int_0^\infty P_i^2(r)e_c^i(r_s)\,dr , \tag{7.72}$$

where e_c^i is given by (7.71), and r_s is in turn given as a function of r by

$$r_s = \left[\frac{4\pi}{3}\sum_{j=1}^{i-1}\rho_j(r)\right]^{-1/3} = \left[\frac{1}{3r^2}\sum_{j=1}^{i-1}P_j^2(r)\right]^{-1/3} . \tag{7.73}$$

This final method (7.71)-(7.73) of computing E_c is more laborious than (7.66)-(7.67) because an integral has to be evaluated for N−1 electrons rather than for only q subshells. However, the computing time required to obtain E_c is still small compared with the time required to obtain the SCF radial wavefunctions, and the gain in accuracy of computed energies is great enough to make the effort worthwhile.

A correlation contribution to the potential-energy function $V^i(r)$ of the one-electron differential equation for $P_i(r)$ can readily be derived[26,28,41] by taking the variation with respect to P_i of the correlation energy (7.72). The result is just $e_c^i(r_s)$, where r_s is given as a function of r by (7.73), except with the summation carried over all orbitals other than i. However, inclusion of such a term in $V^i(r)$ has only a negligible effect on the computed function $P_i(r)$, and we shall not consider this option further.

[41]P. Gombás, Acta Phys. Hung. **13**, 233 (1961), **14**, 83 (1962).

*7-16. APPENDIX: THE THOMAS-FERMI AND THOMAS-FERMI-DIRAC ATOMS

For a free electron (the particle-in-a-box problem) there exist two quantum states, one for each possible spin orientation, for each volume h^3 of phase space. Thus using spherical coordinates in momentum space, there are

$$dN = 2(d\Omega)\, 4\pi p^2\, dp/h^3 \tag{7.74}$$

possible states for an electron in volume $d\Omega$ with momentum $|\mathbf{p}|$ between p and p + dp. For a system of electrons in its ground state, there will be one electron in each one-electron quantum state for all momenta from zero up to some maximum value p (the Fermi momentum), and so the density of electrons will be

$$\rho = dN/d\Omega = (8\pi/h^3)\int_0^p p^2\, dp = 8\pi p^3/3h^3 \ . \tag{7.75}$$

The kinetic-energy density of the electrons is

$$\begin{aligned} dE_k/d\Omega &= \int_0^p (p^2/2m)\, 8\pi p^2\, dp/h^3 \\ &= (8\pi/10mh^3)p^5 \\ &= (3h^2/10m)(3\pi^2)^{2/3}\rho^{5/3} \\ &= (3/10)e^2a_0(3\pi^2)^{2/3}\rho^{5/3} \ . \end{aligned}$$

From now on we shall measure distances and energies in our usual units of a_0 and rydbergs, respectively, and thus have

$$dE_k/d\Omega = c_k\rho^{5/3} \tag{7.76}$$

with

$$c_k = (3/5)(3\pi^2)^{2/3} \ . \tag{7.77}$$

Let us now consider a spherically symmetric atom in which the electron density at a distance r_2 from the nucleus is $\rho(r_2)$; we suppose this electron distribution to extend out to some radius r_0, which may be infinite. We assume the kinetic energy of the electrons in the atom to be given by the volume integral of the free-electron expression (7.76), where $\rho = \rho(r_2)$ is the local electron density in the atom. We calculate the potential energy of the system from the classical expression

$$V(r_1) = -\frac{2Z}{r_1} + \int_0^{r_o} \frac{2}{r_>} \rho(r_2)\, 4\pi r_2^2\, dr_2$$

$$= -\frac{2Z}{r_1} + \left\{\frac{2}{r_1}\int_0^{r_1} + \int_{r_1}^{r_o} \frac{2}{r_2}\right\} \rho(r_2)\, 4\pi r_2^2\, dr_2 \qquad (7.78)$$

for the potential energy of a test-charge electron at a distance r_1 from the nucleus; see (7.15) and (7.16). The total kinetic-plus-potential energy of the atom is thus

$$E = \int_0^{r_o} \left[c_k \rho^{2/3}(r_1) - \frac{2Z}{r_1} + \frac{1}{2}\int_0^{r_o} \frac{2}{r_>} \rho(r_2)\, 4\pi r_2^2\, dr_2\right] \rho(r_1)\, 4\pi r_1^2\, dr_1 \ , \qquad (7.79)$$

where $r_>$ is the greater of r_1 and r_2, and we have introduced a factor 1/2 in the electron-electron potential-energy term in order to avoid double counting of each pair of charge volume elements.

We shall determine the form of the distribution function $\rho(r_2)$ by applying the variational principle—adjusting the electron distribution so as to minimize the total energy. The variation must be carried out subject to the condition that the total number of electrons

$$N = \int_0^{r_o} \rho(r_1)\, 4\pi r_1^2\, dr_1 \qquad (7.80)$$

be constant. Similarly to the procedure followed in the derivation of the HF equations (Sec. 7-1), we allow for this condition by introducing a Lagrangian multiplier $-V_o$ and thus have the condition

$$\delta\left(E - V_o \int_0^{r_o} \rho(r_1)\, 4\pi r_1^2\, dr_1\right) = 0$$

or

$$\int_0^{r_o} [\delta\rho(r_1)] \left[\frac{5}{3} c_k \rho^{2/3}(r_1) - \frac{2Z}{r_1} + \int_0^{r_o} \frac{2}{r_>} \rho(r_2)\, 4\pi r_2^2\, dr_2 - V_o\right] 4\pi r_1^2\, dr_1 = 0 \ ,$$

where a factor 2 has been obtained in the variation of the electron-electron term, much as in (7.9) and (7.10). Since the variation $\delta\rho(r_1)$ is arbitrary in form, we take it to be zero everywhere except in the immediate vicinity of the point $r_1 = r$, and thereby see that E will be a minimum provided ρ satisfies the relation

$$(5/3)c_k \rho^{2/3}(r) + V(r) - V_o = 0$$

or

$$[3\pi^2\rho(r)]^{2/3} + V(r) - V_o = 0 \ , \qquad r \leqq r_o \ . \tag{7.81}$$

The value of the Lagrangian multiplier V_o is to be determined from the condition (7.80); i.e., r_o must have such a value that the integral in (7.80) gives the correct number of electrons N. For an isolated atom (i.e., an atom under no external compressional forces) we may expect the electron density to be zero at the surface of the atom $r = r_o$, and from (7.81), (7.78), and (7.80) we therefore obtain

$$V_o = V(r_o) = -\ 2(Z - N)/r_o \ . \tag{7.82}$$

Differentiation of (7.78) leads to the expressions

$$\frac{d}{dr}rV(r) = \int_r^{r_o} \frac{2}{r_2}\rho(r_2)\, 4\pi r_2{}^2\, dr_2 \tag{7.83}$$

and[42]

$$\frac{1}{r}\frac{d^2}{dr^2}rV(r) = -8\pi\rho(r) \ , \qquad r \leqq r_o \ . \tag{7.84}$$

Using (7.81) and (7.78) we thus find that V(r) is given by that solution of the differential equation

$$\frac{d^2}{dr^2}[r(V_o - V)] = \frac{8}{3\pi\, r^{1/2}}[r(V_o - V)]^{3/2} \tag{7.85}$$

which satisties the boundary conditions (7.82)

$$\Big\{r[V_o - V(r)]\Big\}_{r\,=\,0} = 2Z \ , \tag{7.86}$$

and [from (7.83)]

$$\Big\{d[rV(r)]/dr\Big\}_{r\,=\,r_o} = 0 \ . \tag{7.87}$$

With the definition

$$\varphi(x) = r[V_o - V(r)]/2Z \tag{7.88}$$

[42]This is just Laplace's equation $\nabla^2(-V/e) = -\ 4\pi(-e\rho)$ written in our present units—the coefficient 8π appearing in place of 4π because we are measuring V in rydbergs rather than in hartrees.

and the value (7.82) for V_o, Eqs. (7.85)-(7.87) are equivalent to (7.32)-(7.34).

The exchange energy of a free-electron gas may be found by a method given by Gombás.[17] The wavefunction for a free electron with momentum $\mathbf{p}_j$ and wavevector $\mathbf{k}_j = \mathbf{p}_j/\hbar$ may be written

$$\varphi_j(\mathbf{r}) = \Omega^{-1/2} e^{i\mathbf{k}_j \cdot \mathbf{r}} \; ; \tag{7.89}$$

this function is obviously properly normalized in volume Ω:

$$\int |\varphi_j(\mathbf{r})|^2 \, d\Omega = 1 \; .$$

For the exchange energy between two electrons of parallel spin, we find from (6.11) [cf. (6.29) regarding the spin limitation]

$$\begin{aligned} E_{ex}{}^{jl} \equiv - \langle jl | 2/r_{12} | lj \rangle &= -2\Omega^{-2} \iint e^{i(\mathbf{k}_l - \mathbf{k}_j)\cdot \mathbf{r}_1} e^{i(\mathbf{k}_j - \mathbf{k}_l)\cdot \mathbf{r}_2} (1/r_{12}) \, d\Omega_1 \, d\Omega_2 \\ &= -2 \iint \rho_{jl}{}^*(\mathbf{r}_1) \rho_{jl}(\mathbf{r}_2)(1/r_{12}) \, d\Omega_1 \, d\Omega_2 \; , \end{aligned} \tag{7.90}$$

where

$$\rho_{jl}(\mathbf{r}) \equiv \Omega^{-1} e^{i(\mathbf{k}_j - \mathbf{k}_l)\cdot \mathbf{r}} \; .$$

The quantity

$$V_{jl}(\mathbf{r}_1) \equiv 2 \int \rho_{jl}(\mathbf{r}_2)(1/r_{12}) \, d\Omega_2 \tag{7.91}$$

has the nature of a potential function, which must satisfy Laplace's equation[42]

$$\nabla^2 V_{jl}(\mathbf{r}_1) = -8\pi \rho_{jl}(\mathbf{r}_1) \; .$$

To within an additive constant, the solution of this equation is

$$V_{jl}(\mathbf{r}_1) = \frac{8\pi}{|\mathbf{k}_j - \mathbf{k}_l|^2} \rho_{jl}(\mathbf{r}_1) \; , \tag{7.92}$$

as may readily be seen by letting $\mathbf{K} = \mathbf{k}_j - \mathbf{k}_l$ and noting that

$$\frac{d^2}{dx_1{}^2} e^{i(K_x x_1 + K_y y_1 + K_z z_1)} = -K_x{}^2 e^{i\mathbf{K}\cdot \mathbf{r}_1} \; , \qquad \text{etc.} \; ;$$

the additive constant must be zero, since from (7.91) $V_{jl} = 0$ if $\rho_{jl} = 0$. We thus see that (7.90) simplifies to

$$E_{ex}{}^{jl} = -\frac{8\pi}{|k_j - k_l|^2}\int |\rho_{jl}(r_1)|^2 \, d\Omega_1 = -\frac{8\pi}{\Omega|k_j - k_l|^2} = -\frac{2h^2}{\pi\Omega|p_j - p_l|^2} \cdot$$

It is convenient to express $|\mathbf{p}_j - \mathbf{p}_l|^2$ in terms of spherical coordinates in momentum space, with the z-axis oriented parallel to $\mathbf{p}_j$. Then using (1/2) dN from (7.74) for the number of electrons l having spin parallel to electron j, we find the total exchange energy between electron j and all electrons l having $|\mathbf{p}_l| \leqq p$ to be

$$\begin{aligned}\sum_{l=1}^{N} \delta_{m_{s_j} m_{s_l}} E_{ex}{}^{jl} &= -\frac{2}{\pi h}\int_0^p\int_0^\pi \frac{2\pi p_l^2 \, dp_l \sin\theta \, d\theta}{p_j^2 + p_l^2 - 2p_j p_l \cos\theta} \\ &= -\frac{4}{hp_j}\int_0^p p_l \, ln\left|\frac{p_j + p_l}{p_j - p_l}\right| dp_l \\ &= -\frac{2}{h}\left[2p + \frac{p^2 - p_j^2}{p_j} \, ln\left|\frac{p + p_j}{p - p_j}\right|\right] . \end{aligned} \qquad (7.93)$$

Finally, the total exchange energy for all electrons in volumn Ω is (including a factor 1/2 to avoid double counting of electron pairs)

$$\begin{aligned} E_{ex} &= \frac{1}{2}\sum_j \sum_l \delta_{m_{s_j} m_{s_l}} E_{ex}{}^{jl} \\ &= -\frac{8\pi\Omega}{h^4}\int_0^p\left[2pp_j^2 + (p^2 p_j - p_j^3) \, ln\left|\frac{p + p_j}{p - p_j}\right|\right] dp_j = -8\pi\Omega(p/h)^4 ; \end{aligned}$$

evaluation of the logarithmic integral is carried out by integrating by parts several times.

Using (7.75), we may then write the exchange energy per unit volume in the form

$$\frac{dE_{ex}}{d\Omega} = -8\pi\left(\frac{3\rho}{8\pi}\right)^{4/3} = -\frac{3}{2}\left(\frac{3}{\pi}\right)^{1/3}\rho^{4/3} , \qquad (7.94)$$

which is the expression used in (7.36).

If the total exchange energy of the atom is added to the expression (7.79), application of the variational principle gives in place of (7.81)

$$[3\pi^2\rho(r)]^{2/3} - \frac{2}{\pi}[3\pi^2\rho(r)]^{1/3} + V(r) - V_o = 0 ,$$

or

$$\rho(r) = \frac{1}{3\pi^2}\left\{\frac{1}{\pi} \pm \left[\frac{1}{\pi^2} + V_o - V(r)\right]^{1/2}\right\}^3 , \tag{7.95}$$

where the plus sign must be chosen because ρ cannot be negative. Laplace's equation (7.84) then leads to the differential equation

$$\begin{aligned}\frac{d^2}{dr^2}\left\{r\left(\frac{1}{\pi^2} + V_o - V\right)\right\} &= 8\pi r\rho \\ &= \frac{8}{3\pi r^{1/2}}\left\{\frac{r^{1/2}}{\pi} + \left[r\left(\frac{1}{\pi^2} + V_o - V\right)\right]^{1/2}\right\}^3\end{aligned} \tag{7.96}$$

for the potential-energy function V(r) in the TFD atom. Equations (7.95) and (7.96) differ from the corresponding TF expressions (7.81) and (7.85) in the presence of the new terms $\pi^{-1}\,(\mathrm{Ry})^{1/2}$ and π^{-2} Ry which appear inside the braces, and which arose from the exchange term.

The exchange energy causes the electron density $\rho_o = \rho(r_o)$ at the boundary of the atom to be greater than zero, and causes r_o to be finite even for neutral atoms. The value of the Lagrangian multiplier V_o is correspondingly more difficult to evaluate than in the TF case. The proper procedure (for an isolated atom, under zero external pressure) is to minimize the total energy of the atom with respect to r_o, subject to the restriction (7.80). However, this involves a somewhat involved calculation,[17] and the same result can be obtained by applying the virial theorem (3.61) to a free-electron gas of density ρ_o: Since the only potential energy is the exchange energy, the virial theorem for zero pressure gives

$$\langle dE_k/d\Omega\rangle = -\frac{1}{2}\langle dE_{ex}/d\Omega\rangle ,$$

or from (7.76) and (7.94)

$$\frac{3}{5}(3\pi^2)^{2/3}\rho_o^{5/3} = \frac{3}{4\pi}(3\pi^2)^{1/3}\rho_o^{4/3} .$$

Thus

$$(3\pi^2\rho_o)^{1/3} = \frac{5}{4\pi} \tag{7.97}$$

and from (7.95)

$$V_o = V(r_o) - \frac{15}{16\pi^2} = -\frac{2(Z-N)}{r_o} - \frac{15}{16\pi^2}\quad \mathrm{Ry} . \tag{7.98}$$

With the definition

$$\psi(x) = \frac{r}{2Z}\left[\frac{1}{\pi^2} + V_o - V(r)\right] \tag{7.99}$$

and with the boundary conditions (7.87), (7.98), and

$$\left\{r\left[\frac{1}{\pi^2} + V_o - V(r)\right]\right\}_{r=0} = \left\{-rV(r)\right\}_{r=0} = 2Z ,$$

the differential equation (7.96) is equivalent to (7.37)-(7.40).

*7-17. APPENDIX: SMALL-r SOLUTION FOR THE HXR METHOD

In trying to find a series solution of the differential equation (7.60), we run into difficulty with the Darwin factor

$$\delta_{l0}\frac{\alpha^2/4}{1+(\varepsilon-V)\alpha^2/4} \cong \frac{r}{2Z}\left[\frac{\delta_{l0}}{1+(\varepsilon+4/\alpha^2)r/2Z}\right] , \tag{7.100}$$

where we have taken $V \cong -2Z/r$ for small r. The quantity $\varepsilon + 4/\alpha^2 \geqq -Z^2 + 274^2 \cong 274^2$ is so large that a series expansion of the factor in brackets in (7.100) is invalid at r_2 (on an integration mesh $r_j = jh$) for any customary value of h; for example, on a mesh commonly used for non-relativistic problems,[43] $h \cong 0.0022\ Z^{-1/3}$, the limitation $2r_2/\alpha^2 Z \ll 1$ requires $Z \gg (0.0022 \cdot 274^2)^{3/4} = 46$. In order to be able to handle all values of Z (including Z as small as unity), we try using an r-independent value

$$b \equiv b_{jh} = \frac{\delta_{l0}}{1+(\varepsilon+4/\alpha^2)jh/2Z} . \tag{7.101}$$

Because of this approximation, there is no point in carrying series expansions beyond a couple of terms and so we assume

$$P \cong r^{\lambda+1} + a_1 r^{\lambda+2} \tag{7.102}$$

for the (unnormalized) radial function. Differentiating this expression we find

$$\frac{P'}{P} - \frac{1}{r} \cong \frac{1}{r}(\lambda + a_1 r) .$$

In the approximation $V \cong -2Z/r$, the Darwin term is

$$-\frac{rb}{2Z}\left(\frac{2Z}{r^2}\right)\frac{1}{r}(\lambda + a_1 r) = -\frac{b}{r^2}(\lambda + a_1 r) ,$$

and the differential equation (7.60) becomes to second order in r^{-1}

$$P'' = (\lambda + 1)\lambda r^{\lambda-1} + a_1(\lambda + 2)(\lambda + 1)r^{\lambda}$$
$$= \left[\frac{l(l+1) - \alpha^2 Z^2 - b\lambda}{r^2} - \frac{2Z + \alpha^2\varepsilon Z + ba_1}{r}\right]\left[r^{\lambda+1} + a_1 r^{\lambda+2}\right] .$$

[43]More precisely, $h = 0.0025\ \mu$, where μ is the TF scale factor defined by (7.31). See F. Herman and S. Skillman, *Atomic Structure Calculations* (Prentice-Hall, Englewood Cliffs, N.J., 1963).

Equating coefficients of $r^{\lambda-1}$ gives

$$(\lambda + 1)\lambda = l(l + 1) - \alpha^2Z^2 - b\lambda$$

or

$$\lambda + 1 = \frac{1-b}{2} + \left[l(l + 1) + \left(\frac{1+b}{2}\right)^2 - \alpha^2Z^2\right]^{1/2} , \tag{7.103}$$

and equating coefficients of r^{λ} gives

$$a_1 = -\frac{(2 + \alpha^2\varepsilon)Z}{(\lambda + 2 + b)(\lambda + 1) - l(l + 1) + \alpha^2Z^2} . \tag{7.104}$$

In the small-Z limit, $b \cong 0$ and $|\alpha^2\varepsilon| \leqq \alpha^2Z^2 \cong 0$, and these expressions reduce to the usual non-relativistic results

$$\lambda + 1 = l + 1 , \qquad a_1 = -\frac{2Z}{2(l + 1)} . \tag{7.105}$$

On the other hand, for large Z and $l = 0$ (s electrons), $b \cong 1$ and (7.103) becomes

$$\lambda + 1 = [1 - \alpha^2Z^2]^{1/2} , \tag{7.106}$$

showing firstly that $\lambda < 0$ $(=l)$ and secondly that solutions are possible only for $Z \leqq \alpha^{-1} \cong 137$—which is however not a serious restriction for our purposes.[44]

There still remains the question of what value to use for b_{jh}. This was settled empirically by using $j = 0$, 1, and 2, and using $h = 1$, 0.4, and 0.16 times the Herman-Skillman value,[43] to make calculations for uranium; in all cases, the series solution (7.102)–(7.104) was employed at r_1 and r_2, and the integration carried forward numerically to r_3, r_4, $\cdots$ in the usual manner. [The derivatives P′ and V′ required in the Darwin term were evaluated via the simple two-interval formula, $f_i' = (f_{i+1} - f_{i-1})/2h$; the value of the Darwin term at $r = h$ was taken to be four times its value at $r = 2h$.] The computed results were found to be fairly insensitive to j and h; $j = 2$ with the Herman-Skillman h was chosen as the most appropriate combination.

It is of course somewhat less than satisfactory to use a calculational method whose numerical results depend on the numerical procedures employed. However, the HXR and HFR methods are at best only approximations to Dirac-Hartree-Fock, and computed results are a sufficiently great improvement over HX and HF that this further approximation is not serious.

[44]With $-1 < \lambda < 0$, the radial factor $r^{-1}P$ of the spin-orbital becomes infinite as r tends to zero, though P itself tends to zero. The kinetic-energy radial integral (6.18) will remain finite provided $\lambda > -1/2$, or $Z \leqq 118$. These high-Z problems are normally surmounted by using a finite-sized nucleus;[34] here, the $118 < Z \leqq 137$ problem is sloughed over by our use of a rather coarse integration mesh, and by indirect evaluation of the kinetic energy from the eigenvalue ε (see Sec. 8-1).

8

RADIAL WAVEFUNCTIONS AND RADIAL INTEGRALS

8-1. COMPUTER CALCULATION OF RADIAL FUNCTIONS

Before the development of modern digital electronic computers, self-consistent solutions of the H and HF equations were obtained with the aid of a desk calculator and weeks or months of heroic labor.[1] Nowadays the calculation of radial functions $P_{nl}(r)$ is vastly simplified by the wide availability of appropriate computer programs. The essence of such a program for use with homogeneous-equation methods is shown in Fig. 8-1.

The calculation is begun with the aid of an appropriate potential-energy function[2] $V_0(r)$, scaled to $-2Z/r$ at small radii and to $-2(Z - N + 1)/r$ at large radii. This provides an initial potential $V^i(r)$ in which to calculate trial functions $P_i^{(1)}(r)$ for all orbitals n_il_i on the first cycle of the SCF iteration ($m = 1$). Thereafter, the potential-energy function $V^i(r)$ for cycle m is computed for the desired method (H, HFS, HX, or HS), using the functions $P_j(r)$ from cycle $m - 1$. In order to speed the convergence, each trial function P_j used to compute V^i on the third and subsequent cycles is taken to be a linear combination of the trial input function and the integral of the differential equation on the preceding cycle:

$$P_j^{(m)}(\text{input}) = c_j\, P_j^{(m-1)}(\text{output}) + (1 - c_j)\, P_j^{(m-1)}(\text{input}) \ . \tag{8.1}$$

The best value of c_j usually lies between 0.3 and 1.0; the value 0.4 can be used at the start, and adjusted appropriately as the iteration proceeds. (This adjustment is not indicated in Fig. 8-1.) The inner iteration on ε_i to find that integral of the differential equation which satisfies the boundary conditions has already been discussed in Secs. 3-4 and 7-7, and the nature of the procedure is illustrated by Fig. 3-2 and the right-hand column of Fig. 7-2.

The flow diagram of Fig. 8-1 is appropriate to the TF or TFD problem, except that $V_0(r)$ is the TF or TFD potential for the Z and N in question and no SCF iteration is involved. It

[1]See, for example, D. R. Hartree, Proc. Roy. Soc. (London) **141**, 282 (1933); D. R. Hartree and W. Hartree, Proc. Roy. Soc. (London) **157**, 490 (1936).

[2]The computer program [R. D. Cowan, Phys. Rev. **163**, 54 (1967)] is based on that described by F. Herman and S. Skillman, *Atomic Structure Calculations* (Prentice-Hall, Englewood Cliffs, N.J., 1963), which book may be consulted for basic numerical methods. The function V_0 used is obtained from their HFS function U(x) for argon.

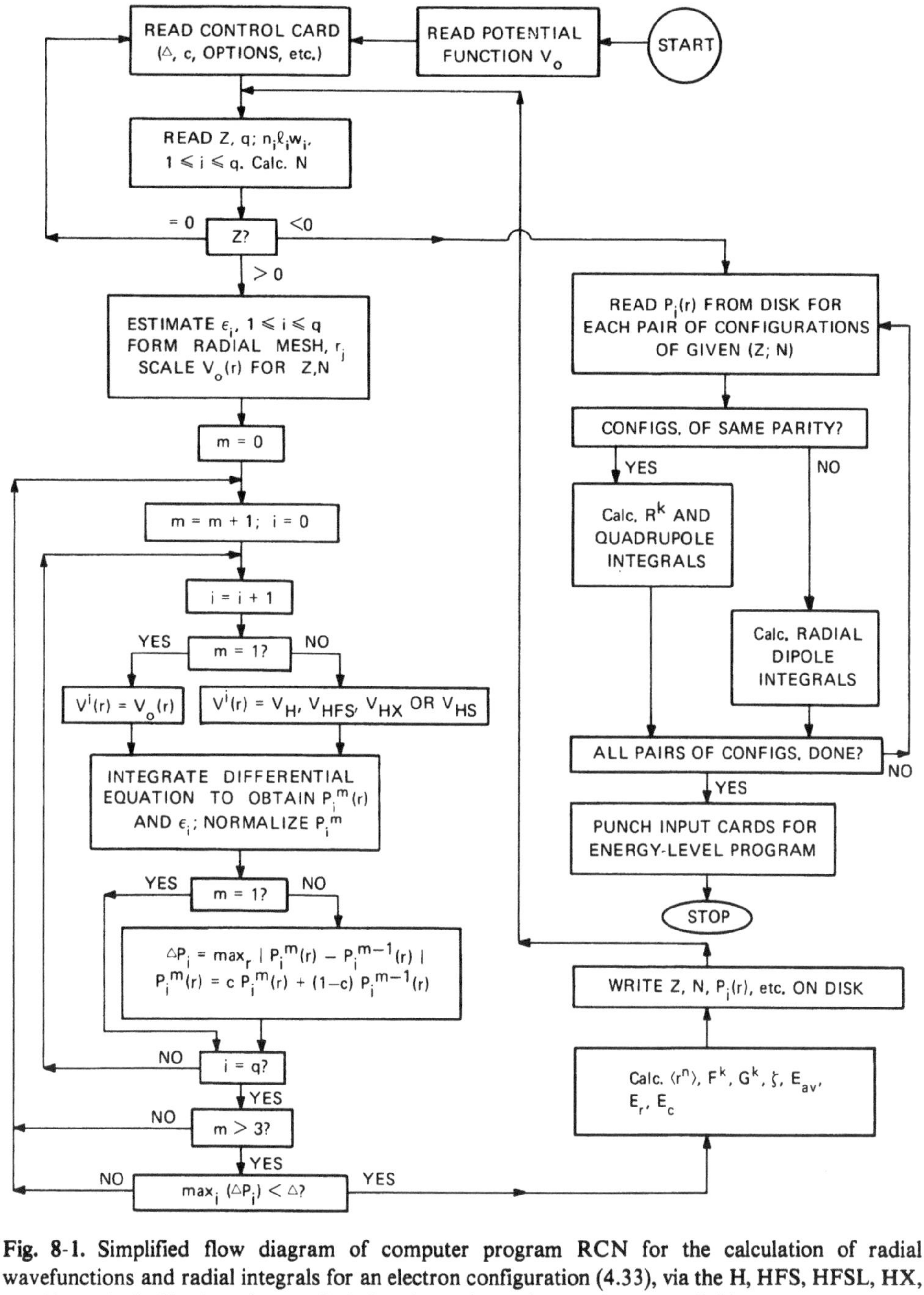

Fig. 8-1. Simplified flow diagram of computer program RCN for the calculation of radial wavefunctions and radial integrals for an electron configuration (4.33), via the H, HFS, HFSL, HX, or HS method. The iteration to find the eigenvalue ε_i is not shown explicitly.

is also basically appropriate to the HF problem[3] except for the much more complex procedures involved in the ε_i iteration, as described in Secs. 7-5 and 7-6.

In any case, once the iteration(s) have converged to within the desired tolerances, the radial functions $P_i(r)$ can be used to evaluate numerically the Coulomb radial integrals F^k and G^k defined by (6.28) and (6.30), and the integral of the radial portion of the spin-orbit operator (4.2)

$$\zeta_i \equiv \int_0^\infty \xi(r)|P_i(r)|^2\,dr = \frac{\alpha^2}{2}\int_0^\infty \frac{1}{r}\left(\frac{dV}{dr}\right)|P_i(r)|^2\,dr\ , \tag{8.2}$$

as well as other radial integrals of interest, such as expectation values of r^n

$$\langle r^n\rangle_i \equiv \int_0^\infty r^n|P_i(r)|^2\,dr\ , \tag{8.3}$$

the kinetic-energy and electron-nuclear-energy integrals (6.18) and (6.19), respectively, and the relativistic and correlation energy corrections defined in Secs. 7-14 and 7-15.

With these radial integrals available, it is then a straightforward matter to calculate the configuration-average Coulomb interaction energies E^{ij}, one-electron binding energies E^i, and total binding energies E_{av} from (6.13)-(6.39). For convenience of reference, we repeat the important relations here:

$$E^{ii} = F^0(ii) - \frac{2l_i+1}{4l_i+1}\sum_{k>0}\begin{pmatrix} l_i & k & l_i \\ 0 & 0 & 0\end{pmatrix}^2 F^k(ii)\ , \tag{8.4}$$

$$E^{ij} = F^0(ij) - \frac{1}{2}\sum_k \begin{pmatrix} l_i & k & l_j \\ 0 & 0 & 0\end{pmatrix}^2 G^k(ij)\ , \qquad n_il_i \neq n_jl_j\ , \tag{8.5}$$

$$F^k(ij) = \int_0^\infty\!\!\int_0^\infty \frac{2r_<^k}{r_>^{k+1}} P_i^*(r_1)P_j^*(r_2)P_i(r_1)P_j(r_2)\,dr_1\,dr_2\ , \tag{8.6}$$

$$G^k(ij) = \int_0^\infty\!\!\int_0^\infty \frac{2r_<^k}{r_>^{k+1}} P_i^*(r_1)P_j^*(r_2)P_j(r_1)P_i(r_2)\,dr_1\,dr_2\ , \tag{8.7}$$

$$E^i = E_k^i + E_n^i + \sum_{j\neq i} E^{ij}\ , \tag{8.8}$$

[3]Actually, all HF results quoted in this book were computed with an entirely different program, modified by D. C. Griffin [Phys. Rev. A 3, 1233 (1971)] from the well-known program of C. Froese Fischer, Comput. Phys. Commun. 1, 151 (1969).

$$E_{av} = \sum_{i=1}^{N}\left(E^i - \frac{1}{2}\sum_{j\neq i} E^{ij}\right) . \tag{8.9}$$

For the application of these equations, it is not actually necessary to evaluate the second derivative of P_i involved in the integral for E_k^i. For any of the homogeneous-equation methods, we find by multiplying both sides of (7.28) by $P_i^*(r)$ and integrating that

$$\varepsilon_i = E_k^i + \int_0^{\infty} V^i(r)|P_i(r)|^2\, dr . \tag{8.10}$$

Thus E_k^i can be found more simply by evaluating the expectation value of V^i, and we may write[4]

$$E^i = \varepsilon_i - \langle V^i\rangle_i + E_n^i + \sum_{j\neq i} E^{ij} . \tag{8.11}$$

In the HF case, the problem is even simpler: we saw in Eq. (7.14) that E^i is given precisely by the Lagrangian multiplier ε_i, and so the one-electron binding energies are already known, and the total binding energy may be found from

$$E_{av} = \sum_{i=1}^{N}\left(\varepsilon_i - \frac{1}{2}\sum_{j\neq i} E^{ij}\right)_{HF} ; \tag{8.12}$$

if the kinetic energy is desired for some purpose, it may be found from

$$E_k^i = \left(\varepsilon_i - E_n^i - \sum_{j\neq i} E^{ij}\right)_{HF} . \tag{8.13}$$

Finally, if radial functions have been found for several configurations of the same atom or ion, radial functions from two configurations of the same parity can be combined to compute the generalized Coulomb integrals R^k defined by (6.23), which are involved in configuration-interaction calculations. Certain radial integrals that are involved in the calculation of radiative transition probabilities can be evaluated similarly. The use of these two-configuration integrals will be discussed in Chapters 13-16 and 18. All pertinent quantities are punched on cards to provide input for the energy level and spectrum program outlined in Fig. 16-1. Computing times for each configuration on CDC 7600 or IBM 370/195 computers run about 1 or 2 seconds per orbital for HX or HXR (1 to 3 seconds for HFR).

[4]If the relativistically corrected differential equation (7.60) is used, then $\varepsilon_i = E_k^i + E_m^i + E_D^i + \langle V^i\rangle_i$; thus use of (8.11) and (8.9) results in relativistically corrected values of E^i and E_{av}, so that calculation of mass-velocity and Darwin energies from (7.58) and (7.59) is unnecessary.

8-2. COMPARISON OF METHODS

In Fig. 8-2 we show the radial functions $P_{nl}(r)$ for the two outermost orbitals of the excited configuration $3p^5 3d$ of neutral argon, as computed by the HX method. Also shown are various approximations to the effective potential-energy function

$$V^i_{eff}(r) = \frac{l_i(l_i+1)}{r^2} - \frac{2Z}{r} + V_{ee}(r) , \tag{8.14}$$

with the electron-electron term V_{ee} calculated from (7.17), (7.18), (7.48), and (7.54) for the classical, Hartree, HFS, and HX approximations, respectively, using in all cases the HX radial functions.

For the 3p orbital, which concerns an inner, multiply occupied subshell, the 3p electron is incompletely shielded from the nucleus by the other electrons, and would therefore be fairly strongly bound even in the classical potential. Correction for the self-shielding of the

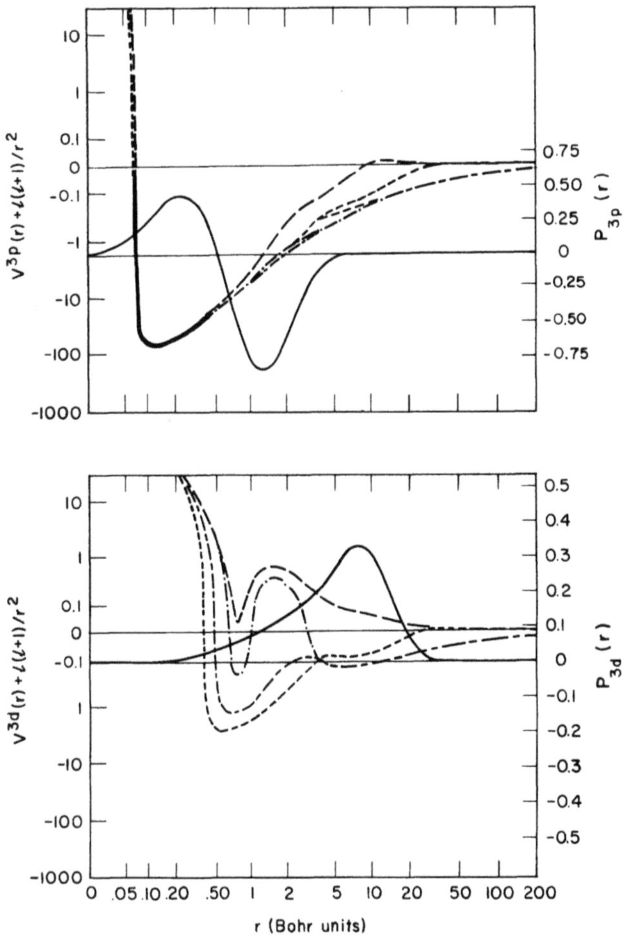

Fig. 8-2. HX radial wavefunctions (solid lines) and effective potentials for the 3p (top) and 3d (bottom) electrons of Ar I $3p^5 3d$, using various approximations for the electron-electron potential energy: — — — — the classical potential V_c (including self-interaction); — · — · — · — the Hartree potential V_H (excluding self-interaction); — - — - — - the HX potential (V_H plus exchange correction); - - - - - - - - - - the HFS potential (V_c plus exchange correction) with $\alpha = 1$, both without and with the Latter tail cutoff (for 3d, the cut-off curve is essentially identical with V_H). The HX radial function $P_{nl}(r)$ is plotted with respect to a horizontal axis placed at the position of the HX eigenvalue ε_{nl} on the V_{eff} scale; the points at which this axis intersects the V_{HX} curve are the classical turning points, at which $d^2P/dr^2 = 0$.

3p electron in question (the Hartree approximation) gives, however, a considerably lower potential curve at large radii and correspondingly stronger binding. The HX exchange allowance provides a relatively minor additional lowering of the potential curve, but nonetheless contributes significantly to the binding, as may be seen from the eigenvalues ε_{3p} listed in Table 8-1. The HFS approximation has the difficult task of correcting for both the self-shielding effect and the exchange effect. With $\alpha = 1$, it overcorrects at small r and undercorrects at large r; with $\alpha = 2/3$, it is considerably better at small r, but is even poorer at large r unless the Latter tail cutoff is used (see the final paragraph of Sec. 7-11).

Much the same comments can be made regarding the 3d electron, except that (1) the shielding by the other electrons is much more thorough and the repulsive centrifugal term $l(l + 1)/r^2$ is considerably larger than before, so that with only the classical term the effective potential is everywhere positive and provides no possibility of binding whatever; and (2) at small r the exchange contribution is much more important than the Hartree self-shielding term, and has a strong effect on the form of P_{3d} where it overlaps the 3p function—and therefore also a strong effect on the computed values of the radial integrals $F^k(3p,3d)$ and $G^k(3p,3d)$, which are a measure of the Coulomb interaction energy between the 3d electron and each 3p electron.

The above remarks are illustrated quantitatively by the various energies and radial integrals given in Table 8-1, where HFSL indicates the HFS method with incorporation of the Latter potential cutoff. It may be seen that the HX method gives much the best overall agreement with Hartree-Fock;[5] in particular, it does the best job of satisfying Koopmans' theorem (7.14), $\varepsilon^l = E^l$. It should be emphasized that the configuration considered here is somewhat unusual, the 3d orbital in elements near the beginning of the first transition series being abnormally sensitive to the assumed analytical form of the central-field potential, for reasons that will be discussed in Sec. 8-6; for most configurations, especially those in which all subshells are multiply occupied, the various methods give more nearly equal results. Because of the existence of difficult cases like Ar I $3p^5 3d$, however, HX is the only approximation method that we shall consider further.

8-3. ACCURACY OF COMPUTED CONFIGURATION-AVERAGE ENERGIES

In Sec. 6-4, details of the calculation of E_{av} from theoretical values of the various radial integrals were indicated by means of a numerical example for N I. Calculation of corresponding experimental values was outlined in Table 6-5, and comparisons between theory and experiment drawn in Tables 6-5 and 6-6. Further comparisons for total binding energies are given in Table 8-2. It is evident that correlation-energy corrections are important in all cases, and that relativistic corrections are significant for Z as low as 5 or 6. The most important relativistic effects of course occur for the inner-shell electrons, which have the highest velocities; an example may be seen in the value $E_r = -0.097$ Ry for the two 1s electrons of O^{6+}, compared with $E_r = -0.111$ for all eight electrons of neutral oxygen.

[5]As discussed in Sec. 7-4, the HF method does not usually provide a suitable potential for the evaluation of spin-orbit parameters via (8.2). All values of ζ given in the table were computed by the more accurate, but much more complicated, Blume-Watson method [M. Blume and R. E. Watson, Proc. Roy. Soc. (London) **A270**, 127 (1962), **A271**, 565 (1963)].

TABLE 8-1. NUMERICAL RESULTS FOR Ar I $3p^5 3d$, COMPUTED BY VARIOUS SELF-CONSISTENT-FIELD METHODS (all results in Ry except ζ in cm^{-1} and $\langle r^k \rangle$ in units of a_o^k)

		HFS		HFSL				
	H	$\alpha = 1$	$\alpha = 2/3$	$\alpha = 1$	$\alpha = 2/3$	HS	HX	HF
E_{av}	−1052.526	−1052.327	−1052.614	−1052.363	−1052.622	−1052.647	−1052.716	−1052.664
ε_{3p}	−1.440	−1.430	−1.228	−1.507	−1.252	−1.653	−1.757	−1.831
E^{3p}	−2.075	−1.381	−1.801	−1.456	−1.823	−1.761	−1.835	−1.831
ε_{3d}	−0.113	−0.039	−0.023	−0.136	−0.120	−0.127	−0.118	−0.117
E^{3d}	−0.117	+0.008	−0.107	−0.041	−0.116	−0.109	−0.117	−0.117
$\langle r^{-2} \rangle_{3p}$	1.598	1.874	1.742	1.887	1.744	1.764	1.786	1.633
$\langle r^2 \rangle_{3p}$	3.393	2.653	2.952	2.636	2.965	2.936	2.885	2.871
$\langle r^{-2} \rangle_{3d}$	0.018	0.198	0.060	0.140	0.041	0.064	0.028	0.027
$\langle r^2 \rangle_{3d}$	118.5	44.60	115.6	53.28	94.15	74.29	102.0	104.11
$F^0(3p,3d)$	0.234	0.575	0.320	0.481	0.295	0.356	0.269	0.264
$F^2(3p,3d)$	0.020	0.193	0.064	0.137	0.044	0.071	0.031	0.029
$G^1(3p,3d)$	0.008	0.222	0.064	0.151	0.036	0.066	0.020	0.017
$G^3(3p,3d)$	0.004	0.129	0.036	0.087	0.020	0.037	0.011	0.010
ζ_{3p}	875	1067	977	1076	978	991	1001	893
ζ_{3d}	0.07	7.61	1.62	5.04	0.83	1.54	0.32	0.27

TABLE 8-2. HF AND HX TOTAL BINDING ENERGIES (FOR CENTER-OF-GRAVITY OF GROUND CONFIGURATION, IN RY)

	HF	HX			
Atom	E_{av}	E_{av}	$E_{av} + E_r$	$E_{av} + E_r + E_c$	Exp[a]
He $1s^2$	−5.7234	−5.7234	−5.7236	−5.7924	−5.807
Li 2s	−14.8655	−14.8723	−14.8739	−14.9612	−14.956
Be $2s^2$	−29.1461	−29.1586	−29.1643	−29.3145	−29.337
B 2p	−49.0581	−49.0694	−49.0844	−49.3071	−49.316
C $2p^2$	−75.3194	−75.3290	−75.3617	−75.6609	−75.667
N $2p^3$	−108.592	−108.599	−108.662	−109.041	−109.057
O^{6+} $1s^2$	−118.222	−118.222	−118.319	−118.399	−118.388
O $2p^4$	−149.538	−149.541	−149.652	−150.113	−150.151
F $2p^5$	−198.819	−198.816	−199.000	−199.542	−199.612
Ne $2p^6$	−257.094	−257.085	−257.374	−257.999	−258.102
Na 3s	−323.719	−323.712	−324.150	−324.810	−324.857
Mg $3s^2$	−399.229	−399.231	−399.870	−400.599	−400.621
Al 3p	−483.753	−483.765	−484.671	−485.471	−485.424
Ar $3p^6$	−1053.635	−1053.689	−1057.394	−1058.599	$-1058._{23}$
Kr $4p^6$	−5504.111	−5504.127	−5575.333	−5577.964	---
Xe $5p^6$	−14464.28	−14464.364	−14862.132	−14866.178	---
Pb $6p^2$	−39047.99	−39047.900	−41346.268	−41352.584	---

[a]C. E. Moore, *Ionization Potentials and Ionization Limits Derived from the Analyses of Optical Spectra,* NSRDS-NBS 34 (1970).

Some comparisons between theory and experiment for ionization energies are given in Table 8-3. The theoretical results, calculated by differencing values of E_{av} for the ground configurations of atom and ion, involve a loss of as many as five significant figures; it is evident, therefore, that the iterations on the ε_l and the SCF iteration on the P_l must be carried out to very tight tolerances, using a computer with a floating-point word length of at least 9 or 10 significant figures.[6] (The numerical methods employed need not be nearly this accurate, however: errors inherent in numerical quadratures largely cancel each other when values of E_{av} are differenced.) Correlation-energy corrections are nearly always important, but relativistic corrections (since they pertain to the outer-shell electrons) are comparatively small except for highly penetrating s electrons in heavy atoms. With both corrections included, computed ionization energies are usually accurate to within 2 to 5 percent, although errors occasionally rise to 10 or 15 percent as in the cases of Cu, Ag,

[6]Calculations reported in this book were done on CDC 6600 and 7600 computers, which have a 60-bit word length, with a 48-bit (14^+-decimal-place) floating-point fraction. For the ε_l iterations, fractional changes in each eigenvalue were required to be less than 10^{-11}; from one SCF cycle to the next, fractional changes in $P_l(r)$ and absolute changes in $V^l(r)$ were required to be less than $5 \cdot 10^{-6}$ and $5 \cdot 10^{-8}$, respectively.

TABLE 8-3. HF AND HX CONFIGURATION-AVERAGE IONIZATION ENERGIES (RY)

	HF	HX				
Atom	ΔE_{av}	ΔE_{av}	ΔE_r	ΔE_c	ΔE^t_{av}	Exp[a]
O $2p^4$	1.072	1.068	−0.001	0.064	1.132	1.164
Al 3p	0.404	0.412	−0.001	0.048	0.460	0.439
Si $3p^2$	0.523	0.527	−0.001	0.052	0.578	0.572
S $3p^4$	0.786	0.790	−0.001	0.059	0.849	0.854
Ar $3p^6$	1.086	1.089	−0.001	0.063	1.151	1.163
Ca $4s^2$	0.376	0.380	0.002	0.043	0.424	0.449
Sc 3d $4s^2$	0.403	0.407	0.003	0.046	0.456	0.486
Sc^+ 3d 4s	0.883	0.890	0.005	0.043	0.937	0.942
Sc^{++}3d	1.755	1.748	−0.011	0.066	1.803	1.819
Cu $3d^{10}4s$	0.471	0.471	0.015	0.048	0.534	0.568
Cu^+ $3d^{10}$	1.287	1.284	−0.011	0.066	1.340	1.499
Ge $4p^2$	0.508	0.509	−0.003	0.052	0.559	0.555
Ag $4d^{10}5s$	0.434	0.429	0.036	0.049	0.514	0.557
Sn $5p^2$	0.472	0.473	−0.004	0.051	0.520	0.516
Xe $5p^6$	0.859	0.859	−0.002	0.060	0.918	0.924
Cs 6s	0.247	0.248	0.010	0.026	0.284	0.286
Au $5d^{10}6s$	0.437	0.428	0.107	0.050	0.586	0.678
Pb $6p^2$	0.456	0.458	−0.008	0.051	0.501	0.501

[a]C. E. Moore, NSRDS-NBS 34 (1970).

and Au. (Energies calculated via the HXR method, instead of by adding relativistic corrections to the HX results, differ little from the values in Table 8-3 except for Au, where the value 0.586 increases to 0.613 Ry.)

A further illustration of the importance of both corrections is given in Table 8-4, where it is seen that errors in computed excitation energies of doubly ionized yttrium are reduced by about an order of magnitude for the low-lying configurations where relativistic and correlation effects are appreciable. In general, relativistic corrections are more important in ions than in neutral atoms, because the valence electrons move in stronger electric fields. At the same time, relative calculational errors tend to decrease with an increase in ionization stage[7] because of the greater importance of the exactly-known nuclear potential as compared with the approximately-treated inter-electronic interactions.

For configurations that consist of a single non-penetrating electron outside a tightly-bound core (such as we have in Y III $4p^6nl$, $l > 1$), the correlation between the position of this outer electron and those of the others amounts to a polarization of the core by the outer electron, and the correlation energy is frequently discussed in terms of core

[7]R. D. Cowan, J. Opt. Soc. Am. 58, 924 (1968).

TABLE 8-4. CONFIGURATION-AVERAGE ENERGIES OF $_{39}$Y III, RELATIVE TO Y IV $4p^6$ (Ry)

Config.		Exp.[a]	HX computed error (Theory − Exp.)			
			ΔE_{av}	$\Delta(E_{av} + E_r)$	$\Delta(E_{av} + E_c)$	ΔE_{av}^t
Y IV	$4p^6$	0	0	0	0	0
Y III	6f	−0.261	0.008	0.008	0.004	0.004
	7p	−0.376	0.006	0.006	0.002	0.002
	5f	−0.377	0.012	0.013	0.004	0.005
	6d	−0.424	0.007	0.009	0.001	0.003
	7s	−0.434	0.011	0.007	0.006	0.003
	4f	−0.587	0.017	0.018	0.001	0.002
	6p	−0.600	0.013	0.012	0.005	0.004
	5d	−0.702	0.014	0.018	0.000	0.004
	6s	−0.718	0.023	0.016	0.013	0.005
	5p	−1.122	0.034	0.032	0.001	−0.001
	5s	−1.441	0.072	0.050	0.030	0.008
	4d	−1.505	0.038	0.062	−0.023	0.001

[a]G. L. Epstein and J. Reader, J. Opt. Soc. Am. **65**, 310 (1975).

polarizabilities.[8] The transitions $np^5 4f - np^5 5g$ in the noble gases, which occur close to the position of the first line of the Brackett series in hydrogen at 4.05 μm (Table 3-3), provide an interesting example in which the polarization effects are very small, but nevertheless provide an essential contribution to the Z-dependence of observed spectra.[9]

8-4. RELATIVISTIC EFFECTS

For Z greater than about 30, relativistic effects on the radial wavefunctions become appreciable, and first-order perturbation energy corrections are no longer always sufficient. For example, Table 8-5 shows the improvement in computed total binding energies (assuming DHF values as a standard) that the second-order HXR method provides, as compared with the non-relativistic HX method with first-order energy corrections. As illustrated by Table 8-6, the principal effect on the radial functions is a contraction of the s functions, which always penetrate close to the nucleus (Secs. 3-4 and 3-5); to a lesser extent, p and d electrons with small n are also pulled in toward the nucleus. As an indirect effect of these contractions, outer-shell electrons with $l > 0$ become more fully shielded

[8]See, for example, B. Edlén, "Atomic Spectra," in S. Flügge, ed., *Handbuch der Physik* (Springer-Verlag, Berlin, 1964), Vol. XXVII, pp. 88, 125-29.

[9]C. J. Humphreys, E. Paul, Jr., R. D. Cowan, and K. L. Andrew, J. Opt. Soc. Am. 57, 855 (1967). (The agreement between experimental and theoretical values of ΔE_{av} shown in this paper is significantly improved by using the revised method of calculating E_c described in Sec. 7-15.)

TABLE 8-5. COMPARISON OF NON-RELATIVISTIC AND RELATIVISTIC TOTAL BINDING ENERGIES (Ry)

	E_{av}^{HF}	E_{av}^{HX}	$E_{av}^{HX} + E_r$	E_{av}^{HXR}	$E_{av}^{DHF(a)}$
$_{2}$He $1s^2$	−5.72336	−5.72336	−5.72363	−5.72363	−5.72363
$_{4}$Be $2s^2$	−29.14605	−29.15857	−29.16428	−29.16421	−29.15178
$_{10}$Ne $2p^6$	−257.0942	−257.0848	−257.3740	−257.3697	−257.3838
$_{12}$Mg $3s^2$	−399.2293	−399.2305	−399.8698	−399.8615	−399.8701
$_{18}$Ar $3p^6$	−1053.635	−1053.689	−1057.394	−1057.377	−1057.367
$_{20}$Ca $4s^2$	−1353.517	−1353.584	−1359.433	−1359.425	−1359.420
$_{30}$Zn $3d^{10}4s^2$	−3555.697	−3555.690	−3588.476	−3589.053	−3589.226
$_{36}$Kr $4p^6$	−5504.112	−5504.127	−5575.333	−5577.533	−5577.723
$_{38}$Sr $5s^2$	−6263.092	−6263.120	−6352.792	−6355.981	−6356.162
$_{48}$Cd $4d^{10}5s^2$	−10930.27	−10930.31	−11171.74	−11186.64	−11186.65
$_{54}$Xe $5p^6$	−14464.28	−14464.36	−14862.13	−14894.11	−14893.82
$_{56}$Ba $6s^2$	−15767.09	−15767.19	−16231.30	−16271.74	−16271.32
$_{70}$Yb $4f^{14}6s^2$	−26782.92	−26782.84	−27966.14	−28137.21	−28135.53
$_{80}$Hg $5d^{10}6s^2$	−36817.99	−36817.88	−38889.85	−39302.07	−39298.30
$_{86}$Rn $6p^6$	−43733.56	−43733.49	−46541.04	−47211.91	−47205.11
$_{88}$Ra $7s^2$	−46188.62	−46188.57	−49280.98	−50065.84	−50057.47
$_{102}$No $5f^{14}7s^2$	−65579.05	−65578.92	−71312.91	−73538.57	−73487.29

[a] J. B. Mann, private communication; see also J. B. Mann and J. T. Waber, At. Data 5, 201 (1973).

from the nucleus and therefore move out to larger radii,[10] with correspondingly *smaller* binding than given by non-relativistic calculations (Table 8-7).

The examples shown in Tables 8-2 to 8-7 establish the importance of both correlation and relativistic effects. Except as stated explicitly to the contrary, all further numerical results quoted in this book may be understood to include both correlation and relativistic corrections (calculated via the HXR method for $Z > 30$). The simple symbol E_{av} will henceforth be used to represent the total binding energy, including corrections, which has previously been designated E_{av}^{t}.

8-5. VARIATION WITH THE PRINCIPAL QUANTUM NUMBER

Of great importance is the concept of a Rydberg series of configurations (Sec. 4-7), in which the various configurations are identical except for the value of the principal quantum number n of a single excited electron (the "running" electron). Computed radial functions $P_{np}(r)$ for a few members of the Rydberg series Ar I $3p^5np$ are shown in Fig. 8-3. As is to be expected physically, the radial density distribution of the excited electron, $\rho_{np}(r)4\pi r^2\,dr = P_{np}^2(r)\,dr$, is seen to move to greater r with increasing n (i.e., with increasing energy).

[10] D. F. Mayers, Proc. Roy. Soc. (London) A241, 93 (1957); R. G. Boyd, A. C. Larson, and J. T. Waber, Phys. Rev. 129, 1629 (1963).

TABLE 8-6. EXPECTATION VALUES OF r^k FOR U I $5f^3 6d 7s^2$ (atomic units)[a]

	HF		HX		HXR		DHF[b]	
	$\langle r^{-2} \rangle$	$\langle r^{2} \rangle$	$\langle r^{-2} \rangle$	$\langle r^{2} \rangle$	$\langle r^{-2} \rangle$	$\langle r^{2} \rangle$	$\langle r^{-2} \rangle$	$\langle r^{2} \rangle$
1s	1.68^{4}	3.61^{-4}	1.68^{4}	3.61^{-4}	4.09^{4}	2.58^{-4}	4.10^{4}	2.62^{-4}
2s	1.90^{3}	5.54^{-3}	1.90^{3}	5.54^{-3}	5.73^{3}	3.87^{-3}	6.52^{3}	3.86^{-3}
$2p_-$							2.08^{3}	2.66^{-3}
	6.35^{2}	4.07^{-3}	6.37^{2}	4.05^{-3}	7.92^{2}	3.51^{-3}		
$2p_+$							7.13^{2}	3.83^{-3}
3s	4.47^{2}	3.33^{-2}	4.44^{2}	3.34^{-2}	1.32^{3}	2.49^{-2}	1.54^{3}	2.49^{-2}
$3p_-$							5.09^{2}	2.24^{-2}
	1.50^{2}	3.05^{-2}	1.49^{2}	3.05^{-2}	1.83^{2}	2.68^{-2}		
$3p_+$							1.78^{2}	2.85^{-2}
$3d_-$							1.00^{2}	2.15^{-2}
	8.74^{1}	2.31^{-2}	8.76^{1}	2.30^{-2}	9.17^{1}	2.24^{-2}		
$3d_+$							8.88^{1}	2.31^{-2}
4s	1.22^{2}	1.49^{-1}	1.20^{2}	1.51^{-1}	3.58^{2}	1.16^{-1}	4.16^{2}	1.16^{-1}
5s	3.03^{1}	6.38^{-1}	3.14^{1}	6.43^{-1}	9.23^{1}	4.97^{-1}	1.03^{2}	5.04^{-1}
6s	5.58^{0}	3.25^{0}	6.16^{0}	3.23^{0}	1.87^{1}	2.44^{0}	2.04^{1}	2.51^{0}
5f	1.79^{0}	1.94^{0}	1.75^{0}	2.07^{0}	1.43^{0}	2.78^{0}	1.53^{0}	2.53^{0}
6d	4.89^{-1}	9.63^{0}	4.72^{-1}	1.04^{1}	3.83^{-1}	1.31^{1}	4.92^{-1}	1.11^{1}
7s	4.73^{-1}	2.88^{1}	6.56^{-1}	2.67^{1}	1.97^{0}	1.97^{1}	1.87^{0}	2.18^{1}

[a]The exponents indicate the powers of ten by which the associated numbers are to be multiplied.
[b]J. B. Mann, private communication (for the configuration $5f_-^3\, 6d_-\, 7s^2$). The DHF values labeled 5f and 6d are actually for $5f_-$ and $6d_-$. It should be recalled from Sec. 7-14 that for non-s electrons the HXR values are to be compared with the $(2j+1)$-weighted averages of the DHF values.

In all cases, the excited electron lies almost completely outside the region occupied by the core electrons $1s^2 \cdots 3p^5$, which may be seen from Fig. 8-2 to extend out only to about $r = 4$. In this region outside the core, the potential is essentially hydrogenic—$V(r) \cong -2(Z-N+1)/r$—and so we may expect the binding energy of the excited electron to be given approximately by the hydrogenic expression

$$E_H^j = -\frac{(Z-N+1)^2}{n_j^2}\ \mathrm{Ry}\ , \tag{8.15}$$

with the approximation improving as n_j increases. Indeed, it was discovered by Rydberg in the early days of empirical spectrum analysis that (in modern notation) if an effective quantum number n^* is defined such that[11]

[11]In principle, it would be better to define n^* in terms of an ionization energy calculated by differencing appropriate values of E_{av}. However, the relation $-E^j \cong \Delta E_{av}$, Eq. (6.48), is quite accurate for the excited electron in a Rydberg series.

TABLE 8-7. ONE-ELECTRON AND TOTAL ENERGIES FOR U I $5f^3 6d7s^2$ (Ry)

	HF	HX		HXR	DHF[a]	Exp.[b]
	E^i	E^i	$E^i + E^i_r$	E^i	E^i	(I.E.)
1s	−7433.	−7433	−8370	−8644	−8559	−8497
2s	−1302.	−1302	−1560	−1630	−1612	−1599
$2p_-$					−1553	−1540
	−1258.	−1258	−1374	−1356		
$2p_+$					−1271	−1262
3s	−333.3	−333.8	−403.0	−417.9	−413.2	−407.8
$3p_-$					−386.2	−380.8
	−311.9	−312.4	−348.1	−341.1		
$3p_+$					−320.6	−316.4
$3d_-$					−278.0	−274.0
	−272.3	−272.8	−283.8	−272.0		
$3d_+$					−264.8	−261.1
4s	−87.04	−87.43	−106.8	−110.2	−108.6	−106.0
5s	−20.12	−20.34	−25.38	−25.71	−25.15	−23.8
6s	−3.365	−3.422	−4.382	−4.404	−4.248	−5.2
5f	−1.269	−1.356	−1.450	−0.740	−0.704	---
6d	−0.533	−0.528	−0.577	−0.372	−0.416	---
7s	−0.333	−0.338	−0.435	−0.414	−0.398	---
E_{av}	−51328.4	−51328.4	−55054.2	−56120.8	−56107.5	---

[a] J. B. Mann, private communication (for the configuration $5f_-^3 6d_- 7s^2$).
[b] K. Siegbahn et al., *ESCA—Atomic, Molecular, and Solid State Structure Studied by Means of Electron Spectroscopy* (Almqvist and Wiksells, Uppsala, 1967).

$$E^j \equiv -\frac{(Z-N+1)^2}{(n_j^*)^2}\ \text{Ry}\ , \tag{8.16}$$

then the quantum defect

$$\delta \equiv n - n^* \tag{8.17}$$

is nearly constant for a given Rydberg series. Because the excited electron does penetrate somewhat into the ion core, the magnitude of the binding energy is greater than the hydrogenic value, and so δ is always positive (for configuration-average energies, to which our present discussion is limited). However, the penetration is smaller the greater the value of l, because of the effect of the centrifugal term $l(l+1)/r^2$ on the form of the effective potential (8.14); consequently δ decreases with increasing l.

In the region of the ion core, the *shape* of $P_{nl}(r)$ for the excited electron may be seen from Fig. 8-3 to be nearly independent of n. This is readily understandable, because V_{eff}^{nl} is

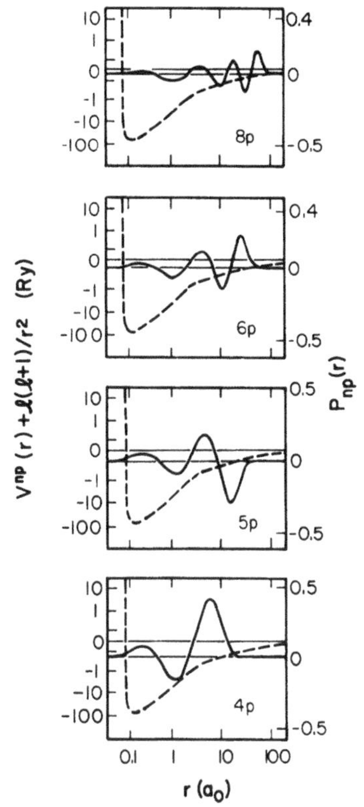

Fig. 8-3. HX computed effective potentials and radial wavefunctions for the excited electron in the configurations Ar I $3p^5 4p$, 5p, 6p, and 8p.

nearly independent of n, and in most of the core region $|V_{eff}(r)|$ is large compared with $|\varepsilon_{nl}|$, so that

$$P''_{nl}(r)/P_{nl}(r) = V_{eff}(r) - \varepsilon_{nl}$$

is roughly independent of n. Thus in this region the main variation is just a decrease in amplitude with increasing n—resulting from the greater radial extent of $P_{nl}(r)$ and the fact that P_{nl} has unit norm for all n. Since contributions to the spin-orbit radial integral (8.2) come primarily from the small-r portion of the wavefunction, we may expect the value of this integral to decrease monotonically with increasing n. Indeed, calculations in specific cases show that for a running electron $n_j l_j$ and large n_j,

$$\zeta_j \propto (n_j^*)^{-3} ; \tag{8.18}$$

this is consistent with the hydrogenic result (3.52).[12] Similarly, the radial integrals $F^k(ij)$ and $G^k(ij)$ associated with the Coulomb interaction between the running electron and a core electron $n_i l_i$ are of such a form that only the small-r portion of P_j is important. Since P_i is nearly independent of n_j it follows that these integrals must also decrease monotonically with increasing n_j, and in fact one finds that[12]

[12]Several experimental examples of these relationships are discussed by B. Edlén, ref. 8, pp. 121, 129-36.

$$F^0(ij) \propto (n^*_j)^{-2} , \tag{8.19}$$

$$F^k(ij) \propto (n^*_j)^{-3} , \qquad k > 0 , \tag{8.20}$$

$$G^k(ij) \propto (n^*_j)^{-3} , \qquad \text{all } k . \tag{8.21}$$

Examples for the configurations Ar I $3p^5np$ are shown in Fig. 8-4.

As a consequence of these relations, it follows from (8.4) and (8.5) that for large n_j

$$E^{ij} \cong F^0(ij) \propto (n^*_j)^{-2} . \tag{8.22}$$

Physically, this result is simply an expression of the classical Coulomb interaction energy of two concentric, spherically-symmetric, non-overlapping charge distributions because

$$\langle r_j^{-1} \rangle \propto (n^*_j)^{-2} , \tag{8.23}$$

as illustrated by the example in Fig. 8-4. (Compare also the hydrogenic result $\langle r^{-1} \rangle = Z/n^2$ in Table 3-2.) F^k ($k > 0$) and G^k decrease (with increasing n_j) at a faster

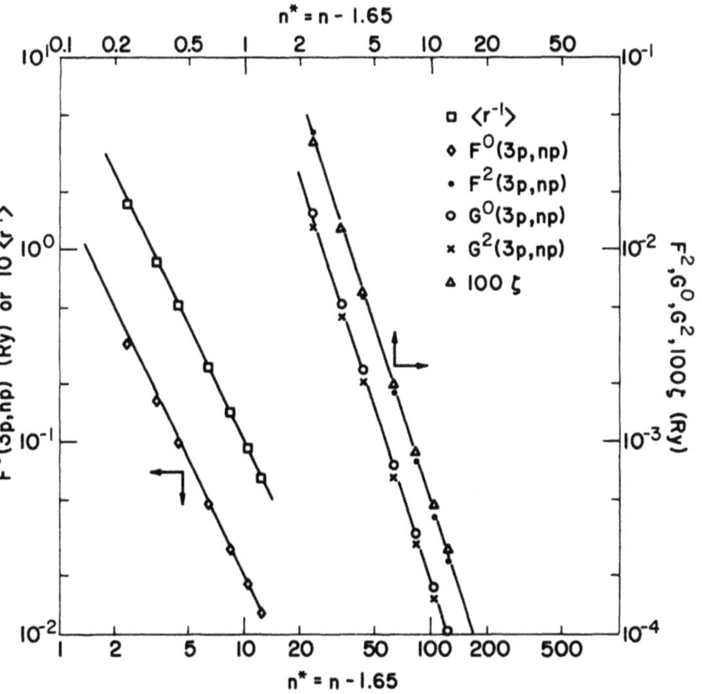

Fig. 8-4. Variation with n* of HX radial integrals for the Rydberg series Ar I $3p^5np$ ($4 \leq n \leq 14$). The straight lines at the left have been drawn with a slope of −2; those at the right, with a slope of −3.

rate than does F^0: the interaction energy between classical multipoles always decreases with increasing separation at a faster rate than the r^{-1} dependence for two monopoles.

A second consequence of (8.18)-(8.21) is that (so far as the spin-orbit interaction of electron j and the Coulomb interaction between electrons i and j are concerned) the energy spread of the levels within a configuration is proportional to $(n_j^*)^{-3}$ since, as we shall see, this spread depends linearly on ζ_j and on the F^k and G^k other than F^0. From (8.16)-(8.17) the spacing between successive configurations is

$$\frac{dE^j}{dn_j} \propto (n_j^*)^{-3} \; ; \qquad (8.24)$$

it follows that the ratio of the energy width of a configuration to the distance between centers of gravity is essentially constant for all members of a Rydberg series, so far as the above interactions are concerned. (This is *not* true of the spin-orbit effect of the electron i, nor of the contribution of Coulomb interactions between i and other core electrons, which are of course nearly independent of the far-distant electron j.)

Problem

8-5(1). Using the experimental energies for Y III given in Table 8-4, calculate configuration-average quantum defects, and verify the strong dependence of δ on l and its near independence of n.

8-6. VARIATION WITH Z OF ONE-ELECTRON BINDING ENERGIES

With increasing Z, the effective potential for an ns electron in neutral atoms varies roughly in the manner shown in Fig. 8-5, because V_{eff} must tend to $-2Z/r$ at small r where the electron sees the full nuclear potential, and to $-2/r$ at large r where it is fully shielded from the nucleus by the other $Z - 1$ electrons. Since the radial function $P_{ns}(r)$ has appreciable amplitude at small r, the s electron feels strongly the increasing value of Z, the binding energy $|E^i|$ increases rapidly with Z, and correspondingly n_i^* decreases rapidly (and relatively smoothly) with increasing Z.

For an np electron, the centrifugal term $l_i(l_i + 1)/r^2$ in V_{eff} prevents the electron from penetrating so far into the small-r region; thus the value of n_i^* is not as small as for an s electron. However, the variation with Z is qualitatively much the same.

For a d electron, the situation is drastically different: the centrifugal term in V_{eff} is so important that (at small Z) it keeps the d electron at rather large radii; the electron penetrates only slightly into the core region, and therefore sees mainly a hydrogenic potential. This is illustrated in Fig. 8-6, where in each frame the upper dotted curve is the hydrogenic effective potential $-2/r + 6/r^2$ for a d electron in a neutral atom; this potential is negative only in the region $r > 3$, which lies largely outside the ion core. The lower curve is the computed (HX) effective potential $V(r) + 6/r^2$ for a 3d electron in the atom in question. With increasing Z, the lowering of V(r) at small r sketched in Fig. 8-5 produces a corresponding lowering of $V_{eff}(r)$; however, even in $_{19}$K I $3p^63d$, where this lowering has resulted in the development of a pronounced inner potential well, the 3d electron remains concentrated

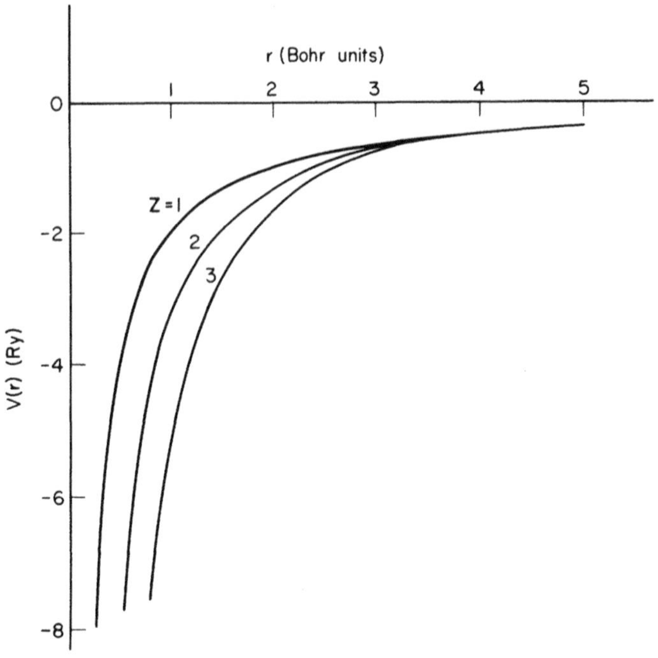

Fig. 8-5. Schematic form of central potentials in neutral atoms, showing the constant form $-2/r$ at large r, but the increasingly more negative potential (with increasing Z) in the region within the electron cloud.

almost entirely in the outer well, which is still closely hydrogenic. The eigenvalue is therefore almost the same as in hydrogen, and since $E^l \cong \varepsilon_l$ for the HX method (see Sec. 7-12 and Table 8-1) it follows from (8.16) that $n^*_{3d} \cong n_{3d} = 3$.

In $_{20}$Ca I 3d4s, however, the 3d electron is not completely shielded from the added unit of nuclear charge by the 4s electron, and the 3d electron is pulled strongly inward toward the nucleus; the 3d wavefunction and eigenvalue are no longer close to hydrogenic, and $n^*_{3d} \ll 3$. This computed collapse of the 3d function at $Z = 20$ is well verified experimentally, as shown in Fig. 8-7. It should be noted, however, that the 3d electron is not as tightly bound as the 4s electron, and that no 3d electron appears in the ground configuration Ca I $4s^2$. It is only at $Z = 21$ that a ground-configuration 3d electron appears (in Sc I $3d4s^2$), and the first transition series of elements begins.

As Z increases still further, the inner-well region becomes still deeper and wider, eventually resulting in the collapse of the 4d, 5d, ··· functions, and the formation of the second, third, ··· transition series of elements (Fig. 8-7).

The f-wavefunction collapse is even more pronounced than that for d electrons, because the hydrogenic effective potential $-2/r + 12/r^2$ becomes negative only for $r > 6$, and the hydrogenic well therefore hardly overlaps the ion core at all. In the core region, the centripetal repulsion is so strong that it is overcome by V(r) only at rather large Z; the inner well of V_{eff} does not become sufficiently wide and deep to hold the 4f function until $_{57}$La

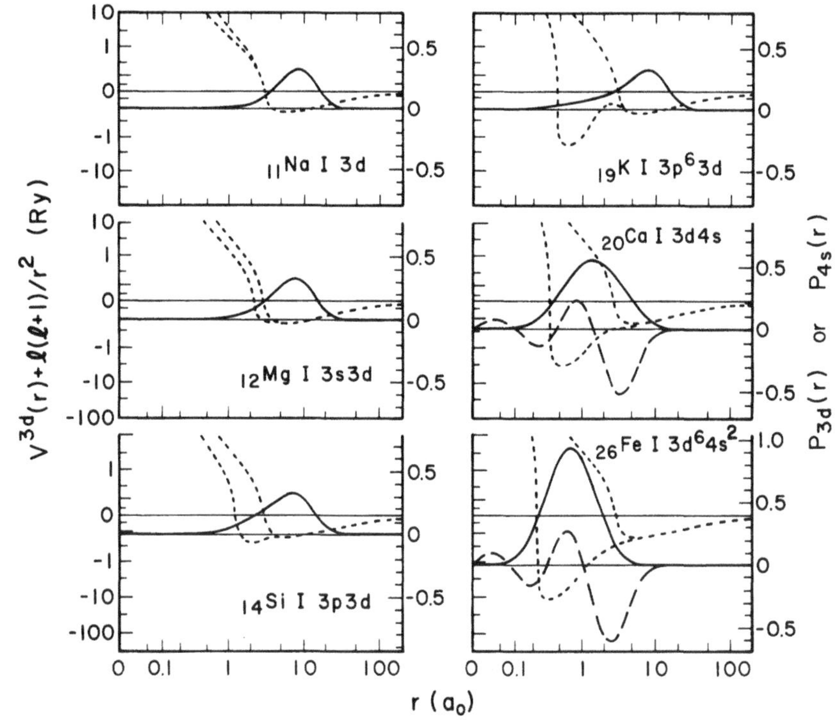

Fig. 8-6. The effective potential (lower dotted curve) and radial wavefunction (solid curve) for a 3d electron in neutral atoms near the beginning of the first transition series. Also shown are the hydrogenic potential for d electrons (upper dotted curve) and the radial function for 4s electrons (where present, dashed curve). In each case the radial functions are plotted (on a linear vertical scale) with respect to a base line drawn at the position of the 3d eigenvalue on the potential scale.

(Fig. 8-8). The lanthanide series of rare-earth elements may be considered to begin at this point, though it is actually only at $_{58}$Ce that a 4f electron becomes sufficiently tightly bound to appear in the ground configuration (Table 4-3).

The collapse, into the inner well, of the 4f function at Z = 57 and of the 5f function at Z = 89 or 90 (the 5f collapse initiating the actinide series of elements) is further illustrated in Fig. 8-9. A striking feature of this figure is that when the 4f collapse occurs, with its associated sudden decrease in the value of n^* there is a simultaneous decrease of almost exactly unity in n^* for an excited 5f, 6f, 7f, $\cdots$ electron. The reason for this[13,14] is that the innermost loop of the nf function drops into the inner well, the first node of the function occurs in the potential-barrier region between the inner and outer wells where the wavefunction must have very small amplitude, and the portion of the wavefunction in the outer hydrogenic well must then look essentially like a hydrogenic function with one less loop

[13]M. Goeppert Mayer, Phys. Rev. **60**, 184 (1941).

[14]D. C. Griffin, K. L. Andrew, and R. D. Cowan, Phys. Rev. **177**, 62 (1969).

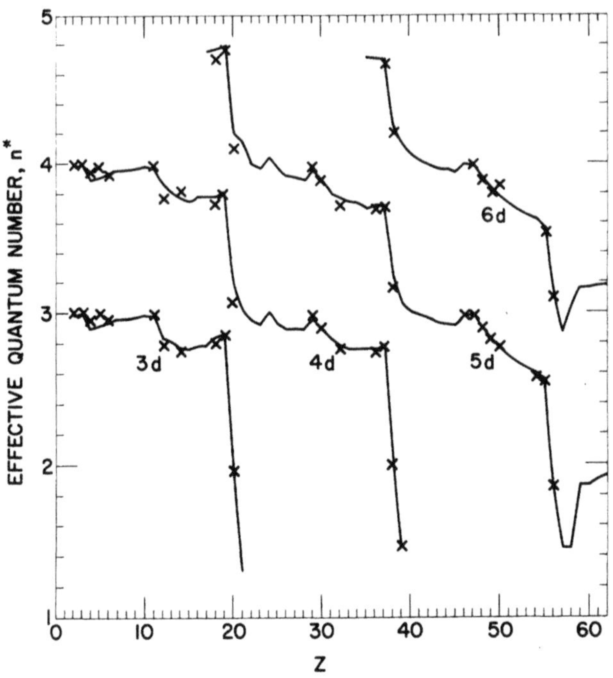

Fig. 8-7. Effective quantum numbers for d electrons in neutral atoms, $n^* = |E^{nd}|^{-1/2}$. The theoretical values (solid lines) were calculated from $E^{nd} = \Delta E^t_{av} = \Delta(E_{av} + E_r + E_c)$. The experimental values (XXX) were calculated from the difference between the centers of gravity of the pertinent configurations of the atom and ion.

than normal; the eigenvalue of an excited nf function must thus be essentially equal to the hydrogenic eigenvalue for an (n − 1)f function, or $\delta = 1$. Similarly, a further unit decrease in n* occurs for 6f, 7f, ⋯ upon the collapse of the 5f function at $Z \cong 90$.

A similar effect occurs for excited d electrons, as shown in Fig. 8-7. The sudden decreases in n* are, however, not quite so close to unity because the outer hydrogenic well overlaps the ion core somewhat, and the potential barrier between inner and outer well is low or absent. Minor irregularities in between drops occur because of shell effects; for example, the decrease in n* between $_{11}$Na I 3d and $_{12}$Mg I 3s3d is due to the 3s electron partially overlapping the hydrogenic well region (i.e., to imperfect shielding of the 3d electron from the added unit of nuclear charge by the 3s electron).

The values of Z at which d collapse occurs are clearly more closely related to the shell structure of the atom than is the case for f electrons. Accurate calculation of the d collapse can therefore be expected only for a self-consistent-field method, in which the computed effective potential takes full account of shell structure. At the same time, the computed width and depth of the inner well must be reasonably accurate. This is illustrated by the results

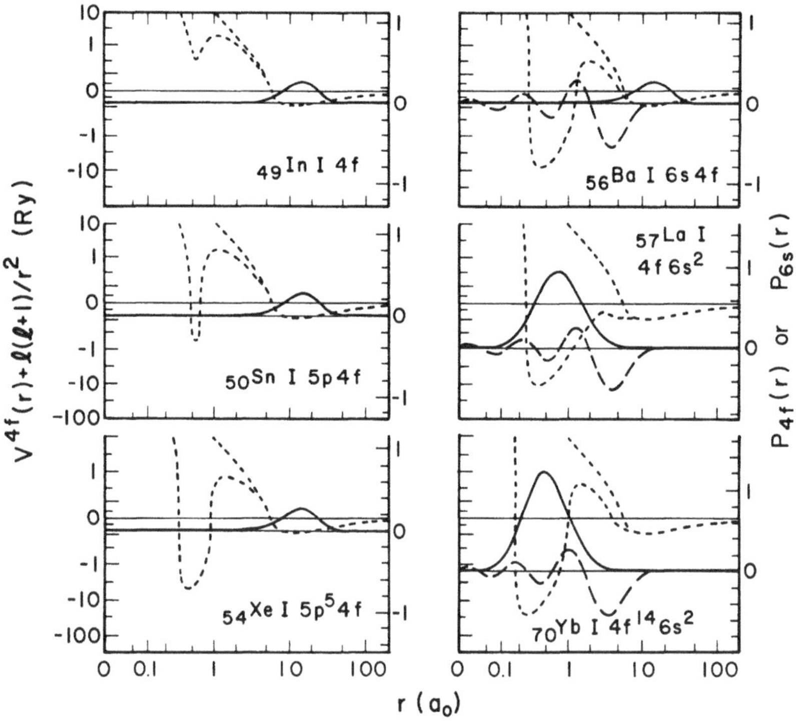

Fig. 8-8. The 4f effective potential (lower dotted curve) and radial wavefunctions for 4f (solid curve) and 6s (dashed curve) electrons in neutral atoms near the beginning of the lanthanide series of rare-earth elements. The upper dotted curve is the hydrogenic effective potential.

given in Table 8-8, where the observed values of Z at which the binding-energy jump occurs are compared with values computed[15] by methods in which the effective potential includes: (1) neither exchange nor shell effects (TF), (2) the deeper inner well produced by the effect of exchange (TFD), (3) shell structure only (Hartree), and (4) both exchange and shell effects (HF and HX). The errors in the Hartree results for 3d and 4f show that the shell structure alone is not the entire story, whereas the presence of exchange effects in the TFD method gives results fairly close to experiment even without the inclusion of shell effects. It should be noted that the indirect relativistic effect (Sec. 8-4) results in a significant shift to higher Z for the predicted point of collapse of 6f and 5g functions for relativistic calculations (HXR and DHF) as compared with non-relativistic ones (HX). The 5f collapse similarly is computed to be marginally postponed from Z = 89 to Z = 90; unfortunately, the spectrum of $_{89}$Ac has not been analyzed with sufficient completeness to indicate whether the collapse does indeed not occur till $_{90}$Th or has already occurred in Ac.

[15]In the TF and TFD methods, the change in value of n* is somewhat gradual. The criterion used is n* (computed from the pertinent eigenvalue) less than 2.5, 3.5, or 4.5 for d, f, or g electrons, respectively.

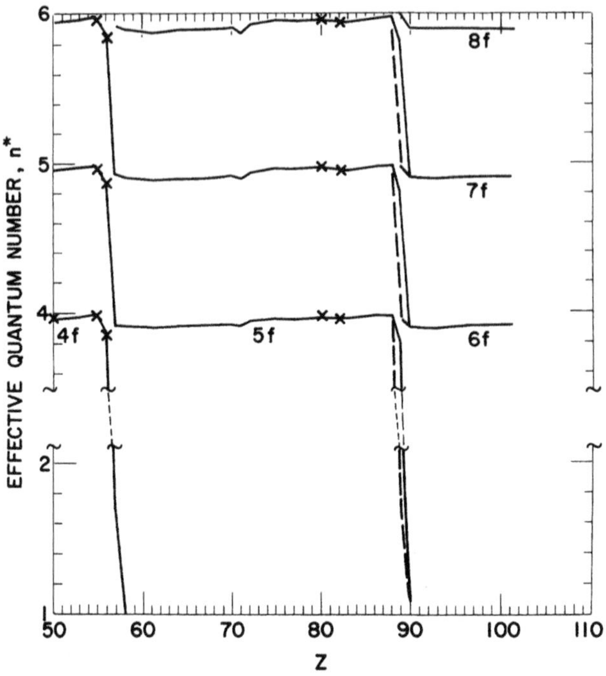

Fig. 8-9. Variation with Z of the effective quantum number for f electrons in neutral atoms: dashed curves, HX; solid curves, HXR; ×××, experiment.

A survey of the Z dependence of radial wavefunctions and binding energies for the entire periodic table is given in Figs. 8-10 and 8-11. Minor shell-structure effects are visible at $Z = 2, 10, 18, 30, 36 \cdots$, where the various s and p subshells become filled, as are the gross changes in d, f, and g functions upon collapse into the inner potential well. It is worth noting that upon collapse of the 4f function at $Z = 57$, the value of $\langle r \rangle_{4f}$ immediately becomes semi-hydrogenic in the sense that it is comparable with $\langle r \rangle_{5s}$. The binding energy $|E^{4f}|$, on the other hand, remains abnormally small until $Z \cong 84$. Indeed, the 4f binding energy is initially about the same as that for 5d and 6s electrons. All three kinds of electrons thus compete for positions in the ground and low-excited configurations of the lanthanides. A similar competition among 5f, 6d, and 7s electrons exists in the actinide elements, as does competition between nd and (n + 1)s electrons in the various d transition series. These competitions are the cause of the numerous irregularities in the ground configurations of the elements, listed in Table 4-3 and discussed in Sec. 4-13.

Ions. All of the above discussion concerns neutral atoms. In a $(Z - N)$-fold ionized atom, the net charge seen by an electron far distant from the nucleus is $Z_c \equiv Z - N + 1$ units, and the hydrogenic potential is

$$-\frac{2Z_c}{r} + \frac{l(l+1)}{r^2} . \tag{8.25}$$

TABLE 8-8. ATOMIC NUMBERS AT WHICH d, f, AND g ELECTRONS COLLAPSE INTO THE CORE FOR NEUTRAL ATOMS

	A priori		*Self-consistent-field*					
	TF[a]	TFD[a]	H[b]	HF[c]	HX[b]	HXR	DHF[d]	Exp.[e]
3d	27	22[f]	21	20	20	20	20	20
4d	45	37	38	38	38	38	38	38
5d	71[f]	59	56	56	56	56	56	56
6d	>92	89	88	88	88	88	88	88?
7d	---	---	120	---	120	120	120	---
4f	69	60	58	57	57	57	57	57
5f	>92	89	90	89	89	90	90	89 or 90
6f	---	---	121	---	120	122	122	---
5g	---	---	122	---	121	124	124	---

[a] R. Latter, Phys. Rev. **99**, 510 (1955).
[b] D. C. Griffin, K. L. Andrew, and R. D. Cowan, Phys. Rev. **177**, 62 (1969).
[c] D. C. Griffin, R. D. Cowan, and K. L. Andrew, Phys. Rev. A **3**, 1233 (1971).
[d] R. D. Cowan and J. B. Mann, in G. H. Woodgate and P. G. H. Sandars, eds., *Atomic Physics 2* (Plenum, London, 1971), p. 215; J. B. Mann, private communication.
[e] C. E. Moore, *Atomic Energy Levels*.
[f] These are borderline cases; the arbitrary criterion on n* (see fn. 15) is very nearly satisfied for Z = 21 and 70.

The hydrogenic well thus begins at

$$r = \frac{l(l+1)}{2Z_c} , \qquad (8.26)$$

and extends well into the electron-core region for d electrons in singly ionized atoms ($Z_c = 2$) and for f electrons in 2- or 3-fold ionized atoms. In such ions, therefore, d and f electrons behave much like p electrons in neutral atoms; that is, for a given ionization stage the radial wavefunctions contract gradually with increasing Z, rather than collapsing suddenly. Similarly, the value of the effective quantum number n* decreases gradually, and tends to reach any given value at a considerably lower value of Z than for neutral atoms.

Problems

8-6(1). Using experimental energy levels from AEL for K I and Ca I, verify that n^*_{3d} decreases suddenly from close to the hydrogenic value n = 3 in K I 3d to approximately n* = 2 in Ca I 3d4s

8-6(2). Using experimental data from AEL (for, say, the even-Z elements), verify that n^*_{3d} in singly ionized atoms decreases gradually from about 3 in Mg II to about 2.3 in Ca II, passing the value n* = 2.5 at about Z = 16 (S II). [Note: In configurations $3p^2 3d$ and

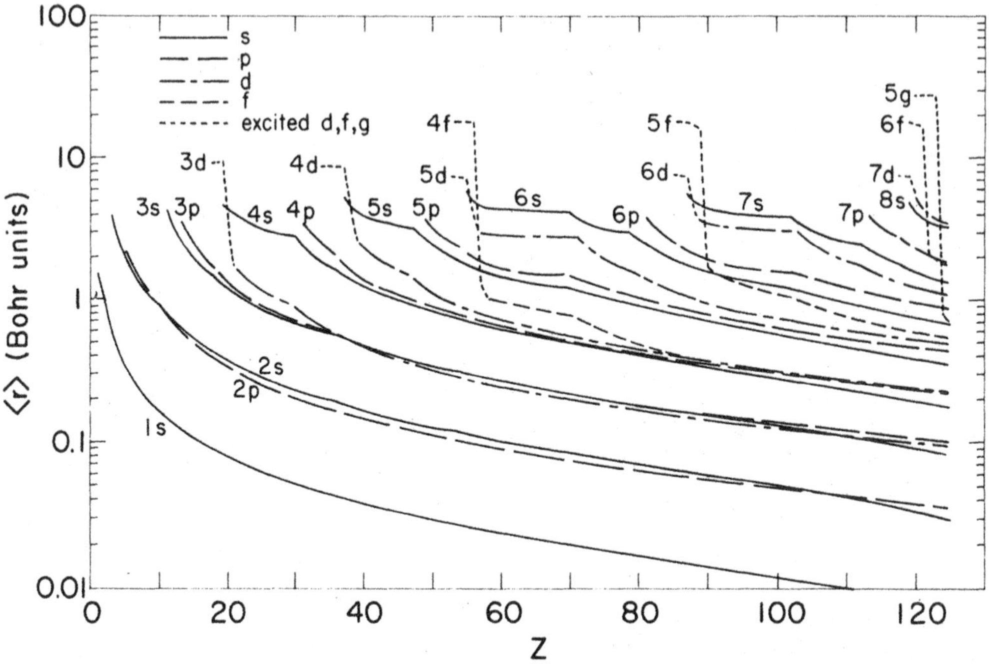

Fig. 8-10. Expectation values of r for the various orbitals of neutral atoms, computed by the HXR method. (The 3s and 3p curves are indistinguishable for $36 \leqq Z \leqq 80$.)

$3p^4 3d$ it is sufficiently accurate for present purposes to use only the levels based on the parent term $3p^w(^3P)$, evaluated relative to the ground term $3p^w\ ^3P$ of the doubly ionized atom.]

8-6(3). Calculate n^*_{nf} for the configurations Ba II nf ($4 \leqq n \leqq 11$), using data from AEL. The theoretical forms of V_{eff} and P(r) for the 4f electron are shown in Fig. 8-12. Discuss qualitatively the form that $P_{nf}(r)$ ($n \geqq 5$) must have to explain the experimentally observed variation of n^*, and the properties of V_{eff} that give rise to this abnormal behavior. [The large variation of δ_{nf} is so unusual—perhaps unique to Ba II nf and Ra II nf—that for a time it was suspected to be a spurious result of misinterpretation of the observed spectrum; cf. V. Kaufman and J. Sugar, J. Res. Natl. Bur. Stand. **71A**, 583 (1967).]

8-7. VARIATION WITH Z OF COULOMB AND SPIN-ORBIT INTEGRALS

The energy-level structure of a configuration involving a single open subshell nl^w ($l > 0$, $2 \leqq w \leqq 4l$) is determined primarily by the Coulomb radial integral(s) $F^k(nl,nl)$ ($k > 0$) and the spin-orbit integral ζ_{nl}. Theoretical values of these integrals are shown in

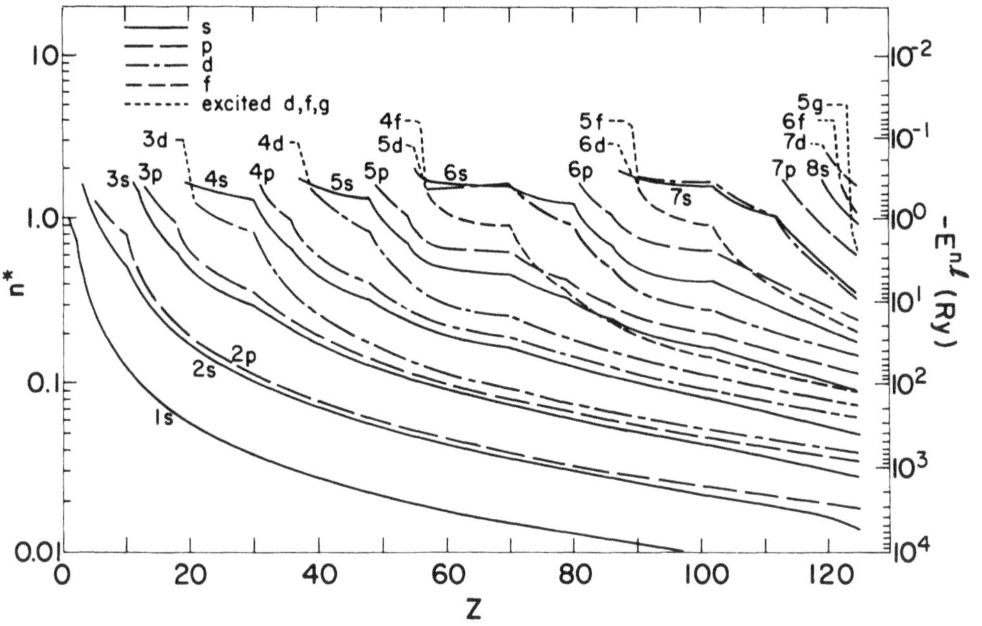

Fig. 8-11. One-electron binding energies $|E^{nl}| \cong |\varepsilon_{nl} + E_c^{nl}|$ and effective quantum numbers $n^*_{nl} = |E^{nl}|^{-1/2}$ for the various orbitals of neutral atoms, computed by the HXR method.

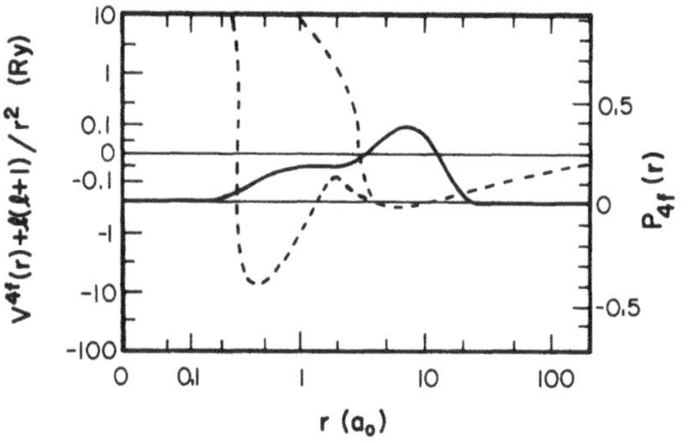

Fig. 8-12. HXR effective potentials and radial wavefunction for the 4f electron in Ba II $5p^64f$. (See the caption to Fig. 8-6 for details.)

Fig. 8-13 for the valence shell in ground configurations of neutral atoms.[16] Several characteristics of this figure are worth noting.

(1) There is an overall trend for the values of ζ to increase significantly with Z (or with n), whereas the values of F^k tend to decrease somewhat with increasing n. This results in a gradual change from nearly pure LS-coupling conditions at low Z toward approximate jj coupling at high Z (cf. Figs. 4-5 and 4-6).

[16]Values of ζ have been computed using the Blume-Watson formulation rather than the simple integral (8.2). These values usually agree with experiment within about 5%. Theoretical values of F^k are usually 15% or more greater than experimental values inferred by least-squares fitting of observed energy levels; see Sec. 16-2. For HF and DHF values, see the references in Sec. XIV of the bibliography.

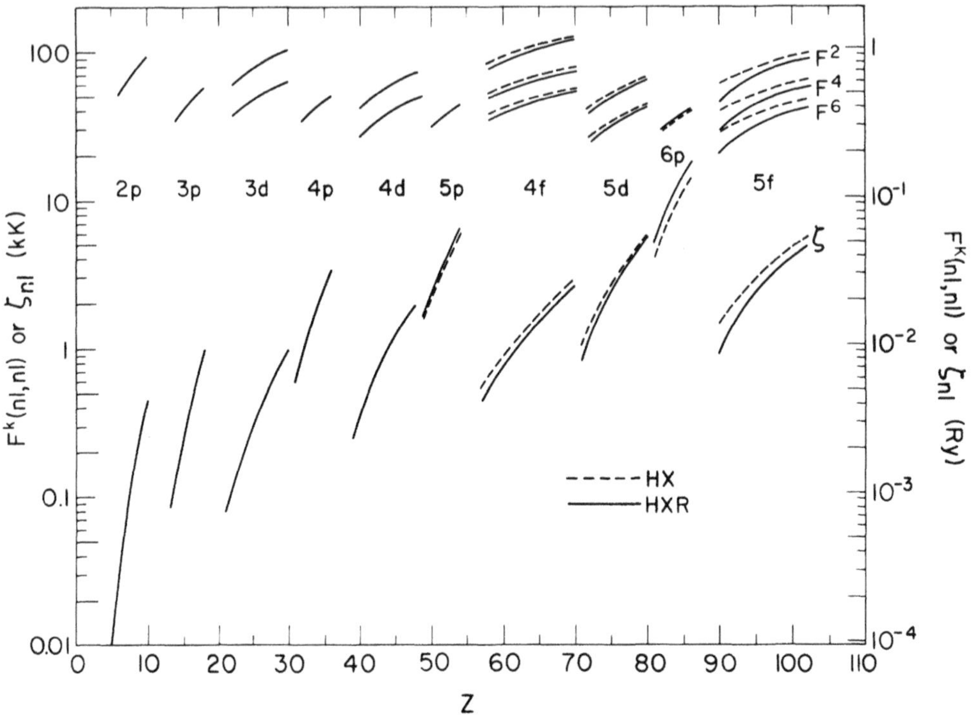

Fig. 8-13. Values of the Coulomb integrals $F^k(nl,nl)$ and of the spin-orbit parameters ζ_{nl} for single-open-subshell configurations in neutral atoms; where dashed curves are not shown, relativistic effects are negligibly small. (In practically all cases, Hartree-Fock values differ from those shown by amounts too small to be plotted.)

(2) Within a given subshell, ζ increases much more rapidly with Z than do the F^k. The coupling conditions therefore shift in the direction from LS toward jj in going from the two-electron configuration nl^2 to the two-hole configuration nl^{4l}.

(3) The Coulomb integrals F^k tend to increase with increasing l. This is primarily because each open p shell shown in the figure is always the outermost shell of the atom; the electron density is comparatively low so that the electrons are far apart and the Coulomb repulsions are comparatively weak. The d and (especially) the f shells, however, lie buried inside the atom as discussed in the preceding section (see Fig. 8-10), so that the electrons are closer together and the Coulomb repulsions are much stronger.

For these same reasons, the spin-orbit parameters ζ might also be expected to increase with l. However, the effects of the centripetal term of V_{eff} in preventing high-l electrons from penetrating close to the nucleus largely compensate [cf. the approximate l^{-3} hydrogenic dependence (3.52)], and ζ tends to decrease with increasing l. As a consequence, none of the transition nor rare-earth (d^n and f^n) configurations show anything approaching pure jj-coupling conditions, even in the heaviest elements; such conditions (in the ground configurations of neutral atoms) are approximated only in the 6p subshell ($_{82}$Pb to $_{84}$Po).

Ions. Coupling conditions usually change rapidly in going from neutral atoms to higher-Z members of an isoelectronic sequence. Electrons in a given subshell see a central field characterized by an effective nuclear charge

$$Z^* = Z - S \; , \tag{8.27}$$

where the screening constant S is approximately independent of Z. The spin-orbit energies increase with Z^* much faster than do the Coulomb energies, as may be surmised from the hydrogenic expressions (3.52) and $\langle r^{-1} \rangle \propto Z$ (Table 3-2):[17]

$$\zeta \propto (Z^*)^4 \tag{8.28}$$

and

$$F^k \propto Z^* \; , \qquad G^k \propto Z^* \; . \tag{8.29}$$

Therefore, the coupling approaches pure jj conditions for ground configurations[18] of ions with sufficiently high Z^*. These topics will be discussed in greater detail in Chapter 19.

As a function of Z (rather than Z^*), it has been noted both empirically[19] and theoretically[20] that the spin-orbit parameter for p electrons in excited subshells of neutral atoms increases as about the 7/3 power of Z:

$$\zeta_p = 0.450 \, Z^{2.33}/(n^*)^3 \; . \tag{8.30}$$

[17]Empirical examples of these relations are discussed by B. Edlén, "Atomic Spectra," in S. Flügge, ed., *Handbuch der Physik* (Springer-Verlag, Berlin, 1964), Vol. XXVII, pp. 165-85.

[18]For *excited* configurations of elements near the beginnings of the transition and rare-earth series, the collapse of the d and f functions may lead to changes in coupling conditions that for small Z^* are greatly different from the usual LS → jj shift; R. D. Cowan, J. Opt. Soc. Am. **58**, 924 (1968).

[19]U. Fano and W. C. Martin, in W. E. Brittin and H. Odabasi, eds., *Topics in Modern Physics* (Colorado Associated University Press, Boulder, 1971).

[20]J. L. Dehmer, Phys. Rev. A **7**, 4 (1973); U. Fano, C. E. Theodosiou, and J. L. Dehmer, Rev. Mod. Phys. **48**, 49 (1976).

9

COUPLED ANTISYMMETRIC BASIS FUNCTIONS

9-1. COUPLING OF TWO ANGULAR MOMENTA

We have now completed discussion of the first phase of the calculation of the energy states of atoms—namely, the determination of the radial factors of the one-electron wavefunctions

$$\varphi_i(r) = \frac{1}{r}P_{nl}(r)\cdot Y_{lm_l}(\theta,\phi)\cdot\sigma_{m_s}(s_z) \tag{9.1}$$

for a given electron configuration, and the evaluation of the center-of-gravity energy of this configuration. For this evaluation, it was sufficient to consider antisymmetrized but uncoupled (determinantal) basis functions for the atom.

We come now to the second phase of the calculation—the determination of the detailed energy-level structure; i.e., the energy-level splittings about the center of gravity energy E_{av}, such as those illustrated in Figs. 4-5 to 4-8. For the reasons discussed in Sec. 4-23, this portion of the calculation is best done using coupled basis functions. It was pointed out in Sec. 4-6 that construction of such functions by angular-momentum coupling of antisymmetric (determinantal) functions may lead to certain awkward complications; these will be considered shortly. For the present, we consider the coupling of non-antisymmetrized (product) functions (4.21). The coupling problem was considered qualitatively in Secs. 2-8 and 2-10, and a numerical illustration was given there. Here we consider the quantitative details.

Suppose we are given functions $\psi_{j_1m_1}(\mathbf{q}_1) \equiv |j_1m_1\rangle$ that are eigenfunctions of angular-momentum operators $\mathbf{J}_1^2$ and J_{1z} operating in the vector space $\mathbf{q}_1$, with eigenvalues $j_1(j_1 + 1)$ and m_1 (in units of $\hbar^2$ and $\hbar$, respectively); also, that we are given functions $\psi_{j_2m_2}(\mathbf{q}_2) \equiv |j_2m_2\rangle$ that are eigenfunctions of operators $\mathbf{J}_2^2$ and J_{2z} operating in the space $\mathbf{q}_2$. Then the product functions

$$\psi_{j_1m_1}(q_1)\psi_{j_2m_2}(q_2) = |j_1m_1\rangle\,|j_2m_2\rangle \equiv |j_1j_2m_1m_2\rangle \tag{9.2}$$

are eigenfunctions of all four operators $\mathbf{J}_1^2$, J_{1z}, $\mathbf{J}_2^2$, and J_{2z}. Coupled functions $|j_1j_2jm\rangle$ that are eigenfunctions of the four operators

$$\mathbf{J}_1^2, \quad \mathbf{J}_2^2, \quad \mathbf{J}^2 \equiv (\mathbf{J}_1 + \mathbf{J}_2)^2, \quad \mathbf{J}_z \equiv \mathbf{J}_{1z} + \mathbf{J}_{2z} \tag{9.3}$$

can be formed by taking appropriate linear combinations of the uncoupled functions (9.2):

$$|j_1j_2jm\rangle = \sum_{m_1=-j_1}^{j_1} \sum_{m_2=-j_2}^{j_2} C(j_1j_2m_1m_2;jm)\,|j_1j_2m_1m_2\rangle\ . \tag{9.4}$$

For our purposes, the operators $\mathbf{J}_1$ and $\mathbf{J}_2$ will be either orbital or spin angular-momentum operators, or combinations thereof. Simple examples are the orbital momenta of two electrons (in which case the coordinates $\mathbf{q}_1$ and $\mathbf{q}_2$ are the position coordinates $\mathbf{r}_1$ and $\mathbf{r}_2$ of the electrons), the spin momenta of two electrons (corresponding to coordinates s_1 and s_2), or an orbital and a spin momentum for either the same or two different electrons; the momenta may also be compound momenta such as **L** and **S**. In any case, the coordinates associated with the quantum numbers j_1 and j_2 of the coupled function $|j_1j_2jm\rangle$ are of course the same as those of $|j_1m_1\rangle$ and $|j_2m_2\rangle$, respectively.

The coefficients C in (9.4) are known as Clebsch-Gordon (CG), vector-coupling, or Wigner coefficients. Derivation of a general algebraic expression for these coefficients is somewhat long and involved[1,2] and we postpone it to Sec. 9-11; the result can be very simply expressed in terms of a 3-j symbol:

$$C(j_1j_2m_1m_2;jm) = (-1)^{j_1-j_2+m}[j]^{1/2}\begin{pmatrix} j_1 & j_2 & j \\ m_1 & m_2 & -m \end{pmatrix}, \tag{9.5}$$

where as usual $[j] \equiv 2j + 1$. Because 3-j symbols are real and $j_1 - j_2 + m$ is integral, the CG coefficients are also real. The δ-factor present in the definition (5.1) of the 3-j symbol ensures that all terms of the linear combination (9.4) have the same value (m) for the sum $m_1 + m_2$, which is necessary for (9.4) to be an eigenfunction of J_z, and effectively reduces the double sum to a single sum:

$$|j_1j_2jm\rangle = \sum_{m_1=-j_1}^{j_1} C(j_1j_2m_1, m-m_1\,;jm)\,|j_1j_2m_1, m-m_1\rangle\ . \tag{9.4'}$$

Various other properties of the 3-j symbol—namely,

$$j_1 + j_2 + j \quad \text{integral}$$

with

$$j = |j_1 - j_2|,\ |j_1 - j_2| + 1,\ \cdots\ j_1 + j_2 - 1,\ j_1 + j_2\ ,$$

and

$$j + m \quad \text{integral}$$

[1]G. Racah, II, Phys. Rev. **62**, 438 (1942).

[2]R. T. Sharp, Am. J. Phys. **28**, 116 (1960); B. R. Judd, *Operator Techniques in Atomic Spectroscopy* (McGraw-Hill, New York, 1963).

with

$$m = -j, -j + 1, \cdots j - 1, j$$

—provide the mathematical formulation of the vector model of the atom.

Recursion relations. Multiplying both sides of (9.4) with an uncoupled function $\langle j_1j_2m_1'm_2'|$, integrating, and making use of the orthonormal property of these functions, we find that the CG coefficients are just overlap integrals between coupled and uncoupled functions:

$$\langle j_1j_2m_1'm_2'|j_1j_2jm\rangle = C(j_1j_2m_1'm_2';jm) \ .$$

If we operate on both sides of (9.4) with the step-up or step-down operators $J_\pm \equiv J_x \pm iJ_y$ before integrating with the uncoupled function, we obtain

$$\langle j_1j_2m_1m_2|J_\pm|j_1j_2jm\rangle = \sum_{m'_1m'_2} \langle j_1j_2m_1m_2|J_{1\pm}+J_{2\pm}|j_1j_2m_1'm_2'\rangle \ C(j_1j_2m_1'm_2';jm) \ .$$

Using (2.19) we then find

$$[(j\mp m)(j\pm m+1)]^{1/2}C(j_1j_2m_1m_2;j,m\pm 1)$$

$$= \sum_{m'_1m'_2} [(j_1\mp m_1')(j_1\pm m_1'+1)]^{1/2}\langle j_1j_2m_1m_2|j_1j_2(m_1'\pm 1)m_2'\rangle \ C(j_1j_2m_1'm_2';jm)$$

$$+ \sum_{m'_1m'_2} [(j_2\mp m_2')(j_2\pm m_2'+1)]^{1/2}\langle j_1j_2m_1m_2|j_1j_2m_1'(m_2'\pm 1)\rangle \ C(j_1j_2m_1'm_2';jm)$$

$$= [(j_1\pm m_1)(j_1\mp m_1+1)]^{1/2} \ C(j_1j_2,m_1\mp 1,m_2;jm)$$

$$+ [(j_2\pm m_2)(j_2\mp m_2+1)]^{1/2}C(j_1j_2m_1,m_2\mp 1;jm) \ . \tag{9.6}$$

These recursion relations for the CG coefficients are useful in the derivation of (9.5), and will be needed in Secs. 11-4 and 11-6.

Phase relations. It is easily seen from (5.1) and (5.3) that for the extreme values $j = j_1 + j_2$ and $m = \pm j$, then the coupled function is identical with a product function in both magnitude and phase:

$$|j_1j_2, j_1 + j_2, \pm(j_1 + j_2)\rangle = |j_1, \pm j_1\rangle|j_2, \pm j_2\rangle \ . \tag{9.7}$$

For any given j, the relative phases of the coupled functions for different m [as obtained from (9.4)-(9.5)] are consistent with (9.6), and therefore with the choice made in deriving (2.19).

It is, however, very important to note that the phase conventions involved in (9.5) are such that the coupling (j_2j_1) is *not* the same as the coupling (j_1j_2): Using the symmetry relation (5.4) for the interchange of two columns of a 3-j symbol, we find

$$|j_2j_1jm\rangle = \sum_{m_1m_2} (-1)^{j_2-j_1+m}[j]^{1/2}\begin{pmatrix} j_2 & j_1 & j \\ m_2 & m_1 & -m \end{pmatrix}|j_2m_2\rangle|j_1m_1\rangle$$

$$= (-1)^{-3j_1+j_2-j} \sum_{m_1m_2} (-1)^{j_1-j_2+m}[j]^{1/2}\begin{pmatrix} j_1 & j_2 & j \\ m_1 & m_2 & -m \end{pmatrix}|j_2m_2\rangle|j_1m_1\rangle$$

$$= (-1)^{j_1+j_2-j}|j_1j_2jm\rangle \; ; \qquad (9.8)$$

in the last step, use has been made of the fact that $4j_1$ is an even integer. As a consequence of this phase relation, the LS-coupled wavefunctions used in this book differ by a factor

$$(-1)^{L+S-J} \qquad (9.9)$$

from the SL-coupled functions used in Condon and Shortley and most of the other literature,[3] and matrix elements of an operator O in the two coupling schemes are related as follows:

$$\langle(SL)J|O|(S'L')J'\rangle = (-1)^{L'-L+S'-S-J'+J}\langle(LS)J|O|(L'S')J'\rangle \; . \qquad (9.10)$$

(In the special case of the electron-electron Coulomb operator, the two matrix elements are equal because the Coulomb matrix is diagonal in L, S, and J, as discussed in Sec. 2-14.)

Orthonormality. The coupled functions (9.4) are properly normalized, and two such functions for given j_1 and j_2 but different values of j and/or m are orthogonal. This follows directly from the assumed orthonormality of the uncoupled functions $|j_1m_1\rangle$ and $|j_2m_2\rangle$ —see, for example, (2.42) and (2.54)—and from the 3-j orthonormality relation (5.15):

$$\langle j_1j_2jm|j_1j_2j'm'\rangle$$

$$= \sum_{m_1m_2}\sum_{m'_1m'_2} C(j_1j_2m_1m_2;jm)\, C(j_1j_2m'_1m'_2;j'm') \times \langle j_1m_1|j_1m'_1\rangle\langle j_2m_2|j_2m'_2\rangle$$

$$= \sum_{m_1m_2} C(j_1j_2m_1m_2;jm)\, C(j_1j_2m_1m_2;j'm')$$

$$= (-1)^{m-m'}\sum_{m_1m_2}[j]\begin{pmatrix} j_1 & j_2 & j \\ m_1 & m_2 & -m \end{pmatrix}\begin{pmatrix} j_1 & j_2 & j' \\ m_1 & m_2 & -m' \end{pmatrix}$$

$$= \delta_{jj'}\delta_{mm'} \; . \qquad (9.11)$$

[3]However, LS coupling is used by R. I. Karaziya, Ya. I. Vizbaraite, Z. B. Rudzikas, and A. P. Yutsis, *Tables for the Calculation of Matrix Elements of Operators of Atomic Quantities* (Moscow, 1967; English transl. ANL-TRANS-563, National Technical Information Service, Springfield, Va., 1968).

Because of this result, the CG coefficients for given j_1j_2 may be thought of as the elements of an orthogonal matrix[4] representing the transformation between the set of uncoupled product functions $|j_1m_1\rangle\,|j_2m_2\rangle$ and the set of coupled functions $|j_1j_2jm\rangle$; the columns of this matrix are labeled by the values of m_1m_2, and the rows are labeled by the possible values of jm.

Example. As an example, let us consider the CG transformation matrix for the coupling of the l's of two (non-equivalent) p electrons. Abbreviating the notation in (9.4)-(9.5) slightly we may write

$$|LM_L\rangle_c = \sum_{m_1m_2} (-1)^{M_L} (2L + 1)^{1/2} \begin{pmatrix} 1 & 1 & L \\ m_1 & m_2 & -M_L \end{pmatrix} |m_1m_2\rangle \,.$$

Using values of the 3-j symbols from Appendix C we find

$$|2\ 2\rangle_c = (5)^{1/2}\begin{pmatrix} 1 & 1 & 2 \\ 1 & 1 & -2 \end{pmatrix} |1\ 1\rangle = |1\ 1\rangle \,,$$

$$|2\ 1\rangle_c = -(5)^{1/2}\begin{pmatrix} 1 & 1 & 2 \\ 1 & 0 & -1 \end{pmatrix} |1\ 0\rangle - (5)^{1/2}\begin{pmatrix} 1 & 1 & 2 \\ 0 & 1 & -1 \end{pmatrix} |0\ 1\rangle$$

$$= (2)^{-1/2}|1\ 0\rangle + (2)^{-1/2}|0\ 1\rangle \,,$$

$$|2\ 0\rangle_c = (5)^{1/2}\begin{pmatrix} 1 & 1 & 2 \\ 1 & -1 & 0 \end{pmatrix} |1\ -1\rangle + (5)^{1/2}\begin{pmatrix} 1 & 1 & 2 \\ 0 & 0 & 0 \end{pmatrix} |0\ 0\rangle$$

$$+ (5)^{1/2}\begin{pmatrix} 1 & 1 & 2 \\ -1 & 1 & 0 \end{pmatrix} |-1\ 1\rangle$$

$$= (6)^{-1/2}|1\ -1\rangle + (2/3)^{1/2}|0\ 0\rangle + (6)^{-1/2}|-1\ 1\rangle \,,$$

etc. The complete C-G transformation matrix is given in Table 9-1. This matrix not only gives the transformation from the uncoupled to the coupled set of functions; since the matrix is orthogonal, its transpose is the transformation matrix from the coupled set back to the uncoupled set. We therefore have the equality

$$C(j_1j_2m_1m_2;jm) = \langle j_1j_2m_1m_2|j_1j_2jm\rangle = \langle j_1j_2jm|j_1j_2m_1m_2\rangle \,, \tag{9.12}$$

and the inverse transformation may be written

[4]An orthogonal matrix is a square matrix whose rows consitute a set of orthonormal real vectors. This property is a necessary and sufficient condition that the columns also form a set of orthonormal vectors, and that the inverse of the matrix be equal to its transpose, $T^{-1} = \tilde{T}$—see (4.14) to (4.16).

TABLE 9-1. THE CLEBSCH-GORDON TRANSFORMATION MATRIX FOR THE ORBITAL-ANGULAR-MOMENTUM COUPLING OF TWO NON-EQUIVALENT p ELECTRONS

	m_1m_2								
LM_L	1 1	1 0	0 1	1 −1	0 0	−1 1	−1 0	0 −1	−1 −1
2 2	1	0	0	0	0	0	0	0	0
2 1	0	$1/\sqrt{2}$	$1/\sqrt{2}$	0	0	0	0	0	0
1 1	0	$-1/\sqrt{2}$	$1/\sqrt{2}$	0	0	0	0	0	0
2 0	0	0	0	$1/\sqrt{6}$	$\sqrt{2/3}$	$1/\sqrt{6}$	0	0	0
1 0	0	0	0	$1/\sqrt{2}$	0	$-1/\sqrt{2}$	0	0	0
0 0	0	0	0	$1/\sqrt{3}$	$-1/\sqrt{3}$	$1/\sqrt{3}$	0	0	0
1 −1	0	0	0	0	0	0	$-1/\sqrt{2}$	$1/\sqrt{2}$	0
2 −1	0	0	0	0	0	0	$1/\sqrt{2}$	$1/\sqrt{2}$	0
2 −2	0	0	0	0	0	0	0	0	1

$$|j_1j_2m_1m_2\rangle = \sum_{j=|j_1-j_2|}^{j_1+j_2} \sum_{m=-j}^{j} \langle j_1j_2jm|j_1j_2m_1m_2\rangle |j_1j_2jm\rangle$$

$$= \sum_j \sum_m C(j_1j_2m_1m_2;jm)|j_1j_2jm\rangle \ . \tag{9.13}$$

Problems

9-1(1). Verify the values of those matrix elements in Table 9-1 that were not calculated in the text, using values of 3-j symbols from Appendix C. Verify that the matrix is an orthogonal one.

9-1(2). For a pp′ configuration (consisting of two non-equivalent p electrons), calculate the M = +2 CG transformation matrix from the product functions $|l_1m_{l_1}\rangle\ |l_2m_{l_2}\rangle\ |m_{s_1}\rangle\ |m_{s_2}\rangle$ to LS-coupled functions.

9-2. RECOUPLING OF THREE ANGULAR MOMENTA

When we have only two angular momenta to couple together, the coupling can be done in only two different ways—either the second momentum can be coupled onto the first, or the first can be coupled onto the second. These two couplings differ only by a phase factor, as given by (9.8).

When we have more than two momenta to be coupled, the coupling can be performed in a number of different ways.[5] For example, if we have three different momenta j_1, j_2, and j_3,

[5]The treatment given here is essentially that of G. Racah, III, Phys. Rev. 63, 367 (1943).

we can first couple j_2 to j_1 to give a resultant J', and then couple j_3 to J' to give a total momentum J. Using (9.4)-(9.5) twice, the coupled wavefunction according to this coupling scheme is

$$|[(j_1j_2)J',j_3]JM\rangle = \sum_{M'm_3} (-1)^{J'-j_3+M}[J]^{1/2}\begin{pmatrix} J' & j_3 & J \\ M' & m_3 & -M \end{pmatrix}$$

$$\times \sum_{m_1m_2} (-1)^{j_1-j_2+M'}[J']^{1/2}\begin{pmatrix} j_1 & j_2 & J' \\ m_1 & m_2 & -M' \end{pmatrix}|j_1m_1\rangle|j_2m_2\rangle|j_3m_3\rangle \; . \qquad (9.14)$$

Another possibility is to couple j_3 to j_2 to give a resultant J'', and then couple J'' to j_1 to give J; for this coupling, the CG expansion is

$$|[j_1,(j_2j_3)J'']JM\rangle = \sum_{m'_1M''} (-1)^{j_1-J''+M}[J]^{1/2}\begin{pmatrix} j_1 & J'' & J \\ m'_1 & M'' & -M \end{pmatrix}$$

$$\times \sum_{m'_2m'_3} (-1)^{j_2-j_3+M''}[J'']^{1/2}\begin{pmatrix} j_2 & j_3 & J'' \\ m'_2 & m'_3 & -M'' \end{pmatrix}|j_1m'_1\rangle|j_2m'_2\rangle|j_3m'_3\rangle \; . \qquad (9.15)$$

These two couplings are not the same, and the expressions (9.14) and (9.15) are not equivalent. The recoupling coefficients are the overlap integrals between (9.14) and (9.15); each such integral (one for each possible set of values of J′J″JM) involves an eight-fold summation over the magnetic quantum numbers $m_1m_2m_3M'm'_1m'_2m'_3M''$, except that we obtain the delta-factors $\delta_{m_1m'_1}\delta_{m_2m'_2}\delta_{m_3m'_3}$ from the orthonormality of the uncoupled functions. Noting that the exponent of each power of (-1) is integral so that we can change its sign at will, making use of the symmetry properties (5.4)-(5.6) of the 3-j symbol, and changing the sign of the summation index m_1, we find

$$\langle[(j_1j_2)J',j_3]JM|[j_1,(j_2j_3)J'']JM\rangle = [J',J'']^{1/2}[J]$$

$$\times \sum_{\substack{m_1m_2m_3 \\ M'M''}} (-1)^{-J'+j_3-M-j_1+j_2-M'+j_1-J''+M-j_2+j_3-M''}\begin{pmatrix} J' & j_3 & J \\ M' & m_3 & -M \end{pmatrix}$$

$$\times (-1)^{j_1+j_2+J'}\begin{pmatrix} J' & j_1 & j_2 \\ M' & m_1 & -m_2 \end{pmatrix}$$

$$\times (-1)^{-j_2-j_3-J''}\begin{pmatrix} J'' & j_3 & j_2 \\ -M'' & m_3 & m_2 \end{pmatrix}(-1)^{j_1+J''+J}\begin{pmatrix} J'' & j_1 & J \\ M'' & -m_1 & -M \end{pmatrix} . \qquad (9.16)$$

Since the terms of this sum are zero unless $-M' = m_1 - m_2$, and since $j_2 + m_2$ and $J'' + M''$ are integral, we may write the exponent of the phase factor in the above sum as

$$(j_3 + m_1 - m_2 - J'' - M'' + 2j_1 + J) + (2j_2 + 2m_2 + 2J'' + 2M'')$$
$$= (j_1 + j_2 + j_3 + J) + (j_1 + j_2 + J'' + m_1 + m_2 + M'') \ .$$

By comparison with (5.25) we thus find[6]

$$\langle[(j_1j_2)J',j_3]JM\,|\,[j_1,(j_2j_3)J'']JM\rangle = (-1)^{j_1+j_2+j_3+J}[J',J'']^{1/2}\begin{Bmatrix} J' & j_3 & J \\ J'' & j_1 & j_2 \end{Bmatrix}$$
$$= (-1)^{j_1+j_2+j_3+J}[J',J'']^{1/2}\begin{Bmatrix} j_1 & j_2 & J' \\ j_3 & J & J'' \end{Bmatrix} \ . \tag{9.17}$$

Note that the recoupling coefficient is independent of M.

Various other recoupling coefficients can be calculated as above, but are more easily derived directly from (9.17) with the use of (9.8). For example,

$$\langle[(j_1j_2)J',j_3]JM\,|\,[j_1,(j_3j_2)J'']JM\rangle = (-1)^{j_1+J+J''}[J',J'']^{1/2}\begin{Bmatrix} j_1 & j_2 & J' \\ j_3 & J & J'' \end{Bmatrix} \ , \tag{9.18}$$

and

$$\langle[(j_1j_2)J',j_3]JM\,|\,[(j_1j_3)J'',j_2]JM\rangle = (-1)^{J'+j_3-J-j_1-j_3+J''}\langle[j_3,(j_1j_2)J']JM\,|\,[(j_3j_1)J'',j_2]JM\rangle$$
$$= (-1)^{J'-J-j_1+J''+j_3+j_1+j_2+J}[J',J'']^{1/2}\begin{Bmatrix} j_3 & j_1 & J'' \\ j_2 & J & J' \end{Bmatrix}$$
$$= (-1)^{j_2+j_3+J'+J''}[J',J'']^{1/2}\begin{Bmatrix} j_2 & j_1 & J' \\ j_3 & J & J'' \end{Bmatrix} \ . \tag{9.19}$$

Orthogonality. The set of coupled functions $|(j_1j_2)J'M'\rangle$ is an orthonormal set, from (9.11). A second application of (9.11) then shows that the functions (9.14), $|(J'j_3)JM\rangle$, also form an orthonormal set:

$$\langle(J'j_3)JM\,|\,(J'''j_3)J''''M''''\rangle = \delta_{J'J'''}\delta_{JJ''''}\delta_{MM''''} \ . \tag{9.20}$$

Similarly, the set of functions (9.15) is orthonormal. The linear transformation

$$|\,[(j_1j_2)J',j_3]JM\rangle = \sum_{J''}\langle[(j_1j_2)J',j_3]JM\,|\,[j_1,(j_2j_3)J'']JM\rangle\,|\,[j_1,(j_2j_3)J'']JM\rangle \tag{9.21}$$

[6]This remarkable simplification, due to Racah II,[1] makes it possible to evaluate the recoupling coefficients by means of the single summation implicit in a 6-j symbol, in place of the sixfold summation of (9.16). (The explicit fivefold summation is reduced to a twofold one by three independent conditions arising from the fact that the magnetic quantum numbers of each 3-j symbol must sum to zero, but is augmented by the implicit summation in each 3-j symbol.) The derivation involves manipulation of multiple summations of factorials, similar to the procedures used in Sec. 9-11.

from one set of functions to the other is therefore an orthogonal one, and the recoupling coefficients (9.17) form the elements of an orthogonal matrix—or rather a set of orthogonal matrices, one matrix for each value of JM—in which the rows are labeled by J′ and the columns by J″. Orthonormality of the rows of the transformation matrix then gives from (9.17)

$$\sum_{J''}\langle[(j_1j_2)J',j_3]JM\,|\,[j_1,(j_2j_3)J'']JM\rangle\langle[j_1,(j_2j_3)J'']JM\,|\,[(j_1j_2)J''',j_3]JM\rangle$$

$$= \delta_{J'J'''} = \sum_{J''}[J',J'']\begin{Bmatrix} j_1 & j_2 & J' \\ j_3 & J & J'' \end{Bmatrix}\begin{Bmatrix} j_1 & j_2 & J''' \\ j_3 & J & J'' \end{Bmatrix}, \tag{9.22}$$

which provides a proof of the orthogonality relation (5.31) for 6-j symbols.

Similarly, the recoupling coefficients (9.18) and (9.19) are elements of orthogonal transformation matrices, and in each case the transposed matrix is the matrix of the inverse transformation.

9-3. TRANSFORMATIONS BETWEEN COUPLING SCHEMES

The results of the preceding section may be readily applied to calculate the basis-function transformation between any two coupling schemes of physical interest, such as those described in Secs. 2-11 and 4-17 to 4-20. We illustrate the procedure for the four important couplings (LS, LK, jK, jj) of the four angular momenta $l_1s_1l_2s_2$ of two electrons. More complicated cases can be handled similarly.[7]

The transformation matrix elements between LS- and LK-coupled functions are given directly by (9.17), with only notational changes:[8]

$$T_{LS,L'K'} \equiv \langle[(l_1l_2)L,(s_1s_2)S]J\,|\,[(l_1l_2)L',s_1]K',s_2,J\rangle$$

$$= \delta_{LL'}(-1)^{L+s_1+s_2+J}[K',S]^{1/2}\begin{Bmatrix} L & s_1 & K' \\ s_2 & J & S \end{Bmatrix}; \tag{9.23}$$

the δ-factor arises from the orthonormality of the partially coupled functions $|(l_1l_2)LM_L\rangle$, and we have omitted the unnecessary quantum number M from the notation. Similarly, the LK-jK and jK-jj transformation matrix elements follow directly from (9.19) and (9.17), respectively:

$$T_{L'K',j_1K} \equiv \langle[(l_1l_2)L',s_1]K',s_2,J\,|\,[(l_1s_1)j_1,l_2]K,s_2,J\rangle$$

$$= \delta_{KK'}(-1)^{l_2+s_1+L'+j_1}[L',j_1]^{1/2}\begin{Bmatrix} l_2 & l_1 & L' \\ s_1 & K & j_1 \end{Bmatrix}, \tag{9.24}$$

[7]R. D. Cowan, J. Opt. Soc. Am. **58**, 808 (1968).

[8]Note that the transformation matrix element between the SL and LK schemes, $\langle[(s_1s_2)S,(l_1l_2)L]J\,\|[(l_1l_2)L,s_1]K,s_2,J\rangle$, involves from (9.9) an additional factor $(-1)^{L+S-J}$. See also fn. 32 of Chap. 4.

$$\begin{aligned} T_{j_1K,j'_1j_2} &\equiv \langle[(l_1s_1)j_1,l_2]K,s_2,J\,|\,[(l_1s_1)j'_1,(l_2s_2)j_2]J\rangle \\ &= \delta_{j_1j'_1}(-1)^{j_1+l_2+s_2+J}[K,j_2]^{1/2}\begin{Bmatrix} j_1 & l_2 & K \\ s_2 & J & j_2 \end{Bmatrix} . \end{aligned} \tag{9.25}$$

The transformation matrix between LS- and jK-coupled functions is given by the matrix product of the LS-LK and LK-jK transformation matrices. Actually, the summation (over L'K') that is normally involved in calculating each element of the product matrix here reduces to a single term because of the δ-factors in (9.23) and (9.24):

$$\begin{aligned} T_{LS,JK} &\equiv \langle[(l_1l_2)L,(s_1s_2)S]J\,|\,[(l_1s_1)j_1,l_2]K,s_2,J\rangle \\ &= (-1)^{s_2+J-l_2-j_1}[L,S,j_1,K]^{1/2}\begin{Bmatrix} L & s_1 & K \\ s_2 & J & S \end{Bmatrix}\begin{Bmatrix} l_2 & l_1 & L \\ s_1 & K & j_1 \end{Bmatrix} . \end{aligned} \tag{9.26}$$

Similarly, matrix elements for the LK-jj transformation are given essentially by the simple product of (9.24) and (9.25):

$$\begin{aligned} T_{LK,JJ} &\equiv \langle[(l_1l_2)L,s_1]K,s_2,J\,|\,[(l_1s_1)j_1,(l_2s_2)j_2]J\rangle \\ &= (-1)^{s_1+L-s_2-J}[L,K,j_1,j_2]^{1/2}\begin{Bmatrix} l_2 & l_1 & L \\ s_1 & K & j_1 \end{Bmatrix}\begin{Bmatrix} j_1 & l_2 & K \\ s_2 & J & j_2 \end{Bmatrix} . \end{aligned} \tag{9.27}$$

The transformation matrix between LS- and jj-coupled basis functions is given by the double matrix product of the elements (9.23), (9.24), and (9.25). The matrix elements involve only a single summation over K; for example, using (9.23) and (9.27) we obtain

$$\begin{aligned} T_{LS,JJ} &= \sum_K T_{LS,LK}\cdot T_{LK,JK}\cdot T_{JK,JJ} \\ &= [L,S,j_1,j_2]^{1/2}\sum_K(-1)^{2(L+s_1)}[K]\begin{Bmatrix} L & s_1 & K \\ s_2 & J & S \end{Bmatrix}\begin{Bmatrix} l_2 & l_1 & L \\ s_1 & K & j_1 \end{Bmatrix}\begin{Bmatrix} j_1 & l_2 & K \\ s_2 & J & j_2 \end{Bmatrix} . \end{aligned}$$

Noting that $2(L + s_1 + K)$ must be an even integer, we may write this with the aid of (5.37) as

$$T_{LS,JJ} = [L,S,j_1,j_2]^{1/2}\sum_K(-1)^{2K}[K]\begin{Bmatrix} l_2 & s_2 & j_2 \\ J & j_1 & K \end{Bmatrix}\begin{Bmatrix} L & S & J \\ s_2 & K & s_1 \end{Bmatrix}\begin{Bmatrix} l_1 & s_1 & j_1 \\ K & l_2 & L \end{Bmatrix} \tag{9.28}$$

$$\begin{aligned} &= [L,S,j_1,j_2]^{1/2}\begin{Bmatrix} l_2 & L & l_1 \\ s_2 & S & s_1 \\ j_2 & J & j_1 \end{Bmatrix} \\ &= [L,S,j_1,j_2]^{1/2}\begin{Bmatrix} l_1 & l_2 & L \\ s_1 & s_2 & S \\ j_1 & j_2 & J \end{Bmatrix} . \end{aligned} \tag{9.29}$$

Thus the LS-jj transformation matrix elements are proportional to 9-j symbols. Since 9-j symbols can be evaluated only as a sum over products of 6-j symbols, it follows that (unless 9-j tables are available) the matrix elements have to be evaluated from (9.28), or an equivalent expression; the 9-j symbol form (9.29) is then of only notational convenience.

The fact that the LS-jj transformation is an orthogonal one provides a proof of the orthonormality property (5.40) of 9-j symbols. The results (9.23) and (9.29) confirm the expressions (5.44) and (5.45) inferred from graphical considerations.

As examples, Table 9-2 gives the transformation matrices for the J = 1 basis functions of a pp′ configuration. It can be easily verified that each matrix is an orthogonal one, so that in each case the transposed matrix describes the inverse transformation.

Problems

9-3(1). Verify the elements of the LS-LK, LK-jK, and jK-jj transformation matrices in Table 9-2 with the aid of values of 6-j symbols from Appendix D. What are the physical reasons for the various null elements?

9-3(2). For J = 1 functions of a pp′ configuration, calculate the LS-jK, LK-jj, and LS-jj transformation matrices by matrix multiplication of the first three matrices in Table 9-2. Also calculate the LS-jK matrix by multiplication of the jK-jj and LS-jj matrices. Verify that all matrices are orthogonal ones.

9-3(3). Write down an expression for transformation matrix elements between LS basis functions and functions corresponding to the (physically uninteresting) coupling scheme

$$\{[l_1,(s_1s_2)S]Q,l_2\}J \ .$$

9-3(4). With the aid of (9.29), derive an expression for matrix elements of the transformation between the $\mathfrak{L}\mathfrak{S}$- and $\mathfrak{J}\mathfrak{J}$-coupled functions (4.37) and (4.44), respectively.

9-4. ANTISYMMETRIZATION DIFFICULTIES FOR EQUIVALENT ELECTRONS

So far, we have discussed angular-momentum coupling only for simple product basis functions. We have yet to consider the problem of constructing coupled functions that are antisymmetric under interchange of any two electron coordinates. One possible procedure is to start with antisymmetrized (determinantal) functions, and to then couple the angular momenta by means of the same CG formula (9.4)-(9.5) that we used for product functions. So long as the electrons involved are non-equivalent, all values of the magnetic quantum numbers m_l and m_s are allowed under the Pauli exclusion principle; the summations over these quantum numbers that are involved in the CG sum can therefore be carried out with the expectation that no complications will arise. No matter what coupling scheme we use, we may anticipate that all sets of quantum numbers (LS, LK, jK, or j_1j_2) that are predicted by the pertinent triangle relations (i.e., by the vector model) are physically allowed.

However, when two equivalent electrons ($n_1l_1 = n_2l_2$) are involved, we saw in Sec. 4-5 that the combination $m_{l_1}m_{s_1} = m_{l_2}m_{s_2}$ is forbidden by the Pauli principle—or, more precisely, that the antisymmetrized (determinantal) function (4.26) is identically zero for

TABLE 9-2. RECOUPLING TRANSFORMATION MATRICES FOR J = 1 BASIS FUNCTIONS OF A pp′ CONFIGURATION

__LS-LK__

	$D\left[\frac{3}{2}\right]_1$	$P\left[\frac{1}{2}\right]_1$	$P\left[\frac{3}{2}\right]_1$	$S\left[\frac{1}{2}\right]_1$
$^3D_1$	1	0	0	0
$^3P_1$	0	$\sqrt{\frac{2}{3}}$	$\sqrt{\frac{1}{3}}$	0
$^1P_1$	0	$-\sqrt{\frac{1}{3}}$	$\sqrt{\frac{2}{3}}$	0
$^3S_1$	0	0	0	1

__LK-jK__

	$\frac{3}{2}\left[\frac{3}{2}\right]_1$	$\frac{3}{2}\left[\frac{1}{2}\right]_1$	$\frac{1}{2}\left[\frac{3}{2}\right]_1$	$\frac{1}{2}\left[\frac{1}{2}\right]_1$
$D\left[\frac{3}{2}\right]_1$	$-\sqrt{\frac{1}{6}}$	0	$\sqrt{\frac{5}{6}}$	0
$P\left[\frac{1}{2}\right]_1$	0	$-\sqrt{\frac{1}{3}}$	0	$\sqrt{\frac{2}{3}}$
$P\left[\frac{3}{2}\right]_1$	$\sqrt{\frac{5}{6}}$	0	$\sqrt{\frac{1}{6}}$	0
$S\left[\frac{1}{2}\right]_1$	0	$\sqrt{\frac{2}{3}}$	0	$\sqrt{\frac{1}{3}}$

__jK-jj__

	$\left(\frac{3}{2}\frac{3}{2}\right)_1$	$\left(\frac{3}{2}\frac{1}{2}\right)_1$	$\left(\frac{1}{2}\frac{3}{2}\right)_1$	$\left(\frac{1}{2}\frac{1}{2}\right)_1$
$\frac{3}{2}\left[\frac{3}{2}\right]_1$	$\frac{2}{3}$	$\frac{\sqrt{5}}{3}$	0	0
$\frac{3}{2}\left[\frac{1}{2}\right]_1$	$\frac{\sqrt{5}}{3}$	$-\frac{2}{3}$	0	0
$\frac{1}{2}\left[\frac{3}{2}\right]_1$	0	0	$\frac{1}{3}$	$\frac{2}{3}\sqrt{2}$
$\frac{1}{2}\left[\frac{1}{2}\right]_1$	0	0	$\frac{2}{3}\sqrt{2}$	$-\frac{1}{3}$

__LS-jj__

	$\left(\frac{3}{2}\frac{3}{2}\right)_1$	$\left(\frac{3}{2}\frac{1}{2}\right)_1$	$\left(\frac{1}{2}\frac{3}{2}\right)_1$	$\left(\frac{1}{2}\frac{1}{2}\right)_1$
$^3D_1$	$-\frac{1}{3}\sqrt{\frac{2}{3}}$	$-\frac{1}{3}\sqrt{\frac{5}{6}}$	$\frac{1}{3}\sqrt{\frac{5}{6}}$	$\frac{2}{3}\sqrt{\frac{5}{3}}$
$^3P_1$	0	$\frac{1}{\sqrt{2}}$	$\frac{1}{\sqrt{2}}$	0
$^1P_1$	$\frac{1}{3}\sqrt{5}$	$\frac{1}{3}$	$-\frac{1}{3}$	$\frac{1}{3}\sqrt{2}$
$^3S_1$	$\frac{1}{3}\sqrt{\frac{10}{3}}$	$-\frac{2}{3}\sqrt{\frac{2}{3}}$	$\frac{2}{3}\sqrt{\frac{2}{3}}$	$-\frac{1}{3\sqrt{3}}$

this combination. When we proceed to couple the angular momenta of determinantal functions via the CG formula (9.4)-(9.5), the magnetic-quantum-number summations will in certain cases involve these null uncoupled functions. We may therefore anticipate that for equivalent electrons the CG method will fail to give us the desired set of orthonormal coupled functions.

We will be no better off if we apply the CG formula to simple product functions, and then attempt to antisymmetrize the coupled product functions by means of coordinate permutations: because both sums are finite, the result is the same whether the summation over permutations is carried out before the CG sum, or vice versa. In order to discuss the difficulties further, however, it is convenient to follow the latter procedure. For equivalent electrons, the LK- and jK-coupling schemes are not physically appropriate, as they treat the two electrons non-symmetrically; we therefore need consider only the LS and jj schemes.

LS coupling. Let us introduce the notation

$$\psi^{ik}_{12\beta}$$

to represent an LS-coupled two-electron product function, characterized by quantum numbers $12\beta = n_1 l_1 n_2 l_2 LSJM$, and formed by taking the appropriate triple sum of CG coefficients times the product of two spin-orbitals

$$\varphi_{n_1 l_1 m_{l_1} m_{s_1}}(r_i)\cdot\varphi_{n_2 l_2 m_{l_2} m_{s_2}}(r_k) \ ;$$

specifically,

$$\psi^{ik}_{12\beta} = \sum_{M_L M_S} C(LSM_L M_S;JM) \sum_{m_{s_1} m_{s_2}} C(s_1 s_2 m_{s_1} m_{s_2};SM_S) \times \sum_{m_{l_1} m_{l_2}} C(l_1 l_2 m_{l_1} m_{l_2};LM_L)\varphi_1(r_i)\varphi_2(r_k) \ .$$

Antisymmetrizing this by means of a coordinate interchange gives

$$\Psi^{ik}_{12\beta} = 2^{-1/2}\,(\psi^{ik}_{12\beta} - \psi^{ki}_{12\beta}) \ . \tag{9.30}$$

The second term is an eigenfunction of $L = l_1 + l_2$, $S = s_1 + s_2$, J, and J_z and has the same values of the quantum numbers LSJM as does the first term, because all four operators are symmetric in the coordinates 1 and 2; therefore Ψ is a proper coupled function. However, the overlap integral between (9.30) and a second function $\Psi^{ik}_{12\beta'}$ characterized by quantum numbers $n_1 l_2 n_2 l_2 L'S'J'M'$ is

$$\begin{aligned}\langle\Psi^{ik}_{12\beta}|\Psi^{ik}_{12\beta'}\rangle &= \frac{1}{2}\Big[\langle\psi^{ik}_{12\beta} - \psi^{ki}_{12\beta}|\psi^{ik}_{12\beta'} - \psi^{ki}_{12\beta'}\rangle\Big] \\ &= \frac{1}{2}\Big[\langle\psi^{ik}_{12\beta}|\psi^{ik}_{12\beta'}\rangle + \langle\psi^{ki}_{12\beta}|\psi^{ki}_{12\beta'}\rangle - \langle\psi^{ik}_{12\beta}|\psi^{ki}_{12\beta'}\rangle - \langle\psi^{ki}_{12\beta}|\psi^{ik}_{12\beta'}\rangle\Big] \\ &= \langle\psi^{ik}_{12\beta}|\psi^{ik}_{12\beta'}\rangle - \langle\psi^{ki}_{12\beta}|\psi^{ik}_{12\beta'}\rangle \\ &= \langle\psi^{ik}_{12\beta}|\psi^{ik}_{12\beta'}\rangle + (-1)^{l_1+l_2+L+S}\langle\psi^{ik}_{21\beta}|\psi^{ik}_{12\beta'}\rangle\ ,\end{aligned} \tag{9.31}$$

where the final step follows by using the CG symmetry relation (9.8) twice, thereby obtaining a phase factor

$$(-1)^{-l_1-l_2+L-s_1-s_2+S} = (-1)^{l_1+l_2+1+L+S}\ .$$

Following the same procedure that we used to establish the orthonormality (9.11) of singly coupled functions, we find the first overlap integral on the right-hand side of (9.31) to be just $\delta_{LL'}\delta_{SS'}\delta_{JJ'}\delta_{MM'}$. For the second overlap integral we have these same delta-factors, and obtain also $\delta_{n_1n_2}\delta_{l_1l_2}$ from the orthonormality (4.19) of the spin orbitals. Thus

$$\langle\Psi^{ik}_{12\beta}|\Psi^{ik}_{12\beta'}\rangle = \delta_{LL'}\delta_{SS'}\delta_{JJ'}\delta_{MM'}[1 + \delta_{n_1n_2}\delta_{l_1l_2}(-1)^{L+S}]\ . \tag{9.32}$$

For non-equivalent electrons, this indicates that the Ψ are properly orthonormalized functions. For equivalent electrons, however, we have

$$\langle\Psi^{ik}_{12\beta}|\Psi^{ik}_{12\beta'}\rangle = \delta_{LL'}\delta_{SS'}\delta_{JJ'}\delta_{MM'}[1 + (-1)^{L+S}]\ . \tag{9.33}$$

This shows firstly that antisymmetrized LS-coupled functions for nl^2 are null if $L + S$ is odd; thus ^3S, ^1P, ^3D, ^1F, $\cdots$ states cannot exist physically (as we saw also in Sec. 4-9 in the case p^2). The result (9.33) shows secondly that when $L + S$ is even, then an additional factor $1/\sqrt{2}$ is required in (9.30) in order to produce a properly normalized function.

jj coupling. In a similar way, the straightforward antisymmetrization via (9.30) of a jj-coupled function ($12\beta \equiv n_1l_1j_1n_2l_2j_2JM$) leads with the use of

$$\Psi^{ki}_{12\beta} = (-1)^{j_1+j_2-J}\ \Psi^{ik}_{21\beta} \tag{9.34}$$

to

$$\begin{aligned}\langle\Psi^{ik}_{12\beta}|\Psi^{ik}_{1'2'\beta'}\rangle &= \langle\psi^{ik}_{12\beta}|\psi^{ik}_{1'2'\beta'}\rangle - (-1)^{j_1+j_2-J}\langle\psi^{ik}_{21\beta}|\psi^{ik}_{1'2'\beta'}\rangle \\ &= \delta_{JJ'}\delta_{MM'}[\delta_{j_1j'_1}\delta_{j_2j'_2} - \delta_{n_1n_2}\delta_{l_1l_2}\delta_{j_1j'_2}\delta_{j_2j'_1}(-1)^{j_1+j_2-J}]\ .\end{aligned} \tag{9.35}$$

For non-equivalent electrons, the second term in this expression is zero, and the Ψ are always properly orthonormalized functions.

For equivalent electrons, we still have orthogonality with respect to J and M, and for the remainder of the discussion we simplify the notation by considering only the case JM = J′M′ (i.e., β = β′):

$$\langle \Psi^{ik}_{12\beta} | \Psi^{ik}_{1'2'\beta} \rangle = [\delta_{j_1 j'_1} \delta_{j_2 j'_2} - \delta_{j_1 j'_2} \delta_{j_2 j'_1} (-1)^{j_1 + j_2 - J}] \ . \tag{9.36}$$

So far as normalization is concerned ($j_1 = j'_1$, $j_2 = j'_2$), this becomes

$$\langle \Psi^{ik}_{12\beta} | \Psi^{ik}_{12\beta} \rangle = 1 - \delta_{j_1 j_2} (-1)^{2j_1 - J} = 1 + \delta_{j_1 j_2} (-1)^J \ , \tag{9.37}$$

because $2j_1 = 2l_1 \pm 1$ is odd. We see that functions for $j_1 = j_2$ and J odd are null, and that therefore such states are forbidden by the Pauli principle—in agreement with the results found in Sec. 4-17 for p^2. In addition, functions for $j_1 = j_2$ and even J require an extra normalization factor $1/\sqrt{2}$, similarly to the LS case.

Functions with $j_1 \neq j_2$ are seen from (9.37) to be properly normalized, but the two functions ($j_1 j_2$) and ($j_2 j_1$) differ at most by a phase factor: if we let

$$j_1 = j_- \equiv l - \frac{1}{2}$$

$$j_2 = j_+ \equiv l + \frac{1}{2} \ ,$$

then we find from (9.30) and (9.34) that for β = β′ (with appropriate simplifications in the notation)

$$\begin{aligned} \Psi^{ik}_{+-} &\equiv 2^{-1/2} (\psi^{ik}_{+-} - \psi^{ki}_{+-}) \\ &= 2^{-1/2} (-1)^{j_+ + j_- - J} (\psi^{kl}_{-+} - \psi^{ik}_{-+}) \\ &= -(-1)^{2l - J} \Psi^{ik}_{-+} \\ &= -(-1)^J \Psi^{ik}_{-+} \ . \end{aligned} \tag{9.38}$$

Thus, as is to be expected for two equivalent electrons, the functions ($j_- j_+$) and ($j_+ j_-$) are physically equivalent; our convention will be to use the former and discard the latter.

Transformation matrices. In spite of the above complications, the LS-jj transformation matrices for nl^2 can be easily obtained from those for $nln'l$. We need only delete all rows for LS terms with L + S odd, delete all columns for $j_1 = j_2$ and J odd, and delete all columns for $j_1 > j_2$—and then multiply the matrix elements for $j_1 < j_2$ by $\sqrt{2}$ to restore the row normalization lost when the elements for $j_1 > j_2$ were deleted. Matrix elements for L + S even and $j_1 = j_2$ (J even) are the same as for non-equivalent electrons, because the only required basis-function change in going to equivalent electrons was to divide both the LS and the jj functions by $\sqrt{2}$. As an example, the LS-jj matrix in Table 9-2 for pp′ becomes a one-by-one unit matrix for p^2.

TABLE 9-3. LS-jj TRANSFORMATION MATRICES FOR THE CONFIGURATIONS p^2 and p^4

$$\begin{array}{c} \\ {}^3P_0 \\ {}^1S_0 \end{array} \begin{array}{c} \begin{array}{cc} \left(\frac{1}{2}\frac{1}{2}\right)_0 & \left(\frac{3}{2}\frac{3}{2}\right)_0 \end{array} \\ \begin{pmatrix} \sqrt{\frac{2}{3}} & -\sqrt{\frac{1}{3}} \\ \sqrt{\frac{1}{3}} & \sqrt{\frac{2}{3}} \end{pmatrix} \end{array} \qquad \begin{array}{c} \\ {}^3P_1 \end{array} \begin{array}{c} \left(\frac{1}{2}\frac{3}{2}\right)_1 \\ \begin{pmatrix} 1 \end{pmatrix} \end{array} \qquad \begin{array}{c} \\ {}^3P_2 \\ {}^1D_2 \end{array} \begin{array}{c} \begin{array}{cc} \left(\frac{1}{2}\frac{3}{2}\right)_2 & \left(\frac{3}{2}\frac{3}{2}\right)_2 \end{array} \\ \begin{pmatrix} \sqrt{\frac{1}{3}} & \sqrt{\frac{2}{3}} \\ -\sqrt{\frac{2}{3}} & \sqrt{\frac{1}{3}} \end{pmatrix} \end{array}$$

TABLE 9-4. LS-jj TRANSFORMATION MATRICES FOR THE CONFIGURATION p^3

$$\begin{array}{c} \\ {}^2P^{\circ}_{1/2} \end{array} \begin{array}{c} \left(\frac{1}{2}\left(\frac{3}{2}\frac{3}{2}\right)_0\right)^{\circ} \\ \begin{pmatrix} 1 \end{pmatrix} \end{array} \qquad\qquad \begin{array}{c} \\ {}^2D^{\circ}_{5/2} \end{array} \begin{array}{c} \left(\frac{1}{2}\left(\frac{3}{2}\frac{3}{2}\right)_2\right)^{\circ} \\ \begin{pmatrix} 1 \end{pmatrix} \end{array}$$

$$\begin{array}{c} \\ {}^4S^{\circ}_{3/2} \\ {}^2D^{\circ}_{3/2} \\ {}^2P^{\circ}_{3/2} \end{array} \begin{array}{c} \begin{array}{ccc} \left(\left(\frac{1}{2}\frac{1}{2}\right)_0\frac{3}{2}\right)^{\circ} & \left(\frac{1}{2}\left(\frac{3}{2}\frac{3}{2}\right)_2\right)^{\circ} & \left(\frac{3}{2}\frac{3}{2}\frac{3}{2}\right)^{\circ} \end{array} \\ \begin{pmatrix} \frac{\sqrt{2}}{3} & \frac{\sqrt{5}}{3} & -\frac{\sqrt{2}}{3} \\ -\sqrt{\frac{5}{18}} & \frac{2}{3} & \sqrt{\frac{5}{18}} \\ 1/\sqrt{2} & 0 & 1/\sqrt{2} \end{pmatrix} \end{array}$$

The complete set of p^2 transformation matrices is given in Table 9-3. These matrices serve also for p^4 if the j_1j_2 labels are considered to represent holes rather than electrons; i.e., (1/2 1/2) means $(3/2)^4$, (1/2 3/2) means $(1/2)\,(3/2)^3$, and (3/2 3/2) means $(1/2)^2(3/2)^2$. For completeness, matrices for p^3 are given in Table 9-4. As discussed in connection with Fig. 8-13, d^w and f^w subshells never show good jj-coupling conditions, and so transformation matrices for these cases are of little interest in atomic spectroscopy.

9-5. COEFFICIENTS OF FRACTIONAL PARENTAGE

We have seen that attempts to antisymmetrize coupled product functions for equivalent electrons by means of coordinate permutations lead to very awkward complications. We therefore look for some other way of accomplishing the antisymmetrization. The method to be discussed shortly is called the method of fractional parentage;[5] it can be applied to jj coupling,[9] but we shall consider only the LS case.

[9] A. de-Shalit and I. Talmi, *Nuclear Shell Theory* (Academic Press, New York, 1963); B. F. Bayman and A. Landé, Nucl. Phys. 77, 1 (1966).

For the LS coupling of two equivalent electrons, the above-mentioned complications arise from the fact that the coupled product function is *already* antisymmetric (for L + S even), so that no explicit antisymmetrization is required. This may be seen from the CG symmetry relation (9.8)

$$|\, l_1^{(k)} l_2^{(l)} \mathrm{LSJM}\rangle = (-1)^{-l_1-l_2+\mathrm{L}-s_1-s_2+\mathrm{S}}\, |\, l_2^{(l)} l_1^{(k)} \mathrm{LSJM}\rangle ,$$

which for equivalent electrons becomes

$$|\, l^{(k)} l^{(l)} \mathrm{LSJM}\rangle = -\,(-1)^{\mathrm{L+S}}\, |\, l^{(l)} l^{(k)} \mathrm{LSJM}\rangle .$$

The antisymmetry arises in the following way: Because the CG summation involves all possible values of the magnetic quantum numbers, for every term in the sum that involves

$$\varphi_{nlm_{l_a}m_{s_a}}(\mathrm{r}_l)\cdot\varphi_{nlm_{l_b}m_{s_b}}(\mathrm{r}_k) , \tag{9.39}$$

there exists another term containing

$$\varphi_{nlm_{l_b}m_{s_b}}(\mathrm{r}_l)\cdot\varphi_{nlm_{l_a}m_{s_a}}(\mathrm{r}_k) . \tag{9.40}$$

This effective interchange of magnetic quantum numbers, together with the equality of the nl values, has the appearance of an interchange of *all* quantum numbers. It is easily seen that the CG coefficients of the two products (9.39) and (9.40) are equal in magnitude but (for even L + S) opposite in sign—and this is just what is required to produce a function that is antisymmetric upon electron-coordinate interchange.

We thus know how, very simply, to form an antisymmetric coupled function for any allowed (even L + S) term of l^2. The question is, how do we form functions[10]

$$|\, l^{\mathrm{w}}\alpha\mathrm{LS}\rangle$$

(w > 2) that are antisymmetric with respect to interchange of any two of the w electron coordinates?

We begin with the case w = 3, and form a function by coupling an electron l (with coordinate $\mathbf{r}_3$) onto the angular momenta $\overline{\mathrm{L}}\overline{\mathrm{S}}$ of a two-electron antisymmetric function (with electron coordinates $\mathbf{r}_1$ and $\mathbf{r}_2$), to give overall angular momenta LS:

$$|\,(l^2\overline{\mathrm{L}}\overline{\mathrm{S}},l)\mathrm{LS}\rangle . \tag{9.41}$$

By twice using the recoupling formula (9.17), we may write (9.41) as

[10] In general, there may be more than one term of l^{w} with the same values of LS; α is then any appropriate quantity (such as a serial number) that serves to distinguish these terms from each other. The considerations of this section are independent of J and M (or independent of $\mathrm{M_L}$ and $\mathrm{M_S}$ if—as is usually assumed to be the case—L and S are left uncoupled), so these quantum numbers are deleted from the notation.

$$|(l^2\bar{L}\bar{S},l)LS\rangle = (-1)^{3l+L+3s+S}[\bar{L},\bar{S}]^{1/2} \times \sum_{L'S'} [L',S']^{1/2} \begin{Bmatrix} l & l & \bar{L} \\ l & L & L' \end{Bmatrix} \begin{Bmatrix} s & s & \bar{S} \\ s & S & S' \end{Bmatrix} \cdot |(l,llL'S')LS\rangle \ , \quad (9.42)$$

where the wavefunction on the right is one formed by coupling a two-electron coupled function $|llL'S'\rangle$ (with electron coordinates $\mathbf{r}_2$ and $\mathbf{r}_3$) onto the one-electron function $\varphi_l(\mathbf{r}_1)$.

The summation in (9.42) in general involves terms with $L' + S'$ odd, and so $|(l^2\bar{L}\bar{S},l)LS\rangle$ is not antisymmetric with respect to interchange of $\mathbf{r}_2$ and $\mathbf{r}_3$. However, it may be possible to find some linear combination of functions (9.42) of the form

$$|l^3\alpha LS\rangle = \sum_{\bar{L}\bar{S}} |(l^2\bar{L}\bar{S},l)LS\rangle(l^2\bar{L}\bar{S}|\}l^3\alpha LS) \quad (9.43)$$

($\bar{L} + \bar{S}$ even) in which the coefficients $(l^2\bar{L}\bar{S}|\}l^3\alpha LS)$ are such that all functions $|(l,llL'S')LS\rangle$ with $L' + S'$ odd disappear. The function (9.43) is then antisymmetric in $\mathbf{r}_2$ and $\mathbf{r}_3$ as well as in $\mathbf{r}_1$ and $\mathbf{r}_2$, and is therefore antisymmetric in $\mathbf{r}_1$ and $\mathbf{r}_3$ as well. The function (9.43) is then completely antisymmetric, and the term αLS is an allowed term of l^3. The terms $\bar{L}\bar{S}$ of l^2 are said to be parents of the term $l^3\alpha LS$, and the coefficients

$$(l^2\bar{L}\bar{S}|\}l^3\alpha LS) \quad (9.44)$$

are called coefficients of fractional parentage (cfp).[11]

Normalization of the function (9.43) requires that

$$\sum_{\bar{L}\bar{S}} (l^2\bar{L}\bar{S}|\}l^3\alpha LS)^2 = 1 \ , \quad (9.45)$$

since the functions $|(l^2\bar{L}\bar{S},l)LS\rangle$ are orthonormal and the cfp are real. More generally, if there are terms of given LS for more than one value of α, orthogonality requires that

$$\sum_{\bar{\alpha}\bar{L}\bar{S}} (l^w\alpha LS\{|l^{w-1}\bar{\alpha}\bar{L}\bar{S})(l^{w-1}\bar{\alpha}\bar{L}\bar{S}|\}l^w\alpha' LS) = \delta_{\alpha\alpha'} \ . \quad (9.46)$$

Example (p^3). As a numerical example, we calculate a few of the cfp required to form antisymmetric LS-coupled basis functions for p^3 from the functions for p^2. The expression (9.42) specializes to

$$|(p^2\bar{L}\bar{S},p)LS\rangle = (-1)^{s+L+S}[\bar{L},\bar{S}]^{1/2} \times \sum_{L'S'} [L',S']^{1/2} \begin{Bmatrix} 1 & 1 & \bar{L} \\ 1 & L & L' \end{Bmatrix} \begin{Bmatrix} \frac{1}{2} & \frac{1}{2} & \bar{S} \\ \frac{1}{2} & S & S' \end{Bmatrix} \cdot |(p,ppL'S')LS\rangle \ . \quad (9.47)$$

[11]Racah's notation[5] $([l^2\bar{L}\bar{S},l]LS|\}l^3\alpha LS)$ is redundant, and we have abbreviated it for simplicity.

The allowed terms of p^2 are ^1S, ^1D, ^3P (Table 4-2) and so the possible terms of $p^2 + p$ (obtained by using the vector model) are $^2P^\circ$, $^2P^\circ D^\circ F^\circ$, $^{2,4}S^\circ P^\circ D^\circ$.

Let us consider first the quartet terms, which can be formed only from the ^3P parent term of p^2. Then both $\bar{S}$ and S' must be 1, and (9.47) simplifies further to

$$|(p^2\,{}^3P,p)\,{}^4L^\circ\rangle = (-1)^L\cdot 3\sum_{L'}(3[L'])^{1/2}\begin{Bmatrix}1 & 1 & 1\\ 1 & L & L'\end{Bmatrix}\begin{Bmatrix}\frac{1}{2} & \frac{1}{2} & 1\\ \frac{1}{2} & \frac{3}{2} & 1\end{Bmatrix}|(p,pp\,{}^3L')\,{}^4L^\circ\rangle$$

$$= (-1)^{L+1}\sum_{L'}(3[L'])^{1/2}\begin{Bmatrix}1 & 1 & 1\\ 1 & L & L'\end{Bmatrix}|(p,pp\,{}^3L')\,{}^4L^\circ\rangle\ . \qquad (9.48)$$

For the case $L = 0$, the 6-j symbol is zero unless $L' = 1$, and we obtain

$$|(p^2\,{}^3P,p)\,{}^4S^\circ\rangle = -3\begin{Bmatrix}1 & 1 & 1\\ 1 & 0 & 1\end{Bmatrix}|(p,pp\,{}^3P)\,{}^4S^\circ\rangle = |(p,pp\,{}^3P)\,{}^4S^\circ\rangle\ .$$

Thus the function $|(p^2\,{}^3P,p)\,{}^4S^\circ\rangle$, which is antisymmetric in $\mathbf{r}_1$ and $\mathbf{r}_2$, is equal to the recoupled function $|(p,pp\,{}^3P)\,{}^4S^\circ\rangle$ which (since $L' + S'$ is even) is antisymmetric in $\mathbf{r}_2$ and $\mathbf{r}_3$; it is therefore a completely antisymmetric function $|p^3\,{}^4S^\circ\rangle$, $^4S^\circ$ is an allowed term of p^3, and the cfp

$$(p^2\overline{LS}|\}p^3\,{}^4S^\circ)$$

are unity for $\overline{LS} = {}^3P$ and zero for $\overline{LS} = {}^1S$ or ^1D.

For $^4D^\circ$, on the other hand, (9.48) gives

$$|(p^2\,{}^3P,p)\,{}^4D^\circ\rangle = -3\begin{Bmatrix}1 & 1 & 1\\ 1 & 2 & 1\end{Bmatrix}|(p,pp\,{}^3P)\,{}^4D^\circ\rangle - 15^{1/2}\begin{Bmatrix}1 & 1 & 1\\ 1 & 2 & 2\end{Bmatrix}|(p,pp\,{}^3D)\,{}^4D^\circ\rangle$$

$$= -\frac{1}{2}|(p,pp\,{}^3P)\,{}^4D^\circ\rangle + \frac{3^{1/2}}{2}|(p,pp\,{}^3D)\,{}^4D^\circ\rangle\ .$$

Here we have a contribution from $L'S' = {}^3D$, which is not antisymmetric; hence, $^4D^\circ$ is not an allowed term of p^3. Similarly, we find that $^4P^\circ$ is not allowed.

In the case of $^2D^\circ$, there are two possible such terms, one based on the ^1D parent of p^2, and one based on the ^3P parent. From (9.47), numerical evaluation of the 6-j symbols shows that

$$|(p^2\,{}^1D,p)\,{}^2D^\circ\rangle = -\frac{3^{1/2}}{4}|(p,pp\,{}^1P)\,{}^2D^\circ\rangle - \frac{1}{4}|(p,pp\,{}^1D)\,{}^2D^\circ\rangle$$

$$+\frac{3}{4}|(p,pp\,{}^3P)\,{}^2D^\circ\rangle + \frac{3^{1/2}}{4}|(p,pp\,{}^3D)\,{}^2D^\circ\rangle\ ,$$

$$|(p^2\,{}^3P,p)\,{}^2D^\circ\rangle = -\frac{3^{1/2}}{4}|(p,pp\,{}^1P)\,{}^2D^\circ\rangle + \frac{3}{4}|(p,pp\,{}^1D)\,{}^2D^\circ\rangle - \frac{1}{4}|(p,pp\,{}^3P)\,{}^2D^\circ\rangle + \frac{3^{1/2}}{4}|(p,pp\,{}^3D)\,{}^2D^\circ\rangle\,.$$

Neither of these expressions is antisymmetric in $\mathbf{r}_2$ and $\mathbf{r}_3$, because both contain $p^2\,{}^1P$ and $p^2\,{}^3D$ contributions. However, the linear combination

$$|p^3\,{}^2D^\circ\rangle \equiv 2^{-1/2}|(p^2\,{}^3P,p)\,{}^2D^\circ\rangle - 2^{-1/2}|(p^2\,{}^1D,p)\,{}^2D^\circ\rangle = -\,2^{-1/2}|(p,pp\,{}^3P)\,{}^2D^\circ\rangle + 2^{-1/2}|(p,pp\,{}^1D)\,{}^2D^\circ\rangle$$

is completely antisymmetric (and normalized); thus p^3 has one (and only one) allowed ${}^2D^\circ$ term, with cfp

$$(p^2\overline{LS}\{|p^3\,{}^2D^\circ)$$

of $1/\sqrt{2}$, $-1/\sqrt{2}$, and 0 for $\overline{LS} = {}^3P, {}^1D, {}^1S$, respectively. (Note that the antisymmetry requirement fixes only the relative signs of the cfp; the overall phase is not fundamentally important, and has been chosen arbitrarily.)

Similarly, none of the three ${}^2P^\circ$ terms based on the ${}^1S$, ${}^1D$, and ${}^3P$ parents is antisymmetric by itself. However, there exists one linear combination

$$|p^3\,{}^2P^\circ\rangle = \frac{2}{18^{1/2}}|(p^2\,{}^1S,p)\,{}^2P^\circ\rangle - \frac{3}{18^{1/2}}|(p^2\,{}^3P,p)\,{}^2P^\circ\rangle - \frac{5^{1/2}}{18^{1/2}}|(p^2\,{}^1D,p)\,{}^2P^\circ\rangle$$

that is completely antisymmetric, and therefore one allowed ${}^2P^\circ$ term of p^3, defined by cfp of $2/\sqrt{18}$, $-3/\sqrt{18}$, and $-\sqrt{5/18}$ for the ${}^1S$, ${}^3P$, and ${}^1D$ parents, respectively.

The ${}^2S^\circ$ and ${}^2F^\circ$ terms of p^2p prove to be non-antisymmetric, and are therefore not allowed terms of p^3.

In summary, out of the possible terms $({}^1S){}^2P^\circ$, $({}^1D){}^2P^\circ D^\circ F^\circ$, and $({}^3P){}^{2,4}S^\circ P^\circ D^\circ$ of $p^2(\overline{LS})p$, the only allowed terms of p^3 are the ${}^4S^\circ$, a ${}^2D^\circ$, and a ${}^2P^\circ$ [cf. Table 4-2] with cfp given in Table 9-5. Note that the squares of the elements in each row of the array sum to unity, in agreement with (9.45). It can be shown[5] that the elements of the columns of a cfp table obey the relationship

$$\sum_{\alpha LS} [L,S]\,(l^{w-1}\bar{\alpha}\overline{LS}\}|l^w\alpha LS)\,(l^w\alpha LS\{|l^{w-1}\bar{\alpha}'\overline{LS}) = \frac{1}{w}(4l + 3 - w)[\bar{L},\bar{S}]\delta_{\bar{\alpha}\bar{\alpha}'}\;; \qquad (9.49)$$

this expression is useful for checking the correctness of calculated values of the cfp. Additional cfp tables are given in Appendix H.

9-6. COEFFICIENTS OF FRACTIONAL GRANDPARENTAGE

In the preceding section we discussed only the cfp required to produce antisymmetric functions of l^3 from those for l^2. The generalization of (9.43) is

TABLE 9-5. COEFFICIENTS OF FRACTIONAL PARENTAGE FOR p^w SUBSHELLS

$$(p\ {}^2P\{|p^0\ {}^1S) = 1$$

$$(p^2\ LS\{|p\ {}^2P) = 1 \qquad (L + S \text{ even})$$

p^3	p^2: ^3P	^1D	^1S
^4S	1	0	0
^2D	$1/\sqrt{2}$	$-1/\sqrt{2}$	0
^2P	$-3/\sqrt{18}$	$-\sqrt{5/18}$	$2/\sqrt{18}$

p^4	p^3: ^4S	^2D	^2P
^3P	$-1/\sqrt{3}$	$\sqrt{5/12}$	$-1/2$
^1D	0	$-\sqrt{3}/2$	$-1/2$
^1S	0	0	1

p^5	p^4: ^3P	^1D	^1S
^2P	$\sqrt{3/5}$	$1/\sqrt{3}$	$1/\sqrt{15}$

$$(p^6\ {}^1S\{|p^5\ {}^2P) = +1$$

$$|l^w\alpha LS\rangle = \sum_{\bar{\alpha}\bar{L}\bar{S}} |(l^{w-1}\bar{\alpha}\bar{L}\bar{S},l)LS\rangle\,(l^{w-1}\bar{\alpha}\bar{L}\bar{S}|\}l^w\alpha LS)\ , \qquad (9.50)$$

where $|l^{w-1}\bar{\alpha}\bar{L}\bar{S}\rangle$ is a function that is antisymmetric in the electron coordinates $r_1, r_2, \cdots r_{w-1}$, and $|l^w\alpha LS\rangle$ is a function antisymmetric in the coordinates $r_1, r_2, \cdots r_w$. In principle, basis functions $|l^w\alpha LS\rangle$ are constructed from one-electron functions $|l\rangle$ by using a succession of $w-2$ cfp expansions of the form (9.50), for $w = m, m-1, \cdots, 4, 3$. In practice, we shall seldom need to use more than the formal expression (9.50) for the single value $w = m$ of interest.

The cfp in (9.50) for $w > 3$ are determined in a manner somewhat different from those for $w = 3$. Making a further cfp expansion in (9.50), and then a recoupling similar to that used in (9.42), we obtain

$$|l^w \alpha LS\rangle = \sum_{\overline{\alpha LS}} \sum_{\overline{\overline{\alpha LS}}} |[(l^{w-2}\overline{\overline{\alpha LS}},l)\overline{LS},l]LS\rangle (l^{w-2}\overline{\overline{\alpha LS}}|\}l^{w-1}\overline{\alpha LS}) (l^{w-1}\overline{\alpha LS}|\}l^w \alpha LS)$$

$$= \sum_{\overline{\alpha LS}} \sum_{\overline{\overline{\alpha LS}}} \sum_{L'S'} |(l^{w-2}\overline{\overline{\alpha LS}},l l L'S')LS\rangle (-1)^{\overline{L}+2l+L+\overline{S}+2s+S}[\overline{L},\overline{S},L',S']^{1/2}$$

$$\times \begin{Bmatrix} \overline{\overline{L}} & l & \overline{L} \\ l & L & L' \end{Bmatrix} \begin{Bmatrix} \overline{\overline{S}} & s & \overline{S} \\ s & S & S' \end{Bmatrix} (l^{w-2}\overline{\overline{\alpha LS}}|\}l^{w-1}\overline{\alpha LS}) (l^{w-1}\overline{\alpha LS}|\}l^w \alpha LS)$$

$$= \sum_{\overline{\overline{\alpha LS}}} \sum_{L'S'} |(l^{w-2}\overline{\overline{\alpha LS}},l l L'S')LS\rangle (l^{w-2}\overline{\overline{\alpha LS}},l^2 L'S'|\}l^w \alpha LS) , \tag{9.51}$$

where

$$(l^{w-2}\overline{\overline{\alpha LS}},l^2 L'S'|\}l^w \alpha LS) \equiv (-1)^{\overline{L}+\overline{S}+L+S+1} \sum_{\overline{\alpha LS}} [\overline{L},\overline{S},L',S']^{1/2}$$

$$\times \begin{Bmatrix} \overline{\overline{L}} & l & \overline{L} \\ l & L & L' \end{Bmatrix} \begin{Bmatrix} \overline{\overline{S}} & s & \overline{S} \\ s & S & S' \end{Bmatrix} (l^{w-2}\overline{\overline{\alpha LS}}|\}l^{w-1}\overline{\alpha LS}) (l^{w-1}\overline{\alpha LS}|\}l^w \alpha LS) . \tag{9.52}$$

It is assumed that the cfp $(l^{w-2}|\}l^{w-1})$ are known; the unknown cfp $(l^{w-1}|\}l^w)$ are then determined by the requirement that (9.52) must be zero for all cases in which $L' + S'$ is odd. If (and only if) this is possible, then (9.51) is antisymmetric and αLS is an allowed term of l^w. The term $\overline{\overline{\alpha LS}}$ of l^{w-2} is a grandparent of the term $l^w \alpha LS$ (if $L' + S'$ is even), and the quantities (9.52) are called coefficients of fractional grandparentage (cfgp). Values of cfgp for a few simple subshells are tabulated in Appendix H.

It may be seen that the procedure outlined in Sec. 9-5 for determining the coefficients of fractional parentage of l^3 is just the special case of (9.52) for which $w = 3$, $\overline{\overline{L}} = l$, $\overline{\overline{S}} = 1/2$, and $(l^{w-2}\overline{\overline{\alpha LS}}|\}l^{w-1}\overline{\alpha LS}) = 1$ for all even $\overline{L} + \overline{S}$.

9-7. THE SENIORITY NUMBER

When there are two or more allowed terms of l^w with the same values of LS, the additional quantity α is required to distinguish the terms from each other. The "seniority number" v is a quantity introduced by Racah[5] to serve this purpose (among others). The definition of v given by Racah is rather complex, but its significance is essentially that given by the following:

Consider the succession of configurations l^0, l^1, l^2, l^3, $\cdots$. For any given term LS, the *first* time that LS appears as an allowed term of l^w, it is assigned the seniority number $v = w$. Thus ^1S is an allowed term of l^0 and has $v = 0$; ^2L ($L = l$) is an allowed term of l^1 and has $v = 1$; in l^2, the allowed term ^1S again has $v = 0$, whereas all other allowed terms ^3P, ^1D, ^3F, $\cdots$ have $v = 2$.

As we go on to successively larger values of w, we may reach a point at which there are *two* allowed terms with the same LS. The choice of values for the cfp for these two terms is not fixed uniquely by (9.46); there are an infinite number of different possible orthonormal choices. To resolve this arbitrariness, we choose the cfp such that for one of the two terms the cfgp (9.52) is zero for L′ = S′ = 0:

$$(l^{w-2}\bar{\bar{\alpha}}LS, l^2\,{}^1S \| l^w \alpha_2 LS) = 0 \;. \tag{9.53}$$

This "new" term α_2LS is given the seniority v = w, and the term α_1LS that is connected to the term $l^{w-2}\bar{\bar{\alpha}}$LS by a non-zero cfgp,

$$(l^{w-2}\bar{\bar{\alpha}}LS, l^2\,{}^1S \| l^w \alpha_1 LS) \neq 0 \;, \tag{9.54}$$

is given the same seniority number as that of $\bar{\bar{\alpha}}$LS (which may be v = w − 2, or may be smaller).

Going on to l^{w+2}, there may exist *three* allowed terms with the same LS as above. We then choose the cfp $(l^{w+1} \| l^{w+2})$ such that of the three terms α_aLS, α_bLS, and α_cLS,

$$(l^w \alpha_1 LS, l^2\,{}^1S \| l^{w+2}\alpha LS) \neq 0 \qquad \text{for } \alpha_a LS \textit{ only} \;,$$

$$(l^w \alpha_2 LS, l^2\,{}^1S \| l^{w+2}\alpha LS) \neq 0 \qquad \text{for } \alpha_b LS \textit{ only} \;.$$

Then α_aLS has the seniority number of $l^w\alpha_1$LS (and hence of $l^{w-2}\bar{\bar{\alpha}}$LS), α_bLS has the seniority number of $l^w\alpha_2$LS (namely, v = w), and the "new" term α_cLS has the seniority number v = w + 2.

In this way it proves to be possible to form unbroken chains of terms of fixed LS in l^w, l^{w+2}, l^{w+4}, ···, with the terms of each chain being connected by non-zero values of the cfgp

$$(l^n LS, l^2\,{}^1S \| l^{n+2} LS) \;, \qquad n = w,\ w+2,\ \cdots \tag{9.55}$$

and with the LS terms in each chain being labeled by a value of the seniority number equal to the value of w at which the chain begins. For example, in the configurations d^w there are three chains of terms ${}^2_vD$, with seniority numbers v = 1, 3, and 5:

$$\begin{array}{ccccc} d^1 & d^3 & d^5 & d^7 & d^9 \\ {}^2_1D & —\,{}^2_1D & —\,{}^2_1D & —\,{}^2_1D & —\,{}^2_1D \\ & {}^2_3D & —\,{}^2_3D & —\,{}^2_3D & \\ & & {}^2_5D & & \end{array}$$

Similarly, there are two chains each of ${}^1_vS$ and ${}^3_vP$ terms:

$$\begin{array}{cccccc} d^0 & d^2 & d^4 & d^6 & d^8 & d^{10} \\ {}^1_0S & —\,{}^1_0S & —\,{}^1_0S & —\,{}^1_0S & —\,{}^1_0S & —\,{}^1_0S \\ & & {}^1_4S & —\,{}^1_4S & & \\ & {}^3_2P & —\,{}^3_2P & —\,{}^3_2P & —\,{}^3_2P & \\ & & {}^3_4P & —\,{}^3_4P & & \end{array}$$

The above considerations of cfgp of the type (9.55), and the seniority number that results, are sufficient to determine uniquely (except for overall phase) all cfp of d^w; the seniority number can therefore be used as the value of α required to distinguish the different terms of d^w that have the same values of LS.

This is not true of the much more complex subshells f^w, because it frequently happens that (for $3 \leqq w \leqq 7$) more than one "new" allowed term with given LS appears in f^w. The terms of f^w then have to be designated by means of values of αvLS, where α is some distinguishing quantity over and above the seniority number. The quantity α is usually (partially) described in terms of representations of the groups G_2 and R_7, as discussed by Racah and others.[12] Even this is not sufficient for all terms of f^5, f^6, and f^7, and in any case the group-theoretical classification has no strict physical significance; it will therefore not be discussed further here. For our purposes, we may consider the cfp problem solved, and simply make use of the tables given by Racah[5] for p^w and d^w, and by Nielson and Koster[13] for p^w, d^w, and f^w ($w \leqq 2l + 1$).

Phase conventions; more than half-filled shells. Fundamentally, the phase of each basis function $|l^w\alpha vLS\rangle$ is arbitrary. Even though we choose the cfp to be real, the phase is still determined only to within a factor ± 1 (corresponding to multiplication of the cfp for all parents of $l^w\alpha vLS$ by this factor). It is customary to remove much of this arbitrariness by choosing phases such that

$$(l^{w-2}\alpha vLS, l^2\,{}^1S|\}l^w\alpha vLS) > 0 \,. \tag{9.56}$$

The relative phases of all functions within a chain (i.e., all functions having given $l\alpha$vLS) are then fixed, and only the phase of the first member of the chain, $|l^v\alpha vLS\rangle$, is left to be chosen arbitrarily. With this convention it can be shown[5] that

$$(l^{w-2}\bar{\bar{\alpha}}\bar{\bar{v}}LS, l^2\,{}^1S|\}l^w\alpha vLS) = \delta_{\bar{\bar{\alpha}}\alpha}\delta_{\bar{\bar{v}}v}\left[\frac{(w-v)(4l+4-w-v)}{2w(w-1)(2l+1)}\right]^{1/2} , \tag{9.57}$$

$$(l^{w-1}\bar{\alpha}\bar{v}\bar{L}\bar{S}|\}l^w\alpha vLS) = 0 \,, \qquad \bar{v} \neq v \pm 1 \,, \tag{9.58}$$

$$(l^{w-1}\bar{\alpha}(v-1)\bar{L}\bar{S}|\}l^w\alpha vLS) = \left[\frac{v(4l+4-w-v)}{w(4l+4-2v)}\right]^{1/2}(l^{v-1}\bar{\alpha}(v-1)\bar{L}\bar{S}|\}l^v\alpha vLS) \,, \tag{9.59}$$

$$(l^{w-1}\bar{\alpha}(v+1)\bar{L}\bar{S}|\}l^w\alpha vLS) = \left[\frac{(v+2)(w-v)}{2w}\right]^{1/2}(l^{v+1}\bar{\alpha}(v+1)\bar{L}\bar{S}|\}l^{v+2}\alpha vLS) \,. \tag{9.60}$$

The expressions (9.58)-(9.60) make it very easy to obtain most cfp; only for the "new" terms ($v = w$) and some of the cfp for $w = v + 2$ is it necessary to use the methods of Secs. 9-5 and 9-6. The arbitrary phases of the new terms for $v = 0$, 1, and 2 are always chosen as follows:

[12] G. Racah, IV, Phys. Rev. **76**, 1352 (1949); B. R. Judd, *Operator Techniques in Atomic Spectroscopy* (McGraw-Hill, New York, 1963); B. G. Wybourne, *Symmetry Principles and Atomic Spectroscopy* (Wiley-Interscience, New York, 1970).

[13] C. W. Nielson and G. F. Koster, *Spectroscopic Coefficients for the p^n, d^n, and f^n Configurations* (The M. I. T. Press, Cambridge, Mass., 1963).

$$|l^0\,{}^1_0\mathrm{S}\rangle \equiv +1\ , \qquad (l^0\,{}^1_0\mathrm{S}\{|l\,{}^2_1\mathrm{L}) = (l\,{}^2_1\mathrm{L}\{|l^2\,\mathrm{LS}) = +1\ ; \tag{9.61}$$

(9.60) and (9.61) lead also to

$$(l^{4l+1}\,{}^2_1\mathrm{L}\{|l^{4l+2}\,{}^1_0\mathrm{S}) = (l\,{}^2_1\mathrm{L}\{|l^2\,{}^1_0\mathrm{S}) = +1\ . \tag{9.62}$$

The phase conventions involved in (9.56)-(9.61), together with any particular choice of phase for the chains v > 2 (specifically the choice made in refs. 5 and 13), define a set of antisymmetric coupled functions that Racah[1,5] calls

$$|l^{\mathrm{w}}\alpha\mathrm{vLSM_LM_S}\rangle_{\mathfrak{L}}\ . \tag{9.63}$$

It is convenient to consider also a set of functions $\mathfrak{R}$ defined by

$$|l^{\mathrm{w}}\alpha\mathrm{vLSM_LM_S}\rangle_{\mathfrak{R}} = \mathrm{c}\,|l^{\mathrm{w}}\alpha\mathrm{vLSM_LM_S}\rangle_{\mathfrak{L}}\ , \tag{9.64}$$

where

$$\mathrm{c} = \begin{cases} +1, & \mathrm{v} = 0,1,4,5,8,9,\cdots, \\ -1, & \mathrm{v} = 2,3,6,7,10,11,\cdots. \end{cases} \tag{9.65}$$

The phase relations between these two sets of functions are such that if we form coupled functions

$$\begin{aligned} \psi_{\alpha\mathrm{vLS}} &\equiv |(l^{\mathrm{w}}\alpha\mathrm{vLS}_{\mathfrak{L}}, l^{4l+2-\mathrm{w}}\alpha\mathrm{vLS}_{\mathfrak{R}})^1\mathrm{S}\rangle \\ &\equiv \sum_{\mathrm{M_LM_S}} \mathrm{C(LLM_L\,-M_L;00)\,C(SSM_S\,-M_S;00)} \\ &\qquad\qquad \times\ |l^{\mathrm{w}}\alpha\mathrm{vLSM_LM_S}\rangle_{\mathfrak{L}}\ |l^{4l+2-\mathrm{w}}\alpha\mathrm{vLS}\,-\mathrm{M_L}\,-\mathrm{M_S}\rangle_{\mathfrak{R}}\ , \end{aligned}$$

and a completely antisymmetric function

$$|l^{4l+2}\,{}^1_0\mathrm{S}\rangle_{\mathfrak{L}} \equiv |l^{4l+2}\,{}^1_0\mathrm{S}\rangle_{\mathfrak{R}} = \sum_{\alpha\mathrm{vLS}} \mathrm{Q}(\alpha\mathrm{vLS})\,\psi_{\alpha\mathrm{vLS}}\ , \tag{9.66}$$

then the coefficients Q (which are generalizations of cfp and cfgp) are positive for all v, having the value[1]

$$\mathrm{Q}(\alpha\mathrm{vLS}) = \left[\frac{(2\mathrm{L}+1)(2\mathrm{S}+1)\mathrm{w}!(4l+2-\mathrm{w})!}{(4l+2)!}\right]^{1/2}\ . \tag{9.67}$$

Also, the relation

$$(l^{4l+2-w}\bar{\alpha}\bar{v}\bar{L}\bar{S}\{|l^{4l+3-w}\alpha vLS)_{\Re} = (-1)^{L+S+\bar{L}+\bar{S}-l-1/2}$$

$$\times \left[\frac{w(2\bar{L}+1)(2\bar{S}+1)}{(4l+3-w)(2L+1)(2S+1)}\right]^{1/2} (l^{w-1}\alpha vLS\{|l^{w}\bar{\alpha}\bar{v}\bar{L}\bar{S})_{\mathfrak{L}} \qquad (9.68)$$

may be shown[5] to hold for all v and $\bar{v}$.

It is convenient [see Eqns. (11.58) and (11.72)] to choose for our standard set of basis functions the set $|l^w\rangle_{\mathfrak{L}}$ for $w \leqq 2l + 1$ and the set $|l^w\rangle_{\Re}$ for $w > 2l + 1$. Except as specifically indicated to the contrary, this convention will be assumed in all that follows. The relation (9.68) then provides appropriate values of cfp for more-than-half-filled subshells from tabulated cfp[5,13] for less-than-half-filled subshells, except that for $(l^{2l+1}\{|l^{2l+2})$ the required quantities are

$$(l^{2l+1}\bar{\alpha}\bar{v}\bar{L}\bar{S}_{\mathfrak{L}}\{|l^{2l+2}\alpha vLS_{\Re}) = (-1)^{(\bar{v}-1)/2+L+S+\bar{L}+\bar{S}-l-1/2}$$

$$\times \left[\frac{(2l+1)(2\bar{L}+1)(2\bar{S}+1)}{(2l+2)(2L+1)(2S+1)}\right]^{1/2} (l^{2l}\alpha vLS\{|l^{2l+1}\bar{\alpha}\bar{v}\bar{L}\bar{S})_{\mathfrak{L}}\ , \qquad (9.69)$$

from (9.65). [It should be noted that the relations (9.56) and (9.57) can be used neither for $w = 2l + 2$ nor for $w = 2l + 3$, nor can the relations (9.59) and (9.60) be used for $w > 2l + 1$, unless appropriate modifications are made in line with (9.64)-(9.65).]

Problems

9-7(1). Using (9.68)-(9.69), verify that the cfp for p^6, p^5, and p^4 in Table 9-5 follow from the tabulated values for p^1, p^2, and p^3, respectively.

9-7(2). Show that (9.49) follows from (9.68) and the orthonormality relation (9.46).

9-7(3). Calculate values of the cfgp $(d^4\,{}^3_vP,\ d^2\,{}^1S\{|d^6\,{}^3_vP)$ for $v = 2$ and 4, using (a) the (unmodified) relation (9.57), (b) the relations (9.52) and (9.69) with cfp for d^5 from Nielson and Koster,[13] and (c)—for $v = 2$ only—(9.52) and the (unmodified) relations (9.59)-(9.60) with cfp for d^5 and d^4 from Nielson and Koster. [Hint: The expression (5.29) is helpful.] Discuss the phase discrepancies.

9-7(4). Show from (9.52) and (9.68) that

$$(l^{4l}LS,l^2LS\{|l^{4l+2}\,{}^1S) = (l^{4l}LS\{|l^{4l+1}\,{}^2l) = \left[\frac{(2L+1)(2S+1)}{(2l+1)(4l+1)}\right]^{1/2},$$

thereby verifying (9.67) for the case $w = 2$.

9-8. ANTISYMMETRIZED FUNCTIONS FOR AN ARBITRARY CONFIGURATION

The last three sections have shown us how, with the aid of coefficients of fractional parentage, we can obtain basis functions

$$|l^w\alpha LS\rangle \ ,$$

or more precisely functions

$$|l^w\alpha LSM_LM_S\rangle \ , \tag{9.70}$$

that are eigenfunctions of L^2, L_z, S^2, and S_z and are antisymmetric with respect to interchange of any two of the electron coordinates $r_1, r_2, \cdots r_w$. For the general configuration

$$l_1^{w_1}l_2^{w_2} \cdots l_q^{w_q}, \qquad \sum_{j=1}^{q} w_j = N \ , \tag{9.71}$$

we can write down a function consisting of a product of q functions like (9.70):

$$|l_1^{w_1}\alpha_1L_1S_1\rangle\,|l_2^{w_2}\alpha_2L_2S_2\rangle \cdots |l_q^{w_q}\alpha_qL_qS_q\rangle \ , \tag{9.72}$$

where the first factor is a function of the coordinates $r_1, r_2, \cdots r_{w_1}$, the second factor is a function of the coordinates $r_{w_1+1} \cdots r_{w_1+w_2}$, etc., with the coordinates for the final factor being $r_{N-w_q+1}, \cdots r_N$ We can then couple the various quantum numbers $L_1S_1, L_2S_2, \cdots$ according to some suitable coupling scheme, such as the $\mathfrak{L}\mathfrak{S}$ and $J\mathfrak{J}$ schemes (4.37) and (4.44); this coupling is straightforward, because the cfp used to form the functions (9.70) depend only on L and S, and are independent of M_L and M_S.

Whether the quantum numbers L_iS_i in (9.72) are coupled together or not, this function (which we shall call ψ_b) is not antisymmetric with respect to exchange of pairs of electron coordinates between subshells—for example, the coordinates r_{w_1} and r_{w_1+1}. The necessary additional antisymmetrization can be accomplished through the use of mixed-shell coefficients of fractional parentage.[14] However, we shall here use a modificaton of the coordinate permutation scheme employed with one-electron spin-orbitals in (4.26).

In the functions (9.70) we already have antisymmetry with respect to electron-coordinate interchanges within a subshell $l_i^{w_i}$, and we have seen that additional antisymmetrization, done by summing over permutations of coordinates within the subshell, leads to very awkward complications. Hence we sum only over permutations of electron coordinates among different subshells; within a given subshell, the ordering of electron coordinates must be a single standard ordering, which we take to be that of numerically increasing value of the subscript k of the coordinates r_k. Thus in the configuration $l_1^3l_2^2$, for example, the basic ordering of coordinates is

$$l_1(r_1)l_1(r_2)l_1(r_3)\,|\,l_2(r_4)l_2(r_5) \ , \tag{9.73}$$

or for short simply 123|45. The various possible coordinate permutations over which we sum, and the parity of each permutation, are

[14] L. Armstrong, Jr., Phys. Rev. 172, 18 (1968) and references quoted therein.

123|45	even
124|35	odd
125|34	even
134|25	even
135|24	odd
145|23	even
234|15	odd
235|14	even
245|13	odd
345|12	even

Permutations such as 132|45, 123|54, 214|35, 124|53, etc. are *not* to be included.

Returning to the general configuration (9.71), the ***number*** of different permutations to be included is the total number N! of possible permutations of all N coordinates, divided by the product (over all subshells) of the number of unallowed permutations within each subshell:

$$\frac{N!}{w_1!w_2!\cdots w_q!} = \frac{N!}{\prod_j w_j!} \tag{9.74}$$

Thus the normalized, completely antisymmetrized basis function is, similarly to (4.26),

$$\Psi_b = \left[\frac{\prod_j w_j!}{N!}\right]^{1/2} \sum_P (-1)^p \psi_b^{(P)} , \tag{9.75}$$

where $\psi_b^{(P)}$ is a partially antisymmetrized coupled function formed from products like (9.72), except with electron coordinates permuted according to the permutation P with parity p, and the sum is over only those permutations [(9.74) is number] that are of the type just described.

9-9. SINGLE-CONFIGURATION MATRIX ELEMENTS OF SYMMETRIC OPERATORS

The antisymmetrizing permutations of electron coordinates among different subshells appreciably complicate the appearance of the basis functions (9.75). However, these complications largely disappear in the evaluation of matrix elements of symmetric operators. We here discuss these simplifications for the case in which the bra and ket functions belong to the same configuration. Non-configuration-diagonal matrix elements will be treated in Chapter 13 for the Hamiltonian operator, and in Chapter 14 for radiative transitions.

One-electron operators. Because of the antisymmetry of the basis functions Ψ_b, the value of the matrix element

$$\langle \Psi_b | f_i | \Psi_{b'} \rangle$$

is independent of which electron coordinate $\mathbf{r}_i$ is considered. This follows from the same argument used for determinantal functions in Sec. 6-1: interchanging the name of the integration variable $\mathbf{r}_i$ with that of $\mathbf{r}_j$ (say) does not affect the value of the matrix element, but gives a matrix element of the operator f_j; interchanging the electron coordinates i and j in both the bra and ket functions then gives in each case -1 times the original basis function. Using (9.75), matrix elements of a one-electron symmetric operator may thus be written

$$\langle\Psi_b|\sum_{i=1}^{N} f_i|\Psi_{b'}\rangle = N\langle\Psi_b|f_N|\Psi_{b'}\rangle$$

$$= \frac{\prod_k w_k!}{(N-1)!}\sum_P\sum_{P'}(-1)^{p+p'}\langle\psi_b^{(P)}|f_N|\psi_{b'}^{(P')}\rangle \ . \tag{9.76}$$

For the present, we consider some fixed permutation P. In the evaluation of the matrix element in the right-hand side of (9.76), integration over any coordinate $\mathbf{r}_i$ other than $\mathbf{r}_N$ will involve simple overlap integrals of one-electron spin-orbitals

$$\langle\varphi_{n_jl_jm_{l_j}m_{s_j}}(i)|\varphi_{n'_jl'_jm'_{l_j}m'_{s_j}}(i)\rangle \ . \tag{9.77}$$

Because of the orthonormality of these spin-orbitals, (9.77) will be zero unless

$$n_jl_jm_{l_j}m_{s_j} = n'_jl'_jm'_{l_j}m'_{s_j} \ .$$

Equality of the magnetic quantum numbers does not tell us anything simple about (9.76), because many different values of these quantum numbers are involved in the various terms of the Clebsch-Gordon sums involved in forming coupled functions. However, it is certainly true that (9.77)—and therefore the matrix element in (9.76)—will be zero unless the permutation P′ is such that

$$n_jl_j = n'_jl'_j \ ;$$

i.e., unless $\mathbf{r}_i$ occurs in the same subshell $l_j^{w_j}$ in both $\psi_b^{(P)}$ and $\psi_{b'}^{(P')}$. Since such a correspondence of subshells must hold for every coordinate $i \neq N$, it must by a process of elimination hold also for $i = N$. But this means that the matrix element in (9.76) will be zero unless $P' \equiv P$, because the permutations involved do not include coordinate permutations within a subshell. Therefore

$$\langle\Psi_b|\sum f_i|\Psi_{b'}\rangle = \frac{\prod_k w_k!}{(N-1)!}\sum_P\langle\psi_b^{(P)}|f_N|\psi_{b'}^{(P)}\rangle \ . \tag{9.78}$$

Out of all the permutations P, let us first consider only those permutations for which the coordinate N occurs in the subshell $l_j^{w_j}$ for some specific j; because of the nature of the permutations P, this means that $\mathbf{r}_N$ will always be the coordinate of the *last* (w_jth) electron of the subshell. With the position of this coordinate thus fixed, the number of permutations in question will be the number of permutations $(N - 1)!$ of *all* coordinates other than $\mathbf{r}_N$, divided by the numbers of *un*allowed permutations within each subshell:

$$\frac{(N-1)!}{w_1!w_2! \cdots (w_j - 1)! \cdots w_q!} \ . \tag{9.79}$$

Since all such permutations give the same numerical value of the matrix element in (9.78), the partial sum over these permutations gives

$$\frac{\prod_k w_k!}{(N-1)!} \cdot \frac{(N-1)!}{w_1!w_2! \cdots (w_j - 1)! \cdots w_q!} \langle \psi_b^{(P_j)} | f_N | \psi_{b'}^{(P_j)} \rangle = w_j \langle \psi_b^{(P_j)} | f_N | \psi_{b'}^{(P_j)} \rangle \ ,$$

where P_j is any one of these permutations [of total number (9.79)], for which the coordinate $\mathbf{r}_N$ is the last coordinate of $|l_j^{w_j}\rangle$.

Summing over all possible values of j, we may write the final result in the form

$$\langle \Psi_b | \sum_{i=1}^{N} f_i | \Psi_{b'} \rangle = \sum_{j=1}^{q} w_j \langle \psi_b^{(P_j)} | f_N | \psi_{b'}^{(P_j)} \rangle = \sum_{j=1}^{q} w_j \langle \psi_b | f_{(j)} | \psi_{b'} \rangle \ , \tag{9.80}$$

where in the final expression ψ_b and $\psi_{b'}$ are coupled product functions with basic unpermuted coordinate ordering [e.g., (9.73)], antisymmetric only with respect to coordinate interchanges within each subshell, and $f_{(j)}$ operates on the last electron coordinate of the jth subshell.

Two-electron operators. The number of pairs of electron coordinates $i < j$ is $N(N - 1)/2$. Thus analogously to (9.76) we may write

$$\langle \Psi_b | \sum_{i<j} g_{ij} | \Psi_{b'} \rangle = \frac{N(N-1)}{2} \langle \Psi_b | g_{N-1,N} | \Psi_{b'} \rangle$$

$$= \frac{\prod_k w_k!}{2(N-2)!} \sum_P \sum_{P'} (-1)^{p+p'} \langle \psi_b^{(P)} | g_{N-1,N} | \psi_{b'}^{(P')} \rangle \ . \tag{9.81}$$

We consider some fixed permutation P, and use much the same argument as before: Upon integration over any coordinate $\mathbf{r}_i$ with $i < N - 1$, the matrix element in the right-hand side of (9.81) will be zero unless the permutation P′ is such that the coordinate $\mathbf{r}_i$ occurs in the same subshell $l_j^{w_j}$ in both $\psi_b^{(P)}$ and $\psi_{b'}^{(P')}$. This leaves two different classes of permutations P that must be considered.

(1) Suppose P is such that both $\mathbf{r}_{N-1}$ and $\mathbf{r}_N$ occur in the same subshell ($l_j^{w_j}$, say). Then it follows that P′ must be such that these two coordinates both occur in this same subshell $l_j^{w_j}$ in $\psi_{b'}^{(P')}$; this means that P′ is identical with P, because permutations within a subshell are not allowed. The number of permutations P for which $\mathbf{r}_{N-1}$ and $\mathbf{r}_N$ occur in $l_j^{w_j}$ is

$$\frac{(N-2)!}{w_1!w_2! \cdots (w_j - 2)! \cdots w_q!} \ , \tag{9.82}$$

and the terms of (9.81) with $P' = P$ that fit this class (for all possible values of j) give us

$$\sum_{j=1}^{q} \frac{w_j(w_j-1)}{2} \langle \psi_b^{(P_{jj})} | g_{N-1,N} | \psi_{b'}^{(P_{jj})} \rangle \ , \tag{9.83}$$

where P_{jj} is any one of the permutations [of total number (9.82)] for which the coordinates $N-1$ and N occur in $l_j^{w_j}$.

(2) If P is such that $\mathbf{r}_{N-1}$ occurs in one subshell $l_i^{w_i}$ and $\mathbf{r}_N$ occurs in a different subshell $l_j^{w_j}$ of $\psi_b^{(P)}$, then there are two different permutations P′ that can give non-zero matrix elements in (9.81): namely, the permutation $P' = P$, for which $\mathbf{r}_{N-1}$ and $\mathbf{r}_N$ occur in the same subshells i and j of $\psi_{b'}^{(P')}$ as they do in $\psi_b^{(P)}$, and the permutation P′ which is the same as P except that in $\psi_{b'}^{(P')}$ $\mathbf{r}_{N-1}$ occurs in the subshell j and $\mathbf{r}_N$ occurs in the subshell i. For the direct case $(-1)^{p+p'} = +1$, whereas for the exchange case $(-1)^{p+p'} = -1$. For either the direct or the exchange case, and for given values of i and j ($i < j$ or $j < i$), there are

$$\frac{(N-2)!}{w_1!w_2!\cdots(w_i-1)!\cdots(w_j-1)!\cdots w_q!} \tag{9.84}$$

different permutations P with $\mathbf{r}_{N-1}$ in i and $\mathbf{r}_N$ in j; the matrix element in (9.81) has the same value for all of these. Thus the matrix elements in (9.81) for which P is of this second class may be written

$$\sum_{i \neq j} \frac{w_i w_j}{2} \left[\langle \psi_b^{(P_{ij})} | g_{N-1,N} | \psi_{b'}^{(P_{ij})} \rangle - \langle \psi_b^{(P_{ij})} | g_{N-1,N} | \psi_{b'}^{(P_{ji})} \rangle \right] \ ,$$

where P_{ij} is any one of the permutations [of total number (9.84)] for which $\mathbf{r}_{N-1}$ occurs in $l_i^{w_i}$ and $\mathbf{r}_N$ occurs in $l_j^{w_j}$. Combining this result with (9.83) we may write altogether

$$\langle \Psi_b | \sum_{i<j} g_{ij} | \Psi_{b'} \rangle = \sum_{j=1}^{q} \frac{w_j(w_j-1)}{2} \langle \psi_b | g_{(jj)} | \psi_{b'} \rangle$$
$$+ \sum_{i<j} w_i w_j [\langle \psi_b | g_{(ij)} | \psi_{b'} \rangle - \langle \psi_b | g_{(ij)} | \psi_{b'}^{(ex)} \rangle] \ , \tag{9.85}$$

where ψ_b and $\psi_{b'}$ are partially antisymmetrized coupled functions with basic unpermuted coordinate ordering, $\psi_{b'}^{(ex)}$ is the same as $\psi_{b'}$ except with the final electron coordinates of the ith and j^{th} subshells exchanged, $g_{(jj)}$ operates on the last two electron coordinates of the jth subshell, and $g_{(ij)}$ operates on the final coordinates of the ith and j^{th} subshells.

9-10. UNCOUPLING OF SPECTATOR SUBSHELLS

In the evaluation of matrix elements of symmetric operators, we have seen that the antisymmetrizing permutations in (9.75) do not have to be considered explicitly. We now show that, in certain cases, complexities due to angular-momentum coupling also largely disappear.

Let us consider two functions $|j_1m_1(i)\rangle$ and $|j_2m_2(k)\rangle$ that are functions of two sets of coordinates i and k, respectively, and that are coupled together, but not antisymmetrized by means of permutations of coordinates between the two sets i and k. Let O_i be some operator that operates only on the coordinate set i, and which is such that the matrix with elements

$$\langle j_1m_1|O_i|j_1'm_1'\rangle \tag{9.86}$$

is diagonal in j_1 and m_1, and such that (9.86) is independent of m_1. Then from the CG expansion (9.4)

$$\begin{aligned}
&\langle j_1j_2JM|O_i|j_1'j_2'J'M'\rangle \\
&= \sum_{m_1m_2}\sum_{m'_1m'_2} C(j_1j_2m_1m_2;JM)\,C(j_1'j_2'm_1'm_2';J'M')\,\langle j_1m_1j_2m_2|O_i|j_1'm_1'j_2'm_2'\rangle \\
&= \sum_{m_1m_2}\sum_{m'_1m'_2} C(j_1j_2m_1m_2;JM)\,C(j_1'j_2'm_1'm_2';J'M')\,\langle j_1m_1|O_i|j_1'm_1'\rangle\langle j_2m_2|j_2'm_2'\rangle \\
&= \delta_{j_1j'_1}\delta_{j_2j'_2}\langle j_1|O_i|j_1\rangle \sum_{m_1m_2} C(j_1j_2m_1m_2;JM)\,C(j_1j_2m_1m_2;J'M')\ ,
\end{aligned}$$

where the matrix element $\langle j_1m_1|O_i|j_1m_1\rangle$ has been abbreviated to $\langle j_1|O_i|j_1\rangle$ because of its assumed independence of m_1. From the orthonormality of the CG coefficients, (9.11), this result reduces to

$$\langle j_1j_2JM|O_i|j_1'j_2'J'M'\rangle = \delta_{j_1j'_1}\delta_{j_2j'_2}\delta_{JJ'}\delta_{MM'}\langle j_1|O_i|j_1\rangle\ . \tag{9.87}$$

A similar simplification holds for matrix elements of an operator O_k if the matrix $\langle j_2m_2|O_k|j_2'm_2'\rangle$ is diagonal in j_2 and m_2, and is independent of m_2. This can be shown in the same manner as above, or with the aid of (9.87) and the CG symmetry relation (9.8):

$$\begin{aligned}
\langle j_1j_2JM|O_k|j_1'j_2'J'M'\rangle &= (-1)^{j_1+j_2-J-j'_1-j'_2+J'}\langle j_2j_1JM|O_k|j_2'j_1'J'M'\rangle \\
&= \delta_{j_1j'_1}\delta_{j_2j'_2}\delta_{JJ'}\delta_{MM'}\langle j_2|O_k|j_2'\rangle\ .
\end{aligned} \tag{9.88}$$

Thus for operators satisfying the restrictions named above, it is trivially easy to simplify matrix elements by eliminating subshells of "spectator" electrons, even though these are coupled onto the subshell of "active" electrons upon which the operator acts. Application to matrix elements of $f_{(i)}$ and $g_{(jj)}$ in (9.80) and (9.85) is obvious; partial simplification of matrix elements of $g_{(ij)}$ is usually also possible, depending on the subshells i and j involved and the coupling scheme employed.

*9-11. APPENDIX: VECTOR-COUPLING COEFFICIENTS

In order to prove the relation (9.5) between Clebsch-Gordon coefficients and 3-j symbols, we first establish several useful relations involving factorials of non-negative integers.

The binomial theorem

$$(x + y)^a = \sum_{i=0}^{\infty} x^{a-i}y^i \binom{a}{i} = x^a \sum_{i=0}^{\infty} (y/x)^i \binom{a}{i} \tag{9.89}$$

is valid for any $y/x < 1$, where the binomial coefficient is

$$\binom{a}{i} = \frac{a(a-1)(a-2)\cdots(a-i+1)}{1\cdot 2\cdot 3\cdots i} ,$$

with the convention

$$\binom{a}{0} = 1 .$$

We shall be interested only in integral values of a. If a is a positive integer, then

$$\binom{a}{i} = \frac{a!}{i!(a-i)!} ; \tag{9.90}$$

the summation in (9.89) then terminates at the term $i = a$, and (9.89) is valid for any value of y/x. If a is a negative integer, then

$$\begin{aligned}\binom{a}{i} &= (-1)^i \frac{(-a)(1-a)(2-a)\cdots(i-1-a)}{i!} \\ &= (-1)^i \frac{(i-1-a)!}{i!(-a-1)!} = (-1)^i \binom{i-1-a}{i} .\end{aligned} \tag{9.91}$$

From (9.89) and the equality

$$(x + y)^{a+b} = (x + y)^a (x + y)^b$$

we obtain

$$\sum_{i=0}^{\infty} (y/x)^i \binom{a+b}{i} = \left[\sum_{l=0}^{\infty} (y/x)^l \binom{a}{l}\right]\left[\sum_{k=0}^{\infty} (y/x)^k \binom{b}{k}\right] ,$$

whence equating like powers of y/x on the two sides of this expression we find

$$\binom{a+b}{i} = \sum_{l=0}^{i} \binom{a}{l}\binom{b}{i-l} . \tag{9.92}$$

This may be written in the equivalent form

$$\binom{d}{d-c} = \sum_{l} \binom{d-b}{l}\binom{b}{d-c-l} ,$$

or using (9.90) and mutliplying both sides by (d − c)!/b!,

$$\frac{d!}{b!c!} = \sum_l \frac{(d-b)!(d-c)!}{l!(d-b-l)!(d-c-l)!(b+c-d+l)!} \ . \tag{9.93}$$

If a and b are negative integers, and c + 1 = −a and d + 1 = −b are therefore positive integers, then from (9.91) and (9.92)

$$\frac{c!d!(c+d+1+i)!}{i!(c+d+1)!} = \sum_l \frac{(c+l)!(d+i-l)!}{l!(i-l)!} \ . \tag{9.94}$$

On the other hand, if b is positive but a = i − c − 1 and a + b are negative, then (9.92) may be written

$$(-1)^i \frac{(c-b)!(c-i)!}{i!b!(c-b-i)!} = \sum_l (-1)^l \frac{(c-i+l)!}{l!(i-l)!(b-i+l)!} \ . \tag{9.95}$$

We now begin the derivation of the general form of the CG coefficient by showing that in the special case m = j it has a form such that (9.4′) becomes

$$A\,|j_1j_2jj\rangle = \sum_{\mu_1=j-j_2}^{j_1} \delta_{j,\mu_1+\mu_2}(-1)^{j_1-\mu_1}\left[\frac{(j_1+\mu_1)!(j_2+\mu_2)!}{(j_1-\mu_1)!(j_2-\mu_2)!}\right]^{1/2} |j_1\mu_1\rangle\,|j_2\mu_2\rangle \ , \tag{9.96}$$

where A is an appropriate normalizing constant, and μ_1 cannot be smaller than the minimum of $-j_1$ and $j - j_2$. That this form is correct may be seen by showing that operating on (9.96) by $J_+ \equiv J_{1+} + J_{2+}$ leads to a zero result. From (2.19), the result of operating on the right-hand side of (9.96) is

$$\sum_{\mu_1=j-j_2}^{j_1} \delta_{j,\mu_1+\mu_2}(-1)^{j_1-\mu_1}\left[\frac{(j_1+\mu_1)!(j_2+\mu_2)!}{(j_1-\mu_1)!(j_2-\mu_2)!}\right]^{1/2}$$

$$\times\ [(j_1-\mu_1)^{1/2}(j_1+\mu_1+1)^{1/2}\,|j_1,\mu_1+1\rangle\,|j_2\mu_2\rangle + (j_2-\mu_2)^{1/2}(j_2+\mu_2+1)^{1/2}\,|j_1\mu_1\rangle\,|j_2,\mu_2+1\rangle]$$

$$= \sum_{\mu_1=j-j_2}^{j_1} \delta_{j,\mu_1+\mu_2}(-1)^{j_1-\mu_1}\left[\frac{(j_1+\mu_1+1)!(j_2+\mu_2)!}{(j_1-\mu_1-1)!(j_2-\mu_2)!}\right]^{1/2} |j_1,\mu_1+1\rangle\,|j_2\mu_2\rangle$$

$$- \sum_{m_1=j-j_2-1}^{j_1-1} \delta_{j,m_1+m_2}(-1)^{j_1-m_1}\left[\frac{(j_1+m_1+1)!(j_2+m_2)!}{(j_1-m_1-1)!(j_2-m_2)!}\right]^{1/2} |j_1,m_1+1\rangle\,|j_2m_2\rangle$$

where $m_1 = \mu_1 - 1$ and $m_2 = \mu_2 + 1$. But in this last expression, the term $\mu_1 = j_1$ is zero because $|j_1, j_1 + 1\rangle = 0$, and similarly the term $m_1 = j - j_2 - 1$ (or $m_2 = j_2 + 1$) is zero; hence the result is identically zero, as claimed.

The normalization constant in (9.96) can be evaluated with the aid of (9.94):

$$A^2 = \sum_{\mu_1=j-j_2}^{j_1} \frac{(j_1+\mu_1)!(j_2+j-\mu_1)!}{(j_1-\mu_1)!(j_2-j+\mu_1)!}$$

$$= \sum_{l=0}^{j_1+j_2-j} \frac{(j_1+j-j_2+l)!(2j_2-l)!}{(j_1+j_2-j-l)!l!}$$

$$= \frac{(j_1-j_2+j)!(-j_1+j_2+j)!(j_1+j_2+j+1)!}{(j_1+j_2-j)!(2j+1)!} \;. \tag{9.97}$$

We shall choose A to be positive, so that (9.96) will agree with the phase convention (9.7) for the case $j = j_1 + j_2$ (where the summation reduces to the single term $\mu_1 = j_1$).

To obtain the CG coefficient for values of m other than j, we make use of (2.19) again:

$$(J_-)^n\,|jm\rangle = \left[\frac{(j+m)!(j-m+n)!}{(j-m)!(j+m-n)!}\right]^{1/2} |j,\, m-n\rangle \;.$$

Applying

$$(J_-)^{j-m} \equiv (J_{1-} + J_{2-})^{j-m}$$

$$= \sum_l \binom{j-m}{l} (J_{1-})^l (J_{2-})^{j-m-l}$$

to (9.96), we obtain

$$A\left[\frac{(2j)!(j-m)!}{(j+m)!}\right]^{1/2} |j_1 j_2 jm\rangle$$

$$= \sum_{\mu_1} \delta_{j,\mu_1+\mu_2} (-1)^{j_1-\mu_1} \left[\frac{(j_1+\mu_1)!(j_2+\mu_2)!}{(j_1-\mu_1)!(j_2-\mu_2)!}\right]^{1/2}$$

$$\times \sum_l \binom{j-m}{l} \left[\frac{(j_1+\mu_1)!(j_1-\mu_1+l)!(j_2+\mu_2)!(j_2-\mu_2+j-m-l)!}{(j_1-\mu_1)!(j_1+\mu_1-l)!(j_2-\mu_2)!(j_2+\mu_2-j+m+l)!}\right]^{1/2}$$

$$\times\, |j_1,\mu_1-l\rangle\, |j_2,\mu_2-j+m+l\rangle$$

$$= \sum_{m_1} \delta_{m,m_1+m_2} (-1)^{j_1-m_1} \left[\frac{(j_1-m_1)!(j_2-m_2)!}{(j_1+m_1)!(j_2+m_2)!}\right]^{1/2} B\, |j_1 m_1\rangle\, |j_2 m_2\rangle \;, \tag{9.98}$$

where we have substituted

$$\mu_1 = m_1 + l, \qquad \mu_2 = j - m - l + m_2 \;.$$

Using (9.93)—with the summation index l replaced by n—and then (9.95) with (9.4), we find

$$i = j_1 - j_2 - m + n, \qquad b = j_1 - j - m_2 + n, \qquad c = 2j_1 - j_2 - m_2 + n,$$

$$B \equiv \sum_l (-1)^l \frac{(j-m)!(j_1+m_1+l)!}{l!(j_2-j+m_1+l)!} \frac{(j_2+j-m_1-l)!}{(j-m-l)!(j_1-m_1-l)!}$$

$$= (j-m)!(j_2+m_2)!(-j_1+j_2+j)! \sum_n \frac{1}{n!(j_2+m_2-n)!(-j_1+j_2+j-n)!}$$

$$\times \sum_l (-1)^l \frac{(j_1+m_1+l)!}{l!(j_1-j_2-m+n-l)!(j_2-j+m_1+l)!}$$

$$= (j-m)!(j_2+m_2)!(-j_1+j_2+j)!(j_1-j_2+j)!(j_1+m_1)!(-1)^{j_1-j_2-m}$$

$$\times \sum_n \frac{(-1)^n}{n!(j_2+m_2-n)!(-j_1+j_2+j-n)!(j_1-j_2-m+n)!(j_1-j-m_2+n)!(j+m-n)!}$$

$$= (j-m)!(j_2+m_2)!(-j_1+j_2+j)!(j_1-j_2+j)!(j_1+m_1)!(-1)^{j_1-m_1}$$

$$\times \sum_k \frac{(-1)^k}{(j_2+m_2-k)!k!(j-j_1-m_2+k)!(j_1-m_1-k)!(j_1+j_2-j-k)!(j-j_2+m_1+k)!}$$

where in the last step we have substituted $n = j_2 + m_2 - k$. Substituting this and (9.97) into (9.98), and comparing with (9.4), we find

$$C(j_1j_2m_1m_2;jm) = \delta_{m,m_1+m_2}\Big[(j_1-m_1)!(j_1+m_1)!(j_2-m_2)!(j_2+m_2)!(j-m)!(j+m)!$$

$$\times \frac{(j_1+j_2-j)!(j_1-j_2+j)!(-j_1+j_2+j)!(2j+1)}{(j_1+j_2+j+1)!}\Big]^{1/2}$$

$$\times \sum_k \frac{(-1)^k}{k!(j_1+j_2-j-k)!(j_1-m_1-k)!(j_2+m_2-k)!(j-j_2+m_1+k)!(j-j_1-m_2+k)!} . \qquad (9.99)$$

From the definition (5.1) of the 3-j symbol, this is seen to be equivalent to (9.5).

10

ENERGY LEVEL STRUCTURE (SIMPLE CONFIGURATIONS)

10-1. KINETIC AND ELECTRON-NUCLEAR ENERGIES

We are now ready to consider the problem of calculating the energy-level structure of an atom. As discussed in Secs. 4-2 and 4-23, the procedure is to find the eigenvalues of the matrix of the Hamiltonian operator

$$H = -\sum_i \nabla_i^2 - \sum_i \frac{2Z}{r_i} + \sum_i \xi_i(r_i)(l_i \cdot s_i) + \sum_{i>j}\sum \frac{2}{r_{ij}} . \qquad (10.1)$$

The matrix elements of this operator are to be evaluated using the coupled, antisymmetric basis functions (9.75).

For the present we shall consider only the single-configuration approximation, in which the set of basis functions used corresponds to only one of the infinity of possible electron configurations of the atom. The configuration in question will as usual be written in the general form

$$(n_1 l_1)^{w_1}(n_2 l_2)^{w_2} \cdots (n_q l_q)^{w_q} = \prod_{j=1}^{q} (n_j l_j)^{w_j} , \qquad (10.2)$$

where

$$\sum_{j=1}^{q} w_j = N \qquad (10.3)$$

is the number of electrons in the atom. It is convenient to write the energy matrix in the form

$$(H) \equiv \begin{pmatrix} H_{11} & H_{12} & H_{13} & \cdots \\ H_{21} & H_{22} & H_{23} & \cdots \\ H_{31} & H_{32} & H_{33} & \cdots \\ \cdot & \cdot & \cdot & \\ \cdot & \cdot & \cdot & \\ \cdot & \cdot & \cdot & \end{pmatrix}$$

$$= \begin{pmatrix} E_{av} + \Delta H_{11} & H_{12} & H_{13} & \cdots \\ H_{21} & E_{av} + \Delta H_{22} & H_{23} & \cdots \\ H_{31} & H_{32} & E_{av} + \Delta H_{33} & \cdots \\ \cdot & \cdot & \cdot & \\ \cdot & \cdot & \cdot & \\ \cdot & \cdot & \cdot & \end{pmatrix} \tag{10.4}$$

with the configuration center-of-gravity energy E_{av} explicitly written as a separate term of each diagonal matrix element. Thus one aspect of the problem of evaluating the matrix elements is to determine which portion of each diagonal element H_{bb} belongs to E_{av} and which to ΔH_{bb}.

Kinetic energy. The kinetic energy term of the Hamiltonian operator constitutes a one-electron symmetric operator of the form (6.3), and so from the general result (9.80) we have

$$\langle \Psi_b | \sum_i -\nabla_i^2 | \Psi_{b'} \rangle = \sum_j w_j \langle \psi_b | -\nabla_{(j)}^2 | \psi_{b'} \rangle \ .$$

Now the operator $-\nabla_{(j)}^2$ operates on an electron coordinate in the subshell $n_j l_j^{w_j}$, and we shall see shortly that it possesses the necessary properties to be an operator of the type O_j (or O_i or O_k) discussed in Sec. 9-10. Regardless of the coupling scheme involved, repeated application of (9.87) and/or (9.88) then makes it possible to eliminate all spectator subshells, and to obtain

$$\langle \Psi_b | \sum_i -\nabla_i^2 | \Psi_{b'} \rangle = \sum_j w_j \delta_{bb'} \langle l_j^{w_j} \alpha LS | -\nabla_{(j)}^2 | l_j^{w_j} \alpha' L'S' \rangle \ . \tag{10.5}$$

The matrix element on the right-hand side of this expression may be written, with the aid of the cfp expansion (9.50),

$$\langle l^w \alpha LS | -\nabla_{(j)}^2 | l^w \alpha' L'S' \rangle = \sum_{\overline{\alpha LS}} \sum_{\overline{\alpha' L' S'}} (l^w \alpha LS \{| l^{w-1} \overline{\alpha LS}) \\ \times \langle (l^{w-1} \overline{\alpha LS}, l) LS | -\nabla_{(j)}^2 | (l^{w-1} \overline{\alpha' L' S'}, l) L'S' \rangle (l^{w-1} \overline{\alpha' L' S'} |\} l^w \alpha' L'S') \ . \tag{10.6}$$

Since $-\nabla_{(j)}^2$ operates on the *last* coordinate of l^w, then further use of (9.88) gives for the matrix element in (10.6)

$$\delta_{LS,L'S'}\delta_{\overline{\alpha LS},\overline{\alpha' L'S'}}\langle l\,|-\nabla^2|\,l\rangle \ .$$

We saw in Sec. 6-2 that $\langle l\|-\nabla^2\|\,l'\rangle$ is diagonal in all angular-momentum quantum numbers and is independent of the magnetic quantum numbers $m_l m_s$; this verifies that $-\nabla^2$ indeed satisfies the conditions required for this last use of (9.88). Since $\langle l_j\|-\nabla^2\|\,l_j\rangle = E_k^{\,j}$, then from (9.46) we find that (10.6) simplifies to

$$\langle l_j^{\,w_j}\alpha LS\,|-\nabla_{(j)}^{\;2}\,|\,l_j^{\,w_j}\alpha' L'S'\rangle = \delta_{LS,L'S'}E_k^{\,j}\sum_{\overline{\alpha LS}}(l_j^{\,w_j}\alpha LS\{|l_j^{\,w_j-1}\overline{\alpha LS})(l_j^{\,w_j-1}\overline{\alpha LS}|\}l_j^{\,w_j}\alpha' LS)$$

$$= \delta_{\alpha LS,\alpha' L'S'}E_k^{\,j} \ ;$$

this verifies the applicability of (9.87) and (9.88) in the derivation of (10.5). Finally we see that (10.5) may be written

$$\langle\Psi_b\,|\sum_i -\nabla_i^{\,2}\,|\,\Psi_{b'}\rangle = \delta_{bb'}\sum_j w_j E_k^{\,j} = \delta_{bb'}E_k \ ; \qquad (10.7)$$

i.e., the kinetic energy term of the Hamiltonian operator contributes only to the diagonal matrix elements—and in fact contributes precisely the total kinetic energy portion of E_{av}, with zero contribution to ΔH_{bb}.

Electron-nuclear energy. Exactly as for the kinetic energy, the electron-nuclear term of the Hamiltonian operator contributes the total electron-nuclear potential energy of the atom to the E_{av} portion of each diagonal matrix element, and makes zero contribution to ΔH_{bb} and to each off-diagonal matrix element.

10-2. EFFECTS OF CLOSED SUBSHELLS

The spin-orbit and electron-electron terms of the Hamiltonian operator depend on the angular, as well as the radial, components of the electron coordinates; evaluation of their matrix elements is therefore much more complicated than for the kinetic and electron-nuclear energies. However, one important simplification is possible. We have seen that closed electron subshells are spherically symmetrical and possess zero net angular momentum. They therefore contribute nothing to the total angular momentum of the atom, and we may correspondingly anticipate that they make no angularly-dependent contributions to the Hamiltonian matrix elements. That is, closed subshells must contribute only to the value of E_{av}, and not to ΔH_{bb} nor to any off-diagonal matrix element. Indeed, we have already noted this effect in calculations with uncoupled (determinantal) basis functions: In summing the Coulomb interaction between electrons i and j over all electrons in the filled subshell j, we found the results (6.36) and (6.37) to be independent of $m_{l_i}m_{s_i}$; the magnitude of the contribution to E_{av}, when one or both of the electrons lies in a filled subshell, will be given by (6.38) or (6.39) even when using coupled basis functions. Similarly, the argument that led to (6.12) shows that spin-orbit contributions from filled shells are nil, even for E_{av}.

All of this will be proved in detail in Secs. 12-2 to 12-5; for the present we simply accept the above arguments and, for purposes of calculating level splittings about the center of gravity, ignore all filled subshells. For the remainder of this chapter, we consider only simple configurations containing no more than two electrons outside of filled subshells.

10-3. ONE-ELECTRON CONFIGURATIONS

With only one electron other than those contained in closed subshells, there are no Coulomb interactions to be considered in the calculation of level structure. There is only the spin-orbit interaction of the electron in question, and the problem is basically identical with that of the hydrogenic atom treated in Sec. 3-7. The principal difference is that in the expression

$$\xi_i(r_i) = \frac{\alpha^2}{2}\left[\frac{1}{r_i}\frac{dV^i(r_i)}{dr_i}\right] \tag{10.8}$$

for the radial factor of the spin-orbit operator, the central-field potential-energy function V^i for an electron in the spin-orbital i is not a simple Coulomb (r^{-1}) function. The radial integral of ξ_i,

$$\zeta_i \equiv \zeta_{n_i l_i} = \frac{\alpha^2}{2}\int_0^\infty \frac{1}{r}\left(\frac{dV^i}{dr}\right)|P_{n_i l_i}(r)|^2\,dr\ , \tag{10.9}$$

is commonly referred to as the *spin-orbit parameter*. It can be evaluated numerically using, for example, the HX central-field potential (7.54) for V^i, and is obviously positive since dV^i/dr is everywhere greater than zero. (α is the fine-structure constant 1/137.036, and ζ_i has the same units as V^i if r is in Bohr units.)

In evaluating matrix elements of the spin-orbit operator with the aid of (9.80), we can—according to the arguments of the preceding section—ignore all terms in the summation except that corresponding to the singly occupied subshell i. In evaluating the remaining term

$$\langle\Psi_b|\sum_{k=1}^{N}\xi_k(r_k)\cdot(\boldsymbol{l}_k\cdot\mathbf{s}_k)|\Psi_{b'}\rangle = \langle\psi_b|\xi_i(r_i)\cdot(\boldsymbol{l}_i\cdot\mathbf{s}_i)|\psi_{b'}\rangle\ , \tag{10.10}$$

we may note that filled shells offer no coupling complications; that is, when one of the angular momenta being coupled is zero, the CG sum contains only one term and the CG coefficient is unity. The coupled function is therefore a simple product function except for the coupling of the angular momenta within subshell i, integration over the electron coordinates within each closed subshell produces a simple normalization integral, and (10.10) reduces to[1]

$$\langle n_i l_i j_i m_i|\xi_i\cdot(\boldsymbol{l}_i\cdot\mathbf{s}_i)|n_i l_i j_i' m_i'\rangle = \zeta_i\langle l_i j_i m_i|\boldsymbol{l}_i\cdot\mathbf{s}_i|l_i j_i' m_i'\rangle \equiv d_i\zeta_i\ ; \tag{10.11}$$

[1]Alternatively, we could use jj-coupled basis functions and make use of the general expressions (9.87) and/or (9.88), since (10.12) shows the matrix elements (10.11) to possess the required properties.

the total angular momenta of the atom LSJM are identical in the present case with the values $l_i s_i j_i m_i$ for the electron in shell i.

Evaluation of the angular factor d_i of the matrix element is easily accomplished with the aid of the relation (3.41), $\boldsymbol{l} \cdot \boldsymbol{s} = (j^2 - l^2 - s^2)/2$, giving

$$d_i = \delta_{j_i j_i'} \delta_{m_i m_i'} \frac{1}{2} [j_i(j_i + 1) - l_i(l_i + 1) - s_i(s_i + 1)] \ . \qquad (10.12)$$

For $l = 0$, d is obviously zero; for $l > 0$, there are two possible values $j = l \pm 1/2$, and the diagonal values of d are

$$d = \frac{1}{2}\left[\left(l \pm \frac{1}{2}\right)\left(l + 1 \pm \frac{1}{2}\right) - l^2 - l - \frac{3}{4}\right]$$

$$= \begin{cases} \dfrac{l}{2} \ , & j = l + \dfrac{1}{2} \ , \\ -\dfrac{l+1}{2} \ , & j = l - \dfrac{1}{2} \ . \end{cases} \qquad (10.13)$$

Since there are no non-zero off-diagonal matrix elements, the eigenvalues of the Hamiltonian matrix are equal to the diagonal elements. Thus the two energy levels of the one-electron configuration nl are equal to $E_{av} + l\zeta/2$ and $E_{av} - (l + 1)\zeta/2$, and their separation is

$$\Delta E = \frac{2l+1}{2} \zeta_{nl} \ . \qquad (10.14)$$

The center-of-gravity spin-orbit energy is proportional to

$$(2j_+ + 1)d_+\zeta + (2j_- + 1)d_-\zeta = (2l+2)\frac{l}{2}\zeta + (2l)\left(-\frac{l+1}{2}\right)\zeta = 0 \ ,$$

in agreement with our earlier remarks regarding the nil spin-orbit contribution to E_{av}.

Problem

10-3(1). Using experimental energy levels from AEL, infer empirical values of the spin-orbit parameter ζ_{nl} (in cm^{-1}) for the lowest two or three values of n in the np, nd, and nf configurations of K I, Ca II, Cs I, and Ba II. What are the physical reasons for the observed dependence of ζ_{nl} on n, l, Z, and ionization stage—especially the change in ζ_{nd} in going from K I to Ca II, and in ζ_{nf} in going from Cs I to Ba II? [Hint: See Secs. 8-5 to 8-7.]

10-4. TWO-ELECTRON CONFIGURATIONS; SPIN-ORBIT MATRIX ELEMENTS

For a configuration involving two (or more) non-equivalent electrons not in closed subshells, the spin-orbit matrix elements are easily evaluated if jj-coupled basis functions are used. From (9.80) and (9.87) and/or (9.88),

$$\langle \Psi_b | \sum_i \xi_i(r_i) \cdot (l_i \cdot s_i) | \Psi_{b'} \rangle = \sum_t \langle \psi_b | f_{(t)} | \psi_{b'} \rangle$$

$$= \delta_{bb'} \sum_t d_{tb} \zeta_t$$

$$\equiv \delta_{bb'} \sum_{nl} d_{lj} \zeta_{nl} , \tag{10.15}$$

where the summations are over the open subshells. Thus the spin-orbit matrix is completely diagonal, and the coefficients of ζ are given by the expression (10.13). It should be noted that ζ_{nl} has the same value for all basis states of the configuration (but depends on the subshell and atom in question through the quantum number n), whereas d_{lj} depends on the angular basis state through the quantum number j but is independent of the principal quantum number and the particular atom in question.

If the Coulomb interaction between the two electrons is neglected, the eigenvalues of the Hamiltonian matrix are equal to the diagonal matrix elements, and the four energy levels are those given in Table 10-1. A weak Coulomb interaction will make additional small contributions to the diagonal (and off-diagonal) matrix elements, resulting in eigenvalues slightly different from those in the table, and in corresponding energy-level splittings according to the possible values of J. An example is shown in Fig. 4-6.

If the Coulomb interaction is strong, the off-diagonal matrix elements are large and the above picture is not valid at all. It is better to consider the level structure primarily from the point of view of an LS-coupling scheme, which we consider next.

Problem

10-4(1). Show that the level structure in Table 10-1 is valid also for two equivalent electrons (l^2), except that there are only three levels because $l_1 = l_2$ and $\zeta_1 = \zeta_2$ (and of course the number of basis states is smaller because of the Pauli principle). [Hint: Using a relation analogous to (9.34), show that jj-coupled product functions $|ll\rangle$ are correct antisymmetric basis functions $|l^2\rangle$ when $j_1 = j_2$ and J is even.]

TABLE 10-1. ENERGY LEVELS FOR TWO-ELECTRON CONFIGURATIONS
(Spin-Orbit Interaction Only)

$b = (j_1 j_2)J$	E
$\left(l_1 + \frac{1}{2}, l_2 + \frac{1}{2}\right)J$	$E_{av} + \frac{l_1}{2}\zeta_1 + \frac{l_2}{2}\zeta_2$
$\left(l_1 + \frac{1}{2}, l_2 - \frac{1}{2}\right)J$	$E_{av} + \frac{l_1}{2}\zeta_1 - \frac{l_2 + 1}{2}\zeta_2$
$\left(l_1 - \frac{1}{2}, l_2 + \frac{1}{2}\right)J$	$E_{av} - \frac{l_1 + 1}{2}\zeta_1 + \frac{l_2}{2}\zeta_2$
$\left(l_1 - \frac{1}{2}, l_1 - \frac{1}{2}\right)J$	$E_{av} - \frac{l_1 + 1}{2}\zeta_1 - \frac{l_2 + 1}{2}\zeta_2$

10-5. TWO-ELECTRON CONFIGURATIONS; COULOMB MATRIX ELEMENTS

For a configuration containing two non-equivalent electrons (other than closed subshells, which we ignore), the general expression (9.85) in the case of the Coulomb interaction may be written, using LS-coupled basis functions,

$$\begin{aligned}
\langle\Psi_b|\sum_{i<j}\frac{2}{r_{ij}}|\Psi_{b'}\rangle &= \langle l_1(i)l_2(j)LSJM|\frac{2}{r_{ij}}|l_1(i)l_2(j)L'S'J'M'\rangle \\
&\quad - \langle l_1(i)l_2(j)LSJM|\frac{2}{r_{ij}}|l_1(j)l_2(i)L'S'J'M'\rangle \\
&= \langle l_1(i)l_2(j)LSJM|\frac{2}{r_{ij}}|l_1(i)l_2(j)L'S'J'M'\rangle \\
&\quad - (-1)^{l_1+l_2-L'-s_1-s_2+S'}\langle l_1(i)l_2(j)LSJM|\frac{2}{r_{ij}}|l_2(i)l_1(j)L'S'J'M'\rangle \\
&= \delta_{LSJM,L'S'J'M'}\Big[\langle l_1(i)l_2(j)L|\frac{2}{r_{ij}}|l_1(i)l_2(j)L\rangle \\
&\quad + (-1)^{l_1+l_2-L+S}\langle l_1(i)l_2(j)L|\frac{2}{r_{ij}}|l_2(i)l_1(j)L\rangle\Big] \; . \qquad (10.16)
\end{aligned}$$

In the first of the above steps, the CG symmetry relation (9.8) has been used (once for the orbital quantum numbers and once for the spins), and in the second step the relation (9.87) has been employed: the Coulcmb operator possesses the required properties because it does not act on the spins and represents forces internal to the atom, which cannot alter L and cannot depend on M_L (the orientation of L). (For a mathematical proof, see Sec. 11-8.)

Either of the matrix elements on the right-hand side of (10.16) can be evaluated by writing both the bra and ket functions in terms of uncoupled product functions by means of the CG formula (9.4), and by then using (6.22):

$$\begin{aligned}
&\langle l_1l_2LM_L|\frac{2}{r_{ij}}|l_3l_4LM_L\rangle \\
&= \sum_{\substack{m_1m_2\\m_3m_4}} C(l_1l_2m_1m_2;LM_L)\langle l_1m_1l_2m_2|\frac{2}{r_{ij}}|l_3m_3l_4m_4\rangle C(l_3l_4m_3m_4;LM_L) \\
&= \sum_k r_k R^k(l_1l_2,l_3l_4) \; , \qquad (10.17)
\end{aligned}$$

where

$$\begin{aligned}
r_k = &\sum_{\substack{m_1m_2\\m_3m_4}} C(l_1l_2m_1m_2;LM_L)C(l_3l_4m_3m_4;LM_L) \\
&\times \sum_q \delta_{q,m_3-m_1}\delta_{q,m_2-m_4}(-1)^q c^k(l_1m_1,l_3m_3)c^k(l_2m_2,l_4m_4) \; . \qquad (10.18)
\end{aligned}$$

The arguments l_i of the radial integral R^k are abbreviations for $n_i l_i$, and the angular coefficients r_k depend on the basis state through the quantum number L. [The delta-factors on the spin quantum numbers present in (6.22) are absent here because we have already eliminated spins in the final form of (10.16), so that the functions $|l_i m_i\rangle$ in (10.17) are orbitals rather than spin-orbitals.]

The expression (10.18) for r_k actually represents only a two-fold summation (over m_1 and m_3, say) because of the limitations $m_2 = M_L - m_1$ and $m_4 = M_L - m_3$, and the delta-factor on q. All the same, hand evaluation of this expression is onerous, even using tabulated values of the c^k and CG coefficients. If evaluation of r_k is carried out on an electronic computer, table look-up of the coefficients is not feasible because of the large size tables required to cover all argument values of potential interest. It is therefore necessary to evaluate everything numerically in terms of 3-j symbols; using (5.19) and (9.5) we obtain

$$r_k = (-1)^{l_1-l_2-l_3+l_4}[l_1, l_2, l_3, l_4]^{1/2}\begin{pmatrix} l_1 & k & l_3 \\ 0 & 0 & 0 \end{pmatrix}\begin{pmatrix} l_2 & k & l_4 \\ 0 & 0 & 0 \end{pmatrix}$$

$$\times [L] \sum_{\substack{qm_1m_2\\ m_3m_4}} (-1)^{q-m_1-m_2}\begin{pmatrix} l_1 & l_2 & L \\ m_1 & m_2 & -M_L \end{pmatrix}\begin{pmatrix} l_1 & k & l_3 \\ -m_1 & -q & m_3 \end{pmatrix}$$

$$\times \begin{pmatrix} l_2 & k & l_4 \\ -m_2 & q & m_4 \end{pmatrix}\begin{pmatrix} l_3 & l_4 & L \\ m_3 & m_4 & -M_L \end{pmatrix}. \tag{10.19}$$

Evaluation of the summation in this expression involves a six-fold summation (because of the implicit summation in each 3-j symbol), which is very time consuming even with a computer. However, a tremendous simplification became available when Racah found it possible to analytically reduce this six-fold sum to the single summation present in the expression (5.23) for a 6-j symbol.[2] Conversion to 6-j form can be accomplished by using the relation $m_1 + m_2 = M_L = m_3 + m_4$ and the 3-j symmetry relations (5.5)-(5.6), changing the sign of the summation index m_3, and noting that $2(l_4 + m_4)$ must be an even integer; the summation in (10.19) may thus be written

$$[L]\sum(-1)^{2l_4+2m_4+q+m_3-m_4}\begin{pmatrix} l_1 & l_2 & L \\ m_1 & m_2 & -M_L \end{pmatrix}\begin{pmatrix} l_1 & l_3 & k \\ m_1 & m_3 & q \end{pmatrix}$$

$$\times (-1)^{l_2+k+l_4}\begin{pmatrix} l_4 & l_2 & k \\ -m_4 & m_2 & -q \end{pmatrix}(-1)^{l_3+l_4+L}\begin{pmatrix} l_4 & l_3 & L \\ m_4 & -m_3 & -M_L \end{pmatrix}$$

$$= (-1)^{l_2-l_4+L}[L]\sum(-1)^{l_4+l_3+k+m_4+m_3+q}\begin{pmatrix} \\ \end{pmatrix}\begin{pmatrix} \\ \end{pmatrix}\begin{pmatrix} \\ \end{pmatrix}\begin{pmatrix} \\ \end{pmatrix}$$

$$= (-1)^{l_2-l_4+L}\begin{Bmatrix} l_1 & l_2 & L \\ l_4 & l_3 & k \end{Bmatrix},$$

[2]G. Racah, II. Phys. Rev. 62, 438 (1942); cf. also fn. 6 of Chap. 9. The reduction from a six-fold to a single summation is greatly simplified by the knowledge (from the physical argument given above) that r_k must be independent of M_L, so that it is sufficient to consider only the case $M_L = L$.

where the final step follows from (5.25). We therefore obtain

$$r_k = (-1)^{l_1-l_3+L}[l_1, l_2, l_3, l_4]^{1/2}\begin{pmatrix} l_1 & k & l_3 \\ 0 & 0 & 0 \end{pmatrix}\begin{pmatrix} l_2 & k & l_4 \\ 0 & 0 & 0 \end{pmatrix}\begin{Bmatrix} l_1 & l_2 & L \\ l_4 & l_3 & k \end{Bmatrix} . \tag{10.20}$$

By combining the results (10.17) and (10.20), and recalling from (6.28), (6.30) the abbreviations

$$R^k(l_1l_2, l_1l_2) = F^k(l_1l_2)$$

$$R^k(l_1l_2, l_2l_1) = G^k(l_1l_2) ,$$

we find that the Coulomb-interaction matrix element (10.16) between LS-coupled basis functions of a two-electron configuration may be written

$$\langle\Psi_b|\sum_{i<j}\frac{2}{r_{ij}}|\Psi_{b'}\rangle = \delta_{LSJM,L'S'J'M'}\sum_k[f_k'F^k(l_1l_2) + g_k'G^k(l_1l_2)] , \tag{10.21}$$

where

$$f_k' = (-1)^L[l_1, l_2]\begin{pmatrix} l_1 & k & l_1 \\ 0 & 0 & 0 \end{pmatrix}\begin{pmatrix} l_2 & k & l_2 \\ 0 & 0 & 0 \end{pmatrix}\begin{Bmatrix} l_1 & l_2 & L \\ l_2 & l_1 & k \end{Bmatrix} \tag{10.22}$$

and

$$g_k' = (-1)^S[l_1, l_2]\begin{pmatrix} l_1 & k & l_2 \\ 0 & 0 & 0 \end{pmatrix}^2\begin{Bmatrix} l_1 & l_2 & L \\ l_1 & l_2 & k \end{Bmatrix} ; \tag{10.23}$$

in the expression for g_k', use has been made of the fact that the columns of the 3-j symbol can be freely permuted, because from (5.6) this symbol is zero unless $l_1 + k + l_2$ is even. Taking into account the triangle relations for the 3-j (or the 6-j) symbols, the only values of k that lead to non-zero values of the angular integrals f_k and g_k are

$$\begin{aligned} f_k&: \quad k = 0, 2, 4, \cdots \min(2l_1, 2l_2) , \\ g_k&: \quad k = |l_1 - l_2|, |l_1 - l_2| + 2, \cdots l_1 + l_2 . \end{aligned} \tag{10.24}$$

We note from (10.21) that for a simple two-electron configuration—where there exists no more than one basis function with given LSJM—the Coulomb matrix is completely diagonal. If we are to write the diagonal matrix elements of H in the form $E_{av} + \Delta H_{bb}$, we must subtract from (10.22)-(10.23) the portions of these quantities that belong to E_{av}. From (5.11) and (5.29) we easily find $f_0'F^0(l_1l_2) = F^0(l_1l_2)$, which from (6.14) and (6.38) is just the contribution to E_{av} of the direct Coulomb interaction between l_1 and l_2. Noting also the form of the exchange contribution in (6.38), we may write the matrix elements (excluding spin-orbit contributions) in the form

$$\langle\Psi_b|H - H_{so}|\Psi_{b'}\rangle = \delta_{bb'}\left[E_{av} + \sum_{k>0} f_k F^k(l_1l_2) + \sum_k g_k G^k(l_1l_2)\right] , \tag{10.25}$$

where the direct coefficient for k > 0 remains[3]

$$f_k = (-1)^L[l_1, l_2]\begin{pmatrix} l_1 & k & l_1 \\ 0 & 0 & 0 \end{pmatrix}\begin{pmatrix} l_2 & k & l_2 \\ 0 & 0 & 0 \end{pmatrix}\begin{Bmatrix} l_1 & l_2 & L \\ l_2 & l_1 & k \end{Bmatrix}, \tag{10.26}$$

and the exchange coefficient is now[3]

$$g_k = \begin{pmatrix} l_1 & k & l_2 \\ 0 & 0 & 0 \end{pmatrix}^2 \left[\frac{1}{2} + (-1)^S[l_1, l_2]\begin{Bmatrix} l_1 & l_2 & L \\ l_1 & l_2 & k \end{Bmatrix}\right]. \tag{10.27}$$

Energy level structure. If spin-orbit interactions are negligibly small, the Hamiltonian matrix for a two-electron configuration is completely diagonal in the LS representation, the eigenvalues are equal to the diagonal elements (10.25)-(10.27), and the various quantum states of the atom are accurately described by the corresponding basis functions—i.e., the energy levels can be accurately labeled by means of the quantum numbers LS.

The direct Coulomb energy $\sum f_k F^k$ is independent of the total spin S. Its dependence on L is not easily describable in an arbitrary case. However, the general form of the individual contributions $f_k F^k$ may be visualized with the aid of the specific cases shown in Fig. 10-1. Although the magnitude of the angular coefficient f_k is roughly independent of k, the decrease (6.31) in F^k with increasing k indicates that the principal variation of energy with L will be the same as that of f_2: thus the energy tends to be high for the extreme values $L_{max} = l_1 + l_2$ and $L_{min} = |l_1 - l_2|$, and lowest for L equal to or somewhat less than $L_{max} - 1$. The term f_2F^2 represents the interaction energy between the quadrupole components of the charge distribution of the two electrons (or rather, of the corresponding orbitals), and a qualitative picture of the reason why the energy is lowest at intermediate L was given in Sec. 4-15. The terms f_4F^4, f_6F^6, $\cdots$ (if present) represent interactions between 16-pole, 64-pole, $\cdots$ components, and correspondingly show increasingly rapid energy fluctuations as a function of L (i.e., as a function of the angle between l_1 and l_2). It may be noted that these higher-order terms give a negative contribution for $L = L_{max} - 1$ and a positive contribution for $L = L_{max} - 2$, usually sufficient in magnitude to make the total direct energy lowest for $L = L_{max} - 1$ even in the ff case; at the same time, it is evident that these terms make their greatest contribution to the level structure at the smaller values of L.

The spin (singlet-triplet) degeneracy of the direct terms is seen from (10.27) to be removed by the exchange contribution $\sum g_k G^k$. Although the Coulomb operator $1/r_{ij}$ is spin independent, the exchange energy depends on spin as a result of the partial electron correlation (and consequent energy decrease) which exists for parallel-spin orbitals in determinantal functions [see Sec. 4-5 and Eq. (6.29)]. With coupled functions, parallel spins can occur only for triplet terms; however, this does not imply that for each L the ^3L lies below the ^1L—merely that the $(2L + 1)$-weighted average of all triplets lies below the

[3]Values of f_k and g_k for p*l* (and numerous other) configurations are tabulated by J. C. Slater, *Quantum Theory of Atomic Structure* (McGraw-Hill, New York, 1960), Vol. II, App. 21, and by E. U. Condon and H. Odabasi, *Atomic Structure* (Cambridge University Press, 1980), App. 3.

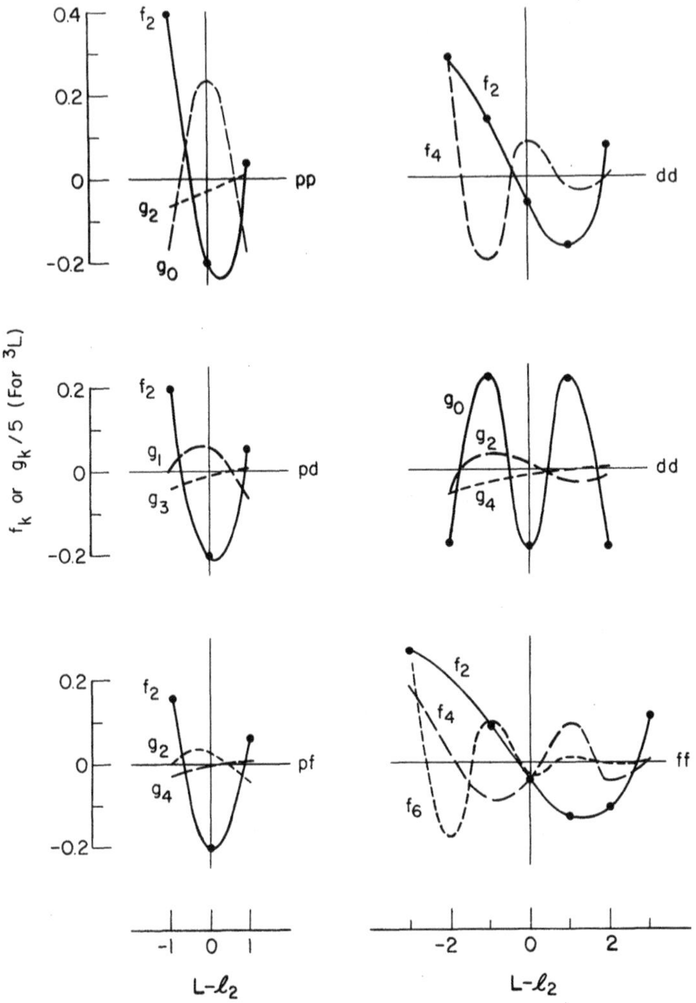

Fig. 10-1. L-dependence of the angular coefficients f_k and g_k of the direct (F^k) and exchange (G^k) Coulomb radial integrals for some two-electron configurations l_1l_2. The g_k are plotted for triplet terms only; values for singlet terms may be obtained by a sign change plus a small upward displacement. (The g_k for ff have been omitted from the figure.)

weighted average of all singlets.[4] Indeed, on the assumption that the dominant exchange term g_kG^k is that with minimum k [cf. (6.32)-(6.33) and Fig. 10-1] it is easily seen that the spin-dependent part of (10.27) has the same sign as

[4]The result (6.29) pertained only to *diagonal* matrix elements in the uncoupled representation; the present result pertains to diagonal elements in the coupled representation, which involve off-diagonal uncoupled elements. We can therefore compare only values of the matrix trace, which is invariant under the orthogonal transformation between the two representations.

$$(-1)^{S+L+L_{max}} = (-1)^{S+L+L_{min}} \ . \tag{10.28}$$

Thus the triplet lies lower than the singlet for $L = L_{max}$, but for successively smaller values of L the triplet lies alternately above and below the corresponding singlet. This singlet-triplet alternation can be seen from data in AEL to exist experimentally in most cases, and an example is shown in Fig. 4-4 for Ti III 3d4p; exceptions can be traced to perturbations from levels in other configurations (i.e., to inadequacy of the single-configuration approximation).

By contrast with the direct terms, the "frequency" of the L-variation of the exchange coefficients g_k is seen from Fig. 10-1 to decrease with increasing k—the coefficient with maximum k showing a monotonic dependence on L.

Weak spin-orbit interactions produce additional small J-dependent contributions to the diagonal (and off-diagonal) matrix elements, resulting in energy-level splittings of each triplet term (except ^3S) according to the three possible values of J, as illustrated in Figs. 4-4 and 4-5.

Special case: s*l* and *l*s configurations. Two-electron configurations in which one (or both) electrons have zero orbital angular momentum constitute a degenerate special case. There is only one possible value of L ($=l$), there are no direct terms with $k > 0$, and there is only one exchange term $g_kG^k(sl)$—with $k = l$. From (10.27), (5.11), and (5.29), the Coulomb energies are found to be

$$E = E_{av} + \frac{1/2 + (-1)^S}{2l+1}G^l$$

$$= \begin{cases} E_{av} + \dfrac{3/2}{2l+1}G^l \ , & \text{singlet} \ , \\ E_{av} - \dfrac{1/2}{2l+1}G^l \ , & \text{triplet} \ . \end{cases} \tag{10.29}$$

The energy difference ΔE between the singlet and the (center of gravity of the) triplet is thus $2G^l/(2l + 1)$, and empirical values of G^l are readily inferred from

$$G^l(sl) = \frac{2l+1}{2}(\Delta E)_{exp} \ . \tag{10.30}$$

Equivalent electrons. For a configuration l^2 consisting of two equivalent electrons, we have seen (Secs. 9-4 and 9-5) that LS-coupled product functions are antisymmetric without any coordinate permutation. Consequently there is no exchange contribution g_kG^k to the energy; except for this, the Hamiltonian matrix elements are still given by (10.21)-(10.22) with $l_1 = l_2 = l$.[5] However, only those basis states for which $L + S$ is even are allowed and the contribution to E_{av} is therefore different from what it is for non-equivalent electrons. Using (6.39) we obtain

[5]Alternatively, we can use the re-antisymmetrized basis functions (9.30), with the required additional renormalization factor $1/\sqrt{2}$ for each, thereby obtaining the result (10.21), including exchange terms, except with f_k' and g_k' half as large as given by (10.22) and (10.23). Since $(-1)^L = (-1)^S$ for $L + S$ even, then $f_k' = g_k'$ for $l_1 = l_2$; since also $F^k(ll) = G^k(ll)$ from (6.28) and (6.30), the result reduces to the form indicated in the text.

$$\langle \Psi_b | H - H_{so} | \Psi_{b'} \rangle = \delta_{bb'} \left[E_{av} + \sum_{k>0} f_k F^k(ll) \right] , \tag{10.31}$$

where

$$f_k = \begin{pmatrix} l & k & l \\ 0 & 0 & 0 \end{pmatrix}^2 \left[\frac{2l+1}{4l+1} + (-1)^L (2l+1)^2 \begin{Bmatrix} l & l & L \\ l & l & k \end{Bmatrix} \right] . \tag{10.32}$$

Except for the constant term, the variation of f_k with L is the same as illustrated in Fig. 10-1 for pp, dd, and ff. Since triplet terms (S = 1) occur only for odd L ($L = L_{max} - 1$, $L_{max} - 3$, $\cdots$), it is clear from the figure (and the earlier discussion regarding the contributions of f_4 and f_6 in the case of ff) that the lowest-energy term ($L = L_{max} - 1$) is that triplet term which has the largest value of L. Hund's rule (Sec. 4-16) is therefore satisfied for l^2, even though it is not for $l_1 l_2$ nor for $nl\,n'l$.

Problems

10-5(1). Verify (10.28) by showing that for minimum k, the sign of the 6-j symbol in (10.27) is given by

$$\begin{Bmatrix} l_1 & l_2 & L \\ l_1 & l_2 & |l_1 - l_2| \end{Bmatrix} \propto (-1)^{L+l_1+l_2} . \tag{10.33}$$

[Hint: Let b be the larger, and a the smaller, of l_1 and l_2. Then show that for

$$\begin{Bmatrix} b & a & L \\ b & a & b-a \end{Bmatrix}$$

the summation in (5.23) contains only the single term in which the value of the summation index is b + a + L.]

10-5(2). Show that for maximum k, the 6-j symbol in (10.27) is always positive but decreases monotonically with increasing L. [Hint: Show that this symbol is proportional to $F(L) \equiv \{(b+a-L)!(a+b+L+1)!\}^{-1}$, and calculate $F(L+1) - F(L)$.]

10-5(3). Using energy-level data from AEL, calculate empirical values of $G^l(sl)$ (in cm^{-1}) for the lowest two or three 3sns, 3snp, 3snd, and 3snf configurations of Mg I, Al II, and P IV. Discuss the dependence on n, *l*, and Z. What is the source of the negative values of "$G^2(sd)$" in Mg I? [Hint: Note the relative energies of $3p^2$ ^3P and ^1D in Al II and in P IV. See also G. Risberg, Ark. Fys. **28**, 381 (1965).]

10-6. INTERMEDIATE COUPLING

When neither the Coulomb nor the spin-orbit interaction is small compared with the other, the Hamiltonian matrix is not even approximately diagonal in either the jj- or LS-coupling representation. The eigenvalues are then not given simply by the diagonal matrix elements in the appropriate representation. The only possible procedure is to write down the complete Hamiltonian matrix, in some convenient representation; this may be accomplished, for example, by transforming the spin-orbit matrix with elements (10.15) from the jj to the LS representation by means of the transformation matrix (9.29) (or its appropriate modification for equivalent electrons, Sec. 9-4), or by similarly transforming the Coulomb matrix (10.25)-(10.27) [or (10.31)-(10.32)] from the LS to the jj representation. The eigenvalues of the Hamiltonian matrix are then found either by numerical diagonalization, or (in simple cases) by analytical or numerical solution of the secular equation (Sec. 4-2).

There is one simplifying feature of the calculation. As explained in Sec. 4-23, the Hamiltonian matrix for an atom free of external fields breaks up into separate matrices, one for each value of J. If for some J the matrix has only one row and column, the problem is of course trivial: the single matrix element is also the eigenvalue, the single eigenvector component is unity, and the atomic wavefunction is (in the approximation being employed) identical with the basis function. An example occurs in the calculation of levels of the configuration p^2 (in the single-configuration approximation), where there is only one basis function with $J = 1$. This is the $^3P_1$ function in the LS-coupling representation, and the $(1/2\ 3/2)_1$ function in jj coupling. (Clearly, these two basis functions must be identical, in spite of the completely different appearance of the quantum-number labeling and the corresponding difference in the CG sums that are involved in coupling the angular momenta.)

If for some value of J there are only two basis functions (of given M)—for example, the $^1S_0$ and $^3P_0$ functions of p^2—then the corresponding eigenvalue problem

$$\begin{pmatrix} A & C \\ C & B \end{pmatrix} \begin{pmatrix} y_1 \\ y_2 \end{pmatrix} = E \begin{pmatrix} y_1 \\ y_2 \end{pmatrix} \tag{10.34}$$

is still simple enough to discuss analytically. The eigenvalues are found from the secular equation

$$\begin{vmatrix} A - E & C \\ C & B - E \end{vmatrix} = E^2 - (A + B)E + AB - C^2 = 0$$

to be

$$E^{\pm} = \frac{A + B}{2} \pm C\left[\left(\frac{A - B}{2C}\right)^2 + 1\right]^{1/2} . \tag{10.35}$$

One eigenvalue is larger than max(A,B) and the other is smaller than min(A,B), so that the effect of C is to produce an apparent mutual repulsion of the two energy levels.

The off-diagonal matrix element C also causes each eigenstate to be a mixture of the two basis states: Substituting (10.35) into (say) the upper of the two equations (10.34),

$$Ay_1^{\pm} + Cy_2^{\pm} = E^{\pm}y_1^{\pm} ,$$

we obtain

$$\frac{y_2^{\pm}}{y_1^{\pm}} = \frac{E^{\pm} - A}{C} = \frac{B - A}{2C} \pm \left[\left(\frac{A - B}{2C}\right)^2 + 1\right]^{1/2} \equiv r_{\pm} \; . \tag{10.36}$$

The normalization requirement

$$y_1^2 + y_2^2 = y_1^2(1 + r^2) = 1$$

(together with an arbitrary choice of the overall eigenvector phase, and the relation $r_+r_- = -1$) then gives

$$Y^{\pm} = \begin{pmatrix} \dfrac{1}{(1 + r_{\pm}^2)^{1/2}} \\ \dfrac{r_{\pm}}{(1 + r_{\pm}^2)^{1/2}} \end{pmatrix} = \begin{pmatrix} \dfrac{r_{\mp}}{(1 + r_{\mp}^2)^{1/2}} \\ \dfrac{-1}{(1 + r_{\mp}^2)^{1/2}} \end{pmatrix} . \tag{10.37}$$

If $\Delta \equiv C/(A - B)$ has a magnitude much less than unity, then the eigenvalues simplify approximately to

$$A + \frac{C^2}{A - B} \; , \qquad B - \frac{C^2}{A - B} \; , \tag{10.38}$$

with corresponding eigenvectors

$$\begin{pmatrix} 1 - \Delta^2/2 \\ \Delta \end{pmatrix} , \qquad \begin{pmatrix} \Delta \\ -1 + \Delta^2/2 \end{pmatrix} ; \tag{10.39}$$

the two eigenstates are therefore nearly pure basis states. On the other hand, if $|\Delta| \gg 1$ then the eigenvalues and vectors become

$$E^{\pm} = \frac{A + B}{2} \pm C \; , \qquad Y^{\pm} = \begin{pmatrix} 1/\sqrt{2} \\ \pm 1/\sqrt{2} \end{pmatrix} ; \tag{10.40}$$

in this limit, the two eigenstates are 50-50 mixtures of the two basis states.

When the J-matrix is larger than 2-by-2, the problem is sufficiently complex that little can be said in general,[6] except that one eigenvalue is greater than the largest diagonal element, and one is smaller than the smallest diagonal element. The relative magnitudes and phases of the various eigenvector components depend in complex ways on the magnitudes and signs of the various matrix elements. One can in general only calculate numerical values of the Hamiltonian matrix elements, and then numerically diagonalize the matrix (usually with the aid of an electronic computer) to obtain the eigenvalues and eigenvectors.

[6]An important special case is discussed in detail by E. U. Condon and G. H. Shortley, *The Theory of Atomic Spectra* (University Press, Cambridge, 1935), pp. 37-42. Application of this case to discrete-continuum interactions is discussed in Sec. 18-7 below.

Problem

10-6(1). Consider a two-level atom assumed to be adequately described by linear combinations $\Psi = y_1\psi_1 + y_2\psi_2$ of two given basis functions ψ_1 and ψ_2. Show that varying y_1 and y_2 in order to minimize the expectation value of the Hamiltonian, $\langle\Psi|H|\Psi\rangle$, leads exactly to the matrix equation (10.34). [Hint: Introduce a Lagrangian-multiplier term $-E(y_1^2 + y_2^2)$ to take account of the normalization condition on Ψ.] Thus diagonalization of the Hamiltonian matrix provides an eigenvalue that is a variational minimum energy obtainable by any linear combination of the given basis functions (and also an eigenvalue that is a variational maximum). [It may be recalled that in Sec. 7-1 the radial parts of the basis functions were themselves variationally adjusted to minimize the mean energy of all levels.] Generalize the above considerations to an N-level system.

10-7. EIGENVECTOR PURITIES

Because neither the Coulomb nor the spin-orbit interactions are ever exactly zero, one never sees 100 percent pure jj- nor LS-coupling conditions. Strictly speaking, one always has to deal with an intermediate-coupling problem, and no component of an eigenvector of a J-matrix is ever exactly zero in a pure-coupling representation. Thus (except in the simple case of a 1-by-1 matrix) an eigenstate of an atom is always a mixture of several pure-coupling basis functions. In principle, the nature of this eigenstate can be properly specified only by giving the complete basis-function expansion; i.e., the numerical values, including signs, of all eigenvector components (together, of course, with the detailed nature and phase of the corresponding basis functions). In practice, it is convenient to be able to label (or "designate") each eigenstate (and its corresponding energy level) by means of some simple symbol that gives some indication of the nature of the state. If, in some representation, one component of an eigenvector is much greater in magnitude than all other components, then the eigenstate may be designated by the name of the dominant contributing basis function. However, it is essential to always keep in mind the distinction between the use of such a name (e.g., "$^3P_1$") for the designation of an energy level, and its use as a label for a pure-coupling basis function: the latter is a rigorously correct description of a mathematical abstraction, whereas the former is a more or less rough description of the approximate angular-momentum characteristics of a real quantum state of the atom.

An indication of the degree of approximation involved in the use of a pure-coupling level designation is provided by the *purity* of the quantum state—the square of the corresponding eigenvector component (usually expressed in percent).[7] Similarly, a general indication of the appropriateness of a given coupling scheme for use in designating the energy levels of some configuration of an atom is afforded by the average purity for all levels of that configuration (excluding those belonging to 1-by-1 matrices, which are 100 percent pure in the approximation being used). Some care must be taken, however, in the use and interpretation of such numbers:

[7]G. H. Shortley, Phys. Rev. **44**, 666 (1933).

(a) Purities have to be rather high before pure-coupling designations have any great physical significance. In the case of a 2-by-2 matrix, for example, 50% purity of the eigenstates indicates complete mixing [Eq. (10.40)] and thus zero significance of such designations. As another example, the J = 1 levels of a pp′ configuration could be essentially pure JK-coupled states, and yet would show average purities of 72% and 75% in the jj and LK representations, respectively. (The eigenvectors are given respectively by the rows of the jK-jj, and the columns of the LK-jK, transformation matrices in Table 9-2). Similarly, the jK-jj transformation matrices[8] of pf show that the states of this configuration can have an average jj purity of 95%, and yet really be (essentially) purely jK coupled.

(b) When an inappropriate representation is employed, blind use of purities (or rather, of dominant eigenvector components) as a basis for deciding on energy-level designations can lead to incongruous results. For example, there is a strong tendency in the literature to use LS notation whenever possible, because it is the most compact, the best known, and (for low-lying configurations) the most nearly universally appropriate. However, if the J = 1 states of a pp′ configuration are actually jj-coupled, the columns of the LS-jj matrix in Table 9-2 would indicate that both the $(3/2\ 1/2)_1$ and $(1/2\ 3/2)_1$ levels should be designated $^3P_1$ and that no level would be given a $^3S_1$ designation. Conversely, if these states were purely LS-coupled but an attempt was made to use jj designations, both the $^1P_1$ and $^3S_1$ levels would have to be called $(3/2\ 3/2)_1$, and there would be no way to decide whether to designate $^3P_1$ as $(3/2\ 1/2)_1$ or as $(1/2\ 3/2)_1$. In either case, calculation of an average purity by uncritically using the dominant component of each eigenvector would give a misleadingly large result. In intermediate-coupling cases and large matrices, ambiguities of the above type are frequent no matter what representation is used.

(c) Attempts to decide on a "best" representation by calculating eigenvectors and average purities in each are sometimes ambiguous. For example, average purities for levels of given J are frequently greater in the LS representation for small J and greater in the jj representation for large J.

Some of the above points are illustrated in Table 10-2.

One further important aspect of eigenvector compositions and purities may be illustrated by means of the 2-by-2 problem (10.34). If the off-diagonal element C is zero, the eigenvalues are equal to A and B and the corresponding eigenvectors are

$$\begin{pmatrix} 1 \\ 0 \end{pmatrix} \quad \text{and} \quad \begin{pmatrix} 0 \\ 1 \end{pmatrix} .$$

If the two basis states for the matrix are b_1 and b_2, then the eigenstate with eigenvalue A is a pure $b_1$state and the eigenstate with eigenvalue B is a pure b_2 state. If the matrix elements A and B are continuous (and monotonic) functions of some parameter χ which varies over the range 0 to 1 (say), and if $A > B$ for $\chi = 0$ and $A < B$ for $\chi = 1$, then the energy levels cross as a function of χ. However, the two eigenstates A and B are always pure b_1 and b_2, respectively (except perhaps at the single point where A = B), as indicated in the top half of Fig. 10-2.

[8]R. D. Cowan and K. L. Andrew, J. Opt. Soc. Am. 55, 502 (1965).

TABLE 10-2. THEORETICAL EIGENVALUES (in kK) AND EIGENVECTORS FOR THE J = 1/2 AND 7/2 LEVELS OF Mo XXIV $3p^5 3d4s$ IN THE $\mathfrak{LS}$, $J\mathfrak{J}$, AND $J_1(L_{23}S_{23}J_{23})J$ REPRESENTATIONS. Purities (squares of eigenvector components) are tabulated, negative components being indicated by asterisks.

	J = 1/2					J = 7/2				All J
Energy:	4425.3	4458.6	4607.0	4862.6		4481.1	4511.9	4556.9	4708.7	
$(^3P^\circ)^4P^\circ$	0.900	0.065	0.032	0.003	$(^3F^\circ)^4F^\circ$	0.655	0.126	*0.001	0.219	
$(^3P^\circ)^2P^\circ$	*0.091	0.821	0.080	0.008	$(^3F^\circ)^2F^\circ$	*0.048	0.847	0.006	*0.099	
$(^3D^\circ)^4D^\circ$	*0.009	*0.114	0.818	0.060	$(^3D^\circ)^4D^\circ$	*0.081	0.001	0.663	0.255	
$(^1P^\circ)^2P^\circ$	*0.000	*0.000	*0.070	0.930	$(^1F^\circ)^2F^\circ$	0.217	*0.026	0.331	*0.427	
Ave. purity:	0.867					0.648				0.575
$\left(\frac{3}{2}\,\frac{3}{2}\right)_0 \frac{1}{2}$	*0.900	0.099	0.000	0.000	$\left(\frac{3}{2}\,\frac{3}{2}\right)_3 \frac{1}{2}$	0.921	*0.050	0.012	*0.018	
$\left(\frac{3}{2}\,\frac{3}{2}\right)_1 \frac{1}{2}$	0.078	0.737	*0.157	0.028	$\left(\frac{3}{2}\,\frac{5}{2}\right)_4 \frac{1}{2}$	0.046	0.950	0.003	*0.002	
$\left(\frac{3}{2}\,\frac{5}{2}\right)_1 \frac{1}{2}$	*0.018	*0.150	*0.454	0.377	$\left(\frac{3}{2}\,\frac{5}{2}\right)_3 \frac{1}{2}$	*0.014	*0.001	0.985	0.000	
$\left(\frac{1}{2}\,\frac{3}{2}\right)_1 \frac{1}{2}$	0.003	0.014	0.388	0.595	$\left(\frac{1}{2}\,\frac{5}{2}\right)_3 \frac{1}{2}$	0.019	0.000	0.000	0.981	
Ave. purity:	0.671					0.959				0.780
$\frac{3}{2}(^3D_1)$	0.643	0.236	*0.105	0.017	$\frac{3}{2}(^3D_2)$	0.799	0.184	*0.001	*0.016	
$\frac{3}{2}(^1D_2)$	0.068	*0.624	*0.142	0.165	$\frac{3}{2}(^1D_2)$	*0.178	0.769	*0.049	0.003	
$\frac{3}{2}(^3D_2)$	*0.286	0.126	*0.365	0.223	$\frac{3}{2}(^3D_3)$	*0.004	0.047	0.949	*0.000	
$\frac{1}{2}(^3D_1)$	0.003	0.014	0.388	0.595	$\frac{1}{2}(^3D_3)$	0.019	0.000	0.000	0.981	
Ave. purity:	0.557					0.875				0.613

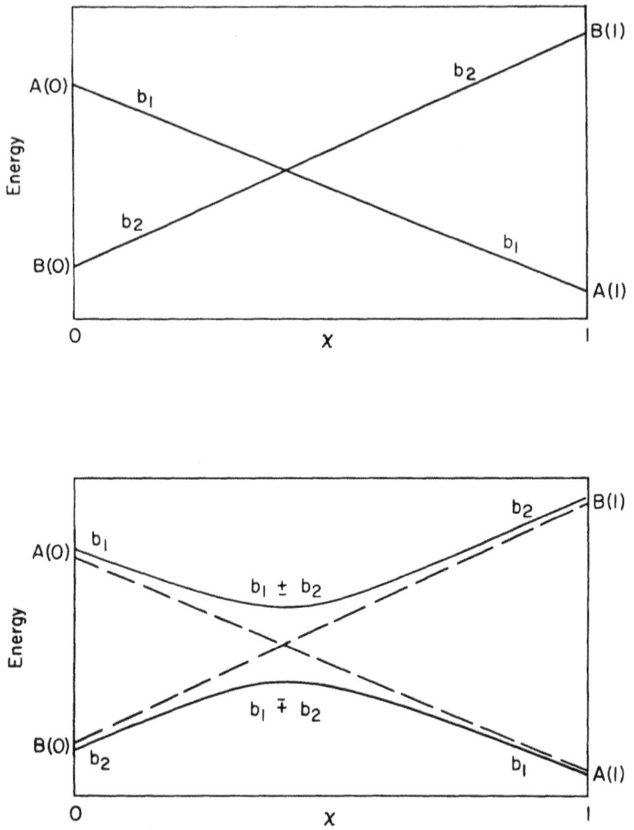

Fig. 10-2. Top: Energies of two non-interacting levels A and B (one being pure basis state b_1, and one pure b_2) as functions of a parameter χ. Bottom: Energies of the two levels when they interact with each other; the levels do not cross, but their basis-state natures for extreme values of χ are the same as though they did cross.

If the off-diagonal element C is non-zero (but is, say, independent of χ), then it may be seen from (10.35) that the eigenvalues no longer cross as χ is varied (Fig. 10-2, bottom). At the point of closest approach of the eigenvalues (where A = B), the eigenvectors are 50-50 mixtures of the basis functions, from (10.40). The eigenvector components along either eigenvalue curve vary smoothly and monotonically as functions of χ, and at either extreme the compositions are nearly the same as for C = O (if $|C|$ is small with respect to $|A - B|$ at these points). Thus, though the eigenvalue curves do not cross, the eigenvectors at the extremes are essentially the same as though the curves *did* cross. This point is very important for the proper quantum-state labeling of experimental energy levels, but is not always fully appreciated.

The above situation not infrequently occurs in considering the variation of energy levels along an isoelectronic sequence (χ then being the charge state of the various ions). It is usually considered desirable to use the same designation in all ions for (say) the lowest energy level in order to indicate which levels "correspond" to each other throughout the sequence; it is clear, however, that such designations will result in an incorrect indication of the nature of the quantum state at one end of the sequence. On the other hand, if correct designations are used at both ends of the sequence, then an apparent discontinuity in designation will appear at some intermediate ionization stage. Clearly there is no simple solution to this difficulty, except to use some sort of innocuous serial number to indicate continuity of energy, and to supplement this with the appropriate angular-momentum symbol to indicate the nature of the quantum state—including an indication of all important contributing basis states when no single one is clearly dominant.

10-8. EXAMPLES: ps AND p^5s

As an example of the intermediate-coupling transition between the LS and jj limits, we consider the simple configuration ps. The level structure is determined by the Coulomb parameter $G^1(ps)$ and the spin-orbit parameter ζ_p. In order to obtain a universal diagram, energies will be plotted in units of $2G^1/3 + 3\zeta/2$—so chosen because the two terms represent the energy range of the levels in the LS and jj limits, respectively. An appropriate variable to represent the type of coupling is

$$\chi = \frac{3\zeta/2}{2G^1/3 + 3\zeta/2} , \tag{10.41}$$

which varies from zero in the LS limit to unity in the jj limit.

The variation with χ of the energy-level structure is shown in the upper portion of Fig. 10-3.[9] Also shown are experimental levels for ps configurations of several atoms and ions; the required empirical values of G^1 and ζ have been obtained by least-squares fitting of the experimental data, as described in Chapter 16. The figure illustrates the degree to which the single-configuration approximation can give agreement with experiment, the discrepancies being attributable to configuration-interaction perturbations. The experimental data illustrate well the shift from LS- toward jj-coupling conditions with increasing principal quantum number of the s electron (Si I 3p4s to 3p7s; see also Fig. 4-9), with increasing ionization stage in an isoelectronic sequence (Si I, S III, Ar V 3p4s), and with increasing Z for a given ionization stage (Si I 3p4s, Ge I 4p5s, Sn I 5p6s, Pb I 6p7s); reasons for such behavior were discussed at length in Secs. 8-5 and 8-7.

The $J = 0$ and $J = 2$ quantum states are of course always 100 percent pure in either the LS or the jj representation. The two $J = 1$ states, on the other hand, are appreciably mixed, as shown in the lower part of the figure. The mixing is small enough that "$^3P^o_1$" and "$^1P^o_1$" are reasonably valid designations for the lower and upper levels, respectively, even in the jj limit; however jj-coupling designations are obviously preferable for $\chi > 0.5$.

The situation is significantly different in the configuration p^5s, for which the energy matrix differs from that of ps only in that the sign of the spin-orbit terms is reversed (see Sec. 12-9). Just as for ps, a pure $(1/2\ 1/2)^o_1$ state of p^5s is 67% $^3P^o_1$ and 33% $^1P^o_1$.

[9] The figure is similar to that given by B. Edlén, "Atomic Spectra," in S. Flügge, ed., *Handbuch der Physik* (Springer-Verlag, Berlin, 1964), Vol. XXVII, p. 113.

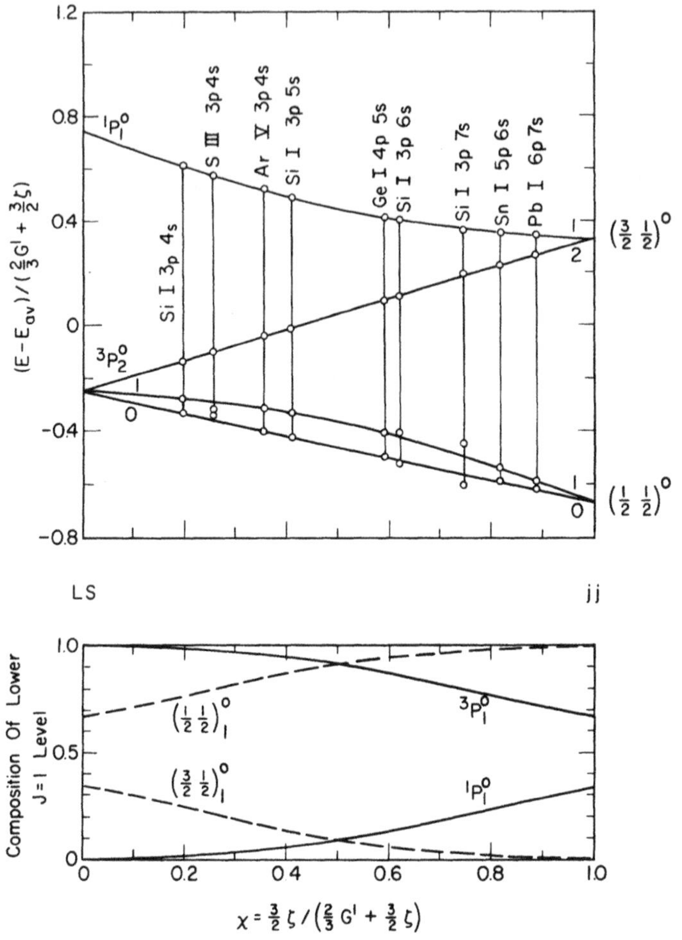

Fig. 10-3. The transition from pure LS- to pure jj-coupling conditions for the configuration ps or sp. The upper portion shows energy levels, and the lower portion indicates the composition of the intermediate-coupling eigenvectors for the J=1 levels.

However, in p^5s the lower J = 1 state in the jj limit is a pure $(3/2\ 1/2)_1^o$ state rather than pure $(1/2\ 1/2)_1^o$—see Fig. 10-4. Consequently, the lower J = 1 state, which is pure $^3P_1^o$ in the LS limit, is more $^1P_1^o$ than $^3P_1^o$ near the jj limit; universal use of the designation $^3P_1^o$ for the lower J = 1 level would therefore give a very misleading indication of the true nature of the quantum state in cases for which χ was large. The situation is the same as that discussed in connection with Fig. 10-2.

Problems

10-8(1). In Fig. 10-4, why does Na II $2p^53s$ show more nearly pure LS coupling than

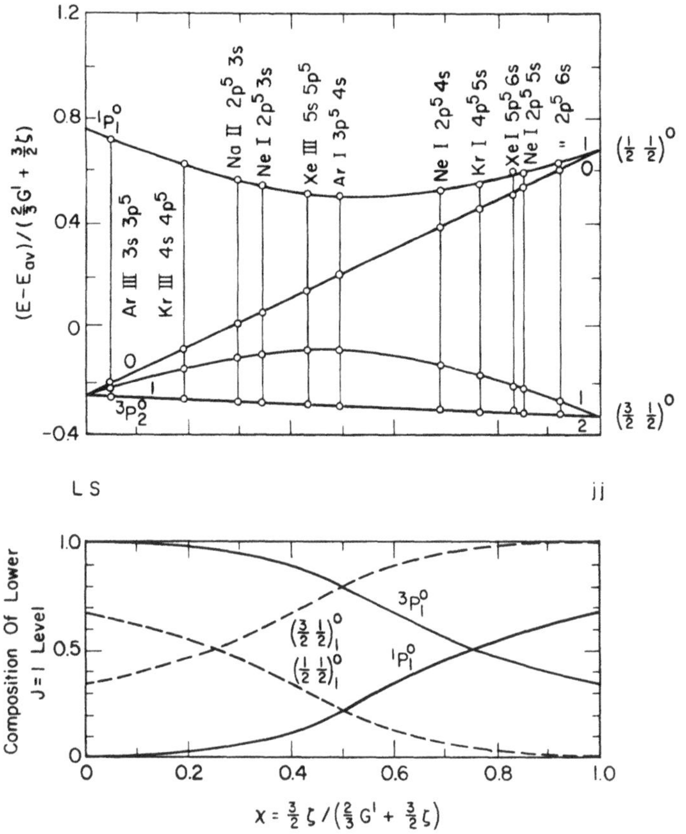

Fig. 10-4. The transition from pure LS- to pure jj-coupling conditions for the configuration p^{5}s.

does Ne I 2p^{5}3s, contrary to the isoelectronic trend for 3p4s in Fig. 10-3? Would you expect this trend toward LS to continue for Mg III, Al IV, etc?

10-8(2). Hartree-Fock calculations, including relativistic and correlation corrections, give the following results (in cm^{-1}; E_{av} relative to the center of gravity of Si II 3p):

		E_{av}	$F^2(3p,3p)$	ζ_{3p}	$G^1(3p,ns)$
Si II	3p	0	---	168	---
Si I	3p7s	−3786	---	168	128
	3p6s	−6056	---	168	248
	3p5s	−10895	---	167	592
	3p4s	−24997	---	164	2266
	3p^2	−62338	35808	129	---

(a) Discuss qualitatively the reasons for the variation in values of ζ_{3p} and $G^1(3p,ns)$ from one configuration to another.

(b) Calculate the energy coefficient-matrices (d_l) and (g_k) for the configurations p and ps, using an LS-coupling representation for ps.

(c) Using HF values from the above table, calculate energy levels (rounded to the nearest cm^{-1}) of the configuration Si II 3p and Si I 3p5s. Also, calculate the eigenvectors in both the LS- and jj-coupled representations. As an intermediate check, the J = 1 matrix for Si I 3p5s is

$$\begin{array}{c c} & \begin{array}{cc} {}^3P^o_1 & {}^1P^o_1 \end{array} \\ \begin{array}{c} {}^3P^o_1 \\ {}^1P^o_1 \end{array} & \begin{pmatrix} -11077 & 118 \\ 118 & -10599 \end{pmatrix} \end{array}.$$

(d) Compare the calculated energies with experimental values from AEL.

10-9. PAIR COUPLING

When both electrons have non-zero orbital angular momentum, LK and jK as well as LS and jj couplings are of physical interest. Table 10-3 summarizes the relative parameter values for which the various pure couplings are approximated;[10] the relationships for the pair couplings are easily inferred from the physical discussion of Sec. 4-18.

TABLE 10-3. PARAMETER-VALUE RELATIONSHIPS FOR VARIOUS TYPES OF COUPLING IN TWO-ELECTRON CONFIGURATIONS

LS coupling	$F^k, G^k \gg \zeta_1, \zeta_2$
Pair coupling	$F^k, \zeta_1 \gg G^k, \zeta_2$
LK coupling	$F^k \gg \zeta_1 \gg G^k, \zeta_2$
jK coupling	$\zeta_1 \gg F^k \gg G^k, \zeta_2$
jj coupling	$\zeta_1, \zeta_2 \gg F^k, G^k$

[10]Strictly speaking, the inequalities should be written in terms of the contributions of the various parameters to the energy spread of the levels of the configuration, rather than in terms of the parameter values themselves. [cf. A. M. Gutman and I. B. Levinson, Astron. Zh. 37, 86 (1960); English transl., Soviet Astron.-AJ 4, 83 (1960).] Because most coefficients f_k, g_k, and d are of the order unity, and cover a range of the order unity, the inequalities reduce approximately to those in the table. Note, however, from (10.14) that the energy range for ζ is $(2l + 1)\zeta/2$, and that the coefficient becomes appreciably greater than unity for large l. Note also (Tables 6-2 and 6-3) that the subscripted Coulomb parameters F_k and G_k differ by orders of magnitude from the F^k and G^k, so that inequalities written in terms of the former would be very misleading.

Because more than two interaction parameters are involved, the intermediate-coupling transitions[11] from one limiting form to another cannot readily be discussed in full detail. The general nature of the simpler transitions is, however, illustrated in Fig. 10-5 for appropriate special parameter values in a pf configuration.[8] In all cases, the relation

$$\frac{9}{25}F^2(pf) + \frac{3}{2}\zeta_p = 1$$

has been assumed, analogous to the plotting of energies in units of $2G^1/3 + 3\zeta/2$ in Figs. 10-3 and 10-4. $G^4(pf)$ has been taken to be zero throughout, as the exchange Coulomb effects can be adequately illustrated by G^2 only. For the LS-LK and jK-jj transitions, it has been assumed respectively that

$$G^2 + 3\zeta_p = 0.3$$

and

$$\frac{3}{25}F^2 + \zeta_f = 0.1 \ ,$$

where the various numerical constants are arbitrary numbers, chosen so as to give splittings of a convenient magnitude.

In the central portion of the figure, a universal diagram has been obtained by omitting the splitting into pairs. The plotted experimental data for pf configurations of P II and N II (including the pair splitting) and of Ge I (pair splitting too small to plot) illustrates the degree of agreement between theory and experiment typically observed. The author knows of no configuration that shows a closer approach to pure LK coupling than P II 3p4f; however, many cases exist with very nearly pure jK coupling. [It may be noted that completely pure LK and jK couplings are physically not even conceivable: the interactions required to remove the K degeneracy (and thereby define the quantum number K) simultaneously destroy the absolute purity of the quantum number L or j, respectively; this is in contrast to the LS case, where L and S can in principle be simultaneously well-defined, and to the jj case, where both j's can simultaneously be well-defined.]

The left-hand portion of Fig. 10-5 shows the transition from pure LS to nearly pure LK coupling. (The limiting value $\chi_1 = 1$ corresponds to the case $\chi_2 = 0.15$ in the center portion of the figure.) The transition from a singlet-triplet to a paired level structure is obvious. Because the figure is drawn for special values of the energy parameters, experimental comparisons cannot be shown; however, configurations can easily be found to cover more or less completely the range between pure LS and pure pair level structures.

Similarly, the right-hand portion of the figure illustrates the transition from nearly pure jK to pure jj level structures. (The limit $\chi_3 = 0$ corresponds to the case $\chi_2 = 0.7$.) The two

[11]The term "intermediate coupling" is commonly used to signify any coupling conditions other than those in which either the spin-orbit or the Coulomb interactions are negligible (pure LS and pure jj coupling, respectively). We here use the term to mean coupling conditions other than those corresponding to any possible mathematical pure-coupling scheme (LS, LK, jK, jj, etc.).

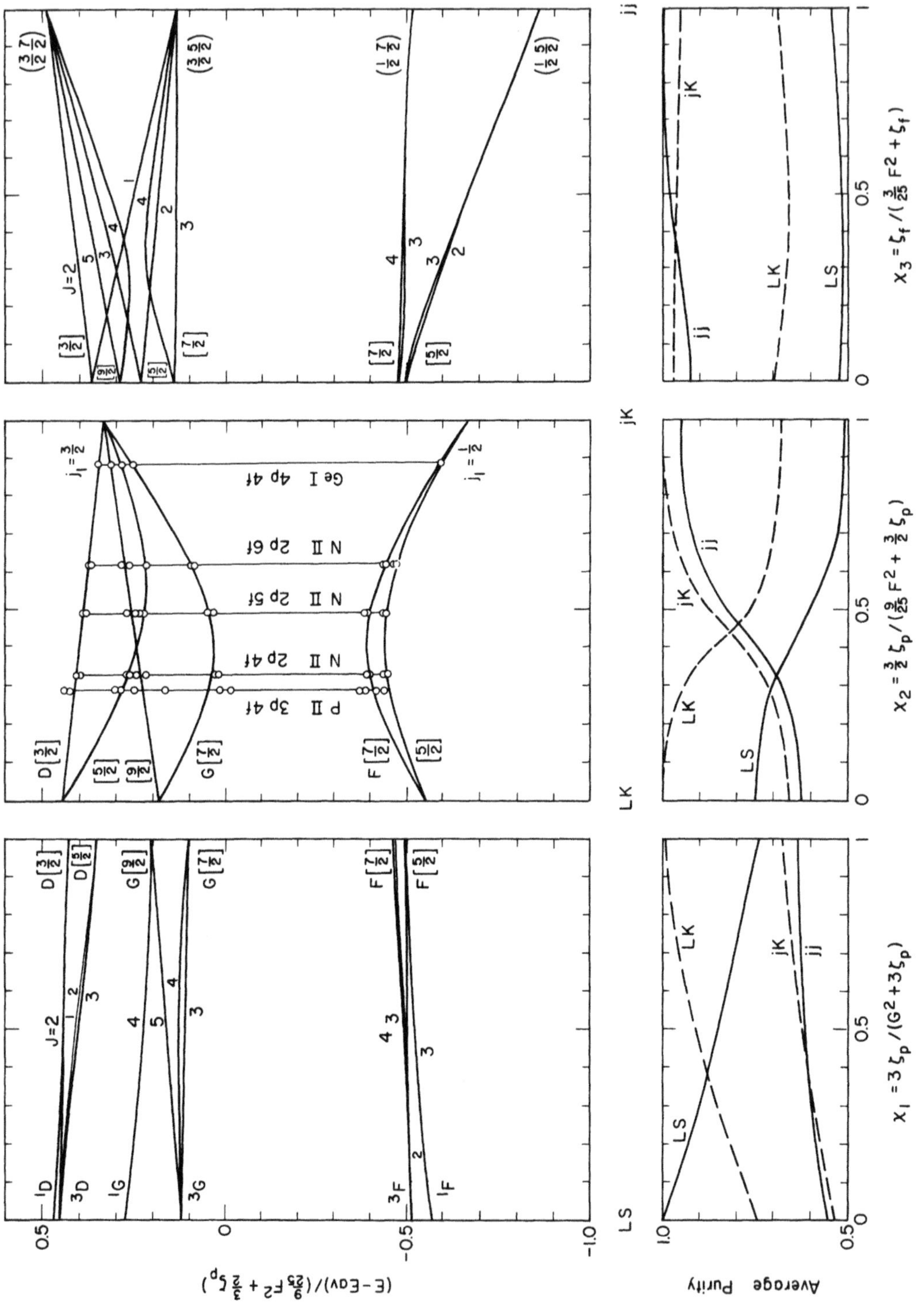
(E−Eav)/(9/25 F2 + 3/2 ζp)
Average Purity
LS
LK
jK
jj
X1 = 3ζp/(G2+3ζp)
X2 = 3/2 ζp/(9/25 F2 + 3/2 ζp)
X3 = ζf/(3/25 F2 + ζf)
P II 3p 4f
N II 2p 4f
N II 2p 5f
N II 2p 6f
Ge I 4p 4f

Fig. 10-5. Upper portions: The changes in energy-level structure in going from pure-LS to near-LK, pure-LK to pure-jK, and near-jK to pure-jj coupling conditions for a pf configuration. In the left-most and right-most sections, the curves are plotted for the special parameter values given in the text; in the center section, the pair splitting is not shown for the curves, but is included for the experimental levels. Bottom portions: The corresponding average eigenvector purities (for all except the J=0 and 5 levels), in each of the four pure-coupling representations.

highest J = 4 levels provide a particularly nice example of the behavior discussed in connection with the lower part of Fig. 10-2: the upper curve shows a continuous energy variation between the pure 3/2 $[9/2]_4$ level at $\chi_3 = 0$ and the pure $(3/2\ 7/2)_4$ level at $\chi_3 = 1$, but nevertheless is 95% pure $(3/2\ 5/2)_4$ in the first limit and 95% pure 3/2 $[7/2]_4$ in the second limit. (See the jK-jj transformation matrix in ref. 8.)

The order of J values in pair coupling. Empirically, it is usually observed that the higher-energy level of every pair either is that having even J (configurations with $l_1 + l_2$ even) or is that of odd J ($l_1 + l_2$ odd). An example may be seen for $\chi_1 \cong 1$ in Fig. 10-5. These relationships can be explained on the assumption that the principal cause of the pair splitting is the Coulomb exchange interaction (rather than ζ_2), and that this interaction is dominated by the term $g_k G^k$ of minimum k.

To prove this, let us first consider the case of good LK coupling. Then in the LK representation the off-diagonal energy matrix elements must be negligible, and the energy levels are given essentially by the diagonal elements:

$$E_{LKJ} = \langle LKJ|\frac{2}{r_{ij}}|LKJ\rangle = \sum_{SS'}\langle LKJ|LSJ\rangle\langle LSJ|\frac{2}{r_{ij}}|LS'J\rangle\langle LS'J|LKJ\rangle . \qquad (10.42)$$

From (9.23), (10.21), and (10.23), and since Coulomb matrix elements are diagonal in S, the coefficient of G^k is

$$[l_1, l_2]\begin{pmatrix} l_1 & k & l_2 \\ 0 & 0 & 0 \end{pmatrix}^2 \begin{Bmatrix} l_1 & l_2 & L \\ l_1 & l_2 & k \end{Bmatrix} \sum_S (-1)^S [K,S] \begin{Bmatrix} L & s_1 & K \\ s_2 & J & S \end{Bmatrix}^2 . \qquad (10.43)$$

The summation is over the two values S = 0 and 1 and so may be written

$$2[K]\begin{Bmatrix} L & s_1 & K \\ s_2 & J & 0 \end{Bmatrix}^2 - \sum_S [K,S]\begin{Bmatrix} L & s_1 & K \\ s_2 & J & S \end{Bmatrix}^2 = \frac{[K]}{[L]}\delta_{JL} - 1 ,$$

from (5.29) and (5.31). Thus the larger value of the summation in (10.43) is obtained for the value J = L. The sign of the remaining factor of (10.43) is, for minimum k, given by (10.33). Thus we see that for given L and K, the higher energy occurs for the level J = L if $L + l_1 + l_2$ is even, or for the level $J = L \pm 1$ if $L + l_1 + l_2$ is odd; i.e., in all cases, the higher energy level is that for which

$$J + l_1 + l_2 \quad \text{is even} . \qquad (10.44)$$

This result must hold in any case of good pair coupling, whether good LK coupling or good jK coupling or something in between. Physically, this is because the J dependence of energy for given K depends only on the coupling of s_2 to K, and not on the details of how the angular momentum K was formed. Mathematically, an example may be seen in that the LK-jK transformation matrix (9.24) is independent of J.

Problem

10-9(1). Show that if good pair coupling exists and the pair splitting is produced primarily by the spin-orbit interaction of the second electron (rather than by the Coulomb exchange interaction), then the higher energy level of each pair is that having the larger value of J (provided $l_2 > l_1$). An example may be seen in Fig. 10-5 for $\chi_3 \cong 0$, and this level ordering has been observed[12] in Cs II $5p^5(^2P^{\circ}_{3/2})nh$ (n=6,7,8). [Hint: Show that the coefficient of ζ_2 for diagonal matrix elements in the jK representation is equal to

$$\frac{1}{2}\sum_{j_2}[K, j_2]\begin{Bmatrix} j_1 & l_2 & K \\ s_2 & J & j_2 \end{Bmatrix}^2 \left[(-1)^{j_2-l_2-1/2}\left(j_2 + \frac{1}{2}\right) - 1 \right]$$

$$= \frac{1}{2}(2l_2 + 1)\left[K, l_2+\frac{1}{2}\right]\begin{Bmatrix} j_1 & l_2 & K \\ s_2 & J & l_2+\frac{1}{2} \end{Bmatrix}^2 - \frac{1}{2}(l_2 + 1)\,,$$

and—using Eqns. (2.13) and (2.16) of Rotenberg, et al[13]—show that the difference between the values of this expression for $J = K + 1/2$ and $J = K - 1/2$ is positive.]

[12] C. J. Sansonetti and K. L. Andrew, private communication (1979).

[13] M. Rotenberg, R. Bivins, N. Metropolis, and J. K. Wooten, Jr., *The 3-j and 6-j Symbols* (The Technology Press, Massachusetts Institute of Technology, Cambridge, 1959).

11

THE ALGEBRA OF IRREDUCIBLE TENSOR OPERATORS

11-1. CALCULATION OF COMPLEX MATRIX ELEMENTS

For configurations containing only one or two electrons outside closed subshells, we have seen how we may calculate spin-orbit and Coulomb matrix elements by comparatively elementary methods. For more complex configurations, the calculation of matrix elements by similar means can become much more difficult. As an example, we consider the calculation of the direct Coulomb matrix elements for a configuration containing three non-equivalent electrons. From the general result (9.85), we need consider only matrix elements between non-antisymmetrized (simple-product) basis functions

$$|l_1(\mathrm{r}_1)\rangle\cdot|l_2(\mathrm{r}_2)\rangle\cdot|l_3(\mathrm{r}_3)\rangle \ .$$

However, there are actually three direct matrix elements to be calculated, corresponding to the operators $2/\mathrm{r}_{12}$, $2/\mathrm{r}_{23}$, and $2/\mathrm{r}_{13}$, and involving the radial integrals $\mathrm{F}^k(l_1l_2)$, $\mathrm{F}^k(l_2l_3)$, and $\mathrm{F}^k(l_1l_3)$, respectively. We shall use basis functions that have been coupled LS-wise from left to right, as in (4.37). From the general results (9.87) and (9.88), we may [as in the derivation of (10.16)] ignore the coupling $(\mathfrak{L}_3\mathfrak{S}_3)\mathfrak{J}_3\mathfrak{M}_3$ and for direct matrix elements ignore spins entirely, except for the delta-factors

$$\delta_{\mathfrak{S}_2\mathfrak{S}'_2}\delta_{\mathfrak{S}_3\mathfrak{S}'_3}\delta_{\mathfrak{J}_3\mathfrak{M}_3,\mathfrak{J}'_3\mathfrak{M}'_3} \ .$$

The first matrix element

$$\langle[(l_1l_2)\mathfrak{L}_2,l_3]\mathfrak{L}_3|2/\mathrm{r}_{12}|[(l_1l_2)\mathfrak{L}'_2,l_3]\mathfrak{L}'_3\rangle$$

presents no difficulties. From the general expression (9.87), we obtain

$$\delta_{\mathfrak{L}_2\mathfrak{L}'_2}\delta_{\mathfrak{L}_3\mathfrak{L}'_3}\langle(l_1l_2)\mathfrak{L}_2|2/\mathrm{r}_{12}|(l_1l_2)\mathfrak{L}'_2\rangle \ ,$$

and this remaining two-electron matrix element is given immediately by (10.21) and (10.22).

For the operator $2/r_{23}$, on the other hand, the matrix is still diagonal in $\mathfrak{L}_3$ but the simplification from a three-electron to a two-electron problem can be accomplished only after first making appropriate recouplings of the orbital angular momenta:

$$\langle[(l_1l_2)\mathfrak{L}_2,l_3]\mathfrak{L}_3\,|\,2/r_{23}\,|\,[(l_1l_2)\mathfrak{L}_2',l_3]\mathfrak{L}_3\rangle = \sum_{\mathfrak{L}_{23}} \langle[(l_1l_2)\mathfrak{L}_2,l_3]\mathfrak{L}_3\,|\,[l_1,(l_2l_3)\mathfrak{L}_{23}]\mathfrak{L}_3\rangle$$
$$\times\ \langle[l_1,(l_2l_3)\mathfrak{L}_{23}]\mathfrak{L}_3\,|\,2/r_{23}\,|\,[l_1,(l_2l_3)\mathfrak{L}_{23}]\mathfrak{L}_3\rangle\langle[l_1,(l_2l_3)\mathfrak{L}_{23}]\mathfrak{L}_3\,|\,[(l_1l_2)\mathfrak{L}_2',l_3]\mathfrak{L}_3\rangle\ ; \qquad (11.1)$$

the double summation over $\mathfrak{L}_{23}\mathfrak{L}_{23}'$ has been reduced to a single summation because the recoupled matrix is diagonal in $\mathfrak{L}_{23}$. Simplifying to a two-electron matrix element with the aid of (9.88), and using (9.17) and (10.22), we find for the coefficient of $F^k(l_2l_3)$

$$f_k'(23) = \delta_{\mathfrak{L}_3\mathfrak{L}_3'}[l_2,l_3]\begin{pmatrix} l_2 & k & l_2 \\ 0 & 0 & 0 \end{pmatrix}\begin{pmatrix} l_3 & k & l_3 \\ 0 & 0 & 0 \end{pmatrix}[\mathfrak{L}_2,\mathfrak{L}_2']^{1/2}\,X\ , \qquad (11.2)$$

where with the use of (5.33)

$$X = \sum_{\mathfrak{L}_{23}} (-1)^{\mathfrak{L}_{23}}[\mathfrak{L}_{23}]\begin{Bmatrix} l_1 & l_2 & \mathfrak{L}_2 \\ l_3 & \mathfrak{L}_3 & \mathfrak{L}_{23} \end{Bmatrix}\begin{Bmatrix} l_2 & l_2 & k \\ l_3 & l_3 & \mathfrak{L}_{23} \end{Bmatrix}\begin{Bmatrix} l_1 & l_2 & \mathfrak{L}_2' \\ l_3 & \mathfrak{L}_3 & \mathfrak{L}_{23} \end{Bmatrix}$$
$$= (-1)^{k+l_2+l_2+l_1+\mathfrak{L}_2'+\mathfrak{L}_2+l_3+\mathfrak{L}_3+l_3}\begin{Bmatrix} k & l_2 & l_2 \\ l_1 & \mathfrak{L}_2' & \mathfrak{L}_2 \end{Bmatrix}\begin{Bmatrix} \mathfrak{L}_2 & k & \mathfrak{L}_2' \\ l_3 & \mathfrak{L}_3 & l_3 \end{Bmatrix}$$
$$= (-1)^{l_1+k+\mathfrak{L}_2+\mathfrak{L}_2'+\mathfrak{L}_3}\begin{Bmatrix} \mathfrak{L}_2 & l_3 & \mathfrak{L}_3 \\ l_3 & \mathfrak{L}_2' & k \end{Bmatrix}\begin{Bmatrix} l_1 & l_2 & \mathfrak{L}_2 \\ k & \mathfrak{L}_2' & l_2 \end{Bmatrix}\ . \qquad (11.3)$$

Although the final result (11.2)-(11.3) is fairly lengthy, it involves no explicit summation and its numerical evaluation is perfectly straightforward.

A similar calculation for $2/r_{13}$, using the recoupling formula (9.19) instead of (9.17), leads to a result similar in form to (11.2)-(11.3).

Evaluation of exchange matrix elements can be carried out by recoupling procedures similar to those used for the direct elements. However, it is necessary to take the spins into account explicitly, and so with the operator $2/r_{23}$ one obtains a double summation over $\mathfrak{L}_{23}$ and $\mathfrak{S}_{23}$. It does not appear to be possible to evaluate these sums analytically, though the double summation can be reduced to a single summation.

11-2. RACAH ALGEBRA

The recoupling methods described in the preceding section can in principle be used for still more complex configurations. However, one obtains summations over dummy quantum numbers (analogous to $\mathfrak{L}_{23}$ and $\mathfrak{S}_{23}$) that are greater in both number and complexity. Simplification by analytical evaluation of some of the summations is correspondingly more cumbersome and difficult, though the graphical methods mentioned in Sec. 5-4 can be used as a guide.

One may well ask whether some alternative method of calculation is possible in which a final result is obtained directly in its simplest possible form, without the necessity of ever introducing superfluous dummy summation variables in the first place. It turns out that such a direct derivation is indeed possible, using the formalism provided by the algebra of irreducible tensor operators (popularly known as Racah-Wigner algebra or, in the atomic spectroscopy literature, simply as Racah algebra).[1-4] This algebra also formalizes calculations such as (10.17)-(10.20) in such a way that we need never again concern ourselves explicitly with multiple sums of products of 3-j symbols, but immediately obtain all results directly in terms of 6-j symbols. In short, Racah algebra provides notational and calculational simplifications of great elegance, which are practically indispensable for the evaluation of matrix elements for complex configurations. In the remainder of this chapter we develop the necessary formalism, which we then apply in subsequent chapters not only to energy-level calculations but also to the calculation of matrix elements for radiative transitions.

It should be emphasized that Racah algebra is concerned with the evaluation of the *angular* portion of a matrix element; it is particularly concerned with the complications that arise from the couplings of the various angular-momentum quantum numbers. The evaluation of the radial integral is always straightforward and unaffected by angular-momentum coupling; it will therefore be ignored in the development of the formalism.

In addition, Racah's methods are concerned strictly with basis functions that have been antisymmetrized (with respect to electron-coordinate interchange) only within a subshell of equivalent electrons, and this antisymmetrization is assumed to have been accomplished via the method of coefficients of fractional parentage discussed in Secs. 9-5 to 9-7. Complications arising from the coordinate permutations used in Sec. 9-8 to produce antisymmetrization between two different subshells are assumed to have already been handled by the method used in Sec. 9-9. Thus the Racah formalism is to be applied only for matrix elements between functions ψ_b (*not* Ψ_b) that are simple products (9.72) of single-subshell functions

$$|l_i^{w_i}\alpha_i L_i S_i\rangle$$

of the type (9.50), except that the angular momenta $L_i S_i$ have been coupled together according to some specified coupling scheme.

11-3. IRREDUCIBLE TENSOR OPERATORS

It can be shown that any classical tensor can be decomposed into irreducible parts, each of which transforms under a rotation of the coordinate axes in the same way as does one of the spherical harmonics $Y_{lm}(\theta,\phi)$. In quantum mechanics, the analogues of these parts are the components of an irreducible tensor operator. Such an operator, of (integral) rank k, may be defined as an operator $T^{(k)}$ whose 2k + 1 components $T_q^{(k)}$ $(q = -k, -k + 1, \cdots$

[1]G. Racah, II, Phys. Rev. 62, 438 (1942).

[2]G. Racah, III, Phys. Rev. 63, 367 (1943).

[3]A. R. Edmonds, *Angular Momentum in Quantum Mechanics* (Princeton University Press, Princeton, N.J., 1960), 2nd ed.

[4]U. Fano and G. Racah, *Irreducible Tensorial Sets* (Academic Press, New York, 1959).

k − 1, k) transform in the same way as do the spherical harmonic operators $Y_{kq}(\theta,\phi)$—or, equivalently, that they satisfy the same commutation rule with respect to J as do the Y_{kq}. Identifying the eigenfunctions ψ_{jm} in (2.19) and in the expression $J_z\psi_{jm} = m\hbar\psi_{jm}$ with spherical harmonics, we see that the defining commutation relations are

$$[J_{\pm},T_q^{(k)}] = [(k \mp q)(k \pm q + 1)]^{1/2}\,T_{q\pm 1}^{(k)} , \tag{11.4}$$

$$[J_z,T_q^{(k)}] = qT_q^{(k)} , \tag{11.5}$$

where the abbreviation

$$J_{\pm} \equiv J_x \pm iJ_y \tag{11.6}$$

has been used for the step-up and step-down operators, and angular momenta are here measured in units of ħ. In line with (2.38)-(2.40), we shall say that an irreducible tensor operator is Hermitian if

$$T_q^{(k)\dagger} = (-1)^q T_{-q}^{(k)} . \tag{11.7}$$

For an operator that does not act on spins but involves only the spatial coordinates (r,θ,ϕ), the above definition reduces to the statement that $T_q^{(k)}$ has the same angular form as Y_{kq}; i.e., that

$$T_q^{(k)} = A_k Y_{kq} , \tag{11.8}$$

where A_k may be a function of $r = |\mathbf{r}|$ but is independent of θ, ϕ and q. An example that we have already encountered in (5.18) is the renormalized spherical harmonic

$$C_q^{(k)}(\theta,\phi) \equiv \left(\frac{4\pi}{2k+1}\right)^{1/2} Y_{kq}(\theta,\phi) . \tag{11.9}$$

Another important example is provided by a vector operator **V**, with ordinary vector components

$$\begin{aligned} V_x &= V \sin\theta \cos\phi , \\ V_y &= V \sin\theta \sin\phi , \\ V_z &= V \cos\theta . \end{aligned} \tag{11.10}$$

Noting that

$$V_x \pm iV_y = V \sin\theta\, e^{\pm i\phi} , \tag{11.11}$$

we see that a vector operator can be written as an irreducible tensor operator of rank one whose components are

$$
\begin{aligned}
T_1^{(1)} &= -\frac{1}{2^{1/2}}(V_x + iV_y) = -\frac{V}{2^{1/2}}\sin\theta\, e^{i\phi} \,, \\
T_0^{(1)} &= V_z = V\cos\theta \,, \qquad (11.12) \\
T_{-1}^{(1)} &= \frac{1}{2^{1/2}}(V_x - iV_y) = \frac{V}{2^{1/2}}\sin\theta\, e^{-i\phi} \,,
\end{aligned}
$$

because these three functions are proportional to the three spherical harmonics Y_{11}, Y_{10}, Y_{1-1} (Table 2-1). The particular normalization ($1/\sqrt{2}$, 1, $1/\sqrt{2}$) has been chosen so as to make $T_0^{(1)}$ identical with V_z, and will be discussed further in Sec. 11-8. A vector of particular importance is the position vector **r**, which may be written in the tensor operator form

$$
r^{(1)} = r\,C^{(1)} \,. \qquad (11.13)
$$

11-4. THE WIGNER-ECKART THEOREM

Suppose that $T^{(k)}$ is an irreducible tensor operator acting in a space that is spanned by a set of basis functions $|\alpha jm\rangle$, each of which is an eigenfunction of the operators J^2 and J_z, with α representing any quantities that (together with the quantum numbers j and m) are required to completely specify the basis function. Evaluating matrix elements of both sides of (11.4), we have

$$
\langle\alpha jm|J_\mp T_q^{(k)} - T_q^{(k)} J_\mp|\alpha' j' m'\rangle = [(k \pm q)(k \mp q + 1)]^{1/2}\langle\alpha jm|T_{q\mp 1}^{(k)}|\alpha' j' m'\rangle \,.
$$

In the first term on the left-hand side we operate to the left with the Hermitian conjugate $J_\mp^\dagger = J_\pm$ [see (11.6)], and in both terms we use (2.19) to obtain

$$
\begin{aligned}
&[(j \mp m)(j \pm m + 1)]^{1/2}\langle\alpha j(m\pm 1)|T_q^{(k)}|\alpha' j' m'\rangle \\
&\quad = [(j' \pm m')(j' \mp m' + 1)]^{1/2}\langle\alpha jm|T_q^{(k)}|\alpha' j'(m'\mp 1)\rangle \\
&\qquad + [(k \pm q)(k \mp q + 1)]^{1/2}\langle\alpha jm|T_{q\mp 1}^{(k)}|\alpha' j' m'\rangle \,. \qquad (11.14)
\end{aligned}
$$

Comparing this with (9.6), we see that matrix elements of an irreducible tensor operator show the same dependence on (m,q,m′) as do the CG coefficients for the angular-momentum coupling (j′k)j [or for (kj′)j]; this result is known as the Wigner-Eckart theorem.[5] Using the standard phase and normalization conventions introduced by Racah,[1] we may thus write from (9.5)

[5]E. Wigner, Z. Physik 43, 624 (1927); C. Eckart, Rev. Mod. Phys. 2, 305 (1930).

$$\begin{aligned}\langle\alpha jm|T^{(k)}_q|\alpha'j'm'\rangle &= [j]^{-1/2}C(j'km'q;jm)\langle\alpha j\|T^{(k)}\|\alpha'j'\rangle \\ &= (-1)^{-j'+k-m}\begin{pmatrix} j' & k & j \\ m' & q & -m\end{pmatrix}\langle\alpha j\|T^{(k)}\|\alpha'j'\rangle \\ &= (-1)^{j-m}\begin{pmatrix} j & k & j' \\ -m & q & m'\end{pmatrix}\langle\alpha j\|T^{(k)}\|\alpha'j'\rangle\ , \end{aligned} \tag{11.15}$$

where in the last step the symmetry relation (5.5) has been used, together with the fact that k is integral. It may be noted from (5.1)-(5.2) that the matrix element is zero unless

$$m = q + m' \tag{11.16}$$

and unless (jkj′) satisfy the triangle relations; these two properties together constitute the matrix-element *selection rules.*

The 3-j symbol in (11.15) involves only purely geometrical properties of the tensor operator—its rank k, and the component q (which depends on the orientation of the coordinate system). The physical nature of the operator is contained entirely in the reduced matrix element $\langle\alpha j\|T^{(k)}\|\alpha'j'\rangle$; this quantity includes any dependence on the quantum numbers $\alpha j\alpha'j'$ other than that involved in the phase factor and 3-j symbol, including the result of integration over r, any constant multiplicative factors, etc.[6] The reduced matrix element is found by evaluating $\langle\alpha jm|T^{(k)}_q|\alpha'j'm'\rangle$ for special values of mqm′ and comparing with (11.15).

For example, in the case of $C^{(0)}_0 \equiv 1$,

$$\langle\alpha jm|C^{(0)}_0|\alpha'j'm'\rangle = \langle\alpha jm|\alpha'j'm'\rangle = \delta_{\alpha jm,\alpha'j'm'}\ ;$$

from (5.11),

$$(-1)^{j-m}\begin{pmatrix} j & 0 & j \\ -m & 0 & m'\end{pmatrix} = \delta_{mm'}[j]^{-1/2}\ , \tag{11.17}$$

and thus

$$\langle\alpha j\|C^{(0)}\|\alpha'j'\rangle = \delta_{\alpha j,\alpha'j'}[j]^{1/2}\ . \tag{11.18}$$

Similarly, for the q = 0 component of the vector operator **J** we have from (11.12) that

$$\langle\alpha jm|J^{(1)}_0|\alpha'j'm'\rangle = \langle\alpha jm|J_z|\alpha'j'm'\rangle = m\,\delta_{\alpha jm,\alpha'j'm'}\ ;$$

from (5.1) we find easily that

[6]Racah's reduced matrix elements (for k = 1) differ from the reduced matrix elements $\langle\alpha j\vdots T\vdots\alpha'j'\rangle$ used by Condon and Shortley, by factors depending on j and j′. The relation between the two is given by Racah II,[1] Eq. (30).

$$(-1)^{j-m}\begin{pmatrix} j & 1 & j \\ -m & 0 & m' \end{pmatrix} = \delta_{mm'}(-1)^{j-m}\begin{pmatrix} j & j & 1 \\ m & -m & 0 \end{pmatrix}$$
$$= \delta_{mm'}\, m\,[j(j+1)(2j+1)]^{-1/2} , \tag{11.19}$$

so that (11.15) leads to

$$\langle \alpha j \| J^{(1)} \| \alpha' j' \rangle = \delta_{\alpha j, \alpha' j'}[j(j+1)(2j+1)]^{1/2} . \tag{11.20}$$

Special cases of importance are

$$\langle l \| l^{(1)} \| l \rangle = [l(l+1)(2l+1)]^{1/2} \tag{11.21}$$

and

$$\langle s \| s^{(1)} \| s \rangle = [\tfrac{1}{2}\cdot\tfrac{3}{2}\cdot 2]^{1/2} = (3/2)^{1/2} . \tag{11.22}$$

Another important example of (11.15) has already been met in (5.19), from which[7]

$$\langle l \| C^{(k)} \| l' \rangle = (-1)^{l}[l,l']^{1/2}\begin{pmatrix} l & k & l' \\ 0 & 0 & 0 \end{pmatrix} . \tag{11.23}$$

From the symmetry properties of the 3-j symbol it follows that

$$\langle l' \| C^{(k)} \| l \rangle = (-1)^{l'-l}\langle l \| C^{(k)} \| l' \rangle = (-1)^{k}\langle l \| C^{(k)} \| l' \rangle , \tag{11.24}$$

which shows that reduced matrix elements are not necessarily symmetric. From (11.17) it may be seen that

$$\langle l \| C^{(0)} \| l' \rangle = \delta_{ll'}[l]^{1/2} \tag{11.25}$$

[which is just a special case of (11.18) with integral j], and that

$$\langle l \| C^{(k)} \| 0 \rangle = \delta_{kl} . \tag{11.26}$$

Numerical values of the matrix elements (11.23) are tabulated in Appendix G.

Sum rules. The Wigner-Eckart theorem (11.15) leads, with the aid of the orthonormality relation (5.15) for 3-j symbols, directly to the important sum rule

[7]It should be noted that Racah's original phase conventions, used here, differ from those employed by Fano and Racah.[4] The latter conventions (generally used in the Israeli literature) employ an additional factor i^{l} in the definition of Y_{lm}, and a corresponding additional factor i^{k} in the definition of $C^{(k)}$; hence the factor $(-1)^{l}$ in (11.23) becomes $(-1)^{l+(-l+k+l')/2} = (-1)^{(l+k+l')/2}$, and from (5.13) their reduced matrix elements $\langle l \| C^{(k)} \| l' \rangle$ are all positive.

$$\sum_{mm'} |\langle\alpha jm|T_q^{(k)}|\alpha'j'm'\rangle|^2 = |\langle\alpha j\|T^{(k)}\|\alpha'j'\rangle|^2 \sum_{mm'} \begin{pmatrix} j & k & j' \\ -m & q & m' \end{pmatrix}^2$$

$$= \frac{1}{2k+1} |\langle\alpha j\|T^{(k)}\|\alpha'j'\rangle|^2 , \tag{11.27}$$

and this in turn gives the additional sum rule

$$\sum_{qmm'} |\langle\alpha jm|T_q^{(k)}|\alpha'j'm'\rangle|^2 = |\langle\alpha j\|T^{(k)}\|\alpha'j'\rangle|^2 . \tag{11.28}$$

[The triangle-relation factor, $\delta(jkj')$, present in matrix elements of $T^{(k)}$ appears implicitly not only in the 3-j symbol in (11.15), but also in the reduced matrix element—as, for example, in (11.23); thus we need not write it explicitly in (11.27) and (11.28).]

Problem

11-4(1). Show with the aid of (5.16) that

$$\sum_{mm'} \langle\alpha jm|T_0^{(k)}|\alpha jm'\rangle$$

is zero if k is nonzero.

11-5. MATRIX ELEMENTS OF THE PRODUCT OF TWO OPERATORS

If $W^{(k_2)}$ operates in a space spanned by an orthonormal set of basis functions $|\alpha jm\rangle$, then

$$|\beta\rangle \equiv W_{q_2}^{(k_2)}|\alpha'j'm'\rangle \tag{11.29}$$

is a function lying in this space, and it can therefore be expanded in terms of the basis functions:

$$|\beta\rangle = \sum_{\alpha''j''m''} |\alpha''j''m''\rangle\langle\alpha''j''m''|\beta\rangle$$

$$= \sum_{\alpha''j''m''} |\alpha''j''m''\rangle\langle\alpha''j''m''|W_{q_2}^{(k_2)}|\alpha'j'm'\rangle . \tag{11.30}$$

If $T^{(k_1)}$ operates in this same space, then it follows directly from the above expressions that

$$\langle\alpha jm|T_{q_1}^{(k_1)}W_{q_2}^{(k_2)}|\alpha'j'm'\rangle = \langle\alpha jm|T_{q_1}^{(k_1)}|\beta\rangle$$

$$= \sum_{\alpha''j''m''} \langle\alpha jm|T_{q_1}^{(k_1)}|\alpha''j''m''\rangle\langle\alpha''j''m''|W_{q_2}^{(k_2)}|\alpha'j'm'\rangle . \tag{11.31}$$

Thus the matrix of a product of two operators is equal to the product of the matrices of the individual operators. (This result is valid even if T and W are not irreducible tensor operators.)

11-6. TENSOR PRODUCT OF TWO TENSOR OPERATORS

Of particular importance is a special linear combination of products of irreducible tensor operators:

$$V_Q^{(K)} \equiv \left[T^{(k_1)} \times W^{(k_2)}\right]_Q^{(K)} \equiv \sum_{q_1 q_2} C(k_1 k_2 q_1 q_2; KQ)\, T_{q_1}^{(k_1)} W_{q_2}^{(k_2)} \,. \tag{11.32}$$

As implied by the notation, $V^{(K)}$ is itself an irreducible tensor operator; this we may verify by showing that the right-hand side of (11.32) satisfies the defining relations (11.4) and (11.5). Firstly, from (11.5),

$$\begin{aligned}\left[J_z, T_{q_1}^{(k_1)} W_{q_2}^{(k_2)}\right] &= J_z T_{q_1}^{(k_1)} W_{q_2}^{(k_2)} - T_{q_1}^{(k_1)} J_z W_{q_2}^{(k_2)} + T_{q_1}^{(k_1)} J_z W_{q_2}^{(k_2)} - T_{q_1}^{(k_1)} W_{q_2}^{(k_2)} J_z \\ &= \left[J_z, T_{q_1}^{(k_1)}\right] W_{q_2}^{(k_2)} + T_{q_1}^{(k_1)} \left[J_z, W_{q_2}^{(k_2)}\right] \\ &= (q_1 + q_2) T_{q_1}^{(k_1)} W_{q_2}^{(k_2)} \,.\end{aligned}$$

Because the CG coefficient in (11.32) is zero unless $q_1 + q_2 = Q$, it follows that

$$\begin{aligned}\left[J_z, V_Q^{(K)}\right] &= Q \sum_{q_1 q_2} C(k_1 k_2 q_1 q_2; KQ)\, T_{q_1}^{(k_1)} W_{q_2}^{(k_2)} \\ &= Q V_Q^{(K)} \,.\end{aligned}$$

Secondly, using (11.4) and (9.6) we find

$$\begin{aligned}\left[J_\pm, V_Q^{(K)}\right] &= \sum_{q_1 q_2} C(k_1 k_2 q_1 q_2; KQ) \Big\{ [(k_1 \mp q_1)(k_1 \pm q_1 + 1)]^{1/2} T_{q_1 \pm 1}^{(k_1)} W_{q_2}^{(k_2)} \\ &\qquad + [(k_2 \mp q_2)(k_2 \pm q_2 + 1)]^{1/2} T_{q_1}^{(k_1)} W_{q_2 \pm 1}^{(k_2)} \Big\} \\ &= \sum_{q_1 q_2} \Big\{ C(k_1 k_2, q_1 \mp 1, q_2; KQ)\, [(k_1 \pm q_1)(k_1 \mp q_1 + 1)]^{1/2} \\ &\qquad + C(k_1 k_2 q_1, q_2 \mp 1; KQ)\, [(k_2 \pm q_2)(k_2 \mp q_2 + 1)]^{1/2} \Big\} T_{q_1}^{(k_1)} W_{q_2}^{(k_2)} \\ &= [(K \mp Q)(K \pm Q + 1)]^{1/2} \sum_{q_1 q_2} C(k_1 k_2 q_1 q_2; K, Q \pm 1) T_{q_1}^{(k_1)} W_{q_2}^{(k_2)} \\ &= [(K \mp Q)(K \pm Q + 1)]^{1/2} V_{Q \pm 1}^{(K)} \,.\end{aligned}$$

Matrix elements of the tensor product are evaluated by first applying the Wigner-Eckart theorem (11.15):

$$\begin{aligned}&\langle \alpha j m | [T^{(k_1)} \times W^{(k_2)}]_Q^{(K)} | \alpha' j' m' \rangle \\ &\qquad = (-1)^{j-m} \begin{pmatrix} j & K & j' \\ -m & Q & m' \end{pmatrix} \langle \alpha j \| [T^{(k_1)} \times W^{(k_2)}]^{(K)} \| \alpha' j' \rangle \,.\end{aligned} \tag{11.33}$$

The reduced matrix element can be evaluated by multiplying both sides of (11.33) by the phase factor and 3-j symbol, summing over mm′, and using (5.15) to give

$$\langle \alpha j \| [T^{(k_1)} \times W^{(k_2)}]^{(K)} \| \alpha' j' \rangle = [K] \sum_{mm'} (-1)^{j-m} \begin{pmatrix} j & K & j' \\ -m & Q & m' \end{pmatrix} \langle \alpha j m | V_Q^{(K)} | \alpha' j' m' \rangle . \tag{11.34}$$

In the right-hand side of this expression we then use (11.32), (9.5), (11.31), the Wigner-Eckart theorem, the facts that $Q = q_1 + q_2 = m - m'$ and that $2j'' + 2m$ is an even integer, and (5.25) to obtain

$$\begin{aligned} \langle \alpha j \| [T^{(k_1)} \times W^{(k_2)}]^{(K)} \| \alpha' j' \rangle &= [K]^{3/2} \sum_{\alpha'' j''} \langle \alpha j \| T^{(k_1)} \| \alpha'' j'' \rangle \langle \alpha'' j'' \| W^{(k_2)} \| \alpha' j' \rangle \\ &\times \sum_{\substack{q_1 q_2 \\ mm'm''}} (-1)^{k_1 - k_2 + 2m - m - m'} \begin{pmatrix} k_1 & k_2 & K \\ q_1 & q_2 & -Q \end{pmatrix} (-1)^{k_1 + j + j''} \begin{pmatrix} k_1 & j & j'' \\ q_1 & -m & m'' \end{pmatrix} \\ &\times (-1)^{j' + k_2 + 2j'' - m''} \begin{pmatrix} j' & k_2 & j'' \\ m' & q_2 & -m'' \end{pmatrix} (-1)^{j' + j + K} \begin{pmatrix} j' & j & K \\ -m' & m & -Q \end{pmatrix} \\ &= (-1)^{j+K+j'} [K]^{1/2} \sum_{\alpha'' j''} \begin{Bmatrix} k_1 & k_2 & K \\ j' & j & j'' \end{Bmatrix} \langle \alpha j \| T^{(k_1)} \| \alpha'' j'' \rangle \langle \alpha'' j'' \| W^{(k_2)} \| \alpha' j' \rangle . \end{aligned} \tag{11.35}$$

In practice, the basis functions $|\alpha jm\rangle$ are usually composite functions that have been formed by coupling together the angular momenta j_1 and j_2 of the product function $|\alpha_1 j_1 m_1\rangle$ $|\alpha_2 j_2 m_2\rangle$, so that

$$|\alpha jm\rangle = |(\alpha_1 j_1, \alpha_2 j_2) jm\rangle ; \tag{11.36}$$

also, the tensor operators $T^{(k_1)}$ and $W^{(k_2)}$ operate only within the subspaces spanned by $|\alpha_1 j_1 m_1\rangle$ and $|\alpha_2 j_2 m_2\rangle$, respectively. A more useful expression than (11.35) can then be obtained by making CG expansions on the right side of (11.34) [instead of using (11.31)]:

$$\begin{aligned} X &\equiv \langle \alpha_1 j_1 \alpha_2 j_2 jm | T_{q_1}^{(k_1)} W_{q_2}^{(k_2)} | \alpha_1' j_1' \alpha_2' j_2' j' m' \rangle \\ &= [j, j']^{1/2} \langle \alpha_1 j_1 \| T^{(k_1)} \| \alpha_1' j_1' \rangle \langle \alpha_2 j_2 \| W^{(k_2)} \| \alpha_2' j_2' \rangle \\ &\times \sum_{\substack{m_1 m_2 \\ m_1' m_2'}} (-1)^{j_1 - j_2 + m_1 + m_2} \begin{pmatrix} j_1 & j_2 & j \\ m_1 & m_2 & -m \end{pmatrix} (-1)^{2j'_1 + m' + j'} \begin{pmatrix} j_1' & j_2' & j' \\ -m_1' & -m_2' & m' \end{pmatrix} \\ &\times (-1)^{j_1 - m_1} \begin{pmatrix} j_1 & j_1' & k_1 \\ m_1 & -m_1' & -q_1 \end{pmatrix} (-1)^{j_2 - m_2} \begin{pmatrix} j_2 & j_2' & k_2 \\ m_2 & -m_2' & -q_2 \end{pmatrix} . \end{aligned}$$

We also sum both sides of (11.34) over all $2K + 1$ values of Q (thereby obtaining a factor [K] on the left which cancels that on the right), and obtain from (5.37)

$$\langle \alpha_1 j_1 \alpha_2 j_2 j \| [T^{(k_1)} \times W^{(k_2)}]^{(K)} \| \alpha_1' j_1' \alpha_2' j_2' j' \rangle$$

$$= \sum_{\substack{q_1 q_2 \\ m m' Q}} X (-1)^{2k_1+m-m'+K} [K]^{1/2} \begin{pmatrix} k_1 & k_2 & K \\ -q_1 & -q_2 & Q \end{pmatrix} (-1)^{-m-j'-K} \begin{pmatrix} j & j' & K \\ -m & m' & Q \end{pmatrix}$$

$$= [j,j',K]^{1/2} \langle \alpha_1 j_1 \| T^{(k_1)} \| \alpha_1' j_1' \rangle \langle \alpha_2 j_2 \| W^{(k_2)} \| \alpha_2' j_2' \rangle \begin{Bmatrix} j_1 & j_2 & j \\ j_1' & j_2' & j' \\ k_1 & k_2 & K \end{Bmatrix} . \tag{11.37}$$

We note that this result, in contrast to (11.35), expresses reduced matrix elements of the tensor product in terms of individual reduced matrix elements for the *uncoupled* (subspace) basis functions.

Problem

11-6(1). In the special case $k_1 = k_2 = 1$, show that the first-rank tensor product $\mathbf{V}^{(1)}$ is equal to $i/\sqrt{2}$ times the tensor-operator form of the ordinary vector product $\mathbf{A} \equiv \mathbf{T} \times \mathbf{W}$; i.e.,

$$[T^{(1)} \times W^{(1)}]_0^{(1)} = \frac{i}{2^{1/2}}(T_x W_y - T_y W_x) = \frac{i}{2^{1/2}} A_z ,$$

$$[T^{(1)} \times W^{(1)}]_{\pm 1}^{(1)} = \frac{i}{2^{1/2}} \left[\mp \frac{1}{2^{1/2}} (A_x \pm i A_y) \right] .$$

11-7. UNCOUPLING FORMULAE FOR REDUCED MATRIX ELEMENTS

In the case of the identity operator $C_0^{(0)} \equiv 1$, it follows from (9.5) and (11.17) that

$$[T^{(k)} \times C^{(0)}]_q^{(k)} = \sum_{q_1} C(k0q_1 0; kq) T_{q_1}^{(k)} = T_q^{(k)} .$$

If $T^{(k)}$ operates only on $|\alpha_1 j_1 m_1\rangle$ then we find from (11.37), (11.18), and (5.38) that

$$\langle \alpha_1 j_1 \alpha_2 j_2 j \| T^{(k)} \| \alpha_1' j_1' \alpha_2' j_2' j' \rangle$$

$$= [j,j',k]^{1/2} \langle \alpha_1 j_1 \| T^{(k)} \| \alpha_1' j_1' \rangle \delta_{\alpha_2 j_2, \alpha_2' j_2'} [j_2]^{1/2} \begin{Bmatrix} j & j_1 & j_2 \\ j' & j_1' & j_2' \\ k & k & 0 \end{Bmatrix}$$

$$= \delta_{\alpha_2 j_2, \alpha_2' j_2'} (-1)^{j_1+j_2+j'+k} [j,j']^{1/2} \begin{Bmatrix} j_1 & j_2 & j \\ j' & k & j_1' \end{Bmatrix} \langle \alpha_1 j_1 \| T^{(k)} \| \alpha_1' j_1' \rangle . \tag{11.38}$$

If $\mathbf{W}^{(k)}$ operates only on $|\alpha_2 j_2 m_2\rangle$, then we can obtain similarly—or, more simply, directly by use of the CG symmetry relation (9.8) and use of (11.38)—

$$\begin{aligned}
&\langle \alpha_1 j_1 \alpha_2 j_2 j \| \mathbf{W}^{(k)} \| \alpha'_1 j'_1 \alpha'_2 j'_2 j' \rangle \\
&\qquad = (-1)^{-j_1-j_2+j+j'_1+j'_2-j'} \langle \alpha_2 j_2 \alpha_1 j_1 j \| \mathbf{W}^{(k)} \| \alpha'_2 j'_2 \alpha'_1 j'_1 j' \rangle \\
&\qquad = \delta_{\alpha_1 j_1, \alpha'_1 j'_1} (-1)^{j_1+j'_2+j+k} [j,j']^{1/2} \begin{Bmatrix} j_1 & j_2 & j \\ k & j' & j'_2 \end{Bmatrix} \langle \alpha_2 j_2 \| \mathbf{W}^{(k)} \| \alpha'_2 j'_2 \rangle \ . \qquad (11.39)
\end{aligned}$$

The expressions (11.38) and (11.39) provide quite an easy means for evaluating reduced matrix elements for coupled basis functions in terms of reduced matrix elements for subspace functions. This is a very important simplification, and is extremely valuable in the evaluation of matrix elements for basis functions of complex configurations. For example, if the reduced matrix elements on the right-hand side of (11.37) involve basis functions that are still multiply-coupled, (11.38) and/or (11.39) (with appropriate notational changes) can be used repeatedly until one finally obtains matrix elements for completely uncoupled functions.

11-8. SCALAR PRODUCT OF TWO TENSOR OPERATORS

Of particular importance is the special case of a tensor product with $k_1 = k_2$ and $K = Q = 0$. From (9.5) and (11.17), the definition (11.32) simplifies to

$$\begin{aligned}
V_0^{(0)} = [\mathbf{T}^{(k)} \times \mathbf{W}^{(k)}]_0^{(0)} &= \sum_q \begin{pmatrix} k & k & 0 \\ -q & q & 0 \end{pmatrix} T_{-q}^{(k)} W_q^{(k)} \\
&= (-1)^k [k]^{-1/2} \sum_q (-1)^q T_{-q}^{(k)} W_q^{(k)} \ . \qquad (11.40)
\end{aligned}$$

From (11.4)-(11.5), such a quantity commutes with J and is therefore a scalar. It is customary to modify the phase and normalization, and to define the scalar product of two irreducible tensor operators $\mathbf{T}^{(k)}$ and $\mathbf{W}^{(k)}$ as[8]

$$\begin{aligned}
Q \equiv \mathbf{T}^{(k)} \cdot \mathbf{W}^{(k)} &\equiv \sum_q (-1)^q T_{-q}^{(k)} W_q^{(k)} \\
&= (-1)^k [k]^{1/2} [\mathbf{T}^{(k)} \times \mathbf{W}^{(k)}]_0^{(0)} \ . \qquad (11.41)
\end{aligned}$$

In the case $k = 1$, this reduces from (11.12) to the ordinary scalar product of two vectors:

[8]In the case of Hermitian operators (11.7), the scalar product can be written in the more familiar-looking form $\Sigma_q T_q^{(k)\dagger} W_q^{(k)}$.

$$Q = -T_1^{(1)}W_{-1}^{(1)} + T_0^{(1)}W_0^{(1)} - T_{-1}^{(1)}W_1^{(1)}$$
$$= \frac{1}{2}(T_x + iT_y)(W_x - iW_y) + T_zW_z + \frac{1}{2}(T_x - iT_y)(W_x + iW_y)$$
$$= T_xW_x + T_yW_y + T_zW_z \,. \qquad (11.42)$$

A scalar product of great importance for our purpose is the expansion (6.21) for the electron-electron Coulomb energy:

$$\frac{2}{r_{tu}} = \sum_{k=0}^{\infty} \frac{2r_<^{\,k}}{r_>^{\,k+1}} C_{(t)}^{(k)} \cdot C_{(u)}^{(k)} \,. \qquad (11.43)$$

The subscripts on the tensor operators of course here refer to the variables (θ_t,ϕ_t) and (θ_u,ϕ_u) upon which the operators act, and not to the components q of the operators; if we need to indicate both the component and the coordinate variable, we will use the notation $C_q^{(k)}(t)$.

Because of (11.42), the spin-orbit operator in (10.1) can also be written as a scalar product of irreducible tensor operators,

$$\xi(r)\,[\boldsymbol{\ell}^{(1)} \cdot \mathbf{s}^{(1)}] \,. \qquad (11.44)$$

Matrix elements of a scalar product may be found with the aid of (11.31), the Wigner-Eckart theorem (11.15), and the orthonormality property (5.15) of the 3-j symbols:

$$\langle \alpha jm | \mathbf{T}^{(k)} \cdot \mathbf{W}^{(k)} | \alpha' j' m' \rangle = \sum_{\alpha'' j''} \langle \alpha j \| \mathbf{T}^{(k)} \| \alpha'' j'' \rangle \langle \alpha'' j'' \| \mathbf{W}^{(k)} \| \alpha' j' \rangle$$
$$\times \sum_{m'' q} (-1)^{-q+j-m} \begin{pmatrix} j'' & k & j \\ -m'' & q & m \end{pmatrix} (-1)^{-j''+m''} \begin{pmatrix} j'' & k & j' \\ -m'' & q & m' \end{pmatrix}$$
$$= \delta_{jm,j'm'} [j]^{-1} \sum_{\alpha'' j''} (-1)^{j-j''} \langle \alpha j \| \mathbf{T}^{(k)} \| \alpha'' j'' \rangle \langle \alpha'' j'' \| \mathbf{W}^{(k)} \| \alpha' j' \rangle \,. \qquad (11.45)$$

This result shows that the matrix of any scalar product is diagonal in j and m, and independent of m; an example is provided by the spin-orbit matrix elements (10.11)-(10.12). Application to the Coulomb operator (11.43) provides the mathematical proof of the physical arguments used in deriving (10.16).

If, as in (11.36), the basis functions $|\alpha jm\rangle$ have been formed by the angular-momentum coupling $(j_1j_2)j$ of subspace functions, and if $\mathbf{T}^{(k)}$ and $\mathbf{W}^{(k)}$ operate only on $|\alpha_1 j_1 m_1\rangle$ and $|\alpha_2 j_2 m_2\rangle$, respectively, then the result (11.45) can be simplified with the aid of (11.38) and (11.39). If we write

$$|\alpha'' j'' m''\rangle = |\alpha_1'' j_1'' \alpha_2'' j_2'' j'' m''\rangle \qquad (11.46)$$

so that the summation over α'' in (11.45) becomes a multiple summation over $\alpha_1'' j_1'' \alpha_2'' j_2''$, then

$$\langle \alpha_1 j_1 \alpha_2 j_2 jm | \mathbf{T}^{(k)} \cdot \mathbf{W}^{(k)} | \alpha'_1 j'_1 \alpha'_2 j'_2 j'm' \rangle$$

$$= \frac{\delta_{jm,j'm'}}{[j]} \sum_{\substack{\alpha''_1\alpha''_2 \\ j''_1 j''_2 j''}} (-1)^{j-j''} \delta_{\alpha_2 j_2, \alpha''_2 j''_2} (-1)^{j_1+j_2+j''+k} [j,j'']^{1/2} \begin{Bmatrix} j_1 & j_2 & j \\ j'' & k & j''_1 \end{Bmatrix}$$

$$\times \delta_{\alpha''_1 j''_1, \alpha'_1 j'_1} (-1)^{j''_1+j'_2+j''+k} [j'',j']^{1/2} \begin{Bmatrix} j''_2 & j''_1 & j'' \\ j' & k & j'_2 \end{Bmatrix}$$

$$\times \langle \alpha_1 j_1 \| \mathbf{T}^{(k)} \| \alpha''_1 j''_1 \rangle \langle \alpha''_2 j''_2 \| \mathbf{W}^{(k)} \| \alpha'_2 j'_2 \rangle$$

$$= \delta_{jm,j'm'} (-1)^{j+j_2+j'_1} \langle \alpha_1 j_1 \| \mathbf{T}^{(k)} \| \alpha'_1 j'_1 \rangle \langle \alpha_2 j_2 \| \mathbf{W}^{(k)} \| \alpha'_2 j'_2 \rangle$$

$$\times \sum_{j''} (-1)^{j_1+j'_2+j''} [j''] \begin{Bmatrix} j_2 & j & j_1 \\ k & j'_1 & j'' \end{Bmatrix} \begin{Bmatrix} j_2 & k & j'_2 \\ j & j'_1 & j'' \end{Bmatrix} .$$

Using (5.32), we obtain finally

$$\langle \alpha_1 j_1 \alpha_2 j_2 jm | \mathbf{T}^{(k)} \cdot \mathbf{W}^{(k)} | \alpha'_1 j'_1 \alpha'_2 j'_2 j'm' \rangle$$

$$= \delta_{jm,j'm'} (-1)^{j'_1+j_2+j} \begin{Bmatrix} j_1 & j_2 & j \\ j'_2 & j'_1 & k \end{Bmatrix} \langle \alpha_1 j_1 \| \mathbf{T}^{(k)} \| \alpha'_1 j'_1 \rangle \langle \alpha_2 j_2 \| \mathbf{W}^{(k)} \| \alpha'_2 j'_2 \rangle . \qquad (11.47)$$

This equation[9] provides the first step in the evaluation of matrix elements of a scalar product for coupled basis functions, and does so in a very condensed form—accomplishing in one simple step the separation (11.45) into reduced matrix elements of $\mathbf{T}^{(k)}$ and $\mathbf{W}^{(k)}$ separately, and also the uncoupling of the angular momenta $(j_1 j_2)$ according to (11.38)-(11.39). If the basis functions $|\alpha_1 j_1 m_1\rangle$ and/or $|\alpha_2 j_2 m_2\rangle$ are themselves coupled functions, then as indicated earlier the expressions (11.38) and (11.39) (with suitably modified notation) can be used for further reduction of the matrix elements.

Problems

11-8(1). Derive (11.45) and (11.47) directly from (11.41), (11.33), (11.35), and (11.37).

11-8(2). Use the scalar-product form (11.43) for the Coulomb operator, and the Racah-algebra equations of Secs. 11-7 and 11-8 to derive the result (11.2)-(11.3). Similarly derive the analogous expressions for the coefficients of $F^k(l_1 l_3)$ and $F^k(l_1 l_2)$.

11-9. UNIT TENSOR OPERATORS

It is convenient to define a unit irreducible tensor operator $\mathbf{u}^k$, which operates on the spatial coordinates and is normalized such that[10]

[9]Racah did not break α'' down into $\alpha''_1\alpha''_2$, and therefore could not make use of the δ factors in α_1 and α_2; thus in his version of (11.47) [ref. 1, Eq. (38)] he was still left with a summation over α''.

[10]This is a generalization of the $u^{(k)}$ defined by Racah II,[1] Eq. (58), which he defined only for $l = l'$; i.e., $\langle l \| u^{(k)} \| l' \rangle = \delta_{ll'}$. For simplicity, we have omitted the triangle-relation factor $\delta(lkl')$ from the definition (11.48); it will be implicitly understood that all equations involving these reduced matrix elements are valid only for $|l - l'| \leq k \leq l + l'$.

$$\langle l \| u^{(k)} \| l' \rangle = 1 \tag{11.48}$$

for all lkl' that satisfy the triangle relations.

An example valid for $l + k + l'$ even is

$$u_q^{(k)} = \frac{C_q^{(k)}}{\langle l \| C^{(k)} \| l' \rangle} . \tag{11.49}$$

In this definition, the values of l and l' are not indicated by the notation $\mathbf{u}^{(k)}$, but are simply to be inferred from the context in which $\mathbf{u}^{(k)}$ is used in the calculation of a matrix element:

$$\langle \cdots l(i) \cdots | C_q^{(k)}(i) | \cdots l'(i) \cdots \rangle = \langle l \| C^{(k)} \| l' \rangle \langle \cdots l(i) \cdots | u_q^{(k)}(i) | \cdots l'(i) \cdots \rangle . \tag{11.50}$$

Another example, for $k = 1$, is

$$u_q^{(1)} = \frac{J_q^{(1)}}{\langle l \| J^{(1)} \| l' \rangle} . \tag{11.51}$$

For a subshell l^w of equivalent electrons, we may define a symmetric unit tensor operator

$$U^{(k)} \equiv \sum_{i=1}^{w} u_{(i)}^{(k)} . \tag{11.52}$$

Using the cfp expansion (9.50), the uncoupling formula (11.39), and (11.48), the reduced matrix elements of $U^{(k)}$ may be calculated from

$$\begin{aligned} \langle l^w \alpha LS \| U^{(k)} \| l^w \alpha' L'S' \rangle &= w \langle l^w \alpha LS \| u_{(w)}^{(k)} \| l^w \alpha' L'S' \rangle \\ &= \delta_{SS'} w \sum_{\bar{\alpha}\bar{L}\bar{S}} \sum_{\bar{\alpha}'\bar{L}'\bar{S}'} (l^w \alpha LS \{\!| l^{w-1} \bar{\alpha}\bar{L}\bar{S}) \\ &\quad \times \langle (l^{w-1} \bar{\alpha}\bar{L}\bar{S}, l) L \| u_{(w)}^{(k)} \| (l^{w-1} \bar{\alpha}'\bar{L}'\bar{S}', l) L' \rangle (l^{w-1} \bar{\alpha}'\bar{L}'\bar{S}' |\!\} l^w \alpha' L'S') \\ &= \delta_{SS'} w (-1)^{l+L+k} [L,L']^{1/2} \sum_{\bar{\alpha}\bar{L}\bar{S}} (-1)^{\bar{L}} \begin{Bmatrix} l & k & l \\ L & \bar{L} & L' \end{Bmatrix} \\ &\quad \times (l^w \alpha LS \{\!| l^{w-1} \bar{\alpha}\bar{L}\bar{S}) (l^{w-1} \bar{\alpha}\bar{L}\bar{S} |\!\} l^w \alpha' L'S') . \end{aligned} \tag{11.53}$$

We note from the 6-j symbol in this expression that the matrix elements are zero unless $0 \leqq k \leqq 2l$, and unless LkL' satisfy the triangle relations. We see also (since the cfp are symmetric) that

$$\langle l^w \alpha' L'S' \| U^{(k)} \| l^w \alpha LS \rangle = (-1)^{L'-L} \langle l^w \alpha LS \| U^{(k)} \| l^w \alpha' L'S' \rangle . \tag{11.54}$$

In the special case $k = 0$, we find by using (5.29) and the orthogonality property (9.46) of the cfp that

$$\langle l^{w}\alpha LS\|U^{(0)}\|l^{w}\alpha'L'S'\rangle = \delta_{LS,L'S'}\,w\,\frac{[L]^{1/2}}{[l]^{1/2}}\sum_{\bar{\alpha}\bar{L}\bar{S}}(l^{w}\alpha LS\{|l^{w-1}\bar{\alpha}\bar{L}\bar{S})(l^{w-1}\bar{\alpha}\bar{L}\bar{S}|\}l^{w}\alpha'LS)$$

$$= \delta_{\alpha LS,\alpha'L'S'}\,w\left(\frac{2L+1}{2l+1}\right)^{1/2} . \tag{11.55}$$

For k = 1, we find from (11.51) and (11.20) that

$$\langle l^{w}\alpha LS\|U^{(1)}\|l^{w}\alpha'L'S'\rangle = \delta_{\alpha LS,\alpha'L'S'}\left[\frac{L(L+1)(2L+1)}{l(l+1)(2l+1)}\right]^{1/2} . \tag{11.56}$$

For singly occupied subshells (w = 1), the cfp are unity and $\bar{L} = 0$, and (11.53) reduces with the aid of (5.29) to

$$\langle l\|U^{(k)}\|l\rangle = (-1)^{k}[l]\begin{Bmatrix} l & k & l \\ l & 0 & l \end{Bmatrix} = 1 , \tag{11.57}$$

in agreement with (11.48).

It can be shown[11] that

$$\langle l^{4l+2-w}\alpha LS\|U^{(k)}\|l^{4l+2-w}\alpha'L'S'\rangle_{\mathfrak{R}}$$

$$= -(-1)^{k}\langle l^{w}\alpha LS\|U^{(k)}\|l^{w}\alpha'L'S'\rangle_{\mathfrak{L}} , \qquad k > 0 , \tag{11.58}$$

where the subscripts $\mathfrak{R}$ and $\mathfrak{L}$ refer to the basis-function phase conventions defined in Sec. 9-7. With our standard phase conventions ($\mathfrak{L}$ for $w \leqq 2l + 1$, $\mathfrak{R}$ otherwise), the subscripts can be dropped if (11.58) is used only for $w < 2l + 1$. However, for $w = 2l + 1$ we see from (9.64)-(9.65) that

$$|\,l^{2l+1}\alpha vLS\rangle_{\mathfrak{R}} = (-1)^{(v-1)/2}\,|\,l^{2l+1}\alpha vLS\rangle_{\mathfrak{L}} , \tag{11.59}$$

so that (11.58) becomes

$$\langle l^{2l+1}\alpha vLS\|U^{(k)}\|l^{2l+1}\alpha'v'L'S'\rangle$$

$$= -(-1)^{k+(v-v')/2}\langle l^{2l+1}\alpha vLS\|U^{(k)}\|l^{2l+1}\alpha'v'L'S'\rangle , \qquad k > 0 . \tag{11.60}$$

It follows that for half-filled subshells, reduced matrix elements of $U^{(k)}$ are zero for $v = v'$ (and in particular for diagonal elements) if k is even and non-zero, and are zero for $v = v' \pm 2$ if k is odd. [For $|v - v'| > 2$, matrix elements are zero for all k because of (9.58).]

From (11.58) and (11.57) we see that

$$\langle l^{4l+1}\,{}^{2}l\|U^{(k)}\|l^{4l+1}\,{}^{2}l\rangle = -(-1)^{k} , \qquad k > 0 , \tag{11.61}$$

and from the triangle restrictions for the 6-j symbol in (11.53) it follows that

[11]G. Racah II,[1] Eq. (74). (The brackets ⟨ ⟩ on the right side of his equation indicate the complex conjugate.)

$$\langle l^0\ {}^1S\|U^{(k)}\|l^0\ {}^1S\rangle = \langle l^{4l+2}\ {}^1S\|U^{(k)}\|l^{4l+2}\ {}^1S\rangle = 0\ , \qquad k > 0\ . \tag{11.62}$$

The relationship between matrix elements of l^w and l^{4l+2-w} for k = 0 is easily found from (11.55).

From the results (11.54)-(11.62), we see that matrix elements of $U^{(k)}$ need be computed from (11.53) only for $2 \leqq k \leqq 2l$ and $2 \leqq w \leqq 2l + 1$, and then only for elements on and above the diagonal. Values for all these cases have been tabulated for p, d, and f subshells by Nielson and Koster;[12] illustrative values are given for simple subshells in Appendix H.

Problems

11-9(1). Verify (11.56) and (11.60) in a few cases for p^3 by numerical evaluation of (11.53), using cfp from Table 9-5 or Appendix H. Similarly verify (11.61) for p^5.

11-9(2). From (11.53), (9.68), and (9.62), verify (11.61), and also (11.55) for $w = 4l + 1$.

$$\left[\text{Hint: } \sum_{\bar{L}\bar{S}}(-1)^{\bar{L}}[\bar{S}] = 3\sum_{\bar{L}\ \text{odd}}(-1)^{\bar{L}} + \sum_{\bar{L}\ \text{even}}(+1) = 2\sum_{\bar{L}}(-1)^{\bar{L}} - \sum_{\bar{L}}(+1)\ .\right]$$

11-10. DOUBLE TENSOR OPERATORS

The notation that we used in (11.47) for uncoupling two angular momenta j_1 and j_2 which are acted on separately by two different operators $T^{(k)}$ and $W^{(k)}$ is perfectly satisfactory when j_1 and j_2 are angular momenta associated with two different electrons. However, in evaluating matrix elements of the spin-orbit operator $l^{(1)} \cdot s^{(1)}$ for functions $|l^w\alpha LS\rangle$, j_1 and j_2 will be the quantum numbers L and S, and it is rather unsatisfactory to separate L from S when α, L, and S are inherently linked together as the label of a single basis function.

In order to be able to consider mathematically distinct reduced matrix elements in a single-matrix-element notation, and thereby keep intact the label αLS, it is convenient to define a double tensor operator

$$T^{(k\kappa)} \tag{11.63}$$

of rank (k,κ), which behaves as an irreducible tensor operator of rank k with respect to L and as a tensor of rank κ with respect to S. Of particular importance is the unit double tensor[13]

$$v^{(k1)} \equiv u^{(k)}s^{(1)}\ , \tag{11.64}$$

[12] C. W. Nielson and G. F. Koster, *Spectroscopic Coefficients for the p^n, d^n, and f^n Configurations* (The M.I.T. Press, Cambridge, Mass., 1963).

[13] Racah used the notation $T^{(\kappa k)}$ and $v^{(1k)}$ in line with his use of SL rather than LS coupling. Some of the Russian-language literature uses LS coupling and the notation $v^{(k1)}$, but in some cases considers the spin as well as the orbital part of $v^{(k1)}$ to be a unit operator; then $\langle ls\|v^{(k1)}\|l's\rangle = \delta(lkl')$, and an explicit extra factor $\sqrt{3/2}$ appears in all equations that contain this reduced matrix element.

whose reduced matrix elements

$$\langle ls\|v^{(k1)}\|l's\rangle \tag{11.65}$$

are to be considered equivalent, from (11.22) and (11.48), to

$$\langle l\|u^{(k)}\|l'\rangle\langle s\|s^{(1)}\|s\rangle = (3/2)^{1/2} . \tag{11.66}$$

For a subshell l^w of equivalent electrons, the symmetric unit double tensor operator

$$V^{(k1)} \equiv \sum_{i=1}^{w} v_{(i)}^{(k1)} = \sum_{i=1}^{w} u_{(i)}^{(k)} s_{(i)}^{(1)} \tag{11.67}$$

has reduced matrix elements whose values we find similarly to (11.53) to be

$$\begin{aligned}\langle l^w\alpha LS\|V^{(k1)}\|l^w\alpha'L'S'\rangle &= w\sum_{\bar\alpha\bar L\bar S}(l^w\alpha LS\{|l^{w-1}\bar\alpha\bar L\bar S)\\ &\quad\times\langle(l^{w-1}\bar\alpha\bar L\bar S,ls)LS\|u_{(w)}^{(k)}s_{(w)}^{(1)}\|(l^{w-1}\bar\alpha\bar L\bar S,ls)L'S'\rangle(l^{w-1}\bar\alpha\bar L\bar S|\}l^w\alpha'L'S')\\ &= (3/2)^{1/2}w(-1)^{l+L+k}[L,L',S,S']^{1/2}\sum_{\bar\alpha\bar L\bar S}(-1)^{\bar L+\bar S+S+3/2}\\ &\quad\times\begin{Bmatrix} l & l & k\\ L & L' & \bar L\end{Bmatrix}\begin{Bmatrix} s & s & 1\\ S & S' & \bar S\end{Bmatrix}(l^w\alpha LS\{|l^{w-1}\bar\alpha\bar L\bar S)(l^{w-1}\bar\alpha\bar L\bar S|\}l^w\alpha'L'S') .\end{aligned} \tag{11.68}$$

We see that the possible values of k range from 0 to $2l$, and that both LkL′ and S1S′ must satisfy the triangle relations. We also have the symmetry relation

$$\langle l^w\alpha'L'S'\|V^{(k1)}\|l^w\alpha LS\rangle = (-1)^{L'-L+S'-S}\langle l^w\alpha LS\|V^{(k1)}\|l^w\alpha'L'S'\rangle . \tag{11.69}$$

In the special case k = 0, it follows from (11.49), (11.25), and (11.20) that

$$\langle l^w\alpha LS\|V^{(01)}\|l^w\alpha'L'S'\rangle = \delta_{\alpha LS,\alpha'L'S'}\left[\frac{(2L+1)S(S+1)(2S+1)}{(2l+1)}\right]^{1/2} . \tag{11.70}$$

For singly occupied subshells (w = 1), $\bar L = \bar S = 0$, and (11.68) reduces to

$$\langle ls\|V^{(k1)}\|ls\rangle = (3/2)^{1/2}(-1)^k[l,s]\begin{Bmatrix} l & l & k\\ l & l & 0\end{Bmatrix}\begin{Bmatrix} s & s & 1\\ s & s & 0\end{Bmatrix} = (3/2)^{1/2} , \tag{11.71}$$

in agreement with (11.65)-(11.66).

For any k and κ (k + κ > 0), the generalization of (11.58) is, for real matrix elements,[11]

$$\begin{aligned}&\langle l^{4l+2-w}\alpha LS\|T^{(k\kappa)}\|l^{4l+2-w}\alpha'L'S'\rangle_{\mathfrak{R}}\\ &\quad = -(-1)^{k+\kappa}\langle l^w\alpha LS\|T^{(k\kappa)}\|l^w\alpha'L'S'\rangle_{\mathfrak{L}} ,\end{aligned} \tag{11.72}$$

from which it follows that (for the standard phase conventions adopted in Sec. 9-7)

$$\langle l^{4l+2-w}\alpha LS\|V^{(k1)}\|l^{4l+2-w}\alpha' L'S'\rangle$$
$$= (-1)^k \langle l^w \alpha LS\|V^{(k1)}\|l^w \alpha' L'S'\rangle , \qquad w < 2l + 1 , \tag{11.73}$$

and

$$\langle l^{2l+1}\alpha vLS\|V^{(k1)}\|l^{2l+1}\alpha' v' L'S'\rangle$$
$$= (-1)^{k+(v-v')/2} \langle l^{2l+1}\alpha vLS\|V^{(k1)}\|l^{2l+1}\alpha' v' L'S'\rangle . \tag{11.74}$$

Thus for half-filled subshells, reduced matrix elements of $V^{(k1)}$ are zero for $v = v'$ if k is odd, and are zero for $v = v' \pm 2$ if k is even.

From (11.73) and (11.71) we see that

$$\langle l^{4l+1}\,{}^2l\|V^{(k1)}\|l^{4l+1}\,{}^2l\rangle = (-1)^k (3/2)^{1/2} , \tag{11.75}$$

and from the triangle restrictions on the second 6-j symbol in (11.68) it follows that

$$\langle l^0\,{}^1S\|V^{(k1)}\|l^0\,{}^1S\rangle = \langle l^{4l+2}\,{}^1S\|V^{(k1)}\|l^{4l+2}\,{}^1S\rangle = 0 . \tag{11.76}$$

From the results (11.69)-(11.76), we see that matrix elements of $V^{(k1)}$ need be computed from (11.68) only for $1 \leqq k \leqq 2l$ and $2 \leqq w \leqq 2l + 1$, and then only for elements on and above the diagonal. Values for k = 1 and l = p, d, or f have been tabulated by Nielson and Koster;[12] values for all k, but only through f^4, are given by Karaziya, et al.[14] Illustrative values for simple subshells are given in Appendix H.

Problem

11-10(1). Verify (11.75) directly from (11.68), (9.68), and (9.62). [Hint: for $\bar{S}$ equal either 0 or 1,

$$[\bar{S}]\begin{Bmatrix} \frac{1}{2} & \frac{1}{2} & 1 \\ \frac{1}{2} & \frac{1}{2} & \bar{S} \end{Bmatrix} = 1/2 . \qquad \sum_{\bar{L}\ \mathrm{even}} \begin{Bmatrix} \frac{1}{2} & \frac{1}{2} & 1 \\ \frac{1}{2} & \frac{1}{2} & 0 \end{Bmatrix} = \frac{1}{2}\sum_{\bar{L}}(+1) .$$

11-11. SPIN-ORBIT MATRIX ELEMENTS (ONE-ELECTRON CONFIGURATIONS)

In order to illustrate the use of Racah's methods, we now recalculate some of the matrix elements that we evaluated earlier by other means.

[14]R. I. Karaziya, Ya. I. Vizbaraite, Z. B. Rudzikas, and A. P. Jucys [A. P. Yutsis], *Tables for the Calculation of Matrix Elements of Atomic Quantities* (Moscow, 1967); English transl. by E. K. Wilip, ANL-Trans-563 (National Technical Information Service, Springfield, Va., 1968). These authors consider $U^{(k)}$ to be equivalent to $V^{(k0)}$; therefore, their tabulated matrix elements of $U^{(k)}$ contain an extra factor $(2S + 1)^{1/2}$—see (11.55) for an example involving orbital momenta. This is true also of the tables in A. P. Jucys and A. J. Savukynas, *Mathematical Foundations of the Atomic Theory* (Vilnius, 1973) (in Russian).

For the spin-orbit matrix element of a one-electron configuration, we find from (11.44), (11.51), (11.47), (11.66), and (11.21)

$$\begin{aligned} d &= \langle l s j m | \boldsymbol{l}^{(1)} \cdot s^{(1)} | l s j' m' \rangle \\ &= \langle l \| \boldsymbol{l}^{(1)} \| l \rangle \langle l s j m | u^{(1)} \cdot s^{(1)} | l s j' m' \rangle \\ &= \langle l \| \boldsymbol{l}^{(1)} \| l \rangle \delta_{jm,j'm'} (-1)^{l+s+j} \begin{Bmatrix} l & s & j \\ s & l & 1 \end{Bmatrix} \langle l s \| v^{(11)} \| l s \rangle \\ &= \delta_{jm,j'm'} [(3/2) l(l+1)(2l+1)]^{1/2} (-1)^{l+s+j} \begin{Bmatrix} l & s & j \\ s & l & 1 \end{Bmatrix} . \end{aligned} \tag{11.77}$$

It may be seen directly from the definition (5.23) of the 6-j symbol that

$$(-1)^{l+s+j} \begin{Bmatrix} l & s & j \\ s & l & 1 \end{Bmatrix} = \frac{j(j+1) - l(l+1) - s(s+1)}{2[l(l+1)(2l+1)s(s+1)(2s+1)]^{1/2}} , \tag{11.78}$$

from which we see that (11.77) is equivalent to (10.12).

The above derivation is actually somewhat more complicated than that used to obtain (10.12), but at least it is straightforward and does not depend on the trick (3.41). More importantly, it is easily extendable to subshells l^w, whereas use of the previous method would first require an expansion in cfp and then a transformation from the LS to a JJ representation.

11-12. DIRECT COULOMB MATRIX ELEMENTS (TWO-ELECTRON CONFIGURATIONS)

For the direct Coulomb matrix elements in a configuration of two non-equivalent electrons, we find from (11.43), (11.49), (11.47), and (11.48)

$$f'_k = \langle l_1(i) l_2(j) L M_L | C^{(k)}_{(i)} \cdot C^{(k)}_{(j)} | l_1(i) l_2(j) L' M'_L \rangle \tag{11.79}$$

$$= \langle l_1 \| C^{(k)} \| l_1 \rangle \langle l_2 \| C^{(k)} \| l_2 \rangle \langle l_1(i) l_2(j) L M_L | u^{(k)}_{(i)} \cdot u^{(k)}_{(j)} | l_1(i) l_2(j) L' M'_L \rangle \tag{11.80}$$

$$= \delta_{LM_L,L'M'_L} (-1)^{l_1+l_2+L} \langle l_1 \| C^{(k)} \| l_1 \rangle \langle l_2 \| C^{(k)} \| l_2 \rangle \begin{Bmatrix} l_1 & l_2 & L \\ l_2 & l_1 & k \end{Bmatrix} . \tag{11.81}$$

[In deriving this result, it was not actually necessary to go through the step (11.80) of replacing the operators $\mathbf{C}^{(k)}$ by the corresponding unit operators $\mathbf{u}^{(k)}$; using only (11.47), one can go directly from (11.79) to (11.81) in one step. The indirect procedure was used here to illustrate the use of the $\mathbf{u}^{(k)}$ in order to clarify the discussion of the effective exchange operator in Sec. 11-14.]

From (11.23) it is obvious that (11.81) is equivalent to (10.22). However, the result has here been obtained much more simply, because the multiple summation met in (10.19) has already (in effect) been taken care of in the calculation that led to the formula (11.47).

For the three-electron direct matrix element (11.1), we find by using (11.47) and then (11.39) that

$$\langle[(l_1l_2)\mathfrak{L}_2,l_3]\mathfrak{L}_3\,|\,\mathbf{C}^{(k)}_{(2)}\cdot\mathbf{C}^{(k)}_{(3)}\,|\,[(l_1l_2)\mathfrak{L}'_2,l_3]\mathfrak{L}'_3\rangle$$

$$= \delta_{\mathfrak{L}_3,\mathfrak{L}'_3}(-1)^{\mathfrak{L}'_2+l_3+\mathfrak{L}_3}\begin{Bmatrix}\mathfrak{L}_2 & l_3 & \mathfrak{L}_3\\ l_3 & \mathfrak{L}'_2 & k\end{Bmatrix}\langle(l_1l_2)\mathfrak{L}_2\|\mathbf{C}^{(k)}_{(2)}\|(l_1l_2)\mathfrak{L}'_2\rangle\langle l_3\|\mathbf{C}^{(k)}\|l_3\rangle$$

$$= \delta_{\mathfrak{L}_3,\mathfrak{L}'_3}(-1)^{\mathfrak{L}'_2+l_3+\mathfrak{L}_3+l_1+l_2+\mathfrak{L}_2+k}[\mathfrak{L}_2,\mathfrak{L}'_2]^{1/2}$$

$$\times\begin{Bmatrix}\mathfrak{L}_2 & l_3 & \mathfrak{L}_3\\ l_3 & \mathfrak{L}'_2 & k\end{Bmatrix}\begin{Bmatrix}l_1 & l_2 & \mathfrak{L}_2\\ k & \mathfrak{L}'_2 & l_2\end{Bmatrix}\langle l_2\|\mathbf{C}^{(k)}\|l_2\rangle\langle l_3\|\mathbf{C}^{(k)}\|l_3\rangle\ . \tag{11.82}$$

(For simplicity, we have ignored the magnetic quantum numbers $\mathfrak{M}_3,\mathfrak{M}'_3$.) From (11.23) we see that this is equivalent to (11.2)-(11.3); it has been obtained quite simply, without recourse to angular-momentum recouplings, and without introduction of the summation over $\mathfrak{L}_{23}$ and its subsequent analytical evaluation.

11-13. EXCHANGE COULOMB MATRIX ELEMENTS (TWO-ELECTRON CONFIGURATIONS)

For the exchange matrix element in a configuration of two non-equivalent electrons, we find with the aid of (9.8), (11.47), and (11.24)

$$g'_k = -\langle l_1(i)l_2(j)LS\,|\,\mathbf{C}^{(k)}_{(i)}\cdot\mathbf{C}^{(k)}_{(j)}\,|\,l_1(j)l_2(i)L'S'\rangle \tag{11.83}$$

$$= (-1)^{l_1+l_2-L'+S'}\langle l_1(i)l_2(j)LS\,|\,\mathbf{C}^{(k)}_{(i)}\cdot\mathbf{C}^{(k)}_{(j)}\,|\,l_2(i)l_1(j)L'S'\rangle \tag{11.84}$$

$$= \delta_{LS,L'S'}(-1)^{l_1+l_2+S}\begin{Bmatrix}l_1 & l_2 & L\\ l_1 & l_2 & k\end{Bmatrix}\langle l_1\|\mathbf{C}^{(k)}\|l_2\rangle\langle l_2\|\mathbf{C}^{(k)}\|l_1\rangle$$

$$= \delta_{LS,L'S'}(-1)^{l_1+l_2+k+S}\langle l_1\|\mathbf{C}^{(k)}\|l_2\rangle^2\begin{Bmatrix}l_1 & l_2 & L\\ l_1 & l_2 & k\end{Bmatrix}\ . \tag{11.85}$$

From (11.23) and the fact that $l_1 + l_2 + k$ is even, we see that this is equivalent to (10.23).

11-14. THE EFFECTIVE OPERATOR FOR EXCHANGE MATRIX ELEMENTS

The derivation of the expression (11.85) for g'_k in a two-electron configuration involved first of all the use of the CG symmetry relation (9.8) in order to interchange the order of coupling of $(l_1l_2)L'$ in the ket function. This interchange effectively converted the exchange matrix element (11.83), for which the scalar-product formula (11.47) is *not* applicable, into the "direct" matrix element (11.84) for which the electron coordinates occur in the same order in both basis functions so that (11.47) *is* applicable.

In more complex cases, for example in the evaluation of the exchange matrix element

$$-\langle\{[l_1(\mathrm{i})l_2(\mathrm{j})]\mathfrak{L}_2\mathfrak{S}_2,l_3(\mathrm{t})\}\mathfrak{L}_3\mathfrak{S}_3\,|\,2/r_{\mathrm{jt}}\,|\,\{[l_1(\mathrm{i})l_2(\mathrm{t})]\mathfrak{L}_2'\mathfrak{S}_2',l_3(\mathrm{j})\}\mathfrak{L}_3'\mathfrak{S}_3'\rangle\ , \tag{11.86}$$

conversion to a direct integral cannot be accomplished by means of a simple interchange introducing no more than a phase factor. Rather, the recoupling coefficient (9.19) must be employed (once for orbital momenta and once for spins), together with summations over the new quantum numbers $\mathfrak{L}_{13}'\mathfrak{S}_{13}'$ that are introduced into the recoupled ket function

$$|\{[l_1(\mathrm{i})l_3(\mathrm{j})]\mathfrak{L}_{13}'\mathfrak{S}_{13}',l_2(\mathrm{t})\}\mathfrak{L}_3'\mathfrak{S}_3'\rangle\ . \tag{11.87}$$

For exchange matrix elements involving l_1l_3 instead of l_2l_3, a different recoupling is involved. Thus each case must be handled differently. In still more complex cases, where multiply occupied subshells are involved, expansions via cfp must be carried out before any recoupling can be done, and this complicates evaluation of the matrix elements still further.

All of these complications can be avoided and the calculation of exchange matrix elements systematized to a universal procedure (at comparatively small cost), by replacing the Coulomb operator

$$\mathrm{C}^{(k)}_{(\mathrm{i})}\cdot\mathrm{C}^{(k)}_{(\mathrm{j})} \tag{11.88}$$

by an effective exchange operator; the latter is to be of such a form that when used for the evaluation of a *direct* matrix element, it gives the results that (11.88) would give in an *exchange* matrix element. This effective operator is derived as follows:[1]

By using (5.32) we may write the result (11.85)—deleting the δ-factor for simplicity—in the form

$$g_k' = (-1)^{l_1+l_2+L+S}\langle l_1\|\mathrm{C}^{(k)}\|l_2\rangle^2\sum_r(-1)^r[r]\begin{Bmatrix} l_1 & l_2 & L\\ l_2 & l_1 & r\end{Bmatrix}\begin{Bmatrix} l_1 & l_2 & k\\ l_2 & l_1 & r\end{Bmatrix}\ . \tag{11.89}$$

But from (11.80)-(11.81) we see that the first 6-j symbol can be written in terms of a *direct* matrix element of the scalar product of two unit tensors, so that

$$g_k' = (-1)^S\langle l_1\|\mathrm{C}^{(k)}\|l_2\rangle^2\sum_r(-1)^r[r]\begin{Bmatrix} l_1 & l_2 & k\\ l_2 & l_1 & r\end{Bmatrix}\langle(l_1l_2)LS\,|\,u^{(r)}_{(\mathrm{i})}\cdot u^{(r)}_{(\mathrm{j})}\,|\,(l_1l_2)LS\rangle_{\mathrm{dir}}\ . \tag{11.90}$$

We see here the primary reason for introducing the unit tensor operator $u^{(k)}$—namely, to be able to factor out the exchange reduced matrix elements $\langle l_1\|\mathrm{C}^{(k)}\|l_2\rangle^2$, as opposed to the direct elements $\langle l_1\|\mathrm{C}^{(k)}\|l_1\rangle\,\langle l_2\|\mathrm{C}^{(k)}\|l_2\rangle$ present in (11.81), and thus leave the matrix elements of $u^{(r)}\cdot u^{(r)}$ to keep track only of coupling relationships rather than the magnitudes of one-electron angular integrals.

We next note that if $\mathbf{S} = \mathbf{s}_1 + \mathbf{s}_2$, then

$$\mathbf{S}^2 = \mathbf{s}_1{}^2 + 2\mathbf{s}_1\cdot\mathbf{s}_2 + \mathbf{s}_2{}^2$$

so that

$$\langle (s_1s_2)S \,|\, -1/2 - 2(s_1^{(1)}\cdot s_2^{(1)}) \,|\, (s_1s_2)S\rangle$$
$$= \langle (s_1s_2)S \,|\, -1/2 + s_1^{\,2} + s_2^{\,2} - S^2 \,|\, (s_1s_2)S\rangle$$
$$= -1/2 + 2s(s+1) - S(S+1)$$
$$= 1 - S(S+1) = (-1)^S\,, \qquad S = 0 \text{ or } 1\,. \tag{11.91}$$

Using this result to replace the $(-1)^S$ in (11.90), we obtain

$$g_k' = \langle l_1\|C^{(k)}\|l_2\rangle^2 \sum_r (-1)^r [r] \begin{Bmatrix} l_1 & l_1 & r \\ l_2 & l_2 & k \end{Bmatrix}$$
$$\times \langle (l_1l_2)LS \,|\, -(1/2)[u_{(i)}^{(r)}\cdot u_{(j)}^{(r)}] - 2[v_{(i)}^{(r1)}\cdot v_{(j)}^{(r1)}] \,|\, (l_1l_2)LS\rangle_{dir}\,, \tag{11.92}$$

where $v^{(r1)} = u^{(r)}s^{(1)}$ is the unit double tensor operator defined in Sec. 11-10.

This result may be generalized straightforwardly: For basis functions of a configuration involving the two subshells $l_i^{\,w_i}$ and $l_j^{\,w_j}$, instead of evaluating the *exchange* matrix element of the operator $C_{(i)}^{(k)}\cdot C_{(j)}^{(k)}$, we can obtain the coefficient g_k' of the exchange radial integral $G^k(l_il_j)$ by evaluating the *direct* matrix element of the effective operator

$$-\frac{1}{2}\langle l_i\|C^{(k)}\|l_j\rangle^2 \sum_r (-1)^r [r] \begin{Bmatrix} l_i & l_i & r \\ l_j & l_j & k \end{Bmatrix} [U_{(i)}^{(r)}\cdot U_{(j)}^{(r)} + 4V_{(i)}^{(r1)}\cdot V_{(j)}^{(r1)}]\,. \tag{11.93}$$

[The subscripts i and j on the operators **U** and **V** mean of course that these tensors operate on the coordinates associated with the spin-orbitals of the subshells i and j, respectively. It should be noted that, because **U** and **V** are sums of **u** and **v** over all coordinates of a subshell, the explicit factors w_i and w_j in (9.85) will now appear only implicitly within the expressions (11.53) and (11.68) for the reduced matrix elements of **U** and **V**.]

As indicated previously, the factor $\langle l_i\|C^{(k)}\|l_j\rangle^2$ represents the proper reduced matrix elements for an exchange interaction; the complex form of the remainder of the operator is mainly concerned with providing (as the result of the evaluation of a direct integral) the angular-momentum recoupling relationships appropriate to the evaluation of an exchange integral. The whole is somewhat complicated, and requires evaluation of two matrix elements (of **U·U** and **V·V**), as well as a summation over r. However, in complex configurations this proves to be a small price to pay for a universally applicable formalism that eliminates cfp expansions [or, rather, relegates them to the expressions (11.53) and (11.68), which can be evaluated once and for all], and also eliminates assorted angular-momentum recouplings which would have been different for different pairs of values i,j.

Problems

11-14(1). Outline the mathematical proof (involving cfp expansions and assorted recouplings) that justifies the generalization from the two-electron problem (11.92) to the general case (11.93).

11-14(2). Use Racah's methods to obtain an expression for the exchange matrix element (11.86) by (a) recoupling the ket function to the form (11.87) and working with the resulting "direct" matrix element, and (b) using the effective exchange operator in the original direct matrix element [i.e., in (11.86) without the minus sign, and with coordinate arrangement $l_1(\mathrm{i})l_2(\mathrm{j})l_3(\mathrm{t})$ in the ket function]. [Hints: Be careful with the various spin delta-factors; use (5.37) to evaluate the sum over $\mathfrak{L}_{13}'$ in (a) and the sum over r in (b); compare the final spin factors obtained in (a) and (b) by numerical evaluation for each possible set of values $(\mathfrak{S}_2\mathfrak{S}_2'\mathfrak{S}_3)$.]

11-15. SUMMARY

For convenience of reference, Appendix F summarizes the most important equations involved in the coupling and recoupling of angular-momentum eigenfunctions and in the application of Racah algebra. Included in Appendices G and H are short tables of cfp, cfgp, and reduced matrix elements, for illustrative purposes; for more complete tables, see Nielson and Koster[12] or [for $V^{(k1)}$ only] Karaziya et al.[14]

12

ENERGY LEVEL STRUCTURE (COMPLEX CONFIGURATIONS)

12-1. INTRODUCTION

The techniques of Racah algebra make it easy to derive closed algebraic expressions for Coulomb and spin-orbit matrix elements between basis functions of the general configuration (10.2). For purposes of numerical calculation of energy levels by numerical diagonalization of the Hamiltonian matrix (10.4), it is immaterial what pure-coupling representation is used. We shall use mainly the genealogical $\mathfrak{L}\mathfrak{S}$ scheme (4.37), with the slightly modified notation

$$\{[(l_1^{w_1}\alpha_1 L_1 S_1 \mathfrak{L}_1 \mathfrak{S}_1, l_2^{w_2}\alpha_2 L_2 S_2)\mathfrak{L}_2\mathfrak{S}_2, \cdots l_q^{w_q}\alpha_q L_q S_q]\mathfrak{L}_q\mathfrak{S}_q\}\mathfrak{J}_q\mathfrak{M}_q \ ; \tag{12.1}$$

here $\mathfrak{L}_1 \equiv L_1$ and $\mathfrak{S}_1 \equiv S_1$ are redundant quantum numbers introduced for convenience in later notation. This coupling scheme is not only mathematically convenient, but also provides the best single approximation to actual coupling conditions in atoms—i.e., it provides high-purity eigenvectors for a larger proportion of physically interesting configurations than any other single coupling scheme. However, once eigenvectors $Y_{\mathfrak{L}\mathfrak{S}}$ have been obtained in this representation, the corresponding eigenvectors in any other desired pure-coupling representation can readily be found with the aid of the appropriate transformation matrix.

Consider, for example, the $J\mathfrak{J}$-coupling scheme (4.44)

$$\{[(l_1^{w_1}\alpha_1 L_1 S_1)J_1\mathfrak{J}_1, (l_2^{w_2}\alpha_2 L_2 S_2)J_2]\mathfrak{J}_2, \cdots (l_q^{w_q}\alpha_q L_q S_q)J_q\}\mathfrak{J}_q\mathfrak{M}_q \ , \tag{12.2}$$

where $\mathfrak{J}_1 \equiv J_1$ is again a redundant quantum number introduced for notational convenience. Then from (9.29), the elements of the transformation matrix from the $\mathfrak{L}\mathfrak{S}$ to the $J\mathfrak{J}$ representation are

$$
\begin{aligned}
T_{J\mathfrak{J},\mathfrak{L}\mathfrak{S}} &\equiv \langle J\mathfrak{J}\,|\,\mathfrak{L}\mathfrak{S}\rangle \\
&= \langle[(\mathfrak{L}_1\mathfrak{S}_1)\mathfrak{J}_1,(L_2S_2)J_2]\mathfrak{J}_2\,|\,[(\mathfrak{L}_1L_2)\mathfrak{L}_2,(\mathfrak{S}_1S_2)\mathfrak{S}_2]\mathfrak{J}_2\rangle \\
&\quad\times \langle[(\mathfrak{L}_2\mathfrak{S}_2)\mathfrak{J}_2,(L_3S_3)J_3]\mathfrak{J}_3\,|\,[(\mathfrak{L}_2L_3)\mathfrak{L}_3,(\mathfrak{S}_2S_3)\mathfrak{S}_3]\mathfrak{J}_3\rangle \\
&\quad\times \cdots \times \langle[(\mathfrak{L}_{q-1}\mathfrak{S}_{q-1})\mathfrak{J}_{q-1},(L_qS_q)J_q]\mathfrak{J}_q\,|\,(\mathfrak{L}_{q-1}L_q)\mathfrak{L}_q,(\mathfrak{S}_{q-1}S_q)\mathfrak{S}_q]\mathfrak{J}_q\rangle \\
&= \prod_{i=2}^{q} [\mathfrak{L}_i,\mathfrak{S}_i,\mathfrak{J}_{i-1},J_i]^{1/2} \begin{Bmatrix} \mathfrak{L}_{i-1} & L_i & \mathfrak{L}_i \\ \mathfrak{S}_{i-1} & S_i & \mathfrak{S}_i \\ \mathfrak{J}_{i-1} & J_i & \mathfrak{J}_i \end{Bmatrix} .
\end{aligned} \tag{12.3}
$$

Eigenvectors in the $J\mathfrak{J}$ representation are then given by the matrix product

$$
Y_{J\mathfrak{J}} = T_{J\mathfrak{J},\mathfrak{L}\mathfrak{S}} \cdot Y_{\mathfrak{L}\mathfrak{S}} . \tag{12.4}
$$

If the Hamiltonian matrix is desired in the $J\mathfrak{J}$ representation, it can of course be found by calculating the double matrix product

$$
H_{J\mathfrak{J},J'\mathfrak{J}'} = T_{J\mathfrak{J},\mathfrak{L}\mathfrak{S}} \cdot H_{\mathfrak{L}\mathfrak{S},\mathfrak{L}'\mathfrak{S}'} \cdot T_{\mathfrak{L}'\mathfrak{S}',J'\mathfrak{J}'} . \tag{12.5}
$$

12-2. COULOMB MATRIX ELEMENTS: EQUIVALENT ELECTRONS

In the $\mathfrak{L}\mathfrak{S}$ representation, it follows from (11.43) and (11.45) that the Coulomb-interaction portion of the Hamiltonian matrix is diagonal in $\mathfrak{J}_q$ and $\mathfrak{M}_q$, and independent of $\mathfrak{M}_q$. Moreover, because the Coulomb operator acts only on $\mathfrak{L}$ and not on $\mathfrak{S}$, it follows from (9.87) that for the Coulomb matrix the coupling $(\mathfrak{L}_q\mathfrak{S}_q)\mathfrak{J}_q$ can be ignored, and the matrix is independent of $\mathfrak{J}_q$ and diagonal in $\mathfrak{L}_q$ and $\mathfrak{S}_q$. These facts will be implicitly understood throughout all that follows.

Let us consider first that portion of the Coulomb matrix element (9.85) in which the operator $g_{(jj)} = 2/r_{(jj)}$ involves the last two coordinates (m and n, say) of the subshell $l_j^{w_j}$. Using (11.43) and carrying out the radial integrations, we obtain

$$
\frac{w_j(w_j-1)}{2}\langle\psi_b\,|\sum_k \frac{2r_<^k}{r_>^{k+1}} C_{(m)}^{(k)}\cdot C_{(n)}^{(k)}\,|\,\psi_{b'}\rangle = \sum_k f_k'(l_jl_j)F^k(l_jl_j) , \tag{12.6}
$$

where F^k is the radial integral (6.28), and the angular integral is[1]

$$
f_k'(l_jl_j) = \frac{1}{2}w_j(w_j - 1)\langle\psi_b\,|\,C_{(m)}^{(k)}\cdot C_{(n)}^{(k)}\,|\,\psi_{b'}\rangle . \tag{12.7}
$$

The operator in this matrix element operates only on that portion of ψ_b involving $|l_j^{w_j}\rangle$; because the operator is a scalar, it follows from (11.45) that the matrix elements

[1]In (12.7), ψ_b represents only the angular portion of the basis function, the radial portions having already been integrated to give the two-electron radial integral F^k and $N - 2$ radial overlap integrals.

$$\langle l_j{}^{w_j}\alpha_j L_j S_j | C^{(k)}_{(m)} \cdot C^{(k)}_{(n)} | l_j{}^{w_j}\alpha'_j L'_j S'_j \rangle$$

satisfy the conditions on (9.86) such that (9.87) and (9.88) are valid. Using (9.87) q − j times, and then (9.88) once if j > 1, we immediately obtain

$$f'_k(l_j l_j) = \delta_j \frac{1}{2} w_j(w_j - 1)\langle l_j{}^{w_j}\alpha_j L_j S_j | C^{(k)}_{(m)} \cdot C^{(k)}_{(n)} | l_j{}^{w_j}\alpha'_j L_j S_j \rangle , \tag{12.8}$$

where δ_j is an abbreviation for δ-factors in all quantum numbers $\mathfrak{J}_q \mathfrak{M}_q$, $\alpha_i L_i S_i \mathfrak{L}_i \mathfrak{S}_i$ (all i) except α_j.

The matrix element in (12.8) could be evaluated with the aid of cfgp expansions (9.51), together with the two-electron result (11.81); however, this would involve a double summation over the terms of $l_j{}^{w_j-2}$ and those of $l_j{}^2$. The following alternative method[2] gives a more convenient result, involving a single summation of matrix elements of the symmetric unit operator $U^{(k)}$ defined by (11.52).

Because all w_j electron coordinates in $|l_j{}^{w_j}\rangle$ are equivalent, the value of (12.8) is independent of which two coordinates m and n are involved. Thus we may write, using also (11.49),

$$f'_k(l_j l_j) = \delta_j \frac{1}{2} \langle l_j \| C^{(k)} \| l_j \rangle^2 \langle l_j{}^{w_j}\alpha_j L_j S_j | \sum_{s \neq t}\sum u^{(k)}_{(s)} \cdot u^{(k)}_{(t)} | l_j{}^{w_j}\alpha'_j L_j S_j \rangle . \tag{12.9}$$

The operator may be written in the form

$$\sum_{s=1}^{w_j} \sum_{t=1}^{w_j} u^{(k)}_{(s)} \cdot u^{(k)}_{(t)} - \sum_{t=1}^{w_j} u^{(k)}_{(t)} \cdot u^{(k)}_{(t)} = U^{(k)} \cdot U^{(k)} - \sum_t u^{(k)}_{(t)} \cdot u^{(k)}_{(t)} . \tag{12.10}$$

The matrix element of the first term of (12.10) is seen from (11.45) and the δ-factor in (11.53) to be equal to

$$[L_j]^{-1} \sum_{\alpha'' L''} (-1)^{L_j - L''} \langle l_j{}^{w_j}\alpha_j L_j S_j \| U^{(k)} \| l_j{}^{w_j}\alpha'' L'' S_j \rangle \langle l_j{}^{w_j}\alpha'' L'' S_j \| U^{(k)} \| l_j{}^{w_j}\alpha'_j L_j S_j \rangle . \tag{12.11}$$

The matrix element of the second term of (12.10) may be evaluated with the aid of cfp expansions, (9.88), (11.45), (9.46), and (11.48):

$$\begin{aligned} -w_j \sum_{\bar\alpha \bar L \bar S} \sum_{\bar\alpha' \bar L' \bar S'} & (\alpha_j L_j S_j \{| \bar\alpha \bar L \bar S) \langle (l_j{}^{w_j-1}\bar\alpha \bar L \bar S, l_j) L_j S_j | u^{(k)}_{(n)} \cdot u^{(k)}_{(n)} | (l_j{}^{w_j-1}\bar\alpha' \bar L' \bar S', l_j) L_j S_j \rangle (\bar\alpha' \bar L' \bar S' |\} \alpha'_j L_j S_j) \\ &= - w_j \sum_{\bar\alpha \bar L \bar S} (\alpha_j L_j S_j \{| \bar\alpha \bar L \bar S) \langle l_j | u^{(k)} \cdot u^{(k)} | l_j \rangle (\bar\alpha \bar L \bar S |\} \alpha'_j L_j S_j) \\ &= - w_j \delta_{\alpha_j \alpha'_j} [l_j]^{-1} \langle l_j \| u^{(k)} \| l_j \rangle^2 \\ &= - w_j \delta_{\alpha_j \alpha'_j} [l_j]^{-1} . \end{aligned} \tag{12.12}$$

[2]F. R. Innes, Phys. Rev. 91, 31 (1953), Eq. (13).

Combining (12.9)-(12.12) and using (11.54), we obtain finally

$$f_k'(l_jl_j) = \delta_j \tfrac{1}{2} \langle l_j \| C^{(k)} \| l_j \rangle^2 \Big\{ [L_j]^{-1} \sum_{\alpha''L''} \langle l_j^{w_j}\alpha''L''S_j \| U^{(k)} \| l_j^{w_j}\alpha_j L_j S_j \rangle$$
$$\times \langle l_j^{w_j}\alpha''L''S_j \| U^{(k)} \| l_j^{w_j}\alpha_j' L_j S_j \rangle - \delta_{\alpha_j\alpha_j'} w_j [l_j]^{-1} \Big\} \ . \qquad (12.13)$$

If we wish to write the diagonal matrix elements in the form $E_{av} + \Delta H_{bb}$, as in (10.4), then we have yet to subtract that portion of (12.6) which belongs in E_{av}; i.e., we have yet to convert the diagonal values of f_k' to f_k by subtracting the average value of f_k'. For $k = 0$, we find from (11.25) and (11.55) that

$$f_0'(l_jl_j) = \delta_j \frac{1}{2} [l_j] \Big\{ w_j^2 [l_j]^{-1} \delta_{\alpha_j\alpha_j'} - \delta_{\alpha_j\alpha_j'} w_j [l_j]^{-1} \Big\}$$
$$= \delta_{bb'} \frac{1}{2} w_j (w_j - 1)$$
$$= \delta_{bb'} \cdot (\text{number of electron pairs}) \ . \qquad (12.14)$$

Thus the term $f_0'(l_jl_j)F^0(l_jl_j)$ in (12.6) is non-zero only for diagonal elements, and (being independent of b) is also its own average; i.e., the $k = 0$ term belongs entirely to E_{av} and makes no other contribution to the matrix elements, either on or off the diagonal. [This is in agreement with the first term of (6.39), which shows that each pair of equivalent electrons contributes F^0 to E_{av}.]

For $k > 0$, it is not feasible to find the average diagonal value of f_k' by a direct calculation. However, the correct value per pair of interacting electrons for any value of w_j should be equal to the value per pair in the case of a filled subshell, because the latter is spherically symmetric from Unsöld's theorem (4.35) and thus automatically provides an average over all angles. Using (11.62), we see that the value of (12.13) for $w_j = 4l_j + 2$ is

$$-\delta_{bb'} \frac{1}{2} \langle l_j \| C^{(k)} \| l_j \rangle^2 w_j [l_j]^{-1} = -\delta_{bb'} \frac{1}{2} w_j (w_j - 1) \frac{\langle l_j \| C^{(k)} \| l_j \rangle^2}{(2l_j + 1)(4l_j + 1)} \ ; \qquad (12.15)$$

from (11.23) it may be seen that this result (per electron pair) is indeed equivalent to the corresponding portion of (6.39).

Subtracting the average value (12.15) from (12.13), we find the equivalent-electron Coulomb contribution to the Hamiltonian matrix elements (over and above E_{av}) to be

$$\sum_j \sum_{k>0} f_k(l_jl_j) F^k(l_jl_j) \ , \qquad (12.16)$$

where

$$f_k(l_jl_j) = \delta_j \frac{1}{2} \langle l_j \| C^{(k)} \| l_j \rangle^2 \Big\{ [L_j]^{-1} \sum_{\alpha''L''} \langle l_j^{w_j}\alpha''L''S_j \| U^{(k)} \| l_j^{w_j}\alpha_j L_j S_j \rangle$$
$$\times \langle l_j^{w_j}\alpha''L''S_j \| U^{(k)} \| l_j^{w_j}\alpha_j' L_j S_j \rangle - \delta_{\alpha_j\alpha_j'} \frac{w_j(4l_j + 2 - w_j)}{(2l_j + 1)(4l_j + 1)} \Big\} \ ; \qquad (12.17)$$

as before, δ_j represents δ-factors in all quantum numbers except α_j. It follows directly from (11.58) that the values of $f_k(l_jl_j)$ are the same for a subshell with w_j electrons as for a subshell with w_j holes ($4l_j + 2 - w_j$ electrons). Thus, to the extent that the ratios of the various F^k are not greatly different in the two cases, the equivalent-electron contributions to the energy-level structure are basically the same for conjugate subshells.

It is easily seen from (11.57), (11.61), and (11.62) that (12.17) is zero for $w_j = 0$, 1, $4l_j + 1$, and $4l_j + 2$; i.e., that the Coulomb interactions among a set of equivalent electrons $l_j^{w_j}$ contribute nothing to the level structure if there are less than two electrons or less than two holes in the subshell. Of course, this is not a very profound result, because in each of those four cases there exists only one LS term so that there *is* no structure; also, for $w_j < 2$ there are not enough electrons to produce any interaction at all—not even a contribution to E_{av}. However, this does indicate that the summation over j in (12.16) need extend only over those subshells for which $2 \leqq w_j \leqq 4l_j$; in particular, filled subshells can be disregarded, as inferred from qualitative arguments in Sec. 10-2.

12-3. NON-EQUIVALENT ELECTRONS: DIRECT INTEGRAL

Substituting the Coulomb operator (11.43) into the direct portion of (9.85) for non-equivalent electrons, we find this contribution to the Hamiltonian matrix elements to be

$$\sum_{i<j} w_i w_j \langle \psi_b | \sum_k \frac{2r_<^k}{r_>^{k+1}} \mathbf{C}_{(m)}^{(k)} \cdot \mathbf{C}_{(n)}^{(k)} | \psi_{b'} \rangle = \sum_{i<j} \sum_k f_k(l_i l_j) F^k(l_i l_j) , \tag{12.18}$$

m and n being the final coordinates of the subshells $l_i^{w_i}$ and $l_j^{w_j}$, respectively. The coefficient of the radial integral $F^k(l_il_j)$ is the angular matrix element

$$f_k(l_i l_j) = \langle l_i \| \mathbf{C}^{(k)} \| l_i \rangle \langle l_j \| \mathbf{C}^{(k)} \| l_j \rangle \, I_{ij}^{(k)} , \tag{12.19}$$

where

$$I_{ij}^{(k)} \equiv w_i w_j \langle \psi_b | \mathbf{u}_{(m)}^{(k)} \cdot \mathbf{u}_{(n)}^{(k)} | \psi_{b'} \rangle = \langle \psi_b | \mathbf{U}_{(i)}^{(k)} \cdot \mathbf{U}_{(j)}^{(k)} | \psi_{b'} \rangle \tag{12.20}$$

and $\mathbf{U}_{(i)}^{(k)}$ means the sum of $\mathbf{u}_{(t)}^{(k)}$ over all w_i coordinates t involved in $|l_i^{w_i}\rangle$. The operator in (12.20) is much simpler than that of (12.9)-(12.10) because there are no self-interaction terms s = t to be avoided.

Because the operator in (12.20) acts only on the orbital angular momentum, the matrix will be diagonal in all spins—i.e., there will be δ-factors

$$\delta_{\mathfrak{J}_q \mathfrak{M}_q, \mathfrak{J}'_q \mathfrak{M}'_q} \prod_{m=1}^{q} \delta_{S_m \mathfrak{S}_m, S'_m \mathfrak{S}'_m} \tag{12.21}$$

—and we can ignore all spin couplings in the evaluation of (12.20). In addition, the operator has the properties of the operator O_i in (9.86) with respect to orbital angular momentum; we can therefore apply (9.87) q − j times, thereby eliminating the subshells q, q − 1, ⋯ j + 1 and obtaining the further δ-factors

$$\delta_{\mathfrak{L}_j\mathfrak{L}'_j} \prod_{m=j+1}^{q} \delta_{\alpha_m L_m,\alpha'_m L'_m} \delta_{\mathfrak{L}_m \mathfrak{L}'_m} \, . \tag{12.22}$$

The evaluation of the remaining matrix element

$$\langle (\cdots \mathfrak{L}_{j-1}, L_j)\mathfrak{L}_j | U^{(k)}_{(i)} \cdot U^{(k)}_{(j)} | (\cdots \mathfrak{L}'_{j-1}, L'_j)\mathfrak{L}_j \rangle \tag{12.23}$$

involves use of the uncoupling formulae (11.38) and (11.39). In order to simplify the notation, we introduce the abbreviations

$$\langle \alpha_1 j_1 \alpha_2 j_2 j \| T^{(k)} \| \alpha'_1 j'_1 \alpha'_2 j'_2 j' \rangle = \delta_{\alpha_2 j_2, \alpha'_2 j'_2} U_a(j_1 j_2 j; k; j'_1 j') \langle \alpha_1 j_1 \| T^{(k)} \| \alpha'_1 j'_1 \rangle \tag{12.24}$$

and

$$\langle \alpha_1 j_1 \alpha_2 j_2 j \| W^{(k)} \| \alpha'_1 j'_1 \alpha'_2 j'_2 j' \rangle = \delta_{\alpha_1 j_1, \alpha'_1 j'_1} U_b(j_1 j_2 j; k; j'_2 j') \langle \alpha_2 j_2 \| W^{(k)} \| \alpha'_2 j'_2 \rangle \, , \tag{12.25}$$

where[3]

$$U_a(j_1 j_2 j; k; j'_1 j') = (-1)^{j_1+j_2+j'+k} [j, j']^{1/2} \begin{Bmatrix} j_1 & j_2 & j \\ j' & k & j'_1 \end{Bmatrix} \, , \tag{12.26}$$

$$U_b(j_1 j_2 j; k; j'_2 j') = (-1)^{j_1+j'_2+j+k} [j, j']^{1/2} \begin{Bmatrix} j_1 & j_2 & j \\ k & j' & j'_2 \end{Bmatrix} \, . \tag{12.27}$$

Evaluation of (12.23) is easily accomplished by first using (11.47) to uncouple L_j from $\mathfrak{L}_{j-1}$; then (if $i < j-1$) using (12.24) $j-i-1$ times to successively uncouple $L_{j-1}, L_{j-2}, \cdots L_{i+1}$; and then finally (if $i > 1$) using (12.25) to uncouple L_i from $\mathfrak{L}_{i-1}$. The result of all this is δ-factors in all quantum numbers except $\alpha_i L_i$, $\alpha_j L_j$, and $\mathfrak{L}_m$ ($i \leq m < j$), and the non-zero values of $I^{(k)}_{ij}$ are[4]

$$\begin{aligned} I^{(k)}_{ij} &= (-1)^{\mathfrak{L}'_{j-1}+L_j+\mathfrak{L}_j} \begin{Bmatrix} \mathfrak{L}_{j-1} & L_j & \mathfrak{L}_j \\ L'_j & \mathfrak{L}'_{j-1} & k \end{Bmatrix} \\ &\quad \times \langle \cdots \mathfrak{L}_{j-1} \| U^{(k)}_{(i)} \| \cdots \mathfrak{L}'_{j-1} \rangle \langle l_j^{w_j} \alpha_j L_j S_j \| U^{(k)} \| l_j^{w_j} \alpha'_j L'_j S_j \rangle \\ &= (-1)^{\mathfrak{L}'_{j-1}+L_j+\mathfrak{L}_j} \begin{Bmatrix} \mathfrak{L}_{j-1} & L_j & \mathfrak{L}_j \\ L'_j & \mathfrak{L}'_{j-1} & k \end{Bmatrix} \left[\prod_{m=i+1}^{j-1} U_a(\mathfrak{L}_{m-1} L_m \mathfrak{L}_m; k; \mathfrak{L}'_{m-1} \mathfrak{L}'_m) \right] \\ &\quad \times \langle \cdots \mathfrak{L}_i \| U^{(k)}_{(i)} \| \cdots \mathfrak{L}'_i \rangle \langle l_j^{w_j} \alpha_j L_j S_j \| U^{(k)} \| l_j^{w_j} \alpha'_j L'_j S_j \rangle \\ &= (-1)^{\mathfrak{L}'_{j-1}+L_j+\mathfrak{L}_j} \begin{Bmatrix} \mathfrak{L}_{j-1} & L_j & \mathfrak{L}_j \\ L'_j & \mathfrak{L}'_{j-1} & k \end{Bmatrix} \left[\prod_{m=i+1}^{j-1} U_a(\mathfrak{L}_{m-1} L_m \mathfrak{L}_m; k; \mathfrak{L}'_{m-1} \mathfrak{L}'_m) \right] \\ &\quad \times [\delta_{i1} + (1-\delta_{i1}) U_b(\mathfrak{L}_{i-1} L_i \mathfrak{L}_i; k; L'_i \mathfrak{L}'_i)] \\ &\quad \times \langle l_i^{w_i} \alpha_i L_i S_i \| U^{(k)} \| l_i^{w_i} \alpha'_i L'_i S_i \rangle \langle l_j^{w_j} \alpha_j L_j S_j \| U^{(k)} \| l_j^{w_j} \alpha'_j L'_j S_j \rangle \, . \end{aligned} \tag{12.28}$$

[3]The uncoupling coefficients U_a and U_b are of course not to be confused with the unit tensor operators $U^{(k)}$. In programming a computer to calculate angular matrix elements, it is convenient to have not only a subroutine to calculate 6-j symbols but also individual subroutines for U_a and U_b.

[4]R. D. Cowan, J. Opt. Soc. Am. 58, 808 (1968).

For k = 0, this expression reduces with the aid of (5.29) and (11.55) to

$$I_{ij}^{(0)} = \delta_{\mathcal{L}_{j-1}\mathcal{L}'_{j-1}}\delta_{L_jL'_j}\frac{1}{[\mathcal{L}_{j-1},L_j]^{1/2}}\left[\prod_{m=i+1}^{j-1}\delta_{\mathcal{L}_{m-1}\mathcal{L}'_{m-1}}\delta_{\mathcal{L}_m\mathcal{L}'_m}\frac{[\mathcal{L}_m]^{1/2}}{[\mathcal{L}_{m-1}]^{1/2}}\right]$$

$$\times\left[\delta_{i1} + (1-\delta_{i1})\delta_{\mathcal{L}_i\mathcal{L}'_i}\frac{[\mathcal{L}_i]^{1/2}}{[L_i]^{1/2}}\right]\delta_{\alpha_iL_i,\alpha'_iL'_i}\delta_{\alpha_j\alpha'_j}w_iw_j\frac{[L_i,L_j]^{1/2}}{[l_i,l_j]^{1/2}} \qquad (12.29)$$

$$= \delta_{bb'}w_iw_j[l_i,l_j]^{-1/2} \; ; \qquad (12.30)$$

the result (12.30) follows from (12.29) because $\mathcal{L}_1 \equiv L_1$, and because (12.21), (12.22), and (12.29) together include δ-factor in all quantum numbers. Substituting (12.30) into (12.19) and using (11.25), we find

$$f_0(l_il_j) = \delta_{bb'}w_iw_j = \delta_{bb'}\cdot(\text{number of electron pairs}) \; . \qquad (12.31)$$

Thus the k = 0 terms in (12.18) represent contributions only to E_{av} [see also (6.38)], and make no other contribution to the Hamiltonian matrix elements, either on or off the diagonal.

For k > 0 and either $w_i = 4l_i + 2$ or $w_j = 4l_j + 2$, we see from (11.62) that $f_k = 0$; from this we conclude that the average value of f_k for any values of w_i and w_j is zero, so that (12.19), (12.28) requires no correction for contributions to E_{av}; this agrees with the absence from (6.38) of any terms in F^k for k > 0. Likewise, if either the subshell $l_i^{w_i}$ or the subshell $l_j^{w_j}$ is actually filled in the configuration in question, then f_k is zero for k > 0. Thus the contribution to matrix elements over and above E_{av} is

$$\sum_{i<j}\sum_{k>0} f_k(l_il_j)F^k(l_il_j) \; , \qquad (12.32)$$

where the summation over i and j need be carried only over non-filled subshells, $1 \leqq w \leqq 4l + 1$.

12-4. NON-EQUIVALENT ELECTRONS: EXCHANGE INTEGRAL

Similarly to (12.18), the exchange portion of (9.85) for non-equivalent electrons is of the form

$$\sum_{i<j}\sum_{k} g'_k(l_il_j)G^k(l_il_j) \; . \qquad (12.33)$$

However, rather than attempt to evaluate the angular coefficient g'_k as an exchange integral of the operator $U_{(i)}^{(k)}\cdot U_{(j)}^{(k)}$ analogously to (12.20), we shall evaluate the direct integral of the effective operator (11.93):

$$g'_k(l_il_j) = -\frac{1}{2}\langle l_i\|C^{(k)}\|l_j\rangle^2\sum_r(-1)^r[r]\begin{Bmatrix} l_i & l_i & r \\ l_j & l_j & k \end{Bmatrix}\langle\psi_b|U_{(i)}^{(r)}\cdot U_{(j)}^{(r)} + 4V_{(i)}^{(r1)}\cdot V_{(j)}^{(r1)}|\psi_{b'}\rangle$$

$$= -\frac{1}{2}\langle l_i\|C^{(k)}\|l_j\rangle^2\sum_r(-1)^r[r]\begin{Bmatrix} l_i & l_i & r \\ l_j & l_j & k \end{Bmatrix}[I_{ij}^{(r)} + 4I_{ij}^{(r1)}] \; . \qquad (12.34)$$

Here $I^{(r)}_{ij}$ is identical with the quantity (12.28) except with k replaced by r; it is still diagonal in all spins, and in all other quantum numbers except $\alpha_i L_i$, $\alpha_j L_j$, and $\mathfrak{L}_m$ ($i \leqq m < j$). The operator $V^{(r1)}_{(i)} \cdot V^{(r1)}_{(j)}$ operates on spins as well as orbital momenta, and therefore $I^{(r1)}_{ij}$ is diagonal only in quantum numbers other than $\alpha_i L_i S_i$, $\alpha_j L_j S_j$, and $\mathfrak{L}_m \mathfrak{S}_m$ ($i \leqq m < j$). $I^{(r1)}$ is evaluated by the same procedure that was used for $I^{(r)}$ except that the uncoupling of spins as well as orbital momenta must be taken into account; the result is of exactly the same form as $I^{(r)}$ except that for every factor involving the orbital momenta there is a similar factor that involves the corresponding spins, and the reduced matrix elements are those of $V^{(r1)}$ instead of $U^{(r)}$. Thus the non-zero values are

$$I^{(r1)}_{ij} = (-1)^{\mathfrak{L}'_{j-1}+\mathfrak{S}'_{j-1}+L_j+S_j+\mathfrak{L}_j+\mathfrak{S}_j} \begin{Bmatrix} \mathfrak{L}_{j-1} & L_j & \mathfrak{L}_j \\ L'_j & \mathfrak{L}'_{j-1} & r \end{Bmatrix} \begin{Bmatrix} \mathfrak{S}_{j-1} & S_j & \mathfrak{S}_j \\ S'_j & \mathfrak{S}'_{j-1} & 1 \end{Bmatrix}$$
$$\times \left[\prod_{m=i+1}^{j-1} U_a(\mathfrak{L}_{m-1} L_m \mathfrak{L}_m; r; \mathfrak{L}'_{m-1} \mathfrak{L}'_m) U_a(\mathfrak{S}_{m-1} S_m \mathfrak{S}_m; 1; \mathfrak{S}'_{m-1} \mathfrak{S}'_m) \right]$$
$$\times [\delta_{i1} + (1-\delta_{i1}) U_b(\mathfrak{L}_{i-1} L_i \mathfrak{L}_i; r; L'_i \mathfrak{L}'_i) U_b(\mathfrak{S}_{i-1} S_i \mathfrak{S}_i; 1; S'_i \mathfrak{S}'_i)]$$
$$\times \langle l_i^{w_i} \alpha_i L_i S_i \| V^{(r1)} \| l_i^{w_i} \alpha'_i L'_i S'_i \rangle \langle l_j^{w_j} \alpha_j L_j S_j \| V^{(r1)} \| l_j^{w_j} \alpha'_j L'_j S'_j \rangle \, . \quad (12.35)$$

As in Sec. 12-2, we can find the configuration-average value of g'_k by considering its value for filled subshells. For either $w_i = 4l_i + 2$ or $w_j = 4l_j + 2$, it follows from (11.62) and (11.76) that $I^{(r)} = 0$ ($r > 0$) and $I^{(r1)} = 0$ (all r), so that from (12.30) and (5.29)

$$\langle g'_k(l_i l_j) \rangle_{av} = -\frac{1}{2} \langle l_i \| C^{(k)} \| l_j \rangle^2 \begin{Bmatrix} l_i & l_j & k \\ l_j & l_i & 0 \end{Bmatrix} \delta_{bb'} w_i w_j [l_i, l_j]^{-1/2}$$
$$= -\delta_{bb'} \frac{1}{2} \langle l_i \| C^{(k)} \| l_j \rangle^2 w_i w_j [l_i, l_j]^{-1} \, , \quad (12.36)$$

where the last step follows from the fact that $l_i + l_j + k$ must be even from (11.23) and (5.13). Subtracting this result from (12.34), we find the contribution to the Hamiltonian matrix elements over and above E_{av} to be

$$\sum_{i<j} \sum_k g_k(l_i l_j) G^k(l_i l_j) \, , \quad (12.37)$$

where

$$g_k(l_i l_j) = \frac{1}{2} \langle l_i \| C^{(k)} \| l_j \rangle^2 \left[\frac{\delta_{bb'} w_i w_j}{[l_i, l_j]} - \sum_r (-1)^r [r] \begin{Bmatrix} l_i & l_i & r \\ l_j & l_j & k \end{Bmatrix} [I^{(r)}_{ij} + 4 I^{(r1)}_{ij}] \right] \, , \quad (12.38)$$

$I^{(r)}$ and $I^{(r1)}$ being given by (12.28) and (12.35), respectively, with δ-factors as listed in the text. Because of the manner in which (12.38) was constructed from (12.34) and (12.36) it is clear that $g_k(l_i l_j)$ is zero if either $l_i^{w_i}$ or $l_j^{w_j}$ is a filled subshell; hence the ij summation in (12.37) need run over only those subshells for which $1 \leqq w_i \leqq 4l_i + 1$ and $1 \leqq w_j \leqq 4l_j + 1$.

12-5. SPIN-ORBIT MATRIX ELEMENTS

From (9.80), matrix elements of the spin-orbit operator (11.44) are of the form

$$\langle \Psi_b | \sum_{i=1}^{N} \xi_i(r) l_i^{(1)} \cdot s_i^{(1)} | \Psi_{b'} \rangle = \sum_{j=1}^{q} d_j \zeta_j , \qquad (12.39)$$

where ζ_j is the radial integral (10.9) and the angular coefficient for the jth subshell may be written in terms of the unit operator (11.51) as

$$d_j = w_j \langle l_j \| l^{(1)} \| l_j \rangle \langle \psi_b | u_{(j)}^{(1)} \cdot s_{(j)}^{(1)} | \psi_{b'} \rangle . \qquad (12.40)$$

$\mathfrak{LS}$ representation. In the $\mathfrak{LS}$ representation we evaluate the matrix element in (12.40) by first applying the uncoupling relation (11.47) for scalar products to the coupling $(\mathfrak{L}_q \mathfrak{S}_q) \mathfrak{J}_q$. In addition to the factor $\delta_{\mathfrak{J}_q \mathfrak{M}_q, \mathfrak{J}'_q \mathfrak{M}'_q}$ we obtain

$$w_j \langle \cdots (\mathfrak{L}_q \mathfrak{S}_q) \mathfrak{J}_q | u_{(j)}^{(1)} \cdot s_{(j)}^{(1)} | \cdots (\mathfrak{L}'_q \mathfrak{S}'_q) \mathfrak{J}_q \rangle$$

$$= w_j (-1)^{\mathfrak{L}'_q + \mathfrak{S}_q + \mathfrak{J}_q} \begin{Bmatrix} \mathfrak{L}_q & \mathfrak{S}_q & \mathfrak{J}_q \\ \mathfrak{S}'_q & \mathfrak{L}'_q & 1 \end{Bmatrix} \langle \cdots \mathfrak{L}_q \| u_{(j)}^{(1)} \| \cdots \mathfrak{L}'_q \rangle \langle \cdots \mathfrak{S}_q \| s_{(j)}^{(1)} \| \cdots \mathfrak{S}'_q \rangle$$

$$= (-1)^{\mathfrak{L}'_q + \mathfrak{S}_q + \mathfrak{J}_q} \begin{Bmatrix} \mathfrak{L}_q & \mathfrak{S}_q & \mathfrak{J}_q \\ \mathfrak{S}'_q & \mathfrak{L}'_q & 1 \end{Bmatrix} \langle \cdots \mathfrak{L}_q \mathfrak{S}_q \| V_{(j)}^{(11)} \| \cdots \mathfrak{L}'_q \mathfrak{S}'_q \rangle , \qquad (12.41)$$

where the symmetric double-tensor (11.67) has been introduced to make the notation both more compact and more logical. Simplification of the reduced matrix element in (12.41) follows much the same procedure as that used for $I^{(r1)}$ in (12.35): if $j < q$, we use (12.24) repeatedly to successively uncouple $L_q S_q$, $L_{q-1} S_{q-1}$, $\cdots$ $L_{j+1} S_{j+1}$; then if $j > 1$ we use (12.25) once to uncouple $L_j S_j$ from $\mathfrak{L}_{j-1} \mathfrak{S}_{j-1}$. The final result, using also (11.21), is

$$d_j = \delta_{\mathfrak{J}_q \mathfrak{M}_q, \mathfrak{J}'_q \mathfrak{M}'_q} \Big(\prod_{m \neq j} \delta_{\alpha_m L_m S_m, \alpha'_m L'_m S'_m} \Big) \Big(\prod_{m<j} \delta_{\mathfrak{L}_m \mathfrak{S}_m, \mathfrak{L}'_m \mathfrak{S}'_m} \Big)$$

$$\times (-1)^{\mathfrak{L}'_q + \mathfrak{S}_q + \mathfrak{J}_q} \begin{Bmatrix} \mathfrak{L}_q & \mathfrak{S}_q & \mathfrak{J}_q \\ \mathfrak{S}'_q & \mathfrak{L}'_q & 1 \end{Bmatrix}$$

$$\times \Big[\prod_{m=j+1}^{q} U_a(\mathfrak{L}_{m-1} L_m \mathfrak{L}_m ; 1 ; \mathfrak{L}'_{m-1} \mathfrak{L}'_m) U_a(\mathfrak{S}_{m-1} S_m \mathfrak{S}_m ; 1 ; \mathfrak{S}'_{m-1} \mathfrak{S}'_m) \Big]$$

$$\times [\delta_{j1} + (1 - \delta_{j1}) U_b(\mathfrak{L}_{j-1} L_j \mathfrak{L}_j ; 1 ; L'_j \mathfrak{L}'_j) U_b(\mathfrak{S}_{j-1} S_j \mathfrak{S}_j ; 1 ; S'_j \mathfrak{S}'_j)]$$

$$\times [l_j(l_j + 1)(2l_j + 1)]^{1/2} \langle l_j^{w_j} \alpha_j L_j S_j \| V^{(11)} \| l_j^{w_j} \alpha'_j L'_j S'_j \rangle . \qquad (12.42)$$

[It should be noted that if $\mathfrak{SL}$- instead of $\mathfrak{LS}$-coupled basis functions had been used, then the phase factor $(-1)^{\mathfrak{L}'_q + \mathfrak{S}_q + \mathfrak{J}_q}$ in (12.41)-(12.42) would have been $(-1)^{\mathfrak{S}'_q + \mathfrak{L}_q + \mathfrak{J}_q}$. This is consistent with (9.10), since $\mathfrak{L}_q$ and $\mathfrak{L}'_q$ are integral.]

$J\mathfrak{J}$ representation. If the spin-orbit coefficient (12.40) is evaluated for basis functions constructed according to the $J\mathfrak{J}$-coupling scheme (12.2), the result is quite simple because all uncouplings of the various J_i can be accomplished with the aid of (9.87) and (9.88), to give

$$d_j = \left(\prod_{m=1}^{q} \delta_{J_m\mathfrak{J}_m, J'_m\mathfrak{J}'_m}\right)\left(\prod_{m\neq j} \delta_{\alpha_m L_m S_m, \alpha'_m L'_m S'_m}\right)\langle l_j \| \boldsymbol{l}^{(1)} \| l_j\rangle$$
$$\times\; w_j \langle l_j^{w_j}\alpha_j L_j S_j J_j \,|\, u^{(1)}\cdot s^{(1)} \,|\, l_j^{w_j}\alpha'_j L'_j S'_j J_j\rangle$$
$$= \left(\prod_{m=1}^{q} \delta_{J_m\mathfrak{J}_m, J'_m\mathfrak{J}'_m}\right)\left(\prod_{m\neq j} \delta_{\alpha_m L_m S_m, \alpha'_m L'_m S'_m}\right)(-1)^{L'_j+S_j+J_j}\begin{Bmatrix} L_j & S_j & J_j \\ S'_j & L'_j & 1 \end{Bmatrix}$$
$$\times\; [l_j(l_j+1)(2l_j+1)]^{1/2}\langle l_j^{w_j}\alpha_j L_j S_j \| V^{(11)} \| l_j^{w_j}\alpha'_j L'_j S'_j\rangle \;, \tag{12.43}$$

where (11.47) has been used to undo the coupling $(L_jS_j)J_j$. [The phase factor $(-1)^{L'_j+S_j+J_j}$ would of course have been $(-1)^{S'_j+L_j+J_j}$ if the coupling $(S_jL_j)J_j$ had been used in forming the basis functions.]

The result (12.43) is much simpler than (12.42). However, the matrix elements (12.43) cannot be used as they stand with the Coulomb matrix elements (12.16)-(12.17), (12.19), (12.28), (12.32), (12.35), and (12.37)-(12.38); they must first be transformed from the $J\mathfrak{J}$ to the $\mathfrak{L}\mathfrak{S}$ representation with the aid of the $\mathfrak{L}\mathfrak{S}$-$J\mathfrak{J}$ transformation matrix (12.3).

If the $\mathfrak{L}\mathfrak{S}$-$J\mathfrak{J}$ transformation matrices (one for each value of $\mathfrak{J}_q$) are going to be computed anyway in order to obtain eigenvectors in both representations from (12.4), then for small matrices there is not a great deal to choose between the two approaches. However, for very large matrices (larger than about 100-by-100), the computing time involved in transforming each d_j matrix (12.43) to the $\mathfrak{L}\mathfrak{S}$ representation more than makes up for the added complexity of using (12.42) directly.

Regardless of the representation used, the reduced matrix element of $V^{(11)}$ is zero for a closed subshell from (11.76), so that d_j is also zero; the summation over j in (12.39) need therefore be carried only over partially filled subshells, $1 \leq w_j \leq 4l_j + 1$. (The fact that d_j is zero for filled subshells also indicates that the configuration-~~average~~ value of d_j is zero for an open subshell, so that the spin-orbit interaction makes no contribution to E_{av}.) This result, together with similar conclusions in each of the last three sections, shows that closed subshells can be completely ignored so far as their contributions to energy-level structures are concerned (in the single-configuration approximation), as was discussed qualitatively in Sec. 10-2.

Problems

12-5(1). Simplify the expressions derived above for $f_k(l_jl_j)$, $f_k(l_il_j)$, $g_k(l_il_j)$, and d_j (both representations) to the special case of the configuration $l_1^{w_1}l_2$. Simplify further to the case l_1l_2, and verify that the results agree with (11.81), (11.85), and (11.77).

12-5(2). For the configuration $l_1^{w_1}l_2$, derive expressions for the coefficients of $F^{(k)}(l_1l_1)$, $F^k(l_1l_2)$, and ζ_1, using J_1K-coupled basis functions

$|\{[(\alpha_1L_1S_1)J_1,l_2]K,s_2\}J\rangle$.

12-5(3). From the results of the preceding problem, obtain expressions for the energy levels of a pf configuration under pure pair-coupling conditions—i.e., neglecting the (Ks_2) interaction. [See K. B. S. Eriksson, Phys. Rev. **102**, 102 (1956).]

12-6. EXAMPLE: f^3sd^2

To the approximation in which we are working (i.e., neglecting magnetic spin-spin, orbit-orbit, and spin-other-orbit interactions), the energy-level structure of the various quantum states of a configuration is determined only by the electron-electron Coulomb interactions and the spin-orbit magnetic interactions.

As a specific example of the radial parameters and angular coefficients involved, we consider the low-lying configuration of neutral uranium

$$\text{U I} \quad 1s^22s^22p^6 \cdots 6s^26p^65f^36d^27s \ , \tag{12.44}$$

which is obtained from the ground configuration $\cdots$ $5f^36d7s^2$ by exciting a 7s electron to the 6d subshell. For the numerical calculation of E_{av} using (8.4)-(8.9), all 92 electrons in the 18 subshells of (12.44) must of course be taken into account. However, for the energy-level *structure* (relative to E_{av}), only the open subshells need be considered, and we may abbreviate (12.44) to

$$\text{U I} \quad 5f^37s6d^2 \ , \tag{12.45}$$

or for a strictly qualitative example, simply

$$f^3sd^2 \ . \tag{12.46}$$

[The coupling representation $(f^3d^2)s$ implied by (12.44) gives a closer approximation to physical reality than the representation $(f^3s)d^2$ implied by (12.45), but we consider the latter because it illustrates better the matrix properties of the spin-orbit coefficients.]

For any given value of $\mathfrak{J}_q$, we may write the energy matrix in the form

$$\begin{aligned}(H_{bb'}) = (\delta_{bb'})E_{av} &+ \sum_{j=1}^{q}\Big[\sum_{k>0}\big(f_k(l_jl_j)\big)F^k(l_jl_j) + (d_j)\zeta_j\Big] \\ &+ \sum_{i=1}^{q-1}\sum_{j=i+1}^{q}\Big[\sum_{k>0}\big(f_k(l_il_j)\big)F^k(l_il_j) + \sum_k\big(g_k(l_il_j)\big)G^k(l_il_j)\Big] \ ,\end{aligned} \tag{12.47}$$

where in the present example q is equal to 3. The parenthetical notation $(H_{bb'})$ means a matrix with elements $H_{bb'}$, etc;[5] that is, the Hamiltonian matrix may be thought of as a linear combination of terms, each consisting of an angular coefficient matrix multiplied by its corresponding radial integral. The rows and columns of each matrix are labeled by the

[5]For simplicity, the basis-state subscripts bb′ have been omitted from the Coulomb and spin-orbit angular coefficients f_k, g_k, and d.

various possible sets of values of the quantum numbers—for the $\mathfrak{L}\mathfrak{S}$ representation (12.1), these are [in addition to the configuration labels (12.46)]

$$b = \alpha_1 L_1 S_1 \, \alpha_2 L_2 S_2 \mathfrak{L}_2 \mathfrak{S}_2 \, \alpha_3 L_3 S_3 \mathfrak{L}_3 \mathfrak{S}_3 \, \mathfrak{J}_3 \mathfrak{M}_3 \; . \tag{12.48}$$

In the present example, α_2 and α_3 are unnecessary (because for s and for d^2 there is no more than one term with given $L_j S_j$), and L_2 and S_2 are necessarily equal to 0 and 1/2, respectively; however, we shall retain the full set of quantum numbers for greater generality.

Each parameter that contributes to the level calculation, and the quantum numbers in which the corresponding coefficient matrix is diagonal, are listed in Table 12-1. The Coulomb terms collectively produce an energy matrix that is diagonal only in $\mathfrak{L}_3\mathfrak{S}_3$ (and $\mathfrak{J}_3\mathfrak{M}_3$). Therefore, even if spin-orbit terms are neglected, each eigenvector will be a mixture of all those basis states having some common value of $\mathfrak{L}_3\mathfrak{S}_3\mathfrak{J}_3\mathfrak{M}_3$; the only "good" quantum numbers for an energy state will be these four numbers. When the spin-orbit terms are added, each matrix of given $\mathfrak{J}_3$ (and $\mathfrak{M}_3$) is in general not diagonal in *any* other quantum number; each eigenvector will be a mixture of *all* basis states of the matrix, and the only remaining good quantum numbers will be $\mathfrak{J}_3\mathfrak{M}_3$.

In spite of this mixing of basis states, it is customary to attempt to designate each eigenstate with the quantum numbers of one of this (or of some other pure-coupling) set of basis functions. As discussed in Sec. 10-7, such designations will be physically meaningful only if there exists (and is used!) a pure-coupling representation for which the eigenvectors have fairly high purities.

TABLE 12-1. CONTRIBUTORS TO THE LEVEL STRUCTURE OF f^3sd^2

Parameter	k	Coeff. matrix diagonal in:[a]
E_{av}	---	everything (unit matrix)
$F^k(ff)$	2,4,6	everything except α_1
$F^k(dd)$	2,4	everything except α_3
ζ_f	---	$\alpha_2 L_2 S_2$, $\alpha_3 L_3 S_3$
ζ_d	---	$\alpha_1 L_1 S_1$, $\alpha_2 L_2 S_2$, $\mathfrak{L}_2\mathfrak{S}_2$
[$F^k(fs)$	0	contributes to E_{av} only]
$F^k(fd)$	2,4	all spins, $\mathfrak{L}_3$, $\alpha_2 L_2$
[$F^k(sd)$	0	contributes to E_{av} only]
$G^k(fs)$	3	$\mathfrak{L}_3\mathfrak{S}_3$, $\mathfrak{L}_2\mathfrak{S}_2$, $\alpha_3 L_3 S_3$
$G^k(fd)$	1,3,5	$\mathfrak{L}_3\mathfrak{S}_3$, $\alpha_2 L_2 S_2$
$G^k(sd)$	2	$\mathfrak{L}_3\mathfrak{S}_3$, $\alpha_1 L_1 S_1$

[a]In addition to $\mathfrak{J}_3\mathfrak{M}_3$.

12-7. LEVEL STRUCTURE: SPIN-ORBIT EFFECTS UNDER NEAR-$\mathfrak{LS}$ COUPLING CONDITIONS

The Coulomb matrix elements depend on the various quantum numbers $\alpha_j L_j S_j \mathfrak{L}_j \mathfrak{S}_j$, but not on $\mathfrak{J}_q$ (nor $\mathfrak{M}_q$); thus, so far as Coulomb interactions are concerned, the energy levels are completely independent of $\mathfrak{J}_q$ (and $\mathfrak{M}_q$). A dependence of the energy on $\mathfrak{J}_q$—i.e., a splitting of the levels of an energy term $\mathfrak{L}_q\mathfrak{S}_q$ according to the value of $\mathfrak{J}_q$—arises only as a result of spin-orbit interactions (cf. Fig. 4-5). The discussion of spin-orbit effects involves primarily the total-angular-momentum quantum numbers $\mathfrak{L}_q\mathfrak{S}_q\mathfrak{J}_q$; to simplify the notation we shall replace these by the unsubscripted Roman letters LSJ. Under good LS-coupling conditions, we shall use γ to distinguish different levels having the same values of LSJ; thus an energy term will be designated γLS, and a level will be designated γLSJ.

In the LS representation it may be seen from (12.42) that the entire J dependence of the spin-orbit coefficient matrix element d_j is contained in the factor

$$(-1)^{L'+S+J}\begin{Bmatrix} L & S & J \\ S' & L' & 1 \end{Bmatrix} . \qquad (12.49)$$

This factor is common to d_j for all subshells $l_j^{w_j}$, and so the complete spin-orbit matrix element (12.39) is also proportional to (12.49).

For configurations in which pure LS-coupling conditions are closely approximated, appreciable Coulomb mixing of basis states occurs only among states having common values of LS. Provided spin-orbit splittings are small compared with term separations, the energy dependence on J (relative to the energy $E_{\gamma LS}$ of the unsplit term) must then be proportional to (12.49) with L′S′ = LS:

$$\begin{aligned} E_{\gamma LSJ} - E_{\gamma LS} &\propto (-1)^{L+S+J}\begin{Bmatrix} L & S & J \\ S & L & 1 \end{Bmatrix} \\ &= \frac{1}{2}\frac{[J(J+1) - L(L+1) - S(S+1)]}{[L(L+1)(2L+1)S(S+1)(2S+1)]^{1/2}} , \end{aligned} \qquad (12.50)$$

where the final expression follows from (11.78). This result may by analogy with (10.11)-(10.12) be written in the form

$$E_{\gamma LSJ} - E_{\gamma LS} = \frac{1}{2}[J(J+1) - L(L+1) - S(S+1)]\cdot\zeta(\gamma LS) , \qquad (12.51)$$

where $\zeta(\gamma LS)$ is an effective spin-orbit splitting factor for the term γLS.

It follows directly from (12.51) that

$$E_{\gamma LSJ} - E_{\gamma LS,J-1} = \frac{1}{2}[J(J+1) - (J-1)J]\cdot\zeta(\gamma LS) = J\cdot\zeta(\gamma LS) , \qquad (12.52)$$

so that the energy interval between two levels of a term with consecutive values of J is proportional to the larger of the two values of J. This is known as the *Landé interval rule.*[6]

[6]A Landé, Z. Physik 15, 189 (1923), 19, 112 (1923).

It is of considerable utility in the empirical analysis of atomic spectra: when a regularly spaced group of (three or more) levels is found in a case where LS-coupling conditions are expected, (12.52) makes possible tentative assignment of J values and therefore of probable values of L and S.

The quantity $\zeta(\gamma LS)$ may be either positive (in which case the energy increases with J, and the term is said to be *normal*) or negative (in which case the energy decreases with increasing J, and the term is said to be *inverted*). For less-than-half-filled subshells, $\zeta(\gamma LS)$ is usually positive, especially for the lowest term of a configuration; however, negative values of $\zeta(\gamma LS)$ are not infrequent, and the inverted structure is not by any means to be considered abnormal.

It may be verified by straightforward algebraic evaluation using (12.51) that

$$\sum_J (2J + 1)(E_{\gamma LSJ} - E_{\gamma LS}) = 0 \; ; \qquad (12.53)$$

thus the implied energy $E_{\gamma LS}$ of the un-split term can be calculated as the center of gravity of the observed levels γLSJ. [The result (12.53) will be obtained in a more sophisticated manner in Sec. 12-11.]

Large departures from the interval rule (12.52)—and, by inference, from (12.53)—are frequently observed as the result of large departures from pure LS-coupling conditions; examples may be seen in Figs. 10-3 to 10-5.

Problems

12-7(1). When there is only one basis state of given LS, then (under good LS-coupling conditions) the energy perturbation (12.51) is given by the diagonal s-o matrix element. Show from (12.39) and (12.42) that for the configuration nl^w

$$\zeta(\alpha LS) = \frac{[l(l+1)(2l+1)]^{1/2}\langle l^w \alpha LS \| V^{(11)} \| l^w \alpha LS \rangle}{[L(L+1)(2L+1)S(S+1)(2S+1)]^{1/2}} \zeta_{nl} \; . \qquad (12.54)$$

Show from this that $\zeta(LS) = \zeta_l$ for $w = 1$ and that $\zeta(LS) = -\zeta_l$ for $w = 4l + 1$. From energy-level data in AEL, derive experimental values of ζ_{2p} in C I $2p^2$ and O I $2p^4$, of ζ_{3p} in Si I $3p^2$ and S I $3p^4$, and of ζ_{3d} in Ti I $3d^2 4s^2$; compare with Fig. 8-13.

12-7(2). Show that for the configuration $l_1 l_2$

$$\zeta(LS) = \frac{1}{4L(L+1)} \Big\{ [L(L+1) + l_1(l_1+1) - l_2(l_2+1)]\zeta_1 + [L(L+1) + l_2(l_2+1) - l_1(l_1+1)]\zeta_2 \Big\} \; , \qquad (12.55)$$

and that for $l_1^{4l_1+1} l_2$ the same expression holds except for a change in sign of the term in ζ_1. Show that for $l_2 > l_1$ and $\zeta_2 \ll \zeta_1$, (12.55) is negative for small L. Using data from AEL, estimate values of ζ_{2p} and ζ_{3d} in N II 2p3d and of ζ_{2p} and ζ_{3p} in N II 2p3p. Why can a similar analysis not be applied to N II 2p4f?

12-7(3). Assume that for some given LS there are only two basis states α_1LS and α_2LS. Show that under conditions of good LS coupling the effective spin-orbit splitting factors for the eigenstates with eigenvectors $\gamma_a = (y_1,y_2)$ and $\gamma_b = (y_2, -y_1)$ are

$$\begin{aligned} \zeta(\gamma_a LS) &= y_1{}^2\zeta(\alpha_1 LS) + y_2{}^2\zeta(\alpha_2 LS) + 2y_1y_2\zeta(\alpha_{12}LS) \;, \\ \zeta(\gamma_b LS) &= y_2{}^2\zeta(\alpha_1 LS) + y_1{}^2\zeta(\alpha_2 LS) - 2y_1y_2\zeta(\alpha_{12}LS) \;, \end{aligned} \qquad (12.56)$$

where $\zeta(\alpha_{12}LS)$ is a quantity related to the off-diagonal spin-orbit matrix element in the same way that $\zeta(\alpha_1 LS)$ and $\zeta(\alpha_2 LS)$ are related to the two diagonal elements. [In l^w, for example, $\zeta(\alpha_{12}LS)$ would be given by substituting $\langle l^w\alpha_1 LS\|V^{(11)}\|l^w\alpha_2 LS\rangle$ for the reduced matrix element in (12.54).] Thus if the off-diagonal spin-orbit matrix element is zero (as is commonly true if the two basis states belong to different configurations), the eigenstate splitting factors are simple weighted averages of the basis-state splitting factors.

12-8. LEVEL STRUCTURE: CONFIGURATIONS l^w

The simplest configurations are those involving only a single open subshell l^w. These are also of great physical importance because in most atoms and ions the ground configuration is of this type—p^w for the boron- through fluorine-group elements, d^w for the transition elements, and f^w for the lanthanides and actinides.

For l^w the only basis-state quantum numbers are αLSJ (and M). The level structure depends only on the Coulomb parameters $F^k(ll)$, with coefficients given by (12.17) (which are diagonal in all quantum numbers except α), and on the single spin-orbit parameter ζ_l, with coefficients given essentially by (12.43) [which are diagonal only in J (and M)]. Because the electrons are all equivalent and therefore lie spatially close together—at least in the radial direction—their mutual Coulomb interactions are large and the physical coupling conditions tend to lie closer to the LS than to the jj limit, except for very heavy atoms (especially those with p^w configurations) where the spin-orbit interactions are also very large: see Fig. 8-13. It is therefore appropriate to begin the discussion of level structures by considering only the Coulomb contributions.

Unfortunately, the form of the expression (12.17) for the Coulomb coefficients f_k is rather complicated, because the reduced matrix elements of $U^{(k)}$ involve coefficients of fractional parentage with rather obscure properties. As a consequence, it is difficult or impossible to discuss the level structure in any generality. We have already considered the special case l^2 in Sec. 10-5, using the simpler two-electron expression (10.32) for f_k. The configurations p^3 and p^4 are also simple and will be discussed shortly. Little else can be said beyond Hund's rule: the lowest-energy term of l^w is that term of maximum S which has the largest value of L; qualitative reasons for the validity of this rule were given in Sec. 4-16. The relative positions of the remaining terms of l^w ($2 < w < 4l$) can be predicted only on the basis of quantitative numerical calculations. Some results of such calculations for d^w and f^w configurations will be discussed in Chapters 20 and 21.

Under good LS-coupling conditions, the spin-orbit splitting of a term is given by (12.51), and by (12.54) if no other term of like LS is present. For terms of maximum possible S (including the Hund's-rule term), a much simpler expression for ζ(LS) can be derived. We consider first only less-than-half-filled subshells, $w < 2l + 1$, so that $S = w/2$. We also consider that basis function for which J and M have their maximum values, $J = M = L + S$, so that $M_L = L$ and $M_S = S$. It follows from (9.7) that the coupled basis function $|l^w LSJM\rangle$ is actually a simple product of an orbital and a spin function, $|LM_L\rangle|SM_S\rangle$, and that the spin function is a simple product of one-electron spin functions, each with $m_{s_i} = +1/2$.

In evaluating diagonal matrix elements of the spin-orbit operator

$$\sum_i \xi_i(r_i) l^{(1)} \cdot s^{(1)} = \sum_i \xi_i(r_i) \sum_q (-1)^q l^{(1)}_{-q} s^{(1)}_q ,$$

we will therefore obtain spin factors

$$\langle m_s | s^{(1)}_q | m'_s \rangle$$

only with $m_s = m'_s = +1/2$. The terms $q = \pm 1$ contribute nothing, because of the Wigner-Eckart theorem (11.15) and the fact that the three magnetic quantum numbers of a 3-j symbol must sum to zero. Thus the diagonal spin-orbit matrix element for $J = M = L + S$, $M_L = L$, $M_S = S = w/2$ is

$$\begin{aligned} \zeta_l \langle LSJM | \sum_i l^{(1)}_0 s^{(1)}_0 | LSJM \rangle &= \zeta_l \sum_i \langle LM_L | l^{(1)}_0(i) | LM_L \rangle \langle m_s | s^{(1)}_0(i) | m_s \rangle \\ &= \zeta_l \frac{1}{2} \langle LM_L | \sum_i l^{(1)}_0(i) | LM_L \rangle \\ &= \frac{1}{2} M_L \zeta_l = \frac{1}{2} L \zeta_l . \end{aligned}$$

But from (12.51), this same matrix element is, for $J = L + S$, equal to $L \cdot S \cdot \zeta(LS)$. Equating these two results gives (for maximum S)

$$\zeta(LS) = \frac{1}{2S} \zeta_l = \frac{1}{w} \zeta_l , \qquad 1 < w < 2l + 1 . \tag{12.57}$$

(The case $w = 2l + 1$ is excluded because the only term of maximum S is one with $L = 0$.) From (12.54) and (11.73), it follows that for maximum-S terms of more-than-half-filled subshells

$$\zeta(LS) = -\frac{1}{2S} \zeta_l = -\frac{1}{4l + 2 - w} \zeta_l , \qquad 2l + 1 < w < 4l + 1 . \tag{12.58}$$

Example: p^2. For the configuration p^2, there is only one term of each LS, so that including only Coulomb interactions the term energies are equal to the diagonal matrix elements. From (12.16)-(12.17), or more simply from (10.31)-(10.32), we find

$$E(^1S) = E_{av} + \frac{12}{25}F^2(pp) ,$$
$$E(^1D) = E_{av} + \frac{3}{25}F^2(pp) , \qquad (12.59)$$
$$E(^3P) = E_{av} - \frac{3}{25}F^2(pp) .$$

The term structure is particularly simple because it depends on the single parameter $F^2(pp)$, and the theory predicts that

$$\frac{E(^1S) - E(^1D)}{E(^1D) - E(^3P)} = \frac{3}{2} . \qquad (12.60)$$

From energy-level data for p^2 configurations, we can calculate experimental values for this ratio by using for $E(^3P)$ the center of gravity of the observed triplet levels [see (12.53)]. The experimental ratios are only about 1.14 in $2p^2$ configurations of the C I isoelectronic sequence because of strong configuration-interaction effects,[7] but are fairly close to the theoretical value 1.5 in $3p^2$, $4p^2$, $\cdots$ configurations.

When spin-orbit effects are small, the splitting of the ^3P term can be described by (12.51), where $\zeta(p^2\,{}^3P) = \zeta_p/2$ from (12.57). When the departures from pure LS coupling are large, the theoretical energy-level relationships can be found only by numerical calculations involving the complete Coulomb plus spin-orbit Hamiltonian matrix. The transition from pure LS to pure jj coupling is illustrated in Fig. 12-1, which has been plotted in a manner analogous to Figs. 10-3 and 10-4. Experimental energies for several p^2 configurations are shown, the abscissae and ordinates having been calculated from empirical values of F^2 and ζ obtained by least-squares fitting of the observed levels (Sec. 16-3). It may be seen that $6p^2$ configurations lie closer to the jj than to the LS limit, as do also highly ionized members of the other isoelectronic sequences.

The $J = 1$ level is 100% pure in either the LS or the jj representation. The $J = 0$ and $J = 2$ levels each arise from 2-by-2 matrices and so show more or less strong basis function mixing; the compositions of all four levels in either representation can be inferred from the curves in the lower half of Fig. 12-1. The $J = 0$ levels lie rather far apart and interact fairly weakly; consequently they never show more than 33% mixing. The $J = 2$ levels, on the other hand, would cross if it were not for the off-diagonal spin-orbit matrix element that connects the two $J = 2$ basis states. Correspondingly, the lower $J = 2$ level is more $^1D_2$ than $^3P_2$ for $\chi > 0.7$ (see Fig. 10-2). It is clear that jj-coupling level designations are physically more realistic than LS designations for χ greater than about 0.47.

More-than-half-filled subshells: p^4. It follows immediately from (12.16)-(12.17) and (11.58) that Coulomb contributions to the level structure of l^{4l+2-w} are identical with those of the conjugate configuration l^w. Thus (12.59) is appropriate to p^4 as well as to p^2. Experimental data agree well with (12.60) in most cases, though observed term-interval ratios

[7]P. S. Bagus and C. M. Moser, Phys. Rev. **167**, 13 (1968); A. W. Weiss, Phys. Rev. **162**, 71 (1967). For a survey of level structures in low-lying configurations of light atoms, see B. Edlén, Phys. Scr. **7**, 93 (1973).

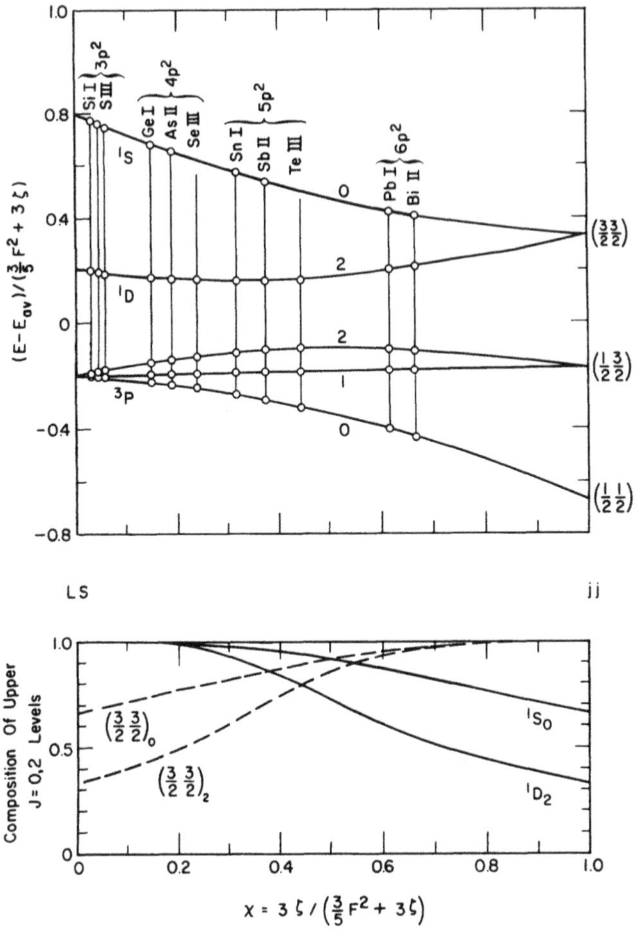

Fig. 12-1. The transition from pure LS to pure jj coupling for p^2 configurations. Upper portion: energy levels; lower portion: squares of eigenvector components.

are only about 1.15 in the O I $2p^4$ isoelectronic sequence because of strong configuration-interaction effects.

For the spin-orbit interaction, (12.39)-(12.42), (12.54), and (11.73) show that d_J and $\zeta(\gamma LS)$ have values in l^{4l+2-w} that are equal in magnitude to, but opposite in sign from, their values in l^w. Thus under near LS-coupling conditions, any term that has normal s-o splitting in l^w is inverted in l^{4l+2-w}, and vice versa. In particular, the ground term (given by Hund's rule) is always normal in less-than-half-filled subshells, and inverted in more-than-half-filled subshells [(12.57)-(12.58)].

Because of the change in sign of the spin-orbit matrix elements, the transition from pure LS to pure jj coupling in p^4 (Fig. 12-2) is rather different from that shown in Fig. 12-1 for p^2. For example, it is now the $J = 0$ rather than the $J = 2$ basis states that mix together

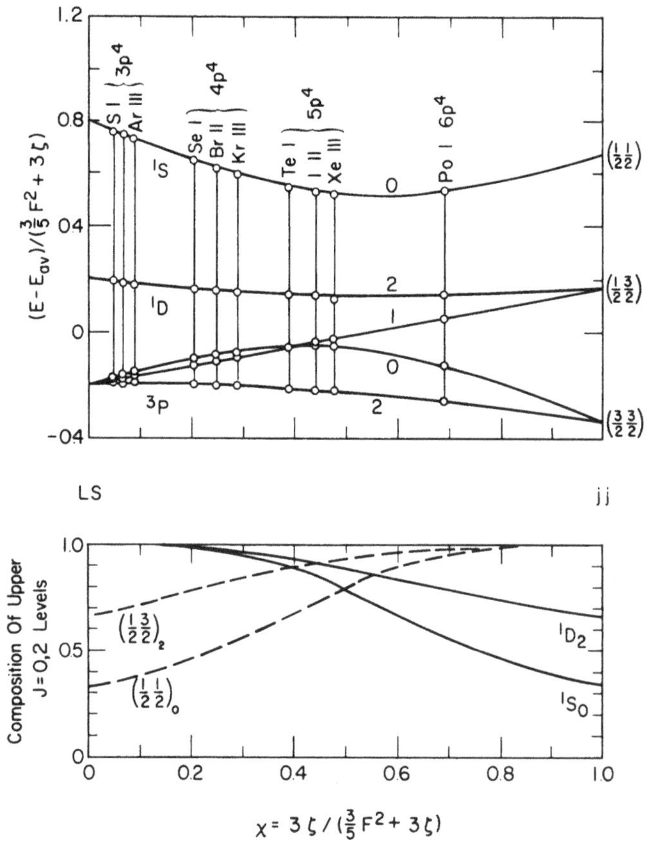

Fig. 12-2. The transition from pure LS to pure jj coupling for p^4 configurations. Upper portion: energy levels; lower portion: squares of eigenvector components.

more strongly; also, for $\chi > 0.39$ the energies of the lowest three levels no longer depend monotonically on J. However, jj-coupling level designations are still preferable only for χ greater than about 0.47.

Half-filled subshells: p^3. For l^{2l+1}, (11.74) shows that diagonal spin-orbit matrix elements are all zero. Thus under near LS-coupling conditions, spin-orbit interactions give zero term splittings to first order. Small spin-spin and spin-other-orbit interactions (whose diagonal matrix elements are in general non-zero) then make important contributions to the term splittings.[8] When departures from pure LS coupling are sizable, however, the off-diagonal s-o matrix elements are sufficiently important to produce the dominant effects.

As an example, we consider the configuration p^3. The two-electron expression (10.32) that we used for p^2 is not applicable, and we are forced to use the general expression (12.17). From (11.23),

[8] L. H. Aller, C. W. Ufford, and J. H. Van Vleck, Astrophys. J. **109**, 42 (1949); R. H. Garstang, Astrophys. J. **115**, 506 (1952).

$$\langle p\|C^{(2)}\|p\rangle^2 = 9\begin{pmatrix} 1 & 2 & 1 \\ 0 & 0 & 0 \end{pmatrix}^2 = \frac{6}{5} .$$

Because there is only one term of p^3 for any given LS, there is no α with which to be concerned; therefore, (12.17) becomes

$$f_2(pp) = \delta_{bb'}\frac{3}{5}\left\{[L]^{-1}\sum_{L''}\langle p^3L''S\|U^{(2)}\|p^3LS\rangle^2 - \frac{3}{5}\right\}$$

and the Coulomb matrix elements are completely diagonal. From Appendix H we find the only non-zero matrix elements of $U^{(2)}$ to be[9]

$$\langle p^3\,{}^2D^\circ\|U^{(2)}\|p^3\,{}^2P^\circ\rangle = -\langle p^3\,{}^2P^\circ\|U^{(2)}\|p^3\,{}^2D^\circ\rangle = \sqrt{3} .$$

Thus the diagonal Coulomb matrix elements, and corresponding term energies, are

$$\begin{aligned} E({}^2P^\circ) &= E_{av} + \frac{3}{5}\left\{\frac{1}{3}\cdot 3 - \frac{3}{5}\right\}F^2(pp) = E_{av} + \frac{6}{25}F^2(pp) , \\ E({}^2D^\circ) &= E_{av} + \frac{3}{5}\left\{\frac{1}{5}\cdot 3 - \frac{3}{5}\right\}F^2(pp) = E_{av} , \\ E({}^4S^\circ) &= E_{av} + \frac{3}{5}\left\{\frac{1}{1}\cdot 0 - \frac{3}{5}\right\}F^2(pp) = E_{av} - \frac{9}{25}F^2(pp) . \end{aligned} \tag{12.61}$$

Under good LS coupling conditions, these expressions predict that

$$\frac{E({}^2P^\circ) - E({}^2D^\circ)}{E({}^2D^\circ) - E({}^4S^\circ)} = \frac{2}{3} . \tag{12.62}$$

Experimentally, $2p^3$ configurations show term-interval ratios more like 0.5 than 2/3. However, most other p^3 configurations show good agreement with the single-configuration theory, as may be seen from Fig. 12-3. The zero term splitting in the LS-coupling limit may be seen, as well as the appearance of sizable spin-orbit effects for χ as small as 0.1.

There are here three levels with $J = 3/2$; the lower half of the figure shows the basis-state compositions for only the lowest level. Though this quantum state is pure ${}^4S^\circ$ in the LS limit, it contains in the jj limit less ${}^4S^\circ$ nature than either ${}^2P^\circ$ or ${}^2D^\circ$. The compositions for the other two levels in the LS and jj limits may be found from the rows and columns, respectively, of the LS-jj transformation matrix in Table 9-4.

12-9. LEVEL STRUCTURE: CONFIGURATIONS $l_1^{w_1}l_2$

Next to the case l^w, the simplest configurations are those that contain two non-filled subshells, with the second one being singly occupied. Such configurations are of great importance because they may be derived from ground configurations $l_1^{w_1+1}$ by the excitation of

[9]Diagonal matrix elements of $U^{(2)}$ are, like those of $V^{(11)}$, necessarily zero. However, this does not imply zero diagonal Coulomb elements, because the summation over $\alpha''L''$ in (12.17) generally involves non-zero off-diagonal matrix elements of $U^{(2)}$.

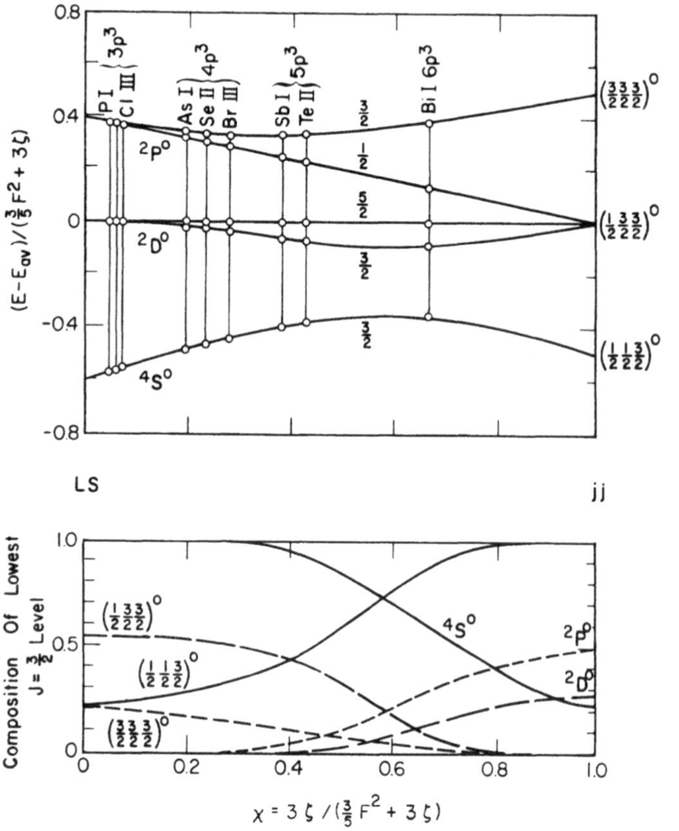

Fig. 12-3. The transition from pure LS to pure jj coupling for p^3 configurations. Upper portion: energy levels; lower portion: squares of eigenvector components.

a single electron, and are therefore a commonly observed type of excited configuration. In addition, ground configurations are of this type in a number of neutral atoms (e.g., Cr I $3d^54s$, Mo I $4d^55s$, Gd I $4f^75d6s^2$, U I $5f^36d7s^2$, Cm I $5f^76d7s^2$), and also in numerous singly ionized atoms (e.g., Ti II $3d^24s$).

The basis states for the LS representation are completely described by the quantum numbers

$$b = \alpha_1 L_1 S_1 LSJM \ . \tag{12.63}$$

The usual notation (with M ignored) is

$$(\alpha_1 \, {}^{2S_1+1}L_1) \, {}^{2S+1}L_J \ , \tag{12.64}$$

and $(\alpha_1 \, {}^{2S_1+1}L_1)$ is called the *parent* of the term ${}^{2S+1}L$.

The matrix elements are of the form

$$H_{bb'} = \delta_{bb'}E_{av} + \sum_{k>0} f_k(l_1l_1)F^k(l_1l_1) + \left[\sum_{k>0} f_k(l_1l_2)F^k(l_1l_2) + \sum_k g_k(l_1l_2)G^k(l_1l_2)\right] + \sum_{j=1}^{2} d_j\zeta_j \ . \qquad (12.65)$$

The l_1l_2 interaction coefficients are given by the expressions (12.19) and (12.38):

$$f_k(l_1l_2) = \langle l_1\|C^{(k)}\|l_1\rangle\langle l_2\|C^{(k)}\|l_2\rangle I_{12}^{(k)} \ , \qquad (12.66)$$

$$g_k(l_1l_2) = \frac{1}{2}\langle l_1\|C^{(k)}\|l_2\rangle^2\left[\frac{\delta_{bb'}w_1}{[l_1,l_2]} - \sum_r(-1)^r[r]\begin{Bmatrix} l_1 & l_1 & r \\ l_2 & l_2 & k \end{Bmatrix}[I_{12}^{(r)} + 4I_{12}^{(r1)}]\right] \ , \qquad (12.67)$$

but the general expressions (12.28) and (12.35) for $I^{(r)}$ and $I^{(r1)}$ simplify greatly because the product over intermediate shells m is absent and because $i = 1$. Using the results (11.57) and (11.71) we find the simplified expressions to be

$$I_{12}^{(r)} = \delta_{spins}(-1)^{L'_1+l_2+L}\begin{Bmatrix} L_1 & l_2 & L \\ l_2 & L'_1 & r \end{Bmatrix}\langle l_1{}^{w_1}\alpha_1L_1S_1\|U^{(r)}\|l_1{}^{w_1}\alpha'_1L'_1S_1\rangle \ , \qquad (12.68)$$

$$I_{12}^{(r1)} = (-1)^{L'_1+S'_1+l_2+s+L+S}\begin{Bmatrix} L_1 & l_2 & L \\ l_2 & L'_1 & r \end{Bmatrix}\begin{Bmatrix} S_1 & s_2 & S \\ s_2 & S'_1 & 1 \end{Bmatrix} \times (3/2)^{1/2}\langle l_1{}^{w_1}\alpha_1L_1S_1\|V^{(r1)}\|l_1{}^{w_1}\alpha'_1L'_1S'_1\rangle \ . \qquad (12.69)$$

The coefficients $f_k(l_1l_1)$ are diagonal in all quantum numbers except α_1, and are independent of all quantum numbers except $\alpha_1\alpha'_1L_1S_1$; thus they are essentially the same for $l_1{}^{w_1}l_2$ as they were for $l_1{}^{w_1}$ alone. Consequently, if the interactions of l_2 with $l_1{}^{w_1}$ are weak, and if spin-orbit effects are also small, then the term structure is essentially that of $l_1{}^{w_1}$ alone $[F^k(l_1l_2) = G^k(l_1l_2) = 0]$, with superimposed small splittings about each parent term $\alpha_1L_1S_1$. An example for the simple case p^3s is shown in Fig. 12-4. Provided the value of E_{av} is held fixed as we add the l_1l_2 interaction, and provided the latter is small, not only is the weighted *total* of all splittings due to l_1l_2 equal to zero, but the weighted average splitting about *each* parent term individually is also zero—for example, the unsplit $(^2P^\circ)$ in the figure lies at the center of gravity of the $(^2P^\circ)^1P^\circ$ and $(^2P^\circ)^3P^\circ$; this will be proved in the appendix to this chapter.

If the strength of the l_1l_2 interaction is gradually increased, the splittings about each parent term increase in size. A physical example may be obtained by starting with the ground configuration $l_1{}^{w_1}$ of an ion, and then adding another electron to the ion in an orbit n_2l_2—first with n_2 very large, and then with successively smaller values of n_2. As n_2 decreases, the l_2 electron moves closer to the ion core so that the l_1l_2 interaction increases. At the same time, however, the binding energy of the valence electron increases, so that the value of E_{av} decreases; thus the effective position of each unsplit parent term decreases by this same amount. The resulting energy-level scheme for p^3s configurations is shown schematically in Fig. 12-5. In this example, there are six Rydberg series of terms: the $^3S^\circ$ and $^5S^\circ$ Rydberg series approach as a limit the $(^4S^\circ)$ term of the ion, which is the normal

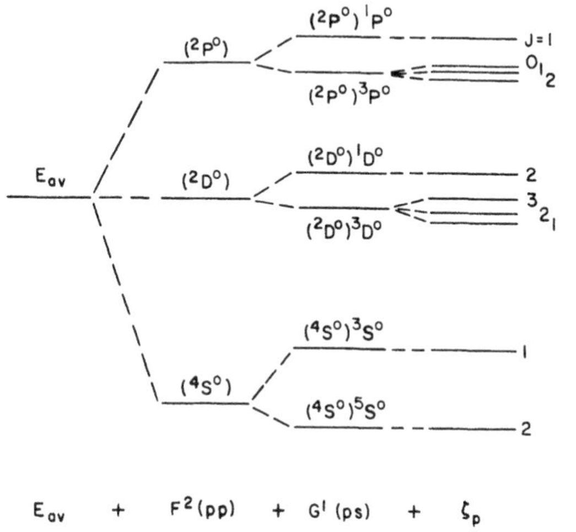

Fig. 12-4. Development of the level structure of a p^3s configuration under good LS-coupling conditions, showing the Coulomb splitting that the (weak) ps interaction produces about each parent term of p^3.

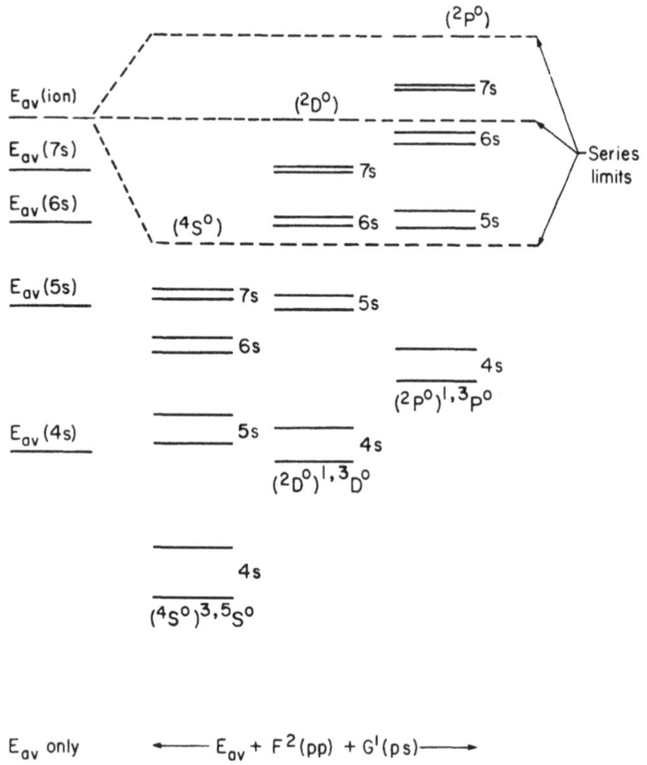

Fig. 12-5. The LS term structure of a Rydberg series of p^3s configurations. Also shown are the three ionization limits, formed by the three terms of the ground configuration p^3 of the ion.

(or "first") ionization limit of the atom; the $^{1,3}D^\circ$ series approach as a limit the excited $(^2D^\circ)$ term of the ion; and the $^{1,3}P^\circ$ series approach the still-more-highly excited $(^2P^\circ)$ term of the ion.

High members of the $^{1,3}D^\circ$ and $^{1,3}P^\circ$ series obviously lie above the normal ionization limit, and may "autoionize" by interaction with continuum states above the $(^4S^\circ)$ series limit. For example $(p^3\ ^2D^\circ)ns\ ^3D^\circ$ may interact with $(p^3\ ^4S^\circ)\varepsilon d\ ^3D^\circ$ states; the excited ns discrete state may thereby make a radiationless transition to the continuum εd state, with the εd electron escaping as a free electron (with kinetic energy ε) and leaving the ion in the $p^3\ ^4S^\circ$ ground state. This will be treated in detail in Chapter 18.

If the $l_1 l_2$ interaction becomes sufficiently large, the resulting splittings about each parent term become comparable in size with the separations of the parents themselves, and the structure of $l_1^{w_1} l_2$ no longer shows an obvious relationship to that of $l_1^{w_1}$. This is particularly true if terms with a given LS can be derived from more than one parent. These terms will in general interact with each other because of non-zero off-diagonal Coulomb matrix elements, producing distortions in the energy-level structure. At the same time, the eigenvectors will show strong mixing of the various basis states, and one can no longer say that each LS term corresponds to a particular parent state.

As an example, the configuration p^4d includes three ^2D basis states, one derived from each of the p^4 parents (^3P), (^1D), and (^1S). The energies of the three corresponding ^2D terms are shown in Fig. 12-6,[10] as computed for the case of fixed E_{av} and F^2 (pp), with increasing G^1(pd) and fixed values of the ratios F^2(pd):G^1(pd):G^3(pd). For small G^1, both the diagonal and off-diagonal pd-interaction terms are small, the mixing of the three ^2D basis states is small, the ^2D terms have essentially the same energies as the parents, and the compositions are essentially pure $(^3P)^2D$, $(^1D)^2D$, and $(^1S)^2D$, in order of increasing energy. This is the situation present in Cl I $3p^43d$, where the 3d electron is essentially hydrogenic and comparatively weakly bound.

By the time G^1 has increased to 10 kK, the two lower ^2D's have interacted sufficiently that the eigenvectors show about a 15% mixing of the two basis states, the lowest ^2D being about 85% (^3P) and 15% (^1D). At $G^1 \cong 25$ kK, the two lowest ^2D's are mixed about 50-50, and both have also begun to show mixing-in of the $(^1S)^2D$ basis state. Beyond $G^1 \cong 35$ kK the eigenvector compositions have changed so drastically that the three ^2D terms should really be called (in order of increasing energy) $(^1D)^2D$, $(^1S)^2D$, and $(^3P)^2D$—though in the limit of large G^1 the purities are only 40%, 56%, and 77%, respectively; this is still another example of the non-crossing of interacting levels, but effective crossing of quantum-state compositions, that was discussed in Sec. 10-7.

As shown in the figure, conditions in Ar II and higher ions lie on the right-hand side of the figure where these composition changes have occurred. The large pd interaction in these ions is a result of a strong contraction of the 3d wavefunction, similar to that associated with the beginning of the first transition series of elements (Sec. 8-6).

The three ^2D terms are frequently designated $3p^4nd\ ^2D$, $3p^4nd'\ ^2D$, and $3p^4nd''\ ^2D$, in order of increasing energy.[11] The primes serve as a noncommittal serial-numbering device, and this notation is therefore quite unobjectionable. The parentage notation $3p^4(^3P)nd\ ^2D$,

[10] R. D. Cowan, J. Opt. Soc. Am. 58, 924 (1968).

[11] See, for example, AEL, Vol. I.

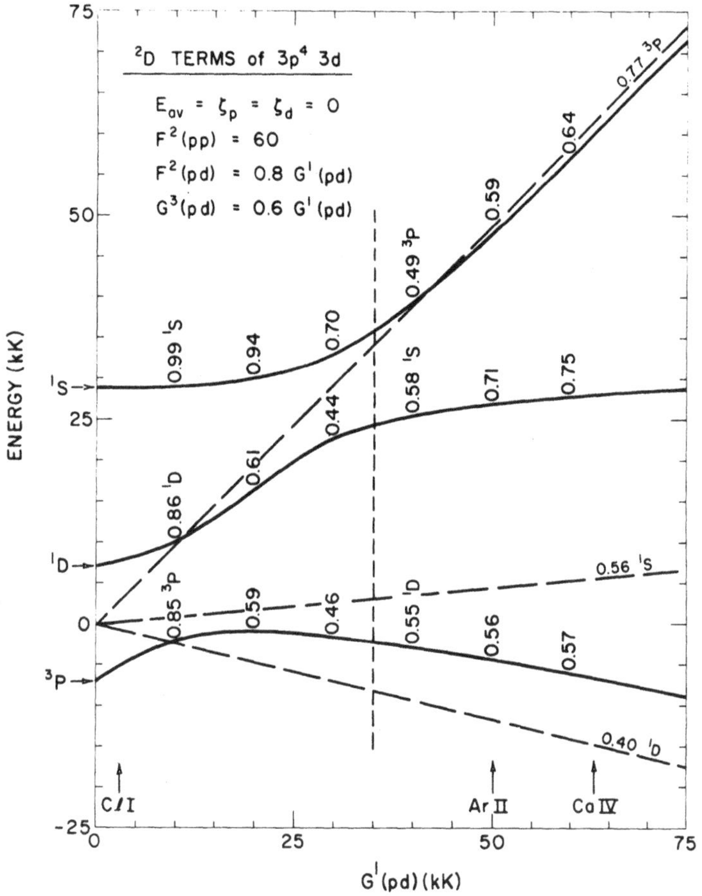

Fig. 12-6. Solid curves: Energies of the three ^{2}D terms of p^{4}d as a function of the strength of the pd interaction. The numbers at various points along the curves give the parentage composition (square of the largest eigenvector component) for the quantum state at that point. Dashed lines: Energies in the limiting case in which the pp interactions are negligible. Conditions in Cl I 3p^{4}3d, and in configurations 3p^{4}nd (n > 4) for all members of the Cl I isoelectronic sequence, lie to the left of the vertical dotted line; conditions in 3p^{4}3d for all ions lie to the right of this line.

3p^4(^{1}D)nd ^{2}D, and 3p^4(^{1}S)nd ^{2}D is also used for energy-level designation, again in order of increasing energy.[11] These designations are satisfactory in all cases, provided they are interpreted *only* as indicating the ^{2}D terms in the same energy order as the parent terms in the ion; they may be construed as indicating the dominant eigenvector component only when the $l_1 l_2$ interaction is small (e.g., in Cl I for all n, or in ions if n is sufficiently large).

Special case: $l_1^{w_1}$s. The special case $l_2 = 0$ is particularly important because most ground configurations of singly ionized transition and rare-earth elements are of this type.

For the l_1s interaction, there are no direct terms $f_k F^k$ (except for k = 0, which contributes only to E_{av}), and the only exchange term is that for which $k = l_1$. The coefficient $g_{l_1}(l_1 s)$ simplifies sufficiently from the expressions (12.67)-(12.69) that it is worth working it out in detail. With $l_2 = 0$, it is necessarily true that $L = L_1$ and $L' = L_1'$; since Coulomb matrices are diagonal in LL′, it follows that in the present case they will also be diagonal in $L_1 L_1'$. From (12.68), (5.29), and (11.55) we find

$$\begin{aligned} I_{12}^{(r)} &= \delta_{(\text{all but } \alpha_1)} (-1)^{L_1+L} \begin{Bmatrix} L_1 & 0 & L \\ 0 & L_1 & r \end{Bmatrix} \langle l_1{}^{w_1} \alpha_1 L_1 S_1 \| U^{(r)} \| l_1{}^{w_1} \alpha_1' L_1 S_1 \rangle \\ &= \delta_{r0} \delta_{(\text{all but } \alpha_1)} [L_1]^{-1/2} \langle l_1{}^{w_1} \alpha_1 L_1 S_1 \| U^{(0)} \| l_1{}^{w_1} \alpha_1' L_1 S_1 \rangle \\ &= \delta_{r0} \delta_{bb'} w_1 [l_1]^{-1/2} . \end{aligned} \tag{12.70}$$

Similarly we find from (12.69), (11.70), and (11.78) that

$$\begin{aligned} I_{12}^{(r1)} &= \delta_{r0} \delta_{(\text{all but } \alpha_1 S_1)} [L_1]^{-1/2} (-1)^{S_1'+s_2+S} \begin{Bmatrix} S_1 & s_2 & S \\ s_2 & S_1' & 1 \end{Bmatrix} \\ &\qquad \times (3/2)^{1/2} \langle l_1{}^{w_1} \alpha_1 L_1 S_1 \| V^{(01)} \| l_1{}^{w_1} \alpha_1' L_1 S_1' \rangle \\ &= \delta_{r0} \delta_{bb'} (3/2)^{1/2} \left[\frac{S_1(S_1+1)(2S_1+1)}{[l_1]} \right]^{1/2} (-1)^{S_1+s_2+S} \begin{Bmatrix} S_1 & s_2 & S \\ s_2 & S_1 & 1 \end{Bmatrix} \\ &= \delta_{r0} \delta_{bb'} (3/2)^{1/2} \frac{[S(S+1) - S_1(S_1+1) - s_2(s_2+1)]}{2[(2l_1+1) s_2(s_2+1)(2s_2+1)]^{1/2}} \\ &= \delta_{r0} \delta_{bb'} \frac{S(S+1) - S_1(S_1+1) - 3/4}{2[l_1]^{1/2}} . \end{aligned} \tag{12.71}$$

Using (11.26), we find that (12.67) reduces to

$$\begin{aligned} g_{l_1}(l_1 s) &= \frac{1}{2} \delta_{bb'} \left\{ \frac{w_1}{[l_1]} - \begin{Bmatrix} l_1 & l_1 & 0 \\ 0 & 0 & l_1 \end{Bmatrix} [I_{12}^{(0)} + 4 I_{12}^{(01)}] \right\} \\ &= \frac{1}{2} \delta_{bb'} \left\{ \frac{w_1}{[l_1]} - \frac{1}{[l_1]^{1/2}} \left[\frac{w_1}{[l_1]^{1/2}} + \frac{2\{S(S+1) - S_1(S_1+1) - 3/4\}}{[l_1]^{1/2}} \right] \right\} \\ &= -\delta_{bb'} \frac{\{S(S+1) - S_1(S_1+1) - 3/4\}}{[l_1]} \\ &= \begin{cases} \delta_{bb'} \dfrac{S_1+1}{2l_1+1} , & S = S_1 - \dfrac{1}{2} , \\ -\delta_{bb'} \dfrac{S_1}{2l_1+1} , & S = S_1 + \dfrac{1}{2} . \end{cases} \end{aligned} \tag{12.72}$$

[If $S_1 = 0$, then only the case $S = S_1 + 1/2$ is possible and $g_{l_1}(l_1 s) = 0$.] We note that the l_1s interaction is completely diagonal, even though the $l_1 l_2$ interaction would not have been diagonal in $\alpha_1L_1S_1$ for non-zero l_2. We note also that the resulting splitting is independent of α_1L_1 (and w_1) and dependent only on l_1 and S_1, and that the weighted-average splitting for given $\alpha_1L_1S_1$ is zero since

$$\sum_S (2S + 1)g_{l_1}(l_1 s) = 0 ,$$

in agreement with our earlier remarks.

Special case: $w_1 = 4l_1 + 1$. The case in which the shell $l_1^{w_1}$ is almost full (one hole) is also of special interest, because the configurations

$$l_1^{4l_1+1}l_2 \tag{12.73}$$

constitute the most easily excited configurations of the noble gases (ground configurations p^6) and of atoms and ions such as Cu II ($3d^{10}$), Pd I ($4d^{10}$), Yb III ($4f^{14}$), and Au II ($5d^{10}$), and also form the ground configurations in Tm II $4f^{13}6s$) and Pt I ($5d^9 6s$).

This type of configuration is similar to the two-electron configuration l_1l_2 in that both $l_1^{4l_1+1}$ and l_1 have only the single term $^2L_1$, with $L_1 = l_1$. However, there are several important differences in the level structures of the two configurations.

We see from (12.43), (11.71), and (11.73) that the spin-orbit splitting of the $l_1^{4l_1+1}\ {}^2L_1$ parent is inverted, while that of $l_1\ {}^2L_1$ is normal. From (12.66), (12.68), and (11.58), we see that the coefficients of $F^k(l_1l_2)$ in $l_1^{4l_1+1}l_2$ are just the negatives of those for l_1l_2; thus the term structure resulting from the direct electrostatic interaction is also inverted in $l_1^{4l_1+1}l_2$ relative to l_1l_2. [We note also that since $L_1 = L_1' = l_1$, and since the reduced matrix elements $\langle l_1^w \| U^{(k)} \| l_1^w \rangle$ for $k > 0$ are independent of l_1 (and w) for $w = 1$ or $4l_1 + 1$, then the direct Coulomb effects are the same in $l_1l_2^{4l_2+1}$ as in $l_1^{4l_1+1}l_2$.]

The situation for the exchange coefficients $g_k(l_1l_2)$ is very different in $l_1^{4l_1+1}l_2$ from that in l_1l_2, and is quite interesting. We note first that $L_1 = L_1' = l_1$ and $S_1 = S_1' = s_1$, so that the Coulomb matrix is completely diagonal. Using (11.61), (11.55), and (11.75) in (12.68) and (12.69) we find

$$I_{12}^{(r)} = -(-1)^{l_1+l_2+L+r} \begin{Bmatrix} l_1 & l_2 & L \\ l_2 & l_1 & r \end{Bmatrix} , \qquad r > 0 , \tag{12.74}$$

$$I_{12}^{(0)} = (-1)^{l_1+l_2+L} \begin{Bmatrix} l_1 & l_2 & L \\ l_2 & l_1 & 0 \end{Bmatrix} w_1 , \tag{12.75}$$

$$I_{12}^{(r1)} = \frac{3}{2}(-1)^{l_1+l_2+L+S+r+1} \begin{Bmatrix} l_1 & l_2 & L \\ l_2 & l_1 & r \end{Bmatrix} \begin{Bmatrix} s_1 & s_2 & S \\ s_2 & s_1 & 1 \end{Bmatrix} , \qquad \text{all } r . \tag{12.76}$$

From these we find, with the aid of (11.78) and (5.31), that the summation in (12.67) may be written

$$- \sum_r (-1)^r [r] \begin{Bmatrix} l_1 & l_1 & r \\ l_2 & l_2 & k \end{Bmatrix} [I_{12}^{(r)} + 4I_{12}^{(r1)}]$$

$$= - \begin{Bmatrix} l_1 & l_1 & 0 \\ l_2 & l_2 & k \end{Bmatrix} (-1)^{l_1+l_2+L} \begin{Bmatrix} l_1 & l_2 & L \\ l_2 & l_1 & 0 \end{Bmatrix} (w_1 + 1)$$

$$+ (-1)^{l_1+l_2+L} \left[1 - 6(-1)^{1+S} \begin{Bmatrix} s_1 & s_2 & S \\ s_2 & s_1 & 1 \end{Bmatrix} \right] \sum_r [r] \begin{Bmatrix} l_1 & l_2 & k \\ l_2 & l_1 & r \end{Bmatrix} \begin{Bmatrix} l_1 & l_2 & L \\ l_2 & l_1 & r \end{Bmatrix}$$

$$= -(-1)^{l_1+l_2+k} \frac{(w_1 + 1)}{[l_1,l_2]} + (-1)^{l_1+l_2+L} \left[1 - \frac{3\{S(S+1) - 3/2\}}{s(s+1)(2s+1)} \right] \frac{\delta_{kL}}{[L]}$$

$$= (-1)^{l_1+l_2+k} \left[- \frac{(w_1 + 1)}{[l_1,l_2]} + \delta_{kL} \delta_{S0} \frac{4}{[L]} \right] ,$$

where the last step follows because the only possible values of S are 0 and 1. Remembering that the Coulomb matrix is diagonal, and that the reduced matrix element in (12.67) is zero unless $l_1 + k + l_2$ is even, we obtain finally

$$g_k(l_1 l_2) = \delta_{bb'} \frac{1}{2} \langle l_1 \| C^{(k)} \| l_2 \rangle^2 \left[- \frac{1}{[l_1,l_2]} + \delta_{kL} \delta_{S0} \frac{4}{[L]} \right] . \qquad (12.77)$$

From (11.24) we see that this, like $f_k(l_1 l_2)$, is symmetric in l_1 and l_2 and independent of w_1; hence the entire Coulomb-interaction structure is the same for $l_1 l_2^{4l_2+1}$ as for $l_1^{4l_1+1} l_2$.

The unusual feature of the exchange terms, which make them of special interest, is that if we ignore the center-of-gravity correction term (which does not contribute to the level *structure*) then the coefficients g_k are non-zero only for those singlet terms for which L equals one of the possible values of k:

$$L = |l_1 - l_2|, \; |l_1 - l_2| + 2, \; \cdots, \; l_1 + l_2 . \qquad (12.78)$$

For each such term, g_k (k = L) is positive, and so the effect of the exchange interaction is always such as to make these singlet levels tend to lie unusually high in energy. This is particularly true of the singlet having smallest L, because this is the case for which both g_k and G^k are largest [see (12.77) and (6.32), respectively].

Example. For the configuration p^5d the diagonal matrix elements (neglecting s-o terms) are computed from (12.66), (12.74), and (12.77) to be

$$\begin{aligned}
E(^1P^\circ) &= E_{av} - 0.200F^2 + 1.267G^1 - 0.043G^3 , \\
E(^3P^\circ) &= E_{av} - 0.200F^2 - 0.067G^1 - 0.043G^3 , \\
E(^1D^\circ) &= E_{av} + 0.200F^2 - 0.067G^1 - 0.043G^3 , \\
E(^3D^\circ) &= E_{av} + 0.200F^2 - 0.067G^1 - 0.043G^3 , \\
E(^1F^\circ) &= E_{av} - 0.057F^2 - 0.067G^1 + 0.325G^3 , \\
E(^3F^\circ) &= E_{av} - 0.057F^2 - 0.067G^1 - 0.043G^3 ,
\end{aligned} \qquad (12.79)$$

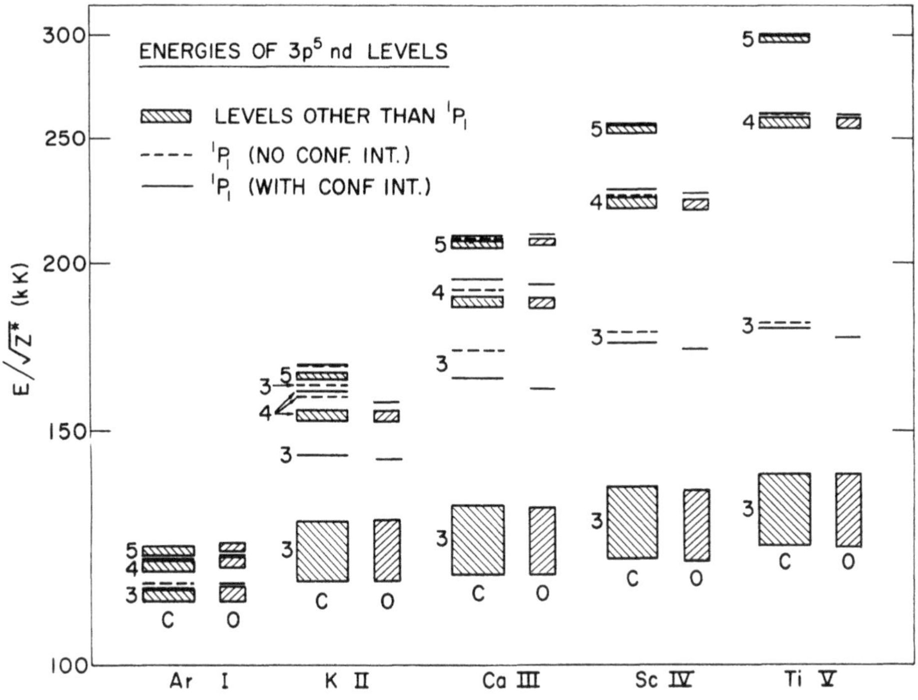

Fig. 12-7. Computed (C) and observed (O) energy levels of p^5d configurations of the Ar I isoelectronic sequence. Each shaded block represents the energy range of the eleven levels other than $^1P_1^\circ$.

where $F^k = F^k(pd)$ and $G^k = G^k(pd)$.

The contribution $1.267\,G^1$ relative to the center of gravity (or $1.333\,G^1$ relative to all other levels) tends to make the $^1P_1^\circ$ level lie higher than the other eleven levels of the configuration, in spite of the negative direct-interaction contribution $-0.2\,F^2$. Figure 12-7[12] shows this to be confirmed by experimental data in the Ar I isoelectronic sequence[13]—especially for the ions, where the 3d-wavefunction collapse (Sec. 8-6) has caused a large increase in the value of G^1 (larger even than for F^2, as it happens).[10] Configuration-interaction effects are correspondingly large for the $^1P_1^\circ$ levels; this will be discussed in detail in Sec. 13-5. Because of these large effects, the $3p^53d$ $^1P_1^\circ$ levels in ions of the Ar I sequence were not found experimentally until fairly recently.[14]

[12]R. D. Cowan, J. phys., Colloq. **31**, C4-191 (1970).

[13]Ar I, AEL; K II, L. Minnhagen and K. Hallén, private communication (1971); Ca III, A. Borgström, Phys. Scr. **3**, 157 (1971); Sc IV, R. Smitt, Phys. Scr. **8**, 292 (1973); Ti V, L. Å. Svensson, Phys. Scr. **13**, 235 (1976).

[14]Ca III—Ni XI, A. H. Gabriel, B. C. Fawcett, and C. Jordan, Nature **206**, 390 (1965), Proc. Phys. Soc. (London) **87**, 825 (1966); K II, ref. 13.

TABLE 12-2. COEFFICIENTS[a] FOR THE MINIMUM-k EXCHANGE TERM IN CONFIGURATIONS $l_1^{4l_1+1}l_2$

Configuration	k	$\langle l_1 \| C^{(k)} \| l_2 \rangle^2$	Level	g''_k
ss′	0	1	$^1S_0$	2
p^5p'	0	3	$^1S_0$	6
d^9d'	0	5	$^1S_0$	10
$f^{13}f'$	0	7	$^1S_0$	14
sp,p^5s	1	1	$^1P^o_1$	2/3
p^5d,d^9p	1	2	$^1P^o_1$	4/3
$d^9f,f^{13}d$	1	3	$^1P^o_1$	2
sd,d^9s	2	1	$^1D_2$	2/5
$p^5f,f^{13}p$	2	9/5	$^1D_2$	18/25

[a]Relative to $g''_k = 0$ for all other levels.

It may be inferred from Table 12-2 that similar effects might be seen in configurations other than p^5d. Examples can be found in Na II $2p^53p$ $^1S_0$, K II and Ca III $3p^54p$ $^1S_0$, and Cu II $3d^94d$ $^1S_0$.[15] In most such configurations, however, the effect is obscured by small values of the exchange parameter G^k or by large spin-orbit interactions. A large effect in Ba I $4d^94f$ $^1P^o_1$ will be discussed in Sec. 18-8.

*12-10. LEVEL STRUCTURE: MORE-COMPLEX CONFIGURATIONS

For configurations more complex than $l_1^{w_1}l_2$, the principles outlined in the preceding section are still applicable.

If in $l_1^{w_1}l_2^{w_2}$, the interactions within $l_2^{w_2}$ and between $l_2^{w_2}$ and $l_1^{w_1}$ are small compared to those within $l_1^{w_1}$, the basic structure is that of $l_1^{w_1}$, with superimposed additional structure due to the l_1l_2 and l_2l_2 interactions.

In $l_1^{w_1}l_2^{w_2}l_3$, we may start with the structure of $l_1^{w_1}l_2^{w_2}$, and add on additional structure due to the l_1l_3 and l_2l_3 interactions; if these interactions are small, the basic $l_1^{w_1}l_2^{w_2}$ structure will be discernible—otherwise it probably will not be.

The detailed level structure quickly becomes very complex, and there is little point in trying to discuss it here. Numerical calculations, especially when spin-orbit effects are included, generally require the services of a large computer.[16]

[15]AEL, except ref. 13 for Ca III.

[16]Numerical values of the Coulomb angular coefficients f_k and g_k are, however, tabulated for numerous moderately complex configurations in J. C. Slater, *Quantum Theory of Atomic Structure* (McGraw-Hill, New York, 1960), Vol. II, App. 21 and in E. U. Condon and H. Odabasi, *Atomic Structure* (Cambridge University Press, Cambridge, 1980), App. 3.

*12-11. APPENDIX: CENTER-OF-GRAVITY RELATIONS

We here prove various center-of-gravity relations, such as that mentioned after (12.69). These all follow from the relation (5.34).

(1) Applying (5.34) to the spin-orbit splitting expression (12.50) for good LS coupling, we find the weighted-average level separation from $E_{\gamma LS}$ to be

$$\sum_J [J](E_{\gamma LSJ} - E_{\gamma LS}) \propto \sum_J (-1)^{L+S+J}[J]\begin{Bmatrix} L & L & 1 \\ S & S & J \end{Bmatrix} = 0 \; ; \tag{12.80}$$

i.e., $E_{\gamma LS}$ is the center of gravity of the levels comprising the term γLS—a result obtained by detailed algebraic evaluation in (12.53).

(2) If, in the configuration $l_1^{w_1} l_2$, the $l_1 l_2$ interaction is sufficiently small compared with the $l_1 l_1$ interactions that the effect of the former is given by the diagonal matrix elements $L_1' = L_1$, $S_1' = S_1$, then applying (5.34) to the couplings $(L_1 l_2)L$ and $(S_1 s_2)S$ in (12.68) and (12.69), we obtain, with the aid of (11.55):

$$\sum_L [L] I_{12}^{(r)} = \delta_{r0}[L_1,l_2]^{1/2}\langle l_1^{w_1}\alpha_1 L_1 S_1 \| U^{(0)} \| l_1^{w_1}\alpha_1 L_1 S_1\rangle = \delta_{r0}[L_1] w_1 \frac{[l_2]^{1/2}}{[l_1]^{1/2}} \tag{12.81}$$

and

$$\sum_{LS} [L,S] I_{12}^{(r1)} \propto \delta_{r0}\delta_{10} = 0 \; . \tag{12.82}$$

Thus the direct coefficients $f_k(l_1 l_2)$ are seen from (12.66) to give a weighted average splitting of zero about the parent term L_1S_1, because we have only terms with $k > 0$. The exchange contributions (12.67) are found with the aid of (12.81), (12.82), and (5.29) to also be zero:

$$\begin{aligned}\sum_{LS} [L,S] g_k &\propto \frac{w_1}{[l_1,l_2]} \sum_{LS} [L,S] - \begin{Bmatrix} l_1 & l_2 & k \\ l_2 & l_1 & 0 \end{Bmatrix} [L_1] w_1 \frac{[l_2]^{1/2}}{[l_1]^{1/2}} \sum_S [S] \\ &= \frac{w_1}{[l_1,l_2]}[L_1,l_2,S_1,s_2] - (-1)^{l_1+l_2+k} \frac{[L_1,S_1,s_2] w_1}{[l_1,l_2]^{1/2}} \frac{[l_2]^{1/2}}{[l_1]^{1/2}} \\ &= 0 \end{aligned} \tag{12.83}$$

since the reduced matrix element in (12.67) requires that $l_1 + k + l_2$ be even.

The above methods can evidently be applied similarly to the general case of any small splitting associated with a weak interaction $(j_1 j_2) j_3$.

13

CONFIGURATION INTERACTION

13-1. GENERAL REMARKS

Our entire theory of atomic structure is based on expanding the wavefunction Ψ^k for an atomic state k in terms of a set of basis functions Ψ_b, as in (4.4):

$$\Psi^k = \sum_b y_b^k \Psi_b \ . \qquad (13.1)$$

So far, we have limited ourselves to the approximation in which the summation has been truncated to include basis functions for only a single configuration. We now wish to improve the accuracy of Ψ^k by including basis functions from more than one configuration.

Let us consider briefly the form of the Hamiltonian matrix for the case in which we include basis states of two configurations (Fig. 13-1). Each row and column label b of the matrix includes the set of angular-momentum quantum numbers that we had in the single-configuration case; in addition, we must now also explicitly state the configuration—which previously could be implicitly understood, as it was common to all rows and columns. The matrix may be thought of as being made up of four different submatrix blocks. The two diagonal blocks (c_1 and c_2) are always square, and are identical with the energy matrices that we would have had for the two configurations c_1 and c_2 when considered separately. In either block, each matrix element consists of sums of terms E_{av}, $f_kF^k(ij)$, $g_kG^k(ij)$, and $d_j\zeta_j$. However, the numerical value of E_{av} that appears in the diagonal elements of the block c_1 will be different from that in c_2. Likewise, the radial integrals $F^k(ij)$, etc., for the block c_1 will normally be computed from radial wavefunctions $P_i(r)$ and $P_j(r)$ obtained from a Hartree-Fock (or HX, HFS, etc.) calculation for the configuration c_1, whereas the radial integrals appearing within c_2 will be computed via a HF calculation for c_2.

The off-diagonal ("configuration-interaction") blocks c_1-c_2 and c_2-c_1 are in general rectangular rather than square, and one block is just the transpose of the other because the total matrix is symmetric. The configuration-interaction (CI) matrix elements are in principle calculated exactly as were the the single-configuration matrix elements, using the

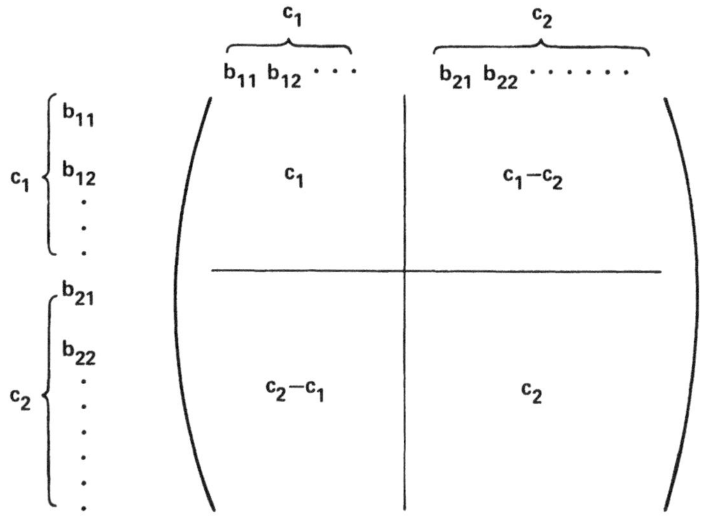

Fig. 13-1. General form of the Hamiltonian matrix (for any given value of J) in the two-configuration approximation. The basis states have been denoted by the labels b_{ij}, where i is the configuration number and j represents the serial number of the basis state within a configuration.

Hamiltonian operator (10.1), except that the bra function is a basis function belonging to one configuration, and the ket function is a basis function from a different configuration. Correspondingly, in place of the single-configuration radial integrals F^k and G^k, the electron-electron Coulomb operator leads to CI radial integrals of the general form (6.23); namely,

$$R_d^k(l_i l_j, l_{i'} l_{j'}) \equiv \int_0^\infty \int_0^\infty \frac{2r_<^k}{r_>^{k+1}} P_i^*(r_1) P_j^*(r_2) P_{i'}(r_1) P_{j'}(r_2)\, dr_1\, dr_2 \tag{13.2}$$

and

$$R_e^k(l_i l_j, l_{i'} l_{j'}) \equiv R_d^k(l_i l_j, l_{j'} l_{i'})$$

$$= \int_0^\infty \int_0^\infty \frac{2r_<^k}{r_>^{k+1}} P_i^*(r_1) P_j^*(r_2) P_{j'}(r_1) P_{i'}(r_2)\, dr_1\, dr_2 \ , \tag{13.3}$$

where P_i and P_j are radial functions for the bra configuration, and $P_{i'}$ and $P_{j'}$ correspond to the ket configuration. (Note that unlike F^k and G^k, R^k is not necessarily positive.) We shall denote the angular coefficients of these radial integrals by r_d^k and r_e^k, analogously to the single-configuration direct and exchange coefficients f_k and g_k, respectively.

There is never any exact analog of the single-configuration parameter E_{av}. In some types of CI (namely, between two configurations belonging to the same Rydberg series) there do exist non-zero contributions from the kinetic-energy and electron-nuclear potential-energy operators; however, we shall see in Sec. 13-5 that these contributions are analogous only to the center-of-gravity corrections to the coefficients f_0' and g_k', with nothing left over analogous to E_{av}.

Whenever there are non-zero KE and e-n contributions, there are also non-zero spin-orbit matrix elements of the form $d_{jj'}\zeta_{jj'}$ $(l_j = l_{j'})$, where analogously to (10.9)

$$\zeta_{jj'} = \frac{\alpha^2}{2}\int_0^\infty P_j^*(r)\left[\frac{1}{r}\frac{dV}{dr}\right]P_{j'}(r)\,dr \ . \qquad (13.4)$$

[The potential V(r) is essentially the same for both P_j and $P_{j'}$ because j and j′ usually are both excited electrons, and core relaxations are small upon excitation of j to j′.] From Schwarz's inequality,

$$\zeta_{jj'} \leqq [\zeta_j\zeta_{j'}]^{1/2} \ , \qquad (13.5)$$

where ζ_j and $\zeta_{j'}$, are the single-configuration spin-orbit parameters computed from $|P_j|^2$ and $|P_{j'}|^2$, respectively; in practice, the equality sign usually gives a very good approximation for $\zeta_{jj'}$, because the major contribution to each radial integral arises at small r, where $P_j(r)$ and $P_{j'}(r)$ are nearly identical except for a scale factor (Fig. 8-3). The effect of the CI spin-orbit matrix elements is usually small (and commonly neglected) except for p electrons in heavy elements.[1]

Selection rules. Because of practical limitations, the basis set employed for the expansion (13.1) must be kept manageably small, and the set of configurations to be included in any calculation must therefore be chosen judiciously. Several guiding principles may be employed. Firstly, the Hamiltonian operator (10.1) has even parity, and so the CI matrix elements are zero unless the bra and ket functions have a common parity (Sec. 2-14). Thus we need consider only those configurations that have the same parity as the configuration of primary interest. Secondly, the Hamiltonian involves only one- and two-electron operators, so that interactions can occur only between two configurations that differ in at most two orbitals; for example, interactions occur between any two of the configurations d^3s^2, d^4s, d^5, d^3p^2, and d^2sp^2, except for d^5—d^2sp^2. Finally, the matrix of the Coulomb operator in an LS representation is (just as in the single-configuration case) diagonal in the quantum numbers LSJM. Therefore, non-zero Coulomb CI matrix elements exist only if each configuration contains a basis state with some common value of LS; thus sp and sf do not interact, because the former contains only $^{1,3}P^\circ$ terms whereas the latter contains only $^{1,3}F^\circ$. This selection rule applies also to the KE, electron-nuclear, and spin-orbit operators: Each is a one-electron operator so that non-zero matrix elements exist only if the two configurations differ in only one orbital; moreover, this orbital must have the same l in both configurations, because the operators are (in effect) functions of $|\mathbf{r}|$ only [Sec. 6-2 and Eq. (3.41)]. The two configurations can therefore only be of the form $\cdots n_il_i^{\,w_i}n_jl_i^{\,w_j}$ — $\cdots n_il_i^{\,w_i-1}n_jl_i^{\,w_j+1}$, and they necessarily possess a common term. (The

[1]W. C. Martin, J. Sugar, and J. L. Tech, Phys. Rev. A 6, 2022 (1972).

subshell l^w always contains the term ^1S if w is even, or ^2l if w is odd; therefore the subshells i and j combined always contribute ^1S if w_i and w_j are both even or both odd, or contribute ^2l if one is even and the other odd.)

The non-zero Coulomb matrix element that connects basis functions $|c\gamma LSJ\rangle$ and $|c'\gamma' LSJ\rangle$ of two configurations c and c′ causes each eigenstate to be a mixture of the two basis states. Because of the spin-orbit mixing present within each configuration separately, each eigenstate of the two-configuration calculation indirectly becomes a mixture of *all* basis states (of given J) of both configurations. Similarly, a three-configuration calculation for $d^5 + d^3s^2 + d^2sp^2$ will give eigenstates each of which is a mixture of basis functions of all three configurations, even though d^5 and d^2sp^2 do not interact directly. (Of course the indirect effect may be small enough that some eigenstates contain very little d^2sp^2 nature, and others contain very little d^5.)

Strong interactions. In addition to fundamental selection rules, a few qualitative remarks can be made concerning those interactions that may be most important. From (10.38)-(10.39), CI energy perturbations and configuration mixings are largest when the magnitude of the CI matrix element is large compared with the energy difference between the unperturbed levels. Consequently, CI effects tend to be largest between configurations whose center-of-gravity energies E_{av} are not greatly different, and/or for cases in which the Coulomb matrix elements $r^kR^k(ij,i'j')$ are large in magnitude. Large values of $|R^k|$ tend to occur particularly when $n_in_j = n_{i'}n_{j'}$ (i.e., when the two configurations belong to the same complex[2]), because the various radial wavefunctions then tend to have maximum overlap. Thus, particularly in high ionization stages where E_{av} depends primarily on the principal quantum numbers and relatively weakly on the orbital angular momenta, the largest interactions tend to occur among configurations within a complex, such as $3s^23p^w - 3p^{w+2}$ or $3s3p^{w+1} - 3s^23p^{w-1}3d$. However, the interaction $nsnp^{w+1} - ns^2np^{w-1}n'd$ is notoriously strong in neutral atoms (especially for $n = 3$) for all values of n′. The interactions $3d^n4s^2 + 3d^{n+1}4s + 3d^{n+2}$ [frequently written $(3d+4s)^{n+2}$] are usually important in elements of the first transition series, as are the corresponding interactions in the later transition series.

Interactions of the type $nl^wn'l' - nl^wn''l'$ among members of a Rydberg series are also sometimes quite strong (see, for example, Sec. 13-5), though more commonly they produce noticeable perturbations only between levels that accidentally have nearly equal energies.[3]

*13-2. WAVEFUNCTION ORTHOGONALITY PROBLEMS

We have assumed from the beginning that the basis functions Ψ_b employed for the wavefunction expansion (13.1) form an orthonormal set,

[2]A complex is the set of all configurations of given parity that can be formed with a given set of N values of the principal quantum numbers, $\{n_i\}$; see D. Layzer, Ann. Phys. 8, 271 (1959). As examples in the case of two electrons, 2s3s+2p3p+2s3d form an even-parity complex, and 2s3p+2p3s+2p3d form the corresponding odd-parity complex. In neutral or weakly ionized atoms, the important interactions are more nearly those between configurations having (approximately) the same set of values of the effective quantum numbers, $\{n^*_i\}$.

[3]Examples are discussed by B. Edlén, "Atomic Spectra," in S. Flügge, ed., *Handbuch der Physik* (Springer-Verlag, Berlin, 1964), Vol. XXVII, pp. 140 ff. See also K. T. Lu and U. Fano, Phys. Rev. A 2, 81 (1970) and C. M. Lee and K. T. Lu, Phys. Rev. A 8, 1241 (1973).

$$\langle \Psi_b | \Psi_{b'} \rangle = \delta_{bb'} . \tag{13.6}$$

When all basis functions belonged to a single configuration, we encountered no difficulty in constructing such a set. It was necessary only that the one-electron spin-orbitals (4.17) be mutually orthonormal. Orthogonality with respect to l, m_l, and m_s was automatic because of the properties of the spherical harmonics and of the spin functions. For given l, m_l, and m_s, we assured orthogonality with respect to n [i.e., orthogonality of the radial functions $P_{nl}(r)$, $P_{n'l}(r)$] by introducing the Lagrange multipliers $\varepsilon_{nl,n'l}$ into the Hartree-Fock method (Secs. 7-1 and 7-2); in the HFS method (Sec. 7-11), orthogonality was automatic because the same effective radial potential $V^i(r) + l_i(l_i + 1)/r^2$ was used for all orbitals i having given l_i.

In constructing a multi-configuration basis set, we have the additional problem of producing orthogonality between basis functions belonging to different configurations. Again, there is no problem regarding the angular factors. Orthogonality of the radial factors may be accomplished by the use of virtual or "unoccupied" orbitals. In the HFS method, for example, radial functions for all excited orbitals may be calculated using the potential function for the ground configuration (in which the excited orbital is not occupied); the fact that a single potential is used for calculating all radial functions (of given l) of all configurations assures the desired orthogonality. The difficulty with this approach is that the basis functions for excited configurations, calculated from a potential inappropriate to those configurations, give a poor first-order (single-configuration) approximation to the wavefunctions and energy levels of those configurations. This is of little consequence if one is interested in computed results for only the ground configuration.

However, we shall usually be interested simultaneously in all (or at least most of) the configurations included in the calculation. In order to obtain good first-order approximations for all levels, we shall use a separate SCF calculation for each configuration. Complete orthogonality does not then exist between basis functions of two configurations that differ only in one or more values of n_i: because of relaxation effects upon exciting (for example) a 3p electron of Si I $3p^2$ to give Si I 3p4p, the overlaps of P_{3p} in the ground configuration with P_{3p} and P_{4p} in the excited configuration will not be exactly one and zero, respectively. We now investigate the consequence of such non-orthogonalities.

We first define an overlap matrix (Gram's matrix) M whose elements are the overlap integrals of the basis functions Ψ_b:

$$M_{ij} \equiv \langle \Psi_i | \Psi_j \rangle = M_{ji} . \tag{13.7}$$

The appearance of this matrix in the two-configuration case is shown in Fig. 13-2. (Many elements of the off-diagonal block m are zero, because of orthogonalities with respect to various coupled angular-momentum quantum numbers.) When the expansion (13.1) is substituted into the Schrödinger equation $H\Psi = E\Psi$ (omitting superscript k's for brevity), the final simplifying step of (4.6) no longer follows, because of the lack of complete orthogonality; in place of the matrix equation

$$HY = EY \tag{13.8}$$

we now obtain

$$M = \left(\begin{array}{cc|ccc} 1 & 0 & M_{13} & M_{14} & M_{15} \\ 0 & 1 & M_{23} & M_{24} & M_{25} \\ \hline M_{31} & M_{32} & 1 & 0 & 0 \\ M_{41} & M_{42} & 0 & 1 & 0 \\ M_{51} & M_{52} & 0 & 0 & 1 \end{array}\right) \equiv \left(\begin{array}{c|c} I_a & m \\ \hline \tilde{m} & I_b \end{array}\right)$$

$$M^{-1/2} = \left(\begin{array}{cc|ccc} 1 & 0 & -\tfrac{1}{2}M_{13} & -\tfrac{1}{2}M_{14} & -\tfrac{1}{2}M_{15} \\ 0 & 1 & -\tfrac{1}{2}M_{23} & -\tfrac{1}{2}M_{24} & -\tfrac{1}{2}M_{25} \\ \hline -\tfrac{1}{2}M_{31} & -\tfrac{1}{2}M_{32} & 1 & 0 & 0 \\ -\tfrac{1}{2}M_{41} & -\tfrac{1}{2}M_{42} & 0 & 1 & 0 \\ -\tfrac{1}{2}M_{51} & -\tfrac{1}{2}M_{52} & 0 & 0 & 1 \end{array}\right) = \left(\begin{array}{c|c} I_a & -\tfrac{1}{2}m \\ \hline -\tfrac{1}{2}\tilde{m} & I_b \end{array}\right)$$

Fig. 13-2. Upper: Form of the overlap matrix M when (for given J) there are two basis states in one configuration and three basis states in another. Lower: Approximate form of $M^{-1/2}$ (to first order), assuming the overlap between basis functions of different configurations to be small in magnitude ($|M_{ij}| \ll 1$ for elements in the off-diagonal blocks). For a Rydberg-series interaction, m is square and equal to a constant times I_a.

$$HY = EMY \ , \tag{13.9}$$

which may be written in the form

$$(M^{-1}H)Y = EY$$

or in the form

$$(M^{-1/2}HM^{-1/2})(M^{1/2}Y) = E(M^{1/2}Y) \ . \tag{13.10}$$

Thus we have to find not eigenvalues and eigenvectors of H, but rather eigenvalues E and eigenvectors Y of $M^{-1}H$, or (equivalently) eigenvalues E and eigenvectors $M^{1/2}Y$ of $M^{-1/2}HM^{-1/2}$.

To see the significance of this latter modified Hamiltonian matrix, we construct a new set of basis functions Φ_b defined by

$$\Phi_i = \sum_j (M^{-1/2})_{ij}\Psi_j = \Psi_i + \sum_j{}'(M^{-1/2})_{ij}\Psi_j \cong \Psi_i - \frac{1}{2}\sum_j{}' M_{ij}\Psi_j \,, \tag{13.11}$$

where the primed summation runs only over basis states of configurations other than that to which Ψ_i belongs, and the final approximate expression is valid if $|M_{ij}| \ll 1$ (see Fig. 13-2). If we construct column vectors of basis functions

$$\Psi \equiv \begin{pmatrix} \Psi_1 \\ \Psi_2 \\ \cdot \\ \cdot \\ \cdot \end{pmatrix}, \qquad \Phi \equiv \begin{pmatrix} \Phi_1 \\ \Phi_2 \\ \cdot \\ \cdot \\ \cdot \end{pmatrix} \tag{13.12}$$

(*not* column vectors of numbers, such as **Y**), (13.11) may be written in the matrix form

$$\Phi = M^{-1/2}\Psi \,. \tag{13.13}$$

It follows from (13.7) that **M** is real symmetric, and consequently that $\mathbf{M}^{1/2}$ and $\mathbf{M}^{-1/2}$ are also real symmetric. Therefore the Hamiltonian matrix in the Φ representation is

$$H_\Phi \equiv \langle\Phi|H|\tilde{\Phi}\rangle = M^{-1/2}\langle\Psi|H|\tilde{\Psi}\rangle(M^{-1/2})^{tr} = M^{-1/2}HM^{-1/2} \,. \tag{13.14}$$

This is precisely the modified Hamiltonian matrix that appears in (13.10). Moreover, the wavefunction (13.1) may be written in terms of the basis functions Φ_i as

$$\Psi^k = \tilde{\Psi}Y = (M^{1/2}\Phi)^{tr}Y = \tilde{\Phi}(M^{1/2}Y) \,, \tag{13.15}$$

and the eigenvector $\mathbf{Y}_\Phi \equiv \mathbf{M}^{1/2}\mathbf{Y}$ is precisely that appearing in (13.10). Thus the awkward-looking equation (13.9) that we obtained in the non-orthogonal basis Ψ becomes, in the basis Φ,

$$H_\Phi Y_\Phi = EY_\Phi \,, \tag{13.16}$$

which is of the familiar form (13.8). This is not surprising, because the basis Φ is an orthonormal one:[4] from (13.13) and (13.7)

$$\langle\Phi|\tilde{\Phi}\rangle = \langle M^{-1/2}\Psi|(M^{-1/2}\Psi)^{tr}\rangle = M^{-1/2}\langle\Psi|\tilde{\Psi}\rangle(M^{-1/2})^{tr} = M^{-1/2}MM^{-1/2} = I \,,$$

where **I** is the identity matrix.

The point of all this is the following: From (13.11), the basis functions Ψ_i are very nearly like the Φ_i except for their small lack of orthogonality. Hence if we calculate Hamiltonian matrix elements, using the Ψ_i but *always treating them as though they were exactly orthonormal*, we must obtain very nearly the matrix $\mathbf{H}_\Phi$ instead of **H**, and may therefore work with the standard equation (13.16) instead of with the more complicated equation (13.9). To the approximation that we obtain a Hamiltonian matrix identical with $\mathbf{H}_\Phi$, we

[4]R. Landshoff, Z. Physik **102**, 201 (1936); H. C. Schweinler and E. P. Wigner, J. Math. Phys. **11**, 1693 (1970). See also S. Perlis, *Theory of Matrices* (Addison-Wesley, Reading, Mass., 1952), Sec. 9-15.

obtain exactly the same eigenvalues as though we had taken the non-orthogonalities into account and used (13.9) [or its equivalent form (13.10)]. The eigenvector components will correspond to a wavefunction expansion in terms of the Φ_l instead of the Ψ_l. However, this just means that if we use these eigenvector components in (13.1), then it will be appropriate to make all further calculations (of transition probabilities, etc.) treating the basis functions Ψ_b in (13.1) as though they again were exactly orthonormal.

We shall therefore henceforth always assume that

$$\langle P_{nl} | P_{n'l} \rangle = \delta_{nn'} , \tag{13.17}$$

even when $P_{nl}(r)$ and $P_{n'l}(r)$ have been obtained via separate SCF calculations for two different configurations.[5] We thereby retain the advantage of having the best possible single-configuration approximation for each and every configuration, without formally introducing any non-orthogonality complications into multi-configuration calculations.

Problem

13-2(1). For two configurations a and b containing d_a and d_b basis states, respectively, write the Hamiltonian matrix for the Ψ representation in block form

$$H = \begin{pmatrix} H_{aa} & H_{ab} \\ H_{ba} & H_{bb} \end{pmatrix} ,$$

where H_{aa} and H_{bb} are the single-configuration Hamiltonian matrices, H_{ab} is the d_a-by-d_b CI matrix, and H_{ba} is the transpose of H_{ab}. Using the block form of the overlap matrix M given in Fig. 13-2, verify (by evaluating $M^{-1/2}M^{-1/2}M$) that to second order in m

$$M^{-1/2} = \begin{pmatrix} I_a + \frac{3}{8}m\tilde{m} & -\frac{1}{2}m \\ -\frac{1}{2}\tilde{m} & I_b + \frac{3}{8}\tilde{m}m \end{pmatrix} ,$$

and that to a corresponding order

$$H_\Phi = \begin{pmatrix} H_{aa} - \frac{1}{2}(h_1\tilde{m} + m\tilde{h}_1) - (h_2\tilde{m} + m\tilde{h}_2) & h_1 \\ \tilde{h}_1 & H_{bb} - \frac{1}{2}(\tilde{h}_1 m + \tilde{m}h_1) + (\tilde{h}_2 m + \tilde{m}h_2) \end{pmatrix} ,$$

where

$$h_1 = H_{ab} - \frac{1}{2}(H_{aa}m + mH_{bb}) , \qquad h_2 = \frac{1}{8}(-H_{aa}m + mH_{bb}) .$$

[5]Typically, actual values of $\langle P_{nl}|P_{nl}\rangle$ are no smaller than 0.98, and values of $|\langle P_{nl}|P_{n'l}\rangle|$ for $n \neq n'$ are less than 0.02 to 0.05. Although the detailed form of M is somewhat different, the above discussion also provides the justification for ignoring non-orthogonalities among basis functions within a single configuration, as for example in using the HX method (Sec. 7-12).

[The final terms of $\mathbf{h}_1$ represent corrections that cancel the non-orthogonality contributions to $\mathbf{H}_{ab}$. These contributions (especially those from the 1s electrons) are very large, and it is therefore essential that the corrections be computed very accurately. Our approximate method of calculating $\mathbf{H}_{\Phi}$, using the Ψ_l but assuming exact orthogonality, in effect makes certain that the cancellation is exact. In this approximate method, the diagonal blocks of $\mathbf{H}_{\Phi}$ are just $\mathbf{H}_{aa}$ and $\mathbf{H}_{bb}$; we are thus neglecting energy-level-*spread* contributions of the order of 2 to 5 percent,[5] as the magnitudes of the matrix elements of $\mathbf{h}_1$ and $\mathbf{h}_2$ are comparable with those of $\mathbf{H}_{aa}$ and $\mathbf{H}_{bb}$ (excluding E_{av}).]

13-3. TWO-ELECTRON CONFIGURATIONS

There are so many different types of pairs of interacting configurations that any sort of semi-general treatment is necessarily very involved and abstract. We therefore first consider a few comparatively simple special cases. One of the simplest is that in which each configuration contains only two electrons outside a common core of closed subshells:

$$\cdots n_\rho l_\rho n_\sigma l_\sigma \; - \; \cdots n'_\rho l'_\rho n'_\sigma l'_\sigma \; ; \tag{13.18}$$

a specific example is Si I $\cdots 3s^2 3p4s - \cdots 3s^2 3p3d$. For the present, we assume that in each configuration the two electrons are non-equivalent, and we also assume that the two configurations are not members of a single Rydberg series (e.g., $\cdots$ 3p4s — $\cdots$ 3p5s); these excluded cases will be considered later in this section, and in Sec. 13-5, respectively. We may also assume, without loss of generality, that if one of the electrons in each configuration lies in a common subshell, then this subshell is $n_\rho l_\rho = n'_\rho l'_\rho$; it then follows that $l_\sigma \neq l'_\sigma$, since we have excluded members of a Rydberg series.

Because the two configurations have the same basic structure [i.e., the same total number of occupied subshells, $q = q'$, and the same occupation number for corresponding subshells, $w_i = w'_i$ ($1 \leq i \leq q$)], complications due to coordinate permutations can be handled exactly as for configuration-diagonal matrix elements, and we need to consider only matrix elements of the form (9.80),(9.85) between non-antisymmetrized basis functions; the only difference from before is that ψ_b is a basis function from one configuration, and ψ'_b is a basis function from the other.

For any of the one-electron operators of the Hamiltonian (KE, e-n energy, or s-o energy), each term of (9.80) with $j < q$ is zero because integration over the coordinate of the qth subshell gives simply the overlap integral

$$\langle n_\sigma l_\sigma | n'_\sigma l'_\sigma \rangle = 0 \; . \tag{13.19}$$

The term $j = q$ also gives zero because either $\langle n_\rho l_\rho | n'_\rho l'_\rho \rangle = 0$ or $l_\sigma \neq l'_\sigma$, and if the latter is true then $\langle n_\sigma l_\sigma | f | n'_\sigma l'_\sigma \rangle = 0$ by the selection rules discussed in Sec. 13-1.

For the two-electron Coulomb matrix elements (9.85) the single summation over j gives zero, because if $j < q$ then we have the zero overlap integral (13.19), and if $j = q$ then $w_j - 1 = 0$. Similarly, all terms of the double summation over i and j will be zero unless $j = \sigma$. Even then, the result will be zero if i is one of the closed subshells; this may be easily seen as follows: We recouple the basis functions so that the subshell σ comes before the subshell ρ, and then from (9.87) find

$$\langle[(\cdots{}^1S,l_\sigma)\mathfrak{L}_\sigma\mathfrak{S}_\sigma,l_\rho]LS|2/r_{ij}|[(\cdots{}^1S,l'_\sigma)\mathfrak{L}'_\sigma\mathfrak{S}'_\sigma,l'_\rho]L'S'\rangle$$

$$= \delta_{\mathfrak{L}_\sigma\mathfrak{S}_\sigma,\mathfrak{L}'_\sigma\mathfrak{S}'_\sigma}\,\delta_{n_\rho l_\rho,n'_\rho l'_\rho}\,\delta_{LS,L'S'}\,\langle(\cdots{}^1S,l_\sigma)\mathfrak{L}_\sigma\mathfrak{S}_\sigma|2/r_{ij}|(\cdots{}^1S,l'_\sigma)\mathfrak{L}'_\sigma\mathfrak{S}'_\sigma\rangle\ ; \qquad (13.20)$$

but if $n_\rho l_\rho = n'_\rho l'_\rho$ then $l_\sigma \neq l'_\sigma$, so that $\mathfrak{L}_\sigma$ cannot be equal to $\mathfrak{L}'_\sigma$. Hence the closed subshells contribute nothing whatever to the matrix elements.[6]

Thus we see that the only contribution to the CI matrix element is the two-electron Coulomb element

$$\langle l_\rho(i)l_\sigma(j)LS|\sum_k \frac{2r_<^k}{r_>^{k+1}}C^{(k)}_{(i)}\cdot C^{(k)}_{(j)}|[l'_\rho(i)l'_\sigma(j) - l'_\rho(j)l'_\sigma(i)]L'S'\rangle$$

$$= \sum_k [r_d^k R_d^k(l_\rho l_\sigma,l'_\rho l'_\sigma) + r_e^k R_e^k(l_\rho l_\sigma,l'_\rho l'_\sigma)]\ . \qquad (13.21)$$

The angular coefficients can be evaluated in the same way as were (11.81) and (11.85), using (11.47) and—for the exchange term—(9.8):

$$r_d^k = \delta_{LS,L'S'}(-1)^{l'_\rho+l_\sigma+L}\langle l_\rho\|C^{(k)}\|l'_\rho\rangle\langle l_\sigma\|C^{(k)}\|l'_\sigma\rangle\begin{Bmatrix} l_\rho & l_\sigma & L \\ l'_\sigma & l'_\rho & k \end{Bmatrix}\ , \qquad (13.22)$$

$$r_e^k = -(-1)^{l'_\rho+l'_\sigma-L'-1+S}\langle l_\rho l_\sigma LS\|C^{(k)}_{(i)}\cdot C^{(k)}_{(j)}\|l'_\sigma l'_\rho L'S'\rangle$$

$$= \delta_{LS,L'S'}(-1)^{l'_\rho+l_\sigma+S}\langle l_\rho\|C^{(k)}\|l'_\sigma\rangle\langle l_\sigma\|C^{(k)}\|l'_\rho\rangle\begin{Bmatrix} l_\rho & l_\sigma & L \\ l'_\rho & l'_\sigma & k \end{Bmatrix}$$

$$= \delta_{LS,L'S'}(-1)^S\langle l_\rho\|C^{(k)}\|l'_\sigma\rangle\langle l'_\rho\|C^{(k)}\|l_\sigma\rangle\begin{Bmatrix} l_\rho & l_\sigma & L \\ l'_\rho & l'_\sigma & k \end{Bmatrix}\ . \qquad (13.23)$$

It follows from the expression (11.23) for the reduced matrix elements of $C^{(k)}$ that the summation over k for the direct elements runs over those values that are common to the two sequences

$$k_d = \begin{cases} |l_\rho - l'_\rho|, & |l_\rho - l'_\rho| + 2, \quad \cdots \quad l_\rho + l'_\rho\ , \\ |l_\sigma - l'_\sigma|, & |l_\sigma - l'_\sigma| + 2, \quad \cdots \quad l_\sigma + l'_\sigma\ , \end{cases} \qquad (13.24)$$

and that the possible values of k for the exchange terms are those common to

$$k_e = \begin{cases} |l_\rho - l'_\sigma|, & |l_\rho - l'_\sigma| + 2, \quad \cdots \quad l_\rho + l'_\sigma\ , \\ |l_\sigma - l'_\rho|, & |l_\sigma - l'_\rho| + 2, \quad \cdots \quad l_\sigma + l'_\rho\ . \end{cases} \qquad (13.25)$$

[6]For single-configuration matrix elements, we saw that closed subshells contributed nothing to the level structure, but they *did* contribute to E_{av}.

It is evident that these restrictions embody the parity selection rule discussed in Sec. 13-1. They are also necessary (but not sufficient) conditions that the two configurations contain at least one common LS term; for example, $f^{14} - f^{13}p$ has possible values $k_d = 2$ and 4, but $f^{13}p$ has no ^1S term.

We can now relax the restriction against equivalent electrons. If the electrons are equivalent in either configuration or in both, then $R_d^k = R_e^k$ and the two terms in (13.21) can be lumped into a single one. The coefficient r_d^k is then effectively doubled, except that from (9.33) we must introduce an additional renormalization factor $2^{-1/2}$ for each pair of equivalent electrons. Thus we have

$$\langle l_\rho^2 LS | 2/r_{12} | l'_\rho l'_\sigma L'S' \rangle = (2)^{1/2} r_d^k R^k(l_\rho^2, l'_\rho l'_\sigma) , \qquad (n'_\rho l'_\rho \neq n'_\sigma l'_\sigma) , \tag{13.26}$$

where $n'_\rho l'_\rho$ may be identical with $n_\rho l_\rho$, and

$$\langle l_\rho^2 LS | 2/r_{12} | l_\sigma^2 L'S' \rangle = r_d^k R^k(l_\rho^2, l_\sigma^2) , \tag{13.27}$$

where in each case r_d^k is given by (13.22). Specific examples are Si III $3p^2 - 3s4d$, Sc II $3d^2 - 3d4s$, and Sc II $3d^2 - 4s^2$; the last of these illustrates that a filled subshell may be involved, provided that it is not filled in both configurations.

13-4. SINGLE-CONFIGURATION-LIKE INTERACTIONS

In some interactions involving more than two electrons outside closed subshells, the two configurations still have the same basic structure—differing only in the nl values of corresponding singly occupied (or doubly occupied) subshells. Examples that arise in rare-earth elements are

$$fsp - fdp \tag{13.28}$$

and

$$f^w s^2 - f^w d^2 . \tag{13.29}$$

Such cases can be treated analogously to the single-configuration case, following the same approach used in Sec. 13-3; provided we exclude the Rydberg-series case (e.g., $fsp - fsp'$), contributions from closed subshells and from one-electron operators in the Hamiltonian are all zero.

As an example, if we use the genealogical $\mathfrak{LS}$ coupling

$$(fs)p - (fd)p , \tag{13.30}$$

then (9.87) shows that the matrix elements for the $fs - fd$ Coulomb interaction are basically just those for the two-electron problem

$$fs - fd ,$$

to which (13.21)-(13.23) are directly applicable. Similarly, use of (9.88) shows that the $s^2 - d^2$ interaction in (13.29) involves nothing more than delta-factors (on the quantum numbers $\alpha_1L_1S_1$ of f^w) times the matrix elements of $s^2 - d^2$, which are given by (13.27). Likewise, (13.26) can be applied directly to $f^wd^2 - f^wds$ provided we first recouple from $(f^wd)s$ to $f^w(ds)$.

For the sp − dp interaction in (13.28) we can either recouple (13.30) to f(sp) − f(dp) and use (13.21)-(13.23), or (better) use appropriate generalizations of (12.19), (12.28), (12.34), and (12.35) so as to automatically eliminate the recoupling summations (cf. Sec. 11-12). This latter approach serves also for

$$f^wsp - f^wdp \tag{13.31}$$

(or even $f^{w1}sp^{w3} - f^{w1}dp^{w3}$, though such a case is of little practical interest). The required generalizations are

$$\langle l_1^{w_1}l_2l_3^{w_3}|2/r_{\rho\sigma}|l_1^{w_1}l_2'l_3^{w_3}\rangle = \sum_k [r_d^k R_d^k(l_\rho l_\sigma, l_\rho' l_\sigma') + r_e^k R_e^k(l_\rho l_\sigma, l_\rho' l_\sigma')] , \tag{13.32}$$

$$r_d^k = \langle l_\rho\|C^{(k)}\|l_\rho'\rangle\langle l_\sigma\|C^{(k)}\|l_\sigma'\rangle\, I_{\rho\sigma}^{(k)} , \tag{13.33}$$

$$r_e^k = -\frac{1}{2}\langle l_\rho\|C^{(k)}\|l_\sigma'\rangle\langle l_\rho'\|C^{(k)}\|l_\sigma\rangle \sum_r (-1)^r[r] \begin{Bmatrix} l_\rho & l_\rho' & r \\ l_\sigma & l_\sigma' & k \end{Bmatrix} [I_{\rho\sigma}^{(r)} + 4I_{\rho\sigma}^{(r1)}] . \tag{13.34}$$

It can be readily verified [using (5.32) in the case of r_e^k] that these expressions reduce to (13.22) and (13.23) for two-electron configurations. For application to (13.31), $(l_\rho l_\sigma, l_\rho' l_\sigma')$ is equal to (fs,fd) for the fs − fd interaction and is equal to (sp,dp) for the sp − dp interaction.

In the evaluation of $I^{(r)}$ and $I^{(r1)}$ from (12.28) and (12.35), we encounter the reduced matrix elements

$$\langle l\|u^{(r)}\|l'\rangle = \delta(lrl')$$

and

$$\langle l\|v^{(r1)}\|l'\rangle = (3/2)^{1/2}\,\delta(lrl')$$

from (11.48) and (11.71), respectively. It was for this application to the calculation of CI matrix elements that we generalized Racah's definition $\langle l\|u^{(r)}\|l'\rangle = \delta_{ll'}$—see fn. 10 in Sec. 11-9.

13-5. RYDBERG-SERIES INTERACTIONS

We now consider the case excluded in the two preceding sections; namely, that in which the two configurations are members of a single Rydberg series, and so are of the form

$$\cdots n_m l_m^{w_m} n_\sigma l_\sigma - \cdots n_m l_m^{w_m} n_\sigma' l_\sigma . \tag{13.35}$$

So far as angular momenta are concerned, the two configurations are completely identical, and the basis-function sets of angular-momentum quantum numbers b and b′ are the same for both configurations. For given total angular momentum J, the combined Hamiltonian matrix is just twice as large as for either configuration separately; the off-diagonal block is square, with quantum-number labels just like those of each diagonal block, except for the configuration labels. The CI parameters R_d^k and R_e^k will have the same values of k as do the diagonal-block parameters F^k and G^k, respectively.

Calculation of Hamiltonian matrix elements proceeds exactly as in Secs. 13-3 and 13-4, with two exceptions. Firstly, the Coulomb terms of (9.85) will still be non-zero only for the $i < j$ summation with $j = \sigma$, but contributions when i represents a closed subshell are no longer zero. These closed-subshell terms will be exactly the same as in the single-configuration case, except with F^0 and G^k replaced by R_d^0 and R_e^k, respectively; from (12.31) and (12.36) these are

$$\delta_{bb'} \sum_i w_i \left[R_d^0(i\sigma,i'\sigma') - \frac{1}{2[l_i,l_\sigma]} \sum_k \langle l_i \| C^{(k)} \| l_\sigma \rangle^2 R_e^k(i\sigma,i'\sigma') \right] , \qquad (13.36)$$

where the notation i′ signifies a radial function P_i for the ith subshell of the primed (ket) configuration. The summation is over only closed subshells i, and $\delta_{bb'}$ refers only to the angular-momentum basis-state labels (not to the configuration labels).

Secondly, contributions of the one-electron operators are non-zero for the $j = \sigma$ term of (9.80). Assuming the radial wavefunctions to be solutions of the HF equations, evaluation of the KE plus electron-nuclear contribution

$$\langle \cdots l_\sigma | -\nabla_\sigma^2 - 2Z/r_\sigma | \cdots l'_{\sigma'} \rangle \qquad (13.37)$$

can be carried out with the use of (7.11) for $i = \sigma$. The terms of (7.11) involving ε_σ and $\varepsilon_{\sigma j}$ give zero because of orthogonality of P_σ and $P_{\sigma'}$. Since $w_\sigma = 1$, the remaining terms of (13.37) are

$$\delta_{bb'} \sum_{j<\sigma} w_j \left[- R_d^0(j'\sigma,j'\sigma') + \int P_\sigma(r) B_{\sigma'j'}(r) P_{j'}(r)\, dr \right]$$

$$= \delta_{bb'} \sum_{i<\sigma} w_i \left[- R_d^0(i\sigma,i'\sigma') + \frac{1}{2[l_i,l_\sigma]} \sum_k \langle l_i \| C^{(k)} \| l_\sigma \rangle^2 R_e^k(i\sigma,i'\sigma') \right] , \qquad (13.38)$$

where use has been made of (11.23) and also of the fact that $P_j(r) \cong P_{j'}(r)$. We thus see that, because of the special form of the HF radial functions, the KE and e-n contributions for closed subshells exactly cancel the closed-subshell Coulomb terms (13.36); in addition, for non-closed shells i they cancel the Coulomb term $r_d^0 R_d^0(i\sigma,i'\sigma')$ [since $r_d^0 = w_i$] and also make corrections to the "diagonal" terms $r_e^k R_e^k(i\sigma,i'\sigma')$. In short, comparison of (13.33), (13.34), and (13.38) with (12.32), (12.19), and (12.37)-(12.38) shows that the CI coefficients r_d^k ($k > 0$) and r_e^k are identical with the corresponding configuration-diagonal coefficients f_k and g_k. There is no CI analog of E_{av}, and if there are more than two unfilled subshells there are likewise no analogs of parameters F^k, G^k that do not involve the final Rydberg-series electron.

The only CI analog of the single-configuration spin-orbit terms is that involving $\zeta_{\sigma\sigma'}$; the angular coefficients are identical with the coefficients of the single-configuration parameters ζ_σ and $\zeta_{\sigma'}$.

Example: $p^5d - p^5d'$. For the interaction between configurations p^5nd and $p^5n'd$, the CI block is square and is diagonal so far as the the Coulomb terms are concerned. Analogous to the single-configuration matrix elements (12.79), the CI matrix elements are

$$\begin{aligned}
&{}^1P^\circ - {}^1P^\circ: && -0.200R_d^2 + 1.267R_e^1 - 0.043R_e^3\,,\\
&{}^3P^\circ - {}^3P^\circ: && -0.200R_d^2 - 0.067R_e^1 - 0.043R_e^3\,,\\
&{}^1D^\circ - {}^1D^\circ: && +0.200R_d^2 - 0.067R_e^1 - 0.043R_e^3\,,\\
&{}^3D^\circ - {}^3D^\circ: && +0.200R_d^2 - 0.067R_e^1 - 0.043R_e^3\,,\\
&{}^1F^\circ - {}^1F^\circ: && -0.057R_d^2 - 0.067R_e^1 + 0.325R_e^3\,,\\
&{}^3F^\circ - {}^3F^\circ: && -0.057R_d^2 - 0.067R_e^1 - 0.043R_e^3\,,
\end{aligned} \tag{13.39}$$

where $R_d^2 = R_d^2(pd,pd')$ and $R_e^k = R_d^k(pd,d'p)$. In the same way that large values of 1.267 G^1 gave high-lying ${}^1P_1^\circ$ levels in the single-configuration approximation (Fig. 12-7), large values of $1.267R_e^1$ produce large perturbations among the ${}^1P_1^\circ$ levels of neighboring configurations. This produces the appreciable CI depression of the computed 3d ${}^1P_1^\circ$ energy level shown in the figure (together with a raising of most higher nd ${}^1P_1^\circ$ levels), with a corresponding improvement in the agreement with experiment.

The matrix elements (13.39) for terms other than ${}^1P^\circ$ are much smaller, so that all other levels of the p^5d configurations show only comparatively small CI perturbations.

13-6. ONE-ELECTRON CONFIGURATIONS

Consider two configurations of the same parity, each of which contains a single electron in addition to a common core of closed subshells:

$$\cdots l - \cdots l'\ . \tag{13.40}$$

All such pairs belong to one of two cases.

(1) If $l = l'$ (but $n \neq n'$) we have a special case of the Rydberg-series interaction treated in the preceding section. Except for spin-orbit contributions, the CI matrix elements are zero because the summation over closed subshells i in (13.36) is identical with the summation over $i < \sigma$ in (13.38). That is, in the HF approximation the kinetic-energy and electron-nuclear contributions exactly cancel the electron-electron Coulomb terms.

(2) If $l \neq l'$, then the KE, e-n, and s-o terms are all zero. All Coulomb terms are likewise zero, because Coulomb matrices are diagonal in the total orbital quantum number L, whereas in the present case $L = l$ and $L' = l'$ so that $L \neq L'$.

Thus in either case, CI effects among the one-electron configurations of an alkali-like atom or ion are effectively nil; there exist only small spin-orbit interactions, involving radial integrals (13.4), among members of each Rydberg series. Explanation of the observed inverted ${}^2D$ and ${}^2F^\circ$ terms in alkali atoms requires consideration of perturbations by configurations in which an electron has been excited out of a closed core subshell (for example, Rb I $4p^6 4f\ ^2F^o - 4p^5 4d5s\ ^2F^o$), or of relativistic effects.[7]

13-7. BRILLOUIN'S THEOREM

Consider two configurations c_1 and c_2 that differ only in the principal quantum number of one electron:

$$\begin{aligned} c_1 &= \cdots n_i l_i^{w_i} n_j l_j^{w_j - 1} , \\ c_2 &= \cdots n_i l_i^{w_i - 1} n_j l_j^{w_j} , \end{aligned} \qquad (13.41)$$

where $l_i = l_j$ and the dots represent a common core of (closed or open) subshells. Two orthogonal functions $|c_1\gamma_1 LS\rangle$ and $|c_2\gamma_2 LS\rangle$ are said to obey Brillouin's theorem[8] if

$$\langle c_1\gamma_1 LS | H' | c_2\gamma_2 LS\rangle = 0 , \qquad (13.42)$$

where the non-relativistic Hamiltonian H′ includes only kinetic-energy and electrostatic-interaction terms. Basis functions usually possess this property[9] if they are constructed from radial wavefunctions obtained by solving the Hartree-Fock equations for the LS term in question (see Sec. 16-8). We have already seen one example in the alkali-configuration case of the preceding section ($w_i = w_j = 1$, all core subshells filled), where our spherically averaged HF method is identical with the LS-dependent HF method because there is only one LS term per configuration.

That HF functions possess this special property follows from the fact that the HF equations were derived from an energy variational principle:[9] Suppose that $|c_1\gamma_1 LS\rangle$ is the lowest-energy single-configuration state having a particular angular symmetry. Then if the CI matrix element in (13.42) were not zero, the resulting CI perturbation would produce an energy lower than the single-configuration value. But if $|c_2\gamma_2 LS\rangle$ has the same angular form as does $|c_1\gamma_1 LS\rangle$, then the mixed state can differ from $|c_1\gamma_1 LS\rangle$ only in its radial dependence—which is contradictory to the derivation of the radial factors of $|c_1\gamma_1 LS\rangle$ as those giving the lowest possible energy.

The form (13.42) of Brillouin's theorem holds only when the HF equations are derived by minimizing the energy of the term $\gamma_1 LS$. For the spherically averaged HF method, which we have used exclusively, Brillouin's theorem takes the form that the *configuration-average* interaction (so to speak) is zero. In the Rydberg-series case, we saw in Sec. 13-5 that this consisted of a zero coefficient of R_d^0, and spherical-average corrections to the other terms on the diagonal of the CI block. The form of Brillouin's theorem in the general case (13.41) will be discussed in Sec. 13-11.

[7]E. Luc-Koenig, Phys. Rev. A 13, 2114 (1976). See also R. M. Sternheimer, J. E. Rodgers, and T. P. Das, Phys. Rev. A 17, 505 (1978) and references therein.

[8]L. Brillouin, J. phys. radium 3, 373 (1932).

[9]J. Bauche and M. Klapisch, J. Phys. B. 5, 29 (1972); J. J. Labarthe, J. Phys. B. 5, L181 (1972); C. Froese Fischer, J. Phys. B. 6, 1933 (1973).

*13-8. ARBITRARY CONFIGURATION INTERACTIONS

In Secs. 13-3 to 13-6 we considered pairs of configurations such that CI matrix elements could be evaluated by a relatively simple extension of single-configuration methods. In general, however, two interacting configurations will not have the same basic shell structure, so that those methods are not directly applicable. Conceptually, the simplest and most straightforward approach is to first make appropriate cfp or cfgp expansions and/or appropriate recouplings to cast the two configurations into a form such that the above methods do apply.[10]

For example, in the case

$$sp^3 \; - \; s^2pd \tag{13.43}$$

the interacting (active) electrons must be p^2 from the first configuration and sd from the second, leaving spectator electrons sp in each. One possible procedure is to make a cfp expansion $p^3 \rightarrow p^2p$ (and the trivial expansion $s^2 \rightarrow$ ss), followed by a series of recouplings:

$$\left\{ \begin{matrix} [sp^3] \\ [s^2p]d \end{matrix} \right\} \rightarrow \left\{ \begin{matrix} [s(p^2p)] \\ [(ss)p]d \end{matrix} \right\} \rightarrow \left\{ \begin{matrix} [s(p^2)]p \\ [(ss)d]p \end{matrix} \right\} \rightarrow \left\{ \begin{matrix} [s(p^2)]p \\ [s(sd)]p \end{matrix} \right\} . \tag{13.44}$$

An alternative is to make a cfgp expansion $p^3 \rightarrow pp^2$ plus a series of recouplings:

$$\left\{ \begin{matrix} [sp^3] \\ [s^2p]d \end{matrix} \right\} \rightarrow \left\{ \begin{matrix} [s(pp^2)] \\ [(ss)p]d \end{matrix} \right\} \rightarrow \left\{ \begin{matrix} [(sp)(p^2)] \\ [(sp)s]d \end{matrix} \right\} \rightarrow \left\{ \begin{matrix} [(sp)(p^2)] \\ [(sp)(sd)] \end{matrix} \right\} . \tag{13.45}$$

In either case, we finish by using (13.26) for the two-electron interaction p^2 – sd. [It must be noted that weighting factors from the multiple occupancies of the p^3 and s^2 subshells have been ignored in the above qualitative outline. Also ignored has been a factor -1, which arises from the fact that from s^2pd we pick electrons 2 and 4, whereas from sp^3 we pick electrons 2 and 3 in (13.44) or electrons 3 and 4 in (13.45).]

In principle, the above procedure of converting to a two-electron problem can be applied to any pair of interacting configurations. In practice, however, it appears to be very difficult to construct a simple and efficient computer program capable of handling all possible types of interaction. In part, this stems from the fact that different cases involve different numbers and different types of recouplings. In cases such as the fs – fd interaction of f^wsp – f^wdp, there is also the inefficiency of making the cfp expansion $f^w \rightarrow f^{w-1}f$, when this expansion can be avoided by making use of the (already-computed) reduced matrix elements of $U^{(r)}$ and $V^{(r1)}$. In order to minimize computer time, it appears to be necessary to divide all possible interactions up into about a dozen different classes, and to handle each one separately; these will be described later. First we consider those aspects of the calculation that can be treated in general.

In order to minimize notational and bookkeeping problems we assume the two configurations of interest to be written initially in the form

[10]N. Rosenzweig, Phys. Rev. **88**, 580 (1952).

$$(n_1l_1)^{w_1}(n_2l_2)^{w_2}\cdots(n_ql_q)^{w_q} ,$$

and (13.46)

$$(n_1l_1)^{w'_1}(n_2l_2)^{w'_2}\cdots(n_ql_q)^{w'_q} .$$

In these two expressions, corresponding values of n_jl_j are identical; the only differences lie in the occupation numbers, where w_j may or may not be equal to w'_j. Some subshells may be full ($w = 4l + 2$) or empty ($w = 0$), but of course

$$\sum_j w_j = \sum_j w'_j = N , \tag{13.47}$$

where N is the number of electrons in the atom. It is assumed that

$$\sum_j w_jl_j = \sum_j w'_jl_j \pmod 2 , \tag{13.48}$$

so that the configurations are of the same parity.

A preview of the basic calculational procedure is as follows:

(1) Ignore all subshells that are filled in both configurations.

Two-electron operators.

(2) Pick out a set of four interacting electrons, two (l_ρ and l_σ, $\rho \leqq \sigma$) from the first configuration, and two ($l_{\rho'}$ and $l_{\sigma'}$, $\rho' \leqq \sigma'$) from the second. [The change from the notation of Secs. 13-3 to 13-6 must be kept clearly in mind. In those sections, the active electrons came from subshells with the same pair of index numbers in the two configurations ($\rho = \rho'$ and $\sigma = \sigma'$), but the subshells themselves could be different ($n_\rho l_\rho \neq n'_\rho l'_\rho$ and/or $n_\sigma l_\sigma \neq n'_\sigma l'_\sigma$). Now the subshells are identical ($n_il_i = n'_il'_i$, all i), but active electrons must come from different pairs of subshells ($\rho \neq \rho'$ and/or $\sigma \neq \sigma'$.]

- (a) If the remaining $N - 2$ (spectator) electrons are not identical in the two configurations, the interaction is null; try a different set of four active electrons.
- (b) If no value of k_d can be found that is common to the two sequences (13.24) [with appropriately modified notation], and no value of k_e can be found that is common to the two sequences (13.25), then the interaction is null; try a different set of four active electrons.

(3) Calculate the phase factor and the occupation-number weighting factor that arise from the coordinate permutations among different subshells.

(4) Calculate the remaining factor of each matrix element by making appropriate cfp and/or cfgp expansions, and by then performing suitable recouplings so as to make the two configurations appear identical (differing at most in nl-values for corresponding singly occupied subshells); then, with the subshells appropriately renumbered, use either the two-electron equations (13.21)-(13.23) or the more general equations (13.33)-(13.34), as appropriate.

One-electron operators.

(5) If two interacting electrons of equal l (one from each configuration) can be chosen such that the remaining $N - 1$ spectator electrons are identical in the two configurations,

then there are non-zero KE, e-n, and s-o contributions. The KE and e-n terms must be calculated, as outlined in Sec. 13-5 for the Rydberg-series case. The s-o terms are usually neglected, but are comparatively simple to evaluate if one wishes to include them. (In the Rydberg-series case, they are identical with the single-configuration values.)

*13-9. MATRIX ELEMENTS OF SYMMETRIC OPERATORS

The first non-trivial step in the above procedure is the simplification of matrix elements between completely antisymmetric basis functions Ψ, to give an expression that involves functions ψ which are antisymmetric only with respect to electron-coordinate interchange within a subshell of equivalent electrons. The procedure is similar to that followed in Sec. 9-9 for the single-configuration case.

One-electron operators. If Ψ_b and $\Psi_{b'}$ are fully antisymmetrized basis functions of the form (9.75) belonging respectively to the first and second of the configurations (13.46), then similarly to (9.76)

$$\langle \Psi_b | \sum_{i=1}^{N} f_i | \Psi_{b'} \rangle = [(N-1)!]^{-1} \left[\left(\prod_k w_k! \right) \left(\prod_k w_k'! \right) \right]^{1/2} \sum_P \sum_{P'} (-1)^{p+p'} \langle \psi_b^{(P)} | f_N | \psi_{b'}^{(P')} \rangle . \tag{13.49}$$

Let σ and σ' be subshell indices such that

$$w_k - \delta_{k\sigma} = w_k' - \delta_{k\sigma'} , \qquad 1 \leqq k \leqq q . \tag{13.50}$$

[If no such σ and σ' exist, then all terms of (13.49) are zero regardless of the permutations P and P′, because integration over at least one electron coordinate r_i ($i < N$) will involve an overlap integral $\langle nl|n'l' \rangle$ for which $nl \neq n'l'$.] The matrix element on the right-hand side of (13.49) will be zero unless the permutation P is such that the coordinate r_N appears in the subshell σ of $\psi_b^{(P)}$ (and is therefore the *final* coordinate of this subshell, because of the nature of the permutations, as defined in Sec. 9-8), and unless the permutation P′ is such that r_N is the final coordinate of the subshell σ' of $\psi_{b'}^{(P)}$ but is otherwise identical with P. The number of coordinate interchanges required to move r_N from its position in the subshell σ (for the permutation P) to its standard position as the final coordinate is evidently

$$\sum_{j=\sigma+1}^{q} w_j ,$$

and the difference in parity of P and P′ is thus

$$\Delta p = p - p' = \sum_{j=\sigma+1}^{q} w_j - \sum_{j=\sigma'+1}^{q} w_j' . \tag{13.51}$$

The number of permutations P for which r_N lies in σ is, similarly to (9.79),

$$\frac{(N-1)!}{w_1!\cdots(w_\sigma-1)!\cdots w_q!} = \frac{(N-1)!}{w_1'!\cdots(w_{\sigma'}'-1)!\cdots w_q'!}\,, \tag{13.52}$$

where use has been made of (13.50). The values of the matrix element and of p − p′ are identical for all these permutations, and so we obtain finally

$$\langle\Psi_b|\sum_{i=1}^{N} f_i|\Psi_{b'}\rangle = (w_\sigma w_{\sigma'}')^{1/2}(-1)^{\Delta p}\langle\psi_b^{(P_\sigma)}|f_N|\psi_{b'}^{(P_{\sigma'})}\rangle\,, \tag{13.53}$$

where P_σ is any permutation for which r_N is the last coordinate of $l_\sigma^{\;w_\sigma}$, and $P_{\sigma'}$ is identical with P_σ except that r_N is the final coordinate of $l_{\sigma'}^{\;w'_{\sigma'}}$. [The single-configuration result (9.80) is essentially the special case $\sigma = \sigma' = j$, $w_\sigma = w_{\sigma'}'$ except for the summation over j.]

Two-electron operators. Similarly to the single-configuration expression (9.81), we find in the CI case

$$\begin{aligned}\langle\Psi_b|\sum_{i<j} g_{ij}|\Psi_b'\rangle &= \frac{N(N-1)}{2}\langle\Psi_b|g_{N-1,N}|\Psi_{b'}\rangle\\ &= [2(N-2)!]^{-1}\left[\left(\prod_k w_k!\right)\left(\prod_k w_k'!\right)\right]^{1/2}\sum_P\sum_{P'}(-1)^{p+p'}\langle\psi_b^{(P)}|g_{N-1,N}|\psi_{b'}^{(P')}\rangle\,.\end{aligned} \tag{13.54}$$

We consider two subshell indices ρ and σ (ρ ≦ σ) in the first configuration and two indices ρ′ and σ′ (ρ′ ≦ σ′) in the second, such that

$$w_k - \delta_{k\rho} - \delta_{k\sigma} = w_k' - \delta_{k\rho'} - \delta_{k\sigma'}\,, \qquad 1 \leqq k \leqq q\,. \tag{13.55}$$

(Clearly we cannot have both $\rho = \rho'$ and $\sigma = \sigma'$, because this would be the single-configuration case.) If it is not possible to choose ρσρ′σ′ in such a way as to satisfy (13.55), then all terms of (13.54) are necessarily zero regardless of the permutations P and P′ and we have no interaction between the two configurations; if (13.55) can be satisfied for more than one choice of the four indices ρσρ′σ′—as, for example, in (13.28)—then we have to consider each possible set in turn.

We consider first a permutation $P \equiv P_{\rho\sigma}$ such that r_N is the final coordinate in the subshell σ and r_{N-1} is the final coordinate in the subshell ρ (or the next-to-last coordinate in ρ, if ρ = σ); at the same time, we consider a permutation $P' \equiv P_{\rho'\sigma'}$ such that r_N is the last coordinate in σ′ and r_{N-1} is the last coordinate in ρ′ (or the next-to-last, if ρ′ = σ′). The matrix element in the right-hand side of (13.54) will be zero unless P and P′ are such that all coordinates r_1 to r_{N-2} match up exactly in the two configurations. All non-zero terms in the double summation over P and P′ (for different permutations of the coordinates r_1 to r_{N-2}) have the same numerical value as to both matrix element and phase factor.

To move the coordinate r_N in P to its standard position as the final coordinate of the subshell $l_q^{\;w_q}$ requires

$$\sum_{j=\sigma+1}^{q} w_j$$

coordinate interchanges; to then move r_{N-1} to its standard position as next-to-last coordinate of the configuration then requires

$$\delta_{\rho\sigma} - 1 + \sum_{j=\rho+1}^{q} w_j$$

additional interchanges. The total number of interchanges required is

$$\delta_{\rho\sigma} - 1 + \sum_{j=\rho+1}^{q} w_j + \sum_{j=\sigma+1}^{q} w_j ,$$

or to within modulo 2

$$\delta_{\rho\sigma} - 1 + \sum_{j=\rho+1}^{\sigma} w_j .$$

The difference in parity between P and P′ is thus (modulo 2)

$$\Delta p = p - p' = \delta_{\rho\sigma} - \delta_{\rho'\sigma'} + \sum_{j=\rho+1}^{\sigma} w_j - \sum_{j=\rho'+1}^{\sigma'} w_j' . \qquad (13.56)$$

The number of different permutations of the coordinates r_1 to r_{N-2} may be written, because of (13.55),

$$\frac{(N-2)!}{\prod_k (w_k - \delta_{k\rho} - \delta_{k\sigma})!} = \frac{(N-2)!}{\prod_k (w_k' - \delta_{k\rho'} - \delta_{k\sigma'})!}$$

$$= \frac{(N-2)!}{\left[\prod_k (w_k - \delta_{k\rho} - \delta_{k\sigma})!\right]^{1/2}\left[\prod_k (w_k' - \delta_{k\rho'} - \delta_{k\sigma'})!\right]^{1/2}} .$$

Thus the contribution to (13.54) of all permutations $P_{\rho\sigma}$ and $P_{\rho'\sigma'}$ is

$$\frac{1}{2}(-1)^{\Delta p} [w_\rho(w_\sigma - \delta_{\rho\sigma})w_{\rho'}'(w_{\sigma'}' - \delta_{\rho'\sigma'})]^{1/2} \langle \psi_b^{(P_{\rho\sigma})} | g_{N-1,N} | \psi_{b'}^{(P_{\rho'\sigma'})} \rangle , \qquad (13.57)$$

where Δp is given by (13.56).

So far we have considered only permutations of the type $P = P_{\rho\sigma}$ and $P' = P_{\rho'\sigma'}$. If $\rho = \sigma$ and $\rho' = \sigma'$, this includes all the permutations allowed (because within a subshell the coordinates may be arranged only in numerical order). However, if $\rho \neq \sigma$ and/or $\rho' \neq \sigma'$, then we have yet to consider permutations $P = P_{\sigma\rho}$ and/or $P' = P_{\sigma'\rho'}$. We have to consider two different cases:

(1) If $\rho \neq \sigma$ and $\rho' \neq \sigma'$, then the matrix elements for $P_{\sigma\rho} - P_{\sigma'\rho'}$ will be numerically equal to those included in (13.57), and can be accounted for by simply adding an additional factor 2 to (13.57). There will also be matrix elements with permutations $P_{\rho\sigma} - P_{\sigma'\rho'}$ and $P_{\sigma\rho} - P_{\rho'\sigma'}$, which may be lumped together in the form of a single exchange term

similar to the direct term, except for an extra factor -1 resulting from the one additional coordinate exchange over those included in (13.56).

(2) If $\rho = \sigma$ but $\rho' \neq \sigma'$, then the only matrix elements arising from the double sum in (13.54) besides those already included in (13.57) are of the form

$$- \langle \psi_b^{(P_{\rho\sigma})} | g_{N-1,N} | \psi_{b'}^{(P_{\sigma'\rho'})} \rangle \ .$$

But because the coordinates $\mathbf{r}_{N-1}$ and $\mathbf{r}_N$ lie in the same subshell in the bra function, and because this subshell is still antisymmetric with respect to coordinate interchanges, this may be written

$$+ \langle \psi_b^{(P_{\sigma\rho})} | g_{N-1,N} | \psi_{b'}^{(P_{\sigma'\rho'})} \rangle = + \langle \psi_b^{(P_{\rho\sigma})} | g_{N-1,N} | \psi_{b'}^{(P_{\rho'\sigma'})} \rangle \ .$$

Thus these terms multiply (13.57) by a factor 2, but introduce no explicit exchange terms. An identical result is obtained if $\rho \neq \sigma$ but $\rho' = \sigma'$.

All the above cases, together with corresponding terms from all possible sets of indices $\rho\sigma\rho'\sigma'$, may be summarized into the final result

$$\langle \Psi_b | \sum_{i<j} g_{ij} | \Psi_{b'} \rangle = \sum_{\rho\sigma\rho'\sigma'} (-1)^{\Delta p} \frac{[w_\rho(w_\sigma - \delta_{\rho\sigma})w'_{\rho'}(w'_{\sigma'} - \delta_{\rho'\sigma'})]^{1/2}}{(1 + \delta_{\rho\sigma}\delta_{\rho'\sigma'})}$$
$$\times \left[\langle \psi_b^{(P_{\rho\sigma})} | g_{N-1,N} | \psi_{b'}^{(P_{\rho'\sigma'})} \rangle - (1 - \delta_{\rho\sigma})(1 - \delta_{\rho'\sigma'}) \langle \psi_b^{(P_{\rho\sigma})} | g_{N-1,N} | \psi_{b'}^{(P_{\sigma'\rho'})} \rangle \right] , \quad (13.58)$$

where Δp is given by (13.56), and the summation is over subshells $\rho\sigma\rho'\sigma'$ satisfying (13.55). This expression is equivalent to that given by Fano;[11] it is easily seen to reduce to (9.85) in the single-configuration case $\rho = \rho'$, $\sigma = \sigma'$.

The expressions (13.53) and (13.58) provide the phase and weight factors that arise as a result of coordinate permutations, and leave us with matrix elements only between non-antisymmetrized wavefunctions to be evaluated.[12]

*13-10. COULOMB MATRIX ELEMENTS; GENERAL CASE

The electron-electron Coulomb matrix elements (13.58) are of the form

$$\langle \Psi_b | \sum_{i<j} \sum_k \frac{2r_<^k}{r_>^{k+1}} \mathbf{C}_{(i)}^{(k)} \cdot \mathbf{C}_{(j)}^{(k)} | \Psi_{b'} \rangle = \sum_{\rho\sigma\rho'\sigma'} \sum_k [r_d^k R_d^k(l_\rho l_\sigma, l_{\rho'} l_{\sigma'}) + r_e^k R_e^k(l_\rho l_\sigma, l_{\rho'} l_{\sigma'})] \ . \quad (13.59)$$

[11]U. Fano, Phys. Rev. **140**, A67 (1965). It is, however, important to remember that Fano's phase conventions for the matrix elements remaining in (13.58) are different from ours: Fano includes an additional factor i^l in each orbital $\varphi_{nlm_lm_s}$, and therefore his matrix elements differ from ours by a factor $(-1)^{(l_{\rho'}+l_{\sigma'}-l_\rho-l_\sigma)/2}$. [In addition, it must be remembered that matrix elements $\langle l \| C^{(k)} \| l' \rangle$ involved in equations that employ the Fano conventions differ from the quantities that we use here by a factor $(-1)^{(k+l'-l)/2}$; see fn. 7 in Sec. 11-4.]

[12]This makes it unnecessary to consider the mixed-shell coefficients of fractional parentage used for antisymmetrization purposes by P. S. Kelly, Off. of Ordnance Research Report OOR-1784.6 (June, 1959), and by L. Armstrong, Jr., Phys. Rev. 172, 12, 18 (1968).

For each possible set of interacting electrons $\rho\sigma\rho'\sigma'$ ($\rho \leqq \sigma$ and $\rho' \leqq \sigma'$), we shall always assume that the primed (ket) configuration has been chosen to be that which makes $\rho < \rho'$, of if $\rho = \rho'$ then $\sigma < \sigma'$; this is permissible because the energy matrix is always symmetric.

We shall compute each angular coefficient r_d^k or r_e^k as a product of several factors, assuming the genealogical $\mathfrak{L}\mathfrak{S}$ coupling (12.1). First, similarly to the single-configuration case, we have delta-factors

$$B_1 = \delta_{\mathfrak{J}_q\mathfrak{J}'_q}\left[\prod_m \delta_{\alpha_m L_m S_m,\alpha'_m L'_m S'_m}\right]_{m\neq\rho\sigma\rho'\sigma'} \times \left[\prod_m \delta_{\mathfrak{L}_m\mathfrak{S}_m,\mathfrak{L}'_m\mathfrak{S}'_m}\right]_{m<\rho,\ m\geqq\max(\sigma,\sigma')} \tag{13.60}$$

Each coefficient is independent of $\mathfrak{J}_q$, and we may in the usual way ignore the coupling $(\mathfrak{L}_q\mathfrak{S}_q)\mathfrak{J}_q$.

Second, there are the phase and occupation-number weighting factors given by (13.58),

$$B_2 = (-1)^{\Delta p}\frac{[w_\rho(w_\sigma - \delta_{\rho\sigma})w'_{\rho'}(w'_{\sigma'} - \delta_{\rho'\sigma'})]^{1/2}}{(1 + \delta_{\rho\sigma}\delta_{\rho'\sigma'})} , \tag{13.61}$$

with Δp given by (13.56).

Our guiding principle in evaluating the remaining factors is to make cfp and/or cfgp expansions and suitable recouplings so as to reduce the problem to what is effectively the evaluation of single-configuration matrix elements, and to do this in a way that will introduce the fewest possible summations over parents, grandparents, and recoupling quantum numbers.[13] We first consider all those steps that can be performed without introducing any summations at all.

For each subshell i such that i is equal to one *and only one* of the four indices $\rho\sigma\rho'\sigma'$, we make a cfp expansion of the form (9.50) for the unprimed configuration (if $i = \rho$ or σ) or for the primed configuration (if $i = \rho'$ or σ'). In evaluation of the matrix element for r^k, integration over the coordinates of the electrons in the parent subshell gives a zero result except for that parent in the summation (9.50) which matches the $\alpha_1 L_1 S_1$ term in the basis function of the opposite configuration. Thus for each pertinent i we obtain (in both r_d^k and r_e^k) no summation, but simply a multiplying factor

$$B_{3i} = (l_i^{w_i}\alpha_i L_i S_i \{| l_i^{w_i-1}\alpha'_i L'_i S'_i) , \qquad i = \rho \text{ and/or } \sigma , \tag{13.62}$$

or

$$B_{4i} = (l_i^{w'_i-1}\alpha_i L_i S_i |\} l_i^{w'_i}\alpha'_i L'_i S'_i) , \qquad i = \rho' \text{ and/or } \sigma' . \tag{13.63}$$

[13]Fano (ref. 11) evaluates the matrix elements by converting them to recoupling coefficients by introducing the fictitious "orbitons" of U. Fano, F. Prats, and Z. Goldschmidt, Phys. Rev. 129, 2643 (1963). This method introduces additional quantum numbers, with corresponding additional recouplings and summations. Moreover, in cases such as the fs—fd interaction of (13.31) it introduces an unnecessary cfp expansion and sum over parents. B. W. Shore, Phys. Rev. 139, A1042 (1965) employs partially uncoupled basis functions; this avoids recoupling problems, but necessitates very large basis sets, and extensive summations associated with Clebsch-Gordon expansions.

Next, we have the possibility of two series of recouplings that will introduce no summations. In discussing these it is convenient to introduce abbreviations for the recoupling coefficients (9.17), (9.18), and (9.19), which will be referred to as shift, jump, and exchange recouplings, respectively:[14]

$$R_s(j_1j_2J',j_3J,J'') \equiv \langle[(j_1j_2)J',j_3]J\,|\,[j_1,(j_2j_3)J'']J\rangle$$

$$= (-1)^{j_1+j_2+j_3+J}[J',J'']^{1/2}\begin{Bmatrix} j_1 & j_2 & J' \\ j_3 & J & J'' \end{Bmatrix} , \qquad (13.64)$$

$$R_j(j_1j_2J',j_3J,J'') \equiv \langle[(j_1j_2)J',j_3]J\,|\,[j_1,(j_3j_2)J'']J\rangle$$

$$= (-1)^{j_1+J+J''}[J',J'']^{1/2}\begin{Bmatrix} j_1 & j_2 & J' \\ j_3 & J & J'' \end{Bmatrix} , \qquad (13.65)$$

$$R_x(j_1j_2J',j_3J,J'') \equiv \langle[(j_1j_2)J',j_3]J\,|\,[(j_1j_3)J'',j_2]J\rangle$$

$$= (-1)^{j_2+j_3+J'+J''}[J',J'']^{1/2}\begin{Bmatrix} j_2 & j_1 & J' \\ j_3 & J & J'' \end{Bmatrix} . \qquad (13.66)$$

We note that, provided the arguments form a vectorially possible set, all three recoupling coefficients are unity if either j_2 or j_3 is zero, and that R_s is unity if j_1 is zero.

The first possibility is a recoupling of the electron ρ toward the right, and is illustrated schematically in Fig. 13-3: a dash represents a subshell $l_i^{w_i}$ containing only spectator electrons; (—ρ) represents a subshell that includes the active electron ρ, with the electron ρ split off from the spectator electrons via a cfp, but still coupled thereto; and the coupling is always assumed to proceed from left to right in the manner (12.1) except as parentheses indicate otherwise. Setting $i = \min(\sigma,\rho')$, then if $\rho < i$ we can make recouplings like those in Fig. 13-3, which (taking into account pertinent basis-function orthogonalities) add to r^k no summations but only the factor

$$B_5 = [\delta_{\rho 1} + (1-\delta_{\rho 1})R_s(\mathfrak{L}_{\rho-1}L'_\rho\mathfrak{L}'_\rho,l_\rho\mathfrak{L}_\rho,L_\rho)]$$

$$\times \left[\prod_{m=\rho+1}^{i-1} R_x(\mathfrak{L}'_{m-1}l_\rho\mathfrak{L}_{m-1},L_m\mathfrak{L}_m,\mathfrak{L}'_m)\right] \times \text{spins} , \qquad (13.67)$$

where the factor "× spins" means factors for the spin quantum numbers sS𝔖 analogous to those written for $lL\mathfrak{L}$.

The second possiblity is that of recoupling the rightmost of the active electrons (which for the moment we assume to be σ') toward the left to the position $j = \max(\sigma,\rho')$, as illustrated in Fig. 13-4. If $j < \sigma'$, this gives a factor

[14]The coefficient $R_s(j_1j_2J',j_3J,J'')$ is identical with the coefficient $U(j_1j_2Jj_3;J'J'')$ used by H. A. Jahn, Proc. Roy. Soc. (London) **A205**, 192 (1951), his ordering of the arguments being that for the Racah coefficient (5.28). For ease in remembering, we use an argument ordering that is the same for all three recoupling coefficients, this being the order in which the various quantum numbers first appear in the corresponding matrix elements. In a computer program, it is convenient to have a special subroutine to compute each of the three recoupling coefficients.

Schematic

$$\left\{ \begin{array}{l} \cdots\ -\overset{\frown}{(}-\ \rho)\ -\ -\ -\ \quad -\ \cdots \\ \cdots\ -\ -\ \quad -\ -\ (-\ \rho')\ -\ \cdots \end{array} \right\}$$

$$\longrightarrow \left\{ \begin{array}{l} \cdots\ -\ -\ \rho\overset{\frown}{}\overset{\frown}{}-\ \quad -\ \cdots \\ \cdots\ -\ -\ \quad -\ -\ (-\ \rho')\ -\ \cdots \end{array} \right\}$$

$$\longrightarrow \left\{ \begin{array}{l} \cdots\ -\ -\ -\ -\ \rho\ -\ \quad -\ \cdots \\ \cdots\ -\ -\ -\ -\ \quad (-\ \rho')\ -\ \cdots \end{array} \right\}$$

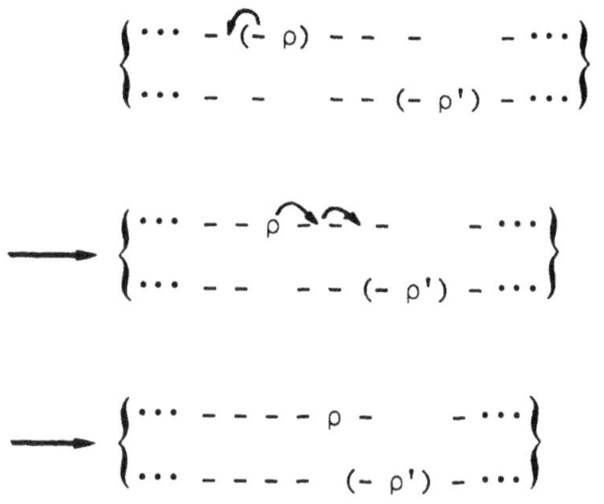

Example (including the coefficient of fractional parentage expansions)

$$\left\{ \begin{array}{l} s\ p^4\ d\ f\ g^2\ h \\ s\ p^3\ d\ f\ g^3\ h \end{array} \right\}$$

$$\longrightarrow \left\{ \begin{array}{l} s\ (p^3 p)\ d\ f\ g^2\ h \\ s\ p^3\ \ d\ f\ (g^2 g)\ h \end{array} \right\}$$

$$\longrightarrow \left\{ \begin{array}{l} s\ p^3\ p\ d\ f\ g^2\ h \\ s\ p^3\ \ d\ f\ (g^2 g)\ h \end{array} \right\}$$

$$\longrightarrow \left\{ \begin{array}{l} s\ p^3\ d\ p\ f\ g^2\ h \\ s\ p^3\ d\ \ f\ (g^2 g)\ h \end{array} \right\}$$

$$\longrightarrow \left\{ \begin{array}{l} s\ p^3\ d\ f\ p\ g^2\ h \\ s\ p^3\ d\ f\ \ (g^2 g)\ h \end{array} \right\}$$

Fig. 13-3. Recouplings of the electron l_ρ that introduce no summations (for the case $\rho <$ $\rho' \leqq \sigma$). The upper line of each pair represents the bra configuration and the lower represents the ket configuration. In all cases, the coupling is assumed to proceed from left to right except as otherwise indicated by parentheses.

Schematic

$$\left\{ \begin{array}{l} \cdots \;\; (-\;\sigma) \; - \; - \quad - \qquad - \; \cdots \\ \cdots \quad - \qquad - \; - \overset{\frown}{(-\;\sigma')} \; - \; \cdots \end{array} \right\}$$

$$\longrightarrow \left\{ \begin{array}{l} \cdots \;\; (-\;\sigma) \; - \; - \qquad - \; - \quad \cdots \\ \cdots \quad - \quad \overset{\frown}{\;\;}\overset{\frown}{-\;}{}_{\sigma'} \; - \; - \quad \cdots \end{array} \right\}$$

$$\longrightarrow \left\{ \begin{array}{l} \cdots \;\; (-\;\sigma) \qquad - \; - \; - \; \cdots \\ \cdots \quad - \quad \sigma' \; - \; - \; - \; \cdots \end{array} \right\}$$

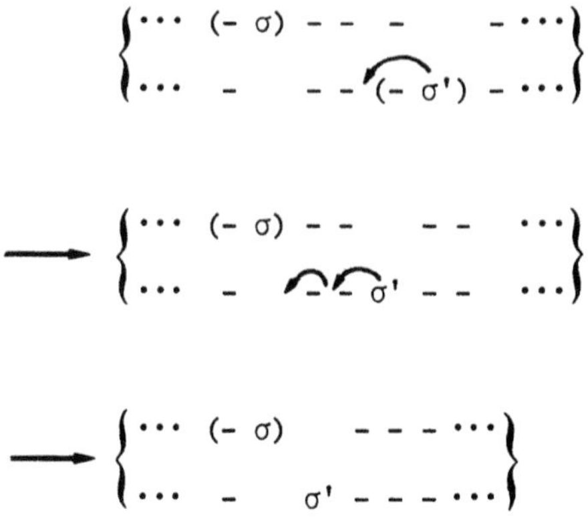

Example (including the coefficient of fractional parentage expansions)

$$\left\{ \begin{array}{l} i^5 \; k \; \ell \; m^3 \; n \\ i^4 \; k \; \ell \; m^4 \; n \end{array} \right\}$$

$$\longrightarrow \left\{ \begin{array}{l} (i^4 i) \; k \; \ell \quad m^3 \quad n \\ i^4 \quad k \; \ell \; (m^3 m) \; n \end{array} \right\}$$

$$\longrightarrow \left\{ \begin{array}{l} (i^4 i) \; k \; \ell \quad m^3 \; n \\ i^4 \quad k \; \ell \; m \; m^3 \; n \end{array} \right\}$$

$$\longrightarrow \left\{ \begin{array}{l} (i^4 i) \; k \quad \ell \; m^3 \; n \\ i^4 \quad k \; m \; \ell \; m^3 \; n \end{array} \right\}$$

$$\longrightarrow \left\{ \begin{array}{l} (i^4 i) \quad k \; \ell \; m^3 \; n \\ i^4 \quad m \; k \; \ell \; m^3 \; n \end{array} \right\}$$

Fig. 13-4. Recouplings of the electron $l_{\sigma'}$ that introduce no summations (for the case $\sigma' > \sigma \geqq \rho'$).

$$B_6' = R_j(\mathcal{L}'_{\sigma'-1}l_{\sigma'}\mathcal{L}_{\sigma'-1},L_{\sigma'}\mathcal{L}'_{\sigma'},L'_{\sigma'})\left[\prod_{m=j+1}^{\sigma'-1} R_x(\mathcal{L}'_{m-1}L_m\mathcal{L}'_m,l_{\sigma'}\mathcal{L}_m,\mathcal{L}_{m-1})\right] \times \text{spins} . \tag{13.68}$$

If the rightmost of the four active electrons is in the subshell σ instead of in σ', then primed quantities in (13.68) are to be replaced by the corresponding unprimed ones and vice versa, giving a factor B_6.

The above operations include everything that can be done in a general way without introducing any summations. They leave the pair of configurations in one of the first twelve forms shown in Fig. 13-5 (except that in classes 2 to 5, the cfgp expansions have not yet been made). For most efficient use of computer time, each of these classes must be handled differently.

Class 1. There is no coefficient r_e^k. In class 1a, for r_d^k we first make a shift recoupling to obtain $(x^w y)$. From (9.87) and (9.88), we can then completely ignore all subshells preceding and following x^{w+1}, make a cfp expansion in x^{w+1}, and obtain

$$r_d^k = (w_\rho')^{-1}B_1B_2B_{4\sigma'}B_6'C_{1a} \tag{13.69}$$

with

$$C_{1a} = R_s(\mathcal{L}_{\rho-1}L_\rho'\mathcal{L}'_\rho,l_{\sigma'}\mathcal{L}_\rho,L_\rho) \times \text{spins}$$
$$\times \sum_{\bar{\alpha}\bar{L}\bar{S}} (l_\rho{}^{w_\rho}\alpha_\rho L_\rho S_\rho\{|l_\rho{}^{w'_\rho}\bar{\alpha}\bar{L}\bar{S})\langle l_\rho{}^{w'_\rho}\bar{\alpha}\bar{L}\bar{S},l_\rho L_\rho S_\rho|g^{(k)}|l_\rho{}^{w'_\rho}\alpha_\rho' L_\rho' S_\rho',l_{\sigma'}L_\rho S_\rho\rangle , \tag{13.70}$$

where $g^{(k)}$ is the two-electron angular operator $C^{(k)}_{(w'_\rho)} \cdot C^{(k)}_{(w'_\rho+1)}$. The final matrix element is given by (13.33) except for minor obvious modifications in notation; the factor $(w_\rho')^{-1}$ is required in (13.69) to compensate for the factor w_ρ' being included implicitly in $I^{(k)}$ as well as explicitly in B_2.

Class 1b is handled basically the same way as 1a, except that we have a jump recoupling instead of a shift:

$$r_d^k = (w_\sigma)^{-1}B_1B_2B_{3\rho}B_5C_{1b} , \tag{13.71}$$

where

$$C_{1b} = -R_j(\mathcal{L}'_{\sigma-1}l_\rho\mathcal{L}_{\sigma-1},L_\sigma\mathcal{L}_\sigma,L_\sigma') \times \text{spins}$$
$$\times \sum_{\bar{\alpha}\bar{L}\bar{S}} \langle l_\sigma{}^{w_\sigma}\alpha_\sigma L_\sigma S_\sigma,l_\rho L_\sigma' S_\sigma'|g^{(k)}|l_\sigma{}^{w_\sigma}\bar{\alpha}\bar{L}\bar{S},l_\sigma L_\sigma' S_\sigma'\rangle(l_\sigma{}^{w_\sigma}\bar{\alpha}\bar{L}\bar{S}|\}l_\sigma{}^{w'_\sigma}\alpha_\sigma' L_\sigma' S_\sigma') . \tag{13.72}$$

The minus sign arises upon interchanging the coordinates of the last two electrons in the subshell x^{w+1} to make these coordinates match those of the final two electrons in the bra function (after recoupling to $x^w y$), so that the direct expression (13.33) can be used in place of the more complicated exchange expression (13.34).

Class 2. In order to simplify the notation slightly, we let $\tau = \rho' = \sigma'$. We make two cfgp expansions of the type (9.51), in which (because of orthogonalities) only a single

Class	Definition		Example	Method (recouple to)
1a	$\rho=\sigma=\rho'<\sigma'$	$\left\{\begin{matrix}(x^{w+1}) \\ (x^{w})y\end{matrix}\right\}$	$\left\{\begin{matrix}f^{3} \\ f^{2}p\end{matrix}\right\}$	$\left\{\begin{matrix}(x^{w}x) \\ (x^{w}y)\end{matrix}\right\}$
1b	$\rho<\sigma=\rho'=\sigma'$	$\left\{\begin{matrix}y(x^{w}) \\ (x^{w+1})\end{matrix}\right\}$	$\left\{\begin{matrix}fp \\ p^{2}\end{matrix}\right\}$	$\left\{\begin{matrix}(x^{w}y) \\ (x^{w}x)\end{matrix}\right\}$
2	$\rho=\sigma, \rho'=\sigma'$	$\left\{\begin{matrix}(-x^{2})\cdots - \\ - \cdots(-y^{2})\end{matrix}\right\}$	$\left\{\begin{matrix}d^{5}s \\ d^{3}sp^{2}\end{matrix}\right\}$	$\left\{\begin{matrix}- x^{2} \cdots - \\ - y^{2} \cdots -\end{matrix}\right\}$
3	$\rho=\sigma<\rho'<\sigma'$	$\left\{\begin{matrix}(-x^{2})\cdots - \\ - \cdots(-y)y'\end{matrix}\right\}$	$\left\{\begin{matrix}f^{4}d \\ f^{2}d^{2}s\end{matrix}\right\}$	$\left\{\begin{matrix}- \cdots (x^{2}\ ) - \\ - \cdots (yy') -\end{matrix}\right\}$
4	$\rho<\rho'=\sigma'<\sigma$	$\left\{\begin{matrix}y - y' \\ (-x^{2})\end{matrix}\right\}$	$\left\{\begin{matrix}s^{2}p\,d \\ s\,p^{3}\end{matrix}\right\}$	$\left\{\begin{matrix}- (yy') \\ - (x^{2}\ )\end{matrix}\right\}$
5	$\rho<\sigma<\rho'=\sigma'$	$\left\{\begin{matrix}y(-y')\cdots - \\ - \cdots(-x^{2})\end{matrix}\right\}$	$\left\{\begin{matrix}d^{10}s^{2}p \\ d^{9}\ s\ p^{3}\end{matrix}\right\}$	$\left\{\begin{matrix}- (yy') \cdots - \\ - (x^{2}\ ) \cdots -\end{matrix}\right\}$
6	$\rho=\rho'<\sigma<\sigma'$	$\left\{\begin{matrix}(x^{w})\cdots(-y) \\ (x^{w})\cdots - y'\end{matrix}\right\}$	$\left\{\begin{matrix}f^{w}s^{2} \\ f^{w}s\ d\end{matrix}\right\}$	$\left\{\begin{matrix}(x^{w}) \cdots y - \\ (x^{w}) \cdots y' -\end{matrix}\right\}$
7	$\rho<\sigma=\rho'<\sigma'$	$\left\{\begin{matrix}y(x^{w}) \\ (x^{w})y'\end{matrix}\right\}$	$\left\{\begin{matrix}sp^{w} \\ p^{w}d\end{matrix}\right\}$	$\left\{\begin{matrix}(x^{w})\ y \\ (x^{w})\ y'\end{matrix}\right\}$
8	$\rho<\rho'<\sigma=\sigma'$	$\left\{\begin{matrix}y - \cdots(x^{w}) \\ (-y')\cdots(x^{w})\end{matrix}\right\}$	$\left\{\begin{matrix}ds\ p^{w} \\ s^{2}p^{w}\end{matrix}\right\}$	$\left\{\begin{matrix}- y \cdots (x^{w}) \\ - y' \cdots (x^{w})\end{matrix}\right\}$
9	$\rho<\sigma<\rho'<\sigma'$	$\left\{\begin{matrix}x(-y)\cdots - \\ - \cdots(-x')y'\end{matrix}\right\}$	$\left\{\begin{matrix}f^{3}d^{2}s \\ f^{2}d\ s^{2}p\end{matrix}\right\}$	$\left\{\begin{matrix}- \cdots (x\ y\ ) - \\ - \cdots (x'y') -\end{matrix}\right\}$
10a	$\rho<\rho'<\sigma<\sigma'$	$\left\{\begin{matrix}x - \cdots(-y) \\ (-x')\cdots - y'\end{matrix}\right\}$	$\left\{\begin{matrix}f^{3}d^{2}s^{2} \\ f^{2}d^{3}s\ p\end{matrix}\right\}$	$\left\{\begin{matrix}- x \cdots y - \\ - x' \cdots y' -\end{matrix}\right\}$
10b	$\rho<\rho'<\sigma'<\sigma$	$\left\{\begin{matrix}x - \cdots - y \\ (-x')\cdots(-y')\end{matrix}\right\}$	$\left\{\begin{matrix}f^{3}d^{2}s\,p \\ f^{2}d^{3}s^{2}\end{matrix}\right\}$	$\left\{\begin{matrix}- x \cdots y - \\ - x' \cdots y' -\end{matrix}\right\}$
11	$\rho=\rho', \sigma=\sigma'$	$\left\{\begin{matrix}(x^{w})\cdots y \\ (x^{w})\cdots y'\end{matrix}\right\}$	$\left\{\begin{matrix}d^{3}s^{2}p \\ d^{3}s^{2}p'\end{matrix}\right\}$	no recoupling required

Fig. 13-5. The possible classes of configuration interaction that remain after the operations (13.62)-(13.63) and (13.67)-(13.68) have been performed. The active electrons are denoted by x, y, x′, and y′; spectator electrons are indicated by — in an active-electron subshell, or by · in other subshells. There may exist any number of additional spectator subshells preceding and/or following the subshells shown. Though not indicated in the figure, it must be remembered that electrons are coupled together from left to right, as in (12.1), except as explicitly indicated by parentheses. [The example for class 5 shows a multiply-occupied d subshell before d^9 has been separated off via operations (13.62) and (13.67), the d^9 subsequently becoming one of the spectator subshells; similar remarks apply to examples 2, 4, 9, and 10.]

summation over $\widetilde{L}\widetilde{S}$ remains; for l_τ^2 we make one jump recoupling and a series of exchange recouplings to the left, and for l_ρ^2 make one shift recoupling to the right if $\rho > 1$. The final result is

$$r_d^k = B_1 B_2 C_2 , \tag{13.73}$$

where

$$\begin{aligned} C_2 = \sum_{\widetilde{L}\widetilde{S}} & (l_\rho^{w_\rho}\alpha_\rho L_\rho S_\rho \{| l_\rho^{w_\rho-2}\alpha'_\rho L'_\rho S'_\rho, l_\rho^2 \widetilde{L}\widetilde{S})(l_\tau^{w_\tau}\alpha_\tau L_\tau S_\tau, l_\tau^2 \widetilde{L}\widetilde{S} |\} l_\tau^{w_\tau+2}\alpha'_\tau L'_\tau S'_\tau) \\ & \times R_j(\mathfrak{L}'_{\tau-1}\widetilde{L}\mathfrak{L}_{\tau-1}, L_\tau \mathfrak{L}_\tau, L'_\tau)\left[\prod_{m=\rho+1}^{\tau-1} R_x(\mathfrak{L}'_{m-1}\widetilde{L}\mathfrak{L}_{m-1}, L_m \mathfrak{L}_m, \mathfrak{L}'_m)\right] \\ & \times [\delta_{\rho 1} + (1-\delta_{\rho 1}) R_s(\mathfrak{L}_{\rho-1} L'_\rho \mathfrak{L}'_\rho, \widetilde{L}\mathfrak{L}_\rho, L_\rho)] \times \text{spins} \\ & \times \langle l_\rho^2 \widetilde{L}\widetilde{S} | g^{(k)} | l_\tau^2 \widetilde{L}\widetilde{S}\rangle ; \end{aligned} \tag{13.74}$$

the value of the two-electron matrix element is given by (13.27) and (13.22) as

$$\langle l_\rho^2 \widetilde{L}\widetilde{S} | g^{(k)} | l_\tau^2 \widetilde{L}\widetilde{S}\rangle = (-1)^{l_\rho + l_\tau + \widetilde{L}} \langle l_\rho \| C^{(k)} \| l_\tau\rangle^2 \begin{Bmatrix} l_\rho & l_\rho & \widetilde{L} \\ l_\tau & l_\tau & k \end{Bmatrix} \times R^k(l_\rho^2, l_\tau^2) \tag{13.75}$$

Class 3. We let $\tau = \rho'$ and $\upsilon = \sigma'$. In the bra function, we expand in cfgp (one summation over $\widetilde{L}\widetilde{S}$), and perform for $l_\rho^2\widetilde{L}\widetilde{S}$ a shift recoupling (if $\rho > 1$) and a series of $\tau - \rho - 1$ exchange recouplings to the right (no additional summations, because of basis function orthogonalities). We then convert to a two-electron problem by making in the ket function a shift recoupling of $L'_\tau S'_\tau$ ($= L_\tau S_\tau$) to the left (involving a summation over $\mathfrak{L}\mathfrak{S}$), a shift recoupling of l_τ to the right, and then an exchange recoupling of $(l_\tau l_\upsilon)\widetilde{L}\widetilde{S}$ to the left, giving

$$\sum_{\mathfrak{L}} R_s(\mathfrak{L}'_{\tau-1} L_\tau \mathfrak{L}, l_\tau \mathfrak{L}'_\tau, L'_\tau) R_s(\mathfrak{L} l_\tau \mathfrak{L}'_\tau, l_\upsilon \mathfrak{L}_\tau, \widetilde{L}) R_x(\mathfrak{L}'_{\tau-1}\widetilde{L}\mathfrak{L}_{\tau-1}, L_\tau \mathfrak{L}_\tau, \mathfrak{L}) \tag{13.76}$$

and a similar factor for spins. The final result is

$$r_d^k = B_1 B_2 B_{4\rho'} B_{4\sigma'} B'_6 C_3 , \tag{13.77}$$

where

$$\begin{aligned} C_3 = \sum_{\widetilde{L}\widetilde{S}} & (l_\rho^{w_\rho}\alpha_\rho L_\rho S_\rho \{| l_\rho^{w_\rho-2}\alpha'_\rho L'_\rho S'_\rho, l_\rho^2 \widetilde{L}\widetilde{S}) \langle l_\rho^2 \widetilde{L}\widetilde{S} | g^{(k)} | l_\tau l_\upsilon \widetilde{L}\widetilde{S}\rangle \\ & \times [\delta_{\rho 1} + (1-\delta_{\rho 1}) R_s(\mathfrak{L}_{\rho-1} L'_\rho \mathfrak{L}'_\rho, \widetilde{L}\mathfrak{L}_\rho, L_\rho)] \left[\prod_{m=\rho+1}^{\tau-1} R_x(\mathfrak{L}'_{m-1}\widetilde{L}\mathfrak{L}_{m-1}, L_m \mathfrak{L}_m, \mathfrak{L}'_m)\right] \\ & \times (-1)^{\widetilde{L} + l_\upsilon + \mathfrak{L}_{\tau-1} + \mathfrak{L}_\tau - \mathfrak{L}'_{\tau-1} - \mathfrak{L}'_\tau} [\widetilde{L}, \mathfrak{L}_{\tau-1}, L'_\tau, \mathfrak{L}'_\tau]^{1/2} \begin{Bmatrix} \widetilde{L} & l_\tau & l_\upsilon \\ \mathfrak{L}'_{\tau-1} & L'_\tau & \mathfrak{L}'_\tau \\ \mathfrak{L}_{\tau-1} & L_\tau & \mathfrak{L}_\tau \end{Bmatrix} \times \text{spins} . \end{aligned} \tag{13.78}$$

The phase factor, square root, and 9-j symbol represent the value of (13.76) obtained with the aid of (5.37), and the two-electron matrix element remaining in (13.78) may be evaluated from (13.26) with appropriate notational modifications.

Class 4. We let $\tau = \rho' = \sigma'$, expand in cfgp (one summation), perform a shift recoupling for l_τ^2 (one more summation), and then an exchange and a shift recoupling for l_ρ (no summations) to convert to a two-electron problem. The result is

$$r_d^k = B_1 B_2 B_{3\rho} B_{3\sigma} B_5 B_6 C_4 \ , \tag{13.79}$$

where

$$\begin{aligned} C_4 = \sum_{\widetilde{L}\widetilde{S}} & (l_\tau{}^{w_\tau}\alpha_\tau L_\tau S_\tau, l_\tau{}^2\widetilde{L}\widetilde{S} \| l_\tau{}^{w_\tau+2}\alpha'_\tau L'_\tau S'_\tau) \langle l_\rho l_\sigma \widetilde{L}\widetilde{S} | g^{(k)} | l_\tau{}^2 \widetilde{L}\widetilde{S} \rangle \\ & \times (-1)^{2l_\rho + l_\sigma + \mathfrak{L}_{\tau-1} - \mathfrak{L}'_{\tau-1} - \widetilde{L}} [\mathfrak{L}_{\tau-1}, \mathfrak{L}_\tau, L'_\tau, \widetilde{L}]^{1/2} \begin{Bmatrix} \mathfrak{L}'_{\tau-1} & l_\rho & \mathfrak{L}_{\tau-1} \\ \mathfrak{L}'_\tau & l_\sigma & \mathfrak{L}_\tau \\ L'_\tau & \widetilde{L} & L_\tau \end{Bmatrix} \times \text{spins} \ . \end{aligned} \tag{13.80}$$

Similarly to class 3, the last three factors of (13.80) represent the value of

$$\sum_{\mathfrak{L}} R_x(\mathfrak{L}'_{\tau-1} l_\rho \mathfrak{L}_{\tau-1}, L_\tau \mathfrak{L}_\tau, \mathfrak{L}) R_s(\mathfrak{L} l_\rho \mathfrak{L}_\tau, l_\sigma \mathfrak{L}'_\tau, \widetilde{L}) R_s(\mathfrak{L}'_{\tau-1} L_\tau \mathfrak{L}, \widetilde{L} \mathfrak{L}'_\tau, L'_\tau) \ ,$$

and the two-electron matrix element may be evaluated from (13.26).

Class 5. We set $\tau = \rho' = \sigma'$, expand in cfgp (one summation), apply a jump recoupling and a series of exchange recouplings to $l_\tau{}^2$, followed by a four-momentum (9-j) recoupling from $[(\mathfrak{L}'_{\sigma-1} l_\rho)\mathfrak{L}_{\sigma-1}, (L'_\sigma l_\sigma)L_\sigma]\mathfrak{L}_\sigma$ to $[(\mathfrak{L}'_{\sigma-1} L'_\sigma)\mathfrak{L}'_\sigma, (l_\rho l_\sigma)\widetilde{L}]\mathfrak{L}_\sigma$ using (9.29). The result is

$$r_d^k = B_1 B_2 B_{3\rho} B_{3\sigma} B_5 C_5 \ , \tag{13.81}$$

where

$$\begin{aligned} C_5 = \sum_{\widetilde{L}\widetilde{S}} & (l_\tau{}^{w_\tau}\alpha_\tau L_\tau S_\tau, l_\tau{}^2\widetilde{L}\widetilde{S} \| l_\tau{}^{w_\tau+2}\alpha'_\tau L'_\tau S'_\tau) \langle l_\rho l_\sigma \widetilde{L}\widetilde{S} | g^{(k)} | l_\tau{}^2 \widetilde{L}\widetilde{S} \rangle \\ & \times R_j(\mathfrak{L}'_{\tau-1} \widetilde{L} \mathfrak{L}_{\tau-1}, L_\tau \mathfrak{L}_\tau, L'_\tau) \left[\prod_{m=\sigma+1}^{\tau-1} R_x(\mathfrak{L}'_{m-1} L'_m \mathfrak{L}'_m, \widetilde{L} \mathfrak{L}_m, \mathfrak{L}_{m-1}) \right] \\ & \times [\mathfrak{L}_{\sigma-1}, L_\sigma, \mathfrak{L}'_\sigma, \widetilde{L}]^{1/2} \begin{Bmatrix} \mathfrak{L}'_{\sigma-1} & l_\rho & \mathfrak{L}_{\sigma-1} \\ L'_\sigma & l_\sigma & L_\sigma \\ \mathfrak{L}'_\sigma & \widetilde{L} & \mathfrak{L}_\sigma \end{Bmatrix} \times \text{spins} \ . \end{aligned} \tag{13.82}$$

The value of the remaining two-electron matrix element may be obtained from (13.26).

Class 6. We let $\upsilon = \sigma'$, perform a jump recoupling on l_σ (one summation) and an exchange recoupling on l_υ (no summation), and obtain [omitting spin quantum numbers in the first matrix element for brevity]

$$C_6 = \langle((l_\rho{}^{w_\rho}\cdots)\mathfrak{L}_{\sigma-1},[(l_\sigma{}^{w_\sigma-1}L'_\sigma,l_\sigma)L_\sigma]\mathfrak{L}_\sigma|g^{(k)}|[(l_\rho{}^{w_\rho}\cdots)\mathfrak{L}'_{\sigma-1},l_\sigma{}^{w_\sigma-1}L'_\sigma]\mathfrak{L}'_\sigma,l_\upsilon\mathfrak{L}_\sigma\rangle$$

$$= \sum_{\mathfrak{L}\mathfrak{S}}[R_j(\mathfrak{L}_{\sigma-1}l_\sigma\mathfrak{L},L'_\sigma\mathfrak{L}_\sigma,L_\sigma)R_x(\mathfrak{L}'_{\sigma-1}l_\upsilon\mathfrak{L},L'_\sigma\mathfrak{L}_\sigma,\mathfrak{L}'_\sigma) \times \text{spins}]$$

$$\times \langle((l_\rho{}^{w_\rho}\cdots)\mathfrak{L}_{\sigma-1}\mathfrak{S}_{\sigma-1},l_\sigma\mathfrak{L}\mathfrak{S}|g^{(k)}|(l_\rho{}^{w_\rho}\cdots)\mathfrak{L}'_{\sigma-1}\mathfrak{S}'_{\sigma-1},l_\upsilon\mathfrak{L}\mathfrak{S}\rangle\ . \qquad (13.83)$$

The angular matrix elements are then

$$r^k = (w_\rho)^{-1}B_1B_2B_{3\sigma}B_{4\sigma'}B'_6C_6\ . \qquad (13.84)$$

The matrix element in (13.83) is given by (13.33) and (13.34) for r^k_d and r^k_e, respectively (with l'_ρ replaced by l_ρ and l'_σ by l_υ); the additional factor $(w_\rho)^{-1}$ is required to compensate for the factor w_ρ being included implicitly in $I^{(r)}_{\rho\sigma}$ and $I^{(r1)}_{\rho\sigma}$ as well as explicitly in B_2.

Class 7. We let $\upsilon = \sigma'$ and perform an exchange recoupling for the electron l_ρ, giving

$$C_7 = -\langle[(\mathfrak{L}'_{\sigma-1}l_\rho)\mathfrak{L}_{\sigma-1},l_\sigma{}^{w_\sigma}L_\sigma]\mathfrak{L}_\sigma|g^{(k)}|[\mathfrak{L}'_{\sigma-1},l_\sigma{}^{w_\sigma}L'_\sigma]\mathfrak{L}'_\sigma,l_\upsilon\mathfrak{L}_\sigma\rangle$$

$$= -\sum_{\mathfrak{L}\mathfrak{S}}[R_x(\mathfrak{L}'_{\sigma-1}l_\rho\mathfrak{L}_{\sigma-1},L_\sigma\mathfrak{L}_\sigma,\mathfrak{L}) \times \text{spins}]$$

$$\times \langle[(\mathfrak{L}'_{\sigma-1},l_\sigma{}^{w_\sigma}L_\sigma)\mathfrak{L},l_\rho]\mathfrak{L}_\sigma|g^{(k)}|[\mathfrak{L}'_{\sigma-1},l_\sigma{}^{w_\sigma}L'_\sigma]\mathfrak{L}'_\sigma,l_\upsilon\mathfrak{L}_\sigma\rangle\ , \qquad (13.85)$$

where spin quantum numbers have been omitted from the matrix elements for brevity. In order to obtain in the final matrix element in (13.85) a standard-order coordinate numbering for the active electrons $l_\sigma l_\rho$ in the bra function, we have (as in class 1b) taken a non-standard order in the original basis function; the minus sign represents the corresponding correction to the standard phase factor given by (13.56) and (13.61). The final matrix element in (13.85) can be evaluated with the aid of (13.33) and (13.34) provided the notation is carefully modified: the coefficient of $R^k_d(l_\rho l_\sigma,l_{\rho'}l_{\sigma'})$ is to be obtained by using the $r^k_e(l_\sigma l_\rho,l_\sigma l_\upsilon)$ of (13.34), and the coefficient of $R^k_e(l_\rho l_\sigma,l_{\rho'}l_{\sigma'})$ by using the $r^k_d(l_\sigma l_\rho,l_\sigma l_\upsilon)$ of (13.33). As in class 6, an additional factor $(w_\sigma)^{-1}$ is also required:

$$r^k = (w_\sigma)^{-1}B_1B_2B_{3\rho}B_{4\sigma'}B_5B'_6C_7\ . \qquad (13.86)$$

Class 8. We let $\tau = \rho'$, perform an exchange recoupling on l_ρ and a shift recoupling on l_τ, giving

$$C_8 = \langle[(\mathfrak{L}'_{\tau-1},l_\rho)\mathfrak{L}_{\tau-1},l_\tau{}^{w_\tau}L_\tau]\mathfrak{L}_\tau,\cdots(l_\sigma{}^{w_\sigma}L_\sigma)\mathfrak{L}_\sigma|g^{(k)}|[\mathfrak{L}'_{\tau-1},(l_\tau{}^{w_\tau}L_\tau,l_\tau)L'_\tau]\mathfrak{L}'_\tau,\cdots(l_\sigma{}^{w_\sigma}L'_\sigma)\mathfrak{L}_\sigma\rangle$$

$$= \sum_{\mathfrak{L}\mathfrak{S}}[R_x(\mathfrak{L}'_{\tau-1}l_\rho\mathfrak{L}_{\tau-1},L_\tau\mathfrak{L}_\tau,\mathfrak{L})R_s(\mathfrak{L}'_{\tau-1}L_\tau\mathfrak{L},l_\tau\mathfrak{L}'_\tau,L'_\tau) \times \text{spins}]$$

$$\times \langle[(\mathfrak{L}'_{\tau-1},L_\tau)\mathfrak{L},l_\rho]\mathfrak{L}_\tau,\cdots(l_\sigma{}^{w_\sigma}L_\sigma)\mathfrak{L}_\sigma|g^{(k)}|[(\mathfrak{L}'_{\tau-1},L_\tau)\mathfrak{L},l_\tau]\mathfrak{L}'_\tau,\cdots(l_\sigma{}^{w_\sigma}L'_\sigma)\mathfrak{L}_\sigma\rangle\ . \qquad (13.87)$$

The coefficients r^k_d and r^k_e are obtainable straightforwardly from

$$r^k = (w_\sigma)^{-1}B_1B_2B_{3\rho}B_{4\rho'}B_5C_8\ , \qquad (13.88)$$

using (13.33)-(13.34).

Class 9. In the bra function we make a four-angular-momentum recoupling

$$[(\mathfrak{L}'_{\sigma-1}l_\rho)\mathfrak{L}_{\sigma-1},(L'_\sigma l_\sigma)L_\sigma]\mathfrak{L}_\sigma \rightarrow [(\mathfrak{L}'_{\sigma-1}L'_\sigma)\mathfrak{L}'_\sigma,(l_\rho l_\sigma)\tilde{L}]\mathfrak{L}_\sigma \ ,$$

which from (9.29) may be written in terms of 9-j symbols and a summation over $\widetilde{LS}$ This leaves the problem in a form almost like that of class 3, and we obtain

$$r^k = B_1B_2B_{3\rho}B_{3\sigma}B_{4\rho'}B_{4\sigma'}B_5B'_6C_9 \ , \tag{13.89}$$

where by comparison with (13.78), and with $\tau = \rho'$ and $\upsilon = \sigma'$,

$$C_9 = \sum_{\widetilde{LS}}\langle l_\rho l_\sigma\widetilde{LS}|g^{(k)}|l_\tau l_\upsilon\widetilde{LS}\rangle\left[\prod_{m=\sigma+1}^{\tau-1} R_x(\mathfrak{L}'_{m-1}\tilde{L}\mathfrak{L}_{m-1},L_m\mathfrak{L}_m,\mathfrak{L}'_m)\right]$$
$$\times (-1)^{\tilde{L}+l_\upsilon+\mathfrak{L}_{\tau-1}+\mathfrak{L}_\tau-\mathfrak{L}'_{\tau-1}-\mathfrak{L}'_\tau}[\tilde{L},\mathfrak{L}_{\tau-1},L'_\tau,\mathfrak{L}'_\tau,\mathfrak{L}_{\sigma-1},L_\sigma,\mathfrak{L}'_\sigma,\tilde{L}]^{1/2}$$
$$\times \begin{Bmatrix} \tilde{L} & l_\tau & l_\upsilon \\ \mathfrak{L}'_{\tau-1} & L'_\tau & \mathfrak{L}'_\tau \\ \mathfrak{L}_{\tau-1} & L_\tau & \mathfrak{L}_\tau \end{Bmatrix}\begin{Bmatrix} \mathfrak{L}'_{\sigma-1} & l_\rho & \mathfrak{L}_{\sigma-1} \\ L'_\sigma & l_\sigma & L_\sigma \\ \mathfrak{L}'_\sigma & \tilde{L} & \mathfrak{L}_\sigma \end{Bmatrix} \times \text{spins} \ . \tag{13.90}$$

Values of the remaining two-electron matrix element are given by (13.22)-(13.23).

Class 10. We let $m = \rho'$, $i = \min(\sigma,\sigma')$, and $j = \max(\sigma,\sigma')$; then perform an exchange recoupling on l_ρ (one summation), a shift recoupling on l_m, a jump recoupling on l_i (one summation), and an exchange recoupling on l_j. Using superscripts a and b to denote unprimed and primed quantum numbers, respectively, for class 10a, and to denote primed and unprimed quantum numbers, respectively, for class 10b, we obtain a factor

$$C_{10} = \sum_{\mathfrak{L}\mathfrak{S}}[R_x(\mathfrak{L}'_{m-1}l_\rho\mathfrak{L}_{m-1},L_m\mathfrak{L}_m,\mathfrak{L})R_s(\mathfrak{L}'_{m-1}L_m\mathfrak{L},l_m\mathfrak{L}'_m,L'_m) \times \text{spins}]$$
$$\times \sum_{\mathfrak{L}'\mathfrak{S}'}[R_j(\mathfrak{L}^a_{i-1}l_i\mathfrak{L}',L^b_i\mathfrak{L}^a_iL^a_i)R_x(\mathfrak{L}^b_{i-1}L^b_i\mathfrak{L}^b_i,l_j\mathfrak{L}^a_i,\mathfrak{L}') \times \text{spins}]$$
$$\times \langle[(\mathfrak{L}\mathfrak{S},l_\rho)\mathfrak{L}_m\mathfrak{S}_m,\cdots\mathfrak{L}_{i-1}\mathfrak{S}_{i-1},l_\sigma]\mathfrak{L}'\mathfrak{S}'|g^{(k)}|[(\mathfrak{L}\mathfrak{S},l_m)\mathfrak{L}'_m\mathfrak{S}'_m,\cdots\mathfrak{L}'_{i-1}\mathfrak{S}'_{i-1},l_{\sigma'}]\mathfrak{L}'\mathfrak{S}'\rangle \ ; \tag{13.91}$$

the remaining matrix element may be found from (13.33) and (13.34) for the direct and exchange terms, respectively. The angular matrix elements are then given by

$$r^k = B_1B_2B_{3\rho}B_{3\sigma}B_{4\rho'}B_{4\sigma'}B_5B^b_6C_{10} \ . \tag{13.92}$$

Class 11. This is a special case of class 6 in which σ and σ' are the final occupied subshells in the two configurations, with $l_\sigma = l_{\sigma'}$ and $w_\sigma = w_{\sigma'} = 1$. This class has been introduced in order to permit handling Rydberg-series type interactions (such as $3p^w3d$–$3p^w4d$–$3p^w5d$–$\cdots$) without requiring separate subshells for 3d, 4d, 5d, $\cdots$, which would necessitate an inordinately large value of q in (13.46) and correspondingly large amounts of computer storage. This represents a relaxation of the restrictions mentioned following (13.46), in that for class 11 we require only that $l_\sigma = l_{\sigma'}$, but not that $n_\sigma = n_{\sigma'}$. The coefficients r_d^k and r_e^k are given directly by (13.33) and (13.34), respectively. [The factor B_1 is included in the definition of $I_{\rho\sigma}^{(k)}$ and $I_{\rho\sigma}^{(k1)}$; $B_{3\sigma} = B_{4\sigma'} = 1$; B_5 and B_6 are inappropriate; if $B_2 = w_\rho$ is included, then a compensating additional factor $(w_\rho)^{-1}$ is required just as in class 6.]

Summary. The expressions (13.69)-(13.92) for the angular factors of CI Coulomb matrix elements are rather complicated in appearance. This has been unavoidable in order to obtain closed expressions for the general pair of interacting configurations (13.46). In practice, however, the various factors frequently simplify considerably. For example, when an active electron comes from a singly or doubly occupied subshell or from a filled subshell, the corresponding cfp factor B_{3l} or B_{4l} is unity. In the singly occupied case the parent subshell is empty, and certain recoupling coefficients [for example, the R_J and R_x in (13.83)] are unity [see textual comments following (13.66)]—in (13.83) the summation over $\mathfrak{L}\mathfrak{S}$ also reduces to a single term. Similarly, the intervening subshells indicated by "..." in Fig. 13-5 may be unoccupied, filled, or absent; the factor $\prod R_x$ in (13.74), (13.78), (13.82), or (13.90) then becomes unity. These and other simplifications are illustrated by the problem that follows. All possibilities are easily tested for in a computer program, so that no computing time need be wasted in calculating factors that will turn out to be unity.

Problems

13-10(1). Some of the possible interacting even configurations of Ce I or Th I, written according to the convention (13.46), are

(1)	$f^4d^0s^0p^0$
(2)	$f^3d^0s^0p$
(3)	$f^2d^0s^2p^0$
(4)	$f\ d\ s\ p$
(5)	$f\ d^0s^2p$
(6)	$f^0d^2s^2p^0$
(7)	$f^0d^0s^2p^2$

For each pair of configurations and for each possible $\rho\sigma\rho'\sigma'$, determine the interaction class with the aid of Fig. 13-5. [For example: for configurations (1) and (2), the f^2—fp interaction is class 1a.]

For the interaction (5)-(6), carry out all possible algebraic simplifications of the general expression (13.79), and verify that the result is that appropriate to the interaction between the two-electron configurations fp and d^2.

13-10(2). In Sec. 12-9, we saw that the Coulomb matrix elements for $l_1{}^{4l_1+1}l_2$ were the same as those for $l_1l_2{}^{4l_2+1}$ Show that, analogously, the CI matrix elements for

$$A: \begin{cases} l_1{}^{4l_1+2}l_2{}^{4l_2+1}l_3{}^0l_4{}^1 \\ l_1{}^{4l_1+1}l_2{}^{4l_2+2}l_3{}^1l_4{}^0 \end{cases}$$

are equal to those for the conjugate configurations

$$\text{B:}\quad \begin{cases} l_1^{\,0} l_2^{\,1} l_3^{\,4l_3+2} l_4^{\,4l_4+1} \\ l_1^{\,1} l_2^{\,0} l_3^{\,4l_3+1} l_4^{\,4l_4+2} \,. \end{cases}$$

[Hint: For A, use (13.64) to recouple the wavefunctions in the matrix element given by (13.91), so as to obtain the same matrix element that (13.91) gives for B.]

Thus, for example, the matrix elements for sp^6f^{13}–dp^5f^{14} are the same as those for $d^{10}sf$–d^9s^2p. How do these values compare with those for $f^{13}p^6s$–$f^{14}p^5d$?

*13-11. ONE-ELECTRON MATRIX ELEMENTS; GENERAL CASE

As discussed in Secs. 13-1 and 13-5, there will in certain cases be non-zero matrix-element contributions from the one-electron terms of the Hamiltonian. In Fig. 13-5, these cases include class 1 if $l_y = l_x$, and classes 6-8 and 11 if $l_y = l_{y'}$. For the present discussion we may represent all possible cases in the form

$$\Psi \equiv \Psi_b: \qquad \cdots l_x^{\,w_x+1} \cdots l_y^{\,w_y} \cdots \,, \tag{13.93}$$

$$\Psi' \equiv \Psi_{b'}: \qquad \cdots l_x^{\,w_x} \cdots l_y^{\,w_y+1} \cdots \,, \tag{13.94}$$

where

$$l_x = l_y \equiv l \tag{13.95}$$

but $n_x \neq n_y$. The case $w_x = w_y = 0$ is the Rydberg series case (class 11) already discussed in Sec. 13-5. The only other case of much practical importance is the case $w_y = 0$ (class 1a), as for example the interaction np^{w+1}–$np^wn'p$ $(n' > n)$; however, we shall carry through the algebra for the general case.

We assume that our basis functions have been constructed using Hartree-Fock radial wavefunctions, so that P_y (for example) is a solution of the HF equation (7.11), which we may write in the form

$$\left[-\frac{d^2}{dr^2} + \frac{l_y(l_y+1)}{r^2} - \frac{2Z}{r}\right]P_y = [\varepsilon_y - V_y]P_y + \sum_{j\neq y} \delta_{l_y l_j} w_j \varepsilon_{yj} P_j \,, \tag{13.96}$$

where V_y includes the exchange terms $-\sum w_j B_{yj} P_j/P_y$. Matrix elements of the Hamiltonian operator (excluding spin-orbit terms) may thus be written in the form

$$\begin{aligned} M &\equiv \langle\Psi| -\sum_i \nabla_i^2 - \sum_i \frac{2Z}{r_i} + \sum_{j<i} \frac{2}{r_{ij}} |\Psi'\rangle \\ &= \left[\varepsilon_y' \langle P_x|P_y'\rangle - \langle P_x|V_y'|P_y'\rangle + \sum_{j\neq y} \delta_{l_y l_j} w_j \varepsilon_{yj}' \langle P_x|P_j'\rangle\right] \langle\Psi|\Psi'\rangle_{ang} \\ &\qquad + \langle\Psi| \sum_{j<i} \frac{2}{r_{ij}} |\Psi'\rangle \end{aligned} \tag{13.97}$$

or

$$M \cong [-\langle P_x|V_y'|P_y'\rangle + w_x\varepsilon_{yx}']\langle\Psi|\Psi'\rangle_{ang} + \langle\Psi|\sum_{j<i}\frac{2}{r_{ij}}|\Psi'\rangle , \tag{13.98}$$

where the primes on the P's mean radial wavefunctions for the primed configuration (13.94), and $\langle\Psi|\Psi'\rangle_{ang}$ includes the factor

$$(-1)^{\Delta p}[(w_x + 1)(w_y + 1)]^{1/2} \tag{13.99}$$

in (13.53) that arises from coordinate permutations, as well as the angular integrals proper; of course $\langle\Psi|\Psi'\rangle_{ang}$ is zero unless $l_x = l_y$. In going from (13.97) to (13.98) we have made the approximation $\langle P_x|P_j'\rangle = \delta_{xj}$ in line with (13.17).

We could equally well have computed the complex-conjugate matrix element

$$\begin{aligned} M &\equiv \langle\Psi'|-\sum_i\nabla_i^2 - \sum_i\frac{2Z}{r_i} + \sum_{j<i}\frac{2}{r_{ij}}|\Psi\rangle \\ &\cong [-\langle P_y'|V_x|P_x\rangle + w_y\varepsilon_{xy}]\langle\Psi|\Psi'\rangle_{ang} + \langle\Psi'|\sum_{j<i}\frac{2}{r_{ij}}|\Psi\rangle . \end{aligned} \tag{13.100}$$

Since the energy matrix is symmetric, (13.98) and (13.100) should be equal, and we shall use the average of the two:

$$M = M_1 + M_2 + M_3 , \tag{13.101}$$

where

$$M_1 = \frac{1}{2}(w_y\varepsilon_{xy} + w_x\varepsilon_{yx}')\langle\Psi|\Psi'\rangle_{ang} , \tag{13.102}$$

$$M_2 = -\frac{1}{2}[\langle P_y'|V_x|P_x\rangle + \langle P_x|V_y'|P_y'\rangle]\langle\Psi|\Psi'\rangle_{ang} , \tag{13.103}$$

and

$$M_3 = \langle\Psi|\sum_{j<i}\frac{2}{r_{ij}}|\Psi'\rangle ; \tag{13.104}$$

M_3 is the two-electron Coulomb term that we discussed in detail in Sec. 13-10.

Introducing the abbreviation

$$X_{ji}^k(r_1) \equiv \int_0^\infty \frac{2r_<^k}{r_>^{k+1}}P_j(r_2)P_i(r_2)\,dr_2 , \tag{13.105}$$

remembering (13.95), and noting the occupation number $w_x + 1$ in (13.93), we may write from (7.11)-(7.13)

$$V_x = \sum_j w_j X^0_{jj} - w_x \frac{2l+1}{4l+1} \sum_{k>0} \begin{pmatrix} l & k & l \\ 0 & 0 & 0 \end{pmatrix}^2 X^k_{xx}$$
$$- \frac{1}{2} \sum_{j \neq x} w_j \sum_k \begin{pmatrix} l & k & l_j \\ 0 & 0 & 0 \end{pmatrix}^2 X^k_{jx} \frac{P_j}{P_x} , \qquad (13.106)$$

whence

$$\langle P'_y | V_x | P_x \rangle = \sum_j w_j R^0_d(jy',jx) - w_x \frac{2l+1}{4l+1} \sum_{k>0} \begin{pmatrix} l & k & l \\ 0 & 0 & 0 \end{pmatrix}^2 R^k_d(xy',xx)$$
$$- \frac{1}{2} \sum_{j \neq x} w_j \sum_k \begin{pmatrix} l & k & l_j \\ 0 & 0 & 0 \end{pmatrix}^2 R^k_d(jy',xj) ; \qquad (13.107)$$

similarly,

$$\langle P_x | V'_y | P'_y \rangle = \sum_j w_j R^0_d(j'x,j'y') - w_y \frac{2l+1}{4l+1} \sum_{k>0} \begin{pmatrix} l & k & l \\ 0 & 0 & 0 \end{pmatrix}^2 R^k_d(y'x,y'y')$$
$$- \frac{1}{2} \sum_{j \neq y} w_j \sum_k \begin{pmatrix} l & k & l_j \\ 0 & 0 & 0 \end{pmatrix}^2 R^k_d(j'x,y'j') . \qquad (13.108)$$

Firstly, let us consider those terms of (13.103) that involve core subshells j that are closed in both configurations. These are

$$M_{2c} = - \frac{1}{2} \sum_{j=j_c} w_j \Bigg[R^0_d(jx,jy') + R^0_d(j'x,j'y')$$
$$- \frac{1}{2} \sum_k \begin{pmatrix} l & k & l_j \\ 0 & 0 & 0 \end{pmatrix}^2 [R^k_d(jx,y'j) + R^k_d(j'x,y'j')] \Bigg] \langle \Psi | \Psi' \rangle_{ang}$$
$$\cong - \sum_{j=j_c} w_j \Bigg[R^0_d(jx,j'y') - \frac{1}{2} \sum_k \begin{pmatrix} l & k & l_j \\ 0 & 0 & 0 \end{pmatrix}^2 R^k_e(jx,j'y') \Bigg] \langle \Psi | \Psi' \rangle_{ang} , \qquad (13.109)$$

the last step following because the core functions P_j and $P_{j'}$ show very little relaxation upon excitation from (13.93) to (13.94) and hence are very nearly equal (and what little difference there is tends to be removed by the averaging process), and because $R^k_d(jx,y'j') = R^k_e(jx,j'y')$. But (13.109) is just the negative of the core-valence terms in (13.104); this may be seen by comparison with (6.38) when we note that, for the pair of configurations (13.93)-(13.94), the core-valence terms of (13.104) are exactly the same as in the single-configuration case except for the form of $\langle \Psi | \Psi' \rangle_{ang}$, and with R^k in place of F^k and

G^k. Hence we can ignore the terms (13.109) provided we delete core electrons from the sum over j in (13.104).

Secondly, let us consider the following R_d^0 terms of (13.103) for non-closed-subshell interactions:

$$-\frac{1}{2}\sum_{j\neq j_c} w_j[R_d^0(jx,jy') + R_d^0(j'x,j'y')]\langle\Psi|\Psi'\rangle_{ang} .$$

Sample calculations show that even for the terms j = x and j = y this is very closely equal to

$$-\sum_{j\neq j_c} w_j R_d^0(jx,j'y')\langle\Psi|\Psi'\rangle_{ang} . \qquad (13.110)$$

As before, it is evident from (6.38) and (6.39)—where the coefficient of F^0 is unity for each electron in the subshell j—that (13.110) will exactly cancel the corresponding R_d^0 terms of (13.104). This is in agreement with the conclusions of Sec. 13-5.

Finally, the remaining terms of (13.103) are

$$\langle\Psi|\Psi'\rangle_{ang}\left[\frac{2l+1}{4l+1}\sum_{k>0}\begin{pmatrix} l & k & l \\ 0 & 0 & 0\end{pmatrix}^2 \frac{w_x R_d{}^k(xy',xx) + w_y R_d{}^k(y'x,y'y')}{2} + \frac{1}{2}\sum_k \begin{pmatrix} l & k & l \\ 0 & 0 & 0\end{pmatrix}^2 \frac{w_y R_e{}^k(yy',yx) + w_x R_e{}^k(x'x,x'y')}{2} + \frac{1}{2}\sum_j{}^* w_j \sum_k \begin{pmatrix} l & k & l_j \\ 0 & 0 & 0\end{pmatrix}^2 \frac{R_e{}^k(jy',jx) + R_e{}^k(j'x,j'y')}{2}\right] , \qquad (13.111)$$

where the asterisk means a summation over open subshells j other than x and y. The expressions (13.102) and (13.111) provide a means of computing the one-electron contributions in terms of radial integrals R_d^k and R_e^k which are just like those met in the evaluation of the two-electron terms (13.104), except that they involve three radial functions from one configuration and one from the other instead of two from each.

From a comparison of (13.61) with (13.99), it is evident that $\langle\Psi|\Psi'\rangle_{ang}$ is just $1/w_a$ times the coefficient of a normal two-electron parameter

$$R_d^0(ax,a'y') , \qquad (13.112)$$

including the case a = x or a = y. Thus we may handle configuration interactions of the type (13.93)-(13.94) by simply calculating the two-electron terms (13.104) in the usual way, ignoring all subshells that are closed in both configurations, setting *all* parameters of the form (13.112) equal to zero, and then setting the value of one and only one of these parameters equal to

$$\text{“}R_d^0(ax,a'y')\text{”} \equiv \frac{1}{w_a}\Bigg[\frac{w_y\mathcal{E}_{xy} + w_x\mathcal{E}'_{yx}}{2}$$
$$+ \frac{2l+1}{4l+1}\sum_{k>0}\begin{pmatrix} l & k & l \\ 0 & 0 & 0 \end{pmatrix}^2 \frac{w_x R_d^{\,k}(xy',xx) + w_y R_d^{\,k}(y'x,y'y')}{2}$$
$$+ \frac{1}{2}\sum_k \begin{pmatrix} l & k & l \\ 0 & 0 & 0 \end{pmatrix}^2 \frac{w_y R_e^{\,k}(yy',yx) + w_x R_e^{\,k}(x'x,x'y')}{2}$$
$$+ \frac{1}{2}\sum_j{}^{*} w_j \sum_k \begin{pmatrix} l & k & l_j \\ 0 & 0 & 0 \end{pmatrix}^2 \frac{R_e^{\,k}(jy',jx) + R_e^{\,k}(j'x,j'y')}{2}\Bigg] \quad . \qquad (13.113)$$

Special case—Rydberg series. In the Rydberg series case $w_x = w_y = 0$, the orbitals P_x and P'_y both lie mainly outside all other orbitals and see very nearly the same potential; i.e., relaxation effects upon exciting the electron x to y′ are very small. Thus $P_j \cong P'_j$, and (13.113) reduces to

$$\text{“}R_d^0(ax,a'y')\text{”} \cong \frac{1}{w_a}\left[\frac{1}{2}\sum_j{}^{*} w_j \sum_k \begin{pmatrix} l & k & l_j \\ 0 & 0 & 0 \end{pmatrix}^2 R_e^k(jx,j'y')\right] \quad . \qquad (13.114)$$

Instead of using (13.114) for the effective value of one of the R_d^0 parameters and setting all other R_d^0's equal to zero, we could if we wished use for *each* open shell $l_a^{\,w_a}$ (a equal to neither x nor y) an effective parameter value

$$R_d^0(ax,a'y') = \frac{1}{2}\sum_k \begin{pmatrix} l & k & l_a \\ 0 & 0 & 0 \end{pmatrix}^2 R_e^k(ax,a'y') \quad . \qquad (13.115)$$

But for a Rydberg series interaction, the configuration-interaction block is square, and this portion of the coefficient matrix for R_d^0 is just w_a times a unit matrix. Thus still another alternative is to set *all* R_d^0 equal to zero (or ignore them completely), and add

$$\frac{w_a}{2}\begin{pmatrix} l & k & l_a \\ 0 & 0 & 0 \end{pmatrix}^2 \qquad (13.116)$$

to all diagonal elements of the CI block of the coefficient matrix for each genuine two-electron parameter $R_e^k(ax,a'y')$. But from (11.23) this is just the negative of (12.36)—i.e., it is the term required in order to correct the coefficients of the single-configuration parameter $G^k(ax)$ to the center-of-gravity energy E_{av}. Thus the result of this alternative is that the off-diagonal blocks of the coefficient matrices for the R_d^k (k > 0) and R_e^k are exactly the same as the single-configuration matrices for the corresponding F^k (k > 0) and G^k; the sole difference is that there is no configuration-interaction analog R_d^0 of E_{av}. An example for p^5d–p^5d' was given in (13.39).

14

RADIATIVE TRANSITIONS (E1)

14-1. THE EINSTEIN TRANSITION PROBABILITIES

Up to this point, we have considered the various possible quantum states of an isolated atom to be completely stationary states. Actually, this is only an approximation (though usually a very good one) because the atom interacts weakly with electromagnetic radiation. An atom in an excited state j of energy E_j can in general make a spontaneous radiative transition to a state i of lower energy E_i, with emission of a photon of energy

$$h\nu_{ji} = E_j - E_i \; , \tag{14.1}$$

corresponding to a spectrum line of wavenumber

$$\sigma_{ji} = 1/\lambda_{ji} = (E_j - E_i)/hc \; . \tag{14.2}$$

The probability per unit time that an atom in state j will make such a transition to the state i we shall denote by a_{ji}. We recall from (4.58) that for an isolated, field-free atom in a state with total angular momentum J_i, there are

$$g_i = 2J_i + 1 \tag{14.3}$$

degenerate quantum states of energy E_i, corresponding to the $2J_i + 1$ possible values of the magnetic quantum number M_i. The Einstein spontaneous emission transition probability rate[1] is defined to be the total probability per unit time of an atom in a *specific* state j making a transition to *any* of the g_i states of the energy level i:

$$A_{ji} = \sum_{M_i} a_{ji} \; . \tag{14.4}$$

[1]A. Einstein, Physik. Z. 18, 121 (1917). An English translation is available in B. L. van der Waerden, *Sources of Quantum Mechanics* (Dover Publications, New York, 1968).

We shall see in Sec. 14-5 that A_{Ji} is independent of M_J; physically, this corresponds to the fact that the transition probability cannot depend on an arbitrary choice of the orientation of the coordinate axes.

If at time t there are $N_J(t)$ atoms in state j, the rate of change of N_J due to spontaneous transitions to all states of the level i—that is, all transitions from state j that give rise to the spectrum line (14.2)—is

$$\frac{dN_J(t)}{dt} = -A_{Ji}N_J(t) \ . \tag{14.5}$$

Under normal (isotropic) excitation conditions, there is the same number of atoms in each of the states belonging to the level j, and therefore the intensity of the spectrum line (energy radiated per unit time) is

$$I(t) = hc\sigma_{Ji}g_JA_{Ji}N_J(t) \ . \tag{14.6}$$

The quantity

$$g_JA_{Ji} = g_J\sum_{M_i} a_{Ji} = \sum_{M_J}\sum_{M_i} a_{Ji} \tag{14.7}$$

is called the weighted (spontaneous-emission) transition probability, and is evidently more symmetric in the upper and lower energy levels than is the Einstein transition probability itself.

The total rate of change of N_J due to all possible spontaneous transitions is

$$\frac{dN_J(t)}{dt} = -N_J(t)\sum_i A_{Ji} \ , \tag{14.8}$$

where the summation is over all levels of the atom having energies less than E_J. It follows that if no other excitation nor de-excitation processes are involved, then

$$N_J(t) = N_J(0)e^{-t/\tau_J} \ , \tag{14.9}$$

where

$$\tau_J = \left(\sum_i A_{Ji}\right)^{-1} \tag{14.10}$$

is the natural lifetime of the atom in any one of the states of the level j. If this lifetime is not infinite, the uncertainty principle implies a finite width of the level j given by (1.17), and a corresponding natural width of the spectrum line; the quantity σ_{Ji} then represents the central wavenumber of the spectrum line, equivalent to σ_o in (1.18).

Transitions may not only occur spontaneously, but may also be induced by the presence of a radiation field. We assume this radiation field to be isotropic and unpolarized, and to have an energy per unit volume of $\rho(\sigma)d\sigma$ in the wavenumber range $d\sigma$. The Einstein coefficients of absorption B_{iJ} and of stimulated emission B_{Ji} are defined as follows: If $\rho(\sigma)$ is essentially constant over the profile of the spectrum line (i.e., over a wavenumber range of several Γ either side of σ_{ji}—see Fig. 1-8), then absorption by atoms in a state i results in transitions to states of the level j at a rate

$$\frac{dN_i(t)}{dt} = -B_{ij}N_i(t)\rho(\sigma_{ji}) , \tag{14.11}$$

and atoms in a state j are stimulated (or induced) to make radiative transitions to states of the level i at a rate

$$\frac{dN_j(t)}{dt} = -B_{ji}N_j(t)\rho(\sigma_{ji}) . \tag{14.12}$$

Values of the three Einstein coefficients are not independent, and their mutual relationships may be inferred as follows.[1] We suppose the radiation field and the atoms to be in mutual thermodynamic equilibrium at temperature T. Then the radiation energy density per unit wavenumber interval is given by Planck's law

$$\rho(\sigma) = \frac{8\pi hc\sigma^3}{e^{hc\sigma/kT} - 1} , \tag{14.13}$$

and the relative numbers of atoms in different quantum states are given by the Maxwell-Boltzmann law

$$\frac{N_j}{N_i} = e^{-(E_j-E_i)/kT} = e^{-hc\sigma_{ji}/kT} . \tag{14.14}$$

According to the law of detailed balance, the rate of transitions from all states of level i to all states of level j due to absorption from the radiation field must be equal to the rate of spontaneous plus induced emission from level j to level i:

$$g_iB_{ij}N_i\rho(\sigma_{ji}) = g_jA_{ji}N_j + g_jB_{ji}N_j\rho(\sigma_{ji}) .$$

Using (14.14) we find

$$\rho(\sigma_{ji}) = \frac{g_jA_{ji}}{g_iB_{ij}e^{hc\sigma_{ij}/kT} - g_jB_{ji}} , \tag{14.15}$$

which by comparison with (14.13) implies

$$g_iB_{ij} = g_jB_{ji} \tag{14.16a}$$

and

$$g_jA_{ji} = 8\pi hc\sigma_{ji}^3 g_jB_{ji} . \tag{14.16b}$$

Because of the relation (14.16a), we shall usually drop the subscripts and refer simply to gA and gB, where it is to be understood that g is always the statistical weight of the *initial* level—i.e., the upper level for emission, and the lower level for absorption.

The Einstein transition probabilities are physical properties of the atom depending only on the states i and j, and are independent of whether or not a state of thermodynamic equilibrium actually exists. Indeed, though the distribution of atoms in excited states may approach the Maxwell-Boltzmann law (14.14), spectroscopic sources are seldom such that the atoms are in equilibrium with the radiation field—as shown by the fact that radiation is streaming out from the source, and by the fact that the radiation consists of a line spectrum rather than a black-body continuum. Usually the radiation density $\rho(\sigma)$ is low enough that stimulated emission is unimportant compared with spontaneous emission; at the same time, N_i may be so much greater than N_j [as in (14.14) for low temperature] that absorption is appreciable, leading to significant self-absorption effects of the type shown in Fig. 1-9, with corresponding apparent line broadening. In lasers, on the other hand, radiation densities are raised by means of highly reflecting mirrors, but absorption is kept small by using optical pumping to make $N_j \gg N_i$; stimulated emission is thereby made the most important effect, with corresponding effective line narrowing.

Because of the relations (14.16), it is necessary to derive a detailed expression only for the spontaneous emission coefficient gA. The calculation involves making multipole expansions to find the contribution to gA from the interaction between the electromagnetic field and each electric and magnetic multipole moment of the atom. The various contributions are denoted E1, E2, E3, $\cdots$ for the electric dipole, quadrupole, octupole, etc. moments and M1, M2, M3, $\cdots$ for the corresponding magnetic moments.

The complete quantum mechanical derivation is long and involved[2,3] and would take us too far afield. We give here only a classical calculation of the electric dipole contribution, sufficient to make the quantum-mechanical result look reasonable and to provide a classical explanation of the polarization of the emitted radiation.

14-2. ELECTRIC DIPOLE RADIATION (CLASSICAL)

If an electron suffers an instantaneous acceleration $\ddot{\mathbf{r}}$, it can be shown[2] that its classical rate of loss of energy by radiation is

$$\dot{E} = -\frac{2e^2}{3c^3}|\ddot{\mathbf{r}}|^2 . \tag{14.17}$$

We consider the accelerations that an electron experiences under two types of simple harmonic motion (SHM), analogous to the (a) longitudinal and (b) circular SHM of a mass held between two springs, illustrated in Fig. 14-1.

(a) If the electron oscillates parallel to the z-axis with amplitude $|\mathbf{r}_0|$ (where $\mathbf{r}_0$ is parallel to the z-axis), then

$$\begin{aligned} \mathbf{r} &= \mathbf{r}_0 \cos \omega t , \\ \dot{\mathbf{r}} &= -\omega \mathbf{r}_0 \sin \omega t , \\ \ddot{\mathbf{r}} &= -\omega^2 \mathbf{r}_0 \cos \omega t = -\omega^2 \mathbf{r} . \end{aligned} \tag{14.18}$$

[2]W. Heitler, *The Quantum Theory of Radiation* (Clarendon Press, Oxford, 1954), 3rd ed.

[3]B. W. Shore and D. H. Menzel, *Principles of Atomic Spectra* (John Wiley, New York, 1968).

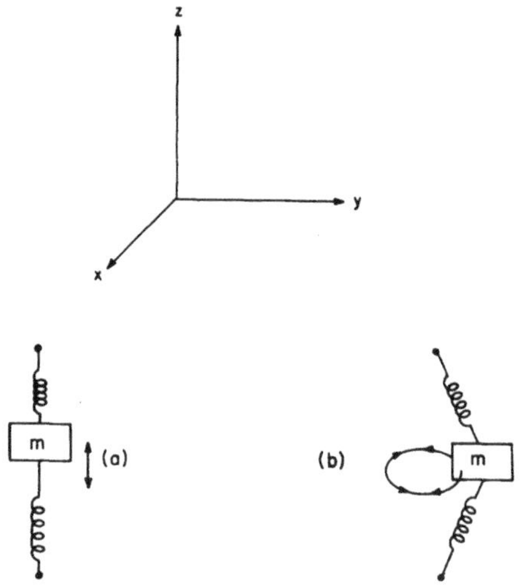

Fig. 14-1. Simple harmonic motion of a mass held between two springs in (a) linear motion parallel to the z-axis, and (b) circular motion either clockwise or counterclockwise in the x-y plane.

From (14.17),

$$\dot{E} = -\frac{2e^2\omega^4}{3c^3}|r|^2 ,$$

or averaged over one cycle of the SHM,

$$\bar{\dot{E}} = -\frac{2e^2\omega^4}{3c^3}|r|^2_{av} . \qquad (14.19)$$

(b) If the electron travels in a circular orbit of radius r_0 in the x,y plane:

$$r = x \pm iy = r_0 e^{\pm i\omega t} , \qquad (14.20)$$

then

$$\ddot{r} = -\omega^2 r_0 e^{\pm i\omega t} = -\omega^2 r .$$

From (14.17) we again obtain the result (14.19), though this time there is no need for averaging.

For a given frequency, the energy of a SH oscillator is proportional to $|r|^2_{av}$. Thus (14.19) indicates that

$$\bar{\dot{E}} \propto -E$$

or that the energy decays exponentially with time:

$$E = E_0 e^{-at} , \qquad (14.21)$$

where E_0 is the energy originally available for radiation and a is the fractional decay rate. (We assume the fractional energy loss per cycle to be small.) The cycle-averaged rate of loss of energy at time t is then

$$\bar{\dot{E}} = -aE_0e^{-at} ,$$

and the weighted-average loss rate over the entire duration of the radiative decay is

$$\bar{\bar{\dot{E}}} = \frac{\int_0^\infty \bar{\dot{E}}\, E\, dt}{\int_0^\infty E\, dt} = -aE_0 \frac{\int_0^\infty e^{-2at}\, dt}{\int_0^\infty e^{-at}\, dt} = -\frac{a}{2}E_0 .$$

The decay rate is therefore

$$a = -\frac{2\bar{\bar{\dot{E}}}}{E_0} = \frac{4e^2\omega^4}{3c^3E_0}|r|^2_{av} , \tag{14.22}$$

where $|r|^2$ is now to be averaged over the entire duration of the radiation loss.

14-3. ELECTRIC DIPOLE RADIATION (QUANTUM MECHANICAL)

The quantum mechanical analog of the average value of **r** for a stationary state Ψ is

$$\langle r\rangle = \langle\Psi|r|\Psi\rangle .$$

In order to deal with radiative transitions between two states Ψ' and Ψ, we must use time-dependent wavefunctions of the form

$$\Psi = \Psi_0 e^{-iEt/h} ,$$

and may thus expect the analog of $\langle r\rangle$ during the time of radiation to be

$$\langle\Psi|r|\Psi'\rangle = \langle\Psi_0|r|\Psi'_0\rangle e^{i(E-E')t/h} .$$

By comparison with (14.20), the analog of the classical angular frequency is $\omega \equiv 2\pi\nu = |E'-E|/h$, in agreement with (14.1).

If we interpret

$$|\langle\Psi|r|\Psi'\rangle|^2 = |\langle\Psi_0|r|\Psi'_0\rangle|^2$$

as the average $|r|^2$ over the entire duration of the radiative process, then we obtain from (14.22) a decay rate

$$a = \frac{4e^2\omega^4}{3c^3h\nu}|\langle\Psi_0|r|\Psi'_0\rangle|^2 = \frac{64\pi^4e^2\nu^3}{3c^3h}|\langle\Psi_0|r|\Psi'_0\rangle|^2 .$$

We rewrite this expression in terms of the wavenumber (14.2) of the spectrum line, include radiation by all N electrons of the atom, and write the vector position r_i as a tensor operator of rank one [cf. (11.12) and (11.42)] measured in units of the Bohr radius a_0:

$$|r_i|^2 = a_o{}^2 \sum_q |r_q^{(1)}(i)|^2 \ .$$

We thereby obtain for the spontaneous-emission transition probability per unit time from an excited state $\gamma'J'M'$ to a state γJM of lower energy

$$a = \frac{64\pi^4 e^2 a_o{}^2 \sigma^3}{3h} \sum_q |\langle \gamma JM | P_q^{(1)} | \gamma'J'M' \rangle|^2 \ , \tag{14.23}$$

where

$$P_q^{(1)} \equiv \sum_{i=1}^{N} r_q^{(1)}(i) = \sum_{i=1}^{N} r_i C_q^{(1)}(i) \tag{14.24}$$

is the (q^{th} component of the) classical dipole moment of the atom measured in units of $-ea_o$.

Three alternative forms of the electric dipole matrix element are sometimes used:[4]

$$\langle \gamma JM | \sum_i \mathbf{r}(i) | \gamma'J'M' \rangle \ , \tag{14.25}$$

$$2(E' - E)^{-1} \langle \gamma JM | \sum_i \nabla_i | \gamma'J'M' \rangle \ , \tag{14.26}$$

and

$$2(E' - E)^{-2} \langle \gamma JM | \Big(\sum_i \nabla_i V \Big) | \gamma'J'M' \rangle \ , \tag{14.27}$$

where E and E′ are the energies (in rydbergs) of the states γJM and $\gamma'J'M'$, V is the central-field potential energy (in rydbergs), and all distances (including those in the gradients) are in Bohr units. The operators in (14.26) and (14.27) correspond to the classical momentum and force, respectively. Hence the three alternatives are called the length, velocity, and acceleration forms; they are analogous to the three classical alternatives (14.18). All are equivalent when exact wavefunctions are used, but usually give rather different results when approximate wavefunctions are used.

The acceleration and velocity forms involve derivatives of approximate functions—always of doubtful accuracy. The acceleration form, particularly, tends to give poor results because the integrand is weighted toward small r, where the wavefunctions usually are poorly known.[5] The velocity form may give the best results when good variational wavefunctions are used, provided the transition energy $|E' - E|$ is not small.

[4]H. A. Bethe and E. E. Salpeter, *Quantum Mechanics of One- and Two-Electron Atoms* (Springer-Verlag, Berlin, 1957), Sec. 59. [The equivalence of (14.25) and (14.26) will be proved in (14.103).]

[5]D. Layzer and R. H. Garstang, Ann. Rev. Astron. Astrophys. 6, 449 (1968); R. J. S. Crossley, Adv. At. Mol. Phys. 5, 237 (1969).

The length form weights the integrand toward large r; however, this is not an inherent disadvantage when using HF-type radial wavefunctions, which have the correct asymptotic form because of their correct eigenvalues (Koopmans' theorem).[5] Since the length form is calculationally the simplest, it is the only one that we shall consider further.

14-4. SELECTION RULES (ELECTRIC DIPOLE RADIATION)

Electric dipole (E1) transitions can occur only when the matrix element in (14.23) is non-zero. Because the electric dipole operator (14.24) has odd parity, it follows that E1 transitions can occur only between states γJM and γ'J'M' that have opposite parity (see Sec. 2-14).

Further selection rules follow by application of the Wigner-Eckart theorem (11.15) to give

$$\langle \gamma JM | P_q^{(1)} | \gamma' J' M' \rangle = (-1)^{J-M} \begin{pmatrix} J & 1 & J' \\ -M & q & M' \end{pmatrix} \langle \gamma J \| \boldsymbol{P}^{(1)} \| \gamma' J' \rangle \; . \qquad (14.28)$$

From the properties of the 3-j symbol we see that transitions can occur only if (J1J') satisfy the triangle relations; that is, we must have

$$\left. \begin{array}{l} \Delta J \equiv J - J' = 0, \ \pm 1 \\ \text{with the restriction that} \\ J = J' = 0 \ \text{ is not allowed.} \end{array} \right\} \qquad (14.29)$$

In addition, it is necessary that $-M + q + M' = 0$, so that the only possible transitions are the three listed in Table 14-1; the indicated polarizations follow from a comparison of the forms of the tensor-operator components (11.12) with the classical motions (14.18) and (14.20).

It may be noted that every matrix met in the calculation of energy levels was diagonal in J and M and independent of M, because the operator was always a scalar. Here at last we have met a vector operator, and the matrix is neither diagonal nor independent of M,M'.

14-5. THE DIPOLE LINE STRENGTH

Substitution of (14.28) into (14.23) leads to

$$a = \frac{64\pi^4 e^2 a_0^2 \sigma^3}{3h} \, \boldsymbol{S} \sum_q \begin{pmatrix} J & 1 & J' \\ -M & q & M' \end{pmatrix}^2 , \qquad (14.30)$$

where

$$\boldsymbol{S} \equiv | \langle \gamma J \| \boldsymbol{P}^{(1)} \| \gamma' J' \rangle |^2 \qquad (14.31)$$

TABLE 14-1. POLARIZATION OF ELECTRIC DIPOLE RADIATION

ΔM	q	polarization[a]
$M = M'$	0	linear, parallel to z-axis
$M = M' + 1$	+1	circular, clockwise in (x,y) plane
$M = M' - 1$	−1	circular, counterclockwise in (x,y) plane

[a]The polarizations for $q = \pm 1$ pertain to radiation seen looking in the negative z direction, if M and M′ refer to the lower and upper energy states, respectively. More generally, the radiation is elliptically polarized, except that when viewed from a point in the (x,y) plane it is plane polarized perpendicular to the z-axis.

is known as the (electric dipole) line strength.[6] Using the sum rule (5.15) we find for the total transition probability from a state $\gamma'J'M'$ to all states M of the level γJ

$$A = \frac{64\pi^4 e^2 a_0{}^2 \sigma^3}{3h} S \sum_{Mq} \begin{pmatrix} J & 1 & J' \\ -M & q & M' \end{pmatrix}^2 = \frac{64\pi^4 e^2 a_0{}^2 \sigma^3}{3h(2J'+1)} S . \tag{14.32}$$

As we anticipated in Sec. 14-1, this quantity is independent of M′.

The weighted transition probability is

$$gA = (2J' + 1)A = \frac{64\pi^4 e^2 a_0{}^2 \sigma^3}{3h} S = 2.0261 \cdot 10^{-6} \sigma^3 S \quad \sec^{-1} , \tag{14.33}$$

where the numerical coefficient applies for σ in kaysers (cm^{-1}) and S in atomic units of $e^2a_0^2$, as indicated in connection with (14.24). For isotropic excitation conditions, the intensity of a spectrum line is seen from (14.6) to be given by

$$I \propto \sigma g A \propto \sigma^4 S . \tag{14.34}$$

Thus the quantity S, as its name implies, is a measure of the total strength of the spectrum line, including all possible transitions M,M′. The manner in which this total strength is divided up among the individual line components is given by the square of that 3-j symbol in (14.30) for which $q = M - M'$. These relative component strengths can be observed experimentally if the various components are separated as to wavelength by placing the light source in an external magnetic field; the details will be discussed in Chapter 17 on the Zeeman effect.

[6]G. H. Shortley, Phys. Rev. 47, 295 (1935).

14-6. OSCILLATOR STRENGTHS

Another quantity much used for the discussion of strengths of spectrum lines, particularly when dealing with absorption from a continuous spectrum, is the (absorption) oscillator strength, which is related to S by the expression

$$f_{ij} = \frac{8\pi^2 mca_o{}^2\sigma}{3h(2J+1)} S = \frac{(E_j - E_i)}{3(2J+1)} S \,, \tag{14.35}$$

where $E_j - E_i$ is the transition energy in rydbergs. It is important to note that this quantity refers to the total probability of absorption from a specific state of the lower level i to all (2J′+1) states of the upper level j. The corresponding quantity for emission (usually taken to be negative) is

$$f_{ji} = -\frac{8\pi^2 mca_o{}^2\sigma}{3h(2J'+1)} S = \frac{(E_i - E_j)}{3(2J'+1)} S \,; \tag{14.36}$$

like the transition probability A_{ji}, f_{ji} refers to the total probability of emission from a specific state of the upper level j to all (2J+1) states of the lower level i.

A quantity symmetric in the upper and lower levels of the transition is the weighted oscillator strength

$$gf = (2J+1)f_{ij} = -(2J'+1)f_{ji}$$

$$= \frac{8\pi^2 mca_o{}^2\sigma}{3h} S = \frac{1}{3}(\Delta E)S = 3.0376 \cdot 10^{-6}\sigma S \,. \tag{14.37}$$

From (14.33) and (14.37),

$$gA = \frac{8\pi^2 e^2\sigma^2}{mc} gf = 0.66702\sigma^2 gf \quad \sec^{-1} \,. \tag{14.38}$$

The numerical coefficients in the last two expressions apply for ΔE in rydbergs, σ in kaysers, and S in atomic units of $e^2a_0^2$. [We emphasize that g is always the statistical weight of the *initial* level (the upper level for A and $f_{emission}$, and the lower level for $f_{absorption}$).]

The oscillator strength is dimensionless, and has the physical significance of the effective number of classical electron simple harmonic oscillators that would absorb radiation of wavenumber σ as strongly as does the atom.[7,8] Thus for strong spectrum lines, f is of the order of unity.

For strong lines of neutral atoms ($\lambda \cong 5000$ Å, $\sigma \cong 20000$ K), it follows from (14.37) and (14.38) that S is of the order of 10 and that

$$A \cong 10^8 \sec^{-1} \,. \tag{14.39}$$

[7] R. Ladenburg and F. Reiche, Naturwissenschaften 11, 584 (1923).

[8] E. W. Foster, Rep. Prog. Phys. 27, 469 (1964).

The natural lifetime (14.10) with respect to radiative decay of an excited state that can decay by one or more strong electric dipole transitions is thus of the order of 10^{-8} sec. The corresponding natural width of the upper level was discussed in Sec. 1-9.

For highly ionized atoms, transition probabilities may be 10^{14} sec^{-1} or even greater, as we shall see in Sec. 14-17.

14-7. THEORETICAL CALCULATION OF LINE STRENGTHS

A number of different methods have been developed for the theoretical calculation of transition probabilities or oscillator strengths[5,9] Some of these are partially or completely empirical. Where experimental energy-level data are unavailable, only *ab initio* theoretical methods can be used; of these, we describe only that which is based on the Slater-Condon theory of atomic structure described in the preceding chapters.

In order to calculate gA or gf, we have to compute the corresponding value of the line strength (14.31), or that of its square root

$$S^{1/2}_{\gamma\gamma'} = \langle \gamma J \| \boldsymbol{P}^{(1)} \| \gamma' J' \rangle \ . \tag{14.40}$$

In order to do this, we expand the wavefunction $|\gamma J\rangle$ in terms of a suitable set of basis functions $|\beta J\rangle$ just as was done in (13.1):

$$|\gamma J\rangle = \sum_{\beta} y^{\gamma}_{\beta J} \, |\beta J\rangle \ . \tag{14.41}$$

We make a similar expansion for $|\gamma' J'\rangle$, except that from Sec. 14-4 the basis set $|\beta' J'\rangle$ will correspond to a configuration (or configurations) having parity opposite to that of the configuration(s) represented by the $|\beta J\rangle$. We may then write (14.40) in the form

$$S^{1/2}_{\gamma\gamma'} = \sum_{\beta} \sum_{\beta'} y^{\gamma}_{\beta J} \langle \beta J \| \boldsymbol{P}^{(1)} \| \beta' J' \rangle \, y^{\gamma'}_{\beta' J'} \ . \tag{14.42}$$

We have seen that the coefficients y are to be found as the components of the energy eigenvector for the state γJ. The double summation in (14.42) may thus be thought of as representing a double matrix product, in which the quantities

$$D_{\beta\beta'} \equiv S^{1/2}_{\beta\beta'} = \langle \beta J \| \boldsymbol{P}^{(1)} \| \beta' J' \rangle \tag{14.43}$$

are elements of a dipole-transition matrix (which is in general rectangular rather than square),[10] multiplied from the right by the column eigenvector for the upper state $\gamma' J'$ (in the β' representation), and from the left by the transposed (row) eigenvector for the lower state γJ (in the β representation); for example, if there are three basis states with total angular momentum J and five with total angular momentum J′, then $S^{1/2}$ is obtained by a matrix product of the form

[9] O. Sinanoğlu, Nucl. Instrum. Methods 110, 193 (1973), and references therein. A. Hibbert, Phys. Scr. 16, 7 (1977).

[10] There will be one such dipole matrix for each pair of values JJ′ that satisfies the selection rule (14.29).

$$S^{1/2}_{\gamma\gamma'} = \begin{pmatrix} \times & \times & \times \end{pmatrix} \begin{pmatrix} \times & \times & \times & \times & \times \\ \times & \times & \times & \times & \times \\ \times & \times & \times & \times & \times \end{pmatrix} \begin{pmatrix} \times \\ \times \\ \times \\ \times \\ \times \end{pmatrix} . \tag{14.44}$$

This matrix-product notation can be extended in the following way. Let **Y** be the matrix of eigenvectors for the states $|\gamma J\rangle$; i.e., a square matrix in which the ith column is the eigenvector for the ith quantum state of specified J, $|\gamma_i J\rangle$. Similarly, let **Y′** be the square matrix of column eigenvectors for the states $|\gamma_j' J'\rangle$. Then the value of $S_{\gamma\gamma'}^{1/2}$, is given by the $\gamma\gamma'$ element of the matrix

$$(S^{1/2}_{\gamma\gamma'}) = \tilde{Y}(D_{\beta\beta'})Y' , \tag{14.45}$$

where $\tilde{Y}$ is the transpose of Y. Thus the $S^{1/2}$ matrix for the actual atom is just the $S^{1/2}$ matrix for the pure-coupling representations $\beta\beta'$, transformed from the left into the actual intermediate-coupling representation γ and from the right into the intermediate-coupling representation γ'.

We may consider the eigenvector matrices **Y** and **Y′** to be known as the result of having made energy-matrix diagonalizations for the corresponding energies. Our problem, then, reduces to that of calculating the value of the pure-coupling dipole matrix element (14.43) between each possible pair of basis functions.

14-8. SELECTION RULES AND RELATIVE INTENSITIES FOR LS COUPLING

Because the electric dipole operator $\boldsymbol{P}^{(1)}$ does not involve spin coordinates, the matrix-element calculation is simplest in LS-coupled bases. From (11.38) it follows that

$$D_{LS} = \langle \cdots LSJ \| \boldsymbol{P}^{(1)} \| \cdots L'S'J' \rangle = \delta_{SS'} D_{line} \langle \cdots LS \| \boldsymbol{P}^{(1)} \| \cdots L'S' \rangle , \tag{14.46}$$

where the "line factor"

$$D_{line} = (-1)^{L+S+J'+1} [J,J']^{1/2} \begin{Bmatrix} L & S & J \\ J' & 1 & L' \end{Bmatrix} \tag{14.47}$$

contains the entire dependence of the matrix element on J and J′. Taking note of the delta factor in (14.46) and of the triangle relations that must be satisfied for the 6-j symbol to be non-zero, we find—in addition to the selection rule (14.29) on J—the following selection rules:

$$\Delta S \equiv S - S' = 0 , \tag{14.48}$$

$$\left.\begin{array}{l} \Delta L \equiv L - L' = 0, \ \pm 1 \\ \text{except that} \\ L = L' = 0 \text{ is not allowed.} \end{array}\right\} \tag{14.49}$$

For conditions that closely approximate pure LS coupling, the eigenvectors in the LS representation are nearly pure basis vectors and the eigenvector matrices **Y** and **Y′** are practically identity matrices. It then follows from (14.45) that $(S_{\gamma\gamma'}^{1/2})$ is essentially equal to $(D_{\beta\beta'})$ and that the strength S of a line is just the square of (14.46). Thus the relative strengths of the various lines of a multiplet (the lines arising from transitions between levels of two specific terms ···LS and ···L′S) are given by

$$S_{LS} \propto (D_{line})^2 = [J,J'] \begin{Bmatrix} L & S & J \\ J' & 1 & L' \end{Bmatrix}^2 . \tag{14.50}$$

For illustrative purposes, values of this quantity for a few simple cases are given in Table 14-2; additional cases are included in Appendix I. As may be seen from the examples in the table, it is usually true (particularly for large LL′) that the strongest lines of a multiplet are those for which

$$\Delta J = \Delta L .$$

These lines are called the principal lines of the multiplet; the principal line of maximum J is always the strongest, and the strengths of the others usually decrease monotonically with J. Lines for which $\Delta J \neq \Delta L$ are sometimes called satellite lines (though this term is also used with other meanings—see Secs. 18-11 and 19-9), and usually are strongest for intermediate values of J; when $\Delta L \neq 0$, the first-order satellite lines ($\Delta J = 0$) are stronger than the second-order satellites ($\Delta J = -\Delta L$).

From the orthonormality relation (5.31) for 6-j symbols, and from (2.65), we easily find the following sum rules:

$$\sum_J S_{LS} \propto \sum_J (D_{line})^2 = \frac{[J']}{[L']} \propto 2J' + 1 , \tag{14.51}$$

$$\sum_{J'} S_{LS} \propto \sum_{J'} (D_{line})^2 = \frac{[J]}{[L]} \propto 2J + 1 , \tag{14.52}$$

$$\sum_{JJ'} S_{LS} \propto \sum_{JJ'} (D_{line})^2 = \frac{[L,S]}{[L]} = 2S + 1 . \tag{14.53}$$

Thus the sum of the strengths of all lines of a multiplet that originate (or end) on a given level J is proportional to the statistical weight 2J+1 of that level, and the total strength of all lines of the multiplet is proportional to 2S+1. In simple cases [e.g., the $^2L - {}^2(L\pm 1)$ and $^3P - {}^3S$ multiplets in Table 14-2], these three relations are sufficient to uniquely determine all values of D_{line}^2—and indeed such statistical-weight arguments provided the earliest theoretical interpretations of observed line strengths; however, they of course provide no information concerning the essential phase factors of the matrix elements.

It cannot be emphasized too strongly that the relations (14.48)-(14.53)—i.e., LS selection rules, and relative "line strengths" and sum rules within a multiplet—apply strictly only to the matrix elements (14.46) for pure-LS-coupled basis functions. All of these relations are invalid to a greater or lesser extent for actual line strengths of a real atom, because of departures from pure LS coupling conditions; i.e., because of the presence in

TABLE 14-2. VALUES OF THE LINE FACTOR D^2_{line}.
[An asterisk indicates that $D_{line}(LSJ - L'S'J')$ is negative.[a]]

$^1L - {}^1L'$: $\delta(L\,1\,L')$

L \ L'	$^2S_{1/2}$	$^2P_{1/2}$	$^2P_{3/2}$	$^2D_{3/2}$	$^2D_{5/2}$
$^2P_{1/2}$	*2/3	4/9	*2/9	10/15	---
$^2P_{3/2}$	4/3	2/9	10/9	2/15	18/15

L \ L'	$^2D_{3/2}$	$^2D_{5/2}$	$^2F_{5/2}$	$^2F_{7/2}$	$^2G_{7/2}$	$^2G_{9/2}$
$^2F_{5/2}$	28/35	*2/35	40/49	*2/49	54/63	---
$^2F_{7/2}$	---	40/35	2/49	54/49	2/63	70/63

L \ L'	$^3S_1$	$^3P_0$	$^3P_1$	$^3P_2$	$^3D_1$	$^3D_2$	$^3D_3$
$^3P_0$	1/3	---	*4/12	---	20/60	---	---
$^3P_1$	*3/3	4/12	3/12	*5/12	15/60	45/60	---
$^3P_2$	5/3	---	5/12	15/12	1/60	15/60	84/60

L \ L'	$^3D_1$	$^3D_2$	$^3D_3$	$^3F_2$	$^3F_3$	$^3F_4$	$^3G_3$	$^3G_4$	$^3G_5$
$^3F_2$	189/315	*35/315	1/315	640/1008	*80/1008	---	720/1008	---	---
$^3F_3$	---	280/315	*35/315	80/1008	847/1008	*81/1008	63/1008	945/1008	---
$^3F_4$	---	---	405/315	---	81/1008	1215/1008	1/1008	63/1008	1232/1008

[a]Note that $D(L'S'J' - LSJ) = (-1)^{L'-L+J-J'} D(LSJ - L'S'J')$. Note also that the extensive tables of $\mathfrak{R}_{line}$ given by Shore and Menzel (refs. 3 and 13) are for SL rather than LS coupling; it follows from (9.10) that their $\mathfrak{R}_{line}(L'J' - LJ)$ is equal to our $D_{line}(LSJ - L'S'J')$.

TABLE 14-3. RELATIVE LINE STRENGTHS IN Ti VI

	J–J′	LS[a]	Obs.[b]	Calc.[c]
$3p^5\ ^2P^o - 3p^4(^1S)3d\ ^2D$	1/2–3/2	5	6	56
	3/2–3/2	1	4	40
	3/2–5/2	9	0	1

[a]From Table 14-2.
[b]L. Å. Svensson and J. O. Ekberg, Ark. Fys. 37, 74 (1968).
[c]Arbitrary normalization.

(14.41)-(14.42) of eigenvector components that represent mixings (in the states γJ and $\gamma' J'$) of basis states with different values of LS and L′S′.[11] Discussion of this topic is made confusing by the standard spectroscopic practice of designating a level γJ by means of those pure-coupling quantum numbers βJ for which $|y_{\beta\gamma}^J|$ is greatest. Thus, so-called *spin-forbidden, intercombination,* or *intersystem* lines (lines for which Δ"S" $\neq$ 0) are never rigorously forbidden in real atoms. They are indeed observed only with difficulty or not at all in light elements ($Z \leqq 5$), where spin-orbit parameters are very small and LS-coupling conditions are closely approximated; however, the larger values of ζ present in heavier atoms produce increasingly great departures from pure LS coupling, and "intercombination" lines rapidly become more the rule than the exception. Thus, for example, the fairly intense resonance line Mg I $3s^2\ ^1S_0 - 3s3p\ ^3P_1^o$ at 4571 Å exists because the level 3s3p "$^3P_1^o$" contains an appreciable admixture of $3s3p\ ^1P_1^o$ character,[12] and the analogous line Hg I $6s^2\ ^1S_0 - 6s6p\ ^3P_1^o$ at 2536 Å is one of the two most intense lines of the mercury arc spectrum.

Even within a single multiplet, relative line strengths are not necessarily close to those predicted by the LS-coupling line factor (14.47). An extreme example appears in Ti VI, which on the whole shows fairly good LS coupling. However, for one particular multiplet the expected relative intensities under pure LS conditions, the observed values, and computed theoretical line strengths in intermediate coupling are as indicated in Table 14-3; the line expected to be strongest has actually almost vanished because of destructive-interference effects among the various terms of the sum in (14.42). Of course, theoretical evaluation of this sum requires values of the reduced matrix elements in (14.46). Before taking up this task in the general case, we first consider some simple special cases.

Problems

14-8(1). Verify the line strength factors in Table 14-2 for $^2F - {}^2D$ from (14.50), and also by using only the sum rules (14.51)-(14.53).

[11]E. U. Condon, Phys. Rev. 36, 1121 (1930).

[12]Particularly in light atoms, where intermediate-coupling mixings are small, relativistic corrections to the transition *operator* can make important contributions to intercombination line strengths. However, there are no such corrections if the length, rather than the velocity or acceleration, form of the electric-dipole operator is used: G. W. F. Drake, J. Phys. B 9, L169 (1976); D. L. Lin, Phys. Rev. A 17, 1939 (1978). See also D. H. Kobe, Phys. Rev. A 19, 205 (1979).

14-8(2). Relative values of D_{LS} for $p^5\ ^2P^\circ - p^4d$ transitions are as follows:

$$\begin{array}{lccccccccc} J - J' & ^4F & ^4D & ^4P & (^3P)^2D & (^1D)^2D & (^1S)^2D & (^3P)^2P & (^1D)^2P \\ 1/2 - 3/2 & \Big(0 & 0 & 0 & (3/2)^{1/2} & (7/18)^{1/2} & (2/9)^{1/2} & (1/6)^{1/2} & (1/6)^{1/2}\Big) \\ 3/2 - 3/2 & \Big(0 & 0 & 0 & (3/10)^{1/2} & (7/90)^{1/2} & (2/45)^{1/2} & -(5/6)^{1/2} & -(5/6)^{1/2}\Big) \end{array}$$

$$\begin{array}{lccccccccc} J - J' & ^4F & ^4D & ^4P & (^3P)^2D & (^1D)^2D & (^1S)^2D & (^3P)^2F & (^1D)^2F \\ 3/2 - 5/2 & \Big(0 & 0 & 0 & (27/10)^{1/2} & (7/10)^{1/2} & (2/5)^{1/2} & 0 & 0\Big) . \end{array}$$

For Ti VI $3p^43d$, theoretically computed eigenvectors (with components arranged in the same order as indicated above) are:

$$\text{“}(^1S)\,^2D_{3/2}\text{”} = (0.05816\ \ 0.01706\ \ 0.02191\ \ -0.10324\ \ -0.49338\ \ 0.85949\ \ -0.01234\ \ -0.05374)\ ,$$

$$\text{“}(^1S)\,^2D_{5/2}\text{”} = (0.04027\ \ 0.02231\ \ 0.05921\ \ -0.10027\ \ -0.45649\ \ 0.87698\ \ -0.07829\ \ \ 0.02665)\ .$$

Verify both the pure-LS and the intermediate-coupling relative line strengths in Table 14-3. Why are the intermediate-coupling ratios so different from the LS values, even though the eigenvector purities are fairly high?

14-9. ONE-ELECTRON CONFIGURATIONS; THE RADIAL DIPOLE INTEGRAL

For a hydrogenic (one-electron) atom, (14.42) and (14.46) reduce to

$$S^{1/2} \equiv D_{LS} = \langle nlsj\|r^{(1)}\|n'l'sj'\rangle = (-1)^{l+j'+3/2}[j,j']^{1/2}\begin{Bmatrix} l & s & j \\ j' & 1 & l' \end{Bmatrix}\langle nl\|r^{(1)}\|n'l'\rangle\ . \tag{14.54}$$

The remaining reduced matrix element may be easily evaluated with the aid of (11.13) and (11.23):

$$\begin{aligned} P^{(1)}_{nl,n'l'} \equiv \langle nl\|r^{(1)}\|n'l'\rangle &= \langle l\|C^{(1)}\|l'\rangle \int_0^\infty P_{nl}(r)\, r\, P_{n'l'}(r)\, dr \\ &= (-1)^l [l,l']^{1/2}\begin{pmatrix} l & 1 & l' \\ 0 & 0 & 0 \end{pmatrix}\int_0^\infty P_{nl}(r)\, r\, P_{n'l'}(r)\, dr\ . \end{aligned} \tag{14.55}$$

(In the future, we shall generally simplify the notation by omitting the principal quantum numbers n and n′, which are however always to be understood.)

Since the 3-j symbol is zero unless $l+1+l'$ is even and the triangle relation $(l1l')$ is satisfied, it follows that $P^{(1)}_{ll'} = 0$ unless

$$l' = l \pm 1 \; ; \tag{14.56}$$

in fact, the 3-j symbol may be easily simplified from (5.13) to give

$$P_{ll'}^{(1)} = \delta_{l',l\pm1} (-1)^{l+l_>} (l_>)^{1/2} \int_0^\infty P_{nl}\, r\, P_{n'l'}\, dr = (-1)^{l-l'} P_{l'l}^{(1)} = - P_{l'l}^{(1)} , \tag{14.57}$$

where $l_>$ is the larger of l and l'. The result (14.56) means that the states $|l\mathrm{sj}\rangle$ and $|l'\mathrm{sj}'\rangle$ have opposite parity, consistent with the general result (Sec. 14-4) that electric dipole transitions involve a change of parity. We shall see shortly that it is true also in multi-electron atoms that non-zero matrix elements $D_{\beta\beta'}$ arise only between two basis functions that belong to configurations differing only in the quantum numbers nl and n$'l'$ of one electron; the matrix element will always involve the factor $P_{ll'}^{(1)}$, and the values of l and l' must always satisfy (14.56).

Notation. The radial portion of the transition matrix element is frequently expressed not in terms of Racah's reduced matrix element (14.55), but in terms simply of the radial integral

$$\int_0^\infty P_{nl}\, r\, P_{n'l'}\, dr = (-1)^{l+l_>} (l_>)^{-1/2} P_{ll'}^{(1)} \tag{14.58}$$

or of Condon and Shortley's reduced matrix element

$$\sigma \equiv \langle l \vdots r \vdots l' \rangle = (4l_>^{\,2} - 1)^{-1/2} \int_0^\infty P_{nl}\, r\, P_{n'l'}\, dr = (-1)^{l+l_>} [l_> (4l_>^{\,2} - 1)]^{-1/2} P_{ll'}^{(1)} . \tag{14.59}$$

(This use of the symbol σ—most commonly appearing as the square, σ^2—is of course not to be confused with its use as a symbol for the wavenumber of a spectrum line.) Tabulated values of these radial integrals for hydrogen may be found in numerous places;[3,4,13] for hydrogenic ions of an element Z, these need only be divided by Z (cf. Table 3-1). For non-hydrogenic atoms whose energy levels are known, values of the radial integrals can be obtained via the Coulomb approximation[14] or the scaled Thomas-Fermi method.[15] When energy-level data are unavailable, these semi-empirical methods cannot be used, and *ab initio* values must be calculated from Hartree-Fock radial wavefunctions (or others such as HX or HS, of adequate accuracy); this is the method that we shall assume henceforth.

It may be noted in passing that for hydrogenic wavefunctions with n ≠ n′, the radial integral (14.58) is always positive. This is decidedly not the case for non-hydrogenic functions; the integral frequently changes sign in going up a Rydberg series (fixed n, variable n′)

[13]L. C. Green, P. P. Rush, and C. D. Chandler, Astrophys. J., Suppl. Ser. 3, 37 (1957) [the radial integral (14.58)], or B. W. Shore and D. H. Menzel, Astrophys. J., Suppl. Ser. 12, 187 (1965) [the reduced matrix element $P_{ll'}^{(1)}$].

[14]D. R. Bates and A. Damgaard, Phil. Trans. Roy. Soc. London A242, 101 (1949); G. K. Oertel and L. P. Shomo, Astrophys. J., Suppl. Ser. 16, 175 (1968).

[15]J. C. Stewart and M. Rotenberg, Phys. Rev. 140, A1508 (1965).

or along an isoelectronic sequence (fixed n, n′, and N, but variable Z). When doing multi-configuration calculations in many-electron atoms, more than one radial dipole integral is involved in the summation (14.42), and it is essential to know the sign of each.

14-10. TRANSITIONS INVOLVING AN ELECTRON IN SINGLY OCCUPIED SUBSHELLS

The set of all possible transitions (for all JJ′) between all levels of one configuration and all levels of some other configuration is called a *transition array*. Before considering the most general type of electric dipole transition array we first consider the special case

$$\cdots l_1^{w_1} l_2 \; - \; \cdots l_1^{w_1} l_2' \, , \tag{14.60}$$

where the dots indicate closed subshells common to the two configurations. This type of transition array is such that all four pure-coupling representations LS, LK, JK, and Jj are of physical interest. In order to evaluate D in any of these representations, the first step is to simplify the effect of the antisymmetrizing coordinate permutations. From (13.53) and (14.24), we find immediately that (14.43) reduces to

$$D \equiv \langle \cdots l_1^{w_1} l_2 \| \sum_i r_i^{(1)} \| \cdots l_1^{w_1} l_2' \rangle = \langle \cdots l_1^{w_1} l_2 \| r_N^{(1)} \| \cdots l_1^{w_1} l_2' \rangle \, , \tag{14.61}$$

where in the final matrix element the basis functions are non-antisymmetrized functions with standard coordinate ordering, and r_N is the coordinate of the spin-orbitals $|l_2\rangle$ and $|l_2'\rangle$.

LS representation. Since the operator $r_N^{(1)}$ acts neither on spins nor on any portion of the subshells $\cdots l_1^{w_1}$, we may use (11.38) to uncouple S from L and then use (11.39) to uncouple l_2 and l_2' from $\cdots L_1$:

$$\begin{aligned} D_{LS} &\equiv \langle [(\cdots \alpha_1 L_1, l_2)L, (\cdots S_1 s_2)S]J \| r_N^{(1)} \| [(\cdots \alpha_1' L_1', l_2')L', (\cdots S_1' s_2)S']J' \rangle \\ &= \delta_{S_1 S_1'} \delta_{SS'} (-1)^{L+S+J'+1} [J,J']^{1/2} \begin{Bmatrix} L & S & J \\ J' & 1 & L' \end{Bmatrix} \langle (\cdots \alpha_1 L_1, l_2)L \| r_N^{(1)} \| (\cdots \alpha_1' L_1', l_2')L' \rangle \\ &= \delta_{\alpha_1 L_1 S_1, \alpha_1' L_1' S_1'} \delta_{SS'} (-1)^{S+J'+L_1+l_2'} [J,J',L,L']^{1/2} \begin{Bmatrix} L & S & J \\ J' & 1 & L' \end{Bmatrix} \begin{Bmatrix} L_1 & l_2 & L \\ 1 & L' & l_2' \end{Bmatrix} P_{l_2 l_2'}^{(1)} \, . \end{aligned} \tag{14.62}$$

The previously discussed one-electron reduced matrix element (14.55) is of course present, and we still have the LS selection rules (14.48)-(14.49). The second 6-j symbol adds nothing new in the way of selection rules beyond those contained in the first 6-j symbol and in $P_{l_2 l_2'}^{(1)}$ [see (14.56)].

It may be noted that the above calculation and result can be directly extended to cases in which there are other open subshells preceding $l_1^{w_1}$, provided l_2 and l_2' are coupled LS-wise onto the resultant quantum numbers for all other subshells: it is necessary only to add further δ-factors for all additional quantum numbers, and to replace L_1 in the second 6-j symbol by the total $\mathfrak{L}_1$ for all subshells preceding l_2, l_2'.

In the special case of a single electron outside closed subshells (e.g., alkali atoms or the simple configurations of boron-group atoms) then $L_1 = 0$, $L = l_2$, and $L' = l_2'$, and (14.62) simplifies with the aid of (5.29) to a result identical with (14.54) except that $P^{(1)}_{l_2 l'_2}$ is to be evaluated using one-electron functions for the N-electron atom in question rather than hydrogenic functions.

LK Representation. Using (11.38) twice and then (11.39) once, we find

$$D_{LK} \equiv \langle[(\cdots \alpha_1 L_1, l_2)L, S_1]K, s_2 J \| r_N^{(1)} \| [(\cdots \alpha_1' L_1', l_2')L', S_1']K', s_2 J'\rangle$$

$$= (-1)^{K+s+J'+1}[J,J']^{1/2}\begin{Bmatrix} K & s & J \\ J' & 1 & K' \end{Bmatrix}\langle[(\alpha_1 L_1, l_2)L, S_1]K \| r_N^{(1)} \| [(\alpha_1' L_1', l_2')L', S_1']K'\rangle$$

$$= \delta_{S_1 S_1'}(-1)^{K+s+J'+L+S_1+K'}[J,J',K,K']^{1/2}$$

$$\times \begin{Bmatrix} K & s & J \\ J' & 1 & K' \end{Bmatrix}\begin{Bmatrix} L & S_1 & K \\ K' & 1 & L' \end{Bmatrix}\langle(\alpha_1 L_1, l_2)L \| r_N^{(1)} \| (\alpha_1' L_1', l_2')L'\rangle$$

$$= \delta_{\alpha_1 L_1 S_1, \alpha_1' L_1' S_1'}(-1)^{K+s+J'+S_1+K'+L_1+l_2'+1}[J,J',K,K',L,L']^{1/2}$$

$$\times \begin{Bmatrix} K & s & J \\ J' & 1 & K' \end{Bmatrix}\begin{Bmatrix} L & S_1 & K \\ K' & 1 & L' \end{Bmatrix}\begin{Bmatrix} L_1 & l_2 & L \\ 1 & L' & l_2' \end{Bmatrix} P^{(1)}_{l_2 l'_2} . \tag{14.63}$$

In addition to the selection rule (14.29) on J, to that on $l_2' = l_2 \pm 1$, and to that on $\alpha_1 L_1 S_1$, the 6-j symbols give also the selection rules

$$\Delta K = 0, \pm 1 \qquad (K = K' = 0 \text{ not allowed}) ,$$
$$\Delta L = 0, \pm 1 \qquad (L = L' = 0 \text{ not allowed}) . \tag{14.64}$$

A comparison of the first 6-j symbol in (14.63) with that in (14.47) shows that the variation of line strength with J and J′ is much like that discussed in Sec. 14-8, except that the roles of L and L′ are now taken by K and K′; thus the strongest lines occur when

$$\Delta J = \Delta K ,$$

and of these the strongest is that of maximum J.

As in the LS case, (14.63) can be extended to the case of additional open subshells preceding $l_1^{w_1}$ by replacing L_1 with the total $\mathfrak{L}_i$ for everything except l_2, l_2'.

JK representation. We use (11.38) once just as in the LK case and then (11.39) once, and obtain

$$D_{JK} \equiv \langle[(\cdots \alpha_1 L_1 S_1)J_1, l_2]K, s_2 J \| r_N^{(1)} \| [(\cdots \alpha_1' L_1' S_1')J_1', l_2']K', s_2 J'\rangle$$

$$= (-1)^{K+s+J'+1}[J,J']^{1/2}\begin{Bmatrix} K & s & J \\ J' & 1 & K' \end{Bmatrix}\langle[(\alpha_1 L_1 S_1)J_1, l_2]K \| r_N^{(1)} \| [(\alpha_1' L_1' S_1')J_1', l_2']K'\rangle$$

$$= \delta_{\alpha_1 L_1 S_1, \alpha_1' L_1' S_1'}\delta_{J_1 J_1'}(-1)^{s+J'-J_1-l_2'}[J,J',K,K']^{1/2}\begin{Bmatrix} K & s & J \\ J' & 1 & K' \end{Bmatrix}\begin{Bmatrix} J_1 & l_2 & K \\ 1 & K' & l_2' \end{Bmatrix} P^{(1)}_{l_2 l'_2} \tag{14.65}$$

Selection rules in addition to those on JJ′, l_2l_2', and $\alpha_1L_1S_1$ are

$$\Delta K = 0, \pm 1 \qquad (K = K' = 0 \text{ not allowed}) \; , \qquad\qquad (14.66)$$
$$\Delta J_1 = 0 \; .$$

Jj representation. Using (11.39) and then (11.38) we find

$$D_{Jj} \equiv \langle [(\cdots \alpha_1L_1S_1)J_1,(l_2s_2)j_2]J \| r_N^{(1)} \| [(\cdots \alpha_1'L_1'S_1')J_1',(l_2's_2)j_2']J' \rangle$$
$$= \delta_{\alpha_1L_1S_1,\alpha_1'L_1'S_1'}\delta_{J_1J_1'}(-1)^{J_1+j_2'+J+1}[J,J']^{1/2}\begin{Bmatrix} J_1 & j_2 & J \\ 1 & J' & j_2' \end{Bmatrix}\langle (l_2s_2)j_2 \| r^{(1)} \| (l_2's_2)j_2' \rangle$$
$$= \delta_{\alpha_1L_1S_1,\alpha_1'L_1'S_1'}\delta_{J_1J_1'}(-1)^{J_1+J-l_2-s}[J,J',j_2,j_2']^{1/2}\begin{Bmatrix} J_1 & j_2 & J \\ 1 & J' & j_2' \end{Bmatrix}\begin{Bmatrix} l_2 & s & j_2 \\ j_2' & 1 & l_2' \end{Bmatrix} P_{l_2l_2'}^{(1)} \; . \qquad (14.67)$$

Selection rules in addition to those on JJ′, l_2l_2', and $\alpha_1L_1S_1$ are

$$\Delta j_2 = 0, \pm 1 \qquad (j_2 = j_2' = 0 \text{ not allowed}) \; , \qquad\qquad (14.68)$$
$$\Delta J_1 = 0 \; .$$

Sum rules and line-strength distributions. To the extent that coupling conditions for the configurations (14.60) in a given atom closely approximate one of the above four pure-coupling schemes, line strengths are given directly by the square of the corresponding pure-coupling matrix element (14.62), (14.63), (14.65), or (14.67). In each case, sum rules similar to (14.51)-(14.53) can be derived with the aid of (5.31) and (2.65). As an example, for the JK-coupling case we find (assuming the appropriate selection rules to be satisfied) that

$$S_{JK} = [J,J',K,K']\begin{Bmatrix} K & s & J \\ J' & 1 & K' \end{Bmatrix}^2 \begin{Bmatrix} J_1 & l_2 & K \\ 1 & K' & l_2' \end{Bmatrix}^2 P^2 \; , \qquad (14.69)$$

$$\sum_{J'} S_{JK} = [J,K']\begin{Bmatrix} J_1 & l_2 & K \\ 1 & K' & l_2' \end{Bmatrix}^2 P^2 \; , \qquad (14.70)$$

$$\sum_{JJ'} S_{JK} = [K,s_2,K']\begin{Bmatrix} J_1 & l_2 & K \\ 1 & K' & l_2' \end{Bmatrix}^2 P^2 \; , \qquad (14.71)$$

$$\sum_{K'JJ'} S_{JK} = \frac{[K,s_2]}{[l_2]} P^2 \; , \qquad (14.72)$$

$$\sum_{KK'JJ'} S_{JK} = [J_1,s_2] P^2 \; , \qquad (14.73)$$

$$\sum_{J_1KK'JJ'} \boldsymbol{S}_{JK} = [L_1,S_1,s_2]\boldsymbol{P}^2 \; . \tag{14.74}$$

We can further sum (14.74) over all terms $\alpha_1L_1S_1$, and say that the total line strength of the transition array (14.60) is proportional to the total statistical weight[16]

$$\sum_{\alpha_1L_1S_1} [L_1,S_1] = \sum_{\alpha_1L_1S_1} (2L_1 + 1)(2S_1 + 1)$$

of all the states of $l_1^{w_1}$. More generally, for the array

$$\cdots l_q^{w_q} l - \cdots l_q^{w_q} l' \; , \tag{14.75}$$

the total line strength is

$$\sum \boldsymbol{S} = g[s]\boldsymbol{P}_{ll'}^2 = 2g\boldsymbol{P}_{ll'}^2 \; , \tag{14.76}$$

where g is the total statistical weight of all states of the core configuration $\cdots l_q^{w_q}$ and is given by (4.63)-(4.65).

If we view the expressions (14.69)-(14.74) in reverse order, we see how the total line strength is divided up into successively smaller parcels as quantum numbers of successively higher order are taken into account:

Equation (14.74) shows that the total line strength of the array is divided up, for all lines between levels based on a given parent term $\alpha_1L_1S_1$, according to the statistical weight $(2L_1+1)(2S_1+1)$ of that term.

Equation (14.73) shows that the total line strength for a given level J_1 of a parent term $l_1^{w_1}\alpha_1L_1S_1$ is divided up in proportion to the statistical weight of the level, $(2J_1+1)$.

Equation (14.71) describes the manner in which the line strength for a given parent level J_1 of $l_1^{w_1}$ is divided up according to the various values of K and K′ (which quantum numbers are related to the interaction of l_2 and l_2', respectively, with J_1).

Finally, the first two factors in (14.69) show how the various line strengths for given K and K′ are split up according to the various values of J and J′ when the interactions $(Ks_2)J$ and $(K's_2)J'$ are included.[17]

Clearly the splitting up of the total line strength is very closely related to the development of energy-level structures such as that illustrated in Fig. 4-7. This relationship is illustrated schematically in Fig. 14-2 for the case pf − pg, where there is only a single parent term L_1S_1 (namely $^2P^o$); the energy splitting is carried through to the JJ′ stage, but the spectrum splitting becomes rather too complex to illustrate beyond the KK′ stage. [In reality, the ζ_1 splitting should be nearly the same in pg as in pf so that the two "lines" in spectrum (b) should fall almost atop one another; the lines have been drawn far apart for

[16]This identical result can be obtained by summing S_{LS}, S_{LK}, or S_{JJ} over all pertinent quantum numbers; this is consistent with a general sum rule to be derived in Sec. 14-13.

[17]In a magnetic field, the line strength is still further divided up into components M,M′, as described by (14.28).

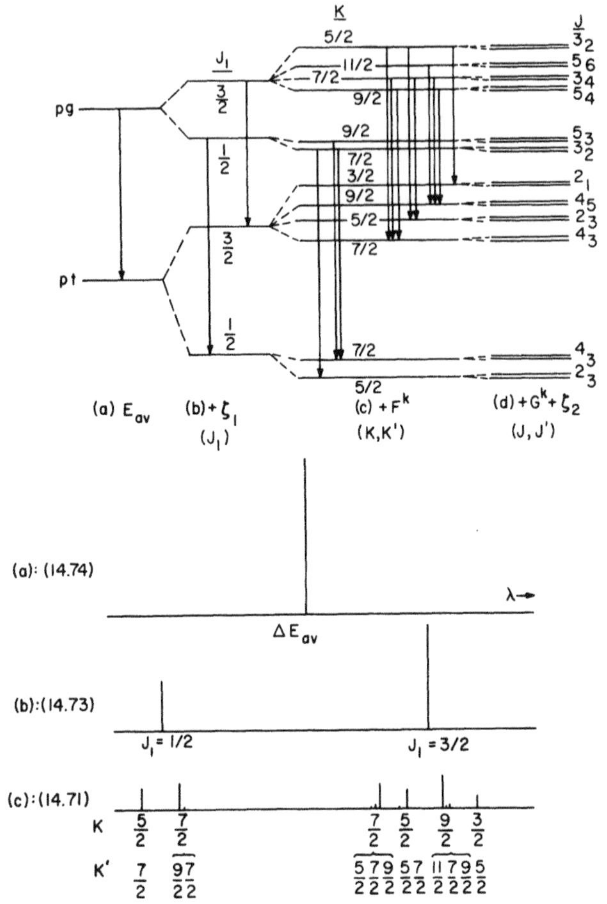

Fig. 14-2. Energy levels and spectra for pf – pg transitions taking into account (a) only E_{av}, (b) spin-orbit splitting for the p electron, (c) direct Coulomb interactions, and (d) other interactions.

greater clarity. Similarly, the spectrum lines for given K have been grouped together, even though the pg levels of different K′ have been drawn rather far apart.]

In the figure, the total line strength of the transition array is represented by the single line in spectrum (a). In spectrum (b) where the J_1 splitting has been taken into account, this total line strength is divided up[18] in the ratio of the statistical weights, $(2\cdot 1/2+1):(2\cdot 3/2+1) = 2:4$. In spectrum (c), the line strengths for transitions with $J_1=1/2$ are divided up (for all possible K′) in the ratio $(2\cdot 5/2+1):(2\cdot 7/2+1) = 6:8$, or (summed instead over K) in the ratio $(2\cdot 7/2+1):(2\cdot 9/2+1) = 8:10$; similarly, for the lines corresponding to $J_1 = 3/2$, the strengths (summed over K′) are in the ratios

[18]Quantitative values (for the LK as well as the JK representation) are tabulated by R. D. Cowan and K. L. Andrew, J. Opt. Soc. Am. 55, 502 (1965).

$(2\cdot 7/2+1):(2\cdot 5/2+1):(2\cdot 9/2+1):(2\cdot 3/2+1) = 8:6:10:4$, or (summed over K) are in the ratios 6:8:10:12. In the case $J_1 = 1/2$, the two ratios plus the known sum is sufficient to determine unambiguously the strengths of all three lines. For $J_1 = 3/2$, the six ratios plus the sum give only seven conditions, which are not sufficient to determine the nine individual line strengths. Nonetheless, it is clear that line strengths basically are simply a matter of statistical weights.

The general rule that the strongest lines usually occur when the various quantum numbers change in the same direction is well illustrated in the figure: the strongest lines are those for which

$$\Delta K \ (\equiv K - K') = \Delta l = -1 \ ,$$

the next strongest are those for which

$$\Delta K = 0 \ ,$$

and the weakest are those for which

$$\Delta K = +1 \ .$$

We remark once again that all the above line strength relations are valid in real atoms only to the extent that the quantum numbers involved are "good" quantum numbers—i.e., to the extent that the physical coupling conditions closely approximate the pure coupling scheme under consideration.

*14-11. LINE STRENGTHS FOR GENERAL TRANSITION ARRAYS

So far, we have considered the calculation of line strengths only for two very simple types of transition arrays, involving either one-electron configurations, or singly occupied outer subshells. Because of the restriction (13.50) that only one electron can be involved in a transition, the most general type array that we must consider is of the form

$$l_1^{w_1} \cdots l_i^{n} \cdots l_j^{k-1} \cdots l_q^{w_q} \ - \ l_1^{w_1} \cdots l_i^{n-1} \cdots l_j^{k} \cdots l_q^{w_q} \ , \qquad (14.77)$$

where $w'_m = w_m$ for all subshells except $m = i$ ($w'_i = w_i - 1$) and $m = j$ ($w'_j = w_j + 1$). We shall evaluate the electric dipole moment matrix elements in the $\mathfrak{LS}$-coupling representation (12.1); if the dipole matrix is desired in some other coupling scheme, it can be found by means of a matrix transformation similar to (12.5), except that the two transformation matrices belong to two different configurations (and could transform to two different coupling schemes).

The calculation of matrix elements involves the same basic principles of Racah algebra used in Chapter 13—namely, cfp expansions, recouplings, and uncouplings. The final result is sufficiently complex that we shall write it in the form

$$S^{1/2}_{\mathfrak{LS}} \equiv D_{\mathfrak{LS}} \equiv \langle \Psi_b \| \sum_m r_m{}^t C_m^{(t)} \| \Psi_{b'} \rangle = D_1 \cdot D_2 \cdot D_3 \cdots D_7 \cdot \langle l_i \| r^t C^{(t)} \| l_j \rangle$$

$$= D_1 \cdot D_2 \cdot D_3 \cdots D_7 \cdot P^{(t)}_{l_i l_j} \ . \qquad (14.78)$$

(The value of t is unity for the present case of electric dipole transitions; we use the general value to make the results applicable to higher-order electric multipole transitions as well—see Chapter 15 and Sec. 18-12.) The D_m are various subfactors encountered in different steps of the calculation, which briefly are respectively: (1) simplify the effect of coordinate permutations, (2) make two cfp expansions, (3) uncouple $\mathfrak{S}$ from $\mathfrak{L}$ and $\mathfrak{L}'$, (4) uncouple the L_m in the shells q, q − 1, ⋯ j + 1, (5)-(6) in the bra function recouple the last electron in subshell i until it is coupled to the subshell l_j^{k-1}, (7) complete the calculation by making one additional recoupling and two uncouplings. In some cases, certain of these steps are unnecessary, and the corresponding factor D_m is then unity.

D_1—Factor arising from coordinate permutations (always present). From (13.53) we obtain directly

$$D = D_1 \langle \psi_b^{(P_N)} \| r_N{}^t C_N^{(t)} \| \psi_{b'}^{(P'_N)} \rangle ,$$

where[19]

$$D_1 = (-1)^{\Delta p} (n \cdot k)^{1/2} \tag{14.79}$$

with

$$\Delta p = k - 1 + \sum_{m=i+1}^{j-1} w_m , \tag{14.80}$$

and where P_N and P'_N are, for example, the identity permutations except with r_N in the ith and j^{th} subshells, respectively.

D_2—If n > 2 and/or k > 2. Coefficient-of-fractional-parentage expansions, together with basis-function orthogonality properties to reduce each summation over parents to a single term, give a factor

$$D_2 = (l_i{}^n \alpha_i L_i S_i \{| l_i{}^{n-1} \alpha'_i L'_i S'_i)(l_j{}^{k-1} \alpha_j L_j S_j |\} l_j{}^k \alpha'_j L'_j S'_j) . \tag{14.81}$$

D_3—Always present. Uncoupling $\mathfrak{S}$ from $\mathfrak{L}$ and $\mathfrak{L}'$ with the aid of (11.38) gives a factor

$$D_3 = \delta_{\mathfrak{S}_q \mathfrak{S}'_q} (-1)^{\mathfrak{L}_q + \mathfrak{S}_q + \mathfrak{J}'_q + t} [\mathfrak{J}_q, \mathfrak{J}'_q]^{1/2} \begin{Bmatrix} \mathfrak{L}_q & \mathfrak{S}_q & \mathfrak{J}_q \\ \mathfrak{J}'_q & t & \mathfrak{L}'_q \end{Bmatrix} ; \tag{14.82}$$

this is basically the line factor D_{line} of (14.47).

D_4—If q > j. Using (11.38) repeatedly to successively uncouple L_q, L_{q-1}, ⋯ L_{j+1}, we obtain

[19]The phase factor $(-1)^{\Delta p}$ was omitted in the less-general derivation of R. D. Cowan, J. Opt. Soc. Am. 58, 808 (1968). It is unimportant in the single-configuration calculations considered there, but is essential when configuration interactions are included.

$$D_4 = \prod_{m=j+1}^{q} \delta_{\alpha_m L_m S_m,\alpha'_m L'_m S'_m} \delta_{\mathfrak{S}_m \mathfrak{S}'_m} (-1)^{\mathcal{L}_{m-1}+L_m+\mathcal{L}'_m+t} [\mathcal{L}_m,\mathcal{L}'_m]^{1/2} \begin{Bmatrix} \mathcal{L}_{m-1} & L_m & \mathcal{L}_m \\ \mathcal{L}'_m & t & \mathcal{L}'_{m-1} \end{Bmatrix} . \tag{14.83}$$

We note the selection-rule triangle requirements $(\mathcal{L}_{m-1}\ t\ \mathcal{L}'_{m-1})$ and $(\mathcal{L}_m\ t\ \mathcal{L}'_m)$, with $t = 1$ for E1 transitions.

D_5—If $i > 1$. We use (9.17) to recouple from $l_{i-1}{}^{w_{i-1}}(l_i{}^{n-1}l_i)$ to $(l_{i-1}{}^{w_{i-1}}l_i{}^{n-1})l_i$, which (noting that the result is zero unless the new recoupled quantum numbers are equal to $\mathcal{L}'_i\mathfrak{S}'_i$) gives a factor (with no summation)

$$D_5 = \left(\prod_{m=1}^{i-1} \delta_{\alpha_m L_m S_m,\alpha'_m L'_m S'_m} \delta_{\mathcal{L}_m \mathfrak{S}_m,\mathcal{L}'_m \mathfrak{S}'_m}\right) (-1)^{\mathcal{L}_{i-1}+L'_i+l_i+\mathcal{L}_i} \times [\mathcal{L}'_i,L_i]^{1/2} \begin{Bmatrix} \mathcal{L}_{i-1} & L'_i & \mathcal{L}'_i \\ l_i & \mathcal{L}_i & L_i \end{Bmatrix} \times \text{spins} ; \tag{14.84}$$

the "$\times$ spins" means another phase factor, square root, and 6-j symbol involving the corresponding spin quantum numbers. If $n = 1$, $L'_i = S'_i = 0$ and D_5 reduces to simply the δ factors.

D_6—If $i < j - 1$. We use (9.19) to jump l_i successively over each of the intervening subshells $i < m < j$, which (with δ-factors between the bra and ket functions to remove summations over the new recoupled quantum numbers) gives

$$D_6 = \left(\prod_{m=i+1}^{j-1} \delta_{\alpha_m L_m S_m,\alpha'_m L'_m S'_m} (-1)^{l_i+L_m+\mathcal{L}_{m-1}+\mathcal{L}'_m} \times [\mathcal{L}_{m-1},\mathcal{L}'_m]^{1/2} \begin{Bmatrix} l_i & \mathcal{L}'_{m-1} & \mathcal{L}_{m-1} \\ L_m & \mathcal{L}_m & \mathcal{L}'_m \end{Bmatrix} \times \text{spins}\right) . \tag{14.85}$$

If any subshell $l_m{}^{w_m}$ is full (or if $L_m = S_m = 0$ for *any* reason), only the δ-factors remain.

D_7—Always present. We use (9.18) to recouple l_i once more, and then use (11.39) twice to obtain

$$\langle[(\mathcal{L}'_{j-1}\mathfrak{S}'_{j-1},l_i)\mathcal{L}_{j-1}\mathfrak{S}_{j-1},L_jS_j]\mathcal{L}_j\mathfrak{S}_j\|r_N{}^t C_N^{(t)}\|[\mathcal{L}'_{j-1}\mathfrak{S}'_{j-1},(l_j{}^{k-1}L_jS_j,l_j)L'_jS'_j]\mathcal{L}'_j\mathfrak{S}'_j\rangle$$

$$= \sum_{LS}(-1)^{\mathcal{L}'_{j-1}+\mathcal{L}_j+L}[\mathcal{L}_{j-1},L]^{1/2}\begin{Bmatrix} \mathcal{L}'_{j-1} & l_i & \mathcal{L}_{j-1} \\ L_j & \mathcal{L}_j & L \end{Bmatrix}(-1)^{\mathfrak{S}'_{j-1}+\mathfrak{S}_j+S}[\mathfrak{S}_{j-1},S]^{1/2}$$

$$\times \begin{Bmatrix} \mathfrak{S}'_{j-1} & s & \mathfrak{S}_{j-1} \\ S_j & \mathfrak{S}_j & S \end{Bmatrix}\delta_{SS'_j}\delta_{\mathfrak{S}_j\mathfrak{S}'_j}\langle[\mathcal{L}'_{j-1},(L_jl_i)L]\mathcal{L}_j\|r_N{}^t C_N^{(t)}\|[\mathcal{L}'_{j-1},(L_jl_j)L'_j]\mathcal{L}'_j\rangle$$

$$= \delta_{\mathfrak{S}_j\mathfrak{S}'_j}(-1)^{\mathfrak{S}'_{j-1}+\mathfrak{S}_j+S'_j}[\mathfrak{S}_{j-1},S'_j]^{1/2}\begin{Bmatrix} \mathfrak{S}'_{j-1} & s & \mathfrak{S}_{j-1} \\ S_j & \mathfrak{S}_j & S'_j \end{Bmatrix}$$

$$\times \sum_{L}(-1)^{\mathcal{L}'_{j-1}+\mathcal{L}_j+L}[\mathcal{L}_{j-1},L]^{1/2}\begin{Bmatrix} \mathcal{L}'_{j-1} & l_i & \mathcal{L}_{j-1} \\ L_j & \mathcal{L}_j & L \end{Bmatrix}(-1)^{\mathcal{L}'_{j-1}+L'_j+\mathcal{L}_j+t}[\mathcal{L}_j,\mathcal{L}'_j]^{1/2}$$

$$\times \begin{Bmatrix} \mathcal{L}'_{j-1} & L & \mathcal{L}_j \\ t & \mathcal{L}'_j & L'_j \end{Bmatrix}(-1)^{L_j+l_j+L+t}[L,L'_j]^{1/2}\begin{Bmatrix} L_j & l_i & L \\ t & L'_j & l_j \end{Bmatrix}\langle l_i\|r^t C^{(t)}\|l_j\rangle .$$

Using (5.37) we may write

$$\sum_{L}(-1)^{2L}[L]\begin{Bmatrix} L_j & \mathfrak{L}_j & \mathfrak{L}_{j-1} \\ \mathfrak{L}'_{j-1} & l_i & L \end{Bmatrix}\begin{Bmatrix} L'_j & \mathfrak{L}'_j & \mathfrak{L}'_{j-1} \\ \mathfrak{L}_j & L & t \end{Bmatrix}\begin{Bmatrix} l_j & t & l_i \\ L & L_j & L'_j \end{Bmatrix} = \begin{Bmatrix} L_j & L'_j & l_j \\ \mathfrak{L}_j & \mathfrak{L}'_j & t \\ \mathfrak{L}_{j-1} & \mathfrak{L}'_{j-1} & l_i \end{Bmatrix}$$

and thus we obtain finally

$$D_7 = \delta_{\mathfrak{S}_j\mathfrak{S}'_j}(-1)^{\mathfrak{S}'_{j-1}+S'_j+\mathfrak{S}_j}[\mathfrak{S}_{j-1},S'_j]^{1/2}\begin{Bmatrix} \mathfrak{S}'_{j-1} & S'_j & \mathfrak{S}_j \\ S_j & \mathfrak{S}_{j-1} & s \end{Bmatrix}$$

$$\times\ (-1)^{L_j+l_j+L'_j}[\mathfrak{L}_{j-1},L'_j,\mathfrak{L}_j,\mathfrak{L}'_j]^{1/2}\begin{Bmatrix} \mathfrak{L}_{j-1} & \mathfrak{L}'_{j-1} & l_i \\ L_j & L'_j & l_j \\ \mathfrak{L}_j & \mathfrak{L}'_j & t \end{Bmatrix}. \tag{14.86}$$

Summary. The final result (14.78)-(14.86) is rather lengthy, but it provides a perfectly general, closed-form expression suitable for use in a computer program. If desired, it can also be readily simplified analytically with the aid of (5.29) and (5.38) to provide a more manageable expression for hand calculations in any special case of interest.[20] In particular, it can be reduced in this way to the expressions (14.62) and (14.54) that we derived earlier.

Similarly, for transition arrays of the form

$$l_1^{n}l_2^{k-1} - l_1^{n-1}l_2^{k} \tag{14.87}$$

(in addition to closed subshells), the general result (14.78)-(14.86) reduces to (in the E1 case, $t = 1$)

$$D_{LS} \equiv \langle(\alpha_1L_1S_1,\alpha_2L_2S_2)LS,J\|r^{(1)}\|(\alpha'_1L'_1S'_1,\alpha'_2L'_2S'_2)L'S',J'\rangle$$

$$= \delta_{SS'}(-1)^{k+l_2+L_2+L'_2+L+S'_1+S'_2-J'}(nk[L_1,S_1,L'_2,S'_2,L,L',J,J'])^{1/2}$$

$$\times\begin{Bmatrix} L & S & J \\ J' & 1 & L' \end{Bmatrix}\begin{Bmatrix} S'_1 & S'_2 & S \\ S_2 & S_1 & s \end{Bmatrix}\begin{Bmatrix} L_1 & L'_1 & l_1 \\ L_2 & L'_2 & l_2 \\ L & L' & 1 \end{Bmatrix}$$

$$\times\ (l_1^{n}\alpha_1L_1S_1\{|l_1^{n-1}\alpha'_1L'_1S'_1)(l_2^{k-1}\alpha_2L_2S_2|\}l_2^{k}\alpha'_2L'_2S'_2)P^{(1)}_{l_1l_2}\,. \tag{14.88}$$

For the array

$$l_1^{n} - l_1^{n-1}l_2 \tag{14.89}$$

(in addition to closed subshells), the expression (14.88) simplifies further to

[20]Results for many special cases are tabulated by Shore and Menzel (refs. 3 and 13). However, they use the SL rather than the LS coupling scheme; the relation between the two schemes is given by (9.9)-(9.10).

$$\begin{aligned} D_{LS} &\equiv \langle \alpha_1 L_1 S_1, J \| r^{(1)} \| (\alpha_1' L_1' S_1', l_2) L' S', J' \rangle \\ &= \delta_{S_1 S'} (-1)^{L_1' + l_2 + S_1 + J'} (n[L_1, L', J, J'])^{1/2} \\ &\times \begin{Bmatrix} L_1 & S & J \\ J' & 1 & L' \end{Bmatrix} \begin{Bmatrix} l_1 & L_1' & L_1 \\ L' & 1 & l_2 \end{Bmatrix} (l_1^n \alpha_1 L_1 S_1 \{| l_1^{n-1} \alpha_1' L_1' S_1') P^{(1)}_{l_1 l_2} \ . \end{aligned} \quad (14.90)$$

Problem

14-11(1). Verify that (14.78)-(14.86) simplifies for the special case (14.87) to the result (14.88), and that the latter simplifies further to (14.90) and then to (14.54) in the appropriate special cases.

14-12. SELECTION-RULE VIOLATIONS

In the preceding sections, we derived a number of different selection rules—on ΔJ (14.29), on Δl (14.56), and on various intermediate quantum numbers (14.48), (14.49), (14.64), (14.66), (14.68), and (14.83)-(14.86). All of these selection rules except for the prohibition of transitions $J = J' = 0$ are observed to be violated with greater or lesser frequency. In most cases, the violation is the result of one type or other of basis-function mixing, leading to the presence of more than one term in the double sum in (14.42). We discuss briefly the various possibilities.

Quantum-number (intermediate-coupling) violations. We have already discussed in Sec. 14-8 the very common apparent violation of the selection rule $\Delta S = 0$, owing to intermediate-coupling mixing of basis states with different values of S. Similarly, apparent violations of selection rules on other quantum numbers L, K, j, etc., are quite common, because conditions in an atom never correspond to an absolutely pure coupling scheme, and there is always at least a small amount of mixing of basis states with different values of L or K or j, etc. Apparent violations will of course be particularly noticeable for configurations in which no pure-coupling scheme is well approximated, or when for reasons of simplicity or familiarity the experimental levels are labeled in LS notation when the coupling conditions actually approximate some other coupling scheme much more closely. It may also happen that the two configurations of a transition array approximate two different pure-coupling schemes; in such a case, the only (approximately) valid selection rules are those involving quantum numbers common to the two schemes. For example, in N II, 2p3d shows approximate LS coupling whereas 2p4f approximates LK coupling; the only pertinent selection rule (other than those on l_2 and J) is $\Delta L = 0, \pm 1$.[21]

Violations of the Δl selection rule: Violations of the relation $\Delta l = \pm 1$ are of several types. In this paragraph we comment only on the case $\Delta l = \pm 3$, such as transitions $p^2 - pg$ observed in Pb I.[22] These non-zero line strengths actually arise as the result of

[21]Numerical examples of relative line strengths for different pairs of pure-coupling schemes (and also for two intermediate-coupling calculations) in the case Ne I $2p^5 3s - 2p^5 3p$ are given by D. R. Shoffstall and D. G. Ellis, J. Opt. Soc. Am. 60, 894 (1970).

[22]D. R. Wood and K. L. Andrew, J. Opt. Soc. Am. 58, 818 (1968).

configuration interaction between pd and pg configurations; the resulting admixture of a small amount of pd into the predominantly pg states makes apparently possible the unallowed p^2 – pg transition via what is really an allowed p^2 – pd transition.

Other types of Δl violations are described in the following two paragraphs.

Two-electron jumps. The electric dipole operator is a one-electron operator, and as a result transitions can theoretically occur only between two configurations that differ in only one electron [see (14.77)]. Nevertheless, transitions such as s^2 – dp are observed that have the appearance of involving two-electron jumps—one jump with $\Delta l = \pm 1$ and another with $\Delta l = \pm 2$. The explanation again lies in configuration-interaction effects[11]—in this case, admixture of some sp nature into the predominantly dp states, which makes possible an allowed one-electron transition of the type s^2 – sp (or admixture of p^2 into s^2, making possible an allowed p^2 – dp transition).

Parity-change violations. One of the most rigorously obeyed restrictions is that transitions occur only between states of opposite parity, but even here violations occur. Most commonly, this can be traced to a Stark-effect mixing of basis states: an external electric field mixes states of opposite parity (for the same reason that absorption of radiation from an electromagnetic field produces a change of parity)—for example,[22] mixing a small amount of 6p7d into Pb I 6p5f—thereby producing an apparently parity-forbidden transition ($6p^2$ – 6p5f) via what is actually an allowed ($6p^2$ – 6p7d) contribution. To the extent that the electric fields are inter-ionic (rather than the direct result of the light-source excitation mechanism), the Stark mixing will of course tend to disappear as the density of atoms in the source is decreased.

Other observed parity violations (for example, $6p^2$ – 6p7p and $6p^2$ – $6p^2$ in Pb I)[22] may be genuine electric-dipole-forbidden transitions, taking place instead by electric quadrupole or magnetic dipole interactions, which require parity conservation, as will be discussed in Chapter 15. Such transitions are relatively infrequently observed in laboratory sources because the transition probabilities are very small, but are quite commonly observed in low-density astrophysical sources (where collisional de-excitation is negligible) for cases in which the upper state of the transition is metastable with respect to electric dipole radiation.

Violations of the ΔJ selection rule. Similarly, violations of the selection rule (14.29) on J may appear as a result of J-mixing caused by external electric or magnetic fields, or caused by interaction of the electronic motions with nuclear magnetic moments (hyperfine-structure effects, Chapter 17).

*14-13. LINE STRENGTH SUM RULES

We have already derived a number of sum rules, (14.51)-(14.53) and (14.69)-(14.74), for pure-coupling line strengths in special types of transition arrays. We now derive some results for arbitrary coupling conditions in a general array.

For any given pair of values JJ′, the total strength of all spectrum lines produced by transitions $\gamma J - \gamma' J'$ is given by the sum of the squares of all elements of the matrix (14.45). But the sum of the squares of all elements of a real matrix A is easily seen to be equal to the trace of $A\widetilde{A}$, where $\widetilde{A} \equiv A^T$ is the transpose of A. Thus, since Y and Y′ are orthogonal matrices ($\widetilde{Y} = Y^{-1}$), the total line strength is

$$\sum_\gamma \sum_{\gamma'} S_{\gamma\gamma'} = \mathrm{tr}[(S_{\gamma\gamma'}{}^{1/2})(S_{\gamma\gamma'}{}^{1/2})^T] = \mathrm{tr}[(\widetilde{Y} D Y')(\widetilde{Y}' \widetilde{D} Y)] = \mathrm{tr}[\widetilde{Y} D \widetilde{D} Y] .$$

But the trace of a matrix is invariant under an orthogonal transformation, and so we find

$$\sum_\gamma \sum_{\gamma'} S_{\gamma\gamma'} = \mathrm{tr}[D\widetilde{D}] = \sum_\beta \sum_{\beta'} (D_{\beta\beta'})^2 = \sum_\beta \sum_{\beta'} S_{\beta\beta'} . \qquad (14.91)$$

Thus for any given JJ′, the total line strength for all transitions $\gamma J - \gamma' J'$ of the atom is the same as the total strength of all transitions between the pure basis states $\beta J - \beta' J'$ of an idealized pure-coupling atom; i.e., the total line strength is independent of the coupling conditions that exist in the actual atom.[23]

In the single-configuration approximation, in which wavefunction expansions of the form (14.41) include basis functions from only one configuration of each parity, the result (14.91) is known as the "J-group" sum rule.[23] This result, summed over all JJ′, indicates that the total line strength of each transition array is conserved; a change in the coupling conditions in either configuration will in general shift strength from one line of the array to others (having the same JJ′), but does not alter the total. In multi-configuration calculations, on the other hand, the mixings that result from configuration interactions in general result in line-strength shifts from one "array" to another[24]—only the total strength of all lines of all arrays included in the calculation is conserved.

The total strength of the lines of an array (or the contribution from each array, in the case of multi-configuration calculations) can be written in a fairly simple form.[19] Because this total is independent of coupling conditions, we may calculate it from the pure-LS-coupling expression (14.78)-(14.86); moreover, we may assume that the subshells in (14.77) have been ordered such that the subshells i and j come last, so that the factors D_4 and D_6 are absent. Then summing $S \equiv D^2$ over $\mathfrak{J}_q'$ and $\mathfrak{J}_q$, we find from (5.31) and (2.65) that the contribution of the factor $D_3{}^2$ reduces to

$$\delta_{\mathfrak{S}_q \mathfrak{S}_q'} \sum_{\mathfrak{J}_q} \frac{[\mathfrak{J}_q]}{[\mathfrak{L}_q]} = \delta_{\mathfrak{S}_q \mathfrak{S}_q'} [\mathfrak{S}_q] \equiv \delta_{\mathfrak{S}_j \mathfrak{S}_j'} [\mathfrak{S}_j] .$$

Multiplying this by $D_7{}^2$ and summing over $\mathfrak{S}_j'$, $\mathfrak{S}_j$, $\mathfrak{L}_j'$, and $\mathfrak{L}_j$ we find similarly [using also (5.40) and remembering that $i = j - 1$] that the result simplifies to

$$\frac{[\mathfrak{L}_i, L_j', \mathfrak{S}_i, S_j']}{[l_i, l_j, s]} .$$

Multiplication by $D_5{}^2$ and summation over $\mathfrak{L}_i'$ and $\mathfrak{S}_i'$ adds nothing but the product of δ-factors in (14.84), and summation over $\mathfrak{L}_i$ and $\mathfrak{S}_i$ then leaves us with

[23]G. R. Harrison and M. H. Johnson, Jr., Phys. Rev. 38, 757 (1931).

[24]The word "array" has been enclosed in quotation marks because the presence of configuration mixing of course physically invalidates the idealized concept of a transition array.

$$\left(\prod_{m=1}^{i-1} \delta_{\alpha_m L_m S_m, \alpha'_m L'_m S'_m} \delta_{\mathfrak{L}_m \mathfrak{S}_m, \mathfrak{L}'_m \mathfrak{S}'_m}\right) \frac{[\mathfrak{L}_{i-1}, L_i, L'_j, \mathfrak{S}_{i-1}, S_i, S'_j]}{[l_i, l_j, s]} .$$

Multiplication by $D_2{}^2$ and summation over $\alpha'_i L'_i S'_i$ and $\alpha_j L_j S_j$ contributes a factor unity from (9.46), $D_1{}^2$ adds the factor nk, and then (14.78) adds the square of $P^{(t)}$. Summation over all $\alpha'_m L'_m S'_m \mathfrak{L}'_m \mathfrak{S}'_m$ ($m < i$) merely removes all the δ-factors, and then summation successively over $\mathfrak{L}_{i-1}\mathfrak{S}_{i-1}$, $\mathfrak{L}_{i-2}\mathfrak{S}_{i-2}$, $\cdots$ leaves us with

$$nk\left(\prod_{m=1}^{i} [L_m, S_m]\right)[L'_j, S'_j] P^2_{l_i l_j} / [l_i, l_j, s] .$$

But [L,S] is just the statistical weight of the term LS. Summing over the remaining quantum numbers $\alpha_m L_m S_m$ ($m \leqq i$) and $\alpha'_j L'_j S'_j$, we thus obtain finally [cf. (4.63)-(4.64)], for the original transition array (14.77),

$$\sum\sum S = 2nk \frac{g(l_1{}^{w_1}) g(l_2{}^{w_2}) \cdots g(l_i{}^{n}) \cdots g(l_j{}^{k}) \cdots}{(4l_i + 2)(4l_j + 2)} P^2_{l_i l_j} ; \tag{14.92}$$

the double sum is over all levels (of both parities) included in the calculation, and $g(l^w)$ is the total statistical weight of all those terms of l^w for which basis functions $|l^w \alpha LS\rangle$ have been included in the calculation. (In complicated configurations, practical limitations frequently require use of a truncated basis set.)

If all terms of l^w are included, we found in (4.65) that g is just the binomial coefficient

$$g(l^w) = \frac{(4l+2)!}{w!(4l+2-w)!} , \tag{14.93}$$

so that (14.92) can be easily evaluated. Because $wg(l^w)/(4l+2)$ as well as $g(l^w)$ is a binomial coefficient, each is integral. Thus the total line strength of a transition array is an even integral multiple of P^2. These results provide useful numerical checks on computed dipole matrices and on computed intermediate-coupling line strengths.

*14-14. OSCILLATOR STRENGTH SUM RULES

The oscillator strength defined by (14.35) is the most commonly used quantity, but there are a number of other less-used variations that differ as to the number of upper states included. In order to clarify the distinctions, we modify and elaborate on the notation previously used.

The component oscillator strength. We first define a line-component oscillator strength, associated with the strength of absorption from a given lower state γJM to a given upper state γ'J'M':

$$f_c(\gamma JM - \gamma' J' M') = \frac{\Delta E}{3} |\langle \gamma JM | \sum_i r_i | \gamma' J' M' \rangle|^2$$

$$= \frac{\Delta E}{3} \boldsymbol{S}(\gamma J - \gamma' J') \sum_q \begin{pmatrix} J & 1 & J' \\ -M & q & M' \end{pmatrix}^2 , \qquad (14.94)$$

where ΔE is the transition energy in rydbergs.

The line oscillator strength. For an atom free of external fields, ΔE is the same for all components of a spectrum line. Therefore, summing the component strength over all states M' of the upper level we obtain from (5.15) the ordinary, or line, oscillator strength defined by (14.35):

$$f_l(\gamma JM - \gamma' J') \equiv \sum_{M'} f_c(\gamma JM - \gamma' J' M') = \frac{\Delta E}{3(2J+1)} \boldsymbol{S}(\gamma J - \gamma' J') . \qquad (14.95)$$

As pointed out previously, this quantity is the same for all states M of the lower level γJ.

The multiplet oscillator strength. In cases that closely approximate pure LS coupling, ΔE is nearly independent of J and J'. We may then define a multiplet oscillator strength by summing (14.95) over all levels of the upper term L'S'. From (14.46) and (5.31), with γ and γ' redefined in an obvious manner,

$$f_m(\gamma LSJM - \gamma' LS) \equiv \sum_{J'} f_l(\gamma LSJM - \gamma' L' S' J')$$

$$= \delta_{SS'} \frac{\langle \Delta E \rangle_{av}}{3(2L+1)} \langle \gamma LS \| \boldsymbol{P}^{(1)} \| \gamma' L' S' \rangle^2 . \qquad (14.96)$$

This quantity has the same value for all states JM of the lower term γLS; it is frequently used in discussing low-Z elements where LS coupling is usually closely approximated.

The array oscillator strength. If the single-configuration approximation is valid, then (regardless of the coupling conditions) it is possible to define an array oscillator strength. This is usually of interest only when ΔE is approximately the same for all lines of the transition array, as in many of the arrays of Ne I (Fig. 4-1) or for $\Delta n \neq 0$ transitions in highly ionized atoms, where energies depend strongly on the configuration but relatively weakly on all other quantum numbers.

Summation of the line oscillator strength (14.95) over all levels $\gamma' J'$ of the upper configuration does not in general give a result that is independent of the lower level γJ, but we can define an averaged array oscillator strength by averaging over all states of the lower configuration.[25] From (14.95) and (14.92), this averaged array oscillator strength for the general transition array (14.77) is (with n and k replaced by w_i and w_j')

[25]Summation over the upper level gives a result independent of γJ, so that averaging over γJ is unnecessary, if the jumping electron in the upper configuration lies in a singly occupied subshell ($k \equiv w_j' = 1$), or if the jumping electron in the lower configuration comes from a full subshell $n \equiv w_i = 4l_i + 2$)—see problem 14-14(2). These two cases cover all arrays for which the lower configuration contains at most one open subshell—which includes the ground configuration of all atoms and ions except some neutral and weakly ionized transition and rare-earth elements, and some fairly strongly ionized heavy elements (Table 4-3 and Sec. 20-6).

$$
\begin{aligned}
\bar{f}_a &\equiv \frac{\sum_{\gamma J}(2J+1)\sum_{\gamma' J'} f_l(\gamma JM - \gamma' J')}{\sum_{\gamma J}(2J+1)} \\
&= \frac{\langle \Delta E \rangle_{av}}{3} \frac{\sum\sum S(\gamma J - \gamma' J')}{g(l_1^{w_1} \ldots l_i^{w_i} \ldots l_j^{w_j - 1} \ldots)} \\
&= \frac{\langle \Delta E \rangle_{av}}{3} \frac{2w_i w_j' g(l_j^{w_j})}{(4l_i+2)(4l_j+2)g(l_j^{w_j'-1})} P^2_{n_i l_i n_j l_j} \\
&= \frac{\langle \Delta E \rangle_{av}}{3} \frac{2w_i(4l_j+3-w_j')}{(4l_i+2)(4l_j+2)} P^2_{n_i l_i n_j l_j} \qquad (14.97) \\
&= \frac{w_i(4l_j+3-w_j')}{(4l_j+2)} \bar{f}_a(n_i l_i - n_j l_j) ,
\end{aligned}
$$

where $\bar{f}_a(n_i l_i - n_j l_j)$ is the averaged array oscillator strength for a hypothetical one-electron atom ($w_i = w_j' = 1$) having the same $\overline{\Delta E}$ and $P_{n_i l_i n_j l_j}$ as the real atom, and is also equal to the multiplet oscillator strength (14.96) for this hypothetical atom.

It should be noted that the array oscillator strength is independent of the presence of subshells other than i and j, except insofar as the other subshells affect the values of $\overline{\Delta E}$ and $P_{n_i l_i n_j l_j}$. The array oscillator strength (and the analogously defined array transition probability) are frequently used for statistical (array-averaged) treatments of dielectronic recombination and collisional excitation, in plasma-physics and astrophysical applications.[26]

The *l*-summed oscillator strength. In one-electron atoms or highly excited levels of multi-electron atoms (especially highly ionized ones), energies depend only slightly on the quantum numbers l_i and l_j of the excited electron, and $\overline{\Delta E}$ therefore depends strongly on only n_i and n_j. We may then consider the l_j-complex oscillator strength; i.e., the array oscillator strength, summed over all l_j for given n_j. This quantity is usually considered only when averaged over all initial states of given n_i, which is appropriate for high-density sources in which collision rates are high enough to maintain a statistical distribution of atoms among the states of different l_i. Thus for $w_j' = w_i = 1$ and $n_j > n_i$ ($\gg 1$ for multi-electron atoms), we have from (14.97) and (4.57)

$$
\begin{aligned}
\bar{f}(n_i - n_j) &\equiv \frac{\sum_{l_i}(4l_i+2)\sum_{l_j} \bar{f}_a(n_i l_i - n_j l_j)}{\sum_{l_i}(4l_i+2)} \\
&= \frac{\langle \Delta E \rangle_{av}}{3 n_i^2} \sum_{l_i l_j} P^2_{n_i l_i n_j l_j} . \qquad (14.98)
\end{aligned}
$$

[26]See, for example, A. Burgess, Astrophys. J. 139, 776 (1964); C. Jordan, Mon. Not. R. Astron. Soc. 142, 501 (1969).

For non-relativistic one-electron (hydrogenic) ions, the double summation can be evaluated approximately via correspondence-principle arguments, leading to the result[27]

$$\bar{f}(n_i - n_j) \cong \frac{32}{3^{3/2}\pi n_i^5 n_j^3}\left(\frac{1}{n_i^2} - \frac{1}{n_j^2}\right)^{-3} = \frac{1.96}{n_i^5 n_j^3}\left(\frac{1}{n_i^2} - \frac{1}{n_j^2}\right)^{-3} ; \qquad (14.99)$$

this expression is accurate for large n_i and n_j, and valid to within +40% for all n_i and n_j. For multi-electron atoms (14.99) is still useful, but its accuracy is in general lower because of the non-hydrogenic values of $\overline{\Delta E}$ and $\mathcal{P}^2$ (cf. Tables 14-4 and 14-5 below).

The total oscillator strength. Finally, we may obtain the total oscillator strength of the atom by summing the component oscillator strength over *all* possible final states of the atom. In this case, the transition energy ΔE cannot of course be considered even approximately constant, and evaluation of the sum is quite different in nature from the cases we have considered so far.

The evaluation can be carried out[28] with the aid of the commutation relations

$$[x_j, p_{xk}] = i\hbar\delta_{jk} , \qquad [x_j, p_{yj}] = 0 , \qquad \text{etc.} , \qquad (14.100)$$

and the identity

$$[a,b^2] = ab^2 - bab + bab - b^2a = [a,b]b + b[a,b] .$$

(In this discussion, we shall use general units, *not* atomic units.)

For a non-relativistic atomic Hamiltonian of the form

$$H = \frac{1}{2m}\sum_{k=1}^{N} p_k^2 + V(r_1, r_2, r_3, \cdots) , \qquad (14.101)$$

it follows that

$$[r_j, H] = \frac{1}{2m}[r_j, p_j^2] = \frac{i\hbar}{m} p_j ; \qquad (14.102)$$

using this, we obtain[29]

$$\langle\alpha|[r_j,H]|\alpha'\rangle = (E_{\alpha'} - E_\alpha)\langle\alpha|r_j|\alpha'\rangle = \frac{i\hbar}{m}\langle\alpha|p_j|\alpha'\rangle , \qquad (14.103)$$

where $|\alpha\rangle \equiv |\gamma JM\rangle$ is an eigenstate of H, and E_α is the corresponding eigenvalue.

With the aid of this last equality, we may [noting that $hc\sigma = E'_\alpha - E_\alpha$, including the appropriate negative sign in (14.36) for emission] write the component oscillator strength (14.94) in either of two alternative forms:

[27] Ya. B. Zel'dovich and Yu. P. Raizer, *Physics of Shock Waves and High-Temperature Hydrodynamic Phenomena* (Academic Press, New York, 1966), Vol. I, p. 296. See also H. A. Bethe and E. E. Salpeter, ref. 4, Sec. 63.

[28] See, for example, H. A. Bethe and E. E. Salpeter, ref. 4, Sec. 62.

[29] The right-hand equality in (14.103) establishes the equivalence of (14.25) and (14.26).

$$f_c(\alpha - \alpha') = \frac{2m}{3h^2}(E_{\alpha'} - E_\alpha)\langle\alpha|\sum r_j|\alpha'\rangle\langle\alpha'|\sum r_j|\alpha\rangle$$
$$= -\frac{2i}{3h}\langle\alpha|\sum r_j|\alpha'\rangle\langle\alpha'|\sum p_j|\alpha\rangle$$
$$= \frac{2i}{3h}\langle\alpha|\sum p_j|\alpha'\rangle\langle\alpha'|\sum r_j|\alpha\rangle \; .$$

We sum the average of the last two expressions over all states α', use the fact that [cf. (11.31)]

$$\sum_{\alpha'}\langle\alpha|a|\alpha'\rangle\langle\alpha'|b|\alpha''\rangle = \langle\alpha|ab|\alpha''\rangle$$

if the sum is carried over a complete set of quantum states, and obtain with the aid of the commutation relations:

$$\sum_{\alpha'} f_c(\alpha - \alpha') = -\frac{i}{3h}\sum_j\sum_k\langle\alpha|[r_j,p_k]|\alpha\rangle = \frac{1}{3}\sum_j\sum_k 3\,\delta_{jk}\langle\alpha|\alpha\rangle \; .$$

Because the wavefunctions are normalized, we thus obtain finally (going back to our earlier notation) the Thomas-Reiche-Kuhn sum rule[30,31]

$$f_t \equiv \sum_{\gamma'J'} f_l(\gamma JM - \gamma'J') = \sum_{\gamma'J'M'} f_c(\gamma JM - \gamma'J'M') = N \; , \qquad (14.104)$$

which is rigorously valid for any N-electron system provided relativistic effects are negligible. The result (14.104) is consistent with the physical concept of oscillator strength: the N electrons of the atom absorb energy as strongly as N classical electron oscillators; however, the absorption by the atom is spread out over an infinite number of different frequencies rather than at most N, and if the initial state is excited some of the individual oscillator strengths are negative (emission). It must be remembered that the summation in (14.104) is to be carried out over *all* energy levels of the atom (or *all* eigenstates of the Hamiltonian H), including continuum states.

We shall consider only atoms with a single electron outside of closed subshells, and consider first the simple case of one-electron atoms, for which a few examples are given in Table 14-4. These illustrate also the Wigner-Kirkwood sum rules[28,31]

[30]W. Thomas, Naturwissenschaften 13, 627 (1925); W. Kuhn, Z. Physik 33, 408 (1925). Historically, the Thomas-Reiche-Kuhn sum rule (derived by classical arguments) was used by W. Heisenberg to derive the commutation relations (14.100) in his first paper on matrix mechanics, Z. Physik 33, 879 (1925).

[31]Numerous other sum rules, involving f times various powers of σ (and in some cases also times $ln\ \sigma$), are discussed by U. Fano and J. W. Cooper, Rev. Mod. Phys. 40, 441 (1968), 41, 724 (1969), and by A. L. Stewart, Adv. At. Mol. Phys. 3, 1 (1967). See also J. L. Dehmer, M. Inokuti, and R. P. Saxon, Phys. Rev. A 12, 102 (1975).

TABLE 14-4. NON-RELATIVISTIC ARRAY (=MULTIPLET) OSCILLATOR STRENGTHS FOR HYDROGENIC ATOMS[a]

Initial	1s	2s	2p		3s	3p		3d	
Final	np	np	ns	nd	np	ns	nd	np	nf
n=1	---	---	−0.139	---	---	−0.026	---	---	---
2	0.4162	---	---	---	−0.041	−0.145	---	−0.417	---
3	0.0791	0.4349	0.014	0.696	---	---	---	---	---
4	0.0290	0.1028	0.0031	0.122	0.484	0.032	0.619	0.011	1.016
5	0.0139	0.0419	0.0012	0.044	0.121	0.007	0.139	0.0022	0.156
6	0.0078	0.0216	0.0006	0.022	0.052	0.003	0.056	0.0009	0.053
7	0.0048	0.0127	0.0003	0.012	0.027	0.002	0.028	0.0004	0.025
8	0.0032	0.0081	0.0002	0.008	0.016	0.001	0.017	0.0002	0.015
n=9 to ∞	0.0109	0.0268	0.0007	0.023	0.048	0.004	0.045	0.0007	0.037
Sum {Discrete	0.5650	0.6489	−0.119	0.928	0.707	−0.121	0.904	−0.402	1.302
Sum {Continuum	0.4350	0.3511	0.008	0.183	0.293	0.010	0.207	0.002	0.098
Total	1.000	1.000	−0.111	1.111	1.000	−0.111	1.111	−0.400	1.400

[a]Values for larger l may be found in H. A. Bethe and E. E. Salpeter, ref. 4, Sec. 63. For Z > 20, oscillator strengths depart increasingly strongly from the above values because of relativistic effects on energies, and at much higher Z because of relativistic effects on the wavefunctions.

$$\sum_{n'} f_a(nlJM - n'[l-1]) = -\frac{l(2l-1)}{3(2l+1)}$$
$$\sum_{n'} f_a(nlJM - n'[l+1]) = \frac{(l+1)(2l+3)}{3(2l+1)} , \tag{14.105}$$

which hold rigorously for non-relativistic one-electron atoms; the sum of these two expressions is equivalent to the Thomas-Reiche-Kuhn sum rule for N=1. Both (14.105) and Table 14-4 indicate that transitions are generally much stronger when Δn and Δl have the same sign than when they have opposite signs; this is usually true also in many-electron atoms.

In the single-configuration approximation, the summation in (14.104) may be broken down into partial sums over excitations out of single subshells. From the fact that the array oscillator strength (14.97) is proportional to the subshell occupation number w_i, we might infer the partial sum rule

$$\sum_{n'l'\gamma'J'} f_l(\cdots l_i^{w_i}\gamma JM - \cdots l_i^{w_i-1}n'l'\gamma'J') = w_i ; \tag{14.106}$$

for example, for an initial state belonging to the ground configuration of neutral fluorine,

$$\sum_{n'l'\gamma'J'} f_l(1s^2 2s^2 2p^5 \gamma JM - 1s2s^2 2p^5 n'l'\gamma'J') = 2 ,$$

$$\sum_{n'l'\gamma'J'} f_l(1s^2 2s^2 2p^5 \gamma JM - 1s^2 2s 2p^5 n'l'\gamma'J') = 2 ,$$

$$\sum_{n'l'\gamma'J'} f_l(1s^2 2s^2 2p^5 \gamma JM - 1s^2 2s^2 2p^4 n'l'\gamma'J') = 5 .$$

Actually, the breakdown (14.106) is valid (even approximately) only if the summation is carried out over a complete set of states $n'l'\gamma'J'$, including those virtual states that are forbidden by the Pauli exclusion principle.[31,32] [Contributions of these virtual transitions add to zero when (14.106) is summed over i to obtain (14.104): in the fluorine example, positive 1s − 2p contributions to the first partial sum are equal in magnitude to negative 2p − 1s contributions to the second partial sum. This can be seen from (14.97), which shows that *allowed* transitions make a contribution proportional to the product of the number of electrons in the initial subshell times the number of vacancies in the final subshell; thus the virtual contribution is proportional to the number of electrons in the initial subshell i times the number of *excluded* positions in subshell j, which is equal to $w_i w_j$ regardless of the direction of the transition.]

Usually, configuration-interaction effects are appreciable. The above breakdown by subshell is then not really even conceptually valid. In addition, we have already noted in the preceding section that configuration mixings do not change the total line strength, but will in general shift line strength from one transition to another. Since oscillator strengths are proportional to σS and the wavenumbers of different lines are different, it follows that configuration interactions in general alter the total oscillator strength from that which would be computed if configuration interactions were neglected. The inference, of course, is that the single-configuration approximation will give erroneous oscillator strengths, and that inclusion of the configuration-interaction effects will improve the agreement with the sum rule (14.104).

In light alkali-like ions, configuration-interaction effects are nil for excitation of the valence electron because of Brillouin's theorem (Secs. 13-6,7) and are small for excitation of inner-shell electrons because of the large energy required. Oscillator strengths associated with excitations of the valence electron are, however, far from hydrogenic (even for fairly high ionization stages), as may be seen by comparing the examples in Table 14-5 with Table 14-4. One may note particularly the great importance of $\Delta n = 0$ transitions [resulting from the large values of the radial integrals (14.57), which arise because of the large overlap of two radial wavefunctions of equal n], and the extent to which the partial sum rules (14.105) and (14.106) are violated (without inclusion of virtual transitions).[32]

Problems

14-14(1). For an orbital angular momentum $\mathbf{l} = \mathbf{r} \times \mathbf{p}$, show from the commutation relations (14.100) that $[x,l_x] = 0$, $[x,l_y] = i\hbar z$, $[x,l_z] = -i\hbar y$, etc., and then that $[\mathbf{r}_j,$

[32] G. A. Martin and W. L. Wiese, Phys. Rev. A 13, 699 (1976).

TABLE 14-5. SINGLE-CONFIGURATION HARTREE-FOCK OSCILLATOR STRENGTHS

	Li I			Na I			Ar VIII		
Initial	2s	2p		3s	3p		3s	3p	
Final	np	ns	nd	np	ns	nd	np	ns	nd
n=2	0.7882	−0.2627							
3	0.0035	0.1197	0.6928	1.0754	−0.3583	0.9401	0.6021	−0.1984	0.5347
4	0.0036	0.0135	0.1285	0.0140	0.2313	0.1023	0.1177	0.0859	0.1309
5	0.0022	0.0045	0.0478	0.0020	0.0151	0.0312	0.0411	0.0154	0.0561
6	0.0014	0.0021	0.0235	0.0006	0.0048	0.0139	0.0195	0.0058	0.0285
7	0.0009	0.0012	0.0135	0.0003	0.0022	0.0075	0.0110	0.0029	0.0165
8	0.0006	0.0007	0.0085	0.0001	0.0012	0.0046	0.0068	0.0017	0.0105
9 − ∞	0.002	0.002	0.025	0.0002	0.003	0.012	0.019	0.005	0.030
Sum: Discrete	0.803	−0.119	0.940	1.093	−0.101	1.112	0.817	−0.082	0.807
Sum: Continuum[a]	0.318	0.023	0.222	0.059	0.021	0.065	0.286	0.027	0.317
Total	1.12	−0.09	1.16	1.15	−0.08	1.18	1.10	−0.05	1.12

[a]For the continuum, HX values are quoted; see Chap. 18.

$l_k \cdot s_k] = -i\hbar(\mathbf{r}_j \times \mathbf{s}_j)\,\delta_{jk}$. From this, show that addition of the spin-orbit term $\sum_k \xi(r_k) l_k \cdot s_k$ to the Hamiltonian (14.101) adds a further term to the right-hand side of (14.103), which, however, does not invalidate the Thomas-Reiche-Kuhn sum rule (14.104).

14-14(2). In the single-configuration, pure-$\mathfrak{LS}$-coupling approximation for the transition array (14.77), show that

$$\frac{\sum_{b'} S_{bb'}}{[\mathfrak{J}_q]\langle l_i \| r^t C^{(t)} \| l_j \rangle^2} = \begin{cases} n/(2l_i+1)\ , & k = 1\ , \\ (4l_j+3-k)/(2l_j+1)\ , & n = 4l_i+2\ , \end{cases} \tag{14.106a}$$

independent of the level b of the first configuration, if the summation is over all levels b′ of the second configuration (see fn. 25). [Hint: Proceed similarly to the derivation of (14.92), except sum over only the primed quantum numbers, and make use of (9.49) as well as (9.46).] That the same results hold for intermediate coupling in the upper configuration follows from the derivation of (14.91); for the lower configuration, a straightforward but messy generalization of the above proof is required, showing that the matrix $D\tilde{D}$ is diagonal (and hence a multiple of the unit matrix).

14-14(3). Tabulated (theoretical) values for Si III 3s4p − 3s4d are (from ref. 45 below):

	λ	gf
$^3P^o_0 - {}^3D_1$	3791.41 Å	1.3
$^3P^o_1 - {}^3D_1$	3796.2	0.96
$^3D_2$	3796.11	2.79
$^3P^o_2 - {}^3D_1$	3806.7	0.060
$^3D_2$	3806.7	0.95
$^3D_3$	3806.54	5.0
$^1P^o_1 - {}^1D_2$	3590.47	3.6

For both absorption and emission, verify that multiplet oscillator strengths are independent of both J and S of the initial level. Also, verify that the total line strengths of the two multiplets are proportional to 2S+1; why is this not true of values tabulated (ref. 45) for the array Si III 3s3p − 3s3d?

14-15. CANCELLATION EFFECTS

In favorable cases, theoretically computed oscillator strengths and transition probabilities are accurate to ten percent or better (depending on the method used). In many cases, however, especially for neutral or weakly ionized atoms having complex configurations, errors can be far greater—ranging from 50 or 100 percent to one or more orders of magnitude. The greatest errors usually arise as a result of cancellation effects of one sort or another.

Intermediate coupling. As we have seen, line strengths are generally computed in terms of single-configuration, pure-coupling basis functions. The quantum states in real atoms (especially those with Z and N greater than 5) do not usually closely approximate pure basis functions; therefore, computed line strengths usually involve evaluating the sum of several terms, because of intermediate-coupling and configuration-interaction mixing of basis states. It must be remembered that this sum, (14.42), represents a mixing of amplitudes rather than of line strengths themselves. Consequently, the effect of mixing is not necessarily a tendency to average out the various line strengths. There frequently are destructive interference effects that cause a weak line to become still weaker, or even cause an otherwise strong line to essentially disappear; an example in Ti VI was discussed in Table 14-3.

In making numerical calculations from (14.42) it is worthwhile evaluating a "cancellation factor"

$$\mathrm{CF} = \left[\frac{\left| \sum \sum y^{\gamma}_{\beta J} \langle \beta J \| \boldsymbol{P}^{(1)} \| \beta' J' \rangle y^{\gamma'}_{\beta' J'} \right|}{\sum \sum \left| y^{\gamma}_{\beta J} \langle \beta J \| \boldsymbol{P}^{(1)} \| \beta' J' \rangle y^{\gamma'}_{\beta' J'} \right|} \right]^2 . \tag{14.107}$$

Very small values of this quantity are by no means unusual; values less than 10^{-4} are found for about one percent of the lines in a complex transition array, just as would be expected on purely statistical grounds. Computed line strengths may be expected to show large percent errors when CF is smaller than about 0.1 or 0.05.[33]

Radial-integral cancellation. Another type of cancellation may occur which is completely different from the angular effects produced by intermediate-coupling mixing of basis states. The reduced matrix element $P^{(1)}_{ll'}$ involves a radial integral

$$\int_0^\infty r\, P_{nl}(r) P_{n'l'}(r)\, dr \tag{14.108}$$

in which the radial wavefunctions have both positive and negative portions if $n > l + 1$ and/or $n' > l' + 1$. It therefore frequently happens that the value of this integral is much smaller in magnitude than is

$$\int_0^\infty r\, |P_{nl}(r) P_{n'l'}(r)|\, dr \ . \tag{14.109}$$

Such a situation represents a destructive-interference effect in the radial portion (rather than the angular portion) of $\mathbf{S}^{1/2}$. In extreme cases, the cancellation effect may be almost 100%; this more or less completely wipes out an entire transition array. Examples have apparently been seen (or, rather, not seen!) in Ti V, where the three possible lines of the array $3p^6 - 3p^54d$ have so far proved to be unobservable,[34,35] and in Mo VII, where the $4p^6 - 4p^56d$ array is extremely weak.[35,36]

Configuration-interaction effects. Mixing of basis functions owing to configuration-interaction effects is particularly likely to produce large changes in computed line strengths. A somewhat extreme example is offered by computed values for the Si I array $3p4s - 3p5p$, for which

$$\langle 4s\|r^{(1)}\|5p\rangle = 0.70 \ . \tag{14.110}$$

Configuration-interaction-induced mixing of the 5p states with 4p basis states is computed to be very small.[37] However, the radial matrix element

[33]C. H. Corliss and R. D. Cowan, unpublished results for Dy I $4f^{10}6s^2 - 4f^{10}6s6p$, found good correlation between observed intensities and computed gf values so long as CF was greater than about 0.02, but large scatter for smaller CF.

[34]L. Å. Svensson and J. O. Ekberg, Ark. Fys. 37, 68 (1968); L. Å. Svensson, Phys. Scr. 13, 235 (1976).

[35]R. D. Cowan, J. Opt. Soc. Am. **58**, 924 (1968). The zero in the radial integral (14.108) is predicted to occur at about Sc IV and Nb VI, rather than Ti V and Mo VII, which illustrates the inaccuracies likely to occur when cancellation effects are serious. For extended treatments of this subject see L. J. Curtis and D. G. Ellis, J. Phys. B **11**, L543 (1978); L. J. Curtis, J. Opt. Soc. Am. **71**, 566 (1981).

[36]J. Reader, G. L. Epstein, and J. O. Ekberg, J. Opt. Soc. Am. 62, 273 (1972).

[37]R. D. Cowan, J. phys., Colloq. 31, C4-191 (1970).

$$\langle 4s \| r^{(1)} \| 4p \rangle = 7.41 \qquad (14.111)$$

is an order of magnitude greater than (14.110), in part because of the equality of n and n′ that gives greater overlap of the two radial wavefunctions in (14.111). As a result, an admixture of only one percent 4p into the 5p states can have a great effect: the eigenvector components are about (±0.1, 0.995), which weighted with the radial integrals (14.111) and (14.110) means a value of $S^{1/2}$ proportional to

$$\pm 0.1 \cdot 7.4 + 0.995 \cdot 0.7 \cong \pm 0.7 + 0.7 \ ,$$

and therefore a line strength S that is either zero or four times the value computed without configuration interaction, depending on the phase relations that exist for any individual transition. Obviously, computed values of S in such a case are completely untrustworthy—particularly since the value (14.110) itself involves a radial cancellation factor [magnitude of the ratio of (14.108) to (14.109)] of only 0.15.

Of course, when configuration mixing is 25 to 50 percent instead of one percent, then pronounced cancellation effects (by factors of 2 to 10 or more) may occur even when the values of the radial integrals for the interacting configurations have about the same magnitude. We point out once again that total line strength is always conserved: cancellation effects in one line are always counterbalanced by increased line strengths elsewhere. In this way, configuration-interaction effects quite commonly result in the net transfer of line strength from one transition array to another.[38]

14-16. EXPERIMENTAL MEASUREMENT OF OSCILLATOR STRENGTHS

Oscillator strengths and transition probabilities can be measured by a variety of different methods,[8,39] involving either absorption from a continuous spectrum [Eq. (14.11)], measurement of intensities of emission lines [Eq. (14.5) or (14.6)], or anomalous dispersion measurements. In the older literature, especially, measurements have been of notoriously low accuracy, in large part because of the difficulty of determining the required values of the atom densities N_i or N_j in the pertinent quantum state; a well-known example is an order of magnitude error in oscillator strengths of Fe I lines, which led to a similar error in the inferred concentration of iron in the solar photosphere. Measurements made since about 1960 tend to be more reliable because of improvements in experimental technique, such as the development of shock-tube sources and wall-stabilized arcs having well-known density and temperature characteristics.[40] Using such methods, improved values of Fe I oscillator strengths have been obtained, which have removed the discrepancy between the

[38] See, for example, ref. 37 and M. Aymar, Nucl. Instrum. Methods 110, 211 (1973).

[39] W. L. Wiese, in *Methods of Experimental Physics* (Academic Press, New York, 1968), 7A, 117.

[40] R. A. Hartunian, in *Methods of Experimental Physics* (Academic Press, New York, 1968), 7B, 141; W. L. Wiese, *ibid.*, 7B, 307.

calculated concentration of iron in the solar photosphere and that in the corona (the latter derived from theoretical transition probabilities for ions such as Fe XIII).[41]

Useful information on transition probabilities can also be obtained by measurements on radiative lifetimes. For example, atoms can be excited to some state j by a mechanism that can suddenly be turned off—e.g., by optical or pulsed-electron-gun excitation.[8,39] After the excitation is cut off, the intensity of each emitted spectrum line originating in the level j decays exponentially, from (14.6) and (14.9), according to

$$I(t) \propto e^{-t/\tau} \ . \tag{14.112}$$

Observation of this intensity decay by means of short-time-constant (nanosecond or faster) detection equipment gives directly an experimental value of the lifetime τ. In cases in which a transition is possible to only one lower level i (or at least in which transition probabilities to all other lower levels are small), this gives from (14.10) the radiative transition probability A_{ji}, provided the physical conditions of the source are such that collisional excitations and de-excitations have a negligible effect on the measured lifetime. Even when several transitions are possible with comparable probabilities, the absolute transition probability for each can be found if relative values ("branching ratios") are known from other experimental measurements or from theoretical calculations.

All the above experimental methods are limited largely to spectrum lines from neutral and one- or two-fold ionized atoms. Lifetime measurements for much higher ionization stages can be made by beam-foil spectroscopy.[42] A well-collimated beam of ions of accurately-known energy (of a few Mev) from a Van de Graaff or other linear accelerator is passed through a thin carbon foil. The foil further ionizes and excites the atoms of the beam without slowing them appreciably. Measurement of spectrum line intensities as a function of distance downstream from the foil, together with the known beam velocity, then provides lifetimes of the upper states involved. The major limitation of this method arises from the very low intensity of the ion-beam spectrum, which necessitates use of a very fast (and hence low-resolution) spectrograph; thus only relatively simple spectra can be examined. Also, it is usually necessary to make sizable (and somewhat uncertain) corrections for repopulation of the upper levels of interest by radiative cascading downward from still higher excited levels. Finally, practical considerations[43] limit the feasibility of lifetime measurements to the range $10^{-5} > \tau > 10^{-12}$ sec, considerably short of what is needed for strong lines of even twenty-fold ionized atoms (Sec. 14-17). Nevertheless, beam-foil experiments have provided a great number of fairly accurate

[41]J. M. Bridges and W. L. Wiese, Astrophys. J. 161, L71 (1970); S. J. Wolnik, R. O. Berthel, and G. W. Wares, Astrophys. J. 162, 1037 (1970) and 166, L31 (1971); J. M. Bridges and R. L. Kornblith, Astrophys. J. 192, 793 (1974).

[42]S. Bashkin, ed., *Beam-Foil Spectroscopy* (Gordon and Breach, New York, 1968), 2 vols. See also Nucl. Instrum. Methods 28, 88 (1964), 90 (1970), and 110 (1973); S. Bashkin, ed., *Beam-Foil Spectroscopy* (Springer-Verlag, Berlin, 1976); I. A. Sellin and D. J. Pegg, *Beam-Foil Spectroscopy* (Plenum, New York, 1976), 2 vols; D. J. Pegg, in *Methods of Experimental Physics* (Academic Press, New-York, 1980), 17, 529.

[43]J. Bromander, Nucl. Instrum. Methods 110, 11 (1973); S. L. Varghese, C. L. Cocke, and B. Curnutte, Phys. Rev. A 14, 1729 (1976); D. J. Pegg, ref. 42.

lifetimes, which have either given transition probabilities directly, or have made possible accurate absolute normalization of relative values obtained by other means.

Wiese and co-workers have tabulated both experimental and theoretical oscillator strengths and transition probabilities for light elements,[44-46] and have published an extensive bibliography on all elements.[47] Few of the available data are accurate to better than 10%, and most are probably in error by 25-50% or more. The general status of the available information is illustrated in Fig. 14-3; in spite of the very great increase in activity in this field during recent years,[47] our knowledge is still very inadequate in both quantity and quality.

*14-17. SYSTEMATICS OF OSCILLATOR STRENGTHS

The low accuracy of, and gaps in, available data have until recently made it difficult to recognize any systematic trends in oscillator strengths. More recently, especially as a result of data on higher ionization stages provided by beam-foil methods, it has become possible to verify theoretically predicted variations with n in Rydberg series, and with Z along isoelectronic sequences.[45,48]

Variation with n. Numerical calculations show that for a running electron $n_j l_j$ and large n_j,

$$\langle n_i l_i \| r^{(1)} \| n_j l_j \rangle^2 \propto (n_j^*)^{-3} , \tag{14.113}$$

analogously to the results for other radial integrals discussed in Sec. 8-5. Thus we may expect that for a Rydberg series of spectrum lines,

$$f_l(n_i l_i \gamma JM - n_j l_j \gamma' J') \propto (n_j^*)^{-3} . \tag{14.114}$$

However, this relation becomes reasonably accurate only at considerably larger n_j^* than was true of the energy integrals discussed in Sec. 8-5, for several reasons:

(a) the wavenumber σ does not become nearly constant till n_j is large;

(b) changes in coupling conditions for small n_j (cf. Fig. 4-9 and Figs. 10-3 to 10-5) in general shift oscillator strength from one line of an array to others, in going up a Rydberg series;

[44]W. L. Wiese, M. W. Smith, and B. M. Glennon, *Atomic Transition Probabilities, Vol. I Hydrogen through Neon,* NSRDS-NBS 4 (U.S. Govt. Printing Off., Washington, D.C., 1966).

[45]W. L. Wiese, M. W. Smith, and B. M. Miles, *Atomic Transition Probabilities, Vol. II Sodium through Calcium,* NSRDS-NBS 22 (U.S. Govt. Printing Off., Washington, D.C., 1969).

[46]B. M. Miles and W. L. Wiese, At. Data 1, 1 (1969); Ba I and II. W. L. Wiese and J. R. Fuhr, J. Phys. Chem. Ref. Data 4, 263 (1975); Sc and Ti. See also additional references in Sec. XI of the bibliography.

[47]J. R. Fuhr, B. J. Miller, and G. A. Martin, *Bibliography on Atomic Transition Probabilities (1914 through October 1977),* NBS Spec. Publ. 505 (U.S. Govt. Printing Off., Washington, D.C., 1978); also Suppl. 1 through March 1980.

[48]M. W. Smith, G. A. Martin, and W. L. Wiese, Nucl. Instrum. Methods 110, 219 (1973).

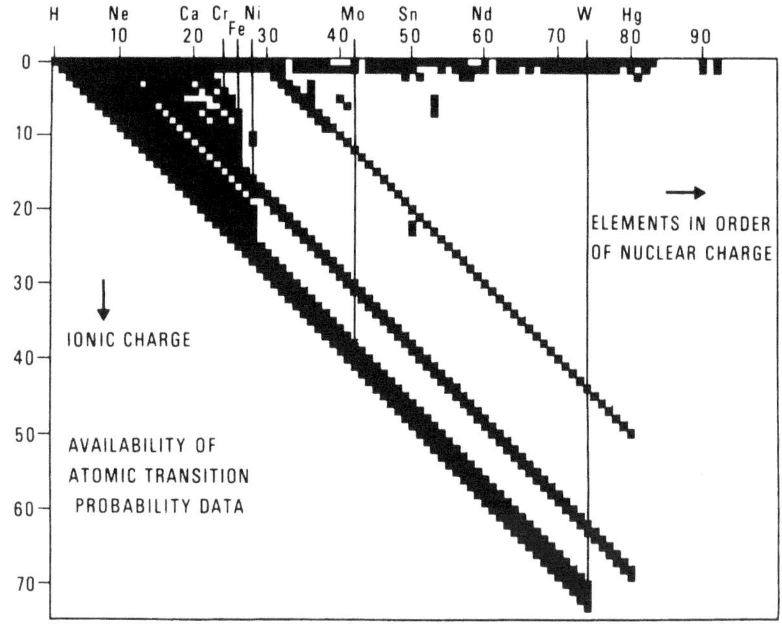

Fig. 14-3. Availability of experimental and theoretical information on atomic transition probabilities. (Courtesy of W. L. Wiese, 1978).

(c) unlike the Coulomb and spin-orbit radial integrals, which are always positive, the radial dipole integral involved in (14.113) can be either positive or negative, and not infrequently changes sign partway up a series; examples in Ti V and Mo VII were mentioned in Sec. 14-15.

For hydrogenic ions, where the complications (b) and (c) above are not present, the dependence (14.114) is shown explicitly by the approximate expression (14.99).

Variation with Z. We consider now variations along an isoelectronic sequence for lines of a given transition array. To the extent that radial wavefunctions in a highly ionized atom are hydrogenic in form, it follows from the footnote to Table 3-1 that the radial transition integral may be written

$$\int_0^\infty Z_c P_i^H(Z_c r)\, r\, P_j^H(Z_c r)\, dr = Z_c^{-1} \int_0^\infty P_i^H(r)\, r\, P_j^H(r)\, dr \ ,$$

where Z_c is the effective core charge seen by the jumping electron, so that

$$P_{ij}^2 = \langle n_i l_i \| r^{(1)} \| n_j l_j \rangle^2 \propto Z_c^{-2} \ . \qquad (14.115)$$

Since $f_{ij} \propto \sigma_{ij} P_{ij}^2$, and since energies are proportional to Z_c^2 from (3.27), it follows that to a first approximation[49]

$$\sigma \propto Z_c^2, \qquad f = \text{constant}, \qquad A \propto \sigma^3 Z_c^{-2} \propto Z_c^4 \qquad (\Delta n \neq 0) \; ; \tag{14.116}$$

from (14.39), it follows that A may be of the order of 10^{14} $\sec^{-1}$ for $\sim$ 30-fold ionized atoms. To a higher order of approximation, (including core penetration and correlation, but not relativistic effects—cf. Chapter 19), it may be shown that one-electron binding energies can be written as a series expansion in powers of Z_c:

$$E = E_0 Z_c^2 + E_1 Z_c + E_2 + E_3 Z_c^{-1} + \cdots , \tag{14.117}$$

where $E_0 = -1/n^2$ is the hydrogenic term, and that f is of the form

$$f = f_0 + f_1 Z_c^{-1} + f_2 Z_c^{-2} + \cdots , \tag{14.118}$$

where f_0 is the hydrogenic oscillator strength.[5] For $\Delta n = 0$ transitions, the leading term (14.117) contributes nothing to σ, and $f_0 = 0$. Thus[49]

$$\sigma \propto Z_c, \qquad f \propto Z_c^{-1}, \qquad A \propto \sigma^3 Z_c^{-2} \propto Z_c \qquad (\Delta n = 0) \; . \tag{14.119}$$

Examples in the Li I isoelectronic sequence are shown in Fig. 14-4. At small Z_c, the $\Delta n \neq 0$ oscillator strength departs greatly from the hydrogenic limit, primarily as a result of non-hydrogenic changes in the relative forms of the s and p radial wavefunctions that produce almost complete cancellation in the radial dipole integral (14.108) for Li I. This illustrates that the Z-expansion formula (14.118) is often not very useful near the beginning of an isoelectronic sequence; this is particularly true when the neutral member is an element near the beginning of a transition or rare-earth series, where violent radial-wavefunction changes may occur upon ionization (cf. Sec. 8-6).

Relativistic effects have been neglected in (14.115)-(14.119). At large Z_c, relativistic energy-level shifts strongly affect the value of σ, expecially for $\Delta n = 0$ transitions, with corresponding effects on f values, as shown in the figure. Additional effects arise from relativistic modifications in the wavefunctions, which in the present cases reduce the magnitude of the radial dipole integral and thereby decrease the f values (except for the $\Delta n = 0$ transitions at very large Z_c, where the increase in the relativistic energy effect more than offsets the effect of the radial integral). It is unlikely that the magnitude of the calculated relativistic effects shown in the figure will ever be checked by direct experimental measurement of oscillator strengths, because experimental difficulties increase even faster with Z_c. The effects themselves, however, are of more than academic interest: for example, 2s − 2p optical transitions in highly ionized Fe have been observed in both

[49]This assumes either LS-allowed transitions or large departures from pure LS coupling. For intercombination lines under nearly pure LS-coupling conditions, the powers of Z_c are greater by 6: R. H. Garstang, Astron. J. 79, 1260 (1974). For discussions of irregularities resulting from level crossings of the sort depicted in Fig. 10-2, see A. W. Weiss in I. A. Sellin and D. J. Pegg, eds., *Beam-Foil Spectroscopy* (Plenum, New York, 1976), p. 51, and C. Froese Fischer, Phys. Rev. A 22, 551 (1980).

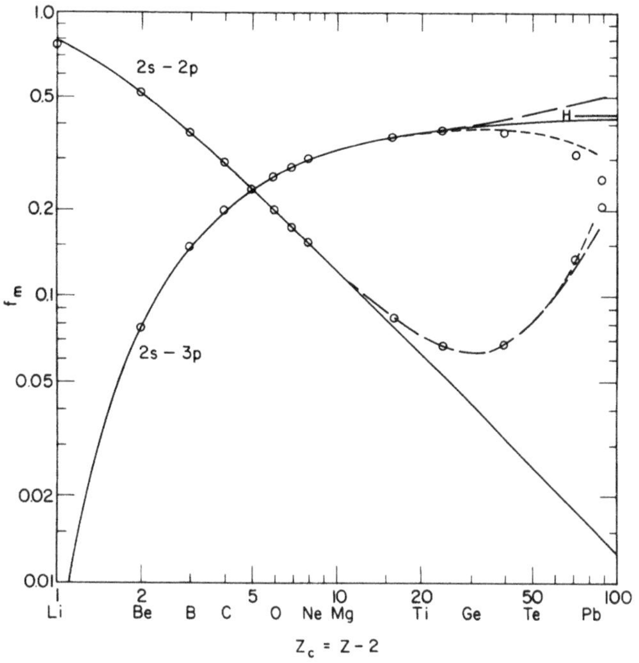

Fig. 14-4. Hartree-Fock multiplet oscillator strengths, $f(s_{1/2} - p_{1/2}) + f(s_{1/2} - p_{3/2})$, for resonance transitions in the Li I isoelectronic sequence: solid curves, non-relativistic; dashed curves, including relativistic energy-level shifts; dotted curves, including also relativistic (HFR) modifications in the radial wavefunctions. The horizontal line marked H indicates the non-relativistic hydrogenic 2s − 3p limit. The circles represent fully relativistic DHF (length-approximation) calculations of L. Armstrong, Jr., W. R. Fielder, and D. L. Lin, Phys. Rev. A **14**, 1114 (1976).

laboratory[50] and solar[51] spectra, and oscillator strengths are of great importance for solar physics diagnostics; also, inner-shell (x-ray) transitions have of course, been observed in (singly ionized) elements of all Z.

Relativistic modifications to oscillator strengths and to the Thomas-Reiche-Kuhn sum rule (14.104) are discussed by Garstang.[52]

[50]U. Feldman, G. A. Doschek, R. D. Cowan, and L. Cohen, Astrophys. J. **196**, 613 (1975).

[51]B. C. Fawcett and R. D. Cowan, Mon. Not. R. Astron. Soc. **171**, 1 (1975).

[52]R. H. Garstang, in W. E. Brittin and H. Odabasi, eds., *Topics in Modern Physics* (Colorado Associated University Press, Boulder, 1971).

*14-18. RADIATIVE DECAY IN LOW-DENSITY PLASMAS

In sources of sufficiently low density, collisional excitation and de-excitation rates are sufficiently low that highly excited atoms decay only by emission of radiation. For most excited states, selection-rule restrictions are such that the ground state cannot be reached by a single transition, but only by a sequence of several successive jumps. The qualitative nature of these radiative cascades can be seen by studying relative transition probabilities in hydrogenic ions—which apply approximately also to highly excited states of multi-electron ions:

The solid curves in Fig. 14-5 give radiative decay probabilities from states $n_j l_j$ with $n_j = 20$ to configurations $n_i l_i$ with $l_i = l_j - 1$. [As discussed in connection with (14.105), decay probabilities for $l_i = l_j + 1$ are much lower.] It is evident that for small l_j, decay is most probable directly to the lowest possible configuration, $n_i = l_i + 1$. For large l_j, however, decay is most probable to a state of large n_i—being, in fact, restricted by selection rules to $n_i = n_j - 1$ in the limiting case $l_j = n_j - 1$. The highest solid curve and the two dashed curves illustrate a typical cascade decay path. From 20m ($n - l = 11$), the most probable transition is to $14l$ ($n - l = 6$); from $14l$, decay is most probable to 11k ($n - l = 4$); from there to 9i ($n - l = 3$); etc. Thus the decay tends toward states with $n - l = 1$, after which the chain necessarily continues in steps of $\Delta n = \Delta l = -1$.

With increasing n_j of the initial state, radiative decay rates decrease but collisional rates increase (because of the larger orbit size and hence larger collision cross-section). For some appropriately large n_j, collisionally induced $\Delta n \neq 0$ transitions are still improbable compared with radiative decay, but collisionally produced $\Delta n = 0$ transitions are much more probable. Collisions then maintain a statistical $(2l_j + 1)$ distribution among all ions in configurations $n_j l_j$ (fixed n_j). The mean rate of decay to states $n_i l_i$ (any l_i) is then given by summing the array transition probability over all l_i and averaging over l_j. From (14.38), (4.57), and (14.99), we obtain the approximate hydrogenic expression

$$\bar{A}(n_j - n_i) \equiv \frac{\sum_{l_j}(2l_j + 1)\sum_{l_i} A(n_j l_j, n_i l_i)}{\sum_{l_j}(2l_j + 1)} \tag{14.120}$$

$$= \frac{8\pi^2 e^2 \bar{\sigma}^2}{mc} \frac{n_i^2}{n_j^2} \bar{f}(n_i - n_j)$$

$$= \frac{1.57 \cdot 10^{10} Z_c^4}{n_i^3 n_j^5} \left(\frac{1}{n_i^2} - \frac{1}{n_j^2} \right)^{-1} \quad \text{sec}^{-1}$$

$$= \frac{1.57 \cdot 10^{10} Z_c^4}{n_i n_j^3 (n_j^2 - n_i^2)} \quad \text{sec}^{-1} . \tag{14.121}$$

This result is illustrated by the dash-dot curve in Fig. 14-5 for the case $n_j = 20$, $n_i = n$. We see that transitions are most probable either directly to very small n_i, or to very large n_i (from which further decays proceed); the former represent small-l_j transitions (which have

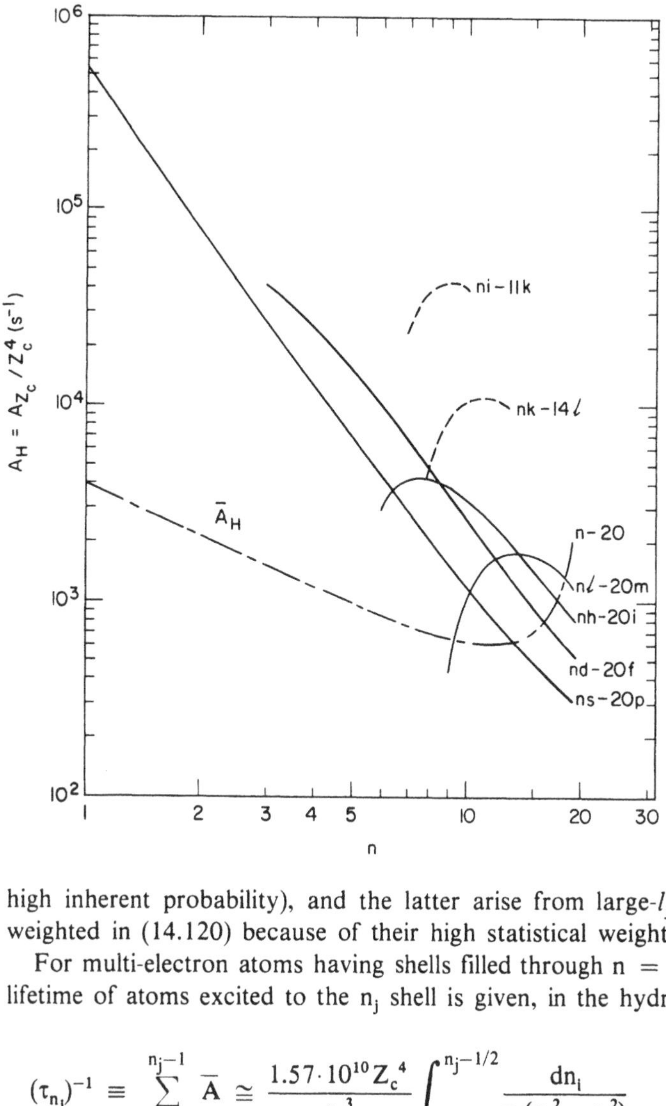

Fig. 14-5. Non-relativistic hydrogenic array transition probabilities for the indicated transitions $nl - n'l'$. Dash-dot curve: total transition probability for $n' = 20$, summed over l and averaged over l'.

high inherent probability), and the latter arise from large-l_j states [which are strongly weighted in (14.120) because of their high statistical weights].[53]

For multi-electron atoms having shells filled through $n = n_o - 1$, the mean radiative lifetime of atoms excited to the n_j shell is given, in the hydrogenic approximation, by

$$(\tau_{n_j})^{-1} \equiv \sum_{n_i=n_o}^{n_j-1} \bar{A} \cong \frac{1.57\cdot 10^{10} Z_c^4}{n_j^3} \int_{n_o-1/2}^{n_j-1/2} \frac{dn_i}{n_i(n_j^2 - n_i^2)}$$

$$= \frac{1.57\cdot 10^{10} Z_c^4}{2n_j^5}\, ln\left[\frac{(n_j - 1/2)^2}{(n_o - 1/2)^2}\left(\frac{n_j^2 - (n_o - 1/2)^2}{n_j - 1/4}\right)\right] \qquad (14.122)$$

$$\cong \frac{1.57\cdot 10^{10} Z_c^4}{2n_j^5}\, ln\left[\frac{n_j^3}{(n_o - 1/2)^2}\right] \ \sec^{-1} \qquad (n_o \ll n_j)\ . \qquad (14.123)$$

[53]Radio-frequency $\Delta n = 1$ transitions in H I have been observed in the upper solar corona ($n_e \cong 10^4$ cm^{-3}) for n as large as 253—see L. Goldberg, in K. B. Gebbie, ed., *The Menzel Symposium on Solar Physics, Atomic Spectra, and Gaseous Nebulae*, NBS Spec. Publ. 353 (U.S. Govt. Printing Off., Washington, D.C., 1971).

15

RADIATIVE TRANSITIONS (M1 AND E2)

The vast majority of observed spectrum lines arise from electric dipole transitions. However, magnetic and higher-order-electric transitions can be observed under special circumstances; they are observed mainly in emission, and correspondingly are described quantitatively in terms of the transition probability rates a_{Jl} and A_{Jl}. We shall discuss only magnetic dipole and electric quadrupole emission, but higher-order transitions are sometimes of interest in highly ionized atoms.[1,2]

15-1. MAGNETIC DIPOLE RADIATION (M1)

Transition probabilities for magnetic dipole radiation are given by exactly the same expression as for electric dipole radiation, except that the electric dipole moment operator $-e\sum r_i^{(1)} = -ea_0 P^{(1)}$ is to be replaced by the magnetic dipole moment operator

$$\begin{aligned} \mu^{(1)} &= -\mu_0 \sum_i [l_i^{(1)} + g_s s_i^{(1)}] \\ &= -\mu_0[\mathbf{L}^{(1)} + g_s \mathbf{S}^{(1)}] \\ &= -\mu_0[\mathbf{J}^{(1)} + (g_s - 1)\mathbf{S}^{(1)}] \ , \end{aligned} \tag{15.1}$$

where

$$g_s \cong 2.0023192 \tag{15.2}$$

is the anomalous gyromagnetic ratio for electron spin. The Bohr magneton may be written

$$\mu_0 = \frac{eh}{2mc} = e\left(\frac{h^2}{me^2}\right)\left(\frac{e^2}{2hc}\right) = ea_0\frac{\alpha}{2} \ , \tag{15.3}$$

[1] R. H. Garstang, in C. de Jaeger, ed., *Highlights of Astronomy* (D. Reidel Publ. Co., Dordrecht, Holland, 1971).

[2] Tensor operators for higher-order transitions are given by B. W. Shore and D. H. Menzel, *Principles of Atomic Spectra* (John Wiley, New York, 1968), p. 440.

where $\alpha \cong 1/137.036$ is the fine structure constant.

For simplicity, we shall henceforth use the approximation $g_s = 2$. By comparison with (14.23), we then have for pure magnetic dipole transitions

$$a_{M1} = \frac{64\pi^4 e^2 a_o^2(\alpha/2)^2\sigma^3}{3h} \sum_q |\langle \gamma JM | J_q^{(1)} + S_q^{(1)} | \gamma' J' M' \rangle|^2$$

$$= \frac{64\pi^4 e^2 a_o^2(\alpha/2)^2\sigma^3}{3h} \sum_q \begin{pmatrix} J & 1 & J' \\ -M & q & M' \end{pmatrix}^2 |\langle \gamma J \| J^{(1)} + S^{(1)} \| \gamma' J' \rangle|^2 \, , \tag{15.4}$$

$$gA_{M1} \equiv (2J' + 1)\sum_M a_{M1} = \frac{64\pi^4 e^2 a_o^2(\alpha/2)^2\sigma^3}{3h} |\langle \gamma J \| J^{(1)} + S^{(1)} \| \gamma' J' \rangle|^2$$

$$= 2.6973 \cdot 10^{-11} \sigma^3 |\langle \gamma J \| J^{(1)} + S^{(1)} \| \gamma' J' \rangle|^2 \quad \text{sec}^{-1} \, , \tag{15.5}$$

where the numerical coefficient applies for σ in kaysers. Thus the total transition probability rate A_{M1} from the upper state $\gamma' J' M'$ to all states of the level γJ is independent of M'. Selection rules for J and M, and relative strengths of line components are exactly the same for magnetic dipole as for electric dipole radiation, because the 3-j symbol in (15.4) is identical with that in (14.30). Viewed from a position on the +z axis, the atom still radiates light of zero intensity for $q = M - M' = 0$, and light circularly polarized clockwise or counterclockwise for $q = \Delta M = \pm 1$, respectively. However, viewed in a direction perpendicular to the z-axis, the magnetic field vector $\mathfrak{H}$ here takes the place of the electric vector $\mathfrak{E}$ in the electric dipole case; since $\mathfrak{H}$ is perpendicular to $\mathfrak{E}$ it follows that polarizations are just opposite to those for electric dipole radiation—namely, linearly polarized perpendicular and parallel to the z-axis for $q = \Delta M = 0$ and for $q = \Delta M = \pm 1$, respectively.

The matrix element in (15.4) and (15.5) is evaluated by making the usual expansions in pure-coupling basis functions:

$$\langle \gamma J \| J^{(1)} + S^{(1)} \| \gamma' J' \rangle = \sum_\beta \sum_{\beta'} y^{\gamma}_{\beta J} \langle \beta J \| J^{(1)} + S^{(1)} \| \beta' J' \rangle y^{\gamma'}_{\beta' J'} \, . \tag{15.6}$$

Because the basis functions are eigenfunctions of the operator $J^{(1)}$ and because $S^{(1)}$ does not involve radii, integration over the radial variables gives delta-factors in all quantum numbers $n_i l_i$—that is, the magnetic dipole matrix is diagonal in the configuration. (Unlike electric dipole transitions, there is no change in parity.)

The reduced matrix element for pure basis functions is easily evaluated in the LS representation by using (11.39) [for the $S^{(1)}$ operator only] and (11.20) [for both $J^{(1)}$ and $S^{(1)}$]:

$$\langle \alpha LSJ \| J^{(1)} + S^{(1)} \| \alpha' L' S' J' \rangle = \delta_{bb'} [J(J+1)(2J+1)]^{1/2}$$

$$+ \delta_{\alpha LS, \alpha' L' S'} (-1)^{L+S+J+1} [J, J']^{1/2} \begin{Bmatrix} L & S & J \\ 1 & J' & S \end{Bmatrix} [S(S+1)(2S+1)]^{1/2} \, ; \tag{15.7}$$

here α represents all quantum numbers other than LSJ, and b ≡ αLSJ (i.e., *all* quantum numbers). Thus, pure LS-coupling matrix elements are non-zero only between two levels of the *same* LS term (of the *same* configuration), differing at most in the values of J. At first glance, a magnetic dipole transition with $\Delta J = 0$ would appear to be a transition from one level to the *same* level, and hence to be no transition at all. However, this does not mean that transitions with $\Delta J = 0$ are impossible in real atoms; so long as two different levels of the same J ($\neq 0$) both contain non-zero amounts of some common basis state αLSJ (by virtue of departures from pure LS coupling), a transition between the two levels can from (15.6) have non-zero probability. Similarly, transitions between levels nominally belonging to different configurations may become possible as a consequence of configuration-interaction mixings. Examples will be discussed in Sec. 15-4.

Where details of intermediate-coupling and configuration-interaction mixings are not important, theoretical magnetic dipole transition probabilities can be calculated with rather high accuracy because no radial integral is involved.

Transition probability magnitudes. For neutral atoms, the magnetic dipole matrix element in (15.4) has about the same magnitude as that of the electric dipole element in (14.23); it follows that

$$A_{M1}/A_{E1} \cong (\alpha/2)^2(\sigma_{M1}/\sigma_{E1})^3 = 1.3\cdot 10^{-5}(\sigma_{M1}/\sigma_{E1})^3 , \tag{15.8}$$

so that magnetic dipole transitions are about five orders of magnitude less probable than electric dipole transitions of the same wavelength. Along an isoelectronic sequence, the M1 matrix element remains constant (rather than decreasing like Z_c^{-1}, as does the E1 matrix element). However, M1 radiation is generally observed only as intra-configuration transitions; therefore, for a specific transition,[3]

$$A_{M1} \propto \sigma^3 \propto Z_c^3 , \qquad A_{M1}/A_{E1} \propto Z_c^2 , \qquad \Delta n = 0 , \tag{15.9}$$

if the energy-level separation is primarily due to electron-electron Coulomb interactions [cf. (14.119)], and

$$A_{M1} \propto \sigma^3 \propto Z_c^{12} \tag{15.10}$$

if primarily due to spin-orbit effects [cf. (3.52)]. Thus magnetic dipole transition probabilities increase considerably faster with Z_c than do electric dipole probabilities of comparable wavelength.

Problem

15-1(1). Show that the magnetic dipole transition probability between the levels $s^2p\ {}^2P^o_{1/2}$ and $p^3\ {}^2P^o_{1/2}$ is identically zero even when CI between s^2p and p^3 is taken into account. [Hint: the CI eigenvectors are of the form (10.37).] Why does the $s^2p\ {}^2P^o_{3/2} - p^3\ {}^2P^o_{3/2}$ transition have non-zero probability?

[3]This assumes large departures from pure LS coupling. When coupling conditions are close to LS, the Z dependence of the mixing of different LS terms results in A varying by an additional factor of Z_c^6; R. H. Garstang, Astron. J. 79, 1260 (1974).

15-2. ELECTRIC QUADRUPOLE RADIATION (E2)

Analogously to the electric dipole moment operator $P_q^{(1)} = \sum_i r_i C_q^{(1)}(i)$ [units of $-ea_o$], the electric quadrupole operator is[2,4]

$$P_q^{(2)} = \sum_i r_i^2 C_q^{(2)}(i) \qquad [\text{units of } -ea_o^2] \ . \qquad (15.11)$$

Because this operator has even parity, electric quadrupole transitions (like magnetic dipole ones) involve no change in parity.

The transition probability rate for pure electric quadrupole transitions is,[2] analogously to (14.23) with an only slightly different coefficient,

$$a_{E2} = \frac{64\pi^6 e^2 a_o^4 \sigma^5}{15h} \sum_q |\langle \gamma JM | P_q^{(2)} | \gamma' J' M' \rangle|^2$$

$$= \frac{64\pi^6 e^2 a_o^4 \sigma^5}{15h} \sum_q \begin{pmatrix} J & 2 & J' \\ -M & q & M' \end{pmatrix}^2 |\langle \gamma J \| P^{(2)} \| \gamma' J' \rangle|^2 \ , \qquad (15.12)$$

$$gA_{E2} = (2J' + 1) \sum_M a_{E2} = \frac{64\pi^6 e^2 a_o^4 \sigma^5}{15h} |\langle \gamma J \| P^{(2)} \| \gamma' J' \rangle|^2$$

$$= 1.1199 \cdot 10^{-22} \sigma^5 |\langle \gamma J \| P^{(2)} \| \gamma' J' \rangle|^2 \quad \sec^{-1} \ , \qquad (15.13)$$

where the numerical coefficient applies for σ in kaysers. As for electric and magnetic dipole radiations, the total transition probability rate from the state $\gamma'J'M'$ to all states of the level γJ is independent of M'.

From the 3-j symbol, the selection rules on J are

$$\Delta J = 0, \pm 1, \pm 2, \qquad J + J' \geqq 2 \ . \qquad (15.14)$$

The selection rules on M, and corresponding polarizations for viewing at an angle θ from the z-axis, are as given in Table 15-1. The polarizations and qualitative intensity variations can be understood on the basis of a rough classical picture similar to that used in Fig. 14-1 for electric dipole radiation. In Fig. 15-1 we consider a classical quadrupole consisting of a stationary positive point charge of two units, together with two unit negative point charges; these negative charges oscillate or rotate with SHM in such a way that they are always diametrically opposite each other and equidistant from the positive charge, so that the instantaneous dipole moment is always zero. Rectilinear motion along the z-axis (part a of the figure) corresponds to $q = \Delta M = 0$, and radiates light linearly polarized parallel to the z-axis; the intensity is zero if viewed at $\theta = 0$ because a point charge does not radiate in the direction of its acceleration, and is zero at $\theta = 90°$ because radiations from the two negative charges cancel each other.

[4]The operator (15.11) is $(3/2)^{1/2}$ times the corresponding dyadic operator. Thus line strengths defined as the square of the reduced matrix element of $P^{(2)}$ are 3/2 times the line strengths tabulated in the older literature.

TABLE 15-1. SELECTION RULES AND POLARIZATION FOR ELECTRIC QUADRUPOLE RADIATION

		Polarization		
ΔM	q	$\theta = 0$	$0 < \theta < 90°$	$\theta = 90°$
$M = M'$	0	(zero intensity)	∥ to z-axis	(zero intensity)
$M = M' \pm 1$	± 1	circular	elliptical	∥ to z-axis
$M = M' \pm 2$	± 2	(zero intensity)	elliptical	⊥ to z-axis

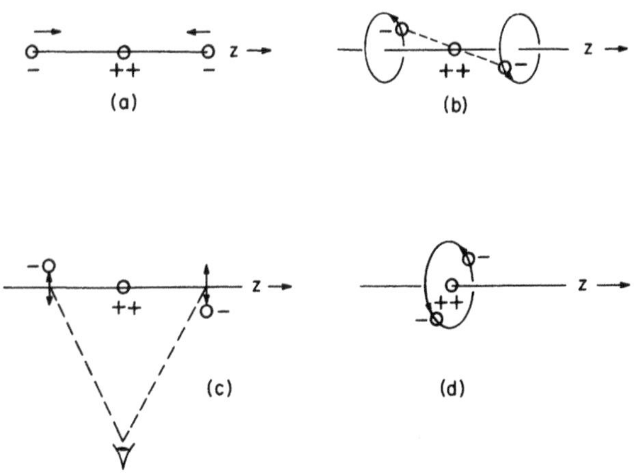

Fig. 15-1. Classical simple harmonic quadrupole motions analogous to quantum-mechanical transitions with (a) $\Delta M = 0$, (b) and (c) $\Delta M = \pm 1$, and (d) $\Delta M = \pm 2$.

Motion of the negative charges in circles parallel to the xy plane and centered symmetrically on the $\pm z$ axis (part b) corresponds to $q = \pm 1$ (each negative charge passes any given point in its circle once per revolution, corresponding to the factor $e^{\pm i\phi}$ in $Y_{2,\pm 1}$, Table 2-1); polarization of radiation corresponding to the direction of electron rotation shown in the figure pertains to $M' > M$ (negative q) if M' refers to the upper energy state. Viewed at $\theta = 90°$ from a point in the plane of the paper, components of the circular motion perpendicular to the paper cancel exactly, whereas the components in the plane of the paper (part c) give a net oscillation parallel to the z-axis.

Motion of the negative charges at opposite ends of the diameter of a circle in the xy plane centered at the origin (part d) corresponds to $q = \pm 2$ (a negative charge passes any given point in the circle twice per revolution of the system, corresponding to the factor $e^{\pm 2i\phi}$ in $Y_{2,\pm 2}$). Viewed at $\theta = 0$, the effects of the two negative charges cancel exactly; for $\theta > 0$, the radiation is elliptically polarized, except at $\theta = 90°$, where the ellipse degenerates into linear polarization perpendicular to the z-axis.

Evaluation of the reduced matrix element in (15.13) begins as usual with basis-function expansions of the wavefunctions,

$$\langle \gamma J \| \boldsymbol{P}^{(2)} \| \gamma' J' \rangle = \sum_{\beta} \sum_{\beta'} y^{\gamma}_{\beta J} \langle \beta J \| \boldsymbol{P}^{(2)} \| \beta' J' \rangle y^{\gamma'}_{\beta' J'} \; . \tag{15.15}$$

Calculation of the basis-function matrix element must be carried out for two different cases.

Inter-configuration transitions. If $|\beta J\rangle$ and $|\beta' J'\rangle$ belong to two different configurations, the calculation is basically the same as that for the electric dipole case. The most general type of transition is again (14.77), except that $l_i + l_j$ must be even if the two configurations are to have the same parity. For the $\mathfrak{LS}$-coupling representation (12.1), the results are again (14.78)-(14.86), except that the value of t is everywhere 2 instead of 1. Selection rules are modified accordingly; for example, the selection rules on total orbital and spin angular momenta are

$$\Delta L = 0, \pm 1, \pm 2, \qquad L + L' \geqq 2 \tag{15.16}$$

and

$$\Delta S = 0 . \tag{15.17}$$

In place of the one-electron radial dipole matrix element (14.55), we have the quadrupole reduced matrix element[4]

$$\begin{aligned} P^{(2)}_{nl,n'l'} &\equiv \langle nl\|r^2 C^{(2)}\|n'l'\rangle \\ &= \langle l\|C^{(2)}\|l'\rangle \int_0^\infty P_{nl}(r)\, r^2\, P_{n'l'}(r)\, dr \\ &= (-1)^l [l, l']^{1/2} \begin{pmatrix} l & 2 & l' \\ 0 & 0 & 0 \end{pmatrix} \int_0^\infty P_{nl}(r)\, r^2\, P_{n'l'}(r)\, dr . \end{aligned} \tag{15.18}$$

[Reduction of the 3-j symbol to an algebraic expression analogous to (14.57) proves to give a result too complex to be worthwhile.] The properties of the 3-j symbol lead to the selection rule

$$\Delta l = 0, \pm 2 , \qquad l + l' \geqq 2 ; \tag{15.19}$$

thus s—s′ transitions are not allowed.

Intra-configuration transitions. If Δl is zero and in addition Δn is also zero, then (as for magnetic dipole radiation) there is no change in configuration at all. The evaluation of basis-function matrix elements is then a single-configuration problem quite different from that just discussed. From (9.80) we may write (using subscripts a and n to denote matrix elements for basis functions that are and are not antisymmetrized for electrons in different subshells)

$$\begin{aligned} \langle \beta J\| \sum_{i=1}^{N} r_i^2 C_i^{(2)} \|\beta' J'\rangle_a &= \sum_{j=1}^{q} w_j \langle \beta J\| r_j^2 C_j^{(2)} \|\beta' J'\rangle_n \\ &= \sum_{j=1}^{q} w_j \langle \beta J\| u_j^{(2)} \|\beta' J'\rangle_n \langle l_j\| r^2 C^{(2)} \|l_j\rangle = \sum_{j=1}^{q} \langle \beta J\| U_j^{(2)} \|\beta' J'\rangle_n P^{(2)}_{jj} ; \end{aligned} \tag{15.20}$$

$U_j^{(2)} \equiv \sum_i u_i^{(2)}$ is the sum of one-electron unit tensor operators defined in (11.52), with the sum extending over all (spatial) coordinates i of the w_j electrons in the subshell $l_j^{w_j}$. For our usual type of $\mathfrak{L}\mathfrak{S}$-coupled basis functions, we may write (15.20) with the aid of (11.38) as

$$\langle \cdots \mathfrak{L}\mathfrak{S}\mathfrak{J} \| \boldsymbol{P}^{(2)} \| \cdots \mathfrak{L}'\mathfrak{S}'\mathfrak{J}' \rangle = \delta_{\text{all spins}} (-1)^{\mathfrak{L}+\mathfrak{S}+\mathfrak{J}'} [\mathfrak{J}, \mathfrak{J}']^{1/2} \times \begin{Bmatrix} \mathfrak{L} & \mathfrak{S} & \mathfrak{J} \\ \mathfrak{J}' & 2 & \mathfrak{L}' \end{Bmatrix} \sum_{j=1}^{q} \langle \cdots \mathfrak{L} \| U_j^{(2)} \| \cdots \mathfrak{L}' \rangle P_{jj}^{(2)} . \tag{15.21}$$

The remaining matrix element may be simplified exactly as was done in the evaluation of (12.28) to give

$$\begin{aligned} &\langle \cdots \mathfrak{L}_q \| U_j^{(2)} \| \cdots \mathfrak{L}_q' \rangle \\ &= \left[\prod_{m=j+1}^{q} \delta_{\alpha_m L_m, \alpha_m' L_m'} (-1)^{\mathfrak{L}_{m-1} + L_m + \mathfrak{L}_m'} [\mathfrak{L}_m, \mathfrak{L}_m']^{1/2} \begin{Bmatrix} \mathfrak{L}_{m-1} & L_m & \mathfrak{L}_m \\ \mathfrak{L}_m' & 2 & \mathfrak{L}_{m-1} \end{Bmatrix} \right] \\ &\times \left[\delta_{j1} + (1-\delta_{j1}) \left(\prod_{m=1}^{j-1} \delta_{\alpha_m L_m, \alpha_m' L_m'} \right) (-1)^{\mathfrak{L}_{j-1} + L_j' + \mathfrak{L}_j} [\mathfrak{L}_j, \mathfrak{L}_j']^{1/2} \begin{Bmatrix} \mathfrak{L}_{j-1} & L_j & \mathfrak{L}_j \\ 2 & \mathfrak{L}_j' & L_j' \end{Bmatrix} \right] \\ &\times \langle l_j^{w_j} \alpha_j L_j S_j \| U^{(2)} \| l_j^{w_j} \alpha_j' L_j' S_j' \rangle ; \end{aligned} \tag{15.22}$$

values of this final matrix element may be obtained from (11.53).

The 6-j symbols give the selection rules

$$\Delta L_j = 0, \pm 1, \pm 2 \qquad (L_j + L_j' \geqq 2) ,$$

and (15.23)

$$\Delta \mathfrak{L}_m = 0, \pm 1, \pm 2 \qquad (\mathfrak{L}_m + \mathfrak{L}_m' \geqq 2), \quad m \geqq j ,$$

in addition to the selection rules

$$\begin{aligned} \Delta S_m &= \Delta \mathfrak{S}_m = 0, \quad \text{all } m , \\ \Delta \alpha_m L_m &= 0, \qquad m \neq j . \end{aligned} \tag{15.24}$$

Because of the triangle rule on $(L_j\ 2\ L_j')$, the final reduced matrix element in (15.22) is zero for a filled subshell [see (11.62)]; hence the summation in (15.21) need be carried out only over partially filled subshells (and any s-electron subshell may also be ignored). For a configuration containing only a single open subshell j, the summation in (15.21) reduces to

$$\langle l_j^{w_j} \alpha_j L_j S_j \| U^{(2)} \| l_j^{w_j} \alpha_j' L_j' S_j \rangle P_{jj}^{(2)} , \tag{15.25}$$

where $L_j S_j = \mathfrak{L}_q \mathfrak{S}_q \equiv \mathfrak{L}\mathfrak{S}$ and $L_q' = \mathfrak{L}_q' = \mathfrak{L}'$.

Sum rules similar to those for electric dipole radiation can be derived.[5]

[5] D. Rosenthal, R. P. McEachran, and M. Cohen, Proc. Roy. Soc. (London) A337, 365 (1974). See also G. H. Shortley, Phys. Rev. 57, 225 (1940).

Transition probability magnitudes. For neutral atoms, the electric quadrupole matrix element in (15.13) has roughly the same magnitude as that of the electric dipole element in (14.33). It follows that

$$\frac{A_{E2}}{A_{E1}} \cong \frac{\pi^2 a_0{}^2(\sigma_{E2})^5}{5(\sigma_{E1})^3} = 0.5\cdot 10^{-16}\frac{(\sigma_{E2})^5}{(\sigma_{E1})^3} \tag{15.26}$$

for σ in kaysers; for radiation in the visible region ($\lambda \equiv 5000$ Å, $\sigma = 20000$ K), electric quadrupole transitions are eight orders of magnitude less probable than electric dipole transitions.

Along an isoelectronic sequence, the E2 radial matrix element decreases like $Z_c{}^{-2}$ (rather than like $Z_c{}^{-1}$, as does the E1 matrix element). For inter-configuration transitions with $\Delta n \neq 0$,

$$\sigma \propto Z_c{}^2 , \qquad A_{E2} \propto \sigma^5 Z_c{}^{-4} \propto Z_c{}^6 , \tag{15.27}$$

so that from (14.116)

$$A_{E2}/A_{E1} \propto Z_c{}^2 , \qquad \Delta n \neq 0 . \tag{15.28}$$

For inter-configuration transitions with $\Delta n = 0$, or for intra-configuration transitions between levels whose energy separation is primarily due to Coulomb interactions,[3]

$$\sigma \propto Z_c , \qquad A_{E2} \propto \sigma^5 Z_c{}^{-4} \propto Z_c , \tag{15.29}$$

so that from (14.119)

$$A_{E2}/A_{E1} \cong \text{constant} , \qquad \Delta n = 0 . \tag{15.30}$$

*15-3. INTERFERENCE BETWEEN M1 AND E2 RADIATION

For a transition γJM—$\gamma' J' M'$ that is allowed for both magnetic dipole and electric quadrupole radiation ($|\Delta J| \leqq 1$ and $J + J' \geqq 2$), the transition probability is equal to $64\pi^4 e^2 a_0^2 \sigma^3/3h$ times the squared magnitude of

$$\begin{aligned}\langle \gamma JM | \frac{\alpha}{2}(J_q^{(1)} + S_q^{(1)}) + \frac{\pi a_0 \sigma}{\sqrt{5}} P_q^{(2)} | \gamma' J' M' \rangle &= (-1)^{J-M}\begin{pmatrix} J & 1 & J' \\ -M & q & M' \end{pmatrix}\frac{\alpha}{2}\langle \gamma J \| J^{(1)} + S^{(1)} \| \gamma' J' \rangle \\ &+ (-1)^{J-M}\begin{pmatrix} J & 2 & J' \\ -M & q & M' \end{pmatrix}\frac{\pi a_0 \sigma}{\sqrt{5}}\langle \gamma J \| \boldsymbol{P}^{(2)} \| \gamma' J' \rangle ,\end{aligned} \tag{15.31}$$

where the Wigner-Eckart theorem (11.15) has been used, and where $q = M - M'$. Squaring this expression, we find that the total transition probability $a(\gamma JM$—$\gamma' J' M')$ is equal to the sum of a_{M1} and a_{E2}, as given by (15.4) and (15.12), plus an interference term that involves the product

$$\begin{pmatrix} J & 1 & J' \\ -M & q & M' \end{pmatrix}\begin{pmatrix} J & 2 & J' \\ -M & q & M' \end{pmatrix} . \qquad (15.32)$$

These interference effects can be observed when the various line components (corresponding to different MM′) are separated as to wavelength by means of the Zeeman effect; they have for example, been investigated for [Pb I] $6p^2\ {}^3P_1{-}{}^1D_2$ and [Pb II] $6p\ {}^2P_{1/2}{-}{}^2P_{3/2}$ by Hults,[6] who found results in good agreement with theory.[7]

For field-free atoms under isotropic excitation conditions (equal numbers of atoms in each state $\gamma'J'M'$ of the level $\gamma'J'$), the total line intensity is from (14.6)-(14.7) proportional to the average transition probability

$$\overline{A} = \frac{\sum_{M'}\sum_{M} a(\gamma JM - \gamma'J'M')}{\sum_{M'} 1} . \qquad (15.33)$$

From the orthogonality property (5.15) of 3-j symbols, the double sum of (15.32) is zero so that the averaged contribution of the interference term is nil, and we have simply

$$g\overline{A} = gA_{M1} + gA_{E2} , \qquad (15.34)$$

where the two terms on the right are given by (15.5) and (15.13).

15-4. EXAMPLES OF FORBIDDEN TRANSITIONS

For neutral atoms, we have seen from (15.8) and (15.26) that magnetic dipole and electric quadrupole transitions are inherently much weaker than electric dipole transitions; for the most part, this remains true even in rather highly ionized atoms. Magnetic and higher-order-electric transitions are therefore commonly referred to as "forbidden" transitions, as opposed to "allowed" electric-dipole transitions. Forbidden spectrum lines are denoted by enclosing the Roman-numeral specification in brackets; e.g., [C I], [Fe II], etc. The term "forbidden" is of course only a relative one; forbidden transitions can be, and are, observed under suitable conditions. Theoretical transition probabilities have been computed for a large number of spectra[8] and a limited number of experimental measurements are available; Wiese and co-workers have prepared extensive bibliographies and tables.[9,10]

The most easily observed forbidden lines are those for which the upper energy state is metastable; i.e., where there are no transitions to lower energy levels that are allowed by electric dipole radiation. In low-density sources in which collisional de-excitation is small,

[6]M. E. Hults, J. Opt. Soc. Am. 56, 1298 (1966). Similar observations in [Te I $5p^4$] have been made by C. Morillon and J. Vergès, Phys. Scr. 12, 145 (1975).

[7]R. H. Garstang, J. Res. Natl. Bur. Stand. 68A, 61 (1964).

[8]R. H. Garstang, in D. R. Bates, ed., *Atomic and Molecular Processes* (Academic Press, New York, 1962); R. H. Garstang, Mem. Soc. Roy. Sci. Liège 17, 35 (1969).

[9]Refs. 44, 45, 47 of Chap. 14; also B. Edlén, Mem. Soc. Roy. Sci. Liege 9, 235 (1976).

[10]For iron-group elements, M. W. Smith and W. L. Wiese, J. Phys. Chem. Ref. Data 2, 85 (1973).

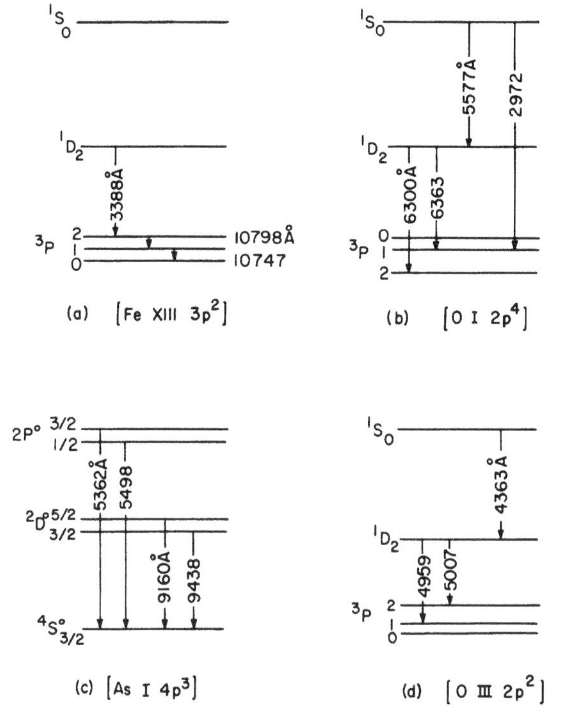

Fig. 15-2. Examples of forbidden transitions within ground configurations p^w.

the density of atoms in metastable states tends to build up to much higher values than for non-metastable states; so far as spectrum line intensitites are concerned, the low transition probabilities are then partially offset by the higher level populations. Looked at from another point of view, line intensities in such sources are determined more by the rate at which the metastable states are populated (via either collisional excitation or radiative decay from still higher levels) than by radiative-decay probabilities. For most atoms and ions, all levels of the ground configuration lie below all levels of excited configurations, and so are all metastable. Most observed forbidden lines are therefore produced by transitions within the ground configuration.

Coronal lines. Under pure LS coupling conditions with no configuration interaction, (15.7) shows that magnetic dipole transitions occur only between levels of a given LS term. This would mean that in ground configurations np^2, for example, the only possible magnetic-dipole transitions would be $^3P_0$—$^3P_1$ and $^3P_1$—$^3P_2$ (Fig. 15-2a). For neutral atoms, these energy intervals are small and do not give rise to readily observable wavelengths except for very high Z. In highly ionized atoms the intervals are much greater (Table 15-2), and in [Fe XIII] these transitions produce prominent lines in the infrared portion of the spectrum of the solar corona. Many other coronal and solar-flare lines have been similarly identified as arising from $2p^n$ and $3p^n$ ground-configuration forbidden transitions[11] of

[11]This explanation for the source of coronal lines was first suggested by W. Grotrian, Naturwissenschaften 27, 214 (1939). The first extensive identifications were made by B. Edlén, Ark. Mat. Astr. Fys. **28B**, No. 1, 1 (1941), Z. Astrophys. 22, 30 (1942).

TABLE 15-2. ^3P INTERVALS AND WAVELENGTHS IN np^2 CONFIGURATIONS

		$^3P_0$–$^3P_1$		$^3P_1$–$^3P_2$	
		σ(K)	λ(Å)	σ	λ
Si I	$3p^2$	77	---	146	---
Pb I	$6p^2$	7819	~12800	2831	~35300
Ca VII	$3p^2$	1627	~61500	2443	~40900
Fe XIII	$3p^2$	9303	10747[a]	9258	10798[a]
Ni XV	$3p^2$	14917	6702[a]	12459	8024[a]

[a]Observed coronal lines, ref. 12.

highly ionized Si, S, Ar, Ca, Fe, Ni, etc.[12] Magnetic-dipole transitions between (nominally) different LS terms (made possible by intermediate-coupling mixings), and electric-quadrupole transitions between such terms, give rise to coronal lines in the visible and ultraviolet regions;[12] one example is included in Fig. 15-2a. Many of these lines are used in the determination of elemental abundances in the solar atmosphere.[13]

Auroral and airglow lines. Atomic oxygen is produced in the earth's upper atmosphere through photo-dissociation of molecular oxygen by ultraviolet light from the sun. In the ground configuration $2p^4$ of O I (Fig. 15-2b), the transition $^1D_2$–$^1S_0$ is allowed only by electric quadrupole radiation (because $\Delta J = 2$), but gives rise to a prominent green line in spectra of aurorae and the airglow. The red auroral/airglow lines $^3P_2$–$^1D_2$ and $^3P_1$–$^1D_2$ are, in pure LS coupling, allowed neither as electric quadrupole ($\Delta S \neq 0$) nor as magnetic dipole ($\Delta S \neq 0$ and $\Delta L \neq 0$) transitions. They become possible for both kinds of radiation as a result of small intermediate-coupling admixtures of $^3P_2$ into the "$^1D_2$" state and (for the 6300 Å line) of $^1D_2$ into the "$^3P_2$" state; calculations indicate both lines to be predominantly magnetic dipole in nature (Table 15-3). Transition probabilities for the red lines are much smaller than for the green line in spite of the estimates (15.8) and (15.26), because A values for the red lines depend on the presence of departures from pure LS coupling, which are quite small.

[As I]. In the ground configuration $4p^3$ of neutral arsenic (Fig. 15-2c), the transition $^4S^\circ_{3/2}$–$^2P^\circ_{3/2}$ (λ5362) has been observed in the laboratory for field-free atoms[14] and in the Zeeman effect.[15] In a pure LS-coupling approximation it is allowed neither as electric quadrupole ($\Delta S \neq 0$) nor magnetic dipole ($\Delta S \neq 0$, $\Delta L \neq 0$) radiation. It occurs because of very small intermediate-coupling admixtures of $^2P^\circ_{3/2}$ (<1%) and $^2D^\circ_{3/2}$ (0.01%) into

[12]See, for example, M. H. L. Pryce, Astrophys. J. **140**, 1192 (1964); J. T. Jefferies, Mem. Soc. Roy. Sci. Liège **17**, 213 (1969); C. Jordan, Nucl. Instrum. Methods **110**, 373 (1973); G. A. Doschek et al., Astrophys. J. **196**, L83 (1975).

[13]See, for example, P. L. Smith, Nucl. Instrum. Methods **110**, 395 (1973).

[14]M. Hults and S. Mrozowski, J. Opt. Soc. Am. **54**, 855 (1964).

[15]Hui Li, unpublished (1971).

TABLE 15-3. TRANSITION PROBABILITIES (sec^{-1}) OF FORBIDDEN LINES OF O I $2p^4$

λ (Å)	Theory[a]		Exp.[b]
	A_{M1}	A_{E2}	A
$5577(^1D_2-^1S_0)$		1.28	1.1 ± 0.3
		1.18	
$6300(^3P_2-^1D_2)$	$6.9 \cdot 10^{-3}$	$2.4 \cdot 10^{-5}$	$(5.2 \pm 1.3) \cdot 10^{-3}$
	$5.7 \cdot 10^{-3}$	$3.5 \cdot 10^{-6}$	
$6363(^3P_1-^1D_2)$	$2.2 \cdot 10^{-3}$	$3.2 \cdot 10^{-6}$	$(1.7 \pm 0.4) \cdot 10^{-3}$
	$1.9 \cdot 10^{-3}$	$4.3 \cdot 10^{-6}$	

[a]R. H. Garstang, Mon. Not. R. Astron. Soc. 111, 115 (1951), except lower number for λ5577 from C. Nicolaides, O. Sinanoğlu, and P. Westhaus, Phys. Rev. A 4, 1400 (1971), and lower numbers for λ6300 and λ6363 from T. Yamanouchi and H. Horie, J. Phys. Soc. Japan 7, 52 (1952).

[b]J. A. Kernahan and P. H.-L. Pang, Can. J. Phys. 53, 455 (1975). See also A. Corney, Nucl. Instrum. Methods 110, 151 (1973).

$^4S^\circ_{3/2}$, and of $^4S^\circ_{3/2}$ (<1%) and $^2D^\circ_{3/2}$ (4%) into $^2P^\circ_{3/2}$. Two other lines made possible by these mixings have also been observed; relative intensitites[14] agree well with theory[16] (Table 15-4).

Nebular lines. The lines [O II] $2p^3$ at 3726 and 3729 Å (analogous to the two long-wavelength lines of [As I]) and [O III] $2p^2$ at 4959 and 5007 Å (Fig. 15-2d) are prominent in the spectra of gaseous nebulae. Before their origin was established,[17] they were known as "nebulium" lines because they were thought to be due to an element not yet discovered on earth—analogous to the discovery of helium lines in the solar spectrum before that element was found on earth.

Excited configurations. Forbidden lines may also arise from metastable levels of excited configurations. For example, in highly ionized atoms having ground configurations $3p^m$ the lowest excited configuration is $3p^{m-1}3d$. For $m > 1$, $p^{m-1}d$ has levels with J' greater by two than the largest J of p^m, and downward electric dipole transitions from those levels are therefore forbidden by the dipole selection rule $|\Delta J| \leqq 1$. Magnetic dipole and electric quadrupole transitions within the configuration Fe IX $3p^5 3d$ and arising from such

[16]R. H. Garstang, J. Res. Natl. Bur. Stand. **68A**, 61 (1964).

[17]I. S. Bowen, Astrophys. J. **67**, 1 (1928).

TABLE 15-4. FORBIDDEN TRANSITIONS IN As I $4p^3$

		Theory[a]		
	λ(Å)	A_{M1} (sec^{-1})	A_{E2} (sec^{-1})	Observed Intensity[b]
$^4S^o_{3/2}-^2P^o_{3/2}$	5362	1.6	0.0001	3
$-^2P^o_{1/2}$	5498	0.69	0.0012	1
$-^2D^o_{5/2}$	9161	0.0020	0.0033	---
$-^2D^o_{3/2}$	9440	0.073	0.0019	v. weak

[a]Ref. 16.
[b]Ref. 14.

metastable levels are responsible for several lines of the solar corona; similar transitions in Fe X-XIII and Ni XI-XV probably account for numerous other unidentified lines.[18]

In Pb I, $6p^2 - 6p7p$ transitions have been observed[19] that are parity-forbidden for electric dipole radiation. The upper levels are not metastable, because E1 transitions to 6p7s are possible; however, these transitions have wavenumbers of only about 9000 K, as opposed to $\sigma \cong 40000$ K for the transitions to $6p^2$. From (15.26),

$$A_{E2}(6p^2 - 6p7p)/A_{E1}(6p7s - 6p7p) \cong 10^{-5} , \tag{15.35}$$

which may be a sufficiently large branching ratio to make the forbidden transitions observable. An alternative explanation is that configuration interaction gives the 6p7p configuration some $6p^2$ character and vice versa, making magnetic dipole transitions possible; from (15.8), $A_{M1}/A_{E1} \cong 10^{-3} \times$ (fractional mixing). For higher excited configurations 6pnp ($n > 7$), frequencies of permitted transitions are comparable with those of the forbidden transitions $6p^2 - 6pnp$ so that the E2 branching ratio is much smaller than (15.35), and no such forbidden transitions have been observed. (M2 transitions would also be weaker because configuration mixing of 6pnp with $6p^2$ decreases with increasing n.)

In absorption, electric quadrupole transitions of the type s − d have been extensively studied in the alkalis (see AEL for references).

Transitions s − s′ are strictly forbidden for isolated atoms, but have been observed by means of Stark-effect-induced mixing of n′s′ with n″p states.[20]

[18]L. Å Svensson, J. O. Ekberg, and B. Edlén, Solar Phys. **34**, 173 (1974).

[19]D. R. Wood and K. L. Andrew, J. Opt. Soc. Am. **58**, 818 (1968).

[20]For absorption and Stark-effect-induced observations in astronomical sources, see papers in Mem. Soc. Roy. Sci. Liège **17** (1969), and H. Ezawa and M. Leventhal, J. Phys. B **8**, 1824 (1975).

Problem

15-4(1). For the ground configuration of O I (Fig. 15-2b), the energy levels and wavefunctions (the latter obtained by least-squares fitting of the former, Sec. 16-3) are

$$\begin{aligned}
&\text{E: } 33793 \text{ cm}^{-1}, && |\text{E}\rangle = 0.00618\,|{}^3\text{P}_0\rangle + 0.99998\,|{}^1\text{S}_0\rangle\,, \\
&\text{D: } 15868, && |\text{D}\rangle = 0.00654\,|{}^3\text{P}_2\rangle - 0.99998\,|{}^1\text{D}_2\rangle\,, \\
&\text{C: } 227, && |\text{C}\rangle = 0.99998\,|{}^3\text{P}_0\rangle - 0.00618\,|{}^1\text{S}_0\rangle\,, \\
&\text{B: } 158, && |\text{B}\rangle = 1.00000\,|{}^3\text{P}_1\rangle\,, \\
&\text{A: } 0, && |\text{A}\rangle = 0.99998\,|{}^3\text{P}_2\rangle + 0.00654\,|{}^1\text{D}_2\rangle\,.
\end{aligned}$$

The single-configuration HF value of the electric-quadrupole radial matrix element $P^{(2)} \equiv \langle 2p\|r^2\|2p\rangle$ is $-2.192\ a_0^2$. Calculate transition probabilities for all possible M1 and E2 transitions, and compare with the values given by Garstang (Table 15-3).

16

NUMERICAL CALCULATION OF ENERGY LEVELS AND SPECTRA

16-1. COMPUTER PROGRAMS

For simple configurations involving only one or two open subshells it is feasible to calculate energy levels with the aid of a desk calculator provided no J-matrix is larger than 2×2, or provided the coupling conditions lie close to some pure-coupling scheme so that the Hamiltonian matrix is nearly diagonal in the appropriate pure-coupling representation. The task is particularly easy in the near-LS case, where extensive tables of the Coulomb angular coefficients f_k and g_k are available,[1,2] and where spin-orbit splittings can be put in as perturbations. Examples were discussed in Secs. 12-7 to 12-9. Calculation of electric-dipole line strengths in an LS representation is similarly facilitated by available tables of the necessary angular factors.[3]

In general, however, routine calculations of energy levels and spectra are practical only with the aid of large electronic computers and elaborate computer programs. By way of summarizing the calculational procedures and the theory developed in Chapters 9-15, we describe briefly the program[4] outlined in Fig. 16-1, which performs all calculations in the $\mathfrak{L}\mathfrak{S}$ representation (12.1). (The letters A, B, C, $\cdots$ labeling the paragraphs that follow are keyed to corresponding blocks in the figure.)

A: For each subshell l^w that will be involved in any of the electron configurations to be considered, an input card deck provides the quantum numbers αLS of l^w, the parents $\overline{\alpha LS}$, and the coefficients of fractional parentage $(l^{w-1}\overline{\alpha LS}|\}l^w\alpha LS)$ defined in Secs. 9-5 to 9-7;

[1]For single subshells, C. W. Nielson and G. F. Koster, *Spectroscopic Coefficients for the p^n, d^n, and f^n Configurations* (The M.I.T. Press, Cambridge, Mass., 1963).

[2]J. C. Slater, *Quantum Theory of Atomic Structure* (McGraw-Hill, New York, 1960), Vol. II, App. 21; E. U. Condon and H. Odabasi, *Atomic Structure* (Cambridge University Press, Cambridge, 1980).

[3]B. W. Shore and D. H. Menzel, Astrophys. J., Suppl. Ser. 12, 187 (1965), or *Principles of Atomic Spectra* (John Wiley, New York, 1968). Note, however, that the tabulated values are for the SL representation; see (9.10).

[4]This is an extension, to include configuration interaction and M1 and E2 spectra, of the program RCG Mod 4 described by R. D. Cowan, J. Opt. Soc. Am. 58, 808 (1968).

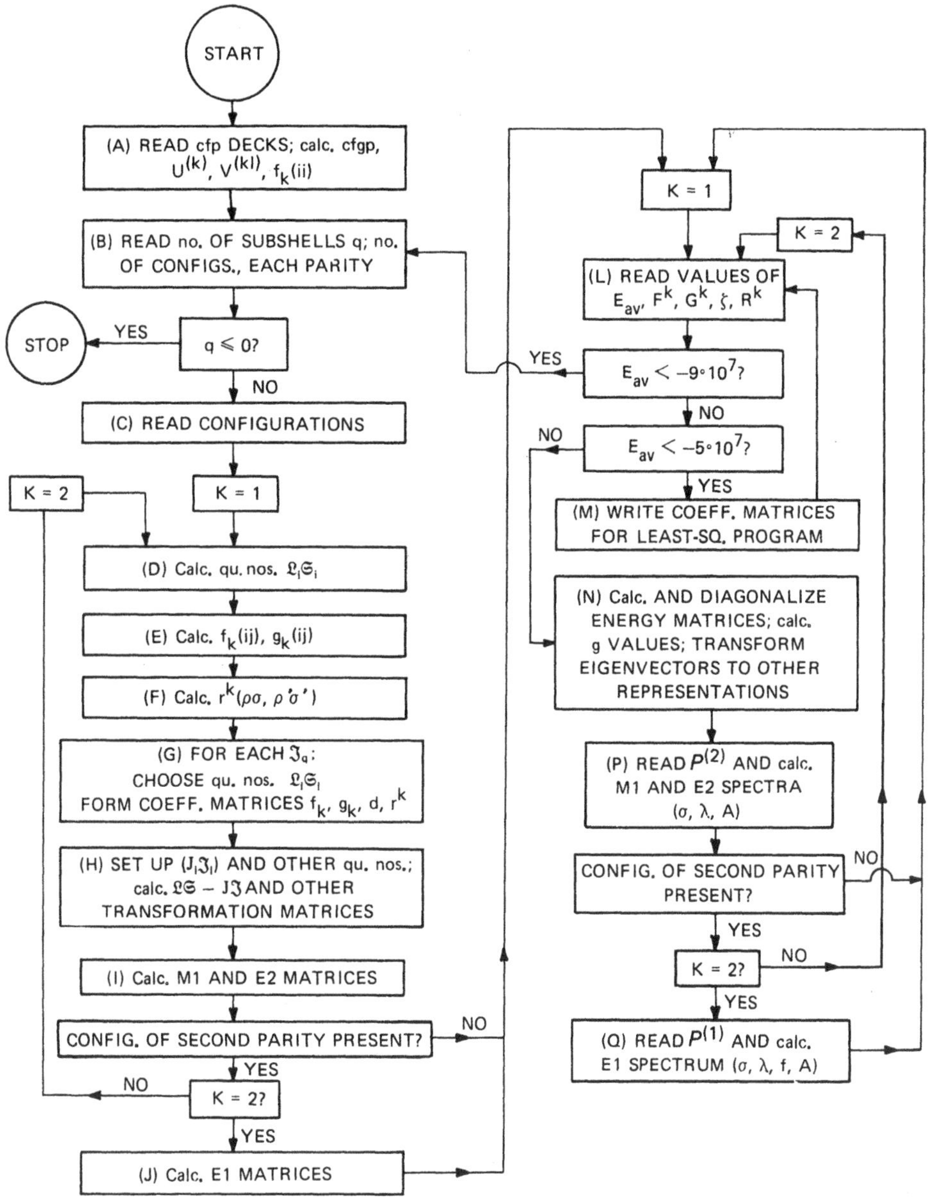

Fig. 16-1. Simplified block diagram of computer program RCG Mod 5. The left-hand column represents the calculation of angular coefficient matrices; the right-hand column represents the calculation of atomic energy levels and spectra for given values of the radial integrals E_{av}, F^k, G^k, ζ, R^k, $P^{(t)}$. K = 1 and 2 represent the first and second parities, respectively. [Following step (L) there is actually a test to see whether the card read contains E_{av}, $\cdots$ or $P^{(t)}$; if the latter, transfer is made to (P) or (Q) according as t = 2 or 1, respectively.]

this information is available for all subshells with $l \leq 3$ from Nielson and Koster[1,5] and (9.68)-(9.69), and for singly and doubly occupied subshells of larger l from (9.61). If decks for both l^w and l^{w-1} are included, then coefficients of fractional grandparentage $(l^{w-2}\alpha\bar{L}\bar{S}, l^2L'S'\|l^w\alpha LS)$ are computed from (9.52) for later use in calculating configuration-interaction matrix elements. All reduced matrix elements of $U^{(k)}$ and $V^{(k1)}$ are then computed from (11.53)-(11.54) and (11.68)-(11.69), and if $2 \leq w \leq 4l$ then coefficients of the Coulomb integrals $F^k(ll)$ are calculated from (12.17) and (11.23). [Each 3n-j symbol is calculated as needed, from the definitions (5.1), (5.13), (5.23), and (5.37) and suitable (floating-point)[5] subroutines.] This information could, of course, be computed once and for all for every l^w and then made available as fundamental program input; however, the required computing time is small except for f^w subshells with $5 \leq w \leq 10$.

B: A calculation for a set of electron configurations (13.46) is begun by reading a control card which (along with numerous options) specifies the number of subshells q (excluding subshells that are filled in every configuration) and the number of configurations of each parity.

C: Configuration-definition cards provide l_i and w_i $(1 \leq i \leq q)$ for each configuration.

D: For the configurations of the first parity, all possible sets of quantum numbers $\alpha_i L_i S_i \mathfrak{L}_i \mathfrak{S}_i$ for the $\mathfrak{L}\mathfrak{S}$ coupling scheme (12.1) are formed, by vector addition of the L_iS_i from A.

E: For each configuration, the angular coefficients of the single-configuration Coulomb integrals $F^k(l_il_j)$ and $G^k(l_il_j)$ $(1 \leq i < j \leq q$, excluding filled or empty subshells) are computed from (12.19), (12.26)-(12.28), (12.38), and (12.35).

F: For each pair of configurations, the angular coefficients of the various configuration-interaction radial integrals $R^k(l_\rho l_\sigma, l_{\rho'} l_{\sigma'})$ are calculated from (13.60)-(13.92).

G: For each possible value of the total angular momentum $\mathfrak{J}_q$, those sets of quantum numbers from D are selected that satisfy $|\mathfrak{L}_q - \mathfrak{S}_q| \leq \mathfrak{J}_q \leq \mathfrak{L}_q + \mathfrak{S}_q$. Coulomb coefficient matrices (f_k), (g_k), (r^k) are set up, using the values of the coefficients calculated in A, E, and F; spin-orbit angular coefficients d_i $(1 \leq i \leq q, l_i > 0, 1 \leq w_i \leq 4l_i + 1)$ are computed from (12.42).

H: The possible sets of quantum numbers $J_i\mathfrak{J}_i$ for the representation (12.2) are formed, and the $\mathfrak{L}\mathfrak{S}$-$J\mathfrak{J}$ transformation matrix is calculated from (12.3). Similarly, quantum number sets for other desired pure-coupling representations are formed, and transformation matrices from the $\mathfrak{L}\mathfrak{S}$ representation are calculated with the aid of equations obtained by the method used for (9.23)-(9.29).[6]

[5]Nielson and Koster tabulate cfp in power-of-prime notation, as do Rotenberg, et al. for 3n-j symbols; however, we here perform all calculations using ordinary floating-point numbers.

[6]A method of finding transformation matrices without having to program the appropriate equations is to diagonalize the $\mathfrak{L}\mathfrak{S}$ energy matrices for artificial values of F^k, G^k, ζ that correspond to the coupling conditions of interest—see text following (4.16), and Table 10-2 (Sec. 10-9). It is necessary to use extreme values (e.g., for $J\mathfrak{J}$ coupling in a two-subshell configuration, $\zeta_1 = 100\zeta_2 = 10000F^2 = \cdots$); a further disadvantage is that it is not always easy to tell from the computed energy-level structure just which eigenvalue (and hence which column of the transformation matrix) belongs to a given set of quantum numbers in the new representation.

I: If desired, then for each permissible pair of values $\mathfrak{J}_q\mathfrak{J}_q'$, angular line-strength matrices for M1 and E2 transitions are calculated from (15.7), and from (15.21)-(15.22) or from (14.78)-(14.86) with operator rank $t = 2$.

If configurations of opposite parity are included, steps D-I are repeated for configurations of the second parity.

J: If configurations of both parities are included, then for each permissible $\mathfrak{J}_q\mathfrak{J}_q'$ angular line-strength matrices for E1 transitions are computed, using (14.78)-(14.86) with operator rank $t = 1$.

The coefficient, transformation, and line-strength matrices computed in G-J are stored on disk for use by the remaining parts of the program; if desired, they can also be written on magnetic tape for use in later runs.

L: For the configurations of the first parity, appropriate values are read from cards for the Coulomb and spin-orbit radial integrals E_{av}, F^k, G^k, ζ_i, R^k. (M: An inappropriately large negative value of E_{av} for the first configuration is a signal that the energy coefficient matrices should be written on disk or tape for use by a least-squares level-fitting program to be described in Sec. 16-4.)

N: For each $\mathfrak{J}_q$, the Hamiltonian matrix is evaluated from (12.47). The matrix is then diagonalized numerically (see fn. 5, Sec. 4-2), to give the various eigenvalues (energy levels) E^k and the eigenvectors $\mathbf{Y}^k$, whose components are the basis-function expansion coefficients y_b^k ($b \equiv \beta\mathfrak{J}_q$) for the corresponding wavefunctions (4.4). The eigenvectors are used to calculate Landé g values (17.13), and are transformed into the $J\mathfrak{J}$ and other pure-coupling representations using transformation matrices from H and relations like (12.4).

P: If desired, the various eigenvalues are differenced to obtain wavelengths (14.1)-(14.2) for forbidden radiative transitions, and the corresponding eigenvectors are used with the $\mathfrak{LS}$-coupling line-strength matrices from I to calculate transition probabilities for magnetic dipole radiation from (15.5)-(15.6), and for electric quadrupole radiation from (15.13) and (15.15) [using values read from cards for the one-electron electric-quadrupole integrals $P^{(2)}_{l_il_j}$ needed in (15.21) or (14.78)].

If configurations of the second parity are present, L-P are repeated to give energy levels, eigenvectors, and forbidden transitions for these configurations.

Q: If both parities are present, values of the one-electron electric-dipole integrals $P^{(1)}_{l_il_j}$ are read to complete the line-strength calculation (14.78), and E1 oscillator strengths and transition probabilities are computed from (14.2), (14.37), (14.38), and (14.42). The same computer subroutine is used for Q as for P. In either case, an option provides for making a microfilm plot of the computed spectrum (gf or gA vs. σ) sufficient to make possible a quick and useful comparison with observed spectra; an example is shown in Fig. 16-2.

Though not shown in Fig. 16-1, the spectrum calculations P and Q can be repeated with other values of $P^{(t)}$, for parameter studies. Control is then returned to L to make further energy-level calculations for the same set of configurations as before, for parameter studies or isoelectronic-sequence calculations. A value less than $-9\cdot 10^7$ for E_{av} of the first configuration is a signal to return to B to start a calculation for a new set of configurations.

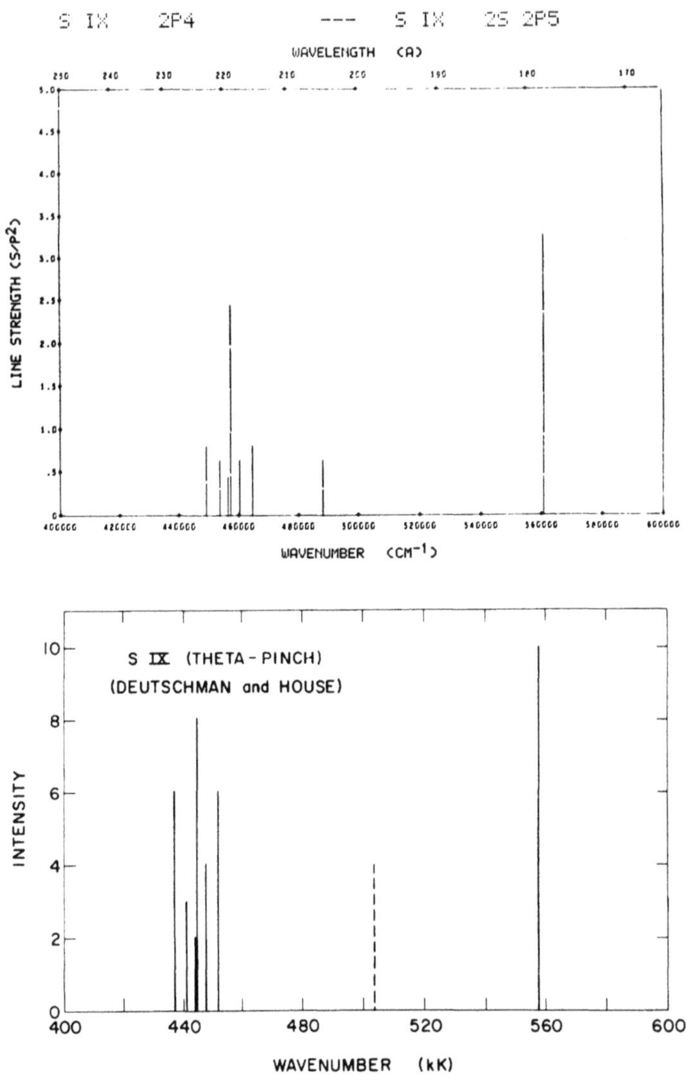

Fig. 16-2. Top: Computer plotted theoretical spectrum for the transition array S IX $2s^2 2p^4 - 2s2p^5$. Bottom: Analogously hand-plotted spectrum lines observed by W. A. Deutschman and L. L. House, Astrophys. J. 144, 435 (1966).

In practice, the complete set of input cards required in blocks B, C, L, P, and Q is punched out by the radial-wavefunction program RCN (Fig. 8-1), thus practically eliminating any hand work.

For illustrative purposes, required computing times are given in Tables 16-1 and 16-2 for several cases. It is evident that times required for A are usually small compared with those for D-J, which in turn are usually small compared with the energy-level-spectrum computing times L-Q; hence it is usually not worthwhile saving the results of A-J for later use, and they are generally recomputed whenever needed.

At the same time, it is obvious that for configurations containing two or more open subshells, matrices can easily become very large, especially for rare-earth configurations involving an f^w subshell. Limitations of storage space and computer time then force drastic simplifications. The usual procedure is to truncate a configuration such as $f^w l$ by including only one or a few of the lowest LS terms of f^w—e.g., $f^5(^6H^\circ,^6F^\circ,^6P^\circ)d$; the resulting set of basis states is frequently referred to as a *sub-configuration.* Another possibility is to truncate by retaining basis states for only one or a few particular values of $\mathfrak{L}_q\mathfrak{S}_q$. Neither procedure is very accurate if the truncation is too severe, because of the almost universal presence in complex configurations of severe basis-state mixing. Still a third possibility is to first diagonalize the matrices for f^w alone, using the appropriate values of $F^k(ff)$ (with $\zeta_f = 0$)—thereby obtaining a transformed (non-Racah) set of $U^{(k)}$ and $V^{(k1)}$ matrices—and to then add the remaining subshells to a truncated subset of the new f^w basis functions. This procedure seems to be the most promising; it has been used for low configurations of first and second neptunium and plutonium spectra with good results.[7]

16-2. *AB INITIO* CALCULATIONS

From a fundamental theoretical point of view, the straightforward way of obtaining values for the radial integrals E_{av}, F^k, G^k, ζ, R^k, and $P^{(t)}$ needed in steps L, P, and Q of the calculation (Fig. 16-1) is to use values computed from Hartree-Fock (or HX, etc.) radial wavefunctions for the configurations in question (cf. Sec. 8-1). We have already seen (Tables 6-6 and 8-4) that theoretical values of E_{av} agree well with experiment (at least in simple cases), provided that relativistic and correlation corrections are included; in many cases, it is necessary also that strong configuration-interaction perturbations be taken into account by performing a multi-configuration rather than single-configuration calculation. Spin-orbit effects are usually predicted with good accuracy, provided that values of ζ are computed via the Blume-Watson theory (see fn. 1 of Chapter 4) using HF radial functions (or HFR functions, for heavy or highly ionized atoms).

On the other hand, computed energy-level splittings resulting from electron-electron Coulomb interactions are generally larger than observed by ten to fifty percent. This is a consequence of the neglect of the LS-term dependence of electron correlation: For a given configuration, levels having a high computed energy must correspond to situations in which the electrons lie relatively close together (the energy being higher because of the stronger Coulomb repulsion). These are therefore the states for which correlation effects will be greatest. Thus if detailed correlation corrections were included, we would expect the computed energy decrease to be greater for the high levels than for the low ones;[8] the result would be a smaller computed energy-spread among the levels of a configuration—in the proper direction to improve the agreement with experiment.

[7]H. M. Crosswhite, H. Crosswhite, and M. Fred, private communication.

[8]The *average* decrease for all levels of the configuration has been taken into account in the correlation energy correction to E_{av}.

TABLE 16-1. CDC 7600 COMPUTING TIMES (in seconds)[a] FOR CALCULATION OF CFGP, $U^{(k)}$, $V^{(k1)}$, $f_k(ll)$, d_l

				time						time[b]	
l^w	No. of LS terms	No. of parents	No. of grandp.	no cfgp	with cfgp	l^w	No. of LS terms	No. of parents	No. of grandp.	no cfgp	with cfgp
p^w	≦3	≦3	≦3	<0.1	<0.1	f^3	17	7	1	1.0	1.2
						f^4	47	17	7	11	15
d^2	5	1	1	<0.1	<0.1	f^5	73	47	17	55	99
d^3	8	5	1	0.2	0.2	f^6	119	73	47	130	
d^4	16	8	5	0.6	1.0	f^7	119	119	73	200	
d^5	16	16	8	1.0	2.1	f^8	119	119	119	170	
d^6	16	16	16	0.8	3.0	f^9	73	119	119	100	
d^7	8	16	16	0.4	1.6	f^{10}	47	73	119	27	
d^8	5	8	16	0.1	0.5	f^{11}	17	47	73	4.9	49
d^9	1	5	8	<0.1	<0.1	f^{12}	7	17	47	0.3	4.5
d^{10}	1	1	5	<0.1	<0.1	f^{13}	1	7	17	<0.1	0.2

[a]Times for the IBM 360/95 or 370/195 would be about the same; times for the CDC 6600 or UNIVAC 1108 would be about 4 times as great, and for the CDC 6400 or IBM 7094 II about 15 times as great.

[b]It is generally impractical to perform multi-configuration calculations for configurations involving untruncated f^w subshells with $5 \leqq w \leqq 10$; hence there is no need to compute cfgp in these cases. If f^w *is* truncated to a small number of terms, then computing times are proportionately smaller.

TABLE 16-2. MATRIX SIZES AND CDC 7600 COMPUTING TIMES (seconds) FOR ENERGY LEVEL AND SPECTRUM CALCULATIONS

	p^3	p^2d	d^4	d^3p	$d^3p + d^2sp$	f^3s^2	f^3sp	f^5	$f^{5(a)}$	$f^5d^{(a)}$	f^7
Number of parameters	3	7	4	8	21	5	11	5	5	11	5
Number of levels	5	28	34	110	200	41	476	198	51	494	327
Maximum matrix size	3	8	8	25	47	7	81	30	9	74	50
D. Calc. quantum numbers	<0.1	0.1	<0.1	0.1	0.1	0.1	0.2	0.1	0.1	0.1	0.4
E. Calc. $f_k(ll')$, $g_k(ll')$	<0.1	0.1	0.1	0.5	0.6	0.1	2.5	0.4	0.2	1.7	0.8
F. Calc. r^k	---	---	---	---	0.3	---	---	---	---	---	---
G-H: Calc. d, $\mathfrak{L}\mathfrak{S}$-$\mathfrak{J}\mathfrak{J}$, f_k, g_k, mx.	0.1	0.3	0.1	1.1	5.8	0.3	26	1.2	0.3	21	5.3
I. Calc. M1 and E2 matrices	0.1	0.4	---	---	---	---	---	---	---	---	---
N. Calc. and diag. energy matrices	<0.1	0.2	0.2	1.0	7.1	0.2	33	2.9	0.4	25	8.6
P. Calc. M1 and E2 spectra	0.1	1.0	---	---	---	---	---	---	---	---	---
Number of spectrum lines	19	544	---	---	---	---	---	---	---	---	---
J. Calc. E1 matrices	0.3		1.4		---	6.9		---	---	---	---
Q. Calc. E1 spectrum	0.3		7		---	16		---	---	---	---
Number of spectrum lines	97		1718			7402[b]					
Total time[c]	10		16		15	90		6.3	2.1	48	17

[a]Subshell f^5 truncated to 12 terms ($^6H°$, $^6F°$, $^6P°$, $^4G°_{124}$, $^4F°_{134}$, $^4I°_3$, $^4L°$, $^4M°$).
[b]Only the 1119 strongest lines included in the calculation of gf and gA, and in the spectrum plot.
[c]Includes overhead time.

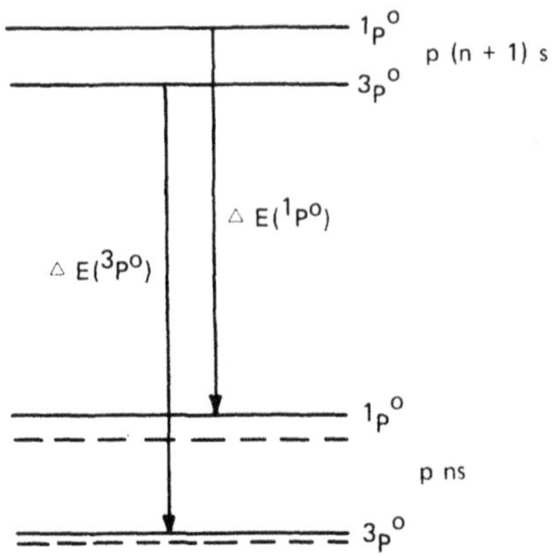

Fig. 16-3. Perturbation of levels of pns by those of p(n+1)s. Because ΔE is smaller for $^1P^\circ$ than for $^3P^\circ$, pns $^1P^\circ$ tends to be perturbed downward a greater distance than does pns $^3P^\circ$, thus reducing the singlet-triplet separation.

Mathematically, we can look at the electron correlation problem as a matter of configuration interaction. For example, in Fig. 16-3 we consider the perturbation of the $^{1,3}P^\circ$ terms of a ps configuration by the corresponding terms of a higher member, ps′, of a Rydberg series of configurations. Because the energy-spread within a configuration decreases as one goes up a Rydberg series, and because the CI matrix elements for $^1P^\circ$ and $^3P^\circ$ are equal (and, for more complex cases, tend to be larger in magnitude for the higher levels), the higher levels of the low configuration tend to be depressed more than the lower levels, thus decreasing the level-spread. (Of course, the level-spread in the upper configuration is correspondingly *increased*, but usually this tendency is more than offset by effects of the infinity of still higher members of the series.)

In making numerical calculations, a few of the most important configuration interactions can be included explicitly, but it is impractical to include more than a very limited number.[9] It is customary in *ab initio* calculations to make a rough allowance for the cumulative effect of the infinity of small perturbations by using scaled-down theoretical values of the single-configuration Coulomb integrals F^k and G^k for the calculation of those parts of energy matrix elements other than E_{av}. Appropriate scale factors range from about 0.7 or 0.8 for neutral atoms to about 0.9 or 0.95 for highly ionized atoms.[10] This semi-empirical correction has been justified qualitatively by the theoretical investigations of Rajnak and Wybourne;[11,12] they have shown that the cumulative perturbations may be represented by adding to the single-configuration energy matrix several new terms, some of which have the same LS dependence as do the f_kF^k and g_kG^k terms except with opposite

[9]Hence it is always appropriate to retain the correlation-energy correction in E_{av}, as this correction represents the cumulative effect of the many weak interactions, rather than of the few strong ones.

[10]See, for example, R. D. Cowan and N. J. Peacock, Astrophys. J. **142**, 390 (1965), **143**, 283 (1966) [Fe VII-XVI]; R. D. Cowan, L. J. Radziemski, Jr., and V. Kaufman, J. Opt. Soc. Am. **64**, 1474 (1974) [Cl I]; V. Kaufman and M.-C. Artru, J. Opt. Soc. Am. **70**, 1135, 1130 (1980) [Mg V and Al VI].

[11]K. Rajnak and B. G. Wybourne, Phys. Rev. **132**, 280 (1963), Phys. Rev. **134**, A596 (1964); B. G. Wybourne, Phys. Rev. **137**, A364 (1965). See also G. Racah and J. Stein, Phys. Rev. **156**, 58 (1967).

[12]B. G. Wybourne, *Spectroscopic Properties of Rare Earths* (Interscience, New York, 1965), Sec. 2-17.

signs. (Some attempts have been made to calculate the magnitude of these contributions theoretically, by means of perturbation theory,[13] but slow convergence has made it difficult to obtain useful quantitative results.)

Assuming that values for the scale factors can be estimated to within five or ten percent (e.g., 0.85 ± 0.05), it follows that theoretically computed wavelengths for intra-configuration (forbidden) transitions should also be accurate to about five or ten percent; the same accuracy would be expected for $\Delta n = 0$ (E1 or E2) transition arrays in very highly ionized atoms, where the level splittings within a configuration may be comparable with the energy difference ΔE_{av} between the centers of gravity of the two configurations (see Sec. 19-2). This accuracy is too low to provide more than highly tentative identifications of observed spectrum lines.

However, for $\Delta n \neq 0$ transitions (and for $\Delta n = 0$ transitions in not too highly ionized atoms), the range of energies within each configuration is usually small compared with the difference between centers of gravity. Wavelengths are then determined primarily by ΔE_{av}. Except for difficult rare-earth and transition-element excitations involving a change in the number of f or d electrons, theoretical values of ΔE_{av} are usually accurate to a couple of percent or so for neutral atoms (Tables 6-5, 6-6, 8-3, and 8-4), and to 0.1 or 0.2 percent for highly ionized atoms. The five to ten percent error in energy-level spreads then affects mainly the wavelength *difference* between lines of a transition array, rather than the absolute wavelengths themselves. Accuracies are thus frequently high enough to make possible fairly reliable identifications of observed spectrum lines, especially for not too complex configurations of highly ionized atoms. An illustrative example in Mn IX and X is given in Fig. 16-4;[14] it may be seen that relative wavelengths and intensities are fairly well predicted. The fact that relative strengths of strong lines are well predicted implies that the intermediate-coupling eigenvectors have been calculated with reasonable accuracy; this is of course to be expected anyway, if the relative values of Coulomb and spin-orbit integrals are accurate to within about ten percent. By transforming the eigenvectors into various pure-coupling representations (block N of Fig. 16-1) to see which representation gives the highest-purity eigenvectors, a decision can be made as to which (if any) pure-coupling set of quantum numbers (in the Mn IX example, LS) provides the most appropriate designations for the energy levels.

Line identifications based on *ab initio* calculations must usually be considered rather tentative and subject to corroboration by other evidence. Nonetheless, such calculations can provide very helpful guidance in the empirical analysis of atomic spectra.[15]

16-3. LEAST-SQUARES CALCULATIONS

Once considerable progress has been made in the interpretation of observed spectrum lines and in the deduction of the associated energy levels—either with the aid of *ab initio*

[13]J. C. Morrison and K. Rajnak, Phys. Rev. A 4, 536 (1971); J. C. Morrison, Phys. Rev. A 6, 643 (1972).

[14]B. C. Fawcett, N. J. Peacock, and R. D. Cowan, J. Phys. B 1, 295 (1968).

[15]See, for example, L. Å. Svensson and J. O. Ekberg, Ark. Fys. 37, 65 (1968); S. Johansson and U. Litzén, Phys. Scr. 10, 121 (1974); B. C. Fawcett, Adv. At. Mol. Phys. 10, 223 (1974).

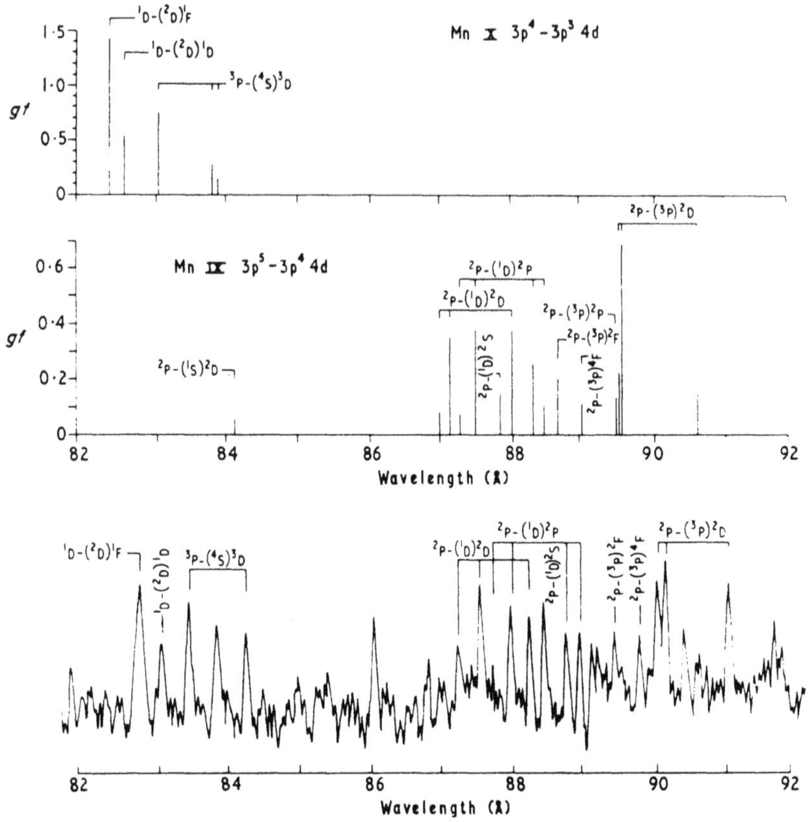

Fig. 16-4. Top: Theoretically calculated spectra of manganese. Bottom: Microdensitometer trace of observed spectrum of manganese, B. C. Fawcett, N. J. Peacock, and R. D. Cowan, J. Phys. B 1, 295 (1968).

calculations, or by the purely empirical methods outlined in Sec. 1-11—the Slater-Condon theory can be applied in quite a different manner, for the semi-empirical checking and interpretation of observed energy levels. The radial integrals E_{av}, F^k, G^k, ζ, and R^k are considered simply as adjustable parameters, whose values are to be determined empirically so as to give the best possible fit between the calculated eigenvalues and the observed energy levels. The accuracy of the fit is measured by means of the standard deviation

$$s = \left[\frac{\sum_k (E^k - T^k)^2}{N_k - N_p} \right]^{1/2} , \tag{16.1}$$

where the T^k are the observed energy levels (or "term values") and the E^k are the computed eigenvalues, N_k is the number of levels being fit, and N_p is the number of adjustable parameters involved in the fit. For cases in which pure LS coupling is closely approximated, a statistically more significant quantity is obtained by using for E^k and T^k the centers of gravity of the LS terms, summing only over the N^k terms instead of over all levels, and using for N_p the number of Coulomb interaction parameters (including E_{av}, but excluding all spin-orbit parameters ζ).[16]

From the point of view of the semi-empirical least-squares approach, the accuracy of the theory is judged by answers to questions such as the following:

(1) Is the accuracy of the fit to the observed energy levels reasonably good? (Experience indicates that a fit may be considered good if the standard deviation s is less than one percent of the energy range ΔE covered by the levels being fit; a value $s/\Delta E$ greater than about two percent usually indicates the presence of strong configuration-interaction perturbations that have been neglected in the calculation.)

(2) Are the least-squares values of the F^k, G^k, and ζ all positive, as required by their physical significance? (Negative values may result from neglect of important interconfiguration or magnetic interactions.)

(3) Do the values of the parameters vary as expected (cf. Figs. 8-4 and 8-13) from one configuration to another, and from one element to another down a column of the periodic table (e.g., C, Si, Ge, Sn, Pb) or along an isoelectronic sequence?

The application of the semi-empirical theory is to problems of the following sort:

(1) Confirmation of, or identification of errors in, a tentative energy-level analysis.

(2) Aid in ascertaining the proper configuration and other quantum numbers for designating experimentally observed but unidentified levels. (This includes determination of the most appropriate pure-coupling representation to use for the designation of the energy levels.)

(3) Prediction of the energies of missing (experimentally unknown) levels.

(4) Aid in determining improved values of series limits (ionization energies) by helping to identify and correct for individual irregularities.

(5) Determination of the ratios of empirical to HF parameter values, to provide estimates of scale factors to be used in *ab initio* calculations for cases where experimental levels are unknown. (Caution is advisable, however. For example, in the configuration p^5d under good LS-coupling conditions, (12.79) shows the empirical value of $G^1(pd)$ to be determined entirely by the position of the ${}^1P^o_1$ level. Fig. 12-7 shows that a physically highly erroneous value of G^1 may be indicated if configuration interactions are not properly taken into account in the least-squares calculation.)

(6) The calculation of more accurate eigenvectors (wavefunctions) than those given by *ab initio* calculations, for use in improved calculations of transition probabilities, Zeeman-effect splittings, etc. (It must be noted, however, that the least-squares eigenvectors can actually be poorer than the *ab initio* ones. This is usually a result of inadvertent forced fitting of perturbed levels, or of other features not adequately accounted for in the theoretical expressions employed. In such cases, it may be necessary to use a judicious mixture of the *ab initio* and least-squares approaches—for example, by using a fixed theoretical value for a parameter whose value is not well determined by least-squares fitting.[17])

[16] G. Racah, Bull. Res. Council Israel **8F**, 1 (1959).

[17] A comparison of different calculational approaches in Ne I is given by R. M. Schectman, D. R. Shoffstall, D. G. Ellis, and D. A. Chojnacki, J. Opt. Soc. Am. 63, 80 (1973).

16-4. BASIC LEAST-SQUARES METHOD

We now describe a straightforward and practical iterative procedure for finding the least-squares values of the radial parameters, and at the same time finding the corresponding eigenvalues and eigenvectors.[18,16]

We denote the various energy parameters (E_{av}, F^k, G^k, ζ, R^k), arranged in some arbitrary order, by the symbols $x_l (1 \leqq l \leqq N_p)$; the parameters may be arranged in the form of a column vector

$$X = \begin{pmatrix} x_1 \\ x_2 \\ x_3 \\ \cdot \\ \cdot \\ \cdot \end{pmatrix} . \tag{16.2}$$

The matrix of coefficients of x_l in any chosen pure-coupling representation will be denoted

$$C^l = (c^l_{ij}) . \tag{16.3}$$

(In the case of a single-configuration calculation, then for the parameter $x_l = E_{av}$, C^l is the identity matrix; for a parameter $x_l = F^k$, C^l is the matrix of coefficients f_k; etc.) Similarly to (12.47), the Hamiltonian matrix may be written

$$H = \sum_l C^l x_l , \tag{16.4}$$

the elements of which are

$$H_{ij} = \sum_l c^l_{ij} x_l ; \tag{16.5}$$

for present purposes, we consider H to be a single matrix that includes basis states of all J, even though H of course breaks up into diagonal blocks according to the different possible values of J.

From the eigenvalue equation (4.11),

$$HY^k = E^k Y^k , \tag{16.6}$$

and from the orthonormality of the eigenvectors, it follows that the kth eigenvalue is

$$\begin{aligned} E^k &= (Y^k)^{tr} H Y^k \\ &= \sum_i \sum_j y^k_i H_{ij} y^k_j \\ &= \sum_l \Big(\sum_i \sum_j y^k_i c^l_{ij} y^k_j \Big) x_l . \end{aligned} \tag{16.7}$$

[18]G. H. Shortley, Phys. Rev. 47, 295 (1935).

We define V to be the $N_k \times N_p$ matrix whose elements are

$$v_{kl} \equiv \sum_i \sum_j y_i^k c_{ij}^l y_j^k \cong \partial E^k / \partial x_l \ ; \tag{16.8}$$

the approximate equality is valid to the extent that v_{kl} remains constant as the x_l are varied.

We arrange the eigenvalues E^k into a column vector

$$E \equiv \begin{pmatrix} E^1 \\ E^2 \\ E^3 \\ \cdot \\ \cdot \\ \cdot \end{pmatrix}, \tag{16.9}$$

and write (16.7) as the matrix equation

$$E = VX \ . \tag{16.10}$$

We write the observed energy levels ("term values") of the atom, T^k, in the form of a vector T analogous to E, and our problem then is to minimize the residual

$$\begin{aligned} R &\equiv \sum_k (E^k - T^k)^2 \\ &= |E - T|^2 \\ &= (VX - T)^{tr}(VX - T) \\ &= \tilde{X}\tilde{V}VX - \tilde{X}\tilde{V}T - \tilde{T}VX - \tilde{T}T \ . \end{aligned} \tag{16.11}$$

Minimization of R with respect to each of the x_l gives the vector equation

$$(\partial R/\partial x_l) = 0 = \tilde{V}VX + (\tilde{X}\tilde{V}V)^{tr} - \tilde{V}T - (\tilde{T}V)^{tr} = 2\tilde{V}VX - 2\tilde{V}T$$

or

$$(\tilde{V}V)X = (\tilde{V}T) \ . \tag{16.12}$$

This represents a set of N_p nonhomogeneous linear equations in the N_p unknowns x_l, and can be solved by standard methods to give the vector **X** of energy parameters—for example, by finding the inverse of the square matrix $\tilde{V}V$:

$$X = (\tilde{V}V)^{-1}(\tilde{V}T) \ . \tag{16.13}$$

[Actually, it is convenient to use, instead, Eq. (16.17) below.]

There are three complications involved in finding the solution (16.13). Firstly, calculation of the matrix V from (16.8) requires knowledge of the eigenvectors $\mathbf{Y}^k$, and finding the eigenvectors from (16.6) requires a knowledge of H and hence from (16.5) of **X**. It is therefore necessary to carry out an iterative process of the sort shown in Fig. 16-5, in

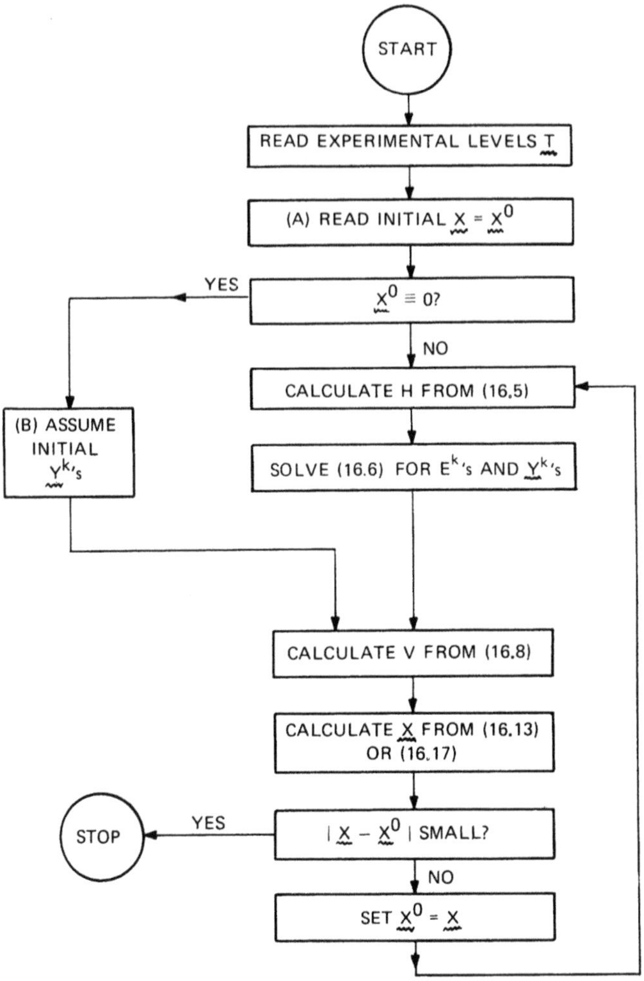

Fig. 16-5. Simplified block diagram of computer program RCE for the least-squares fitting of a set of energy levels T. The coefficient matrices (c^l_{ij}) required in (16.8) are obtained via the option (M) in program RCG (Fig. 16-1).

which one can begin either at point A by assuming non-zero estimates of the parameter values x_l, or at point B by assuming approximate eigenvectors. The second starting method is straightforward for simple cases approaching pure-coupling conditions if the calculation is done in the appropriate pure-coupling representation: one simply assumes basis vectors ($y^k_i = \delta_{ik}$) for the initial eigenvector estimates. In general, however, the calculation is started at point A by assuming, say, *ab initio* scaled-HF values for the parameters x_l, or by extrapolating least-squares values from a neighboring atom or ion (perhaps using HF values as a guide).

The second complication is the problem of deciding how to assign the values of k to the various eigenvalues (and their corresponding eigenvectors)—i.e., how to make the correlation between eigenvalues E^k and term values T^k that is implied in (16.11). In principle this is simple: In the process of calculating the elements of the coefficient matrices C^l, specific basis states b were assigned to the various rows and columns i and j of C^l. The preliminary empirical term analysis assigns a specific value of b to each experimental level T^k, and this then correlates k with i or j as far as the T^k are concerned. The value of k for a given eigenvector $Y^k \equiv (y^k_i)$ is the value of i for which $|y^k_i|$ is largest (Sec. 10-7). Since the correlation of eigenvalues with eigenvectors is always known, we thus obtain a definite correlation between the T^k and E^k through the dominant components of the eigenvectors, provided there *is* an unambiguously dominant component for each vector—i.e., provided the eigenvectors can be assigned values of k such that the matrix (y^k_i) of eigenvectors is somewhere near diagonal. For this purpose the largest component y^k_k may be considered dominant if, say,

$$|y^k_i|^2 < 0.7|y^k_k|^2 , \qquad \text{all } i \neq k . \tag{16.14}$$

If some of the eigenvectors do not have dominant components, the best that can be done is to assign those eigenvectors (and corresponding eigenvalues) that are unambiguous, and then (for each value of J) to correlate the remaining E^k with the remaining T^k by magnitude—the smallest remaining E^k being correlated with the smallest remaining T^k, etc. If the coupling conditions of the atom are not close to any pure-coupling scheme or if the calculation is done in a representation other than that which best approximates the coupling conditions, it may be necessary to make correlations by magnitude for all k; it is then essential that rather good initial estimates be used for the x_l and that the course of the iteration be examined carefully each cycle.

This brings us to the third complication. It may well happen that on one cycle of the iteration two or more eigenvectors of the same J do not have dominant components, so that the corresponding correlations are made by eigenvalue magnitude. On the next cycle, one of these eigenvectors may change sufficiently that it acquires a dominant component according to the criterion (16.14); the change need be only very small in a borderline case. The result may be an (effectively discontinuous) change in the correlation between the E^k and the T^k; this causes a change in the course of the iteration which may produce wild changes in the x_l, which may in turn cause the iteration to blow up.[19] Various solutions to this problem are possible. The difficulty may arise from incorrect term assignments having been made for two or more experimental levels, resulting in the computer attempting the impossible task of fitting the wrong levels; in this case the solution is to correct the assignments,[20] and the calculation has accomplished its goal of detecting such errors. If the difficulty is only accidental, and due to inherent strong basis function mixing, it can sometimes be eliminated by replacing the criterion (16.14) with a somewhat more stringent

[19]For this reason, some workers prefer not to let the computer program perform the automatic iteration shown in Fig. 16-5, but to calculate for only one cycle at a time and to make the eigenvalue-termvalue correlation by hand.

[20]This can be programmed to be done automatically by having the computer (after a preset minimum number of iterations) reorder the *term values* to have the same order by magnitude as do the eigenvalues (rather than vice versa, as done when there are no dominant eigenvector components).

one, or by performing the calculation in a more appropriate pure-coupling representation (in which case, of course, all energy levels have to be given new designations corresponding to the new coupling scheme, and the calculation has accomplished its goal of forcing the use of the proper coupling scheme).

Even with the most appropriate representation and correct term assignments, there may be some parameters x_l (notably configuration-interaction parameters R^k) whose values have such a small effect on the calculated eigenvalues that they tend to fluctuate wildly during the course of the iteration, and may destroy convergence altogether. It may be necessary to hold these parameters fixed (see next section) at, for example, their theoretical values.

Alternatively, the fluctuations can be controlled (perhaps to the point of making true convergence possible) by the following means. If $\mathbf{X}^0$ and $\mathbf{E}^0$ are values from the preceding iteration cycle, then from (16.10)

$$\tilde{V}E^0 = \tilde{V}VX^0 . \qquad (16.15)$$

Subtraction of this from (16.12) gives

$$(\tilde{V}V)(X - X^0) = \tilde{V}(T - E^0) . \qquad (16.16)$$

Thus the predicted changes Δx_l for the current cycle are given by

$$\Delta X \equiv X - X^0 = (\tilde{V}V)^{-1}\tilde{V}(T - E^0) . \qquad (16.17)$$

However, we actually take for the new estimates

$$X = X^0 + c \cdot \Delta X , \qquad (16.18)$$

where c is chosen such that for the largest $|\Delta x_l|$,

$$c|\Delta x_l| = (\Delta x)_{max} , \qquad (16.19)$$

where $(\Delta x)_{max}$ is some value (such as 100 cm^{-1}) that is small enough to control the erratic fluctuations.

Experience indicates that even when the iteration is proceeding smoothly, the indicated change (16.17) tends to overshoot the mark in some cases and to undershoot in others. Convergence can therefore be speeded somewhat by decreasing or increasing the value of c, depending on whether or not $\mathbf{\Delta X}$ tends to change sign each cycle.

On each cycle of the iterative calculation, we have a set of parameter values

$$x_l \pm \delta x_l , \qquad (16.20)$$

where an estimate of the uncertainty in each value is given by[21]

$$\delta x_l = \{[(\tilde{V}V)^{-1}]_{ll}\}^{1/2} s , \qquad (16.21)$$

[21]See, for example, A. Hald, *Statistical Theory with Engineering Applications* (John Wiley, New York, 1952), Eq. (20.3.8).

where s is the standard deviation (16.1). The calculation can be considered to have converged when the changes Δx_l given by (16.17) have become small compared with the uncertainties δx_l.

It is important to note that these uncertainties are strictly statistical estimates, based on the error in the fit and the degree to which the computed eigenvalues are insensitive to changes in the parameter values. They do not necessarily have any physical significance. For example, in an LS-coupled p^5d configuration [Eqs. (12.79) and Fig. 12-7], the parameter $G^1(pd)$ affects only the energy of the $^1P^o_1$ level and will always be determined exactly; however, the least-squares value of G^1 can be legitimately compared to the *ab initio* theoretical value only if all important $p^5d - p^5d'$ configuration interactions have been included in the least-squares calculation.[21a]

*16-5. MODIFICATIONS OF THE BASIC METHOD

Missing levels. If there are experimental levels T^k that are suspected of being incorrect or of being strongly perturbed by interactions not included in the calculation, or if there are values that are completely unknown, then it is necessary to be able to delete these values from consideration in the least-squares fit. To ignore a term value T^m and the corresponding eigenvalue E^m in (16.11), we need only delete the mth component of T and of $E = VX$. Both of these can be accomplished in (16.10), (16.12), and (16.16) by simply setting

$$(\tilde{V})_{lm} = v_{ml} = 0 \ , \qquad \text{each } l \ . \tag{16.22}$$

Fixed parameters. It is frequently necessary to hold certain parameters fixed (at, say, their scaled HF values), at least during early cycles of the iteration. This is particularly true of some configuration-interaction parameters R^k (especially those of large k), which may have little effect on computed eigenvalues, and whose values may therefore fluctuate so wildly as to prevent convergence. It is probably true also for any parameter which, if left free, has a least-squares value much smaller than its uncertainty δx_l; this tends to be the case for spin-orbit parameters of excited electrons, whose values are very small and sometimes come out to be unphysically negative. In two-electron configurations of the form $nln'l$, the coefficient matrices of both spin-orbit parameters are the same, and hence only $\zeta_{nl} + \zeta_{n'l}$ is determined by the least-squares calculation; it is necessary to hold $\zeta_{n'l}$ fixed at its theoretical value to prevent both ζ's from fluctuating wildly, and/or to obtain a reliable value of ζ_{nl}.

To hold fixed a given parameter x_j, it is necessary in (16.12) to delete the jth linear equation (since this came from minimizing R with respect to x_j), and to treat the term $(\tilde{V}V)_{ij}x_j$ of every other equation as a constant by subtracting it from both sides of the equation. The first of these requirements is conveniently met in a computer program by setting $(\tilde{V}V)_{ji} = 0$, all i, and $(\tilde{V}T)_j = 0$ [or $(\tilde{V}T - \tilde{V}E^0)_j = 0$ if (16.16) is being used in place of (16.12)]; the second requirement has already been accomplished in the right side of (16.16), but needs formalizing in the left side by setting $(\tilde{V}V)_{ij} = 0$, all i. All of this can be done at once by setting $v_{kj} = 0$, all k. The resulting singularity in $\tilde{V}V$ can be removed by setting $(\tilde{V}V)_{jj} = 1$;

[21a]For similar situations in other configurations, see M. Wilson, J. Phys. B 2, 514 (1969).

solution of (16.16) will still automatically give the desired result $\delta x_j = 0$. In summary, for each parameter j to be held fixed, we formally set

$$v_{kj} = 0 \ , \qquad \text{all } k \ , \tag{16.23}$$

and after calculating $\tilde{V}V$ set

$$(\tilde{V}V)_{jj} = 1 \ , \tag{16.24}$$

and then use (16.16)-(16.17) rather than (16.12)-(16.13).

Linked parameters. It is sometimes convenient to be able to link several parameters together in such a way that the mutual ratios of their values remain constant during the iteration. In the configuration $f^w l$, for example, if only levels based on one or two terms of f^w are known, then too little information is known to determine $F^2(ff)$, $F^4(ff)$, and $F^6(ff)$ individually, but a least-squares fit might be made in which the ratios F^2:F^4:F^6 were held at (say) their HF values. Or, in any configuration, empirical best scale factors for HF values could be determined by linking together all Coulomb parameters in one group and all spin-orbit parameters in a second group; one would then be doing a least-squares fitting in which (in addition to E_{av}) there were only two adjustable parameters—namely, the scale factors for the two kinds of interaction parameters. To treat such cases mathematically, we replace the single subscript l of Sec. 16-4 by the double subscript ij, representing the ith parameter of the jth linked group. Then in place of (16.7)-(16.8) we have

$$E^k = \sum_j \sum_i v_{k(ij)} x^0_{(ij)} q_j \ , \tag{16.25}$$

where x^0 is the starting parameter value and $x^0 q$ is the parameter value on the current iterative cycle, q being the quantity actually changed during the iteration. Minimizing

$$R \equiv \sum_k \left(\sum_j \sum_i v_{k(ij)} x^0_{(ij)} q_j - T^k \right)^2$$

with respect to each q_j gives

$$\sum_k \left(\sum_j \sum_i v_{k(ij)} x^0_{(ij)} q_j - T^k \right) \left(\sum_{i'} v_{k(i'j')} x^0_{(i'j')} \right) = 0$$

or

$$\tilde{U}UQ = \tilde{U}T \ , \tag{16.26}$$

where Q is a vector of the N_p' ($<N_p$) scale factors q_j, and U is the $N_k \times N'_p$ matrix with elements

$$u_{kj} \equiv \sum_i v_{k(ij)} x^0_{(ij)} \ . \tag{16.27}$$

If E^0 is the vector of eigenvalues obtained using the scale factors Q^0 from the preceding cycle, then the indicated changes Δq_j on the current cycle are given by

$$\Delta Q \equiv Q - Q^0 = (\tilde{U}U)^{-1}\tilde{U}(T - E^0) . \tag{16.28}$$

Fine-structure fitting. When coupling conditions lie close to some pure-coupling scheme, the basic least-squares method may give rather poor values for the small parameters that are responsible for the fine-structure level separations. Under near-LS conditions, for example, the spin-orbit parameters may come out with values determined not so much to properly fit the LS-term splittings as to try to help fit the centers of gravity of the terms (which the Coulomb parameters have been unable to fit exactly, because of inadequacies of the theory). Similarly, under near-jK conditions the values of ζ_2 and G^k (which should affect mainly the pair splittings) may be unduly influenced in an attempt to better fit the pair centers of gravity (which should be accomplished almost entirely by ζ_1 and F^k).

A way of handling this problem is to divide the parameters x_l into two groups, one group *a* (including E_{av}) to be varied to fit a set of center-of-gravity energies T^k_c, and a second group *b* to be varied so as to fit a set of splittings $T^k - T^k_c$ from the appropriate centers of gravity. We define an $N_k \times N_p$ matrix V_c with elements

$$(V_c)_{kl} = \frac{1}{N_j} \sum_i{}' v_{il} , \qquad k \text{ in term } j , \tag{16.29}$$

where v_{kl} is the standard matrix element defined by (16.8), and where the summation is over only those N_j levels i that belong to the jth fine-structure group (LS term, or level pair, etc.). We then define a matrix W with elements

$$\begin{aligned} W_{kl} &= (V_c)_{kl} , \qquad l \text{ in group } a , \\ W_{kl'} &= v_{kl'} - (V_c)_{kl'} , \qquad l' \text{ in group } b . \end{aligned} \tag{16.30}$$

If l is in group *a* and l' in group *b*, then from (16.29)

$$\begin{aligned} (\tilde{W}W)_{ll'} &= \sum_k (V_c)_{lk}[v_{kl} - (V_c)_{kl}] \\ &= \sum_j (V_c)_{lk} \sum_i{}' [v_{kl} - (V_c)_{kl}] \\ &= 0 , \end{aligned} \tag{16.31}$$

so that $\tilde{W}W$ consists of two diagonal blocks corresponding to the two groups of parameters. We define a vector D with elements

$$\begin{aligned} D_l &= \sum_k W_{lk} T^k_c , \qquad l \text{ in group } a , \\ D_{l'} &= \sum_k W_{l'k}(T^k - T^k_c) , \qquad l' \text{ in group } b . \end{aligned} \tag{16.32}$$

Then the set of N_p linear equations

$$(\tilde{W}W)X = D \tag{16.33}$$

really represents two independent sets of equations, one corresponding to least-squares adjustment of the parameter set *a* to fit the centers of gravity

$$T_c^k = \frac{\sum_i'(2J_i + 1)T^i}{\sum_i'(2J_i + 1)}, \tag{16.34}$$

and the other corresponding to least-squares adjustment of the set *b* to fit the fine-structure displacements $T^k - T_c^k$ from the pertinent centers of gravity.

This least-squares method has been little used because of its complexity and its applicability only to rather pure coupling conditions. It has, however, given satisfactory values of G^k and ζ_f in the 4pnf configurations of Ge I, where the basic least-squares method gave unphysical negative values.[22] Nevertheless, it is easier and probably generally about as satisfactory to simply use the more straightforward least-squares methods, with the relatively unimportant parameters held fixed at scaled HF values, or with inclusion of effective-operator parameters (Sec. 16-7) to allow more flexibility in the fitting process and thereby reduce the artificial strain on the fine-structure parameters.

Fitting of g factors. In principle, eigenvectors (and hence the angular nature of the wavefunctions) can be improved by including in the least-squares fitting process not only experimental energy levels, but also data on other physical quantities whose values depend more sensitively on the eigenvector components. Perhaps the most sensitive quantities are line strengths of weak lines; however, experimental data are usually of low accuracy, if available at all—and indeed, one of the major purposes of obtaining accurate eigenvectors is the theoretical calculation of line strengths. The only quantities used in practice are Landé g factors;[23,24,17] thus the parameters are varied to minimize

$$R = \sum_k (E^k - T^k)^2 + c\sum_k (g_{calc}^k - g_{exp}^k)^2 , \tag{16.35}$$

where c is some constant chosen to make the g-factor terms comparable in magnitude with the energy-level terms. The method is useful only when conditions are far from pure LS coupling so that computed g factors are sensitive to the eigenvectors. Even then, results depend on the value assumed for the arbitrary constant c.[17] By and large, one can again probably do as well by using scaled HF values in place of least-squares parameter values that have high uncertainties δx_l or that appear to be obviously out of line, or by introducing effective operators.

[22] R. D. Cowan and K. L. Andrew, J. Opt. Soc. Am. 55, 502 (1965).

[23] G. Racah, paper scheduled to have been given at the Zeeman Centennial Conference, Amsterdam, July, 1965 (cancelled because of his untimely death).

[24] R. Mehlhorn, J. Opt. Soc. Am. 59, 1453 (1969).

16-6. PHASE OF THE CONFIGURATION-INTERACTION PARAMETERS

In making multi-configuration least-squares fits, some ambiguity exists in the signs of the CI parameters R^k: for any given pair of configurations, the relative signs of the various R^k are uniquely determined, but the overall phase of the set of values $\{R^k\}$ may be determined only to within a factor ± 1. In a two-configuration fit, for example, changing the signs of all the R^k does not change the computed eigenvalues, but only changes the signs of all eigenvector components for basis functions of one configuration relative to those for the other configuration; these changes correspond to reversing the phase of (the radial wavefunction for one active electron of) each basis function of one configuration. In a three-configuration fit, changing the phase of each basis function of configuration A corresponds to a change in sign of all R^k for interactions A-B and A-C, but not for the interaction B-C, etc.

These phase ambiguities are of no consequence if only computed eigenvalues are of interest. They are, however, of critical importance in the calculation of line strengths. A phase error for the basis functions of one configuration means that configuration mixing effects that should lead to destructive interference will instead produce constructive interference, and vice versa, resulting in completely incorrect oscillator strengths. It is thus essential that the phases of R^k be consistent with those of the electric dipole matrix elements. The only way that this can be assured is to calculate *ab initio* (HF) phases of the R^k [using phase conventions identical with those used in calculating the $\boldsymbol{P}^{(t)}$], and to alter the signs of the various sets of least-squares R^k (if necessary) to agree with the theoretical signs. One may run into difficulties here if (for some pair of configurations) the relative signs of the least-squares R^k are not the same as the relative signs of the *ab initio* R^k; however, this usually happens only when some R^k are very small in magnitude—as a consequence of a small value of the cancellation factor

$$\frac{\displaystyle\iint \frac{r_<^k}{r_>^{k+1}} P_1(r_1)P_2(r_2)P_3(r_1)P_4(r_2)\, dr_1\, dr_2}{\displaystyle\iint \left| \frac{r_<^k}{r_>^{k+1}} P_1(r_1)P_2(r_2)P_3(r_1)P_4(r_2) \right| dr_1\, dr_2} \tag{16.36}$$

for the *ab initio* R^k in question—in which case these R^k values may be considered inaccurate, and so may be ignored in deciding on the proper phase choice.

*16-7. EFFECTIVE OPERATORS

Even at best—including as many interacting configurations as feasible, as well as small magnetic (spin-spin, orbit-orbit, and spin-other-orbit) interactions.[12,25,26]—it is not in

[25]B. R. Judd, *Operator Techniques in Atomic Spectroscopy* (McGraw-Hill, New York, 1963).

[26]S. Fraga and G. Malli, *Many-Electron Systems: Properties and Interactions* (W. B. Saunders, Philadelphia, 1968).

general possible to fit experimental energy levels with the accuracy one might wish. Difficulties are particularly severe for complicated transition and rare-earth configurations involving partially filled d and f subshells, where the average spacing between energy levels may be so small as to be comparable with the standard deviation of the fit. In an attempt to improve the quality of a fit (and therefore, presumably, the accuracy of the resulting eigenvectors), a bewildering variety of "effective-operator" parameters have been introduced, representing corrections to both the electrostatic[27-35,11,12] and the magnetic[35-37] single-configuration effects. Only a few of the simpler electrostatic operators will be described here.

Rajnak and Wybourne[11,12] have shown that the effective two-body operators have the same angular dependence as do the L- and S-dependent factors of the expressions (12.17), (12.19) and (12.38) for $f_k(l_i l_j)$ and $g_k(l_i l_j)$, including *all* values $|l_i - l_j| \leqq k \leqq l_i + l_j$. For the "legal" values of k ($l_i + k + l_j$ even), these effective operators are therefore identical in form with the normal Coulomb-interaction terms of the Hamiltonian matrix. In the least-squares approach their effects are automatically absorbed into the (smaller-than-theoretical) values obtained for the usual Slater integrals F^k and G^k; in *ab initio* calculations these effects are roughly accounted for by scaling down the theoretical values of F^k and G^k, as discussed in Sec. 16-2.

The simplest way of introducing the "illegal-k" effective operators is to replace the normal Coulomb operator $C_i^{(k)} \cdot C_j^{(k)}$ with the unit operator $u_i^{(k)} \cdot u_j^{(k)}$, and thereby add to the Hamiltonian matrix elements terms of the form

$$\sum_k \sum_j f_{ku}(l_j l_j)\, F^k(l_j l_j) \,, \qquad k \text{ odd} \,, \tag{16.37}$$

$$\sum_k \sum_{i<j} f_{ku}(l_i l_j)\, F^k(l_i l_j) \,, \qquad k \text{ odd} \,, \tag{16.38}$$

[27]R. F. Bacher and S. Goudsmit, Phys. Rev. 46, 948 (1934).

[28]R. E. Trees, Phys. Rev. 83, 756 (1951), 84, 1089 (1951), 85, 382 (1952), J. Res. Natl. Bur. Stand. 53, 35 (1954).

[29]G. Racah, Phys. Rev. 85, 381 (1952).

[30]N. Sack, Phys. Rev. 102, 1302 (1956).

[31]B. R. Judd, Phys. Rev. 141, 4 (1966); *Second Quantization and Atomic Spectroscopy* (The Johns Hopkins Press, Baltimore, 1967).

[32]S. Feneuille, C. R. Acad. Sci. 262B, 23 (1966).

[33]G. Racah and J. Stein, Phys. Rev. 156, 58 (1967).

[34]Y. Shadmi, in F. Bloch, et al., eds., *Spectroscopic and Group Theoretical Methods in Physics* (North-Holland Publ. Co., Amsterdam, 1968).

[35]H. M. Crosswhite, Phys. Rev. A 4, 485 (1971).

[36]K. Rajnak and B. G. Wybourne, Phys. Rev. 134, A596 (1964).

[37]H. Crosswhite, H. M. Crosswhite, and B. R. Judd, Phys. Rev. 174, 89 (1968); H. Crosswhite and B. R. Judd, At. Data 1, 329 (1970); H. M. Crosswhite and A. P. Paszek, J. Opt. Soc. Am. 62, 1343A (1972).

$$\sum_{k}\sum_{i<j} g_{ku}(l_il_j)\, G^k(l_il_j) \;, \qquad l_i + k + l_j \text{ odd} \;. \tag{16.39}$$

The coefficients f_{ku} are identical with the expressions (12.17) and (12.19) except with the reduced matrix elements of $C^{(k)}$ replaced by unity; the coefficient g_{ku} is given by (12.38) except with $\langle l_i\|C^{(k)}\|l_j\rangle\,\langle l_j\|C^{(k)}\|l_i\rangle$ replaced by unity [cf. (11.85)], and hence with $\langle l_i\|C^{(k)}\|l_j\rangle^2$ replaced by $(-1)^k$. The illegal-k parameters F^k and G^k are not radial integrals to be calculated theoretically in the same way as for the analogous legal-k parameters, but are only to be determined empirically by least-squares level fitting.

The inequivalent-electron expressions (16.38) and (16.39) are a generalization of the effective operators used by Crosswhite[35] for two-electron configurations. The coefficients f_{ku} and g_{ku} (and hence also our parameter values F^k and G^k) differ by a phase factor $(-1)^{k+1}$ from the corresponding quantities obtained using the Fano-Racah phase conventions;[38] the parameters using the Fano-Racah convention are denoted F^k and G^k by the Israeli group,[39] and D^k and X^k by the National Bureau of Standards group.[40]

For group-theoretical reasons similar to those that led Racah to replace the equivalent-electron legal-k parameters $F^k(l_jl_j)$ by the linear combinations E^k of (6.44), the equivalent-electron illegal-k expression (16.37) is usually replaced by

$$\delta_{b_jb'_j}[c_1\alpha + c_3\beta + c_5\gamma] \;, \tag{16.40}$$

where α, β, and γ are new radial parameters; the angular coefficients c_p ($p \leqq 2l_j - 1$) are linear combinations of the f_{ku}, and are taken to be non-zero only for matrix elements diagonal in the basis states $b_j \equiv \alpha_jL_jS_j$ of $l_j^{w_j}$. The linear combinations are chosen such that[12,41] $c_1 = L_j(L_j + 1)$, $c_{2l-1} = G(R_{2l+1})$ for $l \equiv l_j > 1$, and $c_3 = G(G_2)$ for $l = 3$; here $G(R_5)$, $G(G_2)$, and $G(R_7)$ are eigenvalues of the Casimir operators[42,25] for the group R_5 and for the groups G_2 and R_7 used to classify the basis states of the d^w and the f^w subshells, respectively.

Tables of these eigenvalues are available,[12] but we can calculate them (and at the same time eliminate any need to keep track of the group theoretical properties of the basis states) from the expressions[25,43]

$$4G(G_2) = \langle \alpha_jL_jS_j|\,3U^{(1)}\cdot U^{(1)} + 11U^{(5)}\cdot U^{(5)}\,|\alpha_jL_jS_j\rangle \;, \tag{16.41}$$

$$(2l - 1)G(R_{2l+1}) = \sum_{t=1}^{l}(4t - 1)\langle \alpha_jL_jS_j|\,U^{(2t-1)}\cdot U^{(2t-1)}\,|\alpha_jL_jS_j\rangle \;. \tag{16.42}$$

[38]This phase difference arises from an explicitly introduced minus sign together with the different phase convention for $\langle l\|C^{(k)}\|l'\rangle$—see fn. 7 of Chap. 11.

[39]Z. B. Goldschmidt and J. Starkand, J. Phys. B. 3, L141 (1970).

[40]J. Sugar and V. Kaufman, J. Opt. Soc. Am. 62, 562 (1972), 64, 1656 (1974).

[41]The angular coefficients are sometimes chosen to be $5G(R_7)$ or $10G(R_7)$ and $12G(G_2)$.

[42]H. Casimir, Proc. K. Akad. Wet. Amsterdam 34, 844 (1931).

[43]G. Racah IV, Phys. Rev. 76, 1352 (1949).

We shall modify the coefficients by subtracting their weighted-average values, so that (16.40) makes no contribution to the center-of-gravity energy of the configuration (the appropriate contribution already having been allowed for in the correlation-energy correction to E_{av}). We then have

$$c_3 = \frac{2}{3}(3f_{1u} + 7f_{3u}) , \qquad l_j = 2 , \tag{16.43}$$

$$c_3 = \frac{1}{2}(3f_{1u} + 11f_{5u}) , \qquad l_j = 3 , \tag{16.44}$$

$$c_5 = \frac{2}{5}(3f_{1u} + 7f_{3u} + 11f_{5u}) , \qquad l_j = 3 , \tag{16.45}$$

where $f_{ku} = f_{ku}(l_j l_j)$. From (2.15) and (11.51)-(11.52) we find that

$$\begin{aligned} L_j(L_j + 1) &= \langle l_j^{w_j}\alpha_j L_j S_j | L^{(1)} \cdot L^{(1)} | l_j^{w_j}\alpha_j L_j S_j \rangle \\ &= \langle l_j \| l^{(1)} \| l_j \rangle^2 \langle \alpha_j L_j S_j | U^{(1)} \cdot U^{(1)} | \alpha_j L_j S_j \rangle . \end{aligned} \tag{16.46}$$

Using (11.21) we find that, including the center-of-gravity correction from (12.17),

$$\begin{aligned} c_1 &= 2[l_j(l_j + 1)(2l_j + 1)]f_{1u} \\ &= L_j(L_j + 1) - \frac{w_j(4l_j + 2 - w_j)l_j(l_j + 1)}{(4l_j + 1)} . \end{aligned} \tag{16.47}$$

The term $c_1\alpha$ is basically the $L_j(L_j + 1)$ correction first introduced empirically by Trees[28] in his investigation of $d^w s$ configurations of iron-group elements—its physical interpretation as representing sums of pair correlations of the form $l_i \cdot l_j$ was first given by Racah.[29]

A few attempts have been made[13] to obtain *ab initio* theoretical values of the parameters α, β, and γ. However, the calculations are extremely laborious and are impractical for routine energy-level calculations, and so these parameters are usually only determined empirically by least-squares fitting of energy levels.

Numerous other empirical operators are sometimes used, such as the Sack[30] operators

$$\delta_{bb'}[L(L + 1)a + S(S + 1)b] \tag{16.48}$$

(where L and S are total angular momenta of the complete multi-subshell configuration) and various effective three-body operators.[31-34]

The total number of operators can easily be greater than the number of energy levels being fit, and so they must be used with considerable care. That these operators have a physical as well as a purely mathematical significance is shown by the fact that not only does their introduction produce a statistically-significant reduction in the standard deviation (16.1), but also the least-squares values of the effective-operator parameters vary smoothly from one element to another through a transition-element or rare-earth series.[33,34,44] Nevertheless, it is important to remember that though these operators may improve an energy-level fit and the angular composition of computed wavefunctions (eigenvectors), they do not in general give a proper indication of radial electron-correlation effects. Specifically, they cannot give any indication of transfers of line strength from one transition array to another; such effects can be treated only by calculating the radial correlation effects in detail—e.g., by the explicit configuration-interaction calculations discussed in Secs. 14-15 and 18-8.

*16-8. LS-DEPENDENT HARTREE-FOCK CALCULATIONS

In all discussions up to this point, we have (for any given electron configuration) considered only basis functions constructed from a set of radial wavefunctions $P_{nl}(r)$ that were the same for all basis functions of the configuration. These radial wavefunctions were obtained by choosing them so as to minimize the computed energy for the center of gravity of the configuration.

For configurations in which conditions approximate pure LS coupling, an alternative approach is to use a different set of radial wavefunctions for each different term γLS of the configuration. Evidently, these radial wavefunctions are to be obtained by choosing them so as to minimize the energy of the term in question. For the present, let us suppose that there is only one term having the value of LS in question (i.e., that there is only one possible value of γ for this LS). Then the center-of-gravity energy of this term (that is, the energy neglecting spin-orbit effects) is given by the corresponding diagonal element of the Hamiltonian matrix:

$$E_{av} + \sum_k \sum_i \left\{ f_k(l_i l_i) F^k(l_i l_i) + \sum_{j>i} [f_k(l_i l_j) F^k(l_i l_j) + g_k(l_i l_j) G^k(l_i l_j)] \right\} , \tag{16.49}$$

where the angular coefficients f_k and g_k are the diagonal quantities (12.17), (12.19), and (12.38) for the LS basis function in question. Taking the variation of (16.49) with respect to $P_i(r)$, we obtain an LS-dependent (LSD) Hartree-Fock equation identical with (7.11) except for the presence of the additional direct terms

$$\frac{1}{w_i} \sum_k \left\{ 2f_k(l_i l_i) \int_0^\infty \frac{2r_<^k}{r_>^{k+1}} P_i^2(r_2)\, dr_2 + \sum_{j\neq i} f_k(l_i l_j) \int_0^\infty \frac{2r_<^k}{r_>^{k+1}} P_j^2(r_2)\, dr_2 \right\} \tag{16.50}$$

within the brackets on the left side of (7.11), and the additional exchange terms

$$-\frac{P_j}{w_i} \sum_k \sum_{j\neq i} g_k(l_i l_j) \int_0^\infty \frac{2r_<^k}{r_>^{k+1}} P_j(r_2) P_i(r_2)\, dr_2 \tag{16.51}$$

[44]C. Roth, J. Res. Natl. Bur. Stand. 73A, 125 (1969); H. M. Crosswhite, in *Spectroscopie des éléments de transition et des éléments lourds dans les solides* (Éditions du Centre National de la Recherche Scientifique, Paris, 1977), p. 65; see also H. M. Crosswhite and W. T. Carnall, J. Opt. Soc. Am. 63, 1316 (1973).

on the right-hand side.

In most cases, the radial functions P_{nl} obtained by the LSD-HF method do not vary greatly from one LS term to another. The center-of-gravity (c-g) HF method is then much preferable because only one HF calculation per configuration is required instead of several, and because a strictly orthogonal set of basis functions is obtained for the entire configuration. Term separations within a configuration can also be calculated with considerably greater assurance (especially for large-N atoms) because the value of E_{av} in (16.49) is the same for all terms, whereas with LSD-HF wavefunctions the value of the very large quantity "E_{av}" may vary significantly from term to term.

There are, however, numerous cases in which the radial functions differ markedly from term to term. A well-known example[45] is Be I 2s2p $^1P^\circ$, $^3P^\circ$. Another example is $3p^53d\ ^1P^\circ$ in ions of the Ar I isoelectronic sequence.[46] The strong LSD effect here is related to the very strong configuration interaction with $3p^5nd\ ^1P^\circ$ $(n > 3)$ found if c-g radial functions are used (Sec. 13-5), and with the fact that (by Brillouin's theorem, Sec. 13-7) the LSD calculation gives immediately the correct radial "3d" function for the lowest $^1P^\circ$ level; the essential equivalence of the LSD and c-g-configuration-interaction approaches is shown in Fig. 16-6.

Vastly larger effects are found in the configuration Ba I $4d^94f \cdots 6s^2$. In a c-g HF calculation, the 4f wavefunction has collapsed into the inner potential well and $G^1(4d,4f) = 115000$ cm^{-1}; however, with LSD-HF calculations[47] the collapse has not occurred for $^1P^\circ$, and for this term G^1 is only 3 cm^{-1}!

When the radial wavefunctions $P_{nl}(r)$ vary appreciably from one LS term to another, it may be necessary to take account of the corresponding variation of the radial integrals F^k and G^k in least-squares energy-level-fitting calculations.[48] One way of doing this is to scale the coefficients f_k and g_k (in the LS representation) in proportion to the LSD-HF values of F^k and G^k for the various LS terms, and then to carry out the fit in the usual manner, using a single value of each radial integral for the entire configuration.[48,49] Only by thus introducing an effective LS-dependence of radial integrals was it found possible to obtain eigenvectors that gave accurate theoretical g values for the levels of N II and homologous ions.[49] In Ne I, on the other hand, the Sack $L(L + 1)$ correction (16.48) has been found adequate to give much-improved g values and transition probabilities.[50]

So far, we have considered only configurations in which there exists only a single term having the value of LS in question. If there exist two or more terms, then the energy of each term is not given by the simple expression (16.49), because of off-diagonal Coulomb terms in the Hamiltonian matrix. LSD-HF calculations are then usually made by an iterative

[45] D. R. Hartree and W. Hartree, Proc. Roy. Soc. (London) A154, 588 (1936).

[46] J. E. Hansen, J. Phys. B 5, 1083, 1096 (1972).

[47] J. E. Hansen, A. W. Fliflet, and H. P. Kelly, J. Phys. B 8, L127 (1975). Somewhat similar shell-collapse LSD effects on isotope shift and hyperfine structure have been discussed by M. Wilson, Phys. Rev. A 3, 45 (1971).

[48] M. Wilson and M. Fred, J. Opt. Soc. Am. 59, 827 (1969).

[49] H. Li and K. L. Andrew, J. Opt. Soc. Am. 62, 1476 (1972).

[50] R. A. Lilly, J. Opt. Soc. Am. 65, 389 (1975).

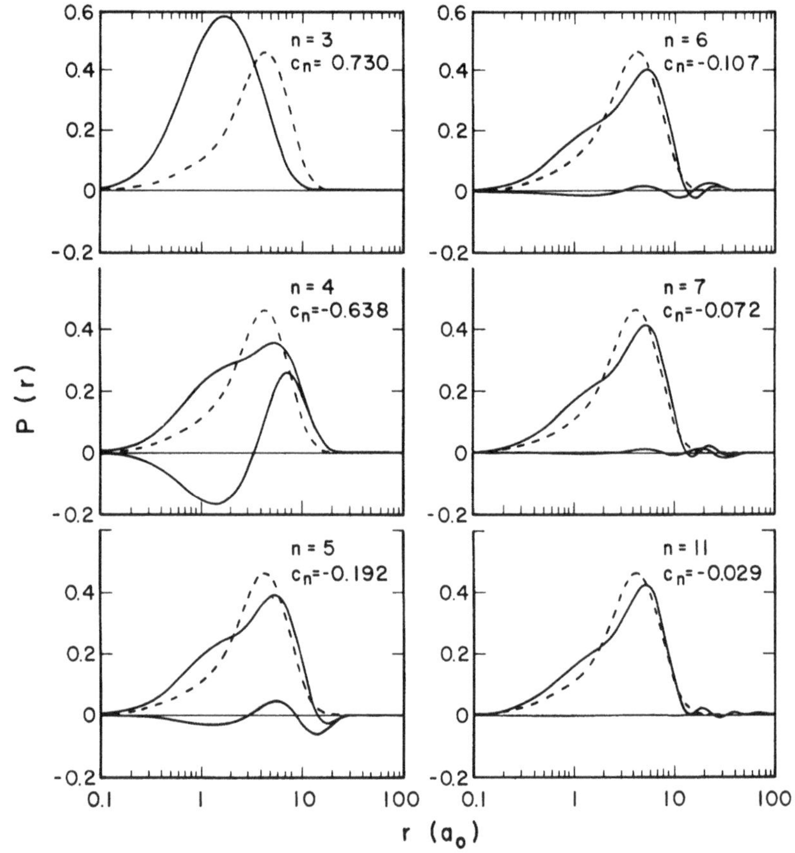

Fig. 16-6. Dotted curves: LSD-HF 3d radial wavefunction for K II $3p^5 3d$ $^1P^o$. Solid curves: (upper) renormalized configuration-interaction "3d" $^1P^o$ radial wavefunction $\Sigma c_{n'} P_{n'd}(r)$ [summed from n′ = 3 to n], and (lower) $c_n P_{nd}(r)$, where the $P_{nd}(r)$ are center-of-gravity HF functions.

procedure in which the off-diagonal terms are treated in the same manner as the configuration-interaction matrix elements of a multiconfiguration HF calculation.[51]

*16-9. HIGHLY EXCITED CONFIGURATIONS

All of the methods described above—either *ab-initio* or least-squares-fitting calculations, even including explicit configuration interactions and/or effective operators and/or LSD-HF effects—work reasonably well only for comparatively low-lying configurations. The more highly excited configurations (e.g., involving an electron with n greater by 6 or 8

[51]C. Froese Fischer, Comput. Phys. Commun. 1, 151 (1969), 4, 107 (1972); *The Hartree-Fock Method for Atoms* (John Wiley, New York, 1977).

than its value in the ground configuration) overlap and mix to such a great extent that perturbations frequently become comparable in magnitude with the distance between successive members of a Rydberg series; i.e., the perturbations produce changes in the effective quantum number n^* that are comparable with unity.[52] The perturbations may become so numerous and irregular that it becomes extremely difficult (if not meaningless) to try to designate an experimental level in terms of any quantum numbers (even configuration) other than J and parity.

A case of special interest is mutual perturbation of two Rydberg series of levels that approach different series limits.[53] At first glance the perturbations appear to vary irregularly with increasing n. However, an interesting (if somewhat complex) regularity appears when the levels are analyzed by a method introduced by Lu and Fano.[54] In this method, two values of n^* are calculated for each level—corresponding to the two series limits; the fractional part of n_1^* is then plotted against the fractional part of n_2^*. This procedure has made it possible to extend the analysis of one of the $J = 1$ Rydberg series in Si I 3pnd from $n = 20$ to $n = 44$.[55] The method becomes increasingly more complex in the case of three or more mutually interacting series, as one then has a multi-dimensional instead of a two-dimensional analysis problem.[56] Zeeman-effect studies of perturbation mixings among these high lying levels have been carried out using multi-photon laser-excitation spectroscopy.[57]

[52]A. G. Shenstone and H. N. Russell, Phys. Rev. 39, 415 (1932); L. J. Radziemski, Jr. and K. L. Andrew, J. Opt. Soc. Am. 55, 474 (1965); L. J. Radziemski, Jr. and V. Kaufman, J. Opt. Soc. Am. 59, 424 (1969).

[53]B. Edlén, in S. Flügge, ed., *Handbuch der Physik* (Springer-Verlag, Berlin, 1964), Vol. XXVII, p. 140.

[54]K. T. Lu and U. Fano, Phys. Rev. A 2, 81 (1970); C.-M. Lee and K. T. Lu, Phys. Rev. A 8, 1241 (1973); A. F. Starace, J. Phys. B. 6, 76 (1973).

[55]C. M. Brown, S. G. Tilford, and M. L. Ginter, J. Opt. Soc. Am. 65, 385 (1975); for similar results in Ge I, Sn I, and Ba I see J. Opt. Soc. Am. 67, 584, 607 (1977), 68, 817 (1978).

[56]J. A. Armstrong, P. Esherick, and J. J. Wynne, Phys. Rev. A 15, 180 (1977).

[57]J. J. Wynne, J. A. Armstrong, and P. Esherick, Phys. Rev. Lett. 39, 1520 (1977).

17

EXTERNAL FIELDS AND NUCLEAR EFFECTS

17-1. THE ZEEMAN EFFECT

Up to this point we have considered energy states only of an isolated atom or ion. We now consider states of an atom or ion that lies in a uniform external magnetic field of flux density **B**. The energy of interaction of the atom with this field is

$$H_{mag} = -\mathbf{B}\cdot\mu \ , \tag{17.1}$$

where the intrinsic magnetic moment of the atom is given by (15.1) as

$$\mu = -\mu_0[J + (g_s - 1)S] \ . \tag{17.2}$$

Here $g_s \cong 2.0023192$ is the anomalous gyromagnetic ratio for the electron spin, J and S are measured in units of h, and the Bohr magneton

$$\mu_0 = \frac{eh}{2mc} = 9.2741\cdot10^{-21} \text{ ergs/gauss} = 4.2543\cdot10^{-10} \text{ Ry/gauss} \tag{17.3}$$

gives in our usual pseudo-energy units

$$\frac{\mu_0}{hc} = \frac{e}{4\pi mc^2} = 4.6686\cdot10^{-5} \text{ cm}^{-1}\text{/gauss} \ . \tag{17.4}$$

The magnetic operator (17.1) must be added to the zero-field Hamiltonian (10.1). Thus in order to find the eigenstates of the atom in the external magnetic field, we have to evaluate matrix elements of (17.1), add these to the matrix elements of (10.1), and diagonalize the combined matrix. The end result of this is that the (2J + 1)-fold degeneracy of each level is completely removed, the level being split into a set of 2J + 1 *sublevels*. Correspondingly, each spectrum line is split into a number of *components* of slightly different wavelengths—a phenomenon first observed by Zeeman.[1]

We shall consider only relatively weak static fields easily producible in the laboratory, with flux densities of 10^4 to 10^5 gauss. However, fields as strong as 10^7 G have been

[1]P. Zeeman, Versl. K. Akad. Wet. Amsterdam 5, 181, 242 (1896), Phil. Mag. 43, 226 (1897).

produced by explosively driven compression of weaker fields[2] and in laser-produced plasmas,[3] and have also been detected in spectra of white dwarfs.[4] It is then necessary to consider a quadratic Zeeman effect,[4] corresponding to diamagnetic effects proportional to B^2.

17-2. MATRIX ELEMENTS OF THE MAGNETIC-ENERGY OPERATOR

Matrix elements of the added term (17.1) of the Hamiltonian are most easily evaluated in an LS representation, using basis functions of the form $|b\rangle = |\alpha LSJM\rangle$. It is customary to choose the coordinate system with the z-axis in the direction of **B**. The magnetic-energy operator may then be written, using the tensor-operator notation (11.12),

$$H_{mag} = -B\mu_z = B\mu_o[J_0^{(1)} + (g_s - 1)S_0^{(1)}] \ . \tag{17.5}$$

Using (11.15) and (11.39), we obtain

$$\frac{1}{B\mu_o}\langle b|H_{mag}|b'\rangle = M\delta_{bb'} + (g_s - 1)(-1)^{M-J}\begin{pmatrix} J & 1 & J' \\ -M & 0 & M' \end{pmatrix}$$

$$\times\ \delta_{\alpha L,\alpha' L'}(-1)^{L+S'+J+1}[J,J']^{1/2}\begin{Bmatrix} L & S & J \\ 1 & J' & S \end{Bmatrix}\langle \alpha S\|S^{(1)}\|\alpha' S'\rangle \ .$$

From (5.1), the 3-j symbol is zero unless $M = M'$. Using (11.20) we obtain finally

$$\frac{1}{B\mu_o}\langle b|H_{mag}|b'\rangle = M\delta_{bb'} - \delta_{\alpha LSM,\alpha' L'S'M'}(g_s - 1)(-1)^{L+S+M}$$

$$\times\ [J,J']^{1/2}\{S(S+1)(2S+1)\}^{1/2}\begin{pmatrix} J & 1 & J' \\ -M & 0 & M \end{pmatrix}\begin{Bmatrix} L & S & J \\ 1 & J' & S \end{Bmatrix} . \tag{17.6}$$

Thus the magnetic matrix is diagonal in all quantum numbers except J; much as for the magnetic dipole operator, (15.7), non-zero matrix elements exist only between levels of the same LS term of the same configuration, and then only for

$$\Delta J = 0,\ \pm 1 \qquad (J = J' = 0 \text{ not allowed}) \ . \tag{17.7}$$

For diagonal matrix elements, the 3-j and 6-j symbols can be evaluated analytically[5] to give

[2] C. M. Fowler, Science **180**, 261 (1973).

[3] J. A. Stamper and B. H. Ripin, Phys. Rev. Lett. **34**, 138 (1975).

[4] R. H. Garstang, Rep. Prog. Phys. **40**, 105 (1977), and references therein. See also A. R. P. Rau, R. O. Mueller, and L. Spruch, Phys. Rev. A **11**, 1865 (1975) regarding fields of 10^{12} G presumed to be present in pulsars.

[5] See for example M. Rotenberg, R. Bivins, N. Metropolis, and J. K. Wooten, Jr., *The 3-j and 6-j Symbols* (The Technology Press, Massachusetts Institute of Technology, Cambridge, 1959).

$$\frac{1}{B\mu_o}\langle\alpha LSJM|H_{mag}|\alpha LSJM\rangle = M\, g_{LSJ} \,, \tag{17.8}$$

where

$$\begin{aligned} g_{LSJ} &= 1 + (g_s - 1)\frac{J(J+1)+S(S+1)-L(L+1)}{2J(J+1)} \\ &= \frac{g_s+1}{2} + \frac{(g_s-1)}{2}\left[\frac{S(S+1)-L(L+1)}{J(J+1)}\right] \end{aligned} \tag{17.9}$$

is known as the Landé g factor (or g value)[6]—usually written in the approximation $g_s = 2$. For singlet terms (S = 0), L is equal to J and so $g_{LSJ} = 1$. For S states (L = 0), $g_{LSJ} = g_s \cong 2.0023$. For terms with L = S ($^3P$, ^5D, ^7F, etc.), $g_{LSJ} = (g_s + 1)/2 \cong 1.5012$ for all J. For all other cases, g_{LSJ} is either greater than 1.5 or less than 1.5 for all J, depending on whether S > L or S < L; the departure from 1.5 is of course greatest for smallest J, and g_{LSJ} may even be negative when S = L − 1/2 or S = L − 1 so that J can be especially small. Extensive tables of values of g_{LSJ} are available.[7]

For the off-diagonal elements J′ = J − 1, analytical evaluation of the 3n-j symbols[5] leads to

$$\frac{1}{B\mu_o}\langle\alpha LSJM|H_{mag}|\alpha'L'S',J-1,M'\rangle$$
$$= -\delta_{\alpha LSM,\alpha'L'S'M'}(g_s-1)\left\{\frac{(J^2-M^2)(J+L+S+1)(J+L-S)(J+S-L)(L+S-J+1)}{4J^2(2J-1)(2J+1)}\right\}^{1/2} . \tag{17.10}$$

As a simple example, we consider the configuration p^2, with terms ^1S, ^3P, and ^1D. The magnetic part of the energy matrix is shown in Fig. 17-1. It should be recalled that for the field-free atom the energy matrix involves no matrix elements connecting states of different J; the matrix thus partitions into a 2 × 2 matrix for J = 0, a 1 × 1 matrix for J = 1, and a 2 × 2 matrix for J = 2, each of which is independent of M. For B ≠ 0, we still have a 2 × 2 matrix for M = ± 2, but have to diagonalize a 3 × 3 matrix for M = ± 1, and a 5 × 5 matrix for M = 0. For M = ± 2 the eigenstates (in the single-configuration approximation) are pure J = 2 states, but for M = ± 1 they are mixtures of J = 2 and J = 1 states, and for M = 0 they are mixtures of basis states of all values of J.

In general, states of any given value of M are mixtures of basis states for all values of J ≧ M. Thus J is no longer a rigorously good quantum number of the atomic state. The semiclassical interpretation in terms of the vector model is that the magnetic field exerts a torque which causes the angular momentum **J** to precess about the direction of the field; hence only the z-component of **J** remains a constant of the motion, and only M remains a good quantum number.

[6]A. Landé, Z. Physik 5, 231 (1921).

[7]W. C. Martin, R. Zalubas, and L. Hagan, *Atomic Energy Levels—The Rare Earth Elements*, NSRDS-NBS 60 (U.S. Govt. Printing Off., Washington, D.C., 1978).

$$
\begin{array}{c|ccccc}
 & {}^1S_0 & {}^3P_0 & {}^3P_1 & {}^3P_2 & {}^1D_2 \\
\hline
{}^1S_0 & \frac{0}{0}M & 0 & 0 & 0 & 0 \\
{}^3P_0 & 0 & \frac{0}{0}M & -\sqrt{\frac{2}{3}}\sqrt{1-M^2} & 0 & 0 \\
{}^3P_1 & 0 & -\sqrt{\frac{2}{3}}\sqrt{1-M^2} & \frac{3}{2}M & -\frac{\sqrt{3}}{6}\sqrt{4-M^2} & 0 \\
{}^3P_2 & 0 & 0 & -\frac{\sqrt{3}}{6}\sqrt{4-M^2} & \frac{3}{2}M & 0 \\
{}^1D_2 & 0 & 0 & 0 & 0 & M
\end{array}
\cdot B\mu_0
$$

M = 0 → M = 1 → M = 2 →

Fig. 17-1. The matrix of the magnetic-field operator $-B\mu_z$ in an LS representation for the states of a p^2 configuration, in the approximation $g_s = 2$. The figure actually represents three different matrices: a 5 × 5 matrix for M = 0, a 3 × 3 matrix for M = ± 1 (because |M| cannot be greater than J), and a 2 × 2 matrix for M = ± 2. (Note: The indeterminate matrix elements (0/0)M are encountered only for M = 0, when of course they become simply zero.)

17-3. THE WEAK-FIELD LIMIT

For sufficiently small values of B, the off-diagonal matrix elements of H_{mag} that connect basis states of different J will be negligible compared with the energy differences provided by the Coulomb and spin-orbit interactions. Mixing of basis states of different J can then be neglected. (In the vector-model picture, the precession of J about the z-axis is so slow that J is almost a constant of the motion.) The energy matrix breaks up into blocks according to J value, just as in the field-free case. For each M, the eigenvectors are essentially the same as for zero field.

The magnetic-field energy contribution can be calculated as a simple perturbation, given by the diagonal elements of the magnetic matrix in the intermediate-coupling representation:

$$E_{mag}(k) = \langle k|H_{mag}|k'\rangle = \sum_{bb'}\langle k|b\rangle\langle b|H_{mag}|b'\rangle\langle b'|k\rangle \, , \qquad (17.11)$$

where the wavefunction-expansion coefficients $\langle k|b\rangle$ are (as mentioned above) the zero-field eigenvector components, and the summation extends over all basis functions of given parity and J. In an LS basis, (H_{mag}) is diagonal and so (17.11) reduces, for a state $k \equiv \gamma JM$, to

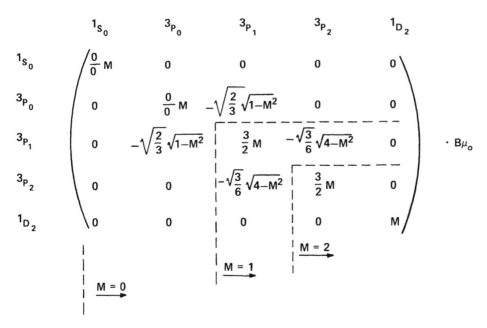

$$E_{mag}(\gamma JM) = \sum_{\alpha LS} \langle \gamma J | \alpha LSJ \rangle^2 \langle \alpha LSJM | H_{mag} | \alpha LSJM \rangle ,$$

or from (17.8)

$$\frac{E_{mag}(\gamma JM)}{B\mu_o} = M\, g_{\gamma J} , \tag{17.12}$$

where

$$g_{\gamma J} = \sum_{\alpha LS} \langle \gamma J | \alpha LSJ \rangle^2 g_{LSJ} . \tag{17.13}$$

Thus the qualitative form of the contribution of the magnetic field to the level structure is independent of the coupling conditions: each zero-field level is broken up into a set of 2J + 1 equally spaced sublevels, symmetrically arranged about the zero-field position; M increases from −J for the lowest sublevel to +J for the highest (for positive $g_{\gamma J}$). The effect of the coupling conditions appears only quantitatively in determining the value of the intermediate-coupling g factor $g_{\gamma J}$, which in turn determines the energy difference $g_{\gamma J}B\mu_o$ between successive sublevels. Conversely, experimental measurement of g values provides important information on the coupling conditions within the atom.[8]

Sum rules. If we sum (17.13) over all levels γ of given J, we find from the fact that the eigenvector matrix is an orthogonal one:

$$\sum_{\gamma} g_{\gamma J} = \sum_{\alpha LS} g_{LSJ} \sum_{\gamma J} \langle \gamma J | \alpha LSJ \rangle^2 = \sum_{\alpha LS} g_{LSJ} . \tag{17.14}$$

Thus the sum of the g factors for any given J is independent of the coupling conditions.[9]

If configuration interactions are unimportant, then the summations in (17.14) need be carried only over levels belonging to the configuration of interest, and we then have the more restrictive rule that the sum of the g factors for all levels of given J within a single configuration is independent of coupling. g-factor sums within a "configuration" remain unchanged even when configuration interactions are strong, if the configurations involved are closely LS coupled; this is because Coulomb interactions occur only between levels of the same L, S, and J, and because g_{LSJ} depends only on these three quantum numbers and not on the configuration. In general, however, two interacting levels will be different mixtures of various LS basis states and would therefore (in the absence of CI) have different g values; the effect of the CI mixing is to produce a partial averaging of the two zero-CI g factors. Consequently the g sum for levels of one "configuration" is reduced from the single-configuration value, and the g sum for the second "configuration" is increased by an equal amount above the single-configuration value. Thus comparison of observed g-sum values for the levels of various "configurations" with the theoretical single-configuration values can frequently pinpoint pairs of strongly-interacting levels.[8]

[8]See, for example, K. L. Andrew, R. D. Cowan, and A. Giacchetti, J. Opt. Soc. Am. 57, 715 (1967).

[9]This result is of course just a specific example of the general property that the trace of a matrix (in this case, the g-factor matrix) is invariant under an orthogonal transformation.

The g-factor matrix in other representations. When conditions in the atom closely approximate some mathematically pure coupling, the g factors are given by the diagonal elements of the g-factor matrix in the corresponding pure-coupling representation. These diagonal elements can be calculated semiclassically from the vector model of the atom, and from the relation

$$\mu_J = -\mu_0 g_J \mathbf{J} \qquad (17.15)$$

that follows from (17.1) and (17.12). Thus in the LS-coupling case, we find with the aid of Fig. 17-2 that

$$\mu_J = (\mu_L \cdot \mathbf{J} + \mu_S \cdot \mathbf{J})\mathbf{J}/\mathbf{J}^2 = -\mu_0 \frac{\mathbf{L}\cdot\mathbf{J} + g_s \mathbf{S}\cdot\mathbf{J}}{\mathbf{J}^2}\mathbf{J} . \qquad (17.16)$$

But

$$\mathbf{S}^2 = (\mathbf{J} - \mathbf{L})^2 = \mathbf{J}^2 + \mathbf{L}^2 - 2\,\mathbf{L}\cdot\mathbf{J}$$

and

$$\mathbf{L}^2 = (\mathbf{J} - \mathbf{S})^2 = \mathbf{J}^2 + \mathbf{S}^2 - 2\,\mathbf{S}\cdot\mathbf{J} .$$

Replacing L^2 by its quantum-mechanical expectation value L(L + 1), and similarly for S^2 and J^2, we find

$$g_{LSJ} = \frac{J(J+1) + L(L+1) - S(S+1)}{2J(J+1)} + g_s \frac{J(J+1) + S(S+1) - L(L+1)}{2J(J+1)} , \qquad (17.17)$$

which is the same as (17.9).

For other coupling schemes, a multi-step calculation is involved. For example, in the jK-coupling case $|(jl)K,s|J$, the strong coupling is that between j and *l*, which vectors therefore precess rapidly about their resultant K. Thus

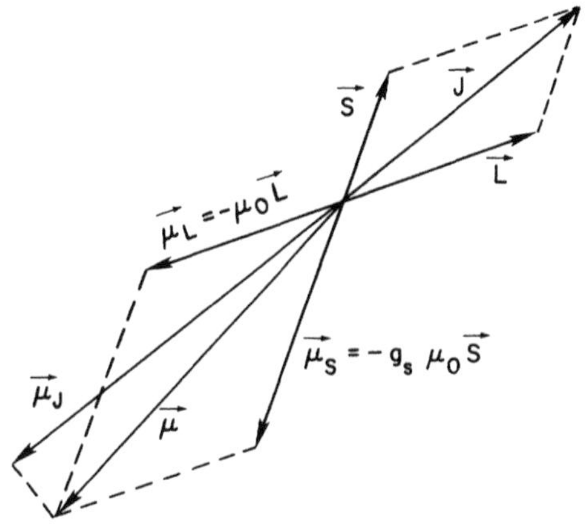

Fig. 17-2. Vector-model calculation of the component of μ in the direction of **J**, under LS-coupling conditions. (The figure has been drawn using units such that $\mu_0 = 1$.)

$$\mu_K = -\mu_0 \frac{g_j \mathbf{j}\cdot\mathbf{K} + \mathbf{l}\cdot\mathbf{K}}{K^2}\mathbf{K} \tag{17.18}$$

and in complete analogy with (17.17) or (17.9) we find

$$g_K = 1 + (g_j - 1)\frac{K(K+1) + j(j+1) - l(l+1)}{2K(K+1)} , \tag{17.19}$$

where g_j is the g factor for the parent-ion state. Secondarily, the weaker interaction between K and s causes these vectors to precess about J, so that

$$\mu_J = -\mu_0 \frac{g_K \mathbf{K}\cdot\mathbf{J} + g_s \mathbf{s}\cdot\mathbf{J}}{J^2}\mathbf{J} \tag{17.20}$$

and

$$g_{jKJ} = \frac{g_K[J(J+1)+K(K+1)-s(s+1)]+g_s[J(J+1)+s(s+1)-K(K+1)]}{2J(J+1)} . \tag{17.21}$$

In view of the fact that J can be equal only to K ± 1/2, it is possible (in the approximation $g_s = 2$) to simplify (17.19), (17.21) to[10]

$$g_{jKJ} = \frac{2J+1}{2K+1} + 2(g_j - 1)\frac{K(K+1) + j(j+1) - l(l+1)}{(2K+1)(2J+1)} . \tag{17.22}$$

Results for pure LK and jj coupling can be derived similarly.[11]

It must be emphasized that if intermediate-coupling g values are to be calculated in a non-LS representation, then it is necessary to carry out the double matrix multiplication (17.11); simplification to a single summation as in (17.13) does *not* follow. The necessary off-diagonal elements of the g-factor matrix cannot be obtained from the semiclassical vector model, and the matrix is most easily obtained simply by a matrix transformation of the LS-representation g-factor matrix. As an example, the g-factor matrix in the jj representation for the J = 2 levels of a p^2 configuration is found from Fig. 17-1 and Table 9-3 to be (in the approximation $g_s = 2$):

$$(g_{jj}) = \begin{pmatrix} (\frac{1}{3})^{1/2} & -(\frac{2}{3})^{1/2} \\ (\frac{2}{3})^{1/2} & (\frac{1}{3})^{1/2} \end{pmatrix} \begin{pmatrix} \frac{3}{2} & 0 \\ 0 & 1 \end{pmatrix} \begin{pmatrix} (\frac{1}{3})^{1/2} & (\frac{2}{3})^{1/2} \\ -(\frac{2}{3})^{1/2} & (\frac{1}{3})^{1/2} \end{pmatrix}$$

$$= \begin{array}{c} \\ (\frac{1}{2}\,\frac{3}{2})_2 \\ (\frac{3}{2}\,\frac{3}{2})_2 \end{array} \begin{array}{c} (\frac{1}{2}\,\frac{3}{2})_2 \quad (\frac{3}{2}\,\frac{3}{2})_2 \\ \begin{pmatrix} \frac{7}{6} & (\frac{1}{18})^{1/2} \\ (\frac{1}{18})^{1/2} & \frac{4}{3} \end{pmatrix} \end{array} . \tag{17.23}$$

[10]G. Racah, Phys. Rev. 61, 537 (1942).

[11]J. C. van den Bosch. "The Zeeman Effect," in S. Flügge, ed., *Handbuch der Physik* (Springer-Verlag, Berlin, 1957), Vol. XXVIII.

Problem

17-3(1). From the vector model, derive an expression for g values under pure jj-coupling conditions. Using this result in the approximation $g_s = 2$, verify the values of the diagonal elements of the matrix (17.23).

17-4. WEAK-FIELD ZEEMAN PATTERNS

Selection rules for electric dipole transitions between the sublevels of two different levels γJ and $\gamma' J'$ are given by (14.29) and Table 14-1. An example of the possible transitions for the case $J = 1$, $J' = 2$ is shown in the upper portion of Fig. 17-3. Under normal excitation conditions, all states (sublevels) of $\gamma' J'$ are equally populated. It follows from (14.30) and a relation analogous to (14.6) that the relative intensities of the various Zeeman components are given by

$$I \propto \begin{pmatrix} J & 1 & J' \\ -M & M-M' & M' \end{pmatrix}^2 . \qquad (17.24)$$

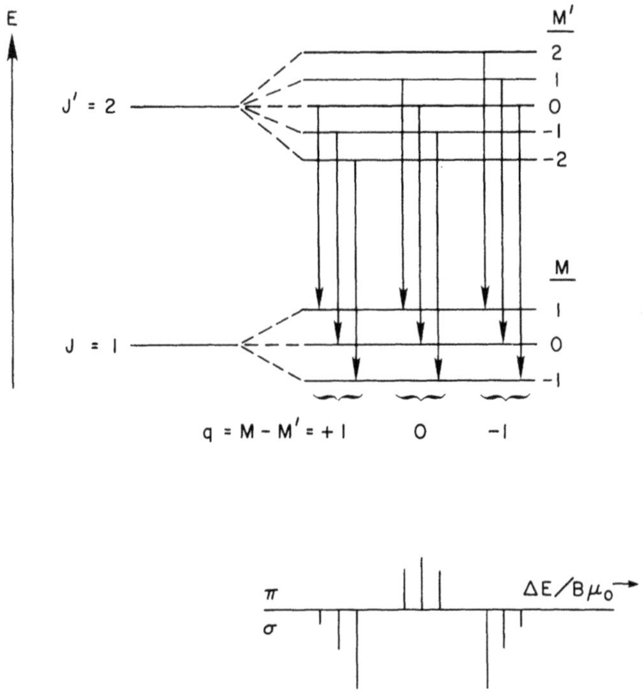

Fig. 17-3. Top: Grotrian diagram for dipole Zeeman transitions in the case $J' > J$, $g' < g$. Bottom: Corresponding transverse electric-dipole spectrum; intensities plotted upward for $\Delta M = 0$ (π components), and downward for $\Delta M = \pm 1$ (σ components).

The 3-j symbol can be evaluated analytically,[5] with the result that relative intensities are proportional to the following quantities:

$$J = J': \begin{cases} I \propto M^2 \,, & \Delta M = 0 \,, \\ I \propto (1/2)(J \pm M)(J \mp M + 1) \,, & M' = M \mp 1 \end{cases} \tag{17.25}$$

$$J_2 = J_1 + 1: \begin{cases} I \propto J_2^2 - M^2 \,, & \Delta M = 0 \,, \\ I \propto (1/2)(J_2 \pm M)(J_2 \pm M - 1) \,, & M_1 = M \mp 1 \end{cases} \tag{17.26}$$

where J_1 and J_2 are, respectively, the smaller and larger of J and J′, and M takes on all values from $-J_2$ to $+J_2$ (in integral steps).

In practice, Zeeman observations are usually made transversely (in a direction perpendicular to the magnetic field). It follows from Table 14-1 that electric dipole radiation arising from $\Delta M = 0$ transitions is polarized parallel to B (π), and that components corresponding to $\Delta M = \pm 1$ transitions are polarized perpendicular to B (σ, from the German *senkrecht*). Experimentally, spectra of the two polarizations are usually selected one at a time by means of a suitable analyzer, and recorded along adjacent sections of the photographic plate. A corresponding graphical representation of these two spectra is shown in the lower portion of Fig. 17-3, using relative intensities obtained from (17.24).

All possible types of dipole Zeeman patterns are illustrated in Fig. 17-4. Except for the pseudotriplet resulting when $g_2 = g_1$ (which is indistinguishable from the normal triplet produced by 0 – 1 transitions), it is evident that a well-resolved, properly exposed Zeeman pattern immediately provides the J values of the two levels involved: the number of π components (including the zero-intensity central component in the even-N, $\Delta J = 0$ case) is always equal to $2J_1 + 1$, where J_1 is the smaller of J and J′. (However, no indication is given as to which level has the higher energy.) Zeeman patterns for a number of lines of neutral tin are reproduced in Fig. 17-5.

It is convenient to measure the photon energies of the Zeeman components in units of $B\mu_0$—or, equivalently, to measure the wavenumbers of the components in Lorentz units $B\mu_0/hc = 4.6686 \cdot 10^{-5}$ B cm^{-1}. Then it is easily verified from (17.12) that the separation of successive components (either in the π spectrum, or in either half of the σ pattern) is $|g_1 - g_2|$, and that the distance between the centers of the two portions of the σ pattern is $g_1 + g_2$ for $\Delta J = 0$ transitions and $2g_2$ for $\Delta J = \pm 1$ transitions (where g_2 is the g value for the level with greater J). Thus the g values of the two levels can be unambiguously determined. For maximum accuracy, least-squares fitting of all measureable components is preferable;[12] this procedure also provides an estimate of the uncertainties of the g values.[8,12] In favorable cases, analysis of Zeeman patterns can give g values accurate to about 0.001; use of the approximation $g_s = 2$ in (17.9) is then inadequate.[13] For Zeeman patterns that are only partially resolved, it may still be possible to determine one g value if the other is already known from other spectrum lines,[14] but the accuracy is of course considerably poorer.

[12] K. L. Vander Sluis, J. Opt. Soc. Am. 46, 605 (1956).

[13] g values of low-lying levels can be measured by magnetic resonance methods with errors of 10^{-5} or less.

[14] A. G. Shenstone and H. A. Blair, Phil. Mag. 8, 765 (1929).

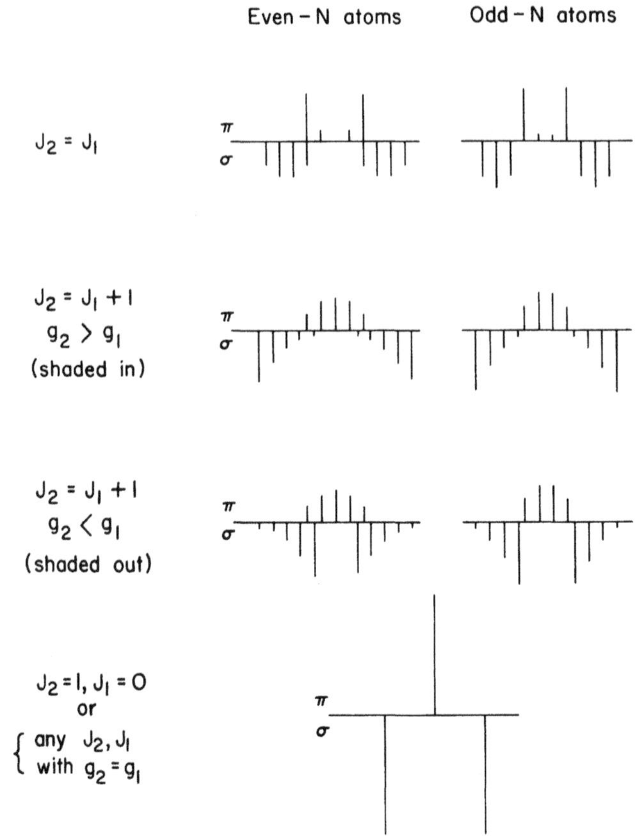

Fig. 17-4. The possible types of weak-field Zeeman patterns for electric dipole transitions. (For magnetic-dipole, interchange the labels π and σ.) Except for the triplet, all plots are for the cases $J_1 = 2$ (even N) or $J_1 = 3/2$ (odd N), and $|g_2 - g_1|$ equals $(g_2 + g_1)/7$ for $\Delta J = 0$ or $2g_2/7$ for $|\Delta J| = 1$.

*17-5. STRONG MAGNETIC FIELDS; THE PASCHEN-BACK EFFECT

When the external field is sufficiently strong, the discussion of the preceding section is no longer appropriate. For the case of near LS coupling, for example, the vector-model picture is no longer that **L** and **S** precess rapidly about each other to give a resultant **J**, which then precesses slowly about the direction of the magnetic field **B**. Instead, the field-induced precessions are so rapid that we must think of **L** and **S** as individually precessing about **B**; i.e., the effect of **B** is to effectively decouple **L** from **S**, and to make **J** meaningless. It is then appropriate to describe the atom in terms of partially coupled basis functions $|\gamma LSM_LM_S\rangle$. In place of (17.15) and (17.12) we have

$$\mu_B = \mu_L \cdot \mathbf{B} + \mu_S \cdot \mathbf{B} \qquad (17.27)$$

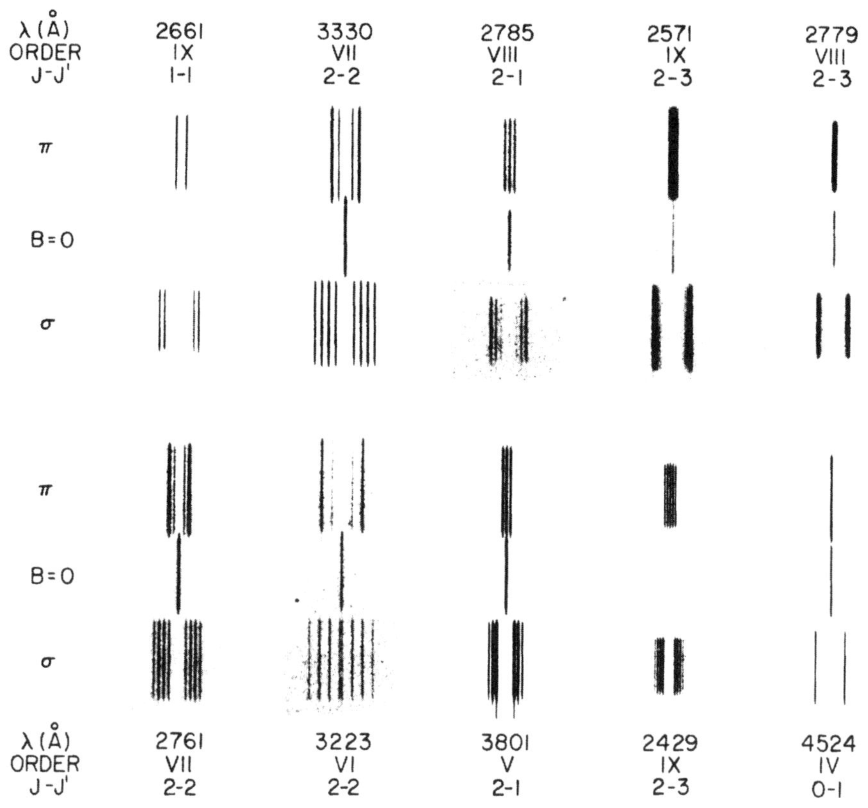

Fig. 17-5. Zeeman patterns of some lines of neutral tin in a magnetic field of 24025 gauss. (Photographic negatives, courtesy of K. L. Andrew.) For the 3223 Å line, Δg is large enough that the two halves of the σ pattern nearly overlap. For the 2779 Å line, Δg is so small as to almost produce a degenerate pseudotriplet, similar to the genuine triplet at 4524 Å.

and

$$\frac{E_{mag}(\gamma LSM_L M_S)}{B\mu_o} = M_L + g_s M_S \cong M_L + 2M_S \ . \tag{17.28}$$

For electric dipole transitions the operator **r** does not act on the spin, and so the selection rules are

$$\begin{aligned} \Delta M_S &= 0, \\ \Delta M_L &= 0, \ \pm 1 \quad (M_L = M'_L = 0 \text{ not allowed}) \ . \end{aligned} \tag{17.29}$$

The resulting Zeeman pattern is a simple triplet like that shown at the bottom of Fig. 17-4 (with a separation of the two σ components that corresponds to g = 1).

Thus in the case of $^3S_1 - {}^3P_{0,1,2}$ transitions, for example, we would see as B is increased from zero a gradual change from a set of three narrow Zeeman patterns (similar to those in the left-hand column of Fig. 17-4) to a single very wide triplet. This is known as the Paschen-Back effect.[15] and the final triplet is referred to as the Paschen-Back limit. (Actually, because of residual effects of the spin-orbit interaction, each of the two σ components is really a set of three closely spaced components.[11])

As a natural extension of the above discussion, one might expect that when the magnetic field is so strong as to produce magnetic energies that are large with respect to all Coulomb and spin-orbit splittings within a configuration, then the field will decouple all the various $\boldsymbol{l}_i$ and s_i so that each individual angular momentum will precess separately about the field direction. The atom is then best described in terms of completely uncoupled basis functions $|\{l_i m_{l_i} s_i m_{s_i}\}\rangle$, and we have

$$\mu_B = \sum_i (\mu_{l_i} \cdot \mathbf{B} + \mu_{s_i} \cdot \mathbf{B}) \ , \tag{17.30}$$

$$\frac{E_{mag}}{B\mu_o} = \sum_i (m_{l_i} + g_s m_{s_i}) \ . \tag{17.31}$$

For electric dipole transitions the selection rules are

$$\begin{aligned} \Delta m_{s_i} &= 0, && \text{all electrons} \ , \\ \Delta m_{l_i} &= 0, && \text{spectator electrons} \ , \\ \Delta m_{l_i} &= 0, \ \pm 1, && \text{jumping electron} \ , \end{aligned} \tag{17.32}$$

and Zeeman patterns again are simple triplets.

This situation is sometimes called the complete Paschen-Back effect. It is not usually of much importance in complex atoms; fields strong enough to produce magnetic splittings large compared with zero-field intra-configuration splittings are usually large enough both to mix neighboring configurations and to introduce diamagnetic (quadratic-Zeeman) effects.

Problems

17-5(1). Using a magnetic moment μ_K such as that defined in (17.18), discuss the analog of the Paschen-Back effect for the case of pure pair coupling.

17-5(2). For transitions such as $^2S_{1/2} - {}^2S_{1/2}$, $^3S_1 - {}^3S_1$, etc., M is necessarily equal to M_S so that σ components are impossible for electric dipole lines. Discuss the reasons why π-only Zeeman patterns are not observed.

[15]F. Paschen and E. Back, Ann. Physik 39, 897 (1912), 40, 960 (1913).

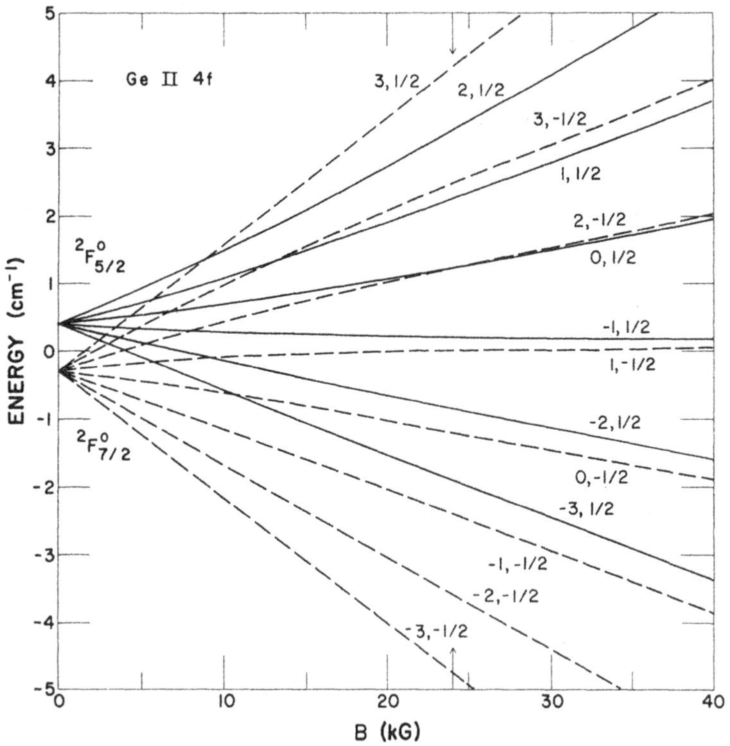

Fig. 17-6. Computed energies of the sublevels of the configuration Ge II 4f, for which the zero-field level separtion $^2F^o_{5/2} - {}^2F^o_{7/2}$ is 0.68 cm^{-1} and the Lande g-values are 6/7 and 8/7. The vertical arrows indicate the flux density that corresponds to Fig. 17-7. The labels on the curves are strong-field values of M_L, M_S; the energy for weak fields is proportional to $M = M_L + M_S$, but at strong fields depends mainly on $M_L + 2M_S$.

*17-6. INTERMEDIATE MAGNETIC FIELDS

In the case of intermediate field strengths, energies of the various magnetic sublevels can be found only by diagonalizing the detailed energy matrices of Sec. 17-2, for each M value and for each field strength of interest. If departures from pure LS coupling are appreciable, it is generally necessary to include all basis states of the configuration of interest; if configuration interaction is important, then basis states must be included for all important interacting configurations.

As an example, results of such calculations[8] are shown in Fig. 17-6[16] for the sublevels of the simple one-electron configuration Ge II 4f. The various curves can be labeled by the

[16]As may be seen from the figure or from (17.4), the intermediate and strong-field regions can be reached at flux densities of 10^4 to 10^5 G only if the interacting levels lie within a few cm^{-1} of each other. Such close-lying levels occur particularly in light elements (small spin-orbit splittings of LS terms) and under pair-coupling conditions (excited electron with $l \geqq 3$).

corresponding values of M, which are good quantum numbers for all B. In this simple case, S and L (^2F) are good quantum numbers over the entire range of B values shown. At the left edge of the figure, where J is a good quantum number, we see the weak-field energy splittings (17.12) proportional to the value of M. At the right edge, we see the approach to the strong-field (Paschen-Back) limit, where M_L and M_S become good quantum numbers and the energy departures (17.28) from E_{av} become proportional to $M_L + 2M_S$. During the transition from one limit to the other, sublevels of equal M never cross, because of the off-diagonal matrix elements in Fig. 17-1 (cf. Fig. 10-2 and the accompanying text).

Although the Landé g factor is inherently a weak-field concept, reasonably accurate g values can be derived even from intermediate- and strong-field Zeeman patterns.[17]

The calculation of component strengths involves an expression completely analogous to (14.42):

$$S^{1/2}_{\gamma\gamma'q} = \sum_{\beta J} \sum_{\beta' J'} y^{\gamma}_{\beta JM} \langle \beta JM | P^{(t)}_q | \beta' J' M' \rangle \, y^{\gamma'}_{\beta' J' M'} \, , \tag{17.33}$$

where $t = 1$ for E1 and $t = 2$ for E2 transitions, and where the eigenvector components y (and the states γ and γ') now depend on M and M' as well as on J and J'. Similarly to (14.28), the basis-state matrix element is given by

$$\langle \beta JM | P^{(t)}_q | \beta' J' M' \rangle = (-1)^{J-M} \begin{pmatrix} J & t & J' \\ -M & q & M' \end{pmatrix} \langle \beta J \| \boldsymbol{P}^{(t)} \| \beta' J' \rangle \; ; \tag{17.34}$$

the reduced matrix element in the LS-coupling representation is given just as in the field-free problem by (14.78) or (15.21)-(15.22) for electric multipole transitions and by (15.7) for magnetic dipole transitions. Because of interference effects among terms in (17.33) arising from different values of J and from different values of J', the relative strengths of different components may differ markedly from the weak-field values (17.24).[17] An example[8] involving the levels of Fig. 17-6 is given in Fig. 17-7. Intensity perturbations may even be pronounced when wavelength perturbations are indiscernably small.[18] Field-induced wavefunction mixings have also made observable the $6p^2\,(3/2,3/2)_2 - 6p6d\ 1/2\ [3/2]_2$ line of Pb I at 4063 Å, which is undetectably weak at zero field because of configuration-interaction-induced cancellation effects.[19]

*17-7. THE STARK EFFECT

We now consider the effects of placing the atom in an external electric field. This field may be a macroscopic one such as that present between two oppositely charged parallel

[17] C. C. Kiess and G. Shortley, J. Res. Natl. Bur. Stand. 42, 183 (1949); M. A. Catalán, J. Res. Natl. Bur. Stand. 47, 502 (1951).

[18] See, for example, D. R. Wood, K. L. Andrew, A. Giacchetti, and R. D. Cowan, J. Opt. Soc. Am. 58, 830 (1968).

[19] Ref. 18. A similar unobservably weak line is Sr III $4p^5(^2P^o_{3/2})\,5s[3/2]^o_2 - 4p^5(^2P^o_{3/2})\,6p[5/2]_3$, W. Persson and S. Valind, Phys. Scr. 5, 187 (1972). In principle, Zeeman-effect mixings should reveal this line also, but experimental problems would be appreciable because the line lies well into the vacuum ultraviolet at 1133 Å.

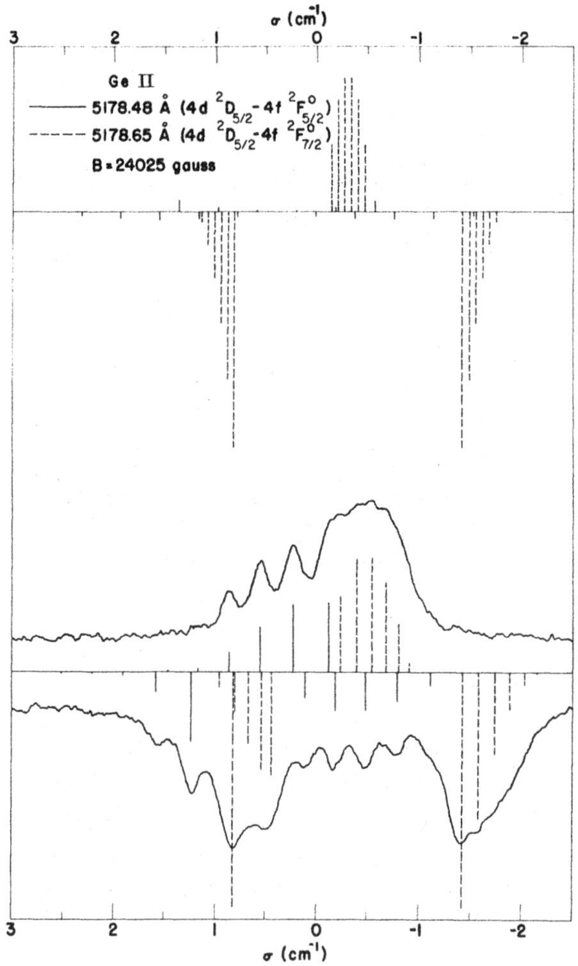

Fig. 17-7. Upper portion: Schematic form of the computed Zeeman patterns of the Ge II 4d $^2D_{5/2}$ – 4f $^2F^{\circ}_{5/2,7/2}$ transitions at 24025 gauss, assuming no interaction between the two $^2F^{\circ}$ states. Lower portion: The computed transitions with interactions included, and superimposed microdensitometer tracings of the spectrum lines photographed in the ninth order. Note the sizable perturbations in position of all components except those involving the $M = \pm 7/2$ sublevels of $^2F^{\circ}_{7/2}$, and the even greater effects on component strengths—especially for the 5/2 – 5/2 transition.

plates, or it may be a microscopic field produced by neighboring electrons and ions in the spectroscopic source. In the latter case, the field will be a randomly fluctuating one; however, the time and spatial variations will normally be sufficiently slow that we may consider the field to be approximately a uniform static field $\mathfrak{E}$.

The interaction between the atom and the external field will add to the Hamiltonian a term

$$H_{elec} = \mathfrak{E} \cdot \sum_i e\, r_i = -\mathfrak{E} \cdot \boldsymbol{P}^{(1)} , \tag{17.35}$$

where $\boldsymbol{P}^{(1)}$ is the electric dipole moment of the atom. As usual, we shall measure energies in rydbergs ($e^2/2a_o$) and $\boldsymbol{P}^{(1)}$ in units of ea_o; we shall measure the electric field strength in units of

$$\frac{e}{a_o^2} = 5.1423 \cdot 10^9 \text{ volts/cm} . \tag{17.36}$$

An additional factor 2 is thereby introduced on the right-hand side of (17.35). Choosing the z-axis in the direction of $\mathfrak{E}$ and using the Wigner-Eckart theorem (11.15), matrix elements of H_{elec} are

$$\langle\gamma JM|H_{elec}|\gamma'J'M'\rangle = -2\mathfrak{E}\langle\gamma JM|P_0^{(1)}|\gamma'J'M'\rangle$$
$$= -2\mathfrak{E}(-1)^{J-M}\begin{pmatrix} J & 1 & J' \\ -M & 0 & M' \end{pmatrix}\langle\gamma J\|\boldsymbol{P}^{(1)}\|\gamma'J'\rangle . \qquad (17.37)$$

Because of the properties of the 3-j symbol, M′ must be equal to M; hence M remains a good quantum number just as in the case of an external magnetic field.

The reduced matrix element in (17.37) is identical with that discussed in connection with electric dipole radiation, and is given by (14.42) and—in the LS basis—(14.78). This matrix element is non-zero only if the levels γJ and γ′J′ have opposite parity and therefore belong to two different configurations. Consequently there are no non-zero diagonal matrix elements of H_{elec}, and the effect of an external electric field is considerably more complicated than that of a magnetic field. Nevertheless, there are in principle still perceptible energy-level shifts, and these are obviously M dependent. Thus each spectrum line is in general broken up into a number of different components—an effect first observed by Stark.[20]

Because of the field-induced wavefunction mixings that result from the off-diagonal matrix elements (17.37), neither J nor parity remains a good quantum number. Consequently, apparent violations of the normal electric-dipole selection rules for radiative transitions ($\Delta J = 0, \pm 1$, and a parity change) are readily observable, as a result either of an externally applied field or of residual inter-ionic fields within the source (see Sec. 14-12).

Weak fields (quadratic Stark effect). When the off-diagonal electric-field matrix elements are small compared with the difference between the corresponding diagonal elements, the field-induced energy shifts can be calculated by perturbation theory. From (10.38), the energy perturbation of a state γJM resulting from all other states γ′J′M′ is

$$\Delta E_{\gamma JM} = 4\mathfrak{E}^2 \sum_{\gamma'J'M'} \begin{pmatrix} J & 1 & J' \\ -M & 0 & M' \end{pmatrix}^2 \frac{\langle\gamma J\|\boldsymbol{P}^{(1)}\|\gamma'J'\rangle^2}{E_{\gamma J} - E_{\gamma'J'}} . \qquad (17.38)$$

[Though the 3-j symbol is zero unless $M' = M$, we retain the general expression for convenience in deriving (17.41).] From (5.6) we see that $\Delta E_{\gamma JM}$ is independent of the sign of M; thus unlike Zeeman splittings, Stark splittings are not symmetrical about the unperturbed level. We saw from (14.50) and (14.62) that the reduced matrix element tends to be largest when ΔJ equals Δ*l* for the active electron, and hence when $|J - J'| = 1$; it may then be seen that the square of the 3-j symbol increases with decreasing M. It follows that $\Delta E_{\gamma JM}$ tends to have a sign that is independent of M (and most commonly negative), and a magnitude that increases with decreasing M.

[20] J. Stark, Sitzber. preuss. Akad. Wiss. Berlin, 932 (1913), Ann. Physik 43, 965 (1914). For a recent treatise on the Stark effect, see N. Ryde, *Atoms and Molecules in Electric Fields* (Almqvist and Wiksells International, Stockholm, 1976).

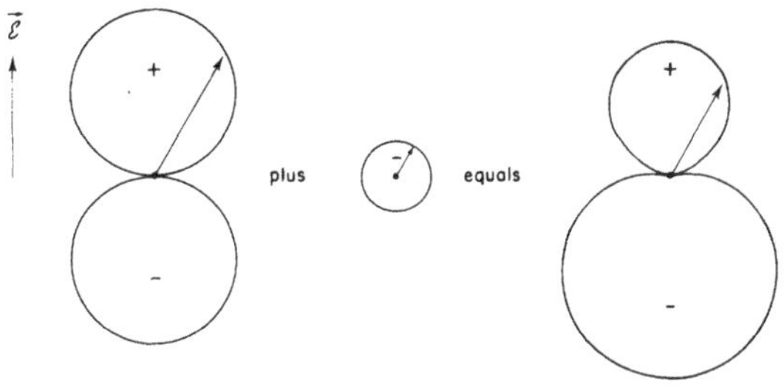

Fig. 17-8. Polar-diagram plots of the asymmetric electron-probability amplitude distribution (and hence non-zero electric dipole moment) that is produced by admixture via $\langle p\|\boldsymbol{P}^{(1)}\|s\rangle$ of some s-wavefunction nature (with negative coefficient) into a predominantly p_z state. The $\pm$ signs represent the signs of the distribution functions in the respective lobes—see Fig. 2-3.

The effect of fluctuating inter-ionic electric fields in a light source is thus not only to broaden each spectrum line (especially those involving highly excited levels, for which the reduced matrix elements are large and the levels are close together) but also usually to shift its wavelength.[21] This effect can be useful for diagnostic purposes in a plasma source (e.g., the measurement of electron density);[22] however, it seriously limits the accuracy of wavelength measurements for use in experimental spectrum analysis, especially in the high-density, strong-field sources used to obtain high ionization stages.

The level shift (17.38) lends itself to a simple physical interpretation. The external electric field exerts forces on the nucleus and on the atomic electrons that are parallel and antiparallel to $\mathfrak{E}$, respectively, thereby polarizing the atom and producing a net non-zero dipole moment $\boldsymbol{P}^{(1)}$ that has a magnitude proportional to the field strength,

$$\boldsymbol{P} = \alpha_d \mathfrak{E} \,, \tag{17.39}$$

where α_d is the dipole polarizability of the atom. (The manner in which the asymmetric electron density distribution arises quantum mechanically is illustrated in Fig. 17-8.) The energy of this induced dipole in the field $\mathfrak{E}$ is (including the factor 2 dictated by our choice of units)

[21]Data on Stark widths and shifts have been tabulated by N. Konjevic and J. R. Roberts, J. Phys. Chem. Ref. Data 5, 209 (1976), and by N. Konjevic and W. L. Wiese, J. Phys. Chem. Ref. Data 5, 259 (1976). See also J. R. Fuhr, W. L. Wiese, and L. J. Roszman, *Bibliography on Atomic Line Shapes and Shifts (1889 through March 1972)*, U.S. Natl. Bur. Stand. Special Publ. 366 (U.S. Govt. Printing Off., Washington D.C., 1972); Suppl. 1, through June 1973 (1974); J. R. Fuhr, G. A. Martin, B. J. Specht, and B. J. Miller, Suppl. 2 and 3 through June 1978 (1975 and 1978).

[22]H. R. Griem, *Plasma Spectroscopy* (McGraw-Hill, New York, 1964); *Spectral Line Broadening by Plasmas* (Academic Press, New York, 1974).

$$\Delta E = -2\int_0^{\mathfrak{E}} P \, d\mathfrak{E} = -\alpha_d \mathfrak{E}^2 \; . \tag{17.40}$$

The polarizability in the level γJ, averaged over all possible orientations M of the atom is, from (17.38) and (5.15),

$$\begin{aligned} \alpha_d(\gamma J) &= -\frac{\mathfrak{E}^{-2}}{(2J+1)} \sum_M \Delta E_{\gamma JM} \\ &= \frac{4}{3(2J+1)} \sum_{\gamma' J'} \frac{\langle \gamma J \| \boldsymbol{P}^{(1)} \| \gamma' J' \rangle^2}{(E_{\gamma' J'} - E_{\gamma J})} \; . \end{aligned} \tag{17.41}$$

Alternatively, this may be written with the aid of (14.35) in terms of oscillator strengths as

$$\alpha_d(\gamma J) = 4 \sum_{\gamma' J'} \frac{f(\gamma J - \gamma' J')}{(E_{\gamma' J'} - E_{\gamma J})^2} \; . \tag{17.42}$$

In both (17.41) and (17.42), $\boldsymbol{P}^{(1)}$ is to be measured in units of ea_o, energies are in rydbergs, and α_d is from (17.39) and (17.36) in units of a_o^3. As in the Thomas-Reiche-Kuhn sum rule (14.104), the summation over $\gamma' J'$ in principle extends over all levels having parity opposite to γJ, including continuum levels. In practice, those levels lying fairly close to γJ contribute most importantly, because of the energy denominator.

The polarization concept is frequently employed in the discussion of hydrogenic levels of an N-electron atom.[23] For a configuration in which the single excited electron has both large n and large l, this electron lies always well outside the ion core and produces an electric field at the nucleus of $1/r^2$ (in units of e/a_o^2). Thus from (17.40) the binding energy of the excited electron is, neglecting core penetration effects,

$$E^{nl} = -\frac{(Z-N+1)^2}{n^2} - \alpha_d \langle r^{-4} \rangle \quad \text{Ry} \; , \tag{17.43}$$

where α_d is the dipole polarizability of the $(N-1)$-electron ion core in units of a_o^3, and r is in Bohr units.

Strong fields (linear Stark effect). The Stark effect is quite different when the electric-field matrix element becomes large compared with the energy difference of the two unperturbed levels that it connects. For example, in the case of two levels of given M, then in the strong-field limit we find from (10.40) that the electric-field energy perturbation is simply $\pm$ the expression (17.37). Physically, the energy shifts are linear instead of quadratic in $\mathfrak{E}$ because of a saturation effect—the two wavefunctions of opposite parity are seen from (10.40) to be mixed 50-50, so that the induced dipole moment has reached a maximum possible value independent of $\mathfrak{E}$.

[23]B. Edlén, "Atomic Spectra," in S. Flügge, ed., *Handbuch der Physik* (Springer-Verlag, Berlin, 1964), Vol. XXVII, Sec. 20.

Of course, if $\mathfrak{E}$ is increased still further, more and more distant levels may successively enter the picture. Stark splittings may thus successively pass from a quadratic regime into a more or less linear regime and then back into a complex nonlinear one.

For hydrogenic (one-electron) ions, where there is zero-field degeneracy between levels of opposite parity (for $n \geqq 2$, neglecting small spin-orbit and relativistic effects), the Stark-effect shifts start out linearly even at very small $\mathfrak{E}$. The matrix elements (17.37) can be evaluated analytically, and the Stark shifts for states with principal quantum number n shown to be given by

$$\begin{aligned} \Delta E &= 3nk\mathfrak{E}/Z \quad \text{Ry} , \qquad [\mathfrak{E} \text{ in the units (17.36)}] , \\ &= 0.0640\, nk\mathfrak{E}/Z \quad \text{cm}^{-1} , \qquad [\mathfrak{E} \text{ in kV/cm}] , \end{aligned} \tag{17.44}$$

where $k = 0, \pm 1, \pm 2, \cdots \pm(n-1)$.

A general quantitative treatment of the Stark effect involves, as usual, numerical diagonalization of the entire energy matrix, including all basis states that are important for the field strengths of interest. The results of such calculations are shown in Fig. 17-9 for two hypothetical atoms possessing one-electron levels for n = 2 and 3 only. The figure illustrates the following general facts:

(1) Stark shifts are independent of the sign of M. This follows from (17.37) and (5.6), which show that changing the sign of M merely changes the sign of the Stark matrix element if

$$(-1)^{2M+J+1+J'} = (-1)^{M+M'+J+1+J'}$$

is negative. This change is equivalent to reversing the phase of each even-parity (say) bra basis function if

$$(-1)^{J+M}$$

is negative, and reversing the phase of each odd-parity ket function if

$$(-1)^{J'+M'+1}$$

is negative. Basis-function phases have no physical significance; thus changing the sign of M changes the signs of certain eigenvector components, but cannot alter the eigenvalues.

(2) Stark shifts really depend only on $|M_L|$ rather than on $|M|$, because the electric field operator (17.35) does not involve electron spins. Thus each Stark level in the figure is characterized by a specific value of $|M_L|$; the Stark state for either sign of M_L is a linear combination of all zero-field basis states that have $M = M_L \pm 1/2$ (provided, of course, that $L \geqq M_L$).

(3) Stark shifts increase quadratically with the electric field strength $\mathfrak{E}$ if the shifts are small compared with distances between levels (of opposite parity) that are connected by electric dipole transitions.

(4) Basically, Stark shifts increase linearly with $\mathfrak{E}$, except for the quadratic regions mentioned above and for distortions resulting from the non-crossing rule for certain sublevels of equal M.

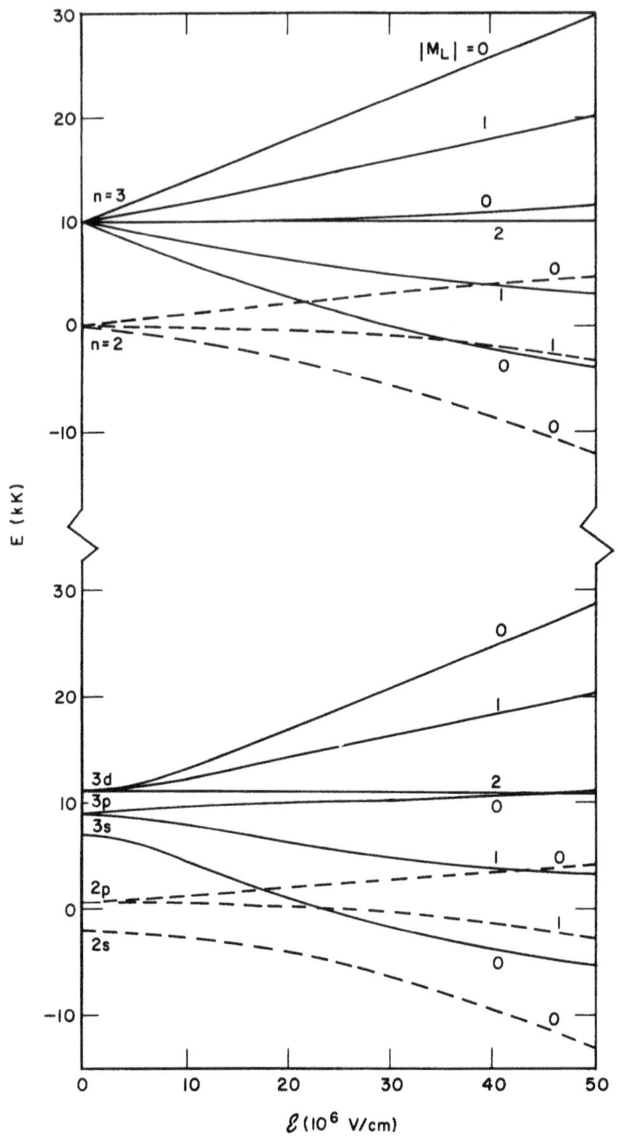

Fig. 17-9. Computed Stark-effect sublevels for atoms having a single valence electron, using Stark matrix elements for hydrogen, but hypothetical zero-field energies. Bottom: Multi-electron atoms, showing a quadratic effect at small $\mathfrak{E}$, and a "linear" behavior at large $\mathfrak{E}$. Top: One-electron ions, showing an initial linear effect, followed by a quadratic effect when n = 2 to 3 mixings become important. (The effects of n > 3 levels have been omitted for simplicity.)

It should be noted that practically obtainable electric field strengths extend only to about 10^6 V/cm, so that extreme Stark mixings in multi-electron atoms appear mainly for highly excited levels where zero-field, opposite-parity level separations are small. Also, Stark mixings rapidly decrease in importance along an isoelectronic sequence, because the Stark matrix elements decrease as 1/Z (cf. Table 3-1) and level separations increase at least as fast as Z (Sec. 19-2).

*17-8. ISOTOPE SHIFTS

The energy levels of two different isotopes of a given element are slightly different, because of two distinctly different effects.[24,25]

The first effect arises strictly from the difference in nuclear masses. The total kinetic energy of the N electrons and the nucleus of mass M is

$$E_k = \sum_{i=1}^{N} \frac{p_i^2}{2m} + \frac{p^2}{2M} . \tag{17.45}$$

Relative to the center of mass of the atom, $p = -\sum p_i$ and so

$$\begin{aligned} E_k &= \sum \frac{p_i^2}{2m} + \frac{\left(\sum p_i\right)^2}{2M} \\ &= \sum \frac{p_i^2}{2m} + \sum \frac{p_i^2}{2M} + \frac{1}{M}\Bigl(\sum_{i<j} p_i \cdot p_j\Bigr) \qquad (17.46) \\ &= \sum \frac{p_i^2}{2\mu} + \frac{1}{M}\Bigl(\sum_{i<j} p_i \cdot p_j\Bigr) , \qquad (17.47) \end{aligned}$$

where $\mu = Mm/(M + m)$ is the reduced mass of the electron-nuclear system.

In the case of a hydrogenic (one-electron) atom, comparison of (17.47) with (17.45) shows that the energy levels are the same as those for an atom with infinitely heavy (stationary) nucleus, except that the rydberg

$$R_\infty = me^4/2h^2$$

has to be replaced by the value

$$R = \mu e^4/2h^2 = R_\infty M/(M + m) = R_\infty (A - Zm)/(A - Zm + m) . \tag{17.48}$$

In a multi-electron atom, there is the additional "coupling effect" of the momentum-correlation terms in (17.47), which in general produces different shifts for different energy levels.[24-27]

The contribution that results from using μ instead of m in (17.47)—i.e., the contribution from the first of the two nuclear terms in (17.46)—is called the *normal mass shift*, and is easily calculable. The momentum-correlation contribution is called the *specific mass shift*;

[24]J. Bauche, thesis (University of Paris, 1969); J. Bauche and R.-J. Champeau, Adv. At. Mol. Phys. 12, 39-86 (1976); D. N. Stacey, Rep. Prog. Phys. XXIX, 171 (1966).

[25]H. Kopfermann, *Nuclear Moments* (Academic Press, New York, 1958).

[26]J. P. Vinti, Phys. Rev. 56, 1120 (1939).

[27]J. Bauche, J. phys. 35, 19 (1974).

it may be an order of magnitude larger than the normal shift,[27,28] but is very difficult to calculate accurately for heavy atoms.[26,27]

The second type of isotope effect arises from the difference in nuclear volume for different isotopes. If we assume the nuclear charge to be distributed throughout the nuclear volume, then the potential energy of an atomic electron inside the nucleus is less negative than for a point nuclear charge. Consequently, electrons are less tightly bound the larger the nuclear volume and hence the heavier the isotope. The effect is appreciable only for orbitals that represent an appreciable electron density at the nucleus—i.e., for s electrons, and to a much smaller degree (taking relativistic effects into account) for $p_{1/2}$ electrons. Thus the *volume isotope shift* (or *field isotope shift*) tends to be the same for all levels of a configuration, and is greater the larger the number of s electrons in the configuration. Measured isotope shifts therefore can be of considerable help in assigning experimental energy levels to specific configurations.[29] They can also sometimes pinpoint pairs of interacting levels: configuration-interaction mixing of two levels of equal J belonging to different configurations leads to a partial averaging of the isotope shifts, so that one level has an abnormally large isotope shift and the other has an abnormally small value.[29,30]

The volume effect increases in importance with increasing nuclear size, whereas (17.46) shows that the mass effect decreases in importance with increasing atomic mass. Thus, observed isotope shifts in light elements ($Z < 30$) are due primarily to the mass effect, whereas in heavy elements ($Z > 50$) they tend to be primarily the result of the volume effect.[31] This is fortunate, because it means that the above mentioned aid to configuration assignments of empirically determined energy levels is available where it is needed most—in the extremely complex spectra of the lanthanide and actinide elements (Chapter 20).

The directly observable effect of isotope shifts is a change in wavelength or wavenumber of each spectrum line. The usual (but not universal) convention is to consider an isotope shift to be positive if the wavenumber increases with increasing atomic mass number A. It is difficult to infer absolute energy-level shifts from the wavenumber data, especially in atoms containing more than a very few electrons. Usually, only relative level shifts are given, with the isotope shift of the ground level arbitrarily chosen as zero. The isotope shift of an excited level is then positive if the excitation energy increases with increasing A.

*17-9. HYPERFINE STRUCTURE

Most nuclei having an odd number of protons and/or an odd number of neutrons possess an intrinsic angular momentum, or nuclear spin, $I\hbar$, where I is integral or half-integral according as the atomic mass number is even or odd, respectively. Corresponding to this spin, the nucleus possesses a magnetic dipole moment, which by analogy with the electronic result (17.15) may be written in the form

[28]W. H. King, J. Opt. Soc. Am. 53, 638 (1963).

[29]K. Rajnak and M. Fred, J. Opt. Soc. Am. 67, 1314 (1977), and references therein.

[30]See, for example, D. C. Griffin, J. S. Ross, and R. D. Cowan, J. Opt. Soc. Am. 62, 571 (1972).

[31]However, even in lanthanide atoms, the mass effect may still be comparable in magnitude with the volume effect because of very large specific mass shifts—see ref. 28.

$$\mu_I = \mu_n g_I \mathbf{I} . \tag{17.49}$$

The nuclear magneton μ_n is smaller than the Bohr magneton (17.3) by the ratio of the electron mass to the proton mass:

$$\mu_n \equiv \frac{eh}{2m_p c} = \mu_o \frac{m}{m_p}$$
$$= 5.0508 \cdot 10^{-24} \text{ ergs/gauss}$$
$$= 2.3170 \cdot 10^{-13} \text{ Ry/gauss} . \tag{17.50}$$

Both the neutron and the proton have spin $I = 1/2$, with g factors of -3.8263 and 5.5856, respectively. Naturally occurring multiple-particle nuclei are known[32] with spins as large as 7 (^{176}Lu), g factors between -4.3 (^{3}He) and $+5.3$ (^{19}F), and magnetic moments μ_I/μ_n of -2.1 (^{3}He) to $+6.2$ (^{93}Nb).

The orientation of the nuclear spin **I** is quantized in the magnetic field produced at the position of the nucleus by the orbital motions and spins of the electrons, resulting in a quantized total (electronic-plus-nuclear) angular momentum

$$\mathbf{F} = \mathbf{J} + \mathbf{I} \tag{17.51}$$

(in units of h). Because of the precession of the $\boldsymbol{l}_i$ and s_i about **J**, the time average of the flux density at the nucleus, **B**(0), lies in the direction of **J**:

$$\bar{\mathbf{B}}_J = \frac{\mathbf{B}(0)\cdot\mathbf{J}}{J^2}\mathbf{J} . \tag{17.52}$$

Provided the interaction of **J** with **I** is weak enough that J remains a good quantum number, the Hamiltonian for the orientation energy of the nuclear magnetic dipole may be written

$$H_M = -\bar{\mathbf{B}}_J \cdot \mu_I = -\mu_n g_I \frac{\mathbf{B}(0)\cdot\mathbf{J}}{J^2}\mathbf{J}\cdot\mathbf{I} \equiv A\,\mathbf{J}\cdot\mathbf{I} . \tag{17.53}$$

The energy of this interaction is always extremely small (generally a fraction of a cm^{-1}), leading to hyperfine splitting of energy levels and hyperfine structure (hfs) of spectrum lines.[33] Contributions to **B**(0) from closed subshells are zero because the magnetic effects of the electrons cancel in pairs. (For every electron with m_l, m_s there is another with $-m_l$, $-m_s$.) Effects will be greatest for those electrons that approach the nucleus most closely; hence there is a tendency for the magnetic-dipole coefficient A to be greatest for levels belonging to a configuration that has a half-filled s subshell.

[32] G. H. Fuller and V. W. Cohen, Nucl. Data Tables A5, 433 (1969).

[33] The name "hyperfine structure" is in contrast to the so-called fine structure that results from the interaction of the electronic orbital and spin momenta, proportional to $\boldsymbol{l}\cdot\mathbf{s}$. The term "fine structure" is, however, something of a misnomer except in very light atoms, as spin-orbit level splittings can become very large, leading to spectrum lines that are quite widely separated in wavelength.

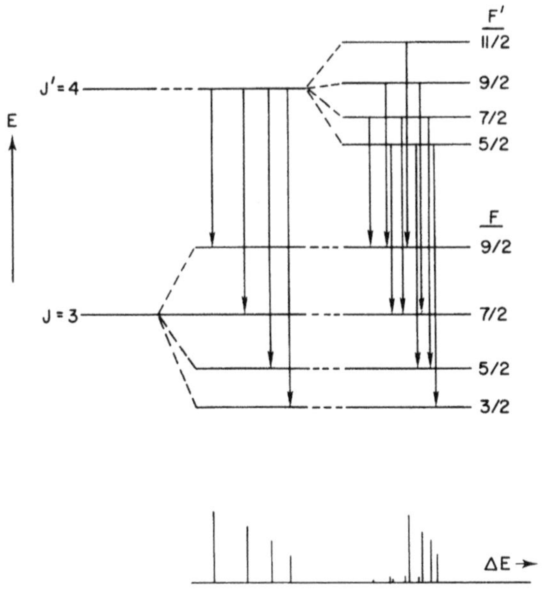

Fig. 17-10. Above: Hyperfine splitting of J = 3 and 4 levels for the case I = 3/2, assuming B = 0 so that the Landé interval rule (17.55) is valid. Below: Relative intensities of the various components of the hfs pattern. In the left-hand side of the figure, the hyperfine splitting of the upper level has been assumed zero, producing a pure flag pattern with component strengths proportional to 2F + 1 for the various sublevels of the J = 3 level.

Because of the small magnitude of (17.53), hfs splittings can usually be calculated as a perturbation energy given by the diagonal matrix elements, which using a relation analogous to (3.42) are

$$\begin{aligned}\Delta E_M &= \langle \gamma JIF | A\, \mathbf{J}\cdot\mathbf{I} | \gamma JIF \rangle \\ &= (1/2)A[F(F+1) - J(J+1) - I(I+1)] \equiv (1/2)AC\ . \qquad (17.54)\end{aligned}$$

The number of different hfs levels is from (17.51) equal to the smaller of 2J + 1 and 2I + 1. Similarly to the Landé interval rule (12.52), the energy spacing between successive levels F − 1 and F is proportional to F:

$$\Delta E_M(F) - \Delta E_M(F-1) = (1/2)A[F(F+1) - (F-1)F] = AF\ ; \qquad (17.55)$$

the center of gravity of the hfs levels lies at the position of the unsplit level J. Illustrative examples are given in Fig. 17-10.

In addition to a magnetic dipole moment, the nucleus may (if I ≧ 1) have an electric quadrupole moment Q. The interaction of Q with the inhomogeneous electric field produced by the electrons results in an additional hfs splitting energy linear in C(C+1), where C is defined in (17.54), giving a total energy of the form[25,34,35]

$$\Delta E_M + \Delta E_Q = (1/2)AC + B[C(C+1) - (4/3)J(J+1)\,I(I+1)]\ , \qquad (17.56)$$

[34] N. F. Ramsey, *Nuclear Moments* (John Wiley, New York, 1953).

[35] I. I. Sobel'man, *Introduction to the Theory of Atomic Spectra* (Pergamon, Oxford, 1972), English translation of the 1963 Russian edition.

where B is proportional[36] to $Q/[J(2J - 1)]$. The quadrupole term introduces departures from the Landé interval rule (17.55), but these tend to be small. Values of the constants A and B can be determined from experimental measurements of hfs level intervals; extensive (though outdated) tables are available.[37]

Determination of μ_I and Q from A and B involves the theoretical calculation of the magnetic and electric fields produced at the nucleus by the atomic electrons. We shall not go into this problem in detail here,[25,34,35] nor shall we discuss various relativistic effects.[35,38] It should, however, be noted that hyperfine splittings may be comparable with fine-structure separations in very light atoms or for complex or highly excited configurations where levels lie close together,[39] or in very heavy atoms where the hfs interaction becomes quite large. The hfs interaction may then produce appreciable mixing of basis states with different J, so that J is no longer a good quantum number. It is then necessary to replace the approximate Hamiltonian (17.53) by a more exact form[35,40] involving the individual r_i, l_i, and s_i, to compute off-diagonal as well as diagonal matrix elements, and to diagonalize the "complete" matrix (including all important interacting basis states) for each value of F. Interactions among the various basis states of given F belonging to different levels produce distortions in the hfs pattern of each level beyond that represented by (17.56). Failure to take these distortions into account can grossly alter the values of A and B that would be inferred using (17.56) alone, and thereby lead to incorrect values of μ_I and Q.

Treatment of the Zeeman effect of hfs would also take us too far afield. The basic principles involved have already been described: it is necessary to numerically diagonalize the matrix of the complete Hamiltonian, including both the hfs and the external-magnetic-field contributions. Examples of the Zeeman effect are given by Kopfermann.[25]

Intensities of hfs components. Strengths of hfs line components for electric dipole transitions are given by a straightforward generalization of (14.42):

$$S_{\gamma\gamma'}^{1/2} = \sum_{\beta J} \sum_{\beta' J'} y_{\beta JIF}^{\gamma} \langle \beta JIF \| \boldsymbol{P}^{(1)} \| \beta' J' IF' \rangle \, y_{\beta' J' IF'}^{\gamma'} \, , \tag{17.57}$$

where the y's are components of the eigenvectors obtained by diagonalizing the hfs matrices mentioned above. We shall discuss only the case of weak interactions in which neighboring levels do not interact, so that J and J′ are good quantum numbers. Strengths of the components of a hfs multiplet are then given from (11.38) by

[36]The strong dependence on J is sometimes allowed for explicitly by introducing a new constant B′ such that $B = (3/8)B'/[I(2I - 1)J(2J - 1)]$.

[37]Landolt-Börnstein, *Zahlenwerte und Funktionen* (Springer-Verlag, Berlin, 1952), 6th ed., Vol. I, part 5.

[38]L. Armstrong, Jr., *Theory of the Hyperfine Structure of Free Atoms* (Wiley-Interscience, New York, 1971).

[39]For example, the hfs splittings are about four times greater than fs splittings in the 1s6h configuration of ^{13}C V: J. L. Subtil, O. Poulsen, P. S. Ramanujam, and D. B. Iversen, J. Phys. B. 10, 1607 (1977).

[40]J. Bauche, "Nuclear Effects in Atomic Spectra," in *Atoms, Molecules, and Lasers* (International Atomic Energy Agency, Vienna, 1974).

$$\boldsymbol{S}_{\text{hfs}} = |\langle\gamma\text{JIF}\|\boldsymbol{P}^{(1)}\|\gamma'\text{J}'\text{IF}'\rangle|^2 \tag{17.58}$$

$$= [\text{F,F}']\begin{Bmatrix} \text{J} & \text{I} & \text{F} \\ \text{F}' & 1 & \text{J}' \end{Bmatrix}^2 |\langle\gamma\text{J}\|\boldsymbol{P}^{(1)}\|\gamma'\text{J}'\rangle|^2 , \tag{17.59}$$

where the final squared matrix element is the total line strength (14.31). The triangle relations in the 6-j symbol lead to the selection rule

$$\Delta\text{F} = 0, \ \pm 1 \qquad (\text{F} = \text{F}' = 0 \text{ not allowed}) , \tag{17.60}$$

in addition to the familiar analogous selection rule on J.

When I is smaller than the other four quantum numbers in the 6-j symbol, the strongest components of the hfs multiplet are those for which

$$\Delta\text{F} = \Delta\text{J} , \tag{17.61}$$

the strength increasing with increasing F. These strong components lead to a "flag pattern" in which the intensities increase monotonically with changing wavelength from one component to the next.

The flag pattern is especially striking in cases where the hfs splitting of one level is very small, so that the weak components of the multiplet are not resolved from the strong ones (Fig. 17-10). Assuming it to be the upper level that is unsplit, the total strengths of the resolved hfs components are from (5.31)

$$\sum_{\text{F}'} \boldsymbol{S}_{\text{hfs}} = \frac{2\text{F} + 1}{2\text{J} + 1} |\langle\gamma\text{J}\|\boldsymbol{P}^{(1)}\|\gamma'\text{J}'\rangle|^2 , \tag{17.62}$$

so that the strengths of the various resolved components are proportional to the values of 2F + 1 for the various resolved hfs sublevels.

From (2.65), the total strength of the entire hfs multiplet is

$$\sum_{\text{F}} \sum_{\text{F}'} \boldsymbol{S}_{\text{hfs}} = (2\text{I} + 1)|\langle\gamma\text{J}\|\boldsymbol{P}^{(1)}\|\gamma'\text{J}'\rangle|^2 , \tag{17.63}$$

which is (2I + 1) times the line strength (14.31) defined without consideration of nuclear effects. However, when the nuclear spin is taken into account there are (2I + 1) times as many quantum states, and consequently in any given light source there are $(2\text{I} + 1)^{-1}$ times as many atoms in each quantum state. The total intensity, summed over all components of a hfs multiplet, is therefore the same as the value we would have obtained for the spectrum line J − J′ ignoring nuclear effects completely.

Analogous to (14.32), the radiative transition probability from the state $|\gamma'\text{J}'\text{IF}'\text{M}'_\text{F}\rangle$ is

$$\text{A}(\gamma'\text{J}'\text{IF}'\text{M}'_\text{F} - \gamma\text{JIF}) = \frac{64\pi^4 e^2 a_0{}^2 \sigma^3}{3h(2\text{F}'+1)} \boldsymbol{S}_{\text{hfs}}$$

$$= [\text{F, J}']\begin{Bmatrix} \text{J} & \text{I} & \text{F} \\ \text{F}' & 1 & \text{J}' \end{Bmatrix}^2 \text{A}(\gamma'\text{J}'\text{M}'_\text{J} - \gamma\text{J}) , \tag{17.64}$$

and the total transition probability to all hfs sublevels of the lower level J is

$$\sum_{F} A(\gamma'J'IF'M'_F - \gamma JIF) = A(\gamma'J'M'_J - \gamma J) , \quad (17.65)$$

in agreement with physical intuition. Similarly,

$$f(\gamma JIFM_F - \gamma'J'IF') = [F',J]\begin{Bmatrix} J & I & F \\ F' & 1 & J' \end{Bmatrix}^2 f(\gamma JM_J - \gamma'J') , \quad (17.66)$$

and the total absorption oscillator strength is

$$\sum_{F'} f(\gamma JIFM_F - \gamma'J'IF') = f(\gamma JM_J - \gamma'J') . \quad (17.67)$$

It may be noted that in the case of small I, only the principal component (17.61) (out of the three possibilities F′) is strong. This component contributes the lion's share to the summation in (17.67), and has a component oscillator strength nearly equal to the total oscillator strength of the parent line. However, in any light source there are only $(2I + 1)^{-1}$ times as many atoms in each hfs quantum state. Thus when hfs splittings are greater than line widths, self-absorption effects for the principal components are (2I + 1) times smaller than when these splittings are negligible (and of course they are much smaller still for the satellite components.)

18

CONTINUUM STATES; IONIZATION AND RECOMBINATION

18-1. INTRODUCTION

Up to this point, we have considered only discrete energy states corresponding to configurations in which all electrons of the atom are more or less tightly bound. That is, we considered only those basis functions that could be constructed from one-electron spin-orbitals

$$\varphi_{nlm_lm_s}(r) = \frac{1}{r}P_{nl}(r)\cdot Y_{lm_l}(\theta,\phi)\cdot\sigma_{m_s}(s_z) \ , \tag{18.1}$$

in which the radial function $P_{nl}(r)$ is bound to the atom in the sense that

$$\lim_{r\to\infty} P_{nl}(r) = 0 \ ; \tag{18.2}$$

in fact, P tends to zero exponentially, which is more than fast enough to make it square-integrable, and to permit the ortho-normalization relation

$$\int_0^\infty P_{nl}(r)P_{n'l}(r)\,dr = \delta_{nn'} \ . \tag{18.3}$$

Among the set of all possible bound-state configurations, there exist subsets each of which constitutes a Rydberg series of bound configurations—that is, a series of configurations of the form

$$\cdots n_il_i^{w_i}nl \ , \tag{18.4}$$

in which the various configurations differ only in the values of

$$n = \cdots 6, 7, 8, \cdots, \infty \ .$$

The center-of-gravity energies of these configurations tend to a finite limit:

$$\cdots E_{av}^{6l},\ E_{av}^{7l},\ E_{av}^{8l}, \cdots, E_{av}^{\infty l} ,$$

where the limit $E_{av}^{\infty l}$ is the c-of-g energy of the corresponding configuration

$$\cdots n_i l_i^{w_i} \tag{18.5}$$

of the ion.

As a natural extension of this Rydberg series of bound configurations we may consider the infinite series of continuum configurations

$$\cdots n_i l_i^{w_i} \varepsilon l , \tag{18.6}$$

in which ε is the kinetic energy of a free electron having an angular momentum $l\hbar$ about the ion core (18.5). Because in the limit $n \rightarrow \infty$ the interaction of the excited electron nl with the ion core becomes zero, the center-of-gravity energy of a continuum configuration is

$$E_{av}^{\infty l} + \varepsilon . \tag{18.7}$$

The c-of-g energies of this extended Rydberg series are illustrated in Fig. 18-1.

Similarly, for each possible Rydberg series of *states*

$$\cdots n_i l_i^{w_i}(\gamma J_c) nlj, JM \tag{18.8}$$

there is a corresponding continuum of states

$$\cdots n_i l_i^{w_i}(\gamma J_c) \varepsilon lj, JM . \tag{18.9}$$

The energy of each such state is evidently

$$E_{\gamma J_c} + \varepsilon , \tag{18.10}$$

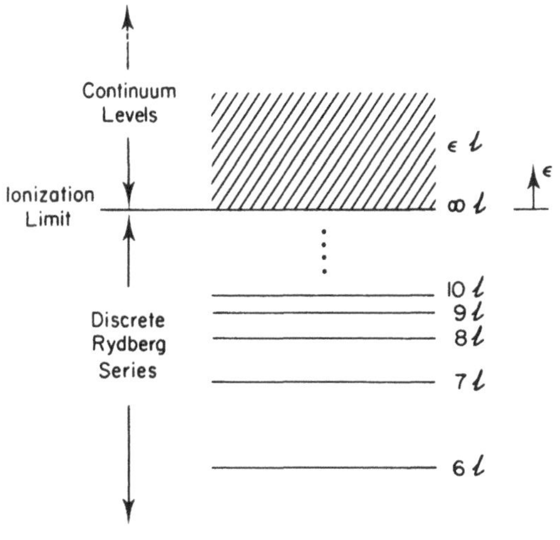

Fig. 18-1. Center-of-gravity energies of an extended Rydberg series of configurations, consisting of an ion core configuration plus an excited bound electron nl ($n = \cdots, 6, 7, 8, 9, 10, \cdots$) or a free electron εl with kinetic energy $0 \leqq \varepsilon < \infty$.

where $E_{\gamma J_c}$ is the energy of the parent state

$$\cdots n_l l_l^{w_l} \gamma J_c \tag{18.11}$$

of the ion—i.e., the limiting energy of the discrete states (18.8) of the atom. [Here γJ_c are all those quantum numbers (other than M_c) required to specify the state of the (N − 1)-electron core $\cdots n_l l_l^{w_l}$. In the notation of (12.2), $J_c jJM$ correspond to $\mathfrak{J}_{q-1} J_q \mathfrak{J}_q \mathfrak{M}_q$.]

Most commonly we are interested only in the discrete bound states (18.11) of the ion, and we ignore the possible presence of a free electron. Nonetheless, we also have a legitimate interest in the continuum states (18.9) of the atom, for several reasons:

(a) When working with absorption spectra of atoms in a suitable absorbing state, photo-absorption excitation into the excited bound states (18.8) of the atom produces a series of absorption lines (e.g., the simple Lyman series 1s − np of H I, or the 3s $^2S_{1/2}$ − np $^2P^o_{3/2}$ absorption series of Na I) whose wavenumbers converge to the threshold limit for photoionization of the atom. Beyond this limit there exists a continuous absorption spectrum, resulting from photoionization of one electron into the state $\cdots \varepsilon l jJM$; this electron is ejected from the atom with kinetic energy ε. The properties of the absorption continuum are a direct extension of the properties of the series of absorption lines, and as such may be of as great interest as the lines themselves.

(b) For maximum accuracy in the calculation of energy levels and eigenvectors of discrete energy states of an atom, the basis-function expansion

$$\Psi_J^k = \sum_\beta y_{\beta J}^k \Psi_{\beta J} \tag{18.12}$$

must usually include basis functions from more than one configuration, including perhaps several bound configurations of a Rydberg series (18.4) of configurations. In cases of strong configuration interaction, a wide energy range of configurations must be included, frequently extending into the continuum:

$$\Psi_J^k = \sum_{\gamma J_c l j} \left[\sum_n y^k_{\gamma J_c n l j J} \Psi_{\gamma J_c n l j J} + \int_0^\infty y^k_{\gamma J_c \varepsilon l j J} \Psi_{\gamma J_c \varepsilon l j J}\, d\varepsilon \right] . \tag{18.13}$$

(c) In the case of bound configurations involving more than one excited electron, or bound states (18.8) involving one excited electron added to an *excited* state γJ_c of the core, there will exist discrete energy levels that may lie higher than the first ionization limit—i.e., higher than the lowest state (18.11) of the ion [cf. Fig. 12-5]. As discussed in connection with that figure, these discrete levels may interact with continuum states having the same parity and J (and, in the case of good LS coupling, having the same L and S), resulting in broadening of the discrete level and autoionization of the atom by a radiationless transition to the continuum state. A theoretical treatment of such high-lying semi-discrete states obviously requires a consideration of the continuum states as well, with basis function expansions similar to (18.13) that include both discrete and continuum functions.

18-2. ONE-ELECTRON CONTINUUM BASIS FUNCTIONS

We assume the one-electron wavefunction for a free electron to be of the same basic form as that for a bound electron,

$$\varphi_{\varepsilon l m_l m_s}(r) = \frac{1}{r} P_{\varepsilon l}(r) \cdot Y_{l m_l}(\theta,\phi) \cdot \sigma_{m_s}(s_z) \ , \tag{18.14}$$

except that the radial factor $P_{\varepsilon l}$ is labeled by the kinetic energy ε (in rydbergs) of a free electron, rather than by the principal quantum number n of a bound electron.

The radial wavefunction is to be found just as before by solving a one-electron differential equation similar to (7.28):

$$\left[-\frac{d^2}{dr^2} + \frac{l(l+1)}{r^2} + V^{\varepsilon l}(r)\right] P_{\varepsilon l}(r) = \varepsilon P_{\varepsilon l}(r) \ . \tag{18.15}$$

We shall assume[1] the potential-energy function $V^{\varepsilon l}(r)$ to be just the limiting form of the HX function $V^{nl}(r)$ [defined by (7.54)] for a bound function P_{nl}, as $n \to \infty$. This limiting function depends only on the $N-1$ electrons of the ion core. Moreover, the core electrons are independent of the continuum electron εl. The problem of solving (18.15) is thus much simpler than for bound configurations: The set of one-electron differential equations (7.28) for the $N-1$ core electrons is solved once by the usual self-consistent iterative procedure. The one-electron functions $P_{n_i l_i}(r)$ thereby obtained provide a potential function $V^{\varepsilon l}(r)$ that is independent of ε. For each desired value of ε, $P_{\varepsilon l}(r)$ is obtained simply by integrating the differential equation (18.15) once; there is no iteration on ε and no self-consistency iteration on the core wavefunctions.

The value of ε is greater than zero (instead of less than zero, as for a bound function), and it is convenient to let

$$q = +\sqrt{\varepsilon} \qquad (\varepsilon \text{ in rydbergs}) \ , \tag{18.16}$$

where q is real. As $r \to \infty$ and the centrifugal and potential terms in (18.15) tend to zero, $P_{\varepsilon l}$ does not tend asymptotically to zero, but rather tends asymptotically to a solution of

$$P''_{\varepsilon l}(r) = -\varepsilon P_{\varepsilon l}(r) = -q^2 P_{\varepsilon l}(r) \ ,$$

which is of the form[2]

$$P_{\varepsilon l}(r) \xrightarrow[r\to\infty]{} \text{const.} \times \cos(qr + \delta) \ , \tag{18.17}$$

[1]Regarding HF continuum functions, see M. J. Seaton, Phil. Trans. Roy. Soc. London A245, 469 (1953). For a discussion of energy-dependent local potential methods, see M. E. Riley and D. G. Truhlar, J. Chem. Phys. 63, 2182 (1975); note that such potentials invalidate the orthogonality proof below—see text following (18.24).

[2]Such a solution corresponds to a standing-wave function, rather than the running-wave solution $P_{\varepsilon l} \propto e^{i(qr+\delta)}$ usually used for scattering problems. We have used the letter q, rather than the letter k generally used in scattering theory, in order to avoid confusion with our multipole expansion index k.

where δ is a phase factor that is in general a function of l and q (and a slowly varying function of r). Sample continuum wavefunctions for $l = 1$ are illustrated in Fig. 18-2 (normalized in a manner to be discussed in the following section). These may be compared with the bound functions shown in Fig. 3-4; it should be noted that at small r, continuum functions are essentially identical in form with bound functions of large n and the same l, the difference being only in amplitude.

18-3. NORMALIZATION OF THE RADIAL FUNCTION

If the radial function $P_{\varepsilon l}(r)$ were of the asymptotic form (18.17) at all values of r, then two such functions corresponding to different values of ε (or of q) would clearly be orthogonal:

$$\int_0^\infty \cos(qr + \delta)\cos(q'r + \delta')\,dr = 0\ , \qquad (q \neq q')\ ; \tag{18.18}$$

however, normalization would present a problem because when $q = q'$

$$\int_0^\infty \cos^2(qr + \delta)\,dr = \infty\ .$$

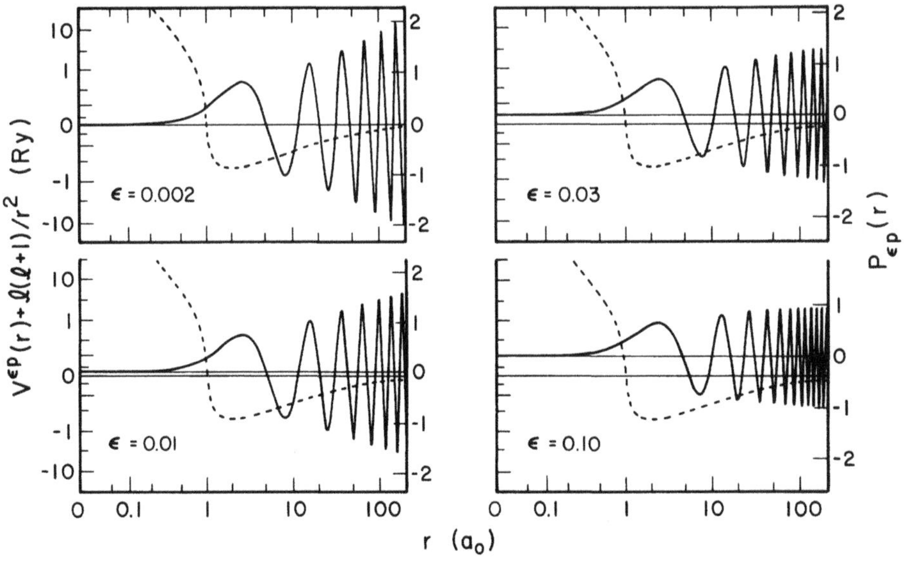

Fig. 18-2. Solid curves: Continuum wavefunctions for εp electrons in the field of a hydrogen nucleus, plotted relative to a horizontal base line drawn at the position of the kinetic energy ε. Dotted curves: The effective potential $V(r) + l(l + 1)/r^2 = -2/r + 2/r^2$.

This singularity is obviously unavoidable even in the correct function $P_{\varepsilon l}$, and so we cannot expect to satisfy an orthonormality condition like (18.3), which we had for discrete states. For continuum functions the best we can hope for is the relation

$$\int_0^\infty P_{\varepsilon l}(r) P_{\varepsilon' l}(r)\, dr = \delta(\varepsilon - \varepsilon') , \tag{18.19}$$

involving a Dirac delta function (which is infinite at $\varepsilon = \varepsilon'$) instead of a Kronecker delta.

The relation (18.19) means that we hope to be able to normalize the functions $P_{\varepsilon l}$ such that for a small but finite energy interval Δ,

$$\int_\Delta \int_0^\infty P_{\varepsilon l}(r) P_{\varepsilon' l}(r)\, dr\, d\varepsilon = \int_\Delta \delta(\varepsilon - \varepsilon')\, d\varepsilon = \begin{cases} 0, & \varepsilon' \text{ outside } \Delta , \\ 1, & \varepsilon' \text{ inside } \Delta . \end{cases} \tag{18.20}$$

Assuming that this can be done, the quantity

$$P_{\varepsilon l}(r) P_{\varepsilon' l}(r)\, d\varepsilon\, dr \tag{18.21}$$

is to be interpreted as the probability that a free electron lies within a shell of radius r and thickness dr, and that the kinetic energy of the electron lies within an energy range $d\varepsilon$ rydbergs centered at ε'.

In order to verify the orthogonality relation in (18.20) and find the proper normalization constant,[3] we write $P_{\varepsilon l}$ in the form

$$P_{\varepsilon l}(r) = c_q r X_{ql}(r) , \tag{18.22}$$

where rX is a radial function normalized to unit amplitude at infinity,

$$r X_{ql}(r) \underset{r \to \infty}{\longrightarrow} \cos(qr + \delta) , \tag{18.23}$$

and c_q is the required normalization constant to make (18.20) hold for ε' within Δ.

Noting that

$$\frac{c_q}{r}\frac{d}{dr} r^2 \frac{d}{dr} X = \frac{1}{r}\frac{d}{dr} r^2 \frac{d}{dr}(r^{-1}P) = \frac{1}{r}\frac{d}{dr} r^2 (r^{-1}P' - r^{-2}P) = P'' ,$$

we may write $X_{q'}$ times (18.15) in the form

$$-\frac{X_{q'}}{r^2}\frac{d}{dr} r^2 \frac{d}{dr} X_q + \left[\frac{l(l+1)}{r^2} + V(r)\right] X_{q'} X_q = q^2 X_{q'} X_q . \tag{18.24}$$

Writing an equation like (18.24) with q and q′ interchanged, and subtracting it from (18.24), we obtain (since V is independent of ε)

[3]E. Fues, Ann. Physik **81**, 281 (1926); J. Hargreaves, Proc. Cambridge Phil. Soc. **25**, 75 (1929).

$$I \equiv \int_0^\infty r^2 X_q X_{q'}\, dr = \frac{1}{q^2 - q'^2}\int_0^\infty \left[X_q \frac{d}{dr} r^2 \frac{d}{dr} X_{q'} - X_{q'} \frac{d}{dr} r^2 \frac{d}{dr} X_q\right] dr .$$

We first carry out an integration by parts to obtain

$$(q^2 - q'^2)I = \lim_{r\to\infty} \left[X_q r^2 \frac{dX_{q'}}{dr} - X_{q'} r^2 \frac{dX_q}{dr}\right]_0^r - \int_0^\infty \left[\left(\frac{dX_q}{dr}\right) r^2 \frac{dX_{q'}}{dr} - \left(\frac{dX_{q'}}{dr}\right) r^2 \frac{dX_q}{dr}\right] dr$$

$$= \lim_{r\to\infty} \left[X_q r^2 \frac{dX_{q'}}{dr} - X_{q'} r^2 \frac{dX_q}{dr}\right]_0^r .$$

But for small r the potential V(r) is Coulombic and (just as in the solution for a hydrogen atom)

$$X = c_1 r^l + c_2 r^{l+1} + \cdots$$

$$r^2 \frac{dX}{dr} = l c_1 r^{l+1} + (l+1) c_2 r^{l+2} + \cdots$$

so that

$$\left[X r^2 \frac{dX}{dr}\right]_{r\to 0} = 0 .$$

For large r we find from (18.23) that

$$\frac{dX}{dr} = -r^{-2}\cos(qr+\delta) - r^{-1} q \sin(qr+\delta)$$

so that

$$(q^2 - q'^2)I = \lim_{r\to\infty} \Big[r^{-1}\cos(qr+\delta)\Big\{-\cos(q'r+\delta') - rq'\sin(q'r+\delta')\Big\} - r^{-1}\cos(q'r+\delta')\Big\{-\cos(qr+\delta) - rq\sin(qr+\delta)\Big\}\Big]$$

$$= \lim_{r\to\infty} \Big[-\frac{q'}{2}\Big\{\sin(q'r+\delta'+qr+\delta) + \sin(q'r+\delta'-qr-\delta)\Big\} + \frac{q}{2}\Big\{\sin(qr+\delta+q'r+\delta') + \sin(qr+\delta-q'r-\delta')\Big\}\Big]$$

$$= \lim_{r\to\infty} \left[\left(\frac{q-q'}{2}\right)\sin\{(q+q')r+\delta+\delta'\} + \left(\frac{q+q'}{2}\right)\sin\{(q-q')r+\delta-\delta'\}\right] .$$

For the integral in (18.20) we thus obtain

$$\int_\Delta \int_0^\infty P_{\varepsilon l} P_{\varepsilon' l}\, dr\, d\varepsilon = \int_\Delta c_q c_{q'} I\, d\varepsilon = \int_\Delta 2 c_q c_{q'} q I\, dq$$

$$= \lim_{r\to\infty} \left[\int_\Delta \frac{c_q c_{q'} q}{(q+q')} \sin\{(q+q')r + \delta + \delta'\}\, dq \right.$$

$$\left. + \int_\Delta \frac{c_q c_{q'} q}{(q-q')} \sin\{(q-q')r + \delta - \delta'\}\, dq \right]$$

$$= \lim_{r\to\infty} \int_\Delta \frac{c_q c_{q'} q}{(q-q')} \sin\{(q-q')r + \delta - \delta'\}\, dq \ ; \qquad (18.25)$$

the integral involving $q + q'$ gives zero contribution because in the limit the sine function oscillates infinitely rapidly. If q' does not lie within the range of integration Δ, then (18.25) is equal essentially to

$$\frac{c_q c_{q'} q}{q - q'} \lim_{r\to\infty} \int_\Delta \sin\{(q-q')r + \delta - \delta'\}\, dq = 0 \ ,$$

which proves the orthogonality portion of (18.20). If q' *does* lie within Δ, then for very large r the entire contribution to (18.25) comes from the integration range $q \cong q'$. We change to the integration variable $x \equiv (q - q')r$, for which the integration range becomes infinite in the limit $r \to \infty$, and obtain

$$\int_\Delta \int_0^\infty P_{\varepsilon l} P_{\varepsilon' l}\, dr\, d\varepsilon = (c_q)^2 q \lim_{r\to\infty} \int_\Delta \frac{\sin\{(q-q')r\}}{q-q'}\, dq$$

$$= (c_q)^2 q \int_{-\infty}^{\infty} \frac{\sin x}{x}\, dx = \pi q (c_q)^2 \ .$$

The normalization condition (18.20) thus gives

$$c_q = \pi^{-1/2} \varepsilon^{-1/4} \qquad (18.26)$$

or

$$P_{\varepsilon l}(r) = \pi^{-1/2} \varepsilon^{-1/4} r X_{\varepsilon l}(r) \ ; \qquad (18.27)$$

i.e., the desired function $P_{\varepsilon l}$ is that which has been normalized such that the asymptotic amplitude of $P_{\varepsilon l}$ is $\pi^{-1/2}\varepsilon^{-1/4}$, where ε is in rydbergs.[4]

[4]Conventions used in the literature are by no means standardized. Other normalizations sometimes used are $c_q = 1$, $c_q = \varepsilon^{-1/4}$, and (when energies are measured in hartrees instead of rydbergs) $c_q = 2^{1/4}\pi^{-1/2}\varepsilon^{-1/4}$.

Certain numerical problems arise in calculating continuum functions from (18.15). Firstly, an (unnormalized) integral $P^u_{\varepsilon l}(r)$ of (18.15) can practicably be evaluated only to some large but finite radius $r = r_o$. In order to properly normalize this function, it is necessary to extrapolate to find its amplitude at infinity. This can be done by noting that for large r the central-field potential is essentially the Coulomb potential $V(r) \cong -2Z_c/r$, where $Z_c = Z - N + 1$ (N being the total number of electrons, including the free electron). Thus for such r, the equation (18.15) may be written

$$P''_{\varepsilon l} + \lambda P_{\varepsilon l} = 0 \ ,$$

where

$$\lambda = \varepsilon + 2Z_c/r - l(l+1)/r^2 \ .$$

It is easily verified that a solution is[5]

$$P_{\varepsilon l}(r) = C\,y \cos\varphi(r) = C\,y \cos(qr + \delta) \ ,$$

where δ is a (slowly varying) function of r, and $y \equiv (\varphi')^{-1/2}$ is given by

$$y^{-4} = \lambda + y^{-1}y'' \ .$$

By using the approximation $y = \lambda^{-1/4}$ in the second term of this expression, it is easily seen that this approximation is accurate to terms in r^{-2} and that asymptotically

$$rX_{\varepsilon l}(r) = A(r,\varepsilon)\cos(qr + \delta) \ ,$$

where the slowly varying amplitude function $A = \varepsilon^{1/4}y$ is

$$A(r,\varepsilon) \cong 1 - \frac{Z_c}{2\varepsilon r}\left[1 - \frac{5Z_c}{4\varepsilon r} - \frac{l(l+1)}{2Z_c r}\right] \ . \qquad (18.28)$$

The properly normalized continuum function is then

$$P_{\varepsilon l}(r) = \frac{A(r_o,\varepsilon)}{\pi^{1/2}\varepsilon^{1/4}} \cdot \frac{P_{\varepsilon l}{}^u(r)}{B} \ , \qquad (18.29)$$

where B is the amplitude of $P^u_{\varepsilon l}$ at $r = r_o$.

Secondly, acceptable accuracy of the numerical integration of (18.15) requires that there be at least half a dozen mesh points per half-cycle of $P_{\varepsilon l}(r)$, and therefore from (18.16)-(18.17) that the mesh size Δr satisfy

$$\Delta r < \frac{\pi}{6\varepsilon^{1/2}} \cong \frac{1}{2\varepsilon^{1/2}} \qquad (\varepsilon \text{ in Ry}) \qquad (18.30)$$

over the entire integration mesh. At the same time, reasonable accuracy of (18.28) requires

[5]See, for example, D. R. Bates and M. J. Seaton, Mon. Not. R. Astron. Soc. 109, 698 (1949).

$$r_o > \max\left[\frac{10Z_c}{\varepsilon}, \frac{5l(l+1)}{Z_c}, r_c\right] , \qquad (18.31)$$

where r_c is the maximum radius to which the $N-1$ core electrons extend with appreciable density. For large ε and l and small Z_c, (18.30) and (18.31) cannot both be satisfied on the logarithmic mesh employed in many HF computer programs without using a very fine mesh (e.g., $\Delta r/r < 10^{-3}$ instead of the usual value 1/16) and hence a prohibitively large number of mesh points (> 12500 instead of 200). The Herman-Skillman linear mesh[6] (used in program RCN, Fig. 8-1) can, however, be readily adapted by discontinuing the periodic mesh-interval doubling at the appropriate point; even so, a rather large number of mesh points is still required. (RCN allows up to 1801 points.)

18-4. THE ENERGY DIMENSIONS OF $P_{\varepsilon l}$

At first glance, the normalization factor (18.26) appears to give $P_{\varepsilon l}$ the dimensions of $\varepsilon^{-1/4}$; however, it must be remembered that ε is really a dimensionless quantity (which *must* be measured in rydbergs). The true dimensions of $P_{\varepsilon l}$ are indicated by (18.20) and (18.21). The integration variable ε is again dimensionless (and measured in rydbergs), but the physical interpretation of (18.21) indicates that $P_{\varepsilon l}$ has in principle the dimensions[7] of $(\text{energy})^{-1/2}$ and may be considered to have units of $(\text{rydbergs})^{-1/2}$.

It is important to keep in mind this inherent difference between discrete and continuum radial wavefunctions. For example, the radial integrals R^k, (13.2) and (13.3), met in the calculation of configuration interaction between two bound-state configurations have the dimensions of energy and are measured in rydbergs. The energy perturbation of a discrete state 1 by another discrete state 2 with unperturbed energies E_1 and E_2, respectively, is to a first approximation indicated by (10.38) to be proportional to

$$\frac{[R^k \ (\text{Ry})]^2}{E_1 - E_2 \ (\text{Ry})} = \frac{(R^k)^2}{E_1 - E_2} \ (\text{Ry}) . \qquad (18.32)$$

In contrast, the corresponding radial integral met in the calculation of configuration interaction between a bound-state and a continuum configuration is of the form[8]

$$R^k(n_1l_1,n_2l_2;n_1'l_1',\varepsilon l_2') = \int_0^\infty\int_0^\infty \frac{2r_<^k}{r_>^{k+1}} P_{n_1l_1}(r_1)P_{n_2l_2}(r_2)P_{n_1'l_1'}(r_1)P_{\varepsilon l_2'}(r_2)\,dr_1\,dr_2 \ ; \qquad (18.33)$$

this clearly has dimensions of $(\text{energy})^{1/2}$ and units of $(\text{rydbergs})^{1/2}$. If E_1 is the energy of the discrete state and ε that of the continuum state, the quantity analogous to (18.32) is

[6]F. Herman and S. Skillman, *Atomic Structure Calculations* (Prentice-Hall, Englewood Cliffs, N. J., 1963).

[7]This is over and above the dimension of $(\text{length})^{-1/2}$, which is in principle present in P_{nl} as well as in $P_{\varepsilon l}$.

[8]Note that no problem results from the fact that $P_{\varepsilon l_2'}(r_2)$ extends to infinity, because the exponential decay of $P_{n_2l_2}(r_2)$ effectively cuts the integral off at finite values of r_2. (Even for the interaction between two continua, where $P_{n_2l_2}$ becomes replaced by $P_{\varepsilon_2 l_2}$, convergence of the integral is assured by the factor $1/r_>^{k+1}$.)

$$\frac{[R^k \ (Ry^{1/2})]^2}{E_1 - \varepsilon \ \ (Ry)} = \frac{(R^k)^2}{E_1 - \varepsilon} \qquad \text{(dimensionless)} \ . \tag{18.34}$$

This represents a quantity proportional to

$$\frac{dE_1}{d\varepsilon} \ , \tag{18.35}$$

the perturbation (in rydbergs) of the discrete energy E_1 *per rydberg energy range of continuum*; the total perturbation of the discrete state produced by a finite energy range Δ of the continuum is obtained only upon integration of (18.34):

$$\int_\Delta \frac{(R^k)^2}{E_1 - \varepsilon} d\varepsilon \ , \tag{18.36}$$

and the integration variable $d\varepsilon$ provides the units (rydbergs) of the perturbation.

Similarly, the electric-multipole radial integral analogous to (14.55) and (15.18) for a radiative transition between a discrete and a continuum state is

$$P^{(t)}_{nl,\varepsilon l'} = (-1)^l \, [l,l']^{1/2} \begin{pmatrix} l & t & l' \\ 0 & 0 & 0 \end{pmatrix} \int_0^\infty P_{nl} \, r^t \, P_{\varepsilon l'} \, dr \tag{18.37}$$

and has energy units of $(\text{rydbergs})^{-1/2}$. The square of this quantity (together with the appropriate angular and proportionality factors) provides a value of

$$\frac{dS}{d\varepsilon} \ , \tag{18.38}$$

the "line" strength per rydberg energy range of continuum.

18-5. RELATION BETWEEN RADIAL INTEGRALS FOR DISCRETE AND CONTINUUM STATES

We have already pointed out that the potential function V(r) in the differential equation (18.15) is essentially the same for continuum states εl as for discrete states nl with large n. Consequently, as a function of the one-electron eigenvalue ε, it must be true that the discrete solutions $P_{nl}(r)$ $(n \rightarrow \infty)$ pass smoothly over into continuum solutions $P_{\varepsilon l}$, provided that proper allowance is made for the difference between the two normalization procedures (18.3) and (18.19). (The similarity in form of P_{nl} and $P_{\varepsilon l}$ for not-too-large r may be seen by a comparison of Figs. 3-4 and 18-2.) Thus values of radial integrals such as R^k and $P^{(t)}$ (which effectively contain contributions from $P_{\varepsilon l}$ for only a finite range of values of r) must also be continuous functions of ε if the normalization differences are taken into account—which can be done in the following way:

From the eigenvalue ε_n for a discrete solution P_{nl}, we can define an effective quantum number n^* such that

$$\varepsilon_n = -\frac{(Z-N+1)^2}{(n^*)^2} \equiv -\left(\frac{Z_c}{n^*}\right)^2 \quad \text{Ry} , \tag{18.39}$$

where according to (8.17) successive values of n^* in a Rydberg series increase by essentially unity just as do successive values of n. Differentiation of this expression gives

$$\frac{d\varepsilon_n}{dn} \cong \frac{d\varepsilon_n}{dn^*} = \frac{2Z_c^{\,2}}{(n^*)^3} = \frac{2}{Z_c} |\varepsilon_n|^{3/2} . \tag{18.40}$$

Thus analogously to $(R^k_{\varepsilon l})^2$, which as shown in Sec. 18-4 represents a configuration-interaction strength per unit $d\varepsilon$, we should have for the discrete case of interaction between a fixed configuration and successive members of a Rydberg series

$$\frac{\text{CI strength per unit } \Delta n}{d\varepsilon_n/dn} = \frac{(n^*)^3}{2Z_c^{\,2}}(R^k_{nl})^2 \equiv \frac{Z_c}{2|\varepsilon_n|^{3/2}}(R^k_{nl})^2 , \tag{18.41}$$

and we should expect the quantities

$$\left(\frac{(n^*)^3}{2Z_c^{\,2}}\right)^{1/2} R^k_{nl} \equiv \left(\frac{Z_c}{2|\varepsilon_n|^{3/2}}\right)^{1/2} R^k_{nl} \qquad \text{and} \qquad R^k_{\varepsilon l} \tag{18.42}$$

to form two functions of ε that join smoothly across the ionization limit $\varepsilon = 0$. Similarly,

$$\left(\frac{(n'^*)^3}{2Z_c^{\,2}}\right)^{1/2} P^{(t)}_{nl,n'l'} \equiv \left(\frac{Z_c}{2|\varepsilon_{n'}|^{3/2}}\right)^{1/2} P^{(t)}_{nl,n'l'} \qquad \text{and} \qquad P^{(t)}_{nl,\varepsilon l'} \tag{18.43}$$

should also join smoothly across the dividing point $\varepsilon = 0$. [The continuity across $\varepsilon = 0$ is not affected if n'^* is replaced by the actual quantum number n', because the percent difference between the two becomes negligible at large n'.]

Examples are shown in Fig. 18-3 for the configuration interactions

$$\begin{cases} 3s3p^2 + 3s^2nd \\ 3s3p^2 + 3s^2\varepsilon d \end{cases}$$

and the dipole transitions

$$\begin{cases} 3s^23p - 3s^2nd \\ 3s^23p - 3s^2\varepsilon d \end{cases}$$

in members of the Al I isoelectronic sequence.

As a consequence of (18.43), the discrete absorption spectrum of a Rydberg series of lines, observed with finite resolution, fades gradually and smoothly into the continuous absorption, as shown in Fig. 18-4.

18-6. PHOTOIONIZATION

The continuous-absorption region of Fig. 18-4 corresponds to bound-free radiative transitions: an atom in a discrete energy state i absorbs a photon (of energy greater than

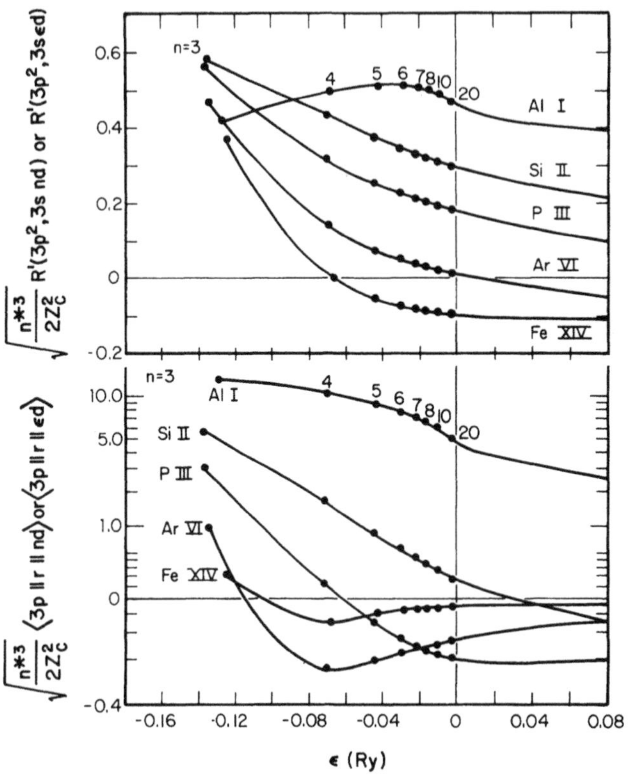

Fig. 18-3. Continuity across the series limit ε=0 of properly weighted configuration-interaction (top) and electric-dipole (bottom) radial integrals involving the outer electron in the Rydberg series $3s^2nd$, $3s^2\varepsilon d$ of the Al I isoelectronic sequence. (Note the nonlinear ordinate scale in the bottom figure—specifically, the logarithm of the dependent variable plus 0.5.)

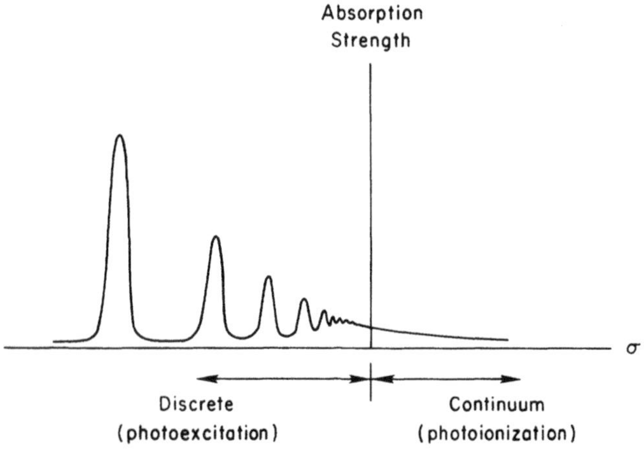

Fig. 18-4. Continuity of absorption strength across the series limit of a Rydberg series of spectrum lines observed at finite resolution—which near the limit effectively averages over a wavenumber range large compared with the interval between successive lines.

the ionization energy from state i), and thereby is caused to make a transition to a continuum state j consisting of an ion plus a free electron (photoelectron). The transition probability for such a process is usually expressed in terms of a photoionization cross section Q.

The relation between Q and oscillator strengths may be found from the differential analogs of the relations (14.16), (14.37), and (14.38), which correspond to integration over a spectrum line. If $\rho(\sigma)d\sigma$ is the photon energy density in the wavenumber range $d\sigma$ at $\sigma = \sigma_{iJ}$, then the number density of photons in this wavenumber range is $\rho\, d\sigma/hc\sigma$ and the probability per unit time that these photons will ionize an atom in state i is

$$\frac{\rho\, d\sigma}{hc\sigma} c Q_{iJ} \ . \qquad (18.44)$$

But from the differential form of (14.11), this probability may also be written $\rho\, dB_{iJ}$. Thus

$$Q_{iJ} = h\sigma(dB_{iJ}/d\sigma) = h^2c\sigma(dB_{iJ}/dE) = \frac{\pi e^2 h}{mc}\frac{df_{iJ}}{dE} \ .$$

If, as usual, we measure the energy interval dE of the continuum in units of one rydberg $(= \hbar^2/2a_o^2m)$, then

$$\begin{aligned} Q_{iJ} &= 4\pi^2\alpha a_o{}^2 \frac{df_{iJ}}{d\varepsilon} = 8.067\cdot 10^{-18}\frac{df_{iJ}}{d\varepsilon} \quad cm^2 \\ &= 2.450\cdot 10^{-23}\frac{\sigma}{g_i}\frac{dS_{iJ}}{d\varepsilon} \quad cm^2 \qquad (\varepsilon \text{ in Ry}) \ , \end{aligned} \qquad (18.45)$$

where $\alpha = e^2/\hbar c$ is the fine-structure constant and (14.37) has been used in the final step; g_i is the statistical weight of the initial level of the atom from which the photoionization occurs. Cross sections are frequently measured in megabarns (1 Mb = 10^{-18} cm²), or in atomic units of $\pi a_o^2 = 87.974\cdot 10^{-18}$ cm².

In simple cases in which the single-configuration approximation is valid, the calculation of $df/d\varepsilon$ (and thereby of Q) at any given energy ε is essentially identical with the calculation of bound-bound oscillator strengths f, except that the radial dipole reduced matrix element $P^{(1)}_{nl,n'l'}$ of (14.55) is replaced by the bound-continuum matrix element $P^{(1)}_{nl,\varepsilon l'}$ defined by (18.37). Because $P_{\varepsilon l'}(r)$—and therefore the radial integral in (18.37)—is a smooth function of ε, a continuous absorption spectrum is basically a smooth and featureless function of wavelength. This smooth function can be extrapolated back into the discrete part of the spectrum by noting that the oscillator strength for a discrete spectrum line corresponds to $\Delta f/\Delta n$, where $\Delta n = 1$; from (18.40), we thus have as the analog of $df_{iJ}/d\varepsilon$ in (18.45),

$$\frac{\Delta f_{iJ}}{\Delta\varepsilon} \doteq \frac{(n_J{}^*)^3}{2Z_c{}^2} f_{iJ} = \frac{Z_c}{2|\varepsilon_{n_J}|^{3/2}} f_{iJ} \ , \qquad (18.46)$$

where $n_J{}^*$ and $|\varepsilon_{n_J}|$ are the effective quantum number and binding energy (in rydbergs) for the excited electron in the upper state of the transition.

For highly ionized atoms, configuration-interaction effects are particularly small. Moreover, for photoionization of a single valence electron nl, hydrogenic approximations may be reasonably accurate. A statistical average over all l can be obtained by combining (18.45)-(18.46) with (14.99) and extrapolating the result into the continuum:

$$\bar{Q}_n^p \equiv \bar{Q}_{n,\epsilon} \cong \frac{64\pi\alpha a_o{}^2 Z_c{}^4}{3^{3/2} n^5 \varepsilon^3}$$

$$= \frac{64\pi\alpha a_o{}^2 n}{3^{3/2} Z_c{}^2}\left(\frac{\varepsilon_n}{\varepsilon}\right)^3 = 7.91\cdot 10^{-18}\frac{n}{Z_c{}^2}\left(\frac{\varepsilon_n}{\varepsilon}\right)^3 \quad \text{cm}^2 , \tag{18.47}$$

where ε is the photon energy in rydbergs and $\varepsilon_n = Z_c^2/n^2$ is the threshold ionization energy for the shell n. It must be remembered that this quasi-classical formula (due to Kramers[9]) is only an approximation, even for non-relativistic one-electron atoms. Accurate quantum-mechnical results[10] show that $\bar{Q}_n^p$ varies as $\varepsilon^{-8/3}$ to $\varepsilon^{-7/2}$, depending on the value of ε and on the shell n.

For many-electron atoms and sufficiently high photon energies, photoionization of inner-shell electrons will also occur, and for these the Kramers formula is not even approximately valid; reasonably accurate results require elaborate numerical calculations,[11,12] such as those involved in the PELEC computer program.[13] The calculation of photoionization contributions to the radiation opacity of a plasma involves not only a summation over all electrons of an ion, but also a summation over all the different ionization stages present at the temperature and density of interest. We shall not discuss details here, nor will we discuss free-free radiation (bremsstrahlung) nor absorption (inverse-bremsstrahlung) processes.[14]

For neutral and weakly ionized atoms, there usually exist important configuration-interaction effects between states belonging either to the same or to different Rydberg series of levels ("channels"). The interactions may produce either localized (resonance) distortions in the cross section or widespread redistributions of the oscillator strength from one spectral region to another,[12,15] which we shall discuss in the following two sections.

18-7. INTERACTION OF A DISCRETE STATE WITH A SINGLE CONTINUUM; AUTOIONIZATION

As an example of configuration-interaction effects in the continuum, we first consider the case of a single (or, at least, an isolated) discrete state that lies within the energy range of the continuum states εl. If transitions are possible from some initial bound state to both the discrete state and the continuum, and if these final states do not interact with each

[9]H. A. Kramers, Phil. Mag. **46**, 836 (1923).

[10]H. A. Bethe and E. E. Salpeter, *Quantum Mechanics of One- and Two-Electron Atoms* (Springer-Verlag, Berlin, 1957), Sec. 71.

[11]A. L. Stewart, Adv. At. Mol. Phys. 3, 1 (1967).

[12]G. V. Marr, *Photoionization Processes in Gases* (Academic Press, New York, 1967).

[13]H. Brysk and C. D. Zerby, Phys. Rev. **171**, 292 (1968); W. D. Barfield, G. D. Koontz, and W. F. Huebner, J. Quant. Spectrosc. Radiat. Transfer 12, 1409 (1972).

[14]Ya. B. Zel'dovich and Yu. P. Raizer, *Physics of Shock Waves and High-Temperature Hydrodynamic Phenomena* (Academic Press, New York, 1966), Vol. I.

[15]U. Fano and J. W. Cooper, Rev. Mod. Phys. **40**, 441 (1968), **41**, 724 (1969).

other, then a discrete absorption line appears superimposed upon the continuum; the total absorption strength S at any given ε is simply the sum of the discrete and continuum absorption strengths at that ε. If on the other hand the discrete state interacts with the continuum, then the wavefunctions for any energy ε are linear combinations (18.13) of the unperturbed discrete and continuum functions, it is the ***amplitudes*** $S^{1/2}$ that are added, and the total absorption strength S shows pronounced interference effects. In addition, the "discrete" state acquires some of the properties of the continuum, and the discrete state may autoionize as described in Sec. 18-1. If the interaction between discrete state and continuum is not too strong, the energy range over which the interaction is appreciable is small and does not include other discrete states. The problem can then be conveniently treated analytically by a method described by Fano.[16]

Let the unperturbed discrete state be described by a wavefunction φ—calculated, in general, as a linear combination (18.12) of discrete basis states, with coefficients determined by diagonalization of the discrete-basis-state portion of the energy matrix. Similarly, we assume a mutually orthogonal set of continuum wavefunctions[17] ψ_ε, constructed such that the continuum-state portion of the energy matrix is diagonal; if we measure all energies relative to the ionization limit $E_{\gamma J_c}$, then the continuum-continuum matrix elements are

$$\langle\psi_{\varepsilon'}|H|\psi_\varepsilon\rangle = \varepsilon\langle\psi_{\varepsilon'}|\psi_\varepsilon\rangle = \varepsilon\,\delta(\varepsilon-\varepsilon') \; . \tag{18.48}$$

For the lone discrete state of interest we have the diagonal element

$$\langle\varphi|H|\varphi\rangle = \varepsilon_\varphi \; , \tag{18.49}$$

where ε_φ is the eigenvalue of φ relative to the ionization limit. We abbreviate the configuration-interaction matrix elements, which involve radial integrals of the form (18.33), by V_ε:

$$\langle\varphi|H|\psi_\varepsilon\rangle = V_\varepsilon \; . \tag{18.50}$$

Eigenfunctions with configuration interaction included. The eigenfunction of the Hamiltonian matrix for eigenvalue ε will be similar to (18.13): with only the one discrete state φ and including only the one continuum $\gamma J_c \varepsilon l$,

$$\Psi^\varepsilon = a^\varepsilon\varphi + \int b^\varepsilon_{\varepsilon'}\psi_{\varepsilon'}\,d\varepsilon' \; . \tag{18.51}$$

The coefficients a^ε and $b^\varepsilon_{\varepsilon'}$ may as usual be found as the components of that eigenvector of the matrix (18.48)-(18.50) which corresponds to the eigenvalue ε. In the present case, the matrix is sufficiently simple in form that its diagonalization may in effect be accomplished analytically.

Substitution of (18.51) into the eigenvalue equation

[16]U. Fano, Phys. Rev. 124, 1866 (1961).

[17]The functions φ and ψ_ε are antisymmetric with respect to electron-coordinate interchange, even though (contrary to our normal convention) they have been designated by lower-case letters.

$$H\Psi^{\epsilon} = \varepsilon\Psi^{\epsilon} \tag{18.52}$$

gives

$$a^{\epsilon}H\varphi + \int b^{\epsilon}_{\epsilon'}(H\psi_{\epsilon'})\,d\varepsilon' = \varepsilon a^{\epsilon}\varphi + \varepsilon\int b^{\epsilon}_{\epsilon'}\psi_{\epsilon'}\,d\varepsilon' \,. \tag{18.53}$$

Since the functions φ and $\psi_{\epsilon'}$ are orthogonal, multiplication of (18.53) by φ^* and integration over all coordinates gives from (18.49)-(18.50)

$$\varepsilon_{\varphi}a^{\epsilon} + \int V_{\epsilon'}b^{\epsilon}_{\epsilon'}\,d\varepsilon' = \varepsilon a^{\epsilon} \;; \tag{18.54}$$

similarly, multiplication of (18.53) by $\psi^*_{\epsilon''}$ and integration gives from (18.19) and (18.48)

$$V_{\epsilon''}a^{\epsilon} + \int b^{\epsilon}_{\epsilon'}\varepsilon'\,\delta(\varepsilon''-\varepsilon')\,d\varepsilon' = \varepsilon\int b^{\epsilon}_{\epsilon'}\delta(\varepsilon''-\varepsilon')\,d\varepsilon'$$

or (changing the notation from ε'' to ε'),

$$V_{\epsilon'}a^{\epsilon} + \varepsilon' b^{\epsilon}_{\epsilon'} = \varepsilon b^{\epsilon}_{\epsilon'} \,. \tag{18.55}$$

(As usual, phases can be chosen such that the matrix elements V_{ϵ} as well as the eigenvector components a^{ϵ} and $b^{\epsilon}_{\epsilon'}$ are real, so that complex conjugates need not be indicated.)

If $\varepsilon' \neq \varepsilon$, then (18.55) may be written in the form

$$b^{\epsilon}_{\epsilon'} = \frac{1}{\varepsilon-\varepsilon'}V_{\epsilon'}a^{\epsilon} \;;$$

the case $\varepsilon' = \varepsilon$ can be included by writing[18]

$$b^{\epsilon}_{\epsilon'} = \left[\frac{1}{\varepsilon-\varepsilon'} + \pi\eta(\varepsilon)\delta(\varepsilon-\varepsilon')\right]V_{\epsilon'}a^{\epsilon} \,, \tag{18.56}$$

with the understanding that one shall take the principal part of integrals over $(\varepsilon-\varepsilon')^{-1}$:

$$\mathcal{P}\int_{\varepsilon_1}^{\varepsilon_2}\frac{f(\varepsilon')}{\varepsilon-\varepsilon'}\,d\varepsilon' \equiv \lim_{\Delta\to 0}\left[\int_{\varepsilon_1}^{\varepsilon-\Delta} + \int_{\varepsilon+\Delta}^{\varepsilon_2}\right]\frac{f(\varepsilon')}{\varepsilon-\varepsilon'}\,d\varepsilon' \,. \tag{18.57}$$

Substitution of (18.56) into (18.54) gives

$$\varepsilon_{\varphi}a^{\epsilon} + a^{\epsilon}\cdot\mathcal{P}\int\frac{V_{\epsilon'}^{2}}{\varepsilon-\varepsilon'}\,d\varepsilon' + \pi\eta(\varepsilon)V_{\epsilon}^{2}a^{\epsilon} = \varepsilon a^{\epsilon}$$

or

[18] P. A. M. Dirac, Z. Physik **44**, 585 (1927).

$$\eta(\varepsilon) = \frac{\varepsilon - \varepsilon_\varphi - F(\varepsilon)}{\pi V_\epsilon^2} , \tag{18.58}$$

where

$$F(\varepsilon) = \mathcal{P}\int \frac{V_{\epsilon'}^2}{\varepsilon - \varepsilon'} d\varepsilon' . \tag{18.59}$$

The physical significance of $F(\varepsilon)$ is evident from a comparison of (18.59) with (18.36): The value of $F(\varepsilon_\varphi)$ is the perturbation of the discrete level by the entire continuum; $F(\varepsilon)$ is the value that the perturbation would assume if the unperturbed level were located at ε instead of ε_φ, and (like V_ϵ) is usually a very slowly varying function of ε. [Indeed, $F(\varepsilon)$ tends to be zero, because the perturbation caused by that portion of the continuum $\varepsilon' < \varepsilon$ tends to cancel the perturbation caused by the portion $\varepsilon' > \varepsilon$.] Thus the function $\eta(\varepsilon)$ is essentially the energy displacement from the *perturbed* position of the discrete level, measured in units of πV_ϵ^2 [which it should be remembered, from (18.34), has the dimensions of energy, not (energy)^2].

The coefficient a^ϵ can now be obtained by substituting (18.56) into (18.51) and requiring that Ψ^ϵ be normalized in the same sense (18.19) as for radial continuum functions:

$$\begin{aligned}\langle \Psi^{\bar\epsilon} | \Psi^\epsilon \rangle = \delta(\bar\varepsilon - \varepsilon) &= a^{\bar\epsilon} a^\epsilon \langle \varphi | \varphi \rangle + \iint b^{\bar\epsilon}_{\epsilon''} b^\epsilon_{\epsilon'} \langle \psi_{\epsilon''} | \psi_{\epsilon'} \rangle d\varepsilon' d\varepsilon'' \\ &= a^{\bar\epsilon} a^\epsilon + \int b^{\bar\epsilon}_{\epsilon'} b^\epsilon_{\epsilon'} d\varepsilon' .\end{aligned} \tag{18.60}$$

Using the expression (18.56) for b, Fano[16] shows that one obtains

$$\begin{aligned}\int b^{\bar\epsilon}_{\epsilon'} b^\epsilon_{\epsilon'} d\varepsilon' &= a^{\bar\epsilon} a^\epsilon \int V_{\epsilon'}^2 \left[\frac{1}{\bar\varepsilon - \varepsilon'} + \pi\eta(\bar\varepsilon)\delta(\bar\varepsilon - \varepsilon')\right]\left[\frac{1}{\varepsilon - \varepsilon'} + \pi\eta(\varepsilon)\delta(\varepsilon - \varepsilon')\right] d\varepsilon' \\
&= a^{\bar\epsilon} a^\epsilon \left\{ V_\epsilon^2 \pi^2 \eta(\varepsilon)^2 \delta(\bar\varepsilon - \varepsilon) + \frac{V_\epsilon^2 \pi\eta(\varepsilon)}{\bar\varepsilon - \varepsilon} - \frac{V_{\bar\epsilon}^2 \pi\eta(\bar\varepsilon)}{\bar\varepsilon - \varepsilon} + \int \frac{V_{\epsilon'}^2}{(\bar\varepsilon - \varepsilon')(\varepsilon - \varepsilon')} d\varepsilon' \right\} \\
&= a^{\bar\epsilon} a^\epsilon \left\{ V_\epsilon^2 \pi^2 \eta(\varepsilon)^2 \delta(\bar\varepsilon - \varepsilon) + \frac{V_\epsilon^2 \pi\eta(\varepsilon)}{\bar\varepsilon - \varepsilon} - \frac{V_{\bar\epsilon}^2 \pi\eta(\bar\varepsilon)}{\bar\varepsilon - \varepsilon} \right. \\
&\qquad \left. + \int \left[\frac{V_{\epsilon'}^2}{\bar\varepsilon - \varepsilon}\left(\frac{1}{\varepsilon - \varepsilon'} - \frac{1}{\bar\varepsilon - \varepsilon'}\right) + \pi^2 V_{\epsilon'}^2 \delta(\bar\varepsilon - \varepsilon)\delta(\varepsilon' - \varepsilon)\right] d\varepsilon' \right\} \\
&= (\pi a^\epsilon V_\epsilon)^2 \left\{ \eta(\varepsilon)^2 + 1 \right\} \delta(\bar\varepsilon - \varepsilon) \\
&\qquad + \frac{a^{\bar\epsilon} a^\epsilon}{\bar\varepsilon - \varepsilon} \left\{ V_\epsilon^2 \pi\eta(\varepsilon) - V_{\bar\epsilon}^2 \pi\eta(\bar\varepsilon) + F(\varepsilon) - F(\bar\varepsilon) \right\} \\
&= (\pi a^\epsilon V_\epsilon)^2 \left\{ \eta(\varepsilon)^2 + 1 \right\} \delta(\bar\varepsilon - \varepsilon) - a^{\bar\epsilon} a^\epsilon ,\end{aligned}$$

where (18.58) has been used in the final step. Use of this result in the normalization condition (18.60) thus gives, with (18.58),

$$(a^\epsilon)^2 = \frac{1}{\pi^2 V_\epsilon^{\,2}[\eta(\varepsilon)^2 + 1]} = \frac{V_\epsilon^{\,2}}{[\varepsilon - \varepsilon_\varphi - F(\varepsilon)]^2 + \pi^2 V_\epsilon^{\,4}} \ . \tag{18.61}$$

Substitution of (18.56) into (18.51) gives for the eigenfunction

$$\begin{aligned} \Psi^\epsilon &= a^\epsilon\varphi + a^\epsilon \pi V_\epsilon \eta(\varepsilon)\psi_\epsilon + a^\epsilon \cdot \mathcal{P}\!\int \frac{V_{\epsilon'}\psi_{\epsilon'}}{\varepsilon - \varepsilon'}\,d\varepsilon' \\ &= a^\epsilon\varphi + \frac{a^\epsilon}{V_\epsilon}[\varepsilon - \varepsilon_\varphi - F(\varepsilon)]\,\psi_\epsilon + a^\epsilon \cdot \mathcal{P}\!\int \frac{V_{\epsilon'}\psi_{\epsilon'}}{\varepsilon - \varepsilon'}\,d\varepsilon' \ . \end{aligned} \tag{18.62}$$

Since Ψ^ϵ is proportional to a^ϵ and the phase of Ψ^ϵ is arbitrary, we may take either sign for the square root of (18.61); we shall choose

$$a^\epsilon = \frac{1}{\pi V_\epsilon[\eta(\varepsilon)^2 + 1]^{1/2}} = \frac{V_\epsilon}{\{[\varepsilon - \varepsilon_\varphi - F(\varepsilon)]^2 + \pi^2 V_\epsilon^{\,4}\}^{1/2}} \ . \tag{18.63}$$

Together with the expression (18.59) for $F(\varepsilon)$, this provides a complete analytical result for the eigenfunction Ψ^ϵ for the mixed discrete state plus continuum.

Autoionization. From the form of (18.51) or (18.62) it may be seen that $(a^\epsilon)^2$ gives the fractional contribution of the pure discrete state φ to the eigenfunction Ψ^ϵ. The result (18.61) shows this contribution to be spread out over a finite energy range; in fact, to the extent that $F(\varepsilon)$ and V_ϵ are constant, the configuration interaction has spread the discrete state out into a resonance line shape having its center at $\varepsilon_0 = \varepsilon_\varphi + F(\varepsilon_\varphi)$ and having a half-width at half-maximum of

$$\Gamma_a = \pi V_\epsilon^{\,2} \quad \text{Ry} \ , \tag{18.64}$$

as illustrated in Fig. 18-5. The half-life τ_a with which an atom prepared initially in the discrete state autoionizes into the continuum state ψ_ϵ ($\varepsilon \cong \varepsilon_0$) can be inferred from the uncertainty principle $(\Delta E)(\Delta t) = \hbar$ [cf. (1.18), (1.20)]. Thus the half-life and the corresponding autoionization transition-probability rate are given by

$$A^a = \frac{1}{\tau_a} = \frac{2\Gamma_a}{\hbar} = \frac{2\pi V_\epsilon^{\,2}}{\hbar} = 1.2988 \cdot 10^{17} V_\epsilon^{\,2} \quad \text{sec}^{-1} \ , \tag{18.65}$$

if $V_\epsilon^{\,2}$ is in rydbergs. In terms of A^a, the full width at half maximum is

$$\begin{aligned} 2\Gamma_a = \hbar A^a &= 4.838 \cdot 10^{-17} A^a \quad \text{Ry} \\ &= 6.582 \cdot 10^{-16} A^a \quad \text{eV} \\ &= 5.309 \cdot 10^{-12} A^a \quad \text{cm}^{-1} \ . \end{aligned} \tag{18.65a}$$

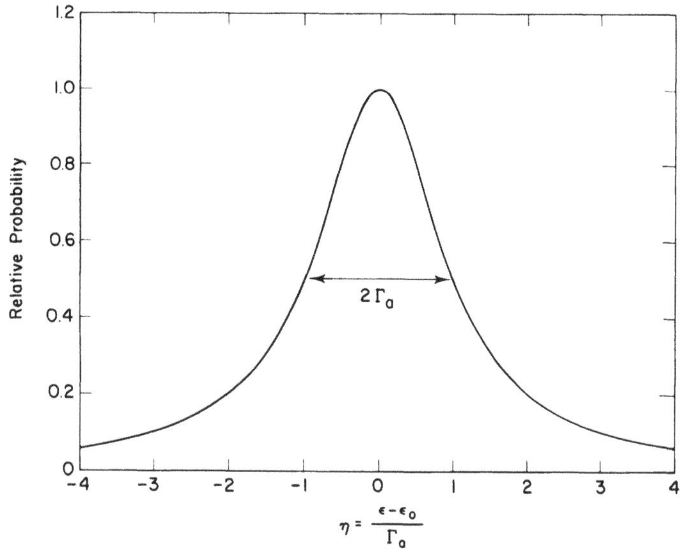

Fig. 18-5. Resonance shape of an autoionizing level at energy ε_0, with half-width at half-maximum $\Gamma_a = \hbar A^a/2 = \pi V_{\varepsilon_0}{}^2$, where A^a is the autoionization transition probability rate resulting from Coulomb interaction with a continuum channel.

Autoionization from a state that involves a hole in an inner subshell of the core is known as the Auger effect,[19] and the ejected electron is called an Auger electron.

Absorption line shapes. Let us suppose that absorption of radiation is possible from some discrete lower state Ψ_l to the discrete state φ, or to the continuum states ψ_ε, or to both. Then in the absence of configuration interaction the absorption spectrum would consist of a discrete spectrum line, or a continous spectrum, or a superposition of both as shown in the upper portion of Fig. 18-6.

With the configuration interaction taken into account, the situation in the third case is not a simple superposition because it is necessary to add values of $S^{1/2}$ rather than of S; i.e., the problem is a coherent one, and there are strong interference effects. From (18.62) we may write

$$S^{1/2} \propto \langle \Psi_l \| \boldsymbol{P} \| \Psi^\varepsilon \rangle = a^\varepsilon \langle \Psi_l \| \boldsymbol{P} \| \Phi^\varepsilon \rangle + a^\varepsilon \pi V_\varepsilon \eta \langle \Psi_l \| \boldsymbol{P} \| \psi_\varepsilon \rangle$$

$$= \frac{1}{\pi V_\varepsilon (\eta^2 + 1)^{1/2}} \langle \Psi_l \| \boldsymbol{P} \| \Phi^\varepsilon \rangle + \frac{\eta}{(\eta^2 + 1)^{1/2}} \langle \Psi_l \| \boldsymbol{P} \| \psi_\varepsilon \rangle \ . \qquad (18.66)$$

The function

$$\Phi^\varepsilon \equiv \varphi + \mathcal{P} \int \frac{V_{\varepsilon'} \psi_{\varepsilon'}}{\varepsilon - \varepsilon'} d\varepsilon' \qquad (18.67)$$

[19]P. Auger, J. phys. radium 6, 205 (1925). The term "autoionization" (applied to levels produced by excitation of loosely bound electrons) was coined by A. G. Shenstone, Phys. Rev. 38, 873 (1931).

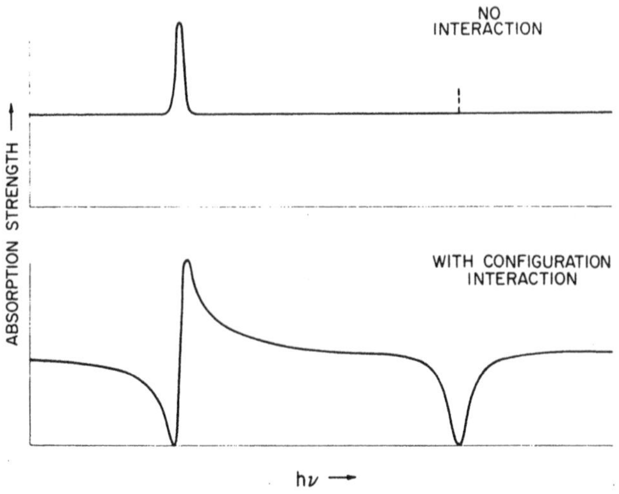

Fig. 18-6. Upper portion: Total absorption strength from a discrete lower level to a continuum channel, plus two discrete levels that do not interact with the continuum. The strength of the higher frequency discrete line is assumed to be zero, and the position of that "line" is indicated by the vertical dashes. Lower portion: Same, except for "discrete" levels that interact with the continuum. The finite-strength discrete line leads to an asymmetric Beutler-Fano absorption profile, whereas the zero-strength line results in a symmetric transmission window in the continuous absorption.

is the discrete-state wavefunction φ modified by an admixture of continuum states; however, the effect of the admixture in (18.66) tends to be small because (for the values of r for which Ψ_l is appreciable) $\psi_{\epsilon'}$ is almost independent of ε, so that contributions for $\varepsilon' < \varepsilon$ nearly cancel out those for $\varepsilon' > \varepsilon$. [This cancellation is incomplete near the "end" of the continuum $\varepsilon = 0$ (i.e., near the ionization limit) unless the discrete Rydberg states are taken into account, in which case these discrete states effectively provide a continuation of the continuum for reasons discussed in Secs. 18-5 and 18-6.] The general forms of the two coefficients in (18.66) are illustrated in Fig. 18-7.

(1) If absorption is possible only to the discrete state (i.e., if $\langle\Psi_l\|P\|\psi_\epsilon\rangle$ is zero for all ε), then the wavenumber dependence of the line strength is proportional to

$$(a^\epsilon)^2 = \frac{1}{\pi^2 V_\epsilon^{\,2}(\eta^2 + 1)}$$

and we see the "discrete" absorption line only, but with center displaced by an energy $F(\varepsilon)$, and broadened into a resonance shape with half-width at half-maximum Γ_a, corresponding to the level shape depicted in Fig. 18-5.

(2) If absorption is possible only to the continuum (i.e., if $\langle\Psi_l\|P\|\Phi^\epsilon\rangle \cong \langle\Psi_l\|P\|\varphi\rangle = 0$), then the line strength is proportional to

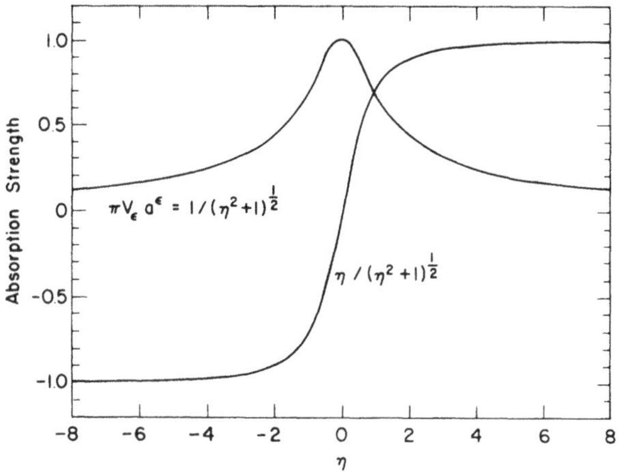

Fig. 18-7. Form of the coefficients, in the final mixed eigenfunction, of the discrete basis function (symmetrical curve) and of the continuum basis functions (asymmetrical curve).

$$\frac{\eta^2}{(\eta^2+1)}$$

and we see the absorption continuum, except with a transparent "window" centered at a wavenumber corresponding to the perturbed position of the discrete line (lower-right portion of Fig. 18-6). It is tempting to think of this window as being the result of the discrete state φ having repelled the continuum states away from the position $\eta = 0$, so that there is essentially a "hole" in the continuum of energy levels. More accurately, however, the window is due to interference effects. The continuum states are not perturbed so much in energy as in radial position; as shown by the last two terms of Eq. (18.62), the unperturbed continuum states (in the vicinity of $\eta = 0$) are mixed together with opposite phases in such a way as to nearly cancel each other out at those radii that contribute to the multipole matrix elements in (18.66), and they interfere constructively only at large r where they cannot contribute to radiative absorption.

(3) If absorption is possible to both the discrete state and the continuum, then neither term in (18.66) is zero. Because the coefficients of the two terms are respectively even and odd functions of η, the terms interfere constructively on one side of the point $\eta = 0$, and interfere destructively on the other side (cf. Fig. 18-7). Since the two reduced matrix elements usually are slowly varying functions of η, then the η dependence of the two coefficients is such that there clearly exists a value of η at which the destructive interference is 100% complete—namely, that value of η which is equal to the negative of

$$q \equiv \frac{\langle\Psi_l\|\boldsymbol{P}\|\Phi^\epsilon\rangle}{\pi V_\epsilon\langle\Psi_l\|\boldsymbol{P}\|\psi_\epsilon\rangle} \cong \frac{\langle\Psi_l\|\boldsymbol{P}\|\varphi\rangle}{\pi V_\epsilon\langle\Psi_l\|\boldsymbol{P}\|\psi_\epsilon\rangle} \quad . \tag{18.68}$$

It may be noted from (18.64) that

$$q^2 = \frac{\langle\Psi_l\|\boldsymbol{P}\|\Phi^\epsilon\rangle^2}{\pi\Gamma_a\langle\Psi_l\|\boldsymbol{P}\|\psi_\epsilon\rangle^2} \cong \frac{\langle\Psi_l\|\boldsymbol{P}\|\varphi\rangle^2}{\pi\Gamma_a\langle\Psi_l\|\boldsymbol{P}\|\psi_\epsilon\rangle^2} \tag{18.69}$$

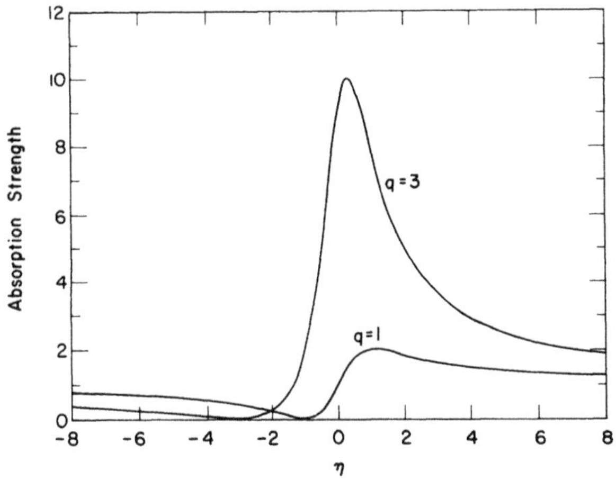

Fig. 18-8. Beutler-Fano profiles. The maximum appears at $\eta = 1/q$, and the zero of absorption at $\eta = -q$.

is approximately the ratio of the line strength for the unperturbed discrete state φ to the line strength for a bandwidth $\pi\Gamma_a$ of the unperturbed continuum.

From (18.66) and (18.68) it follows that the ratio of the line strength with configuration interaction to the line strength of the unperturbed continuum is

$$\frac{\langle\Psi_l\|\boldsymbol{P}\|\Psi^\epsilon\rangle^2}{\langle\Psi_l\|\boldsymbol{P}\|\psi_\epsilon\rangle^2} = \left[\frac{1}{(\eta^2+1)^{1/2}}\frac{\langle\Psi_l\|\boldsymbol{P}\|\Phi^\epsilon\rangle}{\pi V_\epsilon\langle\Psi_l\|\boldsymbol{P}\|\psi_\epsilon\rangle} + \frac{\eta}{(\eta^2+1)^{1/2}}\right]^2 = \frac{(\eta+q)^2}{(\eta^2+1)} . \qquad (18.70)$$

Thus in terms of the scaled energy variable η, the line shape[20] has a very simple form that depends only on the parameter q. The line shape is highly asymmetrical (except when $q = 0$ or ∞) as a result of the interference effects between the discrete state and the continuum (Fig. 18-8); the direction of the asymmetry depends on the sign of q, which in turn depends on the signs of the two dipole matrix elements and on the sign of the configuration-interaction matrix element V_ϵ. [The case $q = 0$ is the symmetric case (2) sketched in the lower-right portion of Fig. 18-6.] Asymmetric autoionization resonances (as well as symmetric absorption windows) have been observed in the photoionization continua of numerous gases, and the profiles generally agree well with the Fano expression (18.70).[12,15,20,21]

The absorption minimum occurring at $\eta = -q$ (the Fano minimum) and arising as a result of interchannel destructive interference should not be confused with an intensity minimum (the Cooper minimum[22]) that may appear in a single continuum channel $\epsilon l'$, as the result of a zero in the radial dipole integral

[20]Usually referred to as a Beutler-Fano profile: H. Beutler, Z. Physik **93**, 177 (1935); H. Beutler and W. Demeter, Z. Physik **91**, 202 (1934); U. Fano, ref. 16.

[21]R. P. Madden and K. Codling, Phys. Rev. Lett. **10**, 516 (1963), Astrophys. J. **141**, 364 (1965); R. P. Madden, D. L. Ederer, and K. Codling, Phys. Rev. **177**, 136 (1969); B. W. Shore, J. Opt. Soc. Am. **57**, 881 (1967); J. A. R. Samson, Phys. Reports **28C**, 303 (1976) and references therein; W. R. S. Garton and M. Wilson, Proc. Phys. Soc. (London) **87**, 841 (1966).

[22]U. Fano and J. W. Cooper, ref. 15, p. 465.

$$P^{(1)}_{nl,\varepsilon l'} = \langle nl \| \boldsymbol{P}^{(1)} \| \varepsilon l' \rangle \ .$$

(This type of radial-integral cancellation, occurring in the discrete part of the spectrum, has been discussed in Sec. 14-15. Examples in both the discrete and continuum region are shown in Fig. 18-3.)

So far, we have discussed only the simple case of an isolated discrete state interacting with a single continuum. The theory can be readily extended to the case of a single discrete state interacting with several continua, or to the case of several discrete states interacting with one continuum.[16] In the former case, the result is simply a Beutler-Fano profile of the form (18.70) with an added slowly varying continuous background; the Fano minimum thus does not actually reach zero. In the latter case, interference effects among closely spaced, overlapping resonances may result in resonance shapes and widths that differ greatly from those that would exist if the resonances were far apart.[23]

*18-8. PSEUDO-DISCRETE TREATMENT OF CONTINUUM PROBLEMS

The analytical treatment of the preceding section is very instructive, but is useful primarily for rather simple cases. For more complicated problems, it is generally necessary to resort to numerical methods, such as the random phase approximation with exchange (RPAE),[24] or numerical solution of the close-coupling equations in Seaton's multi-channel quantum defect theory (MQDT).[25] We shall not discuss these methods, but will only describe an alternative approach that is particularly appropriate when the interaction of a bound state φ with a continuum is so strong as to extend into the discrete Rydberg series (or if φ lies in the midst of the Rydberg series, and the interactions extend into the continuum). The procedure is to mock up the continuum (so far as radial integrals involved in matrix elements are concerned) by means of a set of pseudo-discrete configurations, thereby converting the mixed discrete-continuum problem into what is formally a purely discrete configuration-interaction problem of the sort with which we are already familiar: one (or several) isolated discrete states are combined with the Rydberg series of interacting genuine-discrete (bound) and pseudo-discrete (continuum) states into one giant matrix (for each J), which is then diagonalized in the usual manner to obtain perturbed energy levels and eigenvectors.

In order to mock up the continuum as a set of discrete states we first divide the desired energy range of the continuum up into equal segments, each of some appropriate width Δ

[23]F. H. Mies, Phys. Rev. **175**, 164 (1968); B. W. Shore, Phys. Rev. **171**, 43 (1968).

[24]U. Fano and J. W. Cooper, ref. 15 and references quoted therein; P. L. Altick and A. E. Glassgold, Phys. Rev. 133, A632 (1964); G. Wendin, J. Phys. B 5, 110 (1972), 6, 42 (1973); L. Armstrong, Jr., J. Phys. B 7, 2320 (1974).

[25]M. J. Seaton, Proc. Phys. Soc. (London) **88**, 801, 815 (1966); P. G. Burke and M. J. Seaton, Meth. Comput. Phys. **10**, 1 (1971); K. Smith, *The Calculation of Atomic Collision Processes* (Wiley-Interscience, New York, 1971); P. G. Burke and K. T. Taylor, J. Phys. B. **8**, 2620 (1975); U. Fano, J. Opt. Soc. Am. **65**, 979 (1975). The differential equation (18.15) that we are using to obtain continuum functions is a distorted-wave approximation to the close-coupling equations, in which interchannel coupling is completely neglected (see the article by Burke and Seaton, p. 74).

rydbergs;[26] the value of Δ must be chosen as a compromise between adequate resolution of continuum details and the limitations on feasible matrix size. The jth segment will be characterized by the value ε_j at the center of the segment.

Care must be taken in formulating the proper pseudo-discrete representation of the various energy integrals:

(1) In line with the discussion of Sec. 18-4 concerning the energy dimensions of $P_{\varepsilon l}$, the wavefunction-expansion integral in (18.13) must be represented in the form

$$\int y_\epsilon^k \psi_\epsilon \, d\varepsilon = \sum_j y_{\epsilon j}^k \psi_{\epsilon j} \Delta_j = \sum_j (y_{\epsilon_j}^k \Delta_j^{1/2})(\psi_{\epsilon_j} \Delta_j^{1/2}) \ ; \qquad (18.71)$$

i.e., the weighting factor Δ_j for the jth energy segment is to be divided up square-root wise between the continuum wavefunction and its coefficient. Thus the jth pseudo-discrete wavefunction is

$$\psi_j \equiv \psi_{\epsilon_j} \Delta_j^{1/2} \ , \qquad (18.72)$$

and the square of the corresponding eigenvector component is

$$[y_j^k]^2 \equiv [y_{\epsilon_j}^k]^2 \Delta_j \ , \qquad (18.73)$$

which involves the proper weighting factor Δ_j for the contribution to $(\psi^k)^2$ from the continuum states in the jth energy segment.

(2) Integration of (18.50) over the segment Δ_j, with the assumption that the matrix element is essentially constant over this interval, gives

$$\langle \varphi | H | \psi_{\epsilon_j} \rangle \Delta_j = V_{\epsilon_j} \Delta_j \ ,$$

or from (18.72)

$$\langle \varphi | H | \psi_j \rangle = V_{\epsilon_j} \Delta_j^{1/2} \ . \qquad (18.74)$$

Inasmuch as the *angular* form of the continuum function (18.14) is identical with that of the discrete function (18.1), the angular portion of the matrix element (18.74) is identical with that for corresponding genuine discrete wavefunctions. The difference between continuum and discrete wavefunctions lies only in the radial portion of the wavefunction; hence the matrix element (18.74) is calculated exactly the same as for discrete-function configuration-interaction matrix elements, except that the analog of the discrete-case radial integral

$$R^k(n_1 l_1, n_2 l_2; n_1' l_1', n_2' l_2')$$

is

[26]It is not essential that the segments be of equal width. For example, when ε_φ lies in the discrete portion of the Rydberg series and it is necessary to include a very large portion of the continuum, it is convenient to let Δ_j increase with j in an arithmetic progression.

$$\Delta_j^{1/2}\, R^k(n_1l_1,n_2l_2;n_1'l_1',\varepsilon_j l_2') \tag{18.75}$$

with (as usual) Δ_j measured in rydbergs.

(3) Integration of (18.48) with respect to ε' over the segment Δ_j gives

$$\int_{\Delta_j} \langle \psi_{\varepsilon'} | H | \psi_\varepsilon \rangle \, d\varepsilon' = \int_{\Delta_j} \varepsilon\, \delta(\varepsilon - \varepsilon')\, d\varepsilon' = \varepsilon\, \delta_{(\varepsilon \text{ in } \Delta_j)} \ .$$

The analogous pseudo-discrete expression is

$$\langle \psi_{\varepsilon_j} | H | \psi_{\varepsilon_i} \rangle \Delta_j = \varepsilon_i \delta_{ij} \ ,$$

or from (18.72)

$$\langle \psi_j | H | \psi_i \rangle = \varepsilon_i \delta_{ij} \ .$$

Thus matrix elements between two pseudo-discrete states of the continuum are non-zero only on the diagonal, *provided* the matrix for the unperturbed continuum wavefunctions was previously diagonalized as described in Sec. 18-7. More generally, for $\varepsilon_i = \varepsilon_j$ the matrix elements will be those for the ion core, plus ε_i:

$$\langle \psi_{c_1 i} | H | \psi_{c_2 i} \rangle = \langle \psi_{c_1} | H_c | \psi_{c_2} \rangle + \varepsilon_i \tag{18.76}$$

(energies now being measured with respect to an arbitrary zero); for $\varepsilon_i \neq \varepsilon_j$ the matrix elements will be evaluated as for any other configuration-interaction matrix elements, except that the appropriate radial integral is

$$(\Delta_j \Delta_i)^{1/2}\, R^k(n_1l_1,\varepsilon_j l_2;n_1'l_1',\varepsilon_i l_2') \ . \tag{18.77}$$

(4) Similarly to (18.75), line-strength matrix elements for discrete to pseudo-discrete transitions are computed exactly as for a discrete-discrete case, except that in place of a radial integral

$$\int r^t P_{nl} P_{n'l'}\, dr$$

we must use

$$\Delta_j^{1/2} \int r^t P_{nl} P_{\varepsilon_j l'}\, dr \ . \tag{18.78}$$

With the above modifications, then, the continuum can be handled (with limited resolution) in exactly the same way as a set of discrete quantum states. In order to obtain the necessary radial integrals in (18.75)-(18.78), it is of course not necessary to actually compute values of R^k and $P_{nl,\varepsilon_j l'}$ at every desired value of ε_j; it is sufficient to evaluate these radial integrals at a few judiciously chosen values of ε, and then to interpolate numerically (or graphically, using figures like Fig. 18-3).

It is convenient also to use the above procedure in reverse; the infinity of discrete states for large n ($\rightarrow \infty$) can be thought of as a pseudo-continuum, and then represented as one or a few pseudo-discrete states.

Example. As an example of these methods we show in Fig. 18-9 computed spectra for a hypothetical discrete line plus a continuum. In section (a) of the figure, configuration interaction of the upper states has been neglected, and we see simply a strong discrete line in the middle of a set of pseudo-discrete lines representing a continuum of slowly varying strength; this section is analogous to the upper-left portion of Fig. 18-6. In sections (b), (c), and (d), the configuration interaction has been included, and the radial dipole integral $\langle \psi_i \| P^{(1)} \| \varphi \rangle$ for the discrete upper state has been multiplied respectively by -1, 0, and $+1$, corresponding to values of q equal to $-|q|$, 0, $+|q|$, respectively, and producing line profiles like those in Fig. 18-8 and the lower part of Fig. 18-6. The tick marks in sections (b)-(d) indicate the configuration purity of the corresponding upper state; absence of a tick mark indicates a purity greater than 99%. It is evident that appreciable intensity changes extend into the region of the continuum where configuration mixing is very small.

Fe XXI. As a quantitative example, we consider the case

$$\text{Fe XXI } 2s^2 2p 10s\ ^3P_2^o \rightarrow 2s2p^2 10s\ ^3D_3 + 2s^2 2p\varepsilon p\ ^3D_3 \ .$$

The autoionization half-width at half-maximum, computed via the perturbation formula (18.64), is

$$\Gamma_a = \pi V_\epsilon^{\ 2} = \pi(-0.00829 \text{ Ry}^{1/2})^2 = 2.16 \cdot 10^{-4} \text{ Ry} = 23.7 \text{ cm}^{-1} \ .$$

The value of the Fano parameter q, computed from the approximate form of (18.69), is -74. (Both values take account of the fact that the intermediate-coupled $2s2p^2 10s$ "$^3D_3$" function is computed to be only 92.9% pure $^3D_3$ and 7.1% pure $^5P_3$.)

In order to obtain a more appropriate illustrative example, we arbitrarily reduce the discrete line strength by a factor of 400, thus giving a value $q = -74/20 = -3.7$. The result of a numerical discretized-continuum calculation, using 48 continuum segments each of width $\Delta = 20 \text{ cm}^{-1}$, is indicated by the circles in Fig. 18-10—the ordinates representing the

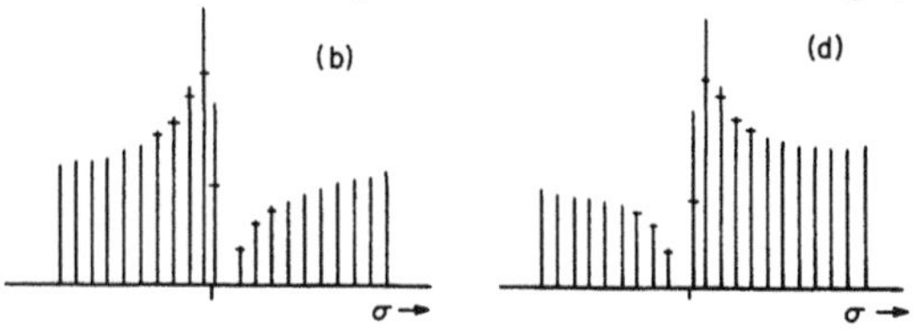

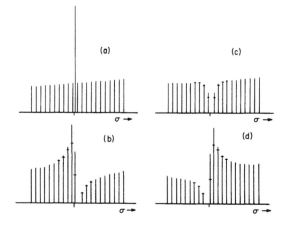

Fig. 18-9. Hypothetical absorption strength for a discrete spectrum line plus a discretized continuum: (a) no configuration interaction; (b)-(d) including CI, and with negative, zero, and positive values of $S^{1/2}$ for the pure discrete line. The tick marks indicate the (pseudo-discrete) configuration purities, if less than 99 percent.

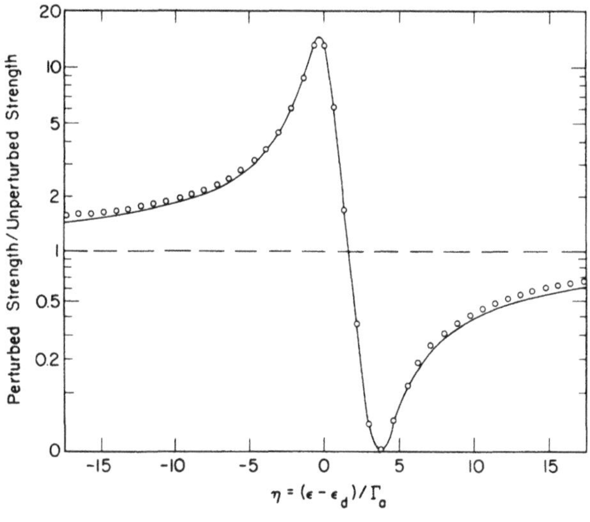

Fig. 18-10. Beutler-Fano autoionization profile for an absorption transition with $\Gamma_a = 23.7\ \text{cm}^{-1}$ and $q = -3.70$, as calculated (circles) via the numerical pseudo-discrete CI method and (solid curve) from the Fano expression (18.70). The horizontal dashed line represents the unperturbed continuum; the unperturbed discrete spectrum line would be a very narrow line at $\eta = 0$ with total area equal to that of the unperturbed continuum for $|\eta| \leq \pi q^2/2 = 21.5$. (Note: For clarity, a nonlinear ordinate scale has been used—specifically, the logarithm of the dependent variable plus 0.1.)

ratios of the various perturbed pseudo-discrete line strengths to the integrated *unperturbed* "line" strengths for the *perturbed* width

$$\Delta_j \cong \frac{1}{2}(E^{j+1} + E^j) - \frac{1}{2}(E^j + E^{j-1}) = \frac{1}{2}(E^{j+1} - E^{j-1}) \tag{18.79}$$

of the jth continuum segment, E^j being the eigenvalue representing the central energy of this segment. The solid curve in the figure is Fano's analytical result (18.70), for $\Gamma_a = 23.7$ cm^{-1} and $q = -3.7$. [The quantity $F(\varepsilon)$ in (18.58) has been taken to be zero because the strength of the discrete-continuum interaction is essentially constant over the energy range considered.]

The discrepancy between the circles and the Fano curve at large $|\eta|$ is partly due to an inherent inaccuracy[16] in the Fano result, but is mainly due to end effects resulting from use of a finite section of the continuum; this is shown by the fact that the discrepancy becomes greater when five segments are discarded from each end of the continuum section. The close agreement between the two curves at small $|\eta|$ shows that the pseudo-discrete method is basically accurate even when the segment width Δ is comparable with the half-width Γ_a.

It should not be forgotten that the low-q Beutler-Fano profile shown in the figure was chosen only for clarity of illustration. Typical profiles for highly ionized atoms correspond to much larger values of q, so that the peak of the profile lies very much higher—at $q^2 + 1 = 74^2 + 1 = 5477$ for the Fe XXI example, rather than at $3.7^2 + 1 = 14.7$ for the plotted modification.

Al I. Beutler-Fano profiles may sometimes be much wider than the (comparatively) narrow example just discussed. Indeed, in neutral and weakly ionized atoms, configuration interactions are occasionally so strong as to extend throughout both the discrete and continuum portions of a Rydberg series. In Al I, for example, the discrete term $3s3p^2\ ^2D$ mixes very strongly with the Rydberg series of terms $3s^2nd\ ^2D$ and $3s^2\varepsilon d\ ^2D$—so strongly

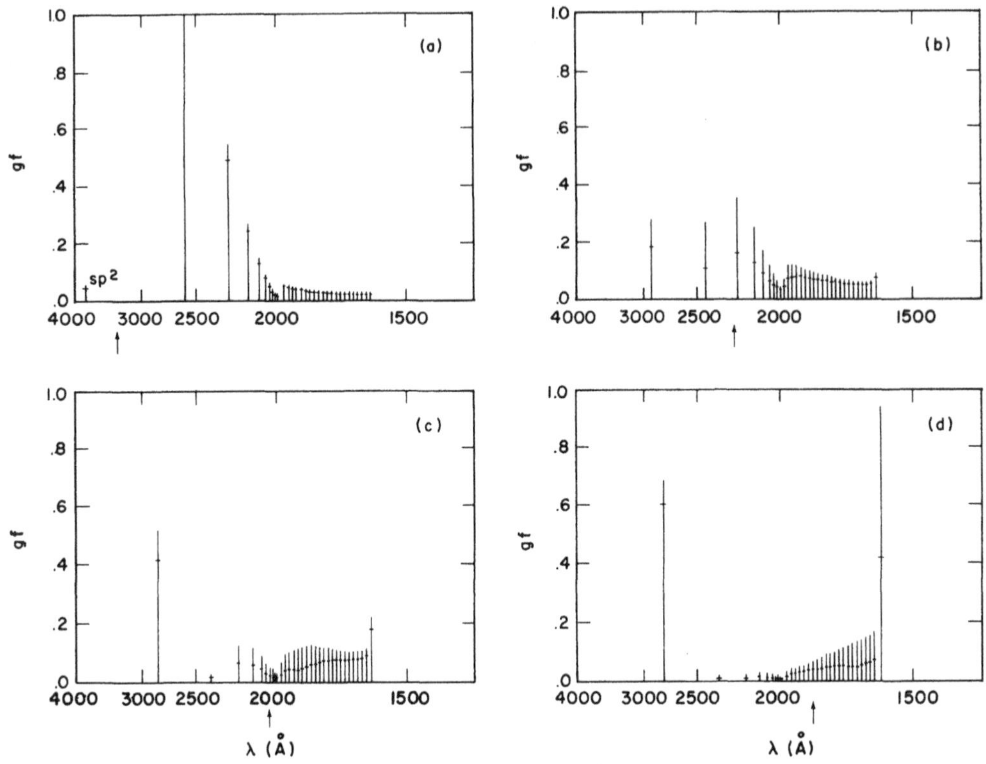

Fig. 18-11. Computed oscillator strengths for the transitions Al I $3s^23p\ ^2P^o_{1/2} - (3s3p^2 + 3s^2d)\ ^2D_{3/2}$, including a discretized continuum. The unperturbed position of the $3s3p^2\ ^2D_{3/2}$ level has arbitrarily been assumed to lie at a position corresponding to the vertical arrow. The tick mark on each line indicates the (s^2d-basis-state) purity of the upper level of the transition.

in fact, that it is not really possible to say which observed ^2D term belongs to sp^2; the unperturbed position of this term can only be estimated theoretically.[27,28]

For illustrative purposes, we shall arbitrarily vary the unperturbed energy of the $sp^2\ ^2D$ term in order to examine the resulting changes in the computed line-strength distribution among the discrete lines and continuous spectrum for the transitions

$$3s^23p\ ^2P^o_{1/2} - 3s^2d\ ^2D_{3/2}\ .$$

In the first section of Fig. 18-11, the unperturbed sp^2 level has been placed below the lowest member of the s^2d series, giving an unperturbed $s^2p - sp^2$ line at the position indicated by the arrow. Perturbations are therefore comparatively small, the $s^2p - sp^2$ line is easily identified, and line strengths in the $s^2p - s^2d$ Rydberg series decrease monotonically with increasing n (and ε) in a straightforward fashion. [The apparent

[27]A. W. Weiss, Phys. Rev. **178**, 82 (1969).

[28]C. D. Lin, Astrophys. J. **187**, 385 (1974).

intensity discontinuity at the discrete series limit (computed at ~1950 Å) is not real, but is only the result of a discontinuity between the spacing of successive discrete levels nd and the spacing Δ of the pseudo-discrete continuum levels.]

In the remaining sections of the figure, the sp^2 level lies in the midst of the s^2nd or $s^2\varepsilon d$ series; configuration mixing extends over such a wide energy range that it is not possible to say which energy state (and therefore which spectrum line) should be associated with sp^2. Intensity perturbations are extensive, and are characterized primarily by a transfer of line strength from the long-wavelength lines to the short-wavelength lines; i.e., from the discrete spectrum to the continuum. (The strong rightmost line in the last two sections is unphysical, and an extreme form of the end effect mentioned in connection with the Fe XXI example; the excess line strength would have been passed on to the shorter-wavelength portion of the continuum if such had been included in the calculation. Omission of the entire continuum in early calculations[27,29] resulted in a mistaken association of such a spurious strong line with a strong absorption line observed just beyond the discrete series limit.[30])

The theoretically computed "correct" position of the unperturbed $3s3p^2\ ^2D$ term corresponds to a case intermediate between those shown in Figs. 18-11b and 18-11c. The computed strength of the second spectrum line for this case is much smaller than that of the first or third lines; this is in agreement with experimental observations,[31] as shown in Table 18-1. The second line is weak because of destructive configuration-mixing interference, and in fact corresponds to a point near the intensity zero of the Beutler-Fano profile at $\eta = -q$, Fig. 18-9d; the only difference is that the point $\eta = 0$ here occurs in the discrete region of the spectrum instead of in the continuum. The photoionization cross section computed here agrees only moderately well with observation[32] (more accurate calculations are available[33]), but the results do illustrate the large effects that configuration interactions can produce on the strength and shape of the photoionization continuum.

Cl I. An analogous case of extensive configuration interaction appears in Cl I $3s3p^6\ ^2S + 3s^23p^4d\ ^2S$. Here the unperturbed sp^6 level lies in the midst of the continuum, but is perturbed to a position below $3s^23p^43d\ ^2S$ by high lying continuum states[34]—a physically correct "end effect" resulting from the fact that the discrete Rydberg series does not extend to $-\infty$, but terminates with the $3s^23p^43d$ configuration. As in Al I, the Beutler-Fano profile in the photoabsorption spectrum is computed to be so wide as to be hardly recognizable,

[29]K. B. S. Eriksson, Ark. Fys. 39, 421 (1969); R. D. Cowan, J. phys., Colloq. 31, C4-191 (1970).

[30]W. R. S. Garton, in H. Maecker, ed., *Proc. Fifth Intern. Conf. Ionization Phenomena in Gases* (North-Holland Publ. Co., Amsterdam, 1962), Vol. 2, p. 1884; J. Quant. Spectrosc. Radiat. Transfer 2, 335 (1962).

[31]N. P. Penkin and L. N. Shabanova, Optics Spectrosc. 18, 504 (1965); B. Budick, Bull. Am. Phys. Soc. 11, 456 (1966).

[32]J. L. Kohl and W. H. Parkinson, Astrophys. J. 184, 641 (1973); R. A. Roig, J. Phys. B 8, 2939 (1975).

[33]A. W. Weiss, Phys. Rev. A 9, 1524 (1974); M. LeDourneuf, Vo Ky Lan, P. G. Burke, and K. T. Taylor, J. Phys. B 8, 2640 (1975).

[34]R. D. Cowan, L. J. Radziemski, Jr., and V. Kaufman, J. Opt. Soc. Am. 64, 1474 (1974); R. D. Cowan and J. E. Hansen, J. Opt. Soc. Am. 71, 60 (1981).

TABLE 18-1. ABSORPTION STRENGTHS IN Al I $3s^2 3p\ ^2P^o_{1/2} - {}^2D_{3/2}$ TRANSITIONS

	Single config.	Fig. 18-11 (a)	(b)	(c)	(d)	Other calcs.[c]	Exp.[d]
f (first line)	0.57[a]	0.03[b]	0.14	0.26	0.34	0.18	0.18
f (second line)	0.17	0.60[a]	0.13	0.01	0.01	0.05	0.04
f (third line)	0.06	0.27	0.18	0.06	0.01	0.12	0.12
f (fourth line)	0.61[b]	0.14	0.13	0.05	0.02	0.10	0.10
f (fifth line)	0.03	0.07	0.08	0.04	0.02	0.07	0.07
$Q^P(\varepsilon = 0)$, Mb	7	48	105	80	35	58	65
$Q^P(\varepsilon = 0.05)$, Mb	4	20	60	98	90	38	38

[a] $3s^2 3p - 3s^2 3d$
[b] $3s^2 3p - 3s3p^2$
[c] Ref. 33
[d] Refs. 31 and 32

the Fano minimum lying at the extreme long-wavelength end of the discrete spectrum, but effects on the magnitude and shape of the continuous spectrum are pronounced. Experimental photoionization cross-section data do not yet exist.[35]

Intrachannel interactions—Ar I and Ba I. The previous examples have all been cases in which the photoionization cross section was distorted by interactions between the continuum and a discrete state. However, widespread redistribution of the absorption strength can also result from intra-channel interactions. A prominent example[36] is that of

$$\text{Ar I}\ 3p^6\ {}^1S_0 - 3p^5(nd + \varepsilon d)\ {}^1P^o_1 : \tag{18.80}$$

as discussed in Sec. 13-5, the angular coefficient of $R^1(3p,d;d',3p)$ is unusually large for the $^1P^o$ terms; this results in large configuration mixings among members of this channel, and in transfer into the continuum of much of the strength of the strong $\Delta n = 0$ 3p–3d transition.

A similar, but even more pronounced, effect appears in the excitation of an inner-shell 4d electron in neutral barium,

$$_{56}\text{Ba I}\ 4d^{10}5s^2 5p^6 6s^2\ {}^1S_0 - 4d^9 5s^2 5p^6 6s^2(nf + \varepsilon f)\ {}^1P^o_1 , \tag{18.81}$$

[35] Experimental cross sections will include contributions from numerous channels other than $3p^4(^1D)\varepsilon d\ ^2S$. Calculations for all channels have been made via the RPAE by A. F. Starace and L. Armstrong, Jr., Phys. Rev. A 13, 1850 (1976); however, the effects of $3s3p^6\ ^2S$ were omitted, and should make a major contribution to the total $3p^4(^1D)\varepsilon l$ cross section. See also E. R. Brown, S. L. Carter, and H. P. Kelly, Phys. Rev. A 21, 1237 (1980).

[36] See Starace and Armstrong, ref. 35, for discussion and references.

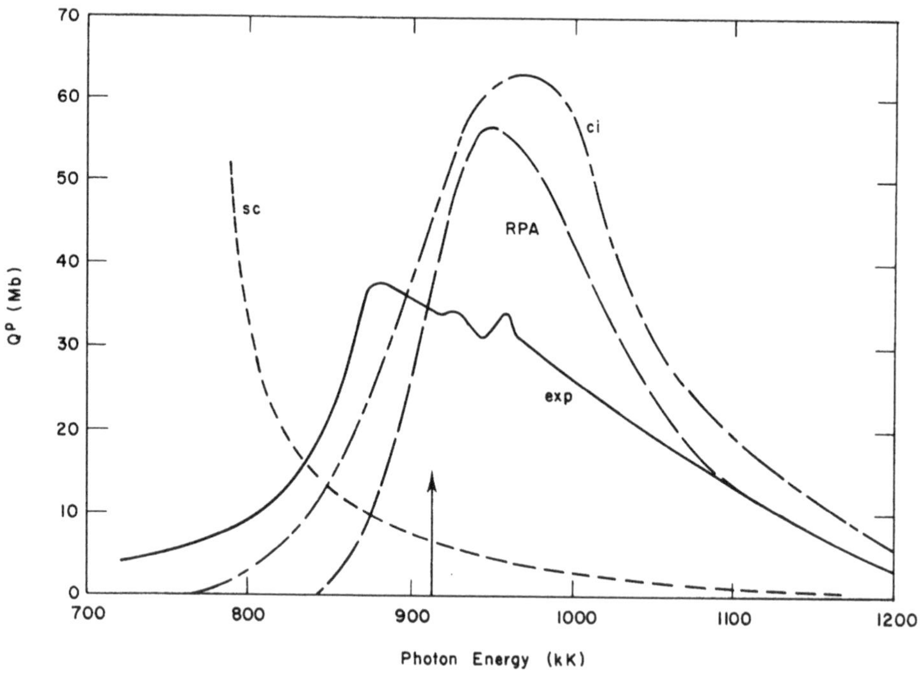

Fig. 18-12. Photoionization cross sections (4d $-$ εf) in Ba I. Solid curve: experiment (Ederer et al.[37]); dotted and dash-dot curves: pseudo-discrete calculation without and with CI; dashed curve: RPAE (Wendin[40]). The vertical arrow indicates the computed position of the unperturbed discrete $4d^{10}\ ^1S_0 - 4d^94f\ ^1P_1^o$ spectrum line.

such that a very strong and broad resonance appears in the photoionization continuum at about 100 to 130 Å.[37] The effect is due firstly to the fact that the hole in the 4d subshell makes the inner potential well as deep as for neutral lanthanum ($Z = 57$); consequently, the 4f orbital is collapsed into the core (Fig. 8-8), close to the 4d electrons, and the $\Delta n = 0$ 4d $-$ 4f transition is extremely strong. Secondly, the large value of $G^1(4d,4f)$ together with its large angular coefficient for $^1P^o$ (larger even than for the Ar I $p^5d\ ^1P^o$ case—see Table 12-2) causes the $4d^94f\ ^1P_1^o$ level to lie far above the other levels of the configuration, and in fact about 10 eV above the ionization limit; the situation is similar to, but even more extreme than, that shown in Fig. 12-7 for K II. The computed unperturbed position of the corresponding 4d $-$ 4f transition (18.81) is shown by the vertical arrow in Fig. 18-12. Because of the unique position of the unperturbed $4f\ ^1P_1^o$ level, the $f\ ^1P_1^o$ intra-channel interactions are in this case very similar to the interaction of an isolated autoionizing level with a continuum. The interaction is very strong, and the Fano profile very broad, because of the large value of the CI radial integral $R^1(4d4f,\varepsilon f4d)$, together with its large angular coefficient.

[37]J. P. Connerade and M. W. D. Mansfield, Proc. Roy. Soc. (London) **A341**, 267 (1974); Papers 3.4-3.6 by J. P. Connerade et al., D. L. Ederer et al., and P. Rabe et al., in E.-E. Koch, R. Haensel, and C. Kunz, eds., *Proc. IV Intern. Conf. Vacuum Ultraviolet Radiation Physics, Hamburg, 1974* (Pergamon-Vieweg, Braunschweig, 1974). D. L. Ederer, et al., J. Phys. B **8**, L21 (1975).

The above interpretation of the Ba I photoionization cross section has been given in terms of our usual CI treatment, using configuration-average basis wavefunctions. An alternative interpretation can be given in terms of LS-dependent HF radial functions (Sec. 16-8) for the $4d^9f\ ^1P^o$ terms. In this treatment, the 4f radial function is *not* collapsed,[38] because the large positive exchange term makes the inner potential well too shallow. At the same time, however, a sizable positive potential barrier is formed between the inner and outer wells—much like that shown for Ba I in Fig. 8-8. Because of this barrier, continuum (positive-ε) f electrons with energy less than the barrier height are partially trapped within the inner well; i.e., there is a broad-resonance range of energies for which the continuum f electrons have an unusually large amplitude within the inner-well region.[39] The resulting large overlap with the 4d radial function produces the large photoionization cross section for the corresponding range of photon energies. The $4d^9f\ ^1P^o$ resonance has also been discussed by means of the RPAE.[40] Analogous effects for $4d^{10}4f^w - 4d^9(4f^{w+1}+4f^w\varepsilon f)$ transitions in rare-earth metals have also been observed and discussed.[41]

In the remainder of this chapter, we turn from the subject of oscillator-strength distributions to the distribution of ionization stages of atoms in a plasma.

*18-9. IONIZATION EQUILIBRIUM

Whether or not the spectrum of a particular ion appears in a given light source containing the element in question depends on the properties of the plasma, such as the electron density and temperature. We shall not go into this subject in detail, but will only discuss briefly two limiting types of equilibrium plasmas.

High density. When the plasma density (more specifically, the electron density) is very high, electron-atom collisional frequencies are much higher than radiative transition probability rates. We then have a condition of *local thermodynamic equilibrium* (LTE) in which the relative occupancies of the various states j of a given ion (Z, N) are simply proportional to the Boltzmann factor $\exp(-E_j/kT)$. Since all *states* of a given energy are equally populated, the density n_N^j of ions N in the energy *level* j (with statistical weight $g_j = 2J_j+1$) is

$$\frac{n^j_N}{n_N} = \frac{g_j e^{-E_j/kT}}{u^i_N}, \tag{18.82}$$

where $n_N = \sum_j n_N^j$ is the total density of ions (Z, N) and

$$u^i_N(T) = \sum_j g_j e^{-E_j/kT} \tag{18.83}$$

[38]J. E. Hansen, A. W. Fliflet, and H. P. Kelly, J. Phys. B **8**, L127, L268 (1975).

[39]The resonance may actually be fairly sharp in the single-configuration approximation, but be greatly broadened by intra-channel configuration mixings.

[40]G. Wendin, Phys. Lett. **46A**, 119 (1973), J. Phys. B **6**, 42 (1973).

[41]J. Sugar, Phys. Rev. B **5**, 1785 (1972) and references therein; A. F. Starace, Phys. Rev. B **5**, 1773 (1972); J. L. Dehmer and A. F. Starace, Phys. Rev. B **5**, 1792 (1972).

is the internal partition function for this ion. The translational partition function for a free particle (electron or ion) of mass M is, with $g = h^{-3}\, dV\, 4\pi p^2 dp$,

$$u^t(T) = \frac{4\pi V}{h^3}\int_0^\infty p^2 e^{-p^2/2MkT}\, dp = \left(\frac{2\pi MkT}{h^2}\right)^{3/2} V \,. \tag{18.84}$$

The equilibrium condition for the ionization reaction

$$(Z, N) \rightleftarrows (Z, N-1) + e \tag{18.85}$$

is[42]

$$\frac{(n_e V)(n_{N-1} V)}{(n_N V)} = \frac{(u^t u^i)_e (u^t u^i)_{N-1}}{(u^t u^i)_N} \,. \tag{18.86}$$

Noting that $u_e^i = 2$ (corresponding to the two possible values of m_s) and that $u_{N-1}^t/u_N^t = (M_{N-1}/M_N)^{3/2} \cong 1$, we see that the relative abundances of the different ions of a given element are given by the Saha equation[42,43]

$$\frac{n_e n_{N-1}}{n_N} = 2\left(\frac{2\pi mkT}{h^2}\right)^{3/2} \frac{u^i_{N-1}}{u^i_N}\, e^{-I_N/kT} \,; \tag{18.87}$$

I_N is the ionization energy of (Z, N), and the exponential arises because u_{N-1} and u_N in (18.86) must be calculated using a common zero of potential energy whereas in (18.87) we wish to calculate the partition functions using excitation energies from the ground state of the ion concerned.

Equations (18.82) and (18.87), together with

$$n_e = \sum_N (Z - N) n_N \,, \tag{18.88}$$

give a complete description of the occupations of all levels of all ions of the element in question. Provided the plasma is optically thin (i.e., shows negligible self-absorption), a complete description of the intensities of all spectrum lines follows from (14.6) if all radiative transition probabilities are known.

Low density. In the opposite limit of low electron density, collisional excitation rates are much smaller than radiative decay rates, and essentially all ions exist in the ground level. The distribution of atoms among the various ionization stages is no longer determined by purely statistical considerations (Saha equilibrium). Instead, it is determined by the balance between the rates at which the various detailed ionization and recombination processes take place. This situation is commonly referred to as *coronal equilibrium*, because it is exemplified by conditions in the solar corona ($n_e \lesssim 10^8$ electrons/cm^3).

[42]Ya. B. Zel'dovich and Yu. P. Raizer, *Physics of Shock Waves and High-Temperature Hydrodynamic Phenomena* (Academic Press, New York, 1966), Vol. I, p. 192.

[43]M. N. Saha, Phil. Mag. **40**, 472 (1920), Proc. Roy. Soc. (London) **A99**, 135 (1921).

The important ionization processes are
(a) collisional ionization

$$(Z, N) + e \rightarrow (Z, N-1) + 2e \ ,$$

(b) photoionization

$$(Z, N) + h\nu \rightarrow (Z, N-1) + e \ ,$$

(c) autoionization

$$(Z, N)^* \rightarrow (Z, N-1) + e \ .$$

The first two processes can in principle take place either directly from the ground state, or from states excited collisionally or by absorption of a photon; the third process can of course occur only for an atom excited to a discrete level above the first ionization limit. The last two processes have already been discussed. The first is an extremely complex theoretical and numerical problem, and little is available beyond hydrogenic and semi-empirical formulae;[44] the subject is not of direct interest for atomic spectroscopy, and will not be treated here.

The important recombination processes are the inverses of the ionization processes:
(a′) three-body recombination

$$(Z, N-1) + 2e \rightarrow (Z, N) + e \ ,$$

(b′) radiative recombination

$$(Z, N-1) + e \rightarrow (Z, N) + h\nu \ ,$$

(c′) dielectronic recombination

$$(Z, N-1) + e \rightarrow (Z, N)^* \ .$$

The first two processes either can produce the recombined atom (Z, N) directly in the ground state, or they can produce an excited state which can then decay to the ground state by collisional or radiative de-excitation; the third process necessarily produces an excited state lying above the ionization limit, which can then either re-ionize by process (c) or decay by collision or radiation. Process (a′) is unimportant in low-density plasmas ($n_e < \sim 10^{20}$ cm^{-3} for temperatures greater than 10^6 °K $\cong$ 100 eV) because the rate is proportional to n_e^2. The other processes will be discussed in the following two sections.

Calculation of the spectrum of a plasma in coronal equilibrium involves (1) simultaneous solution of the rate equations for all important ionization and recombination processes to obtain the relative abundances of the various ionization stages;[45] and (2) calculation of

[44] M. J. Seaton, Planet. Space Sci. 12, 55 (1964); W. Lotz, Z. Physik **206**, 205 (1967); A. Burgess, H. P. Summers, D. M. Cochrane, and R. W. P. McWhirter, Mon. Not. R. Astron. Soc. **179**, 275 (1977). Regarding the relative importance of the first and third processes, see R. D. Cowan and J. B. Mann, Astrophys. J. 232, 940 (1979), and references therein.

[45] See, for example, C. Jordan, Mon. Not. R. Astron. Soc. **142**, 501 (1969), **148**, 17 (1970); H. P. Summers, Mon. Not. R. Astron. Soc. **169**, 663 (1974); C. Breton, C. De Michelis, M. Finkenthal, and M. Mattioli, "Ionization Equilibrium of Selected Elements from Neon to Tungsten of Interest in Tokamak Plasma Research," Association EURATOM-CEA sur la fusion Report EUR-CEA-FC-948 (March 1978).

the rate of population of excited states by collisional excitation from the ground level and by recombination to excited states, together with the calculation of branching ratios for the various possible decay paths.

At intermediate electron densities, ions will exist with appreciable concentrations in other than the ground level. The ionization equilibrium problem then involves not only the calculation of ion abundances, but also calculation of the populations of all important excited states. This is much more complex than the coronal equilibrium problem, involving the calculation and inversion of very large matrices, of size equal to the total number of levels considered for all ions.[46]

*18-10. RADIATIVE RECOMBINATION

The inverse of photoionization is the free-bound transition in which a free electron in the field of an ion is captured into a discrete state of the atom, the excess energy being radiated as a photon of wavenumber σ. If the ion and the atom are respectively in quantum states j and i with energies E_j and E_i, and if the free electron has a velocity v, then σ is given by

$$hc\sigma = mv^2/2 + E_j - E_i \ . \tag{18.89}$$

The radiative-recombination capture cross section $Q^r(v)$ can be calculated in terms of the photoionization cross section $Q^p(\sigma)$ with the aid of the principle of detailed balance:

From the definition of Q^r, the rate of recombinations of the above type (per unit volume) is

$$n_e n^j_{N-1} f(v)\, dv \cdot vQ^r(v) \ , \tag{18.90}$$

where f(v)dv is the fraction of free electrons having velocities between v and v + dv. From (18.44), the rate of inverse (photoionization) transitions is

$$n^i_N \frac{\rho\, d\sigma}{h\sigma} Q^p(\sigma) \ . \tag{18.91}$$

Here Q^p is the *total* photoionization cross section from state i to produce an ion in level j plus a free electron of velocity v; that is, it is the quantity $Q_{ij'}$ given by (18.45) for the level

$$j' = (j,\varepsilon l)J \ ,$$

summed over all possible *l* and J for given j and $\varepsilon = mv^2/2$.

In complete thermodynamic equilibrium, the rate (18.90) is equal to the *net* photoionization rate obtained by multiplying (18.91) by a factor $1 - \exp(-hc\sigma/kT)$ to allow for the effects of induced emission [cf. the denominator of (14.13)]. Thus, using $d\sigma/dv = mv/hc$ from (18.89),

$$Q^r = Q^p \frac{m\rho n^i_N}{h^2 c\sigma f(v) n_e n^j_{N-1}} (1 - e^{-hc\sigma/kT}) \ . \tag{18.92}$$

[46]A. Burgess and H. P. Summers, Astrophys. J. 157, 1007 (1969).

But under equilibrium conditions, ρ is given by (14.13) and f(v) is given by the Maxwell-Boltzmann distribution

$$f(v) = 4\pi\left(\frac{m}{2\pi kT}\right)^{3/2} v^2 e^{-mv^2/2kT} . \qquad (18.93)$$

Also, from (18.82), (18.87), and (18.89) we find

$$\frac{n^i{}_N}{n_e n^j{}_{N-1}} = \frac{g_i}{2g_j}\left(\frac{h^2}{2\pi mkT}\right)^{3/2} e^{hc\sigma/kT - mv^2/2kT} . \qquad (18.94)$$

Combining all this we find[47]

$$Q^r(v) = \frac{g_i h^2 \sigma^2}{g_j m^2 v^2} Q^p(\sigma) = \frac{g_i \alpha^2 \varepsilon^2}{g_j 4(\varepsilon - \varepsilon_i)} Q^p(\varepsilon) , \qquad \varepsilon > \varepsilon_i , \qquad (18.95)$$

where ε is now the *photon* energy in rydbergs and $\varepsilon_i > 0$ is the threshold ionization energy from state i. Like the Einstein-transition-probability relations (14.16), Eq. (18.95) involves only physical properties of the atom and is valid whether thermodynamic equilibrium exists or not.

It is common to define a radiative-recombination rate coefficient α^r_{ji} such that the total rate of recombinations (per unit volume) from state j to level i is

$$R^r_{ji} = n_e n^j{}_{N-1} \alpha^r_{ji} , \qquad (18.96)$$

where from (18.90)

$$\alpha^r_{ji} = \int_0^\infty vQ^r(v)f(v)\,dv = \langle vQ^r(v)\rangle . \qquad (18.97)$$

Usually the free-electron distribution function is assumed to be the thermal distribution (18.93) at some electron temperature T_e (though the ion velocity distribution need not be thermal), and then $\alpha^r_{ji} = \alpha^r_{ji}(T_e)$.

We shall consider detailed results corresponding only to the approximate hydrogenic cross section (18.47) for photoionization from the n shell (averaged over all *l*). The inverse process is capture by an ion in state j of an electron into the (vacant) n shell to give states i = (j,n*l*)JM; this is to be summed over all J, M, and *l*. Then from (4.57) and (2.65) $\sum g_i/g_j = 2n^2$, and from (18.95) and (18.47)

$$\overline{Q}^r_n = \frac{32\pi\alpha^3 a_0{}^2 n\varepsilon_n{}^2}{3^{3/2}\varepsilon(\varepsilon - \varepsilon_n)} , \qquad \varepsilon > \varepsilon_n . \qquad (18.98)$$

Substituting this and (18.93) into (18.97), a straightforward calculation gives

[47] Ya. B. Zel'dovich and Yu. P. Raizer, ref. 42, p. 265. I. I. Sobel'man, *Introduction to the Theory of Atomic Spectra* (Pergamon, Oxford, 1972), Eq. (34.23).

$$\bar{\alpha}_n^r(T) = \frac{64\pi^{1/2}\alpha^4 c a_0{}^2 n\varepsilon_n{}^2}{(3kT)^{3/2}} e^{\varepsilon_n/kT} E_1(\varepsilon_n/kT)$$

$$= 5.20\cdot 10^{-14} \frac{n\varepsilon_n{}^2}{(kT)^{3/2}} e^{\varepsilon_n/kT} E_1(\varepsilon_n/kT) \quad \text{cm}^3/\text{sec} , \tag{18.99}$$

where $\varepsilon_n = Z_c^2/n^2 > 0$ and kT are in rydbergs, and

$$E_1(x) = \int_x^\infty y^{-1} e^{-y}\, dy \tag{18.100}$$

is the exponential integral of index one. The total radiative-recombination rate coefficient is given by

$$\bar{\alpha}^r(T) = \sum_{n=n_0}^{\infty} \alpha_n^r(T) , \tag{18.101}$$

where n_0 is the lowest unoccupied shell of the recombining ion.

More accurate recombination rate coefficients can be obtained by making use of non-hydrogenic photoionization cross sections, such as those calculated numerically with the PELEC computer program.[13]

*18-11. DIELECTRONIC RECOMBINATION

The relationship between the inverse processes autoionization and dielectronic recombination is illustrated in Fig. 18-13, which shows a few of the possible levels of a system consisting of a nucleus of charge Z plus N electrons, at most one of which is free. The state k_0 is one of the states of the ground level of the ion (Z, N). The first ionization limit of this ion is also the ground-level energy of the next-higher ion (Z, N−1); m_0 is one of the states of this level. Excitation of one of the most weakly bound electrons of the ion (Z, N) gives rise to levels (such as k)[48] lying below the ionization limit m_0. Excitation of an inner-subshell electron, or of two or more electrons from any subshells, may lead to a state j that lies above m_0;[49] this state may autoionize—spontaneously making a non-radiative transition to a state i consisting of the ion (Z, N−1) in some bound state m of energy less than $E^i = E^j$, plus a free electron with kinetic energy

$$\varepsilon = E^i - E^m \equiv E^{im} . \tag{18.102}$$

(If the energy E^j is great enough, more than one Auger electron may be ejected, but we shall consider only the single-electron case.)

[48]For simplicity, we shall use k to denote either a level or a specific one of the g_k states of that level (and similarly, for j, i, or m). When necessary, it will be stated explicitly whether the level or a state is meant.

[49]Such a state may also be produced by excitation of a single electron from the outermost subshell, together with excitation within the remaining core configuration (if the latter contains at least one open subshell). However, this type of autoionizing state is of little importance for dielectronic recombination in highly ionized atoms.

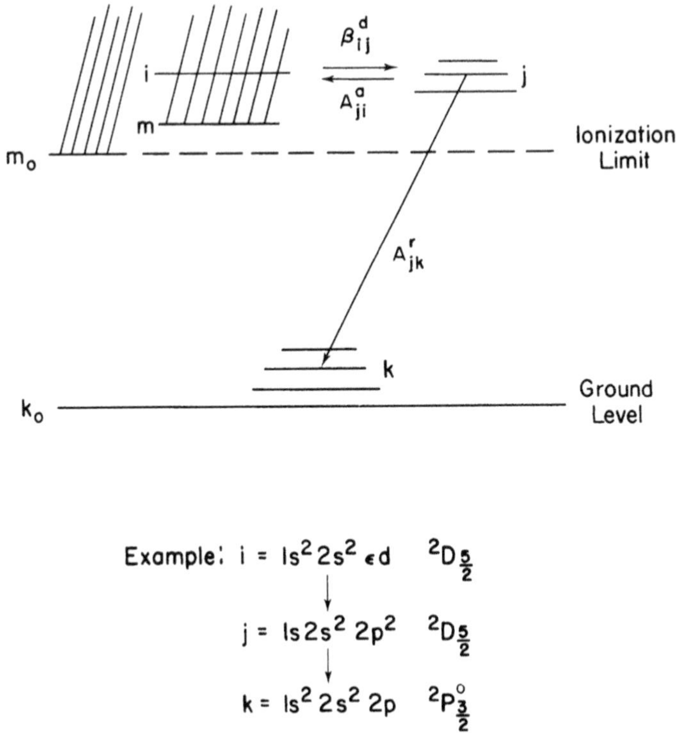

Fig. 18-13. Schematic diagram of a few levels of an N-electron system. The levels k are singly excited bound levels lying below the ionization limit, and the levels j are doubly excited discrete but potentially autoionizing levels lying above the limit. The shaded regions are continua: levels such as i consist of a free electron in the field of an N − 1 electron ion in an excited state m. (In the example, i is a level of the continuum based on the ground state m_o.)

Dielectronic recombination is the inverse of this process. Thus, let i be an N-electron state consisting of the ion (Z, N−1) in some bound state m plus a free electron with kinetic energy $\varepsilon = E^{lm}$. If ε is such that $E^l \cong E^J$, the free electron may excite an electron of the ion (Z, N−1), thereby losing enough energy to be captured into a bound orbit and producing the metastable state j of the ion (Z, N). Thus, dielectronic recombination may be thought of as a resonance in a (radiationless) inelastic scattering process[50] in which the incident free electron loses more than 100% of its kinetic energy, and which is described by an energy-dependent cross section $Q^d_{mJ}(\varepsilon)$ for the capture by the ion (Z, N−1) in state m into (all states of) the level j of the ion (Z, N). This cross-section is closely related to the Beutler-Fano profile (18.70), the unperturbed background continuum of which corresponds to the radiative recombination cross section Q^r.

[50]Both dielectronic recombination and autoionization are frequently treated by the techniques of electron scattering theory; see, for example, refs. 25 and M. J. Seaton and P. J. Storey, in P. G. Burke and B. L. Moiseiwitsch, eds., *Atomic Processes and Applications* (North-Holland Publ. Co., Amsterdam, 1976).

We shall not here be interested in the cross section itself, but only in the dielectronic recombination rate coefficient

$$\beta^d_{mj}(T) = \langle v_\epsilon Q^d_{mj}(\varepsilon)\rangle \tag{18.103}$$

such that

$$n_e n^m_{N-1} \beta^d_{mj}(T) \tag{18.104}$$

is the rate of captures m $\rightarrow$ j (per unit volume of plasma) out of a sea of free electrons with uniform density n_e in thermodynamic (Maxwell-Boltzmann) equilibrium at an electron temperature T; n^m_{N-1} is the density of ions (Z, N−1) in (all g_m states of) the level m.

We shall evaluate the rate coefficient in terms of the inverse (autoionization) process by means of the principle of detailed balance.[51] The rate (per unit volume) at which the inverse transitions take place is

$$n^j_N A^a_{jm} \; , \tag{18.105}$$

where n^j_N is the density of ions (Z, N) in (all g_j states of) the level j, and A^a_{jm} is the probability per unit time that such an ion in any specific state j will autoionize, leaving an ion (Z, N−1) in (all states of) the level m. Under conditions of complete thermodynamic equilibrium, (18.104) and (18.105) must be equal, so that

$$\beta^d_{mj}(T) = \frac{n^j{}_N A^a{}_{jm}}{n_e n^m{}_{N-1}} \; .$$

But under such conditions, the Boltzmann and Saha equations (18.82) and (18.87) are valid, and so

$$\begin{aligned} \beta^d_{mj}(T) &= \frac{h^3}{2(2\pi m k T)^{3/2}} \frac{g_j A^a{}_{jm}}{g_m} e^{-E^{jm}/kT} \\ &= \frac{4\pi^{3/2} a_0{}^3}{T^{3/2}} \frac{g_j A^a{}_{jm}}{g_m} e^{-E^{jm}/T} \; , \end{aligned} \tag{18.106}$$

where in the second expression $E^{jm} \equiv E^j - E^m$ and T are in rydbergs, and a_0 is the Bohr radius. This expression for the dielectronic recombination rate coefficient holds whether or not the ion populations correspond to equilibrium conditions, provided that the free electrons of uniform density n_e in (18.104) have a Maxwell-Boltzmann velocity distribution corresponding to an electron temperature T.

Some clarification of the relationship between the states i and m is appropriate. Before the recombination process begins, the free electron will interact only weakly with the ion, and so the state m will be characterized by a well-defined total angular momentum J_m. The statistical weight g_m in (18.106) will therefore be $2J_m + 1$. The wavefunction of the free

[51]D. R. Bates and A. Dalgarno, "Electronic Recombination," in D. R. Bates, ed., *Atomic and Molecular Processes* (Academic Press, New York, 1962), p. 258.

electron (assumed to be approaching in a direction parallel to the z-axis) is given by the usual partial-wave expansion[52]

$$\psi(r,\theta) = r^{-1} \sum_{l'=0}^{\infty} (2l' + 1) i^{l'} e^{i\eta_{l'}} P_{\varepsilon l'}(r) P_{l'}(\cos\theta) \; ; \qquad (18.107)$$

each term in this expansion corresponds to a definite angular momentum l' of the free electron about the ion (Z, N−1). The corresponding N-electron state can be represented as a superposition of states i of the form

$$(J_m, l's')J_i \; , \qquad (18.108)$$

where s′ is the spin of the free electron, and where dielectronic recombination to the state j can occur only if $J_i = J_j$ because the angular momentum of the total system is conserved in the recombination process. Calculation of the autoionization probability rate A^a_{jm} in (18.106) involves a summation over all possible ways in which l' and s′ can be coupled to J_m to produce $J_i = J_j$, and also a summation over all possible values of l':

$$A^a_{jm} = \left(\sum_i A^a_{ji} \right)_{\text{fixed m and } J_i} \; . \qquad (18.109)$$

[Ultimately, all states i are to be taken into account, each being given the usual statistical weight $2J_i + 1$. For given m and l', the total statistical weight of the possible states i is, from (2.65),

$$\sum_i (2J_i + 1) = (2J_m + 1)(2l' + 1)(2s' + 1) \; . \qquad (18.110)$$

This expression includes the factor $(2l' + 1)$ appearing in the partial-wave expansion (18.107); however, as we are calculating recombination rates in terms of the autoionization process we have not had to concern ourselves explicitly with the details of this expansion.]

Values of A^a_{ji} can be computed from the perturbation expression (18.65), (18.50):

$$A^a_{ji} = \frac{2\pi}{\hbar} |\langle j|H|i\rangle|^2 = \frac{2\pi}{\hbar} | \sum_{bb'} \langle j|b\rangle\langle b|H|b'\rangle\langle b'|i\rangle |^2 \; , \qquad (18.111)$$

where in the final step we have introduced the usual wavefunction expansion in terms of basis functions b and b′. In all cases of importance here,[49] the bound configuration to which b belongs differs from the continuum configuration b′ by two electrons, because in order for the free electron to be captured, an electron of the ion core m must be excited; an example is

$$1s^2 2s^2 2p\varepsilon l' \rightarrow 1s^2 2s 2p^2 nl \; . \qquad (18.112)$$

Thus evaluation of the matrix elements $H_{bb'}$ involves only the two-electron Coulomb portion of the Hamiltonian operator; evaluation of these configuration-interaction matrix

[52]See, for example, N. F. Mott and H. S. W. Massey, *The Theory of Atomic Collisons* (Clarendon Press, Oxford, 1965), 3rd ed., p. 24.

elements has already been discussed in Sec. 13-10, the only difference being that one of the radial functions (l'_σ) involved in the radial integrals R^k in (13.59) is now a continuum function. Because of the parity and angular-momentum restrictions (13.24)-(13.25), it follows that for any given configuration j only a few values of l'_σ are possible; in the example (18.112), these possibilities are $l' = l + 1$ and $l - 1$. Thus (for given j) only a limited number of terms in the partial-wave expansion (18.107) come into play, and the summation over i in (18.109) is strictly finite.

It may be noted that in order to obtain the eigenvector components $\langle j|b\rangle$ and $\langle b'|i\rangle$ in (18.111), it is necessary to diagonalize the corresponding single-configuration energy matrices. It is possible to use the computer program of Fig. 16-1 to obtain both the bound and the continuum eigenvectors, and also the CI matrix elements $H_{bb'}$, in a single "configuration-interaction" calculation; it is sufficient simply to artificially shift the centers of gravity E_{av} of all continuum configurations by large negative amounts, in order to reduce the undesired configuration mixing effectively to zero. This simultaneously provides the computer program with a means of distinguishing continuum eigenvectors from discrete eigenvectors by means of the large negative eigenvalues of the former.

Radiative stabilization. Once the ion (Z, N) has been formed in the state j, there is the possibility that it may reionize at a rate A^a_{jm} (any energetically possible m). We cannot really consider the process of dielectronic recombination to be complete until the ion (Z, N) has been stabilized by de-excitation to a level lying below the ionization limit m_0. In low-density plasmas, collisional de-excitation rates are negligible—at least, for values of n of the highly excited electron that are not too large. The only process of importance is radiative decay to some lower level k, at the rate A^r_{jk} given by (14.32). [Usually, radiative decay of the highly excited electron nl is slow because of the large values of n and/or l; hence the cases contributing most strongly to dielectronic recombination are those that involve an electric-dipole-allowed ion-core excitation: 2s − 2p in the example (18.112).]

The probability that decay will take place via a radiative transition to a specific lower level k is given by the branching ratio

$$B_{jk} = \frac{A^r_{jk}}{\sum_{m'} A^a_{jm'} + \sum_{k'} A^r_{jk'}} , \qquad (18.113)$$

where the summation over m′ extends over all possible levels of the ion (Z, N−1) that lie below j, and the summation over k′ extends over all levels of (Z, N) that lie below m_0 (or that will much more likely decay than autoionize). The *net* recombination rate coefficient via the path m → j → k is thus

$$\alpha^d_{mjk}(T) = \beta^d_{mj}(T)\, B_{jk} \; . \qquad (18.114)$$

The effective net rate coefficient for the path j → k, averaged over all initial states of the ion (Z, N−1), is

$$\alpha^d_{jk}(T) = \frac{1}{n_{N-1}} \sum_m n^m_{N-1}\, \alpha^d_{mjk}(T) , \qquad (18.115)$$

and the total net rate coefficient for all paths is

$$\alpha^d(T) = \sum_j \sum_k \alpha^d_{jk}(T) \ . \tag{18.116}$$

In the low electron-density limit, collisional excitation rates and dielectronic recombination rates will be negligible compared with radiative decay rates, and so essentially all ions will exist in the ground level m_o. Thus

$$\alpha^d_{jk}(T) = \alpha^d_{m_o jk}(T) = \frac{4\pi^{3/2} a_o^{\,3}}{T^{3/2}} e^{-E^{jm_o}/T} \left(\frac{g_j}{g_{m_o}} \right) \frac{A^a_{jm_o} A^r_{jk}}{\sum_{m'} A^a_{jm'} + \sum_{k'} A^r_{jk'}} \ . \tag{18.117}$$

These equations are pertinent to highly ionized atoms (e.g., Fe XV to XXV) in the low-density portions of the solar corona ($n_e \lesssim 10^8$ cm^{-3}).

For highly ionized ions in low-density laboratory plasmas ($n_e \cong 10^{14}$ to 10^{18} cm^{-3}), collisional excitation rates are still too low to appreciably populate levels of excited configurations. However, they may be high enough to produce an approximately statistical population of the metastable levels of the ground configuration, which can decay radiatively only by relatively slow forbidden transitions (M1, E2, etc.). The range of energies within the ground configuration of an ion is so small compared with the temperature required to produce that ion, that the Boltzmann factor in (18.82) is essentially unity, and so for these levels

$$\frac{n^m_{N-1}}{n_{N-1}} = \frac{g_m}{G_m} \ , \tag{18.118}$$

where

$$G_m = \sum g_m \tag{18.119}$$

is the total statistical weight of the ground configuration. Substituting (18.118) into (18.115), we find for this case

$$\alpha^d_{jk}(T) = \frac{4\pi^{3/2} a_o^{\,3}}{T^{3/2}} e^{-E_s/T} \left(\frac{g_j}{G_m} \right) \frac{\sum_m A^a_{jm} A^r_{jk}}{\sum_{m'} A^a_{jm'} + \sum_{k'} A^r_{jk'}} \ , \tag{18.120}$$

where the summation over m extends over all levels of the ground configuration of the ion, and E_s is an average value of $E^j - E^m$, averaged over all levels of this configuration. [This expression is identical with (18.117) for $N - 1 \lesssim 4$, because the ground configuration contains only one level.]

The rate of radiation of photons in the spectrum line $j \rightarrow k$ (per unit volume of an optically thin plasma) as a result of dielectronic recombination into the level j is

$$R(j \rightarrow k) = n_e n_{N-1} \alpha^d_{jk}(T) \ . \tag{18.121}$$

This expression can be used to calculate absolute or relative intensities of spectrum lines resulting from this mechanism,[53] which is a primary one for the production of so-called satellite lines[54]—in the example (18.112) for 20-fold ionized iron, satellites

$$\mathrm{Fe\,XXI}\quad 1s^2 2s^2 2pnl \ - \ 1s^2 2s2p^2 nl$$

lying to the long-wavelength side of the corresponding parent lines

$$\mathrm{Fe\,XXII}\quad 1s^2 2s^2 2p \ - \ 1s^2 2s2p^2 \ .$$

Compared with the intensities of spectrum lines produced by a plasma in LTE, which are given from (18.82) by

$$R(j\rightarrow k) = n^j_N A^r_{jk} = n_N e^{-E_j/T}\left(\frac{g_j}{u^i_N}\right) A^r_{jk} \ , \qquad (18.122)$$

relative intensities of lines from different levels j involve in (18.120) an additional factor

$$\frac{\sum_m A^a{}_{jm}}{\sum_{m'} A^a{}_{jm'} + \sum_{k'} A^r{}_{jk'}} \ . \qquad (18.123)$$

Dielectronic recombination in low density plasmas can therefore produce relative line intensities that are considerably different from what they are in high-density sources. An example is shown in Fig. 18-14. The reasons for the differences in this case can easily be understood qualitatively: Because of parity and angular-momentum restrictions, the only continuum configurations $1s^2 2s^2 \varepsilon l'$ that can recombine to $1s2s^2 2p^2$ are εs and εd, which are necessarily pure ^2S and ^2D, respectively. Thus, because of the Coulomb selection rules LSJ = L′S′J′, recombination can take place into the ^2P and ^4P levels only to the limited extent that these levels contain some admixture of ^2S and ^2D nature.

Approximate expressions for α^d. In order to make the evaluation of (18.116) easier, we make a number of simplifying assumptions. We consider only the expression (18.120) for α^d_{jk}, and in the summation over m′ neglect terms representing autoionization to levels of excited configurations of the ion. This can be a very poor approximation for those terms in (18.116) that involve small n and *l*,[55,56] but does not usually introduce too great an error in the overall value of α^d because much of the contribution comes from the large n and/or *l* terms for which the sum over m′ is small compared with the sum over k′ (see Fig. 18-15 below). Then in both numerator and denominator we have the quantity

[53]A. H. Gabriel, Mon. Not. R. Astron. Soc. 160, 99 (1972); C. P. Bhalla, A. H. Gabriel, and L. P. Presnyakov, Mon. Not. R. Astron. Soc. 172, 359 (1975).

[54]B. Edlén and F. Tyrén, Nature 143, 940 (1939). The "satellite lines" defined here should not be confused with those of Sec. 14-8, which are simply the weak lines of a multiplet.

[55]A. L. Merts, R. D. Cowan, and N. H. Magee, Jr., Los Alamos Scientific Laboratory Report LA-6220-MS (March 1976).

[56]V. L. Jacobs, J. Davis, P. C. Kepple, and M. Blaha, Astrophys. J. 211, 605 (1977), 215, 690 (1977).

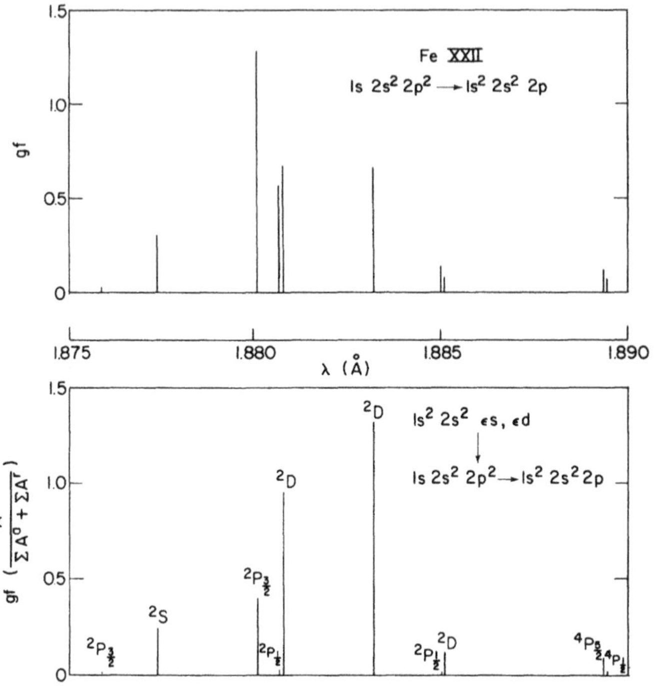

Fig. 18-14. Theoretical relative intensities of 2p → 1s lines in Fe XXII under conditions of (top) collisional excitation (LTE), and (bottom) dielectronic recombination. The LS terms indicated in the bottom figure refer to the configuration $1s2s^22p^2$.

$$A_j^a \equiv \sum_m A_{jm}^a , \tag{18.124}$$

which is the total autoionization rate from a state of level j to all levels of the ground configuration of the ion. The summation over k of A_{jk}^r is identical with the summation over k′, and we abbreviate each by

$$A_j^r \equiv \sum_k A_{jk}^r . \tag{18.125}$$

In practice, the summation is usually limited to levels k of a single configuration, for whatever transition array is the strongest; for example, in the recombination

$$\text{Fe XXI} \quad 1s^22s^22p\varepsilon l \;\rightarrow\; 1s^22s2p3pnl ,$$

3p → 2s decay to $1s^22s^22pnl$ is more than two orders of magnitude faster than 2p → 2s decay to $1s^22s^23pnl$ because of the much greater transition energy.

Next, we divide the summation over all possible autoionizing levels j into a three-fold summation over excited configurations j_c of the *recombining* ion, over the nl of the captured electron, and finally over the various levels of the configuration (j_cnl) of the recombined ion. [In the example (18.112), $j_c = 1s^22s2p^2$, and nl can be either 2p or any orbital with $n \geqq 3$, provided the energy is high enough to lie above the ionization limit $1s^22s^22p$. In Fe XXI, n must be $\geqq$ 10.] In highly ionized atoms, the levels of the configuration $j_c nl$ cover an energy range small compared with kT, and so we obtain

$$\alpha^d(T) = \sum_{j_c} \alpha^d_{j_c} , \tag{18.126}$$

where

$$\alpha^d_{j_c} = \frac{4\pi^{3/2} a_o^{\ 3}}{T^{3/2} G_m} \sum_{nl} e^{-E_s/T} \sum_j \frac{g_j A_j^{\ a} A_j^{\ r}}{A_j^{\ a} + A_j^{\ r}} ; \tag{18.127}$$

$E_s \equiv E_s^{\,j_c nl}$ is now the free-electron kinetic energy averaged over the configuration $j_c nl$ as well as over the configuration m, and the summation over j now extends only over the levels of $j_c nl$.

We now define configuration-average total transition probabilities

$$\bar{A}^a = \frac{1}{G_j} \sum_j g_j A_j^a , \qquad \bar{A}^r = \frac{1}{G_j} \sum_j g_j A_j^r , \tag{18.128}$$

where $G_j = \sum g_j$ is the total statistical weight of the configuration $j_c nl$. Illustrative computed values[55] of $\bar{A}^a$ and $\bar{A}^r$ are shown in Fig. 18-15. These show that $\bar{A}^a$ varies asymptotically with n in the manner

$$\bar{A}^a \propto n^{-3} , \tag{18.129}$$

as is to be expected from (8.20)-(8.21), and in addition decreases very rapidly with increasing l for l greater than about 3.[57] Also, if the radiative decay rate is determined by transition of an excited electron interior to nl, then $\bar{A}^r$ is essentially independent of nl; this behavior follows from (14.38) and (14.97), and the near constancy of the radial dipole integral P for the present type of transition in highly ionized atoms.

If departures from pure coupling conditions are large, no radiative selection rules are effective except for $\Delta J = 0, \pm 1$; then for large nl, $A_j^r \gg A_j^a$ for essentially all j, and the summation in (18.127) is equal essentially to $G_j\bar{A}^a$. Similarly, provided conditions are not close to pure LS coupling, then in the opposite limit $A_j^a \gg A_j^r$, the summation is equal approximately to $G_j\bar{A}^r$. The obvious interpolation formula between these limits is

$$\sum_j \frac{g_j A_j^{\ a} A_j^{\ r}}{A_j^{\ a} + A_j^{\ r}} \cong \frac{G_j \bar{A}^a \bar{A}^r}{\bar{A}^a + \bar{A}^r} . \tag{18.130}$$

[57]Unlike the radial integrals F^k and G^k, which involve P_{nl} twice, the configuration-interaction integrals R^k contain P_{nl} only once; however, A^a is from (18.111) proportional to the square of R^k. The l dependence is approximately l^{-5} or faster, see Merts et al. (ref. 55) or W. E. Cooke and T. F. Gallagher, Phys. Rev. A 19, 2151 (1979).

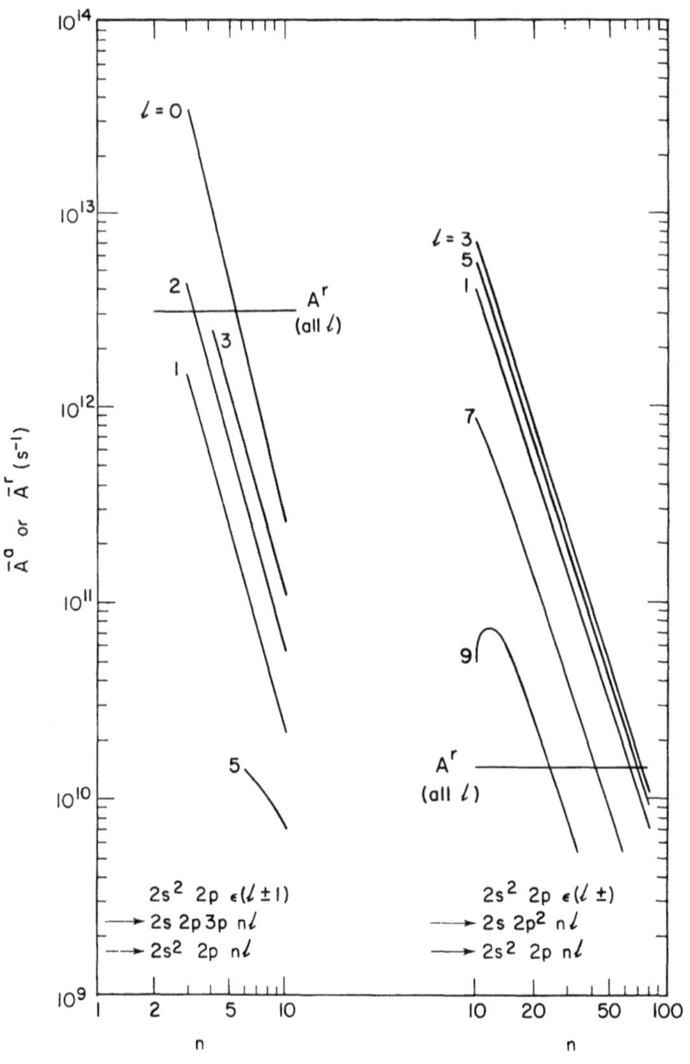

Fig. 18-15. Computed values of $\bar{A}^a$ and $\bar{A}^r$ for the indicated dielectronic recombination processes in Fe XXI.

It is easily seen that the equality would hold exactly if $A_j^a = \bar{A}^a$ and $A_j^r = \bar{A}^r$ for all j, or if $A_j^a = CA_j^r$ where C is some constant; the degree of validity of the approximation in some sample cases[55] is shown in Table 18-2.

For the present, we use the approximation (18.130). Noting that $G_j = (4l + 2)G_{j_c}$ provided that nl is not equivalent to any electrons of the core configuration j_c, then we may write

$$\alpha_{j_c}^d = \frac{8\pi^{3/2} a_o^{\,3} G_{j_c} \bar{A}_\infty^r}{T^{3/2} G_m} e^{-E_s^{\,\infty}/T} \sum_n B_n \ , \tag{18.131}$$

TABLE 18-2. RATIO OF THE LEFT-HAND SIDE OF (18.130) TO THE RIGHT-HAND SIDE

		$\bar{A}^a$	$\bar{A}^r$	Ratio[(a)] Pure LS	Interm.	Pure jj
Fe XXI	$2s^22p10g - 2s2p^210g$	3.27^{12}	1.44^{10}	0.98	0.98	0.98
	$2s^22p40g - 2s2p^240g$	4.71^{10}	1.44^{10}	0.85	0.84	0.84
	$2s^22p150g - 2s2p^2150g$	8.94^{8}	1.44^{10}	0.96	0.96	0.96
Fe XXI	$2s^22pns - 2s2p3pns$[(b)]	1.48^{14}	3.15^{12}	0.56	0.86	0.92
	$2s^22p6s - 2s2p3p6s$	1.48^{12}	3.15^{12}	0.33	0.59	0.68
	$2s^22pns - 2s2p3pns$[(b)]	1.48^{10}	3.15^{12}	0.38	0.79	0.85
Fe XXII	$2s^211d - 2s2p11d$	1.00^{12}	4.61^{9}	0.66	0.90	
	$2s^215d - 2s2p15d$	3.68^{11}	4.61^{9}	0.66	0.85	
	$2s^225d - 2s2p25d$	7.79^{10}	4.61^{9}	0.63	0.75	
	$2s^260d - 2s2p60d$	5.64^{9}	4.61^{9}	0.54	0.55	
	$2s^2150d - 2s2p150d$	3.62^{8}	4.61^{9}	0.77	0.78	

[(a)]The pure LS and jj values were obtained by artificially modifying the ratio of the values of the spin-orbit parameters to the Coulomb parameters.
[(b)]Examples produced by arbitrarily scaling the A^a values for 6s.
Note: Values of A^a and A^r are in sec^{-1}, the exponents representing powers of ten that multiply the corresponding decimal fractions.

where T is in rydbergs and

$$B_n = \sum_l C^g_{nl} C^r_{nl} e^{(E_s^\infty - E_s)/T} \frac{(2l+1)\bar{A}^a}{\bar{A}^a + C^r_{nl}\bar{A}^r_\infty} . \qquad (18.132)$$

Here E^∞_s is the value of E_s for $n = \infty$ (which is also the mean transition energy for the excitation $m \rightarrow j_c$), and $\bar{A}^r_\infty$ is the average total radiative transition probability for large n. The quantities C_{nl} are unity except for that single term nl (if any) for which nl is equivalent to electrons of j_c; non-unit values of the C_{nl} then correct for the fact that the occupation number of the nl subshell is no longer unity so that $G_j \neq (4l+2)G_{j_c}$, and for the fact that $\bar{A}^r$ is not (even approximately) equal to $\bar{A}^r_\infty$.

Some illustrative values[55] of B_n are shown in Fig. 18-16. So far as the value of α^d_{jc} is concerned, we note that though the value of $\bar{A}^r_\infty$ for the $\Delta n = 0$ excitation is much smaller than for $\Delta n > 0$ excitations, the smaller excitation energy E^∞_s and the much greater range of contributing values of n and l (Fig. 18-15) largely compensate, as was first pointed out by Burgess.[58]

The summation over n in (18.131) is to be carried from the minimum energetically allowed value n_0 up to a value n_t at which the highly excited electron nl is collisionally

[58]A. Burgess, Astrophys. J. 139, 776 (1964), Ann. d'astrophys. 28, 774 (1965).

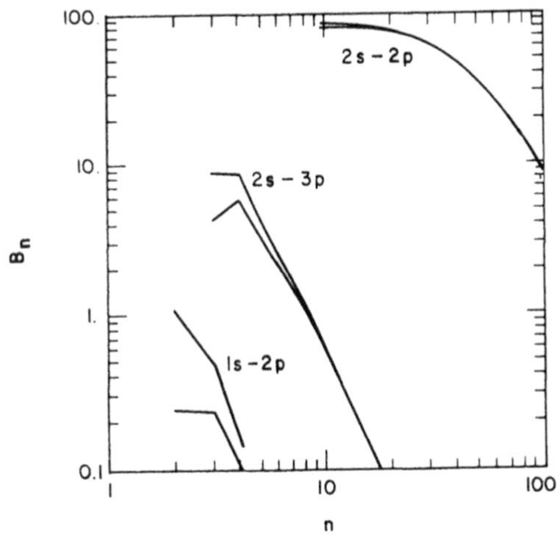

Fig. 18-16. The summation B_n defined by (18.132) for three cases in Fe XXI; the lower of each pair of curves is for $T = \infty$, and the higher is for $T = 1$ keV. The difference in shape between curves for $\Delta n = 0$ and $\Delta n > 0$ excitations is pronounced. For $\Delta n > 0$, B_n becomes negligible at such small values of n that plasma-density effects are unimportant, at least up to $n_e = 10^{18}$ electrons/cm^3.

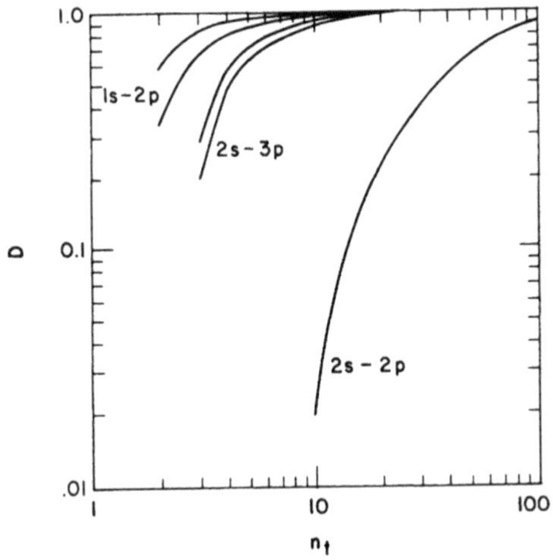

Fig. 18-17. Relative value of $\sum B_n$ for Fe XXI when summed to n_t instead of to ∞. (Lower curve of a pair is for $T = \infty$, the upper curve is for $T = 1$ keV.)

ionized faster than it can decay radiatively. The latter limit can be estimated from the expression[59]

$$n_t^7 = 5.6 \cdot 10^{17} (Z - N + 1)^6 T^{1/2}/n_e \ , \tag{18.133}$$

where the electron temperature T is in rydbergs, and the electron density is in cm^{-3}. The density dependence of n_t provides a means of estimating the reduction of the net dielectronic recombination rate in finite density plasmas. For example, Fig. 18-17 shows the computed fractional reduction from the zero-density value of $\alpha^d_{j_c}$ for three cases of Fe XXI.[55] At $T = 1$ keV (73.5 Ry) and $n_e = 10^{14}$ cm^{-3}, $n_t = 64$. The density reduction is then about 25% for the 2s − 2p contribution to α^d, which constitutes about 30% of the total; the density reduction in α^d is thus only about 8%—which is small compared with other approximations and uncertainties in theoretical values.

[59] R. Wilson, J. Quant. Spectrosc. Radiat. Transfer 2, 477 (1962).

TABLE 18-3. COMPARISON OF PARTIAL DIELECTRONIC-RECOMBINATION RATE COEFFICIENTS FOR T = 1 keV AND $n_e = 0$, CALCULATED IN VARIOUS APPROXIMATIONS

	Fe XXI 2s-2p	Fe XXII 2s-2p	Fe XXI 2s-3p	Fe XXI 1s-2p
G_{j_c}	30	12	72	30
G_m	6	1	6	6
E_s^∞ (Ry)	5.4	4.1	80	483.
$\bar{A}_\infty^r$ (sec^{-1})	$1.44 \cdot 10^{10}$	$4.61 \cdot 10^9$	$3.16 \cdot 10^{12}$	$2.42 \cdot 10^{14}$
f_{mj_c}	0.183	0.187	0.745	0.645
$\alpha_{j_c}^d$ (10^{-14} cm^3/sec)				
(18.127)	250	171	311	3.07
(18.131)-(18.132)	284	272	409	3.27
Burgess, (18.134)	287	287	1290	5.25

One of the major approximations involved in (18.132) is that represented by (18.130). The examples given in Table 18-2 show that the error involved may be as great as a factor of two or three. Fortunately, the errors are greatest for ions showing good LS coupling (not true of highly ionized heavy ions) and for cases with $\bar{A}^a \cong \bar{A}^r$ (true for only a few values of nl). Consequently, the effect on the total α^d for iron ions (Table 18-3) is usually only 10 or 20%, except for ions such as Fe XXII that involve particularly simple configurations so that selection-rule effects are serious in spite of the basic strong departure from pure LS coupling.

Also shown in Table 18-3 are values calculated from a useful approximation formula given by Burgess:[60]

$$\alpha^d = 4.8 \cdot 10^{-11} T^{-3/2} B(z) \sum_{j_c} \bar{f}_{mj_c} A(x) e^{-E_s^\infty/aT} \quad cm^3/sec , \tag{18.134}$$

where E_s^∞ and T are in rydbergs, with $E_s^\infty/aT \leqq 5$,

$$z = Z - N + 1 , \qquad x = E_s^\infty/(z + 1) ,$$

$$B(z) = z^{1/2}(z + 1)^{5/2}/(z^2 + 13.4)^{1/2} , \qquad z \leqq 20 ,$$

$$A(x) = x^{1/2}/(1 + 0.105x + 0.015x^2) , \qquad x > 0.05 ,$$

and

$$a = 1.0 + 0.015 z^3/(z + 1)^2 .$$

[60]A. Burgess, Astrophys. J. **141**, 1588 (1965).

The quantity a is a correction from E_s^{∞} to an appropriately weighted value of E_s in (18.131)-(18.132); $\bar{f}_{mj_c}$ is the average array oscillator strength (14.97), related to $G_{j_c}\bar{A}_{\infty}^{r}/G_m$ similarly to (14.38) via an appropriately averaged value of σ. (Note, however, that $\sigma \cong 109735\ E_s^{\infty}$ is usually a very poor approximation in the case of $\Delta n = 0$ transitions, for reasons discussed in Sec. 19-5; see also Sec. 21-4.)

Judging from the few examples available, the Burgess approximation is quite good for $\Delta n = 0$ excitations [except where the approximation (18.130) is poor], but for $\Delta n > 0$ excitations it appears to give results too high by a factor of roughly two to four.[55,61] It has been tentatively suggested[55] that for $\Delta n > 0$ excitations the function A(x) be modified to

$$A(x) = 0.5x^{1/2}/(1 + 0.210x + 0.030x^2) \ . \tag{18.135}$$

Different modifications have also been suggested.[61a]

Collisional redistribution in *l*. In addition to density effects described above, one further effect should be mentioned. We have seen in Fig. 18-15 that, for given n, dielectronic recombination occurs most rapidly into those outer-electron states that have small *l*. By the same token, these states have a high probability (if n is not too large) of reionizing before stabilizing radiative decay can occur. At sufficiently high density, however, electronic or ionic collisions may rapidly produce an essentially statistical redistribution among the semi-degenerate states of fixed n but different *l*. Because of the high statistical weight of the large-*l* states, together with their low probability of reionizing, the result can be a significant increase in the overall recombination rate.[62] However, available expressions for the *l*-redistribution cross sections (especially for heavy-ion collisions)[63] are of uncertain accuracy, and little quantitative work has been done on this effect.[55,64] In any case, the effect is of consequence only for those terms in (18.126)-(18.127) for which $n \gg 6$ (so that large *l* values exist, and the redistribution rate is large), but for which n is not so large that $A^a < A^r$ even for small *l*; the overall effect on the total α^d is probably no more than about 25 percent.[55]

Summary. In general, the importance of dielectronic recombination relative to radiative recombination increases with increasing degree of ionization and therefore with temperature (provided of course the ions are not completely stripped), because radiative stabilization rates increase much faster than autoionization rates (see Sec. 19-8). Dielectronic recombination therefore has a very significant influence on computed ionization equilibria for high-temperature, heavy-ion plasmas.[45,65] Indeed, it was the discrepancy between observations and ionization equilibrium calculations that first led to

[61] B. W. Shore, Astrophys. J. **158**, 1205 (1969).

[61a] Y. Hahn, Phys. Rev. A **22**, 2896 (1980).

[62] A. Burgess and H. P. Summers, Astrophys. J. **157**, 1007 (1969).

[63] R. M. Pengelly and M. J. Seaton, Mon. Not. R. Astron. Soc. **127**, 165 (1964).

[64] M. J. Seaton and P. J. Storey, ref. 50; J. C. Weisheit, J. Phys. B **8**, 2556 (1975); V. L. Jacobs, J. Davis, and P. C. Kepple, Phys. Rev. Lett. **37**, 1390 (1976).

[65] A. Burgess and M. J. Seaton, Mon. Not. R. Astron. Soc. **127**, 355 (1964); J. W. Allen and A. K. Dupree, Astrophys. J. **155**, 27 (1969); H. P. Summers, Mon. Not. R. Astron. Soc. **158**, 255 (1972).

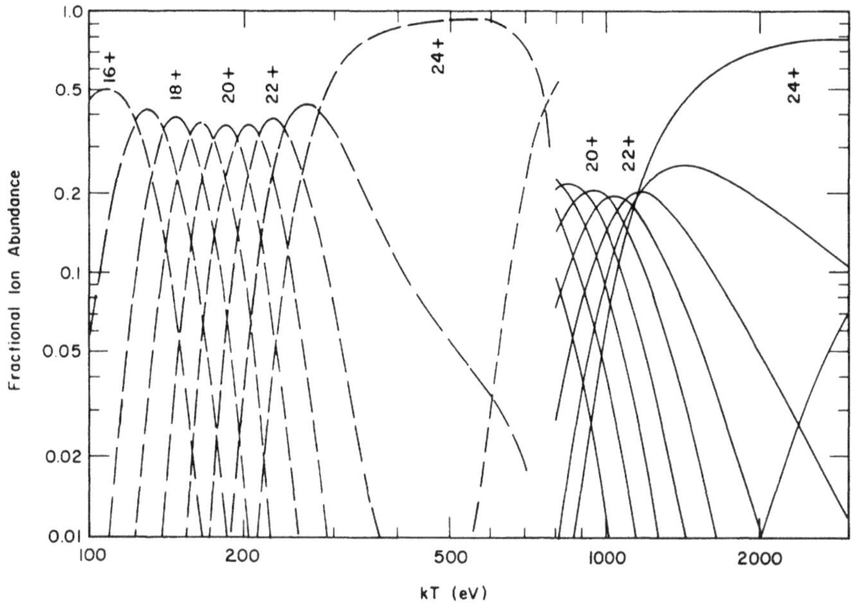

Fig. 18-18. Results of theoretical ionization equilibrium calculations in iron for LTE (dashed curves) and coronal equilibrium (solid curves).

examination of dielectronic recombination rates; a brief historical survey has been given by Seaton and Storey[50] (see also Seaton[44]).

A rough but useful rule of thumb is that under coronal equilibrium conditions the ion stage of maximum abundance is (except for low stages) the ion whose ionization energy is one to three times kT. This may be contrasted with the situation for Saha equilibrium, where the corresponding ratio is about three to ten. An example is shown in Fig. 18-18.[55,66] Approximate theoretical ionization energies with accuracy (±5%) more than adequate for these estimates are available for all ionization stages of all elements.[67]

*18-12. GENERALIZED OSCILLATOR STRENGTHS

In all the preceding sections of this chapter, we have represented the continuum electron by a spherical wave (18.14), in which the "distorted wave" radial factor $P_{\varepsilon l}(r)$ was the solution of a differential equation (18.15) involving the spherically symmetric potential V(r) of the target atom. When the kinetic energy of the free electron is very large compared with V(r), it is sufficient for some purposes to represent the continuum electron by a less accurate wavefunction.

Thus, neglecting V(r) in the Schrödinger equation (3.1)-(3.2) and working in Cartesian coordinates, but as usual measuring distances in Bohr units a_0 and energies in rydbergs ($\hbar^2/2ma_0^2$), we have

[66]N. H. Magee, Jr. and A. L. Merts, private communication.

[67]T. A. Carlson, C. W. Nestor, Jr., N. Wasserman, and J. D. McDowell, At. Data 2, 63 (1970).

$$-\left(\frac{\partial^2}{\partial x^2} + \frac{\partial^2}{\partial y^2} + \frac{\partial^2}{\partial z^2}\right)F = \varepsilon F \ . \tag{18.136}$$

A solution to this equation (with unit-amplitude normalization) is

$$F_k(r) = e^{i(k_x x + k_y y + k_z z)} = e^{i\mathbf{k}\cdot\mathbf{r}} \ , \tag{18.137}$$

where

$$k^2 = \varepsilon \tag{18.138}$$

is the kinetic energy of the free electron in rydbergs. Thus

$$k = \varepsilon^{1/2} = [(p^2/2m)/(\hbar^2/2ma_o^2)]^{1/2} = p/(\hbar/a_o) \ ,$$

so that k is the free-electron momentum in units of $\hbar/a_o$, and (18.137) represents a plane wave aimed in the direction **k** with de Broglie wavelength $\lambda = 2\pi/k = h/a_o p$ (in Bohr units).

We consider the problem of the free electron (18.137) interacting with an N-electron atom (or ion) in a state $a = |\gamma JM\rangle$ so as to excite the atom to a state $a' = |\gamma' J' M'\rangle$. The free electron is thereby inelastically scattered into a different plane-wave state $F_{k'}(r) = \exp(i\mathbf{k}'\cdot\mathbf{r})$ with

$$(k')^2 = k^2 - \Delta E \ , \tag{18.139}$$

where ΔE is the excitation energy from a to a' (in rydbergs). We consider the target-atom states a and a' to be represented by the usual intermediate-coupling linear combination of antisymmetrized pure-coupling bound basis functions. However, we take the wavefunctions of the (N+1)-electron system to be non-antisymmetrized products of the target and free-electron functions:

$$\alpha \equiv |\gamma JM\rangle F_k \qquad \text{and} \qquad \alpha' \equiv |\gamma' J' M'\rangle F_{k'} \ .$$

We shall thus perforce be neglecting effects of exchange of the free electron with one of the bound electrons; these effects are, however, small in the high energy region.

Because of orthogonality of the functions a and a', and orthogonality of the functions F_k and $F_{k'}$, the one-electron (kinetic-energy, electron-nuclear, and spin-orbit) terms of the Hamiltonian operator contribute nothing to matrix elements; likewise, electron-electron terms contribute only if one coordinate is that of the free electron. Thus we have

$$\begin{aligned} H_{\alpha\alpha'} &\equiv \langle \gamma JM e^{i\mathbf{k}\cdot\mathbf{r}} | H | \gamma' J' M' e^{i\mathbf{k}'\cdot\mathbf{r}} \rangle \\ &= \langle \gamma JM e^{i\mathbf{k}\cdot\mathbf{r}} | \sum_{m=1}^{N} \frac{2}{|\mathbf{r}_m - \mathbf{r}|} | \gamma' J' M' e^{i\mathbf{k}'\cdot\mathbf{r}} \rangle \\ &= \frac{8\pi}{K^2} \langle \gamma JM | \sum_m e^{i\mathbf{K}\cdot\mathbf{r}_m} | \gamma' J' M' \rangle \ , \end{aligned} \tag{18.140}$$

where

$$\mathbf{K} = \mathbf{k}' - \mathbf{k} \tag{18.141}$$

is the (negative of the) momentum transferred from the free electron to the atom; the final step in (18.140) follows from (7.91)-(7.92).

To evaluate the matrix element (18.140) we employ the expansion[68] [see (2.44), (2.49), and (11.9)]

$$e^{i\mathbf{K}\cdot\mathbf{r}_m} = \sum_{t=0}^{\infty} (2t + 1)\, i^t\, j_t(Kr_m) P_t(\hat{\mathbf{K}}\cdot\hat{\mathbf{r}}_m)$$

$$= \sum_{t=0}^{\infty} (2t + 1)\, i^t\, j_t(Kr_m) \mathbf{C}^{(t)}(\hat{\mathbf{K}})\cdot\mathbf{C}^{(t)}(\hat{\mathbf{r}}_m) \;, \qquad (18.142)$$

where

$$j_t(Kr) = (\pi/2Kr)^{1/2} J_{t+1/2}(Kr)$$

$$= \frac{(Kr)^t}{1\cdot 3\cdot 5\cdots(2t+1)}\left[1 - \frac{(K^2r^2/2)}{1!(2t+3)} + \frac{(K^2r^2/2)^2}{2!(2t+3)(2t+5)} - \cdots\right] \qquad (18.143)$$

is the spherical Bessel function[69] of order t. Using the definition (11.41) for the scalar product of two tensor operators and then the Wigner-Eckart theorem (11.15), we find

$$\langle\gamma JM|\sum_m e^{i\mathbf{K}\cdot\mathbf{r}_m}|\gamma' J'M'\rangle = \sum_t (2t + 1)\, i^t$$

$$\times \sum_q (-1)^q C_{-q}^{(t)}(\mathbf{K})(-1)^{J-M}\begin{pmatrix} J & t & J' \\ -M & q & M' \end{pmatrix}\langle\gamma J\|\sum_m j_t(Kr_m)\mathbf{C}_m^{(t)}\|\gamma' J'\rangle \;. \qquad (18.144)$$

We shall be interested in the squared magnitude of this matrix element—which is given by the product of the right hand side of (18.144) with an identical expression, except having t and q replaced by (say) t′ and q′, i^t replaced by $(-i)^{t'} = i^{-t'}$, and $C_{-q}^{(t)}$ replaced by $C_{q'}^{(t')}$. If we sum this product over all values of M and M′, then from (5.15) the double sum of the product of 3-j symbols gives

$$(2t + 1)^{-1}\delta_{tt'}\delta_{qq'} \;.$$

[The triangle requirement δ(JtJ′) occurs also in the reduced matrix element, so we need not note it explicitly.] Then from (11.9) and (2.50)

$$\sum_{qq'} \delta_{tt'}\delta_{qq'}(-1)^{q+q'} C_{-q}^{(t)} C_{q'}^{(t')} = \sum_q |C_q^{(t)}|^2 = 1 \;,$$

[68]N. F. Mott and H. S. W. Massey, *The Theory of Atomic Collisions* (Clarendon Press, Oxford, 1965), 3rd ed., p. 22. This plane-wave expansion may be contrasted with the partial-wave expansion (18.107) involving distorted-wave functions.

[69]M. Abramowitz and I. A. Stegun, *Handbook of Mathematical Functions* (U.S. Govt. Printing Off., Washington, D.C., 1964), formula 10.1.2.

and we obtain[70]

$$gf_{JJ'}(K) \equiv \frac{\Delta E}{K^2} \sum_{MM'} |\langle \gamma JM | \sum_m e^{iK\cdot r_m} | \gamma' J'M' \rangle|^2$$

$$= \frac{\Delta E}{K^2} \sum_t (2t + 1) \langle \gamma J \| \sum_m j_t(Kr_m) C_m^{(t)} \| \gamma' J' \rangle^2 , \qquad (18.145)$$

where $g = 2J + 1$ is the statistical weight of the level γJ.

As usual—see, for example, (14.42)—the matrix element is evaluated in terms of eigenvector components and pure-coupling matrix elements:

$$\sum_{\beta\beta'} \langle \gamma J | \beta J \rangle \langle \beta J \| \sum_m j_t(Kr_m) C_m^{(t)} \| \beta' J' \rangle \langle \beta' J' | \gamma' J' \rangle , \qquad (18.146)$$

where $|\beta J\rangle$ and $|\beta' J'\rangle$ are pure-coupling (most conveniently, LS-coupled), single-configuration basis functions. This final reduced matrix element is identical with the electric multipole matrix element (14.78), except with r_m^t replaced by $j_t(Kr_m)$. Thus, evaluation of the reduced matrix element is identical with the evaluation of electric multipole matrix elements discussed in Sec. 14-11, except with the one-electron radial integral

$$\langle l_i \| r^t C^{(t)} \| l_k \rangle$$

for the jumping electron $l_i \rightarrow l_k$ replaced by

$$\langle l_i \| j_t(Kr) C^{(t)} \| l_k \rangle = (-1)^{l_i} [l_i, l_k]^{1/2} \begin{pmatrix} l_i & t & l_k \\ 0 & 0 & 0 \end{pmatrix} \int_0^\infty P_i(r) j_t(Kr) P_k(r) \, dr . \qquad (18.147)$$

Selection rules are identical with those for Et electromagnetic radiation, namely:

(a) $l_i + t + l_k$ must be even (t odd for parity-changing excitations, and t even for parity-conserving excitations), with

$$|l_i - l_k| \leqq t \leqq l_i + l_k , \qquad (18.148)$$

(b) $$|J - J'| \leqq t \leqq J + J' , \qquad (18.149)$$

(c) for matrix elements between pure LS-coupled basis functions,

$$S = S' ,$$
$$|L - L'| \leqq t \leqq L + L' . \qquad (18.150)$$

The conservation of S is a consequence of the absence of exchange effects, resulting from our use of incompletely antisymmetrized wavefunctions, as mentioned earlier. In the

[70]For simplicity, we abbreviate $f_{\gamma J, \gamma' J'}$ to $f_{JJ'}$.

absence of external magnetic and electric fields, the parity and ΔJ selection rules are valid even for the intermediate-coupling matrix element in (18.145).

In the limit $K \to 0$, we find by substituting (18.143) into (18.145) that (for parity-changing excitations)

$$g f_{JJ'}(0) = \frac{\Delta E}{3} \langle \gamma J \| \sum_m r_m C_m^{(1)} \| \gamma' J' \rangle^2 , \tag{18.151}$$

so that $f_{JJ'}(0)$ is just the optical dipole oscillator strength defined by (14.37), (14.31), and (14.24). The quantity $f_{JJ'}(K)$ is therefore called the generalized oscillator strength.[71,72] It satisfies for *any* K the same sum rules as does the optical oscillator strength—for example[71,72]

$$\sum_{\gamma' J'} f_{JJ'}(K) = N , \tag{18.152}$$

where N is the number of electrons in the atom, the optical analog being (14.104).

*18-13. PLANE-WAVE-BORN COLLISION STRENGTHS

The principal application of the generalized oscillator strength is to the calculation of electron-impact excitation cross sections and collision strengths. In the first Born approximation, the differential excitation cross section $I_{aa'}(\theta)$ in units of πa_0^2/steradian is given by[72,73]

$$\pi a_0^2 I_{aa'}(\theta) = \frac{m^2}{4\pi^2 \hbar^4} \frac{k'}{k} \left| H_{\alpha\alpha'} a_0^3 \frac{\hbar^2}{2 m a_0^2} \right|^2 \tag{18.153}$$

or

$$I_{aa'}(\theta) = \frac{1}{16\pi^3} \frac{k'}{k} \left| H_{\alpha\alpha'} \right|^2 , \tag{18.154}$$

where $H_{\alpha\alpha'}$ is the matrix element (18.140), in rydbergs; the factor a_0^3 in (18.153) arises from the volume integral dr^3 for the free electron, which in (18.140) is measured in Bohr units.

The integrated cross section

$$Q_{aa'} = 2\pi \int_0^\pi I_{aa'}(\theta) \sin\theta \, d\theta$$

is conveniently evaluated by replacing the θ integral with one over the magnitude of the momentum transfer K. From (18.141),

[71]H. Bethe, Ann. Physik 5, 325 (1930), Secs. 8 and 11.

[72]M. Inokuti, Rev. Mod. Phys. 43, 297 (1971).

[73]Ref. 68, Secs. XII,2 and XVI,7.

$$K^2 = (k' - k)^2 = (k')^2 + k^2 - 2kk' \cos\theta \tag{18.155}$$

and thus

$$K\,dK = kk' \sin\theta\, d\theta ,$$

whence

$$Q_{aa'} = \frac{1}{8\pi^2}\int_{K_{min}}^{K_{max}} \frac{K}{k^2}\,|H_{\alpha\alpha'}|^2\,dK = \frac{8}{k^2}\int_{K_{min}}^{K_{max}} K^{-3}\langle a|\sum_m e^{iK\cdot r_m}|a'\rangle^2\,dK , \tag{18.156}$$

where $a = |\gamma JM\rangle$ and $a' = |\gamma' J' M'\rangle$.

The total cross section $Q_{JJ'}$ (in units of πa_0^2), obtained by summing $Q_{aa'}$ over M' and averaging over M, and the corresponding collision strength $\Omega_{JJ'}$ are given from (18.145) by

$$\Omega_{JJ'}(\varepsilon) \equiv \varepsilon g Q_{JJ'}(\varepsilon) = \frac{8}{\Delta E}\int_{K_{min}}^{K_{max}} g f_{JJ'}(K)\, d(ln\, K) , \tag{18.157}$$

where ΔE is the excitation energy $\gamma J \to \gamma' J'$, $\varepsilon = k^2 \geqq \Delta E$ is the kinetic energy of the impacting electron in rydbergs, $g = 2J + 1$ is the statistical weight of the initial level γJ, and $f_{JJ'}(K)$ is the generalized oscillator strength (18.145). The limits of integration may be easily obtained from (18.139) and (18.155):

$$\begin{aligned} K_{min} &= k - k' = \varepsilon^{1/2} - (\varepsilon - \Delta E)^{1/2} , \\ K_{max} &= k + k' = \varepsilon^{1/2} + (\varepsilon - \Delta E)^{1/2} . \end{aligned} \tag{18.158}$$

As indicated in the preceding section, generalized oscillator strengths can be evaluated by making minor modifications in the atomic structure and spectra computer programs outlined in Figs. 8-1 and 16-1; evaluation of plane-wave Born (PWB) collision strengths from (18.157) is then a simple matter.

As already mentioned, the results may be expected to be most accurate for neutral or weakly ionized atoms and for high energies—or, more precisely, for large values of the reduced energy

$$X \equiv \frac{\varepsilon}{\Delta E} . \tag{18.159}$$

But high energies are just where the simple PWB calculations are most useful: for high energy (and hence large angular momentum) of the impacting electron, the partial-wave expansion (18.107) employed in distorted-wave (DW) and close-coupling (CC) calculations involves a huge number of terms ($l' = 50$ to 100) for adequate convergence. By contrast, the Bessel-function expansion employed in PWB calculations involves from (18.148) at most 3 or 4 terms for the usual excitations of interest (l_i and $l_k \leqq 2$ or 3).

Moreover, PWB results (for spin-allowed excitations) are surprisingly good even for highly ionized ions and relatively small X. Even though at threshold ($\varepsilon = \Delta E$, or $X = 1$)

$K_{max} = K_{min}$ from (18.158) so that the PWB collision strength is zero, whereas the true threshold value is nonzero, PWB values typically agree with DW and CC values to within 25 or 50 percent for X as small as 3 or 4.[74] Indeed, the empirical modification[75]

$$\Omega^m(X) = \Omega\left(X + \frac{3}{1+X}\right) \tag{18.160}$$

appears to give results Ω^m this accurate for most spin-allowed excitations in atoms more than once or twice ionized at *all* values $X \geqq 1$. Even when PWB absolute values are poor, relative cross sections for excitation of different M′ states can accurately predict the resulting polarization of spontaneous-decay radiation.[76]

[74]K. L. Bell and A. E. Kingston, Adv. At. Mol. Phys. 10, 53 (1974).

[75]Use of some such modification was suggested to the author by W. D. Robb (private communication, 1980). The detailed form of the prescription (18.160), somewhat different from that which he first proposed, was arrived at by comparison with DW and CC results for a considerable variety of ions and transitions. This prescription should be considered as only a tentative proposal, and should certainly not be used for diagnostic applications involving comparison of collision strengths for different individual $\gamma J \rightarrow \gamma' J'$ excitations. However, it may be adequate where only rough or total values (summed over J′ and averaged over J) are required, or to determine whether a given excitation is sufficiently important in a given application to warrant more accurate and elaborate calculations.

[76]W. D. Robb, J. Phys. B 7, 1006 (1974).

19

HIGHLY IONIZED ATOMS

19-1. BINDING ENERGIES IN ISOELECTRONIC SEQUENCES

When dealing with energies of an excited electron nl for a Rydberg series of a given atom, we found it convenient in Sec. 8-5 to discuss binding energies in terms of the hydrogen-like formula

$$E^{nl} = -\frac{Z_c^{\,2}}{(n^*)^2} = -\frac{Z_c^{\,2}}{(n-\delta)^2} \quad \text{Ry} , \tag{19.1}$$

where[1]

$$Z_c = Z - N + 1 \tag{19.2}$$

is the net charge of the nucleus plus the N − 1 electrons of the ion core, and n^* is an effective principal quantum number. This formula is useful because the quantum defect

$$\delta = n - n^* \tag{19.3}$$

is approximately independent of n.

For a given electron nl in an isoelectronic sequence, it is more convenient to employ a different hydrogen-like formula

$$E^{nl} = -\frac{(Z^*)^2}{n^2} = -\frac{(Z-S)^2}{n^2} = -\frac{(Z_c+p)^2}{n^2} \quad \text{Ry} , \tag{19.4}$$

where the effective nuclear charge

$$Z^* = Z - S \equiv Z_c + p \tag{19.5}$$

is equal to the actual nuclear charge minus the effective screening charge of the N − 1 electrons. The screening defect (or penetration)

[1]The net charge is frequently denoted by ζ, but we here use Z_c to avoid confusion with the spin-orbit parameter ζ_{nl}.

$$p = Z - Z_c - S = (N - 1) - S \tag{19.6}$$

is the amount by which the $N - 1$ core electrons fail to perfectly shield the electron nl from the nucleus. Evidently p and S are functions of nl: p tends to zero with increasing n and l (i.e., with decreasing penetration of the ion core by the electron nl).

The utility of the expression (19.4) arises from the empirical observation[2,3] that values of S inferred from experimental binding energies are approximately independent of Z, especially at large Z. The theoretical basis for this fact is semi-obvious from the concept of screening, but added insight may be obtained as follows.

For very large Z, the nuclear field becomes much stronger than the interelectronic fields, and so the electronic orbitals must be approximately hydrogenic in form. Thus, from Table 3-1, we see that the radial extent of the orbitals scales as Z^{-1}. The electron-nuclear energies therefore scale as $-2Z/r \propto Z^2$. Kinetic energies scale, from (6.18), as r^{-2} and thus also go as Z^2. Electron-electron Coulomb energies, on the other hand, go only as r^{-1} and hence scale as Z. Binding energies must therefore involve a term linear in Z as well as a quadratic term. Of course, screening effects cause the radial wavefunctions to be non-hydrogenic in form and so contributions from the various terms of the Hamiltonian do not scale in precisely the above manner. Examples are given in Figs. 19-1 to 19-3 for Hartree-Fock kinetic, electron-nuclear, and electron-electron contributions to one-electron binding energies. The various contributions can be represented fairly accurately in the form

$$\begin{aligned} E^{nl}_{kin} &\propto (Z - s_1)^2 , \\ E^{nl}_{e-n} &\propto -Z(Z - s_2) , \\ E^{nl,n'l'}_{e-e} &\propto (Z - s_3) . \end{aligned} \tag{19.7}$$

The sum of these contributions includes a constant term as well as terms in Z^2 and Z. The expression (19.4) includes all three of these terms,[4] though not with independent coefficients of the two lower-order terms. It should be noted that the term linear in Z—and hence the constant S—represents not only the zero-order electron-electron interactions (direct screening effects), but also first-order corrections to the hydrogenic kinetic and electron-nuclear energies (indirect screening effects).

Relativistic effects. The above discussion has been entirely non-relativistic, and relativistic corrections have yet to be added to the binding energy (19.4). From (7.58)-(7.59), the mass-velocity term is proportional to the expectation value of $(\varepsilon_i - V^i)^2$ and the Darwin term goes as V^i/r^2. Thus from the previous discussion we may expect both terms to increase as the fourth power of Z:

[2]An extensive discussion of empirical relations along isoelectronic sequences is given by B. Edlén, "Atomic Spectra," in S. Flügge, ed., *Handbuch der Physik* (Springer-Verlag, Berlin, 1964), Vol. XXVII, pp. 88-90 and 147-201.

[3]B. Edlén, Phys. Scr. 7, 93 (1973), also "The Term Analysis of Atomic Spectra," in I. A. Sellin and D. J. Pegg, eds., *Beam-Foil Spectroscopy* (Plenum, New York, 1976), Vol. I.

[4]We shall not discuss here the Z-expansion theory, which represents the total atom binding energy by an infinite series $E = Z^2 \sum_i E_i Z^{-i}$; see E. A. Hylleraas, Z. Physik 65, 209 (1930) and D. Layzer, Ann. Phys. 8, 271 (1959).

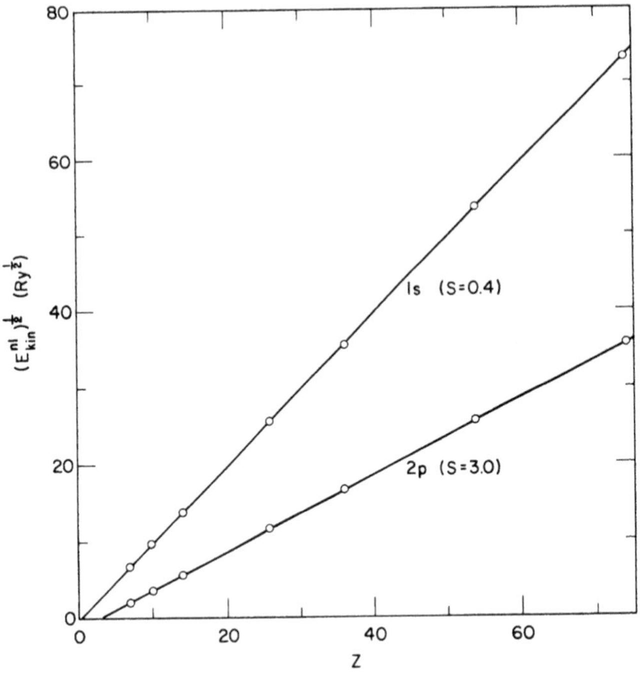

Fig. 19-1. Hartree-Fock values (circles) of the square root of the kinetic energy of a 1s and a 2p electron in the isoelectronic sequence N I $1s^2 2s^2 2p^3$. The solid curves are straight lines drawn (by eye) to best fit the computed points; their x-intercepts are the values of S indicated on the figure.

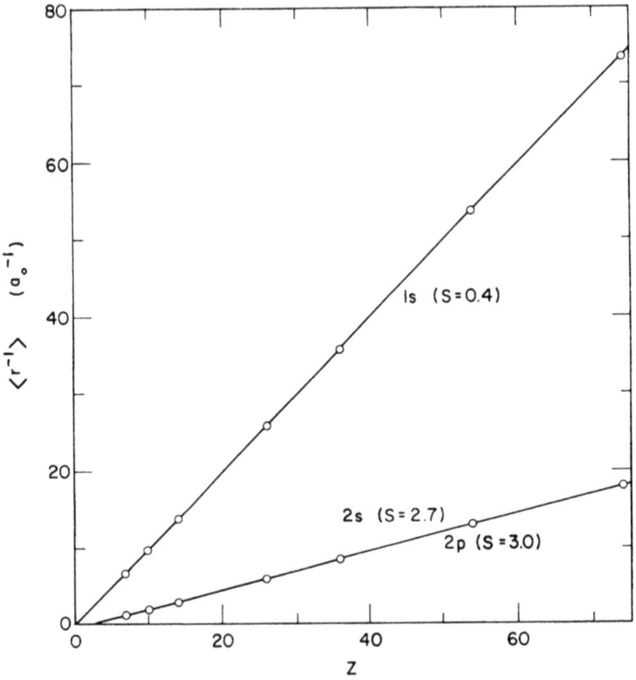

Fig. 19-2. Similar to Fig. 19-1, except expectation values of r^{-1}. Computed values for 2p lie only slightly below those for 2s, and could not be drawn separately on the scale of the figure.

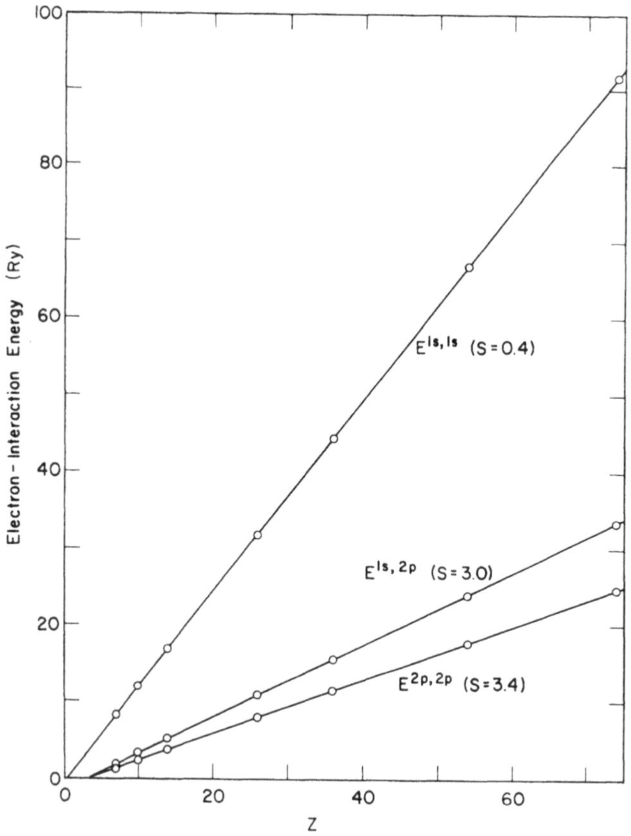

Fig. 19-3. Similar to Fig. 19-1, except electron-electron Coulomb energies for 1s and 2p electrons.

$$E^{nl}_{rel} \propto (Z - S)^4 , \tag{19.8}$$

where S has approximately the same value as in (19.4).[5] At very high Z, where relativistic modifications of the radial wavefunctions become appreciable, the relativistic contributions increase somewhat faster even than given by (19.8).

19-2. ENERGY-LEVEL STRUCTURES

At intermediate Z, where relativistic effects are not yet too important, the basic nature of the energy-level structure of an ion follows directly from (19.4). For two configurations that differ only in the excitation of one electron to a shell of different n, the energy difference between the configurations is[6]

[5]We make no attempt here to indicate more than the approximate magnitude of values of the various screening constants, as we wish to use Eqs. (19.4)-(19.8) only for semi-quantitative discussion of energy level systematics.

[6]In highly ionized atoms, the nuclear attraction is so much stronger than inter-electron forces that orbital relaxations upon electron excitation are relatively unimportant. Thus one-electron binding energies provide a particularly accurate measure of ionization and excitation energies (Sec. 6-3).

$$E^{n_2l_2} - E^{n_1l_1} = \frac{(Z - S_1)^2}{n_1^2} - \frac{(Z - S_2)^2}{n_2^2} . \tag{19.9}$$

This difference increases quadratically with Z, and thus remains roughly a constant fraction of the ionization energy.

On the other hand, the energy difference between two configurations of the same n is

$$E^{nl_2} - E^{nl_1} = \frac{(Z - S_1)^2 - (Z - S_2)^2}{n^2} = \frac{2(S_2 - S_1)}{n^2} [Z - (S_2 + S_1)/2] . \tag{19.10}$$

This increases only linearly with Z, and thus decreases rapidly relative to $\Delta n \neq 0$ excitation energies. The level structure thus tends to become hydrogen-like, energies depending strongly on n but only weakly on l.

This hydrogenic tendency may also be seen by combining (19.4) and (19.1) to give

$$\frac{Z_c + p}{n} = \frac{Z_c}{n - \delta}$$

or

$$\frac{p}{Z_c} = \frac{n}{n - \delta} - 1 = \frac{\delta}{n - \delta} ,$$

whence it follows that

$$E^{nl} = -\frac{(Z - N + 1)^2}{(n - \delta)^2} = -\frac{(Z - S)^2}{n^2} = -\frac{p^2}{\delta^2} \tag{19.11}$$

and also that

$$\delta = \frac{pn}{Z_c + p} = \frac{pn}{Z - S} = \frac{(N - 1 - S)n}{Z - S} ; \tag{19.12}$$

these equations show not only that δ tends to zero along an isoelectronic sequence as $(Z - S)^{-1}$, but also that p tends to zero up a Rydberg series as $(n - \delta)^{-1} = (n^*)^{-1}$.

An example of the tendency toward hydrogenic level structure is given in Fig. 19-4, which shows center-of-gravity energies for some of the low-lying configurations of the N I isoelectronic sequence. Energies are plotted relative to the ionization limit $1s^22s^22p^2$, and have been scaled as $(Z - N + 1)^{-2}$ (i.e., in a manner appropriate to the case $S = N - 1$, or zero penetration p). Consider first the solid curves, representing configurations formed by adding an electron nl to the ion-core ground configuration. For large n and l (3d and 4f) the curves are essentially horizontal straight lines, as would be expected for non-penetrating electrons for which $S = N - 1$ is indeed appropriate at low Z and which do not suffer appreciable relativistic effects at high Z. The semi-hydrogenic level structure at intermediate and high Z is quite clear. At low Z, there are large departures from hydrogenic behavior, because of sizable core penetration by the s and p electrons—especially, of course, by the 2p electron; in particular, it may be noted that in N I the 4s configuration lies (slightly) below 3d. In addition, a slight relativistic depression of the 3s configuration

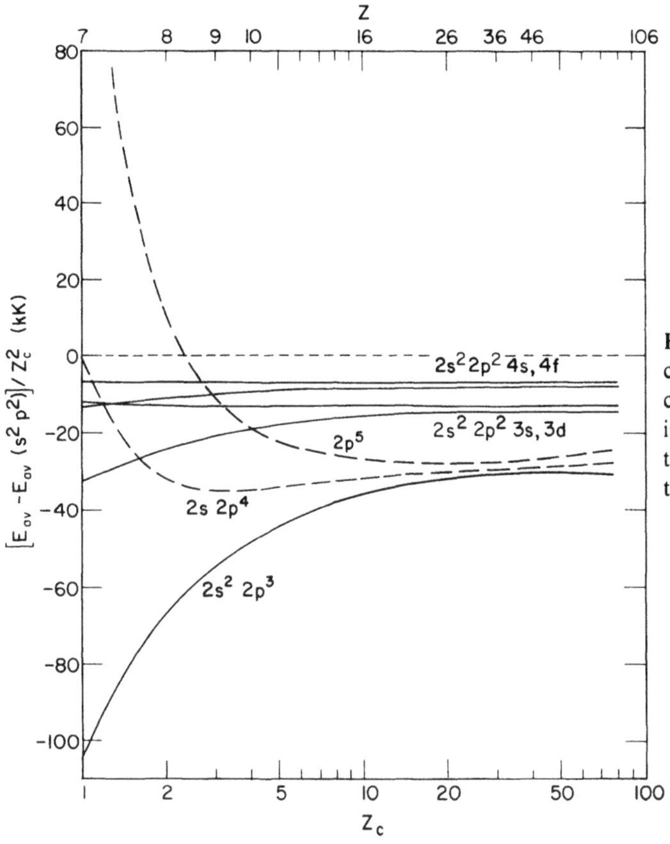

Fig. 19-4. Theoretical (HFR) center-of-gravity energies of low configurations of the N I isoelectronic sequence, relative to the center-of-gravity ionization limit $1s^2 2s^2 2p^2$.

may be noted at large Z; relativistic effects are too small for larger n or *l* to be seen on the scale of the figure.[7]

The dashed curves in Fig. 19-4 represent configurations derived from the ground configuration by excitation of one or both of the 2s electrons to the 2p subshell. At moderate to high Z, the (Z_c^{-2}-scaled) excitation energies are hydrogenically small, but at low Z they become very large because of the large changes in nuclear screening; at the lowest Z, they even lie above the ionization limit. Above Z ≅ 30, the decrease in the magnitude of the relativistic energy upon promoting an electron from 2s to 2p results in pronounced increases in the (scaled) excitation energies.

Energy-level structures within a configuration are determined mainly by the electron-electron Coulomb parameters $F^k(i,j)$ and $G^k(i,j)$, and by the spin-orbit parameters ζ_l. As indicated previously, the Coulomb energies scale linearly with Z, or approximately

$$F^k, G^k \propto Z - S . \tag{19.13}$$

[7]The tendency (with increasing Z) toward a hydrogenic level structure, and the increase in relativistic effects, can also be seen for inner shells of neutral atoms; see Figs. 8-10 and 8-11.

However, the spin-orbit integral (10.9) scales as V^1/r^2, and thus approximately

$$\zeta_l \propto (Z - S)^4 \; ; \tag{19.14}$$

at very high Z where relativistic effects on the wavefunctions become significant, ζ_l may increase even faster than Z^4. At low Z where Coulomb interactions dominate, energy spreads within a configuration decrease as $(Z - S)^{-1}$ relative to the ionization energy or to $\Delta n \neq 0$ excitation energies, but are essentially constant relative to $\Delta n = 0$ excitation energies. Numerous empirical examples of the latter relation are given by Edlén,[3,2] and a theoretical example is shown in Fig. 19-5 (for $Z < 20$). At large Z, the spin-orbit interactions dominate, increasing as they do faster than anything except the relativistic contribution (19.8) to configuration centers of gravity. As a consequence, coupling conditions change rapidly from nearly pure LS at low Z to nearly pure jj at high Z; the dividing line between more nearly LS and more nearly jj coupling occurs at about $Z = 30$ in the $2s^22p^3$ example in Fig. 19-5.

It may be noted that the configurations $2s^22p^3$ and $2p^5$ would overlap beyond about $Z = 54$ were it not for the strong relativistic increase in the center-of-gravity energy of $2p^5$,

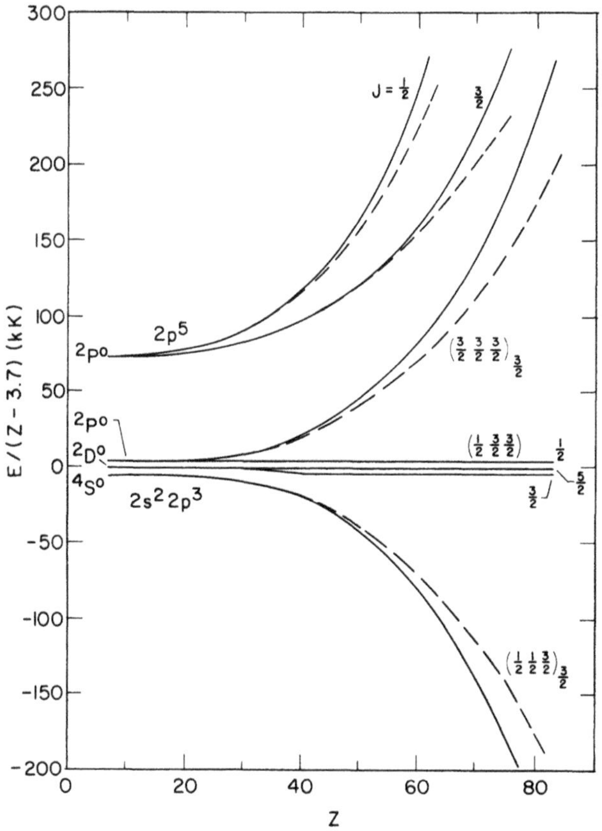

Fig. 19-5. Theoretical energy levels of the complex $2s^22p^3 + 2p^5$ in the N I isoelectronic sequence, plotted relative to the center of gravity of $2s^22p^3$. Dashed curves: HF plus relativistic center-of-gravity energy corrections; solid curves: HFR. Configuration-interaction perturbations are too small to be seen on the scale of the figure.

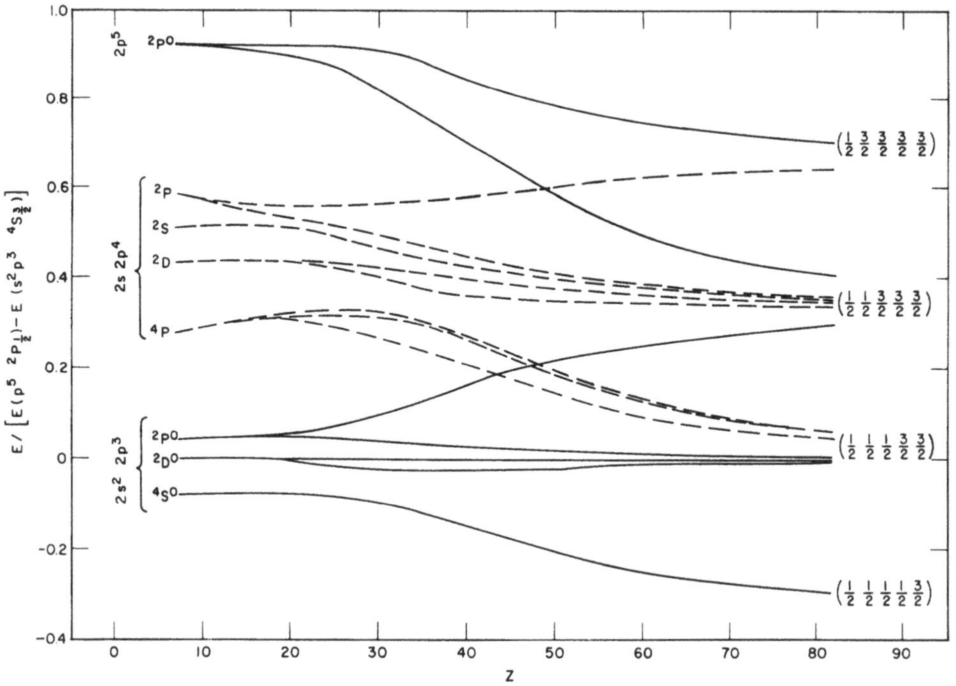

Fig. 19-6. Theoretical (HFR) energy levels of the n = 2 complexes of the N I isoelectronic sequence (solid curves, odd parity; dashed curves, even parity). Note that the Pauli exclusion principle allows at most two equivalent p electrons with j = 1/2 and a maximum of four such electrons with j = 3/2; thus jj limits with all five j values equal to 1/2 or all five equal to 3/2 are not possible.

which begins to become important at about Z = 15. Relativistic effects on the radial wavefunctions themselves do not become significant until Z = 35 or so.

The transition to pure jj coupling is shown even more clearly in Fig. 19-6, where energies have been scaled to make the distance from lowest level of s^2p^3 to highest level of p^5 unity for all Z. Levels of the even-parity configuration sp^4 have also been added. At high Z, energies depend primarily on j and only secondarily on *l*, approaching the hydrogenic situation illustrated in Fig. 3-9.[8] It should be noted that the details of this approach to the hydrogenic limit are completely unverified experimentally. In spite of the great progress made during recent years in the analysis of spectra of highly ionized ions,[9] the N I sequence is known only as far as Ni XXII.[10] It is doubtful that experimental values of the n = 2 levels will be available beyond Z = 50 for decades to come.

[8]Figures similar to Fig. 19-6 for all of the $2s^m2p^n$ isoelectronic sequences are given by E. Ya. Kononov and U. I. Safronova, Opt. Spectrosc. 43, 1 (1977).

[9]B. C. Fawcett, "Recent Progress in the Classification of the Spectra of Highly Ionized Atoms," Adv. At. Mol. Phys. 10, 223 (1974).

[10]B. C. Fawcett, At. Data Nucl. Data Tables 16, 135 (1975); K. D. Lawson and N. J. Peacock, J. Phys. B 13, 3313 (1980).

The N I sequence is of course relatively simple. Energy structures rapidly become vastly more complicated in detail for sequences involving $n = 3, 4, \cdots$ ground-configuration electrons.[3] The basic features are, however, much the same, except that the relativistic nlj limit is never closely approached for other than s and p electrons because of the small values of ζ_{nl} for $l > 1$.[11] [Screening-constant relations like (19.4) are highly inaccurate for excited d and f configurations in the first few members of isoelectronic sequences, such as those of Ar I and Cs I, that begin shortly before the transition and rare-earth elements, because of the sudden collapse of the d and f orbitals upon increasing the nuclear charge (cf. Sec. 8-6).[12] These complications have, however, largely vanished by the fourth or fifth member of the sequence.]

19-3. CONFIGURATION INTERACTIONS

Like all other electron-electron Coulomb energies, configuration-interaction Coulomb radial integrals may be expected to scale linearly with Z: In the hydrogenic-wavefunction approximation $P^Z(r) = Z^{1/2}P^H(Zr)$,

$$R^k(l_1l_2, l_3l_4) = Z^2\int_0^\infty\int_0^\infty \frac{r_<^k}{r_>^{k+1}} P_1^H(Zr_1)P_2^H(Zr_2)P_3^H(Zr_1)P_4^H(Zr_2)\, dr_1\, dr_2$$

$$= Z\int_0^\infty\int_0^\infty \frac{(Zr_<)^k}{(Zr_>)^{k+1}} P_1^H P_2^H P_3^H P_4^H\, d(Zr_1)\, d(Zr_2)$$

$$= Z\,(R^k)^H \,, \tag{19.15a}$$

or somewhat more accurately

$$R^k(l_1l_2, l_3l_4) \propto (Z - S) \,. \tag{19.15b}$$

Some examples showing this behavior are given in Fig. 19-7. [For some interactions, R^k may vary considerably more or less rapidly than indicated by (19.15), because of a strong monotonic increase or decrease with Z in the cancellation factor (16.36); cf. Fig. 18-3. However, this implies small values of the cancellation factor for most values of Z, and values of R^k are then usually so small that configuration-interaction effects are unimportant. Thus (19.15b) is sufficiently accurate for present purposes.]

For an interaction between two configurations belonging to different complexes, it follows from (19.9) that the energy difference between the interacting levels varies at high Z as $(Z - S)^2$—S here being a weighted average of S_1 and S_2. The magnitude of the energy perturbation then varies from (10.38) as

[11] Cf. the high-Z values of x-ray-determined energy levels given, for example, by W. Lotz, J. Opt. Soc. Am. 60, 206 (1970).

[12] R. D. Cowan, J. Opt. Soc. Am. 58, 924 (1968).

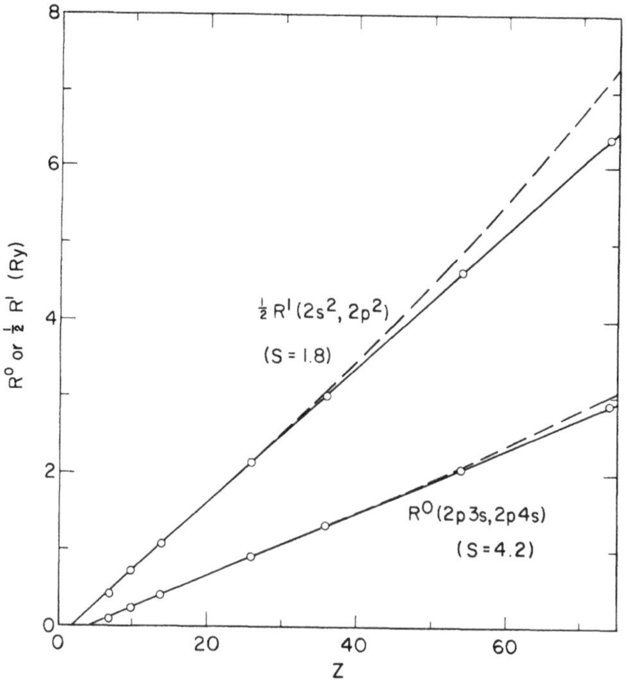

Fig. 19-7. Theoretical values of configuration-interaction radial Coulomb integrals for the interactions $2s^22p^3 + 2p^5$ and $2s^22p^23s + 2s^22p^24s$ in the N I isoelectronic sequence. Circles, HF; dashed curves, HFR.

$$\Delta E \propto \frac{(R^k)^2}{E_2 - E_1} \propto \frac{(Z-S)^2}{(Z-S)^2} \cong \text{constant} . \tag{19.16}$$

The ratio of the perturbation to the energy separation of the interacting levels, and also the amount of configuration mixing Δ^2 from (10.39), go as

$$\frac{(R^k)^2}{(E_2 - E_1)^2} \propto (Z - S)^{-2} , \tag{19.17}$$

and thus decrease rapidly with increasing Z. Even the ratio of the perturbation to the energy spread within each configuration, when governed primarily by Coulomb interactions (19.13), decreases fairly rapidly:

$$\frac{\Delta E}{F^k} \propto (Z - S)^{-1} . \tag{19.18}$$

Thus interactions between configurations belonging to different complexes tend to become unimportant for highly ionized atoms. (Indeed, most interactions are strong only in cases of accidental near-degeneracy, as in the cases of $2p^23d - 2p^24s$ near N I and $2s2p^4 - 2s^22p^23s$ near O II in Fig. 19-4, and these occur mainly for rather low ionization stages.[13]) This is the reason that values of the semi-empirical scaling factors for F^k and G^k discussed in Sec. 16-2 increase toward unity with increasing degree of ionization. The unimportance of interactions between distant configurations shows up also in the fact that

[13]B. Edlén, ref. 2, Sec. 27.

the approximate correlation energy (7.72) becomes essentially constant at high Z, and therefore negligible compared with excitation energies.

For interactions between two configurations of the same complex, energy perturbations are seen from (19.10) to vary as

$$\Delta E \propto \frac{(R^k)^2}{E_2 - E_1} \propto \frac{(Z-S)^2}{(Z-S)} = (Z - S) , \tag{19.19}$$

and the ratio of this to either the distance between configurations or the energy spread within each configuration (where determined mainly by Coulomb interactions) is

$$\frac{\Delta E}{E_2 - E_1} \cong \text{constant} . \tag{19.20}$$

Thus interactions within a complex are about equally important for all members of an isoelectronic sequence. In general, they must be included explicitly in all numerical calculations (though in the N I $2s^22p^3 - 2p^5$ sequence, wavefunction mixings are always less than about one percent, and energy perturbations are too small to be seen on the scale of Figs. 19-5 and 19-6). Appropriate scaling factors for the R^k are equal to or less than those for the F^k and G^k—usually about 0.6 to 0.9, except that values as low as 0.25 seem to be indicated in actinide spectra.[14]

Inasmuch as an increase in Z decreases the energy spread of levels within a configuration (relative to binding energies), and also decreases configuration interactions between distant configurations, we may expect our use of a single set of radial wavefunctions for all states of a configuration to be a particularly good approximation for highly ionized atoms.[15] The use of LS-dependent HF calculations (Sec. 16-8) for such atoms[15] represents an unnecessarily complicated and time-consuming approach; moreover, this procedure is usually physically inappropriate, because coupling conditions tend to lie closer to jj than to LS coupling.

19-4. RADIAL MULTIPOLE INTEGRALS

The isoelectronic-sequence variation in values of radial dipole integrals has already been discussed in Sec. 14-17. In the same manner used there, we see that

$$|P_{ii}^{(t)}|^2 \equiv \langle n_i l_i \| r^t \| n_j l_j \rangle^2 \propto (Z - S)^{-2t} . \tag{19.21}$$

The associated scalings of oscillator strengths and radiative transition probabilities were discussed in Secs. 14-17 and 15-2 for electric dipole and quadrupole transitions, respectively. It must be remembered, however, that the scaling (19.21)—like that for radial integrals R^k—may be quite inaccurate because of large changes with Z in cancellation effects. Examples have been mentioned in Secs. 14-15 and 14-17, and shown in Fig. 18-3.

[14]R. D. Cowan, L. J. Radziemski, Jr., and V. Kaufman, J. Opt. Soc. Am. 64, 1474 (1974); G. E. Bromage, R. D. Cowan, and B. C. Fawcett, Phys. Scr. 15, 177 (1977); K. Rajnak, Phys. Rev. A 14, 1979 (1976).

[15]L. F. Chase, W. C. Jordan, J. D. Perez, and R. R. Johnston, Phys. Rev. A 13, 1497 (1976), especially Sec. III C ii.

19-5. SPECTRA

It is appropriate to divide electric-dipole transitions of highly ionized atoms into two classes—arrays in which the jumping electron changes the value of its principal quantum number, and those in which it does not.

$\Delta n \neq 0$ transitions. At low to intermediate values of Z_c, where relativistic effects are not too important, ΔE_{av} increases as $(Z - S)^2$ whereas intra-configuration splittings increase only as $(Z - S)$. The lines of a transition array therefore tend to form a compact group covering a rather narrow wavelength range about the mean value $\lambda_o = hc/\Delta E_{av}$. As an example, the lines of the array Fe XIX $2p^4 - 2p^3 4d$ are computed to lie within the range 10.4 to 11.1 Å, with $\lambda_o = 10.8$; the *strong* lines of the array are computed (and observed[16]) to lie within the even more limited range 10.5 to 10.9 Å. Thus, the lines of various $\Delta n \neq 0$ transition arrays for one or more ionization stages of a given element tend to be clustered in non-overlapping, fairly easily recognizable groups;[16] the array corresponding to an observed group of lines can be identified with fair reliability simply from the calculated value of $\lambda_o = hc/\Delta E_{av}$.[16,17] On the other hand, the lines within a group lie so close together that the limited resolution available at these short wavelengths usually makes it difficult or impossible to resolve and identify individual spectrum lines. Even at best, experimental difficulties often preclude the observation of the weak lines of an array, thereby making impossible the cross-checks required for unambiguous empirical identification of spectrum lines and energy levels. The establishment of level values for highly ionized atoms therefore relies very heavily on theoretical calculations, and must usually be viewed with considerable reservation: although λ_o (and therefore the wavelengths of individual lines) are predictable with an accuracy of about 0.1-0.5%, wavelength differences within an array are subject to the five or ten percent uncertainty in the values of F^k and G^k.

For very high ionization stages, relativistic effects become more important and nullify the above remarks in some respects. However, the wavelengths become so short and experimental difficulties so great that the matter is at present of mainly academic interest.

$\Delta n = 0$ transitions. For transitions that involve no change in principal quantum number of the jumping electron, Coulomb contributions to ΔE_{av} increase only linearly, Eq. (19.10), instead of quadratically with Z. At medium to high Z, level separations within each configuration are comparable with ΔE_{av} (cf. Fig. 19-6), and so wavelengths within an array cover a range comparable with λ_o. For example, in Fe XX $2s^2 2p^3 - 2s2p^4$ computed wavelengths range from 76 to 231 Å, with $\lambda_o = 122$ Å; even just the strong lines are computed (and observed[18]) to extend over the range 84 to 140 Å. The individual lines lie far enough apart and (except for very high Z_c) at such long wavelengths as to be easily resolvable. On the other hand, the lines of an array cover such a wide wavelength range that different arrays of an ion, or corresponding arrays of different ionization stages of an atom, tend to overlap and make identifications difficult;[18] this is especially true since ΔE_{av} tends to be independent of ionization stage (ref. 12, especially Figs. 2 and 12). Also,

[16] G. E. Bromage, B. C. Fawcett, and R. D. Cowan, Mon. Not. R. Astron. Soc. 178, 599 (1977).

[17] P. G. Burkhalter, U. Feldman, and R. D. Cowan, J. Opt. Soc. Am. 64, 1058 (1974).

[18] G. A. Doschek, U. Feldman, R. D. Cowan, and L. Cohen, Astrophys. J. 188, 417 (1974).

theoretical calculations are of less help because of their lower accuracy: ΔE_{av} is a smaller fraction of the ionization energy than is true for $\Delta n \neq 0$ transitions (and relativistic corrections are much more important), and so the uncertainty in ΔE_{av} may be one or two percent; since energy separations within a configuration are comparable with ΔE_{av}, the accuracies of individual computed wavelengths are not much better than the five to ten percent uncertainties in the F^k and G^k.

In spite of all this, theoretical calculations can be of considerable help in the identification of observed lines, provided the calculations are done in full intermediate coupling, with inclusion of relativistic energy corrections and of configuration interactions within the complex. Calculations simply of ΔE_{av} can be rather misleading (even with relativistic corrections included) because they usually give a poor estimate of the center of gravity of the intensity of the array: Firstly, the high-strength lines tend to be the high-energy (short-wavelength) lines;[12,19] an extreme example is the array $3p^6 - 3p^5 3d$, where in pure LS coupling the only non-zero-strength line is $^1S_0 - {}^1P^o_1$, which in Fe IX has a transition energy 30% greater than ΔE_{av}.[12] Secondly, in high density sources where line intensities are determined by emission transition probabilities, intensity-averaged transition energies are strongly affected by the fact that $A \propto \sigma^3$. Thus in the above Fe XX example, ΔE_{av} lies at 818 kK, the center of gravity of the line strength lies at about 917 kK, and the centers of gravity of gf and gA lie at 931 and 982 kK, respectively. Thus use of ΔE_{av} to predict the intensity center of gravity of the transition array would result in an error of about −16%, and use of ΔE_{av} for $\bar{\sigma}$ in an expression analogous to (14.97) for the mean transition probability $\bar{A}$ would result in an error of about $(818^3 - 953^3)/953^3 = -37\%$. (In some arrays, the latter error is more than −60%; i.e., values of $\bar{A}$ computed using ΔE_{av} for $\bar{\sigma}$ are too small by factors greater than 2.5!)

Forbidden transitions. In low density sources such as the solar corona or tokamaks, spectrum line intensities are determined by collisional-excitation (or by recombination) rates rather than by radiative decay rates. Electric-dipole-forbidden radiative transitions may then become as prominent as dipole-allowed decays. The most important forbidden lines arise from intra-configuration magnetic-dipole transitions. These lines are particularly useful for diagnostic purposes because their relatively long wavelengths (sometimes in the near ultraviolet, or even in the visible or near infrared) make possible high resolution measurements, and therefore determination of ion-temperature measurements via Doppler broadening.[20,21] Such lines, however, provide an extreme example of $\Delta n = 0$ transitions, and it is very difficult to predict their wavelengths accurately. For relatively low Z and intra-LS-term transitions, or for very high Z, transition energies depend almost entirely on values of the spin-orbit parameter ζ, and are therefore in favorable cases accurate to one or

[19] C. Bauche-Arnoult, J. Bauche, and M. Klapisch, J. Opt. Soc. Am. 68, 1136 (1978), Phys. Rev. A 20, 2424 (1979).

[20] S. Suckewer and E. Hinnov, Phys. Rev. Lett. 41, 756 (1978), Phys. Rev. A 20, 578 (1979).

[21] G. A. Doschek and U. Feldman, J. Appl. Phys. 47, 3083 (1976). Doppler-width measurements have also been made in the x-ray region: M. Bitter, S. von Goeler, R. Horton, M. Goldman, K. W. Hill, N. R. Sauthoff, and W. Stodiek, Phys. Rev. Lett. 42, 304 (1979).

two percent. For other cases, transition energies depend appreciably on the Coulomb integrals as well as the spin-orbit integrals, and *ab initio* wavelength predictions cannot be depended on to better than five percent or so. Fortunately, the wavelength density of spectrum lines for highly ionized atoms at the long wavelengths in question is usually small enough that theoretical predictions can still be of some aid in line identification, particularly if time or spatial intensity variations within the source provide clues to the element and ionization stage.[20,22,23] Inter-configuration electric-quadrupole lines, resulting from $3d^w - 3d^{w-1}4s$ transitions in Mo XV-XVI, have also been observed in tokamak spectra; predicted wavelengths for these $\Delta n \neq 0$ transitions are quite accurate.[24]

19-6. X-RAY SPECTRA

Very few experimental results are available for atoms more than about 30-fold ionized, and it is therefore difficult to check the accuracy of theoretical calculations, especially for $\Delta n = 0$ transitions (where relativistic corrections are important and of uncertain accuracy). It is, however, possible to obtain some checks by means of comparison with extensive and fairly accurate data on x-ray spectra. We therefore digress in this section to discuss such spectra.

By x-ray spectra we here mean radiation resulting from inner-shell transitions in (singly ionized) atoms: A single vacancy is produced in one of the tightly bound subshells $n_1 l_1$ of a neutral atom (most commonly as the result of electron-collisional ionization from this subshell); this vacancy is then filled as the result of a radiative transition $n_2 l_2 \rightarrow n_1 l_1$, which then leaves a vacancy in the less tightly bound subshell 2; this second vacancy may then be filled through another radiative transition $n_3 l_3 \rightarrow n_2 l_2$ from a still less tightly bound subshell 3; etc.

Except for the noble gases and mercury, most experimental observations are made on molecules such as F_2 and Cl_2, or on elemental solids. For present purposes, we shall neglect all molecular and solid-state effects, and make theoretical calculations only for an isolated, singly ionized atom. We shall also neglect interactions between the inner-shell vacancy and electrons in any partially filled valence subshells. The levels of such an ion are then characterized solely by the quantum numbers nlj of the vacancy. In x-ray spectroscopy, the vacancies $1s_{1/2}$, $2s_{1/2}$, $2p_{1/2}$, $2p_{3/2}$, $3s_{1/2}$, $\cdots$ are usually denoted by K, L_I, L_{II}, L_{III}, M_I, $\cdots$. (For brevity, Arabic instead of Roman subscripts are sometimes used.) X-ray transitions are denoted by a combination of the above symbols for the two levels involved, or by an empirical symbol consisting of the shell for the upper level plus a single or double subscript; for example, $K_{\alpha_1} = KL_{III} = 1s_{1/2} - 2p_{3/2}$, $K_{\alpha_2} = KL_{II} = 1s_{1/2} - 2p_{1/2}$, K_α = the unresolved KL doublet. A qualitative energy-level diagram is given in Fig. 19-8, and

[22] W. M. Burton, A. Ridgeley, and R. Wilson, Mon. Not. R. Astron. Soc. 135, 207 (1967).

[23] B. Edlén, Solar Phys. 9, 439 (1969); L. Å. Svensson, J. O. Ekberg, and B. Edlén, Solar Phys. 34, 173 (1974).

[24] M. W. D. Mansfield, N. J. Peacock, C. C. Smith, M. G. Hobby, and R. D. Cowan, J. Phys. B 11, 1521 (1978); M. Klapisch, J. L. Schwob, M. Finkenthal, B. S. Fraenkel, S. Egert, A. Bar-Shalom, C. Breton, C. DeMichelis, and M. Mattioli, Phys. Rev. Lett. 41, 403 (1978).

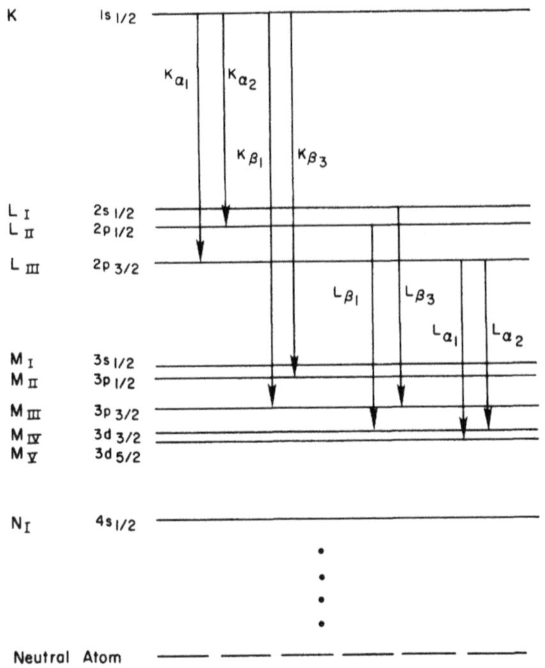

Fig. 19-8. Grotrian diagram for x-ray transitions in singly ionized atoms, neglecting level splittings resulting from outer unfilled subshells and solid-state effects. The x-ray levels are denoted by K, L_I, L_{II}, $\cdots$, and the corresponding inner-shell vacancies by $1s_{1/2}$, $2s_{1/2}$, $2p_{1/2}$, $\cdots$ in order of decreasing energy. Note that for $l < 2$ the energy depends hydrogenically more on j than on l.

a more complete set of transitions[25] is given in Table 19-1. Note that the usual electric-dipole selection rules $\Delta l = \pm 1$ and $\Delta j = 0, \pm 1$ are not necessarily obeyed, because of the influence of environmental (condensed-phase) effects.

Extensive tables of x-ray energy levels are available.[26-28]

19-7. COMPARISON OF HIGHLY IONIZED AND X-RAY SPECTRA

In the upper portion of Fig. 19-9, center-of-gravity energy differences are plotted for the Ni I sequence $3d^{10} - 3d^9 4p$ and for the x-ray transition $M_{IV,V}N_{II,III}$, as computed in various single-configuration Hartree-Fock approximations. The data are presented in the form of Moseley diagrams, since from (19.9) $[\Delta E_{av}]^{1/2}$ should vary approximately linearly with Z in the non-relativistic approximation. Relativistic effects produce (or, rather, increase) a slight downward curvature, but are so small that the HFR method gives results essentially the same as HF plus first-order perturbation energy corrections (the difference

[25] J. A. Bearden, Rev. Mod. Phys. 39, 78 (1967). Reprinted as NSRDS-NBS 14 (U.S. Govt. Printing Off., Washington, D.C., 1967).

[26] J. A. Bearden and A. F. Burr, Rev. Mod. Phys. 39, 125 (1967). Reprinted as NSRDS-NBS 14 (U.S. Govt. Printing Off., Washington, D.C., 1967).

[27] K. Siegbahn et al., *ESCA—Atomic, Molecular, and Solid State Structure Studied by Means of Electron Spectroscopy* (Almqvist and Wiksells, Uppsala, 1967).

[28] W. Lotz, J. Opt. Soc. Am. 60, 206 (1970); data given here have been corrected to correspond to the free ion.

TABLE 19-1. NOMENCLATURE FOR X-RAY TRANSITIONS[a]

		Upper level								
Lower level		K	L_I	L_{II}	L_{III}	M_I	M_{II}	M_{III}	M_{IV}	M_V
K	$1s_{1/2}$									
L_I	$2s_{1/2}$									
L_{II}	$2p_{1/2}$	α_2								
L_{III}	$2p_{3/2}$	α_1								
M_I	$3s_{1/2}$		---	η	l					
M_{II}	$3p_{1/2}$	β_3	β_4	---	t					
M_{III}	$3p_{3/2}$	β_1	β_3	β_{17}	s					
M_{IV}	$3d_{3/2}$	β_5^{II}	β_{10}	β_1	α_2					
M_V	$3d_{5/2}$	β_5^{I}	β_9	---	α_1					
N_I	$4s_{1/2}$		---	γ_5	β_6		---	---		
N_{II}	$4p_{1/2}$	β_2^{II}	γ_2	---	---	---			ζ_2	
N_{III}	$4p_{3/2}$	β_2^{I}	γ_3	---	---	---			---	ζ_1
N_{IV}	$4d_{3/2}$	β_4	---	γ_1	β_{15}		---	γ_2		
N_V	$4d_{5/2}$	β_4	γ_{11}	---	β_2			γ_1		
N_{VI}	$4f_{5/2}$		---	v	u′				β	α_2
N_{VII}	$4f_{7/2}$		---		u					α_1
O_I	$5s_{1/2}$		---	γ_8	β_7			---		
O_{II}	$5p_{1/2}$	---	γ_4'	---	---				---	
O_{III}	$5p_{3/2}$	---	γ_4	---	---	---				
O_{IV}	$5d_{3/2}$		---	γ_6	β_5		---	---		
O_V	$5d_{5/2}$		---		β_5			---		

[a]Dashes indicate observed transitions that have not been given special symbols.

in these values of ΔE_{av} being less than one percent even for $_{82}$Pb). Available experimental data[26,28,29] verify the HFR results to within the experimental uncertainties of about 0.1%.

For the $\Delta n = 0$ transition 4s − 4p, a Moseley diagram (bottom of Fig. 19-9) gives an S-shaped curve because at low atomic number it is ΔE_{av} [not $(\Delta E_{av})^{1/2}$] that increases linearly

[29]Sources of data for the Ni I sequence are: Br VIII, AEL; Mo XV, E. Alexander, M. Even-Zohar, B. S. Fraenkel, and S. Goldsmith, J. Opt. Soc. Am. 61, 508 (1971); Sn XXIII, P. G. Burkhalter, U. Feldman, and R. D. Cowan, J. Opt. Soc. Am. 64, 1058 (1974); I XXVI, C. L. Cocke et al., Phys. Rev. A 12, 2413 (1975); Gd XXXVII, P. G. Burkhalter, D. J. Nagel, and R. R. Whitlock, Phys. Rev. A 9, 2331 (1974); Ag XX, W XLVII, and Au LII, P. G. Burkhalter, C. M. Dozier, and D. J. Nagel, Phys. Rev. A 15, 700 (1977), and private communication. In all cases, only two or three levels of $3d^9 4p$ are known; the value of ΔE_{av} was inferred by comparing the experimental levels with theoretically computed ones.

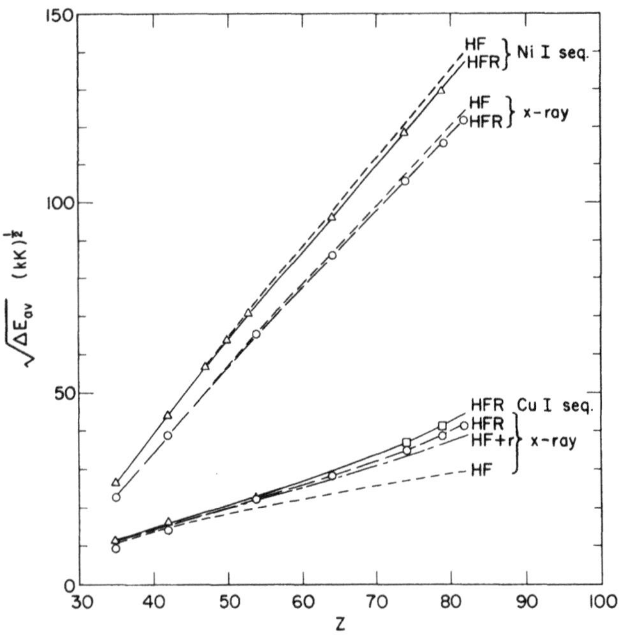

Fig. 19-9. Comparison of optical (valence-shell) and x-ray (inner-shell, singly-ionized-atom) transition energies. Top: The $\Delta n = 1$ 3d – 4p transition, which in the Ni I isoelectronic sequence is $3d^{10} - 3d^9 4p$; bottom: The $\Delta n = 0$ 4s – 4p transition, which in the Cu I sequence is $3d^{10}4s - 3d^{10}4p$. All curves represent theoretical single-configuration Hartree-Fock results, either without (HF) or with (HF + r) relativistic perturbation energy corrections, or including approximate relativistic effects in the radial wavefunctions (HFR). The triangles and circles represent experimental values; the squares are DHF theoretical values, which agree closely with recent experimental results (ref. 30).

with Z [Eq. (19.10)], and because for high atomic number the relativistic contribution is both large and positive. (In the 3d – 4p transition, the relativistic contribution was negative because the upper level had the smaller *l* and hence suffered the greater relativistic depression.) Available data[30] for optical transitions in the Cu I $3d^{10}4s - 3d^{10}4p$ sequence have recently been extended to $Z = 74$, where they agree with both HFR and DHF values of ΔE_{av} within two percent. Thus HFR gives considerably better accuracy than HF + r for both the optical and the x-ray transitions. (The poor agreement between theory and experiment for $_{35}$Br II $4s^2 4p^4 - 4s4p^5$, which is not really an x-ray transition at all, is due to a very strong $4s4p^5 - 4s^2 4p^3 4d$ interaction, neglected in the present calculation. The poor agreement for $_{42}$Mo II $4s^2 4p^5 4d^6 - 4s4p^6 4d^6$ may similarly be due to a neglected $4s4p^6 4d^6 - 4s^2 4p^4 4d^7$ interaction, or to solid state effects on the x-ray transition.)

For the 3s – 3p transitions shown in Fig. 19-10, relativistic effects are all important, constituting 75% of ΔE_{av} at $Z = 80$. Experimental data[31] at present extend only to $_{46}$Pd XXXVI, where there is no appreciable difference between HFR and HF + r. However, the x-ray data show HFR to be much more accurate than HF + r above $Z \cong 50$. Above $Z \cong 70$ even HFR becomes inaccurate, and full DHF calculations are called for; however, such high ionization stages will remain of only academic interest for many years to come.

[30] Br VII from AEL; Mo XIV and Xe XXVI from E. Hinnov, Phys. Rev. 14, 1533 (1976); $_{44}$Ru XVI to $_{74}$W XLVI from J. Reader and G. Luther, Phys. Rev. Lett. 45, 609 (1980).

[31] U. Feldman, L. Katz, W. Behring, and L. Cohen, J. Opt. Soc. Am. 61, 91 (1971) to Cu XIX; Kr XXVI and Mo XXXII from E. Hinnov, ref. 30; Sr XXVIII to Pd XXXVI from J. Reader, private communication (January 1979).

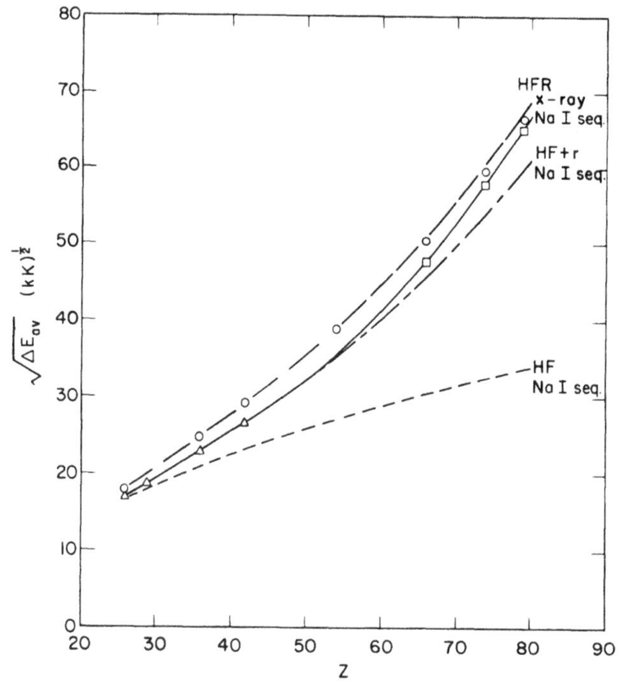

Fig. 19-10. Comparison of $\Delta n = 0$ $3s - 3p$ transitions for x-ray levels and for the Na I $2p^6 3s - 2p^6 3p$ isoelectronic sequence. Note that unlike the Ni I and Cu I cases, the x-ray transition here has an energy greater than the optical transition. The triangles and circles represent experimental values; the squares are DHF results.

For one-electron spectra such as the Cu I and Na I sequences, the calculation of wavelengths of individual spectrum lines involves, in addition to ΔE_{av}, only spin-orbit parameter values. (Configuration interactions are negligible, as shown in Secs. 13-6 and 19-3.) The accuracy of computed values of ζ may be estimated from the results given in Table 19-2; HF and HFR values are computed by the Blume-Watson method, whereas DHF and experimental values are calculated from[32]

$$\zeta_{nl} = |E(nl_+) - E(nl_-)| \cdot \frac{2}{2l+1} . \qquad (19.22)$$

It may be seen that relativistic effects are significant even at $Z \cong 30$, that HFR values are always considerably better than HF, and that errors in HFR values are usually in the range 2 to 5%.

In summary, the HFR method is not greatly inferior to DHF calculations in the accuracy of calculation of ΔE_{av} and ζ_l. For configurations involving more than one electron outside of closed subshells, the HFR method is much simpler than DHF,[33] and accuracies are limited in either approach by uncertainties in the appropriate scaling factors needed for the Coulomb-interaction integrals F^k, G^k, and R^k.

[32]The DHF values of E are total energies calculated with the computer program of J. P. Desclaux, Comput. Phys. Commun. 9, 31 (1975).

[33]See Sec. 7-14; also, R. D. Cowan and D. C. Griffin, J. Opt. Soc. Am. 66, 1010 (1976). The evaluation of relativistic matrix elements is discussed in detail by Z. B. Rudzikas, V. I. Sivcev, I. S. Kickin, and A. A. Slepcov, At. Data Nucl. Data Tables 18, 205, 223 (1976).

TABLE 19-2. SPIN-ORBIT PARAMETER VALUES (kK)

	ζ_{4p} (Cu I $3d^{10}4p$ seq.)				ζ_{4p} (x-ray)		
	HF	HFR	DHF	Exp.	HF	HFR	Exp.[i]
$_{36}$Kr	5.68	5.99	6.33	6.52[a]	3.23	3.40	3.58[f]
$_{42}$Mo	18.70	20.09	20.71	21.03[b]	11.50	12.43	~16[j]
$_{54}$Xe	86.3	97.7	97.7	98.7[c]	57.3	65.2	~54[j]
$_{74}$W	442	567	544		285	373	356
$_{79}$Au	607	810	769		398	543	529
	ζ_{3p} (Na I $2p^63p$ seq.)				ζ_{3p} (x-ray)		
$_{26}$Fe	13.18	13.53	13.88	13.99[d]	7.66	7.87	
$_{29}$Cu	22.91	23.70	24.14	24.25[d]	12.75	13.23	~11[j]
$_{36}$Kr	65.2	68.9	69.2	69.0[c]	39.9	42.2	48
$_{42}$Mo	133.6	144.1	143.6	144[e]	87.6	94.7	94
$_{54}$Xe	414	472	462		296	338	333
$_{74}$W	1636	2122	2008		1265	1645	1580
$_{79}$Au	2165	2925	2743		1693	2295	2177
	ζ_{2p} (Li I $1s^22p$ seq.)				ζ_{2p} (x-ray)		
$_{6}$C	0.0686			0.0718[f]			
$_{8}$O	0.346			0.355[f]			
$_{10}$Ne	1.083			1.100[g]			
$_{14}$Si	5.38	5.42	5.37	5.59[f]			
$_{18}$Ar	16.81	17.03	16.86	17.1[h]	11.32	11.45	11.1
$_{26}$Fe	83.8	86.2	85.2	85.8[h]	63.6	65.3	70
$_{36}$Kr	334	354	347		271	285	281
$_{42}$Mo	638	689	674		529	569	564
$_{54}$Xe	1813	2067	1997		1554	1763	1729
$_{74}$W	6630	8580	8070		5900	7572	7190
$_{79}$Au	8670	11660	10890		7759	10350	9760

[a] B. C. Fawcett, B. B. Jones, and R. Wilson, Proc. Phys. Soc. (London) **78**, 1223 (1961). See also R. L. Kelly and L. J. Palumbo, Naval Research Laboratory Report NRL-7599 (U.S. Govt. Printing Off., Washington, D.C., 1973).
[b] J. Reader and N. Acquista, Phys. Rev. Lett. 39, 184 (1977).
[c] E. Hinnov, ref. 30.
[d] U. Feldman et al., ref. 31.
[e] P. G. Burkhalter, J. Reader, and R. D. Cowan, J. Opt. Soc. Am. 67, 1521 (1977).
[f] AEL.
[g] K. Bockasten, R. Hallin, and T. P. Hughes, Proc. Phys. Soc. (London) **81**, 522 (1963).
[h] J. D. Purcell and K. G. Widing, Astrophys. J. 176, 239 (1972).
[i] J. A. Bearden and A. F. Burr, ref. 26, except as indicated otherwise.
[j] W. Lotz, ref. 28.

*19-8. AUTOIONIZATION AND DIELECTRONIC RECOMBINATION

The manner in which autoionization transition probabilities vary along an isoelectronic sequence is easily determined if we assume hydrogenic wavefunctions. If the autoionizing level involves a $\Delta n \neq 0$ excitation, then the excitation energy—and hence also the free-electron kinetic energy ε—is proportional to Z^2. It follows from the form of the radial equation (18.15) for $P_{\varepsilon l}(r)$ that $P^Z_{\varepsilon l}(r) = cP^H_{\varepsilon l}(Zr)$, just as for bound functions. However, the normalization (18.27) of free-electron functions ($\propto \varepsilon^{-1/4} \propto Z^{-1/2}$) is such that $c = Z^{-1/2}$ rather than $Z^{+1/2}$. Comparison of (18.33) with (19.15a) then shows that R^k is independent of Z.

Since autoionization transition probabilities are from (18.111) proportional to $(R^k)^2$, then A^a is independent of Z for hydrogenic wavefunctions. Numerical calculations show this to be approximately true even for multi-electron wavefunctions. This is in strong contrast to the interaction energy between two bound electrons, which we have seen in (19.7) and (19.15) to increase linearly with Z_c. Physically, the kinetic energy of the free electron (when close to the nucleus) increases with Z_c, so that the time during which the bound-free interaction can take place decreases.

Radiative transition probabilities, on the other hand, increase rapidly with Z_c: linearly for $\Delta n = 0$ excitations, (14.119), and as the fourth power for $\Delta n \neq 0$ excitations, (14.116). As a result, radiative decay rates for doubly excited states of highly ionized atoms may be comparable to or much greater than autoionization decay rates—especially for $\Delta n \neq 0$ excitations of the inner electron, and for large n or *l* of the outer electron (cf. Fig. 18-15). This has several consequences:

(1) Emission lines arising from autoionizing states tend to be much more prominent than in spectra of neutral or weakly ionized atoms, and tend to be little broader than those having non-autoionizing upper states.

(2) The parameter q defined by (18.68) tends to be very large in magnitude ($q = -74$ for the Fe XXI example in Sec. 18-8). Absorption lines tend to be narrow and without obvious Beutler-Fano-profile asymmetries.

(3) Dielectronic recombination is an important process in high temperature plasmas: the branching ratio (18.113) for radiative stabilization tends to be large, the recombination rate coefficient (18.116)-(18.117) being limited by autoionization (first-step-recombination) rates rather than by radiative decay rates. $\Delta n \neq 0$ excitations of the inner electron tend to form the dominant contribution to α^d, in spite of the fact that the exponential factor is smaller than for $\Delta n = 0$ excitations.

(4) Because of the importance of dielectronic recombination, satellite lines of the sort discussed following Eq. (18.121) are more pronounced in spectra of high temperature plasmas than for low temperature sources.

(5) For moderately highly ionized atoms, collisional excitation of an inner shell electron, followed by autoionization, is an ionization process comparable in importance with direct collisional ionization.[34] In very highly ionized systems, however, the indirect process tends to become less important because the highly excited states tend to decay radiatively rather than by autoionization. For example, the importance of the indirect ionization process in

[34] A. Burgess, H. P. Summers, D. M. Cochrane, and R. W. P. McWhirter, Mon. Not. R. Astron. Soc. 179, 275 (1977).

Fe XV and XVI[35] was in early calculations overestimated by a factor of about 1.5 because of neglect of the radiative decay (and by another factor two because of an overestimate in collisional excitation rates).

Quasi-bound autoionizing compound states may produce pronounced resonances in collisional-excitation cross sections, sometimes providing an energy-averaged contribution comparable in importance to the direct excitation process.[36]

*19-9. INNER-SUBSHELL EXCITATIONS

In neutral atoms, most observed emission lines arise from levels that involve excitation of a single electron from the outer, most weakly bound subshell. This is because excitation from an inner subshell, even one belonging to the same shell as the valence subshell, usually requires so much energy as to result in levels near or above the ionization limit—see, for example, the $2s2p^4$ configuration of N I in Fig. 19-4.

In multiply ionized atoms, however, inner-subshell electrons of the valence shell are bound little more strongly than valence-subshell electrons, and levels involving excitation of the former are quite commonly observed. Even the excitation of an electron from a shell interior to the valence shell (producing an autoionizing state) may produce readily observable lines, for reasons discussed in the preceding section. Inner-subshell collisional excitations can contribute significantly to the intensity of satellite lines;[37] a specific example will be discussed in Sec. 19-11.

*19-10. TEMPERATURE AND DENSITY DIAGNOSTICS

We have already alluded briefly to the use of line shape measurements for the determination of plasma temperature and density, through the Doppler effect (Sec. 19-5) or through Stark effect broadening (Sec. 17-7). These methods are difficult to apply to spectra of highly ionized atoms because of the relatively small magnitude of the Doppler width, and because Stark widths depend in a complicated (and not too accurately known) way on both temperature and density. Consequently, considerable attention has been given to intensities of spectrum lines whose ratios depend strongly either on temperature or on density but not on both. We shall not go into great detail here, but only discuss the general principles involved.[38]

[35]L. Goldberg, A. K. Dupree, and J. W. Allen, Ann. d'astrophys. 28, 589 (1965); See also O. Bely, Ann. d'astrophys. 30, 953 (1967); R. D. Cowan and J. B. Mann, Astrophys. J. 232, 940 (1979).

[36]M. J. Seaton, J. Phys. B 2, 5 (1969); M. D. Hershkowitz and M. J. Seaton, J. Phys. B 6, 1176 (1973); L. P. Presnyakov and A. M. Urnov, J. Phys. B 8, 1280 (1975); K. A. Berrington, P. G. Burke, P. L. Dufton, A. E. Kingston, and A. L. Sinfailam, J. Phys. B 10, 1465 (1977), 12, L275 (1979); R. D. Cowan, J. Phys. B 13, 1471 (1980).

[37]A. H. Gabriel, Mon. Not. R. Astron. Soc. 160, 99 (1972); C. P. Bhalla, A. H. Gabriel, and L. P. Presnyakov, Mon. Not. R. Astron. Soc. 172, 359 (1975).

[38]For further details and references to numerous papers, see A. H. Gabriel and C. Jordan, Case Studies Atom. Coll. Phys. 2, 209 (1972); G. A. Doschek, U. Feldman, et al., Phys. Rev. A 12, 980 (1975), Astrophys. J. 212, L147 (1977), 215, 652 (1977), 219, 304 (1978), Astron. Astrophys. 60, L11 (1977); H. E. Mason, G. A. Doschek, U. Feldman, and A. K. Bhatia, Astron. Astrophys. 73, 74 (1979).

As a simple example, consider two emission lines $l \rightarrow j$ and $n \rightarrow m$ arising from different excited states l and n of a given ion. For a plasma in local thermodynamic equilibrium (LTE), it follows from (18.122) that the ratio of line intensities (measured in photons per second) is

$$\frac{I_{nm}}{I_{lj}} = \frac{g_n A^r_{nm}}{g_l A^r_{lj}} e^{-(E_n - E_l)/kT} . \tag{19.23}$$

This ratio is a sensitive function of temperature provided $|E_n - E_l|$ is not small compared with kT.

Under coronal equilibrium conditions, intensities are determined by the rates at which the levels are populated rather than by the radiative transition probabilities A^r. If we neglect population of excited states via downward cascading from higher levels and via dielectronic recombination, then population rates are given by the rates of electron-collisional excitation from the ground level i. These may be calculated with the aid of the appropriate collisional excitation rate coefficients as given (approximately), for example, by the Van Regemorter-Seaton[39] semi-empirical expression (appropriate only for E1-allowed excitations):

$$\begin{aligned} C^e_{il} &= 8(\pi/3)^{1/2} \frac{ha_o}{m} \frac{f_{il}}{(kT_e)^{1/2} E_l} P(E_l/kT_e) e^{-E_l/kT_e} \\ &= 3.151 \cdot 10^{-7} \frac{f_{il}}{(kT_e)^{1/2} E_l} P(E_l/kT_e) e^{-E_l/kT_e} \quad \text{cm}^3/\text{sec} , \end{aligned} \tag{19.24}$$

where E_l and kT_e are in rydbergs, and P is a slowly varying function, with mean value[39,40] about 0.2 for $E_l/kT_e > 1$ and about 1.0 for $E_l/kT_e = 0.02$. The photon intensity ratio is in this case

$$\frac{I_{nm}}{I_{lj}} = \frac{B_{nm} E_l f_{in}}{B_{lj} E_n f_{il}} \frac{P(E_n/kT_e)}{P(E_l/kT_e)} e^{-(E_n - E_l)/kT_e} . \tag{19.25}$$

where B_{nm} is the branching ratio for radiative decay of level n via the transition $n \rightarrow m$. Except for the minor effects of the functions P, this shows the same temperature dependence as for the LTE expression (19.23); however, the indicated temperature for a given value of I_{nm}/I_{lj} is of course different.

In general, dielectronic recombination and/or radiative cascades from higher levels must be taken into account, and the problem becomes much more complicated; to obtain the

[39]H. Van Regemorter, Astrophys. J. **136**, 906 (1962); M. J. Seaton, Planet. Space Sci. **12**, 55 (1964). See also R. Mewe, Solar Phys. **22**, 459 (1972).

[40]O. Bely, Proc. Phys. Soc. (London) **88**, 587 (1966), indicates that P may be a factor two or so larger for $\Delta n = 0$ excitations, and significantly smaller for $\Delta n \neq 0$ excitations. See also ref. 36 regarding contributions from autoionization resonances.

various level populations, it is necessary to solve a set of simultaneous linear equations involving a multitude of different rate coefficients.[41] However, line intensity ratios are still independent of density so long as all important radiative decay rates are large compared with collisional excitation and dielectronic recombination rates. (Calculation of dielectronic recombination contributions of course requires knowledge of the relative abundance of neighboring ionization stages, but the ionization equilibrium is density independent so long as 3-body recombination is unimportant.)

When the electron density is large enough that collisional rates become comparable with radiative decay rates, then line ratios may become strongly density dependent. Consider, for example, densities at which the rate of collisional excitation

$$2s^2 2p^w \rightarrow 2s2p^{w+1} \tag{19.26}$$

is comparable with the inverse radiative decay rate. The population of $2s2p^{w+1}$ states may then become high enough that the collisional excitation

$$2s2p^{w+1} \rightarrow 2p^{w+2} \tag{19.27}$$

contributes significantly to the rate of population of $2p^{w+2}$ levels. The two-step excitation process (19.26)-(19.27) is clearly density dependent, so that intensity ratios of the type

$$\frac{I(2p^{w+2} \rightarrow 2s2p^{w+1})}{I(2s2p^{w+1} \rightarrow 2s^2 2p^w)} \tag{19.28}$$

will increase strongly with density over some finite electron density range.

In highly ionized atoms, the excitation energies of all the above levels are small compared with the ionization energy (Fig. 19-4), and therefore usually small compared with the temperature at which the ion stage in question is abundant (Fig. 18-18). Consequently, the ratio (19.28) is nearly independent of temperature, and is particularly useful as a density diagnostic. An added advantage is that the mean transition energies for (19.26) and (19.27) are nearly equal so that the lines in question tend to have nearly equal wavelengths; this reduces the need for accurate wavelength calibration of spectrograph and detector efficiencies. A problem is that for any given element, each ion has a high abundance over only a limited temperature range, and the range of strong density sensitivity is also limited. It is therefore not necessarily possible to find an element suitable for the temperature and density conditions present in the plasma of interest—especially in astrophysical plasmas, where the naturally occurring abundant elements are strictly limited in variety, and artificial seeding with other elements is difficult. Even when a suitable ion can be found, the reliability of temperature and density determinations is limited by the uncertain accuracy of the required rate coefficients; the situation in this regard is improving rapidly, but much work remains to be done in this field.

[41] Gabriel and Jordan, ref. 38. Under non-steady-state conditions, a set of simultaneous linear first-order differential equations is involved.

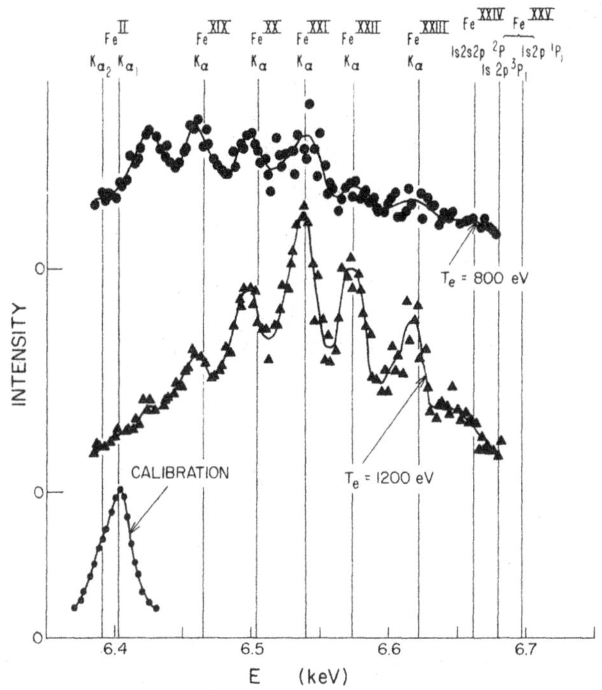

Fig. 19-11. Observed x-ray spectrum of the Princeton ST tokamak in the region of Fe 1s – 2p transitions (1.94 to 1.85 Å).

*19-11. IRON K_α DIAGNOSTICS

The diagnostic methods described in the preceding section involve primarily the use of relative intensities of two lines originating in a single ion. Here we consider the determination of temperature via comparison of intensities of lines from neighboring ionization stages of an element. This involves the added necessity of a calculation of the ionization-equilibrium distribution of ion abundances. On the other hand, it does not necessarily require resolution of individual spectral lines; in fact, use of low resolution has the advantage of averaging over entire transition arrays, thereby eliminating uncertainties arising from inadequate knowledge of relative rate coefficients for different transitions of an array.

As a specific example, we consider 1s – 2p transitions in various ionization stages of iron, which vary in wavelength from the Ly_α line of hydrogenic Fe XXVI at 1.79 Å up to the wavelength 1.94 Å of the K_α lines of singly ionized iron. Actually, the wavelengths differ appreciably from one ion to the next only over the range Fe XXVI to about Fe XVIII, where the number of electrons in the K and L shells is changing. This is illustrated by Fig. 19-11, which shows low-resolution observations of a tokamak spectrum.[42] In such a low

[42]N. Bretz, D. Dimock, A. Greenberger, E. Hinnov, E. Meservey, W. Stodiek, and S. von Goeler, in *Plasma Physics and Controlled Nuclear Fusion Research 1974*, Proc. 5th Intern. Conf., Tokyo, 1974 (International Atomic Energy Agency, Vienna, 1975), Vol. I, p. 55.

density source ($n_e \cong 10^{13}$ cm^{-3}), essentially all ions lie in some level of the ground configuration. Emission of 1s − 2p radiation can arise only if an ion is momentarily raised to an excited configuration that has one or more 2p electrons and a 1s hole. Photoexcitation is unimportant in low density sources. There are then three possible mechanisms by which such an excited configuration can be produced; we illustrate these using the special case of Fe XXII.

(a) Collisional excitation out of the 1s shell:

$$1s^2 2s^2 2p \xrightarrow{C^e} 1s2s^2 2p^2 \xrightarrow{A^r} 1s^2 2s^2 2p \ . \tag{19.29}$$

For ions having (as here) more than four electrons, an additional possibility is

$$1s^2 2s^2 2p \xrightarrow{C^e} 1s2s^2 2pnl \xrightarrow{A^r} 1s^2 2s^2 nl \ , \qquad n > 2 \ . \tag{19.30}$$

(b) Dielectronic recombination involving excitation out of the 1s shell:

$$1s^2 2s^2 2p + \varepsilon(l \pm 1) \xrightarrow{\alpha^d} 1s2s^2 2p^2 nl \xrightarrow{A^r} 1s^2 2s^2 2pnl \ , \qquad n \geqq 2 \tag{19.31}$$

$$1s^2 2s^2 2p + \varepsilon l \xrightarrow{\alpha^d} 1s2s^2 2pn'l'n''l'' \xrightarrow{A^r} 1s^2 2s^2 n'l'n''l'' \ , \qquad n'' \geqq n' > 2 \ . \tag{19.32}$$

(c) For ions with more than four electrons, collisional ionization out of the 1s shell:

$$1s^2 2s^2 2p \xrightarrow{C^i} 1s2s^2 2p + \varepsilon l \xrightarrow{A^r} 1s^2 2s^2 + \varepsilon l \ . \tag{19.33}$$

For iron atoms at the temperatures of interest here ($0.5 \lesssim kT_e \lesssim 5$ keV), the contribution of ionization proves to be unimportant relative to processes (a) and (b) and we shall not consider it further.

Each of the radiative processes in (19.29)-(19.32) involves numerous spectrum lines because each configuration possesses several energy levels. However, the lines of each transition array cover only a relatively small wavelength range. The instrumental resolution corresponding to Fig. 19-11 was low enough ($\Gamma \cong 70$ kK $\cong 9$ eV) that individual lines were not observed—only a single maximum for each transition array. On the other hand, any given observed maximum does not in general arise from a single ionization stage: the labeling Fe XIX, Fe XX, ··· in the figure applies to the "basic" collisionally excited arrays of the type (19.29). For example, the Fe XXII array (19.30) for $n > 2$ lies essentially at the position labeled Fe XXIII for the basic array $1s2s^2 2p \rightarrow 1s^2 2s^2$; similarly, the arrays (19.31) and (19.32) arise from what were initially Fe XXII ions, but lie at the basic positions of Fe XXI for (19.31) with $n = 2$, of Fe XXII for (19.31) with $n > 2$, and of Fe

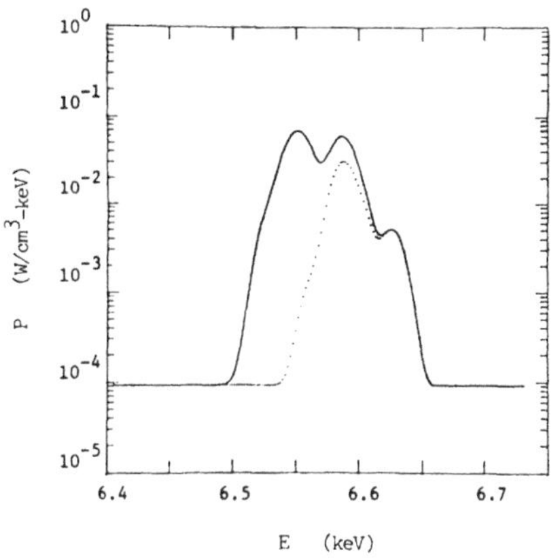

Fig. 19-12. Computed 1s – 2p spectrum resulting from Fe XXII atoms in the plasma: dotted curve, as a result of collisional excitation; solid curve minus dotted curve, from excited Fe XXI ions produced by dielectronic recombination. E is the transition energy, and P the radiated power per unit volume and photon-energy range. The ordinate scale corresponds to 10^{12} Fe XXII ions/cm^3 and 10^{14} electrons/cm^3 with T_e = 1 keV. (A small constant background has been added for computer-plotting convenience.)

XXIII for (19.32) with $n'' \geqq n' > 2$. This is illustrated quantitatively in Fig. 19-12,[43] except that the dielectronic-recombination contributions for $n'' \geqq n' > 2$ have been neglected.

In Fig. 19-13 are shown superpositions of curves like Fig. 19-12 for all ionization stages from Fe XVIII to Fe XXV, using coronal-equilibrium ion abundances from Fig. 18-18. [Included are collisional excitations 1s – 2p for all ions and 1s to 3s, 3p, 3d, 4p, 5p for Fe XXII-XVIII, and dielectronic recombinations (19.31) with nl = 2s (where possible), 2p, 3s, 3p, 3d, 4p, 5p. Recombination rate coefficients were obtained from the modified Burgess approximation (18.134)-(18.135).] Agreement with the experimental curves in Fig. 19-11 is good, except that the temperatures disagree by about 25 percent. This may be the result of errors in the theoretical rate coefficients used, but may also be due to the fact that the experimental observations are time and spatially integrated, and the quoted temperatures are peak values.

Higher-resolution Fe K_α observations have been made on both laboratory plasmas[44] and solar flares;[45] agreement of theory with the observed spectral details is reasonably good considering the complexity of the problem.

Finally, we show in Fig. 19-14 the computed relative contributions of the different excitation processes to the spectrally integrated Fe K_α radiation. It is clear that dielectronic recombination is the dominant excitation mechanism up to $kT_e \cong 2$ keV. In part, this is because the required free-electron kinetic energy [E_s in (18.120) or (18.134)] is about 1.5 keV less than the 1s – 2p excitation energy, so that the collisional-excitation exponential factor in (19.24) is much more unfavorable. At higher temperatures, the exponential factor is less important, and the power of T_e multiplying the exponential is also one higher for collisional excitation than for dielectronic recombination.

[43]A. L. Merts, R. D. Cowan, and N. H. Magee, Jr., Los Alamos Scientific Laboratory Report LA-6220-MS (March 1976).

[44]E. Ya. Kononov, K. N. Koshelev, and Yu. V. Sidel'nikov, Sov. J. Plasma Phys. 3, 375 (1977); Yu. A. Mikhailov et al., Opt. Spectrosc. 42, 469 (1977); V. A. Boiko et al., Mon. Not. R. Astron. Soc. 185, 305 (1978); K. W. Hill, S. von Goeler, M. Bitter, L. Campbell, R. D. Cowan, B. Fraenkel, A. Greenberger, R. Horton, J. Hovey, W. Roney, N. R. Sauthoff, and W. Stodiek, Phys. Rev. A 19, 1770 (1979).

[45]Yu. I. Grineva et al., Solar Phys. 29, 441 (1973); V. V. Korneev et al., Solar Phys. 63, 319, 329 (1979); G. A. Doschek, R. W. Kreplin, and U. Feldman, Astrophys. J. 233, L157 (1979).

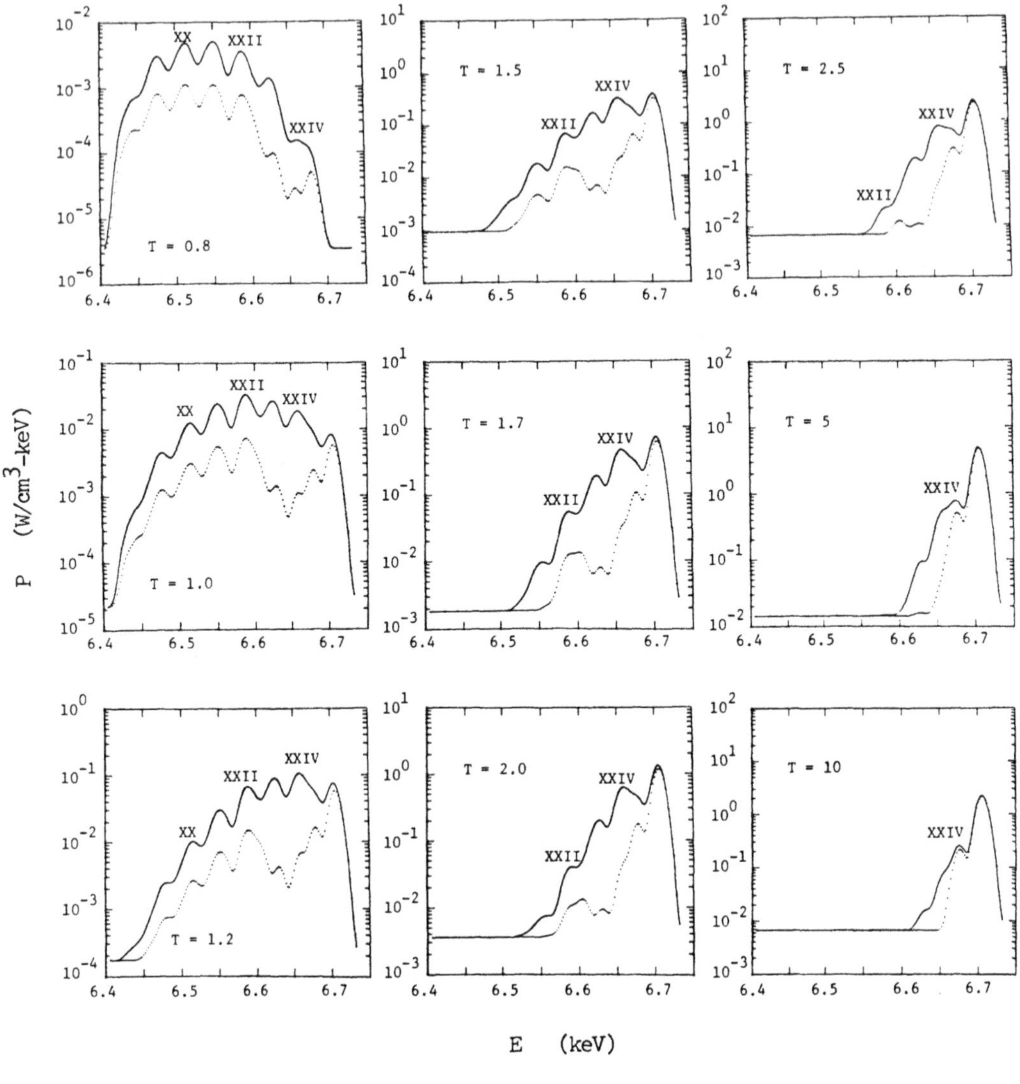

Fig. 19-13. Computed 1s – 2p spectra for coronal-equilibrium distributions of Fe ionization stages at the indicated electron temperatures (in keV). All spectra are plotted as in Fig. 19-12, except that the *total* iron ion density is 10^{12} cm^{-3}. The ionization-stage labeling of the various maxima follows the convention used in Fig. 19-11.

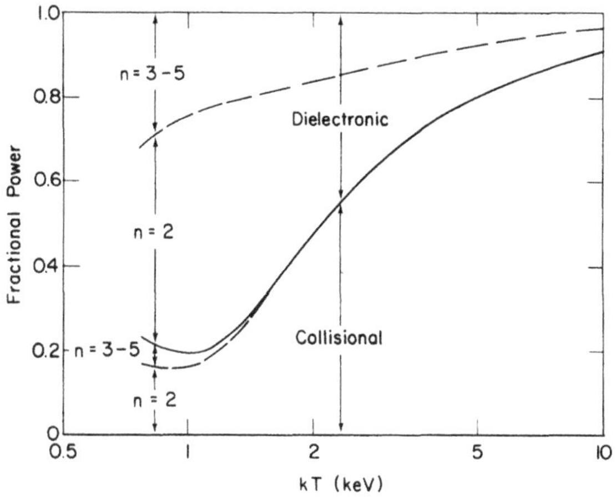

Fig. 19-14. Relative contributions of the various excitation processes to 1s − 2p radiation from Fe ions. The distances between successive curves represent contributions of collisional excitation or of dielectronic recombination, for principal quantum numbers of the (outer) excited electron of n = 2 or of n = 3 to 5.

20

RARE-EARTH AND TRANSITION ELEMENTS

20-1. LANTHANIDE CONFIGURATIONS

The rare earths consist of two series of elements, the lanthanides (Z = 57 to 70) and the actinides (Z = 89 to 102), which roughly speaking involve filling of the 4f and 5f subshells, respectively. For simplicity, we shall first consider only the lanthanide elements, and postpone discussion of the actinides to Sec. 20-7.

The lanthanides are sometimes considered to consist of the elements Z = 58 to 71 (cerium to lutecium), on the basis that in the solid state these atoms are generally triply ionized and hence show valence subshells running from $4f^1$ to $4f^{14}$ (Table 4-3). However, we are here interested primarily in free atoms, and take the lanthanides to start with that neutral atom at which the 4f orbital first collapses from a small-quantum-defect, large-radius hydrogenic orbital to a large-quantum-defect, small-radius non-hydrogenic orbital; this collapse occurs at lanthanum (Z = 57), as indicated in Figs. 8-8 to 8-11 and Table 8-8.

The unique properties of the lanthanide elements are a direct result of the small radii of the 4f orbitals (smaller even than those of 5s electrons, Fig. 8-10) such that the 4f electrons are well shielded from outer valence electrons and environmental effects, together with the fact that their binding energies are relatively small and comparable with those of 5d, 6s, and 6p electrons. (This latter property is even more true than appears from Fig. 8-11 because of relaxation effects that invalidate Koopmans' theorem, which is implicitly involved in the figure.) The low-lying configurations of lanthanide atoms and ions are thus of the form indicated in Table 20-1.[1,2]

Because of the large number of quantum states that may belong to an f^w subshell, the number of levels of a configuration involving additional partially filled subshells may be enormous: some examples are given in Table 20-2.[1,3,4] Correspondingly, the spectra of

[1]J. Blaise and J.-F. Wyart, Rev. chim. miner. 10, 199 (1973).

[2]Z. B. Goldschmidt, in K. A. Gschneidner, Jr. and L. Eyring, eds., *Handbook on the Physics and Chemistry of Rare Earths* (North-Holland Publ. Co., Amsterdam, 1978), Vol. I.

[3]B. G. Wybourne, J. Opt. Soc. Am. 55, 928 (1965).

[4]R. D. Cowan, Nucl. Instrum. Methods 110, 173 (1973).

TABLE 20-1. LOW CONFIGURATIONS OF NEUTRAL AND SINGLY IONIZED LANTHANIDE ATOMS[a]

$4f^{w}6s^{2}$	$4f^{w}6s6p$
$4f^{w-1}5d6s^{2}$	$4f^{w-1}5d6s6p$
$4f^{w}5d6s$	$4f^{w}5d6p$
$4f^{w-1}5d^{2}6s$	$4f^{w-1}5d^{2}6p$
$4f^{w+1}6s$	$4f^{w+1}6p$
$4f^{w+1}5d$	

[a]For multiply ionized atoms, delete a 6s and/or a 6p electron.

TABLE 20-2. NUMBERS OF LS TERMS AND LEVELS OF SOME RARE-EARTH CONFIGURATIONS

Configuration	No. of LS terms	No. of levels	Largest matrix J	Largest matrix rank
f^{14}	1	1	0	1
f^{1}, f^{13}	1	2	5/2, 7/2	1
f^{2}, f^{12}	7	13	2, 4	3
f^{3}, f^{11}	17	41	5/2 - 9/2	7
f^{4}, f^{10}	47	107	4	19
f^{5}, f^{9}	73	198	7/2	30
f^{6}, f^{8}	119	295	4	46
f^{7}	119	327	7/2	50
$f^{3}sp$	160	476	7/2	81
$f^{3}ds$	258	759	7/2	122
$f^{2}d^{2}s$	304	893	7/2	149
$f^{7}dsp$	11168	36262	4	4829
$f^{7}d^{2}p$	24662	78822		

rare-earth atoms by and large are tremendously complex, and one spectrum may contain tens or hundreds of thousands of observable lines; an example is shown in Fig. 20-1. Some beginnings in analysis of the simpler lanthanide spectra were made between 1927 and the mid-1930's. However, major progress was not made in the more complex cases until the 1960's, following development of improved experimental apparatus, the use of computers for data reduction and analysis, and the introduction of adequate theoretical aids involving the use of Racah algebra.[5,6] A brief historical outline of experimental progress has been given by Martin,[7] and the status of analyses in 1976 for the first five spectra of the

[5]B. R. Judd, *Operator Techniques in Atomic Spectroscopy* (McGraw-Hill, New York, 1963).

[6]B. G. Wybourne, *Spectroscopic Properties of Rare Earths* (Interscience, New York, 1965).

[7]W. C. Martin, Opt. Pura y Apl. 5, 181 (1972).

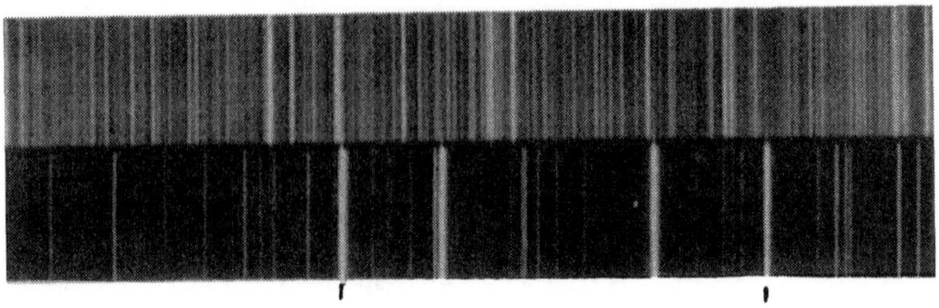

Fig. 20-1. Uranium (upper) and iron (lower) spectra in a small portion of the near ultraviolet region. (Photographic positives, courtesy of D. W. Steinhaus.) The spectra U I + U II average about 15 lines/Å in this region [J. Blaise and L. J. Radziemski, Jr., J. Opt. Soc. Am. 66, 644 (1976)]. This is about 10 times the line density for spectra of the simpler transition element iron [cf. U. Litzén and J. Vergès, Phys. Scr. 13, 240 (1976)], which in turn is an order of magnitude greater than for elements with s^2p^w ground configurations [see for example P I and P II, W. C. Martin, J. Opt. Soc. Am. **49**, 1071 (1959)].

lanthanides has been tabulated by Wyart[8] (see also Table 1-4). Only in the late 1970's—nearly 20 years after completion of the first three volumes of AEL—has our knowledge of lanthanide spectra reached the point where preparation of volume four has become appropriate.[9] Comparisons between theory and observations have been given by Wybourne[6] and by Goldschmidt.[2]

20-2. LEVEL STRUCTURE; GENERAL REMARKS

Because of the small radii of the 4f orbitals, the 4f electrons necessarily lie rather close together. Consequently, their mutual Coulomb repulsions are quite large, and the values of the Slater integrals $F^k(4f,4f)$ are also large—50 to 100 kK (Fig. 8-13). The total energy range covered by the levels of $4f^w$ is therefore unusually large for a valence subshell—over 100 kK ($\cong$ 12 eV) in most cases ($4 \leqq w \leqq 10$). This is shown schematically in Fig. 20-2[10,11] for configurations $4f^w6s^2$, which in most cases are the ground configurations of the neutral lanthanides. It may be noted that:

(1) Because of the gradual contraction of the 4f orbital with increasing Z, the parameter values F^k and ζ increase with Z, so that the level spacings tend to increase also. Thus, for

[8] J.-F. Wyart, J. Opt. Soc. Am. 68, 197 (1978).

[9] W. C. Martin, R. Zalubas, and L. Hagan, *Atomic Energy Levels—The Rare Earth Elements*, NSRDS-NBS 60 (U.S. Govt. Printing Off., Washington, D.C., 1978).

[10] R. D. Cowan, ref. 4. The levels plotted here were computed using HX values of F^k and ζ, all scaled down by the factor $(Z + 14)/100$; essentially the same results are obtained using HFR values with the F^k only scaled by 0.80 for all Z.

[11] A similar figure is given by G. H. Dieke, *Spectra and Energy Levels of Rare Earth Ions in Crystals* (Interscience, New York, 1968), p. 55.

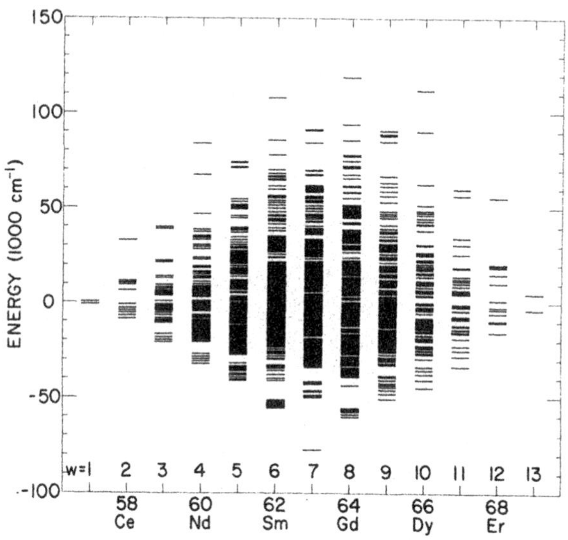

Fig. 20-2. Theoretical energy levels of configurations $4f^w6s^2$ of neutral lanthanide atoms, relative to the center-of-gravity energy E_{av}. The density of levels within each solid block is too great for the levels to be drawn separately.

example, the level structure of Dy I $4f^{10}6s^2$ is very similar to, but considerably more widely spaced than, that of the conjugate configuration Nd I $4f^46s^2$, and the spin-orbit splitting of the $^2F^\circ$ term of Tm I $4f^{13}6s^2$ is much greater than that of La I $4f6s^2$.

(2) The spin-orbit parameters ζ increase much faster with Z than do the Coulomb parameters F^k (Fig. 8-13). Consequently there is a strong tendency to move from fairly good LS coupling in Ce toward moderately good jj coupling in Er. There results also a tendency for the levels to be more uniformly spaced in the high-Z elements; for example, the level distribution in Ho I $4f^{11}6s^2$ is comparatively uniform, whereas the levels of the conjugate configuration Pr I $4f^36s^2$ tend to be clustered into doublet and quartet LS terms.

(3) Ionization of a neutral lanthanide atom involves removal not of one of the 4f electrons, but of one of the comparatively loosely bound 6s electrons, so that ionization energies are only about 50 kK $\cong$ 6 eV (Table 1-1). Thus many of the levels of the *ground* configuration lie above the lowest ionization limit; the same may be true to a lesser extent of the ground-configuration levels of the singly ionized atom, as the binding energy of the remaining 6s electron is only 11 or 12 eV. An example is shown in Fig. 20-3.[4] Levels lying above the ionization limit, arising from the ground configuration as well as from excited configurations, are in general strongly autoionizing. No such levels have been observed in emission, but they can be detected by laser absorption spectroscopy.[12]

Configurations that involve partially filled subshells in addition to the 4f subshell have of course more complicated level structures than those shown in Fig. 20-2. If the additional unfilled subshells all have $n > 5$, then the interactions of these electrons with the 4f electrons are small, and the level structure consists basically of the $4f^w$ structure, except with each level of the latter split into several levels as a result of the additional small interactions (Sec. 12-9).[13] For example, for each level of $4f^w$ with total quantum number J_c, there exist

[12] R. W. Solarz, C. A. May, L. R. Carlson, E. F. Worden, S. A. Johnson, J. A. Paisner, and L. J. Radziemski, Jr., Phys. Rev. A 14, 1129 (1976).

[13] B. G. Wybourne, ref. 6, pp. 49-69, gives numerous illustrations.

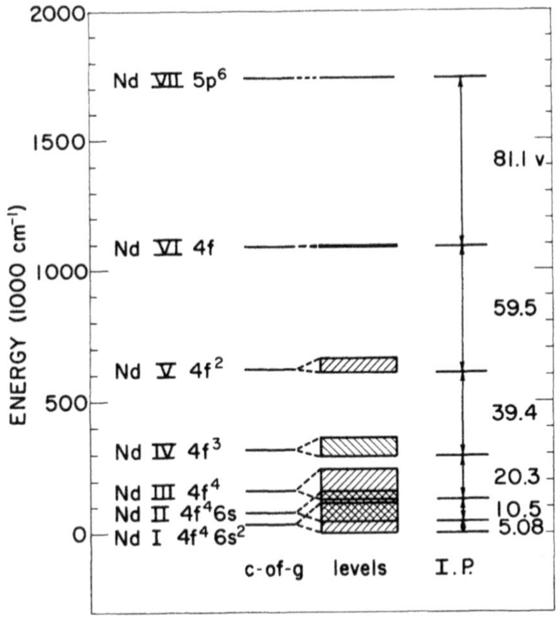

Fig. 20-3. Computed energy range of levels of the ground configurations of neodymium ions; ionization potentials are in volts.

in $4f^w6s$ two levels with $J = J_c \pm 1/2$ whose center-of-gravity energy corresponds to the energy of the parent level J_c; a more complex example will be discussed in Sec. 20-4. When one or more 5d electrons are present, the 5d-4f interactions are sufficiently large that each level of the configuration will in general correspond to a mixture of several parent levels of $4f^w$; it is then not a good approximation to picture the level structure as consisting of a set of levels grouped about the position of each parent level J_c, and the detailed nature of the level structure depends on the magnitude of the 5d-4f interaction.

Because the binding energies of 4f, 5d, 6s, and 6p electrons in neutral lanthanides differ from each other by only a couple of electron volts whereas the spread of energies within a configuration is 5 to 10 eV or more, the various configurations overlap each other to a very high degree. The result is that the density of energy levels becomes very high, even at energies as low as 20 kK (2.5 eV) above the ground level, and the interpretation of experimental spectra and energy levels is extremely difficult. Except for the low levels of each configuration and levels of high J, which are relatively few in number,[3] the computed mixing of theoretical basis functions is usually so great as to make impossible the assignment of meaningful pure-coupling-quantum-number designations, other than parity and J, and perhaps configuration. The distinguishing characteristics of pure basis states are smeared out to such a degree that all levels of given J and configuration tend to have much the same properties (Landé g-values, for instance).[3]

Although the various configurations overlap greatly and configuration interactions are by no means negligible, strong configuration mixings tend to be limited to a comparatively few isolated pairs of levels. The levels of such a pair must not only have the same parity and J and lie relatively close together, but they must belong to two configurations that can interact directly (rather than indirectly through an intermediary third configuration), and must contain large components of common parentage (of $4f^{w-1}$ or $4f^{w-2}$, etc.). Experimental isotope-shift evidence of strong configuration mixing for relatively low-lying levels will be discussed in Sec. 20-5. However, even levels lying very close to the ionization limit do not necessarily show strong mixings; for example, such levels have been shown by laser-probing excitation to have very different radiative-decay lifetimes: short lifetimes if they are simply high-lying levels of low-excited configurations, and long lifetimes if they are low levels of highly excited configurations (Rydberg series members with $n \cong 40$ to 50 for the excited outer electron).[12]

20-3. LOW-LEVEL STRUCTURES AND COUPLING CONDITIONS

It may be seen from Fig. 20-2 that the very lowest portion of each $4f^w$ configuration is comparatively simple, consisting usually of a very few levels that are fairly well isolated from the higher, more complex portion. These levels tend (especially for $w \leqq 8$) to constitute quite high purity LS terms. It is these facts that made it possible to break into the analyses of Sm I, Eu I, and Gd I in the mid-1930's, long before any success was had with neighboring elements.[1,7]

When another electron is present in addition to $4f^w$, the low-level structure is of course made much more complicated. In the case of $4f^w5d$, ($w < 7$), the Coulomb interaction between 5d and 4f is considerably stronger than the spin-orbit interactions, and so the atom still shows some approximation to LS coupling. However, the various terms $(f^w\alpha_1L_1S_1,d)LS$ [fixed $\alpha_1L_1S_1$, different LS] overlap greatly, and so one would expect considerable basis-state mixing and strong departures from the Landé interval rule. Nonetheless, quite regular LS terms are frequently observed for the lowest levels; an example is shown in Fig. 20-4. This surprising result has been explained by Judd[14] as arising from the facts that the low terms have maximum possible spin and that $\zeta_{4f} \cong \zeta_{5d}$ (the higher n for the 5d electron compensating for the lower l).

In the case of $4f^w6p$, the Coulomb interaction between the outer electron and the core is much weaker, and ζ_{6p} is quite large. The coupling is then much closer to

$$(4f^w\alpha_1L_1S_1J_1,\ 6p_j)J\,. \tag{20.1}$$

Correspondingly, the dependence of energy on J bears no resemblance to the monotonic variation of the Landé interval rule, but rather tends to form a characteristic concave-upward (bowl-shaped) or concave-downward (umbrella-shaped) pattern; examples are depicted in Fig. 20-5. Goldschmidt[15] has shown that the pattern depends on whether the reduced matrix element $\langle f^w\alpha_1L_1S_1\|U^{(2)}\|f^w\alpha_1L_1S_1\rangle$ is positive (bowl-shaped) or negative (umbrella-shaped). For $2 < w < 13$, both signs are present in the same configuration; however, bowl shapes tend to predominate for $w < 7$, and umbrella shapes for $w > 7$.

[14]B. R. Judd, J. Opt. Soc. Am. 58, 1311 (1968).

[15]Z. B. Goldschmidt, "Properties and Methods of Interpretation of Rare-Earth Spectra," in F. Bloch, S. G. Cohen, A. de-Shalit, S. Sambursky, and I. Talmi, eds., *Spectroscopic and Group Theoretical Methods in Physics* (North-Holland Publ. Co., Amsterdam, 1968).

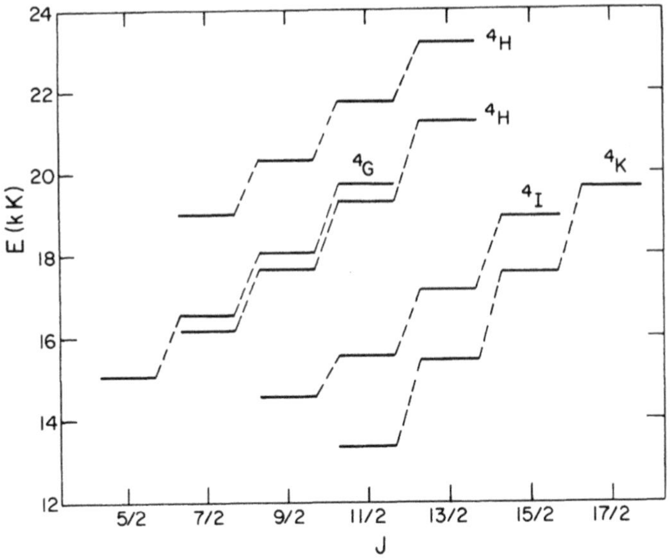

Fig. 20-4. Five low terms of maximum multiplicity for Pr III $4f^2 5d$. In spite of substantial overlapping, the Landé interval rule is obeyed fairly well by all terms.

Configurations $4f^w 6s$ show $J_1 j$ coupling, similarly to $4f^w 6p$. For $w > 7$, $4f^w 5d$ tends toward this type coupling because of the increasing value of ζ_{4f}; this is particularly true for $w = 4l_1 + 1 = 13$ because of the zero contribution of the exchange integrals G^k to the level structure (Sec. 12-9). For all configurations of the type $4f^w nl$ $(n > 6)$, J_1 is a quite good quantum number because of the small interaction of the nl electron with the $4f^w$ core; the coupling may be closely $J_1 j$ (small l) or more nearly $J_1 K$ (large l).

Configurations $4f^w 6s6p$, especially for $w > 7$, tend to show coupling close to[16]

$$[4f^w \alpha_1 L_1 S_1 J_1, (6s6p) L_2 S_2 J_2] J \tag{20.2}$$

because of the strong Coulomb interaction between the 6s and 6p electrons, and because the 6p spin-orbit interaction is large compared with interactions of 6s and 6p with $4f^w$.

20-4. SPECTRA

Although all lanthanide spectra are exceptionally complex, those in the right half of the series tend to be somewhat simpler than those in the left half, in that the former spectra tend to show contrasting strong and weak lines whereas lines of the latter tend to be more uniform in intensity.[8] The reasons for this are associated with the gradual contraction of the 4f orbital with increasing Z, so that the $4f^w$ subshell becomes more and more deeply buried within the atom, and therefore interacts less and less strongly with the outer valence

[16] G. Racah, J. Opt. Soc. Am. 50, 408 (1960); N. Spector, J. Opt. Soc. Am. 57, 308 (1967).

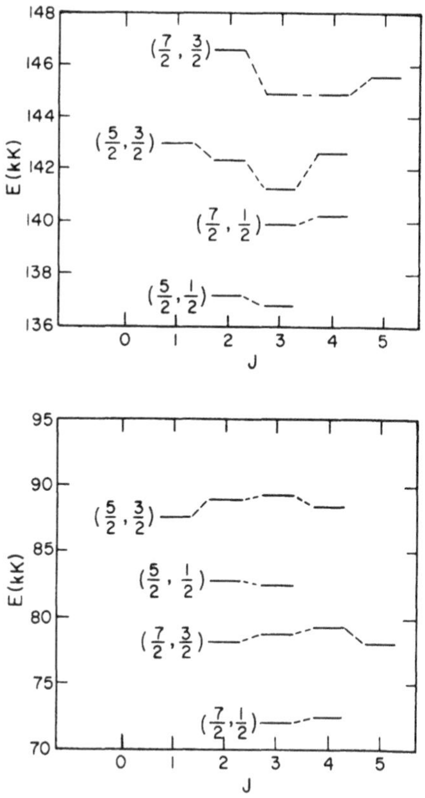

Fig. 20-5. Top: Energy level diagram for Pr IV 4f6p, showing bowl-shaped jj-coupling patterns. Bottom: Energy level diagram for Yb III $4f^{13}6p$, showing umbrella-shaped patterns.

electrons. Thus, for example, coupling conditions for large w approach quite closely the pure-coupling scheme (20.2), so that the neutral-atom transition array

$$4f^w(6s^2) - 4f^w(6s6p) \tag{20.3}$$

reduces essentially to the array $6s^2 - 6s6p$. As illustrated in Fig. 20-6, the only strong transitions are limited to

$$6s^2\ {}^1S_0 - 6s6p\ {}^1P_1^\circ\ ; \tag{20.4}$$

there can be no change in the core quantum numbers $\alpha_1L_1S_1J_1$, and (for given $\alpha_1L_1S_1J_1$) there are just three spectrum lines, corresponding to $J = J_1$ and $J' = J_1, J_1 \pm 1$. The myriads of other spectrum lines, corresponding to $6s6p\ {}^3P^\circ_{0,1,2}$ and/or to a change in $\alpha_1L_1S_1J_1$, are all orders of magnitude weaker.[17]

Similarly, for singly ionized atoms the array

$$4f^w6s - 4f^w6p \tag{20.5}$$

[17]D. C. Griffin, J. S. Ross, and R. D. Cowan, J. Opt. Soc. Am. 62, 571 (1972).

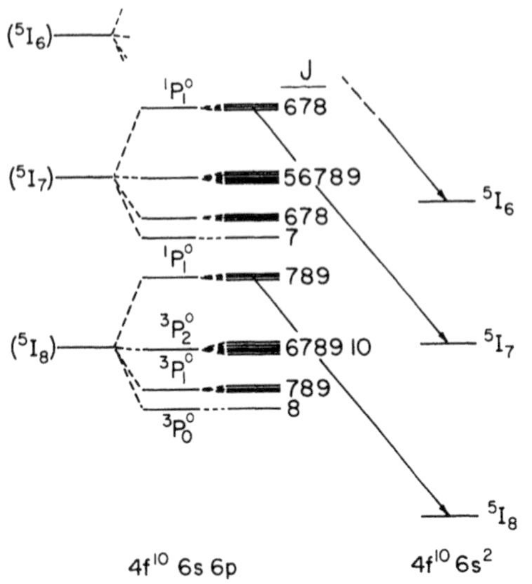

Fig. 20-6. Low energy levels of Dy I $4f^{10}6s^2\ ^5I_J$ and $4f^{10}(^5I_{J_1})$ $(6s6p\ ^{1,3}P^o_{J_2})J'$. Because of the weak interaction between 6s6p and the $4f^{10}$ core, the only strong transitions are the "triplets" corresponding to $J_1 = J$ and 6s6p $^1P^o_1$. (Relative splittings are illustrative only, and not to scale.)

reduces for large w to the array

$$6s\ ^2S_{1/2} - 6p\ ^2P^o_{1/2,3/2}\ , \tag{20.6}$$

and again the only strong lines are those for which $\alpha_1L_1S_1J_1$ does not change. The strong lines are, however, more numerous than in the neutral-atom example because one can have $J = J_1 \pm 1/2$ and $J' = J_1 \pm 1/2, J_1 \pm 3/2$ (subject of course to the usual dipole selection rule $J' = J, J \pm 1$). To a somewhat lesser degree, the array $4f^w5d$–$4f^w6p$ also gives strong lines only when the $4f^w$ core state does not change.

20-5. AIDS TO EMPIRICAL SPECTRUM ANALYSIS

The very high density of energy levels not only makes their theoretical interpretation difficult, but also greatly complicates even the deduction from spectra of the energies of the various levels. The basic empirical procedure (Sec. 1-11) is to search through the list of wavenumbers of observed spectrum lines, looking for multiply repeated wavenumber differences (or sums) that in principle indicate energy differences between pairs of levels. However, the constant-difference (or constant-sum) search can be made only within the limitation set by the accuracy of the experimental wavenumbers (usually, at best, about ± 0.01 to $0.002\ cm^{-1}$—see Table 1-2). The very high level density leads to large numbers of chance coincidences, and thereby a high probability of inference of fictitious energy levels. Elimination of these spurious "levels" requires consistency cross-checks made on the basis of additional types of information other than simply wavenumbers of spectrum lines. These additional data, which simultaneously provide aids to the theoretical interpretation of the real levels, are of several types.[1]

Zeeman effect. Measurement of a resolved Zeeman pattern provides J values and Landé g-values for both energy levels associated with the spectrum line. If spectrum lines that the constant-difference search has associated with a given tentative "level" do not all imply the same value of J and the same value of g for that level, then the level must be discarded as spurious. At the same time, values of J and g obtained for each real level provide valuable clues toward the theoretical interpretation of that level. Deficiencies with Zeeman effect observations include: (1) difficulty in accurate measurement of Zeeman components owing to overlapping of patterns from neighboring lines in these very complex spectra; (2) the tendency for all high lying levels to have much the same g-values (see Sec. 20-2 above and ref. 3); (3) unresolved (pseudo-triplet) Zeeman patterns, resulting from near equality of the two g-values, that give no information about J for either level (and none about either g, unless g is already known for one of the two levels involved).

Isotope shift. When analysis has proceeded far enough to establish a coherent level system, measured isotope shifts of spectrum lines can be converted to isotope shifts of levels. By and large, configuration mixings prove to be relatively small; consequently, levels can (to a limited extent) be grouped into configurations through approximate equality of isotope shift values, and each configuration tentatively identified as to the number of s electrons present (see Sec. 17-8). At the same time, isolated pairs of close-lying, strongly interacting levels can frequently be identified through their abnormal isotope shifts.[17]

Hyperfine structure. The presence of hfs splittings complicates the determination of accurate (center-of-gravity) line positions for use in the constant-difference wavenumber search; the probability of chance coincidences is increased, and this greatly hampers the process of finding energy levels. In addition, it becomes more difficult to make accurate Zeeman-effect and isotope-shift measurements to provide aids to level identifications.[18] Consequently, spectrum analysis of the odd-Z rare earths has lagged considerably behind that for the even-Z elements (even-even isotopes always having zero nuclear spin, and hence no hfs). On the other hand, hfs patterns can themselves provide information on J values and on the number of s electrons in the configuration involved (Sec. 17-9), and the pattern widths may provide invaluable level tags that can be used similarly to Landé g-values to eliminate spurious wavenumber differences.[18,19]

Intensities and temperature classification. Even qualitative observations of relative line intensities (in emission spectra) provide indispensable information. By varying the excitation conditions in the source, one can tentatively separate lines according to the ionization stage responsible, and even (to a limited extent) according to the excitation energy of the upper level within a given stage.[20] By and large, the strongest lines involve the ground level or other very low-lying levels. For large w especially, a group of very strong lines lying within a limited wavelength range ($\Delta\lambda/\lambda \cong 5\%$) very likely belongs to the array $4f^w6s^2 - 4f^w6s6p$ or $4f^w6s - 4f^w6p$, or possibly to the array $4f^w5d - 4f^w6p$ (Sec. 20-4). Arrays involving excitation of a 4f electron, such as $4f^w - 4f^{w-1}5d$ always involve many relatively weak lines, scattered over a very wide wavelength range.

[18]A. Giacchetti, J. Opt. Soc. Am. 57, 728 (1967).

[19]W. J. Childs and L. S. Goodman, J. Opt. Soc. Am. 69, 815 (1979).

[20]A. S. King, Astrophys. J. 68, 194 (1928); A. S. King, J. G. Conway, E. F. Worden, and C. E. Moore, J. Res. Natl. Bur. Stand. 74A, 355 (1970), and references therein.

Absorption spectra; self-reversal. Relatively cool atomic vapors contain mostly neutral atoms in the ground level and perhaps the lowest one or two excited levels. Absorption spectra of such vapors are thus vastly simpler than emission spectra of hot vapors, and can immediately indicate those lines that involve low levels.[20] Even in emission sources—especially in the outer portions of dc arcs—the lowest few levels are highly populated, and self-reversed lines (Sec. 1-9) almost certainly involve such levels.

Laser excitation. Tunable dye lasers can be used to excite individual levels of atoms in a cool vapor. When strong fluorescence is observed, the laser is tuned to one of the spectrum lines involving a low-lying level, and all fluorescent lines observed (having wavelengths less than twice that of the exciting line) involve the same upper level; if the energy of the upper level is known, then each fluorescent wavelength immediately establishes the energy of another level.

Crystal and solution spectra. Lanthanide atoms in molecules, solutions, or crystal matrices are usually triply ionized and (except for lanthanum) exist in a configuration $4f^w$. Because the 4f electrons are so deeply buried within the ion, they are well shielded from environmental influences and show quite sharp energy levels; the quantitative level structure of $4f^w$ is very nearly the same as in the isolated triply ionized atom, except for small Stark-effect broadening or crystal-field splittings.[5,6,11] At the same time, these external electric fields remove the rotational symmetry of the ion's environment and make possible strong intra-$4f^w$ transitions (see Secs. 14-12 and 17-7). Absorption- or fluorescence-spectra observation of these transitions (which conveniently lie in or near the visible region because of the large 4f − 4f Coulomb interactions) make it relatively easy to establish the $4f^w$ level structure. These data greatly aided the spectrum analysis of isolated triply ionized lanthanides, and served as a guide also to analysis of spectra of other ionization stages.

System differences. Because transition arrays that involve a 4f − 5d excitation (such as $4f^w6s^2 - 4f^{w-1}5d6s^2$) consist of large numbers of comparatively weak lines, it is very difficult to identify and analyze these arrays. In addition, the binding energies of 4f and 5d are so nearly equal that these arrays tend to lie in the experimentally difficult infrared region. Consequently, early spectrum analyses frequently resulted in two separate systems of energy levels A and B, the latter containing those low-lying configurations involving a $4f^w$ core, and the former those involving a $4f^{w-1}5d$ core. A major problem has been to identify lines connecting levels of these two systems, thus consolidating the entire set of energy levels and establishing the so-called system difference (SD): the difference between the lowest level of $4f^{w-1}5d6s^m$ and the lowest level of $4f^w6s^m$, where $m = 2$, 1, or 0 for neutral, singly ionized, or multiply ionized atoms, respectively. The utility of the SD lies in the fact that once known, the position of the lowest level of most other configurations of interest can be reliably predicted.[21]

By about 1970, a sufficient number of SD values had been determined to indicate the variation of the SD with Z along the lanthanide series.[7,8,21] Such curves (Fig. 20-7) show a sharp break between europium ($w = 7$) and gadolinium ($w = 8$). This break arises not from any discontinuity in the Z dependence of

$$E_{av}(f^{w-1}ds^m) - E_{av}(f^ws^w) \ , \tag{20.7}$$

[21] W. C. Martin, J. Opt. Soc. Am. 61, 1682 (1971).

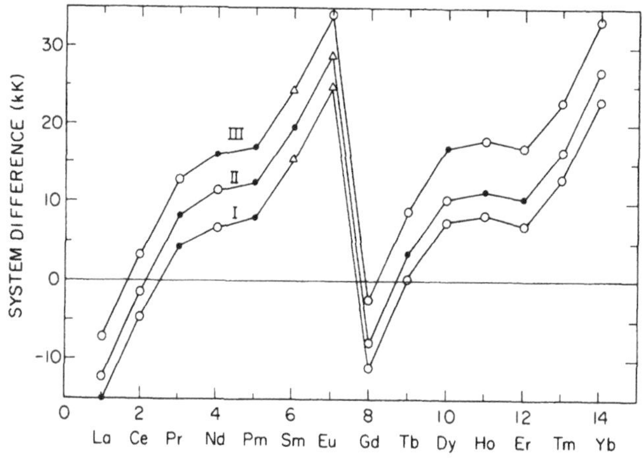

Fig. 20-7. The difference between the lowest level of $4f^{w-1}5d6s^m$ and the lowest level of $4f^w6s^m$ for neutral, singly ionized, and doubly ionized lanthanide atoms (m = 2, 1, and 0, respectively). Open circles, experimental values; triangles, estimates based on incomplete experimental data; closed circles, predicted values.

which is quite smooth, but rather arises from an obvious characteristic of Fig. 20-2: the distance between the lowest level and the center of gravity of f^w,

$$E_{av}(f^w) - E_{min}(f^w) \, , \tag{20.8}$$

increases up to w = 7 and then decreases. To a rough approximation, we can neglect the contribution to $E_{av} - E_{min}$ of adding 5d and 6s electrons, so that

$$SD \equiv E_{min}(f^{w-1}ds^m) - E_{min}(f^w s^m) \cong [E_{av}(f^{w-1}ds^m) - E_{av}(f^w s^m)]$$
$$+ [E_{av}(f^w) - E_{min}(f^w)] - [E_{av}(f^{w-1}) - E_{min}(f^{w-1})] \; ;$$

the first term here varies smoothly (and slowly) with w, but the total of the last two terms changes from about +20 kK at w = 7 to about −20 kK at w = 8. The form of the SD curves can be calculated quantitatively by semi-empirical methods,[22] but only in a rough qualitative form by *ab initio* methods: values of (20.7) are inaccurate because of the extreme relaxation effects on all orbitals produced by exciting an electron out of the deeply buried 4f subshell, and the calculational accuracy of

$$E_{av}(f^{w-1}ds^m) - E_{min}(f^{w-1}ds^m)$$

is poor because of the large size of the energy matrices, necessitating truncation of the f^{w-1} basis and neglect of configuration interaction.

Theoretical aids. In spite of the matrix-size difficulties just mentioned, theoretical calculations are indispensable for the interpretation of observed energy level structures. *Ab initio* calculations are mainly useful only for relatively low-lying levels. The high level density at higher energies necessitates least-squares fitting, with inclusion of two- and three-body effective Coulomb operators and spin-spin and spin-other-orbit magnetic operators in order to make the rms fitting error smaller than the mean energy difference between levels.[2,15] Very little has yet been done in the way of generalized least-squares fitting, in

[22]L. J. Nugent and K. L. Vander Sluis, J. Opt. Soc. Am. 61, 1112 (1971). A somewhat similar discussion of ionization energies has been given by K. Rajnak and B. W. Shore, J. Opt. Soc. Am. 68, 360 (1978).

TABLE 20-3. THEORETICAL GROUND CONFIGURATIONS OF LANTHANIDE AND TUNGSTEN IONS[a]

Spectrum	$_{60}$Nd	$_{64}$Gd	$_{68}$Er	$_{74}$W
V	$5p^64f^2$	$5p^64f^6$	$5p^64f^{10}$	$5p^64f^{14}5d^2$
VI	$5p^64f$	$5p^64f^5$	$5p^64f^9$	$5p^64f^{14}5d$
VII	$5p^6$	$5p^64f^4$	$5p^64f^8$	$5p^64f^{14}$
VIII	$5p^44f$	$5p^54f^4$	$5p^54f^8$	$5p^64f^{13}$
IX	$5p^34f$	$5p^34f^5$	$5p^44f^8$	$5p^44f^{14}$
X	$5p4f^2$	$5p^24f^5$	$5p^24f^9$	$5p^34f^{14}$

[a]R. D. Cowan, refs. 4 and 26.

which levels of all lanthanides are simultaneously fit by varying the coefficients in analytical functions of Z for each Coulomb and magnetic integral.[8]

Summary. The difficulties in analysis and interpretation of rare-earth spectra are so great that as many as possible of the above aids have to be utilized simultaneously. For example, establishment and identification of the low levels of Dy I $4f^{10}6s^2$ and $4f^95d6s^2$ involved intensity, self-reversal, and Zeeman-effect data, together with *ab initio* theoretical calculations and least-squares fitting;[23] identification of $4f^{10}6s6p$ levels involved intensity, Zeeman-effect, and isotope-shift data, combined with *ab initio* and least-squares calculations.[17]

20-6. IONS

The first two ionization stages of the lanthanides involve simply the removal of the two 6s electrons, except that[24] the core of Ce changes from 4f5d to $4f^2$. The third ionization then removes a 4f electron, except for the 5d electron still present in La and Gd, leaving the ground configurations RE IV $5s^25p^64f^w$, where w runs from 0 to 13 for La to Yb. Thereafter, one might expect further ionizations to involve removal of one 4f electron at a time until all are gone and each ion is reduced to the xenon core $5s^25p^6$. No experimental data are available, but theoretical calculations indicate that the above straightforward approach to the xenon core continues only to about the VIIth ionization stage (Table 20-3). At this point the semi-hydrogenic nature of highly ionized atoms (Sec. 19-2) starts to come into evidence, and the binding energy of a 4f electron becomes greater than that of a 5p electron. Thereafter, ionization involves removal of a 5p electron—sometimes accompanied by transfer of an electron from the 5p to the 4f subshell. The presence of two open subshells in the ground configuration means that spectra of these ions will be extremely complex. As yet, analyses have not proceeded beyond the IVth or Vth spectrum.[8]

[23]J. G. Conway and E. F. Worden, J. Opt. Soc. Am. 61, 704 (1971).

[24]See Table 4-3 and W. C. Martin, L. Hagan, J. Reader, and J. Sugar, J. Phys. Chem. Ref. Data 3, 771 (1974).

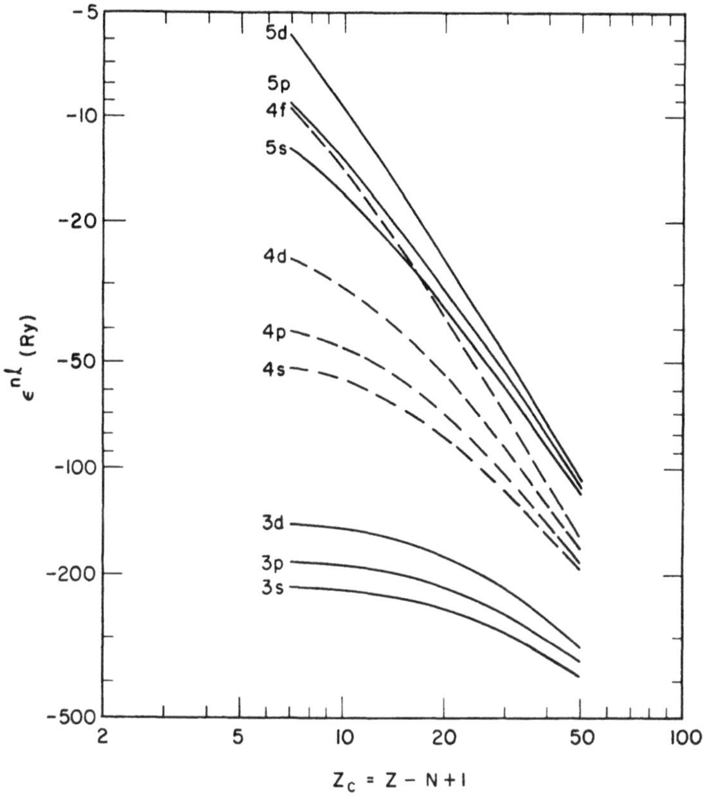

Fig. 20-8. Binding energies of electrons in tungsten ions (Z = 74), as estimated by one-electron HFR eigenvalues and Koopmans' theorem.

Ionization of elements heavier than ytterbium will of course eventually result in ions having ground configurations with open 4f subshells, and hence spectra similar to those of rare-earth ions. Examples for tungsten are included in Table 20-3, which show that the 4f subshell is first broken into at W VIII; this has been tentatively verified experimentally by Sugar and Kaufman.[25] Beyond this point, however, HFR calculations[26] predict that the 4f subshell will be full (except for W XII $5p^2 4f^{13}$) until all 5p electrons have been removed; the 4f electrons then start coming off one after the other, except for 5s electrons at W XVII $4d^{10}5s4f^{11}$ and W XVIII $4d^{10}4f^{11}$. The changes in 4f binding energy from non-hydrogenic to semi-hydrogenic values are illustrated further in Fig. 20-8.[26]

[25] J. Sugar and V. Kaufman, Phys. Rev. A 12, 994 (1975), Tables VI and VII.

[26] R. D. Cowan, Los Alamos Scientific Laboratory Report LA-6679-MS (January 1977).

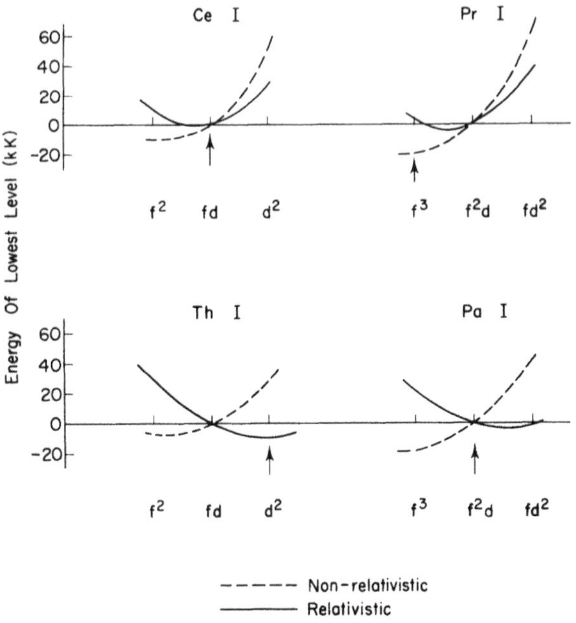

Fig. 20-9. Theoretical energies of the lowest levels of f^w and $f^{w-2}d^2$ configurations, relative to that of $f^{w-1}d$. For each element, the experimental ground configuration is indicated by an arrow.

*20-7. ACTINIDES

All that has been said above about the lanthanide elements ${}_{57}$La to ${}_{70}$Yb is qualitatively true also of the actinides ${}_{89}$Ac to ${}_{102}$No, except with all principal quantum numbers increased by one (5f, 6d, 7s, 7p instead of 4f, 5d, 6s, 6p). The salient features of actinides have been discussed by Fred.[27] The principal quantitative differences from the lanthanides are the following:

(1) It may be seen from Fig. 8-10 that the 5f orbitals tend to be more diffuse and to lie less deeply buried in the core. (On the Bohr orbit picture, the 5f orbit is elliptical rather than circular.) Correspondingly, 5f electrons tend to be relatively weakly bound compared with 6d electrons, and the system-difference curve lies rather lower than the lanthanide curve shown in Fig. 20-7, at least for the left-hand half.[22,27] Thus, many more actinides than lanthanides have a d electron in the ground configuration (Table 4-3). Examples are shown in Fig. 20-9, which illustrates also the importance of relativistic effects in the calculation of nf $\rightarrow$ (n+1)d excitation energies, especially in the actinides. [Indeed, because of relativistic effects the 5f electron has not even collapsed into the core until Z = 90 (Table 8-8); if we take this collapse to define the beginning of the actinides, then we end up with the amusing situation that lanthanum is a lanthanide but actinium is not an actinide.]

(2) Because of the diffuseness of the 5f orbitals, the Coulomb integrals F^k(5f,5f) are smaller than the corresponding lanthanide integrals; at the same time, the spin-orbit radial

[27] M. Fred, "Electronic Structure of the Actinide Elements," in R. F. Gould, ed., *Lanthanide/Actinide Chemistry* (American Chemical Society, Washington, D.C., 1967), p. 180.

integrals are much larger because of the higher Z (see Fig. 8-13). Both of these act to make the coupling conditions closer to jj and farther from LS coupling.

(3) Also because of the greater 5f radii, 5f-7s and 5f-7p interactions tend to be stronger than the corresponding lanthanide interactions. The special coupling conditions discussed in Sec. 20-4 are thus not so appropriate for the actinides, and the spectra tend to be more complicated in that there are fewer prominent very strong lines.

(4) Because of the higher Z and larger nuclei, isotope shifts and hyperfine-structure splittings tend to be much larger in the actinides. This offers both advantages and disadvantages in the empirical analysis of spectra, as discussed in Sec. 20-5.

Finally, the actinide elements are of course all radioactive—many of them extremely so. The associated handling problems and the extremely limited availability of the transplutonium elements compound the already formidable problems of spectrum analysis. As a consequence, much less is known about the structure and spectra of the actinides than of the lanthanides (Table 1-4 and AEL[9]). Very little is known about atoms more than once ionized, except for U V and a few lines of U VI,[28] and for intra-$5f^w$ transitions in crystal and solution spectra of various oxidation states.[29]

20-8. TRANSITION ELEMENTS

The transition-series elements ${}_{21}Sc$—${}_{30}Zn$, ${}_{39}Y$—${}_{48}Cd$, and ${}_{71}Lu$—${}_{80}Hg$ have structures and spectra much less complex than do the rare-earth elements (Fig. 20-1), but there nonetheless are a number of similarities between the two types of series.

Each transition series is preceded by the collapse into the atom core of an nd orbital (n = 3, 4, and 5, respectively—see Figs. 8-6 to 8-11), and involves the subsequent filling of the nd subshell, which in neutral and singly ionized atoms is not the outermost subshell (Table 4-3). Observed levels belong to the configurations

$$nd^w(n+1)s^2 ,$$
$$nd^{w+1}(n+1)s ,$$
$$nd^{w+2} ,$$
$$nd^w(n+1)p^2 ,$$
$$nd^{w+1}n'l' ,$$
$$nd^w n'l'n''l'' .$$

The nd and (n+1)s electrons have similar binding energies and so the first three configurations listed have comparable energies; being all of the same parity, configuration interactions among all three are generally important.

[28] J.-F. Wyart, V. Kaufman, and J. Sugar, Phys. Scr. 22, 389 (1980); V. Kaufman and L. J. Radziemski, Jr., J. Opt. Soc. Am. 66, 599 (1976).

[29] See L. P. Varga, J. D. Brown, M. J. Reisfeld, and R. D. Cowan, J. Chem. Phys. 52, 4233 (1970), and references therein.

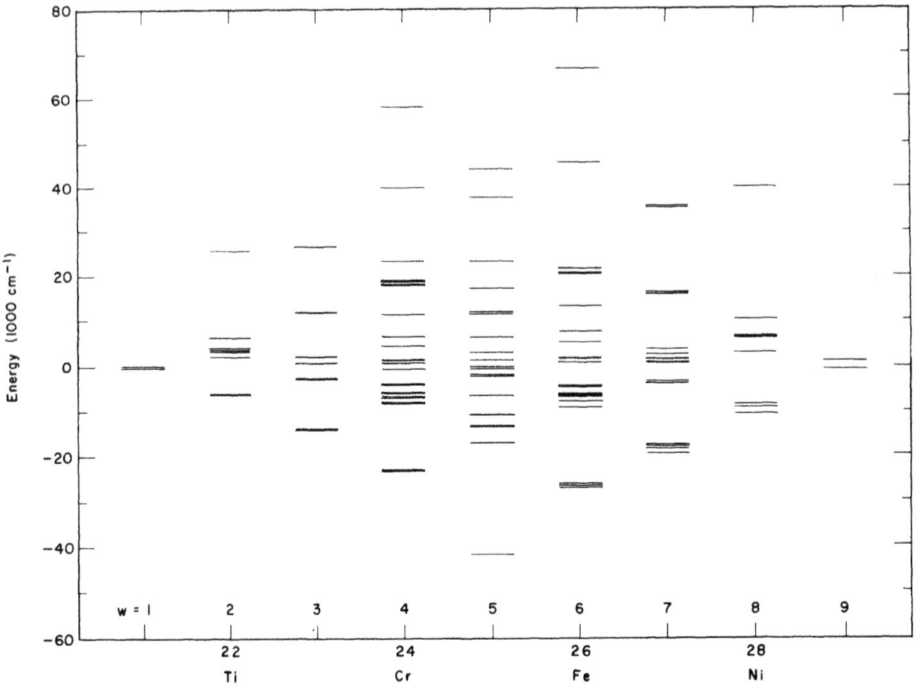

Fig. 20-10. Theoretical energy levels of configurations $3d^w4s^2$ of neutral atoms of the first transition series, relative to the center-of-gravity energy E_{av}.

The basic character of each configuration is established by the level structure of the nd subshell; this structure is shown schematically in Fig. 20-10, analogous to Fig. 20-2 for the rare earths. The interaction of (n + 1)*l* electrons with the nd^w subshell is, however, considerably stronger than the interaction of 5d or 6*l* electrons with the $4f^w$ subshell of the lanthanides. Consequently, the transition elements do not so strongly show simplifications in level structures and spectra analogous to (20.2)-(20.6).

It is evident from Fig. 8-13 that for the first transition series, conditions within the $3d^w$ subshell lie close to pure LS coupling; the same coupling is valid to a good approximation even when one or two 4*l* electrons are added. This circumstance made possible very early work on analysis of these spectra, beginning with Catalán's discovery[30] of multiplets in Mn I and II and continuing through the remarkably complete and accurate work of Kiess, Meggers, Russell, Shenstone, and others,[31] based at most on the vector model of the atom, with no quantitative level structure calculations whatever. These spectra have since been treated theoretically in great detail, using effective operators and generalized least-squares

[30] M. A. Catalán, Phil. Trans. Roy. Soc. London 223, 127 (1923).

[31] See references given in AEL, Vol. II.

fitting techniques,[32] and some of the simpler regularities have been discussed by Edlén.[33] The energy difference between lowest levels of $d^{w-1}s$ and d^w was discussed fairly early by Catalán, Rohrlich, and Shenstone,[34] and the system-difference curve shows a sharp break at the position of the half-filled shell quite analogous to that for the lanthanides.

The 4d and 5d transition elements show coupling conditions progressively farther from LS coupling, and their spectra and level structures are correspondingly more poorly known than for the 3d elements (see Table 1-4). Still less, of course, is known about ions of heavier elements isoelectronic with the transition elements.

[32]See, for example, Y. Shadmi, Phys. Rev. 139, A43 (1965); C. Roth, J. Res. Natl. Bur. Stand. 73A, 125 (1969), and references therein.

[33]B. Edlén, Opt. Pura y Apl. 5, 101 (1972); Phys. Scr. 7, 93 (1973); article in I. A. Sellin and D. J. Pegg, eds., *Beam-Foil Spectroscopy* (Plenum, New York, 1976), Vol. 1.

[34]M. A. Catalán, F. Rohrlich, and A. G. Shenstone, Proc. Roy. Soc. (London) A221, 421 (1954); see also D. Layzer, Astrophys. J. 122, 351 (1955); S. Johansson, U. Litzén, J. Sinzelle, and J.-F. Wyart, Phys. Scr. 21, 40 (1980). An analogous, though less pronounced, break in the middle of p^w subshells appears in the ionization-energy curves of Fig. 4-3; see also B. Edlén, in W. E. Brittin and H. Odabasi, eds., *Topics in Modern Physics* (Colorado Associated University Press, Boulder, 1971).

21

STATISTICAL DISTRIBUTIONS

In order to indicate reasons why statistical distributions of energy levels and spectrum lines can be of interest, we first digress with brief remarks regarding partition functions and radiative opacities.

*21-1. THERMODYNAMIC FUNCTIONS OF ATOMIC GASES

An atomic vapor or plasma may to a first approximation be considered as a system of non-interacting particles (atoms, or ions and free electrons). Under conditions of complete thermal equilibrium, the contribution of the internal structure of the atoms or ions to the thermodynamic properties of the vapor may be derived from the internal partition function (18.83),

$$u(T) = \sum_j g_j e^{-E_j/kT} , \tag{21.1}$$

where E_j is the energy of the jth level, and $g_j = 2J_j + 1$ is its statistical weight. For example, using the Boltzmann distribution function (18.82) for the density n^j of atoms in the level j, we may write the internal energy density of the atoms or ions of the gas in the form

$$\begin{aligned} E/V &= \sum_j n^j E_j \\ &= \frac{n}{u} \sum_j g_j E_j e^{-E_j/kT} \\ &= \frac{nkT^2}{u} \cdot \frac{du}{dT} . \end{aligned} \tag{21.2}$$

We have seen also, in Sec. 18-9, how the partition function enters into the calculation of Saha ionization equilibrium.

Evaluation of the partition function of course requires knowledge of all possible energy levels of the atom, or at least of all levels that contribute significantly to the total.[1] Except at low temperatures, the necessary levels are seldom all known experimentally, even for atoms with relatively simple structures. At temperatures greater than 0.1 or 0.2 of the ionization energy, gross errors may result if unknown energy levels are ignored. However, the missing ones can be computed with reasonable accuracy by the methods described in Chapter 16.

*21-2. OPACITIES OF THICK PLASMAS

An important aspect of the theory of stellar structure and of numerous other problems involving high temperature plasmas is the treatment of the diffusion of radiation as a result of temperature gradients within the plasma. Properly speaking, this treatment requires knowledge of the frequency-dependent absorption coefficient κ_σ of the plasma. However, to a good approximation in optically thick plasmas under conditions of local thermodynamic equilibrium, calculations can be greatly simplified by introducing the Rosseland mean opacity κ defined by[2]

$$\frac{1}{\kappa} = \frac{\displaystyle\int_0^\infty \frac{1}{\kappa'_\sigma}\frac{d\rho(\sigma)}{dT}\,d\sigma}{\displaystyle\int_0^\infty \frac{d\rho(\sigma)}{dT}\,d\sigma}\,, \tag{21.3}$$

where $\rho(\sigma)$ is the black-body distribution function (14.13) and κ'_σ is the monochromatic absorption coefficient [modified as in (18.92) to allow for induced emission] plus scattering coefficient.

Important contributions to κ'_σ and κ may arise from bound-bound (discrete-spectrum-line) transitions within complex ions having half a dozen or more bound electrons. The importance of the lines arises largely because they fill up the regions of low continuous absorption just below the bound-free absorption edges. An example is shown in Fig. 21-1 for an iron plasma at $kT = 60$ eV and density of 0.052 g/cm^3 ($n_e = 7.0 \cdot 10^{21}$ cm^{-3}), where line contributions are computed to constitute about 85 percent of the total Rosseland mean opacity.[3]

[1]The summation in (21.1) is at best semi-convergent, because in going up a Rydberg series of configurations $\cdots nl$, E_j approaches a finite limit whereas the total statistical weight of the configurations of all l increases as n^2. The summation must therefore be cut off at some finite value of n, corresponding roughly to the n at which the orbital nl has a mean radius equal to half the mean distance between atoms.

[2]See, for example, A. N. Cox, in L. H. Aller and D. B. McLaughlin, eds., *Stellar Structure* (University of Chicago Press, Chicago, 1965), pp. 201 and 224.

[3]A. L. Merts and N. H. Magee, Jr., private communication. The most abundant ion stages under these conditions are Fe XII-XV.

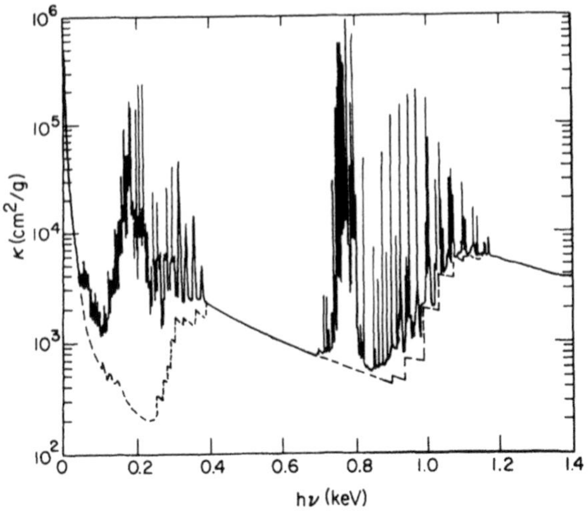

Fig. 21-1. Theoretical monochromatic absorption cross section for an iron plasma at $kT = 60$ eV and a density of 0.052 g/cm^3. Dashed curve: Continuous absorption resulting from bound-free transitions plus (important only at low photon energies) free-free transitions (inverse bremsstrahlung). Solid curve: Total cross section, including bound-bound absorption.

*21-3. ENERGY DISTRIBUTION OF LEVELS OF A CONFIGURATION

Because of the effective averaging of energy levels, wavelengths, and oscillator strengths involved in (21.1) and (21.3), it is not always necessary to use accurate values of these quantities; knowledge of approximate statistical distributions may suffice, except perhaps for the lowest-energy terms in the partition function. In the case of the very complex transition and rare-earth atoms, statistical treatments become almost essential.

A great deal of theoretical work has been done on the distribution of levels with respect to energy. However, most of this has been concerned with the distribution of energy *differences* between successive levels of given J,[4] or with the distribution of levels having a given value of the angular momentum projection M,[5] with applications primarily to nuclear energy levels. A number of empirical studies have been made of the level distribution for all known levels of an atom, particularly those of neutral iron.[6]

We shall here consider the energy distribution of quantum states (or of levels k, each weighted with its statistical weight $g_k = 2J_k + 1$) belonging to a single configuration. The simplest possible distribution, sometimes used for very rough work, is the rectangular distribution shown in the top portion of Fig. 21-2, in which dg/dE is constant between the lowest and highest energies, E_1 and E_2, of the configuration:

$$\frac{dg}{dE} = \left\{ \begin{array}{ll} c\,, & E_1 \leqq E \leqq E_2\,, \\ 0\,, & \text{otherwise}\,. \end{array} \right\} \qquad (21.4)$$

[4]C. E. Porter, *Statistical Theories of Spectra: Fluctuations* (Academic Press, New York, 1965); B. G. Wybourne, J. Opt. Soc. Am. 55, 928 (1965).

[5]J. N. Ginocchio and M. M. Yen, Nucl. Phys. **A239**, 365 (1975).

[6]C. W. Allen, Mon. Not. R. Astron. Soc. **168**, 121 (1974), and references therein.

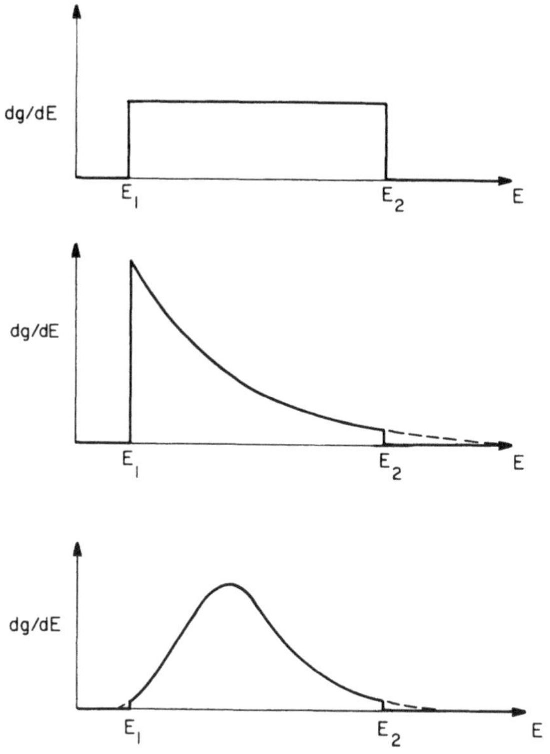

Fig. 21-2. Hypothetical statistical-weight distributions: top, rectangular; middle, exponential; bottom, skewed Gaussian.

A more general, and somewhat better form is the exponential distribution[7] shown in the middle of the figure:

$$\frac{dg}{dE} = \begin{cases} c e^{-a(E-E_1)} , & E_1 \leqq E \leqq E_2 , \\ 0 , & \text{otherwise} . \end{cases} \tag{21.5}$$

However, even a cursory inspection of the level plots for f^w configurations shown in Fig. 20-2 indicates that neither of the above is very realistic. A more appropriate form is the skewed Gaussian[8] shown in the bottom of Fig. 21-2:

$$\frac{dg}{dE} = \begin{cases} c\left[1 - \frac{1}{2}\alpha_3\left(x - \frac{1}{3}x^3\right)\right] e^{-x^2/2} , & E_1 \leqq E \leqq E_2 , \\ 0 , & \text{otherwise} , \end{cases} \tag{21.6}$$

[7]S. A. Moszkowski, Prog. Theor. Phys. **28**, 1 (1962).

[8]F. E. Croxton, D. J. Cowden, and S. Klein, *Applied General Statistics* (Prentice-Hall, Englewood Cliffs, N. J., 1967), 3rd ed., Chap. 10 and p. 533.

where

$$x = (E - E_{av})/d \ , \tag{21.7}$$

$$\begin{aligned} E_{av} &= \langle E_k \rangle , \\ d &= \langle (E_k - E_{av})^2 \rangle^{1/2} , \\ \alpha_3 &= d^{-3} \langle (E_k - E_{av})^3 \rangle ; \end{aligned} \tag{21.8}$$

calculation of each of the three energy moments is to be carried out using the weighting factor g_k for the kth level. The constant c is determined by the fact that the total statistical weight of the configuration is given by (4.63), (4.65):

$$\int_{E_1}^{E_2} \frac{dg}{dE} dE = \prod_{i=1}^{q} \binom{4l_i + 2}{w_i} \ . \tag{21.9}$$

Examples of theoretically computed distributions for the configuration f^4 are shown in Fig. 21-3 in the form of histograms. Shown for comparison are the corresponding skewed Gaussians (21.6); in each case, the required values of E_{av}, d, and α_3 were obtained by numerical evaluation of (21.8) using the computed energy levels E_k. For part (a) of the figure, the values used for $F^k(ff)$ and ζ_f were appropriate to neutral elements near the beginning of the lanthanide series. For (b), (c), and (d), the same F^k were used, but ζ_f was successively increased to illustrate the changes in the level distribution and in the skewness parameter α_3 as conditions change from fairly pure LS coupling to nearly pure jj coupling. In (d), the five groups of levels correspond to the five possible sets of quantum numbers

$$\left(\frac{5}{2}\frac{5}{2}\frac{5}{2}\frac{5}{2}\right) \ \left(\frac{5}{2}\frac{5}{2}\frac{5}{2}\frac{7}{2}\right) \ \left(\frac{5}{2}\frac{5}{2}\frac{7}{2}\frac{7}{2}\right) \ \left(\frac{5}{2}\frac{7}{2}\frac{7}{2}\frac{7}{2}\right) \ \left(\frac{7}{2}\frac{7}{2}\frac{7}{2}\frac{7}{2}\right) \ .$$

The change from positive to negative skewness in going from pure LS to pure jj coupling (for a less-than-half-filled subshell) may appear somewhat surprising at first, but becomes plausible when one considers the following: (1) by Hund's rule (Sec. 4-16), the lowest levels under LS coupling conditions belong to the term of largest S and L, and this term carries the greatest statistical weight; (2) under jj-coupling conditions, electrons with $j = l + 1/2$ have higher energy than those with $j = l - 1/2$ [Eq. (10.13)], and the former of course carry greater statistical weight than the latter.

For more-than-half-filled subshells, the level distribution under pure LS-coupling conditions is identical with that for the conjugate less-than-half-filled subshell [see the text following (12.17)], and hence $\alpha_3 > 0$. Under pure jj coupling conditions, it follows from (12.39)-(12.42) and (11.73) that level distributions for conjugate subshells are mirror images of each other, and hence that $\alpha_3 > 0$ for more-than-half-filled subshells. (Less energy is required to remove a $j = l + 1/2$ electron than a $j = l - 1/2$ electron from a filled subshell, and hence the high-statistical-weight hole has the *lower* energy.)

The detailed variation of α_3 with coupling conditions is shown for f^4 and f^{10} in Fig. 21-4. In general, α_3 is positive for all configurations that are complex enough to justify statistical treatment (i.e., transition and rare-earth elements involving partially filled nd^w or nf^w

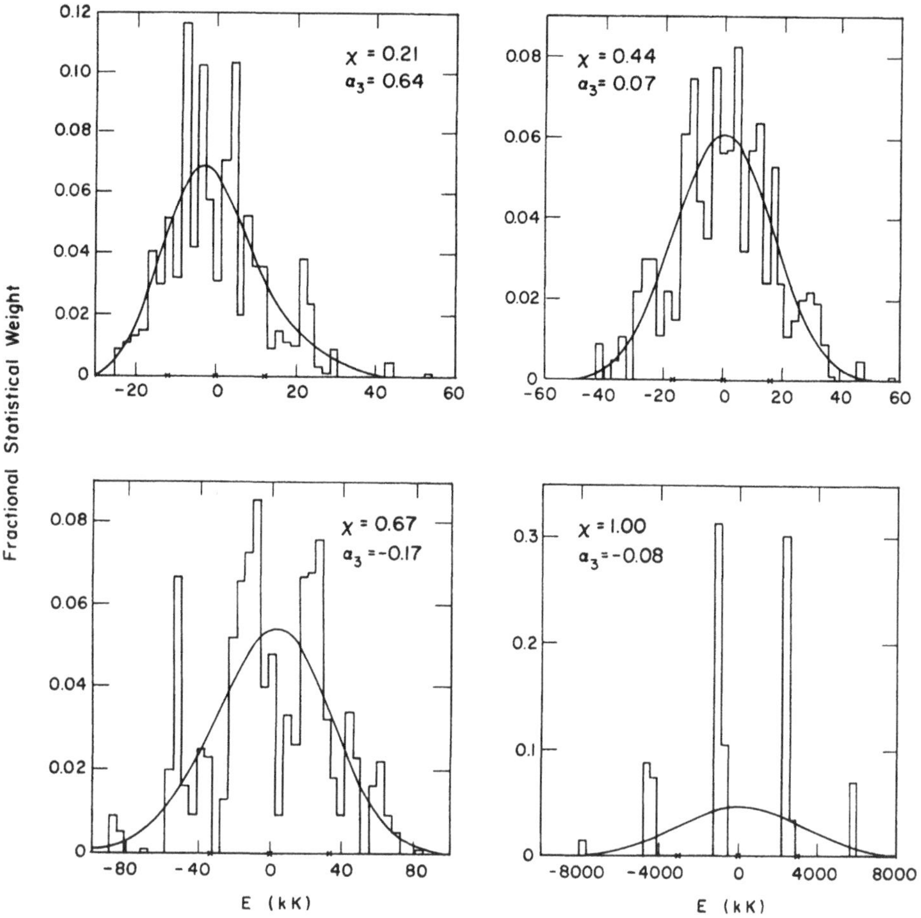

Fig. 21-3. Computer calculated and plotted histograms of the statistical-weight distribution for energy levels of the configuration f^4 (total number of levels 107, total statistical weight 1001), for $F^2 = 46$, $F^4 = 29$, and $F^6 = 21$ kK, with variable ζ_f; $\chi \equiv \zeta_f/(5 + \zeta_f)$. The curves represent the skewed Gaussian function (21.6). The three X's along the abscissa scale represent E_{av} and $E_{av} \pm d$.

subshells), because Coulomb interactions are always stronger than spin-orbit interactions in such subshells. However, α_3 does decrease appreciably with a change in Z that involves an increase in either w or n, for reasons that are evident from Fig. 8-13; similarly, α_3 decreases with increasing Z along an isoelectronic sequence [cf. (19.13)-(19.14)]. Several examples are given in Table 21-1. The detailed distributions for four of the examples are shown in Fig. 21-5.

If facilities for making detailed theoretical calculations are not available, a possible procedure for determining the constants in (21.6) for any configuration of interest is the following: (1) estimate a value of α_3 from Fig. 21-4 or Table 21-1; (2) obtain the lowest energy of the configuration, E_1, from experimental data (AEL, etc.) or from Brewer's

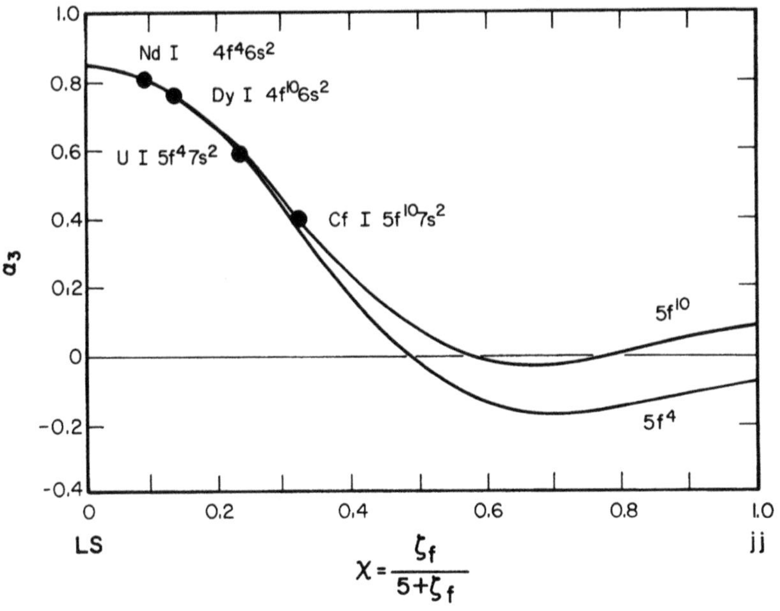

Fig. 21-4. Computed variation with coupling conditions of the skewness parameter α_3, for configurations f^4 and f^{10}. The circles indicate theoretical values for neutral lanthanide and actinide elements.

predictions;[9] (3) estimate the contribution of l^w to the configuration width $E_2 - E_1$ from Figs. 20-2, 20-10, and 8-13 (scaled upward for ions with the aid of Table 21-1), and increase the result by up to ten or twenty percent if there are other unfilled subshells; (4) choose E_{av} and d such that (21.6) has about five percent of its maximum value at $E = E_1$ and $E = E_2$; (5) choose c to satisfy (21.9).

Various modifications of this procedure can easily be constructed. For example, for single-open-subshell configurations Moszkowski[7,10] has derived the Coulomb-interaction relation

$$d(l^w) = \left[\frac{w(w-1)(4l+2-w)(4l+1-w)}{8l(4l-1)}\right]^{1/2} d(l^2) \; . \tag{21.10}$$

[9]L. Brewer, J. Opt. Soc. Am. 61, 1101, 1666 (1971). See references given in the first article for similar results on transition elements.

[10]D. Layzer, Phys. Rev. 132, 2125 (1963).

TABLE 21-1. THEORETICAL SKEWED-GAUSSIAN PARAMETERS[a]

Configuration[b]	No. levels	$\sum g$	$E_2 - E_1$ (kK)	d (kK)	α_3
Fe I $3d^6 4s^2$	34	210	93	16.6	0.73
$3d^6 4s4p$	360	2520	116	18.4	0.59
Ru I $4d^6 5s^2$	34	210	70	12.4	0.66
$4d^6 5s5p$	360	2520	94	14.7	0.46
Os I $5d^6 6s^2$	34	210	70	13.0	0.39
$5d^6 6s6p$	360	2520	97	15.5	0.26
Fe II $3d^7$	19	120	47	12.7	0.82
Fe I $3d^7 4s$	38	240	49	12.6	0.77
V III $3d^3$	19	120	43	11.7	0.82
V II $3d^3 4p$	110	720	50	11.8	0.78
Zn IX $3d^3 4p$	110	720	129	28.6	0.75
Sr XVII $3d^3 4p$	110	720	237	49.5	0.55
Co III $3d^7 4p$	110	720	60	15.2	0.80
Se IX $3d^7 4p$	110	720	139	32.3	0.71
Mo XVII $3d^7 4p$	110	720	261	56.5	0.44
Ho I $4f^{11} 6s^2$	41	364	90	19.0	0.80
$4f^{10} 5d6s^2$	977	10010	172	25.9	0.70
W X $4f^{11}$	41	364	160	33.4	0.62
$4f^{10} 5d$	977	10010	302	44.7	0.53
Ho II $4f^{11} 6s$	82	728	91	19.0	0.80
$4f^{11} 6p$	242	2184	94	19.0	0.80
W IX $4f^{11} 6s$	82	728	161	33.3	0.62
$4f^{11} 6p$	242	2184	191	35.8	0.47
Ce I $4f^2 6s^2$	13	91	43	7.97	1.22
Pr I $4f^3 6s^2$	41	364	63	13.4	0.92
Nd I $4f^4 6s^2$	107	1001	119	18.1	0.81
Sm I $4f^6 6s^2$	295	3003	165	24.8	0.72
Eu I $4f^7 6s^2$	327	3432	168	26.3	0.71
Gd I $4f^8 6s^2$	295	3003	177	26.8	0.70
Dy I $4f^{10} 6s^2$	107	1001	152	23.2	0.74
Er I $4f^{12} 6s^2$	13	91	69	12.9	0.93
U I $5f^4 7s^2$	107	1001	88	13.3	0.60
Cf I $5f^{10} 7s^2$	107	1001	125	19.8	0.40

[a]Computed using HFR values of the Coulomb integrals F^k and G^k, scaled by a factor 0.80 for low ionization stages, 0.85 for IX and X spectra, and 0.88 for high ionization stages.

[b]The ground configurations of W X and W IX are computed to be $4f^{14} 5p^3$ and $4f^{14} 5p^4$, and the experimental ground configurations of Gd I and U I are $4f^7 5d6s^2$ and $5f^3 6d7s^2$, respectively.

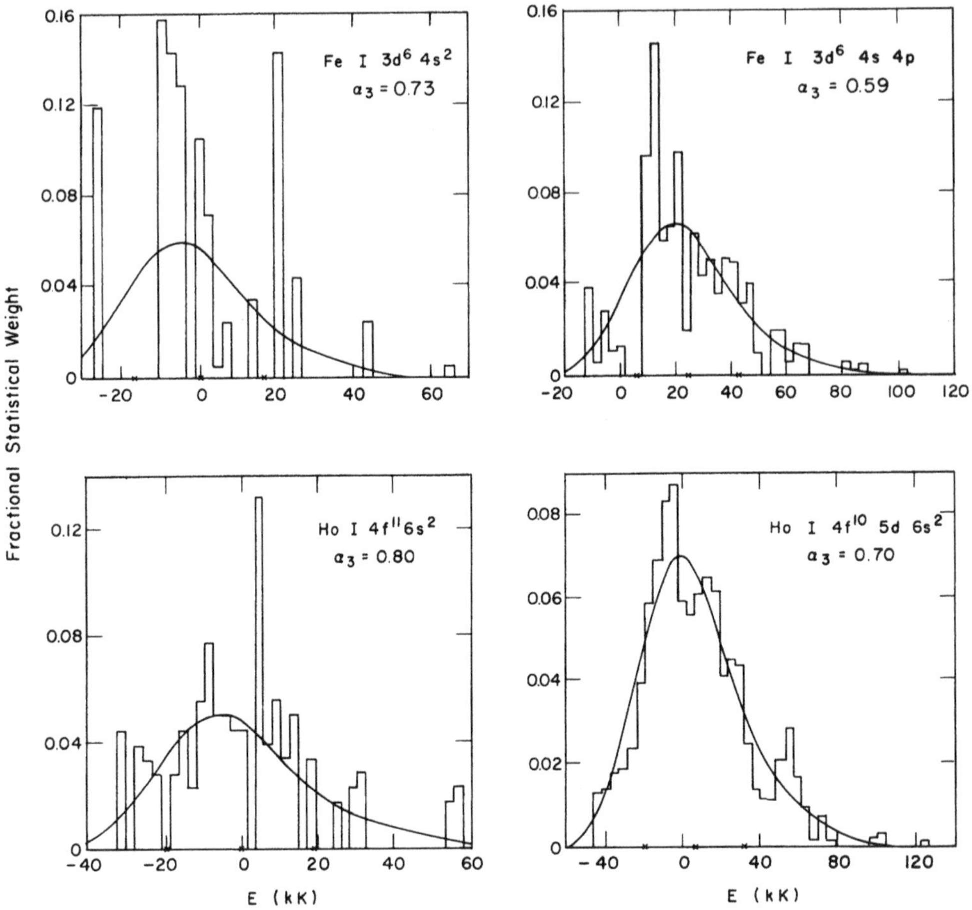

Fig. 21-5. Theoretical statistical-weight distributions for levels of Fe I and Ho I configurations included in Table 21-1.

From this, the variation of d with w can be obtained with good accuracy[11] provided the increase with w of the F^k is taken into account (via Fig. 8-13 or from HF calculations[12]). The value of d for l^2 can usually be obtained from experimental level values; alternatively, if the F^k are obtained as just mentioned, then $d(l^2)$ can be calculated fairly easily from the relatively simple expression (10.31)-(10.32) for the diagonal Coulomb-interaction matrix elements (which are also the energy levels if spin-orbit effects are neglected).

[11]For the neutral-lanthanide configurations $4f^w6s^2$ listed in Table 21-1, relative HFR values of the F^k are 1, 1.07, 1.13, 1.24, 1.28, 1.33, 1.42, 1.46, and 1.50; these indicate (21.10) to be accurate within one percent, except for $w \geq 10$ where the rapidly increasing spin-orbit effects become appreciable (2% for f^{10} and 9% for f^{12}).

[12]C. Froese Fischer, At. Data Nucl. Data Tables 12, 87 (1973); J. B. Mann, At. Data Nucl. Data Tables 12, 1 (1973); J. B. Mann and J. T. Waber, At. Data 5, 201 (1973) [DHF, lanthanides only]; and other references in Sec. XIV of the bibliography.

Detailed expressions are available from which d can be calculated for an arbitrary configuration if suitable values are available for the various F^k, G^k, and ζ.[13]

For application of (21.6) to the evaluation of partition functions, several points should be noted: (1) it is of course necessary to consider a separate skewed Gaussian for each configuration that can make a significant contribution to u(T) at the temperatures of interest; (2) experimental values should be used for low-lying levels, if known, and (21.6) then used only for energies greater than E_1', where

$$\int_{E_1}^{E_1'} \left(\frac{dg}{dE}\right) dE$$

is equal to the statistical weight of the experimental levels used individually in (21.1); (3) the upper limit of the skewed Gaussian should probably be truncated at the first ionization limit in order to delete autoionizing levels, and at an even lower energy for levels involving a highly excited electron.[1]

*21-4. WAVELENGTH AND OSCILLATOR-STRENGTH DISTRIBUTIONS WITHIN A TRANSITION ARRAY

Several empirical studies have been made of wavelength and oscillator-strength distributions, particularly for Fe I and other absorption lines in the solar spectrum;[6,14] all spectrum lines of a given ion were lumped together. Here we shall consider briefly computed distributions within a single transition array. Four examples are depicted in Figs. 21-6 to 21-9, shown in the form of histograms, except in the case of the distribution of number of lines by weighted oscillator strength.

In the case of W X $4f^{11} - 4f^{10}5d$ (Fig. 21-9), the two configurations do not overlap in energy, and the spectrum lines (upper-left portion of the figure) show nearly a symmetrical Gaussian distribution in spite of the asymmetries in the two energy level distributions.[7] The wavenumber distribution of weighted oscillator strength (upper-right section) is also fairly symmetrical, but is considerably narrower and the center is displaced somewhat to higher energies.[13,15] (For $\Delta n = 0$ transitions, especially in highly ionized atoms, the displacement is usually considerably more pronounced; see Sec. 19-5.) For the other three examples shown (Figs. 21-6 to 21-8), the two configurations overlap greatly in energy; the distribution of lines in each case is approximately a Gaussian, with the left-hand portion folded over about the line $\sigma = 0$ onto the right-hand portion.

In the Fe I $3d^64s^2 - 3d^64s4p$ and Ho II $4f^{11}6s - 4f^{11}6p$ examples, the wavenumber distribution of oscillator strength is quite narrow. This is a consequence of the weak interaction between the jumping electron and the core electrons ($3d^6$ or $4f^{11}$), so that to a

[13] C. Bauche-Arnoult, J. Bauche, and M. Klapisch, J. Opt. Soc. Am. **68**, 1136 (1978), Phys. Rev. A **20**, 2424 (1979). Expressions are also given for the width and energy shift (from ΔE_{av}) of line strength distributions in a transition array.

[14] J. P. Mutschlecner and C. F. Keller, Solar Phys. **14**, 294 (1970), **22**, 70 (1972).

[15] Further numerical examples are given by R. D. Cowan, J. Opt. Soc. Am. **58**, 924 (1968); Nucl. Instrum. Methods **110**, 173 (1973).

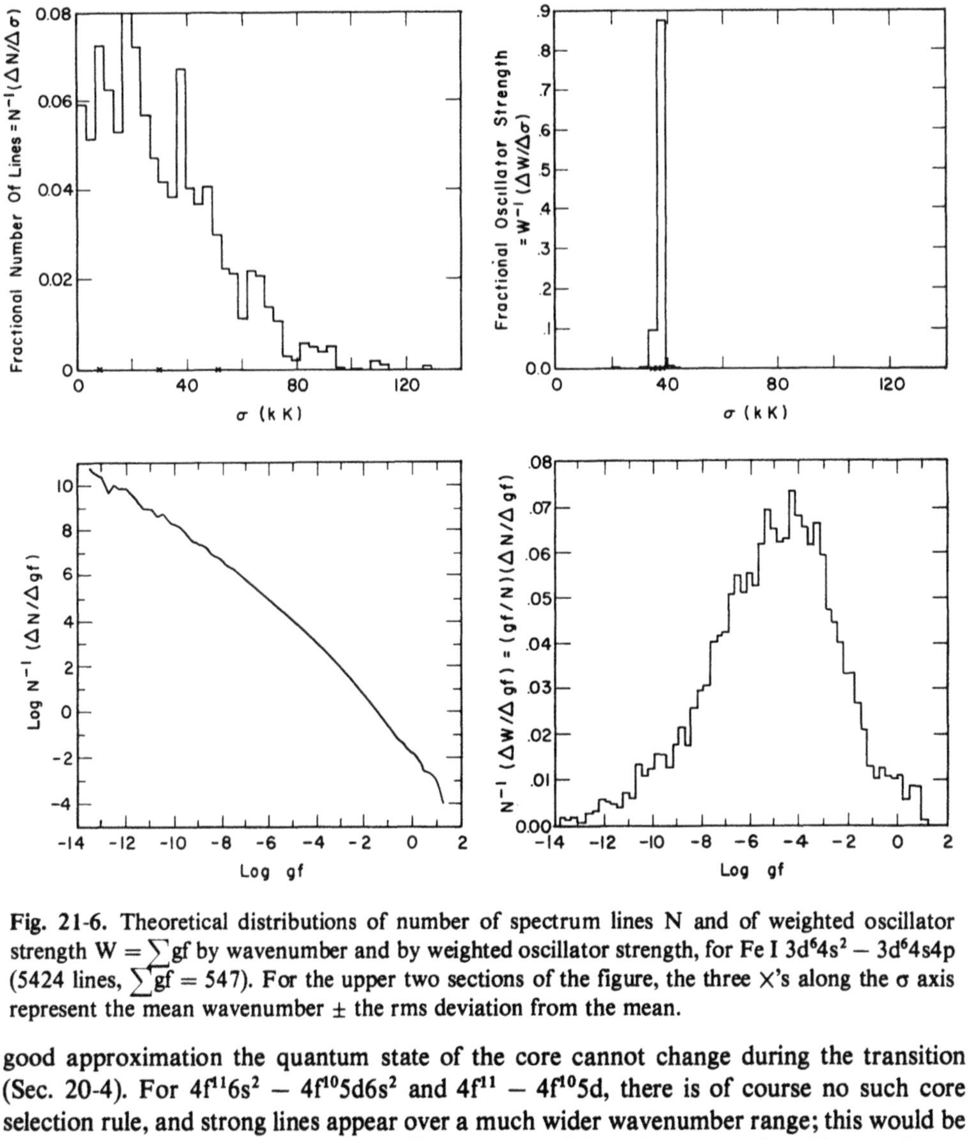

Fig. 21-6. Theoretical distributions of number of spectrum lines N and of weighted oscillator strength $W = \sum gf$ by wavenumber and by weighted oscillator strength, for Fe I $3d^6 4s^2 - 3d^6 4s4p$ (5424 lines, $\sum gf = 547$). For the upper two sections of the figure, the three X's along the σ axis represent the mean wavenumber $\pm$ the rms deviation from the mean.

good approximation the quantum state of the core cannot change during the transition (Sec. 20-4). For $4f^{11}6s^2 - 4f^{10}5d6s^2$ and $4f^{11} - 4f^{10}5d$, there is of course no such core selection rule, and strong lines appear over a much wider wavenumber range; this would be true also of transitions such as $f^w d - f^w p$, because of the fairly strong $f - d$ interaction.

The weighted oscillator strengths of spectrum lines cover a tremendous range—from values of the order unity down to values less than 10^{-10} (lower-left section of each figure). For Fe I, the computed distribution of number of lines by value of gf is, on a log-log plot, approximately a straight line with slope -1. For the strongest lines, in the range $-3 \leq \log gf \leq 0$, the slope is about -1.2; this agrees well with the slope

$$-1.2^{+0.1}_{-0.3}$$

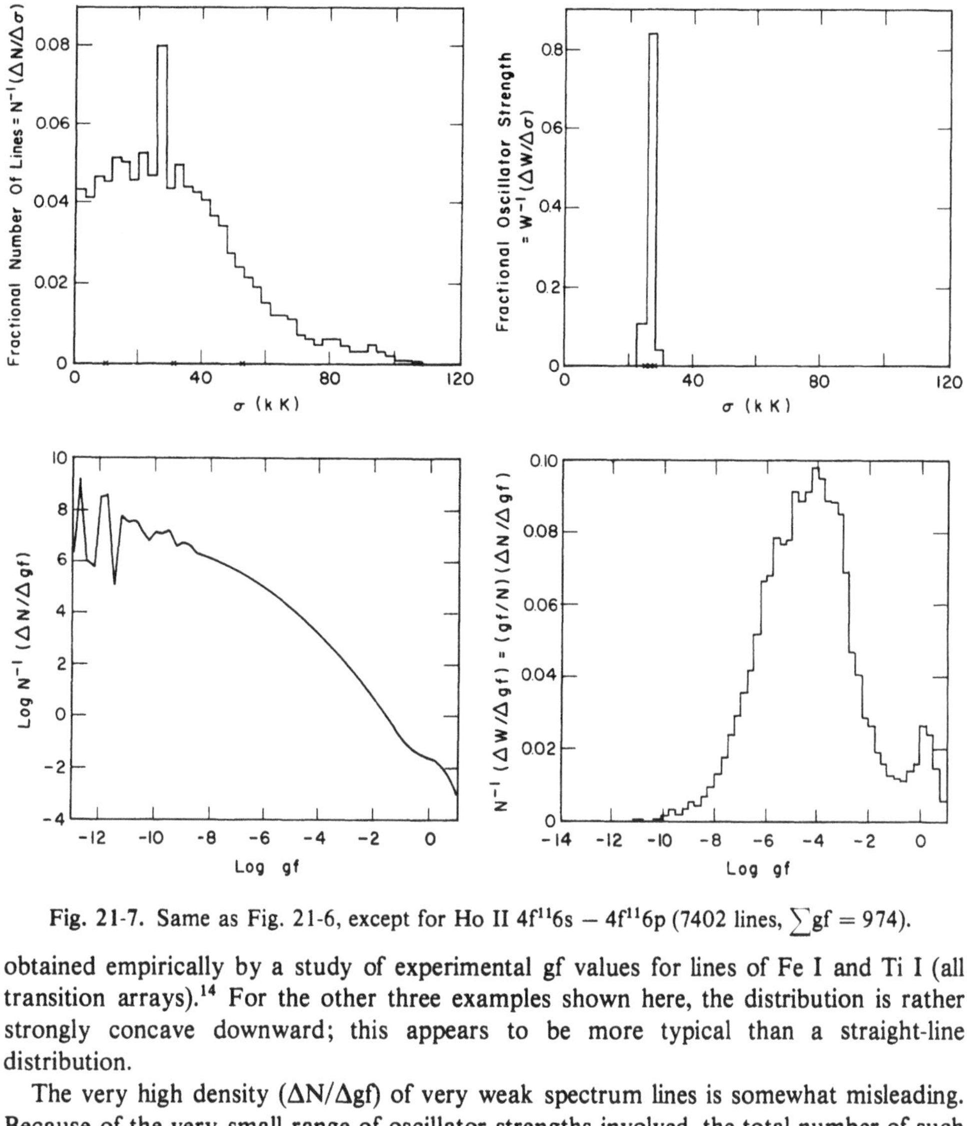

Fig. 21-7. Same as Fig. 21-6, except for Ho II $4f^{11}6s - 4f^{11}6p$ (7402 lines, $\sum gf = 974$).

obtained empirically by a study of experimental gf values for lines of Fe I and Ti I (all transition arrays).[14] For the other three examples shown here, the distribution is rather strongly concave downward; this appears to be more typical than a straight-line distribution.

The very high density ($\Delta N/\Delta gf$) of very weak spectrum lines is somewhat misleading. Because of the very small range of oscillator strengths involved, the total number of such lines is only a small fraction of the total; the oscillator strength carried by these lines is of course a completely negligible fraction of the total. This is illustrated in the lower-right section of Figs. 21-6 to 21-9, where the ordinates are those of the lower-left section weighted by gf. The resulting histogram (with ordinates multiplied by N) may be viewed in either of two ways: as the distribution of weighted oscillator strength $\Delta W/\Delta gf$, or as $\Delta N/\Delta(ln\ gf)$—the area of which gives the total number of lines. Using the latter interpretation, it may be seen that the median value of gf is about 10^{-5} to 10^{-3}, depending on the array; the mean value of gf (obtainable from the numbers given in the figure captions) is much larger—typically 10^{-2} to 10^{-1}.

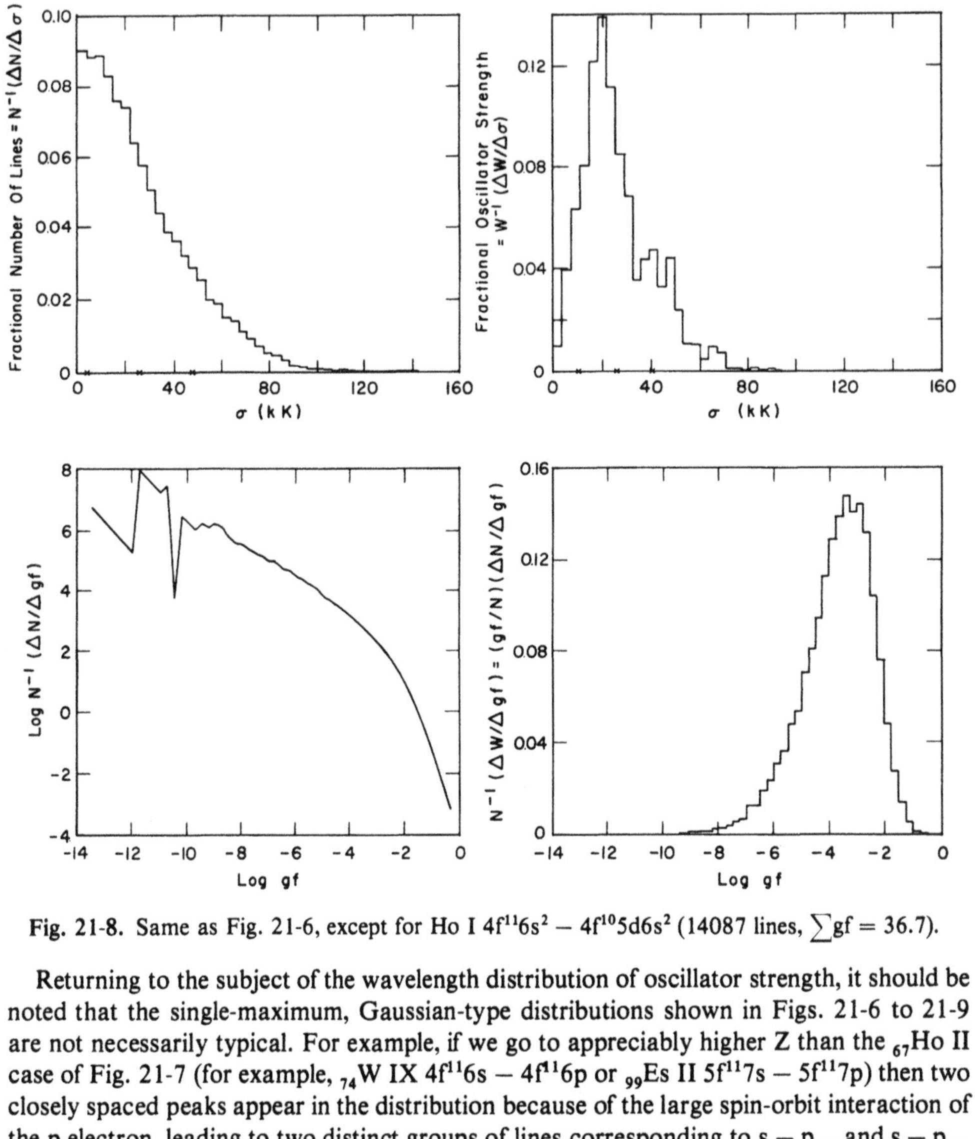

Fig. 21-8. Same as Fig. 21-6, except for Ho I $4f^{11}6s^2 - 4f^{10}5d6s^2$ (14087 lines, $\sum gf = 36.7$).

Returning to the subject of the wavelength distribution of oscillator strength, it should be noted that the single-maximum, Gaussian-type distributions shown in Figs. 21-6 to 21-9 are not necessarily typical. For example, if we go to appreciably higher Z than the $_{67}$Ho II case of Fig. 21-7 (for example, $_{74}$W IX $4f^{11}6s - 4f^{11}6p$ or $_{99}$Es II $5f^{11}7s - 5f^{11}7p$) then two closely spaced peaks appear in the distribution because of the large spin-orbit interaction of the p electron, leading to two distinct groups of lines corresponding to $s - p_{1/2}$ and $s - p_{3/2}$ transitions.

Similarly, in the arrays

$$4p^64d^w \; - \; 4p^54d^{w+1} \tag{21.11}$$

of W XXX-XXXVIII (w = 9 to 1), the very large spin-orbit splitting for $4p^5\ ^2P^\circ$ ($\zeta_{4p} =$ 470 kK, whereas $\zeta_{4d} < 60$ kK) leads to two groups of spectrum lines, separated by more than 25 percent in wavelength (Fig. 21-10, top). In contrast, lines of the array

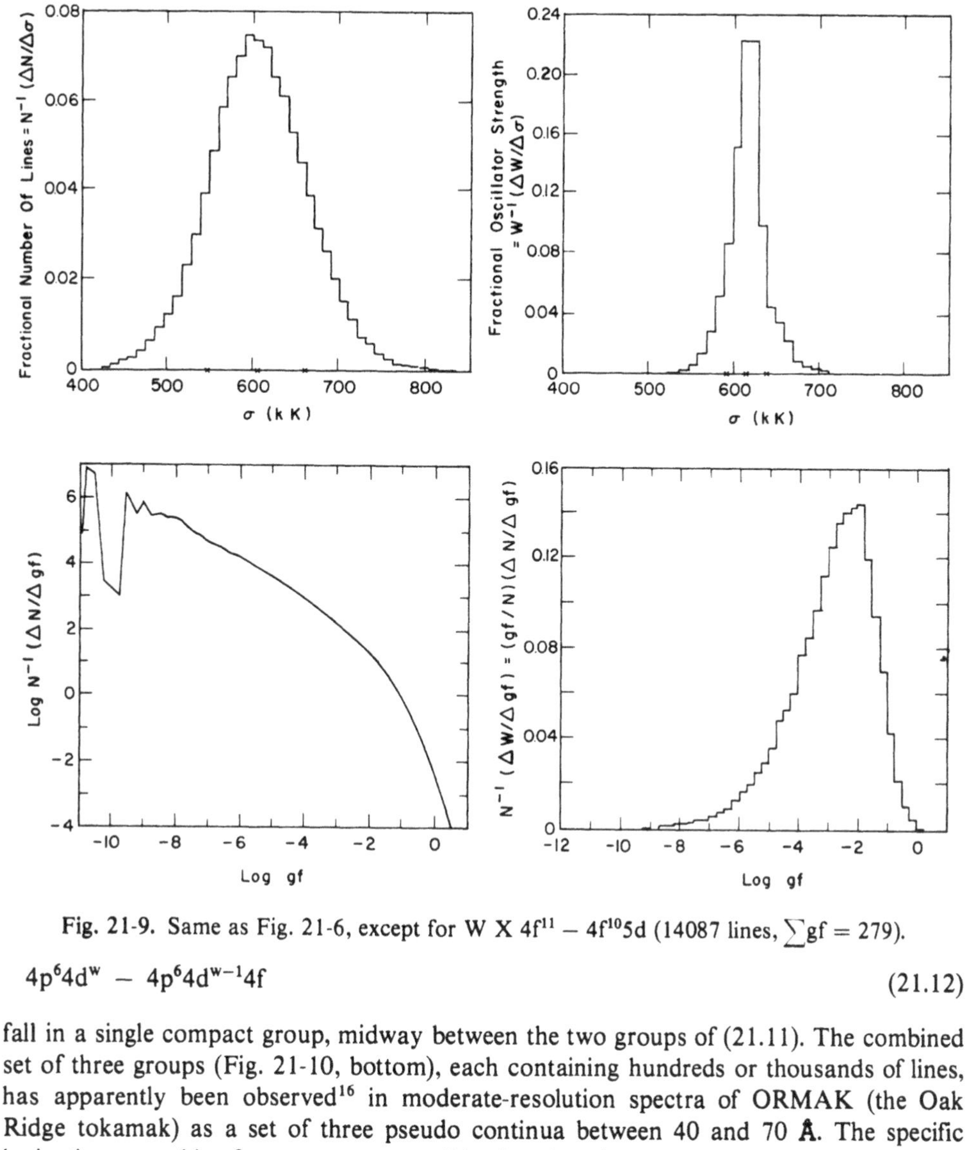

Fig. 21-9. Same as Fig. 21-6, except for W X $4f^{11} - 4f^{10}5d$ (14087 lines, $\sum gf = 279$).

$$4p^64d^w - 4p^64d^{w-1}4f \tag{21.12}$$

fall in a single compact group, midway between the two groups of (21.11). The combined set of three groups (Fig. 21-10, bottom), each containing hundreds or thousands of lines, has apparently been observed[16] in moderate-resolution spectra of ORMAK (the Oak Ridge tokamak) as a set of three pseudo continua between 40 and 70 Å. The specific ionization stage(s) of tungsten responsible for the observations cannot of course be determined by identification of specific spectrum lines, but can be estimated in the following way.

[16]R. C. Isler, R. V. Neidigh, and R. D. Cowan, Phys. Lett. **63A**, 295 (1977). The tungsten impurity ions occur in the tokamak discharge as a result of sputtering from the tungsten limiter. For further information, on gold as well as tungsten, see E. Hinnov and M. Mattioli, Phys. Lett. **66A**, 109 (1978); B. M. Johnson, K. W. Jones, J. L. Cecchi, E. Hinnov, and T. H. Kruse, Phys. Lett. **70A**, 320 (1979); S. Kasai, A. Funahashi, M. Nagami, and T. Sugie, Nucl. Fusion **19**, 195 (1979); J. Sugar and V. Kaufman, Phys. Rev. A **21**, 2096 (1980).

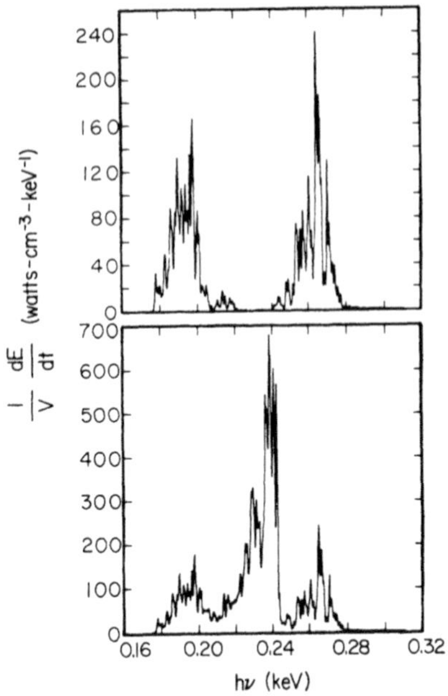

Fig. 21-10. Computed power loss dE/dt from W XXXIII ions ($n_i = 10^{12}$ ions/cm^3) in a plasma with electron density $n_e = 10^{14}$ cm^{-3} and electron temperature 1 keV. Top: Radiation in the transition array $4p^64d^6 - 4p^54d^7$; the 70-eV separation of the two groups of lines corresponds to the spin-orbit splitting of the $4p^5\ ^2P^\circ$ term. Bottom: As above, plus a central group of lines representing the array $4p^64d^6 - 4p^64d^54f$. The integrated power loss (resulting from electron-collisional excitation only) is 3.4 and 11.0 watts/cm^3 for the upper and lower figures, respectively.

For the general transition array (14.77), the integrated value of gf over all lines is (neglecting the dependence of f on wavelength) proportional to the total line strength (14.92)-(14.93). For two-open-subshell arrays

$$l_i^n l_j^{k-1} - l_i^{n-1} l_j^k$$

we have

$$S_{tot} \equiv \sum\sum S = 2\frac{(4l_i+1)!(4l_j+1)!}{(n-1)!(4l_i+2-n)!(k-1)!(4l_j+2-k)!}P^2_{l_il_j} ,$$

or

$$S_{tot} = 2\frac{9!}{w!(9-w)!}P^2_{pd} \quad \text{for (21.11)} ,$$

$$S_{tot} = 2\frac{9!}{(w-1)!(10-w)!}P^2_{df} \quad \text{for (21.12)} .$$

Both of these expressions peak at an intermediate value of w, but their ratio

$$\frac{S_{tot}(p-d)}{S_{tot}(d-f)} = \frac{10-w}{w}\left(\frac{P_{pd}}{P_{df}}\right)^2$$

$$\cong 0.79\frac{10-w}{w} \qquad (21.13)$$

increases monotonically from zero to infinity with increasing ionization stage (w = 10 to w = 0). The observed intensity ratio of the outer bands to the inner one is somewhat lower[17] than that for the case shown in Fig. 21-10, indicating that the ORMAK spectrum may be predominantly about W XXXII (w = 7) or W XXXI (w = 8).

Two-or-more-maximum oscillator strength distributions have also been computed for 7s − 7p, 7s − 8p, and 7p − 8d arrays in U I to U IV; the superimposed total of all U II arrays agrees well with the intensity distribution in observed emission spectra of a high current uranium arc.[18]

[17]Much of the intensity of the lowest-energy observed band probably came from the 4d − 4f peak (21.12) of gold ions, sputtered from the gold-plated inner surface of the vacuum vessel (R. C. Isler, private communication, January 1979; see also the 1978-1980 references in fn. 16).

[18]J. M. Mack, Jr., Los Alamos Scientific Laboratory Report LA-7101-T (January 1978); J. M. Mack, Jr. and L. J. Radziemski, Physica **102B**, 66 (1980).

Appendix A

Physical Constants, Units, and Conversion Factors[a]

Speed of light in vacuum	$c = 2.997925 \cdot 10^{10}$ cm/s
Elementary charge	$e = 4.803242 \cdot 10^{-10}$ esu
Electron rest mass	$m = 9.109534 \cdot 10^{-28}$ g
Planck constant	$h = 6.626176 \cdot 10^{-27}$ erg-s
	$\hbar = 1.054589 \cdot 10^{-27}$ erg-s
	$= 4.837769 \cdot 10^{-17}$ Ry-s
Bohr radius	$a_0 = \hbar^2/me^2 = 5.291771 \cdot 10^{-9}$ cm
	$= 0.5291771$ Å
	$1\ \text{cm}^{-1} \equiv 1$ kayser $= 1$ K
	$1\ \text{eV} = 10^8 e/c$ ergs
	$= 1.602189 \cdot 10^{-12}$ ergs
	$1\ \text{eV}/hc = 10^8 e/hc^2\ \text{cm}^{-1}$
	$= 8065.479\ \text{cm}^{-1}$
	$= 8.065479$ kK
	$T(°K) = 11604.50\ T(\text{eV})$
Photon wavelength	1 Å $= 10^{-8}$ cm
Photon wavenumber	$\sigma \equiv \lambda^{-1} = 10^8/\lambda(\text{Å})\ \text{cm}^{-1}$
Photon energy	$h\nu = hc/\lambda = 12398.52/\lambda(\text{Å})$ eV
	$= 12.39852/\lambda(\text{Å})$ keV
Wavenumber-frequency equivalences	$1\ \text{cm}^{-1} \equiv 1\ \text{K} = 29979.25$ MHz
	$1\ \text{MHz} = 0.03335640$ mK
Rydberg	$\text{Ry} = me^4/2\hbar^2 = \hbar^2/2ma_0^2 = \alpha^2 mc^2/2 = e^2/2a_0$
	$= 2.179907 \cdot 10^{-11}$ ergs
	$= 13.60580$ eV
Rydberg constant	$R_\infty = \text{Ry}/hc = 109737.3\ \text{cm}^{-1}$
	$= 109.7373$ kK

Fine structure constant	$\alpha = e^2/\hbar c = \hbar/mca_0 = 2a_0\ Ry/\hbar c$
	$= 4\pi a_0 R_\infty = 0.007297351$
	$\alpha^{-1} = 137.0360$
Bohr magneton	$\mu_0 = e\hbar/2mc = 9.274079 \cdot 10^{-21}$ ergs/gauss
	$\mu_0/hc = 4.66860 \cdot 10^{-5}\ cm^{-1}$/gauss
	$= 0.466860\ cm^{-1}$/tesla
Atomic unit, electric field strength	$= e/a_0^2$ esu/cm
	$= ec/10^8 a_0^2$ V/cm
	$= 5.142250 \cdot 10^9$ V/cm

[a]E. R. Cohen and B. N. Taylor, J. Phys. Chem. Ref. Data 2, 663 (1973).

Appendix B

Conversion Between Vacuum and Air Wavelengths

It is frequently desirable to determine the wavelength of a spectrum line from the difference between two tabulated energy levels, or conversely from a known wavelength to determine which two tabulated levels could be responsible for the line. The relation between wavelength (in Å) and wavenumber or energy difference (in cm^{-1}) is

$$\lambda_{vac} = \frac{10^8}{\sigma} = \frac{10^8}{E_2 - E_1} .$$

However, wavelengths above 2000 Å are usually given in air. Conversion between λ_{vac} and λ_{air}, sufficiently accurate for most purposes, can be obtained from the following table, computed from the relations (1.6)-(1.7).

λ_{air}	$\lambda_{vac}-\lambda_{air}$	Δ	λ_{air}	$\lambda_{vac}-\lambda_{air}$	Δ
2000 Å	0.648 Å		11000	3.013	
		226			271
3000	0.874		12000	3.284	
		257			271
4000	1.131		13000	3.555	
		264			272
5000	1.395		14000	3.827	
		267			272
6000	1.662		15000	4.099	
		269			272
7000	1.931		16000	4.371	
		270			272
8000	2.201		17000	4.643	
		270			272
9000	2.471		18000	4.915	
		271			272
10000	2.742		19000	5.187	
		271			273
11000	3.013		20000	5.460	

Appendix C

3-j Symbols

The Wigner 3-j symbol

$$\begin{pmatrix} j_1 & j_2 & j_3 \\ m_1 & m_2 & m_3 \end{pmatrix}$$

is defined by (5.1). The table that follows includes values only for $j_1 \geqq j_2 \geqq j_3$. The first section of the table includes 3-j symbols for $m_1 = m_2 = m_3 = 0$ with $j_1 \leqq 10$, $j_2 \leqq 6$; the second section includes values for $j_k \leqq 4$, with the m_k satisfying the restrictions

$$\begin{aligned} m_1 &\geqq 0 \;, & & \\ m_2 &\geqq 0 & &\text{if } m_1 = 0 \;, \\ |m_1| &\geqq |m_2| & &\text{if } j_1 = j_2 \;, \\ m_2 &\geqq m_3 & &\text{if } j_2 = j_3 \;, \\ |m_1| &\geqq |m_3| & &\text{if } j_1 = j_2 = j_3 \;. \end{aligned}$$

3-j symbols for other possible values of the m_k may be found with the aid of the symmetry relations (5.4)-(5.6); namely, the 3-j symbol is multiplied by

$$(-1)^{j_1+j_2+j_3}$$

for any odd permutation of the columns, or for a change in sign of all three m_k.

The value of each 3-j symbol is given as a decimal number, and also in power-of-prime notation

$$c_0 \quad a_1a_2a_3a_4 \quad a_5a_6a_7a_8 \quad a_9$$

such that

$$\text{3-j symbol} = c_0\left[\prod_{i=1}^{9} p_i^{a_i}\right]^{1/2} ,$$

where the first 9 primes p_l are

2,3,5,7, 11,13,17,19, 23 .

Thus, for example,

$$\begin{pmatrix} \frac{3}{2} & \frac{5}{2} & 3 \\ -\frac{1}{2} & \frac{3}{2} & -1 \end{pmatrix} = -\begin{pmatrix} 3 & \frac{5}{2} & \frac{3}{2} \\ -1 & \frac{3}{2} & -\frac{1}{2} \end{pmatrix} = \begin{pmatrix} 3 & \frac{5}{2} & \frac{3}{2} \\ 1 & -\frac{3}{2} & \frac{1}{2} \end{pmatrix} = (-1)\left[2^{-3}3^{-1}5^{-1}7^{1}\right]^{1/2}$$

$$= -0.2415229458 \ .$$

(For the 3-j symbols with $j_k \leqq 4$, it was possible to simplify the appearance and use of the table by using only the first 6 primes and absorbing higher primes into c_0.)

Entries are arranged in odometer order of the arguments

$j_1 j_2 j_3 m_1 m_2$

(with of course $m_3 = -m_1 - m_2$), and are omitted if the 3-j symbol is necessarily zero as a consequence of the symmetry relations (i.e., for $j_1 + j_2 + j_3$ odd, if all m_k are zero, or if $m_k = m_l$ when $j_k = j_l$).

Values of 3-j symbols for j_k up to 8 are given by Rotenberg et al., ref. [109] in the bibliography.

TABLE C-1. 3-j SYMBOLS

j_1	j_2	j_3	m_1	m_2	m_3	3-j symbol				
0	0	0	0	0	0	1.00000000	1	0		
1	1	0	0	0	0	-.57735027	-1	0-1		
2	1	1	0	0	0	.36514837	1	1-1-1		
2	2	0	0	0	0	.44721360	1	0 0-1		
2	2	2	0	0	0	-.23904572	-1	1 0-1-1		
3	2	1	0	0	0	-.29277002	-1	0 1-1-1		
3	3	0	0	0	0	-.37796447	-1	0 0 0-1		
3	3	2	0	0	0	.19518001	1	2-1-1-1		
4	2	2	0	0	0	.23904572	1	1 0-1-1		
4	3	1	0	0	0	.25197632	1	2-2 0-1		
4	3	3	0	0	0	-.16116459	-1	1 0 0-1	-1	
4	4	0	0	0	0	.33333333	1	0-2		
4	4	2	0	0	0	-.16988240	-1	2-2 1-1	-1	
4	4	4	0	0	0	.13409705	1	1 2 0-1	-1-1	
5	3	2	0	0	0	-.20806259	-1	1-1 1-1	-1	
5	4	1	0	0	0	-.22473329	-1	0-2 1 0	-1	
5	4	3	0	0	0	.14135070	1	2 0 1-1	-1-1	
5	5	0	0	0	0	-.30151134	-1	0 0 0 0	-1	
5	5	2	0	0	0	.15267620	1	1-1 1 0	-1-1	
5	5	4	0	0	0	-.11826248	-1	1 0 0 0	-1-1	
6	3	3	0	0	0	.18248297	1	2-1 2-1	-1-1	
6	4	2	0	0	0	.18698940	1	0 0 1 0	-1-1	
6	4	4	0	0	0	-.12465960	-1	2-2 1 0	-1-1	
6	5	1	0	0	0	.20483662	1	1 1 0 0	-1-1	
6	5	3	0	0	0	-.12773808	-1	0-1 0 1	-1-1	
6	5	5	0	0	0	.10473501	1	4-1 1 0	-1-1-1	
6	6	0	0	0	0	.27735010	1	0 0 0 0	0-1	
6	6	2	0	0	0	-.13993005	-1	1 0-1 1	-1-1	
6	6	4	0	0	0	.10732145	1	2 0 0 1	-1-1-1	
6	6	6	0	0	0	-.09305950	-1	4 0 2 0	-1-1-1-1	
7	4	3	0	0	0	-.16490915	-1	0-2 1 1	-1-1	
7	5	2	0	0	0	-.17137861	-1	0 1-1 1	-1-1	
7	5	4	0	0	0	.11312674	1	3-2 1 1	-1-1-1	
7	6	1	0	0	0	-.18946619	-1	0-1-1 1	0-1	
7	6	3	0	0	0	.11756476	1	3 1-1 1	-1-1-1	
7	6	5	0	0	0	-.09535761	-1	2 1 1 1	-1-1-1-1	
8	4	4	0	0	0	.14965261	1	1-2 1 2	-1-1-1	
8	5	3	0	0	0	.15177545	1	3 0 0 1	-1-1-1	
8	5	5	0	0	0	-.10299799	-1	1 0 1 2	-1-1-1-1	
8	6	2	0	0	0	.15918344	1	2 0-1 1	0-1-1	
8	6	4	0	0	0	-.10445903	-1	3 2 0 1	-1-1-1-1	
8	6	6	0	0	0	.08704919	1	1 0 2 1	-1-1-1-1	
9	5	4	0	0	0	-.13818631	-1	1 2 0 2	-1-1-1-1	
9	6	3	0	0	0	-.14143820	-1	2 1 0 1	0-1-1-1	
9	6	5	0	0	0	.09535761	1	2 1 1 1	-1-1-1-1	
10	5	5	0	0	0	.12793566	1	2 3 0 1	-1-1-1-1	
10	6	4	0	0	0	.12911482	1	1 0 1 1	0-1-1-1	
10	6	6	0	0	0	-.08847571	-1	2 3 0 1	0-1-1-1	-1

j1	j2	j3	m1	m2	m3	3-j symbol		
0	0	0	0	0	0	1.00000000	1	0
1/2	1/2	0	1/2	-1/2	0	.70710678	1	-1
1	1/2	1/2	0	1/2	-1/2	.40824829	1	-1-1
1	1/2	1/2	1	-1/2	-1/2	-.57735027	-1	0-1
1	1	0	0	0	0	-.57735027	-1	0-1
1	1	0	1	-1	0	.57735027	1	0-1
1	1	1	1	0	-1	-.40824829	-1	-1-1
3/2	1	1/2	1/2	-1	1/2	.28867513	1	-2-1
3/2	1	1/2	1/2	0	-1/2	.40824829	1	-1-1
3/2	1	1/2	3/2	-1	-1/2	-.50000000	-1	-2
3/2	3/2	0	1/2	-1/2	0	-.50000000	-1	-2
3/2	3/2	0	3/2	-3/2	0	.50000000	1	-2
3/2	3/2	1	1/2	-1/2	0	-.12909944	-1	-2-1-1
3/2	3/2	1	1/2	1/2	-1	.36514837	1	1-1-1
3/2	3/2	1	3/2	-3/2	0	.38729833	1	-2 1-1
3/2	3/2	1	3/2	-1/2	-1	-.31622777	-1	-1 0-1
2	1	1	0	0	0	.36514837	1	1-1-1
2	1	1	0	1	-1	.18257419	1	-1-1-1
2	1	1	1	0	-1	-.31622777	-1	-1 0-1
2	1	1	2	-1	-1	.44721360	1	0 0-1
2	3/2	1/2	0	1/2	-1/2	-.31622777	-1	-1 0-1
2	3/2	1/2	1	-3/2	1/2	.22360680	1	-2 0-1
2	3/2	1/2	1	-1/2	-1/2	.38729833	1	-2 1-1
2	3/2	1/2	2	-3/2	-1/2	-.44721360	-1	0 0-1
2	3/2	3/2	0	1/2	-1/2	.22360680	1	-2 0-1
2	3/2	3/2	0	3/2	-3/2	.22360680	1	-2 0-1
2	3/2	3/2	1	1/2	-3/2	-.31622777	-1	-1 0-1
2	3/2	3/2	2	-1/2	-3/2	.31622777	1	-1 0-1
2	2	0	0	0	0	.44721360	1	0 0-1
2	2	0	1	-1	0	-.44721360	-1	0 0-1
2	2	0	2	-2	0	.44721360	1	0 0-1
2	2	1	1	-1	0	-.18257419	-1	-1-1-1
2	2	1	1	0	-1	.31622777	1	-1 0-1
2	2	1	2	-2	0	.36514837	1	1-1-1
2	2	1	2	-1	-1	-.25819889	-1	0-1-1
2	2	2	0	0	0	-.23904572	-1	1 0-1-1
2	2	2	1	0	-1	.11952286	1	-1 0-1-1
2	2	2	2	-1	-1	-.29277002	-1	0 1-1-1
2	2	2	2	0	-2	.23904572	1	1 0-1-1
5/2	3/2	1	1/2	-3/2	1	.12909944	1	-2-1-1
5/2	3/2	1	1/2	-1/2	0	.31622777	1	-1 0-1
5/2	3/2	1	1/2	1/2	-1	.22360680	1	-2 0-1
5/2	3/2	1	3/2	-3/2	0	-.25819889	-1	0-1-1
5/2	3/2	1	3/2	-1/2	-1	-.31622777	-1	-1 0-1
5/2	3/2	1	5/2	-3/2	-1	.40824829	1	-1-1
5/2	2	1/2	1/2	-1	1/2	-.25819889	-1	0-1-1
5/2	2	1/2	1/2	0	-1/2	-.31622777	-1	-1 0-1
5/2	2	1/2	3/2	-2	1/2	.18257419	1	-1-1-1
5/2	2	1/2	3/2	-1	-1/2	.36514837	1	1-1-1
5/2	2	1/2	5/2	-2	-1/2	-.40824829	-1	-1-1
5/2	2	3/2	1/2	-2	3/2	-.16903085	-1	0 0-1-1
5/2	2	3/2	1/2	-1	1/2	-.24397502	-1	-2-1 1-1
5/2	2	3/2	1/2	0	-1/2	.11952286	1	-1 0-1-1
5/2	2	3/2	1/2	1	-3/2	.25354628	1	-2 2-1-1

j_1	j_2	j_3	m_1	m_2	m_3	3-j symbol		
5/2	2	3/2	3/2	-2	1/2	.27602622	1	3 -1 -1 -1
5/2	2	3/2	3/2	-1	-1/2	.06900656	1	-1 -1 -1 -1
5/2	2	3/2	3/2	0	-3/2	-.29277002	-1	0 1 -1 -1
5/2	2	3/2	5/2	-2	-1/2	-.30860670	-1	1 -1 0 -1
5/2	2	3/2	5/2	-1	-3/2	.26726124	1	-1 0 0 -1
5/2	5/2	0	1/2	-1/2	0	.40824829	1	-1 -1
5/2	5/2	0	3/2	-3/2	0	-.40824829	-1	-1 -1
5/2	5/2	0	5/2	-5/2	0	.40824829	1	-1 -1
5/2	5/2	1	1/2	-1/2	0	.06900656	1	-1 -1 -1 -1
5/2	5/2	1	1/2	1/2	-1	-.29277002	-1	0 1 -1 -1
5/2	5/2	1	3/2	-3/2	0	-.20701967	-1	-1 1 -1 -1
5/2	5/2	1	3/2	-1/2	-1	.27602622	1	3 -1 -1 -1
5/2	5/2	1	5/2	-5/2	0	.34503278	1	-1 -1 1 -1
5/2	5/2	1	5/2	-3/2	-1	-.21821789	-1	0 -1 0 -1
5/2	5/2	2	1/2	-1/2	0	-.19518001	-1	2 -1 -1 -1
5/2	5/2	2	3/2	-3/2	0	.04879500	1	-2 -1 -1 -1
5/2	5/2	2	3/2	-1/2	-1	.16903085	1	0 0 -1 -1
5/2	5/2	2	3/2	1/2	-2	-.25354628	-1	-2 2 -1 -1
5/2	5/2	2	5/2	-5/2	0	.24397502	1	-2 -1 1 -1
5/2	5/2	2	5/2	-3/2	-1	-.26726124	-1	-1 0 0 -1
5/2	5/2	2	5/2	-1/2	-2	.18898224	1	-2 0 0 -1
3	3/2	3/2	0	1/2	-1/2	.25354628	1	-2 2 -1 -1
3	3/2	3/2	0	3/2	-3/2	.08451543	1	-2 0 -1 -1
3	3/2	3/2	1	-1/2	-1/2	-.29277002	-1	0 1 -1 -1
3	3/2	3/2	1	1/2	-3/2	-.16903085	-1	0 0 -1 -1
3	3/2	3/2	2	-1/2	-3/2	.26726124	1	-1 0 0 -1
3	3/2	3/2	3	-3/2	-3/2	-.37796447	-1	0 0 0 -1
3	2	1	0	0	0	-.29277002	-1	0 1 -1 -1
3	2	1	0	1	-1	-.16903085	-1	0 0 -1 -1
3	2	1	1	-2	1	.09759001	1	0 -1 -1 -1
3	2	1	1	-1	0	.27602622	1	3 -1 -1 -1
3	2	1	1	0	-1	.23904572	1	1 0 -1 -1
3	2	1	2	-2	0	-.21821789	-1	0 -1 0 -1
3	2	1	2	-1	-1	-.30860670	-1	1 -1 0 -1
3	2	1	3	-2	-1	.37796447	1	0 0 0 -1
3	2	2	0	1	-1	.23904572	1	1 0 -1 -1
3	2	2	0	2	-2	.11952286	1	-1 0 -1 -1
3	2	2	1	0	-1	-.16903085	-1	0 0 -1 -1
3	2	2	1	1	-2	-.20701967	-1	-1 1 -1 -1
3	2	2	2	0	-2	.26726124	1	-1 0 0 -1
3	2	2	3	-1	-2	-.26726124	-1	-1 0 0 -1
3	5/2	1/2	0	1/2	-1/2	.26726124	1	-1 0 0 -1
3	5/2	1/2	1	-3/2	1/2	-.21821789	-1	0 -1 0 -1
3	5/2	1/2	1	-1/2	-1/2	-.30860670	-1	1 -1 0 -1
3	5/2	1/2	2	-5/2	1/2	.15430335	1	-1 -1 0 -1
3	5/2	1/2	2	-3/2	-1/2	.34503278	1	-1 -1 1 -1
3	5/2	1/2	3	-5/2	-1/2	-.37796447	-1	0 0 0 -1
3	5/2	3/2	0	1/2	-1/2	-.16903085	-1	0 0 -1 -1
3	5/2	3/2	0	3/2	-3/2	-.20701967	-1	-1 1 -1 -1
3	5/2	3/2	1	-5/2	3/2	-.13363062	-1	-3 0 0 -1
3	5/2	3/2	1	-3/2	1/2	-.24152295	-1	-3 -1 -1 1
3	5/2	3/2	1	-1/2	-1/2	.04879500	1	-2 -1 -1 -1
3	5/2	3/2	1	1/2	-3/2	.25354628	1	-2 2 -1 -1
3	5/2	3/2	2	-5/2	1/2	.24397502	1	-2 -1 1 -1

j_1	j_2	j_3	m_1	m_2	m_3	3-j symbol		
3	5/2	3/2	2	-3/2	-1/2	.10910895	1	-2-1 0-1
3	5/2	3/2	2	-1/2	-3/2	-.26726124	-1	-1 0 0-1
3	5/2	3/2	3	-5/2	-1/2	-.29880715	-1	-3 0 1-1
3	5/2	3/2	3	-3/2	-3/2	.23145502	1	-3 1 0-1
3	5/2	5/2	0	1/2	-1/2	-.11268723	-1	2-2-1-1
3	5/2	5/2	0	3/2	-3/2	.19720266	1	-2-2-1 1
3	5/2	5/2	0	5/2	-5/2	.14085904	1	-2-2 1-1
3	5/2	5/2	1	-1/2	-1/2	.19518001	1	2-1-1-1
3	5/2	5/2	1	1/2	-3/2	-.06900656	-1	-1-1-1-1
3	5/2	5/2	1	3/2	-5/2	-.21821789	-1	0-1 0-1
3	5/2	5/2	2	-1/2	-3/2	-.10910895	-1	-2-1 0-1
3	5/2	5/2	2	1/2	-5/2	.24397502	1	-2-1 1-1
3	5/2	5/2	3	-3/2	-3/2	.25197632	1	2-2 0-1
3	5/2	5/2	3	-1/2	-5/2	-.19920477	-1	-1-2 1-1
3	3	0	0	0	0	-.37796447	-1	0 0 0-1
3	3	0	1	-1	0	.37796447	1	0 0 0-1
3	3	0	2	-2	0	-.37796447	-1	0 0 0-1
3	3	0	3	-3	0	.37796447	1	0 0 0-1
3	3	1	1	-1	0	.10910895	1	-2-1 0-1
3	3	1	1	0	-1	-.26726124	-1	-1 0 0-1
3	3	1	2	-2	0	-.21821789	-1	0-1 0-1
3	3	1	2	-1	-1	.24397502	1	-2-1 1-1
3	3	1	3	-3	0	.32732684	1	-2 1 0-1
3	3	1	3	-2	-1	-.18898224	-1	-2 0 0-1
3	3	2	0	0	0	.19518001	1	2-1-1-1
3	3	2	1	-1	0	-.14638501	-1	-2 1-1-1
3	3	2	1	0	-1	-.06900656	-1	-1-1-1-1
3	3	2	1	1	-2	.23904572	1	1 0-1-1
3	3	2	2	-2	0	0.00000000	0	
3	3	2	2	-1	-1	.18898224	1	-2 0 0-1
3	3	2	2	0	-2	-.21821789	-1	0-1 0-1
3	3	2	3	-3	0	.24397502	1	-2-1 1-1
3	3	2	3	-2	-1	-.24397502	-1	-2-1 1-1
3	3	2	3	-1	-2	.15430335	1	-1-1 0-1
3	3	3	1	0	-1	.15430335	1	-1-1 0-1
3	3	3	2	0	-2	-.15430335	-1	-1-1 0-1
3	3	3	3	-1	-2	.21821789	1	0-1 0-1
3	3	3	3	0	-3	-.15430335	-1	-1-1 0-1
7/2	2	3/2	1/2	-2	3/2	.05976143	1	-3 0-1-1
7/2	2	3/2	1/2	-1	1/2	.20701967	1	-1 1-1-1
7/2	2	3/2	1/2	0	-1/2	.25354628	1	-2 2-1-1
7/2	2	3/2	1/2	1	-3/2	.11952286	1	-1 0-1-1
7/2	2	3/2	3/2	-2	1/2	-.13363062	-1	-3 0 0-1
7/2	2	3/2	3/2	-1	-1/2	-.26726124	-1	-1 0 0-1
7/2	2	3/2	3/2	0	-3/2	-.18898224	-1	-2 0 0-1
7/2	2	3/2	5/2	-2	-1/2	.23145502	1	-3 1 0-1
7/2	2	3/2	5/2	-1	-3/2	.26726124	1	-1 0 0-1
7/2	2	3/2	7/2	-2	-3/2	-.35355339	-1	-3
7/2	5/2	1	1/2	-3/2	1	-.13363062	-1	-3 0 0-1
7/2	5/2	1	1/2	-1/2	0	-.26726124	-1	-1 0 0-1
7/2	5/2	1	1/2	1/2	-1	-.18898224	-1	-2 0 0-1
7/2	5/2	1	3/2	-5/2	1	.07715167	1	-3-1 0-1
7/2	5/2	1	3/2	-3/2	0	.24397502	1	-2-1 1-1
7/2	5/2	1	3/2	-1/2	-1	.24397502	1	-2-1 1-1

j_1	j_2	j_3	m_1	m_2	m_3	3-j symbol			
7/2	5/2	1	5/2	-5/2	0	-.18898224	-1	-2 0 0-1	
7/2	5/2	1	5/2	-3/2	-1	-.29880715	-1	-3 0 1-1	
7/2	5/2	1	7/2	-5/2	-1	.35355339	1	-3	
7/2	5/2	2	1/2	-5/2	2	-.08908708	-1	-1-2 0-1	
7/2	5/2	2	1/2	-3/2	1	-.21912525	-1	-3-2-1-1	2
7/2	5/2	2	1/2	-1/2	0	-.06900656	-1	-1-1-1-1	
7/2	5/2	2	1/2	1/2	-1	.19720266	1	-2-2-1 1	
7/2	5/2	2	1/2	3/2	-2	.15936381	1	3-2-1-1	
7/2	5/2	2	3/2	-5/2	1	.17251639	1	-3-1 1-1	
7/2	5/2	2	3/2	-3/2	0	.18898224	1	-2 0 0-1	
7/2	5/2	2	3/2	-1/2	-1	-.10910895	-1	-2-1 0-1	
7/2	5/2	2	3/2	1/2	-2	-.21821789	-1	0-1 0-1	
7/2	5/2	2	5/2	-5/2	0	-.24397502	-1	-2-1 1-1	
7/2	5/2	2	5/2	-3/2	-1	-.04454354	-1	-3-2 0-1	
7/2	5/2	2	5/2	-1/2	-2	.25197632	1	2-2 0-1	
7/2	5/2	2	7/2	-5/2	-1	.26352314	1	-3-2 1	
7/2	5/2	2	7/2	-3/2	-2	-.23570226	-1	-1-2	
7/2	3	1/2	1/2	-1	1/2	.23145502	1	-3 1 0-1	
7/2	3	1/2	1/2	0	-1/2	.26726124	1	-1 0 0-1	
7/2	3	1/2	3/2	-2	1/2	-.18898224	-1	-2 0 0-1	
7/2	3	1/2	3/2	-1	-1/2	-.29880715	-1	-3 0 1-1	
7/2	3	1/2	5/2	-3	1/2	.13363062	1	-3 0 0-1	
7/2	3	1/2	5/2	-2	-1/2	.32732684	1	-2 1 0-1	
7/2	3	1/2	7/2	-3	-1/2	-.35355339	-1	-3	
7/2	3	3/2	1/2	-2	3/2	.17251639	1	-3-1 1-1	
7/2	3	3/2	1/2	-1	1/2	.18898224	1	-2 0 0-1	
7/2	3	3/2	1/2	0	-1/2	-.10910895	-1	-2-1 0-1	
7/2	3	3/2	1/2	1	-3/2	-.21821789	-1	0-1 0-1	
7/2	3	3/2	3/2	-3	3/2	-.10910895	-1	-2-1 0-1	
7/2	3	3/2	3/2	-2	1/2	-.23145502	-1	-3 1 0-1	
7/2	3	3/2	3/2	-1	-1/2	0.00000000	0		
7/2	3	3/2	3/2	0	-3/2	.24397502	1	-2-1 1-1	
7/2	3	3/2	5/2	-3	1/2	.21821789	1	0-1 0-1	
7/2	3	3/2	5/2	-2	-1/2	.13363062	1	-3 0 0-1	
7/2	3	3/2	5/2	-1	-3/2	-.24397502	-1	-2-1 1-1	
7/2	3	3/2	7/2	-3	-1/2	-.28867513	-1	-2-1	
7/2	3	3/2	7/2	-2	-3/2	.20412415	1	-3-1	
7/2	3	5/2	1/2	-3	5/2	.10910895	1	-2-1 0-1	
7/2	3	5/2	1/2	-2	3/2	.19920477	1	-1-2 1-1	
7/2	3	5/2	1/2	-1	1/2	-.04454354	-1	-3-2 0-1	
7/2	3	5/2	1/2	0	-1/2	-.15430335	-1	-1-1 0-1	
7/2	3	5/2	1/2	1	-3/2	.12598816	1	0-2 0-1	
7/2	3	5/2	1/2	2	-5/2	.17817416	1	1-2 0-1	
7/2	3	5/2	3/2	-3	3/2	-.18898224	-1	-2 0 0-1	
7/2	3	5/2	3/2	-2	1/2	-.10910895	-1	-2-1 0-1	
7/2	3	5/2	3/2	-1	-1/2	.17251639	1	-3-1 1-1	
7/2	3	5/2	3/2	0	-3/2	0.00000000	0		
7/2	3	5/2	3/2	1	-5/2	-.21821789	-1	0-1 0-1	
7/2	3	5/2	5/2	-3	1/2	.23145502	1	-3 1 0-1	
7/2	3	5/2	5/2	-2	-1/2	-.06299408	-1	-2-2 0-1	
7/2	3	5/2	5/2	-1	-3/2	-.14085904	-1	-2-2 1-1	
7/2	3	5/2	5/2	0	-5/2	.21821789	1	0-1 0-1	
7/2	3	5/2	7/2	-3	-1/2	-.20412415	-1	-3-1	
7/2	3	5/2	7/2	-2	-3/2	.23570226	1	-1-2	

j_1	j_2	j_3	m_1	m_2	m_3	3-j symbol			
7/2	3	5/2	7/2	-1	-5/2	-.16666667	-1	-2-2	
7/2	7/2	0	1/2	-1/2	0	-.35355339	-1	-3	
7/2	7/2	0	3/2	-3/2	0	.35355339	1	-3	
7/2	7/2	0	5/2	-5/2	0	-.35355339	-1	-3	
7/2	7/2	0	7/2	-7/2	0	.35355339	1	-3	
7/2	7/2	1	1/2	-1/2	0	-.04454354	-1	-3-2 0-1	
7/2	7/2	1	1/2	1/2	-1	.25197632	1	2-2 0-1	
7/2	7/2	1	3/2	-3/2	0	.13363062	1	-3 0 0-1	
7/2	7/2	1	3/2	-1/2	-1	-.24397502	-1	-2-1 1-1	
7/2	7/2	1	5/2	-5/2	0	-.22271770	-1	-3-2 2-1	
7/2	7/2	1	5/2	-3/2	-1	.21821789	1	0-1 0-1	
7/2	7/2	1	7/2	-7/2	0	.31180478	1	-3-2 0 1	
7/2	7/2	1	7/2	-5/2	-1	-.16666667	-1	-2-2	
7/2	7/2	2	1/2	-1/2	0	.17251639	1	-3-1 1-1	
7/2	7/2	2	3/2	-3/2	0	-.10350983	-1	-3 1-1-1	
7/2	7/2	2	3/2	-1/2	-1	-.10910895	-1	-2-1 0-1	
7/2	7/2	2	3/2	1/2	-2	.21821789	1	0-1 0-1	
7/2	7/2	2	5/2	-5/2	0	-.03450328	-1	-3-1-1-1	
7/2	7/2	2	5/2	-3/2	-1	.19518001	1	2-1-1-1	
7/2	7/2	2	5/2	-1/2	-2	-.18898224	-1	-2 0 0-1	
7/2	7/2	2	7/2	-7/2	0	.24152295	1	-3-1-1 1	
7/2	7/2	2	7/2	-5/2	-1	-.22360680	-1	-2 0-1	
7/2	7/2	2	7/2	-3/2	-2	.12909944	1	-2-1-1	
7/2	7/2	3	1/2	-1/2	0	.06978632	1	-3 1 0-1	-1
7/2	7/2	3	1/2	1/2	-1	-.16116459	-1	1 0 0-1	-1
7/2	7/2	3	3/2	-3/2	0	-.16283474	-1	-3-1 0 1	-1
7/2	7/2	3	3/2	-1/2	-1	.10403130	1	-1-1 1-1	-1
7/2	7/2	3	3/2	1/2	-2	.06579517	1	0-1 0-1	-1
7/2	7/2	3	3/2	3/2	-3	-.20806259	-1	1-1 1-1	-1
7/2	7/2	3	5/2	-5/2	0	.11631053	1	-3-1 2-1	-1
7/2	7/2	3	5/2	-3/2	-1	.04652421	1	-1-1 0-1	-1
7/2	7/2	3	5/2	-1/2	-2	-.17094086	-1	-2 2 0-1	-1
7/2	7/2	3	5/2	1/2	-3	.18609684	1	3-1 0-1	-1
7/2	7/2	3	7/2	-7/2	0	.16283474	1	-3-1 0 1	-1
7/2	7/2	3	7/2	-5/2	-1	-.21320072	-1	-1 0 0 0	-1
7/2	7/2	3	7/2	-3/2	-2	.19462474	1	-2-1 1 0	-1
7/2	7/2	3	7/2	-1/2	-3	-.12309149	-1	-1-1 0 0	-1
4	2	2	0	0	0	.23904572	1	1 0-1-1	
4	2	2	0	1	-1	.15936381	1	3-2-1-1	
4	2	2	0	2	-2	.03984095	1	-1-2-1-1	
4	2	2	1	0	-1	-.21821789	-1	0-1 0-1	
4	2	2	1	1	-2	-.08908708	-1	-1-2 0-1	
4	2	2	2	-1	-1	.25197632	1	2-2 0-1	
4	2	2	2	0	-2	.15430335	1	-1-1 0-1	
4	2	2	3	-1	-2	-.23570226	-1	-1-2	
4	2	2	4	-2	-2	.33333333	1	0-2	
4	5/2	3/2	0	1/2	-1/2	-.21821789	-1	0-1 0-1	
4	5/2	3/2	0	3/2	-3/2	-.08908708	-1	-1-2 0-1	
4	5/2	3/2	1	-5/2	3/2	.04454354	1	-3-2 0-1	
4	5/2	3/2	1	-3/2	1/2	.17251639	1	-3-1 1-1	
4	5/2	3/2	1	-1/2	-1/2	.24397502	1	-2-1 1-1	
4	5/2	3/2	1	1/2	-3/2	.14085904	1	-2-2 1-1	
4	5/2	3/2	2	-5/2	1/2	-.10910895	-1	-2-1 0-1	
4	5/2	3/2	2	-3/2	-1/2	-.24397502	-1	-2-1 1-1	

j_1	j_2	j_3	m_1	m_2	m_3	3-j symbol			
4	5/2	3/2	2	-1/2	-3/2	-.19920477	-1	-1 -2 1 -1	
4	5/2	3/2	3	-5/2	-1/2	.20412415	1	-3 -1	
4	5/2	3/2	3	-3/2	-3/2	.26352314	1	-3 -2 1	
4	5/2	3/2	4	-5/2	-3/2	-.33333333	-1	0 -2	
4	5/2	5/2	0	1/2	-1/2	.12598816	1	0 -2 0 -1	
4	5/2	5/2	0	3/2	-3/2	.18898224	1	-2 0 0 -1	
4	5/2	5/2	0	5/2	-5/2	.06299408	1	-2 -2 0 -1	
4	5/2	5/2	1	1/2	-3/2	-.19920477	-1	-1 -2 1 -1	
4	5/2	5/2	1	3/2	-5/2	-.12598816	-1	0 -2 0 -1	
4	5/2	5/2	2	-1/2	-3/2	.14085904	1	-2 -2 1 -1	
4	5/2	5/2	2	1/2	-5/2	.18898224	1	-2 0 0 -1	
4	5/2	5/2	3	-1/2	-5/2	-.23570226	-1	-1 -2	
4	5/2	5/2	4	-3/2	-5/2	.23570226	1	-1 -2	
4	3	1	0	0	0	.25197632	1	2 -2 0 -1	
4	3	1	0	1	-1	.15430335	1	-1 -1 0 -1	
4	3	1	1	-2	1	-.10910895	-1	-2 -1 0 -1	
4	3	1	1	-1	0	-.24397502	-1	-2 -1 1 -1	
4	3	1	1	0	-1	-.19920477	-1	-1 -2 1 -1	
4	3	1	2	-3	1	.06299408	1	-2 -2 0 -1	
4	3	1	2	-2	0	.21821789	1	0 -1 0 -1	
4	3	1	2	-1	-1	.24397502	1	-2 -1 1 -1	
4	3	1	3	-3	0	-.16666667	-1	-2 -2	
4	3	1	3	-2	-1	-.28867513	-1	-2 -1	
4	3	1	4	-3	-1	.33333333	1	0 -2	
4	3	2	0	1	-1	-.19920477	-1	-1 -2 1 -1	
4	3	2	0	2	-2	-.12598816	-1	0 -2 0 -1	
4	3	2	1	-3	2	-.06900656	-1	-1 -1 -1 -1	
4	3	2	1	-2	1	-.19720266	-1	-2 -2 -1 1	
4	3	2	1	-1	0	-.10910895	-1	-2 -1 0 -1	
4	3	2	1	0	-1	.15430335	1	-1 -1 0 -1	
4	3	2	1	1	-2	.17817416	1	1 -2 0 -1	
4	3	2	2	-3	1	.14638501	1	-2 1 -1 -1	
4	3	2	2	-2	0	.19518001	1	2 -1 -1 -1	
4	3	2	2	-1	-1	-.06299408	-1	-2 -2 0 -1	
4	3	2	2	0	-2	-.21821789	-1	0 -1 0 -1	
4	3	2	3	-3	0	-.22360680	-1	-2 0 -1	
4	3	2	3	-2	-1	-.07453560	-1	-2 -2 -1	
4	3	2	3	-1	-2	.23570226	1	-1 -2	
4	3	2	4	-3	-1	.25819889	1	0 -1 -1	
4	3	2	4	-2	-2	-.21081851	-1	1 -2 -1	
4	3	3	0	0	0	-.16116459	-1	1 0 0 -1	-1
4	3	3	0	1	-1	.02686077	1	-1 -2 0 -1	-1
4	3	3	0	2	-2	.18802536	1	-1 -2 0 1	-1
4	3	3	0	3	-3	.08058230	1	-1 0 0 -1	-1
4	3	3	1	0	-1	.10403130	1	-1 -1 1 -1	-1
4	3	3	1	1	-2	-.15194744	-1	4 -2 0 -1	-1
4	3	3	1	2	-3	-.14712247	-1	0 -1 1 -1	-1
4	3	3	2	-1	-1	-.16988240	-1	2 -2 1 -1	-1
4	3	3	2	0	-2	.04652421	1	-1 -1 0 -1	-1
4	3	3	2	1	-3	.19738551	1	0 1 0 -1	-1
4	3	3	3	-1	-2	.10050378	1	0 -2 0 0	-1
4	3	3	3	0	-3	-.21320072	-1	-1 0 0 0	-1
4	3	3	4	-2	-2	-.22473329	-1	0 -2 1 0	-1
4	3	3	4	-1	-3	.17407766	1	0 -1 0 0	-1

j_1	j_2	j_3	m_1	m_2	m_3	3-j symbol				
4	7/2	1/2	0	1/2	-1/2	-.23570226	-1	-1-2		
4	7/2	1/2	1	-3/2	1/2	.20412415	1	-3-1		
4	7/2	1/2	1	-1/2	-1/2	.26352314	1	-3-2 1		
4	7/2	1/2	2	-5/2	1/2	-.16666667	-1	-2-2		
4	7/2	1/2	2	-3/2	-1/2	-.28867513	-1	-2-1		
4	7/2	1/2	3	-7/2	1/2	.11785113	1	-3-2		
4	7/2	1/2	3	-5/2	-1/2	.31180478	1	-3-2 0 1		
4	7/2	1/2	4	-7/2	-1/2	-.33333333	-1	0-2		
4	7/2	3/2	0	1/2	-1/2	.14085904	1	-2-2 1-1		
4	7/2	3/2	0	3/2	-3/2	.18898224	1	-2 0 0-1		
4	7/2	3/2	1	-5/2	3/2	.14638501	1	-2 1-1-1		
4	7/2	3/2	1	-3/2	1/2	.19518001	1	2-1-1-1		
4	7/2	3/2	1	-1/2	-1/2	-.06299408	-1	-2-2 0-1		
4	7/2	3/2	1	1/2	-3/2	-.21821789	-1	0-1 0-1		
4	7/2	3/2	2	-7/2	3/2	-.09128709	-1	-3-1-1		
4	7/2	3/2	2	-5/2	1/2	-.21912525	-1	-3-2-1-1	2	
4	7/2	3/2	2	-3/2	-1/2	-.03450328	-1	-3-1-1-1		
4	7/2	3/2	2	-1/2	-3/2	.23145502	1	-3 1 0-1		
4	7/2	3/2	3	-7/2	1/2	.19720266	1	-2-2-1 1		
4	7/2	3/2	3	-5/2	-1/2	.14907120	1	0-2-1		
4	7/2	3/2	3	-3/2	-3/2	-.22360680	-1	-2 0-1		
4	7/2	3/2	4	-7/2	-1/2	-.27888668	-1	-1-2-1 1		
4	7/2	3/2	4	-5/2	-3/2	.18257419	1	-1-1-1		
4	7/2	5/2	0	1/2	-1/2	.10403130	1	-1-1 1-1	-1	
4	7/2	5/2	0	3/2	-3/2	-.15194744	-1	4-2 0-1	-1	
4	7/2	5/2	0	5/2	-5/2	-.14712247	-1	0-1 1-1	-1	
4	7/2	5/2	1	-7/2	5/2	.08703883	1	-2-1 0 0	-1	
4	7/2	5/2	1	-5/2	3/2	.19125921	1	-2-1-1-1	-1	2
4	7/2	5/2	1	-3/2	1/2	.00600625	1	-3-2-1-1	-1	
4	7/2	5/2	1	-1/2	-1/2	-.16283474	-1	-3-1 0 1	-1	
4	7/2	5/2	1	1/2	-3/2	.06579517	1	0-1 0-1	-1	
4	7/2	5/2	1	3/2	-5/2	.18993429	1	0-2 2-1	-1	
4	7/2	5/2	2	-7/2	3/2	-.16514456	-1	-1 1-1 0	-1	
4	7/2	5/2	2	-5/2	1/2	-.13241022	-1	-2 3-1-1	-1	
4	7/2	5/2	2	-3/2	-1/2	.14440004	17	-2-2-1-1	-1	
4	7/2	5/2	2	-1/2	-3/2	.04652421	1	-1-1 0-1	-1	
4	7/2	5/2	2	1/2	-5/2	-.20806259	-1	1-1 1-1	-1	
4	7/2	5/2	3	-7/2	1/2	.21846572	1	-3 1-1 1	-1	
4	7/2	5/2	3	-5/2	-1/2	-.02752409	-1	-3-1-1 0	-1	
4	7/2	5/2	3	-3/2	-3/2	-.15731330	-1	-2-2-1 2	-1	
4	7/2	5/2	3	-1/2	-5/2	.19462474	1	-2-1 1 0	-1	
4	7/2	5/2	4	-7/2	-1/2	-.20597146	-1	0-1-1 1	-1	
4	7/2	5/2	4	-5/2	-3/2	.22019275	1	3-1-1 0	-1	
4	7/2	5/2	4	-3/2	-5/2	-.14213381	-1	1-2 0 0	-1	
4	7/2	7/2	0	1/2	-1/2	-.12087344	-1	-3 2 0-1	-1	
4	7/2	7/2	0	3/2	-3/2	-.04029115	-1	-3 0 0-1	-1	
4	7/2	7/2	0	5/2	-5/2	.17459498	1	-3-2 0-1	-1	2
4	7/2	7/2	0	7/2	-7/2	.09401268	1	-3-2 0 1	-1	
4	7/2	7/2	1	1/2	-3/2	.13957263	1	-1 1 0-1	-1	
4	7/2	7/2	1	3/2	-5/2	-.10403130	-1	-1-1 1-1	-1	
4	7/2	7/2	1	5/2	-7/2	-.15891043	-1	-1-2 1 0	-1	
4	7/2	7/2	2	-1/2	-3/2	-.13159034	-1	2-1 0-1	-1	
4	7/2	7/2	2	1/2	-5/2	-.01899343	-1	-2-2 0-1	-1	
4	7/2	7/2	2	3/2	-7/2	.19462474	1	-2-1 1 0	-1	

j_1	j_2	j_3	m_1	m_2	m_3	3-j symbol			
4	7/2	7/2	3	-1/2	-5/2	.14213381	1	1-2 0 0	-1
4	7/2	7/2	3	1/2	-7/2	-.18802536	-1	-1-2 0 1	-1
4	7/2	7/2	4	-3/2	-5/2	-.19462474	-1	-2-1 1 0	-1
4	7/2	7/2	4	-1/2	-7/2	.13295401	1	-2-2 0 1	-1
4	4	0	0	0	0	.33333333	1	0-2	
4	4	0	1	-1	0	-.33333333	-1	0-2	
4	4	0	2	-2	0	.33333333	1	0-2	
4	4	0	3	-3	0	-.33333333	-1	0-2	
4	4	0	4	-4	0	.33333333	1	0-2	
4	4	1	1	-1	0	-.07453560	-1	-2-2-1	
4	4	1	1	0	-1	.23570226	1	-1-2	
4	4	1	2	-2	0	.14907120	1	0-2-1	
4	4	1	2	-1	-1	-.22360680	-1	-2 0-1	
4	4	1	3	-3	0	-.22360680	-1	-2 0-1	
4	4	1	3	-2	-1	.19720266	1	-2-2-1 1	
4	4	1	4	-4	0	.29814240	1	2-2-1	
4	4	1	4	-3	-1	-.14907120	-1	0-2-1	
4	4	2	0	0	0	-.16988240	-1	2-2 1-1	-1
4	4	2	1	-1	0	.14440004	17	-2-2-1-1	-1
4	4	2	1	0	-1	.04652421	1	-1-1 0-1	-1
4	4	2	1	1	-2	-.20806259	-1	1-1 1-1	-1
4	4	2	2	-2	0	-.06795296	-1	4-2-1-1	-1
4	4	2	2	-1	-1	-.13241022	-1	-2 3-1-1	-1
4	4	2	2	0	-2	.19738551	1	0 1 0-1	-1
4	4	2	3	-3	0	-.05945884	-1	-2-2-1 1	-1
4	4	2	3	-2	-1	.19462474	1	-2-1 1 0	-1
4	4	2	3	-1	-2	-.16514456	-1	-1 1-1 0	-1
4	4	2	4	-4	0	.23783536	1	2-2-1 1	-1
4	4	2	4	-3	-1	-.20597146	-1	0-1-1 1	-1
4	4	2	4	-2	-2	.11009638	1	1-1-1 0	-1
4	4	3	1	-1	0	.10811250	1	-1 2-1-1	-1
4	4	3	1	0	-1	-.13957263	-1	-1 1 0-1	-1
4	4	3	2	-2	0	-.15616249	-1	-1-2-1-1	-1 2
4	4	3	2	-1	-1	.05884899	1	2-1-1-1	-1
4	4	3	2	0	-2	.10403130	1	-1-1 1-1	-1
4	4	3	2	1	-3	-.18993429	-1	0-2 2-1	-1
4	4	3	3	-3	0	.08408750	1	-1-2-1 1	-1
4	4	3	3	-2	-1	.07784989	1	0-1-1 0	-1
4	4	3	3	-1	-2	-.17407766	-1	0-1 0 0	-1
4	4	3	3	0	-3	.15891043	1	-1-2 1 0	-1
4	4	3	4	-4	0	.16817499	1	1-2-1 1	-1
4	4	3	4	-3	-1	-.20597146	-1	0-1-1 1	-1
4	4	3	4	-2	-2	.17407766	1	0-1 0 0	-1
4	4	3	4	-1	-3	-.10050378	-1	0-2 0 0	-1
4	4	4	0	0	0	.13409705	1	1 2 0-1	-1-1
4	4	4	1	0	-1	-.06704852	-1	-1 2 0-1	-1-1
4	4	4	2	-1	-1	.14135070	1	2 0 1-1	-1-1
4	4	4	2	0	-2	-.08194820	-1	-1-2 0-1	1-1
4	4	4	3	-1	-2	-.06232980	-1	0-2 1 0	-1-1
4	4	4	3	0	-3	.15644655	1	-1 0 0 1	-1-1
4	4	4	4	-2	-2	.18698940	1	0 0 1 0	-1-1
4	4	4	4	-1	-3	-.16490915	-1	0-2 1 1	-1-1
4	4	4	4	0	-4	.10429770	1	1-2 0 1	-1-1

Appendix D

6-j Symbols

The Wigner 6-j symbol

$$\begin{Bmatrix} j_1 & j_2 & j_3 \\ l_1 & l_2 & l_3 \end{Bmatrix}$$

is defined by (5.23)-(5.24), and is unchanged in value by any permutation of columns, or if any *two* columns are inverted. The table that follows assumes that these symmetry operations have been used to make the argument values decrease from left to right in the top row, and also (insofar as possible) down each column and from left to right in the bottom row; specifically,

$$j_1 \geqq j_2 \geqq j_3 \,,$$

$$j_1 \geqq l_1 \,,$$

$$j_2 \geqq l_2 \,,$$

$$j_2 \geqq l_3 \,,$$

$$j_3 \geqq l_2 \qquad \text{if } j_2 = l_3 \,,$$

$$j_3 \geqq l_3 \qquad \text{if } j_1 = l_1 \text{ or } j_2 = l_2 \,,$$

$$l_1 \geqq l_2 \qquad \text{if } j_1 = j_2 \,,$$

$$l_2 \geqq l_3 \qquad \text{if } j_2 = j_3 \,.$$

The value of each 6-j symbol is given as a decimal number, and also in power-of-prime notation similar to that used for 3-j symbols in Appendix C; for 6-j symbols with arguments no greater than 4, all primes greater than 13 occur only with positive even power, and have been absorbed into c_0. As an example,

$$\begin{Bmatrix} 2 & 3 & 4 \\ 3 & 4 & 4 \end{Bmatrix} = \begin{Bmatrix} 4 & 3 & 2 \\ 4 & 4 & 3 \end{Bmatrix} = \begin{Bmatrix} 4 & 4 & 3 \\ 4 & 3 & 2 \end{Bmatrix} = 23\left[2^{-2}3^{-2}5^{-1}7^{-2}11^{-1}\right]^{1/2}$$

$$= 0.0738409373 \,.$$

Entries are arranged in odometer order of the arguments

$j_1 j_2 j_3 l_1 l_2 l_3$.

Values of 6-j symbols for arguments up to 8 are given by Rotenberg et al., ref. [109]. Values of 9-j symbols are tabulated by Koozekanani, ref. [110], and by Smith and Stephenson, ref. [111].

TABLE D-1. 6-j SYMBOLS

j_1	j_2	j_3	l_1	l_2	l_3	6-j symbol		
0	0	0	0	0	0	1.00000000	1	0
1/2	1/2	0	0	0	1/2	-.70710678	-1	-1
1/2	1/2	0	1/2	1/2	0	-.50000000	-1	-2
1	1/2	1/2	0	1/2	1/2	.50000000	1	-2
1	1/2	1/2	1	1/2	1/2	.16666667	1	-2-2
1	1	0	0	0	1	.57735027	1	0-1
1	1	0	1/2	1/2	1/2	.40824829	1	-1-1
1	1	0	1	1	0	.33333333	1	0-2
1	1	1	1/2	1/2	1/2	-.33333333	-1	0-2
1	1	1	1	1	0	-.33333333	-1	0-2
1	1	1	1	1	1	.16666667	1	-2-2
3/2	1	1/2	0	1/2	1	-.40824829	-1	-1-1
3/2	1	1/2	1/2	1	1/2	-.33333333	-1	0-2
3/2	1	1/2	1	1/2	1	-.16666667	-1	-2-2
3/2	1	1/2	3/2	1	1/2	-.08333333	-1	-4-2
3/2	3/2	0	0	0	3/2	-.50000000	-1	-2
3/2	3/2	0	1/2	1/2	1	-.35355339	-1	-3
3/2	3/2	0	1	1	1/2	-.28867513	-1	-2-1
3/2	3/2	0	3/2	3/2	0	-.25000000	-1	-4
3/2	3/2	1	1/2	1/2	1	.26352314	1	-3-2 1
3/2	3/2	1	1	0	3/2	.28867513	1	-2-1
3/2	3/2	1	1	1	1/2	.26352314	1	-3-2 1
3/2	3/2	1	1	1	3/2	-.10540926	-1	-1-2-1
3/2	3/2	1	3/2	1/2	1	.16666667	1	-2-2
3/2	3/2	1	3/2	3/2	0	.25000000	1	-4
3/2	3/2	1	3/2	3/2	1	-.18333333	-1	-4-2-2 0 2
2	1	1	0	1	1	.33333333	1	0-2
2	1	1	1	1	1	.16666667	1	-2-2
2	1	1	2	1	1	.03333333	1	-2-2-2
2	3/2	1/2	0	1/2	3/2	.35355339	1	-3
2	3/2	1/2	1/2	1	1	.28867513	1	-2-1
2	3/2	1/2	1	1/2	3/2	.15811388	1	-3 0-1
2	3/2	1/2	1	3/2	1/2	.25000000	1	-4
2	3/2	1/2	3/2	1	1	.09128709	1	-3-1-1
2	3/2	1/2	2	3/2	1/2	.05000000	1	-4 0-2
2	3/2	3/2	0	3/2	3/2	-.25000000	-1	-4
2	3/2	3/2	1/2	1	1	-.20412415	-1	-3-1
2	3/2	3/2	1	3/2	1/2	-.22360680	-1	-2 0-1
2	3/2	3/2	1	3/2	3/2	.05000000	1	-4 0-2
2	3/2	3/2	3/2	1	1	-.16329932	-1	1-1-2
2	3/2	3/2	2	3/2	1/2	-.10000000	-1	-2 0-2
2	3/2	3/2	2	3/2	3/2	.15000000	1	-4 2-2
2	2	0	0	0	2	.44721360	1	0 0-1
2	2	0	1/2	1/2	3/2	.31622777	1	-1 0-1
2	2	0	1	1	1	.25819889	1	0-1-1
2	2	0	3/2	3/2	1/2	.22360680	1	-2 0-1
2	2	0	3/2	3/2	3/2	-.22360680	-1	-2 0-1
2	2	0	2	2	0	.20000000	1	0 0-2
2	2	1	1/2	1/2	3/2	-.22360680	-1	-2 0-1
2	2	1	1	0	2	-.25819889	-1	0-1-1
2	2	1	1	1	1	-.22360680	-1	-2 0-1
2	2	1	1	1	2	.07453560	1	-2-2-1
2	2	1	3/2	1/2	3/2	-.15811388	-1	-3 0-1
2	2	1	3/2	3/2	1/2	-.21213203	-1	-3 2-2

j_1	j_2	j_3	l_1	l_2	l_3	6-j symbol		
2	2	1	3/2	3/2	3/2	.14142136	1	-1 0-2
2	2	1	2	1	1	-.10000000	-1	-2 0-2
2	2	1	2	2	0	-.20000000	-1	0 0-2
2	2	1	2	2	1	.16666667	1	-2-2
2	2	2	1	1	1	.15275252	1	-2-1-2 1
2	2	2	3/2	3/2	1/2	.18708287	1	-3 0-2 1
2	2	2	3/2	3/2	3/2	0.00000000	0	
2	2	2	2	1	1	.15275252	1	-2-1-2 1
2	2	2	2	2	0	.20000000	1	0 0-2
2	2	2	2	2	1	-.10000000	-1	-2 0-2
2	2	2	2	2	2	-.04285714	-1	-2 2-2-2
5/2	3/2	1	0	1	3/2	-.28867513	-1	-2-1
5/2	3/2	1	1/2	3/2	1	-.25000000	-1	-4
5/2	3/2	1	1	1	3/2	-.15811388	-1	-3 0-1
5/2	3/2	1	3/2	3/2	1	-.10000000	-1	-2 0-2
5/2	3/2	1	2	1	3/2	-.04082483	-1	-3-1-2
5/2	3/2	1	5/2	3/2	1	-.01666667	-1	-4-2-2
5/2	2	1/2	0	1/2	2	-.31622777	-1	-1 0-1
5/2	2	1/2	1/2	1	3/2	-.25819889	-1	0-1-1
5/2	2	1/2	1	1/2	2	-.14907120	-1	0-2-1
5/2	2	1/2	1	3/2	1	-.22360680	-1	-2 0-1
5/2	2	1/2	3/2	1	3/2	-.09128709	-1	-3-1-1
5/2	2	1/2	3/2	2	1/2	-.20000000	-1	0 0-2
5/2	2	1/2	2	3/2	1	-.05773503	-1	-2-1-2
5/2	2	1/2	5/2	2	1/2	-.03333333	-1	-2-2-2
5/2	2	3/2	0	3/2	2	.22360680	1	-2 0-1
5/2	2	3/2	1/2	1	3/2	.17078251	1	-4-1-1 1
5/2	2	3/2	1/2	2	3/2	.15000000	1	-4 2-2
5/2	2	3/2	1	1/2	2	.19720266	1	-2-2-1 1
5/2	2	3/2	1	3/2	1	.18708287	1	-3 0-2 1
5/2	2	3/2	1	3/2	2	-.02357023	-1	-3-2-2
5/2	2	3/2	3/2	1	3/2	.15275252	1	-2-1-2 1
5/2	2	3/2	3/2	2	1/2	.18708287	1	-3 0-2 1
5/2	2	3/2	3/2	2	3/2	-.10000000	-1	-2 0-2
5/2	2	3/2	2	1/2	2	.10000000	1	-2 0-2
5/2	2	3/2	2	3/2	1	.10801234	1	-3-1-2 1
5/2	2	3/2	2	3/2	2	-.13363062	-1	-3 0 0-1
5/2	2	3/2	5/2	1	3/2	.05000000	1	-4 0-2
5/2	2	3/2	5/2	2	1/2	.06666667	1	0-2-2
5/2	2	3/2	5/2	2	3/2	-.11190476	-47	-4-2-2-2
5/2	5/2	0	0	0	5/2	-.40824829	-1	-1-1
5/2	5/2	0	1/2	1/2	2	-.28867513	-1	-2-1
5/2	5/2	0	1	1	3/2	-.23570226	-1	-1-2
5/2	5/2	0	3/2	3/2	1	-.20412415	-1	-3-1
5/2	5/2	0	3/2	3/2	2	.20412415	1	-3-1
5/2	5/2	0	2	2	1/2	-.18257419	-1	-1-1-1
5/2	5/2	0	2	2	3/2	.18257419	1	-1-1-1
5/2	5/2	0	5/2	5/2	0	-.16666667	-1	-2-2
5/2	5/2	1	1/2	1/2	2	.19720266	1	-2-2-1 1
5/2	5/2	1	1	0	5/2	.23570226	1	-1-2
5/2	5/2	1	1	1	3/2	.19720266	1	-2-2-1 1
5/2	5/2	1	1	1	5/2	-.05634362	-1	0-2-1-1
5/2	5/2	1	3/2	1/2	2	.14907120	1	0-2-1
5/2	5/2	1	3/2	3/2	1	.18708287	1	-3 0-2 1

j_1	j_2	j_3	l_1	l_2	l_3	6-j symbol			
5/2	5/2	1	3/2	3/2	2	-.11581320	-1	-3-2-2-1	0 2
5/2	5/2	1	2	1	3/2	.10000000	1	-2 0-2	
5/2	5/2	1	2	1	5/2	-.14253933	-1	5-2-2-1	
5/2	5/2	1	2	2	1/2	.17638342	1	0-2-2 1	
5/2	5/2	1	2	2	3/2	-.13858697	-1	-2-2-2-1	2
5/2	5/2	1	5/2	3/2	1	.06666667	1	0-2-2	
5/2	5/2	1	5/2	5/2	0	.16666667	1	-2-2	
5/2	5/2	1	5/2	5/2	1	-.14761905	-31	-2-2-2-2	
5/2	5/2	2	1	1	3/2	-.12472191	-1	-1-2-2 1	
5/2	5/2	2	1	1	5/2	-.14253933	-1	5-2-2-1	
5/2	5/2	2	3/2	1/2	2	-.16329932	-1	1-1-2	
5/2	5/2	2	3/2	3/2	1	-.15275252	-1	-2-1-2 1	
5/2	5/2	2	3/2	3/2	2	-.02182179	-1	-2-1-2-1	
5/2	5/2	2	2	0	5/2	-.18257419	-1	-1-1-1	
5/2	5/2	2	2	1	3/2	-.14142136	-1	-1 0-2	
5/2	5/2	2	2	1	5/2	.07559289	1	0 0-2-1	
5/2	5/2	2	2	2	1/2	-.16329932	-1	1-1-2	
5/2	5/2	2	2	2	3/2	.05832118	1	-1-1 0-2	
5/2	5/2	2	2	2	5/2	.05832118	1	-1-1 0-2	
5/2	5/2	2	5/2	1/2	2	-.10000000	-1	-2 0-2	
5/2	5/2	2	5/2	3/2	1	-.10690450	-1	1 0-2-1	
5/2	5/2	2	5/2	3/2	2	.11428571	1	4 0-2-2	
5/2	5/2	2	5/2	5/2	0	-.16666667	-1	-2-2	
5/2	5/2	2	5/2	5/2	1	.10952381	23	-2-2-2-2	
5/2	5/2	2	5/2	5/2	2	-.01666667	-1	-4-2-2	
3	3/2	3/2	0	3/2	3/2	.25000000	1	-4	
3	3/2	3/2	1	3/2	3/2	.15000000	1	-4 2-2	
3	3/2	3/2	2	3/2	3/2	.05000000	1	-4 0-2	
3	3/2	3/2	3	3/2	3/2	.00714286	1	-4 0-2-2	
3	2	1	0	1	2	.25819889	1	0-1-1	
3	2	1	1/2	3/2	3/2	.22360680	1	-2 0-1	
3	2	1	1	1	2	.14907120	1	0-2-1	
3	2	1	1	2	1	.20000000	1	0 0-2	
3	2	1	3/2	3/2	3/2	.10000000	1	-2 0-2	
3	2	1	2	1	2	.04364358	1	0-1-2-1	
3	2	1	2	2	1	.06666667	1	0-2-2	
3	2	1	5/2	3/2	3/2	.02182179	1	-2-1-2-1	
3	2	1	3	2	1	.00952381	1	0-2-2-2	
3	2	2	0	2	2	-.20000000	-1	0 0-2	
3	2	2	1/2	3/2	3/2	-.14142136	-1	-1 0-2	
3	2	2	1	2	1	-.16329932	-1	1-1-2	
3	2	2	1	2	2	0.00000000	0		
3	2	2	3/2	3/2	3/2	-.14142136	-1	-1 0-2	
3	2	2	2	2	1	-.10690450	-1	1 0-2-1	
3	2	2	2	2	2	.11428571	1	4 0-2-2	
3	2	2	5/2	3/2	3/2	-.06060915	-1	-1 2-2-2	
3	2	2	3	2	1	-.02857143	-1	0 0-2-2	
3	2	2	3	2	2	.07142857	1	-2 0 0-2	
3	5/2	1/2	0	1/2	5/2	.28867513	1	-2-1	
3	5/2	1/2	1/2	1	2	.23570226	1	-1-2	
3	5/2	1/2	1	1/2	5/2	.14085904	1	-2-2 1-1	
3	5/2	1/2	1	3/2	3/2	.20412415	1	-3-1	
3	5/2	1/2	3/2	1	2	.08908708	1	-1-2 0-1	
3	5/2	1/2	3/2	2	1	.18257419	1	-1-1-1	

j_1	j_2	j_3	l_1	l_2	l_3	6-j symbol			
3	5/2	1/2	3/2	2	2	-.16903085	-1	0 0-1-1	
3	5/2	1/2	2	3/2	3/2	.05976143	1	-3 0-1-1	
3	5/2	1/2	2	5/2	1/2	.16666667	1	-2-2	
3	5/2	1/2	5/2	2	1	.03984095	1	-1-2-1-1	
3	5/2	1/2	5/2	2	2	-.06900656	-1	-1-1-1-1	
3	5/2	1/2	3	5/2	1/2	.02380952	1	-2-2 0-2	
3	5/2	3/2	0	3/2	5/2	-.20412415	-1	-3-1	
3	5/2	3/2	1/2	1	2	-.14907120	-1	0-2-1	
3	5/2	3/2	1/2	2	2	-.14142136	-1	-1 0-2	
3	5/2	3/2	1	1/2	5/2	-.17817416	-1	1-2 0-1	
3	5/2	3/2	1	3/2	3/2	-.16329932	-1	1-1-2	
3	5/2	3/2	1	3/2	5/2	.00890871	1	-3-2-2-1	
3	5/2	3/2	1	5/2	3/2	-.10000000	-1	-2 0-2	
3	5/2	3/2	3/2	1	2	-.14253933	-1	5-2-2-1	
3	5/2	3/2	3/2	2	1	-.16329932	-1	1-1-2	
3	5/2	3/2	3/2	2	2	.07559289	1	0 0-2-1	
3	5/2	3/2	2	1/2	5/2	-.09759001	-1	0-1-1-1	
3	5/2	3/2	2	3/2	3/2	-.10690450	-1	1 0-2-1	
3	5/2	3/2	2	3/2	5/2	.12001984	1	-4-1-2-1	2
3	5/2	3/2	2	5/2	1/2	-.15936381	-1	3-2-1-1	
3	5/2	3/2	2	5/2	3/2	.10952381	23	-2-2-2-2	
3	5/2	3/2	5/2	1	2	-.05345225	-1	-1 0-2-1	
3	5/2	3/2	5/2	2	1	-.07619048	-1	6-2-2-2	
3	5/2	3/2	5/2	2	2	.10722219	1	-2-1-2-2	0 2
3	5/2	3/2	3	3/2	3/2	-.02857143	-1	0 0-2-2	
3	5/2	3/2	3	5/2	1/2	-.04761905	-1	0-2 0-2	
3	5/2	3/2	3	5/2	3/2	.08452381	71	-6-2-2-2	
3	5/2	5/2	0	5/2	5/2	.16666667	1	-2-2	
3	5/2	5/2	1/2	2	2	.14142136	1	-1 0-2	
3	5/2	5/2	1	3/2	3/2	.10000000	1	-2 0-2	
3	5/2	5/2	1	5/2	3/2	.13093073	1	0 1-2-1	
3	5/2	5/2	1	5/2	5/2	-.05238095	-1	-2-2-2-2	2
3	5/2	5/2	3/2	2	1	.13093073	1	0 1-2-1	
3	5/2	5/2	3/2	2	2	.04040610	1	1 0-2-2	
3	5/2	5/2	2	3/2	3/2	.12857143	1	-2 4-2-2	
3	5/2	5/2	2	5/2	1/2	.14638501	1	-2 1-1-1	
3	5/2	5/2	2	5/2	3/2	-.03499271	-1	-1 1-2-2	
3	5/2	5/2	2	5/2	5/2	-.06904762	-29	-4-2-2-2	
3	5/2	5/2	5/2	2	1	.10497813	1	-1 3-2-2	
3	5/2	5/2	5/2	2	2	-.09091373	-1	-3 4-2-2	
3	5/2	5/2	3	3/2	3/2	.06428571	1	-4 4-2-2	
3	5/2	5/2	3	5/2	1/2	.07142857	1	-2 0 0-2	
3	5/2	5/2	3	5/2	3/2	-.10000000	-1	-2 0-2	
3	5/2	5/2	3	5/2	5/2	.06269841	79	-4-4-2-2	
3	3	0	0	0	3	.37796447	1	0 0 0-1	
3	3	0	1/2	1/2	5/2	.26726124	1	-1 0 0-1	
3	3	0	1	1	2	.21821789	1	0-1 0-1	
3	3	0	3/2	3/2	3/2	.18898224	1	-2 0 0-1	
3	3	0	3/2	3/2	5/2	-.18898224	-1	-2 0 0-1	
3	3	0	2	2	1	.16903085	1	0 0-1-1	
3	3	0	2	2	2	-.16903085	-1	0 0-1-1	
3	3	0	5/2	5/2	1/2	.15430335	1	-1-1 0-1	
3	3	0	5/2	5/2	3/2	-.15430335	-1	-1-1 0-1	
3	3	0	5/2	5/2	5/2	.15430335	1	-1-1 0-1	

j_1	j_2	j_3	l_1	l_2	l_3	6-j symbol			
3	3	0	3	3	0	.14285714	1	0 0 0-2	
3	3	1	1/2	1/2	5/2	-.17817416	-1	1-2 0-1	
3	3	1	1	0	3	-.21821789	-1	0-1 0-1	
3	3	1	1	1	2	-.17817416	-1	1-2 0-1	
3	3	1	1	1	3	.04454354	1	-3-2 0-1	
3	3	1	3/2	1/2	5/2	-.14085904	-1	-2-2 1-1	
3	3	1	3/2	3/2	3/2	-.16903085	-1	0 0-1-1	
3	3	1	3/2	3/2	5/2	.09860133	1	-4-2-1 1	
3	3	1	2	1	2	-.09759001	-1	0-1-1-1	
3	3	1	2	1	3	.13363062	1	-3 0 0-1	
3	3	1	2	2	1	-.15936381	-1	3-2-1-1	
3	3	1	2	2	2	.11952286	1	-1 0-1-1	
3	3	1	5/2	3/2	3/2	-.06900656	-1	-1-1-1-1	
3	3	1	5/2	3/2	5/2	.10350983	1	-3 1-1-1	
3	3	1	5/2	5/2	1/2	-.15058465	-1	1-2 1-2	
3	3	1	5/2	5/2	3/2	.12799695	17	-3-2-1-2	
3	3	1	5/2	5/2	5/2	-.09035079	-1	1 0-1-2	
3	3	1	3	2	1	-.04761905	-1	0-2 0-2	
3	3	1	3	3	0	-.14285714	-1	0 0 0-2	
3	3	1	3	3	1	.13095238	1	-4-2 0-2	2
3	3	2	1	1	2	.10690450	1	1 0-2-1	
3	3	2	1	1	3	.13363062	1	-3 0 0-1	
3	3	2	3/2	1/2	5/2	.14638501	1	-2 1-1-1	
3	3	2	3/2	3/2	3/2	.13093073	1	0 1-2-1	
3	3	2	3/2	3/2	5/2	.03273268	1	-4 1-2-1	
3	3	2	2	0	3	.16903085	1	0 0-1-1	
3	3	2	2	1	2	.13093073	1	0 1-2-1	
3	3	2	2	1	3	-.05976143	-1	-3 0-1-1	
3	3	2	2	2	1	.13997084	1	3 1-2-2	
3	3	2	2	2	2	-.03499271	-1	-1 1-2-2	
3	3	2	2	2	3	-.06415330	-1	-3-1-2-2	2
3	3	2	5/2	1/2	5/2	.09759001	1	0-1-1-1	
3	3	2	5/2	3/2	3/2	.10497813	1	-1 3-2-2	
3	3	2	5/2	3/2	5/2	-.09914601	-17	-3-1-2-2	
3	3	2	5/2	5/2	1/2	.14285714	1	0 0 0-2	
3	3	2	5/2	5/2	3/2	-.07857143	-1	-4 0-2-2	2
3	3	2	5/2	5/2	5/2	-.00476190	-1	-2-2-2-2	
3	3	2	3	1	2	.05714286	1	2 0-2-2	
3	3	2	3	2	1	.07824608	1	-1 1-1-2	
3	3	2	3	2	2	-.10000000	-1	-2 0-2	
3	3	2	3	3	0	.14285714	1	0 0 0-2	
3	3	2	3	3	1	-.10714286	-1	-4 2 0-2	
3	3	2	3	3	2	.04523810	19	-4-2-2-2	
3	3	3	3/2	3/2	3/2	-.07824608	-1	-1 1-1-2	
3	3	3	2	2	1	-.11065667	-1	0 1-1-2	
3	3	3	2	2	2	-.05532833	-1	-2 1-1-2	
3	3	3	5/2	3/2	3/2	-.11736912	-1	-3 3-1-2	
3	3	3	5/2	5/2	1/2	-.13041013	-1	-1-1 1-2	
3	3	3	5/2	5/2	3/2	.01304101	1	-3-1-1-2	
3	3	3	5/2	5/2	5/2	.07389908	17	-3-3-1-2	
3	3	3	3	2	1	-.10101525	-1	-1 0 0-2	
3	3	3	3	2	2	.07377111	1	2-1-1-2	
3	3	3	3	3	0	-.14285714	-1	0 0 0-2	
3	3	3	3	3	1	.07142857	1	-2 0 0-2	

j_1	j_2	j_3	l_1	l_2	l_3	6-j symbol			
3	3	3	3	3	2	.02380952	1	-2-2 0-2	
3	3	3	3	3	3	-.07142857	-1	-2 0 0-2	
7/2	2	3/2	0	3/2	2	-.22360680	-1	-2 0-1	
7/2	2	3/2	1/2	2	3/2	-.20000000	-1	0 0-2	
7/2	2	3/2	1	3/2	2	-.14142136	-1	-1 0-2	
7/2	2	3/2	3/2	2	3/2	-.10000000	-1	-2 0-2	
7/2	2	3/2	2	3/2	2	-.05345225	-1	-1 0-2-1	
7/2	2	3/2	5/2	2	3/2	-.02857143	-1	0 0-2-2	
7/2	2	3/2	3	3/2	2	-.01010153	-1	-3 0-2-2	
7/2	2	3/2	7/2	2	3/2	-.00357143	-1	-6 0-2-2	
7/2	5/2	1	0	1	5/2	-.23570226	-1	-1-2	
7/2	5/2	1	1/2	3/2	2	-.20412415	-1	-3-1	
7/2	5/2	1	1	1	5/2	-.14085904	-1	-2-2 1-1	
7/2	5/2	1	1	2	3/2	-.18257419	-1	-1-1-1	
7/2	5/2	1	3/2	3/2	2	-.09759001	-1	0-1-1-1	
7/2	5/2	1	3/2	5/2	1	-.16666667	-1	-2-2	
7/2	5/2	1	2	1	5/2	-.04454354	-1	-3-2 0-1	
7/2	5/2	1	2	2	3/2	-.06900656	-1	-1-1-1-1	
7/2	5/2	1	5/2	3/2	2	-.02439750	-1	-4-1-1-1	
7/2	5/2	1	5/2	5/2	1	-.04761905	-1	0-2 0-2	
7/2	5/2	1	3	2	3/2	-.01304101	-1	-3-1-1-2	
7/2	5/2	1	7/2	5/2	1	-.00595238	-1	-6-2 0-2	
7/2	5/2	2	0	2	5/2	.18257419	1	-1-1-1	
7/2	5/2	2	1/2	3/2	2	.12247449	1	-3 1-2	
7/2	5/2	2	1/2	5/2	2	.13333333	1	2-2-2	
7/2	5/2	2	1	1	5/2	.14638501	1	-2 1-1-1	
7/2	5/2	2	1	2	3/2	.14142136	1	-1 0-2	
7/2	5/2	2	1	2	5/2	.01259882	1	-2-2-2-1	
7/2	5/2	2	3/2	3/2	2	.13093073	1	0 1-2-1	
7/2	5/2	2	3/2	5/2	1	.14638501	1	-2 1-1-1	
7/2	5/2	2	3/2	5/2	2	-.05238095	-1	-2-2-2-2	2
7/2	5/2	2	2	1	5/2	.10350983	1	-3 1-1-1	
7/2	5/2	2	2	2	3/2	.10497813	1	-1 3-2-2	
7/2	5/2	2	2	2	5/2	-.09914601	-17	-3-1-2-2	
7/2	5/2	2	5/2	3/2	2	.06428571	1	-4 4-2-2	
7/2	5/2	2	5/2	5/2	1	.07824608	1	-1 1-1-2	
7/2	5/2	2	5/2	5/2	2	-.10000000	-1	-2 0-2	
7/2	5/2	2	3	1	5/2	.03194383	1	-2 0-1-2	
7/2	5/2	2	3	2	3/2	.03711537	1	-4 3-2-2	
7/2	5/2	2	3	2	5/2	-.07407785	-1	-1-2-2-2	2
7/2	5/2	2	7/2	3/2	2	.01428571	1	-2 0-2-2	
7/2	5/2	2	7/2	5/2	1	.01785714	1	-6 0 0-2	
7/2	5/2	2	7/2	5/2	2	-.04722222	-17	-6-4-2	
7/2	3	1/2	0	1/2	3	-.26726124	-1	-1 0 0-1	
7/2	3	1/2	1/2	1	5/2	-.21821789	-1	0-1 0-1	
7/2	3	1/2	1	1/2	3	-.13363062	-1	-3 0 0-1	
7/2	3	1/2	1	3/2	2	-.18898224	-1	-2 0 0-1	
7/2	3	1/2	3/2	1	5/2	-.08625819	-1	-5-1 1-1	
7/2	3	1/2	3/2	2	3/2	-.16903085	-1	0 0-1-1	
7/2	3	1/2	3/2	2	5/2	.15526475	1	-5 3-1-1	
7/2	3	1/2	2	3/2	2	-.05976143	-1	-3 0-1-1	
7/2	3	1/2	2	5/2	1	-.15430335	-1	-1-1 0-1	
7/2	3	1/2	2	5/2	2	.14638501	1	-2 1-1-1	
7/2	3	1/2	5/2	2	3/2	-.04225771	-1	-4 0-1-1	

j_1	j_2	j_3	l_1	l_2	l_3	6-j symbol			
7/2	3	1/2	5/2	2	5/2	.06900656	1	-1-1-1-1	
7/2	3	1/2	5/2	3	1/2	-.14285714	-1	0 0 0-2	
7/2	3	1/2	3	5/2	1	-.02916059	-1	-3-1 0-2	
7/2	3	1/2	3	5/2	2	.05050763	1	-3 0 0-2	
7/2	3	1/2	7/2	3	1/2	-.01785714	-1	-6 0 0-2	
7/2	3	3/2	0	3/2	3	.18898224	1	-2 0 0-1	
7/2	3	3/2	1/2	1	5/2	.13363062	1	-3 0 0-1	
7/2	3	3/2	1/2	2	5/2	.13363062	1	-3 0 0-1	
7/2	3	3/2	1	1/2	3	.16366342	1	-4 1 0-1	
7/2	3	3/2	1	3/2	2	.14638501	1	-2 1-1-1	
7/2	3	3/2	1	3/2	3	0.00000000	0		
7/2	3	3/2	1	5/2	2	.09759001	1	0-1-1-1	
7/2	3	3/2	3/2	1	5/2	.13363062	1	-3 0 0-1	
7/2	3	3/2	3/2	2	3/2	.14638501	1	-2 1-1-1	
7/2	3	3/2	3/2	2	5/2	-.05976143	-1	-3 0-1-1	
7/2	3	3/2	3/2	3	3/2	.07142857	1	-2 0 0-2	
7/2	3	3/2	2	1/2	3	.09449112	1	-4 0 0-1	
7/2	3	3/2	2	3/2	2	.10350983	1	-3 1-1-1	
7/2	3	3/2	2	3/2	3	-.10910895	-1	-2-1 0-1	
7/2	3	3/2	2	5/2	1	.14285714	1	0 0 0-2	
7/2	3	3/2	2	5/2	2	-.09035079	-1	1 0-1-2	
7/2	3	3/2	5/2	1	5/2	.05455447	1	-4-1 0-1	
7/2	3	3/2	5/2	2	3/2	.07824608	1	-1 1-1-2	
7/2	3	3/2	5/2	2	5/2	-.10115546	-19	-4-2-1-2	
7/2	3	3/2	5/2	3	1/2	.13832083	1	-4 1 1-2	
7/2	3	3/2	5/2	3	3/2	-.10714286	-1	-4 2 0-2	
7/2	3	3/2	3	3/2	2	.03194383	1	-2 0-1-2	
7/2	3	3/2	3	3/2	3	-.06520507	-1	-3-1 1-2	
7/2	3	3/2	3	5/2	1	.05646924	1	-5 0 1-2	
7/2	3	3/2	3	5/2	2	-.08476659	-1	-5-1-1-2	0 2
7/2	3	3/2	7/2	2	3/2	.01785714	1	-6 0 0-2	
7/2	3	3/2	7/2	3	1/2	.03571429	1	-4 0 0-2	
7/2	3	3/2	7/2	3	3/2	-.06547619	-1	-6-2 0-2	2
7/2	3	5/2	0	5/2	3	-.15430335	-1	-1-1 0-1	
7/2	3	5/2	1/2	2	5/2	-.12598816	-1	0-2 0-1	
7/2	3	5/2	1/2	3	5/2	-.09523810	-1	2-2 0-2	
7/2	3	5/2	1	3/2	2	-.08451543	-1	-2 0-1-1	
7/2	3	5/2	1	3/2	3	-.12198751	-1	-4-1 1-1	
7/2	3	5/2	1	5/2	2	-.12046772	-1	5-2-1-2	
7/2	3	5/2	1	5/2	3	.03764616	1	-3-2 1-2	
7/2	3	5/2	3/2	1	5/2	-.11572751	-1	-5 1 0-1	
7/2	3	5/2	3/2	2	3/2	-.11065667	-1	0 1-1-2	
7/2	3	5/2	3/2	2	5/2	-.04894001	-1	-5-2-1-2	0 2
7/2	3	5/2	3/2	3	3/2	-.10101525	-1	-1 0 0-2	
7/2	3	5/2	3/2	3	5/2	.08333333	1	-4-2	
7/2	3	5/2	2	1/2	3	-.13363062	-1	-3 0 0-1	
7/2	3	5/2	2	3/2	2	-.11736912	-1	-3 3-1-2	
7/2	3	5/2	2	3/2	3	.02061965	1	-4-1 0-2	
7/2	3	5/2	2	5/2	1	-.12371791	-1	-2 1 0-2	
7/2	3	5/2	2	5/2	2	.01304101	1	-3-1-1-2	
7/2	3	5/2	2	5/2	3	.07142857	1	-2 0 0-2	
7/2	3	5/2	5/2	1	5/2	-.10101525	-1	-1 0 0-2	
7/2	3	5/2	5/2	2	3/2	-.10164464	-1	-5 4-1-2	
7/2	3	5/2	5/2	2	5/2	.07377111	1	2-1-1-2	

j₁	j₂	j₃	l_1	l_2	l_3	6-j symbol			
7/2	3	5/2	5/2	3	1/2	-.13041013	-1	-1-1 1-2	
7/2	3	5/2	5/2	3	3/2	.05892557	1	-5-2	
7/2	3	5/2	5/2	3	5/2	.02380952	1	-2-2 0-2	
7/2	3	5/2	3	1/2	3	-.07142857	-1	-2 0 0-2	
7/2	3	5/2	3	3/2	2	-.06776309	-1	-3 2-1-2	
7/2	3	5/2	3	3/2	3	.09221389	1	-2-1 1-2	
7/2	3	5/2	3	5/2	1	-.07985957	-1	-4 0 1-2	
7/2	3	5/2	3	5/2	2	.08914009	29	-4-3-1-2	
7/2	3	5/2	3	5/2	3	-.04347004	-1	-1-3 1-2	
7/2	3	5/2	7/2	1	5/2	-.03571429	-1	-4 0 0-2	
7/2	3	5/2	7/2	2	3/2	-.04123930	-1	-2-1 0-2	
7/2	3	5/2	7/2	2	5/2	.07539683	19	-4-4 0-2	
7/2	3	5/2	7/2	3	1/2	-.05357143	-1	-6 2 0-2	
7/2	3	5/2	7/2	3	3/2	.08333333	1	-4-2	
7/2	3	5/2	7/2	3	5/2	-.07341270	-37	-6-4 0-2	
7/2	7/2	0	0	0	7/2	-.35355339	-1	-3	
7/2	7/2	0	1/2	1/2	3	-.25000000	-1	-4	
7/2	7/2	0	1	1	5/2	-.20412415	-1	-3-1	
7/2	7/2	0	3/2	3/2	2	-.17677670	-1	-5	
7/2	7/2	0	3/2	3/2	3	.17677670	1	-5	
7/2	7/2	0	2	2	3/2	-.15811388	-1	-3 0-1	
7/2	7/2	0	2	2	5/2	.15811388	1	-3 0-1	
7/2	7/2	0	5/2	5/2	1	-.14433757	-1	-4-1	
7/2	7/2	0	5/2	5/2	2	.14433757	1	-4-1	
7/2	7/2	0	5/2	5/2	3	-.14433757	-1	-4-1	
7/2	7/2	0	3	3	1/2	-.13363062	-1	-3 0 0-1	
7/2	7/2	0	3	3	3/2	.13363062	1	-3 0 0-1	
7/2	7/2	0	3	3	5/2	-.13363062	-1	-3 0 0-1	
7/2	7/2	0	7/2	7/2	0	-.12500000	-1	-6	
7/2	7/2	1	1/2	1/2	3	.16366342	1	-4 1 0-1	
7/2	7/2	1	1	0	7/2	.20412415	1	-3-1	
7/2	7/2	1	1	1	5/2	.16366342	1	-4 1 0-1	
7/2	7/2	1	1	1	7/2	-.03636965	-1	-2-3 0-1	
7/2	7/2	1	3/2	1/2	3	.13363062	1	-3 0 0-1	
7/2	7/2	1	3/2	3/2	2	.15526475	1	-5 3-1-1	
7/2	7/2	1	3/2	3/2	3	-.08625819	-1	-5-1 1-1	
7/2	7/2	1	2	1	5/2	.09449112	1	-4 0 0-1	
7/2	7/2	1	2	1	7/2	-.12598816	-1	0-2 0-1	
7/2	7/2	1	2	2	3/2	.14638501	1	-2 1-1-1	
7/2	7/2	1	2	2	5/2	-.10572251	-1	-4-3-1-1	0 2
7/2	7/2	1	5/2	3/2	2	.06900656	1	-1-1-1-1	
7/2	7/2	1	5/2	3/2	3	-.09960238	-1	-3-2 1-1	
7/2	7/2	1	5/2	5/2	1	.13832083	1	-4 1 1-2	
7/2	7/2	1	5/2	5/2	2	-.11373046	-37	-4-3-1-2	
7/2	7/2	1	5/2	5/2	3	.07684491	1	-4-3 3-2	
7/2	7/2	1	3	2	3/2	.05050763	1	-3 0 0-2	
7/2	7/2	1	3	2	5/2	-.07776158	-1	3-3 0-2	
7/2	7/2	1	3	3	1/2	.13122266	1	-5 3 0-2	
7/2	7/2	1	3	3	3/2	-.11664237	-1	1-1 0-2	
7/2	7/2	1	3	3	5/2	.09234188	19	-5-3 0-2	
7/2	7/2	1	7/2	5/2	1	.03571429	1	-4 0 0-2	
7/2	7/2	1	7/2	7/2	0	.12500000	1	-6	
7/2	7/2	1	7/2	7/2	1	-.11706349	-59	-6-4 0-2	
7/2	7/2	2	1	1	5/2	-.09449112	-1	-4 0 0-1	

TABLE D-1 (cont)

j_1	j_2	j_3	l_1	l_2	l_3	6-j symbol			
7/2	7/2	2	1	1	7/2	-.12598816	-1	0-2 0-1	
7/2	7/2	2	3/2	1/2	3	-.13363062	-1	-3 0 0-1	
7/2	7/2	2	3/2	3/2	2	-.11572751	-1	-5 1 0-1	
7/2	7/2	2	3/2	3/2	3	-.03857584	-1	-5-1 0-1	
7/2	7/2	2	2	0	7/2	-.15811388	-1	-3 0-1	
7/2	7/2	2	2	1	5/2	-.12198751	-1	-4-1 1-1	
7/2	7/2	2	2	1	7/2	.04879500	1	-2-1-1-1	
7/2	7/2	2	2	2	3/2	-.12371791	-1	-2 1 0-2	
7/2	7/2	2	2	2	5/2	.02061965	1	-4-1 0-2	
7/2	7/2	2	2	2	7/2	.06598289	1	4-1-2-2	
7/2	7/2	2	5/2	1/2	3	-.09449112	-1	-4 0 0-1	
7/2	7/2	2	5/2	3/2	2	-.10101525	-1	-1 0 0-2	
7/2	7/2	2	5/2	3/2	3	.08748178	1	-3 1 0-2	
7/2	7/2	2	5/2	5/2	1	-.12626907	-1	-5 0 2-2	
7/2	7/2	2	5/2	5/2	2	.05892557	1	-5-2	
7/2	7/2	2	5/2	5/2	3	.01683588	1	-3-2 0-2	
7/2	7/2	2	3	1	5/2	-.05832118	-1	-1-1 0-2	
7/2	7/2	2	3	1	7/2	.09671474	1	-3-1 0-2	1
7/2	7/2	2	3	2	3/2	-.07985957	-1	-4 0 1-2	
7/2	7/2	2	3	2	5/2	.09221389	1	-2-1 1-2	
7/2	7/2	2	3	2	7/2	-.06116777	-1	-2-1-1-2	1
7/2	7/2	2	3	3	1/2	-.12626907	-1	-5 0 2-2	
7/2	7/2	2	3	3	3/2	.08417938	1	-3-2 2-2	
7/2	7/2	2	3	3	5/2	-.02525381	-1	-5 0 0-2	
7/2	7/2	2	7/2	3/2	2	-.03571429	-1	-4 0 0-2	
7/2	7/2	2	7/2	5/2	1	-.05952381	-1	-4-2 2-2	
7/2	7/2	2	7/2	5/2	2	.08333333	1	-4-2	
7/2	7/2	2	7/2	7/2	0	-.12500000	-1	-6	
7/2	7/2	2	7/2	7/2	1	.10119048	17	-6-2 0-2	
7/2	7/2	2	7/2	7/2	2	-.05833333	-1	-6-2-2 2	
7/2	7/2	3	3/2	3/2	2	.06487822	1	-5 1-1-2	1
7/2	7/2	3	3/2	3/2	3	.10813037	1	-5-1 1-2	1
7/2	7/2	3	2	1	5/2	.09671474	1	-3-1 0-2	1
7/2	7/2	3	2	1	7/2	.09671474	1	-3-1 0-2	1
7/2	7/2	3	2	2	3/2	.09175166	1	-4 1-1-2	1
7/2	7/2	3	2	2	5/2	.06116777	1	-2-1-1-2	1
7/2	7/2	3	2	2	7/2	-.06116777	-1	-2-1-1-2	1
7/2	7/2	3	5/2	1/2	3	.11845089	1	-4 0 0-2	1
7/2	7/2	3	5/2	3/2	2	.10594569	1	-2 0-1-2	1
7/2	7/2	3	5/2	3/2	3	0.00000000	0		
7/2	7/2	3	5/2	5/2	1	.10813037	1	-5-1 1-2	1
7/2	7/2	3	5/2	5/2	2	.00720869	1	-5-3-1-2	1
7/2	7/2	3	5/2	5/2	3	-.07208691	-1	-3-3 1-2	1
7/2	7/2	3	3	0	7/2	.13363062	1	-3 0 0-1	
7/2	7/2	3	3	1	5/2	.09671474	1	-3-1 0-2	1
7/2	7/2	3	3	1	7/2	-.05832118	-1	-1-1 0-2	
7/2	7/2	3	3	2	3/2	.09671474	1	-3-1 0-2	1
7/2	7/2	3	3	2	5/2	-.05583828	-1	-3-2 0-2	1
7/2	7/2	3	3	2	7/2	-.03367175	-1	-1-2 0-2	
7/2	7/2	3	3	3	1/2	.11845089	1	-4 0 0-2	1
7/2	7/2	3	3	3	3/2	-.03948363	-1	-4-2 0-2	1
7/2	7/2	3	3	3	5/2	-.03948363	-1	-4-2 0-2	1
7/2	7/2	3	3	3	7/2	.06460957	1	-2 2 0-2	-1
7/2	7/2	3	7/2	1/2	3	.07142857	1	-2 0 0-2	

j_1	j_2	j_3	l_1	l_2	l_3	6-j symbol			
7/2	7/2	3	7/2	3/2	2	.06838765	1	-4-1 0-2	1
7/2	7/2	3	7/2	3/2	3	-.08333333	-1	-4-2	
7/2	7/2	3	7/2	5/2	1	.07896726	1	-2-2 0-2	1
7/2	7/2	3	7/2	5/2	2	-.07896726	-1	-2-2 0-2	1
7/2	7/2	3	7/2	5/2	3	.02380952	1	-2-2 0-2	
7/2	7/2	3	7/2	7/2	0	.12500000	1	-6	
7/2	7/2	3	7/2	7/2	1	-.07738095	-1	-6-2 0-2	0 2
7/2	7/2	3	7/2	7/2	2	.00595238	1	-6-2 0-2	
7/2	7/2	3	7/2	7/2	3	.05032468	31	-6 0 0-2	-2
4	2	2	0	2	2	.20000000	1	0 0-2	
4	2	2	1	2	2	.13333333	1	2-2-2	
4	2	2	2	2	2	.05714286	1	2 0-2-2	
4	2	2	3	2	2	.01428571	1	-2 0-2-2	
4	2	2	4	2	2	.00158730	1	-2-4-2-2	
4	5/2	3/2	0	3/2	5/2	.20412415	1	-3-1	
4	5/2	3/2	1/2	2	2	.18257419	1	-1-1-1	
4	5/2	3/2	1	3/2	5/2	.13363062	1	-3 0 0-1	
4	5/2	3/2	1	5/2	3/2	.16666667	1	-2-2	
4	5/2	3/2	3/2	2	2	.09759001	1	0-1-1-1	
4	5/2	3/2	2	3/2	5/2	.05455447	1	-4-1 0-1	
4	5/2	3/2	2	5/2	3/2	.07142857	1	-2 0 0-2	
4	5/2	3/2	5/2	2	2	.03194383	1	-2 0-1-2	
4	5/2	3/2	3	3/2	5/2	.01190476	1	-4-2 0-2	
4	5/2	3/2	3	5/2	3/2	.01785714	1	-6 0 0-2	
4	5/2	3/2	7/2	2	2	.00532397	1	-4-2-1-2	
4	5/2	3/2	4	5/2	3/2	.00198413	1	-6-4 0-2	
4	5/2	5/2	0	5/2	5/2	-.16666667	-1	-2-2	
4	5/2	5/2	1/2	2	2	-.10540926	-1	-1-2-1	
4	5/2	5/2	1	5/2	3/2	-.12598816	-1	0-2 0-1	
4	5/2	5/2	1	5/2	5/2	-.02380952	-1	-2-2 0-2	
4	5/2	5/2	3/2	2	2	-.12046772	-1	5-2-1-2	
4	5/2	5/2	2	5/2	3/2	-.10101525	-1	-1 0 0-2	
4	5/2	5/2	2	5/2	5/2	.08333333	1	-4-2	
4	5/2	5/2	5/2	2	2	-.06776309	-1	-3 2-1-2	
4	5/2	5/2	3	5/2	3/2	-.04123930	-1	-2-1 0-2	
4	5/2	5/2	3	5/2	5/2	.07539683	19	-4-4 0-2	
4	5/2	5/2	7/2	2	2	-.02007795	-1	3-4-1-2	
4	5/2	5/2	4	5/2	3/2	-.00793651	-1	-2-4 0-2	
4	5/2	5/2	4	5/2	5/2	.02777778	1	-4-4	
4	3	1	0	1	3	.21821789	1	0-1 0-1	
4	3	1	1/2	3/2	5/2	.18898224	1	-2 0 0-1	
4	3	1	1	1	3	.13363062	1	-3 0 0-1	
4	3	1	1	2	2	.16903085	1	0 0-1-1	
4	3	1	3/2	3/2	5/2	.09449112	1	-4 0 0-1	
4	3	1	3/2	5/2	3/2	.15430335	1	-1-1 0-1	
4	3	1	3/2	5/2	5/2	-.13363062	-1	-3 0 0-1	
4	3	1	2	1	3	.04454354	1	-3-2 0-1	
4	3	1	2	2	2	.06900656	1	-1-1-1-1	
4	3	1	2	3	1	.14285714	1	0 0 0-2	
4	3	1	5/2	3/2	5/2	.02571722	1	-3-3 0-1	
4	3	1	5/2	5/2	3/2	.05050763	1	-3 0 0-2	
4	3	1	5/2	5/2	5/2	-.07776158	-1	3-3 0-2	
4	3	1	3	2	2	.01505847	1	-1-2-1-2	
4	3	1	3	3	1	.03571429	1	-4 0 0-2	

j_1	j_2	j_3	l_1	l_2	l_3	6-j symbol			
4	3	1	7/2	5/2	3/2	.00841794	1	-5-2 0-2	
4	3	1	7/2	5/2	5/2	-.02061965	-1	-4-1 0-2	
4	3	1	4	3	1	.00396825	1	-4-4 0-2	
4	3	2	0	2	3	-.16903085	-1	0 0-1-1	
4	3	2	1/2	3/2	5/2	-.10910895	-1	-2-1 0-1	
4	3	2	1/2	5/2	5/2	-.12598816	-1	0-2 0-1	
4	3	2	1	1	3	-.13363062	-1	-3 0 0-1	
4	3	2	1	2	2	-.12598816	-1	0-2 0-1	
4	3	2	1	2	3	-.01992048	-1	-3-2-1-1	
4	3	2	1	3	2	-.09523810	-1	2-2 0-2	
4	3	2	3/2	3/2	5/2	-.12198751	-1	-4-1 1-1	
4	3	2	3/2	5/2	3/2	-.13041013	-1	-1-1 1-2	
4	3	2	3/2	5/2	5/2	.03764616	1	-3-2 1-2	
4	3	2	2	1	3	-.09960238	-1	-3-2 1-1	
4	3	2	2	2	2	-.10101525	-1	-1 0 0-2	
4	3	2	2	2	3	.08748178	1	-3 1 0-2	
4	3	2	2	3	1	-.13041013	-1	-1-1 1-2	
4	3	2	2	3	2	.07142857	1	-2 0 0-2	
4	3	2	5/2	3/2	5/2	-.06520507	-1	-3-1 1-2	
4	3	2	5/2	5/2	3/2	-.07985957	-1	-4 0 1-2	
4	3	2	5/2	5/2	5/2	.09221389	1	-2-1 1-2	
4	3	2	3	1	3	-.03367175	-1	-1-2 0-2	
4	3	2	3	2	2	-.04123930	-1	-2-1 0-2	
4	3	2	3	2	3	.07377111	1	2-1-1-2	
4	3	2	3	3	1	-.05952381	-1	-4-2 2-2	
4	3	2	3	3	2	.08333333	1	-4-2	
4	3	2	7/2	3/2	5/2	-.01683588	-1	-3-2 0-2	
4	3	2	7/2	5/2	3/2	-.02430049	-1	-5-3 2-2	
4	3	2	7/2	5/2	5/2	.05158730	1	-4-4 0-2	0 2
4	3	2	4	2	2	-.00793651	-1	-2-4 0-2	
4	3	2	4	3	1	-.01190476	-1	-4-2 0-2	
4	3	2	4	3	2	.03253968	41	-4-4-2-2	
4	3	3	0	3	3	.14285714	1	0 0 0-2	
4	3	3	1/2	5/2	5/2	.11167657	1	-1-2 0-2	1
4	3	3	1	2	2	.07063046	1	0-2-1-2	1
4	3	3	1	3	2	.11167657	1	-1-2 0-2	1
4	3	3	1	3	3	-.02380952	-1	-2-2 0-2	
4	3	3	3/2	5/2	3/2	.09671474	1	-3-1 0-2	1
4	3	3	3/2	5/2	5/2	.05583828	1	-3-2 0-2	1
4	3	3	2	2	2	.10594569	1	-2 0-1-2	1
4	3	3	2	3	1	.11167657	1	-1-2 0-2	1
4	3	3	2	3	2	0.00000000	0		
4	3	3	2	3	3	-.07142857	-1	-2 0 0-2	
4	3	3	5/2	5/2	3/2	.09671474	1	-3-1 0-2	1
4	3	3	5/2	5/2	5/2	-.05583828	-1	-3-2 0-2	1
4	3	3	3	2	2	.07063046	1	0-2-1-2	1
4	3	3	3	3	1	.07896726	1	-2-2 0-2	1
4	3	3	3	3	2	-.07896726	-1	-2-2 0-2	1
4	3	3	3	3	3	.02380952	1	-2-2 0-2	
4	3	3	7/2	5/2	3/2	.04559177	1	-2-3 0-2	1
4	3	3	7/2	5/2	5/2	-.07445104	-1	1-4 0-2	1
4	3	3	4	2	2	.02354349	1	0-4-1-2	1
4	3	3	4	3	1	.02380952	1	-2-2 0-2	
4	3	3	4	3	2	-.05555556	-1	-2-4	

j_1	j_2	j_3	l_1	l_2	l_3	6-j symbol			
4	3	3	4	3	3	.06998557	97	-2-4 0-2	-2
4	7/2	1/2	0	1/2	7/2	.25000000	1	-4	
4	7/2	1/2	1/2	1	3	.20412415	1	-3-1	
4	7/2	1/2	1	1/2	7/2	.12729377	1	-4-3 0 1	
4	7/2	1/2	1	3/2	5/2	.17677670	1	-5	
4	7/2	1/2	3/2	1	3	.08333333	1	-4-2	
4	7/2	1/2	3/2	2	2	.15811388	1	-3 0-1	
4	7/2	1/2	3/2	2	3	-.14433757	-1	-4-1	
4	7/2	1/2	2	3/2	5/2	.05892557	1	-5-2	
4	7/2	1/2	2	5/2	3/2	.14433757	1	-4-1	
4	7/2	1/2	2	5/2	5/2	-.13608276	-1	-1-3	
4	7/2	1/2	5/2	2	2	.04303315	1	-2-3-1	
4	7/2	1/2	5/2	2	3	-.06804138	-1	-3-3	
4	7/2	1/2	5/2	3	1	.13363062	1	-3 0 0-1	
4	7/2	1/2	5/2	3	2	-.12858612	-1	-3-3 2-1	
4	7/2	1/2	5/2	3	3	.12062448	1	-2-3 0-1	1
4	7/2	1/2	3	5/2	3/2	.03149704	1	-4-2 0-1	
4	7/2	1/2	3	5/2	5/2	-.05143445	-1	-1-3 0-1	
4	7/2	1/2	3	7/2	1/2	.12500000	1	-6	
4	7/2	1/2	7/2	3	1	.02227177	1	-5-2 0-1	
4	7/2	1/2	7/2	3	2	-.03857584	-1	-5-1 0-1	
4	7/2	1/2	7/2	3	3	.05455447	1	-4-1 0-1	
4	7/2	1/2	4	7/2	1/2	.01388889	1	-6-4	
4	7/2	3/2	0	3/2	7/2	-.17677670	-1	-5	
4	7/2	3/2	1/2	1	3	-.12198751	-1	-4-1 1-1	
4	7/2	3/2	1/2	2	3	-.12677314	-1	-4 2-1-1	
4	7/2	3/2	1	1/2	7/2	-.15214515	-1	-3-3 1	
4	7/2	3/2	1	3/2	5/2	-.13363062	-1	-3 0 0-1	
4	7/2	3/2	1	3/2	7/2	-.00575055	-1	-5-3-1-1	
4	7/2	3/2	1	5/2	5/2	-.09449112	-1	-4 0 0-1	
4	7/2	3/2	3/2	1	3	-.12598816	-1	0-2 0-1	
4	7/2	3/2	3/2	2	2	-.13363062	-1	-3 0 0-1	
4	7/2	3/2	3/2	2	3	.04879500	1	-2-1-1-1	
4	7/2	3/2	3/2	3	2	-.07142857	-1	-2 0 0-2	
4	7/2	3/2	3/2	3	3	.09671474	1	-3-1 0-2	1
4	7/2	3/2	2	1/2	7/2	-.09128709	-1	-3-1-1	
4	7/2	3/2	2	3/2	5/2	-.09960238	-1	-3-2 1-1	
4	7/2	3/2	2	3/2	7/2	.10029718	1	-5-1-2-1	0 2
4	7/2	3/2	2	5/2	3/2	-.13041013	-1	-1-1 1-2	
4	7/2	3/2	2	5/2	5/2	.07684491	1	-4-3 3-2	
4	7/2	3/2	2	7/2	3/2	-.05357143	-1	-6 2 0-2	
4	7/2	3/2	5/2	1	3	-.05455447	-1	-4-1 0-1	
4	7/2	3/2	5/2	2	2	-.07776158	-1	3-3 0-2	
4	7/2	3/2	5/2	2	3	.09528769	31	-4-3-1-2	
4	7/2	3/2	5/2	3	1	-.12626907	-1	-5 0 2-2	
4	7/2	3/2	5/2	3	2	.09234188	19	-5-3 0-2	
4	7/2	3/2	5/2	3	3	-.04559177	-1	-2-3 0-2	1
4	7/2	3/2	3	3/2	5/2	-.03367175	-1	-1-2 0-2	
4	7/2	3/2	3	3/2	7/2	.06487822	1	-5 1-1-2	1
4	7/2	3/2	3	5/2	3/2	-.05952381	-1	-4-2 2-2	
4	7/2	3/2	3	5/2	5/2	.08262168	17	-5-3 0-2	
4	7/2	3/2	3	7/2	1/2	-.12198751	-1	-4-1 1-1	
4	7/2	3/2	3	7/2	3/2	.10119048	17	-6-2 0-2	
4	7/2	3/2	7/2	2	2	-.02061965	-1	-4-1 0-2	

j_1	j_2	j_3	l_1	l_2	l_3	6-j symbol			
4	7/2	3/2	7/2	2	3	.04325215	1	-3-1-1-2	1
4	7/2	3/2	7/2	3	1	-.04347004	-1	-1-3 1-2	
4	7/2	3/2	7/2	3	2	.06776309	1	-3 2-1-2	
4	7/2	3/2	7/2	3	3	-.07985957	-1	-4 0 1-2	
4	7/2	3/2	4	5/2	3/2	-.01190476	-1	-4-2 0-2	
4	7/2	3/2	4	7/2	1/2	-.02777778	-1	-4-4	
4	7/2	3/2	4	7/2	3/2	.05198413	131	-6-4-2-2	
4	7/2	5/2	0	5/2	7/2	.14433757	1	-4-1	
4	7/2	5/2	1/2	2	3	.11443443	1	-3-1-1-1	1
4	7/2	5/2	1/2	3	3	.09221389	1	-2-1 1-2	
4	7/2	5/2	1	3/2	5/2	.07386711	1	-5-2 0-1	1
4	7/2	5/2	1	3/2	7/2	.11443443	1	-3-1-1-1	1
4	7/2	5/2	1	5/2	5/2	.11167657	1	-1-2 0-2	1
4	7/2	5/2	1	5/2	7/2	-.02766417	-1	-4 1-1-2	
4	7/2	5/2	1	7/2	5/2	.05952381	1	-4-2 2-2	
4	7/2	5/2	3/2	1	3	.10446386	1	-4-2 0-1	1
4	7/2	5/2	3/2	2	2	.09671474	1	-3-1 0-2	1
4	7/2	5/2	3/2	2	3	.05297285	1	-4 0-1-2	1
4	7/2	5/2	3/2	3	2	.09671474	1	-3-1 0-2	1
4	7/2	5/2	3/2	3	3	-.07142857	-1	-2 0 0-2	
4	7/2	5/2	2	1/2	7/2	.12360331	1	-4-2-1 0	1
4	7/2	5/2	2	3/2	5/2	.10813037	1	-5-1 1-2	1
4	7/2	5/2	2	3/2	7/2	-.01116766	-1	-3-2-2-2	1
4	7/2	5/2	2	5/2	3/2	.10813037	1	-5-1 1-2	1
4	7/2	5/2	2	5/2	5/2	0.00000000	0		
4	7/2	5/2	2	5/2	7/2	-.07071068	-1	-3 0-2	
4	7/2	5/2	2	7/2	3/2	.07896726	1	-2-2 0-2	1
4	7/2	5/2	2	7/2	5/2	-.08333333	-1	-4-2	
4	7/2	5/2	5/2	1	3	.09671474	1	-3-1 0-2	1
4	7/2	5/2	5/2	2	2	.09671474	1	-3-1 0-2	1
4	7/2	5/2	5/2	2	3	-.06116777	-1	-2-1-1-2	1
4	7/2	5/2	5/2	3	1	.11397942	1	-4-3 2-2	1
4	7/2	5/2	5/2	3	2	-.03948363	-1	-4-2 0-2	1
4	7/2	5/2	5/2	3	3	-.03367175	-1	-1-2 0-2	
4	7/2	5/2	3	1/2	7/2	.07042952	1	-4-2 1-1	
4	7/2	5/2	3	3/2	5/2	.06838765	1	-4-1 0-2	1
4	7/2	5/2	3	3/2	7/2	-.08518354	-1	4-2-1-2	
4	7/2	5/2	3	5/2	3/2	.08059562	1	-5-3 2-2	1
4	7/2	5/2	3	5/2	5/2	-.07896726	-1	-2-2 0-2	1
4	7/2	5/2	3	5/2	7/2	.03014544	23	-3-3-1-2	-1
4	7/2	5/2	3	7/2	1/2	.11679415	1	-6-2 1-1	1
4	7/2	5/2	3	7/2	3/2	-.06838765	-1	-4-1 0-2	1
4	7/2	5/2	3	7/2	5/2	.00595238	1	-6-2 0-2	
4	7/2	5/2	7/2	1	3	.03764616	1	-3-2 1-2	
4	7/2	5/2	7/2	2	2	.04559177	1	-2-3 0-2	1
4	7/2	5/2	7/2	2	3	-.07389908	-17	-3-3-1-2	
4	7/2	5/2	7/2	3	1	.06242910	1	-5-2 1-2	1
4	7/2	5/2	7/2	3	2	-.07929560	-1	-5-3-1-2	3
4	7/2	5/2	7/2	3	3	.06024099	1	-4-3 1-2	-1 2
4	7/2	5/2	4	3/2	5/2	.01984127	1	-4-4 2-2	
4	7/2	5/2	4	5/2	3/2	.02791914	1	-5-2 0-2	1
4	7/2	5/2	4	5/2	5/2	-.05555556	-1	-2-4	
4	7/2	5/2	4	7/2	1/2	.04166667	1	-6-2	
4	7/2	5/2	4	7/2	3/2	-.06904762	-29	-4-2-2-2	

j_1	j_2	j_3	l_1	l_2	l_3	6-j symbol			
4	7/2	5/2	4	7/2	5/2	.07146465	283	-6-4-2 0	-2
4	7/2	7/2	0	7/2	7/2	-.12500000	-1	-6	
4	7/2	7/2	1/2	3	3	-.10714286	-1	-4 2 0-2	
4	7/2	7/2	1	5/2	5/2	-.08375742	-1	-5 0 0-2	1
4	7/2	7/2	1	7/2	5/2	-.09221389	-1	-2-1 1-2	
4	7/2	7/2	1	7/2	7/2	.04563492	23	-6-4 0-2	
4	7/2	7/2	3/2	2	2	-.05297285	-1	-4 0-1-2	1
4	7/2	7/2	3/2	3	2	-.09671474	-1	-3-1 0-2	1
4	7/2	7/2	3/2	3	3	-.01190476	-1	-4-2 0-2	
4	7/2	7/2	2	5/2	3/2	-.07896726	-1	-2-2 0-2	1
4	7/2	7/2	2	5/2	5/2	-.06514466	-1	-5-4 0 0	1
4	7/2	7/2	2	7/2	3/2	-.09175166	-1	-4 1-1-2	1
4	7/2	7/2	2	7/2	5/2	.04303315	1	-2-3-1	
4	7/2	7/2	2	7/2	7/2	.04166667	1	-6-2	
4	7/2	7/2	5/2	2	2	-.09417395	-1	4-4-1-2	1
4	7/2	7/2	5/2	3	1	-.09671474	-1	-3-1 0-2	1
4	7/2	7/2	5/2	3	2	-.01861276	-1	-3-4 0-2	1
4	7/2	7/2	5/2	3	3	.06746032	17	-4-4 0-2	
4	7/2	7/2	3	5/2	3/2	-.09118353	-1	0-3 0-2	1
4	7/2	7/2	3	5/2	5/2	.03722552	1	-1-4 0-2	1
4	7/2	7/2	3	7/2	1/2	-.10910895	-1	-2-1 0-1	
4	7/2	7/2	3	7/2	3/2	.02661986	1	-4-2 1-2	
4	7/2	7/2	3	7/2	5/2	.04633922	1	-2-3 3-2	-1
4	7/2	7/2	3	7/2	7/2	-.05465368	-101	-6-2 0-2	-2
4	7/2	7/2	7/2	2	2	-.07063046	-1	0-2-1-2	1
4	7/2	7/2	7/2	3	1	-.07776158	-1	3-3 0-2	
4	7/2	7/2	7/2	3	2	.06734350	1	1-2 0-2	
4	7/2	7/2	7/2	3	3	-.00432900	-1	0-2 0-2	-2
4	7/2	7/2	4	5/2	3/2	-.04761905	-1	0-2 0-2	
4	7/2	7/2	4	5/2	5/2	.07106691	1	-1-2 0 0	-1
4	7/2	7/2	4	7/2	1/2	-.05555556	-1	-2-4	
4	7/2	7/2	4	7/2	3/2	.07539683	19	-4-4 0-2	
4	7/2	7/2	4	7/2	5/2	-.04545455	-1	-2 0 0 0	-2
4	7/2	7/2	4	7/2	7/2	-.01388889	-1	-6-4	
4	4	0	0	0	4	.33333333	1	0-2	
4	4	0	1/2	1/2	7/2	.23570226	1	-1-2	
4	4	0	1	1	3	.19245009	1	0-3	
4	4	0	3/2	3/2	5/2	.16666667	1	-2-2	
4	4	0	3/2	3/2	7/2	-.16666667	-1	-2-2	
4	4	0	2	2	2	.14907120	1	0-2-1	
4	4	0	2	2	3	-.14907120	-1	0-2-1	
4	4	0	5/2	5/2	3/2	.13608276	1	-1-3	
4	4	0	5/2	5/2	5/2	-.13608276	-1	-1-3	
4	4	0	5/2	5/2	7/2	.13608276	1	-1-3	
4	4	0	3	3	1	.12598816	1	0-2 0-1	
4	4	0	3	3	2	-.12598816	-1	0-2 0-1	
4	4	0	3	3	3	.12598816	1	0-2 0-1	
4	4	0	7/2	7/2	1/2	.11785113	1	-3-2	
4	4	0	7/2	7/2	3/2	-.11785113	-1	-3-2	
4	4	0	7/2	7/2	5/2	.11785113	1	-3-2	
4	4	0	7/2	7/2	7/2	-.11785113	-1	-3-2	
4	4	0	4	4	0	.11111111	1	0-4	
4	4	1	1/2	1/2	7/2	-.15214515	-1	-3-3 1	
4	4	1	1	0	4	-.19245009	-1	0-3	

j_1	j_2	j_3	l_1	l_2	l_3	6-j symbol			
4	4	1	1	1	3	-.15214515	-1	-3-3 1	
4	4	1	1	1	4	.03042903	1	-3-3-1	
4	4	1	3/2	1/2	7/2	-.12729377	-1	-4-3 0 1	
4	4	1	3/2	3/2	5/2	-.14433757	-1	-4-1	
4	4	1	3/2	3/2	7/2	.07698004	1	2-3-2	
4	4	1	2	1	3	-.09128709	-1	-3-1-1	
4	4	1	2	1	4	.11941214	1	-3-3-2 1	1
4	4	1	2	2	2	-.13608276	-1	-1-3	
4	4	1	2	2	3	.09525793	1	-3-3-2 2	
4	4	1	5/2	3/2	5/2	-.06804138	-1	-3-3	
4	4	1	5/2	3/2	7/2	.09574271	1	-4-1-2 0	1
4	4	1	5/2	5/2	3/2	-.12858612	-1	-3-3 2-1	
4	4	1	5/2	5/2	5/2	.10286890	1	1-3 0-1	
4	4	1	5/2	5/2	7/2	-.06686478	-1	-3-3-2-1	0 2
4	4	1	3	2	2	-.05143445	-1	-1-3 0-1	
4	4	1	3	2	3	.07628962	1	-1-3-1-1	1
4	4	1	3	3	1	-.12198751	-1	-4-1 1-1	
4	4	1	3	3	2	.10572251	1	-4-3-1-1	0 2
4	4	1	3	3	3	-.08132501	-1	-2-3 1-1	
4	4	1	7/2	5/2	3/2	-.03857584	-1	-5-1 0-1	
4	4	1	7/2	5/2	5/2	.06031224	1	-4-3 0-1	1
4	4	1	7/2	5/2	7/2	-.07715167	-1	-3-1 0-1	
4	4	1	7/2	7/2	1/2	-.11620278	-1	-5-4 1 1	
4	4	1	7/2	7/2	3/2	.10624254	1	5-4-1-1	
4	4	1	7/2	7/2	5/2	-.08964215	-1	-5 2-1-1	
4	4	1	7/2	7/2	7/2	.06640159	1	-1-4 1-1	
4	4	1	4	3	1	-.02777778	-1	-4-4	
4	4	1	4	4	0	-.11111111	-1	0-4	
4	4	1	4	4	1	.10555556	19	-4-4-2	
4	4	2	1	1	3	.08529439	1	-3-3 0-1	1
4	4	2	1	1	4	.11941214	1	-3-3-2 1	1
4	4	2	3/2	1/2	7/2	.12360331	1	-4-2-1 0	1
4	4	2	3/2	3/2	5/2	.10446386	1	-4-2 0-1	1
4	4	2	3/2	3/2	7/2	.04178554	1	-2-2-2-1	1
4	4	2	2	0	4	.14907120	1	0-2-1	
4	4	2	2	1	3	.11443443	1	-3-1-1-1	1
4	4	2	2	1	4	-.04082483	-1	-3-1-2	
4	4	2	2	2	2	.11167657	1	-1-2 0-2	1
4	4	2	2	2	3	-.01116766	-1	-3-2-2-2	1
4	4	2	2	2	4	-.06599070	-1	-3-2 0-2	-1 2
4	4	2	5/2	1/2	7/2	.09128709	1	-3-1-1	
4	4	2	5/2	3/2	5/2	.09671474	1	-3-1 0-2	1
4	4	2	5/2	3/2	7/2	-.07835468	-19	-4-1-2-2	
4	4	2	5/2	5/2	3/2	.11397942	1	-4-3 2-2	1
4	4	2	5/2	5/2	5/2	-.04559177	-1	-2-3 0-2	1
4	4	2	5/2	5/2	7/2	-.02403930	-29	-2-3-2-2	-1
4	4	2	3	1	3	.05832118	1	-1-1 0-2	
4	4	2	3	1	4	-.09258201	-1	-1 1-2-1	
4	4	2	3	2	2	.07896726	1	-2-2 0-2	1
4	4	2	3	2	3	-.08518354	-1	4-2-1-2	
4	4	2	3	2	4	.05168766	1	2 2-2-2	-1
4	4	2	3	3	1	.11397942	1	-4-3 2-2	1
4	4	2	3	3	2	-.06838765	-1	-4-1 0-2	1
4	4	2	3	3	3	.01243412	1	-2-1 0-2	-1

j_1	j_2	j_3	l_1	l_2	l_3	6-j symbol			
4	4	2	7/2	3/2	5/2	.03764616	1	-3-2 1-2	
4	4	2	7/2	3/2	7/2	-.06776309	-1	-3 2-1-2	
4	4	2	7/2	5/2	3/2	.06242910	1	-5-2 1-2	1
4	4	2	7/2	5/2	5/2	-.07985957	-1	-4 0 1-2	
4	4	2	7/2	5/2	7/2	.07037462	31	-3-2-1-2	-1
4	4	2	7/2	7/2	1/2	.11283387	1	-5-3 0 0	1
4	4	2	7/2	7/2	3/2	-.08381944	-1	-3-3-2-2	1 2
4	4	2	7/2	7/2	5/2	.04132357	47	-5-1-2-2	-1
4	4	2	7/2	7/2	7/2	.00586150	1	-1-3 0-2	-1
4	4	2	4	2	2	.02380952	1	-2-2 0-2	
4	4	2	4	3	1	.04671766	1	-4-2-1-1	1
4	4	2	4	3	2	-.06904762	-29	-4-2-2-2	
4	4	2	4	4	0	.11111111	1	0-4	
4	4	2	4	4	1	-.09444444	-17	-4-4-2	
4	4	2	4	4	2	.06370851	883	-4-4-2-2	-2
4	4	3	3/2	3/2	5/2	-.05583828	-1	-3-2 0-2	1
4	4	3	3/2	3/2	7/2	-.10050891	-1	-3 2-2-2	1
4	4	3	2	1	3	-.08650430	-1	-1-1-1-2	1
4	4	3	2	1	4	-.09258201	-1	-1 1-2-1	
4	4	3	2	2	2	-.07896726	-1	-2-2 0-2	1
4	4	3	2	2	3	-.06317381	-1	2-2-2-2	1
4	4	3	2	2	4	.05168766	1	2 2-2-2	-1
4	4	3	5/2	1/2	7/2	-.10910895	-1	-2-1 0-1	
4	4	3	5/2	3/2	5/2	-.09671474	-1	-3-1 0-2	1
4	4	3	5/2	3/2	7/2	-.00824786	-1	-2-1-2-2	
4	4	3	5/2	5/2	3/2	-.09306380	-1	-3-4 2-2	1
4	4	3	5/2	5/2	5/2	-.01861276	-1	-3-4 0-2	1
4	4	3	5/2	5/2	7/2	.06835959	101	-1-4-2-2	-1
4	4	3	3	0	4	-.12598816	-1	0-2 0-1	
4	4	3	3	1	3	-.09221389	-1	-2-1 1-2	
4	4	3	3	1	4	.04879500	1	-2-1-1-1	
4	4	3	3	2	2	-.09118353	-1	0-3 0-2	1
4	4	3	3	2	3	.04303315	1	-2-3-1	
4	4	3	3	2	4	.03896024	47	-2-3-2-2	-1
4	4	3	3	3	1	-.10194629	-1	-2-3 1-2	1
4	4	3	3	3	2	.02038926	1	-2-3-1-2	1
4	4	3	3	3	3	.04633922	1	-2-3 3-2	-1
4	4	3	3	3	4	-.05746064	-31	-2-3-1-2	-1
4	4	3	7/2	1/2	7/2	-.07042952	-1	-4-2 1-1	
4	4	3	7/2	3/2	5/2	-.06873217	-1	-2-3 2-2	
4	4	3	7/2	3/2	7/2	.07560539	1	-4-3 0-2	2
4	4	3	7/2	5/2	3/2	-.07896726	-1	-2-2 0-2	1
4	4	3	7/2	5/2	5/2	.06734350	1	1-2 0-2	
4	4	3	7/2	5/2	7/2	-.01076826	-1	-4 0 0-2	-1
4	4	3	7/2	7/2	1/2	-.10758287	-1	-4-3 1	
4	4	3	7/2	7/2	3/2	.05225454	17	-4-3-1-2	
4	4	3	7/2	7/2	5/2	.01201575	43	-4-3-1-2	-2
4	4	3	7/2	7/2	7/2	-.05309285	-19	-2-3 1-2	-2
4	4	3	4	1	3	-.03968254	-1	-2-4 2-2	
4	4	3	4	2	2	-.04761905	-1	0-2 0-2	
4	4	3	4	2	3	.07142857	1	-2 0 0-2	
4	4	3	4	3	1	-.06299408	-1	-2-2 0-1	
4	4	3	4	3	2	.07384094	23	-2-2-1-2	-1
4	4	3	4	3	3	-.04545455	-1	-2 0 0 0	-2

TABLE D-1 (cont)

j_1	j_2	j_3	l_1	l_2	l_3	6-j symbol			
4	4	3	4	4	0	-.11111111	-1	0-4	
4	4	3	4	4	1	.07777778	1	-2-4-2 2	
4	4	3	4	4	2	-.02409812	-167	-2-4-2-2	-2
4	4	3	4	4	3	-.02756133	-191	-2-4-2-2	-2
4	4	4	2	2	2	.04244363	1	-2-4-1-2	1 1
4	4	4	5/2	5/2	3/2	.06710926	1	-3-4 0-2	1 1
4	4	4	5/2	5/2	5/2	.06710926	1	-3-4 0-2	1 1
4	4	4	3	2	2	.08488725	1	0-4-1-2	1 1
4	4	4	3	3	1	.08584646	1	-2-2 0-2	0 1
4	4	4	3	3	2	.02861549	1	-2-4 0-2	0 1
4	4	4	3	3	3	-.05983238	-23	-2-4 0-2	-2 1
4	4	4	7/2	5/2	3/2	.08584646	1	-2-2 0-2	0 1
4	4	4	7/2	5/2	5/2	-.02440337	-1	1-4 0-2	-1 1
4	4	4	7/2	7/2	1/2	.10015420	1	-4-4 0 0	0 1
4	4	4	7/2	7/2	3/2	-.01430774	-1	-4-4 0-2	0 1
4	4	4	7/2	7/2	5/2	-.05072745	-1	-4-2 0-2	-2 3
4	4	4	7/2	7/2	7/2	.04162253	1	6-4 0-2	-2 1
4	4	4	4	2	2	.06945320	1	0 0-1-2	-1 1
4	4	4	4	3	1	.07570946	1	-2-4 0-1	0 1
4	4	4	4	3	2	-.05787767	-1	-2-2 1-2	-1 1
4	4	4	4	3	3	-.00780422	-1	-2-2 0-2	-2 1
4	4	4	4	4	0	.11111111	1	0-4	
4	4	4	4	4	1	-.05555556	-1	-2-4	
4	4	4	4	4	2	-.01659452	-23	-2-4 0-2	-2
4	4	4	4	4	3	.05266955	73	-2-4 0-2	-2
4	4	4	4	4	4	-.02591853	-467	-2-4 0-2	-2-2

Appendix E

One-Electron and Total-Atom Energies

The (negative) spherically averaged binding energy of an electron in an orbital $i = n_i l_i$ of an atom or ion is

$$E^i = E_k{}^i + E_n{}^i + \sum_{j \neq i} E^{ij} , \tag{6.13}$$

and the total spherically averaged (configuration-center-of-gravity) energy of the atom is

$$E_{av} = \sum_i \left(E_k{}^i + E_n{}^i + \frac{1}{2} \sum_{j \neq i} E^{ij} \right) \tag{6.15}$$

$$= \sum_i \left(E^i - \frac{1}{2} \sum_{j \neq i} E^{ij} \right) . \tag{6.16}$$

The summations are over occupied one-electron orbitals (i.e., over all N electrons of the atom, not over subshells).

In rydbergs, the one-electron kinetic energy and electron-nuclear Coulomb energy are

$$E_k{}^i \equiv \langle i | -\nabla^2 | i \rangle = \int_0^\infty P_i{}^*(r) \left[-\frac{d^2}{dr^2} + \frac{l_i(l_i + 1)}{r^2} \right] P_i(r)\, dr , \tag{6.18}$$

$$E_n{}^i \equiv \langle i | -2Z/r | i \rangle = \int_0^\infty (-2Z/r) \, | P_i(r) |^2 \, dr , \tag{6.19}$$

with r in Bohr units. For Hartree-Fock radial wavefunctions $P_i \equiv P_{n_i l_i}(r)$, E^i is identical with the eigenvalue of the one-electron Schrödinger equation, and so E_k^i and E_n^i need not be evaluated if (6.16) is used rather than (6.15).

The spherically averaged electron-electron Coulomb energy (in Ry) for two electrons i and j is

$$E^{ij} = F^0(ij) - \frac{1}{2}\sum_k \begin{pmatrix} l_i & k & l_j \\ 0 & 0 & 0 \end{pmatrix}^2 G^k(ij) \tag{6.38}$$

for non-equivalent electrons ($n_il_i \neq n_jl_j$), and is

$$E^{ii} = F^0(ii) - \frac{2l_i + 1}{4l_i + 1}\sum_{k>0} \begin{pmatrix} l_i & k & l_i \\ 0 & 0 & 0 \end{pmatrix}^2 F^k(ii) \tag{6.39}$$

for equivalent electrons. Numerical evaluation of the 3-j symbols from Table C-1 leads to the expressions given in Table E-1. [Note that these involve the Slater radial integrals F^k and G^k defined by (6.28) and (6.30), and *not* the renormalized Condon-Shortley integrals F_k and G_k defined by (6.41).]

TABLE E-1. AVERAGE COULOMB ENERGY OF ELECTRON PAIRS

Equivalent Electrons

$$E^{ss} = F^0$$

$$E^{pp} = F^0 - \frac{2}{25} F^2$$

$$E^{dd} = F^0 - \frac{2}{63} F^2 - \frac{2}{63} F^4$$

$$E^{ff} = F^0 - \frac{4}{195} F^2 - \frac{2}{143} F^4 - \frac{100}{5577} F^6$$

$$E^{gg} = F^0 - \frac{20}{1309} F^2 - \frac{162}{17017} F^4 - \frac{20}{2431} F^6 - \frac{490}{41327} F^8$$

Non-Equivalent Electrons

$$E^{ss'} = F^0 - \frac{1}{2} G^0$$

$$E^{sp} = F^0 - \frac{1}{6} G^1$$

$$E^{sd} = F^0 - \frac{1}{10} G^2$$

$$E^{sf} = F^0 - \frac{1}{14} G^3$$

$$E^{sg} = F^0 - \frac{1}{18} G^4$$

$$E^{pp'} = F^0 - \frac{1}{6} G^0 - \frac{1}{15} G^2$$

$$E^{pd} = F^0 - \frac{1}{15} G^1 - \frac{3}{70} G^3$$

$$E^{pf} = F^0 - \frac{3}{70} G^2 - \frac{2}{63} G^4$$

$$E^{pg} = F^0 - \frac{2}{63} G^3 - \frac{5}{198} G^5$$

$$E^{dd'} = F^0 - \frac{1}{10} G^0 - \frac{1}{35} G^2 - \frac{1}{35} G^4$$

$$E^{df} = F^0 - \frac{3}{70} G^1 - \frac{2}{105} G^3 - \frac{5}{231} G^5$$

$$E^{dg} = F^0 - \frac{1}{35} G^2 - \frac{10}{693} G^4 - \frac{5}{286} G^6$$

$$E^{ff'} = F^0 - \frac{1}{14} G^0 - \frac{2}{105} G^2 - \frac{1}{77} G^4 - \frac{50}{3003} G^6$$

$$E^{fg} = F^0 - \frac{2}{63} G^1 - \frac{1}{77} G^3 - \frac{10}{1001} G^5 - \frac{35}{2574} G^7$$

$$E^{gg'} = F^0 - \frac{1}{18} G^0 - \frac{10}{693} G^2 - \frac{9}{1001} G^4 - \frac{10}{1287} G^6 - \frac{245}{21879} G^8$$

Appendix F

Basis Functions, Matrix Elements, and Racah Algebra

F-1. ANTISYMMETRIZED COUPLED WAVEFUNCTIONS

(a) Basis wavefunctions for an N-electron atom are constructed from linear combinations of products of N one-electron spin-orbitals (4.17). If angular momenta are not coupled together, antisymmetrization can be accomplished by forming determinantal functions:

$$\Psi = (N!)^{-1/2} \sum_P (-1)^p \prod_{i=1}^{N} |n_i l_i m_{l_i} m_{s_i}(r_j)\rangle , \tag{4.26}$$

where the summation is over all N! permutations of the coordinate subscripts j, and p is the parity of the permutation P.

(b) Two angular momenta j_1 and j_2 may be coupled together to give eigenfunctions of the angular momentum $j = j_1 + j_2$ by using Clebsch-Gordon (vector-coupling) coefficients:

$$|j_1 j_2 jm\rangle = \sum_{m_1 m_2} C(j_1 j_2 m_1 m_2; jm) \, |j_1 m_1\rangle \, |j_2 m_2\rangle \tag{9.4}$$

$$= (-1)^{j_1 - j_2 + m} [j]^{1/2} \sum_{m_1 m_2} \begin{pmatrix} j_1 & j_2 & j \\ m_1 & m_2 & -m \end{pmatrix} |j_1 m_1\rangle \, |j_2 m_2\rangle , \tag{9.5}$$

where

$$[j] \equiv 2j + 1 , \qquad [j_1, j_2, \cdots] \equiv (2j_1 + 1)(2j_2 + 1) \cdots . \tag{5.7}$$

The double summation is really a single one because the Clebsch-Gordon coefficient and 3-j symbol are zero unless $m_1 + m_2 = m$.

(c) Coupling of more than two angular momenta is accomplished by multiple application of (9.4); for example,

$$|nln'l'LSJM\rangle = \sum_{M_L M_S} C(LSM_LM_S;JM) \sum_{m_s m'_s} C(ss'm_sm'_s;SM_S) \times \sum_{m_l m'_l} C(ll'm_lm'_l;LM_L)\, |nlm_lm_s\rangle\, |n'l'm'_lm'_s\rangle \;. \qquad (F.1)$$

F-2. COEFFICIENT-OF-FRACTIONAL-PARENTAGE EXPANSIONS

When two or more electrons are equivalent ($n_il_i = n_kl_k$), normalization difficulties are encountered if Clebsch-Gordon expansions are applied to antisymmetrized product functions (4.26). These difficulties are avoided by using appropriate linear combinations of coupled simple-product (non-antisymmetrized) functions; the linear expansion is independent of J and M (or of M_L and M_S), which we therefore drop from the notation. If $|l^{w-1}\bar{\alpha}\bar{L}\bar{S}\rangle$ is an antisymmetric function for w−1 equivalent electrons in a subshell nl^{w-1}, and if

$$|(l^{w-1}\bar{\alpha}\bar{L}\bar{S},l)LS\rangle \equiv \sum_{\bar{M}_L m_l} C(\bar{L}l\bar{M}_Lm_l;LM_L) \sum_{\bar{M}_S m_s} C(\bar{S}s\bar{M}_Sm_s;SM_S)\, |l^{w-1}\bar{\alpha}\bar{L}\bar{S}\rangle\, |l\rangle$$

is a coupled[1] but not antisymmetrized function, then a completely antisymmetric function for nl^w can be written

$$|l^w\alpha LS\rangle = \sum_{\bar{\alpha}\bar{L}\bar{S}} |(l^{w-1}\bar{\alpha}\bar{L}\bar{S},l)LS\rangle\, (l^{w-1}\bar{\alpha}\bar{L}\bar{S}|\}l^w\alpha LS) \;. \qquad (9.50)$$

Antisymmetric functions for a given w are in principle constructed by repeated application of (9.50), starting with w = 2; in practice, however, it is seldom necessary to use more than the simple expression (9.50) for the particular value of w of interest. Only those values of $\bar{\alpha}\bar{L}\bar{S}$ and αLS are allowed that are given in Table 4-2 (p. 110). The terms $\bar{\alpha}\bar{L}\bar{S}$ are called parents of the terms αLS.

The coefficients of fractional parentage (cfp) $(l^{w-1}|\}l^w)$ satisfy the relations

$$\sum_{\bar{\alpha}\bar{L}\bar{S}} (l^w\alpha LS\{|l^{w-1}\bar{\alpha}\bar{L}\bar{S})\,(l^{w-1}\bar{\alpha}\bar{L}\bar{S}|\}l^w\alpha'LS) = \delta_{\alpha\alpha'} \qquad (9.46)$$

and

$$\sum_{\alpha LS} [L,S]\,(l^{w-1}\bar{\alpha}\bar{L}\bar{S}|\}l^w\alpha LS)\,(l^w\alpha LS\{|l^{w-1}\bar{\alpha}'\bar{L}\bar{S}) = \frac{1}{w}(4l+3-w)\,[\bar{L},\bar{S}]\,\delta_{\bar{\alpha}\bar{\alpha}'} \;. \qquad (9.49)$$

Sample values of the cfp are given in Appendix H for all p^w and for a few d^w and f^w. A complete set of tables for $w \leq 2l+1$ is given in Nielson and Koster, ref. [112]; cfp for $w > 2l+1$ may be calculated from (9.68)-(9.69).

It is sometimes convenient to write basis functions for l^w in terms of antisymmetric functions for l^{w-2} and l^2 via coefficients of fractional grandparentage:

[1]As mentioned above, L and S may be coupled together by adding a third Clebsch-Gordon sum, without affecting the formal expression (9.50).

$$|l^w\alpha LS\rangle = \sum_{\bar{\bar{\alpha}}\bar{\bar{L}}\bar{\bar{S}}} \sum_{L'S'} |(l^{w-2}\bar{\bar{\alpha}}\bar{\bar{L}}\bar{\bar{S}},l^2L'S')LS\rangle (l^{w-2}\bar{\bar{\alpha}}\bar{\bar{L}}\bar{\bar{S}},l^2L'S'|\}l^w\alpha LS) , \qquad (9.51)$$

where

$$(l^{w-2}\bar{\bar{\alpha}}\bar{\bar{L}}\bar{\bar{S}},l^2L'S'|\}l^w\alpha LS) = (-1)^{\bar{\bar{L}}+\bar{\bar{S}}+L+S+1} \sum_{\bar{\alpha}\bar{L}\bar{S}} [\bar{L},\bar{S},L',S']^{1/2}$$

$$\times \begin{Bmatrix} \bar{\bar{L}} & l & \bar{L} \\ l & L & L' \end{Bmatrix} \begin{Bmatrix} \bar{\bar{S}} & s & \bar{S} \\ s & S & S' \end{Bmatrix} (l^{w-2}\bar{\bar{\alpha}}\bar{\bar{L}}\bar{\bar{S}}|\}l^{w-1}\bar{\alpha}\bar{L}\bar{S}) (l^{w-1}\bar{\alpha}\bar{L}\bar{S}|\}l^w\alpha LS) . \qquad (9.52)$$

Sample values of the cfgp are also tabulated in Appendix H.

F-3. ANTISYMMETRIZATION OF MULTI-SUBSHELL FUNCTIONS

For an electron configuration

$$n_1l_1^{w_1}n_2l_2^{w_2}\cdots n_ql_q^{w_q} , \qquad \sum_{j=1}^{q} w_j = N , \qquad (9.71)$$

completely coupled basis functions ψ_b can be constructed by multiple Clebsch-Gordon expansions of the type (F.1), using products of q functions (9.50), one function for each subshell. It is most convenient for numerical calculations to use an LS-coupling scheme such as the genealogical (left-to-right) coupling

$$\{[(l_1^{w_1}\alpha_1L_1S_1\mathfrak{L}_1\mathfrak{S}_1, l_2^{w_2}\alpha_2L_2S_2)\mathfrak{L}_2\mathfrak{S}_2, l_3^{w_3}\alpha_3L_3S_3]\mathfrak{L}_3\mathfrak{S}_3,\cdots\}\mathfrak{L}_q\mathfrak{S}_q\mathfrak{J}_q\mathfrak{M}_q . \qquad (4.37) \; (12.1)$$

[However, the $J\mathfrak{J}$-coupling scheme (4.44) or (12.2) and other coupling schemes (4.45), (4.47), (4.50), (4.51), etc., may be physically more appropriate. Transformation matrices from one coupling scheme to another may be computed with the aid of the recoupling equations of Sec. F-6.]

The coupled functions ψ_b are antisymmetric with respect to interchange of electron coordinates within any one subshell $n_il_i^{w_i}$, but not with respect to interchange of coordinates between two different subshells. The latter type of antisymmetrization is added similarly to (4.26) by summing over limited types of permutations. Our final coupled, completely antisymmetrized basis functions are of the form

$$\Psi_b = \left[\frac{\prod_j (w_j!)}{N!}\right]^{1/2} \sum_P (-1)^p \psi_b^{(P)} , \qquad (9.75)$$

where the permutations P involve coordinate exchanges only between two different subshells; within any one subshell, coordinate serial numbers must be arranged in numerically increasing order.

F-4. MATRIX ELEMENTS OF SYMMETRIC OPERATORS

We have no great interest in the basis functions (9.75) themselves, but require only matrix elements of operators that are symmetric in all electron coordinates. In the evaluation of matrix elements between two basis functions belonging to the same configuration, complications due to the antisymmetrization permutations in (9.75) are easily reduced. For a symmetric one-electron operator,

$$\langle \Psi_b | \sum_{k=1}^{N} f_k | \Psi_{b'} \rangle = \sum_{j=1}^{q} w_j \langle \psi_b | f_{(j)} | \psi_{b'} \rangle \ , \tag{9.80}$$

where f_k operates on any coordinate r_k, and $f_{(j)}$ operates on the *last* coordinate of subshell j. For a symmetric two-electron operator,

$$\langle \Psi_b | \sum_{k<m} g_{km} | \Psi_{b'} \rangle = \sum_{j=1}^{q} \frac{w_j(w_j - 1)}{2} \langle \psi_b | g_{(jj)} | \psi_{b'} \rangle$$
$$+ \sum_{i<j} w_i w_j \left[\langle \psi_b | g_{(ij)} | \psi_{b'} \rangle - \langle \psi_b | g_{(ij)} | \psi_{b'}^{(ex)} \rangle \right] \ , \tag{9.85}$$

where g_{km} operates on any two electron coordinates r_k and r_m, $q_{(jj)}$ operates on the last two electron coordinates of the jth subshell, $g_{(ij)}$ operates on the last coordinate of subshell i and the last coordinate of subshell j, and $\psi_{b'}^{(ex)}$ is the same as $\psi_{b'}$ except with the final electron coordinates of the ith and j^{th} subshells interchanged.

Reductions similar to (9.80) and (9.85) for matrix elements involving basis functions belonging to two different configurations are given by (13.53) and (13.58).

F-5. UNCOUPLING OF SPECTATOR SUBSHELLS

If O_k is a symmetric operator that operates only within the subspace k of two coupled subspaces k and m (each of which may itself involve the coupling of several angular momenta), and if the matrix of O_k for subspace k is diagonal in j_k and independent of m_k, then

$$\langle j_k j_m JM | O_k | j'_k j'_m J'M' \rangle = \delta_{j_k j_m JM, j'_k j'_m J'M'} \langle j_k | O | j_k \rangle \ . \tag{9.87}$$

A similar result (9.88) holds for an operator O_m operating within the subspace m. Because of these relations, (9.80) and the single-subshell matrix elements in (9.85) can (for operators possessing the necessary properties) be simplified immediately to matrix elements involving only the single-subshell functions

$$| l_j^{w_j} \alpha_j L_j S_j \rangle \ .$$

Similarly, for the two-subshell matrix elements in (9.85), subshells can be removed that are coupled onto the subspace that contains both subshells i and j.

F-6. RECOUPLING OF ANGULAR MOMENTA

Useful relations for changing the order of coupling of angular momenta are

$$|j_2j_1jm\rangle = (-1)^{j_1+j_2-j}\,|j_1j_2jm\rangle\ , \tag{9.8}$$

$$R_s(j_1j_2J',j_3J,J'') \equiv \langle[(j_1j_2)J',j_3]J\,|\,[j_1,(j_2j_3)J'']J\rangle$$

$$= (-1)^{j_1+j_2+j_3+J}[J',J'']^{1/2}\begin{Bmatrix} j_1 & j_2 & J' \\ j_3 & J & J'' \end{Bmatrix}\ , \qquad (9.17)\ (13.64)$$

$$R_j(j_1j_2J',j_3J,J'') \equiv \langle[(j_1j_2)J',j_3]J\,|\,[j_1,(j_3j_2)J'']J\rangle$$

$$= (-1)^{j_1+J+J''}[J',J'']^{1/2}\begin{Bmatrix} j_1 & j_2 & J' \\ j_3 & J & J'' \end{Bmatrix}\ , \qquad (9.18)\ (13.65)$$

$$R_x(j_1j_2J',j_3J,J'') \equiv \langle[(j_1j_2)J',j_3]J\,|\,[(j_1j_3)J'',j_2]J\rangle$$

$$= (-1)^{j_2+j_3+J'+J''}[J',J'']^{1/2}\begin{Bmatrix} j_2 & j_1 & J' \\ j_3 & J & J'' \end{Bmatrix}\ , \qquad (9.19)\ (13.66)$$

$$\langle[(l_1l_2)L,(s_1s_2)S]J\,|\,[(l_1s_1)j_1,(l_2s_2)j_2]J\rangle = [L,S,j_1,j_2]^{1/2}\begin{Bmatrix} l_1 & l_2 & L \\ s_1 & s_2 & S \\ j_1 & j_2 & J \end{Bmatrix}\ . \tag{9.29}$$

For operators of the type defined in Sec. F-5, cfp or cfgp expansions (9.50) and (9.51), together with the above recoupling equations and (9.87), can be used to simplify the matrix elements (9.80) and (9.85) to elements involving only one or two orbitals, respectively, in each bra and ket. However, much more powerful methods of evaluating (9.80) and (9.85) are provided by the equations of Racah algebra, which follow.

F-7. SUMMARY OF RACAH ALGEBRA

Irreducible tensor operators. An irreducible tensor operator of rank k is a quantity $\mathbf{T}^{(k)}$ whose 2k + 1 components $T_q^{(k)}$ ($q = -k, -k+1, \cdots k-1, k$) have matrix elements that depend on q in the same way as do those of the renormalized spherical harmonics

$$C_q^{(k)}(\theta,\phi) \equiv \left(\frac{4\pi}{2k+1}\right)^{1/2} Y_{kq}(\theta,\phi)\ . \tag{11.9}$$

A vector **V** with components V_x, V_y, V_z is equivalent to a tensor operator of rank 1 with components

$$T_1^{(1)} = -\frac{1}{\sqrt{2}}(V_x + iV_y) ,$$

$$T_0^{(1)} = V_z , \tag{11.12}$$

$$T_{-1}^{(1)} = \frac{1}{\sqrt{2}}(V_x - iV_y) ;$$

for the position vector **r**,

$$r^{(1)} = r\,C^{(1)} . \tag{11.13}$$

Wigner-Eckart theorem.

$$\langle \alpha jm | T_q^{(k)} | \alpha' j' m' \rangle = (-1)^{j-m} \begin{pmatrix} j & k & j' \\ -m & q & m' \end{pmatrix} \langle \alpha j \| T^{(k)} \| \alpha' j' \rangle . \tag{11.15}$$

Reduced matrix elements.

$$\langle \alpha j \| J^{(1)} \| \alpha' j' \rangle = \delta_{\alpha j, \alpha' j'} [j(j + 1)(2j + 1)]^{1/2} , \tag{11.20}$$

$$\langle l \| \boldsymbol{l}^{(1)} \| l \rangle = [l(l + 1)(2l + 1)]^{1/2} , \tag{11.21}$$

$$\langle s \| s^{(1)} \| s \rangle = [s(s + 1)(2s + 1)]^{1/2} = (3/2)^{1/2} , \tag{11.22}$$

$$\langle l \| C^{(k)} \| l' \rangle = (-1)^l [l,l']^{1/2} \begin{pmatrix} l & k & l' \\ 0 & 0 & 0 \end{pmatrix} , \tag{11.23}$$

$$= (-1)^k \langle l' \| C^{(k)} \| l \rangle , \tag{11.24}$$

$$\boldsymbol{P}^{(1)}_{nl,n'l'} \equiv \langle nl \| r^{(1)} \| n'l' \rangle = \langle l \| C^{(1)} \| l' \rangle \int_0^\infty P_{nl}(r)\, r\, P_{n'l'}(r)\, dr \tag{14.55}$$

$$= (-1)^{l+l_>} (l_>)^{1/2} \int_0^\infty P_{nl}(r)\, r\, P_{n'l'}(r)\, dr , \tag{14.58}$$

where $l' = l \pm 1$ and $l_> = \max(l,l')$.

$$\boldsymbol{P}^{(2)}_{nl,nl} \equiv \langle nl \| r^2 C^{(2)} \| nl \rangle = (-1)^l (2l + 1) \begin{pmatrix} l & 2 & l \\ 0 & 0 & 0 \end{pmatrix} \langle r^2 \rangle_{nl} . \tag{15.18}$$

We define a unit orbital-angular-momentum operator $\mathbf{u}^{(k)}$ such that

$$\langle l \| u^{(k)} \| l' \rangle = 1 \tag{11.48}$$

provided lkl' satisfy the triangle relations, and a corresponding double (orbital + spin) operator $v^{(k1)}$ such that

$$\langle ls\|v^{(k1)}\|l's'\rangle \equiv \langle l\|u^{(k)}\|l'\rangle\langle s\|s^{(1)}\|s\rangle = (3/2)^{1/2} . \tag{11.66}$$

The corresponding symmetric operators (11.52) and (11.67) for a subshell l^w have matrix elements

$$\langle l^w\alpha LS\|U^{(k)}\|l^w\alpha'L'S'\rangle = \delta_{SS'}\, w\,(-1)^{l+L+k}\,[L,L']^{1/2}$$
$$\times \sum_{\bar{\alpha}\bar{L}\bar{S}} (-1)^{\bar{L}} (l^w\alpha LS\{|l^{w-1}\bar{\alpha}\bar{L}\bar{S})\begin{Bmatrix} l & k & l \\ L & \bar{L} & L' \end{Bmatrix}(l^{w-1}\bar{\alpha}\bar{L}\bar{S}|\}l^w\alpha'L'S') \tag{11.53}$$
$$= (-1)^{L'-L}\langle l^w\alpha'L'S'\|U^{(k)}\|l^w\alpha LS\rangle \tag{11.54}$$

and

$$\langle l^w\alpha LS\|V^{(k1)}\|l^w\alpha'L'S'\rangle = (3/2)^{1/2}\, w\,(-1)^{l+L+k}[L,L',S,S']^{1/2}$$
$$\times \sum_{\bar{\alpha}\bar{L}\bar{S}} (-1)^{\bar{L}+\bar{S}+S+3/2} (l^w\alpha LS\{|l^{w-1}\bar{\alpha}\bar{L}\bar{S})\begin{Bmatrix} l & k & l \\ L & \bar{L} & L' \end{Bmatrix}\begin{Bmatrix} s & 1 & s \\ S & \bar{S} & S' \end{Bmatrix}(l^{w-1}\bar{\alpha}\bar{L}\bar{S}|\}l^w\alpha'L'S') . \tag{11.68}$$
$$= (-1)^{L'+S'-L-S}\langle l^w\alpha'L'S'\|V^{(k1)}\|l^w\alpha LS\rangle . \tag{11.69}$$

Corresponding matrix elements for conjugate subshells l^w and l^{4l+2-w} differ by phase factors $(-1)^{k+1}$ and $(-1)^k$ for $U^{(k)}$ and $V^{(k1)}$, respectively. Values of these reduced matrix elements are given in Appendix H for a few of the simpler subshells. Additional values are tabulated in refs. [112]-[115].

Tensor products. The tensor product (11.32) of two irreducible tensor operators has reduced matrix elements

$$\langle \alpha j\|[T^{(k_1)} \times W^{(k_2)}]^{(K)}\|\alpha'j'\rangle$$
$$= (-1)^{j+K+j'}[K]^{1/2}\sum_{\alpha''j''}\langle \alpha j\|T^{(k_1)}\|\alpha''j''\rangle\begin{Bmatrix} k_1 & K & k_2 \\ j' & j'' & j \end{Bmatrix}\langle \alpha''j''\|W^{(k_2)}\|\alpha'j'\rangle . \tag{11.35}$$

If $T^{(k_1)}$ and $W^{(k_2)}$ operate only within subspaces spanned by $|\alpha_1 j_1 m_1\rangle$ and $|\alpha_2 j_2 m_2\rangle$, respectively, that are coupled together, then

$$\langle \alpha_1 j_1 \alpha_2 j_2 j\|[T^{(k_1)} \times W^{(k_2)}]^{(K)}\|\alpha_1' j_1' \alpha_2' j_2' j'\rangle$$
$$= [j,K,j']^{1/2}\langle \alpha_1 j_1\|T^{(k_1)}\|\alpha_1' j_1'\rangle\langle \alpha_2 j_2\|W^{(k_2)}\|\alpha_2' j_2'\rangle\begin{Bmatrix} j_1 & j_2 & j \\ j_1' & j_2' & j' \\ k_1 & k_2 & K \end{Bmatrix} . \tag{11.37}$$

Scalar products. The scalar product

$$T^{(k)}\cdot W^{(k)} \equiv (-1)^k[k]^{1/2}[T^{(k)} \times W^{(k)}]_0^{(0)} = \sum_q (-1)^q T_{-q}^{(k)} W_q^{(k)} \tag{11.41}$$

has matrix elements

$$\langle \alpha jm | \mathbf{T}^{(k)} \cdot \mathbf{W}^{(k)} | \alpha' j' m' \rangle$$

$$= \delta_{jm,j'm'} [j]^{-1} \sum_{\alpha'' j''} (-1)^{j-j''} \langle \alpha j \| \mathbf{T}^{(k)} \| \alpha'' j'' \rangle \langle \alpha'' j'' \| \mathbf{W}^{(k)} \| \alpha' j' \rangle \ . \tag{11.45}$$

If $\mathbf{T}^{(k)}$ and $\mathbf{W}^{(k)}$ operate in two different, but coupled, subspaces 1 and 2, then

$$\langle \alpha_1 j_1 \alpha_2 j_2 jm | \mathbf{T}^{(k)} \cdot \mathbf{W}^{(k)} | \alpha_1' j_1' \alpha_2' j_2' j' m' \rangle$$

$$= \delta_{jm,j'm'} (-1)^{j_1' + j_2 + j} \langle \alpha_1 j_1 \| \mathbf{T}^{(k)} \| \alpha_1' j_1' \rangle \langle \alpha_2 j_2 \| \mathbf{W}^{(k)} \| \alpha_2' j_2' \rangle \begin{Bmatrix} j_1 & j_2 & j \\ j_2' & j_1' & k \end{Bmatrix} \ . \tag{11.47}$$

Examples.

Spin-orbit operator: $f_i \equiv \xi(r_i) \ \boldsymbol{l}_i^{(1)} \cdot \mathbf{s}_i^{(1)}$. (11.44)

The jth term of the matrix element (9.80) [p. 671] is

$$d_j \zeta_j \ ,$$

where ζ_j is the spin-orbit radial integral (10.9), and where use of (11.47) shows the angular coefficient d_j to be basically the reduced matrix element of

$$\langle l_j \| \boldsymbol{l}^{(1)} \| l_j \rangle \ \mathbf{V}_{(j)}^{(11)} = [l_j(l_j + 1)(2l_j + 1)]^{1/2} \ \mathbf{V}_{(j)}^{(11)} \ . \tag{12.40}$$

Coulomb operator: $g_{ij} \equiv \dfrac{2}{r_{ij}} = \displaystyle\sum_{k=0}^{\infty} \frac{2r_<^k}{r_>^{k+1}} \mathbf{C}_{(i)}^{(k)} \cdot \mathbf{C}_{(j)}^{(k)}$. (11.43)

The j term in (9.85) [p. 671] is

$$f_k^{(j)} F^k(l_j l_j) \ , \tag{12.6}$$

where (except for an additive constant in diagonal elements) $f_k^{(j)}$ is the matrix element of

$$\frac{1}{2} \langle l_j \| \mathbf{C}^{(k)} \| l_j \rangle^2 \ \mathbf{U}_{(j)}^{(k)} \cdot \mathbf{U}_{(j)}^{(k)} \ . \tag{12.9}$$

The ij term in (9.85) is

$$f_k^{(ij)} F^k(l_i l_j) + g_k^{(ij)} G^k(l_i l_j) \ , \qquad (12.18)\ (12.33)$$

where $f_k^{(ij)}$ is the matrix element of

$$\langle l_i \| \mathbf{C}^{(k)} \| l_i \rangle \langle l_j \| \mathbf{C}^{(k)} \| l_j \rangle \ \mathbf{U}_{(i)}^{(k)} \cdot \mathbf{U}_{(j)}^{(k)} \ , \tag{12.19}$$

and (except for a constant correction in diagonal elements for the contribution to E_{av}) $g_k^{(ij)}$ is the direct matrix element of

$$-\frac{1}{2}\langle l_i\|C^{(k)}\|l_j\rangle^2 \sum_r (-1)^r [r] \begin{Bmatrix} l_i & r & l_i \\ l_j & k & l_j \end{Bmatrix} \left(\mathbf{U}^{(r)}_{(i)} \cdot \mathbf{U}^{(r)}_{(j)} + 4\mathbf{V}^{(r1)}_{(i)} \cdot \mathbf{V}^{(r1)}_{(j)} \right) . \qquad (12.34)$$

The quantities F^k and G^k are the Coulomb radial integrals (6.28) and (6.30).

Uncoupling formulae. If $\mathbf{T}^{(k)}$ and $\mathbf{W}^{(k)}$ operate only within the subspaces 1 and 2, respectively, then

$$\langle \alpha_1 j_1 \alpha_2 j_2 j \| \mathbf{T}^{(k)} \| \alpha_1' j_1' \alpha_2' j_2' j' \rangle$$
$$= \delta_{\alpha_2 j_2, \alpha_2' j_2'} (-1)^{j_1 + j_2 + j' + k} [j, j']^{1/2} \begin{Bmatrix} j_1 & j_2 & j \\ j' & k & j_1' \end{Bmatrix} \langle \alpha_1 j_1 \| \mathbf{T}^{(k)} \| \alpha_1' j_1' \rangle \qquad (11.38)$$

and

$$\langle \alpha_1 j_1 \alpha_2 j_2 j \| \mathbf{W}^{(k)} \| \alpha_1' j_1' \alpha_2' j_2' j' \rangle$$
$$= \delta_{\alpha_1 j_1, \alpha_1' j_1'} (-1)^{j_1 + j_2' + j + k} [j, j']^{1/2} \begin{Bmatrix} j_1 & j_2 & j \\ k & j' & j_2' \end{Bmatrix} \langle \alpha_2 j_2 \| \mathbf{W}^{(k)} \| \alpha_2' j_2' \rangle . \qquad (11.39)$$

These relations make it easy to simplify reduced matrix elements for coupled multi-subshell functions ψ_b to reduced matrix elements for single-subshell functions $|l^w \alpha LS\rangle$.

The reduction of configuration-interaction Coulomb matrix elements, and of electric multipole transition matrix elements, is usually much more complex; however, the equations of Racah algebra can frequently be used to advantage, after first making appropriate cfp expansions and recouplings. Details are given in Chapters 13-15.

Appendix G

Matrix Elements of Spherical Harmonics

We tabulate here one-electron angular matrix elements of the renormalized spherical harmonic

$$C_q^{(k)}(\theta,\phi) \equiv \left(\frac{4\pi}{2k+1}\right)^{1/2} Y_{kq}(\theta,\phi) \ . \qquad (11.9)$$

Values are tabulated only for $l \leqq l'$; other values may be obtained from the relation

$$\langle l\|C^{(k)}\|l'\rangle = (-1)^l [l,l']^{1/2} \begin{pmatrix} l & k & l' \\ 0 & 0 & 0 \end{pmatrix} \qquad (11.23)$$

$$= (-1)^k \langle l'\|C^{(k)}\|l\rangle \ ; \qquad (11.24)$$

in general, (11.23) is negative if $(k + l' - l)/2$ is odd. Values are given as both decimal fractions and in power-of-prime notation like that used for 3-j symbols in Appendix C.

TABLE G-1. MATRIX ELEMENTS OF $C^{(k)}$

$\langle s\|C^{(0)}\|s\rangle$	=	1.00000000	1	0	
$\langle s\|C^{(1)}\|p\rangle$	=	-1.00000000	-1	0	
$\langle s\|C^{(2)}\|d\rangle$	=	1.00000000	1	0	
$\langle s\|C^{(3)}\|f\rangle$	=	-1.00000000	-1	0	
$\langle s\|C^{(4)}\|g\rangle$	=	1.00000000	1	0	
$\langle s\|C^{(5)}\|h\rangle$	=	-1.00000000	-1	0	
$\langle p\|C^{(0)}\|p\rangle$	=	1.73205081	1	0 1	
$\langle p\|C^{(2)}\|p\rangle$	=	-1.09544512	-1	1 1-1	
$\langle p\|C^{(1)}\|d\rangle$	=	-1.41421356	-1	1	
$\langle p\|C^{(3)}\|d\rangle$	=	1.13389342	1	0 2 0-1	
$\langle p\|C^{(2)}\|f\rangle$	=	1.34164079	1	0 2-1	
$\langle p\|C^{(4)}\|f\rangle$	=	-1.15470054	-1	2-1	
$\langle p\|C^{(3)}\|g\rangle$	=	-1.30930734	-1	2 1 0-1	
$\langle p\|C^{(5)}\|g\rangle$	=	1.16774842	1	0 1 1 0	-1
$\langle p\|C^{(4)}\|h\rangle$	=	1.29099445	1	0-1 1	
$\langle p\|C^{(6)}\|h\rangle$	=	-1.17669681	-1	1 2 0 0	0-1
$\langle d\|C^{(0)}\|d\rangle$	=	2.23606798	1	0 0 1	
$\langle d\|C^{(2)}\|d\rangle$	=	-1.19522861	-1	1 0 1-1	
$\langle d\|C^{(4)}\|d\rangle$	=	1.19522861	1	1 0 1-1	
$\langle d\|C^{(1)}\|f\rangle$	=	-1.73205081	-1	0 1	
$\langle d\|C^{(3)}\|f\rangle$	=	1.15470054	1	2-1	
$\langle d\|C^{(5)}\|f\rangle$	=	-1.23091491	-1	1-1 2 0	-1
$\langle d\|C^{(2)}\|g\rangle$	=	1.60356745	1	1 2 0-1	
$\langle d\|C^{(4)}\|g\rangle$	=	-1.13960576	-1	2 0 2-1	-1
$\langle d\|C^{(6)}\|g\rangle$	=	1.25436302	1	0 2 2 0	-1-1
$\langle d\|C^{(3)}\|h\rangle$	=	-1.54303350	-1	1-1 2-1	
$\langle d\|C^{(5)}\|h\rangle$	=	1.13227703	1	1-1 2 0	0-1
$\langle d\|C^{(7)}\|h\rangle$	=	-1.27097782	-1	0 1 0 1	0-1
$\langle f\|C^{(0)}\|f\rangle$	=	2.64575131	1	0 0 0 1	
$\langle f\|C^{(2)}\|f\rangle$	=	-1.36626010	-1	2-1-1 1	
$\langle f\|C^{(4)}\|f\rangle$	=	1.12815215	1	1 0 0 1	-1
$\langle f\|C^{(6)}\|f\rangle$	=	-1.27738077	-1	2-1 2 1	-1-1
$\langle f\|C^{(1)}\|g\rangle$	=	-2.00000000	-1	2	
$\langle f\|C^{(3)}\|g\rangle$	=	1.27920430	1	1 2 0 0	-1
$\langle f\|C^{(5)}\|g\rangle$	=	-1.12193639	-1	2 2 1 0	-1-1
$\langle f\|C^{(7)}\|g\rangle$	=	1.30892579	1	0 0 1 2	-1-1
$\langle f\|C^{(2)}\|h\rangle$	=	1.82574186	1	1-1 1	
$\langle f\|C^{(4)}\|h\rangle$	=	-1.24034735	-1	2 0 1 0	0-1
$\langle f\|C^{(6)}\|h\rangle$	=	1.12089708	1	0-1 0 2	0-1
$\langle f\|C^{(8)}\|h\rangle$	=	-1.33182418	-1	3 0 0 2	0-1-1
$\langle g\|C^{(0)}\|g\rangle$	=	3.00000000	1	0 2	
$\langle g\|C^{(2)}\|g\rangle$	=	-1.52894157	-1	2 2 1-1	-1
$\langle g\|C^{(4)}\|g\rangle$	=	1.20687342	1	1 6 0-1	-1-1
$\langle g\|C^{(6)}\|g\rangle$	=	-1.12193639	-1	2 2 1 0	-1-1
$\langle g\|C^{(8)}\|g\rangle$	=	1.34687352	1	1 2 1 2	-1-1-1
$\langle g\|C^{(1)}\|h\rangle$	=	-2.23606798	-1	0 0 1	
$\langle g\|C^{(3)}\|h\rangle$	=	1.40642169	1	2 2 1-1	0-1
$\langle g\|C^{(5)}\|h\rangle$	=	-1.17669681	-1	1 2 0 0	0-1
$\langle g\|C^{(7)}\|h\rangle$	=	1.12559688	1	3 0 1 1	0-1-1
$\langle g\|C^{(9)}\|h\rangle$	=	-1.37493640	-1	1 4 0 2	0-1-1-1

Appendix H

Coefficients of Fractional Parentage, $U^{(k)}$, $V^{(k1)}$

Complete tables of cfp, cfgp, and matrix elements of $U^{(k)}$ and $V^{(k1)}$ would be far too voluminous to present here; we include only illustrative values for p^w and the simpler d^w and f^w subshells.

Values of (squares of the) cfp and cfgp are given in rational-fraction form: a least common denominator d is given for each term, and a numerator n is given for each parent or grandparent; the magnitude of the coefficient is then $|n/d|^{1/2}$, and the sign is that of n. The U and V matrix elements are given as decimal fractions, and also in power-of-prime notation, as defined in Appendix C for 3-j symbols. Only the upper triangle of each U and V matrix is given, and zero matrix elements are omitted; from (11.54) and (11.69), elements in the lower triangle differ from the corresponding upper-triangle elements by a phase factor

$$(-1)^{L'+S'-L-S} ,$$

where $S' = S$ for U matrix elements.

Complete cfp tables for p^w, d^w, and f^w ($w \leqq 2l + 1$) are given in ref. [112]; values for $w > 2l + 1$ can be obtained with the aid of (9.68)-(9.69). Matrix elements of $U^{(k)}$ and $V^{(k1)}$ for certain values of k are given in refs. [112]-[115].

TABLE H-1. CFP, CFGP, $U^{(k)}$, $V^{(k1)}$

$\underline{p^1}$

cfp $\langle p^1 \{\!| p^0 \rangle$

	denom.	^1S
^2P	1	1

$\langle p^1\,{}^2P \| U^{(0)} \| p^1\,{}^2P \rangle$	=	1.00000000	1	0	
$\langle p^1\,{}^2P \| U^{(1)} \| p^1\,{}^2P \rangle$	=	1.00000000	1	0	
$\langle p^1\,{}^2P \| U^{(2)} \| p^1\,{}^2P \rangle$	=	1.00000000	1	0	
$\langle p^1\,{}^2P \| V^{(01)} \| p^1\,{}^2P \rangle$	=	1.22474487	1	-1	1
$\langle p^1\,{}^2P \| V^{(11)} \| p^1\,{}^2P \rangle$	=	1.22474487	1	-1	1
$\langle p^1\,{}^2P \| V^{(21)} \| p^1\,{}^2P \rangle$	=	1.22474487	1	-1	1

TABLE H-1 (cont)

p²

cfp $\langle p^2\{|p^1\rangle$

	denom.	^2P
^3P	1	1
^1D	1	1
^1S	1	1

$\langle p^2\,{}^3P\,\|U^{(0)}\|\,p^2\,{}^3P\rangle$	=	2.00000000	1	2
$\langle p^2\,{}^1D\|U^{(0)}\|\,p^2\,{}^1D\rangle$	=	2.58198890	1	2-1 1
$\langle p^2\,{}^1S\,\|U^{(0)}\|\,p^2\,{}^1S\rangle$	=	1.15470054	1	2-1
$\langle p^2\,{}^3P\,\|U^{(1)}\|\,p^2\,{}^3P\rangle$	=	1.00000000	1	0
$\langle p^2\,{}^1D\|U^{(1)}\|\,p^2\,{}^1D\rangle$	=	2.23606798	1	0 0 1
$\langle p^2\,{}^3P\,\|U^{(2)}\|\,p^2\,{}^3P\rangle$	=	-1.00000000	-1	0
$\langle p^2\,{}^1D\|U^{(2)}\|\,p^2\,{}^1D\rangle$	=	1.52752523	1	0-1 0 1
$\langle p^2\,{}^1D\|U^{(2)}\|\,p^2\,{}^1S\rangle$	=	1.15470054	1	2-1
$\langle p^2\,{}^3P\,\|V^{(01)}\|\,p^2\,{}^3P\rangle$	=	2.44948974	1	1 1
$\langle p^2\,{}^3P\,\|V^{(11)}\|\,p^2\,{}^3P\rangle$	=	1.22474487	1	-1 1
$\langle p^2\,{}^3P\,\|V^{(11)}\|\,p^2\,{}^1D\rangle$	=	-1.11803399	-1	-2 0 1
$\langle p^2\,{}^3P\,\|V^{(11)}\|\,p^2\,{}^1S\rangle$	=	1.00000000	1	0
$\langle p^2\,{}^3P\,\|V^{(21)}\|\,p^2\,{}^3P\rangle$	=	-1.22474487	-1	-1 1
$\langle p^2\,{}^3P\,\|V^{(21)}\|\,p^2\,{}^1D\rangle$	=	-1.50000000	-1	-2 2

---

p³

cfp $\langle p^3\{|p^2\rangle$

	denom.	^3P	^1D	^1S
^4S	1	1	0	0
^2D	2	1	-1	0
^2P	18	-9	-5	4

cfgp $\langle p^3\{|p^1,p^2(^1S)\rangle$

	denom.	^2P
^4S	1	0
^2D	1	0
^2P	9	2

cfgp $\langle p^3\{|p^1,p^2(^3P)\rangle$

	denom.	^2P
^4S	1	1
^2D	2	-1
^2P	2	-1

cfgp $\langle p^3\{|p^1,p^2(^1D)\rangle$

	denom.	^2P
^4S	1	0
^2D	2	1
^2P	18	-5

TABLE H-1 (cont)

$\langle p^3\,{}^4S \| U^{(0)} \| p^3\,{}^4S \rangle$	=	1.73205081	1	0 1
$\langle p^3\,{}^2D \| U^{(0)} \| p^3\,{}^2D \rangle$	=	3.87298335	1	0 1 1
$\langle p^3\,{}^2P \| U^{(0)} \| p^3\,{}^2P \rangle$	=	3.00000000	1	0 2
$\langle p^3\,{}^2D \| U^{(1)} \| p^3\,{}^2D \rangle$	=	2.23606798	1	0 0 1
$\langle p^3\,{}^2P \| U^{(1)} \| p^3\,{}^2P \rangle$	=	1.00000000	1	0
$\langle p^3\,{}^2D \| U^{(2)} \| p^3\,{}^2P \rangle$	=	1.73205081	1	0 1
$\langle p^3\,{}^4S \| V^{(01)} \| p^3\,{}^4S \rangle$	=	2.23606798	1	0 0 1
$\langle p^3\,{}^2D \| V^{(01)} \| p^3\,{}^2D \rangle$	=	1.58113883	1	-1 0 1
$\langle p^3\,{}^2P \| V^{(01)} \| p^3\,{}^2P \rangle$	=	1.22474487	1	-1 1
$\langle p^3\,{}^4S \| V^{(11)} \| p^3\,{}^2P \rangle$	=	1.41421356	1	1
$\langle p^3\,{}^2D \| V^{(11)} \| p^3\,{}^2P \rangle$	=	-1.58113883	-1	-1 0 1
$\langle p^3\,{}^4S \| V^{(21)} \| p^3\,{}^2D \rangle$	=	-1.41421356	-1	1
$\langle p^3\,{}^2D \| V^{(21)} \| p^3\,{}^2D \rangle$	=	-1.87082869	-1	-1 0 0 1
$\langle p^3\,{}^2P \| V^{(21)} \| p^3\,{}^2P \rangle$	=	1.22474487	1	-1 1

---

p^4

cfp $\langle p^4 \{| p^3 \rangle$

	denom.	^4S	^2D	^2P
^3P	12	-4	5	-3
^1D	4	0	-3	-1
^1S	1	0	0	1

cfgp $\langle p^4 \{| p^2, p^2(^1S) \rangle$

	denom.	^3P	^1D	^1S
^3P	18	-1	0	0
^1D	18	0	-1	0
^1S	9	0	0	2

cfgp $\langle p^4 \{| p^2, p^2(^3P) \rangle$

	denom.	^3P	^1D	^1S
^3P	18	6	-5	-1
^1D	2	-1	0	0
^1S	2	-1	0	0

cfgp $\langle p^4 \{| p^2, p^2(^1D) \rangle$

	denom.	^3P	^1D	^1S
^3P	18	-5	0	0
^1D	18	0	7	-1
^1S	18	0	-5	0

$\langle p^4\,{}^3P \| U^{(0)} \| p^4\,{}^3P \rangle$	=	4.00000000	1	4
$\langle p^4\,{}^1D \| U^{(0)} \| p^4\,{}^1D \rangle$	=	5.16397779	1	4-1 1
$\langle p^4\,{}^1S \| U^{(0)} \| p^4\,{}^1S \rangle$	=	2.30940108	1	4-1
$\langle p^4\,{}^3P \| U^{(1)} \| p^4\,{}^3P \rangle$	=	1.00000000	1	0
$\langle p^4\,{}^1D \| U^{(1)} \| p^4\,{}^1D \rangle$	=	2.23606798	1	0 0 1
$\langle p^4\,{}^3P \| U^{(2)} \| p^4\,{}^3P \rangle$	=	1.00000000	1	0
$\langle p^4\,{}^1D \| U^{(2)} \| p^4\,{}^1D \rangle$	=	-1.52752523	-1	0-1 0 1
$\langle p^4\,{}^1D \| U^{(2)} \| p^4\,{}^1S \rangle$	=	-1.15470054	-1	2-1

TABLE H-1 (cont)

$\langle p^4\,{}^3P \| V^{(01)} \| p^4\,{}^3P \rangle$	=	2.44948974	1	1 1		
$\langle p^4\,{}^3P \| V^{(11)} \| p^4\,{}^3P \rangle$	=	-1.22474487	-1	-1 1		
$\langle p^4\,{}^3P \| V^{(11)} \| p^4\,{}^1D \rangle$	=	1.11803399	1	-2 0 1		
$\langle p^4\,{}^3P \| V^{(11)} \| p^4\,{}^1S \rangle$	=	-1.00000000	-1	0		
$\langle p^4\,{}^3P \| V^{(21)} \| p^4\,{}^3P \rangle$	=	-1.22474487	-1	-1 1		
$\langle p^4\,{}^3P \| V^{(21)} \| p^4\,{}^1D \rangle$	=	-1.50000000	-1	-2 2		

$\underline{p^5}$

cfp $\langle p^5 \{\!| p^4 \rangle$

	denom.	^3P	^1D	^1S
^2P	15	9	5	1

cfgp $\langle p^5 \{\!| p^3, p^2(^1S) \rangle$

	denom.	^4S	^2D	^2P
^2P	15	0	0	1

cfgp $\langle p^5 \{\!| p^3, p^2(^3P) \rangle$

	denom.	^4S	^2D	^2P
^2P	20	-4	-5	-3

cfgp $\langle p^5 \{\!| p^3, p^2(^1D) \rangle$

	denom.	^4S	^2D	^2P
^2P	12	0	3	-1

$\langle p^5\,{}^2P \| U^{(0)} \| p^5\,{}^2P \rangle$	=	5.00000000	1	0 0 2
$\langle p^5\,{}^2P \| U^{(1)} \| p^5\,{}^2P \rangle$	=	1.00000000	1	0
$\langle p^5\,{}^2P \| U^{(2)} \| p^5\,{}^2P \rangle$	=	-1.00000000	-1	0
$\langle p^5\,{}^2P \| V^{(01)} \| p^5\,{}^2P \rangle$	=	1.22474487	1	-1 1
$\langle p^5\,{}^2P \| V^{(11)} \| p^5\,{}^2P \rangle$	=	-1.22474487	-1	-1 1
$\langle p^5\,{}^2P \| V^{(21)} \| p^5\,{}^2P \rangle$	=	1.22474487	1	-1 1

$\underline{p^6}$

cfp $\langle p^6 \{\!| p^5 \rangle$

	denom.	^2P
^1S	1	1

cfgp $\langle p^6 \{\!| p^4, p^2(^1S) \rangle$

	denom.	^3P	^1D	^1S
^1S	15	0	0	1

cfgp $\langle p^6 \{\!| p^4, p^2(^3P) \rangle$

	denom.	^3P	^1D	^1S
^1S	5	3	0	0

cfgp $\langle p^6 \{\!| p^4, p^2(^1D) \rangle$

	denom.	^3P	^1D	^1S
^1S	3	0	1	0

TABLE H-1 (cont)

$\underline{d^2}$

cfp $\langle d^2 \{\!| d^1 \rangle$

	denom.	^2D
^3F	1	1
^3P	1	1
^1G	1	1
^1D	1	1
^1S	1	1

$\langle d^2\,{}^3F\,\|U^{(0)}\|\,d^2\,{}^3F\rangle$	=	2.36643191	1	2 0-1 1	
$\langle d^2\,{}^3P\,\|U^{(0)}\|\,d^2\,{}^3P\rangle$	=	1.54919334	1	2 1-1	
$\langle d^2\,{}^1G\|U^{(0)}\|\,d^2\,{}^1G\rangle$	=	2.68328157	1	2 2-1	
$\langle d^2\,{}^1D\|U^{(0)}\|\,d^2\,{}^1D\rangle$	=	2.00000000	1	2	
$\langle d^2\,{}^1S\,\|U^{(0)}\|\,d^2\,{}^1S\rangle$	=	.89442719	1	2 0-1	
$\langle d^2\,{}^3F\,\|U^{(1)}\|\,d^2\,{}^3F\rangle$	=	1.67332005	1	1 0-1 1	
$\langle d^2\,{}^3P\,\|U^{(1)}\|\,d^2\,{}^3P\rangle$	=	.44721360	1	0 0-1	
$\langle d^2\,{}^1G\|U^{(1)}\|\,d^2\,{}^1G\rangle$	=	2.44948974	1	1 1	
$\langle d^2\,{}^1D\|U^{(1)}\|\,d^2\,{}^1D\rangle$	=	1.00000000	1	0	
$\langle d^2\,{}^3F\,\|U^{(2)}\|\,d^2\,{}^3F\rangle$	=	.48989795	1	1 1-2	
$\langle d^2\,{}^3F\,\|U^{(2)}\|\,d^2\,{}^3P\rangle$	=	.97979590	1	3 1-2	
$\langle d^2\,{}^3P\,\|U^{(2)}\|\,d^2\,{}^3P\rangle$	=	-.91651514	-1	0 1-2 1	
$\langle d^2\,{}^1G\|U^{(2)}\|\,d^2\,{}^1G\rangle$	=	2.01017818	1	1 2 0-2	1
$\langle d^2\,{}^1G\|U^{(2)}\|\,d^2\,{}^1D\rangle$	=	.76665188	1	4 2-1-2	
$\langle d^2\,{}^1D\|U^{(2)}\|\,d^2\,{}^1D\rangle$	=	-.42857143	-1	0 2 0-2	
$\langle d^2\,{}^1D\|U^{(2)}\|\,d^2\,{}^1S\rangle$	=	.89442719	1	2 0-1	
$\langle d^2\,{}^3F\,\|U^{(3)}\|\,d^2\,{}^3F\rangle$	=	-.77459667	-1	0 1-1	
$\langle d^2\,{}^3F\,\|U^{(3)}\|\,d^2\,{}^3P\rangle$	=	1.09544512	1	1 1-1	
$\langle d^2\,{}^1G\|U^{(3)}\|\,d^2\,{}^1G\rangle$	=	1.42141062	1	0 2 0-2	1
$\langle d^2\,{}^1G\|U^{(3)}\|\,d^2\,{}^1D\rangle$	=	1.35526185	1	1 2 1-2	
$\langle d^2\,{}^1D\|U^{(3)}\|\,d^2\,{}^1D\rangle$	=	-1.14285714	-1	6 0 0-2	
$\langle d^2\,{}^3F\,\|U^{(4)}\|\,d^2\,{}^3F\rangle$	=	-1.48323970	-1	0 0-1 0	1
$\langle d^2\,{}^3F\,\|U^{(4)}\|\,d^2\,{}^3P\rangle$	=	-.63245553	-1	1 0-1	
$\langle d^2\,{}^1G\|U^{(4)}\|\,d^2\,{}^1G\rangle$	=	.76398525	1	0 0-1-2	1 1
$\langle d^2\,{}^1G\|U^{(4)}\|\,d^2\,{}^1D\rangle$	=	1.49829835	1	1 0 1-2	1
$\langle d^2\,{}^1G\|U^{(4)}\|\,d^2\,{}^1S\rangle$	=	.89442719	1	2 0-1	
$\langle d^2\,{}^1D\|U^{(4)}\|\,d^2\,{}^1D\rangle$	=	.57142857	1	4 0 0-2	
$\langle d^2\,{}^3F\,\|V^{(01)}\|\,d^2\,{}^3F\rangle$	=	2.89827535	1	1 1-1 1	
$\langle d^2\,{}^3P\,\|V^{(01)}\|\,d^2\,{}^3P\rangle$	=	1.89736660	1	1 2-1	
$\langle d^2\,{}^3F\,\|V^{(11)}\|\,d^2\,{}^3F\rangle$	=	2.04939015	1	0 1-1 1	
$\langle d^2\,{}^3F\,\|V^{(11)}\|\,d^2\,{}^1G\rangle$	=	-.94868330	-1	-1 2-1	
$\langle d^2\,{}^3F\,\|V^{(11)}\|\,d^2\,{}^1D\rangle$	=	1.09544512	1	1 1-1	
$\langle d^2\,{}^3P\,\|V^{(11)}\|\,d^2\,{}^3P\rangle$	=	.54772256	1	-1 1-1	
$\langle d^2\,{}^3P\,\|V^{(11)}\|\,d^2\,{}^1D\rangle$	=	-1.02469508	-1	-2 1-1 1	
$\langle d^2\,{}^3P\,\|V^{(11)}\|\,d^2\,{}^1S\rangle$	=	.77459667	1	0 1-1	

TABLE H-1 (cont)

$\langle d^2\,{}^3F \| V^{(21)} \| d^2\,{}^3F \rangle$	=	.60000000	1	0	2	-2			
$\langle d^2\,{}^3F \| V^{(21)} \| d^2\,{}^3P \rangle$	=	1.20000000	1	2	2	-2			
$\langle d^2\,{}^3F \| V^{(21)} \| d^2\,{}^1G \rangle$	=	-1.38873015	-1	-1	3	0	-1		
$\langle d^2\,{}^3F \| V^{(21)} \| d^2\,{}^1D \rangle$	=	1.17108009	1	4	1	-1	-1		
$\langle d^2\,{}^3P \| V^{(21)} \| d^2\,{}^3P \rangle$	=	-1.12249722	-1	-1	2	-2	1		
$\langle d^2\,{}^3P \| V^{(21)} \| d^2\,{}^1D \rangle$	=	-.67082039	-1	-2	2	-1			
$\langle d^2\,{}^3F \| V^{(31)} \| d^2\,{}^3F \rangle$	=	-.94868330	-1	-1	2	-1			
$\langle d^2\,{}^3F \| V^{(31)} \| d^2\,{}^3P \rangle$	=	1.34164079	1	0	2	-1			
$\langle d^2\,{}^3F \| V^{(31)} \| d^2\,{}^1G \rangle$	=	-1.45651247	-1	-2	3	-1	-1		1
$\langle d^2\,{}^3F \| V^{(31)} \| d^2\,{}^1D \rangle$	=	.35856858	1	-1	2	-1	-1		
$\langle d^2\,{}^3F \| V^{(31)} \| d^2\,{}^1S \rangle$	=	.77459667	1	0	1	-1			
$\langle d^2\,{}^3P \| V^{(31)} \| d^2\,{}^1G \rangle$	=	-.62105900	-1	-1	3	-1	-1		
$\langle d^2\,{}^3P \| V^{(31)} \| d^2\,{}^1D \rangle$	=	.71713717	1	1	2	-1	-1		
$\langle d^2\,{}^3F \| V^{(41)} \| d^2\,{}^3F \rangle$	=	-1.81659021	-1	-1	1	-1	0		1
$\langle d^2\,{}^3F \| V^{(41)} \| d^2\,{}^3P \rangle$	=	-.77459667	-1	0	1	-1			
$\langle d^2\,{}^3F \| V^{(41)} \| d^2\,{}^1G \rangle$	=	-1.08562030	-1	-2	1	0	-1		1
$\langle d^2\,{}^3F \| V^{(41)} \| d^2\,{}^1D \rangle$	=	-1.03509834	-1	-1	1	1	-1		
$\langle d^2\,{}^3P \| V^{(41)} \| d^2\,{}^1G \rangle$	=	-1.22474487	-1	-1	1				

---

d^3

cfp $\langle d^3 \{\!| d^2 \rangle$

	denom.	^3F	^3P	^1G	^1D	^1S
^4F	5	4	-1	0	0	0
^4P	15	-7	-8	0	0	0
^2H	2	-1	0	1	0	0
^2G	42	21	0	11	-10	0
^2F	70	7	28	-25	-10	0
$^2D_1$	60	-21	-9	-9	-5	16
$^2D_2$	140	21	-49	-25	45	0
^2P	30	-8	7	0	15	0

cfgp $\langle d^3 \{\!| d^1, d^2(^1S) \rangle$

	denom.	^2D
^4F	1	0
^4P	1	0
^2H	1	0
^2G	1	0
^2F	1	0
$^2D_1$	15	4
$^2D_2$	1	0
^2P	1	0

cfgp $\langle d^3 \{\!| d^1, d^2(^3P) \rangle$

	denom.	^2D
^4F	5	-1
^4P	15	-8
^2H	1	0
^2G	1	0
^2F	5	-2
$^2D_1$	20	-3
$^2D_2$	20	-7
^2P	30	-7

TABLE H-1 (cont)

cfgp $\langle d^3 \{\!| d^1, d^2(^1D)\rangle$

	denom.	^2D
^4F	1	0
^4P	1	0
^2H	1	0
^2G	21	-5
^2F	7	1
$^2D_1$	12	-1
$^2D_2$	28	9
^2P	2	-1

cfgp $\langle d^3 \{\!| d^1, d^2(^3F)\rangle$

	denom.	^2D
^4F	5	4
^4P	15	-7
^2H	2	1
^2G	2	1
^2F	10	-1
$^2D_1$	20	-7
$^2D_2$	20	3
^2P	15	4

cfgp $\langle d^3 \{\!| d^1, d^2(^1G)\rangle$

	denom.	^2D
^4F	1	0
^4P	1	0
^2H	2	-1
^2G	42	11
^2F	14	5
$^2D_1$	20	-3
$^2D_2$	28	-5
^2P	1	0

$\langle d^3\,^4F \Vert U^{(0)} \Vert d^3\,^4F\rangle$	=	3.54964787	1	0 2-1 1	
$\langle d^3\,^4P \Vert U^{(0)} \Vert d^3\,^4P\rangle$	=	2.32379001	1	0 3-1	
$\langle d^3\,^2H \Vert U^{(0)} \Vert d^3\,^2H\rangle$	=	4.44971909	1	0 2-1 0	1
$\langle d^3\,^2G \Vert U^{(0)} \Vert d^3\,^2G\rangle$	=	4.02492236	1	0 4-1	
$\langle d^3\,^2F \Vert U^{(0)} \Vert d^3\,^2F\rangle$	=	3.54964787	1	0 2-1 1	
$\langle d^3\,^2D_1 \Vert U^{(0)} \Vert d^3\,^2D_1\rangle$	=	3.00000000	1	0 2	
$\langle d^3\,^2D_2 \Vert U^{(0)} \Vert d^3\,^2D_2\rangle$	=	3.00000000	1	0 2	
$\langle d^3\,^2P \Vert U^{(0)} \Vert d^3\,^2P\rangle$	=	2.32379001	1	0 3-1	
$\langle d^3\,^4F \Vert U^{(1)} \Vert d^3\,^4F\rangle$	=	1.67332005	1	1 0-1 1	
$\langle d^3\,^4P \Vert U^{(1)} \Vert d^3\,^4P\rangle$	=	.44721360	1	0 0-1	
$\langle d^3\,^2H \Vert U^{(1)} \Vert d^3\,^2H\rangle$	=	3.31662479	1	0 0 0 0	1
$\langle d^3\,^2G \Vert U^{(1)} \Vert d^3\,^2G\rangle$	=	2.44948974	1	1 1	
$\langle d^3\,^2F \Vert U^{(1)} \Vert d^3\,^2F\rangle$	=	1.67332005	1	1 0-1 1	
$\langle d^3\,^2D_1 \Vert U^{(1)} \Vert d^3\,^2D_1\rangle$	=	1.00000000	1	0	
$\langle d^3\,^2D_2 \Vert U^{(1)} \Vert d^3\,^2D_2\rangle$	=	1.00000000	1	0	
$\langle d^3\,^2P \Vert U^{(1)} \Vert d^3\,^2P\rangle$	=	.44721360	1	0 0-1	

TABLE H-1 (cont)

$\langle d^3\,{}^4F\|U^{(2)}\|d^3\,{}^4F\rangle$	=	-.48989795	-1	1 1-2	
$\langle d^3\,{}^4F\|U^{(2)}\|d^3\,{}^4P\rangle$	=	-.97979590	-1	3 1-2	
$\langle d^3\,{}^4P\|U^{(2)}\|d^3\,{}^4P\rangle$	=	.91651514	1	0 1-2 1	
$\langle d^3\,{}^2H\|U^{(2)}\|d^3\,{}^2H\rangle$	=	1.56570386	1	0 1-2-1	1 1
$\langle d^3\,{}^2H\|U^{(2)}\|d^3\,{}^2G\rangle$	=	.61411958	1	1 1-2-1	1
$\langle d^3\,{}^2H\|U^{(2)}\|d^3\,{}^2F\rangle$	=	-1.37321312	-1	1 1-1-1	1
$\langle d^3\,{}^2G\|U^{(2)}\|d^3\,{}^2G\rangle$	=	.40203564	1	1 2-2-2	1
$\langle d^3\,{}^2G\|U^{(2)}\|d^3\,{}^2F\rangle$	=	-.82807867	-1	3 1-1-1	
$\langle d^3\,{}^2G\|U^{(2)}\|d^3\,{}^2D_1\rangle$	=	-1.13389342	-1	0 2 0-1	
$\langle d^3\,{}^2G\|U^{(2)}\|d^3\,{}^2D_2\rangle$	=	-.24743583	-1	0 1 0-2	
$\langle d^3\,{}^2F\|U^{(2)}\|d^3\,{}^2F\rangle$	=	.48989795	1	1 1-2	
$\langle d^3\,{}^2F\|U^{(2)}\|d^3\,{}^2D_1\rangle$	=	.77459667	1	0 1-1	
$\langle d^3\,{}^2F\|U^{(2)}\|d^3\,{}^2D_2\rangle$	=	1.52127766	1	0 4-1-1	
$\langle d^3\,{}^2F\|U^{(2)}\|d^3\,{}^2P\rangle$	=	-.52372294	-1	4 1-2-1	
$\langle d^3\,{}^2D_1\|U^{(2)}\|d^3\,{}^2D_1\rangle$	=	.50000000	1	-2	
$\langle d^3\,{}^2D_1\|U^{(2)}\|d^3\,{}^2D_2\rangle$	=	.98198051	1	-2 3 0-1	
$\langle d^3\,{}^2D_1\|U^{(2)}\|d^3\,{}^2P\rangle$	=	.94868330	1	-1 2-1	
$\langle d^3\,{}^2D_2\|U^{(2)}\|d^3\,{}^2D_2\rangle$	=	.21428571	1	-2 2 0-2	
$\langle d^3\,{}^2D_2\|U^{(2)}\|d^3\,{}^2P\rangle$	=	-.20701967	-1	-1 1-1-1	
$\langle d^3\,{}^2P\|U^{(2)}\|d^3\,{}^2P\rangle$	=	-.26186147	-1	2 1-2-1	
$\langle d^3\,{}^4F\|U^{(3)}\|d^3\,{}^4F\rangle$	=	-.77459667	-1	0 1-1	
$\langle d^3\,{}^4F\|U^{(3)}\|d^3\,{}^4P\rangle$	=	1.09544512	1	1 1-1	
$\langle d^3\,{}^2H\|U^{(3)}\|d^3\,{}^2G\rangle$	=	1.47945108	1	-2 1 0-2	1 1
$\langle d^3\,{}^2H\|U^{(3)}\|d^3\,{}^2F\rangle$	=	-1.08562030	-1	-2 1 0-1	1
$\langle d^3\,{}^2H\|U^{(3)}\|d^3\,{}^2D_2\rangle$	=	1.16057691	1	1 1 0-2	1
$\langle d^3\,{}^2G\|U^{(3)}\|d^3\,{}^2G\rangle$	=	-.71070531	-1	-2 2 0-2	1
$\langle d^3\,{}^2G\|U^{(3)}\|d^3\,{}^2F\rangle$	=	-1.08562030	-1	-2 1 0-1	1
$\langle d^3\,{}^2G\|U^{(3)}\|d^3\,{}^2D_2\rangle$	=	.17496355	1	-1 1 0-2	
$\langle d^3\,{}^2G\|U^{(3)}\|d^3\,{}^2P\rangle$	=	1.11346123	1	-2 5 0-2	
$\langle d^3\,{}^2F\|U^{(3)}\|d^3\,{}^2F\rangle$	=	1.16189500	1	-2 3-1	
$\langle d^3\,{}^2F\|U^{(3)}\|d^3\,{}^2D_2\rangle$	=	1.03509834	1	-1 1 1-1	
$\langle d^3\,{}^2F\|U^{(3)}\|d^3\,{}^2P\rangle$	=	.14638501	1	-2 1-1-1	
$\langle d^3\,{}^2D_1\|U^{(3)}\|d^3\,{}^2D_1\rangle$	=	1.00000000	1	0	
$\langle d^3\,{}^2D_2\|U^{(3)}\|d^3\,{}^2D_2\rangle$	=	-.42857143	-1	0 2 0-2	
$\langle d^3\,{}^2D_2\|U^{(3)}\|d^3\,{}^2P\rangle$	=	1.10656667	1	2 1 1-2	
$\langle d^3\,{}^4F\|U^{(4)}\|d^3\,{}^4F\rangle$	=	1.48323970	1	0 0-1 0	1
$\langle d^3\,{}^4F\|U^{(4)}\|d^3\,{}^4P\rangle$	=	.63245553	1	1 0-1	
$\langle d^3\,{}^2H\|U^{(4)}\|d^3\,{}^2H\rangle$	=	-.67377166	-1	0-2-1-1	1 1
$\langle d^3\,{}^2H\|U^{(4)}\|d^3\,{}^2G\rangle$	=	1.75051013	1	-2 1-1-1	1 1
$\langle d^3\,{}^2H\|U^{(4)}\|d^3\,{}^2F\rangle$	=	.33688583	1	-2-2-1-1	1 1
$\langle d^3\,{}^2H\|U^{(4)}\|d^3\,{}^2D_1\rangle$	=	-1.35400640	-1	-1-1 0 0	1
$\langle d^3\,{}^2H\|U^{(4)}\|d^3\,{}^2D_2\rangle$	=	.29546842	1	-1-2 0-1	1
$\langle d^3\,{}^2H\|U^{(4)}\|d^3\,{}^2P\rangle$	=	.79282497	1	1 0-1-1	1
$\langle d^3\,{}^2G\|U^{(4)}\|d^3\,{}^2G\rangle$	=	-.38199263	-1	-2 0-1-2	1 1
$\langle d^3\,{}^2G\|U^{(4)}\|d^3\,{}^2F\rangle$	=	.48550416	1	-2 1-1-1	1
$\langle d^3\,{}^2G\|U^{(4)}\|d^3\,{}^2D_1\rangle$	=	.88640526	1	-1 0 0-1	1

TABLE H-1 (cont)

$\langle d^3\,{}^2G\|U^{(4)}\|d^3\,{}^2D_2\rangle$	=	-1.16057691	-1	1 1 0-2	1
$\langle d^3\,{}^2G\|U^{(4)}\|d^3\,{}^2P\rangle$	=	.43915503	1	-2 3-1-1	
$\langle d^3\,{}^2F\|U^{(4)}\|d^3\,{}^2F\rangle$	=	.24720662	1	-2-2-1 0	1
$\langle d^3\,{}^2F\|U^{(4)}\|d^3\,{}^2D_1\rangle$	=	.91287093	1	-1-1 1	
$\langle d^3\,{}^2F\|U^{(4)}\|d^3\,{}^2D_2\rangle$	=	.39840954	1	1-2 1-1	
$\langle d^3\,{}^2F\|U^{(4)}\|d^3\,{}^2P\rangle$	=	.92966968	1	-2 0-1-1	2
$\langle d^3\,{}^2D_1\|U^{(4)}\|d^3\,{}^2D_1\rangle$	=	.50000000	1	-2	
$\langle d^3\,{}^2D_1\|U^{(4)}\|d^3\,{}^2D_2\rangle$	=	-.54554473	-1	-2-1 2-1	
$\langle d^3\,{}^2D_2\|U^{(4)}\|d^3\,{}^2D_2\rangle$	=	-.45238095	-1	-2-2 0-2	0 0 0 2
$\langle d^3\,{}^4F\|V^{(01)}\|d^3\,{}^4F\rangle$	=	4.58257569	1	0 1 0 1	
$\langle d^3\,{}^4P\|V^{(01)}\|d^3\,{}^4P\rangle$	=	3.00000000	1	0 2	
$\langle d^3\,{}^2H\|V^{(01)}\|d^3\,{}^2H\rangle$	=	1.81659021	1	-1 1-1 0	1
$\langle d^3\,{}^2G\|V^{(01)}\|d^3\,{}^2G\rangle$	=	1.64316767	1	-1 3-1	
$\langle d^3\,{}^2F\|V^{(01)}\|d^3\,{}^2F\rangle$	=	1.44913767	1	-1 1-1 1	
$\langle d^3\,{}^2D_1\|V^{(01)}\|d^3\,{}^2D_1\rangle$	=	1.22474487	1	-1 1	
$\langle d^3\,{}^2D_2\|V^{(01)}\|d^3\,{}^2D_2\rangle$	=	1.22474487	1	-1 1	
$\langle d^3\,{}^2P\|V^{(01)}\|d^3\,{}^2P\rangle$	=	.94868330	1	-1 2-1	
$\langle d^3\,{}^4F\|V^{(11)}\|d^3\,{}^4F\rangle$	=	2.16024690	1	1-1 0 1	
$\langle d^3\,{}^4F\|V^{(11)}\|d^3\,{}^2G\rangle$	=	-1.73205081	-1	0 1	
$\langle d^3\,{}^4F\|V^{(11)}\|d^3\,{}^2F\rangle$	=	.68313005	1	0-1-1 1	
$\langle d^3\,{}^4F\|V^{(11)}\|d^3\,{}^2D_1\rangle$	=	-1.18321596	-1	0 0-1 1	
$\langle d^3\,{}^4F\|V^{(11)}\|d^3\,{}^2D_2\rangle$	=	1.29099445	1	0-1 1	
$\langle d^3\,{}^4P\|V^{(11)}\|d^3\,{}^4P\rangle$	=	.57735027	1	0-1	
$\langle d^3\,{}^4P\|V^{(11)}\|d^3\,{}^2D_1\rangle$	=	-1.26491106	-1	3 0-1	
$\langle d^3\,{}^4P\|V^{(11)}\|d^3\,{}^2P\rangle$	=	1.36626010	1	2-1-1 1	
$\langle d^3\,{}^2H\|V^{(11)}\|d^3\,{}^2H\rangle$	=	.81240384	1	-1 1-2 0	1
$\langle d^3\,{}^2H\|V^{(11)}\|d^3\,{}^2G\rangle$	=	-1.62480768	-1	1 1-2 0	1
$\langle d^3\,{}^2G\|V^{(11)}\|d^3\,{}^2G\rangle$	=	.90000000	1	-2 4-2	
$\langle d^3\,{}^2G\|V^{(11)}\|d^3\,{}^2F\rangle$	=	.86602540	1	-2 1	
$\langle d^3\,{}^2F\|V^{(11)}\|d^3\,{}^2F\rangle$	=	-.34156503	-1	-2-1-1 1	
$\langle d^3\,{}^2F\|V^{(11)}\|d^3\,{}^2D_1\rangle$	=	-1.18321596	-1	0 0-1 1	
$\langle d^3\,{}^2F\|V^{(11)}\|d^3\,{}^2D_2\rangle$	=	-.25819889	-1	0-1-1	
$\langle d^3\,{}^2D_1\|V^{(11)}\|d^3\,{}^2D_1\rangle$	=	.61237244	1	-3 1	
$\langle d^3\,{}^2D_1\|V^{(11)}\|d^3\,{}^2D_2\rangle$	=	-.93541435	-1	-3 0 0 1	
$\langle d^3\,{}^2D_1\|V^{(11)}\|d^3\,{}^2P\rangle$	=	.59160798	1	-2 0-1 1	
$\langle d^3\,{}^2D_2\|V^{(11)}\|d^3\,{}^2D_2\rangle$	=	-.20412415	-1	-3-1	
$\langle d^3\,{}^2D_2\|V^{(11)}\|d^3\,{}^2P\rangle$	=	-1.16189500	-1	-2 3-1	
$\langle d^3\,{}^2P\|V^{(11)}\|d^3\,{}^2P\rangle$	=	.36514837	1	1-1-1	
$\langle d^3\,{}^4F\|V^{(21)}\|d^3\,{}^4F\rangle$	=	-.63245553	-1	1 0-1	
$\langle d^3\,{}^4F\|V^{(21)}\|d^3\,{}^4P\rangle$	=	-1.26491106	-1	3 0-1	
$\langle d^3\,{}^4F\|V^{(21)}\|d^3\,{}^2H\rangle$	=	1.12122382	1	2 0-1-1	1
$\langle d^3\,{}^4F\|V^{(21)}\|d^3\,{}^2G\rangle$	=	-1.52127766	-1	0 4-1-1	
$\langle d^3\,{}^4F\|V^{(21)}\|d^3\,{}^2F\rangle$	=	1.40000000	1	0 0-2 2	
$\langle d^3\,{}^4F\|V^{(21)}\|d^3\,{}^2D_2\rangle$	=	.82807867	1	3 1-1-1	
$\langle d^3\,{}^4F\|V^{(21)}\|d^3\,{}^2P\rangle$	=	1.17594946	1	1 0-2-1	2
$\langle d^3\,{}^4P\|V^{(21)}\|d^3\,{}^4P\rangle$	=	1.18321596	1	0 0-1 1	

TABLE H-1 (cont)

$\langle d^3\,{}^4P\|V^{(21)}\|d^3\,{}^2F\rangle$	=	.80000000	1	4 0-2	
$\langle d^3\,{}^4P\|V^{(21)}\|d^3\,{}^2D_2\rangle$	=	-1.26491106	-1	3 0-1	
$\langle d^3\,{}^4P\|V^{(21)}\|d^3\,{}^2P\rangle$	=	.40000000	1	2 0-2	
$\langle d^3\,{}^2H\|V^{(21)}\|d^3\,{}^2H\rangle$	=	-.63919592	-1	-1 0-2-1	1 1
$\langle d^3\,{}^2H\|V^{(21)}\|d^3\,{}^2G\rangle$	=	-1.50427961	-1	2 2-2-1	1
$\langle d^3\,{}^2H\|V^{(21)}\|d^3\,{}^2F\rangle$	=	-1.12122382	-1	2 0-1-1	1
$\langle d^3\,{}^2G\|V^{(21)}\|d^3\,{}^2G\rangle$	=	.24619554	1	-2 3-2-2	1
$\langle d^3\,{}^2G\|V^{(21)}\|d^3\,{}^2F\rangle$	=	-.25354628	-1	-2 2-1-1	
$\langle d^3\,{}^2G\|V^{(21)}\|d^3\,{}^2D_2\rangle$	=	-1.21218305	-1	3 2 0-2	
$\langle d^3\,{}^2F\|V^{(21)}\|d^3\,{}^2F\rangle$	=	-1.10000000	-1	-2 0-2 0	2
$\langle d^3\,{}^2F\|V^{(21)}\|d^3\,{}^2D_2\rangle$	=	.82807867	1	3 1-1-1	
$\langle d^3\,{}^2F\|V^{(21)}\|d^3\,{}^2P\rangle$	=	-.42761799	-1	5 0-2-1	
$\langle d^3\,{}^2D_1\|V^{(21)}\|d^3\,{}^2D_1\rangle$	=	1.22474487	1	-1 1	
$\langle d^3\,{}^2D_2\|V^{(21)}\|d^3\,{}^2D_2\rangle$	=	-.17496355	-1	-1 1 0-2	
$\langle d^3\,{}^2D_2\|V^{(21)}\|d^3\,{}^2P\rangle$	=	-.67612340	-1	4 0-1-1	
$\langle d^3\,{}^2P\|V^{(21)}\|d^3\,{}^2P\rangle$	=	-1.01559272	-1	-1 0-2-1	0 0 0 2

$\langle d^3\,{}^4F\|V^{(31)}\|d^3\,{}^4F\rangle$	=	-1.00000000	-1	0	
$\langle d^3\,{}^4F\|V^{(31)}\|d^3\,{}^4P\rangle$	=	1.41421356	1	1	
$\langle d^3\,{}^4F\|V^{(31)}\|d^3\,{}^2H\rangle$	=	1.77281052	1	1 0 0-1	1
$\langle d^3\,{}^4F\|V^{(31)}\|d^3\,{}^2F\rangle$	=	1.26491106	1	3 0-1	
$\langle d^3\,{}^4F\|V^{(31)}\|d^3\,{}^2D_1\rangle$	=	1.54919334	1	2 1-1	
$\langle d^3\,{}^4F\|V^{(31)}\|d^3\,{}^2P\rangle$	=	-.23904572	-1	1 0-1-1	
$\langle d^3\,{}^4P\|V^{(31)}\|d^3\,{}^2G\rangle$	=	1.13389342	1	0 2 0-1	
$\langle d^3\,{}^4P\|V^{(31)}\|d^3\,{}^2F\rangle$	=	.44721360	1	0 0-1	
$\langle d^3\,{}^4P\|V^{(31)}\|d^3\,{}^2D_1\rangle$	=	-.77459667	-1	0 1-1	
$\langle d^3\,{}^4P\|V^{(31)}\|d^3\,{}^2D_2\rangle$	=	-.84515425	-1	0 0 1-1	
$\langle d^3\,{}^2H\|V^{(31)}\|d^3\,{}^2H\rangle$	=	-1.69115345	-1	-1 0-2 0	1 1
$\langle d^3\,{}^2H\|V^{(31)}\|d^3\,{}^2G\rangle$	=	-.36239003	-1	-3 2-2-2	1 1
$\langle d^3\,{}^2H\|V^{(31)}\|d^3\,{}^2F\rangle$	=	-.44320263	-1	-3 0 0-1	1
$\langle d^3\,{}^2H\|V^{(31)}\|d^3\,{}^2D_1\rangle$	=	1.08562030	1	-2 1 0-1	1
$\langle d^3\,{}^2H\|V^{(31)}\|d^3\,{}^2D_2\rangle$	=	.71070531	1	-2 2 0-2	1
$\langle d^3\,{}^2G\|V^{(31)}\|d^3\,{}^2G\rangle$	=	.52225961	1	-3 5-2-2	1
$\langle d^3\,{}^2G\|V^{(31)}\|d^3\,{}^2F\rangle$	=	-1.32960789	-1	-3 2 0-1	1
$\langle d^3\,{}^2G\|V^{(31)}\|d^3\,{}^2D_1\rangle$	=	.98198051	1	-2 3 0-1	
$\langle d^3\,{}^2G\|V^{(31)}\|d^3\,{}^2D_2\rangle$	=	-.42857143	-1	0 2 0-2	
$\langle d^3\,{}^2G\|V^{(31)}\|d^3\,{}^2P\rangle$	=	.15152288	1	-3 2 0-2	
$\langle d^3\,{}^2F\|V^{(31)}\|d^3\,{}^2F\rangle$	=	.15811388	1	-3 0-1	
$\langle d^3\,{}^2F\|V^{(31)}\|d^3\,{}^2D_1\rangle$	=	-.38729833	-1	-2 1-1	
$\langle d^3\,{}^2F\|V^{(31)}\|d^3\,{}^2D_2\rangle$	=	-.50709255	-1	0 2-1-1	
$\langle d^3\,{}^2F\|V^{(31)}\|d^3\,{}^2P\rangle$	=	-.77689860	-1	-3 0-1-1	0 2
$\langle d^3\,{}^2D_1\|V^{(31)}\|d^3\,{}^2D_1\rangle$	=	.61237244	1	-3 1	
$\langle d^3\,{}^2D_1\|V^{(31)}\|d^3\,{}^2D_2\rangle$	=	.40089186	1	-3 2 0-1	
$\langle d^3\,{}^2D_1\|V^{(31)}\|d^3\,{}^2P\rangle$	=	-.41403934	-1	1 1-1-1	
$\langle d^3\,{}^2D_2\|V^{(31)}\|d^3\,{}^2D_2\rangle$	=	-.78733599	-1	-3 5 0-2	
$\langle d^3\,{}^2D_2\|V^{(31)}\|d^3\,{}^2P\rangle$	=	.36140316	1	5 0-1-2	

TABLE H-1 (cont)

$\langle d^3\,{}^4F\|V^{(41)}\|d^3\,{}^4F\rangle$	=	1.91485422	1	0-1 0 0	1
$\langle d^3\,{}^4F\|V^{(41)}\|d^3\,{}^4P\rangle$	=	.81649658	1	1-1	
$\langle d^3\,{}^4F\|V^{(41)}\|d^3\,{}^2H\rangle$	=	1.65039678	1	1-1-1-1	1 1
$\langle d^3\,{}^4F\|V^{(41)}\|d^3\,{}^2G\rangle$	=	1.58564993	1	3 0-1-1	1
$\langle d^3\,{}^4F\|V^{(41)}\|d^3\,{}^2D_2\rangle$	=	.97590007	1	2-1 1-1	
$\langle d^3\,{}^4F\|V^{(41)}\|d^3\,{}^2P\rangle$	=	.69006556	1	1-1 1-1	
$\langle d^3\,{}^4P\|V^{(41)}\|d^3\,{}^2H\rangle$	=	-1.21106014	-1	1-1-1 0	1
$\langle d^3\,{}^4P\|V^{(41)}\|d^3\,{}^2G\rangle$	=	-.44721360	-1	0 0-1	
$\langle d^3\,{}^4P\|V^{(41)}\|d^3\,{}^2F\rangle$	=	1.29099445	1	0-1 1	
$\langle d^3\,{}^2H\|V^{(41)}\|d^3\,{}^2H\rangle$	=	-1.65039678	-1	1-1-1-1	1 1
$\langle d^3\,{}^2H\|V^{(41)}\|d^3\,{}^2G\rangle$	=	.71464277	1	-3 0-1-1	1 1
$\langle d^3\,{}^2H\|V^{(41)}\|d^3\,{}^2F\rangle$	=	1.23779758	1	-3 1-1-1	1 1
$\langle d^3\,{}^2H\|V^{(41)}\|d^3\,{}^2D_2\rangle$	=	-.72374686	-1	0-1 0-1	1
$\langle d^3\,{}^2H\|V^{(41)}\|d^3\,{}^2P\rangle$	=	-.32366944	-1	0-1-1-1	1
$\langle d^3\,{}^2G\|V^{(41)}\|d^3\,{}^2G\rangle$	=	1.40353053	1	-3 3-1-2	1 1
$\langle d^3\,{}^2G\|V^{(41)}\|d^3\,{}^2F\rangle$	=	-.19820624	-1	-3 0-1-1	1
$\langle d^3\,{}^2G\|V^{(41)}\|d^3\,{}^2D_2\rangle$	=	-.71070531	-1	-2 2 0-2	1
$\langle d^3\,{}^2G\|V^{(41)}\|d^3\,{}^2P\rangle$	=	1.01594432	1	-3 0-1-1	0 0 2
$\langle d^3\,{}^2F\|V^{(41)}\|d^3\,{}^2F\rangle$	=	-.90829511	-1	-3 1-1 0	1
$\langle d^3\,{}^2F\|V^{(41)}\|d^3\,{}^2D_2\rangle$	=	-.73192505	-1	-2 1 1-1	
$\langle d^3\,{}^2F\|V^{(41)}\|d^3\,{}^2P\rangle$	=	.03450328	1	-3-1-1-1	
$\langle d^3\,{}^2D_1\|V^{(41)}\|d^3\,{}^2D_1\rangle$	=	1.22474487	1	-1 1	
$\langle d^3\,{}^2D_2\|V^{(41)}\|d^3\,{}^2D_2\rangle$	=	.64153303	1	-1-1 0-2	2

---

d^9

cfp $\langle d^9\{\!|d^8\rangle$

	denom.	^3F	^3P	^1G	^1D	^1S
^2D	45	21	9	9	5	1

cfgp $\langle d^9\{\!|d^7,d^2(^1S)\rangle$

	denom.	^4F	^4P	^2H	^2G	^2F	$^2D_1$	$^2D_2$	^2P
^2D	45	0	0	0	0	0	1	-0	0

cfgp $\langle d^9\{\!|d^7,d^2(^3P)\rangle$

	denom.	^4F	^4P	^2H	^2G	^2F	$^2D_1$	$^2D_2$	^2P
^2D	1200	-56	-64	0	0	56	-15	-35	14

cfgp $\langle d^9\{\!|d^7,d^2(^1D)\rangle$

	denom.	^4F	^4P	^2H	^2G	^2F	$^2D_1$	$^2D_2$	^2P
^2D	5040	0	0	0	-180	-84	-35	135	126

cfgp $\langle d^9\{\!|d^7,d^2(^3F)\rangle$

	denom.	^4F	^4P	^2H	^2G	^2F	$^2D_1$	$^2D_2$	^2P
^2D	1200	224	-56	-110	90	14	-35	15	-16

cfgp $\langle d^9\{\!|d^7,d^2(^1G)\rangle$

	denom.	^4F	^4P	^2H	^2G	^2F	$^2D_1$	$^2D_2$	^2P
^2D	1680	0	0	154	66	-70	-21	-25	0

TABLE H-1 (cont)

$\langle d^9\,^2D\|U^{(0)}\|d^9\,^2D\rangle$	=	9.00000000	1	0 4
$\langle d^9\,^2D\|U^{(1)}\|d^9\,^2D\rangle$	=	1.00000000	1	0
$\langle d^9\,^2D\|U^{(2)}\|d^9\,^2D\rangle$	=	-1.00000000	-1	0
$\langle d^9\,^2D\|U^{(3)}\|d^9\,^2D\rangle$	=	1.00000000	1	0
$\langle d^9\,^2D\|U^{(4)}\|d^9\,^2D\rangle$	=	-1.00000000	-1	0
$\langle d^9\,^2D\|V^{(01)}\|d^9\,^2D\rangle$	=	1.22474487	1	-1 1
$\langle d^9\,^2D\|V^{(11)}\|d^9\,^2D\rangle$	=	-1.22474487	-1	-1 1
$\langle d^9\,^2D\|V^{(21)}\|d^9\,^2D\rangle$	=	1.22474487	1	-1 1
$\langle d^9\,^2D\|V^{(31)}\|d^9\,^2D\rangle$	=	-1.22474487	-1	-1 1
$\langle d^9\,^2D\|V^{(41)}\|d^9\,^2D\rangle$	=	1.22474487	1	-1 1

---

d^{10}

cfp $\langle d^{10}\{\!| d^9\rangle$

	denom.	^2D
^1S	1	1

cfgp $\langle d^{10}\{\!| d^8,d^2(^1S)\rangle$

	denom.	^3F	^3P	^1G	^1D	^1S
^1S	45	0	0	0	0	1

cfgp $\langle d^{10}\{\!| d^8,d^2(^3P)\rangle$

	denom.	^3F	^3P	^1G	^1D	^1S
^1S	5	0	1	0	0	0

cfgp $\langle d^{10}\{\!| d^8,d^2(^1D)\rangle$

	denom.	^3F	^3P	^1G	^1D	^1S
^1S	9	0	0	0	1	0

cfgp $\langle d^{10}\{\!| d^8,d^2(^3F)\rangle$

	denom.	^3F	^3P	^1G	^1D	^1S
^1S	15	7	0	0	0	0

cfgp $\langle d^{10}\{\!| d^8,d^2(^1G)\rangle$

	denom.	^3F	^3P	^1G	^1D	^1S
^1S	5	0	0	1	0	0

---

f^2

cfp $\langle f^2\{\!| f^1\rangle$

	denom.	^2F
^3H	1	1
^3F	1	1
^3P	1	1
^1I	1	1
^1G	1	1
^1D	1	1
^1S	1	1

TABLE H-1 (cont)

$\langle f^2\,{}^3H \| U^{(0)} \| f^2\,{}^3H \rangle$	=	2.50713268	1	2 0 0-1	1	
$\langle f^2\,{}^3F \| U^{(0)} \| f^2\,{}^3F \rangle$	=	2.00000000	1	2		
$\langle f^2\,{}^3P \| U^{(0)} \| f^2\,{}^3P \rangle$	=	1.30930734	1	2 1 0-1		
$\langle f^2\,{}^1I \| U^{(0)} \| f^2\,{}^1I \rangle$	=	2.72554058	1	2 0 0-1	0 1	
$\langle f^2\,{}^1G \| U^{(0)} \| f^2\,{}^1G \rangle$	=	2.26778684	1	2 2 0-1		
$\langle f^2\,{}^1D \| U^{(0)} \| f^2\,{}^1D \rangle$	=	1.69030851	1	2 0 1-1		
$\langle f^2\,{}^1S \| U^{(0)} \| f^2\,{}^1S \rangle$	=	.75592895	1	2 0 0-1		
$\langle f^2\,{}^3H \| U^{(1)} \| f^2\,{}^3H \rangle$	=	1.98206242	1	-1 0 1-1	1	
$\langle f^2\,{}^3F \| U^{(1)} \| f^2\,{}^3F \rangle$	=	1.00000000	1	0		
$\langle f^2\,{}^3P \| U^{(1)} \| f^2\,{}^3P \rangle$	=	.26726124	1	-1 0 0-1		
$\langle f^2\,{}^1I \| U^{(1)} \| f^2\,{}^1I \rangle$	=	2.54950976	1	-1 0 0 0	0 1	
$\langle f^2\,{}^1G \| U^{(1)} \| f^2\,{}^1G \rangle$	=	1.46385011	1	0 1 1-1		
$\langle f^2\,{}^1D \| U^{(1)} \| f^2\,{}^1D \rangle$	=	.59761430	1	-1 0 1-1		
$\langle f^2\,{}^3H \| U^{(2)} \| f^2\,{}^3H \rangle$	=	1.06532654	1	-1-2 0-1	1 1	
$\langle f^2\,{}^3H \| U^{(2)} \| f^2\,{}^3F \rangle$	=	.83571089	1	2-2 0-1	1	
$\langle f^2\,{}^3F \| U^{(2)} \| f^2\,{}^3F \rangle$	=	-.33333333	-1	0-2		
$\langle f^2\,{}^3F \| U^{(2)} \| f^2\,{}^3P \rangle$	=	.92582010	1	1 1 0-1		
$\langle f^2\,{}^3P \| U^{(2)} \| f^2\,{}^3P \rangle$	=	-.80178373	-1	-1 2 0-1		
$\langle f^2\,{}^1I \| U^{(2)} \| f^2\,{}^1I \rangle$	=	2.21906341	1	-1-1 2 0	-1 1	
$\langle f^2\,{}^1I \| U^{(2)} \| f^2\,{}^1G \rangle$	=	.58108720	1	1 0 0-1	-1 1	
$\langle f^2\,{}^1G \| U^{(2)} \| f^2\,{}^1G \rangle$	=	.22381413	1	0 3 0-2	-1	
$\langle f^2\,{}^1G \| U^{(2)} \| f^2\,{}^1D \rangle$	=	.94760708	1	2 0 0-2	1	
$\langle f^2\,{}^1D \| U^{(2)} \| f^2\,{}^1D \rangle$	=	-.64153303	-1	-1-1 0-2	2	
$\langle f^2\,{}^1D \| U^{(2)} \| f^2\,{}^1S \rangle$	=	.75592895	1	2 0 0-1		
$\langle f^2\,{}^3H \| U^{(3)} \| f^2\,{}^3F \rangle$	=	1.25356634	1	0 0 0-1	1	
$\langle f^2\,{}^3F \| U^{(3)} \| f^2\,{}^3F \rangle$	=	-1.00000000	-1	0		
$\langle f^2\,{}^3F \| U^{(3)} \| f^2\,{}^3P \rangle$	=	.65465367	1	0 1 0-1		
$\langle f^2\,{}^1I \| U^{(3)} \| f^2\,{}^1I \rangle$	=	1.77525073	1	3-1 0 0	-1 1	
$\langle f^2\,{}^1I \| U^{(3)} \| f^2\,{}^1G \rangle$	=	1.08711461	1	0 0 0 0	-1 1	
$\langle f^2\,{}^1G \| U^{(3)} \| f^2\,{}^1G \rangle$	=	-.83410601	-1	0 1 3-2	-1	
$\langle f^2\,{}^1G \| U^{(3)} \| f^2\,{}^1D \rangle$	=	1.05945693	1	0 0 1-2	1	
$\langle f^2\,{}^1D \| U^{(3)} \| f^2\,{}^1D \rangle$	=	-.73771111	-1	4-1 1-2		
$\langle f^2\,{}^3H \| U^{(4)} \| f^2\,{}^3H \rangle$	=	-.90851353	-1	2-2 0-1	0 1	
$\langle f^2\,{}^3H \| U^{(4)} \| f^2\,{}^3F \rangle$	=	1.01574900	1	0-2 1-1	0 1	
$\langle f^2\,{}^3H \| U^{(4)} \| f^2\,{}^3P \rangle$	=	.69006556	1	1-1 1-1		
$\langle f^2\,{}^3F \| U^{(4)} \| f^2\,{}^3F \rangle$	=	-.33333333	-1	0-2		
$\langle f^2\,{}^3F \| U^{(4)} \| f^2\,{}^3P \rangle$	=	-.72374686	-1	0-1 0-1	1	
$\langle f^2\,{}^1I \| U^{(4)} \| f^2\,{}^1I \rangle$	=	1.27416946	1	3-2 0 0	-2 1 1	
$\langle f^2\,{}^1I \| U^{(4)} \| f^2\,{}^1G \rangle$	=	1.43945090	1	0 3 1-1	-2 1	
$\langle f^2\,{}^1I \| U^{(4)} \| f^2\,{}^1D \rangle$	=	.43311683	1	1-2 1-1	-1 1	
$\langle f^2\,{}^1G \| U^{(4)} \| f^2\,{}^1G \rangle$	=	-1.07698285	-1	0 0 0-2	-2 1 0 0	2
$\langle f^2\,{}^1G \| U^{(4)} \| f^2\,{}^1D \rangle$	=	.16682120	1	0 1 1-2	-1	
$\langle f^2\,{}^1G \| U^{(4)} \| f^2\,{}^1S \rangle$	=	.75592895	1	2 0 0-1		
$\langle f^2\,{}^1D \| U^{(4)} \| f^2\,{}^1D \rangle$	=	.70630462	1	2-2 1-2	1	

TABLE H-1 (cont)

$\langle f^2\,{}^3H\|U^{(5)}\|f^2\,{}^3H\rangle$	=	-1.36277029	-1	0 0 0-1	0 1
$\langle f^2\,{}^3H\|U^{(5)}\|f^2\,{}^3P\rangle$	=	1.03509834	1	-1 1 1-1	
$\langle f^2\,{}^3F\|U^{(5)}\|f^2\,{}^3F\rangle$	=	1.00000000	1	0	
$\langle f^2\,{}^1I\|U^{(5)}\|f^2\,{}^1I\rangle$	=	.78026625	1	0-1 0 0	-2 1 1
$\langle f^2\,{}^1I\|U^{(5)}\|f^2\,{}^1G\rangle$	=	1.46586505	1	2 0 1 0	-2 1
$\langle f^2\,{}^1I\|U^{(5)}\|f^2\,{}^1D\rangle$	=	.99239533	1	-1-1 1 0	-1 1
$\langle f^2\,{}^1G\|U^{(5)}\|f^2\,{}^1G\rangle$	=	-.21458067	-1	0 1 0-1	-2 1
$\langle f^2\,{}^1G\|U^{(5)}\|f^2\,{}^1D\rangle$	=	-1.01929438	-1	4 0 1-1	-1
$\langle f^2\,{}^3H\|U^{(6)}\|f^2\,{}^3H\rangle$	=	-1.16155342	-1	0-2 1-1	0 0 1
$\langle f^2\,{}^3H\|U^{(6)}\|f^2\,{}^3F\rangle$	=	-1.24721913	-1	1-2 0 1	
$\langle f^2\,{}^3H\|U^{(6)}\|f^2\,{}^3P\rangle$	=	-.46291005	-1	-1 1 0-1	
$\langle f^2\,{}^3F\|U^{(6)}\|f^2\,{}^3F\rangle$	=	-.33333333	-1	0-2	
$\langle f^2\,{}^1I\|U^{(6)}\|f^2\,{}^1I\rangle$	=	.35653234	1	0-1 0-1	-2 0 1 1
$\langle f^2\,{}^1I\|U^{(6)}\|f^2\,{}^1G\rangle$	=	1.06017307	1	3 0 0 0	-2 0 1
$\langle f^2\,{}^1I\|U^{(6)}\|f^2\,{}^1D\rangle$	=	1.37620471	1	-1-1 3 0	-1
$\langle f^2\,{}^1I\|U^{(6)}\|f^2\,{}^1S\rangle$	=	.75592895	1	2 0 0-1	
$\langle f^2\,{}^1G\|U^{(6)}\|f^2\,{}^1G\rangle$	=	1.19769554	1	0 5 1-1	-2
$\langle f^2\,{}^1G\|U^{(6)}\|f^2\,{}^1D\rangle$	=	.36037499	1	1 0 1-1	-1
$\langle f^2\,{}^3H\|V^{(01)}\|f^2\,{}^3H\rangle$	=	3.07059789	1	1 1 0-1	1
$\langle f^2\,{}^3F\|V^{(01)}\|f^2\,{}^3F\rangle$	=	2.44948974	1	1 1	
$\langle f^2\,{}^3P\|V^{(01)}\|f^2\,{}^3P\rangle$	=	1.60356745	1	1 2 0-1	
$\langle f^2\,{}^3H\|V^{(11)}\|f^2\,{}^3H\rangle$	=	2.42752078	1	-2 1 1-1	1
$\langle f^2\,{}^3H\|V^{(11)}\|f^2\,{}^1I\rangle$	=	-.83452296	-1	-3 1 0-1	0 1
$\langle f^2\,{}^3H\|V^{(11)}\|f^2\,{}^1G\rangle$	=	1.03509834	1	-1 1 1-1	
$\langle f^2\,{}^3F\|V^{(11)}\|f^2\,{}^3F\rangle$	=	1.22474487	1	-1 1	
$\langle f^2\,{}^3F\|V^{(11)}\|f^2\,{}^1G\rangle$	=	-1.08562030	-1	-2 1 0-1	1
$\langle f^2\,{}^3F\|V^{(11)}\|f^2\,{}^1D\rangle$	=	1.03509834	1	-1 1 1-1	
$\langle f^2\,{}^3P\|V^{(11)}\|f^2\,{}^3P\rangle$	=	.32732684	1	-2 1 0-1	
$\langle f^2\,{}^3P\|V^{(11)}\|f^2\,{}^1D\rangle$	=	-.89642146	-1	-3 2 1-1	
$\langle f^2\,{}^3P\|V^{(11)}\|f^2\,{}^1S\rangle$	=	.65465367	1	0 1 0-1	
$\langle f^2\,{}^3H\|V^{(21)}\|f^2\,{}^3H\rangle$	=	1.30475322	1	-2-1 0-1	1 1
$\langle f^2\,{}^3H\|V^{(21)}\|f^2\,{}^3F\rangle$	=	1.02353263	1	1-1 0-1	1
$\langle f^2\,{}^3H\|V^{(21)}\|f^2\,{}^1I\rangle$	=	-1.27475488	-1	-3 0 0 0	0 1
$\langle f^2\,{}^3H\|V^{(21)}\|f^2\,{}^1G\rangle$	=	1.30930734	1	2 1 0-1	
$\langle f^2\,{}^3F\|V^{(21)}\|f^2\,{}^3F\rangle$	=	-.40824829	-1	-1-1	
$\langle f^2\,{}^3F\|V^{(21)}\|f^2\,{}^3P\rangle$	=	1.13389342	1	0 2 0-1	
$\langle f^2\,{}^3F\|V^{(21)}\|f^2\,{}^1G\rangle$	=	-1.08562030	-1	-2 1 0-1	1
$\langle f^2\,{}^3F\|V^{(21)}\|f^2\,{}^1D\rangle$	=	.75592895	1	2 0 0-1	
$\langle f^2\,{}^3P\|V^{(21)}\|f^2\,{}^3P\rangle$	=	-.98198051	-1	-2 3 0-1	
$\langle f^2\,{}^3P\|V^{(21)}\|f^2\,{}^1D\rangle$	=	-.40089186	-1	-3 2 0-1	
$\langle f^2\,{}^3H\|V^{(31)}\|f^2\,{}^3F\rangle$	=	1.53529895	1	-1 1 0-1	1
$\langle f^2\,{}^3H\|V^{(31)}\|f^2\,{}^1I\rangle$	=	-1.47196014	-1	-1-1 0 0	0 1
$\langle f^2\,{}^3H\|V^{(31)}\|f^2\,{}^1G\rangle$	=	.99744572	1	-2 1 1-2	0 1

TABLE H-1 (cont)

$\langle f^2\,{}^3H\|V^{(31)}\|f^2\,{}^1D\rangle$	=	.61167774	1	0-1 1-2	1
$\langle f^2\,{}^3F\|V^{(31)}\|f^2\,{}^3F\rangle$	=	-1.22474487	-1	-1 1	
$\langle f^2\,{}^3F\|V^{(31)}\|f^2\,{}^3P\rangle$	=	.80178373	1	-1 2 0-1	
$\langle f^2\,{}^3F\|V^{(31)}\|f^2\,{}^1I\rangle$	=	-.39339790	-1	-2-1 0-1	0 1
$\langle f^2\,{}^3F\|V^{(31)}\|f^2\,{}^1G\rangle$	=	-.32732684	-1	-2 1 0-1	
$\langle f^2\,{}^3F\|V^{(31)}\|f^2\,{}^1D\rangle$	=	-.24397502	-1	-2-1 1-1	
$\langle f^2\,{}^3F\|V^{(31)}\|f^2\,{}^1S\rangle$	=	.65465367	1	0 1 0-1	
$\langle f^2\,{}^3P\|V^{(31)}\|f^2\,{}^1G\rangle$	=	-.71070531	-1	-2 2 0-2	1
$\langle f^2\,{}^3P\|V^{(31)}\|f^2\,{}^1D\rangle$	=	.67763093	1	-1 2 1-2	
$\langle f^2\,{}^3H\|V^{(41)}\|f^2\,{}^3H\rangle$	=	-1.11269728	-1	1-1 0-1	0 1
$\langle f^2\,{}^3H\|V^{(41)}\|f^2\,{}^3F\rangle$	=	1.24403338	1	-1-1 1-1	0 1
$\langle f^2\,{}^3H\|V^{(41)}\|f^2\,{}^3P\rangle$	=	.84515425	1	0 0 1-1	
$\langle f^2\,{}^3H\|V^{(41)}\|f^2\,{}^1I\rangle$	=	-1.40345893	-1	0-1 1 0	-1 1
$\langle f^2\,{}^3H\|V^{(41)}\|f^2\,{}^1G\rangle$	=	.20544535	1	-2 0 0-1	-1 1
$\langle f^2\,{}^3H\|V^{(41)}\|f^2\,{}^1D\rangle$	=	.97590007	1	2-1 1-1	
$\langle f^2\,{}^3F\|V^{(41)}\|f^2\,{}^3F\rangle$	=	-.40824829	-1	-1-1	
$\langle f^2\,{}^3F\|V^{(41)}\|f^2\,{}^3P\rangle$	=	-.88640526	-1	-1 0 0-1	1
$\langle f^2\,{}^3F\|V^{(41)}\|f^2\,{}^1I\rangle$	=	-.83029750	-1	-2-1 0 1	-1 1
$\langle f^2\,{}^3F\|V^{(41)}\|f^2\,{}^1G\rangle$	=	.63705899	1	-2 0 3-1	-1
$\langle f^2\,{}^3F\|V^{(41)}\|f^2\,{}^1D\rangle$	=	-.80917359	-1	-2-1 1-1	1
$\langle f^2\,{}^3P\|V^{(41)}\|f^2\,{}^1G\rangle$	=	-.73192505	-1	-2 1 1-1	
$\langle f^2\,{}^3H\|V^{(51)}\|f^2\,{}^3H\rangle$	=	-1.66904592	-1	-1 1 0-1	0 1
$\langle f^2\,{}^3H\|V^{(51)}\|f^2\,{}^3P\rangle$	=	1.26773138	1	-2 2 1-1	
$\langle f^2\,{}^3H\|V^{(51)}\|f^2\,{}^1I\rangle$	=	-1.09356639	-1	-2-1 1-1	-1 1 1
$\langle f^2\,{}^3H\|V^{(51)}\|f^2\,{}^1G\rangle$	=	-.71168357	-1	0 1 0-1	-1 1
$\langle f^2\,{}^3H\|V^{(51)}\|f^2\,{}^1D\rangle$	=	.62201669	1	-3-1 1-1	0 1
$\langle f^2\,{}^3H\|V^{(51)}\|f^2\,{}^1S\rangle$	=	.65465367	1	0 1 0-1	
$\langle f^2\,{}^3F\|V^{(51)}\|f^2\,{}^3F\rangle$	=	1.22474487	1	-1 1	
$\langle f^2\,{}^3F\|V^{(51)}\|f^2\,{}^1I\rangle$	=	-1.17421799	-1	-1-1 0 1	-1 1
$\langle f^2\,{}^3F\|V^{(51)}\|f^2\,{}^1G\rangle$	=	.79568642	1	-2 1 1-1	-1 1
$\langle f^2\,{}^3F\|V^{(51)}\|f^2\,{}^1D\rangle$	=	.48795004	1	0-1 1-1	
$\langle f^2\,{}^3P\|V^{(51)}\|f^2\,{}^1I\rangle$	=	-.43581540	-1	-3 2 0-1	-1 1
$\langle f^2\,{}^3P\|V^{(51)}\|f^2\,{}^1G\rangle$	=	.54056248	1	-1 2 1-1	-1
$\langle f^2\,{}^3H\|V^{(61)}\|f^2\,{}^3H\rangle$	=	-1.42260659	-1	-1-1 1-1	0 0 1
$\langle f^2\,{}^3H\|V^{(61)}\|f^2\,{}^3F\rangle$	=	-1.52752523	-1	0-1 0 1	
$\langle f^2\,{}^3H\|V^{(61)}\|f^2\,{}^3P\rangle$	=	-.56694671	-1	-2 2 0-1	
$\langle f^2\,{}^3H\|V^{(61)}\|f^2\,{}^1I\rangle$	=	-.62158156	-1	-2 0 0 0	-1 0 1
$\langle f^2\,{}^3H\|V^{(61)}\|f^2\,{}^1G\rangle$	=	-1.16774842	-1	0 1 1 0	-1
$\langle f^2\,{}^3H\|V^{(61)}\|f^2\,{}^1D\rangle$	=	-.79056942	-1	-3 0 1	
$\langle f^2\,{}^3F\|V^{(61)}\|f^2\,{}^3F\rangle$	=	-.40824829	-1	-1-1	
$\langle f^2\,{}^3F\|V^{(61)}\|f^2\,{}^1I\rangle$	=	-1.12815215	-1	1 0 0 1	-1
$\langle f^2\,{}^3F\|V^{(61)}\|f^2\,{}^1G\rangle$	=	-.69084928	-1	-2 1 0 1	-1
$\langle f^2\,{}^3P\|V^{(61)}\|f^2\,{}^1I\rangle$	=	-1.06066017	-1	-3 2	

Appendix I

Relative Line Strengths Within an LS Multiplet

The dependence on J and J′ of electric-dipole-transition matrix elements in an LS representation is given by the line factor

$$D_{line}(LSJ - L'S'J') = \delta_{SS'}(-1)^{L+S+J'+1}[J,J']^{1/2}\begin{Bmatrix} L & S & J \\ J' & 1 & L' \end{Bmatrix} . \qquad (14.47)$$

From the symmetry properties of the 6-j symbol it follows that

$$D_{line}(L'S'J' - LSJ) = (-1)^{L'-L+J-J'} D_{line}(LSJ - L'S'J') . \qquad (I.1)$$

From the recoupling relation (9.8), corresponding line factors for LS and SL coupling differ by this same phase factor:

$$D_{line}(SLJ - S'L'J') = (-1)^{L'-L+J-J'} D_{line}(LSJ - L'S'J') . \qquad (I.2)$$

Under conditions of pure LS coupling, relative line strengths within a multiplet are proportional to

$$(D_{line})^2 = \delta_{SS'}[J,J']\begin{Bmatrix} L & S & J \\ J' & 1 & L' \end{Bmatrix}^2 ; \qquad (14.50)$$

the sum of these quantities over all lines of an LS multiplet satisfies the sum rule

$$\sum_{JJ'}(D_{line})^2 = 2S + 1 . \qquad (14.53)$$

The following table gives values of D_{line} (as a decimal fraction) and of D^2_{line} (as both a decimal and a rational fraction) for LS coupling. Only cases $L \leqq L'$, with $J \leqq J'$ if $L = L'$, are tabulated. Other cases may be obtained by interchanging LSJ with L′S′J′, and changing the sign of D_{line} if the value in the table is followed by an asterisk. From (I.1) and (I.2), the values marked by an asterisk are also those that change sign in switching from LS to SL coupling.

The table also gives relative values of D^2_{line} as percentages of the strongest line of the multiplet.

TABLE I-1. LINE STRENGTH MATRIX ELEMENTS

Matrix Element	D_{line}	$(D_{line})^2$			Percent
$\langle {}^2S_{1/2} \| r^{(1)} \| {}^2P_{1/2} \rangle$	.81649658*	.66666667	=	2/3	50.00
$\langle {}^2S_{1/2} \| r^{(1)} \| {}^2P_{3/2} \rangle$	1.15470054	1.33333333	=	4/3	100.00
$\langle {}^2P_{1/2} \| r^{(1)} \| {}^2P_{1/2} \rangle$	.66666667	.44444444	=	4/9	40.00
$\langle {}^2P_{1/2} \| r^{(1)} \| {}^2P_{3/2} \rangle$	-.47140452*	.22222222	=	2/9	20.00
$\langle {}^2P_{3/2} \| r^{(1)} \| {}^2P_{3/2} \rangle$	1.05409255	1.11111111	=	10/9	100.00
$\langle {}^2P_{1/2} \| r^{(1)} \| {}^2D_{3/2} \rangle$	.81649658	.66666667	=	10/15	55.56
$\langle {}^2P_{3/2} \| r^{(1)} \| {}^2D_{3/2} \rangle$	.36514837*	.13333333	=	2/15	11.11
$\langle {}^2P_{3/2} \| r^{(1)} \| {}^2D_{5/2} \rangle$	1.09544512	1.20000000	=	18/15	100.00
$\langle {}^2D_{3/2} \| r^{(1)} \| {}^2D_{3/2} \rangle$	.84852814	.72000000	=	18/25	64.29
$\langle {}^2D_{3/2} \| r^{(1)} \| {}^2D_{5/2} \rangle$	-.28284271*	.08000000	=	2/25	7.14
$\langle {}^2D_{5/2} \| r^{(1)} \| {}^2D_{5/2} \rangle$	1.05830052	1.12000000	=	28/25	100.00
$\langle {}^2D_{3/2} \| r^{(1)} \| {}^2F_{5/2} \rangle$	.89442719	.80000000	=	28/35	70.00
$\langle {}^2D_{5/2} \| r^{(1)} \| {}^2F_{5/2} \rangle$	.23904572*	.05714286	=	2/35	5.00
$\langle {}^2D_{5/2} \| r^{(1)} \| {}^2F_{7/2} \rangle$	1.06904497	1.14285714	=	40/35	100.00
$\langle {}^2F_{5/2} \| r^{(1)} \| {}^2F_{5/2} \rangle$	.90350790	.81632653	=	40/49	74.07
$\langle {}^2F_{5/2} \| r^{(1)} \| {}^2F_{7/2} \rangle$	-.20203051*	.04081633	=	2/49	3.70
$\langle {}^2F_{7/2} \| r^{(1)} \| {}^2F_{7/2} \rangle$	1.04978132	1.10204082	=	54/49	100.00
$\langle {}^2F_{5/2} \| r^{(1)} \| {}^2G_{7/2} \rangle$	.92582010	.85714286	=	54/63	77.14
$\langle {}^2F_{7/2} \| r^{(1)} \| {}^2G_{7/2} \rangle$	.17817416*	.03174603	=	2/63	2.86
$\langle {}^2F_{7/2} \| r^{(1)} \| {}^2G_{9/2} \rangle$	1.05409255	1.11111111	=	70/63	100.00
$\langle {}^2G_{7/2} \| r^{(1)} \| {}^2G_{7/2} \rangle$	.92962225	.86419753	=	70/81	79.55
$\langle {}^2G_{7/2} \| r^{(1)} \| {}^2G_{9/2} \rangle$	-.15713484*	.02469136	=	2/81	2.27
$\langle {}^2G_{9/2} \| r^{(1)} \| {}^2G_{9/2} \rangle$	1.04231461	1.08641975	=	88/81	100.00
$\langle {}^2G_{7/2} \| r^{(1)} \| {}^2H_{9/2} \rangle$	.94280904	.88888889	=	88/99	81.48
$\langle {}^2G_{9/2} \| r^{(1)} \| {}^2H_{9/2} \rangle$	.14213381*	.02020202	=	2/99	1.85
$\langle {}^2G_{9/2} \| r^{(1)} \| {}^2H_{11/2} \rangle$	1.04446594	1.09090909	=	108/99	100.00
$\langle {}^3S_1 \| r^{(1)} \| {}^3P_0 \rangle$	.57735027	.33333333	=	1/3	20.00
$\langle {}^3S_1 \| r^{(1)} \| {}^3P_1 \rangle$	1.00000000*	1.00000000	=	3/3	60.00
$\langle {}^3S_1 \| r^{(1)} \| {}^3P_2 \rangle$	1.29099445	1.66666667	=	5/3	100.00
$\langle {}^3P_0 \| r^{(1)} \| {}^3P_1 \rangle$	-.57735027*	.33333333	=	4/12	26.67
$\langle {}^3P_1 \| r^{(1)} \| {}^3P_1 \rangle$	.50000000	.25000000	=	3/12	20.00
$\langle {}^3P_1 \| r^{(1)} \| {}^3P_2 \rangle$	-.64549722*	.41666667	=	5/12	33.33
$\langle {}^3P_2 \| r^{(1)} \| {}^3P_2 \rangle$	1.11803399	1.25000000	=	15/12	100.00
$\langle {}^3P_0 \| r^{(1)} \| {}^3D_1 \rangle$	.57735027	.33333333	=	20/60	23.81
$\langle {}^3P_1 \| r^{(1)} \| {}^3D_1 \rangle$	.50000000*	.25000000	=	15/60	17.86
$\langle {}^3P_1 \| r^{(1)} \| {}^3D_2 \rangle$	.86602540	.75000000	=	45/60	53.57
$\langle {}^3P_2 \| r^{(1)} \| {}^3D_1 \rangle$	.12909944	.01666667	=	1/60	1.19
$\langle {}^3P_2 \| r^{(1)} \| {}^3D_2 \rangle$	.50000000*	.25000000	=	15/60	17.86
$\langle {}^3P_2 \| r^{(1)} \| {}^3D_3 \rangle$	1.18321596	1.40000000	=	84/60	100.00

TABLE I-1 (cont)

Matrix Element	D_{line}	$(D_{line})^2$			Percent
$\langle ^3D_1\|r^{(1)}\|^3D_1\rangle$	.67082039	.45000000	=	81/180	36.16
$\langle ^3D_1\|r^{(1)}\|^3D_2\rangle$	-.38729833*	.15000000	=	27/180	12.05
$\langle ^3D_2\|r^{(1)}\|^3D_2\rangle$	.83333333	.69444444	=	125/180	55.80
$\langle ^3D_2\|r^{(1)}\|^3D_3\rangle$	-.39440532*	.15555556	=	28/180	12.50
$\langle ^3D_3\|r^{(1)}\|^3D_3\rangle$	1.11554670	1.24444444	=	224/180	100.00
$\langle ^3D_1\|r^{(1)}\|^3F_2\rangle$	.77459667	.60000000	=	189/315	46.67
$\langle ^3D_2\|r^{(1)}\|^3F_2\rangle$	.33333333*	.11111111	=	35/315	8.64
$\langle ^3D_2\|r^{(1)}\|^3F_3\rangle$	.94280904	.88888889	=	280/315	69.14
$\langle ^3D_3\|r^{(1)}\|^3F_2\rangle$	.05634362	.00317460	=	1/315	.25
$\langle ^3D_3\|r^{(1)}\|^3F_3\rangle$	.33333333*	.11111111	=	35/315	8.64
$\langle ^3D_3\|r^{(1)}\|^3F_4\rangle$	1.13389342	1.28571429	=	405/315	100.00
$\langle ^3F_2\|r^{(1)}\|^3F_2\rangle$	.79681907	.63492063	=	640/1008	52.67
$\langle ^3F_2\|r^{(1)}\|^3F_3\rangle$	-.28171808*	.07936508	=	80/1008	6.58
$\langle ^3F_3\|r^{(1)}\|^3F_3\rangle$	.91666667	.84027778	=	847/1008	69.71
$\langle ^3F_3\|r^{(1)}\|^3F_4\rangle$	-.28347335*	.08035714	=	81/1008	6.67
$\langle ^3F_4\|r^{(1)}\|^3F_4\rangle$	1.09788758	1.20535714	=	1215/1008	100.00
$\langle ^3F_2\|r^{(1)}\|^3G_3\rangle$	.84515425	.71428571	=	720/1008	58.44
$\langle ^3F_3\|r^{(1)}\|^3G_3\rangle$	.25000000*	.06250000	=	63/1008	5.11
$\langle ^3F_3\|r^{(1)}\|^3G_4\rangle$	.96824584	.93750000	=	945/1008	76.70
$\langle ^3F_4\|r^{(1)}\|^3G_3\rangle$	.03149704	.00099206	=	1/1008	.08
$\langle ^3F_4\|r^{(1)}\|^3G_4\rangle$	.25000000*	.06250000	=	63/1008	5.11
$\langle ^3F_4\|r^{(1)}\|^3G_5\rangle$	1.10554160	1.22222222	=	1232/1008	100.00
$\langle ^3G_3\|r^{(1)}\|^3G_3\rangle$	.85391256	.72916667	=	2625/3600	62.14
$\langle ^3G_3\|r^{(1)}\|^3G_4\rangle$	-.22047928*	.04861111	=	175/3600	4.14
$\langle ^3G_4\|r^{(1)}\|^3G_4\rangle$	.95000000	.90250000	=	3249/3600	76.92
$\langle ^3G_4\|r^{(1)}\|^3G_5\rangle$	-.22110832*	.04888889	=	176/3600	4.17
$\langle ^3G_5\|r^{(1)}\|^3G_5\rangle$	1.08320512	1.17333333	=	4224/3600	100.00
$\langle ^3G_3\|r^{(1)}\|^3H_4\rangle$	.88191710	.77777778	=	1925/2475	65.81
$\langle ^3G_4\|r^{(1)}\|^3H_4\rangle$	.20000000*	.04000000	=	99/2475	3.38
$\langle ^3G_4\|r^{(1)}\|^3H_5\rangle$	.97979590	.96000000	=	2376/2475	81.23
$\langle ^3G_5\|r^{(1)}\|^3H_4\rangle$	.02010076	.00040404	=	1/2475	.03
$\langle ^3G_5\|r^{(1)}\|^3H_5\rangle$	.20000000*	.04000000	=	99/2475	3.38
$\langle ^3G_5\|r^{(1)}\|^3H_6\rangle$	1.08711461	1.18181818	=	2925/2475	100.00
$\langle ^4S_{3/2}\|r^{(1)}\|^4P_{1/2}\rangle$	.81649658	.66666667	=	2/3	33.33
$\langle ^4S_{3/2}\|r^{(1)}\|^4P_{3/2}\rangle$	1.15470054*	1.33333333	=	4/3	66.67
$\langle ^4S_{3/2}\|r^{(1)}\|^4P_{5/2}\rangle$	1.41421356	2.00000000	=	6/3	100.00
$\langle ^4P_{1/2}\|r^{(1)}\|^4P_{1/2}\rangle$	-.33333333	.11111111	=	5/45	7.94
$\langle ^4P_{1/2}\|r^{(1)}\|^4P_{3/2}\rangle$	-.74535599*	.55555556	=	25/45	39.68
$\langle ^4P_{3/2}\|r^{(1)}\|^4P_{3/2}\rangle$	.42163702	.17777778	=	8/45	12.70
$\langle ^4P_{3/2}\|r^{(1)}\|^4P_{5/2}\rangle$	-.77459667*	.60000000	=	27/45	42.86
$\langle ^4P_{5/2}\|r^{(1)}\|^4P_{5/2}\rangle$	1.18321596	1.40000000	=	63/45	100.00

Matrix Element	D_{line}	$(D_{line})^2$			Percent
$\langle {}^4P_{1/2} \| r^{(1)} \| {}^4D_{1/2} \rangle$	.57735027*	.33333333	=	25/75	20.83
$\langle {}^4P_{1/2} \| r^{(1)} \| {}^4D_{3/2} \rangle$	.57735027	.33333333	=	25/75	20.83
$\langle {}^4P_{3/2} \| r^{(1)} \| {}^4D_{1/2} \rangle$	.25819889	.06666667	=	5/75	4.17
$\langle {}^4P_{3/2} \| r^{(1)} \| {}^4D_{3/2} \rangle$	.65319726*	.42666667	=	32/75	26.67
$\langle {}^4P_{3/2} \| r^{(1)} \| {}^4D_{5/2} \rangle$	.91651514	.84000000	=	63/75	52.50
$\langle {}^4P_{5/2} \| r^{(1)} \| {}^4D_{3/2} \rangle$	.20000000	.04000000	=	3/75	2.50
$\langle {}^4P_{5/2} \| r^{(1)} \| {}^4D_{5/2} \rangle$	.60000000*	.36000000	=	27/75	22.50
$\langle {}^4P_{5/2} \| r^{(1)} \| {}^4D_{7/2} \rangle$	1.26491106	1.60000000	=	120/75	100.00
$\langle {}^4D_{1/2} \| r^{(1)} \| {}^4D_{1/2} \rangle$	.44721360	.20000000	=	35/175	14.58
$\langle {}^4D_{1/2} \| r^{(1)} \| {}^4D_{3/2} \rangle$	-.44721360*	.20000000	=	35/175	14.58
$\langle {}^4D_{3/2} \| r^{(1)} \| {}^4D_{3/2} \rangle$	.56568542	.32000000	=	56/175	23.33
$\langle {}^4D_{3/2} \| r^{(1)} \| {}^4D_{5/2} \rangle$	-.52915026*	.28000000	=	49/175	20.42
$\langle {}^4D_{5/2} \| r^{(1)} \| {}^4D_{5/2} \rangle$	.83152184	.69142857	=	121/175	50.42
$\langle {}^4D_{5/2} \| r^{(1)} \| {}^4D_{7/2} \rangle$	-.47809144*	.22857143	=	40/175	16.67
$\langle {}^4D_{7/2} \| r^{(1)} \| {}^4D_{7/2} \rangle$	1.17108009	1.37142857	=	240/175	100.00
$\langle {}^4D_{1/2} \| r^{(1)} \| {}^4F_{3/2} \rangle$	.63245553	.40000000	=	490/1225	28.00
$\langle {}^4D_{3/2} \| r^{(1)} \| {}^4F_{3/2} \rangle$	.40000000*	.16000000	=	196/1225	11.20
$\langle {}^4D_{3/2} \| r^{(1)} \| {}^4F_{5/2} \rangle$	.80000000	.64000000	=	784/1225	44.80
$\langle {}^4D_{5/2} \| r^{(1)} \| {}^4F_{3/2} \rangle$	.10690450	.01142857	=	14/1225	.80
$\langle {}^4D_{5/2} \| r^{(1)} \| {}^4F_{5/2} \rangle$	.45714286*	.20897959	=	256/1225	14.63
$\langle {}^4D_{5/2} \| r^{(1)} \| {}^4F_{7/2} \rangle$	.98974332	.97959184	=	1200/1225	68.57
$\langle {}^4D_{7/2} \| r^{(1)} \| {}^4F_{5/2} \rangle$	.09035079	.00816327	=	10/1225	.57
$\langle {}^4D_{7/2} \| r^{(1)} \| {}^4F_{7/2} \rangle$	.40406102*	.16326531	=	200/1225	11.43
$\langle {}^4D_{7/2} \| r^{(1)} \| {}^4F_{9/2} \rangle$	1.19522861	1.42857143	=	1750/1225	100.00
$\langle {}^4F_{3/2} \| r^{(1)} \| {}^4F_{3/2} \rangle$	.67612340	.45714286	=	672/1470	34.91
$\langle {}^4F_{3/2} \| r^{(1)} \| {}^4F_{5/2} \rangle$	-.33806170*	.11428571	=	168/1470	8.73
$\langle {}^4F_{5/2} \| r^{(1)} \| {}^4F_{5/2} \rangle$	.76798172	.58979592	=	867/1470	45.04
$\langle {}^4F_{5/2} \| r^{(1)} \| {}^4F_{7/2} \rangle$	-.39123040*	.15306122	=	225/1470	11.69
$\langle {}^4F_{7/2} \| r^{(1)} \| {}^4F_{7/2} \rangle$	.93313895	.87074830	=	1280/1470	66.49
$\langle {}^4F_{7/2} \| r^{(1)} \| {}^4F_{9/2} \rangle$	-.34503278*	.11904762	=	175/1470	9.09
$\langle {}^4F_{9/2} \| r^{(1)} \| {}^4F_{9/2} \rangle$	1.14434427	1.30952381	=	1925/1470	100.00
$\langle {}^4F_{3/2} \| r^{(1)} \| {}^4G_{5/2} \rangle$	.75592895	.57142857	=	1512/2646	42.86
$\langle {}^4F_{5/2} \| r^{(1)} \| {}^4G_{5/2} \rangle$	.30304576*	.09183673	=	243/2646	6.89
$\langle {}^4F_{5/2} \| r^{(1)} \| {}^4G_{7/2} \rangle$	.87481777	.76530612	=	2025/2646	57.40
$\langle {}^4F_{7/2} \| r^{(1)} \| {}^4G_{5/2} \rangle$	.05832118	.00340136	=	9/2646	.26
$\langle {}^4F_{7/2} \| r^{(1)} \| {}^4G_{7/2} \rangle$	.34776035*	.12093726	=	320/2646	9.07
$\langle {}^4F_{7/2} \| r^{(1)} \| {}^4G_{9/2} \rangle$	1.00921678	1.01851852	=	2695/2646	76.39
$\langle {}^4F_{9/2} \| r^{(1)} \| {}^4G_{7/2} \rangle$	.05143445	.00264550	=	7/2646	.20
$\langle {}^4F_{9/2} \| r^{(1)} \| {}^4G_{9/2} \rangle$	.30429031*	.09259259	=	245/2646	6.94
$\langle {}^4F_{9/2} \| r^{(1)} \| {}^4G_{11/2} \rangle$	1.15470054	1.33333333	=	3528/2646	100.00
$\langle {}^4G_{5/2} \| r^{(1)} \| {}^4G_{5/2} \rangle$	.77151675	.59523810	=	37125/62370	47.22
$\langle {}^4G_{5/2} \| r^{(1)} \| {}^4G_{7/2} \rangle$	-.26726124*	.07142857	=	4455/62370	5.67
$\langle {}^4G_{7/2} \| r^{(1)} \| {}^4G_{7/2} \rangle$	.84994034	.72239859	=	45056/62370	57.31
$\langle {}^4G_{7/2} \| r^{(1)} \| {}^4G_{9/2} \rangle$	-.30832082*	.09506173	=	5929/62370	7.54
$\langle {}^4G_{9/2} \| r^{(1)} \| {}^4G_{9/2} \rangle$	.97124771	.94332211	=	58835/62370	74.83
$\langle {}^4G_{9/2} \| r^{(1)} \| {}^4G_{11/2} \rangle$	-.26967994*	.07272727	=	4536/62370	5.77
$\langle {}^4G_{11/2} \| r^{(1)} \| {}^4G_{11/2} \rangle$	1.12276714	1.26060606	=	78624/62370	100.00

TABLE I-1 (cont)

Matrix Element	D_{line}	$(D_{line})^2$			Percent
$\langle {}^4G_{5/2} \Vert r^{(1)} \Vert {}^4H_{7/2} \rangle$	.81649658	.66666667	=	10890/16335	52.38
$\langle {}^4G_{7/2} \Vert r^{(1)} \Vert {}^4H_{7/2} \rangle$	.24343225*	.05925926	=	968/16335	4.66
$\langle {}^4G_{7/2} \Vert r^{(1)} \Vert {}^4H_{9/2} \rangle$	.91084007	.82962963	=	13552/16335	65.19
$\langle {}^4G_{9/2} \Vert r^{(1)} \Vert {}^4H_{7/2} \rangle$	.03669879	.00134680	=	22/16335	.11
$\langle {}^4G_{9/2} \Vert r^{(1)} \Vert {}^4H_{9/2} \rangle$	.27992740*	.07835935	=	1280/16335	6.16
$\langle {}^4G_{9/2} \Vert r^{(1)} \Vert {}^4H_{11/2} \rangle$	1.01558109	1.03140496	=	16848/16335	81.04
$\langle {}^4G_{11/2} \Vert r^{(1)} \Vert {}^4H_{9/2} \rangle$	.03319531	.00110193	=	18/16335	.09
$\langle {}^4G_{11/2} \Vert r^{(1)} \Vert {}^4H_{11/2} \rangle$	.24393469*	.05950413	=	972/16335	4.68
$\langle {}^4G_{11/2} \Vert r^{(1)} \Vert {}^4H_{13/2} \rangle$	1.12815215	1.27272727	=	20790/16335	100.00
$\langle {}^5S_2 \Vert r^{(1)} \Vert {}^5P_1 \rangle$	1.00000000	1.00000000	=	3/3	42.86
$\langle {}^5S_2 \Vert r^{(1)} \Vert {}^5P_2 \rangle$	1.29099445*	1.66666667	=	5/3	71.43
$\langle {}^5S_2 \Vert r^{(1)} \Vert {}^5P_3 \rangle$	1.52752523	2.33333333	=	7/3	100.00
$\langle {}^5P_1 \Vert r^{(1)} \Vert {}^5P_1 \rangle$	-.50000000	.25000000	=	9/36	16.07
$\langle {}^5P_1 \Vert r^{(1)} \Vert {}^5P_2 \rangle$	-.86602540*	.75000000	=	27/36	48.21
$\langle {}^5P_2 \Vert r^{(1)} \Vert {}^5P_2 \rangle$	.37267800	.13888889	=	5/36	8.93
$\langle {}^5P_2 \Vert r^{(1)} \Vert {}^5P_3 \rangle$	-.88191710*	.77777778	=	28/36	50.00
$\langle {}^5P_3 \Vert r^{(1)} \Vert {}^5P_3 \rangle$	1.24721913	1.55555556	=	56/36	100.00
$\langle {}^5P_1 \Vert r^{(1)} \Vert {}^5D_0 \rangle$	.44721360	.20000000	=	12/60	11.11
$\langle {}^5P_1 \Vert r^{(1)} \Vert {}^5D_1 \rangle$	.67082039*	.45000000	=	27/60	25.00
$\langle {}^5P_1 \Vert r^{(1)} \Vert {}^5D_2 \rangle$	.59160798	.35000000	=	21/60	19.44
$\langle {}^5P_2 \Vert r^{(1)} \Vert {}^5D_1 \rangle$	.38729833	.15000000	=	9/60	8.33
$\langle {}^5P_2 \Vert r^{(1)} \Vert {}^5D_2 \rangle$	.76376262*	.58333333	=	35/60	32.41
$\langle {}^5P_2 \Vert r^{(1)} \Vert {}^5D_3 \rangle$	.96609178	.93333333	=	56/60	51.85
$\langle {}^5P_3 \Vert r^{(1)} \Vert {}^5D_2 \rangle$	.25819889	.06666667	=	4/60	3.70
$\langle {}^5P_3 \Vert r^{(1)} \Vert {}^5D_3 \rangle$	.68313005*	.46666667	=	28/60	25.93
$\langle {}^5P_3 \Vert r^{(1)} \Vert {}^5D_4 \rangle$	1.34164079	1.80000000	=	108/60	100.00
$\langle {}^5D_0 \Vert r^{(1)} \Vert {}^5D_1 \rangle$	-.44721360*	.20000000	=	4/20	13.33
$\langle {}^5D_1 \Vert r^{(1)} \Vert {}^5D_1 \rangle$	.22360680	.05000000	=	1/20	3.33
$\langle {}^5D_1 \Vert r^{(1)} \Vert {}^5D_2 \rangle$	-.59160798*	.35000000	=	7/20	23.33
$\langle {}^5D_2 \Vert r^{(1)} \Vert {}^5D_2 \rangle$	.50000000	.25000000	=	5/20	16.67
$\langle {}^5D_2 \Vert r^{(1)} \Vert {}^5D_3 \rangle$	-.63245553*	.40000000	=	8/20	26.67
$\langle {}^5D_3 \Vert r^{(1)} \Vert {}^5D_3 \rangle$	.83666003	.70000000	=	14/20	46.67
$\langle {}^5D_3 \Vert r^{(1)} \Vert {}^5D_4 \rangle$	-.54772256*	.30000000	=	6/20	20.00
$\langle {}^5D_4 \Vert r^{(1)} \Vert {}^5D_4 \rangle$	1.22474487	1.50000000	=	30/20	100.00
$\langle {}^5D_0 \Vert r^{(1)} \Vert {}^5F_1 \rangle$	.44721360	.20000000	=	14/70	12.73
$\langle {}^5D_1 \Vert r^{(1)} \Vert {}^5F_1 \rangle$	.44721360*	.20000000	=	14/70	12.73
$\langle {}^5D_1 \Vert r^{(1)} \Vert {}^5F_2 \rangle$	.63245553	.40000000	=	28/70	25.45
$\langle {}^5D_2 \Vert r^{(1)} \Vert {}^5F_1 \rangle$	.16903085	.02857143	=	2/70	1.82
$\langle {}^5D_2 \Vert r^{(1)} \Vert {}^5F_2 \rangle$	.53452248*	.28571429	=	20/70	18.18
$\langle {}^5D_2 \Vert r^{(1)} \Vert {}^5F_3 \rangle$	.82807867	.68571429	=	48/70	43.64
$\langle {}^5D_3 \Vert r^{(1)} \Vert {}^5F_2 \rangle$	.16903085	.02857143	=	2/70	1.82
$\langle {}^5D_3 \Vert r^{(1)} \Vert {}^5F_3 \rangle$	.54772256*	.30000000	=	21/70	19.09
$\langle {}^5D_3 \Vert r^{(1)} \Vert {}^5F_4 \rangle$	1.03509834	1.07142857	=	75/70	68.18
$\langle {}^5D_4 \Vert r^{(1)} \Vert {}^5F_3 \rangle$	.11952286	.01428571	=	1/70	.91
$\langle {}^5D_4 \Vert r^{(1)} \Vert {}^5F_4 \rangle$	.46291005*	.21428571	=	15/70	13.64
$\langle {}^5D_4 \Vert r^{(1)} \Vert {}^5F_5 \rangle$	1.25356634	1.57142857	=	110/70	100.00

Matrix Element	D_{line}	$(D_{line})^2$			Percent
$\langle {}^5F_1 \| r^{(1)} \| {}^5F_1 \rangle$	.53452248	.28571429	=	160/560	20.20
$\langle {}^5F_1 \| r^{(1)} \| {}^5F_2 \rangle$	-.37796447*	.14285714	=	80/560	10.10
$\langle {}^5F_2 \| r^{(1)} \| {}^5F_2 \rangle$	.59761430	.35714286	=	200/560	25.25
$\langle {}^5F_2 \| r^{(1)} \| {}^5F_3 \rangle$	-.46291005*	.21428571	=	120/560	15.15
$\langle {}^5F_3 \| r^{(1)} \| {}^5F_3 \rangle$	.75000000	.56250000	=	315/560	39.77
$\langle {}^5F_3 \| r^{(1)} \| {}^5F_4 \rangle$	-.47245559*	.22321429	=	125/560	15.78
$\langle {}^5F_4 \| r^{(1)} \| {}^5F_4 \rangle$	.95150257	.90535714	=	507/560	64.02
$\langle {}^5F_4 \| r^{(1)} \| {}^5F_5 \rangle$	-.39641248*	.15714286	=	88/560	11.11
$\langle {}^5F_5 \| r^{(1)} \| {}^5F_5 \rangle$	1.18923745	1.41428571	=	792/560	100.00
$\langle {}^5F_1 \| r^{(1)} \| {}^5G_2 \rangle$	.65465367	.42857143	=	2160/5040	29.67
$\langle {}^5F_2 \| r^{(1)} \| {}^5G_2 \rangle$	.34503278*	.11904762	=	600/5040	8.24
$\langle {}^5F_2 \| r^{(1)} \| {}^5G_3 \rangle$	.77151675	.59523810	=	3000/5040	41.21
$\langle {}^5F_3 \| r^{(1)} \| {}^5G_2 \rangle$	.08908708	.00793651	=	40/5040	.55
$\langle {}^5F_3 \| r^{(1)} \| {}^5G_3 \rangle$	.41666667*	.17361111	=	875/5040	12.02
$\langle {}^5F_3 \| r^{(1)} \| {}^5G_4 \rangle$	.90468358	.81845238	=	4125/5040	56.66
$\langle {}^5F_4 \| r^{(1)} \| {}^5G_3 \rangle$	.09449112	.00892857	=	45/5040	.62
$\langle {}^5F_4 \| r^{(1)} \| {}^5G_4 \rangle$	.42045893*	.17678571	=	891/5040	12.24
$\langle {}^5F_4 \| r^{(1)} \| {}^5G_5 \rangle$	1.04880885	1.10000000	=	5544/5040	76.15
$\langle {}^5F_5 \| r^{(1)} \| {}^5G_4 \rangle$	.06900656	.00476190	=	24/5040	.33
$\langle {}^5F_5 \| r^{(1)} \| {}^5G_5 \rangle$	.34960295*	.12222222	=	616/5040	8.46
$\langle {}^5F_5 \| r^{(1)} \| {}^5G_6 \rangle$	1.20185043	1.44444444	=	7280/5040	100.00
$\langle {}^5G_2 \| r^{(1)} \| {}^5G_2 \rangle$	.68041382	.46296296	=	5000/10800	34.34
$\langle {}^5G_2 \| r^{(1)} \| {}^5G_3 \rangle$	-.30429031*	.09259259	=	1000/10800	6.87
$\langle {}^5G_3 \| r^{(1)} \| {}^5G_3 \rangle$	.74005756	.54768519	=	5915/10800	40.62
$\langle {}^5G_3 \| r^{(1)} \| {}^5G_4 \rangle$	-.37080992*	.13750000	=	1485/10800	10.20
$\langle {}^5G_4 \| r^{(1)} \| {}^5G_4 \rangle$	.85000000	.72250000	=	7803/10800	53.59
$\langle {}^5G_4 \| r^{(1)} \| {}^5G_5 \rangle$	-.37416574*	.14000000	=	1512/10800	10.38
$\langle {}^5G_5 \| r^{(1)} \| {}^5G_5 \rangle$	.99293803	.98592593	=	10648/10800	73.13
$\langle {}^5G_5 \| r^{(1)} \| {}^5G_6 \rangle$	-.31031645*	.09629630	=	1040/10800	7.14
$\langle {}^5G_6 \| r^{(1)} \| {}^5G_6 \rangle$	1.16109782	1.34814815	=	14560/10800	100.00
$\langle {}^5G_2 \| r^{(1)} \| {}^5H_3 \rangle$	.74535599	.55555556	=	2750/4950	40.74
$\langle {}^5G_3 \| r^{(1)} \| {}^5H_3 \rangle$	.27888668*	.07777778	=	385/4950	5.70
$\langle {}^5G_3 \| r^{(1)} \| {}^5H_4 \rangle$	.83666003	.70000000	=	3465/4950	51.33
$\langle {}^5G_4 \| r^{(1)} \| {}^5H_3 \rangle$	.05504819	.00303030	=	15/4950	.22
$\langle {}^5G_4 \| r^{(1)} \| {}^5H_4 \rangle$	.33844564*	.11454545	=	567/4950	8.40
$\langle {}^5G_4 \| r^{(1)} \| {}^5H_5 \rangle$	.93937439	.88242424	=	4368/4950	64.71
$\langle {}^5G_5 \| r^{(1)} \| {}^5H_4 \rangle$	.06030227	.00363636	=	18/4950	.27
$\langle {}^5G_5 \| r^{(1)} \| {}^5H_5 \rangle$	.33993463*	.11555556	=	572/4950	8.47
$\langle {}^5G_5 \| r^{(1)} \| {}^5H_6 \rangle$	1.05025249	1.10303030	=	5460/4950	80.89
$\langle {}^5G_6 \| r^{(1)} \| {}^5H_5 \rangle$	.04494666	.00202020	=	10/4950	.15
$\langle {}^5G_6 \| r^{(1)} \| {}^5H_6 \rangle$	.28069179*	.07878788	=	390/4950	5.78
$\langle {}^5G_6 \| r^{(1)} \| {}^5H_7 \rangle$	1.16774842	1.36363636	=	6750/4950	100.00
$\langle {}^6S_{5/2} \| r^{(1)} \| {}^6P_{3/2} \rangle$	1.15470054	1.33333333	=	4/3	50.00
$\langle {}^6S_{5/2} \| r^{(1)} \| {}^6P_{5/2} \rangle$	1.41421356*	2.00000000	=	6/3	75.00
$\langle {}^6S_{5/2} \| r^{(1)} \| {}^6P_{7/2} \rangle$	1.63299316	2.66666667	=	8/3	100.00

TABLE I-1 (cont)

Matrix Element	D_{line}	$(D_{line})^2$			Percent
$\langle ^6P_{3/2} \Vert r^{(1)} \Vert ^6P_{3/2} \rangle$	-.63245553	.40000000	=	42/105	23.33
$\langle ^6P_{3/2} \Vert r^{(1)} \Vert ^6P_{5/2} \rangle$	-.96609178*	.93333333	=	98/105	54.44
$\langle ^6P_{5/2} \Vert r^{(1)} \Vert ^6P_{5/2} \rangle$	.33806170	.11428571	=	12/105	6.67
$\langle ^6P_{5/2} \Vert r^{(1)} \Vert ^6P_{7/2} \rangle$	-.97590007*	.95238095	=	100/105	55.56
$\langle ^6P_{7/2} \Vert r^{(1)} \Vert ^6P_{7/2} \rangle$	1.30930734	1.71428571	=	180/105	100.00
$\langle ^6P_{3/2} \Vert r^{(1)} \Vert ^6D_{1/2} \rangle$	.63245553	.40000000	=	210/525	20.00
$\langle ^6P_{3/2} \Vert r^{(1)} \Vert ^6D_{3/2} \rangle$	.74833148*	.56000000	=	294/525	28.00
$\langle ^6P_{3/2} \Vert r^{(1)} \Vert ^6D_{5/2} \rangle$	.61101009	.37333333	=	196/525	18.67
$\langle ^6P_{5/2} \Vert r^{(1)} \Vert ^6D_{3/2} \rangle$	.48989795	.24000000	=	126/525	12.00
$\langle ^6P_{5/2} \Vert r^{(1)} \Vert ^6D_{5/2} \rangle$	.85523597*	.73142857	=	384/525	36.57
$\langle ^6P_{5/2} \Vert r^{(1)} \Vert ^6D_{7/2} \rangle$	1.01418511	1.02857143	=	540/525	51.43
$\langle ^6P_{7/2} \Vert r^{(1)} \Vert ^6D_{5/2} \rangle$	.30860670	.09523810	=	50/525	4.76
$\langle ^6P_{7/2} \Vert r^{(1)} \Vert ^6D_{7/2} \rangle$	.75592895*	.57142857	=	300/525	28.57
$\langle ^6P_{7/2} \Vert r^{(1)} \Vert ^6D_{9/2} \rangle$	1.41421356	2.00000000	=	1050/525	100.00
$\langle ^6D_{1/2} \Vert r^{(1)} \Vert ^6D_{1/2} \rangle$	-.29814240	.08888889	=	420/4725	5.45
$\langle ^6D_{1/2} \Vert r^{(1)} \Vert ^6D_{3/2} \rangle$	-.55777335*	.31111111	=	1470/4725	19.09
$\langle ^6D_{3/2} \Vert r^{(1)} \Vert ^6D_{3/2} \rangle$	.09428090	.00888889	=	42/4725	.55
$\langle ^6D_{3/2} \Vert r^{(1)} \Vert ^6D_{5/2} \rangle$	-.69282032*	.48000000	=	2268/4725	29.45
$\langle ^6D_{5/2} \Vert r^{(1)} \Vert ^6D_{5/2} \rangle$	.45355737	.20571429	=	972/4725	12.62
$\langle ^6D_{5/2} \Vert r^{(1)} \Vert ^6D_{7/2} \rangle$	-.71713717*	.51428571	=	2430/4725	31.56
$\langle ^6D_{7/2} \Vert r^{(1)} \Vert ^6D_{7/2} \rangle$	.84578006	.71534392	=	3380/4725	43.90
$\langle ^6D_{7/2} \Vert r^{(1)} \Vert ^6D_{9/2} \rangle$	-.60858062*	.37037037	=	1750/4725	22.73
$\langle ^6D_{9/2} \Vert r^{(1)} \Vert ^6D_{9/2} \rangle$	1.27656948	1.62962963	=	7700/4725	100.00
$\langle ^6D_{1/2} \Vert r^{(1)} \Vert ^6F_{1/2} \rangle$	.47140452*	.22222222	=	7350/33075	12.96
$\langle ^6D_{1/2} \Vert r^{(1)} \Vert ^6F_{3/2} \rangle$	.42163702	.17777778	=	5880/33075	10.37
$\langle ^6D_{3/2} \Vert r^{(1)} \Vert ^6F_{1/2} \rangle$	.25197632	.06349206	=	2100/33075	3.70
$\langle ^6D_{3/2} \Vert r^{(1)} \Vert ^6F_{3/2} \rangle$	.57015732*	.32507937	=	10752/33075	18.96
$\langle ^6D_{3/2} \Vert r^{(1)} \Vert ^6F_{5/2} \rangle$	.64142698	.41142857	=	13608/33075	24.00
$\langle ^6D_{5/2} \Vert r^{(1)} \Vert ^6F_{3/2} \rangle$	.26186147	.06857143	=	2268/33075	4.00
$\langle ^6D_{5/2} \Vert r^{(1)} \Vert ^6F_{5/2} \rangle$	.62986879*	.39673469	=	13122/33075	23.14
$\langle ^6D_{5/2} \Vert r^{(1)} \Vert ^6F_{7/2} \rangle$	.85714286	.73469388	=	24300/33075	42.86
$\langle ^6D_{7/2} \Vert r^{(1)} \Vert ^6F_{5/2} \rangle$	.22131333	.04897959	=	1620/33075	2.86
$\langle ^6D_{7/2} \Vert r^{(1)} \Vert ^6F_{7/2} \rangle$	.62209263*	.38699924	=	12800/33075	22.57
$\langle ^6D_{7/2} \Vert r^{(1)} \Vert ^6F_{9/2} \rangle$	1.07889812	1.16402116	=	38500/33075	67.90
$\langle ^6D_{9/2} \Vert r^{(1)} \Vert ^6F_{7/2} \rangle$	.14547859	.02116402	=	700/33075	1.23
$\langle ^6D_{9/2} \Vert r^{(1)} \Vert ^6F_{9/2} \rangle$	.51434450*	.26455026	=	8750/33075	15.43
$\langle ^6D_{9/2} \Vert r^{(1)} \Vert ^6F_{11/2} \rangle$	1.30930734	1.71428571	=	56700/33075	100.00
$\langle ^6F_{1/2} \Vert r^{(1)} \Vert ^6F_{1/2} \rangle$	.35634832	.12698413	=	9240/72765	8.36
$\langle ^6F_{1/2} \Vert r^{(1)} \Vert ^6F_{3/2} \rangle$	-.39840954*	.15873016	=	11550/72765	10.45
$\langle ^6F_{3/2} \Vert r^{(1)} \Vert ^6F_{3/2} \rangle$	.39440532	.15555556	=	11319/72765	10.24
$\langle ^6F_{3/2} \Vert r^{(1)} \Vert ^6F_{5/2} \rangle$	-.50709255*	.25714286	=	18711/72765	16.92
$\langle ^6F_{5/2} \Vert r^{(1)} \Vert ^6F_{5/2} \rangle$	.54210474	.29387755	=	21384/72765	19.34
$\langle ^6F_{5/2} \Vert r^{(1)} \Vert ^6F_{7/2} \rangle$	-.55328334*	.30612245	=	22275/72765	20.15
$\langle ^6F_{7/2} \Vert r^{(1)} \Vert ^6F_{7/2} \rangle$	.73873500	.54572940	=	39710/72765	35.92
$\langle ^6F_{7/2} \Vert r^{(1)} \Vert ^6F_{9/2} \rangle$	-.53944906*	.29100529	=	21175/72765	19.15
$\langle ^6F_{9/2} \Vert r^{(1)} \Vert ^6F_{9/2} \rangle$	.97095878	.94276094	=	68600/72765	62.04
$\langle ^6F_{9/2} \Vert r^{(1)} \Vert ^6F_{11/2} \rangle$	-.44136741*	.19480519	=	14175/72765	12.82
$\langle ^6F_{11/2} \Vert r^{(1)} \Vert ^6F_{11/2} \rangle$	1.23267211	1.51948052	=	110565/72765	100.00

Matrix Element	D_{line}	$(D_{line})^2$			Percent
$\langle {}^6F_{1/2}\|r^{(1)}\|{}^6G_{3/2}\rangle$	.53452248	.28571429	=	1386/4851	18.37
$\langle {}^6F_{3/2}\|r^{(1)}\|{}^6G_{3/2}\rangle$	.37796447*	.14285714	=	693/4851	9.18
$\langle {}^6F_{3/2}\|r^{(1)}\|{}^6G_{5/2}\rangle$	.65465367	.42857143	=	2079/4851	27.55
$\langle {}^6F_{5/2}\|r^{(1)}\|{}^6G_{3/2}\rangle$	.12598816	.01587302	=	77/4851	1.02
$\langle {}^6F_{5/2}\|r^{(1)}\|{}^6G_{5/2}\rangle$	.46656947*	.21768707	=	1056/4851	13.99
$\langle {}^6F_{5/2}\|r^{(1)}\|{}^6G_{7/2}\rangle$	.78967257	.62358277	=	3025/4851	40.09
$\langle {}^6F_{7/2}\|r^{(1)}\|{}^6G_{5/2}\rangle$	.14285714	.02040816	=	99/4851	1.31
$\langle {}^6F_{7/2}\|r^{(1)}\|{}^6G_{7/2}\rangle$	.49943278*	.24943311	=	1210/4851	16.03
$\langle {}^6F_{7/2}\|r^{(1)}\|{}^6G_{9/2}\rangle$	.93435318	.87301587	=	4235/4851	56.12
$\langle {}^6F_{9/2}\|r^{(1)}\|{}^6G_{7/2}\rangle$	.12598816	.01587302	=	77/4851	1.02
$\langle {}^6F_{9/2}\|r^{(1)}\|{}^6G_{9/2}\rangle$	.48049998*	.23088023	=	1120/4851	14.84
$\langle {}^6F_{9/2}\|r^{(1)}\|{}^6G_{11/2}\rangle$	1.08711461	1.18181818	=	5733/4851	75.97
$\langle {}^6F_{11/2}\|r^{(1)}\|{}^6G_{9/2}\rangle$	.08494120	.00721501	=	35/4851	.46
$\langle {}^6F_{11/2}\|r^{(1)}\|{}^6G_{11/2}\rangle$	.38924947*	.15151515	=	735/4851	9.74
$\langle {}^6F_{11/2}\|r^{(1)}\|{}^6G_{13/2}\rangle$	1.24721913	1.55555556	=	7546/4851	100.00
$\langle {}^6G_{3/2}\|r^{(1)}\|{}^6G_{3/2}\rangle$	.57735027	.33333333	=	15015/45045	23.21
$\langle {}^6G_{3/2}\|r^{(1)}\|{}^6G_{5/2}\rangle$	-.33333333*	.11111111	=	5005/45045	7.74
$\langle {}^6G_{5/2}\|r^{(1)}\|{}^6G_{5/2}\rangle$	.61721340	.38095238	=	17160/45045	26.53
$\langle {}^6G_{5/2}\|r^{(1)}\|{}^6G_{7/2}\rangle$	-.41785545*	.17460317	=	7865/45045	12.16
$\langle {}^6G_{7/2}\|r^{(1)}\|{}^6G_{7/2}\rangle$	.71713717	.51428571	=	23166/45045	35.82
$\langle {}^6G_{7/2}\|r^{(1)}\|{}^6G_{9/2}\rangle$	-.44721360*	.20000000	=	9009/45045	13.93
$\langle {}^6G_{9/2}\|r^{(1)}\|{}^6G_{9/2}\rangle$	.85280287	.72727273	=	32760/45045	50.65
$\langle {}^6G_{9/2}\|r^{(1)}\|{}^6G_{11/2}\rangle$	-.42876379*	.18383838	=	8281/45045	12.80
$\langle {}^6G_{11/2}\|r^{(1)}\|{}^6G_{11/2}\rangle$	1.01480877	1.02983683	=	46389/45045	71.72
$\langle {}^6G_{11/2}\|r^{(1)}\|{}^6G_{13/2}\rangle$	-.34591635*	.11965812	=	5390/45045	8.33
$\langle {}^6G_{13/2}\|r^{(1)}\|{}^6G_{13/2}\rangle$	1.19828938	1.43589744	=	64680/45045	100.00
$\langle {}^6G_{3/2}\|r^{(1)}\|{}^6H_{5/2}\rangle$	.66666667	.44444444	=	220220/495495	30.56
$\langle {}^6G_{5/2}\|r^{(1)}\|{}^6H_{5/2}\rangle$	.30860670*	.09523810	=	47190/495495	6.55
$\langle {}^6G_{5/2}\|r^{(1)}\|{}^6H_{7/2}\rangle$	.75592895	.57142857	=	283140/495495	39.29
$\langle {}^6G_{7/2}\|r^{(1)}\|{}^6H_{5/2}\rangle$	.07597372	.00577201	=	2860/495495	.40
$\langle {}^6G_{7/2}\|r^{(1)}\|{}^6H_{7/2}\rangle$	.38439998*	.14776335	=	73216/495495	10.16
$\langle {}^6G_{7/2}\|r^{(1)}\|{}^6H_{9/2}\rangle$	.85752757	.73535354	=	364364/495495	50.56
$\langle {}^6G_{9/2}\|r^{(1)}\|{}^6H_{7/2}\rangle$	.08989331	.00808081	=	4004/495495	.56
$\langle {}^6G_{9/2}\|r^{(1)}\|{}^6H_{9/2}\rangle$	.40881023*	.16712580	=	82810/495495	11.49
$\langle {}^6G_{9/2}\|r^{(1)}\|{}^6H_{11/2}\rangle$	.96742157	.93590450	=	463736/495495	64.34
$\langle {}^6G_{11/2}\|r^{(1)}\|{}^6H_{9/2}\rangle$	.08131156	.00661157	=	3276/495495	.45
$\langle {}^6G_{11/2}\|r^{(1)}\|{}^6H_{11/2}\rangle$	.38973913*	.15189659	=	75264/495495	10.44
$\langle {}^6G_{11/2}\|r^{(1)}\|{}^6H_{13/2}\rangle$	1.08389353	1.17482517	=	582120/495495	80.77
$\langle {}^6G_{13/2}\|r^{(1)}\|{}^6H_{11/2}\rangle$	.05574947	.00310800	=	1540/495495	.21
$\langle {}^6G_{13/2}\|r^{(1)}\|{}^6H_{13/2}\rangle$	.31289311*	.09790210	=	48510/495495	6.73
$\langle {}^6G_{13/2}\|r^{(1)}\|{}^6H_{15/2}\rangle$	1.20604538	1.45454545	=	720720/495495	100.00

Bibliography

I. EXPERIMENTAL ATOMIC SPECTROSCOPY

[1] R. A. Sawyer, *Experimental Spectroscopy* (Dover Publications, New York, 1963), 3rd ed.
[2] B. Edlén, Rep. Prog. Phys. **26**, 181 (1963).
[3] K. G. Kessler and H. M. Crosswhite, in V. W. Hughes and H. L. Schultz, eds., *Atomic and Electron Physics*, Methods of Experimental Physics, Vol. 4, part B (Academic Press, New York, 1967).
[4] J. A. R. Samson, *Techniques of Vacuum Ultraviolet Spectroscopy* (John Wiley, New York, 1967).
[5] D. Williams, ed., *Spectroscopy*, Methods of Experimental Physics, Vol. 13, parts A and B (Academic Press, New York, 1976).
[6] K. Shimoda, ed., *High-Resolution Laser Spectroscopy*, Topics in Applied Physics, Vol. 13 (Springer-Verlag, Berlin, 1976).
[7] W. Hanle and H. Kleinpoppen, eds., *Progress in Atomic Spectroscopy* (Plenum, New York, 1978), 2 vols.
[7a] B. M. Agarwal, *X-ray Spectroscopy* (Springer-Verlag, Berlin, 1979).

II. EMPIRICAL AND DESCRIPTIVE SPECTROSCOPY

[8] G. K. Woodgate, *Elementary Atomic Structure* (Clarendon Press, Oxford, 1980), 2nd ed.
[9] H. G. Kuhn, *Atomic Spectra* (Longmans, Green, London, 1969), 2nd ed.
[10] B. Edlén, "Atomic Spectra," in S. Flügge, ed., *Handbuch der Physik* (Springer-Verlag, Berlin, 1964), Vol. XXVII.

III. ANGULAR MOMENTUM THEORY

[11] A. R. Edmonds, *Angular Momentum in Quantum Mechanics* (Princeton University Press, Princeton, N. J., 1960), 2nd ed.
[12] U. Fano and G. Racah, *Irreducible Tensorial Sets* (Academic Press, New York, 1959).
[13] L. C. Biedenharn and H. van Dam, eds., *Quantum Theory of Angular Momentum* (Academic Press, New York, 1965).
[14] J. S. Briggs, "Evaluation of Matrix Elements from a Graphical Representation of the Angular Integral," Rev. Mod. Phys. **43**, 189-230 (1971).

[15] E. El-Baz and B. Castel, *Graphical Methods of Spin Algebras in Atomic, Nuclear, and Particle Physics* (Marcel Dekker, New York, 1972).
[16] W. G. Harter and C. W. Patterson, Phys. Rev. A **13**, 1067 (1976); *A Unitary Calculus for Electronic Orbitals* (Springer-Verlag, Berlin, 1976).
[17] L. C. Biedenharn and J. D. Louck, *Angular Momentum in Quantum Physics* and *The Racah-Wigner Algebra in Quantum Theory*, Encyclopedia of Mathematics and Its Applications, Vols. 8 and 9 (Addison-Wesley, Reading, Mass., 1981).

IV. THEORY OF ATOMIC STRUCTURE AND SPECTRA

[18] E. U. Condon and G. H. Shortley, *The Theory of Atomic Spectra* (University Press, Cambridge, 1935).
[19] G. Racah, "Theory of Complex Spectra," II. **62**, 438 (1942); III. **63**, 367 (1943); IV. **76**, 1352 (1949).
[20] J. C. Slater, *Quantum Theory of Atomic Structure* (McGraw-Hill, New York, 1960), 2 vols.
[21] B. R. Judd, *Operator Techniques in Atomic Spectroscopy* (McGraw-Hill, New York, 1963).
[22] B. G. Wybourne, *Spectroscopic Properties of Rare Earths* (Interscience, New York, 1965).
[23] B. R. Judd, *Second Quantization and Atomic Spectroscopy* (The Johns Hopkins Press, Baltimore, 1967).
[24] B. W. Shore and D. H. Menzel, *Principles of Atomic Spectra* (John Wiley, New York, 1968).
[25] H. A. Bethe and R. W. Jackiw, *Intermediate Quantum Mechanics* (W. A. Benjamin, New York, 1968), 2nd ed.
[26] S. Fraga and G. Malli, *Many-Electron Systems: Properties and Interactions* (W. B. Saunders, Philadelphia, 1968).
[27] I. I. Sobel'man, *Introduction to the Theory of Atomic Spectra* (Pergamon, Oxford, 1972).
[28] J. D. Macomber, *The Dynamics of Spectroscopic Transitions* (John Wiley, New York, 1976).
[29] P. G. Burke and B. L. Moiseiwitsch, eds., *Atomic Processes and Applications* (North-Holland Publ. Co., Amsterdam, 1976).
[30] M. Weissbluth, *Atoms and Molecules* (Academic Press, New York, 1978).
[31] I. I. Sobel'man, *Atomic Spectra and Radiative Transitions* (Springer-Verlag, Berlin, 1979).
[32] E. U. Condon and H. Odabasi, *Atomic Structure* (University Press, Cambridge, 1980).
[33] M. R. C. McDowell and A. M. Ferendeci, eds., *Atomic and Molecular Processes in Controlled Thermonuclear Fusion* (Plenum, New York, 1980).
[33a] P. Pyykko, *Relativistic Theory of Atoms and Molecules, A Bibliography 1916-1985* (Springer-Verlag, Berlin, 1986).

V. CALCULATION OF RADIAL WAVEFUNCTIONS

[34] D. R. Hartree, *The Calculation of Atomic Structures* (John Wiley, New York, 1957).
[35] F. Herman and S. Skillman, *Atomic Structure Calculations* (Prentice-Hall, Englewood Cliffs, N. J., 1963).
[36] N. H. March, *Self-Consistent Fields in Atoms* (Pergamon, Oxford, 1975).
[37] C. Froese Fischer, *The Hartree-Fock Method for Atoms* (John Wiley, New York, 1977).
[38] J. P. Desclaux, Comput. Phys. Commun. **9**, 31 (1975).

VI. WAVELENGTH TABLES

[39] G. R. Harrison, *Massachusetts Institute of Technology Wavelength Tables* (The M.I.T. Press, Cambridge, Mass., 1969), revised ed.; most neutral and singly ionized elements, 2000-10000 Å.

[40] A. R. Striganov and N. S. Sventitskii, *Tables of Spectral Lines of Neutral and Ionized Atoms* (IFI/Plenum, New York, 1968); 22 elements, with line classifications.
[41] A. N. Zaidel' et al, *Tables of Spectral Lines* (IFI/Plenum, New York, 1970); all elements, 2000-8000 Å.
[42] R. L. Kelly and L. J. Palumbo, *Atomic and Ionic Emission Lines below 2000 Å*—Hydrogen through Krypton, Naval Research Laboratory Report NRL-7599 (U.S. Govt. Printing Off., Washington, D.C., 1973); extension to 3200 Å and revision for H-Ar in preparation.
[43] W. F. Meggers, C. H. Corliss, and B. F. Scribner, *Tables of Spectral-Line Intensities*, NBS Monograph 145 (U.S. Govt. Printing Off., Washington, D.C., 1975), 2 vols; 70 elements, first and second spectra, 2000-9000 Å.
[44] R. C. Weast, ed., *CRC Handbook of Chemistry and Physics* (CRC Press, West Palm Beach, Fla.), 59th (1978-79) and later editions; I-V spectra.
[45] J. Reader, C. H. Corliss, W. L. Wiese, and G. A. Martin, *Wavelengths and Transition Probabilities for Atoms and Atomic Ions*, NSRDS-NBS 68 (U.S. Govt. Printing Off., Washington, D.C., 1980); I-V spectra of most elements.
[46] M. Outred, "Tables of Atomic Spectral Lines for the 10000 Å to 40000 Å Region," J. Phys. Chem. Ref. Data 7, 1-262 (1978).
[47] V. Kaufman and B. Edlén, J. Phys. Chem. Ref. Data 3, 825-95 (1974); wavelength standards, 15 Å - 25000 Å, 63 spectra of 31 elements.
[48] J. A. Bearden, "X-Ray Wavelengths," Rev. Mod. Phys. 39, 78-124 (1967); reprinted as NSRDS-NBS 14 (U.S. Govt. Printing Off., Washington, D.C., 1967).
See also refs. [70], [81].

VII. ATOMIC ENERGY LEVEL TABLES

[49] C. E. Moore, *Atomic Energy Levels, as Derived from the Analyses of Optical Spectra*, U.S. Natl. Bur. Stands. Circ. 467 (U.S. Govt. Printing Off., Washington, D.C.), Vols I-III, 1949-58. [Reissued 1971 as NSRDS-NBS 35, Vols. I-III.] All elements except the lanthanides and actinides.
[50] W. C. Martin, R. Zalubas, and L. Hagan, *Atomic Energy Levels—The Rare-Earth Elements*, NSRDS-NBS 60 (U.S. Govt. Printing Off., Washington, D.C., 1978); Z = 57 to 71.
[51] C. E. Moore, *Selected Tables of Atomic Spectra*, NSRDS-NBS 3 (U.S. Govt. Printing Off., Washington, D.C.)
Sec. 1 (1965); Si II, III, IV
Sec. 2 (1967); Si I
Sec. 3 (1970); C I-VI
Sec. 4 (1971); N IV-VII
Sec. 5 (1975); N I-III
Sec. 6 (1972); H I, D I, T I
Sec. 7 (1976); O I
Sec. 8 (1979); O VI-VIII
Sec. 9 (1980); O V.
Sec. 10 (1983); O IV
Sec. 11 (1985); O III
[52] G. W. Erickson, J. Phys. Chem. Ref. Data 6, 831-69 (1977); one-electron atoms, Z = 1 to 105.
[53] W. C. Martin, J. Phys. Chem. Ref. Data 2, 257-66 (1973); He I.
[54] C. L. Pekeris, Phys. Rev. 112, 1649 (1958); Y. Accad, C. L. Pekeris, and B. Schiff, Phys. Rev. A 4, 516 (1971); A. M. Ermolaev and M. Jones, J. Phys. B 7, 199 (1974); two-electron atoms, $Z \leq 42$.
[55] G. A. Odintzova and A. R. Striganov, J. Phys. Chem. Ref. Data 8, 63-67 (1979); B I.

[56] W. C. Martin and R. Zalubas, J. Phys. Chem. Ref. Data **10,** 153-195 (1981); Na I-XI.
[57] W. C. Martin and R. Zalubas, J. Phys. Chem. Ref. Data **9,** 1-58 (1980); Mg I-XII.
[58] W. C. Martin and R. Zalubas, J. Phys. Chem. Ref. Data **8,** 817-64 (1979); Al I-XIII.
[59] W. C. Martin and R. Zalubas, J. Phys. Chem. Ref. Data **12,** 323-380 (1983); Si I-XIV.
[60] C. Corliss and J. Sugar, J. Phys. Chem. Ref. Data **8,** 1109-45 (1979); K I-XIX.
[61] J. Sugar and C. Corliss, J. Phys. Chem. Ref. Data **8,** 865-916 (1979); Ca I-XX.
[62] J. Sugar and C. Corliss, J. Phys. Chem. Ref. Data **9,** 473-511 (1980); Sc I-XXI.
[63] C. Corliss and J. Sugar, J. Phys. Chem. Ref. Data **8,** 1-62 (1979); Ti I-XXII.
[64] C. Corliss and J. Sugar, J. Phys. Chem. Ref. Data **7,** 1191-262 (1978); V I-XXIII.
[65] J. Sugar and C. Corliss, J. Phys. Chem. Ref. Data **6,** 317-83 (1977); Cr I-XXIV.
[66] C. Corliss and J. Sugar, J. Phys. Chem. Ref. Data **6,** 1253-330 (1977); Mn I-XXV.
[67] C. Corliss and J. Sugar, J. Phys. Chem. Ref. Data **11,** 135-241 (1982); Fe I-XXVI.
[68] J. Sugar and C. Corliss, J. Phys. Chem. Ref. Data **10,** 1097-1174 (1981); Co I-XXVII.
[69] C. Corliss and J. Sugar, J. Phys. Chem. Ref. Data **10,** 197-289 (1981); Ni I-XXVIII.
Compilations similar to the above are planned for all other elements up to zinc.
[70] B. C. Fawcett, At. Data Nucl. Data Tables **16,** 135-64 (1975); $2s^n2p^m$ configurations, Z = 3 to 28. See also B. Edlén, Phys. Scr. **19,** 255-66 (1979), **20,** 129-37 (1979), **22,** 593-602 (1980); K. D. Lawson and N. J. Peacock, J. Phys. B **13,** 3313-34 (1980).
[71] J. A. Bearden and A. F. Burr, "Reevaluation of X-Ray Atomic Energy Levels," Rev. Mod. Phys. **39,** 125-42 (1967); reprinted as NSRDS-NBS 14 (U.S. Govt. Printing Off., Washington, D.C., 1967).
[72] W. Lotz, J. Opt. Soc. Am. **58,** 915-21 (1968), **60,** 206-10 (1970); binding energies in all subshells of atoms and ions.
[73] D. A. Shirley, R. L. Martin, S. P. Kowalczyk, F. R. McFeely, and L. Ley, Phys. Rev. B **15,** 544-52 (1977); Z = 1 to 30.
[74] F. T. Porter and M. S. Freedman, J. Phys. Chem. Ref. Data **7,** 1267-84 (1978); binding energies of all subshells, $84 \leq Z \leq 103$.
See also ref. [96].

VIII. ATOMIC ENERGY LEVELS—BIBLIOGRAPHIES

[75] C. E. Moore, *Bibliography on the Analyses of Optical Atomic Spectra*, U.S. Natl. Bur. Stand. Special Publ. 306 (U.S. Govt. Printing Off., Washington, D.C.), 4 sections, corresponding to the 4 volumes of refs. [49]-[50], 1968-69.
[76] L. Hagan and W. C. Martin, *Bibliography on Atomic Energy Levels and Spectra, July 1968 through June 1971*, U.S. Natl. Bur. Stand. Special Publ. 363 (1972).
[77] L. Hagan, Supplement 1 to above, July 1971 through June 1975 (1977).
[78] R. Zalubas and A. Albright, Supplement 2, July 1975 through June 1979 (1980).
[79] B. C. Fawcett, Adv. At. Mol. Phys. **10,** 223-93 (1974); highly ionized atoms, and UV solar spectra.
[80] B. Edlén, in I. A. Sellin and D. J. Pegg, eds., *Beam-Foil Spectroscopy* (Plenum, New York, 1976), Vol. I, pp. 1-27; Z = 2 to 28.
[81] References in Trans. Int. Astron. Union **XVIA** (Part 1), 31-50 (1976); **XVIIA** (Part 1), 37-60 (1979); **XVIIIA,** 115-151 (1982).

IX. IONIZATION ENERGIES

[82] C. E. Moore, *Ionization Potentials and Ionization Limits Derived from the Analyses of Optical Spectra*, NSRDS-NBS 34 (U.S. Govt. Printing Off., Washington, D.C., 1970).

[83] R. L. Kelly and D. E. Harrison, Jr., At. Data Nucl. Data Tables **19**, 301 (1977); Ne I isoelectronic sequence.

[84] W. C. Martin, L. Hagan, J. Reader, and J. Sugar, J. Phys. Chem. Ref. Data **3**, 771-79 (1974); I-IV spectra of lanthanides and actinides.

[85] J. Sugar, J. Opt. Soc. Am. **65**, 1366-67 (1975); V^{th} spectra of lanthanides.

[86] T. A. Carlson, C. W. Nestor, Jr., N. Wasserman, and J. D. McDowell, At. Data **2**, 63-99 (1970); theoretical values (accuracy ~5 percent) for all ions of all elements.

[86a] H. Hotop and W. C. Lineberger, "Binding Energies in Atomic Negative Ions: II," J. Phys. Chem. Ref. Data **14**, 731-750 (1985).

See also refs. [42], [49]-[69], and [71]-[74], and Table 1-1 of main text (pp. 12-15).

X. GROTRIAN DIAGRAMS

[87] C. E. Moore and P. W. Merrill, *Partial Grotrian Diagrams of Astrophysical Interest*, NSRDS-NBS 23 (U.S. Govt. Printing Off., Washington, D.C., 1968).

[88] S. Bashkin and J. O. Stoner, Jr., *Atomic Energy Level and Grotrian Diagrams* (North-Holland Publ. Co., Amsterdam), Vol. I, H I - P XV (1975); Vol. I addenda (1978); Vol. II, S I - Ti XXII (1978); Vol. III, V I - Cr XXIV (1981); Vol. IV, Mn I-Mn XXV (1982).

[89] K. Mori, M. Otsuka, and T. Kato, At. Data Nucl. Data Tables 23, 195-294 (1979); Fe VIII-XXVI; also K. Mori, et. al., At. Data Nucl. Data Tables **34**, 79 (1986); Ti V-XXII.

XI. OSCILLATOR STRENGTH TABLES

[90] W. L. Wiese, M. W. Smith, and B. M. Glennon, *Atomic Transition Probabilities, Volume I Hydrogen through Neon*, NSRDS-NBS 4 (U.S. Govt. Printing Off., Washington, D. C., 1966).

[91] W. L. Wiese, M. W. Smith, and B. M. Miles, *Atomic Transition Probabilities, Volume II Sodium through Calcium*, NSRDS-NBS 22 (1969).

[92] G. A. Martin, W. L. Wiese, J. R. Fuhr, and S. M. Younger, *Atomic Transition Probabilities, Volume III Scandium through Nickel*, NSRDS-NBS series (in preparation).

[93] J. Reader, C. H. Corliss, W. L. Wiese, and G. A. Martin, *Wavelengths and Transition Probabilities for Atoms and Atomic Ions*, NSRDS-NBS 68 (U.S. Govt. Printing Off., Washington, D.C., 1980); I-V spectra of most elements.

[94] G. A. Martin and W. L. Wiese, J. Phys. Chem. Ref. Data **5**, 537-70 (1976): Li I isoelectronic sequence to Ni XXVI.

[95] B. C. Fawcett, At. Data Nucl. Data Tables **22**, 473-89 (1978); 2s − 2p and 2p − 2p transitions in Be I to O I isoelectronic sequences, Z = 7 to 26.

[96] M. Crance, At. Data **5**, 185-200 (1973); Ne I isoelectronic sequence.

[97] A. Lindgård and S. E. Nielsen, At. Data Nucl. Data Tables **19**, 533-633 (1977); alkali isoelectronic sequences.

[98] W. L. Wiese and J. R. Fuhr, J. Phys. Chem. Ref. Data **4**, 263-352 (1975); Sc and Ti.

[99] S. M. Younger, J. R. Fuhr, G. A. Martin, and W. L. Wiese, J. Phys. Chem. Ref. Data **7**, 495-629 (1978); V, Cr, and Mn.

[100] J. R. Fuhr, G. A. Martin, W. L. Wiese, and S. M. Younger, J. Phys. Chem. Ref. Data **10**, 305-565 (1981); Fe, Co, Ni.

[101] M. W. Smith and W. L. Wiese, J. Phys. Chem. Ref. Data **2**, 85-120 (1973); forbidden lines of the iron group elements.

[102] B. M. Miles and W. L. Wiese, At. Data **1**, 1-17 (1969); Ba I-II.

[103] M. W. Smith and W. L. Wiese, Astrophys. J., Suppl. Ser. **23**, 103-92 (1971); systematic trends.

[104] C. H. Corliss and W. R. Bozman, *Experimental Transition Probabilities for Spectral Lines of Seventy Elements*, NBS Monograph 53 (U.S. Govt. Printing Off., Washington, D.C., 1962).
[105] R. L. Kurucz and E. Peytremann, "A Table of Semiempirical gf Values," Smithsonian Astrophysical Observatory Special Report 362 (1975), 3 parts.
[106] M. Aymar and M. Coulombe, At. Data Nucl. Data Tables 21, 537-566 (1978); Kr I and Xe I.
See also refs. [44], [81], and [89].

XII. OSCILLATOR STRENGTHS—BIBLIOGRAPHIES

[107] J. R. Fuhr, B. J. Miller, and G. A. Martin, *Bibliography on Atomic Transition Probabilities* (1914 through October 1977), U.S. Natl. Bur. Stands. Special Publ. 505 (U.S. Govt. Printing Off., Washington, D.C., 1978); Suppl. 1, November 1977 through March 1980 (1980).
[108] K. Way, "Atomic Data Related to X and XUV Radiation," At. Data Nucl. Data Tables 22, 125-130 (1978).

XIII. SPECTROSCOPIC COEFFICIENTS (ANGULAR)

[109] M. Rotenberg, R. Bivins, N. Metropolis, and J. K. Wooten, Jr., *The 3-j and 6-j Symbols* (The Technology Press, Massachusetts Institute of Technology, Cambridge, 1959).
[110] S. H. Koozekanani, *9j Symbols* (Engineering Publications, Ohio State University, Columbus, 1972).
[111] K. Smith and J. W. Stephenson, "A Table of Wigner 9j Coefficients," Argonne National Laboratory Reports ANL-5776 (1957) and ANL-5860 (1958).
[112] C. W. Nielson and G. F. Koster, *Spectroscopic Coefficients for the p^n, d^n, and f^n Configurations* (The M.I.T. Press, Massachusetts Institute of Technology, Cambridge, 1963); cfp, matrix elements of $U^{(k)}$ and $V^{(11)}$, and Coulomb coefficients f_k for l^w.
[113] R. I. Karaziya, Ya. I. Vizbaraite, Z. B. Rudzikas, and A. P. Jucys, *Tables for the Calculation of Matrix Elements of Atomic Quantities* (Moscow, 1967); English transl. by E. K. Wilip, ANL-Trans-563 (National Technical Information Service, Springfield, Va., 1968); cfp and matrix elements of $U^{(k)}$ $(k > 1)$ and $V^{(k1)}$ $(k > 0)$ for p^2-p^3, d^2-d^5, and f^2-f^4. [Warning: The values of $U^{(k)}$ contain an additional factor $(2S+1)^{1/2}$ beyond the definition (11.53) used in the present book and in ref. 112.]
[114] W.-K. Li, At. Data 2, 263-71 (1971); $V^{(21)}$, $V^{(31)}$, and $V^{(41)}$ for d^w.
[115] J. C. Slater, ref. [20], Vol. II, appendices; cfp, matrix elements of $U^{(k)}$ and $V^{(k1)}$, and angular coefficients f_k and g_k for Coulomb matrix elements of many configurations. See also Condon and Odabasi, ref. [32].
[116] J. A. Barnes, B. L. Carroll, L. M. Flores, and R. M. Hedges, At. Data 2, 1-43, 101-17 (1970); energy coefficients f_k, g_k, etc. for d^w.
[117] B. W. Shore and D. H. Menzel, "Generalized Tables for the Calculation of Dipole Transition Probabilities," Astrophys. J., Suppl. Ser. 12, 187-214 (1965); also in ref. [24], Chap. 10. [Warning: The table of line factors is for SL coupling; for LS coupling, multiply by $(-1)^{L_1-J_1-L_2+J_2}$.]
[118] I. B. Levinson and A. A. Nikitin, *Handbook for Theoretical Computation of Line Intensities in Atomic Spectra* (Israel Program for Scientific Translations, Jerusalem, 1965); transl. from Russian by Z. Lerman, publ. in USA by Daniel Davey & Co., New York.

XIV. ATOMIC STRUCTURE RESULTS (RADIAL INTEGRALS, ENERGIES, ETC.)

[119] S. Fraga, J. Karwowski, and K. M. S. Saxena, *Handbook of Atomic Data* (Elsevier Scientific Publ. Co., Amsterdam, 1976); HF results for numerous quantities.

[120] S. Fraga, K. M. S. Saxena, and J. Karwowski, *Atomic Energy Levels, Data for Parametric Calculations* (Elsevier, Amsterdam, 1979).

[121] C. Froese Fischer, "Average-Energy-of-Configuration Hartree-Fock Results for the Atoms Helium to Radon," At. Data Nucl. Data Tables **12**, 87-99 (1973). See also ref. [37].

[122] J. B. Mann, "SCF Hartree-Fock Results for Elements with Two Open Shells and for the Elements Francium to Nobelium," At. Data Nucl. Data Tables **12**, 1-86 (1973).

[123] J. P. Desclaux, "Relativistic Dirac-Fock Expectation Values for Atoms with Z = 1 to Z = 120," At. Data Nucl. Data Tables **12**, 311-406 (1973).

[124] L. C. Green, P. P. Rush, and C. D. Chandler, Astrophys. J., Suppl. Ser. **3**, 37-50 (1957); values of the square of the radial dipole integral $\int P_{nl}\, r\, P_{n'l'} dr$ for hydrogen wavefunctions (divide by Z^2 for hydrogenic wavefunctions of other elements).

[125] B. W. Shore and D. H. Menzel, ref. [117]; same as in ref. [124], except the reduced dipole matrix element $P_{ll'}^{(1)}$ (divide by Z).

[126] D. R. Bates and A. Damgaard, Phil. Trans. Roy. Soc. London **A242**, 101-22 (1949); G. K. Oertel and L. P. Shomo, Astrophys. J., Suppl. Ser. **16**, 175-218 (1968); Coulomb-approximation radial multipole integrals.

XV. LINE SHIFTS AND WIDTHS

[127] H. R. Griem, *Plasma Spectroscopy* (McGraw-Hill, New York, 1964); *Spectral Line Broadening by Plasmas* (Academic Press, New York, 1974).

[128] N. Konjevic and J. R. Roberts, J. Phys. Chem. Ref. Data **5**, 209-257 (1976); N. Konjevic and W. L. Wiese, J. Phys. Chem. Ref. Data **5**, 259-308 (1976); experimental Stark widths and shifts. Also, N. Konjevic, M. S. Dimitrijevic, and W. L. Wiese, J. Phys. Chem. Ref. Data **13**, 619-648 and 649-686 (1984); experimental Stark widths and shifts.

[129] J. R. Fuhr, W. L. Wiese, and L. J. Roszman, *Bibliography on Atomic Line Shapes and Shifts* (1889 through March 1972), U.S. Natl. Bur. Stand. Special Publ. 366 (U.S. Govt. Printing Off., Washington, D.C., 1972); Suppls. 1-3 (1974, 1975, 1978).

[130] K. Heilig, Spectrochim. Acta **32B**, 1-57 (1977); bibliography of optical isotope shifts, 1918-1976.

[131] I. I. Sobel'man, L. A. Vainstein and E. A. Yukov, *Excitation of Atoms and Broadening of Spectral Lines* (Springer-Verlag, Berlin, 1981).

[132] S. Buttgenbach, *Hyperfine Structure in 4d- and 5d-shell Atoms* (Springer-Verlag, Berlin, 1982).

Name Index

Abramowitz, M., 565
Accad, Y., 704
Acquista, N., 588
Albright, A., 705
Alexander, E., 585
Allen, C. W., 618
Allen, J. W., 562, 590
Aller, L. H., 345
Altick, P. L., 535
Amaldi, E., 192
Andrew, K. L., 3, 4, 17, 25, 29, 129, 132, 133, 187, 223, 231, 235, 292, 302, 416, 421, 454, 476, 482, 484, 489, 495, 498
Ångström, A. J., 6
Armstrong, B. H., 20
Armstrong, J. A., 484
Armstrong, L., Jr., 266, 378, 439, 509, 535, 542
Artru, M.-C., 464
Auger, P., 531
Aymar, M., 434, 707

Bacher, R. F., 29, 478
Back, E., 496
Bagus, P. S., 343
Balmer, J. J., 81, 83
Barfield, W. D., 526
Barnes, J. A., 707
Bar-Shalom, A., 583
Bashkin, S., 17, 28, 435, 706
Bates, D. R., 411, 520, 551, 708
Bauche, J., 372, 505, 509, 582, 625
Bauche-Arnoult, C., 582, 625
Bearden, J. A., 9, 174, 584, 588, 704, 705
Behring, W., 586
Bell, K. L., 569
Bell, R. J., 24
Bely, O., 590, 591
Berestetskii, V. B., 84
Berrington, K. A., 590
Berthel, R. O., 435
Bethe, H. A., 83, 86, 401, 427, 429, 526, 567, 703
Beutler, H., 534
Bhalla, C. P., 555, 590
Bhatia, A. K., 590
Biedenharn, L. C., 142, 149, 702, 703
Bird, R. B., 88
Bitter, M., 582, 595
Bivins, R., 143, 302, 486, 707
Bjorken, J. D., 84
Blaha, M., 555
Blair, H. A., 493
Blaise, J., 25, 29, 598, 600
Blume, M., 94, 219
Bockasten, K., 73, 588
Bohr, N., 1, 9, 115
Boiko, V. A., 595
Bordarier, Y., 155
Borgström, A., 355
Bowen, I. S., 453
Boyd, R. G., 224
Bozman, W. R., 5, 8, 707
Brackett, F. S., 83
Breton, C., 546, 583
Bretz, N., 593
Brewer, L., 124, 622
Bridges, J. M., 435
Briggs, J. S., 155, 702
Brillouin, L., 372
Bromage, G. E., 580, 581
Bromander, J., 435
Brown, C. M., 484
Brown, E. R., 542

Brown, J. D., 613
Brueckner, K. A., 202, 203
Brysk, H., 526
Budick, B., 541
Burgess, A., 426, 546, 547, 559, 561, 562, 589
Burgess, D. D., 5
Burke, P. G., 535, 541, 590, 703
Burkhalter, P. G., 6, 581, 585, 588
Burr, A. F., 9, 174, 584, 588, 705
Burton, W. M., 583

Campbell, L., 595
Candler, C., 139
Caplan, P. J., 140
Carlson, L. R., 601
Carlson, T. A., 15, 563, 706
Carnall, W. T., 481
Carr, W. J., Jr., 203
Carroll, B. L., 707
Carter, S. L., 542
Casimir, H., 479
Castel, B., 151, 155, 703
Catalán, M. A., 498, 614, 615
Cecchi, J. L., 629
Champeau, R.-J., 505
Chandler, C. D., 411, 708
Chase, L. F., 580
Childs, W. J., 607
Chisholm, J. S. R., 90
Chojnacki, D. A., 467
Chubb, T. A., 24
Clementi, E., 203
Cochrane, D. M., 546, 589
Cocke, C. L., 435, 585
Codling, K., 9, 534
Cohen, E. R., 633
Cohen, L., 5, 6, 439, 581, 586
Cohen, M., 448
Cohen, V. W., 507
Coldwell-Horsfall, R. A., 203
Coleman, C. D., 8
Condon, E. U., 27, 42, 44, 57, 94, 144, 147, 243, 285, 290, 308, 356, 409, 411, 456, 703
Connerade, J. P., 5, 543
Connes, J., 25
Connes, P., 24, 25
Conway, J. G., 607, 610
Cooke, W. E., 60, 557
Cooper, J. W., 428, 526, 534, 535
Corliss, C. H., 5, 433, 704, 705, 706, 707
Corney, A., 453
Coulombe, M., 707
Coulson, C. A., 192
Cowan, R. D., 5, 15, 17, 22, 32, 116, 129, 187, 189, 196, 201, 203, 214, 222, 223, 231, 235, 239, 248, 292, 332, 350, 355, 416, 418, 433, 439, 456, 464, 465, 466, 476, 489, 498, 506, 541, 546, 555, 578, 580, 581, 583, 585, 587, 588, 590, 595, 598, 600, 605, 611, 613, 625, 629
Cowden, D. J., 619
Cowley, C. R., 18
Cox, A. N., 617
Crance, M., 706
Crawford, F. S., 68
Crossley, R. J. S., 401
Crosswhite, H., 461, 478
Crosswhite, H. M., 5, 461, 478, 479, 481, 702
Croxton, F. E., 619
Curl, R. F., Jr., 109
Curnutte, B., 435
Curtis, L. J., 433
Curtiss, C. F., 88
Cusachs, L. C., 196

Dalgarno, A., 551
Damgaard, A., 411, 708
Darwin, C. G., 84
Das, T. P., 372
Davis, J., 555, 562
Dehmer, J. L., 239, 428, 544
Demeter, W., 534
DeMichelis, C., 546, 583
Desclaux, J. P., 200, 201, 587, 703, 708
de-Shalit, A., 127, 255
Deutschman, W. A., 460
Dieke, G. H., 5, 22, 600
Dimock, D., 593
Dirac, P. A. M., 192, 528
Dixon, W. T., 91
Doschek, G. A., 88, 439, 452, 581, 582, 590, 595
Dozier, C. M., 6, 585
Drake, G. W. F., 409
Drell, S. D., 84
Dufton, P. L., 590
Dupree, A. K., 562, 590

Eckart, C., 307
Edelstein, S. A., 60
Ederer, D. L., 9, 534, 543
Edlén, B., 5, 8, 25, 28, 29, 223, 227, 239, 295, 343, 361, 450, 451, 454, 484, 502, 555, 571, 579, 583, 615, 702, 704, 705
Edmonds, A. R., 42, 142, 305, 702
Egert, S., 583
Einstein, A., 395
Ekberg, J. O., 25, 32, 409, 433, 454, 465, 583

El-Baz, E., 151, 155, 703
Elliott, J. P., 149
Ellis, D. G., 28, 421, 433, 467
Epstein, G. L., 223, 433
Erickson, G. W., 87, 704
Eriksson, K. B. S., 130, 337, 541
Ermolaev, A. M., 704
Esherick, P., 484
Even-Zohar, M., 585
Ezawa, H., 454

Fano, U., 42, 144, 149, 239, 305, 309, 361, 378, 379, 428, 484, 526, 527, 529, 534, 535, 702
Fawcett, B. C., 5, 32, 355, 439, 465, 466, 577, 580, 581, 588, 705, 706
Fein, A. E., 203
Feldman, U., 5, 6, 439, 581, 582, 585, 586, 588, 590, 595
Feneuille, S., 478
Ferendeci, A. M., 703
Fermi, E., 191, 192
Fielder, W. R., 439
Finkenthal, M., 546, 583
Fliflet, A. W., 482, 544
Flores, L. M., 707
Fock, V., 180
Foster, E. W., 404
Fowler, A., 29
Fowler, C. M., 486
Fraenkel, B. S., 583, 585, 595
Fraga, S., 477, 703, 708
Fred, M., 461, 482, 506, 612
Freedman, M. S., 705
Fried, B., 147
Friedman, H., 24
Froese Fischer, C., 177, 180, 216, 372, 438, 483, 624, 703, 708
Fues, E., 517
Fuhr, J. R., 436, 501, 706, 707, 708
Fuller, G. H., 507
Funahashi, A., 629

Gabriel, A. H., 5, 355, 555, 590, 592
Gallagher, T. F., 60, 557
Garcia, J. D., 87, 88
Garstang, R. H., 345, 401, 438, 439, 442, 444, 450, 453, 455, 486
Garton, W. R. S., 534, 541
Gáspár, R., 195
Gates, W., 6
Gaunt, J. A., 146
Gell-Mann, M., 203
Giacchetti, A., 17, 25, 489, 498, 607
Gibbs, R. C., 110
Ginocchio, J. N., 618
Ginter, M. L., 484
Glassgold, A. E., 535
Glennon, B. M., 436, 706
Goeppert Mayer, M., 16, 231
Goldberg, L., 441, 590
Goldman, M., 582
Goldschmidt, Z., 379
Goldschmidt, Z. B., 479, 598, 603
Goldsmith, S., 585
Gombás, P., 191, 205, 209
Goodman, L. S., 607
Gopinathan, M. S., 199
Götze, R., 29
Goudsmit, S., 29, 91, 115, 478
Grant, I. P., 200
Green, A. E. S., 193
Green, L. C., 411, 708
Greenberger, A., 593, 595
Griem, H. R., 18, 501, 708
Griffin, D. C., 187, 201, 216, 231, 235, 506, 587, 605
Grineva, Yu. I., 595
Grotrian, W., 17, 451
Gutman, A. M., 134, 298

Hagan, L., 8, 9, 487, 600, 610, 704, 705, 706
Hahn, Y., 562
Hald, A., 472
Hallén, K., 355
Hallin, R., 588
Hamermesh, M., 33
Hanle, W., 702
Hansen, J. E., 482, 541, 544
Hansen, P., 83
Hargreaves, J., 517
Harmon, B. N., 201
Harrison, D. E., Jr., 706
Harrison, G. R., 1, 7, 423, 703
Harrison, R. W., 196
Harter, W. G., 703
Hartree, D. R., 75, 176, 177, 180, 183, 194, 196, 214, 482, 703
Hartree, W., 176, 214, 482
Hartunian, R. A., 434
Hayes, R. W., 32
Hedges, R. M., 707
Hedin, L., 203
Heilig, K., 708
Heisenberg, W., 428
Heitler, W., 398
Herman, F., 197, 212, 214, 521, 703
Hershkowitz, M. D., 590
Herzberg, G., 2, 78, 81, 82

Hibbert, A., 202, 405
Hill, K. W., 582, 595
Hill, R. M., 60
Hindmarsh, W. R., 1
Hinnov, E., 5, 6, 582, 586, 588, 593, 629
Hirschfelder, J. O., 88
Hobby, M. G., 583
Hoory, S., 25
Horie, H., 453
Horton, R., 582, 595
House, L. L., 460
Hovey, J., 595
Hoyt, F., 19
Huebner, W. F., 526
Hughes, T. P., 588
Hults, M. E., 450, 452
Humphreys, C. J., 24, 25, 83, 223
Hund, F., 124
Hylleraas, E. A., 571

Innes, F. R., 329
Inokuti, M., 428, 567
Irons, F. E., 5
Isler, R. C., 629, 631
Iversen, D. B., 509

Jackiw, R. W., 703
Jacobs, V. L., 555, 562
Jahn, H. A., 380
Janet, C., 115
Jefferies, J. T., 452
Johansson, L., 133
Johansson, S., 465, 615
Johnson, B. M., 629
Johnson, M. H., Jr., 423
Johnson, S. A., 601
Johnston, R. R., 580
Jones, B. B., 5, 588
Jones, K. W., 629
Jones, M., 704
Jordan, C., 355, 426, 452, 546, 590, 592
Jordan, W. C., 580
Jucys, A. P., 151, 176, 243, 321, 707
Judd, B. R., 33, 155, 241, 263, 477, 478, 599, 603, 703

Kaminskas, V. A., 176
Karayianis, N., 109
Karaziya, R. I., 243, 321, 326, 707
Karwowski, J., 708
Kasai, S., 629
Kato, T., 706
Katz, L., 586
Kaufman, V., 25, 236, 464, 479, 484, 541, 580, 611, 613, 629, 704
Kay, L., 28
Kayser, H. G. J., 7
Keller, C. F., 625
Kelly, H. P., 482, 542, 544
Kelly, P. S., 378
Kelly, R. L., 7, 588, 704, 706
Kepple, P. C., 555, 562
Kernahan, J. A., 453
Kessler, K. G., 702
Kickin, I. S., 587
Kielkopf, J. F., 20
Kiess, C. C., 498, 614
Kilpatrick, J. E., 109
King, A. S., 607
King, W. H., 506
Kingston, A. E., 569, 590
Kirkwood, J. G., 203, 428
Klapisch, M., 193, 372, 582, 583, 625
Klein, S., 619
Kleinpoppen, H., 702
Kobe, D. H., 409
Koelling, D. D., 201
Kohl, J. L., 541
Kohn, W., 195
Konjevic, N., 501, 708
Kononov, E. Ya., 32, 577, 595
Koontz, G. D., 526
Koopmans, T., 181
Koozekanani, S. H., 150, 647, 707
Kopfermann, H., 505, 509
Kornblith, R. L., 435
Korneev, V. V., 595
Koshelev, K. N., 595
Koster, G. F., 263, 265, 319, 321, 326, 456, 458, 707
Kowalczyk, S. P., 9, 705
Kramers, H. A., 526
Kreplin, R. W., 24, 595
Kroll, N. M., 87
Kruse, T. H., 629
Kuhn, H. G., 2, 58, 122, 139, 702
Kuhn, W., 428
Kurucz, R. L., 707

Labarthe, J. J., 372
Ladenburg, R., 404
Lafoucrière, J., 155
Lamb, W. E., Jr., 87
Lan, Vo Ky, 541
Landé, A., 339, 487
Landolt-Börnstein, 509

Landshoff, R., 364
Laporte, O., 134
Larson, A. C., 196, 224
Larson, H. P., 89
Latter, R., 192, 235
Lawson, K. D., 577, 705
Layzer, D., 361, 401, 571, 615, 622
LeDourneuf, M., 541
Lee, C. M., 361, 484
Lefebvre, R., 202
Legagneur, C. S., 5
Leventhal, M., 454
Levinson, I. B., 134, 151, 298, 707
Ley, L., 9, 705
Li, H., 452, 482
Li, W.-K., 707
Liberman, D., 196
Lifshitz, E. M., 84
Lilly, R. A., 482
Lin, C. D., 540
Lin, D. L., 409, 439
Lindgård, A., 706
Lindgren, I., 170, 199
Litzén, U., 465, 600, 615
Loofbourow, J. R., 1
Lord, R. C., 1
Lotz, W., 9, 546, 578, 584, 588, 705
Louck, J. D., 703
Lu, K. T., 361, 484
Luc-Koenig, E., 193, 372
Luther, G., 586
Lyman, T., 83

McDowell, J. D., 563, 706
McDowell, M. R. C., 703
McEachran, R. P., 448
McFeely, F. R., 9, 705
McGucken, W., 1, 9
Mack, J. E., 87, 88, 134
Mack, J. M., Jr., 631
McLellan, A. G., 88
Macomber, J. D., 703
McWhirter, R. W. P., 546, 589
Madden, R. P., 9, 534
Magee, N. H., Jr., 555, 563, 595, 617
Malli, G., 477, 703
Mann, J. B., 116, 189, 196, 200, 224, 225, 226, 235, 546, 590, 624, 708
Mansfield, M. W. D., 543, 583
Maradudin, A. A., 203
March, N. H., 703
Margenau, H., 44, 162, 177
Marr, G. V., 526
Martin, G. A., 430, 436, 501, 704, 706, 707
Martin, R. L., 9, 705
Martin, W. C., 8, 9, 239, 360, 487, 599, 600, 608, 610, 704, 705, 706
Martinson, I., 28
Mason, H. E., 590
Massey, H. S. W., 552, 565
Massot, J.-N., 155
Mattioli, M., 546, 583, 629
May, C. A., 601
Mayers, D. F., 224
Meekins, J. F., 24
Meggers, W. F., 8, 24, 29, 614, 704
Mehlhorn, R., 476
Meissner, K. W., 133
Menzel, D. H., 398, 408, 411, 420, 442, 456, 703, 707, 708
Merrill, P. W., 17, 18, 706
Merts, A. L., 555, 557, 563, 595, 617
Meservey, E., 593
Metropolis, N., 143, 302, 486, 707
Mewe, R., 591
Mies, F. H., 535
Mikhailov, Yu. A., 595
Miles, B. M., 436, 706
Miller, B. J., 436, 501, 707
Minnhagen, L., 5, 355
Moiseiwitsch, B. L., 703
Moore, C. E., 8, 9, 10, 17, 18, 29, 115, 119, 174, 175, 221, 222, 235, 607, 704, 705, 706
Mori, K., 706
Morillon, C., 450
Morris, R. M., 90
Morrison, J. C., 465
Moser, C. M., 202, 343
Moszkowski, S. A., 619
Mott, N. F., 552, 565
Mrozowski, S., 452
Mueller, R. O., 486
Murphy, G. M., 44, 162, 177
Mutschlecner, J. P., 625

Nagami, M., 629
Nagel, D. J., 6, 585
Neidigh, R. V., 629
Nestor, C. W., Jr., 563, 706
Neubert, D., 115
Neupert, W. M., 6
Nicolaides, C., 453
Nielsen, S. E., 706
Nielson, C. W., 263, 265, 319, 321, 326, 456, 458, 707
Nikitin, A. A., 707
Nugent, L. J., 609
Numerov, B., 75

Odabasi, H., 285, 356, 456, 703
Odintzova, G. A., 704
Oertel, G. K., 411, 708
Osborne, P. J. H., 24
Otsuka, M., 706
Outred, M., 7, 704
Owens, J. C., 8

Paisner, J. A., 601
Palumbo, L. J., 7, 588, 704
Pang, P. H.-L., 453
Parkinson, W. H., 541
Paschen, F., 25, 29, 83, 496
Paszek, A. P., 478
Patterson, C. W., 703
Paul, E., Jr., 24, 223
Pauli, W., 101
Peacock, N. J., 5, 24, 32, 464, 465, 466, 577, 583, 705
Peck, E. R., 8
Pegg, D. J., 28, 435
Pekeris, C. L., 704
Pengelly, R. M., 562
Penkin, N. P., 541
Perez, J. D., 580
Perlis, S., 364
Persson, W., 498
Petrashen, M. J., 180
Peytremann, E., 707
Pfund, A. H., 83
Pines, D., 203
Pitaevskii, L. P., 84
Porter, C. E., 618
Porter, F. T., 705
Poulsen, O., 509
Prats, F., 379
Pratt, G. W., Jr., 176
Presnyakov, L. P., 555, 590
Preston, G. W., 21
Pryce, M. H. L., 452
Purcell, J. D., 588

Rabe, P., 543
Racah, G., 28, 42, 129, 134, 141, 144, 149, 164, 166, 167, 241, 245, 247, 257, 261, 263, 264, 283, 305, 308, 309, 316, 318, 319, 411, 464, 467, 476, 478, 479, 480, 491, 604, 702, 703
Radziemski, L. J., Jr., 25, 29, 132, 464, 484, 541, 580, 600, 601, 613, 631
Raizer, Yu. P., 427, 526, 545, 548
Rajnak, K., 464, 465, 478, 506, 580, 609
Ramanujam, P. S., 509
Ramsey, N. F., 508
Rau, A. R. P., 486
Reader, J., 223, 433, 586, 588, 610, 704, 705, 706
Reeder, K., 8
Reiche, F., 404, 428
Reisfeld, M. J., 613
Reiss, H. R., 16
Retherford, R. C., 87
Richards, P. I., 115
Ridgeley, A., 583
Riley, M. E., 515
Ripin, B. H., 486
Risberg, G., 288
Ritz, W., 17
Robb, W. D., 569
Roberts, J. R., 501, 708
Rodgers, J. E., 372
Rohrlich, F., 615
Roig, R. A., 541
Roney, W., 595
Rosén, A., 170, 199
Rosenthal, D., 448
Rosenzweig, N., 373
Ross, J. S., 506, 605
Roszman, L. J., 501, 708
Rotenberg, M., 143, 193, 302, 411, 458, 486, 636, 647, 707
Roth, C., 481, 615
Rudzikas, Z. B., 243, 321, 587, 707
Rush, P. P., 411, 708
Russell, H. N., 57, 60, 110, 484, 614
Rydberg, J. R., 9
Ryde, N., 500

Sack, N., 478, 480
Safronova, U. I., 577
Saha, M. N., 545
Salpeter, E. E., 83, 86, 401, 427, 429, 526
Samson, J. A. R., 534, 702
Sandars, P. G. H., 155
Sandlin, G. D., 5
Sansonetti, C. J., 302
Saunders, F. A., 57, 60
Saunders, P. A. H., 5
Sauthoff, N. R., 582, 595
Savukynas, A. J., 321
Sawyer, R. A., 1, 702
Saxena, K. M. S., 708
Saxon, R. P., 428
Schectman, R. M., 467
Schiff, B., 704
Schweinler, H. C., 364
Schwob, J. L., 583
Scribner, B. F., 704

Seaborg, G. T., 116
Seaton, M. J., 515, 520, 535, 546, 550, 562, 563, 590, 591
Sellin, I. A., 28, 435
Series, G. W., 81
Shabanova, L. N., 541
Shadmi, Y., 478, 615
Shadwick, W. F., 193
Sham, L. J., 195, 196
Sharma, C. S., 192
Sharp, R. T., 241
Shenstone, A. G., 29, 60, 128, 484, 493, 531, 614, 615
Shimoda, K., 702
Shirley, D. A., 9, 705
Shoffstall, D. R., 28, 421, 467
Shomo, L. P., 411, 708
Shore, B. W., 379, 398, 408, 411, 420, 442, 456, 534, 535, 562, 609, 703, 707, 708
Shortley, G. H., 27, 42, 44, 57, 94, 144, 147, 243, 290, 291, 308, 403, 411, 448, 468, 498, 703
Shudeman, C. L. B., 109
Sidel'nikov, Yu. V., 595
Siegbahn, K., 9, 174, 226, 584
Sinanoğlu, O., 202, 405, 453
Sinfailam, A. L., 590
Sinzelle, J., 615
Sitterly, C. E. Moore. *See* Moore, C. E.
Sivcev, V. I., 587
Skillman, S., 197, 212, 214, 521, 703
Slater, J. C., 9, 94, 108, 147, 163, 174, 195, 196, 197, 285, 356, 456, 703
Slepkov, A. A., 587
Smith, C. C., 583
Smith, K., 150, 535, 647, 707
Smith, M. W., 436, 450, 706
Smith, P. L., 452
Smitt, R., 355
Sobel'man, I. I., 508, 548, 703
Solarz, R. W., 601
Specht, B. J., 501
Spector, N., 604
Speer, R. J., 5, 24, 25
Spruch, L., 486
Stacey, D. N., 505
Stamper, J. A., 486
Stanley, R. W., 20, 24, 25, 89
Starace, A. F., 484, 542, 544
Stark, J., 21, 500
Starkand, J., 479
Steers, E. B. M., 25
Stegun, I. A., 565
Stein, J., 464, 478
Steinhaus, D. W., 600
Stephenson, J. W., 150, 647, 707
Sternheimer, R. M., 372
Stewart, A. L., 428, 526
Stewart, G. W., 96
Stewart, J. C., 193, 411
Stodiek, W., 582, 593, 595
Stoner, E. C., 115
Stoner, J. O., Jr., 17, 706
Storey, P. J., 550, 562, 563
Striganov, A. R., 704
Strong, J., 83
Subtil, J. L., 509
Suckewer, S., 582
Sugar, J., 236, 360, 479, 544, 610, 611, 613, 629, 705, 706
Sugie, T., 629
Summers, H. P., 546, 547, 562, 589
Svensson, L. Å., 25, 32, 355, 409, 433, 454, 465, 583
Sventitskii, N. S., 704
Swartz, M., 5, 6
Szydlik, P. P., 193

Talman, J. D., 193
Talmi, I., 127, 255
Taylor, B. N., 633
Taylor, K. T., 535, 541
Tech, J. L., 360
Theissing, H. H., 140
Theodosiou, C. E., 239
Thomas, L. H., 91, 191
Thomas, W., 428
Tilford, S. G., 484
Tinkham, M., 33
Tolansky, S., 24
Tong, B. Y., 196
Trees, R. E., 478, 480
Truhlar, D. G., 515
Turner, L. A., 60
Tyrén, F., 555

Ufford, C. W., 345
Uhlenbeck, G. E., 91
Underwood, J. H., 6
Unsöld, A., 107
Urnov, A. M., 590

Valind, S., 498
Vanagas, V. V., 151
van Dam, H., 142, 702
van den Bosch, J. C., 491
Vander Sluis, K. L., 493, 609

van der Waerden, B. L., 395
Van Regemorter, H., 591
Van Vleck, J. H., 345
Varga, L. P., 613
Varghese, S. L., 435
Vergès, J., 450, 600
Vinti, J. P., 505
Vizbaraite, Ya. I., 243, 321, 707
Voigt, W., 20
von Goeler, S., 582, 593, 595

Waber, J. T., 116, 196, 200, 224, 624
Waller, W. A., 24
Wares, G. W., 435
Wasserman, N., 563, 706
Watson, R. E., 94, 219
Way, K., 707
Weast, R. C., 7, 704
Weisheit, J. C., 562
Weiss, A. W., 202, 343, 438, 540, 541
Weissbluth, M., 703
Weisskopf, V., 19
Wendin, G., 535, 543, 544
Westfall, F. O., 5
Westhaus, P., 453
White, H. E., 1, 2, 46, 78, 110
Whitlock, R. R., 585
Widing, K. G., 5, 588
Wiese, W. L., 430, 434, 435, 436, 437, 450, 501, 704, 706, 708
Wigner, E., 19, 33, 142, 307, 364, 428
Wilber, D. T., 110
Wilkinson, J. H., 96
Williams, D., 702
Wilson, M., 473, 482, 534
Wilson, R., 5, 560, 583, 588
Wilson, T. M., 196
Wolnik, S. J., 435
Wood, D. R., 17, 133, 421, 454, 498
Wood, J. H., 196
Woodgate, G. K., 2, 139, 702
Wooten, J. K., Jr., 143, 302, 486, 707
Worden, E. F., 601, 607, 610
Wormer, P. E. S., 109
Wyart, J.-F., 29, 598, 600, 613, 615
Wybourne, B. G., 33, 109, 263, 464, 478, 598, 599, 601, 618, 703
Wynne, J. J., 484

Yen, M. M., 618
Yamanouchi, T., 453
Young, R., 6
Younger, S. M., 706
Yutsis, A. P. *See* Jucys, A. P.

Zaidel', A. N., 704
Zalubas, R., 8, 25, 487, 600, 704, 705
Zeeman, P., 21, 485
Zel'dovich, Ya. B., 427, 526, 545, 548
Zerby, C. D., 526

Subject Index

Abbreviations
  cfp (coefficient of fractional parentage), 257
  cfgp (coefficient of fractional grandparentage), 261
  CI (configuration interaction), 358
  E_{av} (configuration-average energy), 112
  see also Notation and special symbols
Absorption coefficient, 617
Accuracy, experimental
  Landé g factors, 493
  oscillator strengths, 434-36
  wavelengths, 25
Accuracy, theoretical
  correlation energy, 221-23
  dielectronic recombination, 559-62
  E1 transitions, 432-34
  interconfiguration energy ($\Delta n = 0$), 465, 581-83
  interconfiguration energy ($\Delta n \neq 0$), 465, 581, 583
  intraconfiguration energy, 461
  ionization energy, 174, 221-26
  M1 transitions, 444
  in rare earth configurations, 609
  relativistic effects, 221-26, 584-88, 612
  spin-orbit parameters, 94, 219, 237, 461, 587-88
  total binding energy (E_{av}), 174, 219-26
Actinide elements, 115, 612-13
Active electrons, 271, 373-74, 384
Airglow, 452
Ångström unit, 6
Angular momentum
  addition of two, 50
  coupling of, 50-60, 102
  diagrams, 152-55
  eigenvalues of, 35-38, 41, 49
  notation, 58-60
  operators, 34
  orbital, 34, 39
  recoupling of, 245-49, 380, 672
  spin, electron, 49
  spin, nuclear, 506-507
  unit of (h), 53, 68
  vector model, 39, 53-60, 102-103, 108-12, 122-34, 242, 490
  z-component, 41, 51, 56
  see also Coupling schemes; Quantum numbers
Antisymmetrization
  vs. coupling, 102-103
  for equivalent electrons, 250-55
    via cfp, 255-59, 669
  multi-subshell configurations, 265-67, 670
  product functions, 99-102
Approximation
  basis-function expansion, 94
  central-field, 97-98
  energy-matrix truncation, 95, 461, 463-64, 609
  energy-parameter scaling, 464-65
  frozen core, 170
  multi-configuration, 105
  single-configuration, 104
Astronomical sources, 5, 21, 24, 27, 451-54, 486, 592
Atomic-beam sources, 20
Auger effect, 531
Aurora, 452
Autoionization, 350, 514, 530-39, 549-53, 589-90
  level broadening, 514, 530
  transition probabilities, 530
    calculation of, 552-53
    vs. radiative rates, 558, 589
    variation with *l* and n, 557-58
    variation with Z, 589

Autoionizing levels
  contribution to collisional excitation, 590
  contribution to collisional ionization, 589
  effect on absorption spectra, 531-44
  lifetime, 530
  in rare earths, 601
  width, 530-31

Balmer formula for hydrogen, 81
Balmer series, 81-83
Baricenter energy, 113
Basis functions, 94
  continuum, 515
  determinantal, 102
    no need for, 140-41
  multi-subshell, 265-67
  single-subshell, 256-60
Basis-function expansions, 94, 358, 405, 443, 446, 514, 527
  size of, 102, 105, 140
Basis-function mixing, 56, 58, 289-301, 338, 421, 461, 506, 509
  configuration-interaction, 105, 361, 433, 444
    discrete-continuum, 514, 530-33
  parentage, 350-51
  spin-orbit, 338, 343-47, 409
  in Zeeman effect, 489
  see also Coupling conditions, intermediate; Purities
Bates-Damgaard (Coulomb) approximation, 411
Beam-foil spectroscopy, 28, 435
Bessel functions, spherical, 565
Beutler-Fano profiles, 532-34, 538-44
  in discrete part of spectrum, 541-42
  in electron scattering, 550
  window, 533, 538
Biedenharn-Elliott sum rule, 149
Binding energy
  one-electron, 160, 168-74, 222-23, 237
  total atom, 160, 168-74, 219-226
  variation with *l*, 113-14
  variation with n, 113-14, 224-29, 570
  variation with Z, 229-37, 570-78
Binomial coefficient, 138, 272, 424
Black-body radiation (Planck's law), 397
Black-body source, 2-3, 22, 398
Blume-Watson spin-orbit operator, 94, 219
  parameters, numerical values, 237-38, 588
Bohr magneton, 485
Bohr orbits, 81
Bohr theory, 1
Bohr unit (a_0), 10
Boltzmann factor, 397, 544

Boundary conditions
  bound radial functions, 69, 184-85
  continuum functions, 515
Bowl-shaped level patterns, 603, 605
Brackett series, 82-83
Branching ratio for radiative decay, 440-41, 553
Brehmsstrahlung, inverse, 526, 618
Brillouin's theorem, 372
Broadening, line
  autoionization, 514, 530-32, 538-39
  collision, 19
  Doppler, 20, 582, 590
  instrumental effects, 22
  natural (radiative), 18-19
  resonance, 19
  self-absorption, 22-23, 398
  Stark, 21, 501, 590, 708
  Zeeman, 21
  see also Spectrum line profiles
Building-up principle, 115

Calculation of energy levels, 138-41, 456-65
  ab initio, 29, 461-65
  least-squares, 29, 465-81
Calculation of radial functions, 139, 214-17
Calculation of spectra, 28, 456-65
Calculation of transition probabilities
  autoionization, 552-53
  radiative, 405-406
Cancellation effects
  in autoionization resonances, 531-35, 538-43
  configuration-interaction integrals, 477, 578
  line strengths, 432-34
  radial transition integrals, 433, 535, 580
Cascades, radiative, 435, 440-41, 591
Casimir operators, 479
Center-of-gravity relations
  E_{av}, experimental, 113
  E_{av}, theoretical, 112, 156-61, 168-75
  energy-level terms, 340, 357
  relative to parent term, 348, 353, 357
  transition array, line intensity *vs.* ΔE_{av}, 581-82, 625-26
  zero contribution of spin-orbit terms, 160
Central-field model, 97-98
Central-field problems, 68-70
Centrifugal term in effective potential, 74, 229-30
Channels, 526
  multi-channel quantum defect theory, 484, 535
Close-coupling calculations, 535, 568-69
Coefficients
  Clebsch-Gordon, 55, 144-45, 241-45, 271-75, 668

fractional parentage, 255-60, 669, 679-90
mixed-shell, 266, 378
fractional grandparentage, 259-61, 669, 680-90
Racah, 148
recoupling, 245-49, 380, 672
vector-addition, 55, 144-45, 241
Wigner, 241
Collapse of d and f electrons, 230-37, 578, 598, 612-13
effect on coupling conditions, 350, 355, 578
in LSD-HF calculations, 482-83, 544
Collision strengths, 568
plane-wave-Born, 567-69
Commutation relations, 34, 427, 430-31
tensor operators, 306
Complex (of configurations), 361
Components, line, 21, 403, 485
hyperfine-structure, 508-511
Zeeman, 403, 492-99
Computer calculations, see Calculation
Computing times, energy levels and spectra, 463
radial wavefunctions, 217
reduced matrix elements, 462-63
Configuration
center-of-gravity energy, 112-13, 156
electron, 103, 276
excited, 104
ground, 104
multi-subshell, 109
pure *vs.* mixed, 109
Configurations, specific
*l*s, or s*l*, 287
l^2, 287
l^w, 341-47, 451-55
l^{2l+1}, 345
n*l*n′*l*, 473
$l^w l'$, 346-56
$l^{4l+1} l'$, 353-56
l^ws, 351-53
ps and p^5s, 133, 295-98, 464
pp′, 251
pd, 123-24, 129, 131-32
pf, 299-301
p^2, 255, 342, 451-52, 488, 491
p^3, 255, 257, 345, 451-54
p^4, 255, 343, 451-55
p^3s, 349
p^wd, 453
p^4d, 350-51
p^5d, 354-55, 371, 473, 482-83
p^5ds, 293
d^w, $d^w l$, 623
f^w, $f^w l$, 621-29
f^3sd^2, 337
Si I-II, 133
Ti III 3d4p, 123
see also Elements, specific; Transition arrays
Configuration interaction (CI), 106, 358
Coulomb integrals (R^k), 359, 521
discrete-continuum, 514, 526-44
effect on g sums, 489
effect on isotope shifts, 506
end effects in calculations, 539-41
in highly ionized atoms, 578
inter-channel, 484
intra-channel, 542-44
see also Rydberg series
Lu-Fano analysis, 484
matrix elements, general, 373-94
one-electron terms, 360, 371, 390-94
perturbations, 106, 355, 464
in rare earths, 602
Rydberg-series, 361, 369-71, 394, 464, 484
alkali-like atoms, 371
p^5d, 355, 371, 473
selection rules for, 360, 367
spin-orbit integrals ($\zeta_{ll'}$), 360, 371
strong, 361
weak, 464
allowed for by scaling, 464-65, 478
Configuration mixing, 105-106, 109, 361, 433, 444, 514, 530-33, 602, 607
Continuum
basis functions, 515
configurations, 513
discretized, 535-43
as extension of bound series, 513, 522-24
radial integrals, 521-24, 536-37
radial wavefunctions, 515-516
close-coupling, 535, 568
distorted-wave, 535, 563, 568
energy dimensions, 521-22
normalization, 516-21
spectra, 2-3, 524, 532-34, 538-43
states, 513
Cooper minimum, 534
Core, electron
collapse of electrons into, 230-35, 598, 612-13
frozen, approximation, 170
penetration, 128, 221, 229-35, 570-71
polarization, 222, 501-502
relaxation, 169-71, 598
Correlation, electron
energy, 170, 202-205
importance of, 219-23
partial, through Pauli principle, 101, 159
relation to configuration interaction, 464
relation to polarization, 222-23, 501

Coulomb approximation for dipole integrals, 411
Coulomb energy, *see* Matrix elements, Coulomb
Coupled wavefunctions, 53-56, 102-103, 240-41, 668
Coupling conditions, 57-58, 64, 122
  effect of electron collapse on, 350, 355, 578
  intermediate, 58, 130-34, 288-301, 343-51, 421
  parameter values for pure, 298
  in rare earths, 603-606, 612-13
  in transition elements, 614
Coupling of antisymmetrized functions, 102-103, 669
Coupling schemes, 56
  genealogical, 112, 327, 670
  hybrid, 134
  intermediate-pair, 130
  pair, 128-31
  jj, 125-28
  J$\mathfrak{J}$, 128, 327
  jK, 129-31
  LK, 130-32
  LS, 57-58, 111, 122-24, 243
    use in numerical calculations, 64
  $\mathfrak{L}\mathfrak{S}$, 111, 327, 670
  Russell-Saunders, 57
  SL, 57, 130, 243
  see also Transformation matrices
Cross section
  collisional, 567
  photoionization, 524-26
  radiative recombination, 548

Daughter, 112
Detailed balance, principle of, 397, 547, 551
Determinantal functions, 102
  no need for, 140-41
Diagnostics, plasma
  electron density, 501, 590-92
  electron temperature, 590-97
  1s − 2p transitions, 593-97
Diagonalization of matrices, 96
Diagrams, *see* Angular momentum; Energy-level; Grotrian
Dielectronic recombination, 549-63
  Burgess formula, 561
  effect of electron density on, 560, 562
  effect of intermediate coupling, 557-61
  effect on relative line intensities, 555-56
    see also Satellite lines
  in highly ionized atoms, 589, 594-97
  importance in ionization equilibrium, 562-63
  inverse of autoionization, 550
  rate coefficient, 551-54
Dipole moment
  electric, 401
  magnetic (electronic), 442, 485
  magnetic (nuclear), 506-507
  Stark-effect-induced, 501
Dirac equation, 83
Dirac-Hartree-Fock method, 200-201
Direct Coulomb integrals (F^k), 163-64
  see also Matrix elements, Coulomb
Direct Coulomb interaction, 123, 159
Distorted-wave continuum functions, 535, 563, 568-69
Doppler profile, 20-21
Doubly excited levels, 514, 549
  see also Autoionization; Dielectronic recombination; Inner-subshell excitations
Dyadic operator, 445

Effective nuclear charge, 239, 570
Effective one-electron potential, 73-78, 218, 229-37
Effective operators, 477-81
  Coulomb exchange, 323-25
  radial-integral scaling as form of, 464-65
Effective principal quantum number, 225-39, 570
  in d- or f-electron collapse, 231-37
Eigenvalues
  of matrix, 96, 289-90
  of radial equation, 75-77
    in Koopmans' theorem, 181, 196, 219, 402
Eigenvectors, 96
  and transformation matrices, 97
  of 2-by-2 matrices, 290
  see also Basis-function mixing; Energy-level designations; Purities
Einstein A coefficient, 395
  see also Transition probabilities
Einstein B coefficients, 397
Electric field strength, atomic unit of, 499
Electrodeless discharge, 5
Electron probability density, 44-46, 78, 99, 107
  at nucleus, 79, 506-507
Electrons
  active, 271, 373-74, 384
  core, 170, 571
  equivalent, 106
  inner-shell, 9, 583, 590
  spectator, 271, 374, 384
  valence, 9
Electrostatic interaction, 93, 161-65, 278
  see also Matrix elements, Coulomb; Radial integrals, Coulomb
Elements
  actinide, 115, 612-13

discovered spectroscopically, 4
lanthanide, 115, 598-610
periodic system, 115-21
rare-earth, 598
super-heavy, 115
transition, 115, 613-15
Elements, specific
Al I, 539-42
Ar I, 218-20, 227-28, 297, 355, 542; III, 297, 345; V, 296; VIII, 431; XVIII, 88
As I, 347, 451-53; II, 344
Ba I, 482, 542-44; II, 236-37
Be I, 482
Bi I, 347; II, 344
Br II, 345; III, 347
C I-VI, 16; II, 10-11, 114; V, 509
Ca III, 355; VII, 452
Cd I, 4
Ce I, 612, 623
Cf I, 622-23
Cl I, 350, 541; III, 347
Co III, 623
Cs II, 302
Dy I, 433, 606, 610, 622-23
Er I, 623; V-X, 610
Es II, 628
Eu I, 623
Fe I, 434-35, 623-26; I-II, 600, 623; IX, 582; IX-XIII, 453-54; X-XVI, 5; XIII, 451-52; XV-XVI, 590; XVIII-XXVI, 593-97; XIX-XX, 581-82; XXI, 538, 556-561; XXI-XXIII, 559; XXII, 556, 561; ionization equilibrium, 563, 596
Gd I, 623; V-X, 610
Ge I, 296, 300, 344; II, 497-99
H I, 82-83, 429
He II, 89
Hg I, 409
Ho I-II, 623-28
I II, 345
K II, 355, 483
Kr I, 297; III, 297, 345
Li I, 431, 438-39
Mg I, 409; IX, 5
Mn IX-X, 466
Mo VII, 433; XV-XVI, 583; XVII, 623; XXIV, 293
N I-VII, 172-74; II, 300
Na I, 431; II, 297
Nd I, 622-23; I-VII, 602; V-X, 610
Ne I, 105, 297
Ni XI-XV, 454; XV, 452; XVIII, 5
O I, 451-53, 455; II, 453; III, 451, 453
Os I, 623
P I, 347, 600; II, 299-300, 600
Pa I, 612
Pb I, 296, 344, 452, 454
Po I, 345
Pr I, 612, 623; III, 604, IV, 605
Ru I, 623
S I, 345; III, 296, 344; IX, 460
Sb I, 347; II, 344
Sc IV, 355
Se I, 345; II, 347; III, 344; IX, 623
Si I, 131-33, 296-98, 344, 433-34, 452; II, 18, 114, 133, 297; III, 431-32; X, 5
Sm I, 623
Sn I, 296, 344; Zeeman effect, 495
Sr XVII, 623
Te I, 345; II, 347; III, 344
Th I, 612
Ti III 3d4p, 123; V, 355, 433; VI, 409-10
U I, 225-26, 622-23; I-II, 600; I-IV, 631; V-VI, 613
V II-III, 623
W V-X, 610; IX, 623, 628; X, 623, 625, 629; XXX-XXXVIII, 628-31; binding energies in ions, 611; ion configurations, 610-11
Xe I, 297; III, 297, 345
Y III-IV, 223
Yb III, 605
Zn IX, 623
see also Accuracy, theoretical; Ionization energy, tables; Isoelectronic sequences
Empirical spectrum analysis, 25-27
aids to, 27-32, 465-67, 606-610
constant-difference (sum) search, 25-26, 606-607
progress in, 28-31
Energy
average, of electron pair, 165, 173, 665-67
configuration-average, 112, 156, 170
correlation, 202-205, 221-23
Darwin, 85-87, 200
dependence on n and *l*, 113-15
dependence on Z, 120-21
electron-nuclear, 161, 172, 278
excitation, 9; in Y III, 223
ionization, 10-16, 120, 168, 174, 222-26
tables, 12-15, 705-706
kinetic, 161, 172, 277
mass-velocity, 85-87, 200
relaxation, 169-71
spin-orbit, 85-87, 91-92, 280-81
zero of, 9, 11
see also Matrix elements
Energy level, 8, 60
excited, 8

ground, 8
metastable. 450-54
statistical weight, 135-36
tabulations, 704-705
Energy-level designations, 33, 134, 291-95, 338, 350-51, 409
configuration, 506, 607
difficulty in providing, 134, 294-95, 350, 484, 602
parentage, 347, 350-51
see also Notation, coupled basis states
Energy-level diagrams, 10-11, 16, 105, 123
see also Energy-level structure; Grotrian diagrams
Energy-level structure
complex, 356
hydrogenic atoms, 79, 82, 87
intermediate coupling, 133, 296-300, 344-51
jj-coupling, 128-29, 343-47
bowl/umbrella patterns, 603, 605
LS-coupling, 122-25, 133, 339
numerical calculation, 140-41, 456-65
one-electron configurations, 82, 87, 279
pair coupling (jK and LK), 130-33, 298-302
rare-earth elements, 600-606
spin-orbit effects (near LS conditions), 339-41
two-electron configurations, 280, 285-88
see also Configurations, specific
Energy matrix, 94-97, 140-41
see also Matrix; Matrix elements
Energy states, 8, 21, 60
statistical weight, 135
Equilibrium
coronal, 545-47, 563
ionization, 544-47
in iron, 563, 596
local thermodynamic (LTE), 544
Saha, 545, 563
Equivalent electrons, 106
antisymmetrization difficulties, 250-54
basis functions for, 255-60
Coulomb matrix elements, 287-88, 328-31
permitted jj states, 127
permitted LS terms, 108-110
Equivalent orbitals, 106
in jj coupling, 126
Exchange
energy, 159, 166, 170, 209
matrix elements, 159, 163, 270, 284-85, 333-34
Exchange Coulomb integrals (G^k), 163-64
see also Matrix elements, Coulomb
Exchange Coulomb interaction, 123, 159
Exchange potential energy, 183
free-electron (Slater approximation), 195-99
local approximations, 190
Excitation conditions, 17, 607
coronal equilibrium, 545
laser excitation, 60, 484, 608
local thermodynamic equilibrium, 544

Fano profiles, see Beutler-Fano profiles
Feynman diagrams, 155
Fine structure (spin-orbit effects), 124, 339
contrasted with hfs, 507
in H I, 87
in He II, 89
in highly ionized atoms, 576-78, 581-83, 587
in least-squares fitting, 475-76
see also Intermediate coupling
Fine-structure constant (α), 84
Flag pattern (in hfs), 508-510
Forbidden lines
electric-dipole, 450-54, 582-83, 608
spin, 409, 421
Fourier transform spectroscopy, 24-25
Free-electron gas
correlation approximation, 203-205
exchange approximation, 194-99
exchange energy, 209-210
kinetic energy, 206
Free-electron wavefunction, 515, 564-65
Frozen-core approximation, 170

Gaussian, skewed, 619-23
Geissler discharges, 3
Gram's matrix, 362
Granddaughter, 112
Grandparent, 112
Graphical methods, angular momentum, 152-54
Grotrian diagrams, 16-18, 26, 82, 451, 706
hyperfine structure, 508
x-ray, 584
Zeeman effect, 492
Group theory, 33, 109
Gyromagnetic ratio of electron, 92, 442, 485

Hamiltonian
Darwin term, 84, 201
electron-electron, 93, 675
electron-nuclear, 70
external electric field, 499
external magnetic field, 485
hyperfine structure, 507, 509
kinetic-energy term, 67-68
magnetic terms, 93-94, 477-78
mass-velocity term, 83, 201
multi-electron atom, 93, 156, 276
one-electron atom, 67

relativistic terms, 83, 201
spin-orbit term, 83, 93, 279, 675
Hansen-Strong series, 82-83
Hartree (unit of energy), 10
Hartree-exchange method, 197-99
HXR method, 202
Hartree-Fock method, 178-89
equations, 180
solution of, 184-89
HFR method, 202
LS-dependent, 481-83, 544
in highly ionized atoms, 580
in least-squares fitting, 482
multi-configuration, 483
restricted, 178
spin-polarized, 178
unrestricted, 178
Hartree-Fock-Slater method, 194-97
Hartree method, 194
Hartree-Slater method, 199
Heisenberg uncertainty principle, 18, 530
Hollow-cathode sources, 5
Humphreys series, 82-83
Hund's rule, 124
Hyperfine structure, 21, 506-511
as aid to spectrum analysis, 607
contrasted with fine structure, 507

Independent-electron model, 116
Index of refraction of air, 8
effect on wavelength, table, 634
Induced emission, 397-98, 547, 617
Infrared, 7, 24-25
Initial slope, 184-88
Inner-subshell excitations, 9, 575, 590
Intensities, spectrum line
in dielectronic recombination, 555-56, 591-97
effect of autoionization, 589
effect of electron density, 590-92
effect of self absorption, 22-23
effect of temperature, 590-97, 607-608
in hyperfine structure, 509-11
range, 17, 626-29
wavelength distribution, 581-83, 625-31
asymmetrical about ΔE_{av}, 582, 625
in Zeeman effect, 492-94
Interactions, coupling of angular momenta by, 55-57
see also Electrostatic; Spin-orbit
Intercombination lines, 409
Interference
between M1 and E2 radiation, 449-50
in bound-bound line strengths, 432-34
in bound-free transitions, 527-44
destructive, 409-10, 432-34
Interferometers, 4, 8, 24-25
Intermediate coupling, 288-301
pair coupling (pf), 298-302
transformation matrix to, 97, 406
ps, p^5s, 295-97
p^2, p^3, p^4, 341-47
in Mo XXIV, 293
in Ti VI, 410
see also Purities
Intersystem lines, 409, 608
Ionization
autoionization, 530-31, 546, 589-90
collisional, 546, 589
photo, 523-26, 531-35, 538-44, 546
Ionization energy, 10
configuration-average, 168-71
correlation effects, 221-23
effect of core relaxation, 169-71
relativistic effects, 221-26
tables, 12-15, 174, 220-26, 705-706
variation with Z, 120-21, 229-37
Ionization equilibrium, 544-47
iron, 563, 596
rule of thumb, 563
Ionization limit, 10-11, 16
Ionization potential, 10
Irreducible tensor operators, *see* Tensor operators
Isoelectronic sequences, 6
binding energies, 570-80
$\Delta n = 0$ transitions, 581-83
$\Delta n \neq 0$ transitions, 581
Li I sequence, 438-39, 588
N I sequence, 572-79
Na I sequence, 587-88
Al I sequence, 524
Cl I sequence, 351
Ar I sequence, 355
Ni I sequence, 586
Cu I sequence, 586-88
see also Z-dependence
Isotope shifts, 21, 505-506
aid to spectrum analysis, 506, 607, 610
bibliography, 708

Kayser (unit of wavenumber), 7
Kinetic energy, 67, 161, 277
angular momentum part of, 74
in configuration interaction, 360, 370, 390
free-electron, 513
in N I-II, 172
in virial theorem, 79, 89-90

Koopmans' theorem, 181, 219, 402
non-validity of, 598
in Xα method, 196

Lagrangian multipliers, 177, 180-81, 185, 207-208, 211, 217, 291
Lamb shift, 87
Landé g factors, 487-95
accuracy of measurement, 493
in intermediate coupling, 489, 491
in lanthanides, 602
in least-squares fitting, 476
in jj coupling, (p^2), 491
in jK coupling, 491
in LS coupling, 487, 490
Landé interval rule, 339
deviations from, 340
for hyperfine structure, 508
in lanthanides, 603
Lanthanide elements, 115, 598-610
aids to spectrum analysis of, 606-10
coupling conditions, 603-604
level structure, 598-610
spectra, 599-600, 604-10
Lanthanide-like ions, 610-11
Laplace's equation, 208-209
Laplacian operator, 67
Laser, 398
production of highly ionized atoms by, 5
Laser-excitation spectroscopy, 60, 484, 608
Laser-produced plasmas, 5, 486
Latter cutoff, 192, 219-20
l-dependence
autoionization rate, 557-58
coupling conditions, 128
radial wavefunctions, 80-81
spin-orbit parameters, 86, 129
Least-squares level fitting, 29, 465-81
distortion of parameter values, 467, 473-75
LS-dependent parameters in, 482-83
Legendre function, associated, 43
Legendre polynomial, 42
Level, *see* Energy level
Level crossings, avoided, 292-97, 300-301, 350
Level structure, *see* Energy-level structure; Configurations, specific
Lifetimes
autoionization, 530
collisional, 19
measurement, 435-36
radiative, 18-19, 396, 405, 441
Line, *see* Spectrum line
Line factor, 406-408, 418, 694
tables, 408, 695-701
Line strength
calculation of, electric dipole, 405-406
and direction of quantum-number changes, 407, 413, 417, 429
electric dipole, 402-403, 405
electric quadrupole, 445
forbidden transitions, 443-49
in hyperfine structure, 509-10
and statistical weights, 415-17
in Zeeman effect, 498
Local-exchange potential, 190-99
energy-dependent, 515
Lorentz unit of Zeeman splitting, 493
Lyman series, 81-83

Madelung's rule, 116
Magnetic flux densities, 485-86
Magnetic interactions, 94, 477-78
see also Energy, spin-orbit; Radial integrals, spin orbit
Magneton, Bohr, 485
nuclear, 507
Matrix
configuration-interaction, 359
diagonalization of, 96, 140, 290, 457-59
eigenvalues, 95-96
eigenvectors, 96-97
external magnetic field, 140-41, 488
Gram's, 362
Hamiltonian, 96, 140-41, 337
Hermitian, 95
inverse, 97
Landé g factor, 486-91
orthogonal, 97, 244
overlap, 362
radiative transition, 405-406
symmetric, 95
transformation, 97, 244-45, 248-51, 255, 328, 458
transpose, 97
unitary, 97
Matrix elements
collisional excitation, 564-66
complex, 303-304, 337-38
configuration-interaction, 358-94
Coulomb, 159, 282-88, 303-304, 322-23, 328-34, 675
of double tensor operators, 320-21, 679-93
effective operator for exchange, 323-25, 676
electron-nuclear energy, 161, 278
evaluation in LS scheme, 64
external electric field, 500
external magnetic field, 486-88
Hamiltonian, 337

hyperfine structure, 508-10
kinetic-energy, 161, 277
of product of two operators, 310
radiative transition, 405-406, 410, 412-14, 417-21, 443, 447-48, 694-701
coupling-scheme transformation, 417
reduced, 308, 673
angular momentum, 309, 673
Condon and Shortley, 308, 411
spherical harmonic, 309, 677-78
uncoupling formulae, 313-14, 676
of scalar product, 314-16, 674
of spherical harmonics, 146, 677-78
spin-orbit, 279-81, 321, 335-36, 675
of symmetric operators, 157-59, 267-70, 375-78, 674
of tensor product, 311-13, 674
uncoupling formulae, 270-71, 671
of unit tensor operators, 317-19, 679-93
Matrix method for atomic structure, 94-97
Maxwell-Boltzmann distribution, 20, 397, 544
Moseley diagrams, 584-85
Multi-configuration approximation, 105, 358
Multi-configuration HF approximation, 483
Multi-photon decay, 16
Multiplet, 407
Multiplets, discovery of, 614
Multiplicity, 60
Multipole expansion
Coulomb interaction, 161-62
radiation field, 398

n-dependence
autoionization rate, 557-58
coupling conditions, 132-33
one-electron energies, 113-14, 224-29
oscillator strength, 436-37
quantum defect, Ba II, 236
radial integrals, 227-28
radial wavefunctions, 77-81, 224-27
Nebulae, 453
Nodes of radial functions, 77, 185-90
Normal mass shift, 505
Normalization conventions
bound radial functions, 70, 98
continuum radial functions, 516-21
determinantal functions, 101
eigenvectors, 95-96
radial Coulomb integrals, 166-68
radiative transition integrals, 411, 445
spherical harmonics, 44
spin-orbitals, 98
tensor operators, 319, 321

Notation
abbreviated, for levels of 1- and 2-electron atoms, 88
angular momentum, 58-60
Coulomb radial integrals, 166-68
double tensor operators, 319
electric dipole integrals, 411
electron configuration, 107
ionization stage, 4
multipole radiation, 398
parentage, 347, 350
parity (odd symbol), 62
seniority, 262
Notation, coupled basis states
jj coupling, 126
$J\mathfrak{J}$ coupling, 128, 327
jK coupling, 130
LK coupling, 130
LS (Russell-Saunders) coupling, 60, 62
$\mathfrak{L}\mathfrak{S}$ coupling, 111, 327
Notational conventions
forbidden transitions, 450
radiative transitions, 88
Notation and special symbols
$C_q^{(k)}$, 146, 672
d_j, 281, 335
f_i, 156, 671
f_k, g_k, 284-85, 330-34
F^k, F_k, G^k, G_k, 163-67
g, 135-38, 487-89
g_{ij}, 156, 671
H_α, H_β, H_γ, $\cdots$, 81
$I^{(r)}$, $I^{(r1)}$, 332, 334
[j], $[j_1, j_2, j_3, \cdots]$, 144, 668
Ly_α, Ly_β, Ly_γ, $\cdots$, 81
P_{nl}, 69, 176; $P_{\varepsilon l}$, 515
$P_{ll'}^{(t)}$, 410-11, 447
r_d^k, r_e^k, 359, 367
R^k, 163, 359, 521
R_s, R_j, R_x, 380
U_a, U_b, 332
$U^{(k)}$, $V^{(k1)}$, 317, 320
Y_{kq}, 43
$\delta(j_1 j_2 j_3)$, 143
ε, 513
ζ, 279-81, 335
π, 493
σ, 7, 411, 493
σ_{m_s}, 49
ψ *vs.* Ψ, 100
Nuclear effects, 505-11
Nuclear magneton, 507
Numerov method, 75

One-electron model, 116
Opacities of thick plasmas, 617-18
Operators
  angular-momentum, 34
  Casimir, 479
  effective, 477-81
    for exchange, 323-25
  Laplacian, 67
  scalar, 64
  step-down, step-up, 38-39
  symmetric, 156
  vector, 64, 306-307
  see also Hamiltonian
Orbitals, equivalent, 106
Orbitons, 379
Orthogonality
  continuum functions, 517-19
  coupled wavefunctions, 243
  determinantal functions, 101
  product functions, 99
  radial functions (bound), 98, 176, 195
  spherical harmonics, 44
  spin functions, 49
  spin-orbitals, 98
  3n-j symbols, 145, 149, 151
Orthogonalization problems
  Hartree radial functions, 194
  HX radial functions, 199
  in multi-configuration bases, 361-65
Oscillator strength, 404
  array, 425-26
  availability, 437
  component, 424-25
  generalized, 563-67
  in hyperfine structure, 511
  line, 404, 425
  l-summed, 426-27
  measurement of, 434-36
  multiplet, 425
  n-dependence of, 436-37
  negative for emission, 404, 430
  range of values, 433, 626-29
  relativistic effects, 438-39
  sum rules, 428-31, 567
  systematics of, 436-39
  tabulations, 436, 706-707
  total, 427-31
  virtual transitions, 430
  weighted, 404
  Z-dependence of, 437-39
Overlap integrals, 194, 199, 362, 365

Pair coupling, 128-33, 298-302
  order of J values in, 301
Parametric potential method, 193
Parametric treatment of level structures, 29, 465-81
Parent, 112
Parity, 61-62
  notation (odd symbol), 62
Parity selection rules, 62-63
  generalized oscillator strengths, 566
  Hamiltonian, 63
    configuration interaction, 360
    dielectronic recombination (autoionization), 553
  radiative, 402, 411, 422, 443, 445
  Stark effect, 63, 500
  Zeeman effect, 63, 486
Partial-wave expansion, 552
Partition function
  internal, 544-45, 616, 625
  translational, 545
Paschen-Back effect, 494
Paschen series, 82-83
Pauli approximation, 201
Pauli exclusion principle, 101, 126
  energy resulting from, 101, 159, 166
  in jj coupling, 126
  in vector model, 111, 128
Penetration, electron-core, 570-71
  see also Core, electron
Periodic system, 115-21
Pfund series, 82-83
Phase, CI parameters (R^k), 477
Phase conventions
  basis functions, 263
  coefficients of fractional parentage, 263
  configuration-interaction integrals, 477
  coupled wavefunctions, 242, 274
  effective operators, 479
  more-than-half-filled subshells, 263-64, 318, 320-21
  radial wavefunctions, 71, 185
  spherical harmonics, 42-43
    matrix elements, 309
  step-up, step-down operations, 39
  vector-coupling coefficients, 144
Phase relationships
  angular-momentum coupling, 242-43
  equivalent-electron functions, 263-65
  in least-squares fitting, 477
  reduced matrix elements, 317-18, 320-21
Photoionization, 514, 523-26, 531-44, 546
  cross sections, 524-26
    in Al I, 540-42
    in Ba I, 543
    in Cl I, 541-42
Planck distribution, 397

Plasmas
  high-density, 544
  intermediate-density, 547
  low-density, 440, 545, 590-94
  optically thick, 22, 617
  optically thin, 22, 398, 545
  see also Diagnostics
Polarizability of atom, 501-502
Polarization
  of atom, 222
  E1 radiation, 403
  E2 radiation, 445
  M1 radiation, 443
Potential energy
  classical, 181-83
  electron-nuclear, 70
  exchange, 183
  Hartree, 183, 194
  Hartree-exchange, 197-99
  Hartree-Fock, 183
  Hartree-Fock-Slater, 194-97
  Hartree-Slater, 199
  local, 190
  parametric, 193
  Thomas-Fermi, 191
  Thomas-Fermi-Dirac, 192
Potential functions, effective, 74, 218, 227-33
  variation with Z, 230-33, 237
Principal quantum number, 71, 77, 185
  effective, 225-37, 570
    in wavefunction collapse, 230-37
  multi-electron atoms, 77, 185
  one-electron atoms, 71, 77
Principal value of integral, 528
Product wavefunctions, 53, 56, 98-99, 240
  multi-subshell, 266
  reduction of matrix elements to, 267-70
Profiles, *see* Broadening; Spectrum line profiles
Pseudo-discrete treatment of continuum problems, 535-43
Purities, eigenvector, 291-301

Quadrupole moment, nuclear, 508
Quantum defect, 226, 230-34, 570
  see also Principal quantum number, effective
Quantum-defect theory, multi-channel, 484, 535
Quantum numbers
  azimuthal (*l*), 41, 71, 77
  electron spin (s), 49
  good, 56-58, 338, 484, 487, 500
  $\mathfrak{L}$, $\mathfrak{S}$, $\mathfrak{J}$, $\mathfrak{M}$, 111, 128, 327
  magnetic (m), 41
  nuclear spin (I), 507
  one-electron (j), 84, 125
  pair (K), 128-30
  principal (n), 71, 77, 185
  total atom (F), 507
  total electronic (J), 56-57, 125
  total magnetic (M), 51, 57-58
  total orbital (L), 57, 111
  total spin (S), 57, 111

Racah algebra, 28, 141, 304-21, 672-76
Radial equations
  continuum functions, 515
  Hartree-Fock, 180
  hydrogenic atoms, 70
  multi-electron atoms, 178-201
  numerical solution, 72, 184-91, 202, 214-17, 515, 520
  one-electron atoms, 69
  relativistic, 85, 200-202
Radial integrals
  configuration-interaction (R^k), 359-60, 521, 537
  for continua, 521-24, 536-37
  Coulomb, (F^k, G^k, R^k), 162-63
  discrete *vs.* continuum, 521-24
  electric dipole, 410
    sign of, 411, 433, 524, 534
  electric multipole, 417-18
  electric quadrupole, 447
  numerical values, 172, 220, 225, 228, 236-38, 524, 572, 579, 588, 600, 708
  scaling *vs.* CI, 461, 464, 478, 580
  scaling factors for, 464-65, 467, 580
  spin-orbit (ζ), 279, 360
    Blume-Watson, 238, 588
  variation with n, 227-28
  variation with Z, 73, 236-39, 437, 571-80, 589, 624
Radial wavefunctions
  collapse at start of d and f series, 230-36, 598, 612-13
  continuum, 515-21, 535
  contraction in f series, 236, 600-601
  effect of configuration interaction, 483
  hydrogenic, 69-78
  multi-configuration atom, 218, 227, 231, 233, 237
  normalization, 70, 98
  numerical calculation of, 72, 139, 184-91, 202, 214, 515, 520, 703
  phase convention, 71, 185
  relativistic effects on, 223-26
  relaxation effects on, 169-70
  variation with n, 77-81, 224-27
  variation with Z, 72, 231-36, 578, 589
  see also Radial equations

Radiationless transitions, 350, 514
  see also Autoionization; Dielectronic recombination
Radiative transitions
  bound-bound, 395-455, 618
    see also Transition arrays
  bound-free, 523-26, 531-44, 618
  cascade decay, 440
  free-free (brehmsstrahlung), 6, 526, 618
Random-phase approximation, 535, 542-44
Rare earth elements, 598-613
Recombination
  dielectronic, 546, 549-63
    Burgess formula, 561-62
    effect of density, 560, 562
    in highly ionized atoms, 562, 589, 594-97
    radiative stabilization, 553
    rate coefficient, 551, 553-54
    relative line intensities, 555-56, 589
  radiative, 546-49, 550
  three-body, 546
Recoupling
  four angular momenta, 249
  three angular momenta, 245-48, 380
  see also Transformations
Recursion relations
  Clebsch-Gordon coefficients, 242
Reduced mass, 68, 81, 505
Relativistic contributions to Hamiltonian, 84-85, 201
Relativistic effects
  in $\Delta n = 0$ transitions, 582
  in $\Delta n \neq 0$ transitions, 581
  on E_{av}, 200, 219-24
  on energy levels, 81-88, 571, 574-76, 584-87, 612
  in hyperfine structure, 509
  on oscillator strengths, 438-39
  on radial integrals, 238, 579, 587-88
  on radial wavefunctions, 200, 223-25
    indirect, 223-25
  on rare-earth ground configurations, 612
  on transition operators, 409
Relaxation of ion core, 169-71, 598
Resonance lines, 17
Resonance (Lorentzian) profile, 19-21, 530-31
Resonances, autoionization, in electron scattering, 550, 589-90
Ritz combination principle, 16-17, 25
Ritz wavelength standards, 25
Rydberg
  constant, 10, 81
  reduced-mass correction, 68, 81, 505
  series, 10, 58, 106
    in continuum, 513, 524, 538-43
    interactions, 369
    perturbations between two, 484
  unit of energy, 7, 9

Saha equation, 545
Satellite lines
  from doubly excited levels, 555, 589-90
    in dielectronic recombination, 555-56, 589, 594-97
  of a multiplet, 407
Scaled Thomas-Fermi method, 193, 411
Schrödinger equation
  multi-electron atoms, 94
  one-electron atoms, 67
  see also Radial equations
Screened nuclear charge, 239, 570
Screening constants, 239, 570-80
Screening defect, 570
Selection rules, 62-64
  configuration-interaction, 360
  electric-field matrix elements, 500
  generalized oscillator strengths, 566
  magnetic-field matrix elements, 486
  in Wigner-Eckart theorem, 308
  in 3n-j symbols, 142, 148, 150
Selection rules, radiative, 17, 64
  in condensed-phase spectra, 608
  in hyperfine structure, 510
  in rare-earth spectra, 605-606
  in x-ray spectra, 584
  in Zeeman effect, 402-403, 492-93, 495-96
  on *l*, 411
  on J and M, 402-403
  jj coupling, 414
  jK coupling, 414
  LK coupling, 413
  LS coupling, 406, 412
  M1 and E2 radiation, 443-48
Selection-rule violations, 407-409, 421-22
Self absorption, 22, 27, 398, 617
Self-consistent field, 184
  convergence criteria, 221
Self reversal, 22
  as aid to spectrum analysis, 608, 610
Seniority number, 261-65
  phase relationships involving, 318, 321
Series, Rydberg, 10, 58, 106, 513
  limit, 10
Series, of spectrum lines
  sharp, principal, diffuse, fundamental, 58
Shell, electron, 106
  see also Subshell
Single-configuration approximation, 104

Slater determinant, 102, 140-41
Slater integrals (parameters), 163
  see also Radial integrals, Coulomb
Solar corona, 435, 582
  electron density, 441, 545
  spectrum lines, 451, 454
Solar flares, 6, 451, 595
Solar spectrum, 5
Specific mass shift, 505
Spectator electrons, 271, 374, 384
Spectator subshells, 270, 671
Spectra, 2-6, 82, 495, 600
  absorption, 9, 608, 618
  arc, 4
  astrophysical, 5, 21, 24, 27, 451-54, 486, 592
  band, 2-3
  complex, 600
  continuous, 2-3, 524, 531-43
  forbidden, 450-54, 582
  high ionization stages, 5-6, 460-66, 581-97, 610-11, 628-31
  hydrogen, 82
  ion-stage notation, 4-6
  line, 2-3
  photoelectron, 8-9
  rare-earth, 598-600, 604-10
  solar, 5-6, 28, 434-35, 451-54, 582, 595
  spark, 4
  tokamak, 5, 582-83, 593-97
  wavelength distribution, 581-83, 625-31
  x-ray, 9, 583-88
  Zeeman, 492-99
  see also Elements, specific; Transition arrays
Spectrochemical analysis, 4, 27
Spectrographs, 2-4, 7, 24-25
Spectrometers, crystal, 24
Spectrum analysis, 25-27
  status of, 28-31
Spectrum analysis, aids to
  condensed-phase spectra, 608
  constant differences/sums, 25-27, 606-607
  hyperfine structure, 607
  intensities, 607
  isotope shift, 506, 607
  laser excitation, 484, 608
  self reversal, 26, 608
  theoretical, 27-32, 460-67, 484, 609-10
  vector model, 614
  Zeeman effect, 489, 493, 606-607
Spectrum line, 4
  formation by spectrograph, 2
  see also Intensities; Line strength; Spectra; Wavelength
Spectrum line profiles, 17-23
  affected by induced emission, 398
  affected by self absorption, 22-23, 398
  autoionization, 530-31
  Beutler-Fano, 532-34, 538-44
  see also Broadening, line
Spherical harmonics, 44-45
  addition theorem of, 44-48
  expansion of product of two, 145-46
  matrix elements of, 146-47, 309, 677-78
  renormalized, 146, 306, 672
  and tensor operators, 305-306
Spin, electron, 49-50
Spin, nuclear, 506-507
Spin-orbit energy, 86, 91, 280-81
  splitting factor, 339-41
Spin-orbit interaction, 91-92, 123
Spin-orbit matrix elements, 279-81, 322, 335-36, 360, 371
Spin-orbit parameter, 279
  Blume-Watson theory, 94, 237
    numerical values, 238, 588
  central-field formula, 93, 279
  variation with *l*, 86, 129
  variation with n, 227-28
  variation with Z, 236-39, 576-77
  see also Radial integrals
Spin-orbitals, 98
  equivalent, 106
Splitting factor, hyperfine structure, 508
Splitting factor, Landé, 339, 341-42
  relation to spin-orbit parameter, 340-41
Stark effect, 21, 498-504
  as measure of electron density, 501
States, basis, *see* Basis functions
States, energy, 8, 21, 60
  statistical weight, 135
Statistical distributions
  energy levels, 618-24
  oscillator strength, 625-31
  statistical weight, 618-24
  wavelength, 581-83, 625-31
Statistical weights, 134
  configuration, 137-38
  energy distribution, 618-24
  jj term, 137
  level, 135-36
  level pair, 137
  LS term, 137
  shell, 136
  state, 135
  subshell, 135, 138
Step-down/up operators, 38
Stimulated emission, 397-98, 547, 617
Sub-configuration, 112, 461

Sublevels, 20, 485
Subshell, 106
  effect of filled, 107, 278, 367, 370
Sum rules
  Biedenharn-Elliott, 149
  Landé g factors, 489
  matrix elements, 309-10
  3n-j symbols, 145, 149-50, 151
Sum rules, line strength, 422-24, 431
  J-file, 407
  J-group, 423
  jK-coupling, 414-15
  LS coupling, 407
Sum rules, oscillator strength, 424-31
  electric quadrupole, 448
  generalized, 567
  Thomas-Reiche-Kuhn, 428-30, 439
  Wigner-Kirkwood, 428-29
Super-heavy elements, 115
Symmetric operators, 156, 317, 320, 671
  matrix elements of, 157-59, 267-70, 375-78, 671
  see also Tensor operators
Synchrotron radiation, 9
System differences, 608-609, 615

Tensor operators, 305-307
  double, 319-21
  Hermitian, 306
  scalar, 402
  scalar product, 314-16, 674
  tensor product, 311-13, 674
  unit, 316-19
  vector, 306-307, 402
Term, energy, 60
  intervals, p^w configurations, 343-47
  inverted, 340
  normal, 340
Theta-pinch discharges, 5
Thomas-Fermi method, 191-92, 206-208
  scaled, 193, 411
Thomas-Fermi-Dirac method, 192-93, 209-12
Three-n-j symbols, 142-55
  tables, 635-64
Tokamaks, 5, 582, 593-97, 629-31
Transformation matrices
  Clebsch-Gordon, 244-45
  between coupling schemes, 247-51, 255, 328, 380, 672
  to intermediate coupling (eigenvector matrices) 97, 406
  by matrix diagonalization, 458
Transformations, matrix
  between coupling schemes, 153-54, 247-51, 328, 336, 380, 672
  of transition matrices, 406, 417
Transition arrays, 412
  concept invalidated by CI, 423
  general, 417-21
  notation (lower-level first), 88
  transfer of line strengths between, 423, 430, 481
  weak because of cancellation, 409, 433-34
  $l - l'$, 410-11
  $l_1{}^n l_2 - l_1{}^n l_2'$, 412-17
  $l_1{}^n - l_1{}^{n-1} l_2$, 420-21
  $l_1{}^n l_2{}^{k-1} - l_1{}^{n-1} l_2{}^k$, 420, 630
  $s^2p^4 - sp^5$, 460
  pf − pg, 415-17
  $p^3 - p^2d$, 463
  d − f, in Zeeman effect, 499
  $d^4 - d^3p$, 463
  $d^6s^2 - d^6sp$, 623-26
  $f^3s^2 - f^3sp$, 463
  $f^{11} - f^{10}d$, 625-29
  $f^{11}s - f^{11}p$, 625-28
  1s − 2p, 593-97
  $1s^22p - 1s2p^2$, 556
  2s − 2p/3p, 438-39
  3s − 3p, 586-87
  $3s^23p - 3s3p^2 + 3s^2nd$, 540
  $3p^6 - 3p^5nd$, 542, 582
  $3d^{10} - 3d^94p$, 584-86
  4s − 4p, 585-86
  $4p^64d^w - 4p^54d^{w+1}$, 628-31
  $4p^64d^w - 4p^64d^{w-1}4f$, 629-31
  $4d^{10}4f^w - 4d^94f^{w+1}$, 542-44
  7s − 7p/8p, 7p − 8d, 631
Transition elements, 115, 613-15
Transition probabilities
  autoionization, 530, 552-53, 558
  radiative (E1), 395, 400-404, 440, 558
    in hyperfine structure, 510-11
  radiative (E2), 445, 449
  radiative (M1), 443-44
  weighted, 403
Triangle delta function, 142-43
Triangle relations, 142-43
  in radiative transitions, 402, 406, 410, 413-14, 447-48
  in reduced matrix elements, 308-10, 317, 320
  in Wigner-Eckart theorem, 308
  in 3n-j symbols, 143, 148, 150
Truncation of energy matrices, 461

Ultraviolet, 7, 24-25
Umbrella-shaped level patterns, 603, 605
Uncoupling, in matrix elements, 270-71, 671
  in reduced matrix elements, 313-14, 676
Unsöld's theorem, 107

Variational principle
  for angular dependence of energy, 291
  for radial functions, 139, 176-77
Vector-coupling coefficients, 241, 271-75
Vector-model, *see* Angular momentum
Virial theorem, 79, 88-91
Virtual orbitals, 362
Virtual transitions, in T-R-K sum rule, 430
Voigt profile, 20

Wavefunctions
  antisymmetrized, 99, 250-61
  basis expansions, 94, 358, 405, 443, 446, 514, 527-30
  coupled, 52-56
  determinantal, 102, 141
  free-electron, 564-65
  mixing of, *see* Basis-function mixing
  for multi-subshell atoms, 265-67
  one-electron, 98
  product, 50, 98-99
  radial, *see* Radial wavefunctions
  for single-subshell atoms, 256-60
  uncoupled, 52-53
Wavelength, 6-8
  accuracy, 24
  air, 8, 634
  infrared, 24-25
  range, 24-25
    within an array, 581-83, 607, 625
  standards, 20, 25
  tables, 703-704
  ultraviolet, 7, 24-25
  vacuum, 6-8, 24-25
  vacuum to air, table, 634
  x-ray, 25
Wavenumber, 7
Wigner coefficients, 241
Wigner-Eckart theorem, 307-308, 673

Xα method, 196
X-ray energy levels, 583-84
X-ray spectra, 583-87
X-unit, 6

Z-dependence
  autoionization rate, 589
  coupling conditions, 237-39
  effective potentials, 229-35
  excitation energies, 574-78
  ionization energies, 120-21, 574-75
  one-electron energies, 229-37
  oscillator strengths, 437-39
  quantum defect, 231-37
  radial integrals, 73, 236-39, 437, 571-80, 589, 624
  radial wavefunctions, 72, 231-36, 578, 589
  radiative transition probabilities, 438-40
Zeeman effect, 21, 123-24, 485-98
  as aid to spectrum analysis, 489, 493, 606-607
  in Ge II, 497-99
  intermediate fields, 497-99
  in Sn I, 495
  in stars, 21, 486
  strong-field limit, 494-96
  sum rules, 489
  weak-field limit, 488-95
  see also Landé g factors
Z-expansion, for energy, 438, 571
  for oscillator strength, 438

R. D. Cowan, *The Theory of Atomic Structure and Spectra*
Errors and additional references to third and fourth printings
October 2001

p. 174: Note that the Moore values in Table 6-6 correspond to ΔE_{av}, whereas those in Table 6-5 are ground-level-to-ground-level values.

pp. 145, 241, 245, 271, 668(2), 669, 670: for "Gordon" read "Gordan"

pp. 403, 443, 445: in Eqs. (14.32), (15.5), and (15.13), add summations over M' in addition to those over M.

p. 421: in the first 6-j symbol in Eq. (14.90), for "S" read "S_1"

p. 477: for "of R^k" read "of the R^k"

p. 534: in fn 21, for "M. Wilson" read "W. R. S. Garten and M. Wilson"
to fn 22, add "systematics: S. T. Manson, PRA <u>31</u>, 3798 (1985)"
also add: "P. F. Bronson, Zeroes in hydrogenic bound-free quadrupole and octupole radial integrals, PRA <u>34</u>, 3749 (1986)"

p. 595: for "from the modified Burgess approximation (18.134)-(18.135)" read "from Eq. (18.120)."

p. 703: in [19], for "II. <u>62</u>" read "II. Phys. Rev. <u>62</u>"

p. 726: under Phase conventions, add "ci matrix elements, 378"

In the Bibliography:

Under III, add Drake and Schlesinger, PRA <u>15</u>, 1990 (1977) and Patterson and Harter, PRA <u>15</u>, 2372 (1977)

Under VII, in [53], add 2nd ref. to Martin, Phys. Rev. A <u>36</u>, 3575 (1987)
add new compilations:
Shirai, Sugar, Musgrove, and Wiese, JPCRD Monograph no. 8 (2000); Ti, V, Cr, Mn, Fe, Co, Ni, Cu, Kr, Mo
subject-index references in JPCRD <u>20</u>, 1343 (1991) (v. 11-20), Mg, Al, Si, P, S, K-Ni, Sc, Fe, Cu, Kr, Mo
subject-index references in JPCRD <u>29</u>, 1645 (2000) (v. 21-29), Be I, O II, S, V, Cr, Mn, Co, Zn, Ge, Kr
G. W. Parker, PRA <u>33</u>, 799 (1986)

Under VIII, in [78], add Musgrove and Zalubas, Suppl. 3, July 1979 to December 1983 (1985)

Under XI, Martin, Furr, and Wiese, JPCRD <u>17</u> Suppls. 3-4 (1988); Sc-Ni
subject-index references in JPCRD <u>29</u>, 1645 (2000) (v. 21-29)
and subject-index references in ADNDT <u>76</u>, 363 (2000)

Milton Keynes UK
Ingram Content Group UK Ltd.
UKHW030920011024
448965UK00001B/4

9 780520 038219